D1296914

Success can be planned!

This text is designed to help you succeed, so use it to your advantage! Learn to work smarter (not necessarily harder!) in your math course by studying more efficiently and by making the most of the helpful learning resources available to you with this text and through your course (see the Preface and Study Tips for more information).

Record Important Contacts
on this page, including your instructor, tutor, and campus math lab. Talk with your classmates and exchange contact information with at least two people so that you stay in touch about class assignments and help each other with study questions, etc.

CONTACT	NAME	EMAIL	PHONE	FAX	OFFICE HOURS	LOCATION
Instructor						
Campus Tutor						
Campus Math Lab						
Classmate						
Classmate						

Supplements recommended by the instructor:

Online resources (Web address, access code, password, etc.):

Schedule the time you need to succeed!

At the start of the course, use the weekly planner on the reverse side to schedule time to study. Decide that success in this math course is a priority and give yourself 2 to 3 hours of study time (including homework) for each hour of class instruction time that you have each week. (See the Study Tip on page 95.)

Study the Study Tips!

Even the best students can learn to study more efficiently. Read ahead, check off the Study Tips on this list that work best for you, and review them often as you progress through the course. One way to use the Study Tips is by category. For example, if you feel that you can make better use of your time, cover all the suggestions on time management. Before you take a test, revisit all the tips on test taking.

Study Tips by Chapter, Section, and Page Number
(learning resources in red, time management in blue, test-taking tips in green, other helpful tips in black)

Chapter 1
- ☐ **1.2** Using This Textbook (p. 17)
- ☐ **1.3** Small Steps Lead to Great Success (p. 21)
- ☐ **1.5** Making Positive Choices (p. 41)
- ☐ **1.7** Video Resources (p. 56)
- ☐ **1.7** Learning Resources (p. 62)
- ☐ **1.8** Preparing for and Taking a Test (p. 71)

Chapter 2
- ☐ **2.2** Time Management (p. 95)
- ☐ **2.4** Highlighting (p. 112)
- ☐ **2.5** Solving Applied Problems (p. 120)
- ☐ **2.6** Problem-Solving Tips (p. 137)
- ☐ **2.7** Using the Supplements (p. 146)
- ☐ **2.8** Checklist (p. 158)

Chapter 3
- ☐ **3.1** Learning Resources on Campus (p. 177)
- ☐ **3.2** Forming a Study Group (p. 186)
- ☐ **3.3** Asking Questions (p. 196)
- ☐ **3.4** Tune Out Distractions (p. 205)

Chapter 4
- ☐ **4.1** Helping Others Helps You (p. 229)
- ☐ **4.2** Writing All the Steps (p. 243)
- ☐ **4.6** Memorizing Formulas (p. 280)
- ☐ **4.6** Checklist (p. 284)

Chapter 5
- ☐ **5.1** Checklist (p. 323)
- ☐ **5.2** Time Management (p. 332)
- ☐ **5.3** Reading Examples (p. 341)
- ☐ **5.6** Worked-Out Solutions (p. 364)
- ☐ **5.7** Learn from Your Mistakes (p. 368)
- ☐ **5.7** Time Management (p. 370)
- ☐ **5.8** Working with a Classmate (p. 377)
- ☐ **5.9** Five Steps for Problem Solving (p. 388)

Chapter 6
- ☐ **6.1** Watch the Videos (p. 412)
- ☐ **6.2** Studying the Examples (p. 422)
- ☐ **6.4** Working with Rational Expressions (p. 430)
- ☐ **6.7** Are You Calculating or Solving? (p. 457)

Chapter 7
- ☐ **7.1** Asking Questions (p. 502)
- ☐ **7.3** Homework Tips (p. 524)
- ☐ **7.5** Highlighting (p. 547)

Chapter 8
- ☐ **8.2** Skill Maintenance Exercises (p. 579)
- ☐ **8.2** Using This Textbook (p. 582)
- ☐ **8.3** Five Steps for Problem Solving (p. 590)
- ☐ **8.4** Making Applications Real (p. 596)
- ☐ **8.5** Worked-Out Solutions (p. 610)
- ☐ **8.5** Learning Resources (p. 614)

Chapter 9
- ☐ **9.1** Video Resources (p. 640)
- ☐ **9.1** Relying on the Answer Section (p. 643)

Chapter 10
- ☐ **10.1** Avoid Distractions (p. 707)
- ☐ **10.3** Test Taking (p. 721)
- ☐ **10.6** Aim for Mastery (p. 746)
- ☐ **10.8** Getting Started (p. 761)

Chapter 11
- ☐ **11.1** Writing All the Steps (p. 790)
- ☐ **11.2** Registering for Future Courses (p. 797)
- ☐ **11.4** Take the Time! (p. 815)
- ☐ **11.5** Key Terms (p. 828)
- ☐ **11.6** Beginning to Study for the Final Exam (p. 838)

Chapter 12
- ☐ **12.1** Budget Your Time (p. 875)
- ☐ **12.6** Beginning to Study for the Final Exam: Three Days to Two Weeks of Study Time (p. 937)
- ☐ **12.7** Beginning to Study for the Final Exam: One or Two Days of Study Time (p. 947)

Plan to succeed!

On this page, plan a typical week. Consider issues such as class time, study time, work time, travel time, family time, and relaxation time.

Important Dates

Mid-Term Exam

_____

Final Exam

_____

Holidays

_____

_____

_____

_____

Other
(Assignments, Quizzes, etc.)

_____

_____

_____

_____

_____

_____

_____

_____

_____

_____

_____

Weekly Planner

TIME	Sun.	Mon.	Tues.	Wed.	Thurs.	Fri.	Sat.
6:00 AM							
6:30							
7:00							
7:30							
8:00							
8:30							
9:00							
9:30							
10:00							
10:30							
11:00							
11:30							
12:00 PM							
12:30							
1:00							
1:30							
2:00							
2:30							
3:00							
3:30							
4:00							
4:30							
5:00							
5:30							
6:00							
6:30							
7:00							
7:30							
8:00							
8:30							
9:00							
9:30							
10:00							
10:30							
11:00							
11:30							
12:00 AM							

Introductory and Intermediate Algebra

FOURTH EDITION

Marvin L. Bittinger

Indiana University Purdue University Indianapolis

Judith A. Beecher

Indiana University Purdue University Indianapolis

PEARSON

Addison
Wesley

Boston   Columbus   Indianapolis   New York   San Francisco   Upper Saddle River
Amsterdam   Cape Town   Dubai   London   Madrid   Milan   Munich   Paris   Montréal   Toronto
Delhi   Mexico City   São Paulo   Sydney   Hong Kong   Seoul   Singapore   Taipei   Tokyo

Editorial Director	Christine Hoag
Editor in Chief	Maureen O'Connor
Executive Editor	Cathy Cantin
Executive Project Manager	Kari Heen
Associate Editors	Joanna Doxey and Christine Whitlock
Editorial Assistant	Jonathan Wooding
Production Manager	Ron Hampton
Composition	Pre-Press PMG
Editorial and Production Services	Sally Lifland/Lifland et al., Bookmakers; Martha K. Morong/Quadrata, Inc.
Art Editor and Photo Researcher	Geri Davis/The Davis Group, Inc.
Senior Media Producer	Ceci Fleming
Associate Media Producer	Nathaniel Koven
Content Development Managers	Rebecca Williams (MathXL), Mary Durnwald (TestGen)
Executive Marketing Manager	Michelle Renda
Marketing Assistants	Margaret Wheeler and Alicia Frankel
Prepress Supervisor	Caroline Fell
Manufacturing Manager	Evelyn Beaton
Senior Manufacturing Buyer	Carol Melville
Senior Media Buyer	Ginny Michaud
Text Designer	Geri Davis/The Davis Group, Inc.
Senior Designer/Cover Design	Beth Paquin
Cover Photograph	© Robert Llewellyn/Age Fotostock, Inc.

Photo Credits

Photo credits appear on p. G-6.

Library of Congress Cataloging-in-Publication Data
Bittinger, Marvin L.
  Introductory and intermediate algebra. — 4th ed. /
    by Marvin L. Bittinger, Judith A. Beecher.
      p. cm.
  Includes index.
  1. Algebra—Textbooks.   I. Beecher, Judith A.   II. Title.
  QA152.3.B58 2011
  512.9—dc22   2009022154

1 2 3 4 5 6 7 8 9 10—CRK—14 13 12 11 10

© 2011, 2007, 2003, 1999. Pearson Education, Inc.

Addison-Wesley
is an imprint of

www.pearsonhighered.com

ISBN-13: 978-0-321-61337-0
ISBN-10: 0-321-61337-6

Contents

Index of Applications

Authors' Note to Students

Welcome to *Introductory and Intermediate Algebra*. Having a solid grasp of the mathematical skills taught in this book will enrich your life in many ways, both personally and professionally, including increasing your earning power and enabling you to make wise decisions about your personal finances.

As we wrote this text, we were guided by the desire to do everything possible to help you learn its concepts and skills. The material in this book has been developed and refined with feedback from users of the three previous editions so that you can benefit from their class-tested strategies for success. Regardless of your past experiences in mathematics courses, we encourage you to consider this course as a fresh start and to approach it with a positive attitude.

One of the most important things you can do to ensure your success in this course is to allow enough time for it. This includes time spent in class and time spent out of class studying and doing homework. To help you derive the greatest benefit from this textbook, from your study time, and from the many other learning resources available to you, we have included an organizer card at the front of the book. This card serves as a handy reference for contact information for your instructor, fellow students, and campus learning resources, as well as a weekly planner. It also includes a list of the Study Tips that appear throughout the text. You might find it helpful to read all of these tips as you begin your course work.

Knowing that your time is both valuable and limited, we have designed this objective-based text to help you learn quickly and efficiently. You are led through the development of each concept, then presented with one or more examples of the corresponding skills, and finally given the opportunity to use these skills by doing the interactive margin exercises that appear on the page beside the examples. For quick assessment of your understanding, you can check your answers with the answers placed at the bottom of the page. This innovative feature, along with illustrations designed to help you visualize mathematical concepts and the extensive exercise sets keyed to section objectives, gives you the support and reinforcement you need to be successful in your math course.

To help apply and retain your knowledge, take advantage of the new Skill to Review exercises when they appear at the beginning of a section and the comprehensive mid-chapter reviews, summary and reviews, and cumulative reviews. Read through the list of supplementary material available to students that appears in the preface to make sure you get the most out of your learning experience, and investigate other learning resources that may be available to you.

Give yourself the best opportunity to succeed by spending the time required to learn. We hope you enjoy learning this material and that you will find it of benefit.

Best wishes for success!
Marv Bittinger
Judy Beecher

Related Bittinger Paperback Titles

- Bittinger: *Fundamental College Mathematics,* 5th Edition
- Bittinger: *Basic College Mathematics,* 11th Edition
- Bittinger/Penna: *Basic College Mathematics with Early Integers,* 2nd Edition
- Bittinger: *Introductory Algebra,* 11th Edition
- Bittinger: *Intermediate Algebra,* 11th Edition

Accuracy
Students rely on accurate textbooks, and our users value the Bittinger reputation for accuracy. All Bittinger titles go through an exhaustive checking process to ensure accuracy in the problem sets, mathematical art, and accompanying supplements.

Preface

New in This Edition

To maximize retention of the concepts and skills presented, five highly effective review features are included in the 4th edition. Student success is increased when review is integrated throughout each chapter.

Five Types of Integrated Review

Skill to Review exercises, found at the beginning of most sections, link to a section objective. These exercises offer a just-in-time review of a previously presented skill that relates to new material in the section. For convenient studying, section and objective references are followed by two practice exercises for immediate review and reinforcement. Exercise answers are given at the bottom of the page for immediate feedback.

Skill Maintenance Exercises, found in each exercise set, review concepts from other sections in the text to prepare students for their final examination. Section and objective references appear next to each Skill Maintenance exercise. All Skill Maintenance answers are included in the text.

A Mid-Chapter Review reinforces understanding of the mathematical concepts and skills just covered before students move on to new material. Section and objective references are included. Exercise types include Concept Reinforcement, Guided Solutions, Mixed Review, and Understanding Through Discussion and Writing. Answers to all exercises in the Mid-Chapter Review are given at the back of the book.

The Chapter Summary and Review at the end of each chapter is expanded to provide more comprehensive in-text practice and review.

- **Key Terms, Properties, and Formulas** are highlighted, with page references for convenient review.
- **Concept Reinforcement** offers true/false questions to enhance students' understanding of mathematical concepts.
- Important Concepts are listed by section objectives, followed by *worked-out examples* for reference and review and *similar practice exercises* for students to solve.
- **Review Exercises**, including Synthesis exercises and two new multiple-choice exercises, are organized by objective and cover the whole chapter.
- **Understanding Through Discussion and Writing** exercises strengthen understanding by giving students a chance to express their thoughts in spoken or written form.

Section and objective references for all exercises are included. Answers to all exercises in the Summary and Review are given at the back of the book.

Chapter Tests, including Synthesis questions and a new multiple-choice question, allow students to review and test their comprehension of chapter skills prior to taking an instructor's exam. Answers to all questions in the Chapter Tests are given at the back of the book. Section and objective references for each question are included with the answers.

A Cumulative Review after every chapter starting with Chapter 2 revisits skills and concepts from all preceding chapters to help students recall previously learned material and prepare for exams. Answers to all Cumulative Review exercises are coded by section and objective at the back of the book to help students identify areas where additional practice is needed.

A new design enhances the Bittinger guided-learning approach. Margin exercises are now located next to examples for easier navigation, and answers for those exercises are given at the bottom of the page for immediate feedback.

Hallmark Features

Revised! The **Bittinger Student Organizer** card at the front of the text helps students keep track of important contacts and dates and provides a weekly planner to help schedule time for classes, studying, and homework. A helpful list of study tips found in each chapter is also included.

New! **Chapter Openers** feature motivating real-world applications that are revisited later in the chapters. This feature engages students and prepares them for the upcoming chapter material. (See pages 85, 409, and 499.)

New! **Real-Data Applications** encourage students to see and interpret the mathematics that appears every day in the world around them. (See pages 185, 246, 497, 509, 590, and 848.) Many applications are drawn from the fields of business and economics, life and physical sciences, social sciences, medicine, and areas of general interest such as sports and daily life.

Study Tips appear throughout the text to give students pointers on how to develop good study habits as they progress through the course, encouraging them to get involved in the learning process. (See pages 62, 177, and 377.) For easy reference, a list of Study Tips by chapter, section, and page number is included in the Bittinger Student Organizer.

Algebraic–Graphical Connections To provide a visual understanding of algebra, algebraic–graphical connections are included in each chapter beginning with Chapter 3. This feature gives the algebra more meaning by connecting it to a graphical interpretation. (See pages 186, 381, and 574–575.)

Caution Boxes are found at relevant points throughout the text. The heading "*Caution!*" alerts students to coverage of a common misconception or an error often made in performing a particular mathematics operation or skill. (See pages 175, 321, and 357.)

Revised! Optional **Calculator Corners** are located where appropriate throughout the text. These streamlined Calculator Corners are written to be accessible to students and to represent current calculators. A calculator icon indicates exercises suitable for calculator use. (See pages 72, 187, 266, and 456.)

Immediate Practice and Assessment in Each Section

OBJECTIVES ➡ SKILL TO REVIEW ➡ EXPOSITION ➡ EXAMPLES WITH DETAILED ANNOTATIONS AND VISUAL ART PIECES ➡ MARGIN EXERCISES ➡ EXERCISE SETS

Objective Boxes begin each section. A boxed list of objectives is keyed by letter not only to section subheadings, but also to the section exercise sets and the Mid-Chapter Review and the Summary and Review exercises, as well as to the answers to the questions in the Chapter Tests and Cumulative Reviews. This correlation enables students to easily find appropriate review material if they need help with a particular exercise or skill at the objective level. (See pages 98, 174, and 353.)

New! **Skill to Review** exercises, found at the beginning of most sections, link to a section objective and offer students a just-in-time review of a previously presented skill that relates to new material in the section. For convenient studying, objective references are followed by two practice exercises for immediate review and reinforcement. Answers to these exercises are given at the bottom of the page for immediate feedback. (See pages 192, 263, and 420.)

Revised! **Annotated Examples** provide annotations and color highlighting to lead students through the structured steps of the examples. The level of detail in these annotations is a significant reason for students' success with this book. This edition contains over 180 new examples. (See pages 100, 275, and 716.)

Revised! The **art and photo program** is designed to help students visualize mathematical concepts and real-data applications. Many applications include source lines and feature graphs and drawings similar to those students see in the media. The use of color is carried out in a methodical and precise manner so that it conveys a consistent meaning, which enhances the readability of the text. For example, the use of both red and blue in mathematical art increases understanding of the concepts. When two lines are graphed using the same set of axes, one is usually red and the other blue. Note that equation labels are the same color as the corresponding line to aid in understanding. (See pages 2, 18, 210, 462, 515, 537, 652, and 826.)

Revised! **Margin Exercises**, now located next to examples for easier navigation, accompany examples throughout the text and give students the opportunity to work similar problems for immediate practice and reinforcement of the concept just learned. Answers are now available at the bottom of the page. (See pages 99, 294, and 723.)

Exercise Sets

To give students ample opportunity to practice what they have learned, each section is followed by an extensive exercise set *keyed by letter to the section objectives* for easy review and remediation. In addition, students also have the opportunity to synthesize the objectives from the current section with those from preceding sections. **For Extra Help** icons, shown at the beginning of each exercise set, indicate supplementary learning resources that students may need. This edition contains over 1600 new exercises.

- **Skill Maintenance Exercises**, found in each exercise set, review concepts from other sections in the text to prepare students for their final examination. Section and objective codes appear next to each Skill Maintenance exercise for easy reference. All Skill Maintenance answers are included in the text. (See pages 180, 428, 734, 794–795.)

- **Vocabulary Reinforcement Exercises** provide an integrated review of key terms that students must know to communicate effectively in the language of mathematics. These appear once per chapter in the Skill Maintenance portion of an exercise set. (See pages 215, 397, and 513.)

- **Synthesis Exercises** help build critical-thinking skills by requiring students to use what they know to synthesize, or combine, learning objectives from the current section with those from previous sections. These are available in most exercise sets. (See pages 126, 290, and 424.)

Mid-Chapter Review

New! A **Mid-Chapter Review** gives students the opportunity to reinforce their understanding of the mathematical skills and concepts just covered before they move on to new material. Section and objective references are included for convenient studying, and answers to all the Mid-Chapter Review exercises are included in the text. The types of exercises are as follows:

- **Concept Reinforcement** are true/false questions that enhance students' understanding of mathematical concepts. These are also available in the Summary and Review at the end of the chapter. (See pages 117, 445, and 608.)

- **Guided Solutions** present worked-out problems with blanks for students to fill in the correct expressions to complete the solution. (See pages 117, 445, and 608.)

- **Mixed Review** provides free-response exercises, similar to those in the preceding sections in the chapter, reinforcing mastery of skills and concepts. (See pages 117, 445, and 608.)

- **Understanding Through Discussion and Writing** lets students demonstrate their understanding of mathematical concepts by expressing their thoughts in spoken and written form. This type of exercise is also found in each Chapter Summary and Review. (See pages 118, 446, and 609.)

Matching Feature

Translating for Success problem sets give extra practice with the important "Translate" step of the process for solving word problems. After translating each of ten problems into its appropriate equation or inequality, students are asked to choose from fifteen possible translations, encouraging them to comprehend the problem before matching. (See pages 138, 470, and 603.)

Visualizing for Success problem sets ask students to match an equation or inequality with its graph by focusing on characteristics of the equation or inequality and the corresponding attributes of the graph. This feature appears at least once in each chapter that contains graphing instruction and reviews graphing skills and concepts with exercises from all preceding chapters. (See pages 197, 286, and 686.)

End-of-Chapter Material

Revised! The **Chapter Summary and Review** at the end of each chapter is expanded to provide more comprehensive in-text practice and review. Section and objective references and answers to all the Chapter Summary and Review exercises are included in the text. (See pages 164, 307, and 863.)

- **Key Terms, Properties, and Formulas** are highlighted, with page references for convenient review. (See pages 164, 307, and 863.)
- **Concept Reinforcement** offers true/false questions to enhance student understanding of mathematical concepts. (See pages 164, 307, and 863.)
- **New!** Important Concepts are listed by section objectives, followed by *a worked-out example* for reference and review and *a similar practice exercise* for students to solve. (See pages 164–166, 307–310, and 863–866.)
- **Review Exercises**, including Synthesis exercises and two new multiple-choice exercises, covering the whole chapter are organized by objective. (See pages 166–168, 310–312, and 866–868.)
- **Understanding Through Discussion and Writing** exercises strengthen understanding by giving students a chance to express their thoughts in spoken or written form. (See pages 168, 312, and 868.)

Chapter Tests, including Synthesis questions and a new multiple-choice question, allow students to review and test their comprehension of chapter skills prior to taking an instructor's exam. Answers to all questions in the Chapter Test are given at the back of the book. Section and objective references for each question are included with the answers. (See pages 169, 495, and 869.)

New! A **Cumulative Review** now follows every chapter starting with Chapter 2; this review revisits skills and concepts from all preceding chapters to help students recall previously learned material and prepare for exams. Answers to all Cumulative Review exercises are coded by section and objective at the back of the book to help students identify areas where additional practice is needed. (See pages 225, 315, and 871.)

For Extra Help

Student Supplements

New! Worksheets for Classroom or Lab Practice
(ISBN: 978-0-321-61368-4)

These classroom- and lab-friendly workbooks offer the following resources for every section of the text: a list of learning objectives, vocabulary practice problems, and extra practice exercises with ample work space.

Student's Solutions Manual (ISBN: 978-0-321-61362-2)
By Judith Penna

Contains completely worked-out annotated solutions for all the odd-numbered exercises in the text. Also includes fully worked-out annotated solutions for all the exercises (odd- and even-numbered) in the Mid-Chapter Reviews, the Summary and Reviews, the Chapter Tests, and the Cumulative Reviews.

Chapter Test Prep Videos

Chapter Tests can serve as practice tests to help you study. Watch instructors work through step-by-step solutions to all the Chapter Test exercises from the textbook. Chapter Test Prep videos are available on YouTube (search using BittingerIntroInter) and in MyMathLab. They are also included on the Video Resources on DVD described below and available for purchase at www.MyPearsonStore.com.

Video Resources on DVD Featuring Chapter Test Prep Videos
(ISBN: 978-0-321-61365-3)

- Complete set of lectures covering every objective of every section in the textbook
- Complete set of Chapter Test Prep videos (see above)
- All videos include optional English and Spanish subtitles.
- Ideal for distance learning or supplemental instruction
- DVD-ROM format for student use at home or on campus

InterAct Math Tutorial Website (www.interactmath.com)

Get practice and tutorial help online! This interactive tutorial website provides algorithmically generated practice exercises that correlate directly to the exercises in the textbook. Students can retry an exercise as many times as they like with new values each time for unlimited practice and mastery. Every exercise is accompanied by an interactive guided solution that provides helpful feedback for incorrect answers, and students can also view a worked-out sample problem that steps them through an exercise similar to the one they're working on.

MathXL® Tutorials on CD (ISBN: 978-0-321-61361-5)

This interactive tutorial CD-ROM provides algorithmically generated practice exercises that are correlated at the objective level to the exercises in the textbook. Every practice exercise is accompanied by an example and a guided solution designed to involve students in the solution process. Selected exercises may also include a video clip to help students visualize concepts. The software provides helpful feedback for incorrect answers and can generate printed summaries of students' progress.

Instructor Supplements

Annotated Instructor's Edition (ISBN: 978-0-321-61360-8)

Includes answers to all exercises printed in blue on the same page as the exercises. Also includes the student answer section, for easy reference.

Instructor's Solutions Manual (ISBN: 978-0-321-61358-5)
By Judith Penna

Contains brief solutions to the even-numbered exercises in the exercise sets. Also includes fully worked-out annotated solutions for all the exercises (odd- and even-numbered) in the Mid-Chapter Reviews, the Summary and Reviews, the Chapter Tests, and the Cumulative Reviews.

Printed Test Forms (ISBN: 978-0-321-61357-8)
By Laurie Hurley

- Contains one diagnostic test and one pretest for each chapter, plus two cumulative tests per chapter, beginning with Chapter 2.
- **New!** Includes two versions of a short mid-chapter quiz.
- Provides eight test forms for every chapter and eight test forms for the final exam.
- For the chapter tests, four free-response tests are modeled after the chapter tests in the main text, two tests are designed for 50-minute class periods and organized so that each objective in the chapter is covered on one of the tests, and two tests consist of multiple-choice questions. Chapter tests also include more challenging Synthesis questions.
- For the final exam, three test forms are organized by chapter, three forms are organized by question type, and two forms are multiple-choice tests.
- Also includes extra practice exercises for select sections.

Instructor's Resource Manual
(ISBN: 978-0-321-61364-6)

- Features resources and teaching tips designed to help both new and adjunct faculty with course preparation and classroom management.
- **New!** Includes a mini-lecture for each section of the text with objectives, key examples, and teaching tips.
- Additional resources include general first-time advice, sample syllabi, teaching tips, collaborative learning activities, correlation guide, video index, and transparency masters.

Additional Media Supplements

MyMathLab **MyMathLab® Online Course (access code required)**

MyMathLab is a series of text-specific, easily customizable online courses for Pearson Education's textbooks in mathematics and statistics. Powered by CourseCompass™ (our online teaching and learning environment) and MathXL® (our online homework, tutorial, and assessment system), MyMathLab gives instructors the tools they need to deliver all or a portion of their course online, whether their students are in a lab setting or working from home. MyMathLab provides a rich and flexible set of course materials, featuring free-response exercises that are algorithmically generated for unlimited practice and mastery. Students can also use online tools, such as video lectures, animations, interactive math games, and a multimedia textbook, to independently improve their understanding and performance. Instructors can use MyMathLab's homework and test managers to select and assign online exercises correlated directly to the textbook, and they can also create and assign their own online exercises and import TestGen tests for added flexibility. MyMathLab's online gradebook—designed specifically for mathematics and statistics—automatically tracks students' homework and test results and gives the instructor control over how to calculate final grades. Instructors can also add offline (paper-and-pencil) grades to the gradebook. MyMathLab also includes access to the **Pearson Tutor Center** (www.pearsontutorservices.com). The Tutor Center is staffed by qualified mathematics instructors who provide textbook-specific tutoring for students via toll-free phone, fax, email, and interactive Web sessions. MyMathLab is available to qualified adopters. For more information, visit our website at www.mymathlab.com or contact your sales representative.

MathXL **MathXL® Online Course (access code required)**

MathXL® is a powerful online homework, tutorial, and assessment system that accompanies Pearson Education's textbooks in mathematics or statistics.

With MathXL, instructors can

- create, edit, and assign online homework and tests using algorithmically generated exercises correlated at the objective level to the textbook.
- create and assign their own online exercises and import TestGen tests for added flexibility.
- maintain records of all student work tracked in MathXL's online gradebook.

With MathXL, students can

- take chapter tests in MathXL and receive personalized study plans based on their test results.
- use the study plan to link directly to tutorial exercises for the objectives they need to study and retest.
- access supplemental animations and video clips directly from selected exercises.

MathXL is available to qualified adopters. For information, visit our website at www.mathxl.com, or contact your Pearson sales representative.

TestGen® (www.pearsoned.com/testgen) enables instructors to build, edit, and print tests using a computerized bank of questions developed to cover all the objectives of the text. TestGen is algorithmically based, allowing instructors to create multiple but equivalent versions of the same question or test with the click of a button. Instructors can also modify test bank questions or add new questions. The software and test bank are available for download from Pearson Education's online catalog.

PowerPoint® Lecture Slides present key concepts and definitions from the text. Slides are available to download from within MyMathLab and from Pearson Education's online catalog.

Pearson Math Adjunct Support Center (http://www.pearsontutorservices.com/math-adjunct. html) is staffed by qualified instructors with more than 100 years of combined experience at both the community college and university levels. Assistance is provided for faculty in the following areas: suggested syllabus consultation, tips on using materials packed with your book, book-specific content assistance, and teaching suggestions, including advice on classroom strategies.

Acknowledgments

Our deepest appreciation to all of you who helped to shape this edition by reviewing and spending time with us on your campuses. In particular, we would like to thank the following reviewers of *Introductory Algebra, Intermediate Algebra,* and *Introductory and Intermediate Algebra*:

Gus Brar, *Delaware County Community College*
Chris Copple, *Northwest State Community College*
Carol Curtis, *Fresno City College*
Shreyas Desai, *Atlanta Metropolitan College*
Hope Essien, *Malcolm X College*
Kimberly I. Fara, *Des Moines Area Community College–Carroll Campus*
Karen S. Hale, *Onondaga Community College*
Dianne Hendrickson, *Becker College*
Vivian Olivia Jones, *Valencia Community College*
Qiana T. Lewis, *Wilbur Wright College*
Susan Meshulam, *Indiana University Purdue University Indianapolis*
Marcia Venzon, *Texas A&M University–Corpus Christi*
Ethel Wheland, *University of Akron*

The endless hours of hard work by Sally Lifland, Martha Morong, and Geri Davis have led to products of which we are immensely proud. We also want to thank Judy Penna for writing the Student's and Instructor's Solutions Manuals and for her strong leadership in the preparation of the printed supplements and video lectures with Barbara Johnson. Other strong support has come from Laurie Hurley for the Printed Test Forms and for accuracy checking, along with checkers Holly Martinez and Barbara Johnson and proofreader Patty LaGree. Michelle Lanosga assisted with applications research. We also wish to recognize Tom Atwater, Margaret Donlan, and Patty Schwarzkopf, who wrote video scripts.

In addition, a number of people at Pearson have contributed in special ways to the development and production of this textbook including the Developmental Math team: Vice President, Executive Director of Development Carol Trueheart, Senior Development Editor Dawn Nuttall, Production Manager Ron Hampton, Senior Designer Beth Paquin, Associate Editors Joanna Doxey and Christine Whitlock, Editorial Assistant Jonathan Wooding, Associate Media Producer Nathaniel Koven, and Media Producer Ceci Fleming. Executive Editor Cathy Cantin and Executive Marketing Manager Michelle Renda encouraged our vision and provided marketing insight. Kari Heen, Executive Project Manager, deserves special recognition for overseeing every phase of the project and keeping it moving.

Introduction to Real Numbers and Algebraic Expressions

Real-World Application

The tallest mountain in the world, when measured from base to peak, is Mauna Kea (White Mountain) in Hawaii. From its base 19,684 ft below sea level in the Hawaiian Trough, it rises 33,480 ft. What is the elevation of the peak above sea level?

Source: The Guinness Book of Records

This problem appears as Exercise 71 in Exercise Set 1.3.

Introduction to Algebra

The study of algebra involves the use of equations to solve problems. Equations are constructed from algebraic expressions. The purpose of this section is to introduce you to the types of expressions encountered in algebra.

a Evaluating Algebraic Expressions

In arithmetic, you have worked with expressions such as

$$49 + 75, \quad 8 \times 6.07, \quad 29 - 14, \quad \text{and} \quad \frac{5}{6}.$$

In algebra, we can use letters to represent numbers and work with *algebraic expressions* such as

$$x + 75, \quad 8 \times y, \quad 29 - t, \quad \text{and} \quad \frac{a}{b}.$$

Sometimes a letter can represent various numbers. In that case, we call the letter a **variable**. Let $a =$ your age. Then a is a variable since a changes from year to year. Sometimes a letter can stand for just one number. In that case, we call the letter a **constant**. Let $b =$ your date of birth. Then b is a constant.

Where do algebraic expressions occur? Most often we encounter them when we are solving applied problems. For example, consider the bar graph shown at left, one that we might find in a book or a magazine. Suppose we want to know how much higher Mt. McKinley is than Mt. Evans. Using arithmetic, we might simply subtract. But let's see how we can determine this using algebra. We translate the problem into a statement of equality, an equation. It could be done as follows:

Height of Mt. Evans	plus	How much more	is	Height of Mt. McKinley
↓	↓	↓	↓	↓
14,264	+	x	=	20,320.

Note that we have an algebraic expression, $14{,}264 + x$, on the left of the equals sign. To find the number x, we can subtract 14,264 on both sides of the equation:

$$14{,}264 + x = 20{,}320$$
$$14{,}264 + x - 14{,}264 = 20{,}320 - 14{,}264$$
$$x = 6056.$$

This value of x gives the answer, 6056 ft.

We call $14{,}264 + x$ an *algebraic expression* and $14{,}264 + x = 20{,}320$ an *algebraic equation*. Note that there is no equals sign, $=$, in an algebraic expression.

In arithmetic, you probably would do this subtraction without ever considering an equation. *In algebra, more complex problems are difficult to solve without first writing an equation.*

Do Exercise 1.

Mountain Peaks in the United States

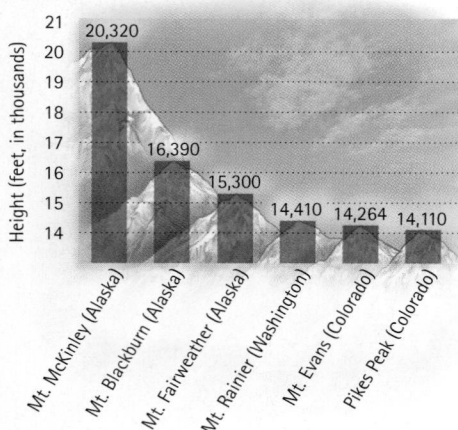

SOURCE: U.S. Department of the Interior, Geological Survey

1. Translate this problem to an equation. Then solve the equation.

Mountain Peaks. There are 92 mountain peaks in the United States that are higher than 14,000 ft. The bar graph above shows data for six of these. How much higher is Mt. Fairweather than Mt. Rainier?

Answer

1. $14{,}410 + x = 15{,}300$; 890 ft

An **algebraic expression** consists of variables, constants, numerals, operation signs, and/or grouping symbols. When we replace a variable with a number, we say that we are **substituting** for the variable. When we replace all of the variables in an expression with numbers and carry out the operations in the expression, we are **evaluating the expression**.

EXAMPLE 1   Evaluate $x + y$ when $x = 37$ and $y = 29$.

We substitute 37 for x and 29 for y and carry out the addition:

$$x + y = 37 + 29 = 66.$$

The number 66 is called the **value** of the expression when $x = 37$ and $y = 29$.

Algebraic expressions involving multiplication can be written in several ways. For example, "8 times a" can be written as

$$8 \times a, \quad 8 \cdot a, \quad 8(a), \quad \text{or simply} \quad 8a.$$

Two letters written together without an operation symbol, such as ab, also indicate a multiplication.

EXAMPLE 2   Evaluate $3y$ when $y = 14$.

$$3y = 3(14) = 42$$

Do Exercises 2–4.

2. Evaluate $a + b$ when $a = 38$ and $b = 26$.

3. Evaluate $x - y$ when $x = 57$ and $y = 29$.

4. Evaluate $4t$ when $t = 15$.

EXAMPLE 3   *Area of a Rectangle.*   The area A of a rectangle of length l and width w is given by the formula $A = lw$. Find the area when l is 24.5 in. and w is 16 in.

We substitute 24.5 in. for l and 16 in. for w and carry out the multiplication:

$$
\begin{aligned}
A = lw &= (24.5 \text{ in.})(16 \text{ in.}) \\
&= (24.5)(16)(\text{in.})(\text{in.}) \\
&= 392 \text{ in}^2, \text{ or } 392 \text{ square inches.}
\end{aligned}
$$

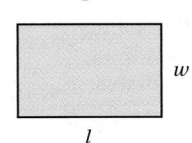

Do Exercise 5.

5. Find the area of a rectangle when l is 24 ft and w is 8 ft.

Algebraic expressions involving division can also be written in several ways. For example, "8 divided by t" can be written as

$$8 \div t, \quad \frac{8}{t}, \quad 8/t, \quad \text{or} \quad 8 \cdot \frac{1}{t},$$

where the fraction bar is a division symbol.

EXAMPLE 4   Evaluate $\dfrac{a}{b}$ when $a = 63$ and $b = 9$.

We substitute 63 for a and 9 for b and carry out the division:

$$\frac{a}{b} = \frac{63}{9} = 7.$$

EXAMPLE 5   Evaluate $\dfrac{12m}{n}$ when $m = 8$ and $n = 16$.

$$\frac{12m}{n} = \frac{12 \cdot 8}{16} = \frac{96}{16} = 6$$

6. Evaluate a/b when $a = 200$ and $b = 8$.

7. Evaluate $10p/q$ when $p = 40$ and $q = 25$.

8. Motorcycle Travel. Find the time it takes to travel 660 mi if the speed is 55 mph.

Do Exercises 6 and 7.

EXAMPLE 6 *Motorcycle Travel.* Ed wants to travel 660 mi on his motorcycle on a particular day. The time t, in hours, that it takes to travel 660 mi is given by

$$t = \frac{660}{r},$$

where r is the speed of Ed's motorcycle. Find the time of travel if the speed r is 60 mph.

We substitute 60 for r and carry out the division:

$$t = \frac{660}{r} = \frac{660}{60} = 11 \text{ hr.}$$

Do Exercise 8.

b Translating to Algebraic Expressions

In algebra, we translate problems to equations. The different parts of an equation are translations of word phrases to algebraic expressions. It is easier to translate if we know that certain words often translate to certain operation symbols.

KEY WORDS, PHRASES, AND CONCEPTS

ADDITION (+)	SUBTRACTION (−)	MULTIPLICATION (·)	DIVISION (÷)
add	subtract	multiply	divide
added to	subtracted from	multiplied by	divided by
sum	difference	product	quotient
total	minus	times	
plus	less than	of	
more than	decreased by		
increased by	take away		

EXAMPLE 7 Translate to an algebraic expression:

Twice (or two times) some number.

Think of some number, say, 8. We can write 2 times 8 as 2×8, or $2 \cdot 8$. We multiplied by 2. Do the same thing using a variable. We can use any variable we wish, such as x, y, m, or n. Let's use y to stand for some number. If we multiply by 2, we get an expression

$$y \times 2, \quad 2 \times y, \quad 2 \cdot y, \quad \text{or} \quad 2y.$$

In algebra, $2y$ is the expression generally used.

EXAMPLE 8 Translate to an algebraic expression:

Thirty-eight percent of some number.

Let $n =$ the number. The word "of" translates to a multiplication symbol, so we could write any of the following expressions as a translation:

$$38\% \cdot n, \quad 0.38 \times n, \quad \text{or} \quad 0.38n.$$

Answers

6. 25 **7.** 16 **8.** 12 hr

EXAMPLE 9 Translate to an algebraic expression:

Seven less than some number.

We let x represent the number. If the number were 10, then 7 less than 10 is $10 - 7$, or 3. If we knew the number to be 34, then 7 less than the number would be $34 - 7$. Thus if the number is x, then the translation is

$x - 7$.

--------- *Caution!* ---------

Note that $7 - x$ is *not* a correct translation of the expression in Example 9. The expression $7 - x$ is a translation of "seven minus some number" or "some number less than seven."

EXAMPLE 10 Translate to an algebraic expression:

Eighteen more than a number.

We let $t =$ the number. Now if the number were 6, then the translation would be $6 + 18$, or $18 + 6$. If we knew the number to be 17, then the translation would be $17 + 18$, or $18 + 17$. If the number is t, then the translation is

$t + 18$,   or $18 + t$.

EXAMPLE 11 Translate to an algebraic expression:

A number divided by 5.

We let $m =$ the number. Now if the number were 7, then the translation would be $7 \div 5$, or 7/5, or $\frac{7}{5}$. If the number were 21, then the translation would be $21 \div 5$, or 21/5, or $\frac{21}{5}$. If the number is m, then the translation is

$m \div 5$, $m/5$,   or $\dfrac{m}{5}$.

EXAMPLE 12 Translate each phrase to an algebraic expression.

PHRASE	ALGEBRAIC EXPRESSION
Five more than some number	$n + 5$, or $5 + n$
Half of a number	$\frac{1}{2}t, \frac{t}{2}$, or $t/2$
Five more than three times some number	$3p + 5$, or $5 + 3p$
The difference of two numbers	$x - y$
Six less than the product of two numbers	$mn - 6$
Seventy-six percent of some number	$76\%z$, or $0.76z$
Four less than twice some number	$2x - 4$

Do Exercises 9–17.

Translate each phrase to an algebraic expression.

9. Eight less than some number

10. Eight more than some number

11. Four less than some number

12. Half of some number

13. Six more than eight times some number

14. The difference of two numbers

15. Fifty-nine percent of some number

16. Two hundred less than the product of two numbers

17. The sum of two numbers

Answers

9. $x - 8$ **10.** $y + 8$, or $8 + y$
11. $m - 4$ **12.** $\frac{1}{2} \cdot p$, or $\frac{p}{2}$
13. $8x + 6$, or $6 + 8x$ **14.** $a - b$
15. $59\%x$, or $0.59x$ **16.** $xy - 200$
17. $p + q$

a Substitute to find values of the expressions in each of the following applied problems.

1. *Commuting Time.* It takes Erin 24 min less time to commute to work than it does George. Suppose that the variable x stands for the time it takes George to get to work. Then $x - 24$ stands for the time it takes Erin to get to work. How long does it take Erin to get to work if it takes George 56 min? 93 min? 105 min?

2. *Enrollment Costs.* At Emmett Community College, it costs $600 to enroll in the 8 A.M. section of Elementary Algebra. Suppose that the variable n stands for the number of students who enroll. Then $600n$ stands for the total amount of money collected for this course. How much is collected if 34 students enroll? 78 students? 250 students?

3. *Area of a Triangle.* The area A of a triangle with base b and height h is given by $A = \frac{1}{2}bh$. Find the area when $b = 45$ m (meters) and $h = 86$ m.

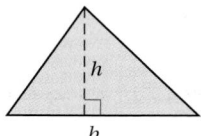

4. *Area of a Parallelogram.* The area A of a parallelogram with base b and height h is given by $A = bh$. Find the area of the parallelogram when the height is 15.4 cm (centimeters) and the base is 6.5 cm.

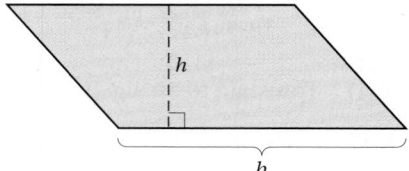

5. *Distance Traveled.* A driver who drives at a constant speed of r miles per hour for t hours will travel a distance of d miles given by $d = rt$ miles. How far will a driver travel at a speed of 65 mph for 4 hr?

6. *Simple Interest.* The simple interest I on a principal of P dollars at interest rate r for time t, in years, is given by $I = Prt$. Find the simple interest on a principal of $4800 at 9% for 2 years. (*Hint*: 9% = 0.09.)

7. *Hockey Goal.* The front of a regulation hockey goal is a rectangle that is 6 ft wide and 4 ft high. Find its area.
Source: National Hockey League

8. *Zoology.* A great white shark has triangular teeth. Each tooth measures about 5 cm across the base and has a height of 6 cm. Find the surface area of one side of one tooth. (See Exercise 3.)

Evaluate.

9. $8x$, when $x = 7$

10. $6y$, when $y = 7$

11. $\dfrac{c}{d}$, when $c = 24$ and $d = 3$

12. $\dfrac{p}{q}$, when $p = 16$ and $q = 2$

13. $\dfrac{3p}{q}$, when $p = 2$ and $q = 6$

14. $\dfrac{5y}{z}$, when $y = 15$ and $z = 25$

15. $\dfrac{x + y}{5}$, when $x = 10$ and $y = 20$

16. $\dfrac{p + q}{2}$, when $p = 2$ and $q = 16$

17. $\dfrac{x - y}{8}$, when $x = 20$ and $y = 4$

18. $\dfrac{m - n}{5}$, when $m = 16$ and $n = 6$

b   Translate each phrase to an algebraic expression. Use any letter for the variable(s) unless directed otherwise.

19. Seven more than some number

20. Nine more than some number

21. Twelve less than some number

22. Fourteen less than some number

23. Some number increased by four

24. Some number increased by thirteen

25. b more than a

26. c more than d

27. x divided by y

28. c divided by h

29. x plus w

30. s added to t

31. m subtracted from n

32. p subtracted from q

33. The sum of two numbers

34. The sum of nine and some number

35. Twice some number

36. Three times some number

37. Three multiplied by some number

38. The product of eight and some number

39. Six more than four times some number

40. Two more than six times some number

41. Eight less than the product of two numbers

42. The product of two numbers minus seven

43. Five less than twice some number

44. Six less than seven times some number

45. Three times some number plus eleven

46. Some number times 8 plus 5

47. The sum of four times a number plus three times another number

48. Five times a number minus eight times another number

49. The product of 89% and your salary

50. 67% of the women attending

51. Your salary after a 5% salary increase if your salary before the increase was s

52. The price of a blouse after a 30% reduction if the price before the reduction was P

53. Danielle drove at a speed of 65 mph for t hours. How far did Danielle travel? (See Exercise 5.)

54. Dino drove his pickup truck at 55 mph for t hours. How far did he travel? (See Exercise 5.)

55. Lisa had $50 before spending x dollars on pizza. How much money remains?

56. Juan has d dollars before spending $29.95 on a DVD of the movie *Chicago*. How much did Juan have after the purchase?

57. Robert's part-time job pays $8.50 per hour. How much does he earn for working n hours?

58. Meredith pays her babysitter $10 per hour. What does it cost her to hire the sitter for m hours?

Synthesis

To the student and the instructor: The Synthesis exercises found at the end of most exercise sets challenge students to combine concepts or skills studied in that section or in preceding parts of the text.

Evaluate.

59. $\dfrac{a - 2b + c}{4b - a}$, when $a = 20$, $b = 10$, and $c = 5$

60. $\dfrac{x}{y} - \dfrac{5}{x} + \dfrac{2}{y}$, when $x = 30$ and $y = 6$

61. $\dfrac{12 - c}{c + 12b}$, when $b = 1$ and $c = 12$

62. $\dfrac{2w - 3z}{7y}$, when $w = 5$, $y = 6$, and $z = 1$

1.2 The Real Numbers

A **set** is a collection of objects. For our purposes, we will most often be considering sets of numbers. One way to name a set uses what is called **roster notation**. For example, roster notation for the set containing the numbers 0, 2, and 5 is $\{0, 2, 5\}$.

Sets that are part of other sets are called **subsets**. In this section, we become acquainted with the set of *real numbers* and its various subsets.

Two important subsets of the real numbers are listed below using roster notation.

OBJECTIVES

a State the integer that corresponds to a real-world situation.

b Graph rational numbers on the number line.

c Convert from fraction notation for a rational number to decimal notation.

d Determine which of two real numbers is greater and indicate which, using < or >. Given an inequality like $a > b$, write another inequality with the same meaning. Determine whether an inequality like $-3 \leq 5$ is true or false.

e Find the absolute value of a real number.

NATURAL NUMBERS

The set of **natural numbers** $= \{1, 2, 3, \ldots\}$. These are the numbers used for counting.

WHOLE NUMBERS

The set of **whole numbers** $= \{0, 1, 2, 3, \ldots\}$. This is the set of natural numbers and 0.

We can represent these sets on the number line. The natural numbers are to the right of zero. The whole numbers are the natural numbers and zero.

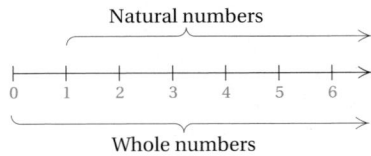

We create a new set, called the *integers*, by starting with the whole numbers, 0, 1, 2, 3, and so on. For each natural number 1, 2, 3, and so on, we obtain a new number to the left of zero on the number line:

For the number 1, there will be an *opposite* number -1 (negative 1).

For the number 2, there will be an *opposite* number -2 (negative 2).

For the number 3, there will be an *opposite* number -3 (negative 3), and so on.

The **integers** consist of the whole numbers and these new numbers.

INTEGERS

The set of **integers** $= \{\ldots, -5, -4, -3, -2, -1, 0, 1, 2, 3, 4, 5, \ldots\}$.

We picture the integers on the number line as follows.

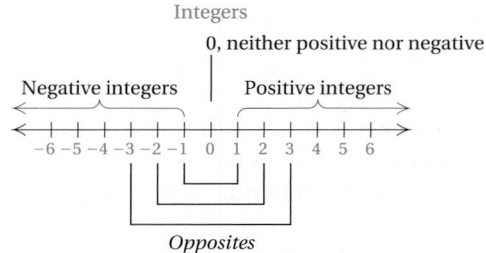

We call the integers to the left of zero **negative integers**. The natural numbers are also called **positive integers**. Zero is neither positive nor negative. We call −1 and 1 **opposites** of each other. Similarly, −2 and 2 are opposites, −3 and 3 are opposites, −100 and 100 are opposites, and 0 is its own opposite. Pairs of opposite numbers like −3 and 3 are the same distance from zero. The integers extend infinitely on the number line to the left and right of zero.

(a) Integers and the Real World

Integers correspond to many real-world problems and situations. The following examples will help you get ready to translate problem situations that involve integers to mathematical language.

EXAMPLE 1 Tell which integer corresponds to this situation: The temperature is 4 degrees below zero.

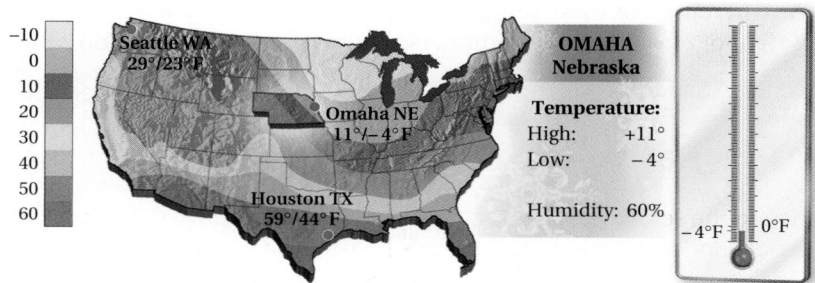

The integer −4 corresponds to the situation. The temperature is −4°.

EXAMPLE 2 *"Jeopardy."* Tell which integer corresponds to this situation: A contestant missed a $600 question on the television game show "Jeopardy."

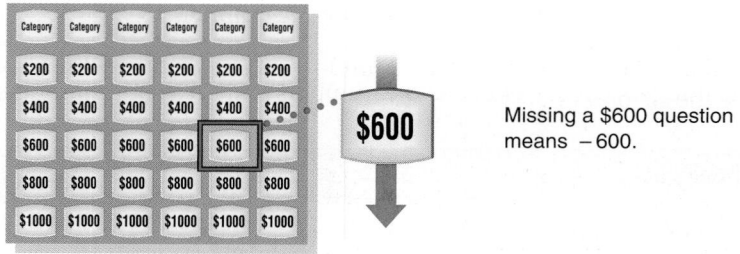

Missing a $600 question causes a $600 loss on the score—that is, the contestant earns −600 dollars.

EXAMPLE 3 *Elevation.* Tell which integer corresponds to this situation: The shores of California's largest lake, the Salton Sea, are 227 ft below sea level.
Source: Salton Sea Authority

The integer −227 corresponds to the situation. The elevation is −227 ft.

EXAMPLE 4 *Stock Price Change.* Tell which integers correspond to this situation: Hal owns a stock whose price decreased $16 per share over a recent period. He owns another stock whose price increased $2 per share over the same period.

The integer −16 corresponds to the decrease in the value of the first stock. The integer 2 represents the increase in the value of the second stock.

Do Exercises 1–5.

Tell which integers correspond to each situation.

1. **Temperature High and Low.** The highest recorded temperature in Nevada is 125°F on June 29, 1994, in Laughlin. The lowest recorded temperature in Nevada is 50°F below zero on January 8, 1937, in San Jacinto.
Source: National Climatic Data Center, Asheville, NC, and Storm Phillips, STORMFAX, INC.

2. **Stock Decrease.** The price of a stock decreased $3 per share over a recent period.

3. At 10 sec before liftoff, ignition occurs. At 148 sec after liftoff, the first stage is detached from the rocket.

4. The halfback gained 8 yd on first down. The quarterback was sacked for a 5-yd loss on second down.

5. A submarine dove 120 ft, rose 50 ft, and then dove 80 ft.

(b) The Rational Numbers

We created the set of integers by obtaining a negative number for each natural number and also including 0. To create a larger number system, called the set of **rational numbers**, we consider quotients of integers with nonzero divisors. The following are some examples of rational numbers:

$$\frac{2}{3}, \quad -\frac{2}{3}, \quad \frac{7}{1}, \quad 4, \quad -3, \quad 0, \quad \frac{23}{-8}, \quad 2.4, \quad -0.17, \quad 10\frac{1}{2}.$$

The number $-\frac{2}{3}$ (read "negative two-thirds") can also be named $\frac{-2}{3}$ or $\frac{2}{-3}$; that is,

$$-\frac{a}{b} = \frac{-a}{b} = \frac{a}{-b}.$$

The number 2.4 can be named $\frac{24}{10}$ or $\frac{12}{5}$, and −0.17 can be named $-\frac{17}{100}$. We can describe the set of rational numbers as follows.

RATIONAL NUMBERS

The set of **rational numbers** = the set of numbers $\frac{a}{b}$, where a and b are integers and b is not equal to 0 ($b \neq 0$).

Answers

1. 125; −50   2. The integer −3 corresponds to the decrease in the stock's value.
3. −10; 148   4. 8; −5   5. −120; 50; −80

Note that this new set of numbers, the rational numbers, contains the whole numbers, the integers, the arithmetic numbers (also called the non-negative rational numbers), and the negative rational numbers.

We picture the rational numbers on the number line as follows.

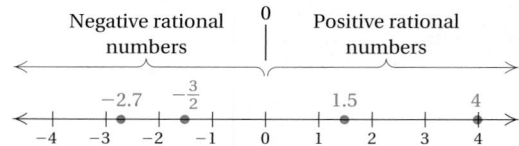

To **graph** a number means to find and mark its point on the number line. Some rational numbers are graphed in the preceding figure.

EXAMPLE 5 Graph: $\frac{5}{2}$.

The number $\frac{5}{2}$ can also be named $2\frac{1}{2}$, or 2.5. Its graph is halfway between 2 and 3.

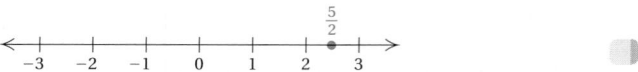

EXAMPLE 6 Graph: -3.2.

The graph of -3.2 is $\frac{2}{10}$ of the way from -3 to -4.

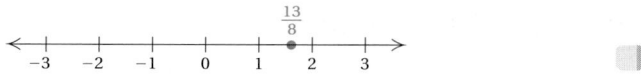

EXAMPLE 7 Graph: $\frac{13}{8}$.

The number $\frac{13}{8}$ can also be named $1\frac{5}{8}$, or 1.625. The graph is $\frac{5}{8}$ of the way from 1 to 2.

Do Exercises 6–8.

(c) Notation for Rational Numbers

Each rational number can be named using fraction notation or decimal notation.

EXAMPLE 8 Convert to decimal notation: $-\frac{5}{8}$.

We first find decimal notation for $\frac{5}{8}$. Since $\frac{5}{8}$ means $5 \div 8$, we divide.

$$
\begin{array}{r}
0.6\,2\,5 \\
8\,\overline{)\,5.0\,0\,0} \\
\underline{4\,8} \\
2\,0 \\
\underline{1\,6} \\
4\,0 \\
\underline{4\,0} \\
0
\end{array}
$$

Thus, $\frac{5}{8} = 0.625$, so $-\frac{5}{8} = -0.625$.

Graph on the number line.

6. $-\dfrac{7}{2}$

-6 -5 -4 -3 -2 -1 0 1 2 3 4 5 6

7. 1.4

-6 -5 -4 -3 -2 -1 0 1 2 3 4 5 6

8. $-\dfrac{11}{4}$

-6 -5 -4 -3 -2 -1 0 1 2 3 4 5 6

Answers

6. $-\dfrac{7}{2}$

-6 -5 -4 -3 -2 -1 0 1 2 3 4 5 6

7. 1.4

-6 -5 -4 -3 -2 -1 0 1 2 3 4 5 6

8. $-\dfrac{11}{4}$

-6 -5 -4 -3 -2 -1 0 1 2 3 4 5 6

Decimal notation for $-\frac{5}{8}$ is -0.625. We consider -0.625 to be a **terminating decimal**. Decimal notation for some numbers repeats.

EXAMPLE 9 Convert to decimal notation: $\frac{7}{11}$.

$$
\begin{array}{r}
0.6\ 3\ 6\ 3\ldots \quad \text{Dividing} \\
1\ 1\ )\overline{7.0\ 0\ 0\ 0} \\
\underline{6\ 6} \\
4\ 0 \\
\underline{3\ 3} \\
7\ 0 \\
\underline{6\ 6} \\
4\ 0 \\
\underline{3\ 3} \\
7
\end{array}
$$

We can abbreviate **repeating decimal** notation by writing a bar over the repeating part—in this case, we write $0.\overline{63}$. Thus, $\frac{7}{11} = 0.\overline{63}$.

> Each rational number can be expressed in either terminating or repeating decimal notation.

The following are other examples showing how rational numbers can be named using fraction notation or decimal notation:

$$
0 = \frac{0}{8}, \qquad \frac{27}{100} = 0.27, \qquad -8\frac{3}{4} = -8.75, \qquad -\frac{13}{6} = -2.1\overline{6}.
$$

> Do Exercises 9–11.

d The Real Numbers and Order

Every rational number has a point on the number line. However, there are some points on the line for which there is no rational number. These points correspond to what are called **irrational numbers**.

What kinds of numbers are irrational? One example is the number π, which is used in finding the area and the circumference of a circle: $A = \pi r^2$ and $C = 2\pi r$.

Another example of an irrational number is the square root of 2, named $\sqrt{2}$. It is the length of the diagonal of a square with sides of length 1. It is also the number that when multiplied by itself gives 2—that is, $\sqrt{2} \cdot \sqrt{2} = 2$. There is no rational number that can be multiplied by itself to get 2. But the following are rational *approximations*:

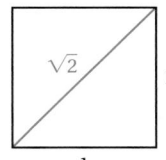

1.4 is an approximation of $\sqrt{2}$ because $(1.4)^2 = 1.96$;

1.41 is a better approximation because $(1.41)^2 = 1.9881$;

1.4142 is an even better approximation because $(1.4142)^2 = 1.99996164$.

We can find rational approximations for square roots using a calculator.

Find decimal notation.

9. $-\dfrac{3}{8}$

10. $-\dfrac{6}{11}$

11. $\dfrac{4}{3}$

Calculator Corner

Approximating Square Roots and π Square roots are found by pressing **2ND** $\sqrt{\ }$. ($\sqrt{\ }$ is the second operation associated with the **x²** key.)

To find an approximation for $\sqrt{48}$, we press **2ND** $\sqrt{\ }$ **4 8** **ENTER**. The approximation 6.92820323 is displayed.

To find $8 \cdot \sqrt{13}$, we press **8** **2ND** $\sqrt{\ }$ **1 3** **ENTER**. The approximation 28.8444102 is displayed.

The number π is used widely enough to have its own key. (π is the second operation associated with the $\frown$ key.)

To approximate π, we press **2ND** π **ENTER**. The approximation 3.141592654 is displayed.

Exercises: Approximate.

1. $\sqrt{76}$
2. $\sqrt{317}$
3. $15 \cdot \sqrt{20}$
4. $29 + \sqrt{42}$
5. π
6. $29 \cdot \pi$
7. $\pi \cdot 13^2$
8. $5 \cdot \pi + 8 \cdot \sqrt{237}$

Answers

9. -0.375 10. $-0.\overline{54}$ 11. $1.\overline{3}$

> Decimal notation for rational numbers *either* terminates *or* repeats.
>
> Decimal notation for irrational numbers *neither* terminates *nor* repeats.

Some other examples of irrational numbers are $\sqrt{3}$, $-\sqrt{8}$, $\sqrt{11}$, and $0.121221222122221\ldots$. Whenever we take the square root of a number that is not a perfect square, we will get an irrational number.

The rational numbers and the irrational numbers together correspond to all the points on the number line and make up what is called the **real-number system**.

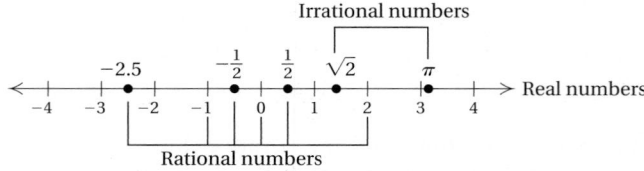

REAL NUMBERS

> The set of **real numbers** = The set of all numbers corresponding to points on the number line.

The real numbers consist of the rational numbers and the irrational numbers. The following figure shows the relationships among various kinds of numbers.

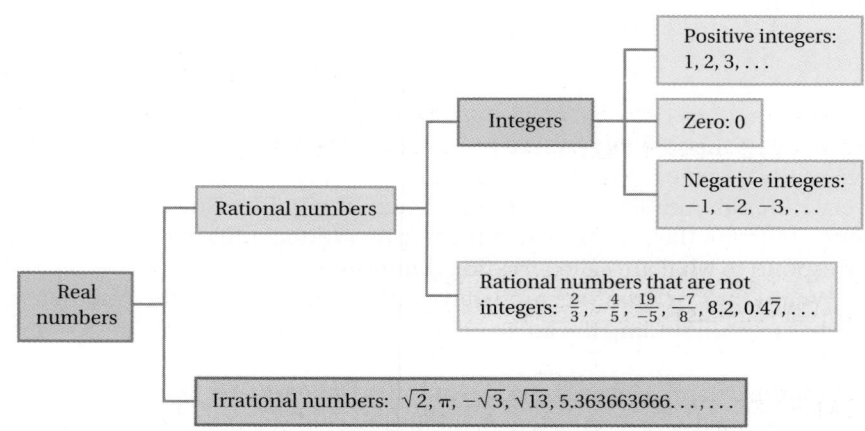

Order

Real numbers are named in order on the number line, increasing as we move from left to right. For any two numbers on the line, the one on the left is less than the one on the right.

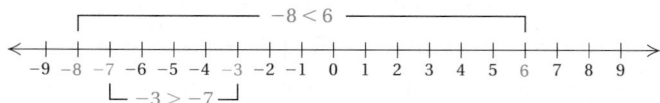

We use the symbol **<** to mean "**is less than.**" The sentence $-8 < 6$ means "-8 is less than 6." The symbol **>** means "**is greater than.**" The sentence $-3 > -7$ means "-3 is greater than -7." The sentences $-8 < 6$ and $-3 > -7$ are **inequalities**.

Calculator Corner

Negative Numbers on a Calculator; Converting to Decimal Notation We use the opposite key (⁻) to enter negative numbers on a graphing calculator. Note that this is different from the (−) key, which is used for the operation of subtraction. To convert $-\frac{5}{8}$ to decimal notation, as in Example 8, we press (⁻) (5) (÷) (8) **ENTER**. The result is -0.625.

Exercises: Convert each of the following negative numbers to decimal notation.

1. $-\dfrac{3}{4}$ 2. $-\dfrac{9}{20}$

3. $-\dfrac{1}{8}$ 4. $-\dfrac{9}{5}$

5. $-\dfrac{27}{40}$ 6. $-\dfrac{11}{16}$

7. $-\dfrac{7}{2}$ 8. $-\dfrac{19}{25}$

EXAMPLES Use either $<$ or $>$ for $\square$ to write a true sentence.

10. $2 \square 9$ Since 2 is to the left of 9, 2 is less than 9, so $2 < 9$.

11. $-7 \square 3$ Since -7 is to the left of 3, we have $-7 < 3$.

12. $6 \square -12$ Since 6 is to the right of -12, then $6 > -12$.

13. $-18 \square -5$ Since -18 is to the left of -5, we have $-18 < -5$.

14. $-2.7 \square -\frac{3}{2}$ The answer is $-2.7 < -\frac{3}{2}$.

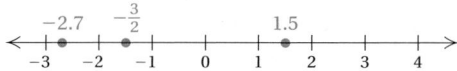

15. $1.5 \square -2.7$ The answer is $1.5 > -2.7$.

16. $1.38 \square 1.83$ The answer is $1.38 < 1.83$.

17. $-3.45 \square 1.32$ The answer is $-3.45 < 1.32$.

18. $-4 \square 0$ The answer is $-4 < 0$.

19. $5.8 \square 0$ The answer is $5.8 > 0$.

20. $\frac{5}{8} \square \frac{7}{11}$ We convert to decimal notation: $\frac{5}{8} = 0.625$ and $\frac{7}{11} = 0.6363\ldots$. Thus, $\frac{5}{8} < \frac{7}{11}$.

21. $-\frac{1}{2} \square -\frac{1}{3}$ The answer is $-\frac{1}{2} < -\frac{1}{3}$.

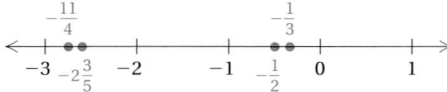

22. $-2\frac{3}{5} \square -\frac{11}{4}$ The answer is $-2\frac{3}{5} > -\frac{11}{4}$.

> Do Exercises 12–19.

Note that both $-8 < 6$ and $6 > -8$ are true. Every true inequality yields another true inequality when we interchange the numbers or variables and reverse the direction of the inequality sign.

ORDER; $>$, $<$

$a < b$ also has the meaning $b > a$.

EXAMPLES Write another inequality with the same meaning.

23. $-3 > -8$ The inequality $-8 < -3$ has the same meaning.

24. $a < -5$ The inequality $-5 > a$ has the same meaning.

A helpful mental device is to think of an inequality sign as an "arrow" with the arrowhead pointing to the smaller number.

> Do Exercises 20 and 21.

Note that all positive real numbers are greater than zero and all negative real numbers are less than zero.

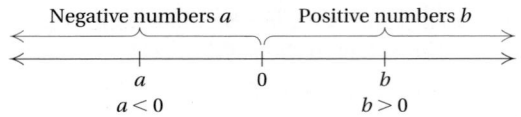

Negative numbers a Positive numbers b

$a < 0$ $b > 0$

> If b is a positive real number, then $b > 0$.
>
> If a is a negative real number, then $a < 0$.

Expressions like $a \le b$ and $b \ge a$ are also inequalities. We read $\boldsymbol{a \le b}$ as "**a is less than or equal to b.**" We read $\boldsymbol{a \ge b}$ as "**a is greater than or equal to b.**"

EXAMPLES   Write true or false for each statement.

25. $-3 \le 5.4$ True since $-3 < 5.4$ is true

26. $-3 \le -3$ True since $-3 = -3$ is true

27. $-5 \ge 1\frac{2}{3}$ False since neither $-5 > 1\frac{2}{3}$ nor $-5 = 1\frac{2}{3}$ is true

Do Exercises 22–24.

(e) Absolute Value

From the number line, we see that numbers like 4 and -4 are the same distance from zero. Distance is always a nonnegative number. We call the distance of a number from zero on the number line the **absolute value** of the number.

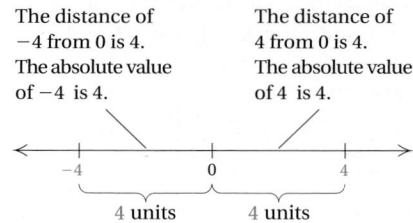

The distance of -4 from 0 is 4. The absolute value of -4 is 4.

The distance of 4 from 0 is 4. The absolute value of 4 is 4.

-4 0   4

4 units   4 units

ABSOLUTE VALUE

The **absolute value** of a number is its distance from zero on the number line. We use the symbol $|x|$ to represent the absolute value of a number x.

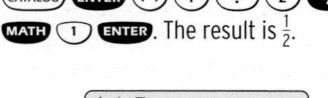

Calculator Corner

Absolute Value   The absolute-value operation is the first item in the Catalog on the T1-84 Plus graphing calculator. To find $|-7|$, as in Example 28 on the following page, we first press **2ND** **CATALOG** **ENTER** to copy "abs(" to the home screen. (CATALOG is the second operation associated with the 0 numeric key.) Then we press **(-)** **7** **)** **ENTER**. The result is 7. To find $\left|-\frac{1}{2}\right|$ and express the result as a fraction, we press **2ND** **CATALOG** **ENTER** **(-)** **1** **÷** **2** **)** **MATH** **1** **ENTER**. The result is $\frac{1}{2}$.

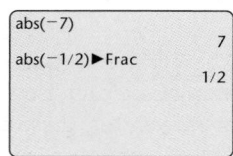

```
abs(-7)
                    7
abs(-1/2)▶Frac
                  1/2
```

Exercises: Find the absolute value.

1. $|-5|$ **2.** $|17|$

3. $|0|$ **4.** $|6.48|$

5. $|-12.7|$ **6.** $|-0.9|$

7. $\left|-\dfrac{5}{7}\right|$ **8.** $\left|\dfrac{4}{3}\right|$

Answers

22. False   **23.** True   **24.** True

FINDING ABSOLUTE VALUE

a) If a number is negative, its absolute value is its opposite.

b) If a number is positive or zero, its absolute value is the same as the number.

EXAMPLES Find the absolute value.

28. $|-7|$ The distance of -7 from 0 is 7, so $|-7| = 7$.

29. $|12|$ The distance of 12 from 0 is 12, so $|12| = 12$.

30. $|0|$ The distance of 0 from 0 is 0, so $|0| = 0$.

31. $\left|\frac{3}{2}\right| = \frac{3}{2}$

32. $|-2.73| = 2.73$

Do Exercises 25–28.

Find the absolute value.

25. $|8|$ **26.** $|-9|$

27. $\left|-\dfrac{2}{3}\right|$ **28.** $|5.6|$

STUDY TIPS

USING THIS TEXTBOOK

You will find many Study Tips throughout the book. An index of all Study Tips can be found on the Bittinger Student Organizer at the front of the book. One of the most important ways to improve your math study skills is to learn the proper use of the textbook. Here we highlight a few points that we consider most helpful.

- **Be sure to note the special symbols ⓐ, ⓑ, ⓒ, and so on, that correspond to the objectives you are to be able to master.** The first time you see them is in the margin at the beginning of each section; the second time is in the subheadings of each section; and the third time is in the exercise set for the section. You will also find them referred to in the skill maintenance exercises in each exercise set, in the mid-chapter review, and in the review exercises at the end of the chapter, as well as in the answers to the chapter tests and the cumulative reviews. These objective symbols allow you to refer to the appropriate place in the text whenever you need to review a topic.

- **Read and study each step of each example.** The examples include important side comments that explain each step. These carefully chosen examples and notes prepare you for success in the exercise set.

- **Stop and do the margin exercises as you study a section.** Doing the margin exercises is one of the most effective ways to enhance your ability to learn mathematics from this text. Don't deprive yourself of its benefits!

- **Note the icons listed at the top of each exercise set.** These refer to the many distinctive multimedia study aids that accompany the book.

- **Odd-numbered exercises.** Usually an instructor assigns some odd-numbered exercises. When you complete these, you can check your answers at the back of the book. If you miss any, check your work in the *Student's Solutions Manual* or ask your instructor for guidance.

- **Even-numbered exercises.** Whether or not your instructor assigns the even-numbered exercises, always do some on your own. Remember, there are no answers given for the class tests, so you need to practice doing exercises without answers. Check your answers later with a friend or your instructor.

Answers

25. 8 **26.** 9 **27.** $\dfrac{2}{3}$ **28.** 5.6

a State the integers that correspond to the situation.

1. *Death Valley.* With an elevation of 282 ft below sea level, Badwater Basin in California's Death Valley has the lowest elevation in the United States.
Source: Desert USA

2. *Pollution Fine.* The Massey Energy Company, the nation's fourth largest coal producer, was fined $20 million for water pollution in 2008.
Source: Environmental Protection Agency

3. On Wednesday, the temperature was 24° above zero. On Thursday, it was 2° below zero.

4. A student deposited her tax refund of $750 in a savings account. Two weeks later, she withdrew $125 to pay technology fees.

5. *Temperature Extremes.* The highest temperature ever created in a lab is 3,600,000,000°F. The lowest temperature ever created is approximately 460°F below zero.
Sources: *Live Science; Guinness Book of World Records*

6. *Extreme Climate.* Verkhoyansk, a river port in northeast Siberia, has the most extreme climate on the planet. Its average monthly winter temperature is 58.5°F below zero, and its average monthly summer temperature is 56.5°F.
Source: *Guinness Book of World Records*

7. In bowling, the Alley Cats are 34 pins behind the Strikers going into the last frame. Describe the situation of each team.

8. During a video game, Maggie intercepted a missile worth 20 points, lost a starship worth 150 points, and captured a landing base worth 300 points.

b Graph the number on the number line.

9. $\frac{10}{3}$

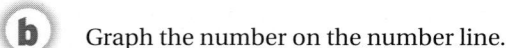

10. $-\frac{17}{4}$

11. -5.2

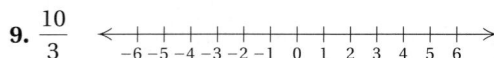

12. 4.78

13. $-4\frac{2}{5}$

14. $2\frac{6}{11}$

c Convert to decimal notation.

15. $-\dfrac{7}{8}$

16. $-\dfrac{3}{16}$

17. $\dfrac{5}{6}$

18. $\dfrac{5}{3}$

19. $-\dfrac{7}{6}$

20. $-\dfrac{5}{12}$

21. $\dfrac{2}{3}$

22. $-\dfrac{11}{9}$

23. $\dfrac{1}{10}$

24. $\dfrac{1}{4}$

25. $-\dfrac{1}{2}$

26. $\dfrac{9}{8}$

27. $\dfrac{4}{25}$

28. $-\dfrac{7}{20}$

d Use either $<$ or $>$ for $\square$ to write a true sentence.

29. $8 \ \square \ 0$

30. $3 \ \square \ 0$

31. $-8 \ \square \ 3$

32. $6 \ \square \ -6$

33. $-8 \ \square \ 8$

34. $0 \ \square \ -9$

35. $-8 \ \square \ -5$

36. $-4 \ \square \ -3$

37. $-5 \ \square \ -11$

38. $-3 \ \square \ -4$

39. $-6 \ \square \ -5$

40. $-10 \ \square \ -14$

41. $2.14 \ \square \ 1.24$

42. $-3.3 \ \square \ -2.2$

43. $-14.5 \ \square \ 0.011$

44. $17.2 \ \square \ -1.67$

45. $-12.88 \ \square \ -6.45$

46. $-14.34 \ \square \ -17.88$

47. $-\dfrac{1}{2} \ \square \ -\dfrac{2}{3}$

48. $-\dfrac{5}{4} \ \square \ -\dfrac{3}{4}$

49. $-\dfrac{2}{3} \ \square \ \dfrac{1}{3}$

50. $\dfrac{3}{4} \ \square \ -\dfrac{5}{4}$

51. $\dfrac{5}{12} \ \square \ \dfrac{11}{25}$

52. $-\dfrac{13}{16} \ \square \ -\dfrac{5}{9}$

Write an inequality with the same meaning.

53. $-6 > x$ **54.** $x < 8$ **55.** $-10 \le y$ **56.** $12 \ge t$

Write true or false.

57. $-5 \le -6$ **58.** $-7 \ge -10$ **59.** $4 \ge 4$ **60.** $7 \le 7$

61. $-3 \ge -11$ **62.** $-1 \le -5$ **63.** $0 \ge 8$ **64.** $-5 \le 7$

 Find the absolute value.

65. $|-3|$ **66.** $|-6|$ **67.** $|10|$ **68.** $|11|$ **69.** $|0|$

70. $|-2.7|$ **71.** $|-30.4|$ **72.** $|325|$ **73.** $\left|-\dfrac{2}{3}\right|$ **74.** $\left|-\dfrac{10}{7}\right|$

75. $\left|\dfrac{0}{4}\right|$ **76.** $|14.8|$ **77.** $|-2.65|$ **78.** $\left|-3\dfrac{5}{8}\right|$ **79.** $\left|-7\dfrac{4}{5}\right|$

Skill Maintenance

This heading indicates that the exercises that follow are Skill Maintenance exercises, which review any skill previously studied in the text. You can expect such exercises in every exercise set. Answers to *all* skill maintenance exercises are found at the back of the book. If you miss an exercise, restudy the objective shown in red.

Evaluate. [1.1a]

80. $15y$, for $y = 7$ **81.** $\dfrac{36}{p}$, for $p = 4$

82. $w - s$, for $w = 23$ and $s = 10$ **83.** $5a - 2b$, for $a = 9$ and $b = 3$

84. $\dfrac{5c}{d}$, for $c = 15$ and $d = 25$ **85.** $\dfrac{2x + y}{3}$, for $x = 12$ and $y = 9$

86. $\dfrac{q - r}{8}$, for $q = 30$ and $r = 6$ **87.** $\dfrac{w}{4y}$, for $w = 52$ and $y = 13$

Synthesis

List in order from the least to the greatest.

88. $-\dfrac{2}{3}, \dfrac{1}{2}, -\dfrac{3}{4}, -\dfrac{5}{6}, \dfrac{3}{8}, \dfrac{1}{6}$ **89.** $\dfrac{2}{3}, -\dfrac{1}{7}, \dfrac{1}{3}, -\dfrac{2}{7}, -\dfrac{2}{3}, \dfrac{2}{5}, -\dfrac{1}{3}, \dfrac{2}{5}, \dfrac{9}{8}$

90. $-5.16, -4.24, -8.76, 5.23, 1.85, -2.13$ **91.** $-8\dfrac{7}{8}, 7^1, -5, |-6|, 4, |3|, -8\dfrac{5}{8}, -100, 0, 1^7, \dfrac{14}{4}, -\dfrac{67}{8}$

Given that $0.\overline{3} = \frac{1}{3}$ and $0.\overline{6} = \frac{2}{3}$, express each of the following as a quotient or a ratio of two integers.

92. $0.\overline{1}$ **93.** $0.\overline{9}$ **94.** $5.\overline{5}$

1.3 Addition of Real Numbers

In this section, we consider addition of real numbers. First, to gain an understanding, we add using the number line. Then we consider rules for addition.

OBJECTIVES

a Add real numbers without using the number line.

b Find the opposite, or additive inverse, of a real number.

c Solve applied problems involving addition of real numbers.

> ### ADDITION ON THE NUMBER LINE
>
> To do the addition $a + b$ on the number line, start at 0, move to a, and then move according to b.
>
> **a)** If b is positive, move from a to the right.
> **b)** If b is negative, move from a to the left.
> **c)** If b is 0, stay at a.

EXAMPLE 1 Add: $3 + (-5)$.

We start at 0 and move to 3. Then we move 5 units left since -5 is negative.

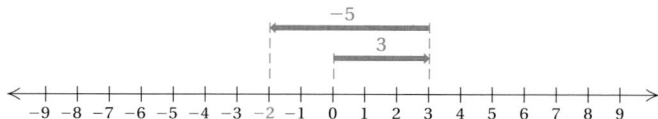

$3 + (-5) = -2$

EXAMPLE 2 Add: $-4 + (-3)$.

We start at 0 and move to -4. Then we move 3 units left since -3 is negative.

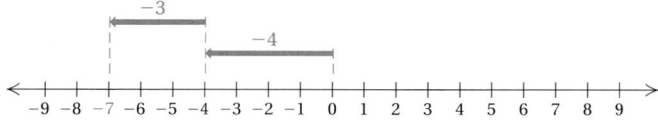

$-4 + (-3) = -7$

EXAMPLE 3 Add: $-4 + 9$.

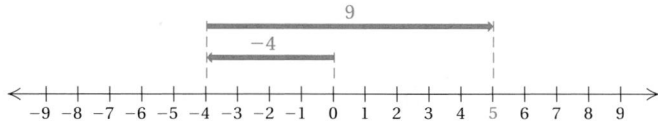

$-4 + 9 = 5$

SKILL TO REVIEW
Objective 1.1a: Evaluate algebraic expressions by substitution.

1. Evaluate $t - h$ when $t = 1$ and $h = 0.05$.

2. Evaluate $44 - 9q$ when $q = 3$.

STUDY TIPS

SMALL STEPS LEAD TO GREAT SUCCESS

What is your long-term goal for getting an education? How does math help you to attain that goal? As you begin this course, approach each short-term task, such as going to class, asking questions, using your time wisely, and doing your homework, as part of the framework of your long-term goal.

Answers

Skill to Review:
1. 0.95   **2.** 17

Add using the number line.

1. $0 + (-3)$

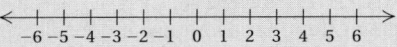

2. $1 + (-4)$

3. $-3 + (-2)$

4. $-3 + 7$

5. $-2.4 + 2.4$

6. $-\dfrac{5}{2} + \dfrac{1}{2}$

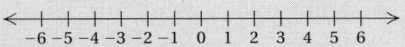

EXAMPLE 4 Add: $-5.2 + 0$.

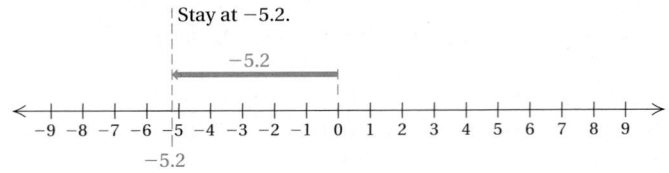

$$-5.2 + 0 = -5.2$$

Do Exercises 1–6.

ⓐ Adding Without the Number Line

You may have noticed some patterns in the preceding examples. These lead us to rules for adding without using the number line that are more efficient for adding larger numbers.

> **RULES FOR ADDITION OF REAL NUMBERS**
>
> 1. *Positive numbers*: Add the same as arithmetic numbers. The answer is positive.
> 2. *Negative numbers*: Add absolute values. The answer is negative.
> 3. *A positive number and a negative number*:
>   • If the numbers have the same absolute value, the answer is 0.
>   • If the numbers have different absolute values, subtract the smaller absolute value from the larger. Then:
>   a) If the positive number has the greater absolute value, the answer is positive.
>   b) If the negative number has the greater absolute value, the answer is negative.
> 4. *One number is zero*: The sum is the other number.

Rule 4 is known as the **identity property of 0.** It says that for any real number a, $a + 0 = a$.

EXAMPLES Add without using the number line.

5. $-12 + (-7) = -19$ Two negatives. Add the absolute values: $|-12| + |-7| = 12 + 7 = 19$. Make the answer *negative*: -19.

6. $-1.4 + 8.5 = 7.1$ One negative, one positive. Find the absolute values: $|-1.4| = 1.4$; $|8.5| = 8.5$. Subtract the smaller absolute value from the larger: $8.5 - 1.4 = 7.1$. The *positive* number, 8.5, has the larger absolute value, so the answer is *positive*: 7.1.

7. $-36 + 21 = -15$ One negative, one positive. Find the absolute values: $|-36| = 36$; $|21| = 21$. Subtract the smaller absolute value from the larger: $36 - 21 = 15$. The *negative* number, -36, has the larger absolute value, so the answer is *negative*: -15.

Answers

1. -3 2. -3 3. -5
4. 4 5. 0 6. -2

8. $1.5 + (-1.5) = 0$ The numbers have the same absolute value. The sum is 0.

9. $-\dfrac{7}{8} + 0 = -\dfrac{7}{8}$ One number is zero. The sum is $-\dfrac{7}{8}$.

10. $-9.2 + 3.1 = -6.1$

11. $-\dfrac{3}{2} + \dfrac{9}{2} = \dfrac{6}{2} = 3$

12. $-\dfrac{2}{3} + \dfrac{5}{8} = -\dfrac{16}{24} + \dfrac{15}{24} = -\dfrac{1}{24}$

> Do Exercises 7-20.

Suppose we want to add several numbers, some positive and some negative, as follows. How can we proceed?

$$15 + (-2) + 7 + 14 + (-5) + (-12)$$

We can change grouping and order as we please when adding. For instance, we can group the positive numbers together and the negative numbers together and add them separately. Then we add the two results.

EXAMPLE 13 Add: $15 + (-2) + 7 + 14 + (-5) + (-12)$.

a) $15 + 7 + 14 = 36$ Adding the positive numbers

b) $-2 + (-5) + (-12) = -19$ Adding the negative numbers

$36 + (-19) = 17$ Adding the results in (a) and (b)

We can also add the numbers in any other order we wish, say, from left to right as follows:

$$
\begin{aligned}
15 + (-2) + 7 + 14 + (-5) + (-12) &= 13 + 7 + 14 + (-5) + (-12) \\
&= 20 + 14 + (-5) + (-12) \\
&= 34 + (-5) + (-12) \\
&= 29 + (-12) \\
&= 17
\end{aligned}
$$

> Do Exercises 21-24.

b) Opposites, or Additive Inverses

Suppose we add two numbers that are **opposites**, such as 6 and −6. The result is 0. When opposites are added, the result is always 0. Opposites are also called **additive inverses**. Every real number has an opposite, or additive inverse.

OPPOSITES, OR ADDITIVE INVERSES

Two numbers whose sum is 0 are called **opposites**, or **additive inverses**, of each other.

Add without using the number line.

7. $-5 + (-6)$ **8.** $-9 + (-3)$

9. $-4 + 6$ **10.** $-7 + 3$

11. $5 + (-7)$ **12.** $-20 + 20$

13. $-11 + (-11)$ **14.** $10 + (-7)$

15. $-0.17 + 0.7$ **16.** $-6.4 + 8.7$

17. $-4.5 + (-3.2)$

18. $-8.6 + 2.4$

19. $\dfrac{5}{9} + \left(-\dfrac{7}{9}\right)$

20. $-\dfrac{1}{5} + \left(-\dfrac{3}{4}\right)$

Add.

21. $(-15) + (-37) + 25 + 42 + (-59) + (-14)$

22. $42 + (-81) + (-28) + 24 + 18 + (-31)$

23. $-2.5 + (-10) + 6 + (-7.5)$

24. $-35 + 17 + 14 + (-27) + 31 + (-12)$

Find the opposite, or additive inverse, of each number.

25. −4   **26.** 8.7

27. −7.74   **28.** $-\dfrac{8}{9}$

29. 0   **30.** 12

EXAMPLES Find the opposite, or additive inverse, of each number.

14. 34   The opposite of 34 is −34 because $34 + (-34) = 0$.

15. −8   The opposite of −8 is 8 because $-8 + 8 = 0$.

16. 0   The opposite of 0 is 0 because $0 + 0 = 0$.

17. $-\dfrac{7}{8}$ The opposite of $-\dfrac{7}{8}$ is $\dfrac{7}{8}$ because $-\dfrac{7}{8} + \dfrac{7}{8} = 0$.

Do Exercises 25–30.

To name the opposite, we use the symbol −, as follows.

> **SYMBOLIZING OPPOSITES**
>
> The opposite, or additive inverse, of a number a can be named $-a$ (read "the opposite of a," or "the additive inverse of a").

Note that if we take a number, say, 8, and find its opposite, −8, and then find the opposite of the result, we will have the original number, 8, again.

> **THE OPPOSITE OF AN OPPOSITE**
>
> The **opposite of the opposite** of a number is the number itself. (The additive inverse of the additive inverse of a number is the number itself.) That is, for any number a,
>
> $$-(-a) = a.$$

EXAMPLE 18 Evaluate $-x$ and $-(-x)$ when $x = 16$.

If $x = 16$, then $-x = -16$.   The opposite of 16 is −16.

If $x = 16$, then $-(-x) = -(-16) = 16$.   The opposite of the opposite of 16 is 16.

EXAMPLE 19 Evaluate $-x$ and $-(-x)$ when $x = -3$.

If $x = -3$, then $-x = -(-3) = 3$.

If $x = -3$, then $-(-x) = -(-(-3)) = -(3) = -3$.

Note that in Example 19 we used a second set of parentheses to show that we are substituting the negative number −3 for x. Symbolism like $--x$ is not considered meaningful.

Do Exercises 31–36.

Evaluate $-x$ and $-(-x)$ when:

31. $x = 14$.   **32.** $x = 1$.

33. $x = -19$.   **34.** $x = -1.6$.

35. $x = \dfrac{2}{3}$.   **36.** $x = -\dfrac{9}{8}$.

A symbol such as −8 is usually read "negative 8." It could be read "the additive inverse of 8," because the additive inverse of 8 is negative 8. It could also be read "the opposite of 8," because the opposite of 8 is −8. Thus a symbol like −8 can be read in more than one way. It is never correct to read −8 as "minus 8."

Answers

25. 4   **26.** −8.7   **27.** 7.74   **28.** $\dfrac{8}{9}$

29. 0   **30.** −12   **31.** −14; 14

32. −1; 1   **33.** 19; −19   **34.** 1.6; −1.6

35. $-\dfrac{2}{3}; \dfrac{2}{3}$ **36.** $\dfrac{9}{8}; -\dfrac{9}{8}$

--------------------------------- *Caution!* ---------------------------------

A symbol like $-x$, which has a variable, should be read "the opposite of x" or "the additive inverse of x" and *not* "negative x," because we do not know whether x represents a positive number, a negative number, or 0. You can check this in Examples 18 and 19.

We can use the symbolism $-a$ to restate the definition of opposite, or additive inverse.

> ### OPPOSITES, OR ADDITIVE INVERSES
>
> For any real number a, the **opposite**, or **additive inverse**, of a, denoted $-a$, is such that
>
> $$a + (-a) = (-a) + a = 0.$$

Signs of Numbers

A negative number is sometimes said to have a "negative sign." A positive number is said to have a "positive sign." When we replace a number with its opposite, we can say that we have "changed its sign."

EXAMPLES   Find the opposite. (Change the sign.)

20. -3 $-(-3) = 3$

21. $-\dfrac{2}{13}$ $-\left(-\dfrac{2}{13}\right) = \dfrac{2}{13}$

22. 0 $-(0) = 0$

23. 14 $-(14) = -14$

Do Exercises 37–40.

Find the opposite. (Change the sign.)

37. -4 **38.** -13.4

39. 0 **40.** $\dfrac{1}{4}$

c Applications and Problem Solving

Addition of real numbers occurs in many real-world situations.

EXAMPLE 24   *Lake Level.*   In the course of one four-month period, the water level of Lake Clearwater went down 2 ft, up 1 ft, down 5 ft, and up 3 ft. By how much had the lake level changed at the end of the four months?

We let $T =$ the total change in the level of the lake. Then the problem translates to a sum:

Total change	is	1st change	plus	2nd change	plus	3rd change	plus	4th change
T	$=$	-2	$+$	1	$+$	(-5)	$+$	$3.$

Adding from left to right, we have

$$
\begin{aligned}
T = -2 + 1 + (-5) + 3 &= -1 + (-5) + 3 \\
&= -6 + 3 \\
&= -3.
\end{aligned}
$$

The lake level had dropped 3 ft at the end of the four-month period.

Do Exercise 41.

41. Change in Class Size.   During the first two weeks of the semester in Jim's algebra class, 4 students withdrew, 8 students enrolled late, and 6 students were dropped as "no shows." By how many students had the class size changed at the end of the first two weeks?

Answers

37. 4   **38.** 13.4   **39.** 0
40. $-\dfrac{1}{4}$ **41.** -2 students

For Extra Help

MyMathLab

Math XL
PRACTICE

WATCH

DOWNLOAD

READ

REVIEW

a Add. Do not use the number line except as a check.

1. $2 + (-9)$ **2.** $-5 + 2$ **3.** $-11 + 5$ **4.** $4 + (-3)$ **5.** $-6 + 6$

6. $8 + (-8)$ **7.** $-3 + (-5)$ **8.** $-4 + (-6)$ **9.** $-7 + 0$ **10.** $-13 + 0$

11. $0 + (-27)$ **12.** $0 + (-35)$ **13.** $17 + (-17)$ **14.** $-15 + 15$ **15.** $-17 + (-25)$

16. $-24 + (-17)$ **17.** $18 + (-18)$ **18.** $-13 + 13$ **19.** $-28 + 28$ **20.** $11 + (-11)$

21. $8 + (-5)$ **22.** $-7 + 8$ **23.** $-4 + (-5)$ **24.** $10 + (-12)$ **25.** $13 + (-6)$

26. $-3 + 14$ **27.** $-25 + 25$ **28.** $50 + (-50)$ **29.** $53 + (-18)$ **30.** $75 + (-45)$

31. $-8.5 + 4.7$ **32.** $-4.6 + 1.9$ **33.** $-2.8 + (-5.3)$ **34.** $-7.9 + (-6.5)$ **35.** $-\dfrac{3}{5} + \dfrac{2}{5}$

36. $-\dfrac{4}{3} + \dfrac{2}{3}$ **37.** $-\dfrac{2}{9} + \left(-\dfrac{5}{9}\right)$ **38.** $-\dfrac{4}{7} + \left(-\dfrac{6}{7}\right)$ **39.** $-\dfrac{5}{8} + \dfrac{1}{4}$ **40.** $-\dfrac{5}{6} + \dfrac{2}{3}$

41. $-\dfrac{5}{8} + \left(-\dfrac{1}{6}\right)$ **42.** $-\dfrac{5}{6} + \left(-\dfrac{2}{9}\right)$ **43.** $-\dfrac{3}{8} + \dfrac{5}{12}$ **44.** $-\dfrac{7}{16} + \dfrac{7}{8}$

45. $-\dfrac{1}{6} + \dfrac{7}{10}$ **46.** $-\dfrac{11}{18} + \left(-\dfrac{3}{4}\right)$ **47.** $\dfrac{7}{15} + \left(-\dfrac{1}{9}\right)$ **48.** $-\dfrac{4}{21} + \dfrac{3}{14}$

49. $76 + (-15) + (-18) + (-6)$

50. $29 + (-45) + 18 + 32 + (-96)$

51. $-44 + \left(-\dfrac{3}{8}\right) + 95 + \left(-\dfrac{5}{8}\right)$

52. $24 + 3.1 + (-44) + (-8.2) + 63$

53. $98 + (-54) + 113 + (-998) + 44 + (-612)$

54. $-458 + (-124) + 1025 + (-917) + 218$

 b Find the opposite, or additive inverse.

55. 24 **56.** -64 **57.** -26.9 **58.** 48.2

Evaluate $-x$ when:

59. $x = 8.$ **60.** $x = -27.$ **61.** $x = -\dfrac{13}{8}.$ **62.** $x = \dfrac{1}{236}.$

Evaluate $-(-x)$ when:

63. $x = -43.$ **64.** $x = 39.$ **65.** $x = \dfrac{4}{3}.$ **66.** $x = -7.1.$

Find the opposite. (Change the sign.)

67. -24 **68.** -12.3 **69.** $-\dfrac{3}{8}$ **70.** 10

c Solve.

71. *Tallest Mountain.* The tallest mountain in the world, when measured from base to peak, is Mauna Kea (White Mountain) in Hawaii. From its base 19,684 ft below sea level in the Hawaiian Trough, it rises 33,480 ft. What is the elevation of the peak above sea level?

Source: *The Guinness Book of Records*

72. *Telephone Bills.* Erika's cell-phone bill for July was $82. She sent a check for $50 and then made $37 worth of calls in August. How much did she then owe on her cell-phone bill?

73. *Temperature Changes.* One day the temperature in Lawrence, Kansas, is 32°F at 6:00 A.M. It rises 15° by noon, but falls 50° by midnight when a cold front moves in. What is the final temperature?

74. *Stock Changes.* On a recent day, the price of a stock opened at a value of $61.38. During the day, it rose $4.75, dropped $7.38, and rose $5.13. Find the value of the stock at the end of the day.

75. *Profits and Losses.* The profit of a business is expressed as a positive number and referred to as operating "in the black." A loss is expressed as a negative number and is referred to as operating "in the red." The profits and losses of Xponent Corporation over various years are shown in the bar graph below. Find the sum of the profits and losses.

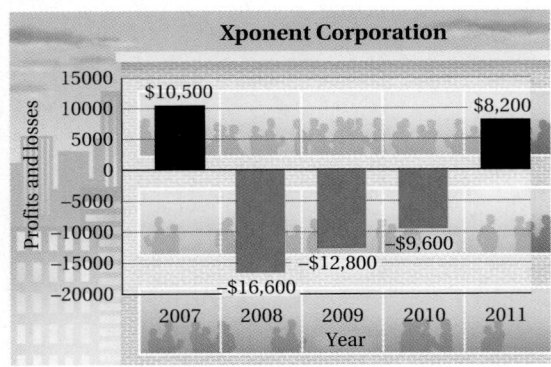

76. *Football Yardage.* In a college football game, the quarterback attempted passes with the following results. Find the total gain or loss.

TRY	GAIN OR LOSS
1st	13-yd gain
2nd	12-yd loss
3rd	21-yd gain

77. *Credit-Card Bills.* On August 1, Lyle's credit-card bill shows that he owes $470. During the month of August, Lyle sends a check for $45 to the credit-card company, charges another $160 in merchandise, and then pays off another $500 of his bill. What is the new amount that Lyle owes at the end of August?

78. *Account Balance.* Leah has $460 in a checking account. She writes a check for $530, makes a deposit of $75, and then writes a check for $90. What is the balance in her account?

Skill Maintenance

Convert to decimal notation. [1.2c]

79. $-\dfrac{5}{8}$

80. $\dfrac{1}{3}$

81. $-\dfrac{1}{12}$

82. $\dfrac{13}{20}$

Find the absolute value. [1.2e]

83. $|2.3|$

84. $|0|$

85. $\left|-\dfrac{4}{5}\right|$

86. $|-21.4|$

Synthesis

87. For what numbers x is $-x$ negative?

88. For what numbers x is $-x$ positive?

89. If a is positive and b is negative, then $-a + b$ is:
  A. Positive.
  C. 0.
  B. Negative.
  D. Cannot be determined without more information

90. If $a = b$ and a and b are negative, then $-a + (-b)$ is:
  A. Positive.
  C. 0.
  B. Negative.
  D. Cannot be determined without more information

1.4 Subtraction of Real Numbers

a Subtraction

We now consider subtraction of real numbers.

> **SUBTRACTION**
>
> The difference $a - b$ is the number c for which $a = b + c$.

Consider, for example, $45 - 17$. *Think*: What number can we add to 17 to get 45? Since $45 = 17 + 28$, we know that $45 - 17 = 28$. Let's consider an example whose answer is a negative number.

EXAMPLE 1 Subtract: $3 - 7$.

Think: What number can we add to 7 to get 3? The number must be negative. Since $7 + (-4) = 3$, we know the number is -4: $3 - 7 = -4$. That is, $3 - 7 = -4$ because $7 + (-4) = 3$.

> Do Exercises 1–3.

The definition above does not provide the most efficient way to do subtraction. We can develop a faster way to subtract. As a rationale for the faster way, let's compare $3 + 7$ and $3 - 7$ on the number line.

To find $3 + 7$ on the number line, we start at 0, move to 3, and then move 7 units farther to the right since 7 is positive.

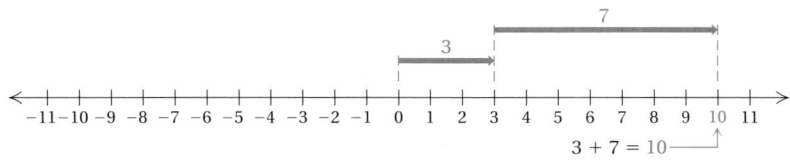

$$3 + 7 = 10$$

To find $3 - 7$, we do the "opposite" of adding 7: We move 7 units to the *left* to do the subtracting. This is the same as *adding* the opposite of 7, -7, to 3.

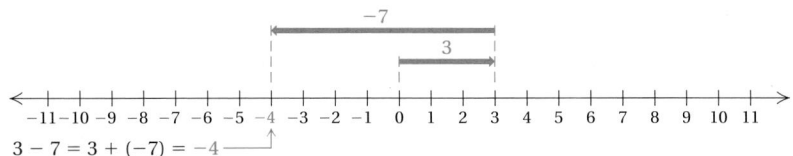

$$3 - 7 = 3 + (-7) = -4$$

> Do Exercises 4–6.

Look for a pattern in the examples shown at right.

SUBTRACTING	ADDING AN OPPOSITE
$5 - 8 = -3$	$5 + (-8) = -3$
$-6 - 4 = -10$	$-6 + (-4) = -10$
$-7 - (-2) = -5$	$-7 + 2 = -5$

OBJECTIVES

a Subtract real numbers and simplify combinations of additions and subtractions.

b Solve applied problems involving subtraction of real numbers.

Subtract.

1. $-6 - 4$

   Think: What number can be added to 4 to get -6:
   $$\square + 4 = -6?$$

2. $-7 - (-10)$

   Think: What number can be added to -10 to get -7:
   $$\square + (-10) = -7?$$

3. $-7 - (-2)$

   Think: What number can be added to -2 to get -7:
   $$\square + (-2) = -7?$$

Subtract. Use the number line, doing the "opposite" of addition.

4. $5 - 9$

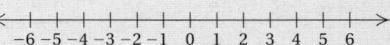

5. $-3 - 2$

6. $-4 - (-3)$

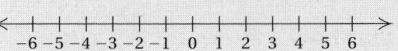

Answers

1. -10 2. 3 3. -5 4. -4
5. -5 6. -1

Do Exercises 7-10.

Complete the addition and compare with the subtraction.

7. $4 - 6 = -2$;
$4 + (-6) =$ _____

8. $-3 - 8 = -11$;
$-3 + (-8) =$ _____

9. $-5 - (-9) = 4$;
$-5 + 9 =$ _____

10. $-5 - (-3) = -2$;
$-5 + 3 =$ _____

Perhaps you have noticed that we can subtract by adding the opposite of the number being subtracted. This can always be done.

SUBTRACTING BY ADDING THE OPPOSITE

For any real numbers a and b,

$$a - b = a + (-b).$$

(To subtract, add the opposite, or additive inverse, of the number being subtracted.)

This is the method generally used for quick subtraction of real numbers.

EXAMPLES Subtract.

2. $2 - 6 = 2 + (-6) = -4$

The opposite of 6 is -6. We change the subtraction to addition and add the opposite. *Check*: $-4 + 6 = 2$.

3. $4 - (-9) = 4 + 9 = 13$

The opposite of -9 is 9. We change the subtraction to addition and add the opposite. *Check*: $13 + (-9) = 4$.

4. $-4.2 - (-3.6) = -4.2 + 3.6 = -0.6$

Adding the opposite. *Check*: $-0.6 + (-3.6) = -4.2$.

5. $-\dfrac{1}{2} - \left(-\dfrac{3}{4}\right) = -\dfrac{1}{2} + \dfrac{3}{4}$

$= -\dfrac{2}{4} + \dfrac{3}{4} = \dfrac{1}{4}$

Adding the opposite. *Check*: $\dfrac{1}{4} + \left(-\dfrac{3}{4}\right) = -\dfrac{1}{2}$.

Subtract.

11. $2 - 8$

12. $-6 - 10$

13. $12.4 - 5.3$

14. $-8 - (-11)$

15. $-8 - (-8)$

16. $\dfrac{2}{3} - \left(-\dfrac{5}{6}\right)$

Do Exercises 11-16.

EXAMPLES Subtract by adding the opposite of the number being subtracted.

6. $3 - 5$ *Think*: "Three minus five is three plus the opposite of five"
$3 - 5 = 3 + (-5) = -2$

7. $\dfrac{1}{8} - \dfrac{7}{8}$ *Think*: "One-eighth minus seven-eighths is one-eighth plus the opposite of seven-eighths"

$\dfrac{1}{8} - \dfrac{7}{8} = \dfrac{1}{8} + \left(-\dfrac{7}{8}\right) = -\dfrac{6}{8}$, or $-\dfrac{3}{4}$

8. $-4.6 - (-9.8)$ *Think*: "Negative four point six minus negative nine point eight is negative four point six plus the opposite of negative nine point eight"

$-4.6 - (-9.8) = -4.6 + 9.8 = 5.2$

9. $-\dfrac{3}{4} - \dfrac{7}{5}$ *Think*: "Negative three-fourths minus seven-fifths is negative three-fourths plus the opposite of seven-fifths"

$-\dfrac{3}{4} - \dfrac{7}{5} = -\dfrac{3}{4} + \left(-\dfrac{7}{5}\right) = -\dfrac{15}{20} + \left(-\dfrac{28}{20}\right) = -\dfrac{43}{20}$

Do Exercises 17-21.

Subtract by adding the opposite of the number being subtracted.

17. $3 - 11$

18. $12 - 5$

19. $-12 - (-9)$

20. $-12.4 - 10.9$

21. $-\dfrac{4}{5} - \left(-\dfrac{4}{5}\right)$

Answers

7. -2 8. -11 9. 4 10. -2 11. -6
12. -16 13. 7.1 14. 3 15. 0 16. $\dfrac{3}{2}$
17. -8 18. 7 19. -3 20. -23.3
21. 0

When several additions and subtractions occur together, we can make them all additions.

EXAMPLES Simplify.

10. $8 - (-4) - 2 - (-4) + 2 = 8 + 4 + (-2) + 4 + 2$ Adding the opposite
$$= 16$$

11. $8.2 - (-6.1) + 2.3 - (-4) = 8.2 + 6.1 + 2.3 + 4 = 20.6$

12. $\dfrac{3}{4} - \left(-\dfrac{1}{12}\right) - \dfrac{5}{6} - \dfrac{2}{3} = \dfrac{9}{12} + \dfrac{1}{12} + \left(-\dfrac{10}{12}\right) + \left(-\dfrac{8}{12}\right)$

$$= \dfrac{9 + 1 + (-10) + (-8)}{12}$$

$$= \dfrac{-8}{12} = -\dfrac{8}{12} = -\dfrac{2}{3}$$

Do Exercises 22–24.

Simplify.

22. $-6 - (-2) - (-4) - 12 + 3$

23. $\dfrac{2}{3} - \dfrac{4}{5} - \left(-\dfrac{11}{15}\right) + \dfrac{7}{10} - \dfrac{5}{2}$

24. $-9.6 + 7.4 - (-3.9) - (-11)$

b   Applications and Problem Solving

Let's now see how we can use subtraction of real numbers to solve applied problems.

EXAMPLE 13   *Surface Temperatures on Mars.*   Surface temperatures on Mars vary from $-128°C$ during polar night to $27°C$ at the equator during mid-day at the closest point in orbit to the sun. Find the difference between the highest value and the lowest value in this temperature range.

Source: Mars Institute

We let $D =$ the difference in the temperatures. Then the problem translates to the following subtraction:

$$\underbrace{\text{Difference in temperature}}_{D} \;\; \underbrace{\text{is}}_{=} \;\; \underbrace{\text{Highest temperature}}_{27} \;\; \underbrace{\text{minus}}_{-} \;\; \underbrace{\text{Lowest temperature}}_{(-128)}$$

$$D = 27 + 128 = 155.$$

The difference in the temperatures is $155°C$.

Do Exercise 25.

25. Temperature Extremes.
The highest temperature ever recorded in the United States is 134°F in Greenland Ranch, California, on July 10, 1913. The lowest temperature ever recorded is −80°F in Prospect Creek, Alaska, on January 23, 1971. How much higher was the temperature in Greenland Ranch than the temperature in Prospect Creek?

Source: National Oceanographic and Atmospheric Administration

Answers

22. -9 **23.** $-\dfrac{6}{5}$ **24.** 12.7   **25.** 214°F

1.4 | Exercise Set

For Extra Help

MyMathLab

Math XL
PRACTICE

WATCH

DOWNLOAD

READ

REVIEW

a Subtract.

1. $2 - 9$

2. $3 - 8$

3. $-8 - (-2)$

4. $-6 - (-8)$

5. $-11 - (-11)$

6. $-6 - (-6)$

7. $12 - 16$

8. $14 - 19$

9. $20 - 27$

10. $30 - 4$

11. $-9 - (-3)$

12. $-7 - (-9)$

13. $-40 - (-40)$

14. $-9 - (-9)$

15. $7 - (-7)$

16. $4 - (-4)$

17. $8 - (-3)$

18. $-7 - 4$

19. $-6 - 8$

20. $6 - (-10)$

21. $-4 - (-9)$

22. $-14 - 2$

23. $-6 - (-5)$

24. $-4 - (-3)$

25. $8 - (-10)$

26. $5 - (-6)$

27. $-5 - (-2)$

28. $-3 - (-1)$

29. $-7 - 14$

30. $-9 - 16$

31. $0 - (-5)$

32. $0 - (-1)$

33. $-8 - 0$

34. $-9 - 0$

35. $7 - (-5)$

36. $7 - (-4)$

37. $2 - 25$

38. $18 - 63$

39. $-42 - 26$

40. $-18 - 63$

41. $-71 - 2$

42. $-49 - 3$

43. $24 - (-92)$

44. $48 - (-73)$

45. $-50 - (-50)$

46. $-70 - (-70)$

47. $-\dfrac{3}{8} - \dfrac{5}{8}$

48. $\dfrac{3}{9} - \dfrac{9}{9}$

49. $\dfrac{3}{4} - \dfrac{2}{3}$

50. $\dfrac{5}{8} - \dfrac{3}{4}$

51. $-\dfrac{3}{4} - \dfrac{2}{3}$

52. $-\dfrac{5}{8} - \dfrac{3}{4}$

53. $-\dfrac{5}{8} - \left(-\dfrac{3}{4}\right)$

54. $-\dfrac{3}{4} - \left(-\dfrac{2}{3}\right)$

55. $6.1 - (-13.8)$

56. $1.5 - (-3.5)$

57. $-2.7 - 5.9$

58. $-3.2 - 5.8$

59. $0.99 - 1$

60. $0.87 - 1$

61. $-79 - 114$

62. $-197 - 216$

63. $0 - (-500)$

64. $500 - (-1000)$

65. $-2.8 - 0$

66. $6.04 - 1.1$

67. $7 - 10.53$

68. $8 - (-9.3)$

69. $\dfrac{1}{6} - \dfrac{2}{3}$

70. $-\dfrac{3}{8} - \left(-\dfrac{1}{2}\right)$

71. $-\dfrac{4}{7} - \left(-\dfrac{10}{7}\right)$

72. $\dfrac{12}{5} - \dfrac{12}{5}$

73. $-\dfrac{7}{10} - \dfrac{10}{15}$

74. $-\dfrac{4}{18} - \left(-\dfrac{2}{9}\right)$

75. $\dfrac{1}{5} - \dfrac{1}{3}$

76. $-\dfrac{1}{7} - \left(-\dfrac{1}{6}\right)$

77. $\dfrac{5}{12} - \dfrac{7}{16}$

78. $-\dfrac{1}{35} - \left(-\dfrac{9}{40}\right)$

79. $-\dfrac{2}{15} - \dfrac{7}{12}$

80. $\dfrac{2}{21} - \dfrac{9}{14}$

Simplify.

81. $18 - (-15) - 3 - (-5) + 2$ **82.** $22 - (-18) + 7 + (-42) - 27$ **83.** $-31 + (-28) - (-14) - 17$

84. $-43 - (-19) - (-21) + 25$ **85.** $-34 - 28 + (-33) - 44$ **86.** $39 + (-88) - 29 - (-83)$

87. $-93 - (-84) - 41 - (-56)$ **88.** $84 + (-99) + 44 - (-18) - 43$

89. $-5.4 - (-30.9) + 30.8 + 40.2 - (-12)$ **90.** $14.9 - (-50.7) + 20 - (-32.8)$

91. $-\dfrac{7}{12} + \dfrac{3}{4} - \left(-\dfrac{5}{8}\right) - \dfrac{13}{24}$ **92.** $-\dfrac{11}{16} + \dfrac{5}{32} - \left(-\dfrac{1}{4}\right) + \dfrac{7}{8}$

 Solve.

93. *Ocean Depth.* The deepest point in the Pacific Ocean is the Marianas Trench, with a depth of 10,924 m. The deepest point in the Atlantic Ocean is the Puerto Rico Trench, with a depth of 8605 m. What is the difference in the elevation of the two trenches?

Source: *The World Almanac and Book of Facts*

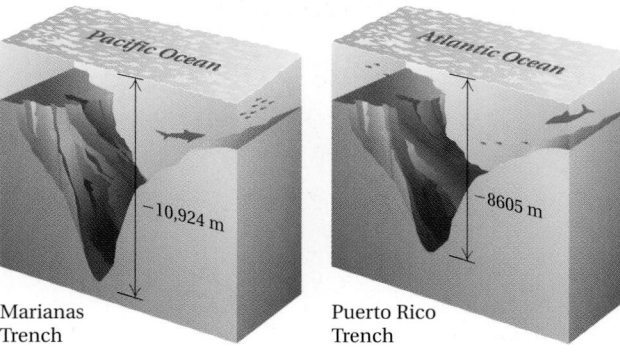

Marianas Trench

Puerto Rico Trench

94. *Elevations in Africa.* The elevation of the highest point in Africa, Mt. Kilimanjaro, Tanzania, is 19,340 ft. The lowest elevation, at Lake Assal, Djibouti, is −512 ft. What is the difference in the elevations of the two locations?

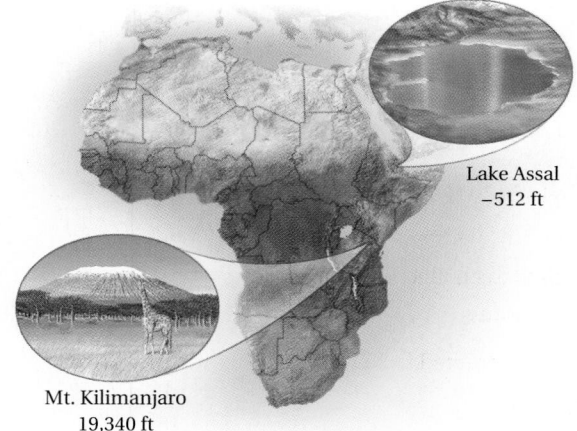

Lake Assal −512 ft

Mt. Kilimanjaro 19,340 ft

95. Claire has a charge of $476.89 on her credit card, but she then returns a sweater that cost $128.95. How much does she now owe on her credit card?

96. Chris has $720 in a checking account. He writes a check for $970 to pay for a sound system. What is the balance in his checking account?

97. *Difference in Elevation.* At its highest point, the elevation of Denver, Colorado, is 5672 ft above sea level. At its lowest point, the elevation of New Orleans, Louisiana, is 4 ft below sea level. Find the difference in the elevations.
Source: *Information Please Almanac*

98. *Difference in Elevation.* The lowest elevation in North America, Death Valley, California, is 282 ft below sea level. The highest elevation in North America, Mount McKinley, Alaska, is 20,320 ft. Find the difference in elevation between the highest point and the lowest point.
Source: National Geographic Society

99. *Low Points on Continents.* The lowest point in Africa is Lake Assal, which is 512 ft below sea level. The lowest point in South America is the Valdes Peninsula, which is 131 ft below sea level. How much lower is Lake Assal than the Valdes Peninsula?
Source: National Geographic Society

100. *Temperature Records.* The greatest recorded temperature change in one 24-hr period occurred between January 23 and January 24, 1916, in Browning, Montana, where the temperature fell from to 44°F to −56°F. By how much did the temperature drop?
Source: *The Guinness Book of Records*

101. *Surface Temperature on Mercury.* Surface temperatures on Mercury vary from 840°F on the equator when the planet is closest to the sun to −290°F at night. Find the difference between these two temperatures.

102. *Run Differential.* In baseball, the difference between the number of runs that a team scores and the number of runs that it allows its opponents to score is called the *run differential*. That is,

$$\text{Run differential} = \frac{\text{Number of}}{\text{runs scored}} - \frac{\text{Number of}}{\text{runs allowed}}.$$

Teams strive for a positive run differential.
Source: Major League Baseball

a) In a recent season, the Chicago White Sox scored 810 runs and allowed 729 runs to be scored on them. Find the run differential.

b) In a recent season, the Pittsburgh Pirates scored 735 runs and allowed 884 runs to be scored on them. Find the run differential.

Skill Maintenance

Translate to an algebraic expression.   [1.1b]

103. 7 more than y

104. 41 less than t

105. h subtracted from a

106. The product of 6 and c

107. r more than s

108. x less than y

Synthesis

Determine whether each statement is true or false for all integers a and b. If false, give an example to show why. Examples may vary.

109. $a - 0 = 0 - a$

110. $0 - a = a$

111. If $a \neq b$, then $a - b \neq 0$.

112. If $a = -b$, then $a + b = 0$.

113. If $a + b = 0$, then a and b are opposites.

114. If $a - b = 0$, then $a = -b$.

Mid-Chapter Review

Concept Reinforcement

Determine whether each statement is true or false.

_____ **1.** All rational numbers can be named using fraction notation. [1.2c]

_____ **2.** If $a > b$, then a lies to the left of b on the number line. [1.2d]

_____ **3.** The absolute value of a number is always nonnegative. [1.2e]

_____ **4.** We can translate "7 less than y" as $7 - y$. [1.1b]

Guided Solutions

Fill in each blank with the number that creates a correct statement or solution.

5. Evaluate $-x$ and $-(-x)$ when $x = -4$. [1.3b]

$$-x = -(\Box) = \Box;$$
$$-(-x) = -(-(\Box)) = -(\Box) = \Box$$

Subtract. [1.4a]

6. $5 - 13 = 5 + (\Box) = \Box$

7. $-6 - 7 = -6 + (\Box) = \Box$

Mixed Review

Evaluate. [1.1a]

8. $\dfrac{3m}{n}$, when $m = 8$ and $n = 6$

9. $\dfrac{a + b}{2}$, when $a = 5$ and $b = 17$

Translate each phrase to an algebraic expression. Use any letter for the variable. [1.1b]

10. Three times some number

11. Five less than some number

12. State the integers that correspond to this situation: Jerilyn deposited $450 in her checking account. Later that week, she wrote a check for $79. [1.2a]

13. Graph -3.5 on the number line. [1.2b]

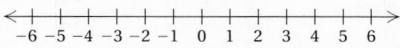

Convert to decimal notation. [1.2c]

14. $-\dfrac{4}{5}$

15. $\dfrac{7}{3}$

Use either $<$ or $>$ for $\Box$ to write a true sentence. [1.2d]

16. $-5 \ \Box \ -3$

17. $-9.9 \ \Box \ -10.1$

Write true or false. [1.2d]

18. $-8 \geq -5$

19. $-4 \leq -4$

Write an inequality with the same meaning. [1.2d]

20. $y < 5$

21. $-3 \geq t$

Find the absolute value. [1.2e]

22. $|15.6|$

23. $|-18|$

24. $|0|$

25. $\left| -\dfrac{12}{5} \right|$

Find the opposite, or additive inverse, of the number. [1.3b]

26. -5.6

27. $\dfrac{7}{4}$

28. 0

29. -49

30. Evaluate $-x$ when x is -19. [1.3b]

31. Evaluate $-(-x)$ when x is 2.3. [1.3b]

Compute and simplify. [1.3a], [1.4a]

32. $7 + (-9)$

33. $-\dfrac{3}{8} + \dfrac{1}{4}$

34. $3.6 + (-3.6)$

35. $-8 + (-9)$

36. $\dfrac{2}{3} + \left(-\dfrac{9}{8} \right)$

37. $-4.2 + (-3.9)$

38. $-14 + 5$

39. $19 + (-21)$

40. $-4.1 - 6.3$

41. $5 - (-11)$

42. $-\dfrac{1}{4} - \left(-\dfrac{3}{5} \right)$

43. $12 - 24$

44. $-8 - (-4)$

45. $-\dfrac{1}{2} - \dfrac{5}{6}$

46. $12.3 - 14.1$

47. $6 - (-7)$

48. $16 - (-9) - 20 - (-4)$

49. $-4 + (-10) - (-3) - 12$

50. $17 - (-25) + 15 - (-18)$

51. $-9 + (-3) + 16 - (-10)$

Solve. [1.3c], [1.4b]

52. *Temperature Change.* In chemistry lab, Ben works with a substance whose initial temperature is 25°C. During an experiment, the temperature falls to -8°C. Find the difference between the two temperatures.

53. *Stock Price Change.* The price of a stock opened at $56.12. During the day, it dropped $1.18, then rose $1.22, and then dropped $1.36. Find the value of the stock at the end of the day.

Understanding Through Discussion and Writing

54. Give three examples of rational numbers that are not integers. Explain. [1.2b]

55. Give three examples of irrational numbers. Explain the difference between an irrational number and a rational number. [1.2b, d]

56. Explain in your own words why the sum of two negative numbers is always negative. [1.3a]

57. If a negative number is subtracted from a positive number, will the result always be positive? Why or why not? [1.4a]

1.5

Multiplication of Real Numbers

OBJECTIVES

a Multiply real numbers.

b Solve applied problems involving multiplication of real numbers.

a Multiplication

Multiplication of real numbers is very much like multiplication of arithmetic numbers. The only difference is that we must determine whether the answer is positive or negative.

Multiplication of a Positive Number and a Negative Number

To see how to multiply a positive number and a negative number, consider the pattern of the following.

This number decreases by 1 each time.　　This number decreases by 5 each time.

$$4 \cdot 5 = 20$$
$$3 \cdot 5 = 15$$
$$2 \cdot 5 = 10$$
$$1 \cdot 5 = 5$$
$$0 \cdot 5 = 0$$
$$-1 \cdot 5 = -5$$
$$-2 \cdot 5 = -10$$
$$-3 \cdot 5 = -15$$

1. Complete, as in the example.

$$4 \cdot 10 = 40$$
$$3 \cdot 10 = 30$$
$$2 \cdot 10 =$$
$$1 \cdot 10 =$$
$$0 \cdot 10 =$$
$$-1 \cdot 10 =$$
$$-2 \cdot 10 =$$
$$-3 \cdot 10 =$$

Do Exercise 1.

According to this pattern, it looks as though the product of a negative number and a positive number is negative. That is the case, and we have the first part of the rule for multiplying real numbers.

> **THE PRODUCT OF A POSITIVE NUMBER AND A NEGATIVE NUMBER**
>
> To multiply a positive number and a negative number, multiply their absolute values. The answer is negative.

Multiply.

2. $-3 \cdot 6$

3. $20 \cdot (-5)$

4. $4 \cdot (-20)$

5. $-\dfrac{2}{3} \cdot \dfrac{5}{6}$

6. $-4.23(7.1)$

7. $\dfrac{7}{8}\left(-\dfrac{4}{5}\right)$

EXAMPLES Multiply.

1. $8(-5) = -40$

2. $-\dfrac{1}{3} \cdot \dfrac{5}{7} = -\dfrac{5}{21}$

3. $(-7.2)5 = -36$

Do Exercises 2–7.

Answers

1. $20; 10; 0; -10; -20; -30$　**2.** -18

3. -100　**4.** -80　**5.** $-\dfrac{5}{9}$

6. -30.033　**7.** $-\dfrac{7}{10}$

Multiplication of Two Negative Numbers

How do we multiply two negative numbers? Again, we look for a pattern.

This number decreases by 1 each time.

$$4 \cdot (-5) = -20$$
$$3 \cdot (-5) = -15$$
$$2 \cdot (-5) = -10$$
$$1 \cdot (-5) = -5$$
$$0 \cdot (-5) = 0$$
$$-1 \cdot (-5) = 5$$
$$-2 \cdot (-5) = 10$$
$$-3 \cdot (-5) = 15$$

This number increases by 5 each time.

Do Exercise 8.

According to the pattern, it appears that the product of two negative numbers is positive. That is actually so, and we have the second part of the rule for multiplying real numbers.

THE PRODUCT OF TWO NEGATIVE NUMBERS

To multiply two negative numbers, multiply their absolute values. The answer is positive.

Do Exercises 9–14.

The following is another way to consider the rules we have for multiplication.

To multiply two nonzero real numbers:

a) Multiply the absolute values.

b) If the signs are the same, the answer is positive.

c) If the signs are different, the answer is negative.

Multiplication by Zero

The only case that we have not considered is multiplying by zero. As with nonnegative numbers, the product of any real number and 0 is 0.

THE MULTIPLICATION PROPERTY OF ZERO

For any real number a,

$$a \cdot 0 = 0 \cdot a = 0.$$

(The product of 0 and any real number is 0.)

EXAMPLES Multiply.

4. $(-3)(-4) = 12$

5. $-1.6(2) = -3.2$

6. $-19 \cdot 0 = 0$

7. $\left(-\frac{5}{6}\right)\left(-\frac{1}{9}\right) = \frac{5}{54}$

8. $0 \cdot (-452) = 0$

9. $23 \cdot 0 \cdot \left(-8\frac{2}{3}\right) = 0$

Do Exercises 15–20.

8. Complete, as in the example.

$$3 \cdot (-10) = -30$$
$$2 \cdot (-10) = -20$$
$$1 \cdot (-10) =$$
$$0 \cdot (-10) =$$
$$-1 \cdot (-10) =$$
$$-2 \cdot (-10) =$$
$$-3 \cdot (-10) =$$

Multiply.

9. $-9 \cdot (-3)$

10. $-16 \cdot (-2)$

11. $-7 \cdot (-5)$

12. $-\frac{4}{7}\left(-\frac{5}{9}\right)$

13. $-\frac{3}{2}\left(-\frac{4}{9}\right)$

14. $-3.25(-4.14)$

Multiply.

15. $5(-6)$

16. $(-5)(-6)$

17. $(-3.2) \cdot 10$

18. $\left(-\frac{4}{5}\right)\left(\frac{10}{3}\right)$

19. $0 \cdot (-34.2)$

20. $-\frac{5}{7} \cdot 0 \cdot \left(-4\frac{2}{3}\right)$

Answers

8. $-10; 0; 10; 20; 30$ **9.** 27 **10.** 32

11. 35 **12.** $\frac{20}{63}$ **13.** $\frac{2}{3}$ **14.** 13.455

15. -30 **16.** 30 **17.** -32 **18.** $-\frac{8}{3}$

19. 0 **20.** 0

Multiplying More Than Two Numbers

When multiplying more than two real numbers, we can choose order and grouping as we please.

EXAMPLES Multiply.

10. $-8 \cdot 2(-3) = -16(-3)$ Multiplying the first two numbers
$$= 48$$

11. $-8 \cdot 2(-3) = 24 \cdot 2$ Multiplying the negatives. Every pair of negative numbers gives a positive product.
$$= 48$$

12. $-3(-2)(-5)(4) = 6(-5)(4)$ Multiplying the first two numbers
$$= (-30)4$$
$$= -120$$

13. $\left(-\dfrac{1}{2}\right)(8)\left(-\dfrac{2}{3}\right)(-6) = (-4)4$ Multiplying the first two numbers and the last two numbers
$$= -16$$

14. $-5 \cdot (-2) \cdot (-3) \cdot (-6) = 10 \cdot 18 = 180$

15. $(-3)(-5)(-2)(-3)(-6) = (-30)(18) = -540$

Considering that the product of a pair of negative numbers is positive, we see the following pattern.

> The product of an even number of negative numbers is positive.
> The product of an odd number of negative numbers is negative.

Do Exercises 21–26.

EXAMPLE 16 Evaluate $2x^2$ when $x = 3$ and when $x = -3$.
$$2x^2 = 2(3)^2 = 2(9) = 18;$$
$$2x^2 = 2(-3)^2 = 2(9) = 18$$

Let's compare the expressions $(-x)^2$ and $-x^2$.

EXAMPLE 17 Evaluate $(-x)^2$ and $-x^2$ when $x = 5$.

$(-x)^2 = (-5)^2 = (-5)(-5) = 25;$ Substitute 5 for x. Then evaluate the power.

$-x^2 = -(5)^2 = -(25) = -25$ Substitute 5 for x. Evaluate the power. Then find the opposite.

In Example 17, we see that the expressions $(-x)^2$ and $-x^2$ are *not* equivalent. That is, they do not have the same value for every allowable replacement of the variable by a real number. To find $(-x)^2$, we take the opposite and then square. To find $-x^2$, we find the square and then take the opposite.

Multiply.

21. $5 \cdot (-3) \cdot 2$

22. $-3 \times (-4.1) \times (-2.5)$

23. $-\dfrac{1}{2} \cdot \left(-\dfrac{4}{3}\right) \cdot \left(-\dfrac{5}{2}\right)$

24. $-2 \cdot (-5) \cdot (-4) \cdot (-3)$

25. $(-4)(-5)(-2)(-3)(-1)$

26. $(-1)(-1)(-2)(-3)(-1)(-1)$

Answers

21. -30 **22.** -30.75 **23.** $-\dfrac{5}{3}$
24. 120 **25.** -120 **26.** 6

EXAMPLE 18 Evaluate $(-a)^2$ and $-a^2$ when $a = -4$.

To make sense of the substitutions and computations, we introduce extra grouping symbols into the expressions.

$$(-a)^2 = [-(-4)]^2 = [4]^2 = 16;$$
$$-a^2 = -(-4)^2 = -(16) = -16$$

Do Exercises 27–29.

Do Exercises 27–29.

27. Evaluate $3x^2$ when $x = 4$ and when $x = -4$.

28. Evaluate $(-x)^2$ and $-x^2$ when $x = 2$.

29. Evaluate $(-x)^2$ and $-x^2$ when $x = -3$.

b Applications and Problem Solving

We now consider multiplication of real numbers in real-world applications.

EXAMPLE 19 *Chemical Reaction.* During a chemical reaction, the temperature in a beaker decreased by 2°C every minute until 10:23 A.M. If the temperature was 17°C at 10:00 A.M., when the reaction began, what was the temperature at 10:23 A.M.?

This is a multistep problem. We first find the total number of degrees that the temperature dropped, using −2° for each minute. Since it dropped 2° for each of the 23 minutes, we know that the total drop d is given by

$$d = 23 \cdot (-2) = -46.$$

To determine the temperature after this time period, we find the sum of 17 and −46, or

$$T = 17 + (-46) = -29.$$

Thus the temperature at 10:23 A.M. was −29°C.

Do Exercise 30.

30. Chemical Reaction. During a chemical reaction, the temperature in a beaker increased by 3°C every minute until 1:34 P.M. If the temperature was −17°C at 1:10 P.M., when the reaction began, what was the temperature at 1:34 P.M.?

STUDY TIPS

MAKING POSITIVE CHOICES

Making the right choices can give you the power to succeed in learning mathematics.

You can choose to improve your attitude and raise the academic goals that you have set for yourself. Projecting a positive attitude toward your study of mathematics and expecting a positive outcome can make it easier for you to learn and to perform well in this course.

Here are some positive choices you can make:

- Choose to make a strong commitment to learning.

- Choose to allocate the proper amount of time to learn.

- Choose to place the primary responsibility for learning on yourself.

Well-known American psychologist William James once said, "The one thing that will guarantee the successful conclusion of a doubtful undertaking is faith in the beginning that you can do it."

Answers
27. 48; 48   **28.** 4; −4
29. 9; −9   **30.** 55°C

1.5 Exercise Set

For Extra Help

MyMathLab

Math XL
PRACTICE

WATCH

DOWNLOAD

READ

REVIEW

a Multiply.

1. $-4 \cdot 2$

2. $-3 \cdot 5$

3. $-8 \cdot 6$

4. $-5 \cdot 2$

5. $8 \cdot (-3)$

6. $9 \cdot (-5)$

7. $-9 \cdot 8$

8. $-10 \cdot 3$

9. $-8 \cdot (-2)$

10. $-2 \cdot (-5)$

11. $-7 \cdot (-6)$

12. $-9 \cdot (-2)$

13. $15 \cdot (-8)$

14. $-12 \cdot (-10)$

15. $-14 \cdot 17$

16. $-13 \cdot (-15)$

17. $-25 \cdot (-48)$

18. $39 \cdot (-43)$

19. $-3.5 \cdot (-28)$

20. $97 \cdot (-2.1)$

21. $9 \cdot (-8)$

22. $7 \cdot (-9)$

23. $4 \cdot (-3.1)$

24. $3 \cdot (-2.2)$

25. $-5 \cdot (-6)$

26. $-6 \cdot (-4)$

27. $-7 \cdot (-3.1)$

28. $-4 \cdot (-3.2)$

29. $\frac{2}{3} \cdot \left(-\frac{3}{5}\right)$

30. $\frac{5}{7} \cdot \left(-\frac{2}{3}\right)$

31. $-\frac{3}{8} \cdot \left(-\frac{2}{9}\right)$

32. $-\frac{5}{8} \cdot \left(-\frac{2}{5}\right)$

33. -6.3×2.7

34. -4.1×9.5

35. $-\frac{5}{9} \cdot \frac{3}{4}$

36. $-\frac{8}{3} \cdot \frac{9}{4}$

37. $7 \cdot (-4) \cdot (-3) \cdot 5$

38. $9 \cdot (-2) \cdot (-6) \cdot 7$

39. $-\frac{2}{3} \cdot \frac{1}{2} \cdot \left(-\frac{6}{7}\right)$

40. $-\frac{1}{8} \cdot \left(-\frac{1}{4}\right) \cdot \left(-\frac{3}{5}\right)$

41. $-3 \cdot (-4) \cdot (-5)$

42. $-2 \cdot (-5) \cdot (-7)$

43. $-2 \cdot (-5) \cdot (-3) \cdot (-5)$

44. $-3 \cdot (-5) \cdot (-2) \cdot (-1)$

45. $\frac{1}{5}\left(-\frac{2}{9}\right)$

46. $-\frac{3}{5}\left(-\frac{2}{7}\right)$

47. $-7 \cdot (-21) \cdot 13$

48. $-14 \cdot (34) \cdot 12$

49. $-4 \cdot (-1.8) \cdot 7$

50. $-8 \cdot (-1.3) \cdot (-5)$

51. $-\frac{1}{9}\left(-\frac{2}{3}\right)\left(\frac{5}{7}\right)$

52. $-\frac{7}{2}\left(-\frac{5}{7}\right)\left(-\frac{2}{5}\right)$

53. $4 \cdot (-4) \cdot (-5) \cdot (-12)$

54. $-2 \cdot (-3) \cdot (-4) \cdot (-5)$

55. $0.07 \cdot (-7) \cdot 6 \cdot (-6)$

56. $80 \cdot (-0.8) \cdot (-90) \cdot (-0.09)$

57. $\left(-\frac{5}{6}\right)\left(\frac{1}{8}\right)\left(-\frac{3}{7}\right)\left(-\frac{1}{7}\right)$

58. $\left(\frac{4}{5}\right)\left(-\frac{2}{3}\right)\left(-\frac{15}{7}\right)\left(\frac{1}{2}\right)$

59. $(-14) \cdot (-27) \cdot 0$

60. $7 \cdot (-6) \cdot 5 \cdot (-4) \cdot 3 \cdot (-2) \cdot 1 \cdot 0$

61. $(-8)(-9)(-10)$

62. $(-7)(-8)(-9)(-10)$

63. $(-6)(-7)(-8)(-9)(-10)$

64. $(-5)(-6)(-7)(-8)(-9)(-10)$

65. $(-1)^{12}$

66. $(-1)^{9}$

67. Evaluate $(-x)^2$ and $-x^2$ when $x = 4$ and when $x = -4$.

68. Evaluate $(-x)^2$ and $-x^2$ when $x = 10$ and when $x = -10$.

69. Evaluate $(-3x)^2$ and $-3x^2$ when $x = 7$.

70. Evaluate $(-2x)^2$ and $-2x^2$ when $x = 3$.

71. Evaluate $5x^2$ when $x = 2$ and when $x = -2$.

72. Evaluate $2x^2$ when $x = 5$ and when $x = -5$.

73. Evaluate $-2x^3$ when $x = 1$ and when $x = -1$.

74. Evaluate $-3x^3$ when $x = 2$ and when $x = -2$.

 Solve.

75. *Weight Loss.* Dave lost 2 lb each week for a period of 10 weeks. Express his total weight change as an integer.

76. *Stock Loss.* Emma lost $3 each day for a period of 5 days in the value of a stock she owned. Express her total loss as an integer.

77. *Chemical Reaction.* The temperature of a chemical compound was 0°C at 11:00 A.M. During a reaction, it dropped 3°C per minute until 11:18 A.M. What was the temperature at 11:18 A.M.?

78. *Chemical Reaction.* The temperature of a chemical compound was −5°C at 3:20 P.M. During a reaction, it increased 2°C per minute until 3:52 P.M. What was the temperature at 3:52 P.M.?

79. *Stock Price.* The price of a stock began the day at $23.75 per share and dropped $1.38 per hour for 8 hr. What was the price of the stock after 8 hr?

80. *Population Decrease.* The population of Bloomtown was 12,500. It decreased 380 each year for 4 yr. What was the population of the town after 4 yr?

81. *Diver's Position.* After diving 95 m below the sea level, a diver rises at a rate of 7 m/min for 9 min. Where is the diver in relation to the surface at the end of the 9-min period?

82. *Checking Account Balance.* Karen had $68 in her checking account. After she had written checks to make seven purchases at $13 each, what was the balance in her checking account?

83. *Drop in Temperature.* The temperature in Osgood was 62°F at 2:00 P.M. It dropped 6°F per hour for the next 4 hr. What was the temperature at the end of the 4-hr period?

84. *Juice Consumption.* Eliza bought a 64-oz container of cranberry juice and drank 8 oz per day for a week. How much juice was left in the container at the end of the week?

Skill Maintenance

85. Evaluate $\dfrac{x - 2y}{3}$ for $x = 20$ and $y = 7$.   [1.1a]

86. Evaluate $\dfrac{d - e}{3d}$ for $d = 5$ and $e = 1$.   [1.1a]

Subtract.   [1.4a]

87. $-\dfrac{1}{2} - \left(-\dfrac{1}{6}\right)$

88. $8 - 12.3$

89. $31 - (-13)$

90. $-\dfrac{5}{12} - \left(-\dfrac{1}{3}\right)$

Write true or false.   [1.2d]

91. $-10 > -12$

92. $0 \leq -1$

93. $4 < -8$

94. $-7 \geq -6$

Synthesis

95. If a is positive and b is negative, then $-ab$ is:

   A. Positive.
   B. Negative.
   C. 0.
   D. Cannot be determined without more information

96. If a is positive and b is negative, then $(-a)(-b)$ is:

   A. Positive.
   B. Negative.
   C. 0.
   D. Cannot be determined without more information

97. Below is a number line showing 0 and two positive numbers x and y. Use a compass or ruler to locate the following as best you can:

   $2x,\quad 3x,\quad 2y,\quad -x,\quad -y,\quad x + y,\quad x - y,\quad x - 2y.$

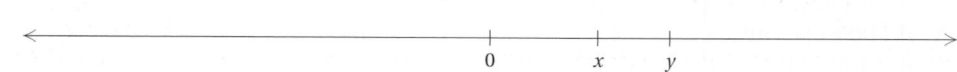

98. Of all possible quotients of the numbers 10, $-\frac{1}{2}$, -5, and $\frac{1}{5}$, which two produce the largest quotient? Which two produce the smallest quotient?

1.6 Division of Real Numbers

We now consider division of real numbers. The definition of division results in rules for division that are the same as those for multiplication.

a Division of Integers

> **DIVISION**
>
> The quotient $a \div b$, or $\dfrac{a}{b}$, where $b \neq 0$, is that unique real number c for which $a = b \cdot c$.

Let's use the definition to divide integers.

EXAMPLES Divide, if possible. Check your answer.

1. $14 \div (-7) = -2$ *Think*: What number multiplied by -7 gives 14? That number is -2. *Check*: $(-2)(-7) = 14$.

2. $\dfrac{-32}{-4} = 8$ *Think*: What number multiplied by -4 gives -32? That number is 8. *Check*: $8(-4) = -32$.

3. $\dfrac{-10}{7} = -\dfrac{10}{7}$ *Think*: What number multiplied by 7 gives -10? That number is $-\frac{10}{7}$. *Check*: $-\frac{10}{7} \cdot 7 = -10$.

4. $\dfrac{-17}{0}$ is **not defined**.   *Think*: What number multiplied by 0 gives -17? There is no such number because the product of 0 and *any* number is 0.

The rules for division are the same as those for multiplication.

> To multiply or divide two real numbers (where the divisor is nonzero):
>
> **a)** Multiply or divide the absolute values.
> **b)** If the signs are the same, the answer is positive.
> **c)** If the signs are different, the answer is negative.

Do Margin Exercises 1–6.

Excluding Division by 0

Example 4 shows why we cannot divide -17 by 0. We can use the same argument to show why we cannot divide any nonzero number b by 0. Consider $b \div 0$. We look for a number that when multiplied by 0 gives b. There is no such number because the product of 0 and any number is 0. Thus we cannot divide a nonzero number b by 0.

On the other hand, if we divide 0 by 0, we look for a number c such that $0 \cdot c = 0$. But $0 \cdot c = 0$ for any number c. Thus it appears that $0 \div 0$ could be any number we choose. Getting any answer we want when we divide 0 by 0 would be very confusing. Thus we agree that division by 0 is not defined.

SKILL TO REVIEW
Objective 1.5a: Multiply real numbers.

Multiply.

1. $-\dfrac{2}{9} \cdot \dfrac{4}{11}$ **2.** $\dfrac{13}{2} \cdot \left(-\dfrac{2}{25}\right)$

Divide.

1. $6 \div (-3)$

Think: What number multiplied by -3 gives 6?

2. $\dfrac{-15}{-3}$

Think: What number multiplied by -3 gives -15?

3. $-24 \div 8$

Think: What number multiplied by 8 gives -24?

4. $\dfrac{-48}{-6}$ **5.** $\dfrac{30}{-5}$

6. $\dfrac{30}{-7}$

Answers

Skill to Review:
1. $-\dfrac{8}{99}$ 2. $-\dfrac{13}{25}$

Margin Exercises:
1. -2 2. 5 3. -3 4. 8
5. -6 6. $-\dfrac{30}{7}$

EXCLUDING DIVISION BY 0

Division by 0 is not defined.

$$a \div 0, \text{ or } \frac{a}{0}, \text{ is not defined for all real numbers } a.$$

Dividing 0 by Other Numbers

Note that

$$0 \div 8 = 0 \text{ because } 0 = 0 \cdot 8; \qquad \frac{0}{-5} = 0 \text{ because } 0 = 0 \cdot (-5).$$

DIVIDENDS OF 0

Zero divided by any nonzero real number is 0:

$$\frac{0}{a} = 0; \quad a \neq 0.$$

EXAMPLES Divide.

5. $0 \div (-6) = 0$ **6.** $\dfrac{0}{12} = 0$ **7.** $\dfrac{-3}{0}$ is not defined.

Do Exercises 7 and 8.

Do Exercises 7 and 8.

Divide, if possible.

7. $\dfrac{-5}{0}$ **8.** $\dfrac{0}{-3}$

b Reciprocals

When two numbers like $\frac{1}{2}$ and 2 are multiplied, the result is 1. Such numbers are called **reciprocals** of each other. Every nonzero real number has a reciprocal, also called a **multiplicative inverse**.

RECIPROCALS

Two numbers whose product is 1 are called **reciprocals**, or **multiplicative inverses**, of each other.

EXAMPLES Find the reciprocal.

8. $\dfrac{7}{8}$ The reciprocal of $\dfrac{7}{8}$ is $\dfrac{8}{7}$ because $\dfrac{7}{8} \cdot \dfrac{8}{7} = 1$.

9. -5 The reciprocal of -5 is $-\dfrac{1}{5}$ because $-5\left(-\dfrac{1}{5}\right) = 1$.

10. 3.9 The reciprocal of 3.9 is $\dfrac{1}{3.9}$ because $3.9\left(\dfrac{1}{3.9}\right) = 1$.

11. $-\dfrac{1}{2}$ The reciprocal of $-\dfrac{1}{2}$ is -2 because $\left(-\dfrac{1}{2}\right)(-2) = 1$.

12. $-\dfrac{2}{3}$ The reciprocal of $-\dfrac{2}{3}$ is $-\dfrac{3}{2}$ because $\left(-\dfrac{2}{3}\right)\left(-\dfrac{3}{2}\right) = 1$.

13. $\dfrac{3y}{8x}$ The reciprocal of $\dfrac{3y}{8x}$ is $\dfrac{8x}{3y}$ because $\left(\dfrac{3y}{8x}\right)\left(\dfrac{8x}{3y}\right) = 1$.

Answers

7. Not defined   **8.** 0

RECIPROCAL PROPERTIES

For $a \neq 0$, the reciprocal of a can be named $\dfrac{1}{a}$ and the reciprocal of $\dfrac{1}{a}$ is a.

The reciprocal of a nonzero number $\dfrac{a}{b}$ can be named $\dfrac{b}{a}$.

The number 0 has no reciprocal.

Do Exercises 9–14.

The reciprocal of a positive number is also a positive number, because the product of the two numbers must be the positive number 1. The reciprocal of a negative number is also a negative number, because the product of the two numbers must be the positive number 1.

THE SIGN OF A RECIPROCAL

The reciprocal of a number has the same sign as the number itself.

Caution!

It is important *not* to confuse *opposite* with *reciprocal*. Keep in mind that the opposite, or additive inverse, of a number is what we add to the number to get 0. The reciprocal, or multiplicative inverse, is what we multiply the number by to get 1.

Compare the following.

NUMBER	OPPOSITE (Change the sign.)	RECIPROCAL (Invert but do not change the sign.)
$-\dfrac{3}{8}$	$\dfrac{3}{8}$	$-\dfrac{8}{3}$
19	-19	$\dfrac{1}{19}$
$\dfrac{18}{7}$	$-\dfrac{18}{7}$	$\dfrac{7}{18}$
-7.9	7.9	$-\dfrac{1}{7.9}$, or $-\dfrac{10}{79}$
0	0	Not defined

$$\left(-\frac{3}{8}\right)\left(-\frac{8}{3}\right) = 1$$

$$-\frac{3}{8} + \frac{3}{8} = 0$$

Do Exercise 15.

Find the reciprocal.

9. $\dfrac{2}{3}$

10. $-\dfrac{5}{4}$

11. -3

12. $-\dfrac{1}{5}$

13. 1.3

14. $\dfrac{a}{6b}$

---

15. Complete the following table.

NUMBER	OPPOSITE	RECIPROCAL
$\dfrac{2}{3}$		
$-\dfrac{5}{4}$		
0		
1		
-8		
-4.7		

---

c Division of Real Numbers

We know that we can subtract by adding an opposite. Similarly, we can divide by multiplying by a reciprocal.

> **RECIPROCALS AND DIVISION**
>
> For any real numbers a and b, $b \neq 0$,
>
> $$a \div b = \frac{a}{b} = a \cdot \frac{1}{b}.$$
>
> (To divide, multiply by the reciprocal of the divisor.)

Rewrite each division as a multiplication.

16. $\dfrac{4}{7} \div \left(-\dfrac{3}{5}\right)$

17. $\dfrac{5}{-8}$

18. $\dfrac{a - b}{7}$

19. $\dfrac{-23}{1/a}$

20. $-5 \div 7$

EXAMPLES Rewrite each division as a multiplication.

14. $-4 \div 3$ $\qquad$ $-4 \div 3$ is the same as $-4 \cdot \dfrac{1}{3}$

15. $\dfrac{6}{-7}$ $\qquad$ $\dfrac{6}{-7} = 6\left(-\dfrac{1}{7}\right)$

16. $\dfrac{3}{5} \div \left(-\dfrac{9}{7}\right)$ $\qquad$ $\dfrac{3}{5} \div \left(-\dfrac{9}{7}\right) = \dfrac{3}{5}\left(-\dfrac{7}{9}\right)$

17. $\dfrac{x + 2}{5}$ $\qquad$ $\dfrac{x + 2}{5} = (x + 2)\dfrac{1}{5}$ $\qquad$ Parentheses are necessary here.

18. $\dfrac{-17}{1/b}$ $\qquad$ $\dfrac{-17}{1/b} = -17 \cdot b$

Do Exercises 16–20.

When actually doing division calculations, we sometimes multiply by a reciprocal and we sometimes divide directly. With fraction notation, it is usually better to multiply by a reciprocal. With decimal notation, it is usually better to divide directly.

EXAMPLES Divide by multiplying by the reciprocal of the divisor.

19. $\dfrac{2}{3} \div \left(-\dfrac{5}{4}\right) = \dfrac{2}{3} \cdot \left(-\dfrac{4}{5}\right) = -\dfrac{8}{15}$

20. $-\dfrac{5}{6} \div \left(-\dfrac{3}{4}\right) = -\dfrac{5}{6} \cdot \left(-\dfrac{4}{3}\right) = \dfrac{20}{18} = \dfrac{10 \cdot 2}{9 \cdot 2} = \dfrac{10}{9} \cdot \dfrac{2}{2} = \dfrac{10}{9}$

--- *Caution!* ---

Be careful *not* to change the sign when taking a reciprocal!

Divide by multiplying by the reciprocal of the divisor.

21. $\dfrac{4}{7} \div \left(-\dfrac{3}{5}\right)$

22. $-\dfrac{12}{7} \div \left(-\dfrac{3}{4}\right)$

21. $-\dfrac{3}{4} \div \dfrac{3}{10} = -\dfrac{3}{4} \cdot \left(\dfrac{10}{3}\right) = -\dfrac{30}{12} = -\dfrac{5 \cdot 6}{2 \cdot 6} = -\dfrac{5}{2} \cdot \dfrac{6}{6} = -\dfrac{5}{2}$

Do Exercises 21 and 22.

Answers

16. $\dfrac{4}{7} \cdot \left(-\dfrac{5}{3}\right)$ **17.** $5 \cdot \left(-\dfrac{1}{8}\right)$

18. $(a - b) \cdot \dfrac{1}{7}$ **19.** $-23 \cdot a$

20. $-5 \cdot \left(\dfrac{1}{7}\right)$ **21.** $-\dfrac{20}{21}$ **22.** $\dfrac{16}{7}$

With decimal notation, it is easier to carry out long division than to multiply by the reciprocal.

EXAMPLES Divide.

22. $-27.9 \div (-3) = \dfrac{-27.9}{-3} = 9.3$ Do the long division $3\overline{)27.9}.$ → 9.3
The answer is positive.

23. $-6.3 \div 2.1 = -3$ Do the long division $2.1_\wedge\overline{)6.3_\wedge}.$ → 3.
The answer is negative.

Do Exercises 23 and 24.

> Divide.
>
> **23.** $21.7 \div (-3.1)$
>
> **24.** $-20.4 \div (-4)$

Consider the following:

1. $\dfrac{2}{3} = \dfrac{2}{3} \cdot 1 = \dfrac{2}{3} \cdot \dfrac{-1}{-1} = \dfrac{2(-1)}{3(-1)} = \dfrac{-2}{-3}.$ Thus, $\dfrac{2}{3} = \dfrac{-2}{-3}.$

(A negative number divided by a negative number is positive.)

2. $-\dfrac{2}{3} = -1 \cdot \dfrac{2}{3} = \dfrac{-1}{1} \cdot \dfrac{2}{3} = \dfrac{-1 \cdot 2}{1 \cdot 3} = \dfrac{-2}{3}.$ Thus, $-\dfrac{2}{3} = \dfrac{-2}{3}.$

(A negative number divided by a positive number is negative.)

3. $\dfrac{-2}{3} = \dfrac{-2}{3} \cdot 1 = \dfrac{-2}{3} \cdot \dfrac{-1}{-1} = \dfrac{-2(-1)}{3(-1)} = \dfrac{2}{-3}.$ Thus, $-\dfrac{2}{3} = \dfrac{2}{-3}.$

(A positive number divided by a negative number is negative.)

We can use the following properties to make sign changes in fraction notation.

SIGN CHANGES IN FRACTION NOTATION

For any numbers a and b, $b \neq 0$:

1. $\dfrac{-a}{-b} = \dfrac{a}{b}$

(The opposite of a number a divided by the opposite of another number b is the same as the quotient of the two numbers a and b.)

2. $\dfrac{-a}{b} = \dfrac{a}{-b} = -\dfrac{a}{b}$

(The opposite of a number a divided by another number b is the same as the number a divided by the opposite of the number b, and both are the same as the opposite of a *divided by* b.)

Do Exercises 25-27.

> Find two equal expressions for each number with negative signs in different places.
>
> **25.** $\dfrac{-5}{6}$
>
> **26.** $-\dfrac{8}{7}$
>
> **27.** $\dfrac{10}{-3}$

Answers

23. -7 **24.** 5.1 **25.** $\dfrac{5}{-6}; -\dfrac{5}{6}$ **26.** $\dfrac{8}{-7}; \dfrac{-8}{7}$

27. $\dfrac{-10}{3}; -\dfrac{10}{3}$

(d) Applications and Problem Solving

EXAMPLE 24 *Chemical Reaction.* During a chemical reaction, the temperature in a beaker decreased every minute by the same number of degrees. The temperature was 56°F at 10:10 A.M. By 10:42 A.M., the temperature had dropped to −12°F. By how many degrees did it change each minute?

We first determine by how many degrees d the temperature changed altogether. We subtract −12 from 56:

$$d = 56 - (-12) = 56 + 12 = 68.$$

The temperature changed a total of 68°. We can express this as −68° since the temperature dropped.

The amount of time t that passed was 42 − 10, or 32 min. Thus the number of degrees T that the temperature dropped each minute is given by

$$T = \frac{d}{t} = \frac{-68}{32} = -2.125.$$

The change was −2.125°F per minute.

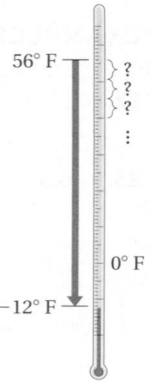

> **Do Exercise 28.**

28. Chemical Reaction. During a chemical reaction, the temperature in a beaker decreased every minute by the same number of degrees. The temperature was 71°F at 2:12 P.M. By 2:37 P.M., the temperature had changed to −14°F. By how many degrees did it change each minute?

Calculator Corner

Operations on the Real Numbers We can perform operations on the real numbers on a graphing calculator. Recall that negative numbers are entered using the opposite key, ⊖, rather than the subtraction operation key, ⊖. Consider the sum −5 + (−3.8). We use parentheses when we write this sum in order to separate the addition symbol and the "opposite of" symbol and thus make the expression more easily read. When we enter this calculation on a graphing calculator, however, the parentheses are not necessary. We can press ⊖ 5 + ⊖ 3 . 8 **ENTER**. The result is −8.8. Note that it is not incorrect to enter the parentheses. The result will be the same if this is done.

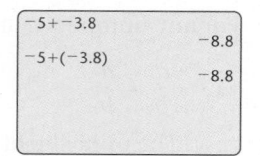

To find the difference 10 − (−17), we press 1 0 ⊖ ⊖ 1 7 **ENTER**. The result is 27. We can also multiply and divide real numbers. To find −5 · (−7), we press ⊖ 5 × ⊖ 7 **ENTER**, and to find 45 ÷ (−9), we press 4 5 ÷ ⊖ 9 **ENTER**. Note that it is not necessary to use parentheses in any of these calculations.

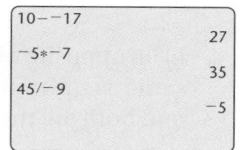

Exercises: Use a calculator to perform each operation.

1. −8 + 4	**2.** 1.2 + (−1.5)	**3.** −7 + (−5)	**4.** −7.6 + (−1.9)
5. −8 − 4	**6.** 1.2 − (−1.5)	**7.** −7 − (−5)	**8.** −7.6 − (−1.9)
9. −8 · 4	**10.** 1.2 · (−1.5)	**11.** −7 · (−5)	**12.** −7.6 · (−1.9)
13. −8 ÷ 4	**14.** 1.2 ÷ (−1.5)	**15.** −7 ÷ (−5)	**16.** −7.6 ÷ (−1.9)

Answer

28. −3.4°F per minute

a Divide, if possible. Check each answer.

1. $48 \div (-6)$

2. $\dfrac{42}{-7}$

3. $\dfrac{28}{-2}$

4. $24 \div (-12)$

5. $\dfrac{-24}{8}$

6. $-18 \div (-2)$

7. $\dfrac{-36}{-12}$

8. $-72 \div (-9)$

9. $\dfrac{-72}{9}$

10. $\dfrac{-50}{25}$

11. $-100 \div (-50)$

12. $\dfrac{-200}{8}$

13. $-108 \div 9$

14. $\dfrac{-63}{-7}$

15. $\dfrac{200}{-25}$

16. $-300 \div (-16)$

17. $\dfrac{75}{0}$

18. $\dfrac{0}{-5}$

19. $\dfrac{0}{-2.6}$

20. $\dfrac{-23}{0}$

b Find the reciprocal.

21. $\dfrac{15}{7}$

22. $\dfrac{3}{8}$

23. $-\dfrac{47}{13}$

24. $-\dfrac{31}{12}$

25. 13

26. -10

27. -32

28. 15

29. $\dfrac{1}{-7.1}$

30. $\dfrac{1}{-4.9}$

31. $\dfrac{1}{9}$

32. $\dfrac{1}{16}$

33. $\dfrac{1}{4y}$

34. $\dfrac{-1}{8a}$

35. $\dfrac{2a}{3b}$

36. $\dfrac{-4y}{3x}$

c Rewrite each division as a multiplication.

37. $4 \div 17$

38. $5 \div (-8)$

39. $\dfrac{8}{-13}$

40. $-\dfrac{13}{47}$

41. $\dfrac{13.9}{-1.5}$

42. $-\dfrac{47.3}{21.4}$

43. $\dfrac{2}{3} \div \left(-\dfrac{4}{5}\right)$

44. $\dfrac{3}{4} \div \left(-\dfrac{7}{10}\right)$

45. $\dfrac{\dfrac{x}{1}}{\dfrac{1}{y}}$

46. $\dfrac{13}{\dfrac{1}{x}}$

47. $\dfrac{3x + 4}{5}$

48. $\dfrac{4y - 8}{-7}$

Divide.

49. $\dfrac{3}{4} \div \left(-\dfrac{2}{3}\right)$

50. $\dfrac{7}{8} \div \left(-\dfrac{1}{2}\right)$

51. $-\dfrac{5}{4} \div \left(-\dfrac{3}{4}\right)$

52. $-\dfrac{5}{9} \div \left(-\dfrac{5}{6}\right)$

53. $-\dfrac{2}{7} \div \left(-\dfrac{4}{9}\right)$

54. $-\dfrac{3}{5} \div \left(-\dfrac{5}{8}\right)$

55. $-\dfrac{3}{8} \div \left(-\dfrac{8}{3}\right)$

56. $-\dfrac{5}{8} \div \left(-\dfrac{6}{5}\right)$

57. $-\dfrac{5}{6} \div \dfrac{2}{3}$

58. $-\dfrac{7}{16} \div \dfrac{3}{8}$

59. $-\dfrac{9}{4} \div \dfrac{5}{12}$

60. $-\dfrac{3}{5} \div \dfrac{7}{10}$

61. $\dfrac{-11}{-13}$

62. $\dfrac{-21}{-25}$

63. $-6.6 \div 3.3$

64. $-44.1 \div (-6.3)$

65. $\dfrac{48.6}{-3}$

66. $\dfrac{-1.9}{20}$

67. $\dfrac{-12.5}{5}$

68. $\dfrac{-17.8}{3.2}$

69. $11.25 \div (-9)$

70. $-9.6 \div (-6.4)$

71. $\dfrac{-9}{17 - 17}$

72. $\dfrac{-8}{-5 + 5}$

 Percent of Increase or Decrease in Employment. A percent of increase is generally positive and a percent of decrease is generally negative. The table below lists estimates of the number of job opportunities for various occupations in 2006 and 2016. In Exercises 73–76, find the missing numbers.

	OCCUPATION	NUMBER OF JOBS IN 2006 (in thousands)	NUMBER OF JOBS IN 2016 (in thousands)	CHANGE	PERCENT OF INCREASE OR DECREASE
	Electrician	705	757	52	7.4%
	File clerk	234	137	−97	−41.5%
73.	Athletic trainer	17	21	4	
74.	Child-care worker	1388	1636	248	
75.	Cashier	3527	3411	−116	
76.	Fisherman	38	32	−6	

SOURCE: U.S. Bureau of Labor Statistics *Occupational Outlook Handbook*

Skill Maintenance

Simplify.

77. $\dfrac{1}{4} - \dfrac{1}{2}$ [1.4a]

78. $-9 - 3 + 17$ [1.4a]

79. $35 \cdot (-1.2)$ [1.5a]

80. $4 \cdot (-6) \cdot (-2) \cdot (-1)$ [1.5a]

81. $13.4 + (-4.9)$ [1.3a]

82. $-\dfrac{3}{8} - \left(-\dfrac{1}{4}\right)$ [1.4a]

Convert to decimal notation. [1.2c]

83. $-\dfrac{1}{11}$

84. $\dfrac{11}{12}$

85. $\dfrac{15}{4}$

86. $-\dfrac{10}{3}$

Synthesis

87. Find the reciprocal of -10.5. What happens if you take the reciprocal of the result?

88. Determine those real numbers a for which the opposite of a is the same as the reciprocal of a.

Determine whether each expression represents a positive number or a negative number when a and b are negative.

89. $\dfrac{-a}{b}$

90. $\dfrac{-a}{-b}$

91. $-\left(\dfrac{a}{-b}\right)$

92. $-\left(\dfrac{-a}{b}\right)$

93. $-\left(\dfrac{-a}{-b}\right)$

1.7

Properties of Real Numbers

OBJECTIVES

a Find equivalent fraction expressions and simplify fraction expressions.

b Use the commutative and associative laws to find equivalent expressions.

c Use the distributive laws to multiply expressions like 8 and $x - y$.

d Use the distributive laws to factor expressions like $4x - 12 + 24y$.

e Collect like terms.

SKILL TO REVIEW

Objective 1.3a: Add real numbers.

Add.

1. $-16 + 5$ **2.** $29 + (-23)$

Complete the table by evaluating each expression for the given values.

1.

Value	$x + x$	$2x$
$x = 3$		
$x = -6$		
$x = 4.8$		

2.

Value	$x + 3x$	$5x$
$x = 2$		
$x = -6$		
$x = 4.8$		

Answers

Skill to Review:
1. -11 **2.** 6

Margin Exercises:
1. 6, 6; $-12, -12$; 9.6, 9.6 **2.** 8, 10; $-24, -30$;
19.2, 24

a Equivalent Expressions

In solving equations and doing other kinds of work in algebra, we manipulate expressions in various ways. For example, instead of $x + x$, we might write $2x$, knowing that the two expressions represent the same number for any allowable replacement of x. In that sense, the expressions $x + x$ and $2x$ are **equivalent**, as are $\dfrac{3}{x}$ and $\dfrac{3x}{x^2}$, even though 0 is not an allowable replacement because division by 0 is not defined.

> **EQUIVALENT EXPRESSIONS**
>
> Two expressions that have the same value for all allowable replacements are called **equivalent**.

The expressions $x + 3x$ and $5x$ are *not* equivalent, as we see in Margin Exercise 2.

Do Exercises 1 and 2.

In this section, we will consider several laws of real numbers that will allow us to find equivalent expressions. The first two laws are the *identity properties of 0 and 1*.

> **THE IDENTITY PROPERTY OF 0**
>
> For any real number a,
>
> $$a + 0 = 0 + a = a.$$
>
> (The number 0 is the *additive identity*.)

> **THE IDENTITY PROPERTY OF 1**
>
> For any real number a,
>
> $$a \cdot 1 = 1 \cdot a = a.$$
>
> (The number 1 is the *multiplicative identity*.)

We often refer to the use of the identity property of 1 as "multiplying by 1." We can use this method to find equivalent fraction expressions. Recall from arithmetic that to multiply with fraction notation, we multiply the numerators and multiply the denominators.

EXAMPLE 1 Write a fraction expression equivalent to $\frac{2}{3}$ with a denominator of $3x$:

$$\frac{2}{3} = \frac{\square}{3x}.$$

Note that $3x = 3 \cdot x$. We want fraction notation for $\frac{2}{3}$ that has a denominator of $3x$, but the denominator 3 is missing a factor of x. Thus we multiply by 1, using x/x as an equivalent expression for 1:

$$\frac{2}{3} = \frac{2}{3} \cdot 1 = \frac{2}{3} \cdot \frac{x}{x} = \frac{2x}{3x}.$$

The expressions $2/3$ and $2x/(3x)$ are equivalent. They have the same value for any allowable replacement. Note that $2x/(3x)$ is not defined for a replacement of 0, but for all nonzero real numbers, the expressions $2/3$ and $2x/(3x)$ have the same value.

Do Exercises 3 and 4.

In algebra, we consider an expression like $2/3$ to be "simplified" from $2x/(3x)$. To find such simplified expressions, we use the identity property of 1 to remove a factor of 1.

EXAMPLE 2 Simplify: $-\dfrac{20x}{12x}$.

$$-\frac{20x}{12x} = -\frac{5 \cdot 4x}{3 \cdot 4x} \qquad \text{We look for the largest factor common to both the numerator and the denominator and factor each.}$$

$$= -\frac{5}{3} \cdot \frac{4x}{4x} \qquad \text{Factoring the fraction expression}$$

$$= -\frac{5}{3} \cdot 1 \qquad \frac{4x}{4x} = 1$$

$$= -\frac{5}{3} \qquad \text{Removing a factor of 1 using the identity property of 1}$$

EXAMPLE 3 Simplify: $\dfrac{14ab}{56a}$.

$$\frac{14ab}{56a} = \frac{14a \cdot b}{14a \cdot 4} = \frac{14a}{14a} \cdot \frac{b}{4} = 1 \cdot \frac{b}{4} = \frac{b}{4}$$

Do Exercises 5–8.

(b) The Commutative and Associative Laws

The Commutative Laws

Let's examine the expressions $x + y$ and $y + x$, as well as xy and yx.

EXAMPLE 4 Evaluate $x + y$ and $y + x$ when $x = 4$ and $y = 3$.

We substitute 4 for x and 3 for y in both expressions:

$$x + y = 4 + 3 = 7; \qquad y + x = 3 + 4 = 7.$$

EXAMPLE 5 Evaluate xy and yx when $x = 3$ and $y = -12$.

We substitute 3 for x and -12 for y in both expressions:

$$xy = 3 \cdot (-12) = -36; \qquad yx = (-12) \cdot 3 = -36.$$

Do Exercises 9 and 10.

3. Write a fraction expression equivalent to $\frac{3}{4}$ with a denominator of 8:
$$\frac{3}{4} = \frac{\square}{8}.$$

4. Write a fraction expression equivalent to $\frac{3}{4}$ with a denominator of $4t$:
$$\frac{3}{4} = \frac{\square}{4t}.$$

Simplify.

5. $\dfrac{3y}{4y}$

6. $-\dfrac{16m}{12m}$

7. $\dfrac{5xy}{40y}$

8. $\dfrac{18p}{24pq}$

9. Evaluate $x + y$ and $y + x$ when $x = -2$ and $y = 3$.

10. Evaluate xy and yx when $x = -2$ and $y = 5$.

Answers

3. $\dfrac{6}{8}$ 4. $\dfrac{3t}{4t}$ 5. $\dfrac{3}{4}$ 6. $-\dfrac{4}{3}$

7. $\dfrac{x}{8}$ 8. $\dfrac{3}{4q}$ 9. $1; 1$ 10. $-10; -10$

The expressions $x + y$ and $y + x$ have the same values no matter what the variables stand for. Thus they are equivalent. Therefore, when we add two numbers, the order in which we add does not matter. Similarly, the expressions xy and yx are equivalent. They also have the same values, no matter what the variables stand for. Therefore, when we multiply two numbers, the order in which we multiply does not matter.

The following are examples of general patterns or laws.

THE COMMUTATIVE LAWS

Addition.   For any numbers a and b,

$$a + b = b + a.$$

(We can change the order when adding without affecting the answer.)

Multiplication.   For any numbers a and b,

$$ab = ba.$$

(We can change the order when multiplying without affecting the answer.)

Using a commutative law, we know that $x + 2$ and $2 + x$ are equivalent. Similarly, $3x$ and $x(3)$ are equivalent. Thus, in an algebraic expression, we can replace one with the other and the result will be equivalent to the original expression.

EXAMPLE 6   Use the commutative laws to write an equivalent expression: **(a)** $y + 5$; **(b)** mn; **(c)** $7 + xy$.

a) An expression equivalent to $y + 5$ is $5 + y$ by the commutative law of addition.

b) An expression equivalent to mn is nm by the commutative law of multiplication.

c) An expression equivalent to $7 + xy$ is $xy + 7$ by the commutative law of addition. Another expression equivalent to $7 + xy$ is $7 + yx$ by the commutative law of multiplication. Another equivalent expression is $yx + 7$.

Use a commutative law to write an equivalent expression.

11. $x + 9$

12. pq

13. $xy + t$

Do Exercises 11–13.

The Associative Laws

Now let's examine the expressions $a + (b + c)$ and $(a + b) + c$. Note that these expressions involve the use of parentheses as *grouping* symbols, and they also involve three numbers. Calculations within parentheses are to be done first.

EXAMPLE 7   Calculate and compare: $3 + (8 + 5)$ and $(3 + 8) + 5$.

$$3 + (8 + 5) = 3 + 13 \qquad \text{Calculating within parentheses first;}$$
$$\text{adding the 8 and the 5}$$
$$= 16;$$
$$(3 + 8) + 5 = 11 + 5 \qquad \text{Calculating within parentheses first;}$$
$$\text{adding the 3 and the 8}$$
$$= 16$$

Answers

11. $9 + x$ **12.** qp
13. $t + xy$, or $yx + t$, or $t + yx$

The two expressions in Example 7 name the same number. Moving the parentheses to group the additions differently does not affect the value of the expression.

EXAMPLE 8 Calculate and compare: $3 \cdot (4 \cdot 2)$ and $(3 \cdot 4) \cdot 2$.

$$3 \cdot (4 \cdot 2) = 3 \cdot 8 = 24; \qquad (3 \cdot 4) \cdot 2 = 12 \cdot 2 = 24$$

Do Exercises 14 and 15.

You may have noted that when only addition is involved, numbers can be grouped any way we please without affecting the answer. When only multiplication is involved, numbers can also be grouped any way we please without affecting the answer.

THE ASSOCIATIVE LAWS

Addition. For any numbers a, b, and c,

$$a + (b + c) = (a + b) + c.$$

(Numbers can be grouped in any manner for addition.)

Multiplication. For any numbers a, b, and c,

$$a \cdot (b \cdot c) = (a \cdot b) \cdot c.$$

(Numbers can be grouped in any manner for multiplication.)

EXAMPLE 9 Use an associative law to write an equivalent expression: **(a)** $(y + z) + 3$; **(b)** $8(xy)$.

a) An expression equivalent to $(y + z) + 3$ is $y + (z + 3)$ by the associative law of addition.

b) An expression equivalent to $8(xy)$ is $(8x)y$ by the associative law of multiplication.

Do Exercises 16 and 17.

The associative laws say that numbers can be grouped any way we please when only additions or only multiplications are involved. Thus we often omit the parentheses. For example,

$$x + (y + 2) \quad \text{means} \quad x + y + 2, \qquad \text{and} \qquad (lw)h \quad \text{means} \quad lwh.$$

Using the Commutative and Associative Laws Together

EXAMPLE 10 Use the commutative and associative laws to write at least three expressions equivalent to $(x + 5) + y$.

a) $(x + 5) + y = x + (5 + y)$ Using the associative law first and then using the commutative law
$= x + (y + 5)$

b) $(x + 5) + y = y + (x + 5)$ Using the commutative law twice
$= y + (5 + x)$

c) $(x + 5) + y = (5 + x) + y$ Using the commutative law first and then the associative law
$= 5 + (x + y)$

---

14. Calculate and compare:

$$8 + (9 + 2) \quad \text{and} \quad (8 + 9) + 2.$$

15. Calculate and compare:

$$10 \cdot (5 \cdot 3) \quad \text{and} \quad (10 \cdot 5) \cdot 3.$$

Use an associative law to write an equivalent expression.

16. $r + (s + 7)$

17. $9(ab)$

Answers

14. 19; 19   **15.** 150; 150   **16.** $(r + s) + 7$

17. $(9a)b$

EXAMPLE 11 Use the commutative and associative laws to write at least three expressions equivalent to $(3x)y$.

a) $(3x)y = 3(xy)$ Using the associative law first and then using the
$\qquad\quad\; = 3(yx)$ commutative law

b) $(3x)y = y(3x)$ Using the commutative law twice
$\qquad\quad\; = y(x \cdot 3)$

c) $(3x)y = (x \cdot 3)y$ Using the commutative law, and then the associative law,
$\qquad\quad\; = x(3y)$ and then the commutative law again
$\qquad\quad\; = x(y \cdot 3)$

Do Exercises 18 and 19.

Use the commutative and associative laws to write at least three equivalent expressions.

18. $4(tu)$

19. $r + (2 + s)$

(c) The Distributive Laws

The *distributive laws* are the basis of many procedures in both arithmetic and algebra. They are probably the most important laws that we use to manipulate algebraic expressions. The distributive law of multiplication over addition involves two operations: addition and multiplication.

Let's begin by considering a multiplication problem from arithmetic:

$$
\begin{array}{r}
4\ 5 \\
7 \\
\hline
3\ 5 \\
2\ 8\ 0 \\
3\ 1\ 5
\end{array}
$$
  ← This is $7 \cdot 5$.
  ← This is $7 \cdot 40$.
  ← This is the sum $7 \cdot 5 + 7 \cdot 40$.

To carry out the multiplication, we actually added two products. That is,

$$7 \cdot 45 = 7(5 + 40) = 7 \cdot 5 + 7 \cdot 40.$$

Let's examine this further. If we wish to multiply a sum of several numbers by a factor, we can either add and then multiply, or multiply and then add.

EXAMPLE 12 Compute in two ways: $5 \cdot (4 + 8)$.

a) $5 \cdot (4 + 8)$ Adding within parentheses first, and then multiplying
$\quad = 5 \cdot \quad 12$
$\quad = 60$

b) $5 \cdot (4 + 8) = (5 \cdot 4) + (5 \cdot 8)$ Distributing the multiplication to terms within parentheses first and then adding
$\qquad\qquad\quad = \quad 20 \;+\; 40$
$\qquad\qquad\quad = \quad 60$

Do Exercises 20–22.

Compute.

20. a) $7 \cdot (3 + 6)$
   b) $(7 \cdot 3) + (7 \cdot 6)$

21. a) $2 \cdot (10 + 30)$
   b) $(2 \cdot 10) + (2 \cdot 30)$

22. a) $(2 + 5) \cdot 4$
   b) $(2 \cdot 4) + (5 \cdot 4)$

THE DISTRIBUTIVE LAW OF MULTIPLICATION OVER ADDITION

For any numbers a, b, and c,

$$a(b + c) = ab + ac.$$

Answers

18. $(4t)u, (tu)4, t(4u)$; answers may vary
19. $(2 + r) + s, (r + s) + 2, s + (r + 2)$;
answers may vary   **20. (a)** $7 \cdot 9 = 63$;
(b) $21 + 42 = 63$ **21. (a)** $2 \cdot 40 = 80$;
(b) $20 + 60 = 80$ **22. (a)** $7 \cdot 4 = 28$;
(b) $8 + 20 = 28$

In the statement of the distributive law, we know that in an expression such as $ab + ac$, the multiplications are to be done first according to the rules for order of operations. So, instead of writing $(4 \cdot 5) + (4 \cdot 7)$, we can write $4 \cdot 5 + 4 \cdot 7$. However, in $a(b + c)$, we cannot omit the parentheses. If we did, we would have $ab + c$, which means $(ab) + c$. For example, $3(4 + 2) = 3(6) = 18$, but $3 \cdot 4 + 2 = 12 + 2 = 14$.

There is another distributive law that relates multiplication and subtraction. This law says that to multiply by a difference, we can either subtract and then multiply, or multiply and then subtract.

> ## THE DISTRIBUTIVE LAW OF MULTIPLICATION OVER SUBTRACTION
>
> For any numbers a, b, and c,
>
> $$a(b - c) = ab - ac.$$

We often refer to "*the* distributive law" when we mean *either* or *both* of these laws.

Do Exercises 23–25.

What do we mean by the *terms* of an expression? **Terms** are separated by addition signs. If there are subtraction signs, we can find an equivalent expression that uses addition signs.

EXAMPLE 13 What are the terms of $3x - 4y + 2z$?

We have

$$3x - 4y + 2z = 3x + (-4y) + 2z. \qquad \text{Separating parts with } + \text{ signs}$$

The terms are $3x$, $-4y$, and $2z$.

Do Exercises 26 and 27.

The distributive laws are a basis for a procedure in algebra called **multiplying**. In an expression like $8(a + 2b - 7)$, we multiply each term inside the parentheses by 8:

$$8(a + 2b - 7) = 8 \cdot a + 8 \cdot 2b - 8 \cdot 7 = 8a + 16b - 56.$$

EXAMPLES Multiply.

14. $9(x - 5) = 9 \cdot x - 9 \cdot 5$ Using the distributive law of multiplication over subtraction

$$= 9x - 45$$

15. $\frac{2}{3}(w + 1) = \frac{2}{3} \cdot w + \frac{2}{3} \cdot 1$ Using the distributive law of multiplication over addition

$$= \frac{2}{3}w + \frac{2}{3}$$

16. $\frac{4}{3}(s - t + w) = \frac{4}{3}s - \frac{4}{3}t + \frac{4}{3}w$ Using both distributive laws

Do Exercises 28–30.

Calculate.

23. a) $4(5 - 3)$

   b) $4 \cdot 5 - 4 \cdot 3$

24. a) $-2 \cdot (5 - 3)$

   b) $-2 \cdot 5 - (-2) \cdot 3$

25. a) $5 \cdot (2 - 7)$

   b) $5 \cdot 2 - 5 \cdot 7$

What are the terms of each expression?

26. $5x - 8y + 3$

27. $-4y - 2x + 3z$

Multiply.

28. $3(x - 5)$

29. $5(x + 1)$

30. $\frac{3}{5}(p + q - t)$

Answers

23. (a) $4 \cdot 2 = 8$; (b) $20 - 12 = 8$
24. (a) $-2 \cdot 2 = -4$; (b) $-10 + 6 = -4$
25. (a) $5(-5) = -25$; (b) $10 - 35 = -25$
26. $5x, -8y, 3$ **27.** $-4y, -2x, 3z$
28. $3x - 15$ **29.** $5x + 5$
30. $\frac{3}{5}p + \frac{3}{5}q - \frac{3}{5}t$

EXAMPLE 17 Multiply: $-4(x - 2y + 3z)$.

$$-4(x - 2y + 3z) = -4 \cdot x - (-4)(2y) + (-4)(3z) \quad \text{Using both distributive laws}$$

$$= -4x - (-8y) + (-12z) \quad \text{Multiplying}$$

$$= -4x + 8y - 12z$$

We can also do this problem by first finding an equivalent expression with all plus signs and then multiplying:

$$-4(x - 2y + 3z) = -4[x + (-2y) + 3z]$$

$$= -4 \cdot x + (-4)(-2y) + (-4)(3z)$$

$$= -4x + 8y - 12z.$$

Do Exercises 31–33.

EXAMPLES Name the property or law illustrated by each equation.

Equation	*Property*
18. $5x = x(5)$	Commutative law of multiplication
19. $a + (8.5 + b) = (a + 8.5) + b$	Associative law of addition
20. $0 + 11 = 11$	Identity property of 0
21. $(-5s)t = -5(st)$	Associative law of multiplication
22. $\frac{3}{4} \cdot 1 = \frac{3}{4}$	Identity property of 1
23. $12.5(w - 3) = 12.5w - 12.5(3)$	Distributive law of multiplication over subtraction
24. $y + \frac{1}{2} = \frac{1}{2} + y$	Commutative law of addition

Do Exercises 34–40.

(d) Factoring

Factoring is the reverse of multiplying. To factor, we can use the distributive laws in reverse:

$$ab + ac = a(b + c) \quad \text{and} \quad ab - ac = a(b - c).$$

> **FACTORING**
>
> To **factor** an expression is to find an equivalent expression that is a product.

To factor $9x - 45$, for example, we find an equivalent expression that is a product: $9(x - 5)$. This reverses the multiplication that we did in Example 14. When all the terms of an expression have a factor in common, we can "factor it out" using the distributive laws. Note the following.

$9x$ has the factors $9, -9, 3, -3, 1, -1, x, -x, 3x, -3x, 9x, -9x$;

-45 has the factors $1, -1, 3, -3, 5, -5, 9, -9, 15, -15, 45, -45$

Multiply.

31. $-2(x - 3)$

32. $5(x - 2y + 4z)$

33. $-5(x - 2y + 4z)$

Name the property or law illustrated by each equation.

34. $(-8a)b = -8(ab)$

35. $p \cdot 1 = p$

36. $m + 34 = 34 + m$

37. $2(t + 5) = 2t + 2(5)$

38. $0 + k = k$

39. $-8x = x(-8)$

40. $x + (4.3 + b) = (x + 4.3) + b$

We generally remove the largest common factor. In this case, that factor is 9. Thus,

$$9x - 45 = 9 \cdot x - 9 \cdot 5$$
$$= 9(x - 5).$$

Remember that an expression has been factored when we have found an equivalent expression that is a product. Above, we note that $9x - 45$ and $9(x - 5)$ are equivalent expressions. The expression $9x - 45$ is the difference of $9x$ and 45; the expression $9(x - 5)$ is the product of 9 and $(x - 5)$.

EXAMPLES Factor.

25. $5x - 10 = 5 \cdot x - 5 \cdot 2$ Try to do this step mentally.
        $= 5(x - 2)$ You can check by multiplying.

26. $ax - ay + az = a(x - y + z)$

27. $9x + 27y - 9 = 9 \cdot x + 9 \cdot 3y - 9 \cdot 1 = 9(x + 3y - 1)$

Note in Example 27 that you might, at first, just factor out a 3, as follows:

$$9x + 27y - 9 = 3 \cdot 3x + 3 \cdot 9y - 3 \cdot 3$$
$$= 3(3x + 9y - 3).$$

At this point, the mathematics is correct, but the answer is not because there is another factor of 3 that can be factored out, as follows:

$$3 \cdot 3x + 3 \cdot 9y - 3 \cdot 3 = 3(3x + 9y - 3)$$
$$= 3(3 \cdot x + 3 \cdot 3y - 3 \cdot 1)$$
$$= 3 \cdot 3(x + 3y - 1)$$
$$= 9(x + 3y - 1).$$

We now have a correct answer, but it took more work than we did in Example 27. Thus it is better to look for the *greatest common factor* at the outset.

EXAMPLES Factor. Try to write just the answer, if you can.

28. $5x - 5y = 5(x - y)$

29. $-3x + 6y - 9z = -3(x - 2y + 3z)$

We usually factor out a negative factor when the first term is negative. The way we factor can depend on the situation in which we are working. We might also factor the expression in Example 29 as follows:

$$-3x + 6y - 9z = 3(-x + 2y - 3z).$$

30. $18z - 12x - 24 = 6(3z - 2x - 4)$

31. $\frac{1}{2}x + \frac{3}{2}y - \frac{1}{2} = \frac{1}{2}(x + 3y - 1)$

Remember that you can always check factoring by multiplying. Keep in mind that an expression is factored when it is written as a product.

Do Exercises 41–46.

Factor.

41. $6x - 12$

42. $3x - 6y + 9$

43. $bx + by - bz$

44. $16a - 36b + 42$

45. $\dfrac{3}{8}x - \dfrac{5}{8}y + \dfrac{7}{8}$

46. $-12x + 32y - 16z$

Answers

41. $6(x - 2)$ **42.** $3(x - 2y + 3)$
43. $b(x + y - z)$ **44.** $2(8a - 18b + 21)$
45. $\dfrac{1}{8}(3x - 5y + 7)$ **46.** $-4(3x - 8y + 4z)$, or $4(-3x + 8y - 4z)$

(e) Collecting Like Terms

Terms such as $5x$ and $-4x$, whose variable factors are exactly the same, are called **like terms**. Similarly, numbers, such as -7 and 13, are like terms. Also, $3y^2$ and $9y^2$ are like terms because the variables are raised to the same power. Terms such as $4y$ and $5y^2$ are not like terms, and $7x$ and $2y$ are not like terms.

The process of **collecting like terms** is also based on the distributive laws. We can apply a distributive law when a factor is on the right because of the commutative law of multiplication.

Later in this text, terminology like "collecting like terms" and "combining like terms" will also be referred to as "simplifying."

EXAMPLES Collect like terms. Try to write just the answer, if you can.

32. $4x + 2x = (4 + 2)x = 6x$ Factoring out the x using a distributive law

33. $2x + 3y - 5x - 2y = 2x - 5x + 3y - 2y$
$$= (2 - 5)x + (3 - 2)y = -3x + 1y = -3x + y$$

34. $3x - x = 3x - 1x = (3 - 1)x = 2x$

35. $x - 0.24x = 1 \cdot x - 0.24x = (1 - 0.24)x = 0.76x$

36. $x - 6x = 1 \cdot x - 6 \cdot x = (1 - 6)x = -5x$

37. $4x - 7y + 9x - 5 + 3y - 8 = 13x - 4y - 13$

38. $\frac{2}{3}a - b + \frac{4}{5}a + \frac{1}{4}b - 10 = \frac{2}{3}a - 1 \cdot b + \frac{4}{5}a + \frac{1}{4}b - 10$
$$= \left(\frac{2}{3} + \frac{4}{5}\right)a + \left(-1 + \frac{1}{4}\right)b - 10$$
$$= \left(\frac{10}{15} + \frac{12}{15}\right)a + \left(-\frac{4}{4} + \frac{1}{4}\right)b - 10$$
$$= \frac{22}{15}a - \frac{3}{4}b - 10$$

Do Exercises 47–53.

Collect like terms.

47. $6x - 3x$

48. $7x - x$

49. $x - 9x$

50. $x - 0.41x$

51. $5x + 4y - 2x - y$

52. $3x - 7x - 11 + 8y + 4 - 13y$

53. $-\frac{2}{3} - \frac{3}{5}x + y + \frac{7}{10}x - \frac{2}{9}y$

STUDY TIPS

LEARNING RESOURCES

Please see the preface for more information on these resources and others. To order any of our products, call (800) 824-7799 in the United States or (201) 767-5021 outside the United States, or visit your campus bookstore.

- The *Student's Solutions Manual* contains fully worked-out solutions to the odd-numbered exercises in the exercise sets, as well as solutions to all exercises in the Mid-Chapter Reviews, end-of-chapter Review Exercises, Chapter Tests, and Cumulative Reviews. (ISBN: 978-0-321-61362-2)

- *Worksheets for Classroom or Lab Practice* provide a list of learning objectives, vocabulary and practice problems, and extra practice problems with ample work space. (ISBN: 978-0-321-61368-4)

- As described on p. 56 and in the Preface, Video Resources on DVD Featuring Chapter Test Prep Videos provide section-level lectures for every objective and step-by-step solutions to all the Chapter Test exercises in this textbook. The Chapter Test videos are also available on YouTube (search using BittingerIntroInter) and in MyMathLab.

- InterAct Math Tutorial Website (www.interactmath.com) provides algorithmically generated practice exercises that correlate directly to the exercises in the textbook.

- MathXL® Tutorials on CD provide practice exercises correlated at the objective level to the exercises in the textbook. Every practice exercise is accompanied by an example and a guided solution, and selected exercises may also include a video clip to help illustrate a concept.

Answers

47. $3x$ **48.** $6x$ **49.** $-8x$ **50.** $0.59x$
51. $3x + 3y$ **52.** $-4x - 5y - 7$
53. $\frac{1}{10}x + \frac{7}{9}y - \frac{2}{3}$

a Find an equivalent expression with the given denominator.

1. $\dfrac{3}{5} = \dfrac{\square}{5y}$
  2. $\dfrac{5}{8} = \dfrac{\square}{8t}$
  3. $\dfrac{2}{3} = \dfrac{\square}{15x}$
  4. $\dfrac{6}{7} = \dfrac{\square}{14y}$
  5. $\dfrac{2}{x} = \dfrac{\square}{x^2}$
  6. $\dfrac{4}{9x} = \dfrac{\square}{9xy}$

Simplify.

7. $-\dfrac{24a}{16a}$
  8. $-\dfrac{42t}{18t}$
  9. $-\dfrac{42ab}{36ab}$
  10. $-\dfrac{64pq}{48pq}$
  11. $\dfrac{20st}{15t}$
  12. $\dfrac{21w}{7wz}$

b Write an equivalent expression. Use a commutative law.

13. $y + 8$
  14. $x + 3$
  15. mn
  16. yz

17. $9 + xy$
  18. $11 + ab$
  19. $ab + c$
  20. $rs + t$

Write an equivalent expression. Use an associative law.

21. $a + (b + 2)$
  22. $3(vw)$
  23. $(8x)y$
  24. $(y + z) + 7$

25. $(a + b) + 3$
  26. $(5 + x) + y$
  27. $3(ab)$
  28. $(6x)y$

Use the commutative and associative laws to write three equivalent expressions.

29. $(a + b) + 2$
  30. $(3 + x) + y$
  31. $5 + (v + w)$
  32. $6 + (x + y)$

33. $(xy)3$
  34. $(ab)5$
  35. $7(ab)$
  36. $5(xy)$

c Multiply.

37. $2(b + 5)$
  38. $4(x + 3)$
  39. $7(1 + t)$
  40. $4(1 + y)$

41. $6(5x + 2)$
  42. $9(6m + 7)$
  43. $7(x + 4 + 6y)$
  44. $4(5x + 8 + 3p)$

45. $7(x - 3)$

46. $15(y - 6)$

47. $-3(x - 7)$

48. $1.2(x - 2.1)$

49. $\dfrac{2}{3}(b - 6)$

50. $\dfrac{5}{8}(y + 16)$

51. $7.3(x - 2)$

52. $5.6(x - 8)$

53. $-\dfrac{3}{5}(x - y + 10)$

54. $-\dfrac{2}{3}(a + b - 12)$

55. $-9(-5x - 6y + 8)$

56. $-7(-2x - 5y + 9)$

57. $-4(x - 3y - 2z)$

58. $8(2x - 5y - 8z)$

59. $3.1(-1.2x + 3.2y - 1.1)$

60. $-2.1(-4.2x - 4.3y - 2.2)$

List the terms of each expression.

61. $4x + 3z$

62. $8x - 1.4y$

63. $7x + 8y - 9z$

64. $8a + 10b - 18c$

d   Factor. Check by multiplying.

65. $2x + 4$

66. $5y + 20$

67. $30 + 5y$

68. $7x + 28$

69. $14x + 21y$

70. $18a + 24b$

71. $14t - 7$

72. $25m - 5$

73. $8x - 24$

74. $10x - 50$

75. $18a - 24b$

76. $32x - 20y$

77. $-4y + 32$

78. $-6m + 24$

79. $5x + 10 + 15y$

80. $9a + 27b + 81$

81. $16m - 32n + 8$

82. $6x + 10y - 2$

83. $12a + 4b - 24$

84. $8m - 4n + 12$

85. $8x + 10y - 22$

86. $9a + 6b - 15$

87. $ax - a$

88. $by - 9b$

89. $ax - ay - az$

90. $cx + cy - cz$

91. $-18x + 12y + 6$

92. $-14x + 21y + 7$

93. $\dfrac{2}{3}x - \dfrac{5}{3}y + \dfrac{1}{3}$

94. $\dfrac{3}{5}a + \dfrac{4}{5}b - \dfrac{1}{5}$

95. $36x - 6y + 18z$

96. $8a - 4b + 20c$

e　Collect like terms.

97. $9a + 10a$

98. $12x + 2x$

99. $10a - a$

100. $-16x + x$

101. $2x + 9z + 6x$

102. $3a - 5b + 7a$

103. $7x + 6y^2 + 9y^2$

104. $12m^2 + 6q + 9m^2$

105. $41a + 90 - 60a - 2$

106. $42x - 6 - 4x + 2$

107. $23 + 5t + 7y - t - y - 27$

108. $45 - 90d - 87 - 9d + 3 + 7d$

109. $\dfrac{1}{2}b + \dfrac{1}{2}b$

110. $\dfrac{2}{3}x + \dfrac{1}{3}x$

111. $2y + \dfrac{1}{4}y + y$

112. $\dfrac{1}{2}a + a + 5a$

113. $11x - 3x$

114. $9t - 17t$

115. $6n - n$

116. $100t - t$

117. $y - 17y$

118. $3m - 9m + 4$

119. $-8 + 11a - 5b + 6a - 7b + 7$

120. $8x - 5x + 6 + 3y - 2y - 4$

121. $9x + 2y - 5x$

122. $8y - 3z + 4y$

123. $11x + 2y - 4x - y$

124. $13a + 9b - 2a - 4b$

125. $2.7x + 2.3y - 1.9x - 1.8y$

126. $6.7a + 4.3b - 4.1a - 2.9b$

127. $\dfrac{13}{2}a + \dfrac{9}{5}b - \dfrac{2}{3}a - \dfrac{3}{10}b - 42$

128. $\dfrac{11}{4}x + \dfrac{2}{3}y - \dfrac{4}{5}x - \dfrac{1}{6}y + 12$

Skill Maintenance

Simplify.

129. $-28 - (-2)$ [1.4a]

130. $-200 + 85$ [1.3a]

131. $-16(-10)$ [1.5a]

132. $-88 \div (-11)$ [1.6a]

133. $\dfrac{400}{-80}$ [1.6a]

134. $38 - (-12)$ [1.4a]

135. Evaluate $9w$ for $w = 20$.   [1.1a]

136. Find the absolute value: $\left| -\dfrac{4}{13} \right|$.   [1.2e]

Write true or false.   [1.2d]

137. $-43 < -40$

138. $-3 \geq 0$

139. $-6 \leq -6$

140. $0 > -4$

Synthesis

Determine whether the expressions are equivalent. Explain why if they are. Give an example if they are not. Examples may vary.

141. $3t + 5$ and $3 \cdot 5 + t$

142. $4x$ and $x + 4$

143. $5m + 6$ and $6 + 5m$

144. $(x + y) + z$ and $z + (x + y)$

145. Factor: $q + qr + qrs + qrst$.

146. Collect like terms:
$$21x + 44xy + 15y - 16x - 8y - 38xy + 2y + xy.$$

1.8 Simplifying Expressions; Order of Operations

We now expand our ability to manipulate expressions by first considering opposites of sums and differences. Then we simplify expressions involving parentheses.

a Opposites of Sums

What happens when we multiply a real number by -1? Consider the following products:

$$-1(7) = -7, \quad -1(-5) = 5, \quad -1(0) = 0.$$

From these examples, it appears that when we multiply a number by -1, we get the opposite, or additive inverse, of that number.

> ### THE PROPERTY OF -1
>
> For any real number a,
>
> $$-1 \cdot a = -a.$$
>
> (Negative one times a is the opposite, or additive inverse, of a.)

The property of -1 enables us to find expressions equivalent to opposites of sums.

EXAMPLES Find an equivalent expression without parentheses.

1. $-(3 + x) = -1(3 + x)$ Using the property of -1

$\qquad\qquad = -1 \cdot 3 + (-1)x$ Using a distributive law, multiplying each term by -1

$\qquad\qquad = -3 + (-x)$ Using the property of -1

$\qquad\qquad = -3 - x$

2. $-(3x + 2y + 4) = -1(3x + 2y + 4)$ Using the property of -1

$\qquad\qquad\qquad = -1(3x) + (-1)(2y) + (-1)4$ Using a distributive law

$\qquad\qquad\qquad = -3x - 2y - 4$ Using the property of -1

Do Exercises 1 and 2.

Suppose we want to remove parentheses in an expression like

$$-(x - 2y + 5).$$

We can first rewrite any subtractions inside the parentheses as additions. Then we take the opposite of each term:

$$-(x - 2y + 5) = -[x + (-2y) + 5]$$
$$= -x + 2y + (-5) = -x + 2y - 5.$$

The most efficient method for removing parentheses is to replace each term in the parentheses with its opposite ("change the sign of every term"). Doing so for $-(x - 2y + 5)$, we obtain $-x + 2y - 5$ as an equivalent expression.

OBJECTIVES

a Find an equivalent expression for an opposite without parentheses, where an expression has several terms.

b Simplify expressions by removing parentheses and collecting like terms.

c Simplify expressions with parentheses inside parentheses.

d Simplify expressions using the rules for order of operations.

SKILL TO REVIEW
Objective 1.6a: Divide integers.

Divide.

  1. $20 \div (-4)$ **2.** $-42 \div (-6)$

Find an equivalent expression without parentheses.

  1. $-(x + 2)$

  2. $-(5x + 2y + 8)$

Find an equivalent expression without parentheses. Try to do this in one step.

3. $-(6 - t)$

4. $-(x - y)$

5. $-(-4a + 3t - 10)$

6. $-(18 - m - 2n + 4z)$

EXAMPLES Find an equivalent expression without parentheses.

3. $-(5 - y) = -5 + y$ Changing the sign of each term

4. $-(2a - 7b - 6) = -2a + 7b + 6$

5. $-(-3x + 4y + z - 7w - 23) = 3x - 4y - z + 7w + 23$

Do Exercises 3–6.

b Removing Parentheses and Simplifying

When a sum is added to another expression, as in $5x + (2x + 3)$, we can simply remove, or drop, the parentheses and collect like terms because of the associative law of addition:

$$5x + (2x + 3) = 5x + 2x + 3 = 7x + 3.$$

On the other hand, when a sum is subtracted from another expression, as in $3x - (4x + 2)$, we cannot simply drop the parentheses. However, we can subtract by adding an opposite. We then remove parentheses by changing the sign of each term inside the parentheses and collecting like terms.

EXAMPLE 6 Remove parentheses and simplify.

$$
\begin{aligned}
3x - (4x + 2) &= 3x + [-(4x + 2)] &&\text{Adding the opposite of } (4x + 2) \\
&= 3x + (-4x - 2) &&\text{Changing the sign of each term} \\
& &&\text{inside the parentheses} \\
&= 3x - 4x - 2 \\
&= -x - 2 &&\text{Collecting like terms}
\end{aligned}
$$

--------- *Caution!* ---------

Note that $3x - (4x + 2) \neq 3x - 4x + 2$. You cannot simply drop the parentheses.

Remove parentheses and simplify.

7. $5x - (3x + 9)$

8. $5y - 2 - (2y - 4)$

Do Exercises 7 and 8.

In practice, the first three steps of Example 6 are usually combined by changing the sign of each term in parentheses and then collecting like terms.

EXAMPLES Remove parentheses and simplify.

7. $5y - (3y + 4) = 5y - 3y - 4$ Removing parentheses by changing the sign of every term inside the parentheses

$$= 2y - 4 \quad\quad \text{Collecting like terms}$$

8. $3x - 2 - (5x - 8) = 3x - 2 - 5x + 8$

$$= -2x + 6$$

9. $(3a + 4b - 5) - (2a - 7b + 4c - 8)$

$$= 3a + 4b - 5 - 2a + 7b - 4c + 8$$

$$= a + 11b - 4c + 3$$

Do Exercises 9–11.

Remove parentheses and simplify.

9. $6x - (4x + 7)$

10. $8y - 3 - (5y - 6)$

11. $(2a + 3b - c) - (4a - 5b + 2c)$

Answers

3. $-6 + t$ **4.** $-x + y$ **5.** $4a - 3t + 10$
6. $-18 + m + 2n - 4z$ **7.** $2x - 9$
8. $3y + 2$ **9.** $2x - 7$ **10.** $3y + 3$
11. $-2a + 8b - 3c$

Next, consider subtracting an expression consisting of several terms multiplied by a number other than 1 or -1.

EXAMPLE 10 Remove parentheses and simplify.

$$
\begin{aligned}
x - 3(x + y) &= x + [-3(x + y)] &&\text{Adding the opposite of } 3(x + y) \\
&= x + [-3x - 3y] &&\text{Multiplying } x + y \text{ by } -3 \\
&= x - 3x - 3y \\
&= -2x - 3y &&\text{Collecting like terms}
\end{aligned}
$$

EXAMPLES Remove parentheses and simplify

11. $3y - 2(4y - 5) = 3y - 8y + 10$ Multiplying each term in the parentheses by -2

$$= -5y + 10$$

12. $(2a + 3b - 7) - 4(-5a - 6b + 12)$

$$= 2a + 3b - 7 + 20a + 24b - 48 = 22a + 27b - 55$$

13. $2y - \frac{1}{3}(9y - 12) = 2y - 3y + 4 = -y + 4$

14. $6(5x - 3y) - 2(8x + y) = 30x - 18y - 16x - 2y = 14x - 20y$

> Do Exercises 12–16.

> **Remove parentheses and simplify.**
> **12.** $y - 9(x + y)$
>
> **13.** $5a - 3(7a - 6)$
>
> **14.** $4a - b - 6(5a - 7b + 8c)$
>
> **15.** $5x - \dfrac{1}{4}(8x + 28)$
>
> **16.** $4.6(5x - 3y) - 5.2(8x + y)$

(c) Parentheses Within Parentheses

In addition to parentheses, some expressions contain other grouping symbols such as brackets $[\,]$ and braces $\{\,\}$.

> When more than one kind of grouping symbol occurs, do the computations in the innermost ones first. Then work from the inside out.

EXAMPLES Simplify.

15. $[3 - (7 + 3)] = [3 - 10] = -7$

16. $\{8 - [9 - (12 + 5)]\} = \{8 - [9 - 17]\}$ Computing $12 + 5$

$$
\begin{aligned}
&= \{8 - [-8]\} &&\text{Computing } 9 - 17 \\
&= 8 + 8 = 16
\end{aligned}
$$

17. $\left[(-4) \div \left(-\frac{1}{4}\right)\right] \div \frac{1}{4} = [(-4) \cdot (-4)] \div \frac{1}{4}$ Working within the brackets; computing $(-4) \div \left(-\frac{1}{4}\right)$

$$
\begin{aligned}
&= 16 \div \frac{1}{4} \\
&= 16 \cdot 4 = 64
\end{aligned}
$$

18. $4(2 + 3) - \{7 - [4 - (8 + 5)]\}$

$$
\begin{aligned}
&= 4 \cdot 5 - \{7 - [4 - 13]\} &&\text{Working with the innermost parentheses first} \\
&= 20 - \{7 - [-9]\} &&\text{Computing } 4 \cdot 5 \text{ and } 4 - 13 \\
&= 20 - 16 &&\text{Computing } 7 - [-9] \\
&= 4
\end{aligned}
$$

> Do Exercises 17–20.

> **Simplify.**
> **17.** $12 - (8 + 2)$
>
> **18.** $9 - [10 - (13 + 6)]$
>
> **19.** $[24 \div (-2)] \div (-2)$
>
> **20.** $5(3 + 4) - \{8 - [5 - (9 + 6)]\}$

Answers

12. $-9x - 8y$ **13.** $-16a + 18$
14. $-26a + 41b - 48c$ **15.** $3x - 7$
16. $-18.6x - 19y$ **17.** 2   **18.** 18
19. 6   **20.** 17

EXAMPLE 19 Simplify.

$$[5(x + 2) - 3x] - [3(y + 2) - 7(y - 3)]$$
$$= [5x + 10 - 3x] - [3y + 6 - 7y + 21] \quad \text{Working with the innermost parentheses first}$$

$$= [2x + 10] - [-4y + 27] \quad \text{Collecting like terms within brackets}$$
$$= 2x + 10 + 4y - 27 \quad \text{Removing brackets}$$
$$= 2x + 4y - 17 \quad \text{Collecting like terms}$$

Do Exercise 21.

21. Simplify:

$$[3(x + 2) + 2x] -$$
$$[4(y + 2) - 3(y - 2)].$$

(d) Order of Operations

When several operations are to be done in a calculation or a problem, we apply the following rules.

RULES FOR ORDER OF OPERATIONS

1. Do all calculations within grouping symbols before operations outside.
2. Evaluate all exponential expressions.
3. Do all multiplications and divisions in order from left to right.
4. Do all additions and subtractions in order from left to right.

These rules are consistent with the way in which most computers and scientific calculators perform calculations.

EXAMPLE 20 Simplify: $-34 \cdot 56 - 17$.

There are no parentheses or powers, so we start with the third step.

$$-34 \cdot 56 - 17 = -1904 - 17 \quad \text{Doing all multiplications and divisions in order from left to right}$$

$$= -1921 \quad \text{Doing all additions and subtractions in order from left to right}$$

EXAMPLE 21 Simplify: $25 \div (-5) + 50 \div (-2)$.

There are no calculations inside parentheses and no powers. The parentheses with (-5) and (-2) are used only to represent the negative numbers. We begin by doing all multiplications and divisions.

$$\underbrace{25 \div (-5)} + \underbrace{50 \div (-2)}$$

$$= -5 + (-25) \quad \text{Doing all multiplications and divisions in order from left to right}$$

$$= -30 \quad \text{Doing all additions and subtractions in order from left to right}$$

Do Exercises 22–24.

Simplify.

22. $23 - 42 \cdot 30$

23. $32 \div 8 \cdot 2$

24. $-24 \div 3 - 48 \div (-4)$

Answers

21. $5x - y - 8$　**22.** -1237　**23.** 8　**24.** 4

EXAMPLE 22 Simplify: $-2^4 + 51 \cdot 4 - (37 + 23 \cdot 2)$.

$$-2^4 + 51 \cdot 4 - (37 + 23 \cdot 2)$$

$= -2^4 + 51 \cdot 4 - (37 + 46)$ Following the rules for order of operations within the parentheses first

$= -2^4 + 51 \cdot 4 - 83$ Completing the addition inside parentheses

$= -16 + 51 \cdot 4 - 83$ Evaluating exponential expressions. Note that $-2^4 \neq (-2)^4$.

$= -16 + 204 - 83$ Doing all multiplications

$= 188 - 83$ Doing all additions and subtractions in order from left to right

$= 105$

A fraction bar can play the role of a grouping symbol, although such a symbol is not as evident as the others.

EXAMPLE 23 Simplify: $\dfrac{-64 \div (-16) \div (-2)}{2^3 - 3^2}$.

An equivalent expression with brackets as grouping symbols is

$$[-64 \div (-16) \div (-2)] \div [2^3 - 3^2].$$

This shows, in effect, that we do the calculations in the numerator and then in the denominator, and divide the results:

$$\frac{-64 \div (-16) \div (-2)}{2^3 - 3^2} = \frac{4 \div (-2)}{8 - 9} = \frac{-2}{-1} = 2.$$

Do Exercises 25 and 26.

Simplify.

25. $-4^3 + 52 \cdot 5 + 5^3 - (4^2 - 48 \div 4)$

26. $\dfrac{5 - 10 - 5 \cdot 23}{2^3 + 3^2 - 7}$

STUDY TIPS

PREPARING FOR AND TAKING A TEST

- **Do a thorough review of the chapter, focusing on the objectives and the examples.** Study the notes that you have taken in class also, as well as any hand-outs that your instructor has prepared for you.

- **Do the review exercises in the Summary and Review at the end of the chapter.** Check your answers using the answers at the back of the book. If you have trouble with an exercise, return to the objective indicated by the objective symbol given with the exercise and study that material further.

- **Do the Chapter Test at the end of the chapter.** Check your answers using the answers at the back of the book.

Use the objective symbols in the answer section to direct yourself to material that requires further study.

- **When taking a test, read each question carefully. Try to answer all the questions the first time through, but be sure to pace yourself.** Don't allow yourself to spend a disproportionate amount of time on any one question. As you answer the questions, mark those to recheck if you have time.

- **Write your test in a neat and orderly manner.** This will make it easier for you to recheck your work and will also allow your instructor to follow your work when grading your test.

Answers

25. 317   **26.** -12

Order of Operations and Grouping Symbols   Parentheses are necessary in some calculations in order to ensure that operations are performed in the desired order. To simplify $-5(3 - 6) - 12$, we press ⊝ ⑤ ❨ ③ ⊝ ⑥ ❩ ⊝ ① ② **ENTER**. The result is 3. Without parentheses, the computation is $-5 \cdot 3 - 6 - 12$, and the result is -33.

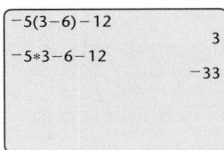

When a negative number is raised to an even power, parentheses must also be used. To find $(-3)^4$, we press ❨ ⊝ ③ ❩ ⌃ ④ **ENTER**. The result is 81. Without parentheses, the computation is $-3^4 = -1 \cdot 3^4 = -1 \cdot 81 = -81$.

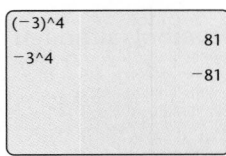

To simplify an expression like $\dfrac{49 - 104}{7 + 4}$, we must enter it as $(49 - 104) \div (7 + 4)$. We press ❨ ④ ⑨ ⊝ ① ⓪ ④ ❩ ÷ ❨ ⑦ ⊕ ④ ❩ **ENTER**. The result is -5.

$$\boxed{\begin{array}{l} (49-104)/(7+4) \\ \hspace{4.5cm} -5 \end{array}}$$

Exercises: Calculate.

1. $-8 + 4(7 - 9) + 5$

2. $-3[2 + (-5)]$

3. $7[4 - (-3)] + 5[3^2 - (-4)]$

4. $(-7)^6$

5. $(-17)^5$

6. $(-104)^3$

7. -7^6

8. -17^5

9. -104^3

10. $\dfrac{38 - 178}{5 + 30}$

11. $\dfrac{311 - 17^2}{2 - 13}$

12. $785 - \dfrac{285 - 5^4}{17 + 3 \cdot 51}$

a Find an equivalent expression without parentheses.

1. $-(2x + 7)$
2. $-(8x + 4)$
3. $-(8 - x)$
4. $-(a - b)$

5. $-(4a - 3b + 7c)$
6. $-(x - 4y - 3z)$
7. $-(6x - 8y + 5)$
8. $-(4x + 9y + 7)$

9. $-(3x - 5y - 6)$
10. $-(6a - 4b - 7)$
11. $-(-8x - 6y - 43)$
12. $-(-2a + 9b - 5c)$

b Remove parentheses and simplify.

13. $9x - (4x + 3)$
14. $4y - (2y + 5)$
15. $2a - (5a - 9)$

16. $12m - (4m - 6)$
17. $2x + 7x - (4x + 6)$
18. $3a + 2a - (4a + 7)$

19. $2x - 4y - 3(7x - 2y)$
20. $3a - 9b - 1(4a - 8b)$
21. $15x - y - 5(3x - 2y + 5z)$

22. $4a - b - 4(5a - 7b + 8c)$
23. $(3x + 2y) - 2(5x - 4y)$
24. $(-6a - b) - 5(2b + a)$

25. $(12a - 3b + 5c) - 5(-5a + 4b - 6c)$
26. $(-8x + 5y - 12) - 6(2x - 4y - 10)$

c Simplify.

27. $9 - 2(5 - 4)$

28. $6 - 5(8 - 4)$

29. $8[7 - 6(4 - 2)]$

30. $10[7 - 4(7 - 5)]$

31. $[4(9 - 6) + 11] - [14 - (6 + 4)]$

32. $[7(8 - 4) + 16] - [15 - (7 + 8)]$

33. $[10(x + 3) - 4] + [2(x - 1) + 6]$

34. $[9(x + 5) - 7] + [4(x - 12) + 9]$

35. $[7(x + 5) - 19] - [4(x - 6) + 10]$

36. $[6(x + 4) - 12] - [5(x - 8) + 14]$

37. $3\{[7(x - 2) + 4] - [2(2x - 5) + 6]\}$

38. $4\{[8(x - 3) + 9] - [4(3x - 2) + 6]\}$

39. $4\{[5(x - 3) + 2] - 3[2(x + 5) - 9]\}$

40. $3\{[6(x - 4) + 5] - 2[5(x + 8) - 3]\}$

d Simplify.

41. $8 - 2 \cdot 3 - 9$

42. $8 - (2 \cdot 3 - 9)$

43. $(8 - 2 \cdot 3) - 9$

44. $(8 - 2)(3 - 9)$

45. $[(-24) \div (-3)] \div \left(-\frac{1}{2}\right)$

46. $[32 \div (-2)] \div \left(-\frac{1}{4}\right)$

47. $16 \cdot (-24) + 50$

48. $10 \cdot 20 - 15 \cdot 24$

49. $2^4 + 2^3 - 10$

50. $40 - 3^2 - 2^3$

51. $5^3 + 26 \cdot 71 - (16 + 25 \cdot 3)$

52. $4^3 + 10 \cdot 20 + 8^2 - 23$

53. $4 \cdot 5 - 2 \cdot 6 + 4$

54. $4 \cdot (6 + 8)/(4 + 3)$

55. $4^3/8$

56. $5^3 - 7^2$

57. $8(-7) + 6(-5)$

58. $10(-5) + 1(-1)$

59. $19 - 5(-3) + 3$

60. $14 - 2(-6) + 7$

61. $9 \div (-3) + 16 \div 8$

62. $-32 - 8 \div 4 - (-2)$

63. $-4^2 + 6$

64. $-5^2 + 7$

65. $-8^2 - 3$

66. $-9^2 - 11$

67. $12 - 20^3$

68. $20 + 4^3 \div (-8)$

69. $2 \cdot 10^3 - 5000$

70. $-7(3^4) + 18$

71. $6[9 - (3 - 4)]$

72. $8[(6 - 13) - 11]$

73. $-1000 \div (-100) \div 10$

74. $256 \div (-32) \div (-4)$

75. $8 - (7 - 9)$

76. $(8 - 7) - 9$

77. $\dfrac{10 - 6^2}{9^2 + 3^2}$

78. $\dfrac{5^2 - 4^3 - 3}{9^2 - 2^2 - 1^5}$

79. $\dfrac{3(6 - 7) - 5 \cdot 4}{6 \cdot 7 - 8(4 - 1)}$

80. $\dfrac{20(8 - 3) - 4(10 - 3)}{10(2 - 6) - 2(5 + 2)}$

81. $\dfrac{|2^3 - 3^2| + |12 \cdot 5|}{-32 \div (-16) \div (-4)}$

82. $\dfrac{|3 - 5|^2 - |7 - 13|}{|12 - 9| + |11 - 14|}$

Skill Maintenance

In each of Exercises 83–90, fill in the blank with the correct term from the given list. Some of the choices may not be used and some may be used more than once.

83. The set of _____ is
$\{\ldots, -5, -4, -3, -2, -1, 0, 1, 2, 3, \ldots\}$.   [1.2a]

84. Two numbers whose sum is 0 are called
_____ of each other.   [1.3b]

85. The _____ of addition says that
$a + b = b + a$ for any real numbers a and b.   [1.7b]

86. The _____ states that for any real number a,
$a \cdot 1 = 1 \cdot a = a$.   [1.7a]

87. The _____ of addition says that
$a + (b + c) = (a + b) + c$ for any real numbers a, b, and c.   [1.7b]

88. The _____ of multiplication says that
$a(bc) = (ab)c$ for any real numbers a, b, and c.   [1.7b]

89. Two numbers whose product is 1 are called
_____ of each other.   [1.6b]

90. The equation $y + 0 = y$ illustrates the _____.
[1.7a]

natural numbers

whole numbers

integers

real numbers

multiplicative inverses

additive inverses

commutative law

associative law

distributive law

identity property of 0

identity property of 1

property of -1

Synthesis

Find an equivalent expression by enclosing the last three terms in parentheses preceded by a minus sign.

91. $6y + 2x - 3a + c$

92. $x - y - a - b$

93. $6m + 3n - 5m + 4b$

Simplify.

94. $z - \{2z - [3z - (4z - 5z) - 6z] - 7z\} - 8z$

95. $\{x - [f - (f - x)] + [x - f]\} - 3x$

96. $x - \{x - 1 - [x - 2 - (x - 3 - \{x - 4 - [x - 5 - (x - 6)]\})]\}$

97. ▦ Use your calculator to do the following.

  a) Evaluate $x^2 + 3$ when $x = 7$, when $x = -7$, and when $x = -5.013$.

  b) Evaluate $1 - x^2$ when $x = 5$, when $x = -5$, and when $x = -10.455$.

98. Express $3^3 + 3^3 + 3^3$ as a power of 3.

Find the average.

99. $-15, \ 20, \ 50, \ -82, \ -7, \ -2$

100. $-1, \ 1, \ 2, \ -2, \ 3, \ -8, \ -10$

Summary and Review

Key Terms and Properties

variable, p. 2
constant, p. 2
algebraic expression, p. 3
substitute, p. 3
evaluate, p. 3
natural numbers, p. 9
whole numbers, p. 9
integers, p. 9

opposites, p. 10
rational numbers, p. 11
terminating decimal, p. 13
repeating decimal, p. 13
irrational numbers, p. 13
real numbers, p. 14
absolute value, p. 16
additive inverse, p. 23

reciprocals, p. 46
multiplicative inverse, p. 46
equivalent expressions, p. 54
factor, p. 60
like terms, p. 62
collect like terms, p. 62

Properties of the Real-Number System

The Commutative Laws: $a + b = b + a$, $ab = ba$

The Associative Laws: $a + (b + c) = (a + b) + c$, $a(bc) = (ab)c$

The Identity Properties: $a + 0 = 0 + a = a$, $a \cdot 1 = 1 \cdot a = a$

The Inverse Properties: For any real number a, there is an opposite $-a$ such that $a + (-a) = (-a) + a = 0$.

For any nonzero real number a, there is a reciprocal $\dfrac{1}{a}$ such that $a \cdot \dfrac{1}{a} = \dfrac{1}{a} \cdot a = 1$.

The Distributive Laws: $a(b + c) = ab + ac$, $a(b - c) = ab - ac$

The Property of -1: $-1 \cdot a = -a$

Concept Reinforcement

Determine whether each statement is true or false.

_____ **1.** Every whole number is also an integer. [1.2d]

_____ **2.** The product of an even number of negative numbers is positive. [1.5a]

_____ **3.** The product of a number and its multiplicative inverse is -1. [1.6b]

_____ **4.** $a < b$ also has the meaning $b \geq a$. [1.2d]

Important Concepts

Objective 1.1a Evaluate algebraic expressions by substitution.

Example Evaluate $y - z$ when $y = 5$ and $z = -7$.
$$y - z = 5 - (-7) = 5 + 7 = 12$$

Practice Exercise

  1. Evaluate $2a + b$ when $a = -1$ and $b = 16$.

Objective 1.2d Determine which of two real numbers is greater and indicate which, using $<$ or $>$.

Example Use $<$ or $>$ for ☐ to write a true sentence:
  -5 ☐ -12.

  Since -5 is to the right of -12 on the number line, we have $-5 > -12$.

Practice Exercise

  2. Use $<$ or $>$ for ☐ to write a true sentence: -6 ☐ -3.

Objective 1.2e Find the absolute value of a real number.

Example Find the absolute value: **(a)** $|21|$; **(b)** $|-3.2|$; **(c)** $|0|$.

a) The number is positive, so the absolute value is the same as the number.
$$|21| = 21$$

b) The number is negative, so we make it positive.
$$|-3.2| = 3.2$$

c) The number is 0, so the absolute value is the same as the number.
$$|0| = 0$$

Practice Exercise

3. Find: $\left| -\dfrac{5}{4} \right|$.

Objective 1.3a Add real numbers without using the number line.

Example Add without using the number line: **(a)** $-13 + 4$; **(b)** $-2 + (-3)$.

a) We have a negative number and a positive number. The absolute values are 13 and 4. The difference is 9. The negative number has the larger absolute value, so the answer is negative.
$$-13 + 4 = -9$$

b) We have two negative numbers. The sum of the absolute values is $2 + 3$, or 5. The answer is negative.
$$-2 + (-3) = -5$$

Practice Exercise

4. Add without using the number line: $-5.6 + (-2.9)$.

Objective 1.4a Subtract real numbers.

Example Subtract: $-4 - (-6)$.
$$-4 - (-6) = -4 + 6 = 2$$

Practice Exercise

5. Subtract: $7 - 9$.

Objective 1.5a Multiply real numbers.

Example Multiply: **(a)** $-1.9(4)$; **(b)** $-7(-6)$.

a) The signs are different, so the answer is negative.
$$-1.9(4) = -7.6$$

b) The signs are the same, so the answer is positive.
$$-7(-6) = 42$$

Practice Exercise

6. Multiply: $-8(-7)$.

Objective 1.6a Divide integers.

Example Divide: **(a)** $15 \div (-3)$; **(b)** $-72 \div (-9)$.

a) The signs are different, so the answer is negative.
$$15 \div (-3) = -5$$

b) The signs are the same, so the answer is positive.
$$-72 \div (-9) = 8$$

Practice Exercise

7. Divide: $-48 \div 6$.

Objective 1.6c Divide real numbers.

Example Divide: **(a)** $-\dfrac{1}{4} \div \dfrac{3}{5}$; **(b)** $-22.4 \div (-4)$.

a) We multiply by the reciprocal of the divisor:

$$-\frac{1}{4} \div \frac{3}{5} = -\frac{1}{4} \cdot \frac{5}{3} = -\frac{5}{12}.$$

b) We carry out the long division:

$$-22.4 \div (-4) = 5.6.$$

Practice Exercise

8. Divide: $-\dfrac{3}{4} \div \left(-\dfrac{5}{3}\right)$.

Objective 1.7a Simplify fraction expressions.

Example Simplify: $-\dfrac{18x}{15x}$.

$$-\frac{18x}{15x} = -\frac{6 \cdot 3x}{5 \cdot 3x} \qquad \text{Factoring the numerator and the denominator}$$

$$= -\frac{6}{5} \cdot \frac{3x}{3x} \qquad \text{Factoring the fraction expression}$$

$$= -\frac{6}{5} \cdot 1 \qquad \frac{3x}{3x} = 1$$

$$= -\frac{6}{5} \qquad \text{Removing a factor of 1}$$

Practice Exercise

9. Simplify: $\dfrac{45y}{27y}$.

Objective 1.7c Use the distributive laws to multiply expressions like 8 and $x - y$.

Example Multiply: $3(4x - y + 2z)$.

$$3(4x - y + 2z)$$
$$= 3 \cdot 4x - 3 \cdot y + 3 \cdot 2z$$
$$= 12x - 3y + 6z$$

Practice Exercise

10. Multiply: $5(x + 3y - 4z)$.

Objective 1.7d Use the distributive laws to factor expressions like $4x - 12 + 24y$.

Example Factor: $12a - 8b + 4c$.

$$12a - 8b + 4c$$
$$= 4 \cdot 3a - 4 \cdot 2b + 4 \cdot c$$
$$= 4(3a - 2b + c)$$

Practice Exercise

11. Factor: $27x + 9y - 36z$.

Objective 1.7e Collect like terms.

Example Collect like terms: $3x - 5y + 8x + y$.

$$3x - 5y + 8x + y$$
$$= 3x + 8x - 5y + y$$
$$= 3x + 8x - 5y + 1 \cdot y$$
$$= (3 + 8)x + (-5 + 1)y$$
$$= 11x - 4y$$

Practice Exercise

12. Collect like terms: $6a - 4b - a + 2b$.

Objective 1.8b   Simplify expressions by removing parentheses and collecting like terms.

Example   Remove parentheses and simplify:
$$5x - 2(3x - y).$$
$$5x - 2(3x - y) = 5x - 6x + 2y = -x + 2y$$

Practice Exercise

13. Remove parentheses and simplify:
$$8a - b - (4a + 3b).$$

Objective 1.8d   Simplify expressions using the rules for order of operations.

Example   Simplify: $12 - (7 - 3 \cdot 6)$.
$$\begin{aligned} 12 - (7 - 3 \cdot 6) &= 12 - (7 - 18) \\ &= 12 - (-11) \\ &= 12 + 11 \\ &= 23 \end{aligned}$$

Practice Exercise

14. Simplify: $75 \div (-15) + 24 \div 8$.

Review Exercises

The review exercises that follow are for practice. Answers are at the back of the book. If you miss an exercise, restudy the objective indicated in red after the exercise or the direction line that precedes it.

1. Evaluate $\dfrac{x - y}{3}$ when $x = 17$ and $y = 5$.   [1.1a]

2. Translate to an algebraic expression: [1.1b]
   Nineteen percent of some number.

3. Tell which integers correspond to this situation: [1.2a]
   David has a debt of $45 and Joe has $72 in his savings account.

Find the absolute value.   [1.2e]

4. $|-38|$

5. $|126|$

Graph the number on the number line.   [1.2b]

6. -2.5

7. $\dfrac{8}{9}$

Use either $<$ or $>$ for $\square$ to write a true sentence.   [1.2d]

8. $-3 \ \square \ 10$

9. $-1 \ \square \ -6$

10. $0.126 \ \square \ -12.6$

11. $-\dfrac{2}{3} \ \square \ -\dfrac{1}{10}$

12. Write another inequality with the same meaning as $-3 < x$.   [1.2d]

Write true or false.   [1.2d]

13. $-9 \leq 11$

14. $-11 \geq -3$

Find the opposite.   [1.3b]

15. 3.8

16. $-\dfrac{3}{4}$

Find the reciprocal.   [1.6b]

17. $\dfrac{3}{8}$

18. -7

19. Evaluate $-x$ when $x = -34$.   [1.3b]

20. Evaluate $-(-x)$ when $x = 5$.   [1.3b]

Compute and simplify.

21. $4 + (-7)$ [1.3a]

22. $6 + (-9) + (-8) + 7$ [1.3a]

23. $-3.8 + 5.1 + (-12) + (-4.3) + 10$ [1.3a]

24. $-3 - (-7) + 7 - 10$ [1.4a]

25. $-\dfrac{9}{10} - \dfrac{1}{2}$ [1.4a]

26. $-3.8 - 4.1$ [1.4a]

27. $-9 \cdot (-6)$ [1.5a]

28. $-2.7(3.4)$ [1.5a]

29. $\dfrac{2}{3} \cdot \left(-\dfrac{3}{7}\right)$ [1.5a]

30. $3 \cdot (-7) \cdot (-2) \cdot (-5)$ [1.5a]

31. $35 \div (-5)$ [1.6a]

32. $-5.1 \div 1.7$ [1.6c]

33. $-\dfrac{3}{11} \div \left(-\dfrac{4}{11}\right)$ [1.6c]

Simplify.   [1.8d]

34. $(-3.4 - 12.2) - 8(-7)$

35. $\dfrac{-12(-3) - 2^3 - (-9)(-10)}{3 \cdot 10 + 1}$

36. $-16 \div 4 - 30 \div (-5)$

37. $\dfrac{-4[7 - (10 - 13)]}{|-2(8) - 4|}$

Solve.

38. On the first, second, and third downs, a football team had these gains and losses: 5-yd gain, 12-yd loss, and 15-yd gain, respectively. Find the total gain (or loss). [1.3c]

39. Kaleb's total assets are $170. He borrows $300. What are his total assets now?   [1.4b]

40. *Stock Price.*   The value of EFX Corp. stock began the day at $17.68 per share and dropped $1.63 per hour for 8 hr. What was the price of the stock after 8 hr? [1.5b]

41. *Checking Account Balance.*   Yuri had $68 in his checking account. After writing a check to buy seven equally priced purchases of DVDs, the balance in his account was −$64.65. What was the price of each DVD? [1.6d]

Multiply.   [1.7c]

42. $5(3x - 7)$ **43.** $-2(4x - 5)$

44. $10(0.4x + 1.5)$ **45.** $-8(3 - 6x)$

Factor.   [1.7d]

46. $2x - 14$ **47.** $-6x + 6$

48. $5x + 10$ **49.** $-3x + 12y - 12$

Collect like terms.   [1.7e]

50. $11a + 2b - 4a - 5b$

51. $7x - 3y - 9x + 8y$

52. $6x + 3y - x - 4y$

53. $-3a + 9b + 2a - b$

Remove parentheses and simplify.

54. $2a - (5a - 9)$ [1.8b]

55. $3(b + 7) - 5b$ [1.8b]

56. $3[11 - 3(4 - 1)]$ [1.8c]

57. $2[6(y - 4) + 7]$ [1.8c]

58. $[8(x + 4) - 10] - [3(x - 2) + 4]$ [1.8c]

59. $5\{[6(x - 1) + 7] - [3(3x - 4) + 8]\}$ [1.8c]

60. Factor out the greatest common factor:
$18x - 6y + 30$.   [1.7d]

    A. $2(9x - 2y + 15)$ **B.** $3(6x - 2y + 10)$
    C. $6(3x + 5)$ **D.** $6(3x - y + 5)$

61. Which expression is *not* equivalent to $mn + 5$?
[1.7b]

    A. $nm + 5$ **B.** $5n + m$
    C. $5 + mn$ **D.** $5 + nm$

Synthesis

Simplify.   [1.2e], [1.4a], [1.6a], [1.8d]

62. $-\left| \dfrac{7}{8} - \left(-\dfrac{1}{2} \right) - \dfrac{3}{4} \right|$

63. $(|2.7 - 3| + 3^2 - |-3|) \div (-3)$

64. $2000 - 1990 + 1980 - 1970 + \cdots + 20 - 10$

65. Find a formula for the perimeter of the figure below.
[1.7e]

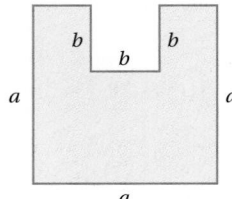

Understanding Through Discussion and Writing

1. Without actually performing the addition, explain why the sum of all integers from -50 to 50 is 0.   [1.3b]

2. What rule have we developed that would tell you the sign of $(-7)^8$ and of $(-7)^{11}$ without doing the computations? Explain.   [1.5a]

3. Explain how multiplication can be used to justify why a negative number divided by a negative number is positive.   [1.6c]

4. Explain how multiplication can be used to justify why a negative number divided by a positive number is negative.   [1.6c]

5. The distributive law was introduced before the discussion on collecting like terms. Why do you think this was done?   [1.7c, e]

6. ▨ Jake keys in $18/2 \cdot 3$ on his calculator and expects the result to be 3. What mistake is he making?   [1.8d]

CHAPTER

1

Test

For Extra Help

Step-by-step test solutions are found on the Chapter Test Prep Videos available via the Video Resources on DVD, in *MyMathLab* , and on You![Tube] (search "BittingerIntroInter" and click on "Channels").

1. Evaluate $\dfrac{3x}{y}$ when $x = 10$ and $y = 5$.

2. Translate to an algebraic expression: Nine less than some number.

Use either $<$ or $>$ for $\square$ to write a true sentence.

3. $-3 \;\square\; -8$

4. $-\dfrac{1}{2} \;\square\; -\dfrac{1}{8}$

5. $-0.78 \;\square\; -0.87$

6. Write an inequality with the same meaning as $x < -2$.

7. Write true or false: $-13 \le -3$.

Simplify.

8. $|-7|$

9. $\left|\dfrac{9}{4}\right|$

10. $|-2.7|$

Find the opposite.

11. $\dfrac{2}{3}$

12. -1.4

Find the reciprocal.

13. -2

14. $\dfrac{4}{7}$

15. Evaluate $-x$ when $x = -8$.

Compute and simplify.

16. $3.1 - (-4.7)$

17. $-8 + 4 + (-7) + 3$

18. $-\dfrac{1}{5} + \dfrac{3}{8}$

19. $2 - (-8)$

20. $3.2 - 5.7$

21. $\dfrac{1}{8} - \left(-\dfrac{3}{4}\right)$

22. $4 \cdot (-12)$

23. $-\dfrac{1}{2} \cdot \left(-\dfrac{3}{8}\right)$

24. $-45 \div 5$

25. $-\dfrac{3}{5} \div \left(-\dfrac{4}{5}\right)$

26. $4.864 \div (-0.5)$

27. $-2(16) - |2(-8) - 5^3|$

28. $-20 \div (-5) + 36 \div (-4)$

29. Maureen kept track of the changes in the stock market over a period of 5 weeks. By how many points had the market risen or fallen over this time?

WEEK 1	WEEK 2	WEEK 3	WEEK 4	WEEK 5
Down 13 pts	Down 16 pts	Up 36 pts	Down 11 pts	Up 19 pts

30. *Antarctica Highs and Lows.* The continent of Antarctica, which lies in the southern hemisphere, experiences winter in July. The average high temperature is −67°F and the average low temperature is −81°F. How much higher is the average high than the average low?

Source: National Climatic Data Center

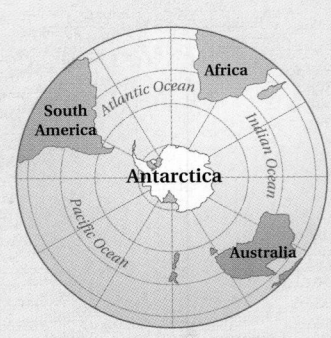

31. *Population Decrease.* The population of Mapleton was 18,600. It dropped 420 each year for 6 yr. What was the population of the city after 6 yr?

32. *Chemical Experiment.* During a chemical reaction, the temperature in a beaker decreased every minute by the same number of degrees. The temperature was 16°C at 11:08 A.M. By 11:52 A.M., the temperature had dropped to −17°C. By how many degrees did it change each minute?

Multiply.

33. $3(6 - x)$

34. $-5(y - 1)$

Factor.

35. $12 - 22x$

36. $7x + 21 + 14y$

Simplify.

37. $6 + 7 - 4 - (-3)$

38. $5x - (3x - 7)$

39. $4(2a - 3b) + a - 7$

40. $4\{3[5(y - 3) + 9] + 2(y + 8)\}$

41. $256 \div (-16) \div 4$

42. $2^3 - 10[4 - (-2 + 18)3]$

43. Which of the following is *not* a true statement?

   A. $-5 \leq -5$ **B.** $-5 < -5$

   C. $-5 \geq -5$ **D.** $-5 = -5$

Synthesis

Simplify.

44. $|-27 - 3(4)| - |-36| + |-12|$

45. $a - \{3a - [4a - (2a - 4a)]\}$

46. Find a formula for the perimeter of the figure shown here.

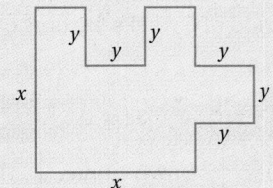

Solving Equations and Inequalities

Real-World Application

The manatee, Florida's state marine mammal, is an endangered species. An aerial wintertime manatee census counted 2817 of these animals in 2007. This was 296 fewer than the number counted in 2006. What was Florida's manatee population in 2006?

Source: Florida Fish and Wildlife Conservation Commission

This problem appears as Exercise 1 in Exercise Set 2.6.

2.1

Solving Equations: The Addition Principle

OBJECTIVES

a Determine whether a given number is a solution of a given equation.

b Solve equations using the addition principle.

SKILL TO REVIEW

Objective 1.1a: Evaluate algebraic expressions by substitution.

1. Evaluate $x - 7$ when $x = 5$.
2. Evaluate $2x + 3$ when $x = -1$.

Determine whether each equation is true, false, or neither.

1. $5 - 8 = -4$

2. $12 + 6 = 18$

3. $x + 6 = 7 - x$

a Equations and Solutions

In order to solve problems, we must learn to solve equations.

> **EQUATION**
>
> An **equation** is a number sentence that says that the expressions on either side of the equals sign, =, represent the same number.

Here are some examples of equations:

$$3 + 2 = 5, \quad 14 - 10 = 1 + 3, \quad x + 6 = 13, \quad 3x - 2 = 7 - x.$$

Equations have expressions on each side of the equals sign. The sentence "$14 - 10 = 1 + 3$" asserts that the expressions $14 - 10$ and $1 + 3$ name the same number.

Some equations are true. Some are false. Some are neither true nor false.

EXAMPLES Determine whether each equation is true, false, or neither.

1. $3 + 2 = 5$ — The equation is *true*.
2. $7 - 2 = 4$ — The equation is *false*.
3. $x + 6 = 13$ — The equation is *neither* true nor false, because we do not know what number x represents.

Do Margin Exercises 1–3.

> **SOLUTION OF AN EQUATION**
>
> Any replacement for the variable that makes an equation true is called a **solution** of the equation. To solve an equation means to find *all* of its solutions.

One way to determine whether a number is a solution of an equation is to evaluate the expression on each side of the equals sign by substitution. If the values are the same, then the number is a solution.

EXAMPLE 4 Determine whether 7 is a solution of $x + 6 = 13$.

We have

$$\begin{array}{ll} x + 6 = 13 & \text{Writing the equation} \\ \overline{7 + 6\ ?\ 13} & \text{Substituting 7 for } x \\ \quad\quad 13\ | & \text{TRUE} \end{array}$$

Since the left-hand and the right-hand sides are the same, 7 is a solution. No other number makes the equation true, so the only solution is the number 7.

Answers

Skill to Review:
1. -2 2. 1

Margin Exercises:
1. False   2. True   3. Neither

EXAMPLE 5 Determine whether 19 is a solution of $7x = 141$.

$$7x = 141 \qquad \text{Writing the equation}$$
$$7(19) \; ? \; 141 \qquad \text{Substituting 19 for } x$$
$$133 \; | \qquad \text{FALSE}$$

Since the left-hand and the right-hand sides are not the same, 19 is not a solution of the equation.

Do Exercises 4–7.

b Using the Addition Principle

Consider the equation

$$x = 7.$$

We can easily see that the solution of this equation is 7. If we replace x with 7, we get

$$7 = 7, \quad \text{which is true.}$$

Now consider the equation of Example 4: $x + 6 = 13$. In Example 4, we discovered that the solution of this equation is also 7, but the fact that 7 is the solution is not as obvious. We now begin to consider principles that allow us to start with an equation like $x + 6 = 13$ and end up with an *equivalent equation*, like $x = 7$, in which the variable is alone on one side and for which the solution is easier to find.

EQUIVALENT EQUATIONS

Equations with the same solutions are called **equivalent equations**.

One of the principles that we use in solving equations involves addition. An equation $a = b$ says that a and b stand for the same number. Suppose this is true, and we add a number c to the number a. We get the same answer if we add c to b, because a and b are the same number.

THE ADDITION PRINCIPLE FOR EQUATIONS

For any real numbers a, b, and c,

$$a = b \quad \text{is equivalent to} \quad a + c = b + c.$$

Let's solve the equation $x + 6 = 13$ using the addition principle. We want to get x alone on one side. To do so, we use the addition principle, choosing to add -6 because $6 + (-6) = 0$:

$$x + 6 = 13$$
$$x + 6 + (-6) = 13 + (-6) \qquad \text{Using the addition principle: adding } -6 \text{ on both sides}$$
$$x + 0 = 7 \qquad \text{Simplifying}$$
$$x = 7. \qquad \text{Identity property of 0: } x + 0 = x$$

The solution of $x + 6 = 13$ is 7.

Do Exercise 8.

Determine whether the given number is a solution of the given equation.

4. 8; $x + 4 = 12$

5. 0; $x + 4 = 12$

6. -3; $7 + x = -4$

7. $-\dfrac{3}{5}$; $-5x = 3$

8. Solve using the addition principle:

$$x + 2 = 11.$$

Answers

4. Yes   **5.** No   **6.** No   **7.** Yes   **8.** 9

When we use the addition principle, we sometimes say that we "add the same number on both sides of the equation." This is also true for subtraction, since we can express every subtraction as an addition. That is, since

$$a - c = b - c \quad \text{is equivalent to} \quad a + (-c) = b + (-c),$$

the addition principle tells us that we can "subtract the same number on both sides of the equation."

EXAMPLE 6 Solve: $x + 5 = -7$.

We have

$$x + 5 = -7$$
$$x + 5 - 5 = -7 - 5 \qquad \text{Using the addition principle: adding } -5 \text{ on both sides or subtracting 5 on both sides}$$
$$x + 0 = -12 \qquad \text{Simplifying}$$
$$x = -12. \qquad \text{Identity property of 0}$$

To check the answer, we substitute -12 in the original equation.

Check: $$\frac{x + 5 = -7}{-12 + 5 \;?\; -7}$$
$$-7 \;\Big|\qquad \text{TRUE}$$

The solution of the original equation is -12.

In Example 6, to get x alone, we used the addition principle and subtracted 5 on both sides. This eliminated the 5 on the left. We started with $x + 5 = -7$, and, using the addition principle, we found a simpler equation $x = -12$ for which it was easy to "see" the solution. The equations $x + 5 = -7$ and $x = -12$ are *equivalent*.

> **9.** Solve using the addition principle, subtracting 5 on both sides:
> $$x + 5 = -8.$$

Do Exercise 9.

Now we use the addition principle to solve an equation that involves a subtraction.

EXAMPLE 7 Solve: $a - 4 = 10$.

We have

$$a - 4 = 10$$
$$a - 4 + 4 = 10 + 4 \qquad \text{Using the addition principle: adding 4 on both sides}$$
$$a + 0 = 14 \qquad \text{Simplifying}$$
$$a = 14. \qquad \text{Identity property of 0}$$

Check: $$\frac{a - 4 = 10}{14 - 4 \;?\; 10}$$
$$10 \;\Big|\qquad \text{TRUE}$$

The solution is 14.

> **10.** Solve: $t - 3 = 19$.

Do Exercise 10.

Answers

9. -13 **10.** 22

88   CHAPTER 2   Solving Equations and Inequalities

EXAMPLE 8 Solve: $-6.5 = y - 8.4$.

We have

$$-6.5 = y - 8.4$$
$$-6.5 + 8.4 = y - 8.4 + 8.4 \qquad \text{Using the addition principle: adding}$$
8.4 on both sides to eliminate -8.4 on the right

$$1.9 = y.$$

Check:
$$\begin{array}{c|c} -6.5 = y - 8.4 \\ \hline -6.5 \;?\; 1.9 - 8.4 \\ \mid \; -6.5 \qquad \text{TRUE} \end{array}$$

The solution is 1.9.

Note that equations are reversible. That is, if $a = b$ is true, then $b = a$ is true. Thus when we solve $-6.5 = y - 8.4$, we can reverse it and solve $y - 8.4 = -6.5$ if we wish.

Do Exercises 11 and 12.

Solve.

11. $8.7 = n - 4.5$

12. $y + 17.4 = 10.9$

EXAMPLE 9 Solve: $-\dfrac{2}{3} + x = \dfrac{5}{2}$.

We have

$$-\frac{2}{3} + x = \frac{5}{2}$$

$$\frac{2}{3} - \frac{2}{3} + x = \frac{2}{3} + \frac{5}{2} \qquad \text{Adding } \tfrac{2}{3} \text{ on both sides}$$

$$x = \frac{2}{3} + \frac{5}{2}$$

$$x = \frac{2}{3} \cdot \frac{2}{2} + \frac{5}{2} \cdot \frac{3}{3} \qquad \begin{array}{l}\text{Multiplying by 1 to obtain equivalent}\\ \text{fraction expressions with the least}\\ \text{common denominator 6}\end{array}$$

$$x = \frac{4}{6} + \frac{15}{6}$$

$$x = \frac{19}{6}.$$

Check:
$$\begin{array}{c|c} -\dfrac{2}{3} + x = \dfrac{5}{2} \\ \hline -\dfrac{2}{3} + \dfrac{19}{6} \;?\; \dfrac{5}{2} \\ -\dfrac{4}{6} + \dfrac{19}{6} \\ \dfrac{15}{6} \\ \dfrac{5}{2} \;\bigm| \text{TRUE} \end{array}$$

The solution is $\dfrac{19}{6}$.

Do Exercises 13 and 14.

Solve.

13. $x + \dfrac{1}{2} = -\dfrac{3}{2}$

14. $t - \dfrac{13}{4} = \dfrac{5}{8}$

Answers

11. 13.2   **12.** -6.5 **13.** -2 **14.** $\dfrac{31}{8}$

a Determine whether the given number is a solution of the given equation.

1. 15; $x + 17 = 32$

2. 35; $t + 17 = 53$

3. 21; $x - 7 = 12$

4. 36; $a - 19 = 17$

5. -7; $6x = 54$

6. -9; $8y = -72$

7. 30; $\dfrac{x}{6} = 5$

8. 49; $\dfrac{y}{8} = 6$

9. 20; $5x + 7 = 107$

10. 9; $9x + 5 = 86$

11. -10; $7(y - 1) = 63$

12. -5; $6(y - 2) = 18$

b Solve using the addition principle. Don't forget to check!

13. $x + 2 = 6$

Check: $\underline{x + 2 = 6}$
$\quad\quad\quad ?$

14. $y + 4 = 11$

Check: $\underline{y + 4 = 11}$
$\quad\quad\quad ?$

15. $x + 15 = -5$

Check: $\underline{x + 15 = -5}$
$\quad\quad\quad ?$

16. $t + 10 = 44$

Check: $\underline{t + 10 = 44}$
$\quad\quad\quad ?$

17. $x + 6 = -8$

Check: $\underline{x + 6 = -8}$
$\quad\quad\quad ?$

18. $z + 9 = -14$

19. $x + 16 = -2$

20. $m + 18 = -13$

21. $x - 9 = 6$

22. $x - 11 = 12$

23. $x - 7 = -21$

24. $x - 3 = -14$

25. $5 + t = 7$

26. $8 + y = 12$

27. $-7 + y = 13$

28. $-8 + y = 17$

29. $-3 + t = -9$

30. $-8 + t = -24$

31. $x + \dfrac{1}{2} = 7$

32. $24 = -\dfrac{7}{10} + r$

33. $12 = a - 7.9$

34. $2.8 + y = 11$

35. $r + \dfrac{1}{3} = \dfrac{8}{3}$

36. $t + \dfrac{3}{8} = \dfrac{5}{8}$

37. $m + \dfrac{5}{6} = -\dfrac{11}{12}$

38. $x + \dfrac{2}{3} = -\dfrac{5}{6}$

39. $x - \dfrac{5}{6} = \dfrac{7}{8}$

40. $y - \dfrac{3}{4} = \dfrac{5}{6}$

41. $-\dfrac{1}{5} + z = -\dfrac{1}{4}$

42. $-\dfrac{1}{8} + y = -\dfrac{3}{4}$

43. $7.4 = x + 2.3$

44. $8.4 = 5.7 + y$

45. $7.6 = x - 4.8$

46. $8.6 = x - 7.4$

47. $-9.7 = -4.7 + y$

48. $-7.8 = 2.8 + x$

49. $5\dfrac{1}{6} + x = 7$

50. $5\dfrac{1}{4} = 4\dfrac{2}{3} + x$

51. $q + \dfrac{1}{3} = -\dfrac{1}{7}$

52. $52\dfrac{3}{8} = -84 + x$

Skill Maintenance

53. Add: $-3 + (-8)$. [1.3a]

54. Subtract: $-3 - (-8)$. [1.4a]

55. Multiply: $-\dfrac{2}{3} \cdot \dfrac{5}{8}$. [1.5a]

56. Divide: $-\dfrac{3}{7} \div \left(-\dfrac{9}{7}\right)$. [1.6c]

57. Divide: $\dfrac{2}{3} \div \left(-\dfrac{4}{9}\right)$. [1.6c]

58. Add: $-8.6 + 3.4$. [1.3a]

59. Subtract: $-\dfrac{2}{3} - \left(-\dfrac{5}{8}\right)$. [1.4a]

60. Multiply: $(-25.4)(-6.8)$. [1.5a]

Translate to an algebraic expression. [1.1b]

61. Jane had $83 before paying x dollars for a pair of tennis shoes. How much does she have left?

62. Justin drove his S-10 pickup truck 65 mph for t hours. How far did he drive?

Synthesis

Solve.

63. $-356.788 = -699.034 + t$

64. $-\dfrac{4}{5} + \dfrac{7}{10} = x - \dfrac{3}{4}$

65. $x + \dfrac{4}{5} = -\dfrac{2}{3} - \dfrac{4}{15}$

66. $8 - 25 = 8 + x - 21$

67. $16 + x - 22 = -16$

68. $x + x = x$

69. $x + 3 = 3 + x$

70. $x + 4 = 5 + x$

71. $-\dfrac{3}{2} + x = -\dfrac{5}{17} - \dfrac{3}{2}$

72. $|x| = 5$

73. $|x| + 6 = 19$

2.2

Solving Equations: The Multiplication Principle

OBJECTIVE

a Solve equations using the multiplication principle.

SKILL TO REVIEW
Objective 1.6b: Find the reciprocal of a real number.

Find the reciprocal.

1. 5
2. $-\dfrac{5}{4}$

a Using the Multiplication Principle

Suppose that $a = b$ is true, and we multiply a by some number c. We get the same number if we multiply b by c, because a and b are the same number.

> **THE MULTIPLICATION PRINCIPLE FOR EQUATIONS**
>
> For any real numbers a, b, and c, $c \neq 0$,
> $$a = b \quad \text{is equivalent to} \quad a \cdot c = b \cdot c.$$

When using the multiplication principle, we sometimes say that we "multiply on both sides of the equation by the same number."

EXAMPLE 1 Solve: $5x = 70$.

To get x alone, we multiply by the *multiplicative inverse*, or *reciprocal*, of 5. Then we get the *multiplicative identity* 1 times x, or $1 \cdot x$, which simplifies to x. This allows us to eliminate 5 on the left.

$$5x = 70 \qquad \text{The reciprocal of 5 is } \tfrac{1}{5}.$$

$$\frac{1}{5} \cdot 5x = \frac{1}{5} \cdot 70 \qquad \text{Multiplying by } \tfrac{1}{5} \text{ to get } 1 \cdot x \text{ and eliminate 5 on the left}$$

$$1 \cdot x = 14 \qquad \text{Simplifying}$$

$$x = 14 \qquad \text{Identity property of 1: } 1 \cdot x = x$$

Check:
$$\begin{array}{c} 5x = 70 \\ \hline 5 \cdot 14 \; ? \; 70 \\ 70 \;\big|\; \quad \text{TRUE} \end{array}$$

The solution is 14.

The multiplication principle also tells us that we can "divide on both sides of the equation by the same nonzero number." This is because dividing is the same as multiplying by a reciprocal. That is,

$$\frac{a}{c} = \frac{b}{c} \quad \text{is equivalent to} \quad a \cdot \frac{1}{c} = b \cdot \frac{1}{c}, \quad \text{when } c \neq 0.$$

In an expression like $5x$ in Example 1, the number 5 is called the **coefficient**. Example 1 could be done as follows, dividing on both sides by 5, the coefficient of x.

EXAMPLE 2 Solve: $5x = 70$.

$$5x = 70$$

$$\frac{5x}{5} = \frac{70}{5} \qquad \text{Dividing by 5 on both sides}$$

$$1 \cdot x = 14 \qquad \text{Simplifying}$$

$$x = 14 \qquad \text{Identity property of 1. The solution is 14.}$$

Answers

Skill to Review:

1. $\dfrac{1}{5}$ 2. $-\dfrac{4}{5}$

Do Exercises 1 and 2.

1. Solve. Multiply on both sides.
$$6x = 90$$

EXAMPLE 3 Solve: $-4x = 92$.

We have

$-4x = 92$

$\dfrac{-4x}{-4} = \dfrac{92}{-4}$ Using the multiplication principle. Dividing by -4 on both sides is the same as multiplying by $-\frac{1}{4}$.

$1 \cdot x = -23$ Simplifying

$x = -23$.   Identity property of 1

2. Solve. Divide on both sides.
$$4x = -7$$

Check: $\dfrac{-4x = 92}{-4(-23) \;?\; 92}$

            $92 \;\big|$ TRUE

The solution is -23.

Do Exercise 3.

3. Solve: $-6x = 108$.

EXAMPLE 4 Solve: $-x = 9$.

We have

$-x = 9$

$-1 \cdot x = 9$ Using the property of -1: $-x = -1 \cdot x$

$\dfrac{-1 \cdot x}{-1} = \dfrac{9}{-1}$ Dividing by -1 on both sides: $-1/(-1) = 1$

$1 \cdot x = -9$

$x = -9$.

Check: $\dfrac{-x = 9}{-(-9) \;?\; 9}$

          $9 \;\big|$ TRUE

The solution is -9.

Do Exercise 4.

4. Solve. Divide on both sides.
$$-x = -10$$

We can also solve the equation $-x = 9$ by multiplying as follows.

EXAMPLE 5 Solve: $-x = 9$.

We have

$-x = 9$

$-1 \cdot (-x) = -1 \cdot 9$ Multiplying by -1 on both sides

$-1 \cdot (-1) \cdot x = -9$ $-x = (-1) \cdot x$

$1 \cdot x = -9$ $-1 \cdot (-1) = 1$

$x = -9$.

The solution is -9.

5. Solve. Multiply on both sides.
$$-x = -10$$

Do Exercise 5.

Answers

1. 15   **2.** $-\dfrac{7}{4}$ **3.** -18 **4.** 10   **5.** 10

In practice, it is generally more convenient to divide on both sides of the equation if the coefficient of the variable is in decimal notation or is an integer. If the coefficient is in fraction notation, it is usually more convenient to multiply by a reciprocal.

EXAMPLE 6 Solve: $\dfrac{3}{8} = -\dfrac{5}{4}x$.

$$\dfrac{3}{8} = -\dfrac{5}{4}x$$

The reciprocal of $-\frac{5}{4}$ is $-\frac{4}{5}$. There is no sign change.

$$-\dfrac{4}{5} \cdot \dfrac{3}{8} = -\dfrac{4}{5} \cdot \left(-\dfrac{5}{4}x\right)$$

Multiplying by $-\frac{4}{5}$ to get $1 \cdot x$ and eliminate $-\frac{5}{4}$ on the right

$$-\dfrac{12}{40} = 1 \cdot x$$

$$-\dfrac{3}{10} = 1 \cdot x$$ Simplifying

$$-\dfrac{3}{10} = x$$ Identity property of 1

Check: $\dfrac{3}{8} = -\dfrac{5}{4}x$

$$\dfrac{3}{8} \;\overset{?}{\vert}\; -\dfrac{5}{4}\left(-\dfrac{3}{10}\right)$$
$$\dfrac{3}{8} \qquad \text{TRUE}$$

The solution is $-\dfrac{3}{10}$.

As noted in Section 2.1, if $a = b$ is true, then $b = a$ is true. Thus we can reverse the equation $\frac{3}{8} = -\frac{5}{4}x$ and solve $-\frac{5}{4}x = \frac{3}{8}$ if we wish.

Do Exercise 6.

6. Solve: $\dfrac{2}{3} = -\dfrac{5}{6}y$.

EXAMPLE 7 Solve: $1.16y = 9744$.

$$1.16y = 9744$$

$$\dfrac{1.16y}{1.16} = \dfrac{9744}{1.16}$$ Dividing by 1.16 on both sides

$$y = \dfrac{9744}{1.16}$$

$$y = 8400$$ Simplifying

Check: $1.16y = 9744$

$$1.16(8400) \;\overset{?}{} \; 9744$$
$$9744 \;\vert\; \qquad \text{TRUE}$$

The solution is 8400.

Do Exercises 7 and 8.

Solve.

7. $1.12x = 8736$

8. $6.3 = -2.1y$

Answers

6. $-\dfrac{4}{5}$ 7. 7800   8. -3

Now we use the multiplication principle to solve an equation that involves division.

EXAMPLE 8 Solve: $\dfrac{-y}{9} = 14$.

$$\dfrac{-y}{9} = 14$$

$$9 \cdot \dfrac{-y}{9} = 9 \cdot 14 \qquad \text{Multiplying by 9 on both sides}$$

$$-y = 126$$

$$-1 \cdot (-y) = -1 \cdot 126 \qquad \text{Multiplying by } -1 \text{ on both sides}$$

$$y = -126$$

Check:
$$\dfrac{-y}{9} = 14$$

$$\dfrac{-(-126)}{9} \;\;?\;\; 14$$

$$\dfrac{126}{9}$$

$$14 \qquad \text{TRUE}$$

The solution is -126.

There are other ways to solve the equation in Example 8. One is by multiplying by -9 on both sides as follows:

$$-9 \cdot \dfrac{-y}{9} = -9 \cdot 14$$

$$\dfrac{(-9)(-y)}{9} = -126$$

$$\dfrac{9y}{9} = -126$$

$$y = -126.$$

Do Exercise 9.

9. Solve: $-14 = \dfrac{-y}{2}$.

Answer

9. 28

a   Solve using the multiplication principle. Don't forget to check!

1. $6x = 36$

Check: $\dfrac{6x = 36}{}$?

2. $3x = 51$

Check: $\dfrac{3x = 51}{}$?

3. $5y = 45$

Check: $\dfrac{5y = 45}{}$?

4. $8y = 72$

Check: $\dfrac{8y = 72}{}$?

5. $84 = 7x$

6. $63 = 9x$

7. $-x = 40$

8. $-x = 53$

9. $-1 = -z$

10. $-47 = -t$

11. $7x = -49$

12. $8x = -56$

13. $-12x = 72$

14. $-15x = 105$

15. $-21w = -126$

16. $-13w = -104$

17. $\dfrac{t}{7} = -9$

18. $\dfrac{y}{5} = -6$

19. $\dfrac{n}{-6} = 8$

20. $\dfrac{y}{-8} = 11$

21. $\dfrac{3}{4}x = 27$

22. $\dfrac{4}{5}x = 16$

23. $-\dfrac{2}{3}x = 6$

24. $-\dfrac{3}{8}x = 12$

25. $\dfrac{-t}{3} = 7$

26. $\dfrac{-x}{6} = 9$

27. $-\dfrac{m}{3} = \dfrac{1}{5}$

28. $\dfrac{1}{8} = -\dfrac{y}{5}$

29. $-\dfrac{3}{5}r = \dfrac{9}{10}$

30. $-\dfrac{2}{5}y = \dfrac{4}{15}$

31. $-\dfrac{3}{2}r = -\dfrac{27}{4}$

32. $-\dfrac{3}{8}x = -\dfrac{15}{16}$

33. $6.3x = 44.1$

34. $2.7y = 54$

35. $-3.1y = 21.7$

36. $-3.3y = 6.6$

37. $38.7m = 309.6$

38. $29.4m = 235.2$

39. $-\dfrac{2}{3}y = -10.6$

40. $-\dfrac{9}{7}y = 12.06$

41. $\dfrac{-x}{5} = 10$

42. $\dfrac{-x}{8} = -16$

43. $-\dfrac{t}{2} = 7$

44. $\dfrac{m}{-3} = 10$

Skill Maintenance

Collect like terms.   [1.7e]

45. $3x + 4x$

46. $6x + 5 - 7x$

47. $-4x + 11 - 6x + 18x$

48. $8y - 16y - 24y$

Remove parentheses and simplify.   [1.8b]

49. $3x - (4 + 2x)$

50. $2 - 5(x + 5)$

51. $8y - 6(3y + 7)$

52. $-2a - 4(5a - 1)$

Translate to an algebraic expression.   [1.1b]

53. Patty drives her van for 8 hr at a speed of r miles per hour. How far does she drive?

54. A triangle has a height of 10 meters and a base of b meters. What is the area of the triangle?

Synthesis

Solve.

55. $-0.2344m = 2028.732$

56. $0 \cdot x = 0$

57. $0 \cdot x = 9$

58. $4|x| = 48$

59. $2|x| = -12$

Solve for x.

60. $ax = 5a$

61. $3x = \dfrac{b}{a}$

62. $cx = a^2 + 1$

63. $\dfrac{a}{b}x = 4$

64. A student makes a calculation and gets an answer of 22.5. On the last step, she multiplies by 0.3 when she should have divided by 0.3. What is the correct answer?

2.3

Using the Principles Together

OBJECTIVES

a) Solve equations using both the addition principle and the multiplication principle.

b) Solve equations in which like terms may need to be collected.

c) Solve equations by first removing parentheses and collecting like terms; solve equations with an infinite number of solutions and equations with no solutions.

a) Applying Both Principles

Consider the equation $3x + 4 = 13$. It is more complicated than those we discussed in the preceding two sections. In order to solve such an equation, we first isolate the x-term, $3x$, using the addition principle. Then we apply the multiplication principle to get x by itself.

EXAMPLE 1 Solve: $3x + 4 = 13$.

$$3x + 4 = 13$$
$$3x + 4 - 4 = 13 - 4 \quad \text{Using the addition principle: subtracting 4 on both sides}$$

> First isolate the x-term. $\rightarrow 3x = 9$ Simplifying

$$\frac{3x}{3} = \frac{9}{3} \quad \text{Using the multiplication principle: dividing by 3 on both sides}$$

> Then isolate x. $\rightarrow x = 3$ Simplifying

Check:
$$\begin{array}{c|c} 3x + 4 = 13 \\ \hline 3 \cdot 3 + 4 \ ? \ 13 \\ 9 + 4 \\ 13 \ | \ \text{TRUE} \end{array}$$

We use the rules for order of operations to carry out the check. We find the product $3 \cdot 3$. Then we add 4.

The solution is 3.

1. Solve: $9x + 6 = 51$.

> Do Exercise 1.

EXAMPLE 2 Solve: $-5x - 6 = 16$.

$$-5x - 6 = 16$$
$$-5x - 6 + 6 = 16 + 6 \quad \text{Adding 6 on both sides}$$
$$-5x = 22$$
$$\frac{-5x}{-5} = \frac{22}{-5} \quad \text{Dividing by } -5 \text{ on both sides}$$
$$x = -\frac{22}{5}, \text{ or } -4\frac{2}{5} \quad \text{Simplifying}$$

Check:
$$\begin{array}{c|c} -5x - 6 = 16 \\ \hline -5\left(-\dfrac{22}{5}\right) - 6 \ ? \ 16 \\ 22 - 6 \\ 16 \ | \ \text{TRUE} \end{array}$$

Solve.

2. $8x - 4 = 28$

3. $-\dfrac{1}{2}x + 3 = 1$

The solution is $-\frac{22}{5}$.

> Do Exercises 2 and 3.

Answers

1. 5   2. 4   3. 4

EXAMPLE 3 Solve: $45 - t = 13$.

$$45 - t = 13$$
$$-45 + 45 - t = -45 + 13 \qquad \text{Adding } -45 \text{ on both sides}$$
$$-t = -32$$
$$-1(-t) = -1(-32) \qquad \text{Multiplying by } -1 \text{ on both sides}$$
$$t = 32$$

The number 32 checks and is the solution.

Do Exercise 4.

4. Solve: $-18 - m = -57$.

EXAMPLE 4 Solve: $16.3 - 7.2y = -8.18$.

$$16.3 - 7.2y = -8.18$$
$$-16.3 + 16.3 - 7.2y = -16.3 + (-8.18) \qquad \text{Adding } -16.3 \text{ on both sides}$$
$$-7.2y = -24.48$$
$$\frac{-7.2y}{-7.2} = \frac{-24.48}{-7.2} \qquad \text{Dividing by } -7.2 \text{ on both sides}$$
$$y = 3.4$$

Check:
$$\begin{array}{c} 16.3 - 7.2y = -8.18 \\ \hline 16.3 - 7.2(3.4) \;?\; -8.18 \\ 16.3 - 24.48 \;\big|\; \\ -8.18 \;\big|\; \qquad \text{TRUE} \end{array}$$

The solution is 3.4.

Do Exercises 5 and 6.

Solve.

5. $-4 - 8x = 8$

6. $41.68 = 4.7 - 8.6y$

(b) Collecting Like Terms

If there are like terms on one side of the equation, we collect them before using the addition principle or the multiplication principle.

EXAMPLE 5 Solve: $3x + 4x = -14$.

$$3x + 4x = -14$$
$$7x = -14 \qquad \text{Collecting like terms}$$
$$\frac{7x}{7} = \frac{-14}{7} \qquad \text{Dividing by 7 on both sides}$$
$$x = -2$$

The number -2 checks, so the solution is -2.

Do Exercises 7 and 8.

Solve.

7. $4x + 3x = -21$

8. $x - 0.09x = 728$

If there are like terms on opposite sides of the equation, we get them on the same side by using the addition principle. Then we collect them. In other words, we get all the terms with a variable on one side of the equation and all the terms without a variable on the other side.

Answers

4. 39 **5.** $-\dfrac{3}{2}$ **6.** -4.3

7. -3 **8.** 800

EXAMPLE 6 Solve: $2x - 2 = -3x + 3$.

$$2x - 2 = -3x + 3$$
$$2x - 2 + 2 = -3x + 3 + 2 \qquad \text{Adding 2}$$
$$2x = -3x + 5 \qquad \text{Collecting like terms}$$
$$2x + 3x = -3x + 3x + 5 \qquad \text{Adding } 3x$$
$$5x = 5 \qquad \text{Simplifying}$$
$$\frac{5x}{5} = \frac{5}{5} \qquad \text{Dividing by 5}$$
$$x = 1 \qquad \text{Simplifying}$$

Check:
$$\begin{array}{c|c} 2x - 2 &= -3x + 3 \\ \hline 2 \cdot 1 - 2 \ ? & -3 \cdot 1 + 3 \qquad \text{Substituting in the original equation} \\ 2 - 2 & -3 + 3 \\ 0 & 0 \qquad \text{TRUE} \end{array}$$

The solution is 1.

Do Exercises 9 and 10.

In Example 6, we used the addition principle to get all the terms with an x on one side of the equation and all the terms without an x on the other side. Then we collected like terms and proceeded as before. If there are like terms on one side at the outset, they should be collected first.

EXAMPLE 7 Solve: $6x + 5 - 7x = 10 - 4x + 3$.

$$6x + 5 - 7x = 10 - 4x + 3$$
$$-x + 5 = 13 - 4x \qquad \text{Collecting like terms}$$
$$4x - x + 5 = 13 - 4x + 4x \qquad \begin{array}{l} \text{Adding } 4x \text{ to get all terms with a} \\ \text{variable on one side} \end{array}$$
$$3x + 5 = 13 \qquad \begin{array}{l} \text{Simplifying; that is, collecting} \\ \text{like terms} \end{array}$$
$$3x + 5 - 5 = 13 - 5 \qquad \text{Subtracting 5}$$
$$3x = 8 \qquad \text{Simplifying}$$
$$\frac{3x}{3} = \frac{8}{3} \qquad \text{Dividing by 3}$$
$$x = \frac{8}{3} \qquad \text{Simplifying}$$

The number $\frac{8}{3}$ checks, so it is the solution.

Do Exercises 11 and 12.

Clearing Fractions and Decimals

In general, equations are easier to solve if they do not contain fractions or decimals. Consider, for example, the equations

$$\frac{1}{2}x + 5 = \frac{3}{4} \quad \text{and} \quad 2.3x + 7 = 5.4.$$

Solve.

9. $7y + 5 = 2y + 10$

10. $5 - 2y = 3y - 5$

Solve.

11. $7x - 17 + 2x = 2 - 8x + 15$

12. $3x - 15 = 5x + 2 - 4x$

Answers

9. 1 **10.** 2 **11.** 2 **12.** $\frac{17}{2}$

If we multiply by 4 on both sides of the first equation and by 10 on both sides of the second equation, we have

$$4\left(\frac{1}{2}x + 5\right) = 4 \cdot \frac{3}{4} \quad \text{and} \quad 10(2.3x + 7) = 10 \cdot 5.4$$

$$4 \cdot \frac{1}{2}x + 4 \cdot 5 = 4 \cdot \frac{3}{4} \quad \text{and} \quad 10 \cdot 2.3x + 10 \cdot 7 = 10 \cdot 5.4$$

$$2x + 20 = 3 \quad \text{and} \quad 23x + 70 = 54.$$

The first equation has been "cleared of fractions" and the second equation has been "cleared of decimals." Both resulting equations are equivalent to the original equations and are easier to solve. *It is your choice* whether to clear fractions or decimals, but doing so often eases computations.

The easiest way to clear an equation of fractions is to multiply *every term on both sides* by the **least common multiple of all the denominators**.

EXAMPLE 8 Solve: $\frac{2}{3}x - \frac{1}{6} + \frac{1}{2}x = \frac{7}{6} + 2x.$

The denominators are 3, 6, and 2. The number 6 is the least common multiple of all the denominators. We multiply by 6 on both sides of the equation.

$$6\left(\frac{2}{3}x - \frac{1}{6} + \frac{1}{2}x\right) = 6\left(\frac{7}{6} + 2x\right) \qquad \text{Multiplying by 6 on both sides}$$

$$6 \cdot \frac{2}{3}x - 6 \cdot \frac{1}{6} + 6 \cdot \frac{1}{2}x = 6 \cdot \frac{7}{6} + 6 \cdot 2x \qquad \begin{array}{l}\text{Using the distributive law} \\ \text{(\textit{Caution}! Be sure to mul-} \\ \text{tiply \textit{all} the terms by 6.)}\end{array}$$

$$4x - 1 + 3x = 7 + 12x \qquad \begin{array}{l}\text{Simplifying. Note that the} \\ \text{fractions are cleared.}\end{array}$$

$$7x - 1 = 7 + 12x \qquad \text{Collecting like terms}$$

$$7x - 1 - 12x = 7 + 12x - 12x \qquad \text{Subtracting } 12x$$

$$-5x - 1 = 7 \qquad \text{Collecting like terms}$$

$$-5x - 1 + 1 = 7 + 1 \qquad \text{Adding 1}$$

$$-5x = 8 \qquad \text{Collecting like terms}$$

$$\frac{-5x}{-5} = \frac{8}{-5} \qquad \text{Dividing by } -5$$

$$x = -\frac{8}{5}$$

Check:

$$\frac{2}{3}x - \frac{1}{6} + \frac{1}{2}x = \frac{7}{6} + 2x$$

$$\frac{2}{3}\left(-\frac{8}{5}\right) - \frac{1}{6} + \frac{1}{2}\left(-\frac{8}{5}\right) \ \overset{?}{\vert} \ \frac{7}{6} + 2\left(-\frac{8}{5}\right)$$

$$-\frac{16}{15} - \frac{1}{6} - \frac{8}{10} \ \vert \ \frac{7}{6} - \frac{16}{5}$$

$$-\frac{32}{30} - \frac{5}{30} - \frac{24}{30} \ \vert \ \frac{35}{30} - \frac{96}{30}$$

$$\frac{-32 - 5 - 24}{30} \ \vert \ \frac{35 - 96}{30}$$

$$-\frac{61}{30} \ \vert \ -\frac{61}{30} \qquad \text{TRUE}$$

The solution is $-\frac{8}{5}$.

Do Exercise 13.

Calculator Corner

Checking Possible Solutions There are several ways to check the possible solutions of an equation on a calculator. One of the most straightforward methods is to substitute and carry out the calculations on each side of the equation just as we do when we check by hand. To check the possible solution, 1, in Example 6, for instance, we first substitute 1 for x in the expression on the left side of the equation. We press ② ⊗ ① ⊖ ② **ENTER**. We get 0. Next, we substitute 1 for x in the expression on the right side of the equation. We then press ⊖ ③ ⊗ ① ⊕ ③ **ENTER**. Again we get 0. Since the two sides of the equation have the same value when x is 1, we know that 1 is the solution of the equation.

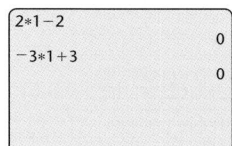

Exercise:

1. Use substitution to check the solutions found in Examples 1–5.

Caution!

Check the possible solution in the *original* equation rather than in the equation that has been cleared of fractions.

13. Solve: $\frac{7}{8}x - \frac{1}{4} + \frac{1}{2}x = \frac{3}{4} + x.$

Answer

13. $\frac{8}{3}$

To illustrate clearing decimals, we repeat Example 4, but this time we clear the equation of decimals first. Compare both methods.

To clear an equation of decimals, we count the greatest number of decimal places in any one number. If the greatest number of decimal places is 1, we multiply every term on both sides by 10; if it is 2, we multiply by 100; and so on.

EXAMPLE 9   Solve: $16.3 - 7.2y = -8.18$.

The greatest number of decimal places in any one number is *two*. Multiplying by 100, which has *two* 0's, will clear all decimals.

$$100(16.3 - 7.2y) = 100(-8.18) \qquad \text{Multiplying by 100 on both sides}$$

$$100(16.3) - 100(7.2y) = 100(-8.18) \qquad \text{Using the distributive law}$$

$$1630 - 720y = -818 \qquad \text{Simplifying}$$

$$1630 - 720y - 1630 = -818 - 1630 \qquad \text{Subtracting 1630}$$

$$-720y = -2448 \qquad \text{Collecting like terms}$$

$$\frac{-720y}{-720} = \frac{-2448}{-720} \qquad \text{Dividing by } -720$$

$$y = \frac{17}{5}, \text{ or } 3.4$$

The number $\frac{17}{5}$, or 3.4, checks, as shown in Example 4, so it is the solution.

14. Solve: $41.68 = 4.7 - 8.6y$.

Do Exercise 14.

(c) Equations Containing Parentheses

To solve certain kinds of equations that contain parentheses, we first use the distributive laws to remove the parentheses. Then we proceed as before.

EXAMPLE 10   Solve: $8x = 2(12 - 2x)$.

$$8x = 2(12 - 2x)$$

$$8x = 24 - 4x \qquad \text{Using the distributive laws to multiply and remove parentheses}$$

$$8x + 4x = 24 - 4x + 4x \qquad \text{Adding } 4x \text{ to get all the } x\text{-terms on one side}$$

$$12x = 24 \qquad \text{Collecting like terms}$$

$$\frac{12x}{12} = \frac{24}{12} \qquad \text{Dividing by 12}$$

$$x = 2$$

The number 2 checks, so the solution is 2.

Do Exercises 15 and 16.

Solve.

15. $2(2y + 3) = 14$

16. $5(3x - 2) = 35$

Answers

14. $-\frac{43}{10}$, or -4.3

15. 2   **16.** 3

Here is a procedure for solving the types of equation discussed in this section.

> **AN EQUATION-SOLVING PROCEDURE**
>
> 1. Multiply on both sides to clear the equation of fractions or decimals. (This is optional, but it can ease computations.)
> 2. If parentheses occur, multiply to remove them using the *distributive laws.*
> 3. Collect like terms on each side, if necessary.
> 4. Get all terms with variables on one side and all numbers (constant terms) on the other side, using the *addition principle.*
> 5. Collect like terms again, if necessary.
> 6. Multiply or divide to solve for the variable, using the *multiplication principle.*
> 7. Check all possible solutions in the original equation.

EXAMPLE 11 Solve: $2 - 5(x + 5) = 3(x - 2) - 1$.

$$2 - 5(x + 5) = 3(x - 2) - 1$$

$2 - 5x - 25 = 3x - 6 - 1$ Using the distributive laws to multiply and remove parentheses

$-5x - 23 = 3x - 7$ Collecting like terms

$-5x - 23 + 5x = 3x - 7 + 5x$ Adding $5x$

$-23 = 8x - 7$ Collecting like terms

$-23 + 7 = 8x - 7 + 7$ Adding 7

$-16 = 8x$ Collecting like terms

$\dfrac{-16}{8} = \dfrac{8x}{8}$ Dividing by 8

$-2 = x$

Check:

$$\dfrac{2 - 5(x + 5) = 3(x - 2) - 1}{2 - 5(-2 + 5) \overset{?}{} 3(-2 - 2) - 1}$$

$$\begin{array}{c|c} 2 - 5(3) & 3(-4) - 1 \\ 2 - 15 & -12 - 1 \\ -13 & -13 \end{array} \quad \text{TRUE}$$

The solution is -2.

Do Exercises 17 and 18.

Equations with Infinitely Many Solutions

The types of equations we have considered thus far in Sections 2.1–2.3 have all had exactly one solution. We now look at two other possibilities.
   Consider

$$3 + x = x + 3.$$

Let's explore the equation and possible solutions in Margin Exercises 19–22.

Do Exercises 19–22.

Solve.

17. $3(7 + 2x) = 30 + 7(x - 1)$

18. $4(3 + 5x) - 4 = 3 + 2(x - 2)$

Determine whether the given number is a solution of the given equation.

19. $10; \ 3 + x = x + 3$

20. $-7; \ 3 + x = x + 3$

21. $\dfrac{1}{2}; \ 3 + x = x + 3$

22. $0; \ 3 + x = x + 3$

Answers

17. -2 **18.** $-\dfrac{1}{2}$ **19.** Yes **20.** Yes
21. Yes **22.** Yes

We know by the commutative law of addition that the equation $3 + x = x + 3$ holds for any replacement of x with a real number. (See Section 1.7.) We have confirmed some of these solutions in Margin Exercises 19–22. Suppose we try to solve this equation using the addition principle:

$$3 + x = x + 3$$
$$-x + 3 + x = -x + x + 3 \qquad \text{Adding } -x$$
$$3 = 3. \qquad\qquad \text{True}$$

We end with a true equation. The original equation holds for all real-number replacements. Every real number is a solution. Thus the number of solutions is **infinite**.

EXAMPLE 12 Solve: $7x - 17 = 4 + 7(x - 3)$.

$$7x - 17 = 4 + 7(x - 3)$$
$$7x - 17 = 4 + 7x - 21 \qquad \text{Using the distributive law to multiply and remove parentheses}$$
$$7x - 17 = 7x - 17 \qquad \text{Collecting like terms}$$
$$-7x + 7x - 17 = -7x + 7x - 17 \qquad \text{Adding } -7x$$
$$-17 = -17 \qquad \text{True for all real numbers}$$

Every real number is a solution. There are infinitely many solutions.

Equations with No Solution

Now consider

$$3 + x = x + 8.$$

Let's explore the equation and possible solutions in Margin Exercises 23–26.

> Do Exercises 23–26.

None of the replacements in Margin Exercises 23–26 is a solution of the given equation. In fact, there are no solutions. Let's try to solve this equation using the addition principle:

$$3 + x = x + 8$$
$$-x + 3 + x = -x + x + 8 \qquad \text{Adding } -x$$
$$3 = 8. \qquad\qquad \text{False}$$

We end with a false equation. The original equation is false for all real-number replacements. Thus it has **no** solution.

EXAMPLE 13 Solve: $3x + 4(x + 2) = 11 + 7x$.

$$3x + 4(x + 2) = 11 + 7x$$
$$3x + 4x + 8 = 11 + 7x \qquad \text{Using the distributive law to multiply and remove parentheses}$$
$$7x + 8 = 11 + 7x \qquad \text{Collecting like terms}$$
$$7x + 8 - 7x = 11 + 7x - 7x \qquad \text{Subtracting } 7x$$
$$8 = 11 \qquad \text{False}$$

There are no solutions.

> Do Exercises 27 and 28.

Answers

23. No   **24.** No   **25.** No   **26.** No
27. All real numbers   **28.** No solution

a Solve. Don't forget to check!

1. $5x + 6 = 31$

Check: $5x + 6 = 31$

2. $7x + 6 = 13$

Check: $7x + 6 = 13$

3. $8x + 4 = 68$

Check: $8x + 4 = 68$

4. $4y + 10 = 46$

Check: $4y + 10 = 46$

5. $4x - 6 = 34$

6. $5y - 2 = 53$

7. $3x - 9 = 33$

8. $4x - 19 = 5$

9. $7x + 2 = -54$

10. $5x + 4 = -41$

11. $-45 = 3 + 6y$

12. $-91 = 9t + 8$

13. $-4x + 7 = 35$

14. $-5x - 7 = 108$

15. $\dfrac{5}{4}x - 18 = -3$

16. $\dfrac{3}{2}x - 24 = -36$

b Solve.

17. $5x + 7x = 72$

Check: $5x + 7x = 72$

18. $8x + 3x = 55$

Check: $8x + 3x = 55$

19. $8x + 7x = 60$

Check: $8x + 7x = 60$

20. $8x + 5x = 104$

Check: $8x + 5x = 104$

21. $4x + 3x = 42$

22. $7x + 18x = 125$

23. $-6y - 3y = 27$

24. $-5y - 7y = 144$

25. $-7y - 8y = -15$

26. $-10y - 3y = -39$

27. $x + \dfrac{1}{3}x = 8$

28. $x + \dfrac{1}{4}x = 10$

29. $10.2y - 7.3y = -58$

30. $6.8y - 2.4y = -88$

31. $8y - 35 = 3y$

32. $4x - 6 = 6x$

33. $8x - 1 = 23 - 4x$

34. $5y - 2 = 28 - y$

35. $2x - 1 = 4 + x$

36. $4 - 3x = 6 - 7x$

37. $6x + 3 = 2x + 11$

38. $14 - 6a = -2a + 3$

39. $5 - 2x = 3x - 7x + 25$

40. $-7z + 2z - 3z - 7 = 17$

41. $4 + 3x - 6 = 3x + 2 - x$

42. $5 + 4x - 7 = 4x - 2 - x$

43. $4y - 4 + y + 24 = 6y + 20 - 4y$

44. $5y - 7 + y = 7y + 21 - 5y$

Solve. Clear fractions or decimals first.

45. $\dfrac{7}{2}x + \dfrac{1}{2}x = 3x + \dfrac{3}{2} + \dfrac{5}{2}x$

46. $\dfrac{7}{8}x - \dfrac{1}{4} + \dfrac{3}{4}x = \dfrac{1}{16} + x$

47. $\dfrac{2}{3} + \dfrac{1}{4}t = \dfrac{1}{3}$

48. $-\dfrac{3}{2} + x = -\dfrac{5}{6} - \dfrac{4}{3}$

49. $\dfrac{2}{3} + 3y = 5y - \dfrac{2}{15}$

50. $\dfrac{1}{2} + 4m = 3m - \dfrac{5}{2}$

51. $\dfrac{5}{3} + \dfrac{2}{3}x = \dfrac{25}{12} + \dfrac{5}{4}x + \dfrac{3}{4}$

52. $1 - \dfrac{2}{3}y = \dfrac{9}{5} - \dfrac{y}{5} + \dfrac{3}{5}$

53. $2.1x + 45.2 = 3.2 - 8.4x$

54. $0.96y - 0.79 = 0.21y + 0.46$

55. $1.03 - 0.62x = 0.71 - 0.22x$

56. $1.7t + 8 - 1.62t = 0.4t - 0.32 + 8$

57. $\frac{2}{7}x - \frac{1}{2}x = \frac{3}{4}x + 1$

58. $\frac{5}{16}y + \frac{3}{8}y = 2 + \frac{1}{4}y$

c Solve.

59. $3(2y - 3) = 27$

60. $8(3x + 2) = 30$

61. $40 = 5(3x + 2)$

62. $9 = 3(5x - 2)$

63. $-23 + y = y + 25$

64. $17 - t = -t + 68$

65. $-23 + x = x - 23$

66. $y - \frac{2}{3} = -\frac{2}{3} + y$

67. $2(3 + 4m) - 9 = 45$

68. $5x + 5(4x - 1) = 20$

69. $5r - (2r + 8) = 16$

70. $6b - (3b + 8) = 16$

71. $6 - 2(3x - 1) = 2$

72. $10 - 3(2x - 1) = 1$

73. $5x + 5 - 7x = 15 - 12x + 10x - 10$

74. $3 - 7x + 10x - 14 = 9 - 6x + 9x - 20$

75. $22x - 5 - 15x + 3 = 10x - 4 - 3x + 11$

76. $11x - 6 - 4x + 1 = 9x - 8 - 2x + 12$

77. $5(d + 4) = 7(d - 2)$

78. $3(t - 2) = 9(t + 2)$

79. $8(2t + 1) = 4(7t + 7)$

80. $7(5x - 2) = 6(6x - 1)$

81. $3(r - 6) + 2 = 4(r + 2) - 21$

82. $5(t + 3) + 9 = 3(t - 2) + 6$

83. $19 - (2x + 3) = 2(x + 3) + x$

84. $13 - (2c + 2) = 2(c + 2) + 3c$

85. $2[4 - 2(3 - x)] - 1 = 4[2(4x - 3) + 7] - 25$

86. $5[3(7 - t) - 4(8 + 2t)] - 20 = -6[2(6 + 3t) - 4]$

87. $11 - 4(x + 1) - 3 = 11 + 2(4 - 2x) - 16$

88. $6(2x - 1) - 12 = 7 + 12(x - 1)$

89. $22x - 1 - 12x = 5(2x - 1) + 4$

90. $2 + 14x - 9 = 7(2x + 1) - 14$

91. $0.7(3x + 6) = 1.1 - (x + 2)$

92. $0.9(2x + 8) = 20 - (x + 5)$

Skill Maintenance

93. Divide: $-22.1 \div 3.4$. [1.6c]

94. Multiply: $-22.1(3.4)$. [1.5a]

95. Factor: $7x - 21 - 14y$. [1.7d]

96. Factor: $8y - 88x + 8$. [1.7d]

Simplify.

97. $-3 + 2(-5)^2(-3) - 7$ [1.8d]

98. $3x + 2[4 - 5(2x - 1)]$ [1.8c]

99. $23(2x - 4) - 15(10 - 3x)$ [1.8b]

100. $256 \div 64 \div 4^2$ [1.8d]

Synthesis

Solve.

101. $\dfrac{2}{3}\left(\dfrac{7}{8} - 4x\right) - \dfrac{5}{8} = \dfrac{3}{8}$

102. $\dfrac{1}{4}(8y + 4) - 17 = -\dfrac{1}{2}(4y - 8)$

103. $\dfrac{4 - 3x}{7} = \dfrac{2 + 5x}{49} - \dfrac{x}{14}$

104. The width of a rectangle is 5 ft, its length is $(3x + 2)$ ft, and its area is 75 ft^2. Find x.

2.4 Formulas

(a) Evaluating Formulas

A **formula** is a "recipe" for doing a certain type of calculation. Formulas are often given as equations. When we replace the variables in an equation with numbers and calculate the result, we are **evaluating** the formula. Evaluating was introduced in Section 1.1.

Let's consider a formula that has to do with weather. Suppose you see a flash of lightning during a storm. Then a few seconds later, you hear the thunder that accompanies that lightning.

Your distance from the place where the lightning struck is given by the formula $M = \frac{1}{5}t$, where t is the number of seconds from the lightning flash to the sound of the thunder and M is in miles.

EXAMPLE 1 *Distance from Lightning.* Consider the formula $M = \frac{1}{5}t$. Suppose it takes 10 sec for the sound of thunder to reach you after you have seen a flash of lightning. How far away did the lightning strike?

$$M = \tfrac{1}{5}t$$

We substitute 10 for t and calculate M:

$$M = \tfrac{1}{5}t = \tfrac{1}{5}(10) = 2.$$

The lightning struck 2 mi away.

EXAMPLE 2 *Socks from Cotton.* Consider the formula $S = 4321x$, where S is the number of socks of average size that can be produced from x bales of cotton. You see a shipment of 300 bales of cotton taken off a ship. How many socks can be made from the cotton?

Source: *Country Woman Magazine*

We substitute 300 for x and calculate S:

$$S = 4321x = 4321(300) = 1{,}296{,}300.$$

Thus, 1,296,300 socks can be made from 300 bales of cotton.

Do Exercises 1 and 2.

1. **Storm Distance.** Refer to Example 1. Suppose that it takes the sound of thunder 14 sec to reach you. How far away is the storm?

2. **Socks from Cotton.** Refer to Example 2. Determine the number of socks that can be made from 65 bales of cotton.

Answers

Skill to Review:
1. 5 2. −6

Margin Exercises:
1. 2.8 mi 2. 280,865 socks

EXAMPLE 3 *Distance, Rate, and Time.* The distance d that a car will travel at a rate, or speed, r in time t is given by

$$d = rt.$$

A car travels at 75 miles per hour (mph) for 4.5 hr. How far will it travel?

We substitute 75 for r and 4.5 for t and calculate d:

$$d = rt = (75)(4.5) = 337.5 \text{ mi.}$$

The car will travel 337.5 mi.

Do Exercise 3.

Do Exercise 3.

3. Distance, Rate, and Time.
A car travels at 55 mph for 6.2 hr. How far will it travel?

(b) Solving Formulas

Refer to Example 2. Suppose a clothing company wants to produce S socks and needs to know how many bales of cotton to order. If this calculation is to be repeated many times, it might be helpful to first solve the formula for x:

$$S = 4321x$$

$$\frac{S}{4321} = x. \qquad \text{Dividing by 4321}$$

Then we can substitute a number for S and calculate x. For example, if the number of socks S to be produced is 432,100, then

$$x = \frac{S}{4321} = \frac{432{,}100}{4321} = 100.$$

The company would need to order 100 bales of cotton.

EXAMPLE 4 Solve for z: $H = \frac{1}{4}z$.

$$H = \frac{1}{4}z \qquad \text{We want this letter alone.}$$

$$4 \cdot H = 4 \cdot \frac{1}{4}z \qquad \text{Multiplying by 4 on both sides}$$

$$4H = z$$

For $H = 2$ in Example 4, $z = 4H = 4(2)$, or 8.

EXAMPLE 5 *Distance, Rate, and Time.* Solve for t: $d = rt$.

$$d = rt \qquad \text{We want this letter alone.}$$

$$\frac{d}{r} = \frac{rt}{r} \qquad \text{Dividing by } r$$

$$\frac{d}{r} = \frac{r}{r} \cdot t$$

$$\frac{d}{r} = t \qquad \text{Simplifying}$$

Do Exercises 4–6.

Do Exercises 4–6.

4. Solve for q: $B = \frac{1}{3}q$.

5. Solve for m: $n = mz$.

6. Electricity. Solve for I: $E = IR$. (This formula relates voltage E, current I, and resistance R.)

Answers

3. 341 mi   **4.** $q = 3B$
5. $m = \dfrac{n}{z}$ **6.** $I = \dfrac{E}{R}$

EXAMPLE 6 Solve for x: $y = x + 3$.

$$y = x + 3 \qquad \text{We want this letter alone.}$$
$$y - 3 = x + 3 - 3 \qquad \text{Subtracting 3}$$
$$y - 3 = x \qquad \text{Simplifying}$$

EXAMPLE 7 Solve for x: $y = x - a$.

$$y = x - a \qquad \text{We want this letter alone.}$$
$$y + a = x - a + a \qquad \text{Adding } a$$
$$y + a = x \qquad \text{Simplifying}$$

Do Exercises 7–9.

Solve for x.

7. $y = x + 5$

8. $y = x - 7$

9. $y = x - b$

EXAMPLE 8 Solve for y: $6y = 3x$.

$$6y = 3x \qquad \text{We want this letter alone.}$$
$$\frac{6y}{6} = \frac{3x}{6} \qquad \text{Dividing by 6}$$
$$y = \frac{x}{2}, \text{ or } \frac{1}{2}x \qquad \text{Simplifying}$$

EXAMPLE 9 Solve for y: $by = ax$.

$$by = ax \qquad \text{We want this letter alone.}$$
$$\frac{by}{b} = \frac{ax}{b} \qquad \text{Dividing by } b$$
$$y = \frac{ax}{b} \qquad \text{Simplifying}$$

Do Exercises 10 and 11.

10. Solve for y: $9y = 5x$.

11. Solve for p: $ap = bt$.

EXAMPLE 10 Solve for x: $ax + b = c$.

$$ax + b = c \qquad \text{We want this letter alone.}$$
$$ax + b - b = c - b \qquad \text{Subtracting } b$$
$$ax = c - b \qquad \text{Simplifying}$$
$$\frac{ax}{a} = \frac{c - b}{a} \qquad \text{Dividing by } a$$
$$x = \frac{c - b}{a} \qquad \text{Simplifying}$$

12. Solve for x: $y = mx + b$.

13. Solve for Q: $tQ - p = a$.

Do Exercises 12 and 13.

Answers

7. $x = y - 5$ **8.** $x = y + 7$

9. $x = y + b$ **10.** $y = \dfrac{5x}{9}$, or $\dfrac{5}{9}x$

11. $p = \dfrac{bt}{a}$ **12.** $x = \dfrac{y - b}{m}$

13. $Q = \dfrac{a + p}{t}$

To solve a formula for a given letter, identify the letter and:

1. Multiply on both sides to clear fractions or decimals, if that is needed.
2. Collect like terms on each side, if necessary.
3. Get all terms with the letter to be solved for on one side of the equation and all other terms on the other side.
4. Collect like terms again, if necessary.
5. Solve for the letter in question.

EXAMPLE 11 *Circumference.* Solve for r: $C = 2\pi r$. This is a formula for the circumference C of a circle of radius r.

$C = 2\pi r$ We want this letter alone.

$\dfrac{C}{2\pi} = \dfrac{2\pi r}{2\pi}$ Dividing by 2π

$\dfrac{C}{2\pi} = r$

EXAMPLE 12 *Averages.* Solve for a: $A = \dfrac{a + b + c}{3}$. This is a formula for the average A of three numbers a, b, and c.

$A = \dfrac{a + b + c}{3}$ We want the letter a alone.

$3 \cdot A = 3 \cdot \dfrac{a + b + c}{3}$ Multiplying by 3 on both sides

$3A = a + b + c$ Simplifying

$3A - b - c = a$ Subtracting b and c

Do Exercises 14 and 15.

14. **Circumference.** Solve for D:

$C = \pi D$.

This is a formula for the circumference C of a circle of diameter D.

15. **Averages.** Solve for c:

$A = \dfrac{a + b + c + d}{4}$.

STUDY TIPS

HIGHLIGHTING

- **Try to keep one section ahead of your syllabus.** Reading and highlighting a section before your instructor lectures on it allows you to listen carefully and concentrate on what is being said in class. Then you can take notes only on special points and on questions related to the lecture.

- **Highlight important points.** You are probably used to highlighting key points as you study. If that works for you, continue to do so. But you will notice many design features throughout this book that already highlight important points. Thus you may not need to highlight as much as you generally do.

- **Highlight points that you do not understand.** Use a special marker to indicate trouble spots that can lead to questions to be asked during class or in a tutoring session.

Answers

14. $D = \dfrac{C}{\pi}$ **15.** $c = 4A - a - b - d$

(a), (b) Solve.

1. *Furnace Output.* The formula
$$B = 30a$$
is used in New England to estimate the minimum furnace output B, in Btu's, for a modern house with a square feet of flooring.
Source: U.S. Department of Energy

a) Determine the minimum furnace output for a 1900-ft^2 modern house.

b) Solve for a. That is, solve $B = 30a$ for a.

2. *Furnace Output.* The formula
$$B = 50a$$
is used in New England to estimate the minimum furnace output B, in Btu's, for an old, poorly insulated house with a square feet of flooring.
Source: U.S. Department of Energy

a) Determine the minimum furnace output for a 3200-ft^2 old, poorly insulated house.

b) Solve for a. That is, solve $B = 50a$ for a.

3. *Distance from Lightning.* The formula
$$M = \tfrac{1}{5}t$$
can be used to determine how far M, in miles, you are from lightning when its thunder takes t seconds to reach your ears.

a) It takes 8 sec for the sound of thunder to reach you after you have seen the lightning. How far away did the lightning strike?

b) Solve for t.

4. *Electrical Power.* The power rating P, in watts, of an electrical appliance is determined by
$$P = I \cdot V,$$
where I is the current, in amperes, and V is measured in volts.

a) A microwave oven requires 12 amps of current and the voltage in the house is 115 volts. What is the wattage of the microwave?

b) Solve for I; for V.

5. *College Enrollment.* At many colleges, the number of "full-time-equivalent" students f is given by
$$f = \frac{n}{15},$$
where n is the total number of credits for which students have enrolled in a given semester.

a) Determine the number of full-time-equivalent students on a campus in which students registered for a total of 21,345 credits.

b) Solve for n.

6. *Surface Area of a Cube.* The surface area A of a cube with side s is given by
$$A = 6s^2.$$

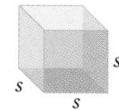

a) Find the surface area of a cube with sides of 3 in.

b) Solve for s^2.

7. *Calorie Density.* The calorie density D, in calories per ounce, of a food that contains c calories and weighs w ounces is given by

$$D = \frac{c}{w}.$$

Eight ounces of fat-free milk contains 84 calories. Find the calorie density of fat-free milk.

Source: *Nutrition Action Healthletter*, March 2000, p. 9. Center for Science in the Public Interest, Suite 300; 1875 Connecticut Ave NW, Washington, D.C. 20008.

8. *Wavelength of a Musical Note.* The wavelength w, in meters per cycle, of a musical note is given by

$$w = \frac{r}{f},$$

where r is the speed of the sound, in meters per second, and f is the frequency, in cycles per second. The speed of sound in air is 344 m/sec. What is the wavelength of a note whose frequency in air is 24 cycles per second?

9. *Size of a League Schedule.* When all n teams in a league play every other team twice, a total of N games are played, where

$$N = n^2 - n.$$

A soccer league has 7 teams and all teams play each other twice. How many games are played?

10. *Size of a League Schedule.* When all n teams in a league play every other team twice, a total of N games are played, where

$$N = n^2 - n.$$

A basketball league has 11 teams and all teams play each other twice. How many games are played?

 Solve for the indicated letter.

11. $y = 5x$, for x

12. $d = 55t$, for t

13. $a = bc$, for c

14. $y = mx$, for x

15. $n = m + 11$, for m

16. $z = t + 21$, for t

17. $y = x - \dfrac{3}{5}$, for x

18. $y = x - \dfrac{2}{3}$, for x

19. $y = 13 + x$, for x

20. $t = 6 + s$, for s

21. $y = x + b$, for x

22. $y = x + A$, for x

23. $y = 5 - x$, for x

24. $y = 10 - x$, for x

25. $y = a - x$, for x

26. $y = q - x$, for x

27. $8y = 5x$, for y

28. $10y = -5x$, for y

29. $By = Ax$, for x

30. $By = Ax$, for y

31. $W = mt + b$, for t

32. $W = mt - b$, for t

33. $y = bx + c$, for x

34. $y = bx - c$, for x

35. *Area of a Parallelogram:*
$A = bh$, for h
(Area A, base b, height h)

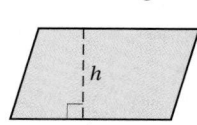

36. *Distance, Rate, Time:*
$d = rt$, for r
(Distance d, speed r, time t)

Speed, r　　　　　Time, t

Distance, d

37. *Perimeter of a Rectangle:*
$P = 2l + 2w$, for w
(Perimeter P, length l, width w)

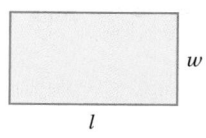

38. *Area of a Circle:*
$A = \pi r^2$, for r^2
(Area A, radius r)

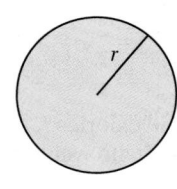

39. *Average of Two Numbers:*
$A = \dfrac{a + b}{2}$, for a

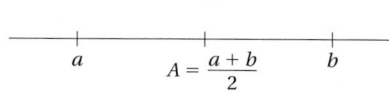

40. *Area of a Triangle:*
$A = \dfrac{1}{2}bh$, for b

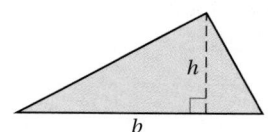

41. $A = \dfrac{a + b + c}{3}$, for b

42. $A = \dfrac{a + b + c}{3}$, for c

43. $A = at + b$, for t

44. $S = rx + s$, for x

45. $Ax + By = c$, for x

46. $Q = \dfrac{p - q}{2}$, for p

47. *Force:*

$$F = ma, \text{ for } a$$

(Force F, mass m, acceleration a)

48. *Simple Interest:*

$$I = Prt, \text{ for } P$$

(Interest I, principal P, interest rate r, time t)

49. *Relativity:*

$$E = mc^2, \text{ for } c^2$$

(Energy E, mass m, speed of light c)

50. $Ax + By = c$, for y

51. $v = \dfrac{3k}{t}$, for t

52. $P = \dfrac{ab}{c}$, for c

Skill Maintenance

53. Evaluate $\dfrac{3x - 2y}{y}$, for $x = 6$ and $y = 2$. [1.1a]

54. Remove parentheses and simplify: [1.8b]
$4a - 8b - 5(5a - 4b)$.

Subtract. [1.4a]

55. $-45.8 - (-32.6)$

56. $-\dfrac{2}{3} - \dfrac{5}{6}$

57. $-123 + 87\dfrac{1}{2}$

Add. [1.3a]

58. $-\dfrac{5}{12} + \dfrac{1}{4}$

59. $0.082 + (-9.407)$

60. $-2\dfrac{1}{2} + 6\dfrac{1}{4}$

Solve.

61. $2y - 3 + y = 8 - 5y$ [2.3b]

62. $10x + 4 = 3x - 2 + x$ [2.3b]

63. $2(5x + 6) = x - 15$ [2.3c]

64. $5a = 3(6 - 3a)$ [2.3c]

Synthesis

65. *Female Caloric Needs.* The number of calories K needed each day by a moderately active woman who weighs w pounds, is h inches tall, and is a years old can be estimated by the formula

$$K = 917 + 6(w + h - a).$$

Source: Parker, M., *She Does Math.* Mathematical Association of America, p. 96

a) Elaine is moderately active, weighs 120 lb, is 67 in. tall, and is 23 yr old. What are her caloric needs?

b) Solve the formula for a; for h; for w.

66. *Male Caloric Needs.* The number of calories K needed each day by a moderately active man who weighs w kilograms, is h centimeters tall, and is a years old can be estimated by the formula

$$K = 19.18w + 7h - 9.52a + 92.4.$$

Source: Parker, M., *She Does Math.* Mathematical Association of America, p. 96

a) Marv is moderately active, weighs 97 kg, is 185 cm tall, and is 55 yr old. What are his caloric needs?

b) Solve the formula for a; for h; for w.

Solve.

67. $H = \dfrac{2}{a - b}$, for b; for a

68. $P = 4m + 7mn$, for m

69. In $A = lw$, if l and w both double, what is the effect on A?

70. In $P = 2a + 2b$, if P doubles, do a and b necessarily both double?

71. In $A = \frac{1}{2}bh$, if b increases by 4 units and h does not change, what happens to A?

72. Solve for F: $D = \dfrac{1}{E + F}$.

Mid-Chapter Review

Concept Reinforcement

Determine whether each statement is true or false.

_____ **1.** $3 - x = 4x$ and $5x = -3$ are equivalent equations. [2.1b]

_____ **2.** For any real numbers a, b, and c, $a = b$ is equivalent to $a + c = b + c$.
[2.1b]

_____ **3.** We can use the multiplication principle to divide on both sides of an
equation by the same nonzero number. [2.2a]

_____ **4.** Every equation has at least one solution. [2.3c]

Guided Solutions

Fill in each blank with the number, variable, or expression that creates a correct statement or solution.

Solve. [2.1b], [2.2a]

5. $x + 5 = -3$

$x + 5 - 5 = -3 - \square$

$x + \square = -8$

$x = \square$

6. $-6x = 42$

$\dfrac{-6x}{-6} = \dfrac{42}{\square}$

$\square \cdot x = -7$

$x = \square$

7. Solve for y: $5y + z = t$. [2.4b]

$5y + z = t$

$5y + z - z = t - \square$

$5y = \square$

$\dfrac{5y}{5} = \dfrac{t - z}{\square}$

$y = \dfrac{\square}{5}$

Mixed Review

Solve. [2.1b], [2.2a], [2.3a, b, c]

8. $x + 5 = 11$

9. $x + 9 = -3$

10. $8 = t + 1$

11. $-7 = y + 3$

12. $x - 6 = 14$

13. $y - 7 = -2$

14. $-\dfrac{3}{2} + z = -\dfrac{3}{4}$

15. $-3.3 = -1.9 + t$

16. $7x = 42$

17. $17 = -t$

18. $6x = -54$

19. $-5y = -85$

20. $\dfrac{x}{7} = 3$

21. $\dfrac{2}{3}x = 12$

22. $-\dfrac{t}{5} = 3$

23. $\dfrac{3}{4}x = -\dfrac{9}{8}$

24. $3x + 2 = 5$

25. $5x + 4 = -11$

26. $6x - 7 = 2$

27. $-4x - 9 = -5$

28. $6x + 5x = 33$ **29.** $-3y - 4y = 49$ **30.** $3x - 4 = 12 - x$ **31.** $5 - 6x = 9 - 8x$

32. $4y - \dfrac{3}{2} = \dfrac{3}{4} + 2y$ **33.** $\dfrac{4}{5} + \dfrac{1}{6}t = \dfrac{1}{10}$ **34.** $0.21n - 1.05 = 2.1 - 0.14n$

35. $5(3y - 1) = -35$ **36.** $7 - 2(5x + 3) = 1$ **37.** $-8 + t = t - 8$

38. $z + 12 = -12 + z$ **39.** $4(3x + 2) = 5(2x - 1)$ **40.** $8x - 6 - 2x = 3(2x - 4) + 6$

Solve for the indicated letter. [2.4b]

41. $A = 4b$, for b **42.** $y = x - 1.5$, for x **43.** $n = s - m$, for m

44. $4t = 9w$, for t **45.** $B = at - c$, for t **46.** $M = \dfrac{x + y + z}{2}$, for y

Understanding Through Discussion and Writing

47. Explain the difference between equivalent expressions and equivalent equations. [1.7a], [2.1b]

48. Are the equations $x = 5$ and $x^2 = 25$ equivalent? Why or why not? [2.1b]

49. When solving an equation using the addition principle, how do you determine which number to add or subtract on both sides of the equation? [2.1b]

50. Explain the following mistake made by a fellow student. [2.1b]

$$x + \frac{1}{3} = -\frac{5}{3}$$

$$x = -\frac{4}{3}$$

51. When solving an equation using the multiplication principle, how do you determine by what number to multiply or divide on both sides of the equation? [2.2a]

52. Devise an application in which it would be useful to solve the equation $d = rt$ for r. [2.4b]

2.5

Applications of Percent

a Translating and Solving

Many applied problems involve percent. Here we begin to see how equation solving can enhance our problem-solving skills.

In solving percent problems, we first *translate* the problem to an equation. Then we *solve* the equation using the techniques discussed in Sections 2.1–2.3. The key words in the translation are as follows.

> **KEY WORDS IN PERCENT TRANSLATIONS**
>
> "**Of**" translates to "·" or "×".
>
> "**Is**" translates to "=".
>
> "**What number**" or "**what percent**" translates to any letter.
>
> "**%**" translates to "$\times \frac{1}{100}$" or "$\times 0.01$".

EXAMPLE 1 Translate:

$$28\% \quad \text{of} \quad 5 \quad \text{is} \quad \text{what number?}$$
$$28\% \quad \cdot \quad 5 \quad = \quad a \qquad \text{This is a percent equation.}$$

EXAMPLE 2 Translate:

$$45\% \quad \text{of} \quad \text{what number} \quad \text{is} \quad 28?$$
$$45\% \quad \times \quad b \quad = \quad 28$$

EXAMPLE 3 Translate:

$$\text{What percent} \quad \text{of} \quad 90 \quad \text{is} \quad 7?$$
$$n \quad \cdot \quad 90 \quad = \quad 7$$

Do Exercises 1–6.

Percent problems are actually of three different types. Although the method we present does *not* require that you be able to identify which type we are studying, it is helpful to know them. Let's begin by using a specific example to find a standard form for a percent problem.

We know that

$$15 \text{ is } 25\% \text{ of } 60, \quad \text{or} \quad 15 = 25\% \times 60.$$

We can think of this as:

> Amount = Percent number × Base.

SKILL TO REVIEW
Objective 2.2a: Solve equations using the multiplication principle.

Solve.

1. $20 = 0.05a$ **2.** $0.3z = 327$

Translate to an equation. Do not solve.

1. 13% of 80 is what number?

2. What number is 60% of 70?

3. 43 is 20% of what number?

4. 110% of what number is 30?

5. 16 is what percent of 80?

6. What percent of 94 is 10.5?

Answers

Skill to Review:
1. 400   **2.** 1090

Margin Exercises:
1. $13\% \cdot 80 = a$ **2.** $a = 60\% \cdot 70$
3. $43 = 20\% \cdot b$ **4.** $110\% \cdot b = 30$
5. $16 = n \cdot 80$ **6.** $n \cdot 94 = 10.5$

Each of the three types of percent problem depends on which of the three pieces of information is missing in the statement

$$\text{Amount} = \text{Percent number} \times \text{Base}.$$

1. Finding the *amount* (the result of taking the percent)

Example: What number is 25% of 60?

Translation: $y = 25\% \cdot 60$

2. Finding the *base* (the number you are taking the percent of)

Example: 15 is 25% of what number?

Translation: $15 = 25\% \cdot y$

3. Finding the *percent number* (the percent itself)

Example: 15 is what percent of 60?

Translation: $15 = y \cdot 60$

Finding the Amount

EXAMPLE 4 What number is 11% of 49?

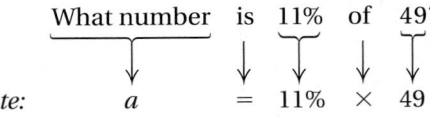

Translate: $a = 11\% \times 49$

Solve: The letter is by itself. To solve the equation, we need only convert 11% to decimal notation and multiply:

$$a = 11\% \times 49 = 0.11 \times 49 = 5.39.$$

Thus, 5.39 is 11% of 49. The answer is 5.39.

Do Exercise 7.

7. What number is 2.4% of 80?

Finding the Base

EXAMPLE 5 3 is 16% of what number?

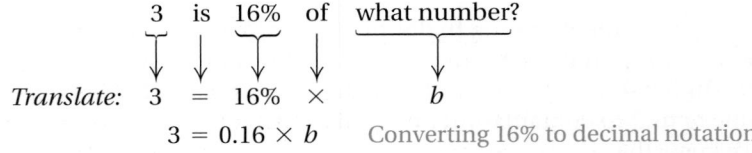

Translate: $3 = 16\% \times b$

$3 = 0.16 \times b$ Converting 16% to decimal notation

Solve: In this case, the letter is not by itself. To solve the equation, we divide by 0.16 on both sides:

$$3 = 0.16 \times b$$

$$\frac{3}{0.16} = \frac{0.16 \times b}{0.16} \qquad \text{Dividing by 0.16}$$

$$18.75 = b. \qquad \text{Simplifying}$$

The answer is 18.75.

Do Exercise 8.

8. 25.3 is 22% of what number?

Answers

7. 1.92   **8.** 115

Finding the Percent Number

In solving these problems, you *must* remember to convert to percent notation after you have solved the equation.

EXAMPLE 6 $32 is what percent of $50?

$$\underbrace{\$32}_{32} \quad \underbrace{is}_{=} \quad \underbrace{what\ percent}_{p} \quad \underbrace{of}_{\times} \quad \underbrace{\$50?}_{50}$$

Translate: $32 = p \times 50$

Solve: To solve the equation, we divide by 50 on both sides and convert the answer to percent notation:

$$32 = p \times 50$$

$$\frac{32}{50} = \frac{p \times 50}{50} \qquad \text{Dividing by 50}$$

$$0.64 = p$$

$$64\% = p. \qquad \text{Converting to percent notation}$$

Thus, $32 is 64% of $50. The answer is 64%.

Do Exercise 9.

9. What percent of $50 is $18?

EXAMPLE 7   *Foreign Visitors to China.*   About 22 million foreign travelers visited China in 2006. Of this number, 9% were from the United States. How many Americans visited China in 2006?

Source: *TIME Magazine*, March 8, 2007

To solve this problem, we first reword and then translate. We let a = the number of Americans, in millions, who visited China in 2006.

Rewording: $\underbrace{What\ number}_{a}$ $\underbrace{is}_{=}$ $\underbrace{9\%}_{9\%}$ $\underbrace{of}_{\times}$ $\underbrace{22?}_{22}$

Translating: $a = 9\% \times 22$

Solve: The letter is by itself. To solve the equation, we need only convert 9% to decimal notation and multiply:

$$a = 9\% \times 22 = 0.09 \times 22 = 1.98.$$

Thus, 1.98 million is 9% of 22 million, so 1.98 million Americans visited China in 2006.

Do Exercise 10.

10. Chinese Visitors to the United States.   About 51 million foreign travelers visited the United States in 2006. Of this number, 1% were from China. How many Chinese travelers visited the United States in 2006?

Source: *TIME Magazine*, March 8, 2007

EXAMPLE 8   *Public School Enrollment.*   In the fall of 2008, 14.9 million students enrolled in grades 9–12 in U.S. public schools. This was 30% of the total enrollment in public schools. What was the total enrollment?

Source: National Center for Educational Statistics

To solve this problem, we first reword and then translate. We let T = the total enrollment, in millions, in U.S. public schools in 2008.

Rewording: $\underbrace{14.9}_{14.9}$ $\underbrace{is}_{=}$ $\underbrace{30\%}_{30\%}$ $\underbrace{of}_{\times}$ $\underbrace{what\ number?}_{T}$

Translating: $14.9 = 30\% \times T$

Answers

9. 36%   **10.** 0.51 million travelers

11. Areas of Texas and Alaska.
The area of the second largest state, Texas, is 268,581 mi². This is about 40.5% of the area of the largest state, Alaska. What is the area of Alaska?

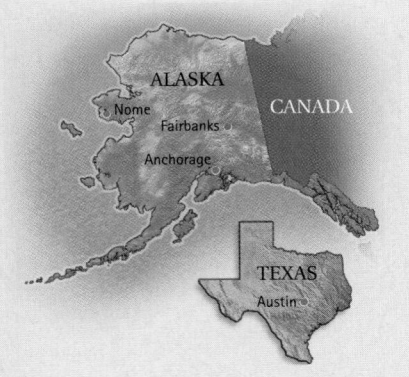

Solve: To solve the equation, we convert 30% to decimal notation and divide by 0.3 on both sides:

$$14.9 = 30\% \times T$$

$$14.9 = 0.3 \times T \qquad \text{Converting to decimal notation}$$

$$\frac{14.9}{0.3} = \frac{0.3 \times T}{0.3} \qquad \text{Dividing by 0.3}$$

$$49.7 \approx T. \qquad \text{Simplifying and rounding to the nearest tenth}$$

About 49.7 million students enrolled in U.S. public schools in 2008.

Do Exercise 11.

EXAMPLE 9 *Employment Outlook.* There were 280 thousand dental assistants in 2006. This number is expected to grow to 362 thousand in 2016. What is the percent of increase?

Source: *Occupational Outlook Handbook*

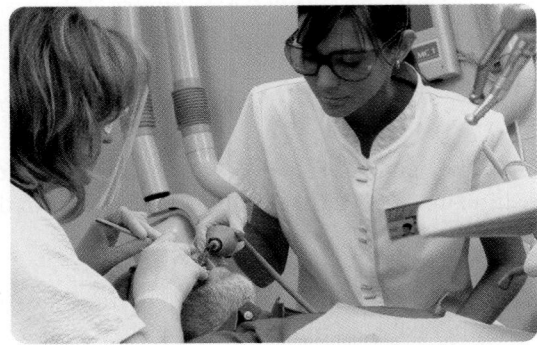

To solve the problem, we must first determine the amount of the increase, in thousands:

Jobs in 2016	minus	Jobs in 2006	=	Increase
362	−	280	=	82.

Using the job increase of 82 thousand, we reword and then translate. We let p = the percent of increase. We want to know, "what percent of the number of jobs in 2006 is 82 thousand?"

Rewording: 82   is   what percent   of   280?

Translating: 82   = p ×   280

Solve: To solve the equation, we divide by 280 on both sides and convert the answer to percent notation:

$$82 = p \times 280$$

$$\frac{82}{280} = \frac{p \times 280}{280} \qquad \text{Dividing by 280}$$

$$0.293 \approx p \qquad \text{Simplifying}$$

$$29.3\% \approx p. \qquad \text{Converting to percent notation}$$

The percent of increase is about 29.3%.

Do Exercise 12.

12. Employment Outlook. There were 234 thousand file clerks in 2006. This number is expected to decrease to 137 thousand in 2016. What is the percent of decrease?

Source: *Occupational Outlook Handbook*

Answers

11. About 663,163 mi²
12. About 41.5%

a   Solve.

1. What percent of 180 is 36?

2. What percent of 76 is 19?

3. 45 is 30% of what number?

4. 20.4 is 24% of what number?

5. What number is 65% of 840?

6. What number is 50% of 50?

7. 30 is what percent of 125?

8. 57 is what percent of 300?

9. 12% of what number is 0.3?

10. 7 is 175% of what number?

11. 2 is what percent of 40?

12. 16 is what percent of 40?

13. What percent of 68 is 17?

14. What percent of 150 is 39?

15. What number is 35% of 240?

16. What number is 1% of one million?

17. What percent of 125 is 30?

18. What percent of 60 is 75?

19. What percent of 300 is 48?

20. What percent of 70 is 70?

21. 14 is 30% of what number?

22. 54 is 24% of what number?

23. What number is 2% of 40?

24. What number is 40% of 2?

25. 0.8 is 16% of what number?

26. 40 is 2% of what number?

27. 54 is 135% of what number?

28. 8 is 2% of what number?

Amount Spent on Pets.   In 2007, $41.2 billion was spent on pets in the United States. The circle graph below shows the breakdown of this spending.

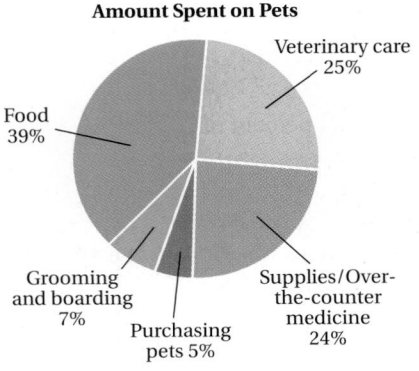

Amount Spent on Pets

Veterinary care 25%

Food 39%

Supplies/Over-the-counter medicine 24%

Grooming and boarding 7%

Purchasing pets 5%

SOURCE: American Pet Products Association

Complete the following table of amounts spent on pets. Round to the nearest tenth.

	CATEGORY	AMOUNT (in billions)		CATEGORY	AMOUNT (in billions)
29.	Food		**30.**	Veterinary care	
31.	Purchasing pets		**32.**	Grooming and boarding	

33. *Smart TV Market.*   Smart TVs, which are designed to be easily connected to the Internet or to a home computer network, are a small but growing part of the TV market. Total TV sales in 2010 are projected to be 209 million units, with smart TVs comprising 25.1 million units. What percent of total sales are projected to be smart TV sales?

Source: IDC

34. *Automobile Sales.*   Sales of cars averaged $26 million per dealership in 2007. Of this amount, new car sales accounted for receipts of $24 million. What percent of total sales are new cars?

Source: U.S. Census Bureau

35. *Graduation Gifts.*   American consumers spent $4.5 billion on graduation gifts in 2008. Cash accounted for 58.8% of this amount. How much cash was given as graduation gifts?

Source: National Retail Federation

36. *Graduation Gifts.*   Refer to Exercise 35. Gift cards accounted for 35.7% of the amount spent on graduation gifts in 2008. What is the total value of these gift cards?

Source: National Retail Federation

37. *Student Loans.* To finance her community college education, Sarah takes out a Stafford loan for $6500. After a year, Sarah decides to pay off the interest, which is 6% of $6500. How much will she pay?

38. *Student Loans.* Paul takes out a PLUS loan for $5400. After a year, Paul decides to pay off the interest, which is 8.5% of $5400. How much will he pay?

39. *Tipping.* Leon left a $4 tip for a meal that cost $25.
   a) What percent of the cost of the meal was the tip?
   b) What was the total cost of the meal including the tip?

40. *Tipping.* Selena left a $12.76 tip for a meal that cost $58.
   a) What percent of the cost of the meal was the tip?
   b) What was the total cost of the meal including the tip?

41. *Tipping.* Leon left a 15% tip for a meal that cost $25.
   a) How much was the tip?
   b) What was the total cost of the meal including the tip?

42. *Tipping.* Sam, Selena, Rachel, and Clement left a 15% tip for a meal that cost $58.
   a) How much was the tip?
   b) What was the total cost of the meal including the tip?

43. *Tipping.* Leon left a 15% tip of $4.50 for a meal.
   a) What was the cost of the meal before the tip?
   b) What was the total cost of the meal including the tip?

44. *Tipping.* Selena left a 15% tip of $8.40 for a meal.
   a) What was the cost of the meal before the tip?
   b) What was the total cost of the meal including the tip?

45. *City Park Space.* Portland, Oregon, has 12,959 acres of park space. This is 15.1% of the acreage of the entire city. What is the total acreage of Portland?
Source: Indy Parks and Recreation master plan

46. *Junk Mail.* About 46.2 billion pieces of unopened junk mail end up in landfills each year. This is about 44% of all the junk mail that is sent annually. How many pieces of junk mail are sent annually?
Source: Globaljunkmailcrisis.org

47. *Size of New Homes.* The median size of a new single-family home grew from 1879 ft^2 in 1997 to 2304 ft^2 in 2008. What is the percent of increase?
Source: U.S. Census Bureau

48. *Health Technology Spending.* With growth in traditional technology markets slowing, many companies are developing products for the health-care market. Worldwide, $68.4 billion was spent on health-care technology in 2005. This amount was expected to increase to $83.6 billion in 2009. What is the percent of increase?
Source: Gartner

49. Renewable Fuel. In 2006, about 4 billion gal of renewable fuels, such as ethanol and other biofuels, were used in the United States. The energy law passed in 2007 requires that 36 billion gal of such fuels be used by 2022. What is the percent of increase?

Source: U.S. Senate Committee on Energy and Natural Resources

50. Accidents at Railroad Crossings. In 1997, 3865 accidents occurred at railroad crossings in the United States. This number dropped to 2918 in 2006. What is the percent of decrease?

Source: Federal Railroad Administration

51. Employment Outlook. In 2006, there were 50 thousand pharmacy aides in the United States. This number is expected to drop to 45 thousand by 2016. What is the percent of decrease?

Source: Occupational Outlook Handbook

52. Employment Outlook. In 2006, there were 767,000 personal and home-care aides in the United States. This number is expected to grow to 1,156,000 by 2016. What is the percent of increase?

Source: Occupational Outlook Handbook

53. Debit IDs. A growing number of colleges are teaming up with banks to issue student ID cards that double as debit cards. There were 52 such partnerships in 2002. This number grew to 127 in 2007. What is the percent of increase?

Source: CR80News

54. Decline in Tuberculosis Cases. The number of cases of tuberculosis in the United States has plunged from 69,895 in 1956 to 13,299 in 2007. What is the percent of decrease?

Source: U.S. Centers for Disease Control and Prevention

Skill Maintenance

Remove parentheses and simplify. [1.8b]

55. $-5a + 3c - 2(c - 3a)$

56. $4(x - 2y) - (y - 3x)$

Add. [1.3a]

57. $-6.5 + 2.6$

58. $-\dfrac{3}{8} + (-5) + \dfrac{1}{4} + (-1)$

Fill in each blank with a word that makes the statement true. [1.8d]

59. To simplify the calculation $18 - 24 \div 3 - 48 \div (-4)$, do all the _____ calculations first, and then the _____ calculations.

60. To simplify the calculation $18 - 24^3 \div 48 \div (-4)^2$, do all the _____ calculations first, and then the _____ calculations, and finally the _____ calculation.

Synthesis

61. It has been determined that at the age of 15, a boy has reached 96.1% of his final adult height. Jaraan is 6 ft 4 in. at the age of 15. What will his final adult height be?

62. It has been determined that at the age of 10, a girl has reached 84.4% of her final adult height. Dana is 4 ft 8 in. at the age of 10. What will her final adult height be?

2.6

Applications and Problem Solving

a Five Steps for Solving Problems

We have discussed many new equation-solving tools in this chapter and used them for applications and problem solving. Here we consider a five-step strategy that can be very helpful in solving problems.

> **FIVE STEPS FOR PROBLEM SOLVING IN ALGEBRA**
>
> 1. *Familiarize* yourself with the problem situation.
> 2. *Translate* the problem to an equation.
> 3. *Solve* the equation.
> 4. *Check* the answer in the original problem.
> 5. *State* the answer to the problem clearly.

Of the five steps, the most important is probably the first one: becoming familiar with the problem situation. The box below lists some hints for familiarization.

> **TO FAMILIARIZE YOURSELF WITH A PROBLEM**
>
> - If a problem is given in words, read it carefully. Reread the problem, perhaps aloud. Try to verbalize the problem as if you were explaining it to someone else.
> - Choose a variable (or variables) to represent the unknown and clearly state what the variable represents. Be descriptive! For example, let $L =$ the length, $d =$ the distance, and so on.
> - Make a drawing and label it with known information, using specific units if given. Also, indicate unknown information.
> - Find further information. Look up formulas or definitions with which you are not familiar. (Geometric formulas appear on the inside back cover of this text.) Consult a reference librarian or the Internet.
> - Create a table that lists all the information you have available. Look for patterns that may help in the translation to an equation.
> - Think of a possible answer and check the guess. Note the manner in which the guess is checked.

EXAMPLE 1 *Knitted Scarf.* Lily knitted a scarf in three shades of blue, starting with a light-blue section, then a medium-blue section, and finally a dark-blue section. The medium-blue section is one-half the length of the light-blue section. The dark-blue section is one-fourth the length of the light-blue section. The scarf is 7 ft long. Find the length of each section of the scarf.

OBJECTIVE

 Solve applied problems by translating to equations.

SKILL TO REVIEW
Objective 1.1b: Translate phrases to algebraic expressions.

Translate each phrase to an algebraic expression.

1. One-third of a number
2. Two more than a number

Answers
Skill to Review:
1. $\frac{1}{3}n$, or $\frac{n}{3}$ **2.** $x + 2$, or $2 + x$

1. **Familiarize.** Because the lengths of the medium-blue section and the dark-blue section are expressed in terms of the length of the light-blue section, we let

x = the length of the light-blue section.

Then $\frac{1}{2}x$ = the length of the medium-blue section

and $\frac{1}{4}x$ = the length of the dark-blue section.

We make a drawing and label it.

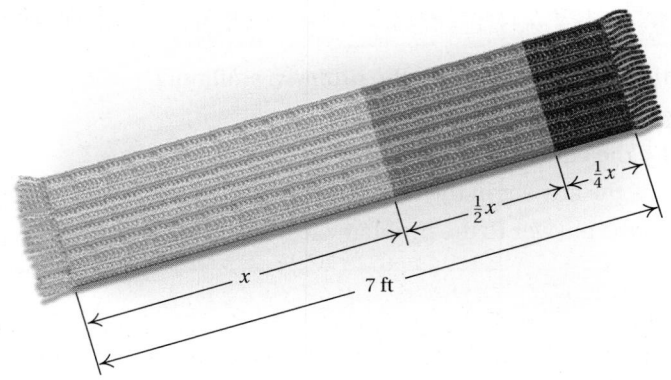

2. **Translate.** From the statement of the problem and the drawing, we know that the lengths add up to 7 ft. This gives us our translation:

Length of light-blue section	plus	Length of medium-blue section	plus	Length of dark-blue section	is	Total length
↓	↓	↓	↓	↓	↓	↓
x	$+$	$\frac{1}{2}x$	$+$	$\frac{1}{4}x$	$=$	7.

3. **Solve.** First, we clear fractions and then carry out the solution as follows:

$$x + \frac{1}{2}x + \frac{1}{4}x = 7 \qquad \text{The LCM of the denominators is 4.}$$

$$4\left(x + \frac{1}{2}x + \frac{1}{4}x\right) = 4 \cdot 7 \qquad \text{Multiplying by the LCM, 4}$$

$$4 \cdot x + 4 \cdot \frac{1}{2}x + 4 \cdot \frac{1}{4}x = 4 \cdot 7 \qquad \text{Using the distributive law}$$

$$4x + 2x + x = 28 \qquad \text{Simplifying}$$

$$7x = 28 \qquad \text{Collecting like terms}$$

$$\frac{7x}{7} = \frac{28}{7} \qquad \text{Dividing by 7}$$

$$x = 4.$$

4. **Check.** Do we have an answer to the *original problem*? If the length of the light-blue section is 4 ft, then the length of the medium-blue section is $\frac{1}{2} \cdot 4$ ft, or 2 ft, and the length of the dark-blue section is $\frac{1}{4} \cdot 4$ ft, or 1 ft. The sum of these lengths is 7 ft, so the answer checks.

5. State. The length of the light-blue section is 4 ft, the length of the medium-blue section is 2 ft, and the length of the dark-blue section is 1 ft. (Note that we must include the unit, feet, in the answer.)

Do Exercise 1.

EXAMPLE 2 *Hiking.* At age 79, Earl Shaffer became the oldest person to through-hike all 2100 miles of the Appalachian Trail—from Springer Mountain, Georgia, to Mount Katahdin, Maine. Shaffer through-hiked the trail three times, in 1948 (Georgia to Maine), in 1965 (Maine to Georgia), and in 1998 (Georgia to Maine) near the 50th anniversary of his first hike. At one point in 1998, Shaffer stood atop Big Walker Mountain, Virginia, which is three times as far from the northern end as from the southern end. How far was Shaffer from each end of the trail?

Source: Appalachian Trail Conference; Earl Shaffer Foundation

1. Familiarize. Let's consider a drawing.

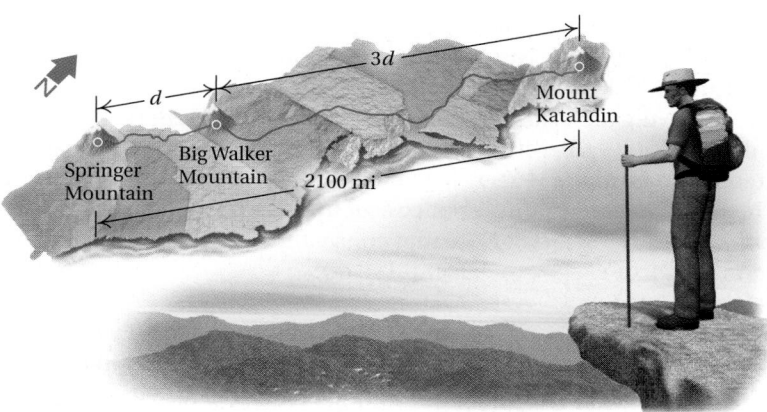

To become familiar with the problem, let's guess a possible distance that Shaffer stood from Springer Mountain—say, 600 mi. Three times 600 mi is 1800 mi. Since 600 mi + 1800 mi = 2400 mi and 2400 mi is greater than 2100 mi, we see that our guess is too large. Rather than guess again, let's use the equation-solving skills that we have learned in this chapter. We let

d = the distance, in miles, to the southern end,   and

$3d$ = the distance, in miles, to the northern end.

(We could also let x = the distance to the northern end and $\frac{1}{3}x$ = the distance to the southern end.)

2. Translate. From the drawing, we see that the lengths of the two parts of the trail must add up to 2100 mi. This leads to our translation:

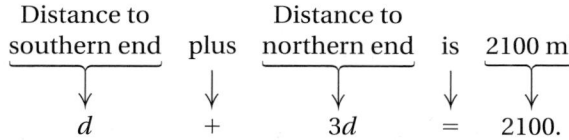

Distance to southern end plus Distance to northern end is 2100 mi

d + $3d$ = 2100.

1. Gourmet Sandwiches. A sandwich shop specializes in sandwiches prepared in buns of length 18 in. Jenny, Emma, and Sarah buy one of these sandwiches and take it back to their apartment. Since they have different appetites, Jenny cuts the sandwich in such a way that Emma gets one-half of what Jenny gets and Sarah gets three-fourths of what Jenny gets. Find the length of each person's sandwich.

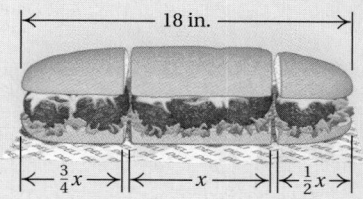

Answer

1. Jenny: 8 in.; Emma: 4 in.; Sarah: 6 in.

3. Solve. We solve the equation:

$$d + 3d = 2100$$

$$4d = 2100 \qquad \text{Collecting like terms}$$

$$\frac{4d}{4} = \frac{2100}{4} \qquad \text{Dividing by 4}$$

$$d = 525.$$

4. Check. As expected, d is less than 600 mi. If $d = 525$ mi, then $3d = 1575$ mi. Since 525 mi + 1575 mi = 2100 mi, we have a check.

5. State. Atop Big Walker Mountain, Shaffer stood 525 mi from Springer Mountain and 1575 mi from Mount Katahdin.

> **Do Exercise 2.**

Recall that the set of integers = $\{\ldots, -5, -4, -3, -2, -1, 0, 1, 2, 3, 4, 5, \ldots\}$. Before we solve the next problem, we need to learn some additional terminology regarding integers.

The following are examples of **consecutive integers:** 16, 17, 18, 19, 20; and $-31, -30, -29, -28$. Note that consecutive integers can be represented in the form $x, x + 1, x + 2$, and so on.

The following are examples of **consecutive even integers:** 16, 18, 20, 22, 24; and $-52, -50, -48, -46$. Note that consecutive even integers can be represented in the form $x, x + 2, x + 4$, and so on.

The following are examples of **consecutive odd integers:** 21, 23, 25, 27, 29; and $-71, -69, -67, -65$. Note that consecutive odd integers can be represented in the form $x, x + 2, x + 4$, and so on.

EXAMPLE 3 *Interstate Mile Markers.* U.S. interstate highways post numbered markers every mile to indicate location in case of an accident or breakdown. In many states, the numbers on the markers increase from west to east. The sum of two consecutive mile markers on I-70 in Kansas is 559. Find the numbers on the markers.

Source: Federal Highway Administration, Ed Rotalewski

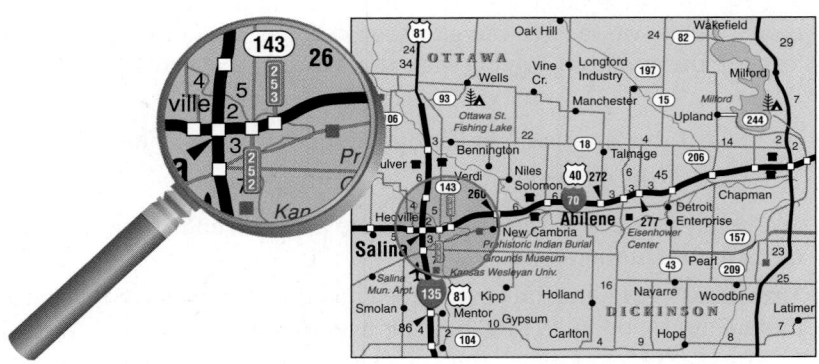

1. Familiarize. The numbers on the mile markers are consecutive positive integers. Thus if we let $x =$ the smaller number, then $x + 1 =$ the larger number.

To become familiar with the problem, we can make a table, as shown at left. First, we guess a value for x; then we find $x + 1$. Finally, we add the two numbers and check the sum.

---

2. Running. Yiannis Kouros of Australia holds the record for the greatest distance run in 24 hr by running 188 mi. After 8 hr, he was approximately twice as far from the finish line as he was from the start. How far had he run?

Source: Australian Ultra Runners Association

x	$x + 1$	Sum of x and $x + 1$
114	115	229
252	253	505
302	303	605

Answer

2. $62\frac{2}{3}$ mi

From the table, we see that the first marker will be between 252 and 302. We could continue guessing and solve the problem this way, but let's work on developing our algebra skills.

2. Translate.   We reword the problem and translate as follows.

Rewording: First integer plus Second integer is 559

Translating: x $+$ $(x + 1)$ $=$ 559

3. Solve. We solve the equation:

$$x + (x + 1) = 559$$
$$2x + 1 = 559 \qquad \text{Collecting like terms}$$
$$2x + 1 - 1 = 559 - 1 \qquad \text{Subtracting 1}$$
$$2x = 558$$
$$\frac{2x}{2} = \frac{558}{2} \qquad \text{Dividing by 2}$$
$$x = 279.$$

If x is 279, then $x + 1$ is 280.

4. Check.   Our possible answers are 279 and 280. These are consecutive positive integers and $279 + 280 = 559$, so the answers check.

5. State.   The mile markers are 279 and 280.

Do Exercise 3.

3. Interstate Mile Markers.   The sum of two consecutive mile markers on I-90 in upstate New York is 627. (On I-90 in New York, the marker numbers increase from east to west.) Find the numbers on the markers.

Source: New York State Department of Transportation

EXAMPLE 4   *Copy Machine Rental.*   It costs the Drake law firm $225 per month plus 1.2¢ per copy to rent a copy machine. The firm needs to lease a machine for use during a special case that they anticipate will take 3 months. If they allot a budget of $1100, how many copies can they make?

Copy Machine Rental
$225 per month
plus 1.2¢ per copy

1. Familiarize.   Suppose that the law firm makes 20,000 copies. Then the cost is given by monthly charges plus copy charges, or

3($225) plus Cost per copy times Number of copies

  $675 $+$ $0.012 $\cdot$ 20,000,

Answer

3. 313 and 314

which is $915. We see that the firm can make more than 20,000 copies. This process familiarizes us with the way in which a calculation is made. Note that we convert 1.2¢ to $0.012 so that all information is in the same unit, dollars. Otherwise, we will not get the correct answer.

We let c = the number of copies that can be made for the budget of $1100.

2. **Translate.** We reword the problem and translate as follows:

$$\underbrace{\text{Monthly cost}}_{3(\$225)} \underset{+}{\text{ plus }} \underbrace{\text{Cost per copy}}_{\$0.012} \underset{\cdot}{\text{ times }} \underbrace{\text{Number of copies}}_{c} \underset{=}{\text{ is }} \underbrace{\text{Budget}}_{\$1100.}$$

3. **Solve.** We solve the equation:

$$3(225) + 0.012c = 1100$$
$$675 + 0.012c = 1100$$
$$0.012c = 425 \qquad \text{Subtracting 675}$$
$$\frac{0.012c}{0.012} = \frac{425}{0.012} \qquad \text{Dividing by 0.012}$$
$$c \approx 35{,}417. \qquad \text{Rounding to the nearest one}$$

4. **Check.** We check in the original problem. The cost for 35,417 pages is $35{,}417(\$0.012) = \$425.004$. The rental for 3 months is $3(\$225) = \$675$. The total cost is then $\$425.004 + \$675 \approx \$1100$, which is the $1100 that was allotted.

5. **State.** The law firm can make 35,417 copies on the copy rental allotment of $1100.

Do Exercise 4.

EXAMPLE 5 *Perimeter of NBA Court.* The perimeter of an NBA basketball court is 288 ft. The length is 44 ft longer than the width. Find the dimensions of the court.
Source: National Basketball Association

1. **Familiarize.** We first make a drawing.

We let w = the width of the rectangle. Then $w + 44$ = the length. The perimeter P of a rectangle is the distance around the rectangle and is given by the formula $2l + 2w = P$, where

l = the length   and w = the width.

Answer
4. 60,417 copies

2. Translate. To translate the problem, we substitute $w + 44$ for l and 288 for P:

Caution!

$$2l + 2w = P$$
$$2(w + 44) + 2w = 288.$$

Parentheses are necessary here.

3. Solve. We solve the equation:

$$2(w + 44) + 2w = 288$$
$$2 \cdot w + 2 \cdot 44 + 2w = 288 \qquad \text{Using the distributive law}$$
$$4w + 88 = 288 \qquad \text{Collecting like terms}$$
$$4w + 88 - 88 = 288 - 88 \qquad \text{Subtracting 88}$$
$$4w = 200$$
$$\frac{4w}{4} = \frac{200}{4} \qquad \text{Dividing by 4}$$
$$w = 50.$$

Thus possible dimensions are

$$w = 50 \text{ ft} \quad \text{and} \quad l = w + 44 = 50 + 44, \text{ or } 94 \text{ ft.}$$

4. Check. If the width is 50 ft and the length is 94 ft, then the perimeter is $2(50 \text{ ft}) + 2(94 \text{ ft})$, or 288 ft. This checks.

5. State. The width is 50 ft and the length is 94 ft.

Do Exercise 5.

5. Perimeter of High School Basketball Court. The perimeter of a standard high school basketball court is 268 ft. The length is 34 ft longer than the width. Find the dimensions of the court.

Source: Indiana High School Athletic Association

Caution!

Always be sure to answer the original problem completely. For instance, in Example 2, we need to find *two* numbers: the distances from *each* end of the trail to the hiker. Similarly, in Example 3, we need to find two mile markers, and in Example 5, we need to find two dimensions, not just the width.

EXAMPLE 6 *Roof Gable.* In a triangular gable end of a roof, the angle of the peak is twice as large as the angle of the back side of the house. The measure of the angle on the front side is 20° greater than the angle on the back side. How large are the angles?

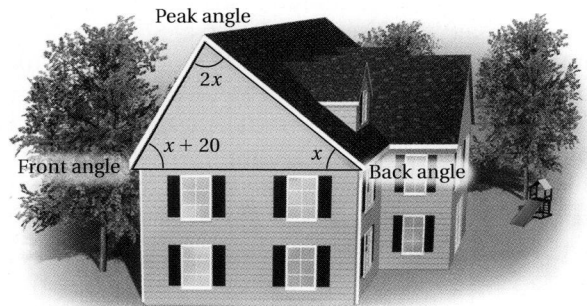

Peak angle

$2x$

$x + 20$

Front angle

x

Back angle

1. Familiarize. We first make a drawing as shown above. We let

measure of back angle $= x$.

Then measure of peak angle $= 2x$

and measure of front angle $= x + 20$.

2. Translate. To translate, we need to know that the sum of the measures of the angles of a triangle is 180°. You might recall this fact from geometry or you can look it up in a geometry book or in the list of formulas inside the back cover of this book. We translate as follows:

Measure of back angle	plus	Measure of peak angle	plus	Measure of front angle	is	180°
x	$+$	$2x$	$+$	$(x + 20)$	$=$	$180°.$

3. Solve. We solve the equation:

$$x + 2x + (x + 20) = 180$$
$$4x + 20 = 180$$
$$4x + 20 - 20 = 180 - 20$$
$$4x = 160$$
$$\frac{4x}{4} = \frac{160}{4}$$
$$x = 40.$$

Possible measures for the angles are as follows:

Back angle: $x = 40°$;
Peak angle: $2x = 2(40) = 80°$;
Front angle: $x + 20 = 40 + 20 = 60°.$

4. Check. Consider our answers: 40°, 80°, and 60°. The peak is twice the back and the front is 20° greater than the back. The sum is 180°. The angles check.

5. State. The measures of the angles are 40°, 80°, and 60°.

Caution!

Units are important in answers. Remember to include them, where appropriate.

> **6.** The second angle of a triangle is three times as large as the first. The third angle measures 30° more than the first angle. Find the measures of the angles.

Do Exercise 6.

EXAMPLE 7 *Fastest Roller Coasters.* The average top speed of the three fastest steel roller coasters in the United States is 116 mph. The third-fastest roller coaster, Superman: The Escape (located at Six Flags Magic Mountain, Valencia, California), reaches a top speed of 28 mph less than the fastest roller coaster, Kingda Ka (located at Six Flags Great Adventure, Jackson, New Jersey). The second-fastest roller coaster, Top Thrill Dragster (located at Cedar Point, Sandusky, Ohio), has a top speed of 120 mph. What is the top speed of the fastest steel roller coaster?

Source: Coaster Grotto

Answer

6. First: 30°; second: 90°; third: 60°

1. Familiarize. The **average** of a set of numbers is the sum of the numbers divided by the number of addends.

We are given that the second-fastest speed is 120 mph. Suppose the three top speeds are 131, 120, and 103. The average is then

$$\frac{131 + 120 + 103}{3} = \frac{354}{3} = 118,$$

which is too high. Instead of continuing to guess, let's use the equation-solving skills we have learned in this chapter. We let x represent the top speed of the fastest roller coaster. Then $x - 28$ is the top speed of the third-fastest roller coaster.

2. Translate. We reword the problem and translate as follows:

$$\frac{\text{Speed of fastest coaster} + \text{Speed of second-fastest coaster} + \text{Speed of third-fastest coaster}}{\text{Number of roller coasters}} = \text{Average speed of three fastest roller coasters}$$

$$\frac{x + 120 + (x - 28)}{3} = 116.$$

3. Solve. We solve as follows:

$$\frac{x + 120 + (x - 28)}{3} = 116$$

$$3 \cdot \frac{x + 120 + (x - 28)}{3} = 3 \cdot 116 \qquad \text{Multiplying by 3 on both sides to clear the fraction}$$

$$x + 120 + (x - 28) = 348$$

$$2x + 92 = 348 \qquad \text{Collecting like terms}$$

$$2x = 256 \qquad \text{Subtracting 92}$$

$$x = 128. \qquad \text{Dividing by 2}$$

4. Check. If the top speed of the fastest roller coaster is 128 mph, then the top speed of the third-fastest is $128 - 28$, or 100 mph. The average of the top speeds of the three fastest is

$$\frac{128 + 120 + 100}{3} = \frac{348}{3} = 116 \text{ mph.}$$

The answer checks.

5. State. The top speed of the fastest steel roller coaster in the United States is 128 mph.

Do Exercise 7.

7. Average Test Score. Sam's average score on his first three math tests is 77. He scored 62 on the first test. On the third test, he scored 9 more than he scored on his second test. What did he score on the second and third tests?

Answer

7. Second: 80; third: 89

EXAMPLE 8 *Simple Interest.* An investment is made at 3% simple interest for 1 year. It grows to $746.75. How much was originally invested (the principal)?

1. **Familiarize.** Suppose that $100 was invested. Recalling the formula for simple interest, $I = Prt$, we know that the interest for 1 year on $100 at 3% simple interest is given by $I = \$100 \cdot 0.03 \cdot 1 = \$3$. Then, at the end of the year, the amount in the account is found by adding the principal and the interest:

$$\begin{array}{ccc}
\text{Principal} & + \quad \text{Interest} & = \quad \text{Amount} \\
\downarrow & \downarrow & \downarrow \\
\$100 & + \qquad \$3 & = \quad \$103.
\end{array}$$

In this problem, we are working backward. We are trying to find the principal, which is the original investment. We let $x =$ the principal. Then the interest earned is $3\%x$.

2. **Translate.** We reword the problem and then translate:

$$\begin{array}{ccc}
\text{Principal} & + \quad \text{Interest} & = \quad \text{Amount} \\
\downarrow & \downarrow & \downarrow \\
x & + \qquad 3\%x & = \quad 746.75.
\end{array}$$

Interest is 3% of the principal.

3. **Solve.** We solve the equation:

$$\begin{aligned}
x + 3\%x &= 746.75 \\
x + 0.03x &= 746.75 \qquad \text{Converting to decimal notation} \\
1x + 0.03x &= 746.75 \qquad \text{Identity property of 1} \\
(1 + 0.03)x &= 746.75 \\
1.03x &= 746.75 \qquad \text{Collecting like terms} \\
\frac{1.03x}{1.03} &= \frac{746.75}{1.03} \qquad \text{Dividing by 1.03} \\
x &= 725.
\end{aligned}$$

4. **Check.** We check by taking 3% of $725 and adding it to $725:

$$3\% \times \$725 = 0.03 \times 725 = \$21.75.$$

Then $725 + $21.75 = $746.75, so $725 checks.

5. **State.** The original investment was $725.

Do Exercise 8.

EXAMPLE 9 *Selling a Home.* The Landers are planning to sell their home. If they want to be left with $117,500 after paying 6% of the selling price to a realtor as a commission, for how much must they sell the house?

1. **Familiarize.** Suppose the Landers sell the house for $120,000. A 6% commission can be determined by finding 6% of $120,000:

$$6\% \text{ of } \$120,000 = 0.06(\$120,000) = \$7200.$$

Subtracting this commission from $120,000 would leave the Landers with

$$\$120,000 - \$7200 = \$112,800.$$

This shows that in order for the Landers to clear $117,500, the house must sell for more than $120,000. Our guess shows us how to translate to an equation. We let $x =$ the selling price, in dollars. With a 6% commission, the realtor would receive $0.06x$.

8. **Simple Interest.** An investment is made at 7% simple interest for 1 year. It grows to $8988. How much was originally invested (the principal)?

Answer

8. $8400

2. Translate. We reword the problem and translate as follows:

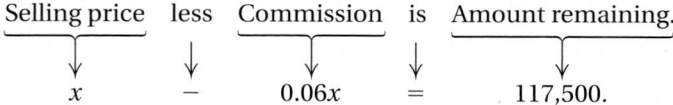

Selling price less Commission is Amount remaining.

$$x - 0.06x = 117{,}500.$$

3. Solve. We solve the equation:

$$x - 0.06x = 117{,}500$$
$$1x - 0.06x = 117{,}500$$
$$(1 - 0.06)x = 117{,}500$$
$$0.94x = 117{,}500$$ Collecting like terms. Had we noted that after the commission has been paid, 94% remains, we could have begun with this equation.

$$\frac{0.94x}{0.94} = \frac{117{,}500}{0.94}$$ Dividing by 0.94

$$x = 125{,}000.$$

4. Check. To check, we first find 6% of $125,000:

$$6\% \text{ of } \$125{,}000 = 0.06(\$125{,}000) = \$7500.$$ This is the commission.

Next, we subtract the commission to find the remaining amount:

$$\$125{,}000 - \$7500 = \$117{,}500.$$

Since, after the commission, the Landers are left with $117,500, our answer checks. Note that the $125,000 selling price is greater than $120,000, as predicted in the *Familiarize* step.

5. State. To be left with $117,500, the Landers must sell the house for $125,000.

Do Exercise 9.

9. **Price Before Sale.** The price of a suit was decreased to a sale price of $526.40. This was a 20% reduction. What was the former price?

Caution!

The problem in Example 9 is easy to solve with algebra. Without algebra, it is not. A common error in such a problem is to take 6% of the price after commission and then subtract or add. Note that 6% of the selling price (6% · $125,000 = $7500) is not equal to 6% of the amount that the Landers want to be left with (6% · $117,500 = $7050).

Answer

9. $658

Translating for Success

1. *Angle Measures.* The measure of the second angle of a triangle is 51° more than that of the first angle. The measure of the third angle is 3° less than twice the first angle. Find the measures of the angles.

2. *Sales Tax.* Tina paid $3976 for a used car. This amount included 5% for sales tax. How much did the car cost before tax?

3. *Perimeter.* The perimeter of a rectangle is 2347 ft. The length is 28 ft greater than the width. Find the length and the width.

4. *Fraternity or Sorority Membership.* At Arches Tech University, 3976 students belong to a fraternity or a sorority. This is 35% of the total enrollment. What is the total enrollment at Arches Tech?

5. *Fraternity or Sorority Membership.* At Moab Tech University, thirty-five percent of the students belong to a fraternity or a sorority. The total enrollment of the university is 11,360 students. How many students belong to either a fraternity or a sorority?

The goal of these matching questions is to practice step (2), *Translate*, of the five-step problem-solving process. Translate each word problem to an equation and select a correct translation from equations A–O.

A. $x + (x - 3) + \frac{4}{5}x = 384$

B. $x + (x + 51) + (2x - 3) = 180$

C. $x + (x + 96) = 180$

D. $2 \cdot 96 + 2x = 3976$

E. $x + (x + 1) + (x + 2) = 384$

F. $3976 = x \cdot 11{,}360$

G. $2x + 2(x + 28) = 2347$

H. $3976 = x + 5\%x$

I. $x + (x + 28) = 2347$

J. $x = 35\% \cdot 11{,}360$

K. $x + 96 = 3976$

L. $x + (x + 3) + \frac{4}{5}x = 384$

M. $x + (x + 2) + (x + 4) = 384$

N. $35\% \cdot x = 3976$

O. $2x + (x + 28) = 2347$

Answers on page A-4

6. *Island Population.* There are 180 thousand people living on a small Caribbean island. The women outnumber the men by 96 thousand. How many men live on the island?

7. *Wire Cutting.* A 384-m wire is cut into three pieces. The second piece is 3 m longer than the first. The third is four-fifths as long as the first. How long is each piece?

8. *Locker Numbers.* The numbers on three adjoining lockers are consecutive integers whose sum is 384. Find the integers.

9. *Fraternity or Sorority Membership.* The total enrollment at Canyonlands Tech University is 11,360 students. Of these, 3976 students belong to a fraternity or a sorority. What percent of the students belong to a fraternity or a sorority?

10. *Width of a Rectangle.* The length of a rectangle is 96 ft. The perimeter of the rectangle is 3976 ft. Find the width.

a Solve. *Although you might find the answer quickly in some other way, practice using the five-step problem-solving strategy.*

1. *Manatee Population.* The manatee, Florida's state marine mammal, is an endangered species. An aerial wintertime manatee census counted 2817 of these animals in 2007. This was 296 fewer than the number counted in 2006. What was Florida's manatee population in 2006?

Source: Florida Fish and Wildlife Conservation Commission

2. *Mass Transit Boom.* Americans took 2.8 billion rides on public transit from April through June in 2008. This was the highest ridership for that period in 50 yr and represented an increase of 0.7 billion rides over the same period in 1998. How many rides were taken from April through June in 1998?

Source: American Public Transportation Association

3. *Pipe Cutting.* A 240-in. pipe is cut into two pieces. One piece is three times the length of the other. Find the lengths of the pieces.

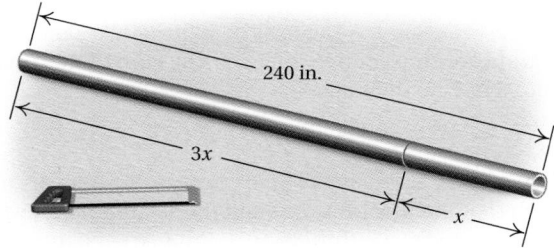

240 in.
3x
x

4. *Board Cutting.* A 72-in. board is cut into two pieces. One piece is 2 in. longer than the other. Find the lengths of the pieces.

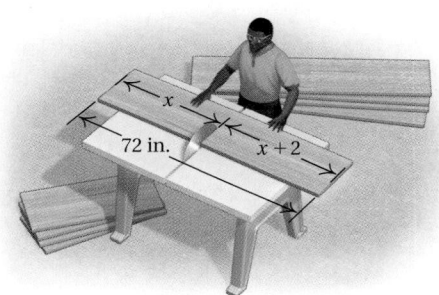

x
72 in.
x + 2

5. *Cost of Movie Tickets.* The average cost of movie tickets for a family of four was $28.32 in 2008. This was $11.76 more than the cost in 1993. What was the average cost of movie tickets for a family of four in 1993? (These prices include senior discounts and children's prices.)

Source: Motion Picture Association of America

6. *Area of Lake Ontario.* The area of Lake Superior is about four times the area of Lake Ontario. The area of Lake Superior is 30,172 mi². What is the area of Lake Ontario?

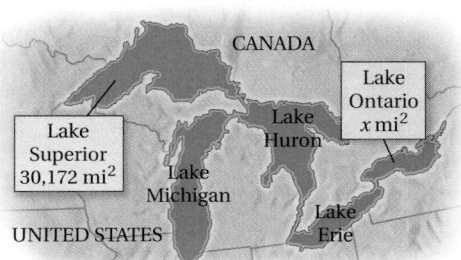

CANADA

Lake Ontario
x mi²

Lake Huron

Lake Superior
30,172 mi²

Lake Michigan

Lake Erie

UNITED STATES

7. *Iditarod Race.* The Iditarod sled dog race in Alaska extends for 1049 mi from Anchorage to Nome. If a musher is twice as far from Anchorage as from Nome, how many miles of the race has the musher completed?

Source: Iditarod Trail Commission

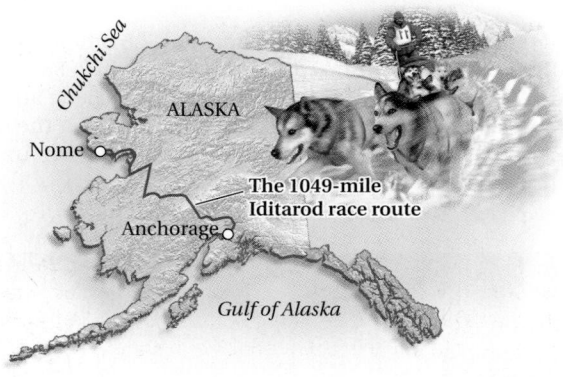

8. *Statue of Liberty.* The height of the Eiffel Tower is 974 ft, which is about 669 ft higher than the Statue of Liberty. What is the height of the Statue of Liberty?

9. *Consecutive Apartment Numbers.* The apartments in Vincent's apartment house are numbered consecutively on each floor. The sum of his number and his next-door neighbor's number is 2409. What are the two numbers?

10. *Consecutive Post Office Box Numbers.* The sum of the numbers on two consecutive post office boxes is 547. What are the numbers?

11. *Consecutive Ticket Numbers.* The numbers on Sam's three raffle tickets are consecutive integers. The sum of the numbers is 126. What are the numbers?

12. *Consecutive Ages.* The ages of Whitney, Wesley, and Wanda are consecutive integers. The sum of their ages is 108. What are their ages?

13. *Consecutive Odd Integers.* The sum of three consecutive odd integers is 189. What are the integers?

14. *Consecutive Integers.* Three consecutive integers are such that the first plus one-half the second plus seven less than twice the third is 2101. What are the integers?

15. *Standard Billboard Sign.* A standard rectangular highway billboard sign has a perimeter of 124 ft. The length is 6 ft more than three times the width. Find the dimensions of the sign.

16. *Two-by-Four.* The perimeter of a cross section or end of a "two-by-four" piece of lumber is 10 in. The length is 2 in. more than the width. Find the actual dimensions of the cross section of a two-by-four.

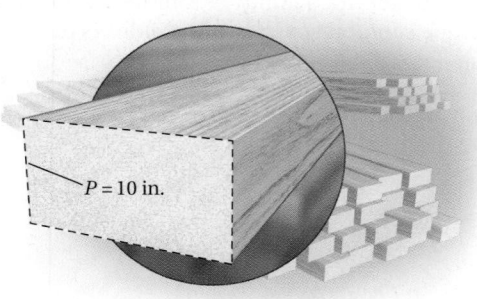

17. *Price of Walking Shoes.* Amy paid $63.75 for a pair of walking shoes during a 15%-off sale. What was the regular price?

18. *Price of a CD Player.* Doug paid $72 for a shockproof portable CD player during a 20%-off sale. What was the regular price?

19. *Price of a Jacket.* Evelyn paid $89.25, including 5% tax, for a jacket. How much did the jacket itself cost?

20. *Price of a Printer.* Jake paid $100.70, including 6% tax, for a color printer. How much did the printer itself cost?

21. *Parking Costs.* A hospital parking lot charges $1.50 for the first hour or part thereof, and $1.00 for each additional hour or part thereof. A weekly pass costs $27.00 and allows unlimited parking for 7 days. Suppose that each visit Ed makes to the hospital lasts $1\frac{1}{2}$ hr. What is the minimum number of times that Ed would have to visit per week to make it worthwhile for him to buy the pass?

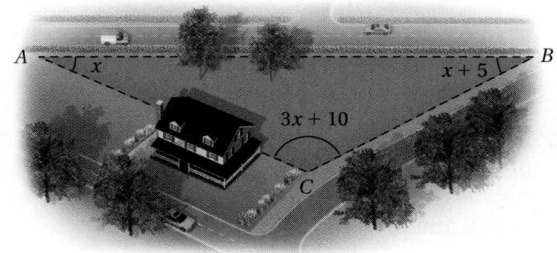

22. *Van Rental.* Value Rent-A-Car rents vans at a daily rate of $84.45 plus 55¢ per mile. Molly rents a van to deliver electrical parts to her customers. She is allotted a daily budget of $250. How many miles can she drive for $250? (*Hint*: 60¢ = $0.60.)

23. *Triangular Field.* The second angle of a triangular field is three times as large as the first angle. The third angle is 40° greater than the first angle. How large are the angles?

24. *Triangular Parking Lot.* The second angle of a triangular parking lot is four times as large as the first angle. The third angle is 45° less than the sum of the other two angles. How large are the angles?

25. *Triangular Backyard.* A home has a triangular backyard. The second angle of the triangle is 5° more than the first angle. The third angle is 10° more than three times the first angle. Find the angles of the triangular yard.

26. *Boarding Stable.* A rancher needs to form a triangular horse pen using ropes next to a stable. The second angle is three times the first angle. The third angle is 15° less than the first angle. Find the angles of the triangular pen.

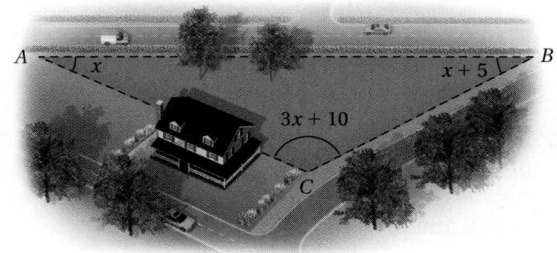

27. *Stock Prices.* Sarah's investment in a technology stock grew 28% to $448. How much did she invest?

28. *Savings Interest.* Sharon invested money in a savings account at a rate of 6% simple interest. After 1 year, she has $6996 in the account. How much did Sharon originally invest?

29. *Credit Cards.* The balance on Will's credit card grew 2%, to $870, in one month. What was his balance at the beginning of the month?

30. *Loan Interest.* Alvin borrowed money from a cousin at a rate of 10% simple interest. After 1 year, $7194 paid off the loan. How much did Alvin borrow?

31. *Taxi Fares.* In Beniford, taxis charge $3 plus 75¢ per mile for an airport pickup. How far from the airport can Courtney travel for $12?

32. *Taxi Fares.* In Cranston, taxis charge $4 plus 90¢ per mile for an airport pickup. How far from the airport can Ralph travel for $17.50?

33. *Tipping.* Leon left a 15% tip for a meal. The total cost of the meal, including the tip, was $41.40. What was the cost of the meal before the tip was added?

34. *Tipping.* Selena left an 18% tip for a meal. The total cost of the meal, including the tip, was $40.71. What was the cost of the meal before the tip was added?

35. *Average Price.* Tom paid an average of $34 per tie for a recent purchase of three ties. The price of one tie was twice as much as another, and the remaining tie cost $27. What were the prices of the other two ties?

36. *Average Test Score.* Jaci averaged 84 on her first three history exams. The first score was 67. The second score was 7 less than the third score. What did she score on the second and third exams?

37. If you double a number and then add 16, you get $\frac{2}{3}$ of the original number. What is the original number?

38. If you double a number and then add 85, you get $\frac{3}{4}$ of the original number. What is the original number?

Skill Maintenance

Calculate.

39. $-\dfrac{4}{5} - \dfrac{3}{8}$ [1.4a]

40. $-\dfrac{4}{5} + \dfrac{3}{8}$ [1.3a]

41. $-\dfrac{4}{5} \cdot \dfrac{3}{8}$ [1.5a]

42. $-\dfrac{4}{5} \div \dfrac{3}{8}$ [1.6c]

43. $\dfrac{1}{10} \div \left(-\dfrac{1}{100}\right)$ [1.6c]

44. $-25.6 \div (-16)$ [1.6c]

45. $-25.6(-16)$ [1.5a]

46. $-25.6 - (-16)$ [1.4a]

47. $-25.6 + (-16)$ [1.3a]

48. $(-0.02) \div (-0.2)$ [1.6c]

Solve.   [2.3c]

49. $2b + 8 - 3b = 10 + 11b - 2 - 12b$

50. $25 - x + 3 = -3 - 2x + x + 10$

51. $-6 + c = c + 6$

52. $w - 43 = -43 + w$

Synthesis

53. Apples are collected in a basket for six people. One-third, one-fourth, one-eighth, and one-fifth are given to four people, respectively. The fifth person gets ten apples, leaving one apple for the sixth person. Find the original number of apples in the basket.

54. *Test Questions.* A student scored 78 on a test that had 4 seven-point fill-ins and 24 three-point multiple-choice questions. The student answered one fill-in incorrectly. How many multiple-choice questions did the student answer correctly?

55. The area of this triangle is 2.9047 in². Find x.

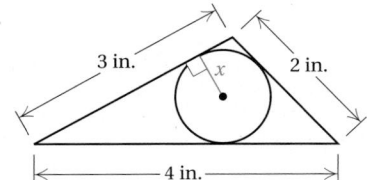

56. Susanne goes to the bank to get $20 in quarters, dimes, and nickels to use to make change at her yard sale. She gets twice as many quarters as dimes and 10 more nickels than dimes. How many of each type of coin does she get?

57. In Connerville, a sales tax of 9% was added to the price of gasoline as registered on the pump. Suppose a driver asked for $10 worth of gas. The attendant filled the tank until the pump read $9.10 and charged the driver $10. Something was wrong. Use algebra to correct the error.

2.7

Solving Inequalities

OBJECTIVES

a Determine whether a given number is a solution of an inequality.

b Graph an inequality on the number line.

c Solve inequalities using the addition principle.

d Solve inequalities using the multiplication principle.

e Solve inequalities using the addition principle and the multiplication principle together.

SKILL TO REVIEW

Objective 1.2d: Determine whether an inequality like $-3 \leq 5$ is true or false.

Write true or false.

1. $-6 \leq -8$ **2.** $1 \geq 1$

Determine whether each number is a solution of the inequality.

1. $x > 3$

   a) 2   **b)** 0

   c) -5 **d)** 15.4

   e) 3   **f)** $-\dfrac{2}{5}$

2. $x \leq 6$

   a) 6   **b)** 0

   c) -4.3 **d)** 25

   e) -6 **f)** $\dfrac{5}{8}$

We now extend our equation-solving principles to the solving of inequalities.

a Solutions of Inequalities

In Section 1.2, we defined the symbols $>$ (is greater than), $<$ (is less than), $\geq$ (is greater than or equal to), and $\leq$ (is less than or equal to).

An **inequality** is a number sentence with $>, <, \geq,$ or $\leq$ as its verb—for example,

$$-4 > t, \quad x < 3, \quad 2x + 5 \geq 0, \quad \text{and} \quad -3y + 7 \leq -8.$$

Some replacements for a variable in an inequality make it true and some make it false. (There are some exceptions to this statement, but we will not consider them here.)

> **SOLUTION**
>
> A replacement that makes an inequality true is called a **solution**. The set of all solutions is called the **solution set**. When we have found the set of all solutions of an inequality, we say that we have **solved** the inequality.

EXAMPLES Determine whether each number is a solution of $x < 2$.

1. -2.7 Since $-2.7 < 2$ is true, -2.7 is a solution.

2. 2   Since $2 < 2$ is false, 2 is not a solution.

EXAMPLES Determine whether each number is a solution of $y \geq 6$.

3. 6   Since $6 \geq 6$ is true, 6 is a solution.

4. $-\dfrac{4}{3}$ Since $-\dfrac{4}{3} \geq 6$ is false, $-\dfrac{4}{3}$ is not a solution.

Do Margin Exercises 1 and 2.

b Graphs of Inequalities

Some solutions of $x < 2$ are $-3, 0, 1, 0.45, -8.9, -\pi, \dfrac{5}{8}$, and so on. In fact, there are infinitely many real numbers that are solutions. Because we cannot list them all individually, it is helpful to make a drawing that represents all the solutions.

A **graph** of an inequality is a drawing that represents its solutions. An inequality in one variable can be graphed on the number line. An inequality in two variables can be graphed on the coordinate plane. We will study such graphs in Chapter 9.

Answers

Skill to Review:
1. False   **2.** True

Margin Exercises:
1. (a) No; (b) no; (c) no; (d) yes; (e) no; (f) no
2. (a) Yes; (b) yes; (c) yes; (d) no; (e) yes; (f) yes

EXAMPLE 5 Graph: $x < 2$.

The solutions of $x < 2$ are all those numbers less than 2. They are shown on the number line by shading all points to the left of 2. The open circle at 2 indicates that 2 is *not* part of the graph.

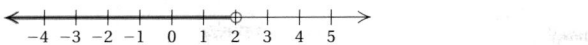

EXAMPLE 6 Graph: $x \geq -3$.

The solutions of $x \geq -3$ are shown on the number line by shading the point for -3 and all points to the right of -3. The closed circle at -3 indicates that -3 *is* part of the graph.

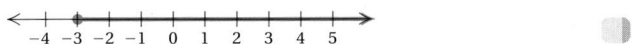

EXAMPLE 7 Graph: $-3 \leq x < 2$.

The inequality $-3 \leq x < 2$ is read "-3 is less than or equal to x *and* x is less than 2," or "x is greater than or equal to -3 *and* x is less than 2." In order to be a solution of this inequality, a number must be a solution of both $-3 \leq x$ and $x < 2$. The number 1 is a solution, as are -1.7, 0, 1.5, and $\frac{3}{8}$. We can see from the graphs below that the solution set consists of the numbers that overlap in the two solution sets in Examples 5 and 6.

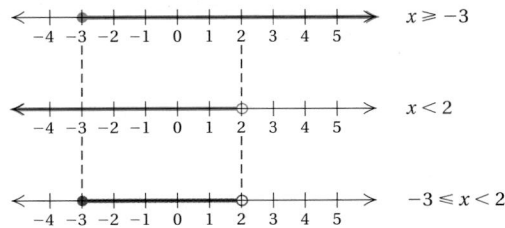

The open circle at 2 means that 2 is *not* part of the graph. The closed circle at -3 means that -3 *is* part of the graph. The other solutions are shaded.

Do Exercises 3–5.

(c) Solving Inequalities Using the Addition Principle

Consider the true inequality $3 < 7$. If we add 2 on both sides, we get another true inequality:

$$3 + 2 < 7 + 2, \quad \text{or} \quad 5 < 9.$$

Similarly, if we add -4 on both sides of $x + 4 < 10$, we get an *equivalent* inequality:

$$x + 4 + (-4) < 10 + (-4),$$

or

$$x < 6.$$

To say that $x + 4 < 10$ and $x < 6$ are **equivalent** is to say that they have the same solution set. For example, the number 3 is a solution of $x + 4 < 10$. It is also a solution of $x < 6$. The number -2 is a solution of $x < 6$. It is also a solution of $x + 4 < 10$. Any solution of one inequality is a solution of the other—they are equivalent.

Graph.

3. $x \leq 4$

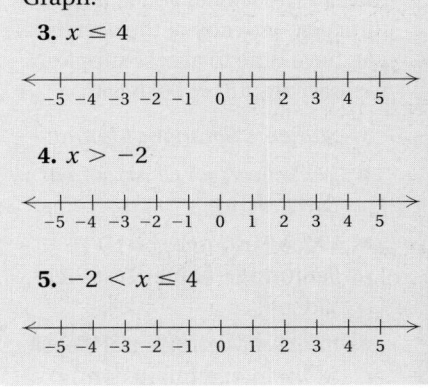

4. $x > -2$

5. $-2 < x \leq 4$

Answers

3.

$x \leq 4$

4.

$x > -2$

5.

$-2 < x \leq 4$

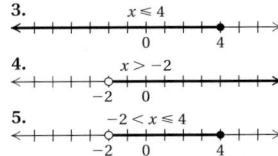

THE ADDITION PRINCIPLE FOR INEQUALITIES

For any real numbers a, b, and c:

$a < b$ is equivalent to $a + c < b + c$;
$a > b$ is equivalent to $a + c > b + c$;
$a \leq b$ is equivalent to $a + c \leq b + c$;
$a \geq b$ is equivalent to $a + c \geq b + c$.

In other words, when we add or subtract the same number on both sides of an inequality, the direction of the inequality symbol is not changed.

As with equation solving, when solving inequalities, our goal is to isolate the variable on one side. Then it is easier to determine the solution set.

EXAMPLE 8   Solve: $x + 2 > 8$. Then graph.

We use the addition principle, subtracting 2 on both sides:

$$x + 2 - 2 > 8 - 2$$
$$x > 6.$$

From the inequality $x > 6$, we can determine the solutions directly. Any number greater than 6 makes the last sentence true and is a solution of that sentence. Any such number is also a solution of the original sentence. Thus the inequality is solved. The graph is as follows:

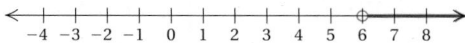

We cannot check all the solutions of an inequality by substitution, as we usually can for an equation, because there are too many of them. A partial check can be done by substituting a number greater than 6—say, 7—into the original inequality:

$$\begin{array}{c} x + 2 > 8 \\ \hline 7 + 2 \; ? \; 8 \\ 9 \; \vert \quad \text{TRUE} \end{array}$$

Since $9 > 8$ is true, 7 is a solution. This is a partial check that any number greater than 6 is a solution.

EXAMPLE 9   Solve: $3x + 1 \leq 2x - 3$. Then graph.

We have

$$\begin{array}{ll} 3x + 1 \leq 2x - 3 & \\ 3x + 1 - 1 \leq 2x - 3 - 1 & \text{Subtracting 1} \\ 3x \leq 2x - 4 & \text{Simplifying} \\ 3x - 2x \leq 2x - 4 - 2x & \text{Subtracting } 2x \\ x \leq -4. & \text{Simplifying} \end{array}$$

Any number less than or equal to -4 is a solution. The graph is as follows:

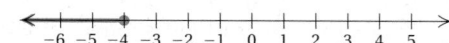

In Example 9, any number less than or equal to -4 is a solution. The following are some solutions:

$$-4, \quad -5, \quad -6, \quad -\frac{13}{3}, \quad -204.5, \quad \text{and} \quad -18\pi.$$

Besides drawing a graph, we can also describe all the solutions of an inequality using **set notation**. We could just begin to list them in a set using roster notation (see p. 9), as follows:

$$\left\{ -4, -5, -6, -\frac{13}{3}, -204.5, -18\pi, \ldots \right\}.$$

We can never list them all this way, however. Seeing this set without knowing the inequality makes it difficult for us to know what real numbers we are considering. There is, however, another kind of notation that we can use. It is

$$\{x \,|\, x \le -4\},$$

which is read

"The set of all x such that x is less than or equal to -4."

This shorter notation for sets is called **set-builder notation**.

From now on, we will use this notation when solving inequalities.

> Do Exercises 6–8.

EXAMPLE 10 Solve: $x + \frac{1}{3} > \frac{5}{4}$.

We have

$$
\begin{aligned}
x + \tfrac{1}{3} &> \tfrac{5}{4} \\
x + \tfrac{1}{3} - \tfrac{1}{3} &> \tfrac{5}{4} - \tfrac{1}{3} && \text{Subtracting } \tfrac{1}{3} \\
x &> \tfrac{5}{4} \cdot \tfrac{3}{3} - \tfrac{1}{3} \cdot \tfrac{4}{4} && \text{Multiplying by 1 to obtain} \\
&&& \text{a common denominator} \\
x &> \tfrac{15}{12} - \tfrac{4}{12} \\
x &> \tfrac{11}{12}.
\end{aligned}
$$

Any number greater than $\frac{11}{12}$ is a solution. The solution set is

$$\left\{ x \,\middle|\, x > \tfrac{11}{12} \right\},$$

which is read

"The set of all x such that x is greater than $\frac{11}{12}$."

When solving inequalities, you may obtain an answer like $\frac{11}{12} < x$. Recall from Chapter 1 that this has the same meaning as $x > \frac{11}{12}$. Thus the solution set in Example 10 can be described as $\left\{ x \,\middle|\, \frac{11}{12} < x \right\}$ or as $\left\{ x \,\middle|\, x > \frac{11}{12} \right\}$. The latter is used most often.

> Do Exercises 9 and 10.

d Solving Inequalities Using the Multiplication Principle

There is a multiplication principle for inequalities that is similar to that for equations, but it must be modified. When we are multiplying on both sides by a negative number, the direction of the inequality symbol must be changed.

Solve. Then graph.

6. $x + 3 > 5$

7. $x - 1 \le 2$

8. $5x + 1 < 4x - 2$

Solve.

9. $x + \dfrac{2}{3} \ge \dfrac{4}{5}$

10. $5y + 2 \le -1 + 4y$

Answers

6. $\{x \,|\, x > 2\}$;

7. $\{x \,|\, x \le 3\}$;

8. $\{x \,|\, x < -3\}$;

9. $\left\{ x \,\middle|\, x \ge \dfrac{2}{15} \right\}$

10. $\{y \,|\, y \le -3\}$

Consider the true inequality $3 < 7$. If we multiply on both sides by a *positive* number, like 2, we get another true inequality:

$$3 \cdot 2 < 7 \cdot 2, \quad \text{or} \quad 6 < 14. \qquad \text{True}$$

If we multiply on both sides by a *negative* number, like -2, and we do not change the direction of the inequality symbol, we get a *false* inequality:

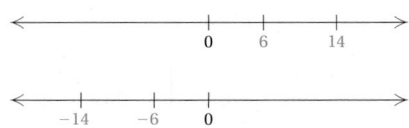

$$3 \cdot (-2) < 7 \cdot (-2), \quad \text{or} \quad -6 < -14. \qquad \text{False}$$

The fact that $6 < 14$ is true but $-6 < -14$ is false stems from the fact that the negative numbers, in a sense, mirror the positive numbers. That is, whereas 14 is to the *right* of 6 on the number line, the number -14 is to the *left* of -6. Thus, if we reverse (change the direction of) the inequality symbol, we get a *true* inequality: $-6 > -14$.

THE MULTIPLICATION PRINCIPLE FOR INEQUALITIES

For any real numbers a and b, and any *positive* number c:

$a < b$ is equivalent to $ac < bc$;

$a > b$ is equivalent to $ac > bc$.

For any real numbers a and b, and any *negative* number c:

$a < b$ is equivalent to $ac > bc$;

$a > b$ is equivalent to $ac < bc$.

Similar statements hold for $\le$ and $\ge$.

   In other words, when we multiply or divide by a positive number on both sides of an inequality, the direction of the inequality symbol stays the same. When we multiply or divide by a negative number on both sides of an inequality, the direction of the inequality symbol is reversed.

EXAMPLE 11   Solve: $4x < 28$. Then graph.

We have

$$4x < 28$$

$$\frac{4x}{4} < \frac{28}{4} \qquad \text{Dividing by 4}$$

The symbol stays the same.

$$x < 7. \qquad \text{Simplifying}$$

The solution set is $\{x \mid x < 7\}$. The graph is as follows:

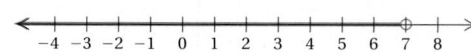

Do Exercises 11 and 12.

Solve. Then graph.

11. $8x < 64$

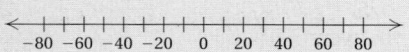

12. $5y \ge 160$

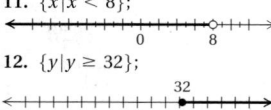

EXAMPLE 12 Solve: $-2y < 18$. Then graph.

$$-2y < 18$$

$$\frac{-2y}{-2} > \frac{18}{-2} \qquad \text{Dividing by } -2$$

The symbol must be reversed!

$$y > -9. \qquad \text{Simplifying}$$

The solution set is $\{y \mid y > -9\}$. The graph is as follows:

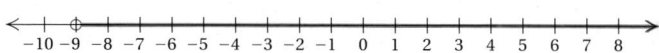

Do Exercises 13 and 14.

Solve.

13. $-4x \le 24$

14. $-5y > 13$

(e) Using the Principles Together

All of the equation-solving techniques used in Sections 2.1–2.3 can be used with inequalities, provided we remember to reverse the inequality symbol when multiplying or dividing on both sides by a negative number.

EXAMPLE 13 Solve: $6 - 5x > 7$.

$$6 - 5x > 7$$

$$-6 + 6 - 5x > -6 + 7 \qquad \text{Adding } -6. \text{ The symbol stays the same.}$$

$$-5x > 1 \qquad \text{Simplifying}$$

$$\frac{-5x}{-5} < \frac{1}{-5} \qquad \text{Dividing by } -5$$

The symbol must be reversed because we are dividing by a *negative* number, -5.

$$x < -\frac{1}{5}. \qquad \text{Simplifying}$$

The solution set is $\left\{ x \mid x < -\frac{1}{5} \right\}$.

Do Exercise 15.

15. Solve: $7 - 4x < 8$.

EXAMPLE 14 Solve: $17 - 5y > 8y - 9$.

$$-17 + 17 - 5y > -17 + 8y - 9 \qquad \text{Adding } -17. \text{ The symbol stays the same.}$$

$$-5y > 8y - 26 \qquad \text{Simplifying}$$

$$-8y - 5y > -8y + 8y - 26 \qquad \text{Adding } -8y$$

$$-13y > -26 \qquad \text{Simplifying}$$

$$\frac{-13y}{-13} < \frac{-26}{-13} \qquad \text{Dividing by } -13$$

The symbol must be reversed because we are dividing by a *negative* number, -13.

$$y < 2$$

The solution set is $\{y \mid y < 2\}$.

Do Exercise 16.

16. Solve. Begin by subtracting 24 on both sides.

$$24 - 7y \le 11y - 14$$

Answers

13. $\{x \mid x \ge -6\}$ **14.** $\left\{ y \mid y < -\frac{13}{5} \right\}$

15. $\left\{ x \mid x > -\frac{1}{4} \right\}$ **16.** $\left\{ y \mid y \ge \frac{19}{9} \right\}$

Typically, we solve an equation or an inequality by isolating the variable on the left side. When we are solving an inequality, however, there are situations in which isolating the variable on the right side will eliminate the need to reverse the inequality symbol. Let's solve the inequality in Example 14 again, but this time we will isolate the variable on the right side.

EXAMPLE 15   Solve: $17 - 5y > 8y - 9$.

Note that if we add $5y$ on both sides, the coefficient of the y-term will be positive after like terms have been collected.

$$17 - 5y + 5y > 8y - 9 + 5y \qquad \text{Adding } 5y$$
$$17 > 13y - 9 \qquad \text{Simplifying}$$
$$17 + 9 > 13y - 9 + 9 \qquad \text{Adding } 9$$
$$26 > 13y \qquad \text{Simplifying}$$
$$\frac{26}{13} > \frac{13y}{13} \qquad \text{Dividing by 13. We leave the inequality symbol the same because we are dividing by a positive number.}$$
$$2 > y$$

The solution set is $\{y \mid 2 > y\}$, or $\{y \mid y < 2\}$.

Do Exercise 17.

17. Solve. Begin by adding $7y$ on both sides.

$$24 - 7y \le 11y - 14$$

EXAMPLE 16   Solve: $3(x - 2) - 1 < 2 - 5(x + 6)$.

First, we use the distributive law to remove parentheses. Next, we collect like terms and then use the addition and multiplication principles for inequalities to get an equivalent inequality with x alone on one side.

$$3(x - 2) - 1 < 2 - 5(x + 6)$$
$$3x - 6 - 1 < 2 - 5x - 30 \qquad \text{Using the distributive law to multiply and remove parentheses}$$
$$3x - 7 < -5x - 28 \qquad \text{Collecting like terms}$$
$$3x + 5x < -28 + 7 \qquad \text{Adding } 5x \text{ and } 7 \text{ to get all } x\text{-terms on one side and all other terms on the other side}$$
$$8x < -21 \qquad \text{Simplifying}$$
$$x < \frac{-21}{8}, \text{ or } -\frac{21}{8}. \qquad \text{Dividing by 8}$$

The solution set is $\left\{x \mid x < -\frac{21}{8}\right\}$.

Do Exercise 18.

18. Solve:

$$3(7 + 2x) \le 30 + 7(x - 1).$$

Answers

17. $\left\{ y \mid y \ge \dfrac{19}{9} \right\}$ **18.** $\{x \mid x \ge -2\}$

EXAMPLE 17 Solve: $16.3 - 7.2p \leq -8.18$.

The greatest number of decimal places in any one number is *two*. Multiplying by 100, which has two 0's, will clear decimals. Then we proceed as before.

$$16.3 - 7.2p \leq -8.18$$

$$100(16.3 - 7.2p) \leq 100(-8.18) \qquad \text{Multiplying by 100}$$

$$100(16.3) - 100(7.2p) \leq 100(-8.18) \qquad \text{Using the distributive law}$$

$$1630 - 720p \leq -818 \qquad \text{Simplifying}$$

$$1630 - 720p - 1630 \leq -818 - 1630 \qquad \text{Subtracting 1630}$$

$$-720p \leq -2448 \qquad \text{Simplifying}$$

$$\frac{-720p}{-720} \geq \frac{-2448}{-720} \qquad \text{Dividing by } -720$$

The symbol must be reversed.

$$p \geq 3.4$$

The solution set is $\{p \mid p \geq 3.4\}$.

Do Exercise 19.

19. Solve:
$$2.1x + 43.2 \geq 1.2 - 8.4x.$$

EXAMPLE 18 Solve: $\frac{2}{3}x - \frac{1}{6} + \frac{1}{2}x > \frac{7}{6} + 2x$.

The number 6 is the least common multiple of all the denominators. Thus we first multiply by 6 on both sides to clear the fractions.

$$\frac{2}{3}x - \frac{1}{6} + \frac{1}{2}x > \frac{7}{6} + 2x$$

$$6\left(\frac{2}{3}x - \frac{1}{6} + \frac{1}{2}x\right) > 6\left(\frac{7}{6} + 2x\right) \qquad \text{Multiplying by 6 on both sides}$$

$$6 \cdot \frac{2}{3}x - 6 \cdot \frac{1}{6} + 6 \cdot \frac{1}{2}x > 6 \cdot \frac{7}{6} + 6 \cdot 2x \qquad \text{Using the distributive law}$$

$$4x - 1 + 3x > 7 + 12x \qquad \text{Simplifying}$$

$$7x - 1 > 7 + 12x \qquad \text{Collecting like terms}$$

$$7x - 1 - 7x > 7 + 12x - 7x \qquad \text{Subtracting } 7x. \text{ The coefficient of the } x\text{-term will be positive.}$$

$$-1 > 7 + 5x \qquad \text{Simplifying}$$

$$-1 - 7 > 7 + 5x - 7 \qquad \text{Subtracting 7}$$

$$-8 > 5x \qquad \text{Simplifying}$$

$$\frac{-8}{5} > \frac{5x}{5} \qquad \text{Dividing by 5}$$

$$-\frac{8}{5} > x$$

The solution set is $\left\{x \mid -\frac{8}{5} > x\right\}$, or $\left\{x \mid x < -\frac{8}{5}\right\}$.

Do Exercise 20.

20. Solve:
$$\frac{3}{4} + x < \frac{7}{8}x - \frac{1}{4} + \frac{1}{2}x.$$

Answers

19. $\{x \mid x \geq -4\}$ **20.** $\left\{x \mid x > \frac{8}{3}\right\}$

a Determine whether each number is a solution of the given inequality.

1. $x > -4$
a) 4
b) 0
c) −4
d) 6
e) 5.6

2. $x \leq 5$
a) 0
b) 5
c) −1
d) −5
e) $7\frac{1}{4}$

3. $x \geq 6.8$
a) −6
b) 0
c) 6
d) 8
e) $-3\frac{1}{2}$

4. $x < 8$
a) 8
b) −10
c) 0
d) 11
e) −4.7

b Graph on the number line.

5. $x > 4$

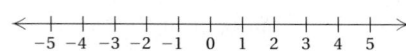

6. $x < 0$

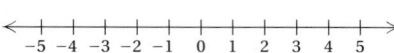

7. $t < -3$

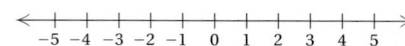

8. $y > 5$

9. $m \geq -1$

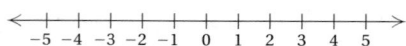

10. $x \leq -2$

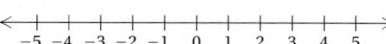

11. $-3 < x \leq 4$

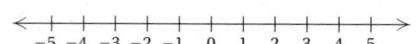

12. $-5 \leq x < 2$

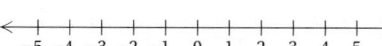

13. $0 < x < 3$

14. $-5 \leq x \leq 0$

c Solve using the addition principle. Then graph.

15. $x + 7 > 2$

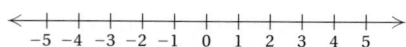

16. $x + 5 > 2$

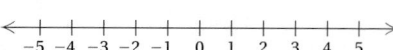

17. $x + 8 \leq -10$

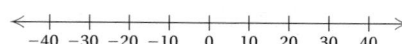

18. $x + 8 \leq -11$

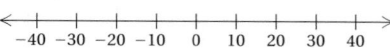

Solve using the addition principle.

19. $y - 7 > -12$

20. $y - 9 > -15$

21. $2x + 3 > x + 5$

22. $2x + 4 > x + 7$

23. $3x + 9 \leq 2x + 6$

24. $3x + 18 \leq 2x + 16$

25. $5x - 6 < 4x - 2$

26. $9x - 8 < 8x - 9$

27. $-9 + t > 5$

28. $-8 + p > 10$

29. $y + \dfrac{1}{4} \leq \dfrac{1}{2}$

30. $x - \dfrac{1}{3} \leq \dfrac{5}{6}$

31. $x - \dfrac{1}{3} > \dfrac{1}{4}$

32. $x + \dfrac{1}{8} > \dfrac{1}{2}$

(d) Solve using the multiplication principle. Then graph.

33. $5x < 35$

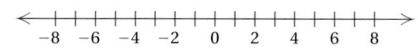

34. $8x \geq 32$

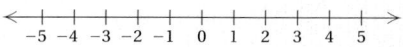

35. $-12x > -36$

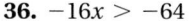

36. $-16x > -64$

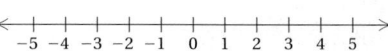

Solve using the multiplication principle.

37. $5y \geq -2$

38. $3x < -4$

39. $-2x \leq 12$

40. $-3x \leq 15$

41. $-4y \geq -16$

42. $-7x < -21$

43. $-3x < -17$

44. $-5y > -23$

45. $-2y > \dfrac{1}{7}$

46. $-4x \leq \dfrac{1}{9}$

47. $-\dfrac{6}{5} \leq -4x$

48. $-\dfrac{7}{9} > 63x$

e Solve using the addition principle and the multiplication principle.

49. $4 + 3x < 28$

50. $3 + 4y < 35$

51. $3x - 5 \leq 13$

52. $5y - 9 \leq 21$

53. $13x - 7 < -46$

54. $8y - 6 < -54$

55. $30 > 3 - 9x$

56. $48 > 13 - 7y$

57. $4x + 2 - 3x \leq 9$

58. $15x + 5 - 14x \leq 9$

59. $-3 < 8x + 7 - 7x$

60. $-8 < 9x + 8 - 8x - 3$

61. $6 - 4y > 4 - 3y$

62. $9 - 8y > 5 - 7y + 2$

63. $5 - 9y \leq 2 - 8y$

64. $6 - 18x \leq 4 - 12x - 5x$

65. $19 - 7y - 3y < 39$

66. $18 - 6y - 4y < 63 + 5y$

67. $0.9x + 19.3 > 5.3 - 2.6x$

68. $0.96y - 0.79 \leq 0.21y + 0.46$

69. $\dfrac{x}{3} - 2 \leq 1$

70. $\dfrac{2}{3} + \dfrac{x}{5} < \dfrac{4}{15}$

71. $\dfrac{y}{5} + 1 \leq \dfrac{2}{5}$

72. $\dfrac{3x}{4} - \dfrac{7}{8} \geq -15$

73. $3(2y - 3) < 27$

74. $4(2y - 3) > 28$

75. $2(3 + 4m) - 9 \geq 45$

76. $3(5 + 3m) - 8 \leq 88$

77. $8(2t + 1) > 4(7t + 7)$

78. $7(5y - 2) > 6(6y - 1)$

79. $3(r - 6) + 2 < 4(r + 2) - 21$

80. $5(x + 3) + 9 \leq 3(x - 2) + 6$

81. $0.8(3x + 6) \geq 1.1 - (x + 2)$

82. $0.4(2x + 8) \geq 20 - (x + 5)$

83. $\dfrac{5}{3} + \dfrac{2}{3}x < \dfrac{25}{12} + \dfrac{5}{4}x + \dfrac{3}{4}$

84. $1 - \dfrac{2}{3}y \geq \dfrac{9}{5} - \dfrac{y}{5} + \dfrac{3}{5}$

Skill Maintenance

Add or subtract.   [1.3a], [1.4a]

85. $-56 + (-18)$

86. $-2.3 + 7.1$

87. $-\dfrac{3}{4} + \dfrac{1}{8}$

88. $8.12 - 9.23$

89. $-56 - (-18)$

90. $-\dfrac{3}{4} - \dfrac{1}{8}$

91. $-2.3 - 7.1$

92. $-8.12 + 9.23$

Simplify.

93. $5 - 3^2 + (8 - 2)^2 \cdot 4$ [1.8d]

94. $10 \div 2 \cdot 5 - 3^2 + (-5)^2$ [1.8d]

95. $5(2x - 4) - 3(4x + 1)$ [1.8b]

96. $9(3 + 5x) - 4(7 + 2x)$ [1.8b]

Synthesis

97. Determine whether each number is a solution of the inequality $|x| < 3$.

    a) 0

    b) −2

    c) −3

    d) 4

    e) 3

    f) 1.7

    g) −2.8

98. Graph $|x| < 3$ on the number line.

$$\xleftarrow{\hspace{1em}} \underset{-5\ -4\ -3\ -2\ -1\ \ 0\ \ 1\ \ 2\ \ 3\ \ 4\ \ 5}{\rule{0pt}{0pt}} \xrightarrow{\hspace{1em}}$$

Solve.

99. $x + 3 < 3 + x$

100. $x + 4 > 3 + x$

2.8

Applications and Problem Solving with Inequalities

OBJECTIVES

a Translate number sentences to inequalities.

b Solve applied problems using inequalities.

The five steps for problem solving can be used for problems involving inequalities.

a Translating to Inequalities

Before solving problems that involve inequalities, we list some important phrases to look for. Sample translations are listed as well.

IMPORTANT WORDS	SAMPLE SENTENCE	TRANSLATION
is at least	Bill is at least 21 years old.	$b \geq 21$
is at most	At most 5 students dropped the course.	$n \leq 5$
cannot exceed	To qualify, earnings cannot exceed $12,000.	$r \leq 12{,}000$
must exceed	The speed must exceed 15 mph.	$s > 15$
is less than	Tucker's weight is less than 50 lb.	$w < 50$
is more than	Boston is more than 200 mi away.	$d > 200$
is between	The film was between 90 and 100 min long.	$90 < t < 100$
no more than	Bing weighs no more than 90 lb.	$w \leq 90$
no less than	Valerie scored no less than 8.3.	$s \geq 8.3$

The following phrases deserve special attention.

> **TRANSLATING "AT LEAST" AND "AT MOST"**
>
> A quantity x is at least some amount q: $x \geq q$.
>   (If x is at least q, it cannot be less than q.)
>
> A quantity x is at most some amount q: $x \leq q$.
>   (If x is at most q, it cannot be more than q.)

Do Exercises 1–8.

Translate.
1. Maggie worked no fewer than 15 hr last week.

2. The price of that PT Cruiser is at most $21,900.

3. The time of the test was between 45 and 55 min.

4. Tania's weight is less than 110 lb.

5. That number is more than -2.

6. The costs of production of that CD-ROM cannot exceed $12,500.

7. At most 1250 people attended the concert.

8. Yesterday, at least 23 people got tickets for speeding.

b Solving Problems

EXAMPLE 1 *Catering Costs.*   To cater a party, Curtis' Barbeque charges a $150 setup fee plus $15.50 per person. The cost of Berry Manufacturing's annual picnic cannot exceed $2100. How many people can attend the picnic?

Source: Curtis' All American Barbeque, Putney, Vermont

1. **Familiarize.**   Suppose that 110 people were to attend the picnic. The cost would then be $150 + $15.50(110), or $1855. This shows that more than 110 people could attend the picnic without exceeding $2100. Instead of making another guess, we let n = the number of people in attendance.

Answers

1. $h \geq 15$ 2. $p \leq 21{,}900$
3. $45 < t < 55$ 4. $w < 110$
5. $n > -2$ 6. $c \leq 12{,}500$
7. $p \leq 1250$ 8. $s \geq 23$

2. Translate. Our guess shows us how to translate. The cost of the picnic will be the $150 setup fee plus $15.50 times the number of people attending. We translate to an inequality:

Rewording:	The setup fee	plus	the cost of the meals	cannot exceed	$2100.
Translating:	150	+	15.50n	≤	2100.

3. Solve. We solve the inequality for n:

$$150 + 15.50n \leq 2100$$
$$150 + 15.50n - 150 \leq 2100 - 150 \qquad \text{Subtracting 150}$$
$$15.50n \leq 1950 \qquad \text{Simplifying}$$
$$\frac{15.50n}{15.50} \leq \frac{1950}{15.50} \qquad \text{Dividing by 15.50}$$
$$n \leq 125.8. \qquad \text{Rounding to the nearest tenth}$$

4. Check. Although the solution set of the inequality is all numbers less than or equal to about 125.8, since n = the number of people in attendance, we round *down* to 125 people. If 125 people attend, the cost will be $150 + $15.50(125), or $2087.50. If 126 attend, the cost will exceed $2100.

5. State. At most, 125 people can attend the picnic.

Do Exercise 9.

Translate to an inequality and solve.

9. Butter Temperatures. Butter stays solid at Fahrenheit temperatures below 88°. The formula

$$F = \tfrac{9}{5}C + 32$$

can be used to convert Celsius temperatures C to Fahrenheit temperatures F. Determine (in terms of an inequality) those Celsius temperatures for which butter stays solid.

------------------------------ *Caution!* ------------------------------

Solutions of problems should always be checked using the original wording of the problem. In some cases, answers might need to be whole numbers or integers or rounded off in a particular direction.

EXAMPLE 2 *Nutrition.* The U.S. Department of Agriculture recommends that for a typical 2000-calorie daily diet, no more than 20 g of saturated fat be consumed. In the first three days of a four-day vacation, Anthony consumed 26 g, 17 g, and 22 g of saturated fat. Determine (in terms of an inequality) how many grams of saturated fat Anthony can consume on the fourth day if he is to average no more than 20 g of saturated fat per day.

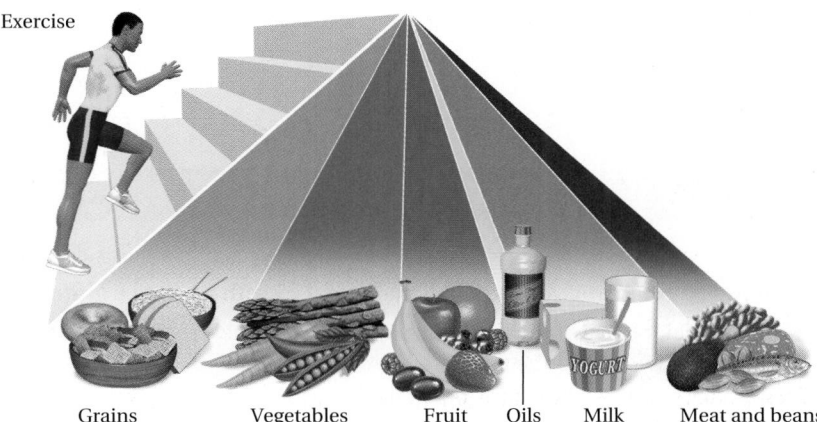

Exercise

Grains   Vegetables   Fruit   Oils   Milk   Meat and beans

SOURCES: U.S. Department of Health and Human Services; U.S. Department of Agriculture

Answer

9. $\dfrac{9}{5}C + 32 < 88;\ \left\{C \mid C < 31\tfrac{1}{9}°\right\}$

1. **Familiarize.** Suppose Anthony consumed 19 g of saturated fat on the fourth day. His daily average for the vacation would then be

$$\frac{26\,g + 17\,g + 22\,g + 19\,g}{4} = \frac{84\,g}{4} = 21\,g.$$

This shows that Anthony cannot consume 19 g of saturated fat on the fourth day, if he is to average no more than 20 g of fat per day. We let $x =$ the number of grams of fat that Anthony consumes on the fourth day.

2. **Translate.** We reword the problem and translate to an inequality as follows:

	The average consumption of saturated fat	should be no more than	20 g.
Rewording:			
Translating:	$\dfrac{26 + 17 + 22 + x}{4}$	$\leq$	20.

3. **Solve.** Because of the fraction expression, it is convenient to use the multiplication principle first to solve the inequality:

$$\frac{26 + 17 + 22 + x}{4} \leq 20$$

$$4\left(\frac{26 + 17 + 22 + x}{4}\right) \leq 4 \cdot 20 \qquad \text{Multiplying by 4}$$

$$26 + 17 + 22 + x \leq 80$$

$$65 + x \leq 80 \qquad \text{Simplifying}$$

$$x \leq 15. \qquad \text{Subtracting 65}$$

4. **Check.** As a partial check, we show that Anthony can consume 15 g of saturated fat on the fourth day and not exceed a 20-g average for the four days:

$$\frac{26 + 17 + 22 + 15}{4} = \frac{80}{4} = 20.$$

5. **State.** Anthony's average intake of saturated fat for the vacation will not exceed 20 g per day if he consumes no more than 15 g of saturated fat on the fourth day.

Do Exercise 10.

Translate to an inequality and solve.

10. Test Scores. A pre-med student is taking a chemistry course in which four tests are given. To get an A, she must average at least 90 on the four tests. The student got scores of 91, 86, and 89 on the first three tests. Determine (in terms of an inequality) what scores on the last test will allow her to get an A.

Answer

10. $\dfrac{91 + 86 + 89 + s}{4} \geq 90$; $\{s | s \geq 94\}$

a Translate to an inequality.

1. A number is at least 7.

2. A number is greater than or equal to 5.

3. The baby weighs more than 2 kilograms (kg).

4. Between 75 and 100 people attended the concert.

5. The speed of the train was between 90 and 110 mph.

6. The attendance was no more than 180.

7. Leah works no more than 20 hr per week.

8. The amount of acid must exceed 40 liters (L).

9. The cost of gasoline is no less than $1.50 per gallon.

10. The temperature is at most $-2°$.

11. A number is greater than 8.

12. A number is less than 5.

13. A number is less than or equal to -4.

14. A number is greater than or equal to 18.

15. The number of people is at least 1300.

16. The cost is at most $4857.95.

17. The amount of water is not to exceed 500 liters.

18. The cost of lettuce is no less than 94 cents per pound.

19. Two more than three times a number is less than 13.

20. Five less than one-half a number is greater than 17.

b Solve.

21. *Test Scores.* James is taking a literature course in which four tests are given. To get a B, he must average at least 80 on the four tests. He got scores of 82, 76, and 78 on the first three tests. Determine (in terms of an inequality) what scores on the last test will allow him to get at least a B.

22. *Test Scores.* Rebecca's quiz grades are 73, 75, 89, and 91. Determine (in terms of an inequality) what scores on the last quiz will allow her to get an average quiz grade of at least 85.

23. *Gold Temperatures.* Gold stays solid at Fahrenheit temperatures below 1945.4°. Determine (in terms of an inequality) those Celsius temperatures for which gold stays solid. Use the formula given in Margin Exercise 9.

24. *Body Temperatures.* The human body is considered to be fevered when its temperature is higher than 98.6°F. Using the formula given in Margin Exercise 9, determine (in terms of an inequality) those Celsius temperatures for which the body is fevered.

25. *World Records in the 1500-m Run.* The formula
$$R = -0.075t + 3.85$$
can be used to predict the world record in the 1500-m run t years after 1930. Determine (in terms of an inequality) those years for which the world record will be less than 3.5 min.

26. *World Records in the 200-m Dash.* The formula
$$R = -0.028t + 20.8$$
can be used to predict the world record in the 200-m dash t years after 1920. Determine (in terms of an inequality) those years for which the world record will be less than 19.0 sec.

27. *Envelope Size.* For a direct-mail campaign, Laramore Advertising determines that any envelope with a fixed width of $3\frac{1}{2}$ in. and an area of at least $17\frac{1}{2}$ in^2 can be used. Determine (in terms of an inequality) those lengths that will satisfy the company constraints.

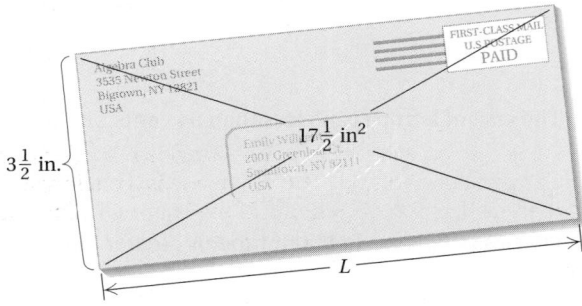

28. *Package Sizes.* Logan Delivery Service accepts packages of up to 165 in. in length and girth combined. (Girth is the distance around the package.) A package has a fixed girth of 53 in. Determine (in terms of an inequality) those lengths for which a package is acceptable.

29. *Blueprints.* To make copies of blueprints, Vantage Reprographics charges a $5 setup fee plus $4 per copy. Myra can spend no more than $65 for copying her blueprints. What numbers of copies will allow her to stay within budget?

30. *Banquet Costs.* The Shepard College women's volleyball team can spend at most $450 for its awards banquet at a local restaurant. If the restaurant charges a $40 setup fee plus $16 per person, at most how many can attend?

31. *Phone Costs.*   Simon claims that it costs him at least $3.00 every time he calls an overseas customer. If his typical call costs 75¢ plus 45¢ for each minute, how long do his calls typically last? (*Hint*: 75¢ = $0.75.)

32. *Parking Costs.*   Laura is certain that every time she parks in the municipal garage it costs her at least $6.75. If the garage charges $1.50 plus 75¢ for each half hour, for how long is Laura's car generally parked?

33. *College Tuition.*   Angelica's financial aid stipulates that her tuition cannot exceed $1000. If her local community college charges a $35 registration fee plus $375 per course, what is the greatest number of courses for which Angelica can register?

34. *Furnace Repairs.*   RJ's Plumbing and Heating charges $45 plus $30 per hour for emergency service. Gary remembers being billed over $150 for an emergency call. How long was RJ's there?

35. *Nutrition.*   Following the guidelines of the Food and Drug Administration, Dale tries to eat at least 5 servings of fruits or vegetables each day. For the first six days of one week, he had 4, 6, 7, 4, 6, and 4 servings. How many servings of fruits or vegetables should Dale eat on Saturday, in order to average at least 5 servings per day for the week?

36. *College Course Load.*   To remain on financial aid, Millie needs to complete an average of at least 7 credits per quarter each year. In the first three quarters of 2009, Millie completed 5, 7, and 8 credits. How many credits of course work must Millie complete in the fourth quarter if she is to remain on financial aid?

37. *Perimeter of a Rectangle.*   The width of a rectangle is fixed at 8 ft. What lengths will make the perimeter at least 200 ft? at most 200 ft?

38. *Perimeter of a Triangle.*   One side of a triangle is 2 cm shorter than the base. The other side is 3 cm longer than the base. What lengths of the base will allow the perimeter to be greater than 19 cm?

39. *Area of a Rectangle.*   The width of a rectangle is fixed at 4 cm. For what lengths will the area be less than 86 cm^2?

40. *Area of a Rectangle.*   The width of a rectangle is fixed at 16 yd. For what lengths will the area be at least 264 yd^2?

L

4 cm $A < 86\ \text{cm}^2$ 4 cm

L

41. Insurance-Covered Repairs. Most insurance companies will replace a vehicle if an estimated repair exceeds 80% of the "blue-book" value of the vehicle. Michelle's insurance company paid $8500 for repairs to her Subaru after an accident. What can be concluded about the blue-book value of the car?

42. Insurance-Covered Repairs. Following an accident, Jeff's Ford pickup was replaced by his insurance company because the damage was so extensive. Before the damage, the blue-book value of the truck was $21,000. How much would it have cost to repair the truck? (See Exercise 41.)

43. Reduced-Fat Foods. In order for a food to be labeled "reduced fat," it must have at least 25% less fat than the regular item. One brand of reduced-fat peanut butter contains 12 g of fat per serving. What can you conclude about the fat content in a serving of the brand's regular peanut butter?

44. Reduced-Fat Foods. One brand of reduced-fat chocolate chip cookies contains 5 g of fat per serving. What can you conclude about the fat content of the brand's regular chocolate chip cookies? (See Exercise 43.)

45. Pond Depth. On July 1, Garrett's Pond was 25 ft deep. Since that date, the water level has dropped $\frac{2}{3}$ ft per week. For what dates will the water level not exceed 21 ft?

46. Weight Gain. A 3-lb puppy is gaining weight at a rate of $\frac{3}{4}$ lb per week. When will the puppy's weight exceed $22\frac{1}{2}$ lb?

47. Area of a Triangular Flag. As part of an outdoor education course, Wendy needs to make a bright-colored triangular flag with an area of at least 3 ft². What heights can the triangle be if the base is $1\frac{1}{2}$ ft?

48. Area of a Triangular Sign. Zoning laws in Harrington prohibit displaying signs with areas exceeding 12 ft². If Flo's Marina is ordering a triangular sign with an 8-ft base, how tall can the sign be?

49. Electrician Visits. Dot's Electric made 17 customer calls last week and 22 calls this week. How many calls must be made next week in order to maintain a weekly average of at least 20 calls for the three-week period?

50. Volunteer Work. George and Joan do volunteer work at a hospital. Joan worked 3 more hr than George, and together they worked more than 27 hr. What possible numbers of hours did each work?

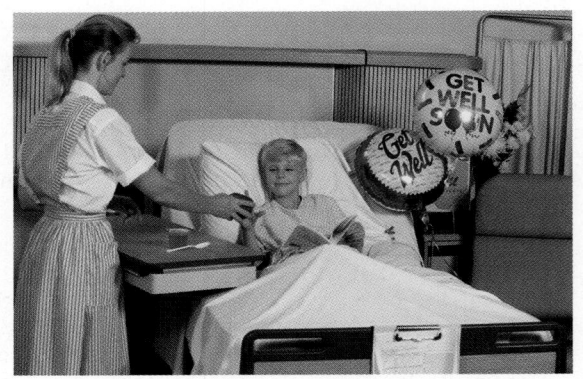

Skill Maintenance

In each of Exercises 51–58, fill in the blank with the correct term from the given list. Some of the choices may not be used.

51. The product of a(n) ——————— number of negative numbers is always positive.   [1.5a]

52. The product of a(n) ——————— number of negative numbers is always negative.   [1.5a]

53. The ——————— inverse of a negative number is always positive.   [1.3b]

54. The ——————— inverse of a negative number is always negative.   [1.6b]

55. Equations with the same solutions are called ——————— equations.   [2.1b]

56. The ——————— for equations asserts that when we add the same number to the expressions on each side of the equation, we get equivalent equations.   [2.1b]

57. The ——————— for inequalities asserts that when we multiply or divide by a negative number on both sides of an inequality, the direction of the inequality symbol ———————.   [2.7d]

58. Any replacement for the variable that makes an equation true is called a(n) ——————— of the equation.   [2.1a]

addition principle

multiplication principle

solution

value

is reversed

stays the same

even

odd

multiplicative

additive

equivalent

Synthesis

59. *Ski Wax.*   Green ski wax works best between 5° and 15° Fahrenheit. Determine those Celsius temperatures for which green ski wax works best. Use the formula given in Margin Exercise 9.

60. *Parking Fees.*   Mack's Parking Garage charges $4.00 for the first hour and $2.50 for each additional hour. For how long has a car been parked when the charge exceeds $16.50?

61. *Low-Fat Foods.*   In order for a food to be labeled "low fat," it must have fewer than 3 g of fat per serving. One brand of reduced-fat tortilla chips contains 60% less fat than regular nacho cheese tortilla chips, but still cannot be labeled low fat. What can you conclude about the fat content of a serving of nacho cheese tortilla chips?

62. *Parking Fees.*   When asked how much the parking charge is for a certain car, Mack replies "between 14 and 24 dollars." For how long has the car been parked? (See Exercise 60.)

Key Terms and Properties

equation, p. 86
solution of an equation, p. 86
equivalent equations, p. 87
clearing fractions, p. 101

clearing decimals, p. 101
formula, p. 109
evaluating a formula, p. 109
inequality, p. 144

solution set, p. 144
graph of an inequality, p. 144
equivalent inequalities, p. 145
set-builder notation, p. 147

The Addition Principle for Equations:	For any real numbers a, b, and c, $a = b$ is equivalent to $a + c = b + c$.
The Multiplication Principle for Equations:	For any real numbers a, b, and c, $c \neq 0$: $a = b$ is equivalent to $a \cdot c = b \cdot c$.
The Addition Principle for Inequalities:	For any real numbers a, b, and c: $a < b$ is equivalent to $a + c < b + c$; $a > b$ is equivalent to $a + c > b + c$. Similar statements hold for $\leq$ and $\geq$.
The Multiplication Principle for Inequalities:	For any real numbers a and b, and any *positive* number c: $a < b$ is equivalent to $ac < bc$; $a > b$ is equivalent to $ac > bc$. For any real numbers a and b, and any *negative* number c: $a < b$ is equivalent to $ac > bc$; $a > b$ is equivalent to $ac < bc$. Similar statements hold for $\leq$ and $\geq$.

Concept Reinforcement

Determine whether each statement is true or false.

_____ **1.** Some equations have no solution.   [2.3c]

_____ **2.** For any number n, $n \geq n$.   [2.7a]

_____ **3.** $2x - 7 < 11$ and $x < 2$ are equivalent inequalities.   [2.7e]

_____ **4.** If $x > y$, then $-x < -y$.   [2.7d]

Important Concepts

Objective 2.3a   Solve equations using both the addition principle and the multiplication principle.

Objective 2.3b   Solve equations in which like terms may need to be collected.

Objective 2.3c   Solve equations by first removing parentheses and collecting like terms.

Example   Solve: $6y - 2(2y - 3) = 12$.

$$6y - 2(2y - 3) = 12$$

$6y - 4y + 6 = 12$ Removing parentheses

$2y + 6 = 12$ Collecting like terms

$2y + 6 - 6 = 12 - 6$ Subtracting 6

$2y = 6$

$\dfrac{2y}{2} = \dfrac{6}{2}$ Dividing by 2

$y = 3$

The solution is 3.

Practice Exercise

1. Solve: $4(x - 3) = 6(x + 2)$.

Objective 2.3c Solve equations with no solutions and equations with an infinite number of solutions.

Example Solve: $8 + 2x - 4 = 6 + 2(x - 1)$.

$$8 + 2x - 4 = 6 + 2(x - 1)$$
$$8 + 2x - 4 = 6 + 2x - 2$$
$$2x + 4 = 2x + 4$$
$$2x + 4 - 2x = 2x + 4 - 2x$$
$$4 = 4$$

Every real number is a solution of the equation $4 = 4$, so all real numbers are solutions of the original equation. The equation has infinitely many solutions.

Example Solve: $2 + 5(x - 1) = -6 + 5x + 7$.

$$2 + 5(x - 1) = -6 + 5x + 7$$
$$2 + 5x - 5 = -6 + 5x + 7$$
$$5x - 3 = 5x + 1$$
$$5x - 3 - 5x = 5x + 1 - 5x$$
$$-3 = 1$$

This is a false equation, so the original equation has no solution.

Practice Exercises

2. Solve: $4 + 3y - 7 = 3 + 3(y - 2)$.

3. Solve: $4(x - 3) + 7 = -5 + 4x + 10$.

Objective 2.4b Solve a formula for a specified letter.

Example Solve for n: $M = \dfrac{m + n}{5}$.

$$M = \frac{m + n}{5}$$
$$5 \cdot M = 5\left(\frac{m + n}{5}\right)$$
$$5M = m + n$$
$$5M - m = m + n - m$$
$$5M - m = n$$

Practice Exercise

4. Solve for b: $A = \dfrac{1}{2}bh$.

Objective 2.7b Graph an inequality on the number line.

Example Graph each inequality: **(a)** $x < 2$; **(b)** $x \geq -3$.

a) The solutions of $x < 2$ are all numbers less than 2. We shade all points to the left of 2, and we use an open circle at 2 to indicate that 2 *is not* part of the graph.

$$x < 2$$
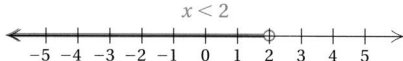

b) The solutions of $x \geq -3$ are all numbers greater than -3 and the number -3 as well. We shade all points to the right of -3, and we use a closed circle at -3 to indicate that -3 *is* part of the graph.

$$x \geq -3$$
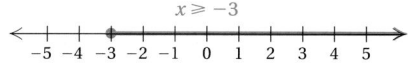

Practice Exercises

5. Graph: $x > 1$.

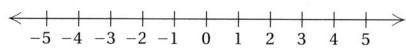

6. Graph: $x \leq -1$.

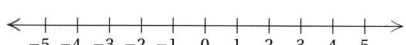

Objective 2.7e Solve inequalities using the addition principle and the multiplication principle together.

Example Solve: $8y - 7 \leq 5y + 2$.

$$8y - 7 \leq 5y + 2$$
$$8y - 7 - 8y \leq 5y + 2 - 8y$$
$$-7 \leq -3y + 2$$
$$-7 - 2 \leq -3y + 2 - 2$$
$$-9 \leq -3y$$
$$\frac{-9}{-3} \geq \frac{-3y}{-3} \quad \text{Reversing the symbol}$$
$$3 \geq y$$

The solution set is $\{y \mid 3 \geq y\}$, or $\{y \mid y \leq 3\}$.

Practice Exercise

7. Solve: $6y + 5 > 3y - 7$.

Review Exercises

Solve. [2.1b]

1. $x + 5 = -17$

2. $n - 7 = -6$

3. $x - 11 = 14$

4. $y - 0.9 = 9.09$

Solve. [2.2a]

5. $-\dfrac{2}{3}x = -\dfrac{1}{6}$

6. $-8x = -56$

7. $-\dfrac{x}{4} = 48$

8. $15x = -35$

9. $\dfrac{4}{5}y = -\dfrac{3}{16}$

Solve. [2.3a]

10. $5 - x = 13$

11. $\dfrac{1}{4}x - \dfrac{5}{8} = \dfrac{3}{8}$

Solve. [2.3b, c]

12. $5t + 9 = 3t - 1$

13. $7x - 6 = 25x$

14. $14y = 23y - 17 - 10$

15. $0.22y - 0.6 = 0.12y + 3 - 0.8y$

16. $\dfrac{1}{4}x - \dfrac{1}{8}x = 3 - \dfrac{1}{16}x$

17. $14y + 17 + 7y = 9 + 21y + 8$

18. $4(x + 3) = 36$

19. $3(5x - 7) = -66$

20. $8(x - 2) - 5(x + 4) = 20 + x$

21. $-5x + 3(x + 8) = 16$

22. $6(x - 2) - 16 = 3(2x - 5) + 11$

Determine whether the given number is a solution of the inequality $x \leq 4$. [2.7a]

23. -3

24. 7

25. 4

Solve. Write set notation for the answers. [2.7c, d, e]

26. $y + \dfrac{2}{3} \geq \dfrac{1}{6}$

27. $9x \geq 63$

28. $2 + 6y > 14$

29. $7 - 3y \geq 27 + 2y$

30. $3x + 5 < 2x - 6$

31. $-4y < 28$

32. $4 - 8x < 13 + 3x$

33. $-4x \leq \dfrac{1}{3}$

Graph on the number line. [2.7b, e]

34. $4x - 6 < x + 3$

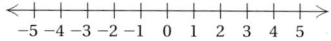

35. $-2 < x \leq 5$

36. $y > 0$

Solve. [2.4b]

37. $C = \pi d$, for d

38. $V = \dfrac{1}{3} Bh$, for B

39. $A = \dfrac{a + b}{2}$, for a

40. $y = mx + b$, for x

Solve. [2.6a]

41. *Dimensions of Wyoming.* The state of Wyoming is roughly in the shape of a rectangle whose perimeter is 1280 mi. The length is 90 mi more than the width. Find the dimensions.

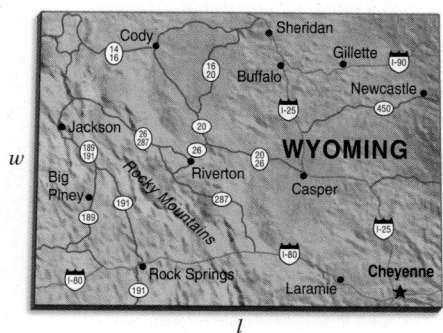

42. *Interstate Mile Markers.* The sum of two consecutive mile markers on I-5 in California is 691. Find the numbers on the markers.

43. An entertainment center sold for $2449 in June. This was $332 more than the cost in February. What was the cost in February?

44. Ty is paid a commission of $4 for each magazine subscription he sells. One week, he received $108 in commissions. How many subscriptions did he sell?

45. The measure of the second angle of a triangle is 50° more than that of the first angle. The measure of the third angle is 10° less than twice the first angle. Find the measures of the angles.

Solve. [2.5a]

46. What number is 20% of 75?

47. Fifteen is what percent of 80?

48. 18 is 3% of what number?

49. *Job Opportunities.* There were 1.388 million child-care workers in 2006. The number of job opportunities in that field is expected to grow to 1.636 million by 2016. What is the percent of increase?

Source: *Occupational Outlook Handbook*

Solve. [2.6a]

50. After a 30% reduction, a bread maker is on sale for $154. What was the marked price (the price before the reduction)?

51. A hotel manager's salary is $61,410, which is a 15% increase over the previous year's salary. What was the previous salary?

52. A tax-exempt organization received a bill of $145.90 for janitorial supplies. The bill incorrectly included sales tax of 5%. How much does the organization actually owe?

Solve. [2.8b]

53. *Test Scores.* Jacinda's test grades are 71, 75, 82, and 86. What is the lowest grade that she can get on the next test and still have an average test score of at least 80?

54. The length of a rectangle is 43 cm. What widths will make the perimeter greater than 120 cm?

55. The solution of the equation $4(3x - 5) + 6 = 8 + x$ is which of the following? [2.3c]
  A. Less than -1
  B. Between -1 and 1
  C. Between 1 and 5
  D. Greater than 5

56. Solve for y: $3x + 4y = P$. [2.4b]
  A. $y = \dfrac{P - 3x}{4}$
  B. $y = \dfrac{P + 3x}{4}$
  C. $y = P - \dfrac{3x}{4}$
  D. $y = \dfrac{P}{4} - 3x$

Synthesis

Solve.

57. $2|x| + 4 = 50$ [1.2e], [2.3a]

58. $|3x| = 60$ [1.2e], [2.2a]

59. $y = 2a - ab + 3$, for a [2.4b]

Understanding Through Discussion and Writing

1. Would it be better to receive a 5% raise and then an 8% raise or the other way around? Why? [2.5a]

2. Erin returns a tent that she bought during a storewide 25%-off sale that has ended. She is offered store credit for 125% of what she paid (not to be used on sale items). Is this fair to Erin? Why or why not? [2.5a]

3. Are the inequalities $x > -5$ and $-x < 5$ equivalent? Why or why not? [2.7d]

4. Explain in your own words why it is necessary to reverse the inequality symbol when multiplying on both sides of an inequality by a negative number. [2.7d]

5. If f represents Fran's age and t represents Todd's age, write a sentence that would translate to $t + 3 < f$. [2.8a]

6. Explain how the meanings of "Five more than a number" and "Five is more than a number" differ. [2.8a]

Test For Extra Help

Solve.

1. $x + 7 = 15$

2. $t - 9 = 17$

3. $3x = -18$

4. $-\frac{4}{7}x = -28$

5. $3t + 7 = 2t - 5$

6. $\frac{1}{2}x - \frac{3}{5} = \frac{2}{5}$

7. $8 - y = 16$

8. $-\frac{2}{5} + x = -\frac{3}{4}$

9. $3(x + 2) = 27$

10. $-3x - 6(x - 4) = 9$

11. $0.4p + 0.2 = 4.2p - 7.8 - 0.6p$

12. $4(3x - 1) + 11 = 2(6x + 5) - 8$

13. $-2 + 7x + 6 = 5x + 4 + 2x$

Solve. Write set notation for the answers.

14. $x + 6 \leq 2$

15. $14x + 9 > 13x - 4$

16. $12x \leq 60$

17. $-2y \geq 26$

18. $-4y \leq -32$

19. $-5x \geq \frac{1}{4}$

20. $4 - 6x > 40$

21. $5 - 9x \geq 19 + 5x$

Graph on the number line.

22. $y \leq 9$

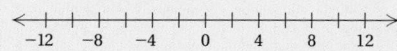

23. $6x - 3 < x + 2$

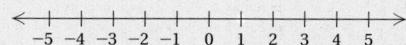

24. $-2 \leq x \leq 2$

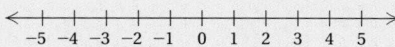

Solve.

25. What number is 24% of 75?

26. 15.84 is what percent of 96?

27. 800 is 2% of what number?

28. *Job Opportunities.* The number of job opportunities for physician's assistants is expected to increase from 66,000 in 2006 to 83,000 in 2016. What is the percent of increase?

Source: *Occupational Outlook Handbook*

29. *Perimeter of a Photograph.* The perimeter of a rectangular photograph is 36 cm. The length is 4 cm greater than the width. Find the width and the length.

30. *Charitable Contributions.* About $102.3 billion was given to religious organizations in 2007. This represents 33% of all charitable donations that year. How much was donated to all charities?

Sources: Giving USA Foundation; Center on Philanthropy at Indiana University

31. *Raffle Tickets.* The numbers on three raffle tickets are consecutive integers whose sum is 7530. Find the integers.

32. *Savings Account.* Money is invested in a savings account at 5% simple interest. After 1 year, there is $924 in the account. How much was originally invested?

33. *Board Cutting.* An 8-m board is cut into two pieces. One piece is 2 m longer than the other. How long are the pieces?

34. *Lengths of a Rectangle.* The width of a rectangle is 96 yd. Find all possible lengths such that the perimeter of the rectangle will be at least 540 yd.

35. *Budgeting.* Jason has budgeted an average of $95 per month for entertainment. For the first five months of the year, he has spent $98, $89, $110, $85, and $83. How much can Jason spend in the sixth month without exceeding his average budget?

36. *Copy Machine Rental.* A catalog publisher needs to lease a copy machine for use during a special project that they anticipate will take 3 months. It costs $225 per month plus 1.2¢ per copy to rent the machine. The company must stay within a budget of $2400 for copies. Determine (in terms of an inequality) the number of copies they can make and still remain within budget.

37. Solve $A = 2\pi rh$ for r.

38. Solve $y = 8x + b$ for x.

39. *Senior Population.* The number of Americans age 65 and older is projected to grow from 40.4 million to 70.3 million between 2011 and 2030. Find the percent of increase.

Source: U.S. Census Bureau

A. 42.5%   B. 47%

C. 57.5%   D. 74%

Synthesis

40. Solve $c = \dfrac{1}{a - d}$ for d.

41. Solve: $3|w| - 8 = 37$.

42. A movie theater had a certain number of tickets to give away. Five people got the tickets. The first got one-third of the tickets, the second got one-fourth of the tickets, and the third got one-fifth of the tickets. The fourth person got eight tickets, and there were five tickets left for the fifth person. Find the total number of tickets given away.

Cumulative Review

Evaluate.

1. $\dfrac{y - x}{4}$, when $y = 12$ and $x = 6$

2. $\dfrac{3x}{y}$, when $x = 5$ and $y = 4$

3. $x - 3$, when $x = 3$

4. Translate to an algebraic expression: Four less than twice w.

Use $<$ or $>$ for $\square$ to write a true sentence.

5. $-4 \ \square \ -6$

6. $0 \ \square \ -5$

7. $-8 \ \square \ 7$

8. Find the opposite and the reciprocal of $\dfrac{2}{5}$.

Find the absolute value.

9. $|3|$

10. $\left| -\dfrac{3}{4} \right|$

11. $|0|$

Compute and simplify.

12. $-6.7 + 2.3$

13. $-\dfrac{1}{6} - \dfrac{7}{3}$

14. $-\dfrac{5}{8}\left(-\dfrac{4}{3} \right)$

15. $(-7)(5)(-6)(-0.5)$

16. $81 \div (-9)$

17. $-10.8 \div 3.6$

18. $-\dfrac{4}{5} \div -\dfrac{25}{8}$

Multiply.

19. $5(3x + 5y + 2z)$

20. $4(-3x - 2)$

21. $-6(2y - 4x)$

Factor.

22. $64 + 18x + 24y$

23. $16y - 56$

24. $5a - 15b + 25$

Collect like terms.

25. $9b + 18y + 6b + 4y$

26. $3y + 4 + 6z + 6y$

27. $-4d - 6a + 3a - 5d + 1$

28. $3.2x + 2.9y - 5.8x - 8.1y$

Simplify.

29. $7 - 2x - (-5x) - 8$

30. $-3x - (-x + y)$

31. $-3(x - 2) - 4x$

32. $10 - 2(5 - 4x)$

33. $[3(x + 6) - 10] - [5 - 2(x - 8)]$

Solve.

34. $x + 1.75 = 6.25$

35. $\dfrac{5}{2}y = \dfrac{2}{5}$

36. $-2.6 + x = 8.3$

37. $4\dfrac{1}{2} + y = 8\dfrac{1}{3}$

38. $-\dfrac{3}{4}x = 36$ **39.** $\dfrac{2}{5}x = -\dfrac{3}{20}$

40. $5.8x = -35.96$ **41.** $-4x + 3 = 15$

42. $-3x + 5 = -8x - 7$

43. $4y - 4 + y = 6y + 20 - 4y$

44. $-3(x - 2) = -15$ **45.** $\dfrac{1}{3}x - \dfrac{5}{6} = \dfrac{1}{2} + 2x$

46. $-3.7x + 6.2 = -7.3x - 5.8$

47. $4(x + 2) = 4(x - 2) + 16$

48. $0(x + 3) + 4 = 0$ **49.** $3x - 1 < 2x + 1$

50. $3y + 7 > 5y + 13$ **51.** $5 - y \le 2y - 7$

52. $H = 65 - m$, for m
(To determine the number of heating degree days H for a day with m degrees Fahrenheit as the average temperature)

53. $I = Prt$, for t
(Simple-interest formula, where I is interest, P is principal, r is interest rate, and t is time)

54. What number is 24% of 105?

55. 39.6 is what percent of 88?

56. $163.35 is 45% of what?

57. *Price Reduction.* After a 25% reduction, a tie is on sale for $18.45. What was the price before reduction?

58. *Rollerblade Costs.* Susan and Melinda purchased rollerblades for a total of $107. Susan paid $17 more for her rollerblades than Melinda did. What did Melinda pay?

59. *Savings Investment.* Money is invested in a savings account at 8% simple interest. After 1 year, there is $1134 in the account. How much was originally invested?

60. *Wire Cutting.* A 143-m wire is cut into three pieces. The second piece is 3 m longer than the first. The third is four-fifths as long as the first. How long is each piece?

61. *Grade Average.* Nadia is taking a literature course in which four tests are given. To get a B, a student must average at least 80 on the four tests. Nadia scored 82, 76, and 78 on the first three tests. What scores on the last test will earn her at least a B?

62. Simplify: $-125 \div 25 \cdot 625 \div 5$.
  A. $-390,625$ **B.** -125
  C. -625 **D.** 25

Synthesis

63. An engineer's salary at the end of a year is $48,418.24. This reflects a 4% salary increase and a later 3% cost-of-living adjustment during the year. What was the salary at the beginning of the year?

64. Grace needs to use a copier to reduce a drawing to fit on a page. The original drawing is 9 in. long and it must fit into a space that is 6.3 in. long. By what percent should she reduce the drawing on the copier?

Solve.

65. $4|x| - 13 = 3$ **66.** $\dfrac{2 + 5x}{4} = \dfrac{11}{28} + \dfrac{8x + 3}{7}$

67. $p = \dfrac{2}{m + Q}$, for Q

Graphs of Linear Equations

Real-World Application

The maximum grade allowed between two stations in a rapid-transit rail system is 3.5%. Between station A and station B, which are 280 ft apart, the tracks rise $8\frac{1}{2}$ ft. What is the grade of the tracks between these two stations? Round the answer to the nearest tenth of a percent. Does this grade meet the rapid-transit rail standards?

Source: Brian Burell, *Merriam Webster's Guide to Everyday Math*, Merriam-Webster, Inc., Springfield MA

This problem appears as Exercise 53 in Section 3.4.

3.1

Introduction to Graphing

OBJECTIVES

a Plot points associated with ordered pairs of numbers; determine the quadrant in which a point lies.

b Find the coordinates of a point on a graph.

c Determine whether an ordered pair is a solution of an equation with two variables.

SKILL TO REVIEW
Objective 2.1a: Determine whether a given number is a solution of a given equation.

Determine whether -3 is a solution of each equation.

1. $8(w - 3) = 0$

2. $15 = -2y + 9$

You probably have seen bar graphs like the following in newspapers and magazines. Note that a straight line can be drawn along the tops of the bars. Such a line is a *graph of a linear equation*. In this chapter, we study how to graph linear equations and consider properties such as slope and intercepts. Many applications of these topics will also be considered.

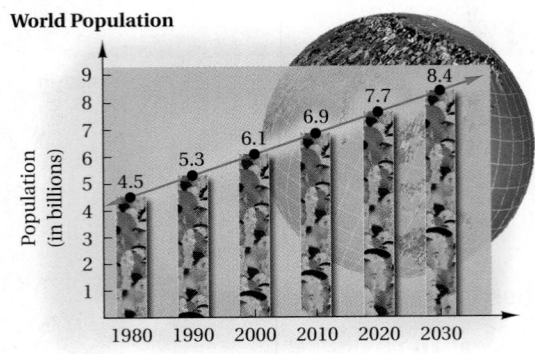

World Population

SOURCE: U.S. Census Bureau; International Data Base

a Plotting Ordered Pairs

In Chapter 2, we graphed numbers and inequalities in one variable on a line. To enable us to graph an equation that contains two variables, we now learn to graph number pairs on a plane.

On the number line, each point is the graph of a number. On a plane, each point is the graph of a number pair. To form the plane, we use two perpendicular number lines called **axes**. They cross at a point called the **origin**. The arrows show the positive directions.

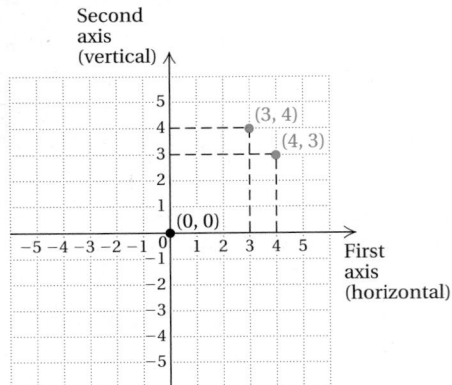

Consider the **ordered pair** (3, 4). The numbers in an ordered pair are called **coordinates**. In (3, 4), the **first coordinate** (the **abscissa**) is 3 and the **second coordinate** (the **ordinate**) is 4. To plot (3, 4), we start at the origin and move horizontally to the 3. Then we move up vertically 4 units and make a "dot."

The point (4, 3) is also plotted above. Note that (3, 4) and (4, 3) represent different points. The order of the numbers in the pair is important. We use the term *ordered* pairs because it makes a difference which number comes first. The coordinates of the origin are (0, 0).

Answers

Skill to Review:
1. -3 is not a solution.　　**2.** -3 is a solution.

EXAMPLE 1 Plot the point $(-5, 2)$.

The first number, -5, is negative. Starting at the origin, we move -5 units in the horizontal direction (5 units to the left). The second number, 2, is positive. We move 2 units in the vertical direction (up).

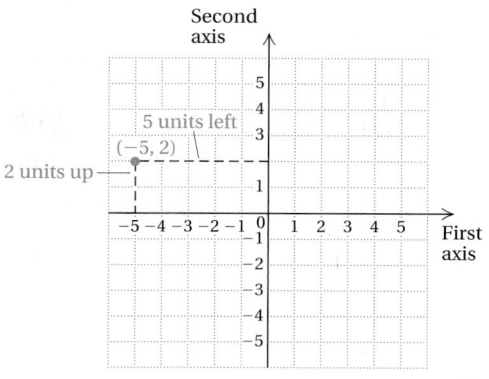

------------ *Caution!* ------------

The *first* coordinate of an ordered pair is always graphed in a *horizontal* direction and the *second* coordinate is always graphed in a *vertical* direction.

Do Exercises 1–8.

The figure below shows some points and their coordinates. In region I (the *first quadrant*), both coordinates of any point are positive. In region II (the *second quadrant*), the first coordinate is negative and the second positive. In region III (the *third quadrant*), both coordinates are negative. In region IV (the *fourth quadrant*), the first coordinate is positive and the second is negative.

EXAMPLE 2 In which quadrant, if any, are the points $(-4, 5)$, $(5, -5)$, $(2, 4)$, $(-2, -5)$, and $(-5, 0)$ located?

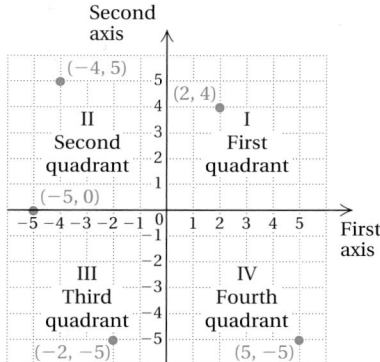

The point $(-4, 5)$ is in the second quadrant. The point $(5, -5)$ is in the fourth quadrant. The point $(2, 4)$ is in the first quadrant. The point $(-2, -5)$ is in the third quadrant. The point $(-5, 0)$ is on an axis and is *not in any quadrant*.

Do Exercises 9–16.

Plot these points on the graph below.

1. $(4, 5)$ **2.** $(5, 4)$

3. $(-2, 5)$ **4.** $(-3, -4)$

5. $(5, -3)$ **6.** $(-2, -1)$

7. $(0, -3)$ **8.** $(2, 0)$

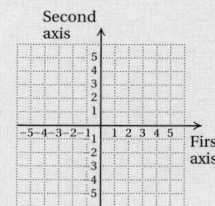

9. What can you say about the coordinates of a point in the third quadrant?

10. What can you say about the coordinates of a point in the fourth quadrant?

In which quadrant, if any, is each point located?

11. $(5, 3)$ **12.** $(-6, -4)$

13. $(10, -14)$ **14.** $(-13, 9)$

15. $(0, -3)$ **16.** $\left(-\dfrac{1}{2}, \dfrac{1}{4}\right)$

Answers

1–8.

Second axis

$(-2, 5)$ $(4, 5)$ $(5, 4)$ $(2, 0)$ $(-2, -1)$ $(0, -3)$ $(5, -3)$ $(-3, -4)$ First axis

9. Both are negative numbers.   **10.** First, positive; second, negative   **11.** I   **12.** III   **13.** IV   **14.** II   **15.** On an axis, not in any quadrant   **16.** II

b Finding Coordinates

To find the coordinates of a point, we see how far to the right or left of the origin it is located and how far up or down from the origin.

EXAMPLE 3 Find the coordinates of points A, B, C, D, E, F, and G.

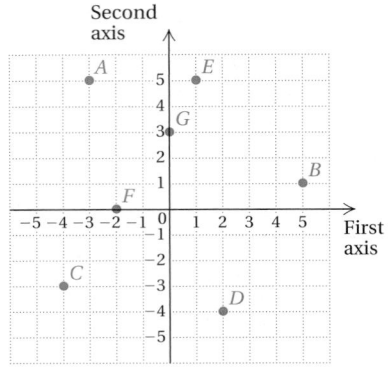

Point A is 3 units to the left (horizontal direction) and 5 units up (vertical direction). Its coordinates are $(-3, 5)$. Point D is 2 units to the right and 4 units down. Its coordinates are $(2, -4)$. The coordinates of the other points are as follows:

B: $(5, 1)$; C: $(-4, -3)$;

E: $(1, 5)$; F: $(-2, 0)$;

G: $(0, 3)$.

17. Find the coordinates of points A, B, C, D, E, F, and G on the graph below.

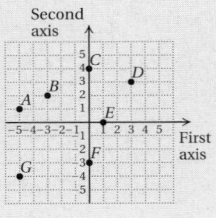

Do Exercise 17.

c Solutions of Equations

Now we begin to learn how graphs can be used to represent solutions of equations. When an equation contains two variables, the solutions of the equation are *ordered pairs* in which each number in the pair corresponds to a letter in the equation. Unless stated otherwise, to determine whether a pair is a solution, we use the first number in each pair to replace the variable that occurs first *alphabetically*.

EXAMPLE 4 Determine whether each of the following pairs is a solution of $4q - 3p = 22$: $(2, 7)$ and $(-1, 6)$.

For $(2, 7)$, we substitute 2 for p and 7 for q (using alphabetical order of variables):

$$\frac{4q - 3p = 22}{4 \cdot 7 - 3 \cdot 2 \overset{?}{\;} 22}$$
$$\begin{array}{c|c} 28 - 6 & \\ 22 & \text{TRUE} \end{array}$$

Thus, $(2, 7)$ is a solution of the equation.

For $(-1, 6)$, we substitute -1 for p and 6 for q:

$$\frac{4q - 3p = 22}{4 \cdot 6 - 3 \cdot (-1) \overset{?}{\;} 22}$$
$$\begin{array}{c|c} 24 + 3 & \\ 27 & \text{FALSE} \end{array}$$

Thus, $(-1, 6)$ is *not* a solution of the equation.

18. Determine whether $(2, -4)$ is a solution of $4q - 3p = 22$.

19. Determine whether $(2, -4)$ is a solution of $7a + 5b = -6$.

Do Exercises 18 and 19.

EXAMPLE 5 Show that the pairs $(3, 7)$, $(0, 1)$, and $(-3, -5)$ are solutions of $y = 2x + 1$. Then graph the three points and use the graph to determine another pair that is a solution.

To show that a pair is a solution, we substitute, replacing x with the first coordinate and y with the second coordinate of each pair:

$$\begin{array}{c|c}
y = 2x + 1 \\ \hline
7 \; ? \; 2 \cdot 3 + 1 \\
\quad \vert \quad 6 + 1 \\
\quad \vert \quad 7 \qquad \text{TRUE}
\end{array} \qquad
\begin{array}{c|c}
y = 2x + 1 \\ \hline
1 \; ? \; 2 \cdot 0 + 1 \\
\quad \vert \quad 0 + 1 \\
\quad \vert \quad 1 \qquad \text{TRUE}
\end{array}$$

$$\begin{array}{c|c}
y = 2x + 1 \\ \hline
-5 \; ? \; 2(-3) + 1 \\
\quad \vert \quad -6 + 1 \\
\quad \vert \quad -5 \qquad \text{TRUE}
\end{array}$$

In each of the three cases, the substitution results in a true equation. Thus the pairs are all solutions.

We plot the points as shown at right. The order of the points follows the alphabetical order of the variables. That is, x comes before y, so x-values are first coordinates and y-values are second coordinates. Similarly, we also label the horizontal axis as the x-axis and the vertical axis as the y-axis.

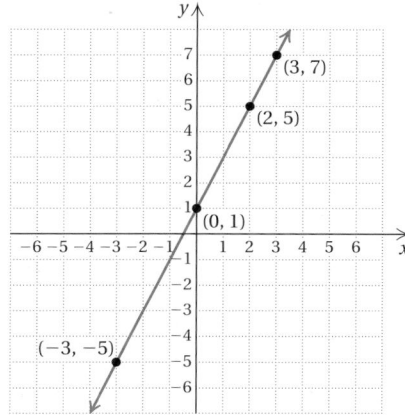

Note that the three points appear to "line up." That is, they appear to be on a straight line. Will other points that line up with these points also represent solutions of $y = 2x + 1$? To find out, we use a straightedge and sketch a line passing through $(3, 7)$, $(0, 1)$, and $(-3, -5)$.

The line appears to pass through $(2, 5)$ as well. Let's see if this pair is a solution of $y = 2x + 1$:

$$\begin{array}{c|c}
y = 2x + 1 \\ \hline
5 \; ? \; 2 \cdot 2 + 1 \\
\quad \vert \quad 4 + 1 \\
\quad \vert \quad 5 \qquad \text{TRUE}
\end{array}$$

Thus, $(2, 5)$ is a solution.

Do Exercise 20.

Example 5 leads us to suspect that any point on the line that passes through $(3, 7)$, $(0, 1)$, and $(-3, -5)$ represents a solution of $y = 2x + 1$. In fact, every solution of $y = 2x + 1$ is represented by a point on that line and every point on that line represents a solution. The line is the *graph* of the equation.

GRAPH OF AN EQUATION

The **graph** of an equation is a drawing that represents all of its solutions.

STUDY TIPS

LEARNING RESOURCES ON CAMPUS

Your college or university probably has resources to support your learning.

1. There may be a learning lab or a tutoring center for drop-in tutoring.

2. There may be group tutoring sessions for this specific course.

3. The math department may have a bulletin board or a network for locating private tutors.

4. Visit your instructor during office hours if you need additional help. Also, many instructors welcome e-mails from students with questions.

20. Use the graph in Example 5 to find at least two more points that are solutions of $y = 2x + 1$.

Answer

20. $(-2, -3)$, $(1, 3)$; answers may vary

ⓐ

1. Plot these points.

$(2, 5)$ $(-1, 3)$ $(3, -2)$ $(-2, -4)$

$(0, 4)$ $(0, -5)$ $(5, 0)$ $(-5, 0)$

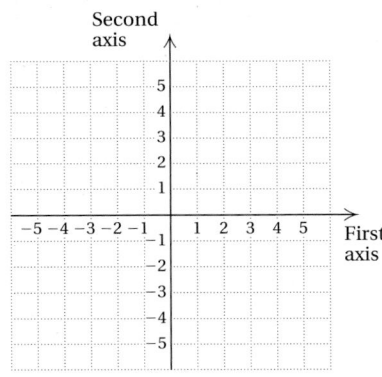

2. Plot these points.

$(4, 4)$ $(-2, 4)$ $(5, -3)$ $(-5, -5)$

$(0, 2)$ $(0, -4)$ $(3, 0)$ $(-4, 0)$

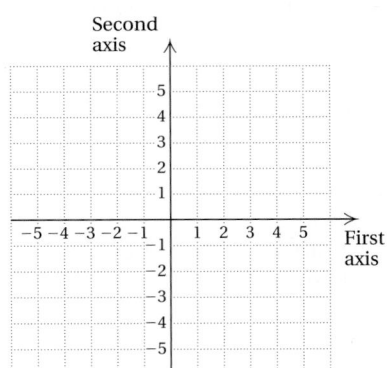

In which quadrant, if any, is each point located?

3. $(-5, 3)$

4. $(1, -12)$

5. $(100, -1)$

6. $(-2.5, 35.6)$

7. $(-6, -29)$

8. $(3.6, 105.9)$

9. $(3.8, 0)$

10. $(0, -492)$

11. $\left(-\dfrac{1}{3}, \dfrac{15}{7}\right)$

12. $\left(-\dfrac{2}{3}, -\dfrac{9}{8}\right)$

13. $\left(12\dfrac{7}{8}, -1\dfrac{1}{2}\right)$

14. $\left(23\dfrac{5}{8}, 81.74\right)$

For each of Exercises 15–18, complete the table regarding the signs of coordinates in certain quadrants.

	QUADRANT	FIRST COORDINATES	SECOND COORDINATES
15.		Positive	Positive
16.	III		Negative
17.	II	Negative	
18.		Positive	Negative

In which quadrant(s) can the point described be located?

19. The first coordinate is negative and the second coordinate is positive.

20. The first and second coordinates are positive.

21. The first coordinate is positive.

22. The second coordinate is negative.

23. The first and second coordinates are equal.

24. The first coordinate is the additive inverse of the second coordinate.

b Find the coordinates of points A, B, C, D, and E.

25.

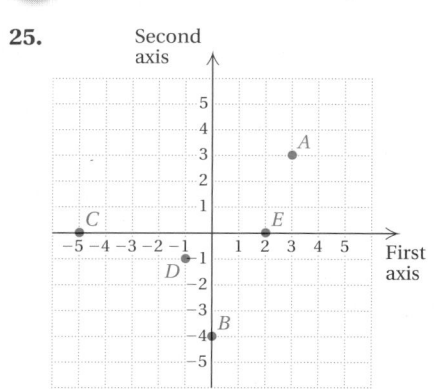

26.

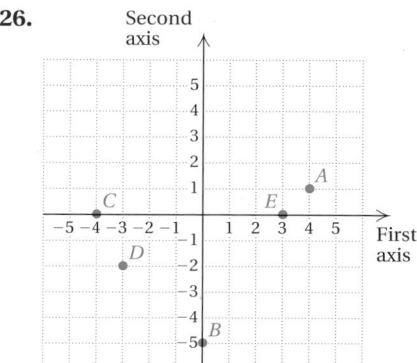

c Determine whether the given ordered pair is a solution of the equation.

27. $(2, 9)$; $y = 3x - 1$

28. $(1, 7)$; $y = 2x + 5$

29. $(4, 2)$; $2x + 3y = 12$

30. $(0, 5)$; $5x - 3y = 15$

31. $(3, -1)$; $3a - 4b = 13$

32. $(-5, 1)$; $2p - 3q = -13$

In Exercises 33–38, an equation and two ordered pairs are given. Show that each pair is a solution of the equation. Then use the graph of the equation to determine another solution. Answers may vary.

33. $y = x - 5$; $(4, -1)$ and $(1, -4)$

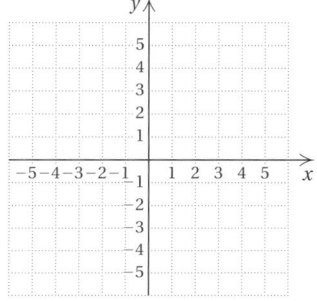

34. $y = x + 3$; $(-1, 2)$ and $(3, 6)$

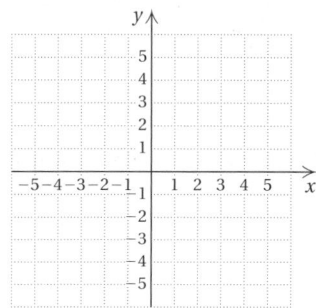

35. $y = \frac{1}{2}x + 3$; $(4, 5)$ and $(-2, 2)$

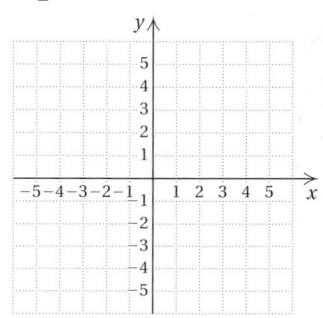

36. $3x + y = 7$; $(2, 1)$ and $(4, -5)$

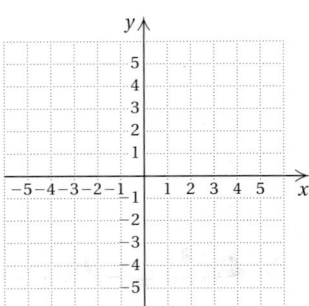

37. $4x - 2y = 10$; $(0, -5)$ and $(4, 3)$

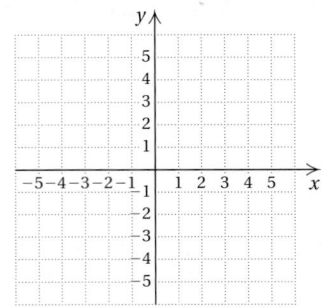

38. $6x - 3y = 3$; $(1, 1)$ and $(-1, -3)$

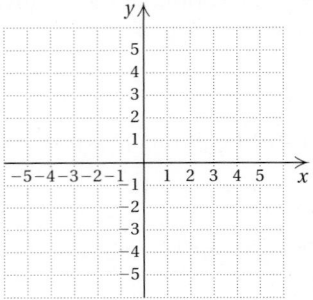

Skill Maintenance

Solve. [2.3c]

39. $6(z - 5) = 2(z + 1)$

40. $8 - 4(q + 2) = -7$

41. $-5 + x = x - 5$

42. $-\frac{3}{4} + x = x + \frac{1}{4}$

43. $4t - 2 = 3(5 - 2t)$

44. $2b - 5 + b = 6(b - 1)$

Solve. [2.5a]

45. What is 15% of $23.80?

46. $7.29 is 15% of what number?

47. Jennifer left an $8.50 tip for a meal that cost $42.50. What percent of the cost of the meal was the tip?

48. Kristen left an 18% tip of $3.24 for a meal. What was the cost of the meal before the tip?

49. Juan left a 15% tip for a meal. The total cost of the meal, including the tip, was $51.92. What was the cost of the meal before the tip was added?

50. After a 25% reduction, a sweater is on sale for $41.25. What was the original price?

Synthesis

51. The points $(-1, 1)$, $(4, 1)$, and $(4, -5)$ are three vertices of a rectangle. Find the coordinates of the fourth vertex.

52. Three parallelograms share the vertices $(-2, -3)$, $(-1, 2)$, and $(4, -3)$. Find the fourth vertex of each parallelogram.

53. Graph eight points such that the sum of the coordinates in each pair is 6. Answers may vary.

54. Graph eight points such that the first coordinate minus the second coordinate is 1. Answers may vary.

55. Find the perimeter of a rectangle whose vertices have coordinates $(5, 3)$, $(5, -2)$, $(-3, -2)$, and $(-3, 3)$.

56. Find the area of a triangle whose vertices have coordinates $(0, 9)$, $(0, -4)$, and $(5, -4)$.

3.2

Graphing Linear Equations

a) Graphs of Linear Equations

Equations like $y = 2x + 1$ and $4q - 3p = 22$ are said to be **linear** because the graph of each equation is a straight line. In general, any equation equivalent to one of the form $y = mx + b$ or $Ax + By = C$, where m, b, A, B, and C are constants (not variables) and A and B are not both 0, is linear.

> To graph a linear equation:
>
> 1. Select a value for one variable and calculate the corresponding value of the other variable. Form an ordered pair using alphabetical order as indicated by the variables.
> 2. Repeat step (1) to obtain at least two other ordered pairs. Two points are essential to determine a straight line. A third point serves as a check.
> 3. Plot the ordered pairs and draw a straight line passing through the points.

In general, calculating three (or more) ordered pairs is not difficult for equations of the form $y = mx + b$. We simply substitute values for x and calculate the corresponding values for y.

EXAMPLE 1 Graph: $y = 2x$.

First, we find some ordered pairs that are solutions. We choose *any* number for x and then determine y by substitution. Since $y = 2x$, we find y by doubling x. Suppose that we choose 3 for x. Then

$$y = 2x = 2 \cdot 3 = 6.$$

We get a solution: the ordered pair $(3, 6)$.

Suppose that we choose 0 for x. Then

$$y = 2x = 2 \cdot 0 = 0.$$

We get another solution: the ordered pair $(0, 0)$.

For a third point, we make a negative choice for x. If x is -3, we have

$$y = 2x = 2 \cdot (-3) = -6.$$

This gives us the ordered pair $(-3, -6)$.

We now have enough points to plot the line, but if we wish, we can compute more. If a number takes us off the graph paper, we either do not use it or we use larger paper or rescale the axes. Continuing in this manner, we create a table like the one shown on the following page.

OBJECTIVES

a) Graph linear equations of the type $y = mx + b$ and $Ax + By = C$, identifying the y-intercept.

b) Solve applied problems involving graphs of linear equations.

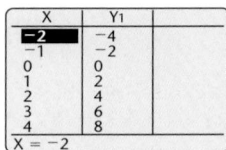

Calculator Corner

Finding Solutions of Equations A table of values representing ordered pairs that are solutions of an equation can be displayed on a graphing calculator. To do this for the equation in Example 1, $y = 2x$, we first press ⟨Y=⟩ to access the equation-editor screen. Next, we enter the equation by positioning the cursor beside "Y1 =" and press ② ⟨X,T,θ,n⟩. Now we press ⟨2ND⟩ ⟨TBLSET⟩ to display the table set-up screen. We then set both INDPNT and DEPEND to AUTO by positioning the cursor over AUTO and pressing ⟨ENTER⟩.

We will display a table of values that starts with $x = -2$ (TBLSTART) and adds 1 (ΔTBL) to the preceding x-value. We press ⟨(-)⟩ ② ⟨▽⟩ ① or ⟨(-)⟩ ② ⟨ENTER⟩ ① to do this. To display the table, we press ⟨2ND⟩ ⟨TABLE⟩.

X	Y1
-2	-4
-1	-2
0	0
1	2
2	4
3	6
4	8
X = -2	

Exercise:

1. Create a table of ordered pairs that are solutions of the equations in Examples 2 and 3.

Complete each table and graph.

1. $y = -2x$

x	y	(x, y)
-3		
-1		
0		
1		
3		

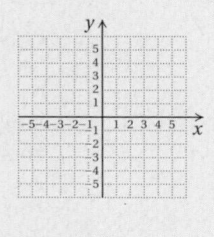

2. $y = \frac{1}{2}x$

x	y	(x, y)
4		
2		
0		
-2		
-4		
-1		

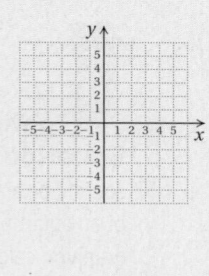

Now we plot these points. Then we draw the line, or graph, with a straightedge and label it $y = 2x$.

x	y $y = 2x$	(x, y)
3	6	$(3, 6)$
1	2	$(1, 2)$
0	0	$(0, 0)$
-2	-4	$(-2, -4)$
-3	-6	$(-3, -6)$

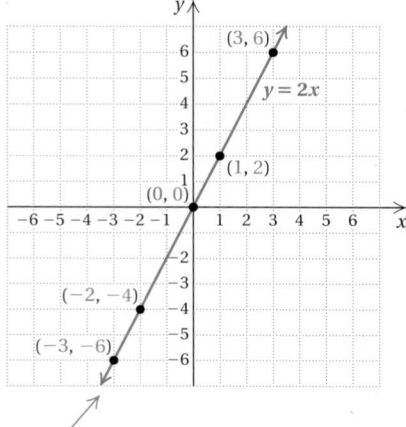

(1) Choose x.
(2) Compute y.
(3) Form the pair (x, y).
(4) Plot the points.

--------- *Caution!* ---------

Keep in mind that you can choose *any* number for x and then compute y. Our choice of certain numbers in the examples does not dictate those that you must choose.

Do Exercises 1 and 2.

EXAMPLE 2 Graph: $y = -3x + 1$.

We select a value for x, compute y, and form an ordered pair. Then we repeat the process for other choices of x.

If $x = 2$,   then $y = -3 \cdot 2 + 1 = -5$,   and $(2, -5)$ is a solution.

If $x = 0$,   then $y = -3 \cdot 0 + 1 = 1$,   and $(0, 1)$ is a solution.

If $x = -1$,   then $y = -3 \cdot (-1) + 1 = 4$,   and $(-1, 4)$ is a solution.

Results are listed in the table below. The points corresponding to each pair are then plotted.

x	y $y = -3x + 1$	(x, y)
2	-5	$(2, -5)$
0	1	$(0, 1)$
-1	4	$(-1, 4)$

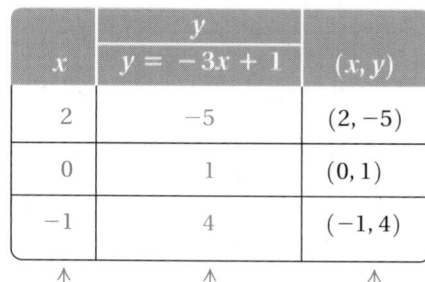

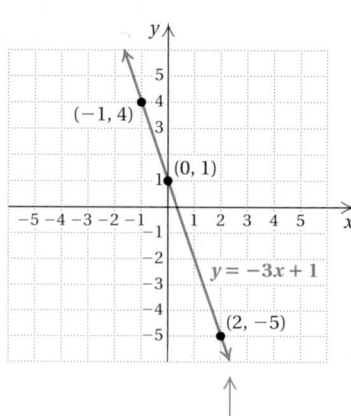

(1) Choose x.
(2) Compute y.
(3) Form the pair (x, y).
(4) Plot the points.

Answers

1.

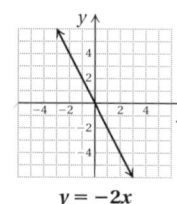

$y = -2x$

2.

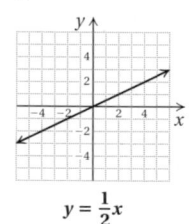

$y = \frac{1}{2}x$

Note that all three points line up. If they did not, we would know that we had made a mistake. When only two points are plotted, a mistake is harder to detect. We use a ruler or other straightedge to draw a line through the points. Every point on the line represents a solution of $y = -3x + 1$.

Do Exercises 3 and 4.

In Example 1, we saw that $(0, 0)$ is a solution of $y = 2x$. It is also the point at which the graph crosses the y-axis. Similarly, in Example 2, we saw that $(0, 1)$ is a solution of $y = -3x + 1$. It is also the point at which the graph crosses the y-axis. A generalization can be made: If x is replaced with 0 in the equation $y = mx + b$, then the corresponding y-value is $m \cdot 0 + b$, or b. Thus any equation of the form $y = mx + b$ has a graph that passes through the point $(0, b)$. Since $(0, b)$ is the point at which the graph crosses the y-axis, it is called the **y-intercept**. Sometimes, for convenience, we simply refer to b as the y-intercept.

y-INTERCEPT

The graph of the equation $y = mx + b$ passes through the **y-intercept** $(0, b)$.

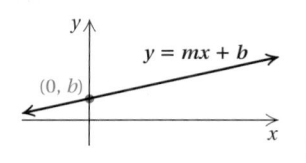

EXAMPLE 3 Graph $y = \frac{2}{5}x + 4$ and identify the y-intercept.

We select a value for x, compute y, and form an ordered pair. Then we repeat the process for other choices of x. In this case, using multiples of 5 avoids fractions. We try to avoid graphing ordered pairs with fractions because they are difficult to graph accurately.

If $x = 0$,   then $y = \dfrac{2}{5} \cdot 0 + 4 = 4$,   and $(0, 4)$ is a solution.

If $x = 5$,   then $y = \dfrac{2}{5} \cdot 5 + 4 = 6$,   and $(5, 6)$ is a solution.

If $x = -5$,   then $y = \dfrac{2}{5} \cdot (-5) + 4 = 2$,   and $(-5, 2)$ is a solution.

The following table lists these solutions. Next, we plot the points and see that they form a line. Finally, we draw and label the line.

x	y $y = \frac{2}{5}x + 4$	(x, y)
0	4	$(0, 4)$
5	6	$(5, 6)$
−5	2	$(-5, 2)$

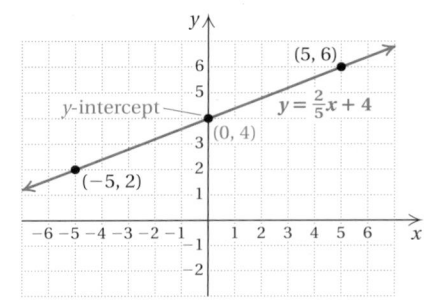

Complete each table and graph.

3. $y = 2x + 3$

x	y	(x, y)

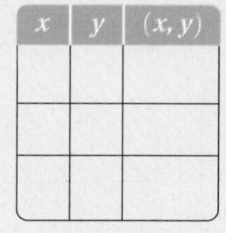

4. $y = -\dfrac{1}{2}x - 3$

x	y	(x, y)

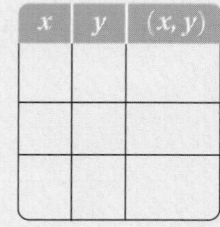

Answers

3.

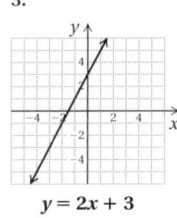

$y = 2x + 3$

4.

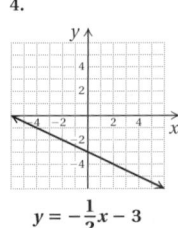

$y = -\dfrac{1}{2}x - 3$

Graph each equation and identify the *y*-intercept.

5. $y = \dfrac{3}{5}x + 2$

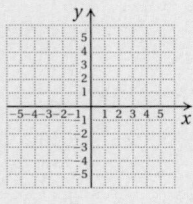

6. $y = -\dfrac{3}{5}x - 1$

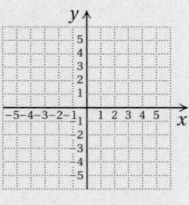

We see that $(0, 4)$ is a solution of $y = \frac{2}{5}x + 4$. It is the *y*-intercept. Because the equation is in the form $y = mx + b$, we can read the *y*-intercept directly from the equation as follows:

$$y = \frac{2}{5}x + 4 \qquad (0, 4) \text{ is the } y\text{-intercept.}$$

Do Exercises 5 and 6.

Calculating ordered pairs is generally easiest when *y* is isolated on one side of the equation, as in $y = mx + b$. To graph an equation in which *y* is not isolated, we can use the addition and multiplication principles to solve for *y*. (See Sections 2.3 and 2.4.)

EXAMPLE 4 Graph $3y + 5x = 0$ and identify the *y*-intercept.

To find an equivalent equation in the form $y = mx + b$, we solve for *y*:

$$3y + 5x = 0$$
$$3y + 5x - 5x = 0 - 5x \qquad \text{Subtracting } 5x$$
$$3y = -5x \qquad \text{Collecting like terms}$$
$$\frac{3y}{3} = \frac{-5x}{3} \qquad \text{Dividing by 3}$$
$$y = -\frac{5}{3}x.$$

Because all the equations above are equivalent, we can use $y = -\frac{5}{3}x$ to draw the graph of $3y + 5x = 0$. To graph $y = -\frac{5}{3}x$, we select *x*-values and compute *y*-values. In this case, if we select multiples of 3, we can avoid fractions.

If $x = 0$, then $y = -\dfrac{5}{3} \cdot 0 = 0$.

If $x = 3$, then $y = -\dfrac{5}{3} \cdot 3 = -5$.

If $x = -3$, then $y = -\dfrac{5}{3} \cdot (-3) = 5$.

We list these solutions in a table. Next, we plot the points and see that they form a line. Finally, we draw and label the line. The *y*-intercept is $(0, 0)$.

x	y	
0	0	← *y*-intercept
3	−5	
−3	5	

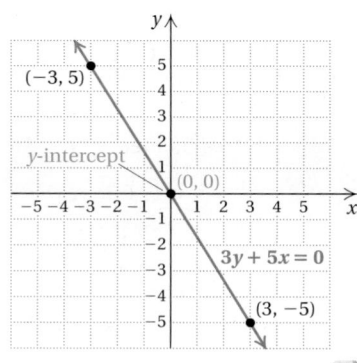

Graph each equation and identify the *y*-intercept.

7. $5y + 4x = 0$

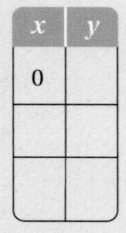
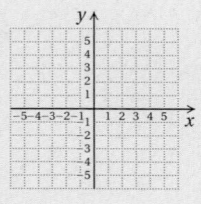

8. $4y = 3x$

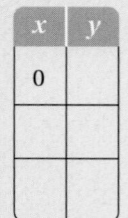
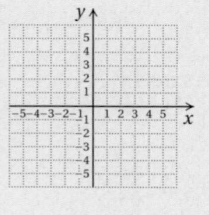

Do Exercises 7 and 8.

Answers

Answers to Margin Exercises 5–8 are on p. 185.

EXAMPLE 5 Graph $4y + 3x = -8$ and identify the y-intercept.

To find an equivalent equation in the form $y = mx + b$, we solve for y:

$$4y + 3x = -8$$
$$4y + 3x - 3x = -8 - 3x \qquad \text{Subtracting } 3x$$
$$4y = -3x - 8 \qquad \text{Simplifying}$$
$$\frac{1}{4} \cdot 4y = \frac{1}{4} \cdot (-3x - 8) \qquad \text{Multiplying by } \tfrac{1}{4} \text{ or dividing by 4}$$
$$y = \frac{1}{4} \cdot (-3x) - \frac{1}{4} \cdot 8 \qquad \text{Using the distributive law}$$
$$y = -\frac{3}{4}x - 2. \qquad \text{Simplifying}$$

Thus, $4y + 3x = -8$ is equivalent to $y = -\frac{3}{4}x - 2$. The y-intercept is $(0, -2)$. We find two other pairs using multiples of 4 for x to avoid fractions. We then complete and label the graph as shown.

x	y
0	-2	$\leftarrow$ y-intercept
4	-5
-4	1

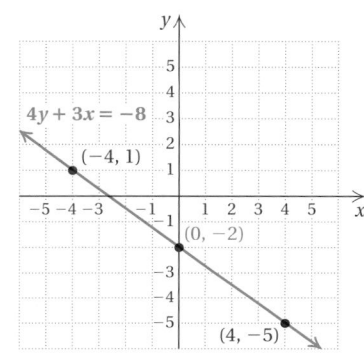

Do Exercises 9 and 10.

Do Exercises 9 and 10.

b Applications of Linear Equations

Mathematical concepts become more understandable through visualization. Throughout this text, you will occasionally see the heading ✳ Algebraic–Graphical Connection, as in Example 6, which follows. In this feature, the algebraic approach is enhanced and expanded with a graphical connection. Relating a solution of an equation to a graph can often give added meaning to the algebraic solution.

EXAMPLE 6 *World Population.* The world population, in billions, is estimated and projected by

$$y = 0.072x + 4.593,$$

where x is the number of years since 1980. That is, $x = 0$ corresponds to 1980, $x = 12$ corresponds to 1992, and so on.

Source: U.S. Census Bureau

Graph each equation and identify the y-intercept.

9. $5y - 3x = -10$

x	y

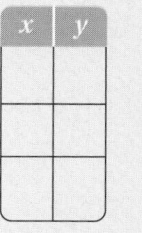

10. $5y + 3x = 20$

x	y

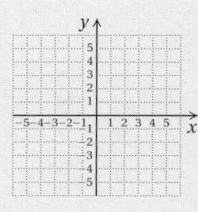

Answers

5.
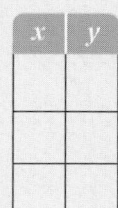
$y = \frac{3}{5}x + 2$

6.
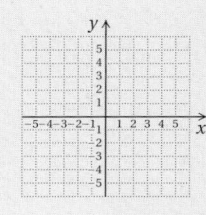
$y = -\frac{3}{5}x - 1$

7.
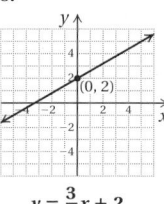
$5y + 4x = 0$

8.
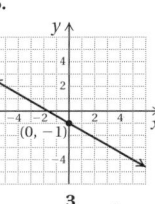
$4y = 3x$

9.
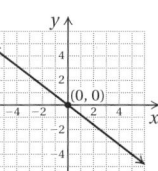
$5y - 3x = -10$

10.
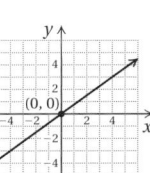
$5y + 3x = 20$

a) Determine the world population in 1980, in 2005, and in 2030.

b) Graph the equation and then use the graph to estimate the world population in 2015.

c) In what year would we estimate the world population to be 7.761 billion?

a) The years 1980, 2005, and 2030 correspond to $x = 0$, $x = 25$, and $x = 50$, respectively. We substitute 0, 25, and 50 for x and then calculate y:

$$y = 0.072(0) + 4.593 = 0 + 4.593 = 4.593;$$
$$y = 0.072(25) + 4.593 = 1.8 + 4.593 = 6.393;$$
$$y = 0.072(50) + 4.593 = 3.6 + 4.593 = 8.193.$$

The world population in 1980, in 2005, and in 2030 is estimated to be 4.593 billion, 6.393 billion, and 8.193 billion, respectively.

※ Algebraic–Graphical Connection

b) We have three ordered pairs from part (a). We plot these points and see that they line up. Thus our calculations are probably correct. Since we are considering only the year 2015 and the number of years since 1980 ($x \geq 0$) and since the population, in billions, for those years will be positive ($y > 0$), we need only the first quadrant for the graph. We use the three points we have plotted to draw a straight line. (See Figure 1.)

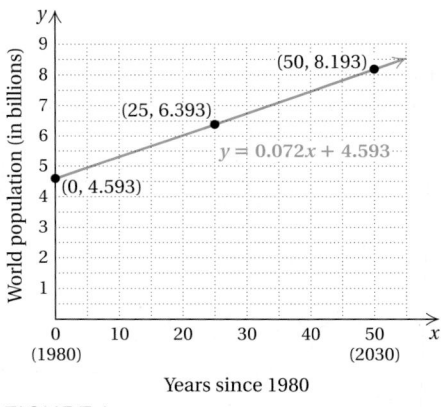

FIGURE 1   FIGURE 2

To use the graph to estimate world population in 2015, we first note in Figure 2 that this year corresponds to $x = 35$. We need to determine which y-value is paired with $x = 35$. We locate the point on the graph by moving up vertically from $x = 35$, and then find the value on the y-axis that corresponds to that point. It appears that the world population in 2015 will be about 7.1 billion.

To find a more accurate value, we can simply substitute into the equation:

$$y = 0.072(35) + 4.593 = 7.113.$$

The world population in 2015 is projected to be 7.113 billion.

c) We substitute 7.761 for y and solve for x:

$$y = 0.072x + 4.593$$
$$7.761 = 0.072x + 4.593$$
$$3.168 = 0.072x$$
$$44 = x.$$

In 44 yr after 1980, or in 2024, the world population will be approximately 7.761 billion.

Do Exercise 11.

Many equations in two variables have graphs that are not straight lines. Three such nonlinear graphs are shown below. We will cover some such graphs in the optional Calculator Corners throughout the text and in Chapter 11.

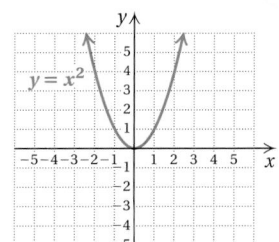 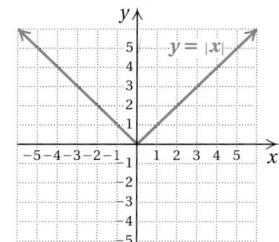 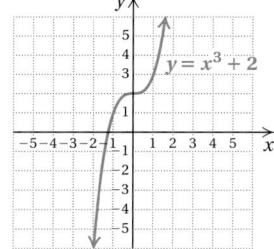

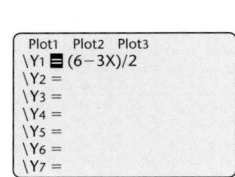

Calculator Corner

Graphing Equations Equations must be solved for y before they can be graphed on the TI-84 Plus. Consider the equation $3x + 2y = 6$. Solving for y, we get $y = \dfrac{6 - 3x}{2}$. We enter this equation as $y_1 = (6 - 3x)/2$ on the equation-editor screen. Then we press ⟨ZOOM⟩ ⟨6⟩ to select the standard viewing window and display the graph.

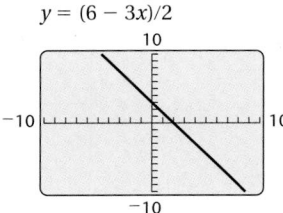

Exercises: Graph each equation in the standard viewing window $[-10, 10, -10, 10]$, with Xscl = 1 and Yscl = 1.

1. $y = 2x + 1$

2. $y = -3x + 1$

3. $y = -5x + 3$

4. $y = 4x - 5$

5. $4x - 5y = -10$

6. $5y + 5 = -3x$

7. $y = 2.085x + 5.08$

8. $y = -3.45x - 1.68$

11. Milk Consumption. Milk consumption per capita (per person) in the United States is given by

$$M = -0.271t + 27.952,$$

where M is the consumption, in gallons, t years from 1980.

Source: U.S. Department of Agriculture

a) Find the per capita consumption of milk in 1980, in 1995, and in 2015.

b) Graph the equation and use the graph to estimate milk consumption in 2010.

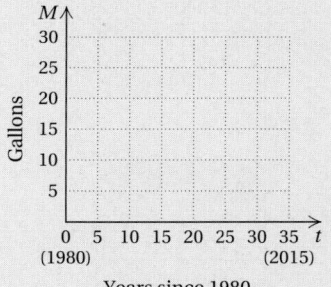

c) In which year would the per capita consumption of milk be 19.28 gal?

Answer

11. (a) 27.952 gal; 23.887 gal; 18.467 gal;
(b) about 19.8 gal;

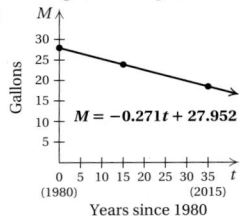

(c) in 32 years, or in 2012

a Graph each equation and identify the *y*-intercept.

✗

y intercept = (0,-1)

1. $y = x + 1$

x	y
−2	−1
−1	0
0	
1	
2	

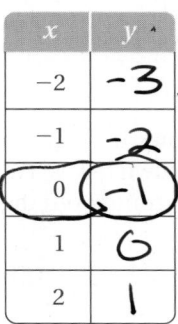

2. $y = x - 1$

x	y
−2	−3
−1	−2
0	−1
1	6
2	1

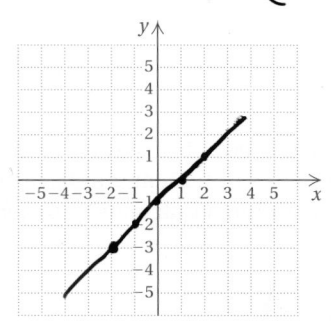

3. $y = x$

x	y
−2	
−1	
0	
1	
2	

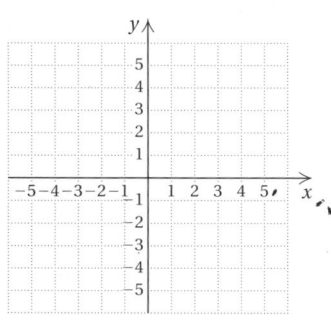

4. $y = -x$

x	y
−2	
−1	
0	
1	
2	

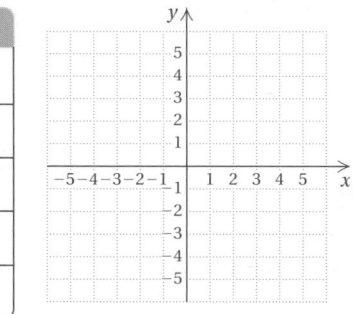

5. $y = \dfrac{1}{2}x$

x	y
−2	
0	
4	

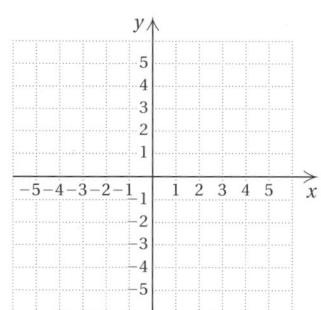

6. $y = \dfrac{1}{3}x$

x	y
−6	
0	
3	

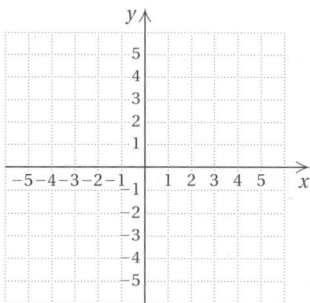

7. $y = x - 3$

x	y

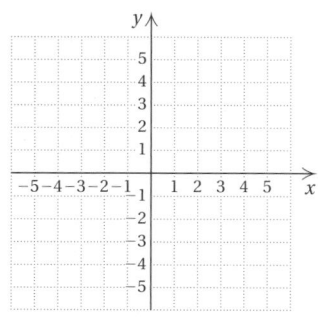

8. $y = x + 3$

x	y

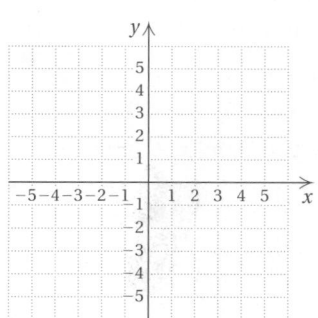

9. $y = 3x - 2$

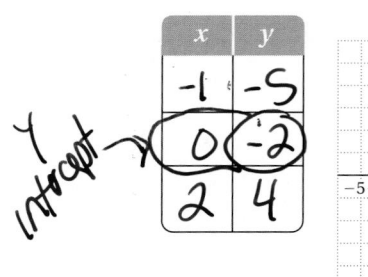

Y intercept →

x	y
-1	-5
0	-2
2	4

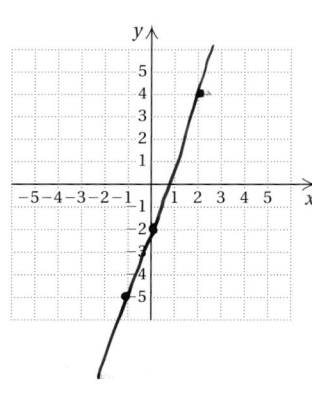

10. $y = 2x + 2$

Y intercept →

x	y
-1	0
0	2
2	6

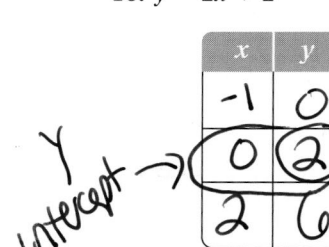

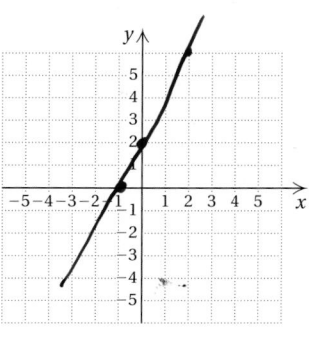

11. $y = \frac{1}{2}x + 1$

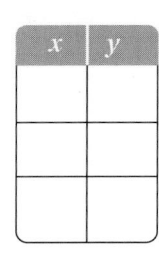

x	y

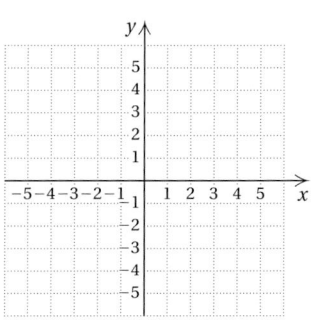

12. $y = \frac{1}{3}x - 4$

x	y

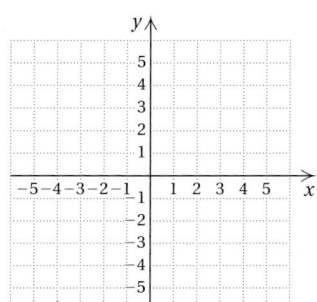

13. $x + y = -5$

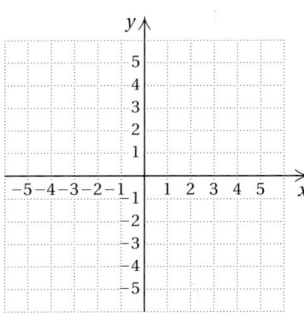

14. $x + y = 4$

$4 - x = y$

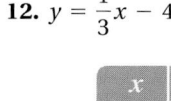

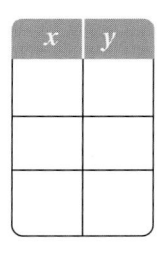

x	y
-1	5
0	4
2	2

Y intercept →

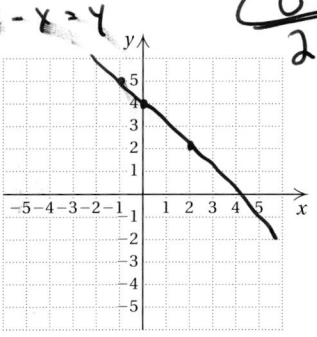

15. $y = \frac{5}{3}x - 2$

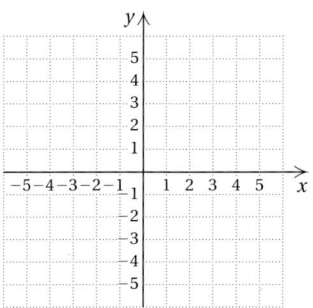

16. $y = \frac{5}{2}x + 3$

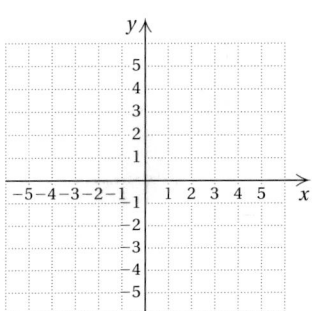

17. $x + 2y = 8$

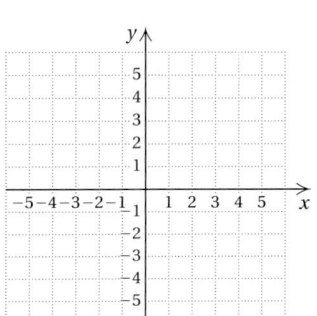

18. $x + 2y = -6$

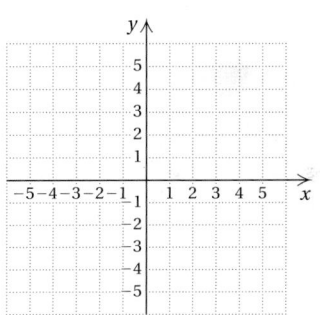

19. $y = \dfrac{3}{2}x + 1$

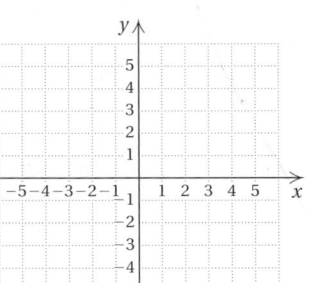

20. $y = -\dfrac{1}{2}x - 3$

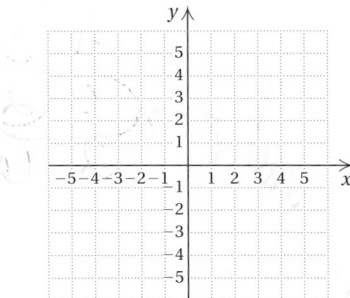

21. $8x - 2y = -10$

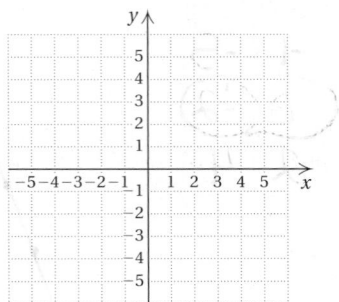

22. $6x - 3y = 9$

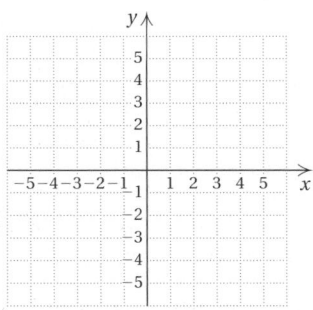

23. $8y + 2x = -4$

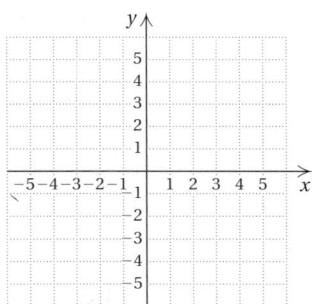

24. $6y + 2x = 8$

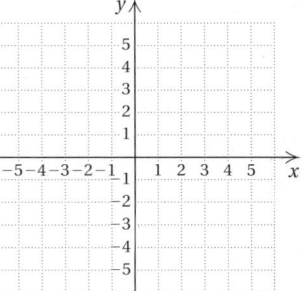

b Solve.

25. *Realtor Income.* The median annual income R, in dollars, for realtors has declined in recent years and can be approximated by

$$R = -1698t + 52{,}620,$$

where t is the number of years since 2002.

Source: National Association of Realtors

a) Find the median income in 2002, in 2007, and in 2010.

b) Graph the equation and then use the graph to estimate the median income in 2005.

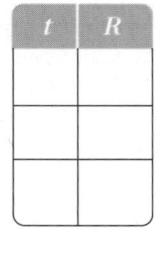
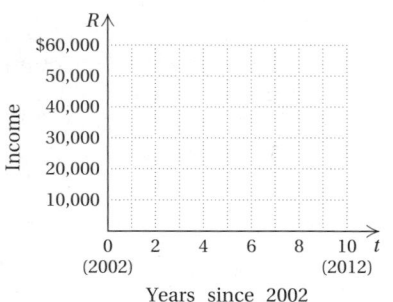

c) At this rate of decline, in what year will the median income be $37,338?

26. *International Visitors.* The number of international visitors V, in millions, to the United States each year can be estimated and projected by

$$V = 2.18t + 46.46,$$

where t is the number of years since 2004.

Sources: TIA's Travel Forecast Model; U.S. Bureau of Labor Statistics, Office of Travel and Tourism Industries

a) Find the number of international visitors in 2004, in 2007, and in 2012.

b) Graph the equation and use the graph to estimate the number of visitors in 2009.

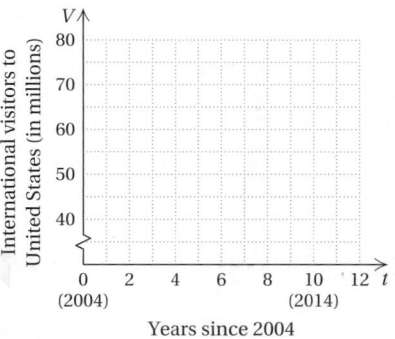

c) In what year will the number of visitors be about 68.3 million?

27. *Bottled Water Consumption.* The average number of gallons W of bottled water consumed each year by the U.S. consumer can be approximated by

$$W = 1.8d + 16.44,$$

where d is the number of years since 2000.

Source: USDA/Economic Research Service

a) Find the average number of gallons of bottled water consumed in 2001 $(d = 1)$, in 2010, and in 2015.

b) Graph the equation and use the graph to estimate what the bottled water consumption was in 2008.

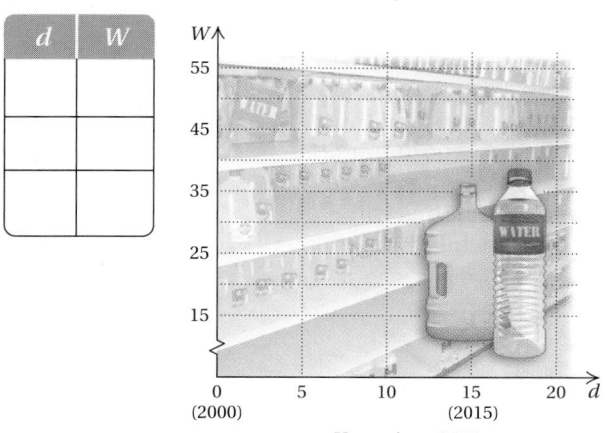

Years since 2000

c) In what year will bottled water consumption be about 36 gal?

28. *Record Temperature Drop.* On 22 January 1943, the temperature T, in degrees Fahrenheit, in Spearfish, South Dakota, could be approximated by

$$T = -2.15m + 54,$$

where m is the number of minutes since 9:00 that morning.

Source: Information Please Almanac

a) Find the temperature at 9:01 A.M., at 9:08 A.M., and at 9:20 A.M.

b) Graph the equation and use the graph to estimate the temperature at 9:15 A.M.

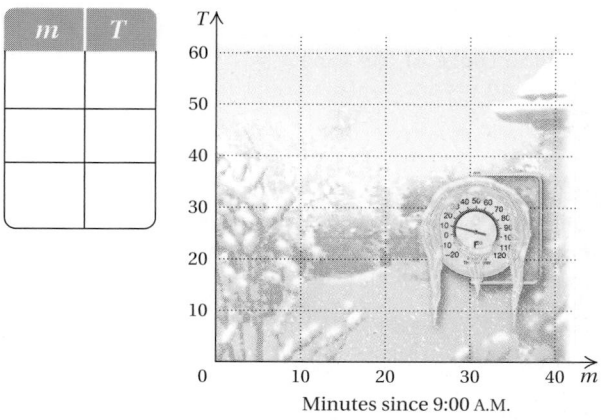

Minutes since 9:00 A.M.

c) The temperature stopped dropping when it reached $-4°F$. At what time did this occur?

Skill Maintenance

Find the absolute value.   [1.2e]

29. $|-12|$

30. $|4.89|$

31. $|0|$

32. $\left|-\frac{4}{5}\right|$

33. $|-3.4|$

34. $|\sqrt{2}|$

35. $\left|\frac{2}{3}\right|$

36. $\left|-\frac{7}{8}\right|$

Solve.   [2.5a]

37. *Older Patients.* In recent years, 24% of U.S. hospital patients are age 75 and older. For a hospital with 200 beds, approximately how many patients will be age 75 and older?

Source: U.S. Centers for Disease Control and Prevention

38. *Under 15 Years Old in India.* It is projected that by 2010, 363.5 million people in India will be under 15 years old. If the projected population in India in 2010 is 1184 million, what percent will be under 15 years old?

Source: U.S. Census Bureau, International Data Base

Calculate.

39. $-\frac{3}{5} \div 5$ [1.6c]

40. $2.8 - (-0.2)$ [1.4a]

41. $-\frac{9}{16} + \left(-\frac{3}{8}\right)$ [1.3a]

42. $4.2 \times (-100)$ [1.5a]

43. $-\frac{8}{7} \div \left(-\frac{1}{4}\right)$ [1.6c]

44. $23.3 - 32.3$ [1.4a]

3.3 More with Graphing and Intercepts

OBJECTIVES

a Find the intercepts of a linear equation, and graph using intercepts.

b Graph equations equivalent to those of the type $x = a$ and $y = b$.

SKILL TO REVIEW
Objective 2.3a: Solve equations using both the addition principle and the multiplication principle.

Solve.

1. $5x - 7 = -10$

2. $-20 = \frac{7}{4}x + 8$

1. Look at the graph shown below.

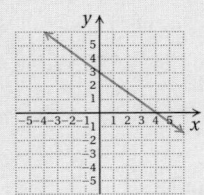

a) Find the coordinates of the y-intercept.

b) Find the coordinates of the x-intercept.

a Graphing Using Intercepts

In Section 3.2, we graphed linear equations of the form $Ax + By = C$ by first solving for y to find an equivalent equation in the form $y = mx + b$. We did so because it is then easier to calculate the y-value that corresponds to a given x-value. Another convenient way to graph $Ax + By = C$ is to use **intercepts**. Look at the graph of $-2x + y = 4$ shown below.

The y-intercept is $(0, 4)$. It occurs where the line crosses the y-axis and thus will always have 0 as the first coordinate. The x-intercept is $(-2, 0)$. It occurs where the line crosses the x-axis and thus will always have 0 as the second coordinate.

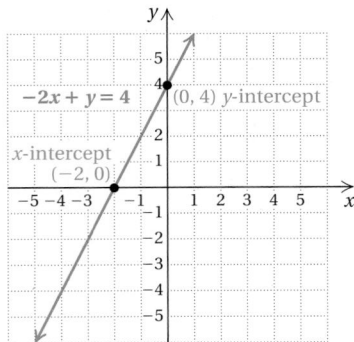

Do Margin Exercise 1.

We find intercepts as follows.

INTERCEPTS

The **y-intercept** is $(0, b)$. To find b, let $x = 0$ and solve the equation for y.

The **x-intercept** is $(a, 0)$. To find a, let $y = 0$ and solve the equation for x.

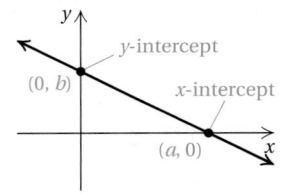

Now let's draw a graph using intercepts.

EXAMPLE 1 Consider $4x + 3y = 12$. Find the intercepts. Then graph the equation using the intercepts.

To find the y-intercept, we let $x = 0$. Then we solve for y:

$$4 \cdot 0 + 3y = 12$$
$$3y = 12$$
$$y = 4.$$

Thus, $(0, 4)$ is the y-intercept. Note that finding this intercept amounts to covering up the x-term and solving the rest of the equation for y.

To find the x-intercept, we let $y = 0$. Then we solve for x:

$$4x + 3 \cdot 0 = 12$$
$$4x = 12$$
$$x = 3.$$

Answers

Skill to Review:

1. $-\frac{3}{5}$ **2.** -16

Margin Exercises:

1. (a) $(0, 3)$; **(b)** $(4, 0)$

Thus, $(3, 0)$ is the x-intercept. Note that finding this intercept amounts to covering up the y-term and solving the rest of the equation for x.

We plot these points and draw the line, or graph.

x	y	
3	0	←x-intercept
0	4	←y-intercept
−2	$6\frac{2}{3}$	←Check point

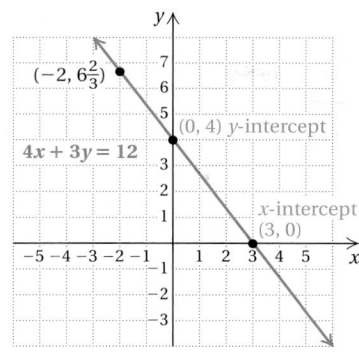

A third point should be used as a check. We substitute any convenient value for x and solve for y. In this case, we choose $x = -2$. Then

$$4(-2) + 3y = 12 \qquad \text{Substituting } -2 \text{ for } x$$
$$-8 + 3y = 12$$
$$3y = 20 \qquad \text{Adding 8 on both sides}$$
$$y = \frac{20}{3}, \text{ or } 6\frac{2}{3}. \qquad \text{Solving for } y$$

It appears that the point $\left(-2, 6\frac{2}{3}\right)$ is on the graph, though graphing fraction values can be inexact. The graph is probably correct.

> Do Exercises 2 and 3.

Graphs of equations of the type $y = mx$ pass through the origin. Thus the x-intercept and the y-intercept are the same, $(0, 0)$. In such cases, we must calculate another point in order to complete the graph. Another point would also need to be calculated if a check is desired.

EXAMPLE 2 Graph: $y = 3x$.

We know that $(0, 0)$ is both the x-intercept and the y-intercept. We calculate values at two other points and complete the graph, knowing that it passes through the origin $(0, 0)$.

x	y	
−1	−3	⎤ x-intercept
0	0	⎬ y-intercept
1	3	⎦

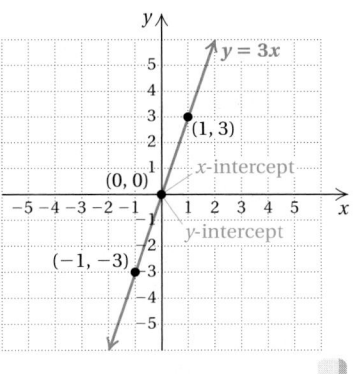

> Do Exercises 4 and 5 on the following page.

For each equation, find the intercepts. Then graph the equation using the intercepts.

2. $2x + 3y = 6$

x	y	
	0	← x-intercept
0		← y-intercept
		← Check point

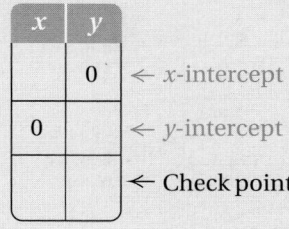

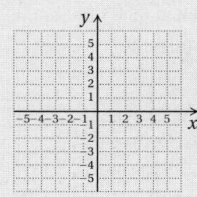

3. $3y - 4x = 12$

x	y	
		← x-intercept
		← y-intercept
		← Check point

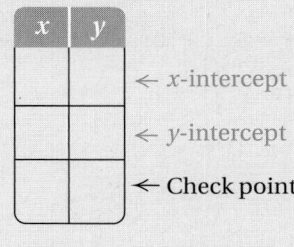

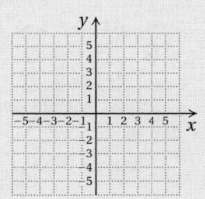

Answers

2.
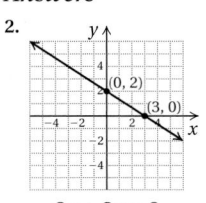
$2x + 3y = 6$

3.
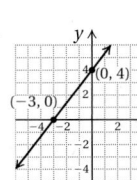
$3y - 4x = 12$

Graph.

4. $y = 2x$

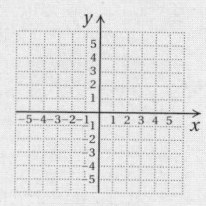

x	y
−1	
0	
1	

5. $y = -\dfrac{2}{3}x$

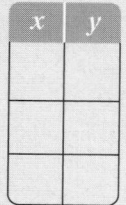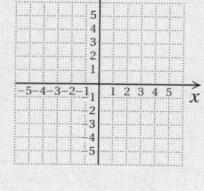

x	y

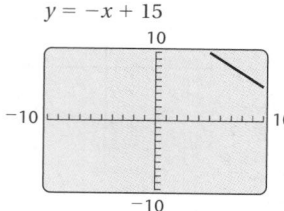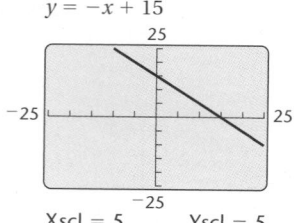
b Equations Whose Graphs Are Horizontal or Vertical Lines

EXAMPLE 3 Graph: $y = 3$.

The equation $y = 3$ tells us that y must be 3, but it doesn't give us any information about x. We can also think of this equation as $0 \cdot x + y = 3$. No matter what number we choose for x, we find that y is 3. We make up a table with all 3's in the y-column.

x	y
	3
	3
	3

Choose any number for x. → y must be 3.

x	y	
−2	3	
0	3	← y-intercept
4	3	

Answers

4.
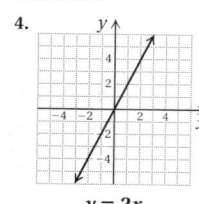
$y = 2x$

5.
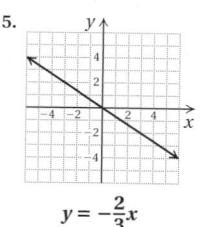
$y = -\dfrac{2}{3}x$

When we plot the ordered pairs $(-2, 3)$, $(0, 3)$, and $(4, 3)$ and connect the points, we obtain a horizontal line. Any ordered pair $(x, 3)$ is a solution. So the line is parallel to the x-axis with y-intercept $(0, 3)$.

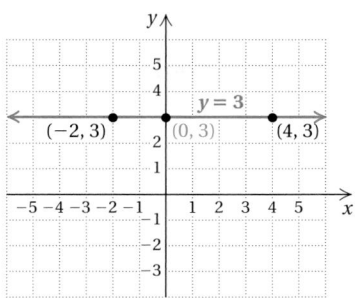

EXAMPLE 4 Graph: $x = -4$.

Consider $x = -4$. We can also think of this equation as $x + 0 \cdot y = -4$. We make up a table with all -4's in the x-column.

x	y
-4	
-4	
-4	
-4	

x must be -4.

x	y
-4	-5
-4	1
-4	3
-4	0

← Choose any number for y.

x-intercept →

When we plot the ordered pairs $(-4, -5)$, $(-4, 1)$, $(-4, 3)$, and $(-4, 0)$ and connect the points, we obtain a vertical line. Any ordered pair $(-4, y)$ is a solution. So the line is parallel to the y-axis with x-intercept $(-4, 0)$.

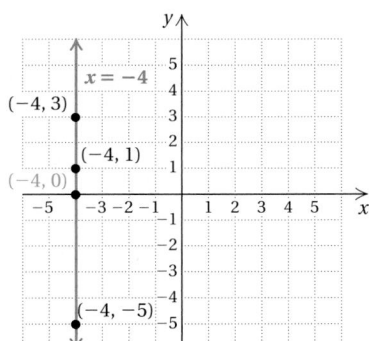

HORIZONTAL AND VERTICAL LINES

The graph of $y = b$ is a **horizontal line**. The y-intercept is $(0, b)$.

The graph of $x = a$ is a **vertical line**. The x-intercept is $(a, 0)$.

Do Exercises 6–9.

Graph.

6. $x = 5$

x	y
5	
5	
5	

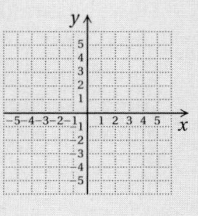

7. $y = -2$

x	y
	-2
	-2
	-2

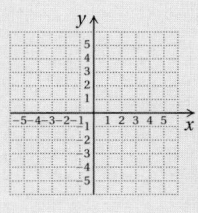

8. $x = -3$

x	y

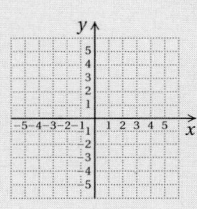

9. $x = 0$

x	y

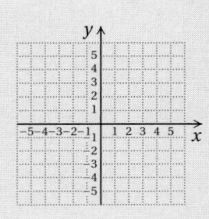

Answers

6.
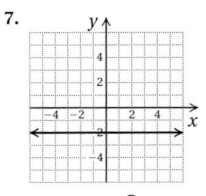
$x = 5$

7.
$y = -2$

8.
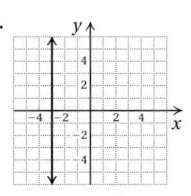
$x = -3$

9.
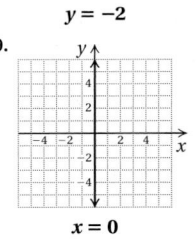
$x = 0$

ASKING QUESTIONS

Don't be afraid to ask questions in class. Most instructors welcome and encourage this. Other students often have the same questions you do.

The following is a general procedure for graphing linear equations.

GRAPHING LINEAR EQUATIONS

1. If the equation is of the type $x = a$ or $y = b$, the graph will be a line parallel to an axis; $x = a$ is vertical and $y = b$ is horizontal.

   Examples.

   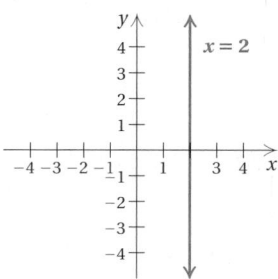   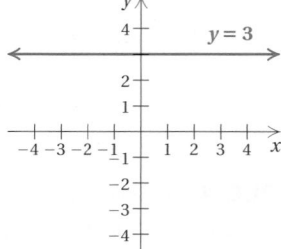

2. If the equation is of the type $y = mx$, both intercepts are the origin, $(0, 0)$. Plot $(0, 0)$ and two other points.

   Example.

3. If the equation is of the type $y = mx + b$, plot the y-intercept $(0, b)$ and two other points.

   Example.

   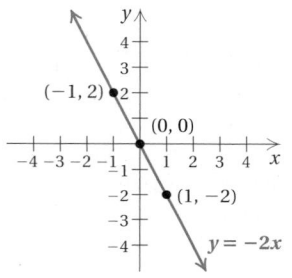   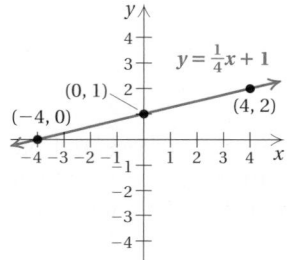

4. If the equation is of the type $Ax + By = C$, but not of the type $x = a$ or $y = b$, then either solve for y and proceed as with the equation $y = mx + b$, or graph using intercepts. If the intercepts are too close together, choose another point or points farther from the origin.

   Examples.

   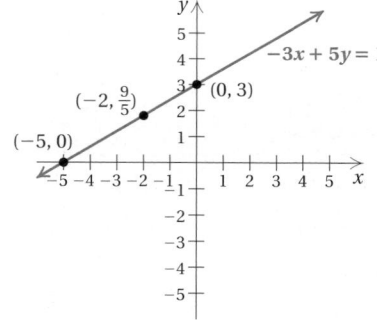   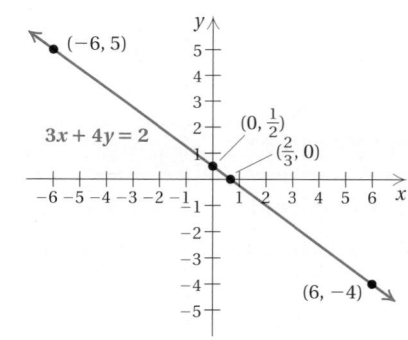

Visualizing for Success

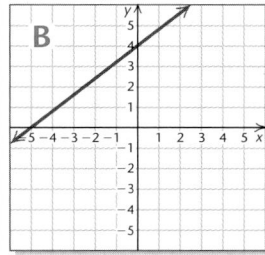

A

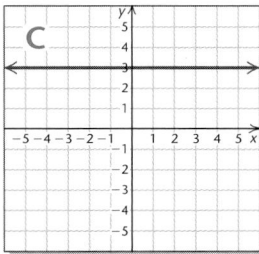

B

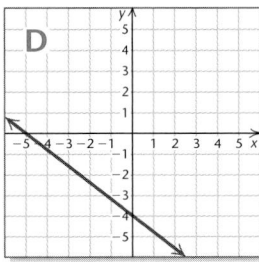

C

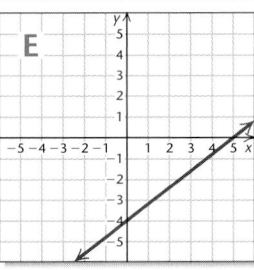

D

E

Match each equation with its graph.

1. $5y + 20 = 4x$

2. $y = 3$

3. $3x + 5y = 15$

4. $5y + 4x = 20$

5. $5y = 10 - 2x$

6. $4x + 5y + 20 = 0$

7. $5x - 4y = 20$

8. $4y + 5x + 20 = 0$

9. $5y - 4x = 20$

10. $x = -3$

Answers on page A-7

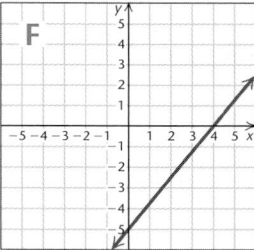

F

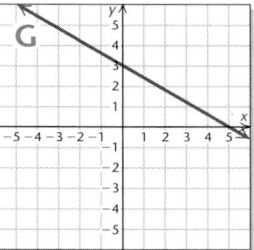

G

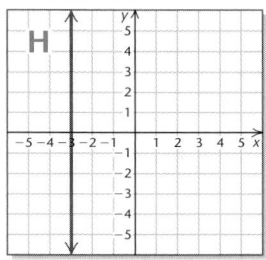

H

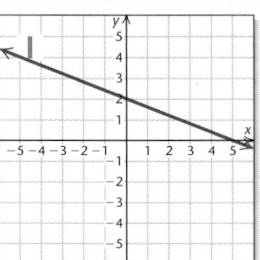

I

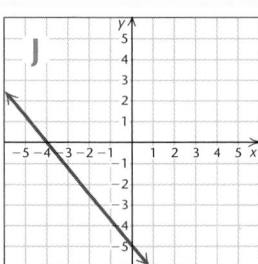

J

a   For Exercises 1–4, find **(a)** the coordinates of the y-intercept and **(b)** the coordinates of the x-intercept.

1.
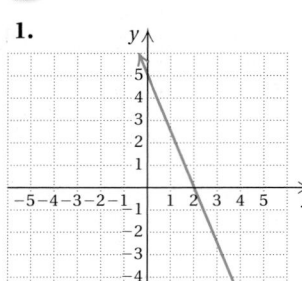

2.
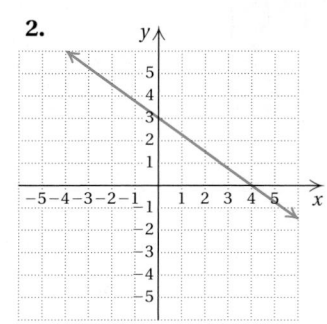

3.
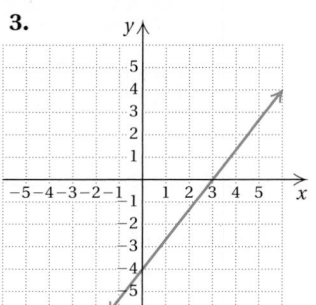

4.
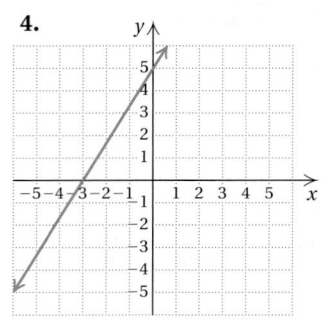

For Exercises 5–12, find **(a)** the coordinates of the y-intercept and **(b)** the coordinates of the x-intercept. Do not graph.

5. $3x + 5y = 15$

6. $5x + 2y = 20$

7. $7x - 2y = 28$

8. $3x - 4y = 24$

9. $-4x + 3y = 10$

10. $-2x + 3y = 7$

11. $6x - 3 = 9y$

12. $4y - 2 = 6x$

For each equation, find the intercepts. Then use the intercepts to graph the equation.

13. $x + 3y = 6$

x	y
0		← y-intercept
	0	← x-intercept

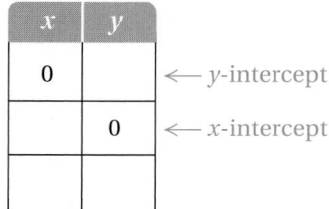

14. $x + 2y = 2$

x	y
0		← y-intercept
	0	← x-intercept

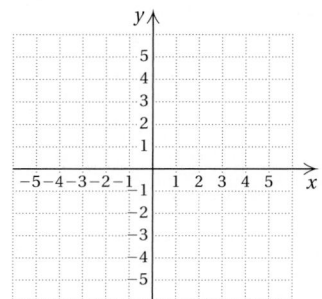

15. $-x + 2y = 4$

x	y
0		← y-intercept
	0	← x-intercept

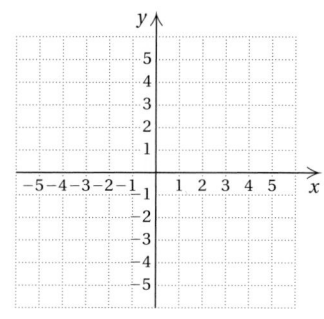

16. $-x + y = 5$

x	y
0		← y-intercept
	0	← x-intercept

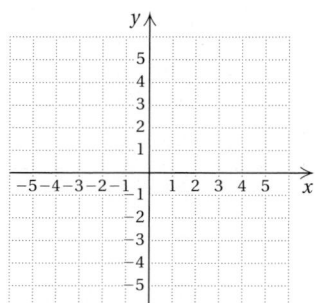

17. $3x + y = 6$

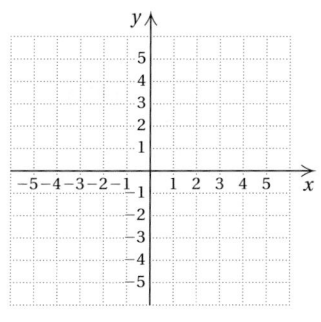

18. $2x + y = 6$

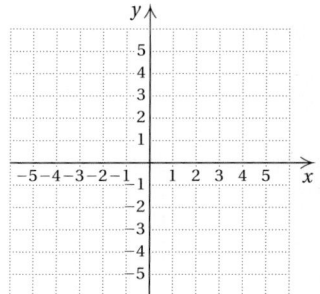

19. $2y - 2 = 6x$

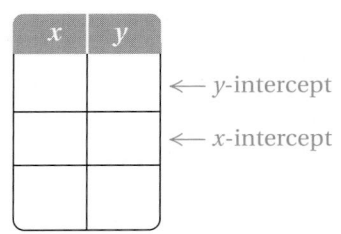

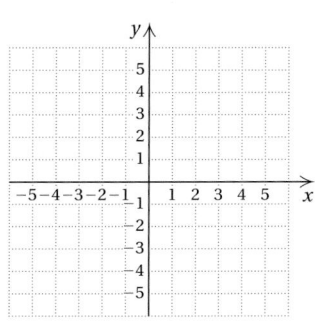

20. $3y - 6 = 9x$

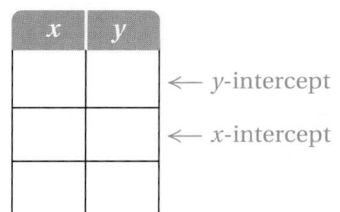

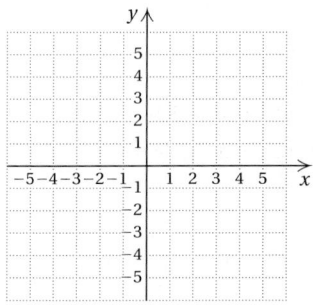

21. $3x - 9 = 3y$

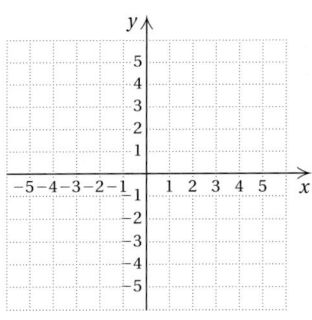

22. $5x - 10 = 5y$

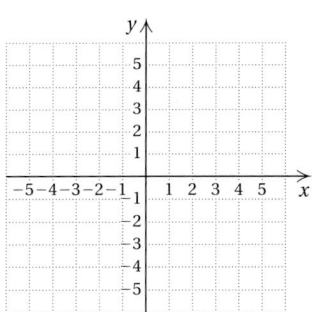

23. $2x - 3y = 6$

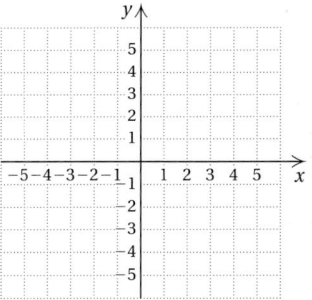

24. $2x - 5y = 10$

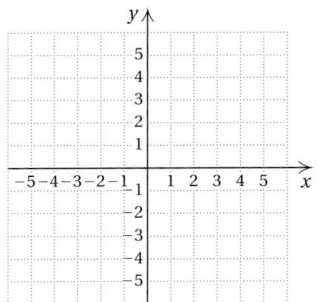

25. $4x + 5y = 20$

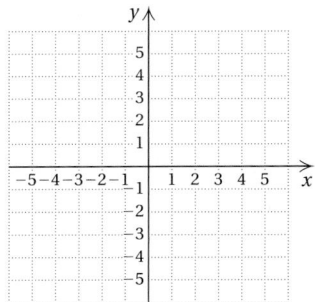

26. $2x + 6y = 12$

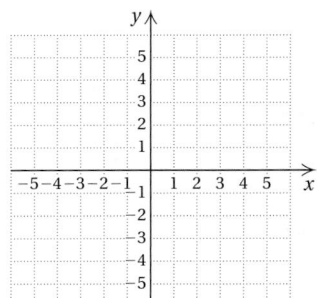

27. $2x + 3y = 8$

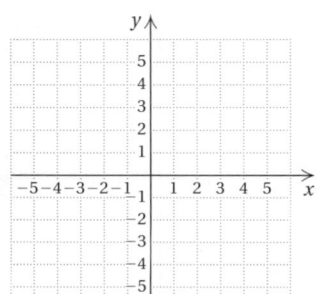

28. $x - 1 = y$

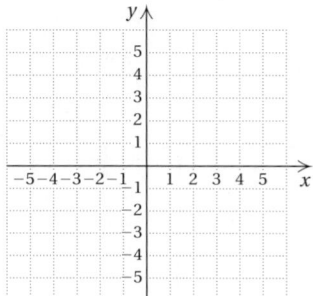

29. $3x + 4y = 5$

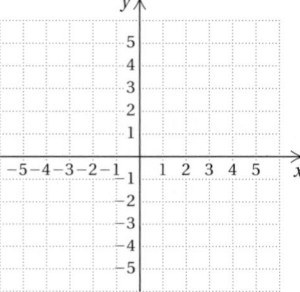

30. $2x - 1 = y$

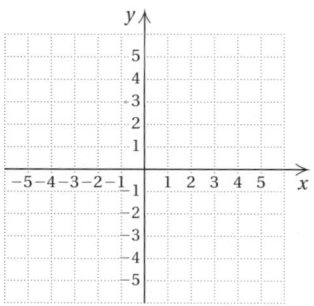

31. $3x - 2 = y$

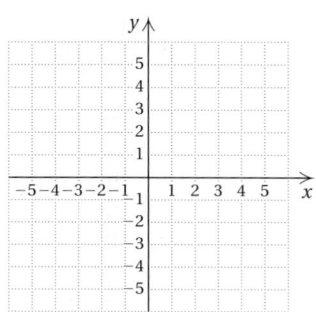

32. $4x - 3y = 12$

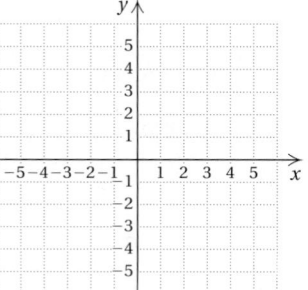

33. $6x - 2y = 12$

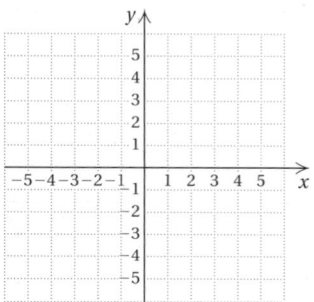

34. $7x + 2y = 6$

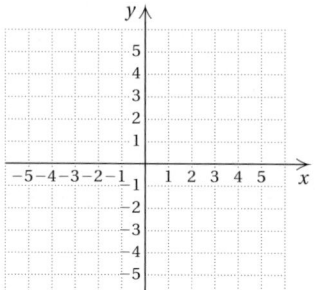

35. $y = -3 - 3x$

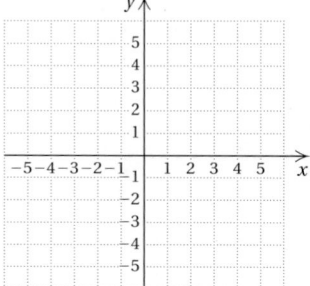

36. $-3x = 6y - 2$

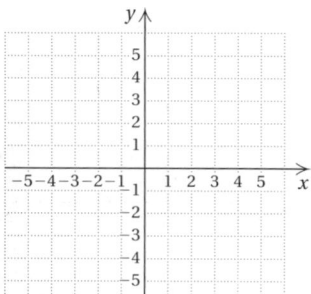

37. $y - 3x = 0$

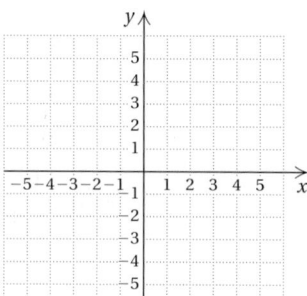

38. $x + 2y = 0$

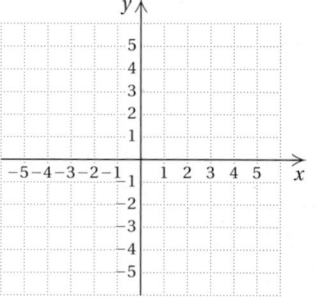

39. $x = -2$

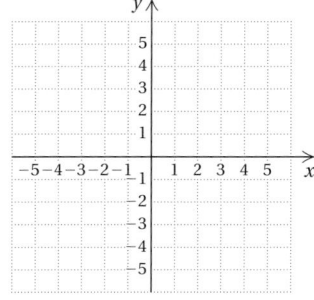

40. $x = 1$

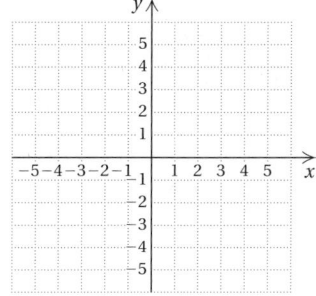

41. $y = 2$

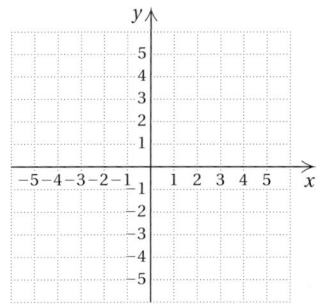
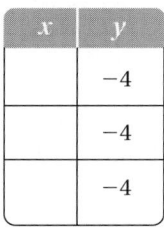

42. $y = -4$

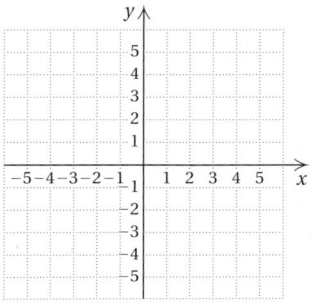

43. $x = 2$

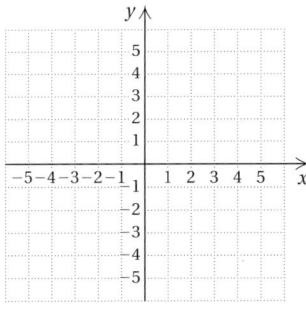

44. $x = 3$

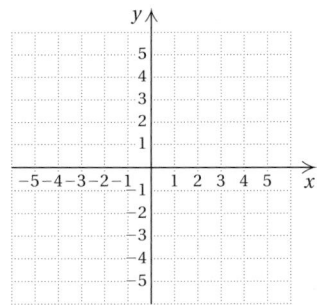

45. $y = 0$

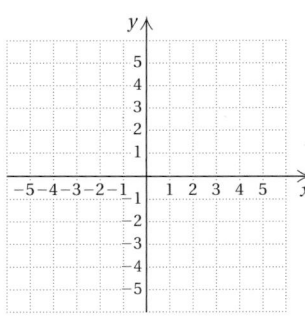

46. $y = -1$

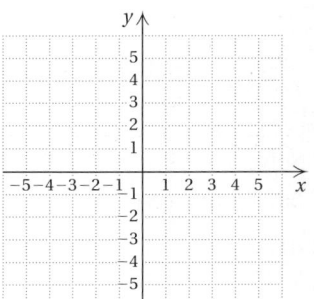

47. $x = \dfrac{3}{2}$

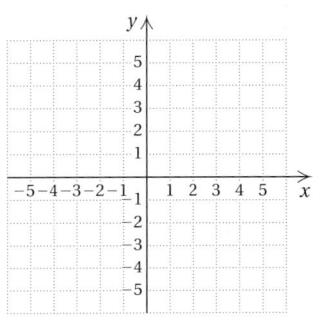

48. $x = -\dfrac{5}{2}$

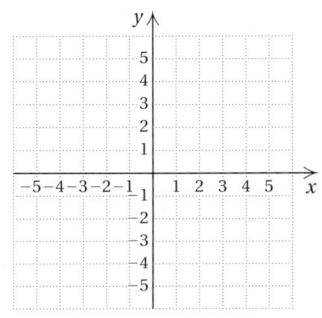

49. $3y = -5$

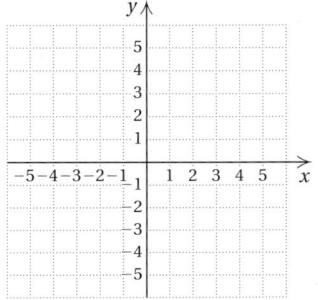

50. $12y = 45$

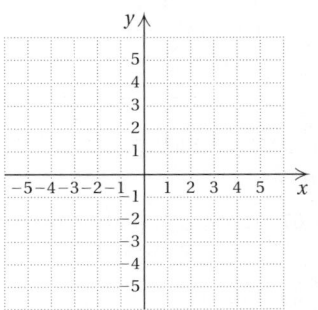

51. $4x + 3 = 0$

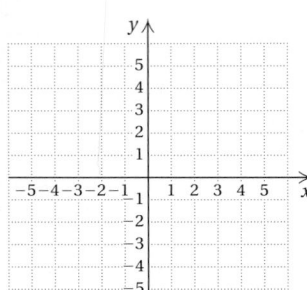

52. $-3x + 12 = 0$

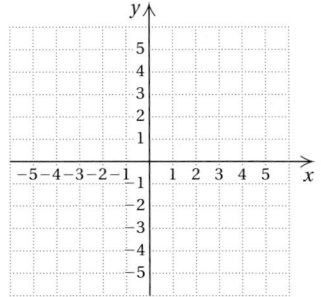

53. $48 - 3y = 0$

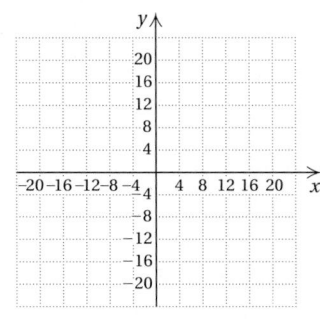

54. $63 + 7y = 0$

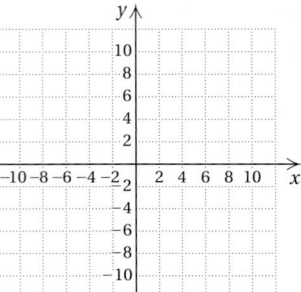

Write an equation for the graph shown.

55.

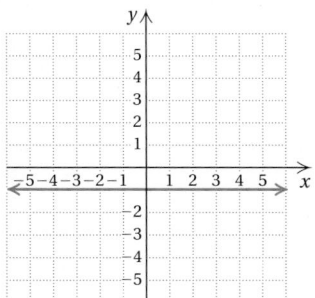

56.

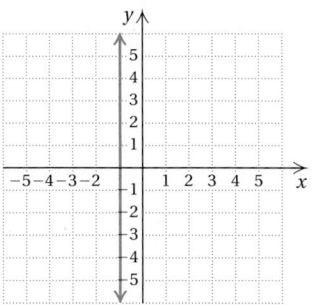

57.

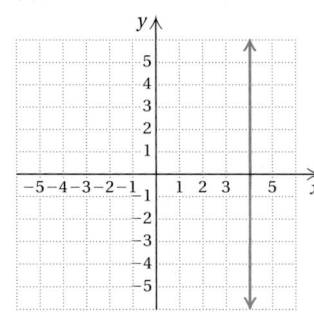

58.

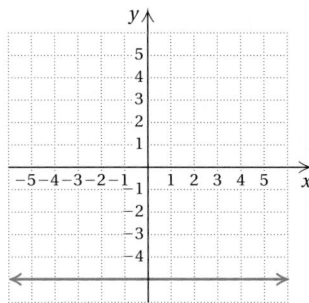

Skill Maintenance

Solve.   [2.7e]

59. $-1.6x < 64$

60. $-12x - 71 \geq 13$

61. $x + (x - 1) < (x + 2) - (x + 1)$

62. $6 - 18x \leq 4 - 12x - 5x$

63. $\dfrac{2x}{7} - 4 \leq -2$

64. $\dfrac{1}{4} + \dfrac{x}{3} > \dfrac{7}{12}$

Solve.   [2.5a]

65. *Foreign-Born Residents.*   The population of Detroit is 1,027,974, and it is estimated that 8.7% are foreign-born. How many of the residents of Detroit are foreign-born?

Source: U.S. Census Bureau

66. *Food Expenditure.*   The average American family spends 7% of its income on food. The Wilsons spent $3024 on food in a year. Estimate their annual income.

Source: "Going Hungry in America," by Peter Jaret. *AARP Bulletin*, Spring 2008

Synthesis

67. Write an equation of a line parallel to the *x*-axis and passing through $(-3, -4)$.

68. Find the value of *m* such that the graph of $y = mx + 6$ has an *x*-intercept of $(2, 0)$.

69. Find the value of *k* such that the graph of $3x + k = 5y$ has an *x*-intercept of $(-4, 0)$.

70. Find the value of *k* such that the graph of $4x = k - 3y$ has a *y*-intercept of $(0, -8)$.

Mid-Chapter Review

Concept Reinforcement

Determine whether each statement is true or false.

_____ **1.** In Quadrant II, the first coordinate of all points is less than the second coordinate.   [3.1a]

_____ **2.** The y-intercept of the graph of $2 - y = 3x$ is $(0, -3)$.   [3.3a]

_____ **3.** The y-intercept of $Ax + By = C$, $B \neq 0$, is $\left(0, \dfrac{C}{B}\right)$.   [3.3a]

_____ **4.** Both coordinates of points in quadrant IV are negative.   [3.1a]

Guided Solutions

5. Given the graph of the line below, fill in the letters and numbers that create correct statements.   [3.3a]

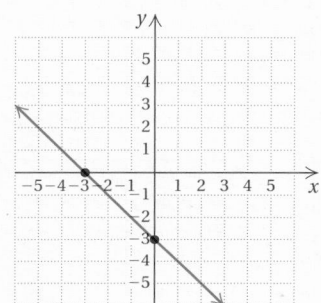

  a) The $\square$-intercept is $(\square, -3)$.
  b) The $\square$-intercept is $(\square, 0)$.

6. Given the graph of the line below, fill in the letters and numbers that create correct statements.   [3.3a]

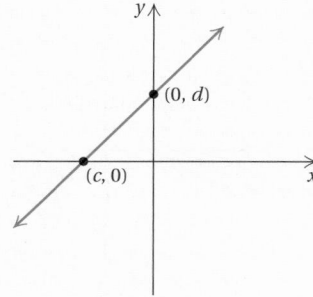

  a) The x-intercept is $(\square, \square)$.
  b) The y-intercept is $(\square, \square)$.

Mixed Review

7. Determine the coordinates of points $A, B, C, D,$ and E.   [3.1a]

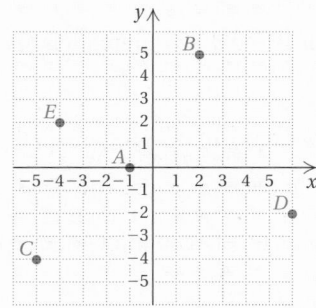

8. Determine the coordinates of points $F, G, H, I,$ and J.   [3.1a]

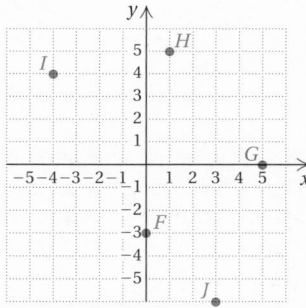

Determine whether the given ordered pair is a solution of the equation.   [3.1c]

9. $(8, -5); \; -2q - 7p = 19$

10. $\left(-1, \dfrac{2}{3}\right); \; 6y = -3x + 1$

Find the coordinates of the *x*-intercept and the *y*-intercept.   [3.3a]

11. $-3x + 2y = 18$

12. $x - \dfrac{1}{2} = 10y$

13. $x - 40 = 20y$

14. $\dfrac{1}{3}y - \dfrac{5}{6}x = 35$

Graph.   [3.2a], [3.3a, b]

15. $-2x + y = -3$

16. $y = -\dfrac{3}{2}$

17. $y = -x + 4$

18. $x = 0$

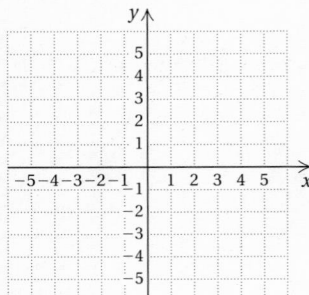
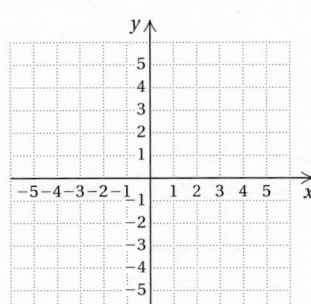
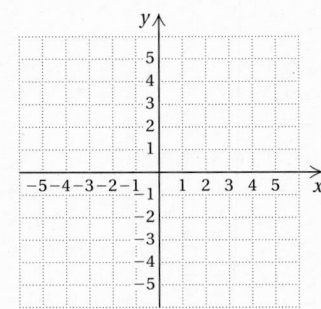
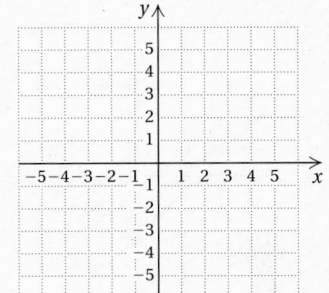

Match each equation with the characteristics listed at the right.   [3.3a, b]

19. $y = -1$

    A. The *y*-intercept is $(0, 1)$ and the *x*-intercept is $(-1, 0)$.

20. $x = 1$

    B. The *x*-intercept is $(-1, 0)$ and the *y*-intercept is $(0, -1)$.

21. $y = -x - 1$

    C. The line is vertical and the *x*-intercept is $(1, 0)$.

22. $y = x - 1$

    D. The line is horizontal and the *y*-intercept is $(0, -1)$.

23. $y = x + 1$

    E. The *y*-intercept is $(0, -1)$ and the *x*-intercept is $(1, 0)$.

Understanding Through Discussion and Writing

24. Do all graphs of linear equations have *y*-intercepts? Why or why not?   [3.3b]

25. The equations $3x + 4y = 8$ and $y = -\frac{3}{4}x + 2$ are equivalent. Which equation is easier to graph and why?   [3.2a]

26. If the graph of the equation $Ax + By = C$ is a horizontal line, what can you conclude about A? Why?   [3.3b]

27. Explain in your own words why the graph of $x = 7$ is a vertical line.   [3.3b]

3.4 Slope and Applications

a Slope

We have considered two forms of a linear equation,

$$Ax + By = C \quad \text{and} \quad y = mx + b.$$

We found that from the form of the equation $y = mx + b$, we know that the y-intercept of the line is $(0, b)$.

$$y = mx + b.$$

?←   The y-intercept is $(0, b)$.

What about the constant m? Does it give us information about the line? Look at the following graphs and see if you can make any connection between the constant m and the "slant" of the line.

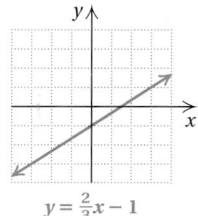
$y = \frac{2}{3}x - 1$

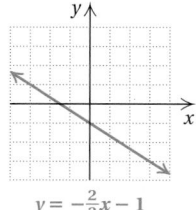
$y = -\frac{2}{3}x - 1$

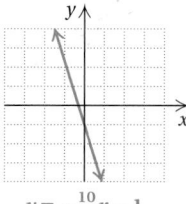
$y = -\frac{10}{3}x - 1$

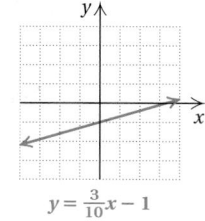
$y = \frac{3}{10}x - 1$

The graphs of some linear equations slant upward from left to right. Others slant downward. Some are vertical and some are horizontal. Some slant more steeply than others. We now look for a way to describe such possibilities with numbers.

Consider a line with two points marked P and Q. As we move from P to Q, the y-coordinate changes from 1 to 3 and the x-coordinate changes from 2 to 6. The change in y is $3 - 1$, or 2. The change in x is $6 - 2$, or 4.

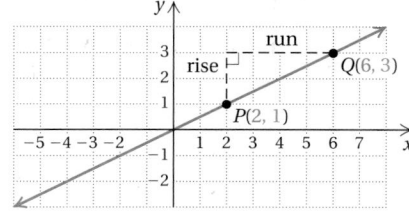

We call the change in y the **rise** and the change in x the **run**. The ratio rise/run is the same for any two points on a line. We call this ratio the **slope** of the line. Slope describes the slant of a line. The slope of the line in the graph above is given by

$$\frac{\text{rise}}{\text{run}} = \frac{\text{the change in } y}{\text{the change in } x}, \text{ or } \frac{2}{4}, \text{ or } \frac{1}{2}.$$

SLOPE

The **slope** of a line containing points (x_1, y_1) and (x_2, y_2) is given by

$$m = \frac{\text{rise}}{\text{run}} = \frac{\text{the change in } y}{\text{the change in } x} = \frac{y_2 - y_1}{x_2 - x_1}.$$

OBJECTIVES

a Given the coordinates of two points on a line, find the slope of the line, if it exists.

b Find the slope of a line from an equation.

c Find the slope, or rate of change, in an applied problem involving slope.

SKILL TO REVIEW
Objective 1.4a: Subtract real numbers.

Subtract.
1. $-4 - 20$
2. $-21 - (-5)$

STUDY TIPS

TUNE OUT DISTRACTIONS

Do you often study in noisy places? If there is constant noise in your home, dorm, or other study area, consider finding a quiet place in the library—preferably a spot that is away from the main traffic areas so that distractions are kept to a minimum.

Answers
Skill to Review:
1. -24 2. -16

In the preceding definition, (x_1, y_1) and (x_2, y_2)—read "x sub-one, y sub-one and x sub-two, y sub-two"—represent two different points on a line. It does not matter which point is considered (x_1, y_1) and which is considered (x_2, y_2) so long as coordinates are subtracted in the same order in both the numerator and the denominator:

$$\frac{y_2 - y_1}{x_2 - x_1} = \frac{y_1 - y_2}{x_1 - x_2}.$$

EXAMPLE 1 Graph the line containing the points $(-4, 3)$ and $(2, -6)$ and find the slope.

The graph is shown below. We consider (x_1, y_1) to be $(-4, 3)$ and (x_2, y_2) to be $(2, -6)$. From $(-4, 3)$ and $(2, -6)$, we see that the change in y, or the rise, is $-6 - 3$, or -9. The change in x, or the run, is $2 - (-4)$, or 6.

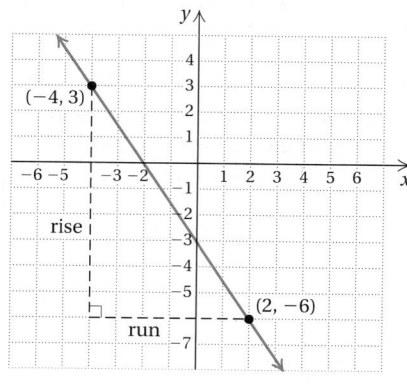

$$\text{Slope} = \frac{\text{rise}}{\text{run}} = \frac{\text{change in } y}{\text{change in } x}$$

$$= \frac{y_2 - y_1}{x_2 - x_1}$$

$$= \frac{-6 - 3}{2 - (-4)}$$

$$= \frac{-9}{6} = -\frac{9}{6}, \text{ or } -\frac{3}{2}.$$

When we use the formula

$$m = \frac{y_2 - y_1}{x_2 - x_1},$$

we must remember to subtract the y-coordinates in the same order that we subtract the x-coordinates. Let's redo Example 1, where we consider (x_1, y_1) to be $(2, -6)$ and (x_2, y_2) to be $(-4, 3)$:

$$\text{Slope} = \frac{\text{change in } y}{\text{change in } x} = \frac{3 - (-6)}{-4 - 2} = \frac{9}{-6} = -\frac{9}{6} = -\frac{3}{2}.$$

Do Exercises 1 and 2.

The slope of a line tells how it slants. A line with positive slope slants up from left to right. The larger the slope, the steeper the slant. A line with negative slope slants downward from left to right.

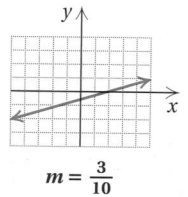

$m = \frac{3}{10}$

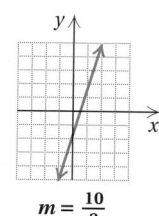

$m = \frac{10}{3}$

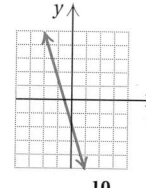

$m = -\frac{10}{3}$

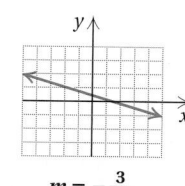

$m = -\frac{3}{10}$

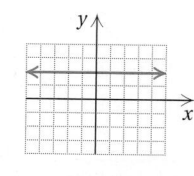

$m = 0$

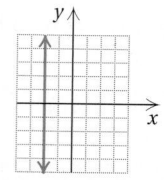

m **is not defined.**

Later in this section, in Examples 7 and 8, we will discuss the slope of a horizontal line and of a vertical line.

Graph the line containing the points and find the slope in two different ways.

1. $(-2, 3)$ and $(3, 5)$

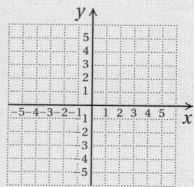

2. $(0, -3)$ and $(-3, 2)$

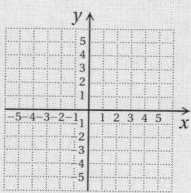

Answers

Answers to Margin Exercises
1 and 2 are on p. 207.

b Finding the Slope from an Equation

It is possible to find the slope of a line from its equation. Let's consider the equation $y = 2x + 3$, which is in the form $y = mx + b$. The graph of this equation is shown at left. We can find two points by choosing convenient values for x—say, 0 and 1—and substituting to find the corresponding y-values. We find the two points on the line to be $(0, 3)$ and $(1, 5)$. The slope of the line is found using the definition of slope:

$$m = \frac{\text{change in } y}{\text{change in } x} = \frac{5 - 3}{1 - 0} = \frac{2}{1} = 2.$$

The slope is 2. Note that this is also the coefficient of the x-term in the equation $y = 2x + 3$.

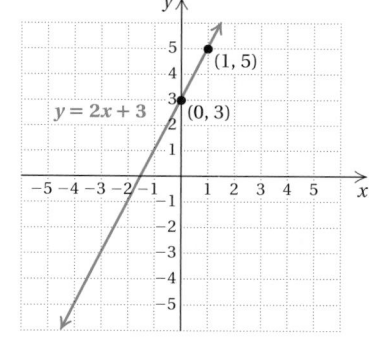

DETERMINING SLOPE FROM THE EQUATION $y = mx + b$

The slope of the line $y = mx + b$ is m. To find the slope of a nonvertical line, solve the linear equation in x and y for y and get the resulting equation in the form $y = mx + b$. The coefficient of the x-term, m, is the slope of the line.

EXAMPLES Find the slope of each line.

2. $y = -3x + \dfrac{2}{9}$

$\longrightarrow m = -3 = \text{Slope}$

3. $y = \dfrac{4}{5}x$

$\longrightarrow m = \dfrac{4}{5} = \text{Slope}$

4. $y = x + 6$

$\longrightarrow m = 1 = \text{Slope}$

5. $y = -0.6x - 3.5$

$\longrightarrow m = -0.6 = \text{Slope}$

> Do Exercises 3–6.

To find slope from an equation, we may need to first find an equivalent form of the equation.

EXAMPLE 6 Find the slope of the line $2x + 3y = 7$.

We solve for y to get the equation in the form $y = mx + b$:

$2x + 3y = 7$

$3y = -2x + 7$

$y = \dfrac{1}{3}(-2x + 7)$

$y = -\dfrac{2}{3}x + \dfrac{7}{3}.$ This is $y = mx + b$.

The slope is $-\frac{2}{3}$.

> Do Exercises 7 and 8.

Find the slope of each line.

3. $y = 4x + 11$

4. $y = -17x + 8$

5. $y = -x + \dfrac{1}{2}$

6. $y = \dfrac{2}{3}x - 1$

Find the slope of each line.

7. $4x + 4y = 7$

8. $5x - 4y = 8$

Answers

1. $\dfrac{2}{5}$ 2. $-\dfrac{5}{3}$

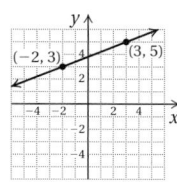   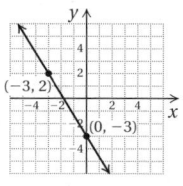

3. 4   4. -17 5. -1 6. $\dfrac{2}{3}$

7. -1 8. $\dfrac{5}{4}$

Calculator Corner

Visualizing Slope

Exercises: Graph each of the following sets of equations using the window settings $[-6, 6, -4, 4]$, with Xscl = 1 and Yscl = 1.

1. $y = x,\ y = 2x,$
   $y = 5x,\ y = 10x$
   What do you think the graph of $y = 123x$ will look like?

2. $y = x,\ y = \frac{3}{4}x,$
   $y = 0.38x,\ y = \frac{5}{32}x$
   What do you think the graph of $y = 0.000043x$ will look like?

3. $y = -x,\ y = -2x,$
   $y = -5x,\ y = -10x$
   What do you think the graph of $y = -123x$ will look like?

4. $y = -x,\ y = -\frac{3}{4}x,$
   $y = -0.38x,\ y = -\frac{5}{32}x$
   What do you think the graph of $y = -0.000043x$ will look like?

What about the slope of a horizontal line or a vertical line?

EXAMPLE 7 Find the slope of the line $y = 5$.

We can think of $y = 5$ as $y = 0x + 5$. Then from this equation, we see that $m = 0$. Consider the points $(-3, 5)$ and $(4, 5)$, which are on the line. The change in $y = 5 - 5$, or 0. The change in $x = -3 - 4$, or -7. We have

$$m = \frac{5 - 5}{-3 - 4}$$

$$= \frac{0}{-7}$$

$$= 0.$$

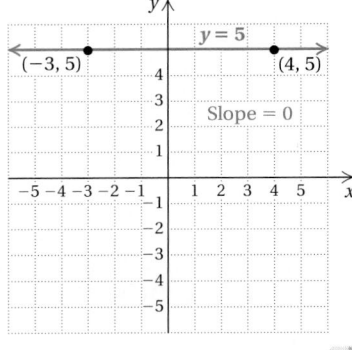

Any two points on a horizontal line have the same y-coordinate. The change in y is 0. Thus the slope of a horizontal line is 0.

EXAMPLE 8 Find the slope of the line $x = -4$.

Consider the points $(-4, 3)$ and $(-4, -2)$, which are on the line. The change in $y = 3 - (-2)$, or 5. The change in $x = -4 - (-4)$, or 0. We have

$$m = \frac{3 - (-2)}{-4 - (-4)}$$

$$= \frac{5}{0}. \qquad \text{Not defined}$$

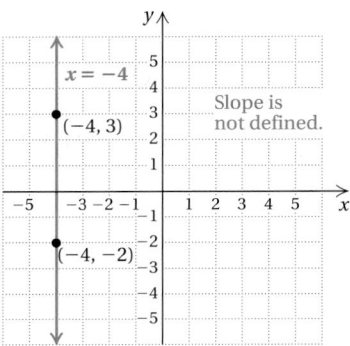

Since division by 0 is not defined, the slope of this line is not defined. The answer in this example is "The slope of this line is not defined."

SLOPE 0; SLOPE NOT DEFINED

The slope of a horizontal line is 0.

The slope of a vertical line is not defined.

Find the slope, if it exists, of each line.

9. $x = 7$

10. $y = -5$

Do Exercises 9 and 10.

Answers

9. Not defined   10. 0

c Applications of Slope; Rates of Change

Slope has many real-world applications. For example, numbers like 2%, 3%, and 6% are often used to represent the *grade* of a road, a measure of how steep a road on a hill or mountain is. For example, a 3% grade $\left(3\% = \frac{3}{100}\right)$ means that for every horizontal distance of 100 ft, the road rises 3 ft, and a −3% grade means that for every horizontal distance of 100 ft, the road drops 3 ft. (Road signs do not include negative signs.) The concept of grade also occurs in skiing or snowboarding, where a 7% grade is considered very tame, but a 70% grade is considered extremely steep. And in cardiology, a physician may change the grade of a treadmill to measure its effect on heart rate (number of beats per minute).

Architects and carpenters use slope when designing and building stairs, ramps, or roof pitches. Another application occurs in hydrology. When a river flows, the strength or force of the river depends on how far the river falls vertically compared to how far it flows horizontally.

EXAMPLE 9 *Skiing.* Among the steepest skiable terrain in North America, the Headwall on Mt. Washington, in New Hampshire, drops 720 ft over a horizontal distance of 900 ft. Find the grade of the Headwall.

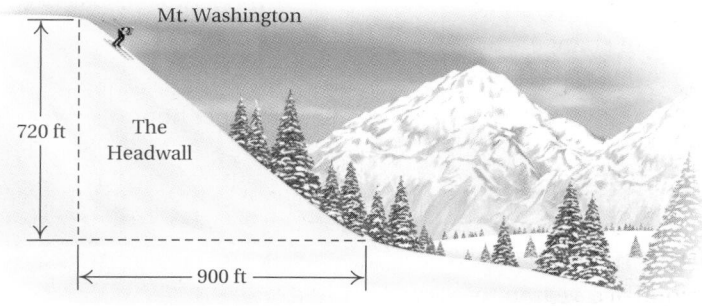

The grade of the Headwall is its slope, expressed as a percent:

$$m = \frac{720}{900} \xleftarrow{} \text{Vertical change}$$
$$ = \frac{720}{900} \xleftarrow{} \text{Horizontal change}$$

$$= \frac{8}{10} = 80\%.$$

Do Exercise 11.

11. **Construction.** Public buildings regularly include steps with 7-in. risers and 11-in. treads. Find the grade of such a stairway.

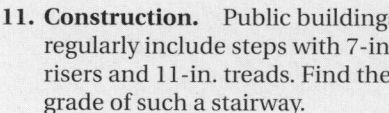

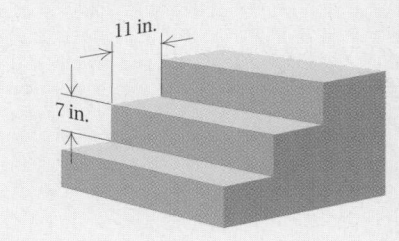

Answer

11. $63\frac{7}{11}\%$, or $63.\overline{63}\%$

Slope can also be considered as a **rate of change**.

EXAMPLE 10 *Masonry.* Jacob, an experienced mason, prepared a graph displaying data from a recent day's work. Use the graph to determine the slope, or the rate of change of the number of bricks he can lay with respect to time.

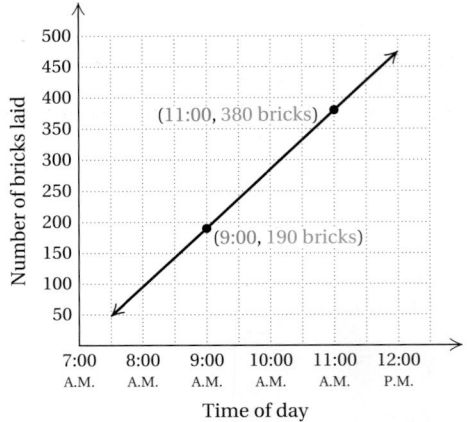

The vertical axis of the graph shows the number of bricks he has laid and the horizontal axis shows the time, in units of one hour. We can describe the rate of change of the number of bricks laid with respect to time as

$$\frac{\text{Bricks}}{\text{Hour}}, \quad \text{or} \quad \text{number of bricks laid per hour.}$$

This value is the slope of the line. We see two ordered pairs on the graph—in this case,

$$(9{:}00, 190 \text{ bricks}) \quad \text{and} \quad (11{:}00, 380 \text{ bricks}).$$

This tells us that in the 2 hr between 9:00 and 11:00, $380 - 190$, or 190, bricks were laid. Thus,

$$\text{Rate of change} = \frac{380 \text{ bricks} - 190 \text{ bricks}}{11{:}00 - 9{:}00}$$

$$= \frac{190 \text{ bricks}}{2 \text{ hours}} = 95 \text{ bricks per hour.}$$

Do Exercise 12.

12. Hair Cutting. Kiddie Kutters has a graph displaying data for a recent day's work. Use the graph to determine the slope, or rate of change of the number of haircuts with respect to time.

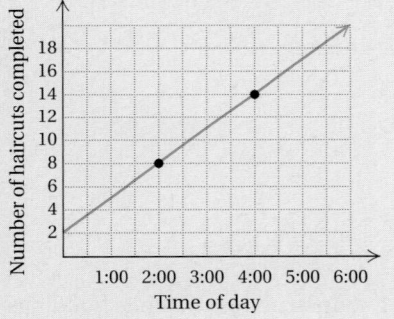

EXAMPLE 11 *Decreased Smoking.* Each year in the United States, the percent of the adult population who smoke declines. Use the graph at right to determine the slope, or rate of change of the percent of the adult population who smoke with respect to time.

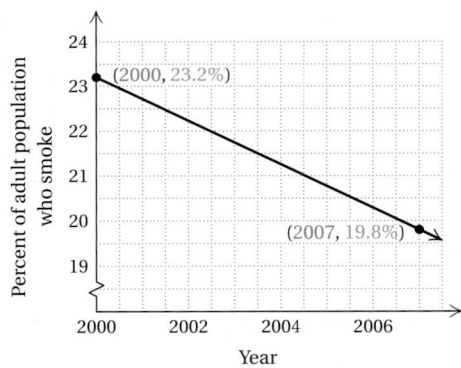

SOURCES: U.S. Centers for Disease Control and Prevention, National Center for Health Statistics; WebMD Health News

Answer

12. 3 haircuts per hour

The vertical axis of the graph shows the percent of the adult population who smoke and the horizontal axis shows the years. We can describe the rate of change of the percent who smoke with respect to time as

$$\frac{\text{Change in percent who smoke}}{\text{Years}}, \quad \text{or} \quad \text{change in percent who smoke per year.}$$

This value is the slope of the line. We determine two ordered pairs on the graph—in this case,

$$(2000, 23.2\%) \quad \text{and} \quad (2007, 19.8\%).$$

This tells us that in the 7 yr from 2000 to 2007, the percent dropped from 23.2% to 19.8%. Thus,

$$\text{Rate of change} = \frac{19.8\% - 23.2\%}{2007 - 2000} = \frac{-3.4\%}{7} \approx -0.5\% \text{ per year.}$$

Do Exercise 13.

13. Farms with Milk Cows. Use the graph below to determine the rate of change of the number of farms in the United States with milk cows.

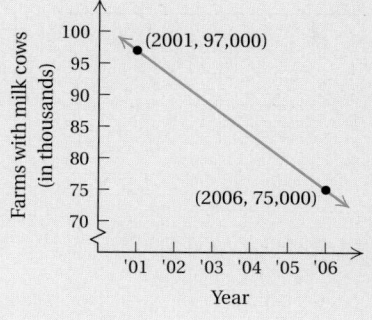

Answer

13. −4400 farms with milk cows per year

3.4 Exercise Set

a Find the slope, if it exists, of each line.

1.

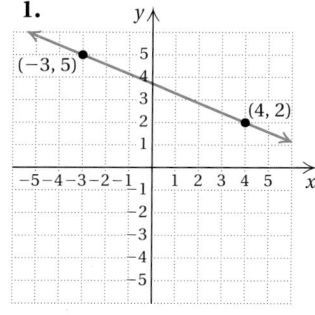

2.

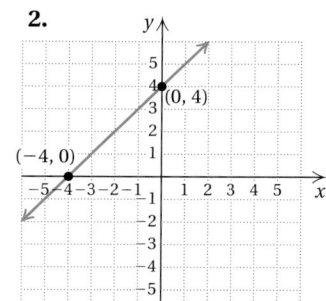

3.

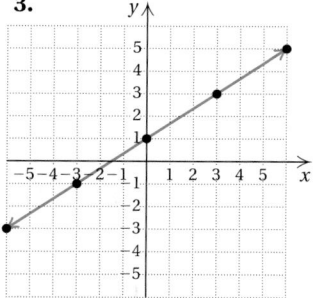

4.

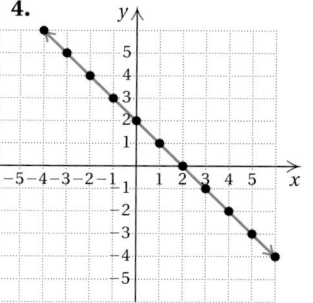

5.

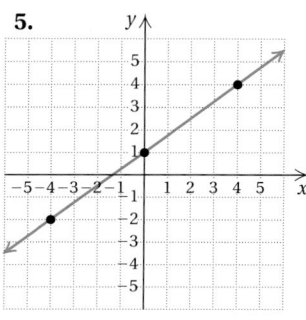

6.

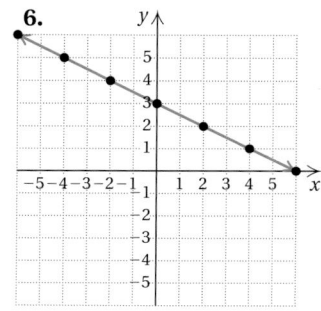

7.

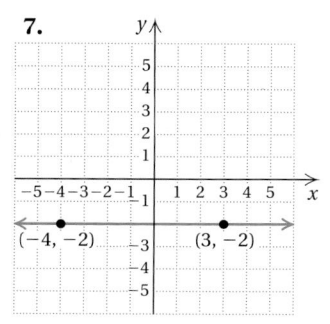

8.

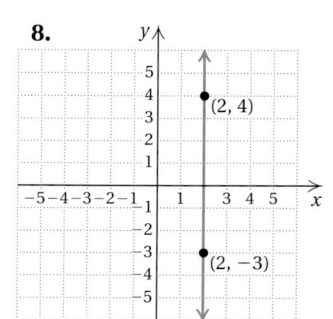

Graph the line containing the given pair of points and find the slope.

9. $(-2, 4), (3, 0)$

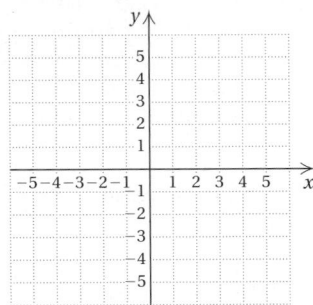

10. $(2, -4), (-3, 2)$

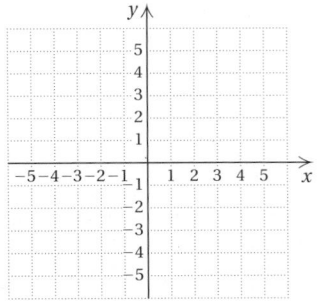

11. $(-4, 0), (-5, -3)$

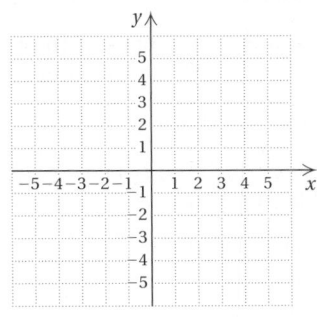

12. $(-3, 0), (-5, -2)$

13. $(-4, 1), (2, -3)$

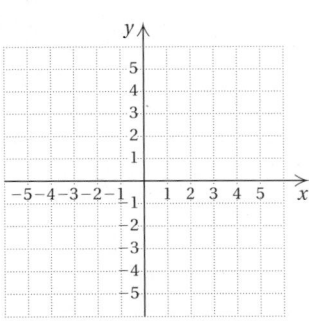

14. $(-3, 5), (4, -3)$

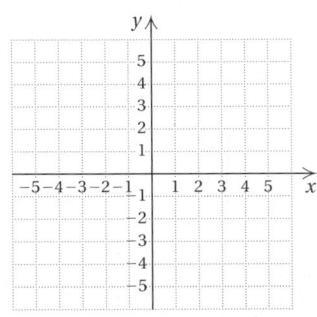

15. $(5, 3), (-3, -4)$

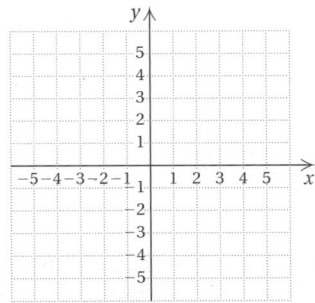

16. $(-4, -3), (2, 5)$

Find the slope, if it exists, of the line containing the given pair of points.

17. $\left(2, -\frac{1}{2}\right), \left(5, \frac{3}{2}\right)$

18. $\left(\frac{2}{3}, -1\right), \left(\frac{5}{3}, 2\right)$

19. $(4, -2), (4, 3)$

20. $(4, -3), (-2, -3)$

21. $(-11, 7), (15, -3)$

22. $(-13, 22), (8, -17)$

23. $\left(-\frac{1}{2}, \frac{3}{11}\right), \left(\frac{5}{4}, \frac{3}{11}\right)$

24. $(0.2, 4), (0.2, -0.04)$

b Find the slope, if it exists, of each line.

25. $y = -10x$

26. $y = \frac{10}{3}x$

27. $y = 3.78x - 4$

28. $y = -\frac{3}{5}x + 28$

29. $3x - y = 4$

30. $-2x + y = 8$

31. $x + 5y = 10$

32. $x - 4y = 8$

33. $3x + 2y = 6$

34. $2x - 4y = 8$

35. $x = \frac{2}{15}$

36. $y = -\frac{1}{3}$

37. $y = 2 - x$

38. $y = \frac{3}{4} + x$

39. $9x = 3y + 5$

40. $4y = 9x - 7$

41. $5x - 4y + 12 = 0$ **42.** $16 + 2x - 8y = 0$ **43.** $y = 4$ **44.** $x = -3$

45. $x = \dfrac{3}{4}y - 2$ **46.** $3x - \dfrac{1}{5}y = -4$ **47.** $\dfrac{2}{3}y = -\dfrac{7}{4}x$ **48.** $-x = \dfrac{2}{11}y$

 c   In Exercises 49–52, find the slope (or rate of change).

49. Find the slope (or pitch) of the roof.

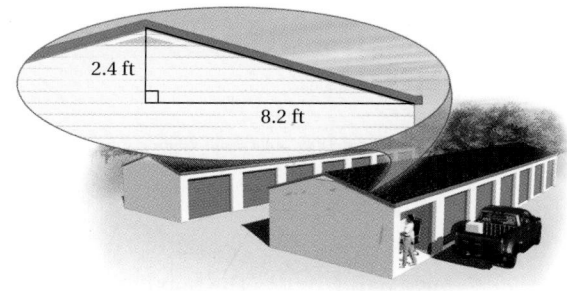

2.4 ft

8.2 ft

50. Find the slope (or grade) of the road.

920.58 m

13,740 m

51. Find the slope of the river.

56 ft

258 ft

52. Find the slope of the treadmill.

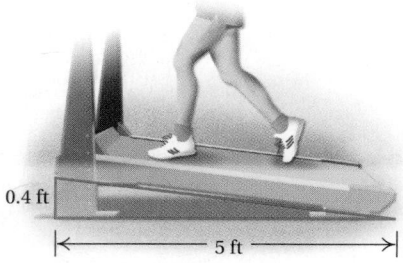

0.4 ft

5 ft

53. *Grade of Transit System.*   The maximum grade allowed between two stations in a rapid-transit rail system is 3.5%. Between station A and station B, which are 280 ft apart, the tracks rise $8\frac{1}{2}$ ft. What is the grade of the tracks between these two stations? Round the answer to the nearest tenth of a percent. Does this grade meet the rapid-transit rail standards?

Source: Brian Burell, *Merriam Webster's Guide to Everyday Math,* Merriam-Webster, Inc., Springfield MA

54. *Slope of Long's Peak.*   From a base elevation of 9600 ft, Long's Peak in Colorado rises to a summit elevation of 14,255 ft over a horizontal distance of 15,840 ft. Find the grade of Long's Peak.

In Exercises 55–58, use the graph to calculate a rate of change in which the units of the horizontal axis are used in the denominator.

55. *Farmland.* The amount of farmland in the United States, in millions of acres, is represented in the following graph. Find the rate of change, rounded to the nearest ten thousand, of the number of acres with respect to time.

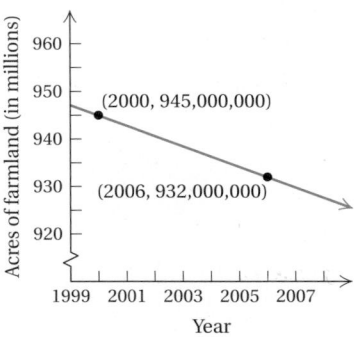

SOURCE: U.S. Department of Agriculture

56. *Movie Cost.* The cost of movie tickets for a family of four is represented in the following graph. Find the rate of change, rounded to the nearest cent, of the cost of four tickets with respect to time.

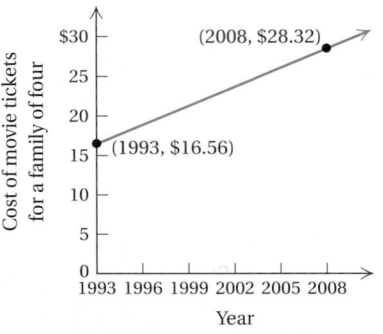

SOURCE: Motion Picture Association of America

57. *Population Growth of Nevada.* The population of Nevada is illustrated in the following graph. Find the rate of change, to the nearest hundred, of the population with respect to time.

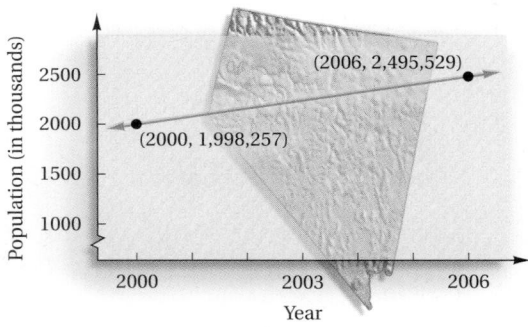

58. *Population Growth of Georgia.* The population of Georgia is illustrated in the following graph. Find the rate of change, to the nearest hundred, of the population with respect to time.

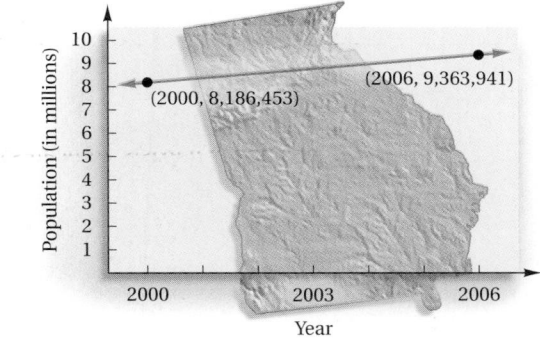

59. *Production of Blueberries.* U.S. production of blueberries is continually increasing. In 2004, 137,000 tons of blueberries were produced. By 2006, this amount had increased to 175,000 tons. Find the rate of change of the production of blueberries with respect to time.

Source: U.S. Department of Agriculture

60. *Manufacturing Jobs.* Employment in manufacturing in the United States has declined for decades. In 1960, approximately 28.5% of all jobs were in manufacturing. By 2007, this number had fallen to 10.1% of all jobs. Find the rate of change, rounded to the nearest tenth of a percent, of the percentage of jobs in manufacturing with respect to time.

Source: U.S. Department of Labor

Skill Maintenance

In each of Exercises 61–68, fill in the blank with the correct term from the given list. Some of the choices may not be used.

61. Equations with the same solutions are called
_____.   [2.1b]

62. The _____ for equations asserts that when we subtract the same number on both sides of an equation, we get an equivalent equation.   [2.1b]

63. The _____ for equations asserts that when we multiply or divide by the same nonzero number on both sides of an equation, we get an equivalent equation.   [2.2a]

64. _____ lines are graphs of equations of the type $y = b$.   [3.3b]

65. _____ lines are graphs of equations of the type $x = a$.   [3.3b]

66. The _____ of a line is a number that indicates how the line slants.   [3.4a]

67. The _____ of a line, if it exists, indicates where the line crosses the x-axis.   [3.3a]

68. The _____ of a line, if it exists, indicates where the line crosses the y-axis.   [3.3a]

vertical

horizontal

variable

addition principle

multiplication principle

coefficient

equivalent equations

slope

x-intercept

y-intercept

Synthesis

In Exercises 69–72, find an equation for the graph shown.

69.
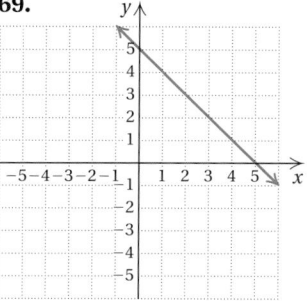

70.
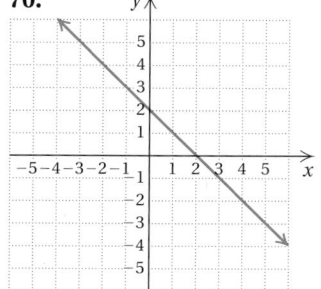

71.
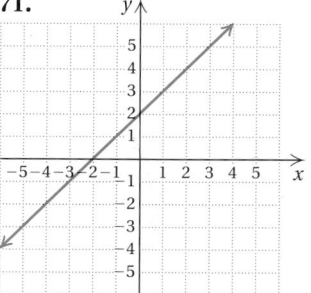

72.
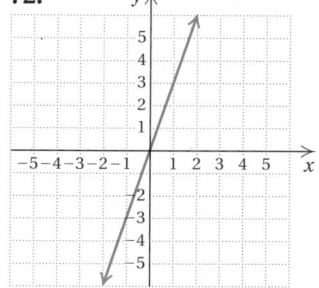

73. A line has slope $-\frac{2}{5}$ and passes through the points $(10, -2)$ and $(-20, x)$. Find x.   [3.4a]

74. A line has slope -6 and passes through the points $\left(\frac{1}{6}, x\right)$ and $(0, -2)$. Find x.   [3.4a]

Summary and Review

Key Terms, Properties, and Formulas

axes, p. 174
origin, p. 174
coordinates, p. 174
first coordinate, p. 174
abscissa, p. 174
second coordinate, p. 174
ordinate, p. 174

ordered pairs, p. 174
first quadrant, p. 175
second quadrant, p. 175
third quadrant, p. 175
fourth quadrant, p. 175
graph, p. 177
x-intercept, p. 192

y-intercept, p. 192
horizontal line, p. 195
vertical line, p. 195
rise, p. 205
run, p. 205
slope, p. 205
rate of change, p. 210

The y-intercept of $y = mx + b$: $(0, b)$

Horizontal line: $y = b$

Vertical line: $x = a$

Slope $= m = \dfrac{y_2 - y_1}{x_2 - x_1}$

Slope of horizontal line: 0

Slope of vertical line: Not defined

Concept Reinforcement

Determine whether each statement is true or false.

_____ **1.** The x- and y-intercepts of $y = mx$ are both $(0, 0)$. [3.3a]

_____ **2.** A slope of $-\frac{3}{4}$ is steeper than a slope of $-\frac{5}{2}$. [3.4a]

_____ **3.** The slope of the line that passes through (a, b) and (c, d) is $\dfrac{d - b}{c - a}$. [3.4a]

_____ **4.** The second coordinate of all points in quadrant III is negative. [3.1a]

_____ **5.** The x-intercept of $Ax + By = C$, $C \neq 0$, is $\left(\dfrac{A}{C}, 0\right)$. [3.3a]

_____ **6.** The slope of the line that passes through $(0, t)$ and $(-t, 0)$ is $\dfrac{1}{t}$. [3.4a]

Important Concepts

Objective 3.1b Find the coordinates of a point on a graph.

Example Find the coordinates of points Q, R, and S.

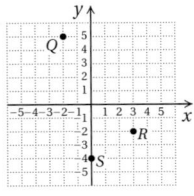

Point Q is 2 units to the left and 5 units up. Its coordinates are $(-2, 5)$.

Point R is 3 units to the right and 2 units down. Its coordinates are $(3, -2)$.

Point S is 0 units to the left or right and 4 units down. Its coordinates are $(0, -4)$.

Practice Exercise

1. Find the coordinates of points F, G, and H.

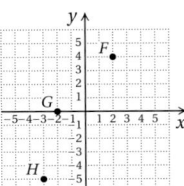

In which quadrant is each point located? [3.1a]

7. $(3, -8)$ **8.** $(-20, -14)$ **9.** $(4.9, 1.3)$

Determine whether each ordered pair is a solution of $2y - x = 10$. [3.1c]

10. $(2, -6)$ **11.** $(0, 5)$

12. Show that the ordered pairs $(0, -3)$ and $(2, 1)$ are solutions of the equation $2x - y = 3$. Then use the graph of the equation to determine another solution. Answers may vary. [3.1c]

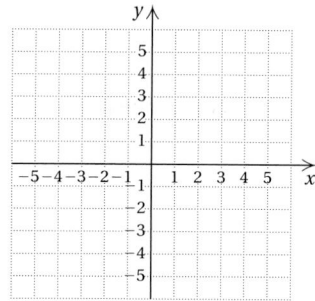

Graph each equation, identifying the y-intercept. [3.2a]

13. $y = 2x - 5$

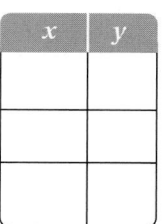
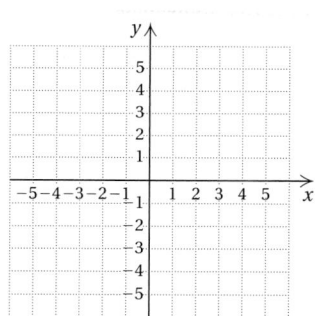

14. $y = -\dfrac{3}{4}x$

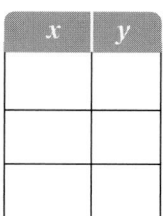
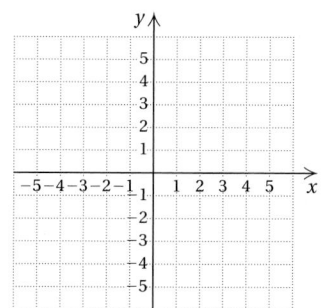

15. $y = -x + 4$

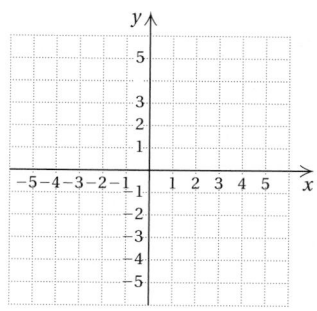

16. $y = 3 - 4x$

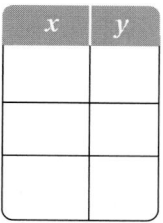
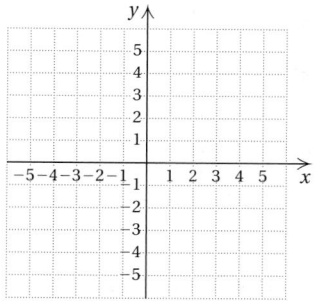

Graph each equation. [3.3b]

17. $y = 3$

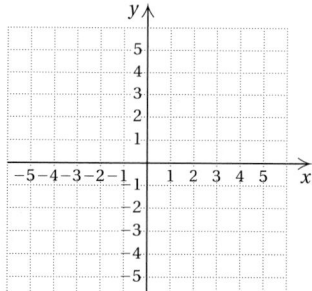

18. $5x - 4 = 0$

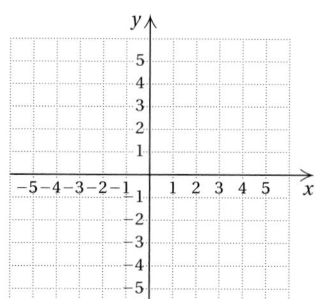

Find the intercepts of each equation. Then graph the equation.   [3.3a]

19. $x - 2y = 6$

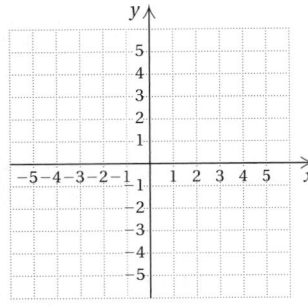

20. $5x - 2y = 10$

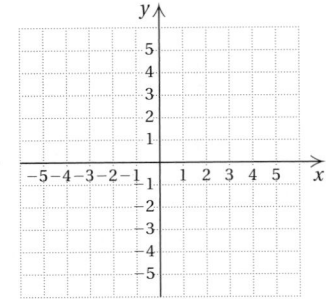

Solve.   [3.2b]

21. *Kitchen Design.*   Kitchen designers recommend that a refrigerator be selected on the basis of the number of people n in the household. The appropriate size S, in cubic feet, is given by

$$S = \frac{3}{2}n + 13.$$

a) Determine the recommended size of a refrigerator if the number of people is 1, 2, 5, and 10.

b) Graph the equation and use the graph to estimate the recommended size of a refrigerator for 4 people sharing an apartment.

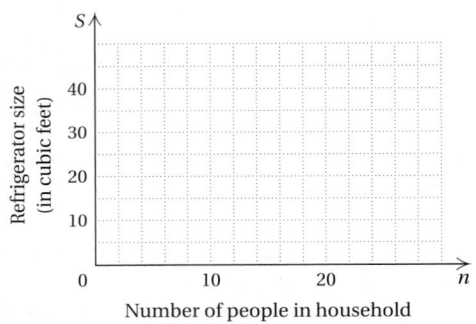

c) A refrigerator is 22 ft³. For how many residents is it the recommended size?

22. *Snow Removal.*   By 3:00 P.M., Erin had plowed 7 driveways and by 5:30 P.M., she had completed 13.

a) Find Erin's plowing rate, in number of driveways per hour.   [3.4c]

b) Find Erin's plowing rate, in minutes per driveway.   [3.4c]

23. *Manicures.*   The following graph shows data from a recent day's work at the O'Hara School of Cosmetology. What is the rate of change, in number of manicures per hour?   [3.4c]

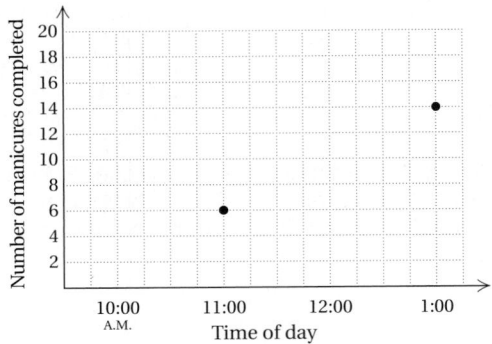

Find the slope.   [3.4a]

24.

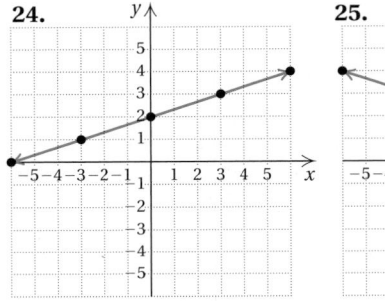

25.

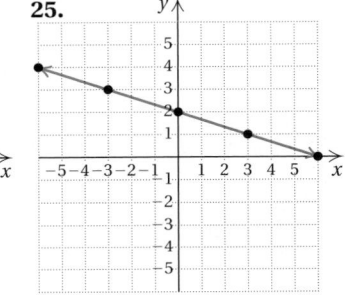

Graph the line containing the given pair of points and find the slope.   [3.4a]

26. $(-5, -2), (5, 4)$

27. $(-5, 5), (4, -4)$

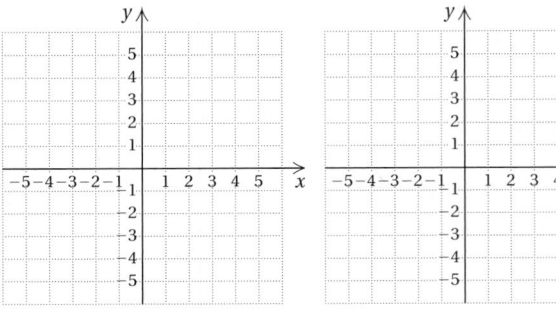

28. *Road Grade.* At one point, Beartooth Highway in Yellowstone National Park rises 315 ft over a horizontal distance of 4500 ft. Find the slope, or grade, of the road. [3.4c]

Find the slope, if it exists. [3.4b]

29. $y = -\dfrac{5}{8}x - 3$

30. $2x - 4y = 8$

31. $x = -2$

32. $y = 9$

33. $x - 10y = 20$

34. $6x - 5 = 4y$

35. Find the x-intercept of the graph of
$5x - y = -15$. [3.3a]

A. $(0, -3)$ **B.** $(-15, 0)$
C. $(0, 15)$ **D.** $(-3, 0)$

36. Find the slope, if it exists, of the line containing the points $(-8, 8)$ and $(8, -8)$ [3.4a]

A. 1 **B.** 0
C. -1 **D.** Not defined

Synthesis

37. Find the area and the perimeter of a rectangle for which $(-2, 2)$, $(7, 2)$, and $(7, -3)$ are three of the vertices. [3.1a]

38. *Gondola Aerial Lift.* In Telluride, Colorado, there is a free gondola ride that provides a spectacular view of the town and the surrounding mountains. The gondolas that begin in the town at an elevation of 8725 ft travel 5750 ft to Station St. Sophia, whose altitude is 10,550 ft. They then continue 3913 ft to Mountain Village, whose elevation is 9500 ft. [3.3c]

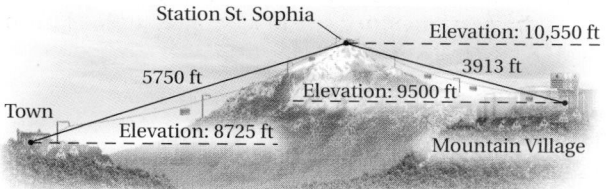

A visitor departs from the town at 11:55 A.M. and with no stop at Station St. Sophia reaches Mountain Village at 12:07 P.M.

a) Find the gondola's average rate of ascent and descent, in feet per minute.
b) Find the gondola's average rate of ascent and descent, in minutes per foot.

Understanding Through Discussion and Writing

1. Explain why the slant of a line with slope $\frac{5}{3}$ is steeper than the slant of a line with slope $\frac{4}{3}$. [3.4a]

2. Do all graphs of linear equations have x-intercepts? Explain. [3.3b]

3. Explain why the first coordinate of the y-intercept is always 0. [3.2a]

4. Explain why the graph of $y = -2$ is a horizontal line. [3.3b]

CHAPTER 3

Test

For Extra Help

CHAPTER Test Prep VIDEOS

Step-by-step test solutions are found on the Chapter Test Prep Videos available via the Video Resources on DVD, in *MyMathLab*, and on YouTube (search "BittingerIntroInter" and click on "Channels").

In which quadrant is each point located?

1. $\left(-\frac{1}{2}, 7\right)$

2. $(-5, -6)$

Find the coordinates of each point.

3. A

4. B

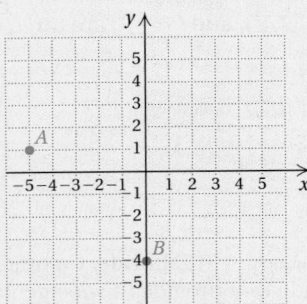

5. Show that the ordered pairs $(-4, -3)$ and $(-1, 3)$ are solutions of the equation $y - 2x = 5$. Then use the graph of the straight line containing the two points to determine another solution. Answers may vary.

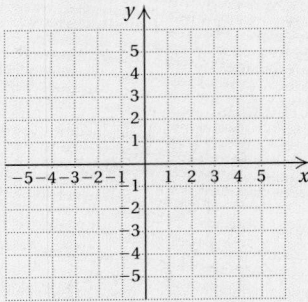

Graph each equation. Identify the y-intercept.

6. $y = 2x - 1$

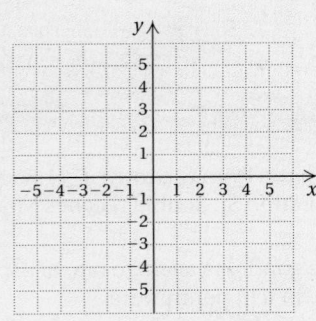

7. $y = -\frac{3}{2}x$

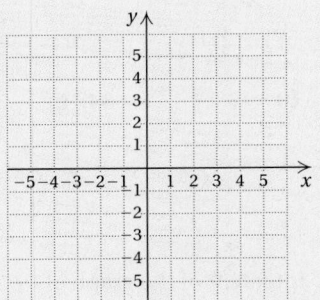

Find the intercepts of each equation. Then graph the equation.

8. $2x - 4y = -8$

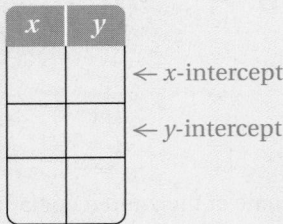
← x-intercept

← y-intercept

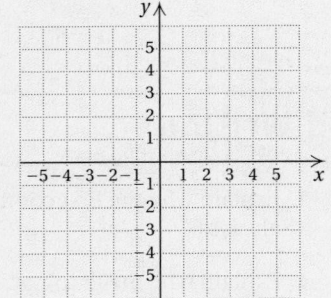

9. $2x - y = 3$

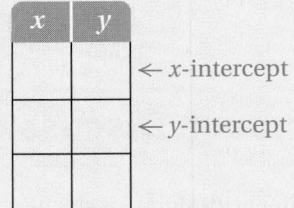
← x-intercept

← y-intercept

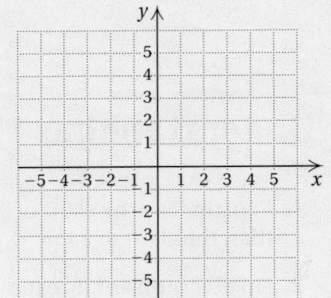

Graph each equation.

10. $2x + 8 = 0$

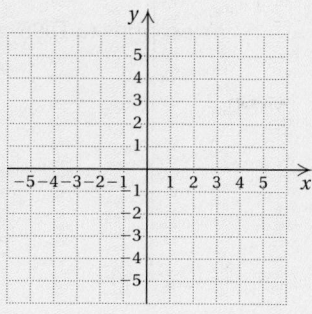

11. $y = 5$

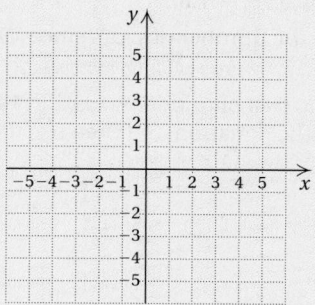

12. Find the slope.

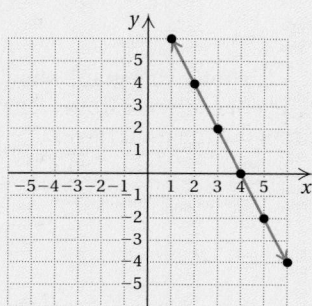

13. Graph the line containing $(-3, 1)$ and $(5, 4)$ and find the slope.

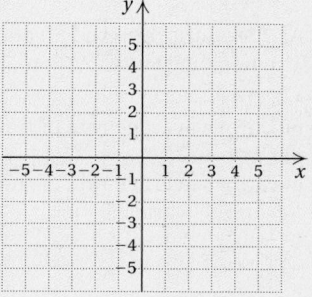

Find the slope, if it exists, of each line.

14. $2x - 5y = 10$

15. $x = -2$

16. $3y = \dfrac{1}{9}$

17. $y = -11x + 6$

18. *Navigation.* Capital Rapids drops 54 ft vertically over a horizontal distance of 1080 ft. What is the slope of the rapids?

19. *Private-College Costs.* The yearly cost T, in thousands of dollars, of tuition and required fees at a private college (includes two- and four-year schools and does not include room and board) can be approximated by

$$T = 0.7n + 7.8,$$

where n is the number of years since 1990. That is, $n = 0$ corresponds to 1990, $n = 5$ corresponds to 1995, and so on.

Source: *Statistical Abstract of the United States,* 2009

a) Find the cost of tuition in 1990, in 1996, in 2005, and in 2010.

b) Graph the equation and then use the graph to estimate the cost of tuition in 2015.

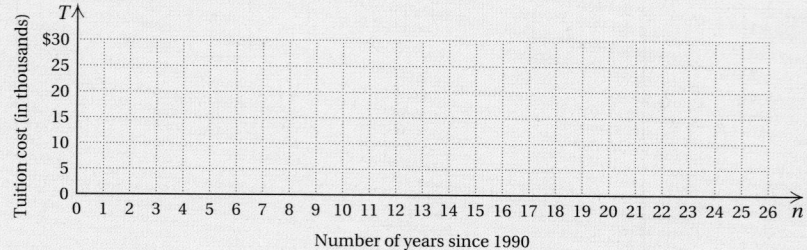

c) Predict the year in which the cost of tuition will be $28,800.

20. *Elevators.* At 2:38, Serge entered an elevator on the 34th floor of the Regency Hotel. At 2:40, he stepped off at the 5th floor.

a) Find the elevator's average rate of travel, in number of floors per minute.

b) Find the elevator's average rate of travel, in seconds per floor.

21. *Train Travel.* The following graph shows data concerning a recent train ride from Denver to Kansas City. At what rate did the train travel?

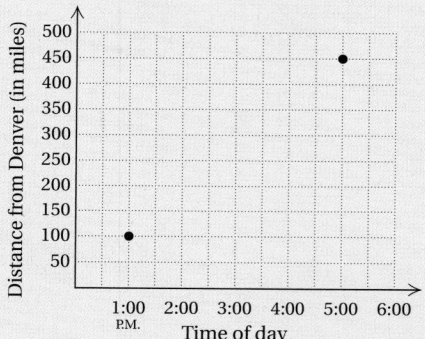

22. Choose the correct description of the line $6x - 1 = 3y + 2$.

A. The slope is -6, and the x-intercept is $\left(-\frac{1}{2}, 0\right)$.

B. The slope is 2, and the y-intercept is $(0, -1)$.

C. The slope is -2, and the x-intercept is $\left(\frac{1}{2}, 0\right)$.

D. The slope is 6, and the y-intercept is $(0, -3)$.

Synthesis

23. Write an equation of a line whose graph is parallel to the x-axis and 3 units above it.

24. A diagonal of a square connects the points $(-3, -1)$ and $(2, 4)$. Find the area and the perimeter of the square.

Cumulative Review

1. Evaluate $\dfrac{2m - n}{5}$ when $m = -1$ and $n = 2$.

2. Multiply: $-\dfrac{2}{3}(x - 6y + 3)$.

3. Factor: $18w - 24 + 9y$.

4. Find decimal notation: $-\dfrac{7}{9}$.

5. Find the absolute value: $\left| -2\dfrac{1}{5} \right|$.

6. Find the opposite of 8.17.

7. Find the reciprocal of $-\dfrac{8}{7}$.

8. Collect like terms: $2x - 5y + (-3x) + 4y$.

Simplify.

9. $-2.6 + (-0.4)$

10. $3 - [81 \div (1 + 2^3)]$

11. $\dfrac{5}{18} \div \left(-\dfrac{5}{12} \right)$

12. $6(x + 4) - 5[x - (2x - 3)]$

13. $\left(-\dfrac{1}{2} \right)(-1.1)(4.8)$

14. $-20 + 30 \div 10 \cdot 6$

Solve.

15. $\dfrac{4}{9}y = -36$

16. $-8 + w = w + 7$

17. $7.5 - 2x = 0.5x$

18. $4(x + 2) = 4(x - 2) + 16$

19. $2(x + 2) \geq 5(2x + 3)$

20. $x - \dfrac{5}{6} = \dfrac{1}{2}$

21. Find the slope, if it exists, of $9x - 12y = -3$.

22. Find the slope, if it exists, of $x = -\dfrac{15}{16}$.

23. Find the slope, if it exists, of $y - 3 = 7$.

24. Solve $7t = sz$ for s.

25. Solve $A = \dfrac{1}{2}h(b + c)$ for h.

26. In which quadrant is the point $(3, -1)$ located?

27. Find the intercepts of $2x - 7y = 21$. Do not graph.

28. Graph on the number line: $-1 < x \leq 2$.

$$\xleftarrow{} \overset{\displaystyle +\ +\ +\ +\ +\ +\ +\ +\ +}{\underset{-4\ -3\ -2\ -1\ \ 0\ \ 1\ \ 2\ \ 3\ \ 4}{}} \xrightarrow{}$$

Graph.

29. $2x + 5y = 10$

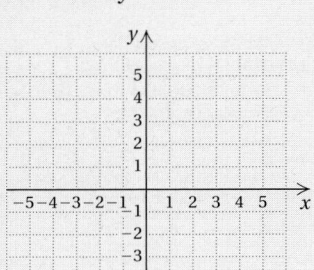

30. $y = -2$

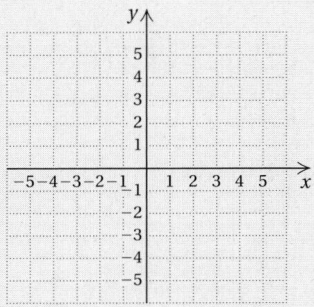

31. $y = -2x + 1$

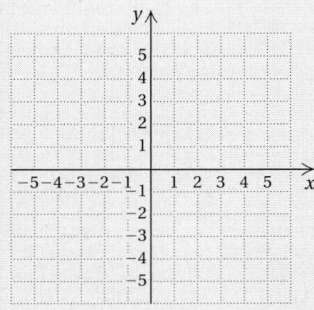

32. $3y + 6 = 2x$

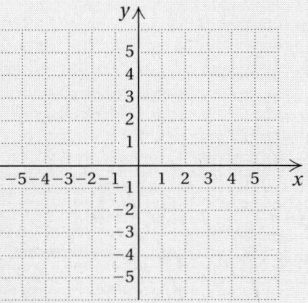

33. $y = -\dfrac{3}{2}x$

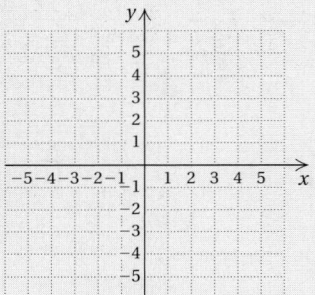

34. $x = 4.5$

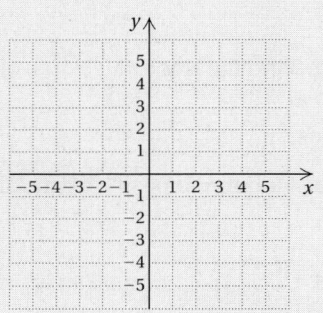

35. Find the slope of the line graphed below.

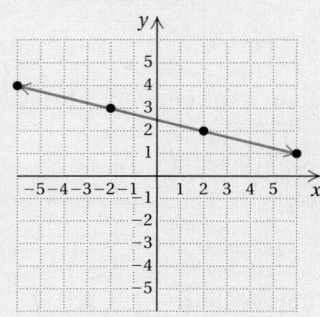

36. *Blood Types.*   There are 134.6 million Americans with either O-positive or O-negative blood. Those with O-positive blood outnumber those with O-negative blood by 94.2 million. How many Americans have O-negative blood?

Source: Stanford University School of Medicine

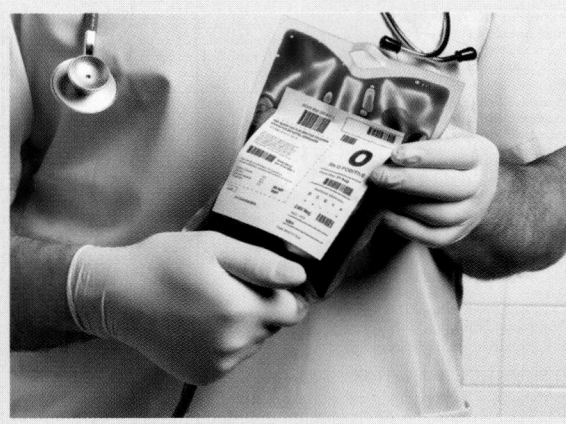

37. *Work Time.*   Cory's contract stipulates that he cannot work more than 40 hr per week. For the first four days of one week, he worked 7, 10, 9, and 6 hr. Determine as an inequality the number of hours he can work on the fifth day without violating his contract.

38. *Wire Cutting.*   A 143-m wire is cut into three pieces. The second piece is 3 m longer than the first. The third is four-fifths as long as the first. How long is each piece?

39. Compute and simplify: $1000 \div 100 \cdot 10 - 10$.
   A. 90
   B. 0
   C. -9
   D. -90

40. The slope of the line containing the points $(2, -7)$ and $(-4, 3)$ is which of the following?
   A. $-\dfrac{5}{2}$
   B. $-\dfrac{3}{5}$
   C. $-\dfrac{2}{5}$
   D. $-\dfrac{5}{3}$

Synthesis

Solve.

41. $4|x| - 13 = 3$

42. $\dfrac{2 + 5x}{4} = \dfrac{11}{28} + \dfrac{8x + 3}{7}$

43. $p = \dfrac{2}{m + Q}$, for Q

Polynomials: Operations

Real-World Application

About 22,750,000 young people, ages 18–29, voted in the 2008 presidential election. Convert the number 22,750,000 to scientific notation.

Source: Center for Information & Research on Civic Learning and Engagement, Tufts University

This problem appears as Exercise 64 in Section 4.2.

4.1

Integers as Exponents

OBJECTIVES

a Tell the meaning of exponential notation.

b Evaluate exponential expressions with exponents of O and 1.

c Evaluate algebraic expressions containing exponents.

d Use the product rule to multiply exponential expressions with like bases.

e Use the quotient rule to divide exponential expressions with like bases.

f Express an exponential expression involving negative exponents with positive exponents.

SKILL TO REVIEW

Objective 1.1a: Evaluate algebraic expressions by substitution.

1. Evaluate $6y$ when $y = 4$.

2. Evaluate $\dfrac{m}{n}$ when $m = 48$ and $n = 8$.

What is the meaning of each of the following?

1. 5^4 **2.** x^5

3. $(3t)^2$ **4.** $3t^2$

5. $(-x)^4$ **6.** $-y^3$

(a) Exponential Notation

An exponent of 2 or greater tells how many times the base is used as a factor. For example,

$$a \cdot a \cdot a \cdot a = a^4.$$

In this case, the **exponent** is 4 and the **base** is a. An expression for a power is called **exponential notation**.

$a^n \longleftarrow$ This is the exponent.

↑

This is the base.

EXAMPLE 1 What is the meaning of 3^5? of n^4? of $(2n)^3$? of $50x^2$? of $(-n)^3$? of $-n^3$?

3^5 means $3 \cdot 3 \cdot 3 \cdot 3 \cdot 3$; n^4 means $n \cdot n \cdot n \cdot n$;

$(2n)^3$ means $2n \cdot 2n \cdot 2n$; $50x^2$ means $50 \cdot x \cdot x$;

$(-n)^3$ means $(-n) \cdot (-n) \cdot (-n)$; $-n^3$ means $-1 \cdot n \cdot n \cdot n$

Do Margin Exercises 1–6.

We read exponential notation as follows: a^n is read the **nth power of a**, or simply **a to the nth**, or **a to the n**. We often read x^2 as "**x-squared**." The reason for this is that the area of a square of side x is $x \cdot x$, or x^2. We often read x^3 as "**x-cubed**." The reason for this is that the volume of a cube with length, width, and height x is $x \cdot x \cdot x$, or x^3.

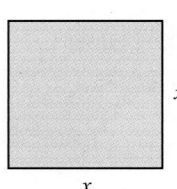
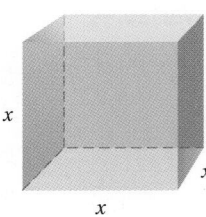

(b) One and Zero as Exponents

Look for a pattern in the following:

On each side, we **divide** by 8 at each step.	$8 \cdot 8 \cdot 8 \cdot 8 = 8^4$	On this side, the exponents **decrease** by 1 at each step.
	$8 \cdot 8 \cdot 8 = 8^3$	
	$8 \cdot 8 = 8^2$	
	$8 = 8^?$	
	$1 = 8^?.$	

To continue the pattern, we would say that

$$8 = 8^1 \quad \text{and} \quad 1 = 8^0.$$

Answers

Skill to Review:
1. 24   **2.** 6

Margin Exercises:
1. $5 \cdot 5 \cdot 5 \cdot 5$ **2.** $x \cdot x \cdot x \cdot x \cdot x$
3. $3t \cdot 3t$ **4.** $3 \cdot t \cdot t$
5. $(-x) \cdot (-x) \cdot (-x) \cdot (-x)$
6. $-1 \cdot y \cdot y \cdot y$

We make the following definition.

> ### EXPONENTS OF 0 AND 1
>
> $a^1 = a$, for any number a;
>
> $a^0 = 1$, for any nonzero number a

We consider 0^0 to be not defined. We will explain why later in this section.

EXAMPLE 2 Evaluate 5^1, $(-8)^1$, 3^0, $(-7.3)^0$, and $(186,892,046)^0$.

$$5^1 = 5; \quad (-8)^1 = -8; \quad 3^0 = 1;$$
$$(-7.3)^0 = 1; \quad (186,892,046)^0 = 1$$

Do Exercises 7–12.

Do Exercises 7–12.

Evaluate.

7. 6^1

8. 7^0

9. $(8.4)^1$

10. 8654^0

11. $(-1.4)^1$

12. 0^1

c Evaluating Algebraic Expressions

Algebraic expressions can involve exponential notation. For example, the following are algebraic expressions:

$$x^4, \quad (3x)^3 - 2, \quad a^2 + 2ab + b^2.$$

We evaluate algebraic expressions by replacing variables with numbers and following the rules for order of operations.

EXAMPLE 3 Evaluate $1000 - x^4$ when $x = 5$.

$$\begin{aligned}
1000 - x^4 &= 1000 - 5^4 &&\text{Substituting} \\
&= 1000 - 5 \cdot 5 \cdot 5 \cdot 5 \\
&= 1000 - 625 \\
&= 375
\end{aligned}$$

EXAMPLE 4 *Area of a Compact Disc.* The standard compact disc used for software and music has a radius of 6 cm. Find the area of such a CD (ignoring the hole in the middle).

$$\begin{aligned}
A &= \pi r^2 \\
&= \pi \cdot (6\,\text{cm})^2 \\
&= \pi \cdot 6\,\text{cm} \cdot 6\,\text{cm} \\
&\approx 3.14 \times 36\,\text{cm}^2 \\
&= 113.04\,\text{cm}^2
\end{aligned}$$

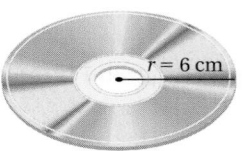
r = 6 cm

In Example 4, "cm^2" means "square centimeters" and "$\approx$" means "is approximately equal to."

EXAMPLE 5 Evaluate $(5x)^3$ when $x = -2$.

When we evaluate with a negative number, we often use extra parentheses to show the substitution.

$$\begin{aligned}
(5x)^3 &= [5 \cdot (-2)]^3 &&\text{Substituting} \\
&= [-10]^3 &&\text{Multiplying within brackets first} \\
&= [-10] \cdot [-10] \cdot [-10] \\
&= -1000 &&\text{Evaluating the power}
\end{aligned}$$

Answers

7. 6   **8.** 1   **9.** 8.4   **10.** 1
11. −1.4   **12.** 0

13. Evaluate t^3 when $t = 5$.

14. Evaluate $-5x^5$ when $x = -2$.

15. Find the area of a circle when $r = 32$ cm. Use 3.14 for π.

16. Evaluate $200 - a^4$ when $a = 3$.

17. Evaluate $t^1 - 4$ and $t^0 - 4$ when $t = 7$.

18. a) Evaluate $(4t)^2$ when $t = -3$.
    b) Evaluate $4t^2$ when $t = -3$.
    c) Determine whether $(4t)^2$ and $4t^2$ are equivalent.

EXAMPLE 6 Evaluate $5x^3$ when $x = -2$.

$$5x^3 = 5 \cdot (-2)^3 \qquad \text{Substituting}$$
$$= 5 \cdot (-2) \cdot (-2) \cdot (-2) \qquad \text{Evaluating the power first}$$
$$= 5(-8) \qquad (-2)(-2)(-2) = -8$$
$$= -40$$

Recall that two expressions are equivalent if they have the same value for all meaningful replacements. Note that Examples 5 and 6 show that $(5x)^3$ and $5x^3$ are *not* equivalent—that is, $(5x)^3 \neq 5x^3$.

Do Exercises 13–18.

(d) Multiplying Powers with Like Bases

There are several rules for manipulating exponential notation to obtain equivalent expressions. We first consider multiplying powers with like bases:

$$a^3 \cdot a^2 = \underbrace{(a \cdot a \cdot a)}_{3 \text{ factors}}\underbrace{(a \cdot a)}_{2 \text{ factors}} = \underbrace{a \cdot a \cdot a \cdot a \cdot a}_{5 \text{ factors}} = a^5.$$

Since an integer exponent greater than 1 tells how many times we use a base as a factor, then $(a \cdot a \cdot a)(a \cdot a) = a \cdot a \cdot a \cdot a \cdot a = a^5$ by the associative law. Note that the exponent in a^5 is the sum of those in $a^3 \cdot a^2$. That is, $3 + 2 = 5$. Likewise,

$$b^4 \cdot b^3 = (b \cdot b \cdot b \cdot b)(b \cdot b \cdot b) = b^7, \quad \text{where} \quad 4 + 3 = 7.$$

Adding the exponents gives the correct result.

> **THE PRODUCT RULE**
>
> For any number a and any positive integers m and n,
>
> $$a^m \cdot a^n = a^{m+n}.$$
>
> (When multiplying with exponential notation, if the bases are the same, keep the base and add the exponents.)

EXAMPLES Multiply and simplify.

Multiply and simplify.

19. $3^5 \cdot 3^5$

20. $x^4 \cdot x^6$

21. $p^4 p^{12} p^8$

22. $x \cdot x^4$

23. $(a^2 b^3)(a^7 b^5)$

7. $5^6 \cdot 5^2 = 5^{6+2}$ Adding exponents: $a^m \cdot a^n = a^{m+n}$
   $= 5^8$

8. $m^5 m^{10} m^3 = m^{5+10+3} = m^{18}$

9. $x \cdot x^8 = x^1 \cdot x^8$ Writing x as x^1
   $= x^{1+8}$
   $= x^9$

10. $(a^3 b^2)(a^3 b^5) = (a^3 a^3)(b^2 b^5)$
    $= a^6 b^7$

11. $(4y)^6 (4y)^3 = (4y)^{6+3} = (4y)^9$

Do Exercises 19–23.

Answers

13. 125   14. 160   15. 3215.36 cm²
16. 119   17. 3; −3   18. (a) 144; (b) 36;
(c) no   19. 3^{10} 20. x^{10} 21. p^{24}
22. x^5 23. $a^9 b^8$

e Dividing Powers with Like Bases

The following suggests a rule for dividing powers with like bases, such as a^5/a^2:

$$\frac{a^5}{a^2} = \frac{a \cdot a \cdot a \cdot a \cdot a}{a \cdot a} = \frac{a \cdot a \cdot a \cdot a \cdot a}{1 \cdot a \cdot a} = \frac{a \cdot a \cdot a}{1} \cdot \frac{a \cdot a}{a \cdot a}$$

$$= \frac{a \cdot a \cdot a}{1} \cdot 1 = a \cdot a \cdot a = a^3.$$

Note that the exponent in a^3 is the difference of those in $a^5 \div a^2$. That is, $5 - 2 = 3$. In a similar way, we have

$$\frac{t^9}{t^4} = \frac{t \cdot t \cdot t \cdot t \cdot t \cdot t \cdot t \cdot t \cdot t}{t \cdot t \cdot t \cdot t} = t^5, \quad \text{where} \quad 9 - 4 = 5.$$

Subtracting exponents gives the correct answer.

THE QUOTIENT RULE

For any nonzero number a and any positive integers m and n,

$$\frac{a^m}{a^n} = a^{m-n}.$$

(When dividing with exponential notation, if the bases are the same, keep the base and subtract the exponent of the denominator from the exponent of the numerator.)

EXAMPLES Divide and simplify.

12. $\dfrac{6^5}{6^3} = 6^{5-3}$ Subtracting exponents

$= 6^2$

13. $\dfrac{x^8}{x^1} = x^{8-1}$

$= x^7$

14. $\dfrac{(3t)^{12}}{(3t)^2} = (3t)^{12-2}$

$= (3t)^{10}$

15. $\dfrac{p^5 q^7}{p^2 q^5} = \dfrac{p^5}{p^2} \cdot \dfrac{q^7}{q^5} = p^{5-2} q^{7-5}$

$= p^3 q^2$

The quotient rule can also be used to explain the definition of 0 as an exponent. Consider the expression a^4/a^4, where a is nonzero:

$$\frac{a^4}{a^4} = \frac{a \cdot a \cdot a \cdot a}{a \cdot a \cdot a \cdot a} = 1.$$

This is true because the numerator and the denominator are the same. Now suppose we apply the rule for dividing powers with the same base:

$$\frac{a^4}{a^4} = a^{4-4} = a^0.$$

Since $a^4/a^4 = 1$ and $a^4/a^4 = a^0$, it follows that $a^0 = 1$, when $a \neq 0$.

We can explain why we do not define 0^0 using the quotient rule. We know that 0^0 is 0^{1-1}. But 0^{1-1} is also equal to $0^1/0^1$, or $0/0$. We have already seen that division by 0 is not defined, so 0^0 is also not defined.

Do Exercises 24–27.

Divide and simplify.

24. $\dfrac{4^5}{4^2}$

25. $\dfrac{y^6}{y^2}$

26. $\dfrac{p^{10}}{p}$

27. $\dfrac{a^7 b^6}{a^3 b^4}$

Answers

24. 4^3 **25.** y^4 **26.** p^9 **27.** $a^4 b^2$

f Negative Integers as Exponents

We can use the rule for dividing powers with like bases to lead us to a definition of exponential notation when the exponent is a negative integer. Consider $5^3/5^7$ and first simplify it using procedures we have learned for working with fractions:

$$\frac{5^3}{5^7} = \frac{5 \cdot 5 \cdot 5}{5 \cdot 5 \cdot 5 \cdot 5 \cdot 5 \cdot 5 \cdot 5} = \frac{5 \cdot 5 \cdot 5 \cdot 1}{5 \cdot 5 \cdot 5 \cdot 5 \cdot 5 \cdot 5 \cdot 5}$$

$$= \frac{5 \cdot 5 \cdot 5}{5 \cdot 5 \cdot 5} \cdot \frac{1}{5 \cdot 5 \cdot 5 \cdot 5} = \frac{1}{5^4}.$$

Now we apply the rule for dividing exponential expressions with the same bases. Then

$$\frac{5^3}{5^7} = 5^{3-7} = 5^{-4}.$$

From these two expressions for $5^3/5^7$, it follows that

$$5^{-4} = \frac{1}{5^4}.$$

This leads to our definition of negative exponents.

> ### NEGATIVE EXPONENT
>
> For any real number a that is nonzero and any integer n,
>
> $$a^{-n} = \frac{1}{a^n}.$$

In fact, the numbers a^n and a^{-n} are reciprocals because

$$a^n \cdot a^{-n} = a^n \cdot \frac{1}{a^n} = \frac{a^n}{a^n} = 1.$$

The following is another way to arrive at the definition of negative exponents.

On each side, we **divide** by 5 at each step.		On this side, the exponents **decrease** by 1 at each step.
	$5 \cdot 5 \cdot 5 \cdot 5 = 5^4$	
	$5 \cdot 5 \cdot 5 = 5^3$	
	$5 \cdot 5 = 5^2$	
	$5 = 5^1$	
	$1 = 5^0$	
	$\dfrac{1}{5} = 5^?$	
	$\dfrac{1}{25} = 5^?$	

To continue the pattern, it should follow that

$$\frac{1}{5} = \frac{1}{5^1} = 5^{-1} \quad \text{and} \quad \frac{1}{25} = \frac{1}{5^2} = 5^{-2}.$$

EXAMPLES Express using positive exponents. Then simplify.

16. $4^{-2} = \dfrac{1}{4^2} = \dfrac{1}{16}$

17. $(-3)^{-2} = \dfrac{1}{(-3)^2} = \dfrac{1}{(-3)(-3)} = \dfrac{1}{9}$

18. $m^{-3} = \dfrac{1}{m^3}$

19. $ab^{-1} = a\left(\dfrac{1}{b^1}\right) = a\left(\dfrac{1}{b}\right) = \dfrac{a}{b}$

20. $\dfrac{1}{x^{-3}} = x^{-(-3)} = x^3$

21. $3c^{-5} = 3\left(\dfrac{1}{c^5}\right) = \dfrac{3}{c^5}$

Example 20 might also be done as follows:

$$\dfrac{1}{x^{-3}} = \dfrac{1}{\dfrac{1}{x^3}} = 1 \cdot \dfrac{x^3}{1} = x^3.$$

-------- *Caution!* --------

As shown in Examples 16 and 17, a negative exponent does not necessarily mean that an expression is negative.

> Do Exercises 28–33.

The rules for multiplying and dividing powers with like bases hold when exponents are 0 or negative.

EXAMPLES Simplify. Write the result using positive exponents.

22. $7^{-3} \cdot 7^6 = 7^{-3+6}$ Adding exponents

$\qquad = 7^3$

23. $x^4 \cdot x^{-3} = x^{4+(-3)} = x^1 = x$

24. $\dfrac{5^4}{5^{-2}} = 5^{4-(-2)}$ Subtracting exponents

$\qquad = 5^{4+2} = 5^6$

25. $\dfrac{x}{x^7} = x^{1-7} = x^{-6} = \dfrac{1}{x^6}$

26. $\dfrac{b^{-4}}{b^{-5}} = b^{-4-(-5)}$

$\qquad = b^{-4+5} = b^1 = b$

27. $y^{-4} \cdot y^{-8} = y^{-4+(-8)}$

$\qquad = y^{-12} = \dfrac{1}{y^{12}}$

> Do Exercises 34–38.

The following is a summary of the definitions and rules for exponents that we have considered in this section.

DEFINITIONS AND RULES FOR EXPONENTS

1 as an exponent:	$a^1 = a$
0 as an exponent:	$a^0 = 1, a \neq 0$
Negative integers as exponents:	$a^{-n} = \dfrac{1}{a^n}, \dfrac{1}{a^{-n}} = a^n; a \neq 0$
Product Rule:	$a^m \cdot a^n = a^{m+n}$
Quotient Rule:	$\dfrac{a^m}{a^n} = a^{m-n}, a \neq 0$

Express with positive exponents. Then simplify.

28. 4^{-3} **29.** 5^{-2}

30. 2^{-4} **31.** $(-2)^{-3}$

32. $4p^{-3}$ **33.** $\dfrac{1}{x^{-2}}$

Simplify.

34. $5^{-2} \cdot 5^4$

35. $x^{-3} \cdot x^{-4}$

36. $\dfrac{7^{-2}}{7^3}$

37. $\dfrac{b^{-2}}{b^{-3}}$

38. $\dfrac{t}{t^{-5}}$

Answers

28. $\dfrac{1}{4^3} = \dfrac{1}{64}$ **29.** $\dfrac{1}{5^2} = \dfrac{1}{25}$ **30.** $\dfrac{1}{2^4} = \dfrac{1}{16}$
31. $\dfrac{1}{(-2)^3} = -\dfrac{1}{8}$ **32.** $\dfrac{4}{p^3}$ **33.** x^2
34. 5^2 **35.** $\dfrac{1}{x^7}$ **36.** $\dfrac{1}{7^5}$ **37.** b **38.** t^6

a What is the meaning of each of the following?

1. 3^4

2. 4^3

3. $(-1.1)^5$

4. $(87.2)^6$

5. $\left(\frac{2}{3}\right)^4$

6. $\left(-\frac{5}{8}\right)^3$

7. $(7p)^2$

8. $(11c)^3$

9. $8k^3$

10. $17x^2$

11. $-6y^4$

12. $-q^5$

b Evaluate.

13. $a^0, a \neq 0$

14. $t^0, t \neq 0$

15. b^1

16. c^1

17. $\left(\frac{2}{3}\right)^0$

18. $\left(-\frac{5}{8}\right)^0$

19. $(-7.03)^1$

20. $\left(\frac{4}{5}\right)^1$

21. 8.38^0

22. 8.38^1

23. $(ab)^1$

24. $(ab)^0, a, b \neq 0$

25. ab^0

26. ab^1

c Evaluate.

27. m^3, when $m = 3$

28. x^6, when $x = 2$

29. p^1, when $p = 19$

30. x^{19}, when $x = 0$

31. $-x^4$, when $x = -3$

32. $-2y^7$, when $x = 2$

33. x^4, when $x = 4$

34. y^{15}, when $y = 1$

35. $y^2 - 7$, when $y = -10$

36. $z^5 + 5$, when $z = -2$

37. $161 - b^2$, when $b = 5$

38. $325 - v^3$, when $v = -3$

39. $x^1 + 3$ and $x^0 + 3$, when $x = 7$

40. $y^0 - 8$ and $y^1 - 8$, when $y = -3$

41. Find the area of a circle when $r = 34$ ft. Use 3.14 for π.

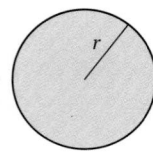

42. The area A of a square with sides of length s is given by $A = s^2$. Find the area of a square with sides of length 24 m.

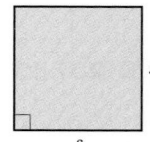

f Express using positive exponents. Then simplify.

43. 3^{-2}

44. 2^{-3}

45. 10^{-3}

46. 5^{-4}

47. a^{-3}

48. x^{-2} **49.** $\dfrac{1}{8^{-2}}$ **50.** $\dfrac{1}{2^{-5}}$ **51.** $\dfrac{1}{y^{-4}}$ **52.** $\dfrac{1}{t^{-7}}$

53. $5z^{-4}$ **54.** $6n^{-5}$ **55.** xy^{-2} **56.** ab^{-3}

Express using negative exponents.

57. $\dfrac{1}{4^3}$ **58.** $\dfrac{1}{5^2}$ **59.** $\dfrac{1}{x^3}$ **60.** $\dfrac{1}{y^2}$ **61.** $\dfrac{1}{a^5}$ **62.** $\dfrac{1}{b^7}$

d , **f** Multiply and simplify.

63. $2^4 \cdot 2^3$ **64.** $3^5 \cdot 3^2$ **65.** $8^5 \cdot 8^9$ **66.** $n^3 \cdot n^{20}$

67. $x^4 \cdot x$ **68.** $y \cdot y^9$ **69.** $9^{17} \cdot 9^{21}$ **70.** $t^0 \cdot t^{16}$

71. $(3y)^4(3y)^8$ **72.** $(2t)^8(2t)^{17}$ **73.** $(7y)^1(7y)^{16}$ **74.** $(8x)^0(8x)^1$

75. $3^{-5} \cdot 3^8$ **76.** $5^{-8} \cdot 5^9$ **77.** $x^{-2} \cdot x^2$ **78.** $x \cdot x^{-1}$

79. $x^{14} \cdot x^3$ **80.** $x^9 \cdot x^4$ **81.** $x^{-7} \cdot x^{-6}$ **82.** $y^{-5} \cdot y^{-8}$

83. $a^{11} \cdot a^{-3} \cdot a^{-18}$ **84.** $a^{-11} \cdot a^{-3} \cdot a^{-7}$ **85.** $(s^2t^3)(st^4)$ **86.** $(m^4n)(m^2n^7)$

(e), (f) Divide and simplify.

87. $\dfrac{7^5}{7^2}$

88. $\dfrac{5^8}{5^6}$

89. $\dfrac{y^9}{y}$

90. $\dfrac{x^{11}}{x}$

91. $\dfrac{16^2}{16^8}$

92. $\dfrac{7^2}{7^9}$

93. $\dfrac{m^6}{m^{12}}$

94. $\dfrac{a^3}{a^4}$

95. $\dfrac{(8x)^6}{(8x)^{10}}$

96. $\dfrac{(8t)^4}{(8t)^{11}}$

97. $\dfrac{(2y)^9}{(2y)^9}$

98. $\dfrac{(6y)^7}{(6y)^7}$

99. $\dfrac{x}{x^{-1}}$

100. $\dfrac{y^8}{y}$

101. $\dfrac{x^7}{x^{-2}}$

102. $\dfrac{t^8}{t^{-3}}$

103. $\dfrac{z^{-6}}{z^{-2}}$

104. $\dfrac{x^{-9}}{x^{-3}}$

105. $\dfrac{x^{-5}}{x^{-8}}$

106. $\dfrac{y^{-2}}{y^{-9}}$

107. $\dfrac{m^{-9}}{m^{-9}}$

108. $\dfrac{x^{-7}}{x^{-7}}$

109. $\dfrac{a^5 b^3}{a^2 b}$

110. $\dfrac{s^8 t^4}{s t^3}$

Matching. In Exercises 111 and 112, match each item in the first column with the appropriate item in the second column by drawing connecting lines. Items in the second column may be used more than once.

111.

5^2	$-\dfrac{1}{10}$
5^{-2}	$\dfrac{1}{10}$
$\left(\dfrac{1}{5}\right)^2$	$-\dfrac{1}{25}$
$\left(\dfrac{1}{5}\right)^{-2}$	10
-5^2	25
$(-5)^2$	-25
$-\left(-\dfrac{1}{5}\right)^2$	$\dfrac{1}{25}$
$\left(-\dfrac{1}{5}\right)^{-2}$	-10

112.

$-\left(\dfrac{1}{8}\right)^2$	16
$\left(\dfrac{1}{8}\right)^{-2}$	-16
8^{-2}	64
8^2	-64
-8^2	$\dfrac{1}{64}$
$(-8)^2$	$-\dfrac{1}{64}$
$\left(-\dfrac{1}{8}\right)^{-2}$	$-\dfrac{1}{16}$
$\left(-\dfrac{1}{8}\right)^2$	$\dfrac{1}{16}$

Skill Maintenance

Solve.   [2.6a]

113. *Cutting a Submarine Sandwich.*   A 12-in. submarine sandwich is cut into two pieces. One piece is twice as long as the other. How long are the pieces?

114. *Book Pages.*   The sum of the page numbers on the facing pages of a book is 457. Find the page numbers.

115. The perimeter of a rectangle is 640 ft. The length is 15 ft more than the width. Find the area of the rectangle.

116. The first angle of a triangle is 24° more than the second. The third angle is twice the first. Find the measures of the angles of the triangle.

Solve.   [2.3c]

117. $-6(2 - x) + 10(5x - 7) = 10$

118. $-10(x - 4) = 5(2x + 5) - 7$

Factor.   [1.7d]

119. $4x - 12 + 24y$

120. $256 - 2a - 4b$

Synthesis

Determine whether each of the following is correct.

121. $(x + 1)^2 = x^2 + 1$

122. $(x - 1)^2 = x^2 - 2x + 1$

123. $(5x)^0 = 5x^0$

124. $\dfrac{x^3}{x^5} = x^2$

Simplify.

125. $(y^{2x})(y^{3x})$

126. $a^{5k} \div a^{3k}$

127. $\dfrac{a^{6t}(a^{7t})}{a^{9t}}$

128. $\dfrac{\left(\frac{1}{2}\right)^4}{\left(\frac{1}{2}\right)^5}$

129. $\dfrac{(0.8)^5}{(0.8)^3(0.8)^2}$

130. $\dfrac{(x - 3)^5}{x - 3}$

Use >, <, or = for ☐ to write a true sentence.

131. 3^5 ☐ 3^4

132. 4^2 ☐ 4^3

133. 4^3 ☐ 5^3

134. 4^3 ☐ 3^4

Evaluate.

135. $\dfrac{1}{-z^4}$, when $z = -10$

136. $\dfrac{1}{-z^5}$, when $z = -0.1$

137. Determine whether $(a + b)^2$ and $a^2 + b^2$ are equivalent. (*Hint*: Choose values for a and b and evaluate.)

4.2

Exponents and Scientific Notation

OBJECTIVES

a Use the power rule to raise powers to powers.

b Raise a product to a power and a quotient to a power.

c Convert between scientific notation and decimal notation.

d Multiply and divide using scientific notation.

e Solve applied problems using scientific notation.

We now add to our ability to work with exponential expressions by considering three more rules. The rules are also applied to a new way to name numbers called *scientific notation*.

a Raising Powers to Powers

Consider an expression like $(3^2)^4$. We are raising 3^2 to the fourth power:

$$(3^2)^4 = (3^2)(3^2)(3^2)(3^2)$$
$$= (3 \cdot 3)(3 \cdot 3)(3 \cdot 3)(3 \cdot 3)$$
$$= 3 \cdot 3 \cdot 3 \cdot 3 \cdot 3 \cdot 3 \cdot 3 \cdot 3$$
$$= 3^8.$$

Note that in this case we could have multiplied the exponents:

$$(3^2)^4 = 3^{2 \cdot 4} = 3^8.$$

Likewise, $(y^8)^3 = (y^8)(y^8)(y^8) = y^{24}$. Once again, we get the same result if we multiply the exponents:

$$(y^8)^3 = y^{8 \cdot 3} = y^{24}.$$

THE POWER RULE

For any real number a and any integers m and n,

$$(a^m)^n = a^{mn}.$$

(To raise a power to a power, multiply the exponents.)

EXAMPLES Simplify. Express the answers using positive exponents.

1. $(3^5)^4 = 3^{5 \cdot 4}$ Multiplying
$= 3^{20}$ exponents

2. $(2^2)^5 = 2^{2 \cdot 5} = 2^{10}$

3. $(y^{-5})^7 = y^{-5 \cdot 7} = y^{-35} = \dfrac{1}{y^{35}}$

4. $(x^4)^{-2} = x^{4(-2)} = x^{-8} = \dfrac{1}{x^8}$

5. $(a^{-4})^{-6} = a^{(-4)(-6)} = a^{24}$

Do Exercises 1–4.

Simplify. Express the answers using positive exponents.

1. $(3^4)^5$

2. $(x^{-3})^4$

3. $(y^{-5})^{-3}$

4. $(x^4)^{-8}$

b Raising a Product or a Quotient to a Power

When an expression inside parentheses is raised to a power, the inside expression is the base. Let's compare $2a^3$ and $(2a)^3$:

$$2a^3 = 2 \cdot a \cdot a \cdot a;$$ The base is a.

$$(2a)^3 = (2a)(2a)(2a)$$ The base is $2a$.
$$= (2 \cdot 2 \cdot 2)(a \cdot a \cdot a)$$ Using the associative and commutative laws of multiplication to regroup the factors
$$= 2^3 a^3$$
$$= 8a^3.$$

Answers

1. 3^{20} 2. $\dfrac{1}{x^{12}}$ 3. y^{15} 4. $\dfrac{1}{x^{32}}$

We see that $2a^3$ and $(2a)^3$ are *not* equivalent. We also see that we can evaluate the power $(2a)^3$ by raising each factor to the power 3. This leads us to the following rule for raising a product to a power.

RAISING A PRODUCT TO A POWER

For any real numbers a and b and any integer n,

$$(ab)^n = a^n b^n.$$

(To raise a product to the nth power, raise each factor to the nth power.)

EXAMPLES Simplify.

6. $(4x^2)^3 = (4^1 x^2)^3$ Since $4 = 4^1$

$\qquad = (4^1)^3 \cdot (x^2)^3$ Raising *each* factor to the third power

$\qquad = 4^3 \cdot x^6 = 64x^6$

7. $(5x^3 y^5 z^2)^4 = 5^4 (x^3)^4 (y^5)^4 (z^2)^4$ Raising *each* factor to the fourth power

$\qquad = 625 x^{12} y^{20} z^8$

8. $(-5x^4 y^3)^3 = (-5)^3 (x^4)^3 (y^3)^3$

$\qquad = -125 x^{12} y^9$

9. $[(-x)^{25}]^2 = (-x)^{50}$ Using the power rule

$\qquad = (-1 \cdot x)^{50}$ Using the property of -1 (Section 1.8)

$\qquad = (-1)^{50} x^{50}$

$\qquad = 1 \cdot x^{50}$ The product of an even number of negative factors is positive.

$\qquad = x^{50}$

10. $(5x^2 y^{-2})^3 = 5^3 (x^2)^3 (y^{-2})^3 = 125 x^6 y^{-6}$ Be sure to raise *each* factor to the third power.

$$= \frac{125 x^6}{y^6}$$

11. $(3x^3 y^{-5} z^2)^4 = 3^4 (x^3)^4 (y^{-5})^4 (z^2)^4 = 81 x^{12} y^{-20} z^8 = \dfrac{81 x^{12} z^8}{y^{20}}$

12. $(-x^4)^{-3} = (-1 \cdot x^4)^{-3} = (-1)^{-3} \cdot x^{4(-3)} = (-1)^{-3} \cdot x^{-12}$

$$= \frac{1}{(-1)^3} \cdot \frac{1}{x^{12}} = \frac{1}{-1} \cdot \frac{1}{x^{12}} = -\frac{1}{x^{12}}$$

13. $(-2x^{-5} y^4)^{-4} = (-2)^{-4} (x^{-5})^{-4} (y^4)^{-4} = \dfrac{1}{(-2)^4} \cdot x^{20} \cdot y^{-16}$

$$= \frac{1}{16} \cdot x^{20} \cdot \frac{1}{y^{16}} = \frac{x^{20}}{16 y^{16}}$$

Do Exercises 5–11.

Simplify.

5. $(2x^5 y^{-3})^4$

6. $(5x^5 y^{-6} z^{-3})^2$

7. $[(-x)^{37}]^2$

8. $(3y^{-2} x^{-5} z^8)^3$

9. $(-y^8)^{-3}$

10. $(-2x^4)^{-2}$

11. $(-3x^2 y^{-5})^{-3}$

Answers

5. $\dfrac{16x^{20}}{y^{12}}$ **6.** $\dfrac{25x^{10}}{y^{12} z^6}$ **7.** x^{74} **8.** $\dfrac{27z^{24}}{y^6 x^{15}}$

9. $-\dfrac{1}{y^{24}}$ **10.** $\dfrac{1}{4x^8}$ **11.** $-\dfrac{y^{15}}{27x^6}$

There is a similar rule for raising a quotient to a power.

> **RAISING A QUOTIENT TO A POWER**
>
> For any real numbers a and b, $b \neq 0$, and any integer n,
>
> $$\left(\frac{a}{b}\right)^n = \frac{a^n}{b^n}.$$
>
> (To raise a quotient to the nth power, raise both the numerator and the denominator to the nth power.) Also,
>
> $$\left(\frac{a}{b}\right)^{-n} = \left(\frac{b}{a}\right)^n = \frac{b^n}{a^n}, \quad a \neq 0.$$

EXAMPLES Simplify.

14. $\left(\dfrac{x^2}{4}\right)^3 = \dfrac{(x^2)^3}{4^3} = \dfrac{x^6}{64}$

15. $\left(\dfrac{3a^4}{b^3}\right)^2 = \dfrac{(3a^4)^2}{(b^3)^2} = \dfrac{3^2(a^4)^2}{b^{3\cdot 2}} = \dfrac{9a^8}{b^6}$

16. $\left(\dfrac{y^2}{2z^{-5}}\right)^4 = \dfrac{(y^2)^4}{(2z^{-5})^4} = \dfrac{(y^2)^4}{2^4(z^{-5})^4} = \dfrac{y^8}{16z^{-20}} = \dfrac{y^8 z^{20}}{16}$

17. $\left(\dfrac{y^3}{5}\right)^{-2} = \dfrac{(y^3)^{-2}}{5^{-2}} = \dfrac{y^{-6}}{5^{-2}} = \dfrac{\dfrac{1}{y^6}}{\dfrac{1}{5^2}} = \dfrac{1}{y^6} \div \dfrac{1}{5^2} = \dfrac{1}{y^6} \cdot \dfrac{5^2}{1} = \dfrac{25}{y^6}$

Example 17 might also be done as follows:

$$\left(\frac{y^3}{5}\right)^{-2} = \left(\frac{5}{y^3}\right)^2 \qquad \left(\frac{a}{b}\right)^{-n} = \left(\frac{b}{a}\right)^n$$
$$= \frac{5^2}{(y^3)^2} = \frac{25}{y^6}.$$

Do Exercises 12–15.

Sidebar (left):

Simplify.

12. $\left(\dfrac{x^6}{5}\right)^2$

13. $\left(\dfrac{2t^5}{w^4}\right)^3$

14. $\left(\dfrac{a^4}{3b^{-2}}\right)^3$

15. $\left(\dfrac{x^4}{3}\right)^{-2}$
Do this two ways.

(c) Scientific Notation

There are many kinds of symbols, or notation, for numbers. You are already familiar with fraction notation, decimal notation, and percent notation. Now we study another, **scientific notation**, which makes use of exponential notation. Scientific notation is especially useful when calculations involve very large or very small numbers. The following are examples of scientific notation.

① *Niagara Falls*: On the Canadian side, the amount of water that spills over the falls in 1 day during the summer is about

$$4.9793 \times 10^{10} \text{ gal} = 49{,}793{,}000{,}000 \text{ gal.}$$

①

② *The mass of the earth:*

6.615×10^{21} tons = 6,615,000,000,000,000,000,000 tons.

③ *The mass of a hydrogen atom:*

1.7×10^{-24} g = 0.0000000000000000000000017 g.

②

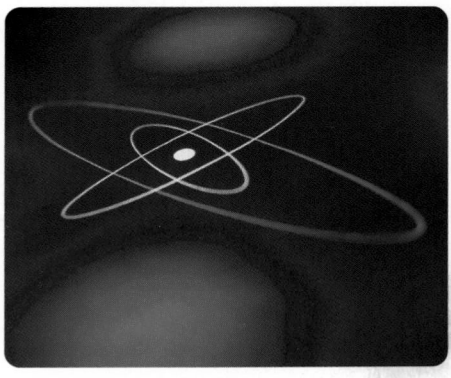

③

SCIENTIFIC NOTATION

Scientific notation for a number is an expression of the type

$$M \times 10^n,$$

where n is an integer, M is greater than or equal to 1 and less than 10 ($1 \le M < 10$), and M is expressed in decimal notation. 10^n is also considered to be scientific notation when $M = 1$.

You should try to make conversions to scientific notation mentally as much as possible. Here is a handy mental device.

A positive exponent in scientific notation indicates a large number (greater than or equal to 10) and a negative exponent indicates a small number (between 0 and 1).

EXAMPLES Convert to scientific notation.

18. $78,000 = 7.8 \times 10^4$

   7.8,000.
   4 places

   Large number, so the exponent is positive.

19. $0.0000057 = 5.7 \times 10^{-6}$

   0.000005.7
   6 places

   Small number, so the exponent is negative.

Do Exercises 16 and 17.

EXAMPLES Convert mentally to decimal notation.

20. $7.893 \times 10^5 = 789,300$

   7.89300.
   5 places

   Positive exponent, so the answer is a large number.

21. $4.7 \times 10^{-8} = 0.000000047$

   .00000004.7
   8 places

   Negative exponent, so the answer is a small number.

Convert to scientific notation.
16. 0.000517

17. 523,000,000

---------- *Caution!* ----------

Each of the following is *not* scientific notation.

$$12.46 \times 10^7$$

This number is greater than 10.

$$0.347 \times 10^{-5}$$

This number is less than 1.

Answers

16. 5.17×10^{-4} **17.** 5.23×10^8

Do Exercises 18 and 19.

Convert to decimal notation.

18. 6.893×10^{11}

19. 5.67×10^{-5}

Multiply and write scientific notation for the result.

20. $(1.12 \times 10^{-8})(5 \times 10^{-7})$

21. $(9.1 \times 10^{-17})(8.2 \times 10^3)$

(d) Multiplying and Dividing Using Scientific Notation

Multiplying

Consider the product

$$400 \cdot 2000 = 800{,}000.$$

In scientific notation, this is

$$(4 \times 10^2) \cdot (2 \times 10^3) = (4 \cdot 2)(10^2 \cdot 10^3) = 8 \times 10^5.$$

By applying the commutative and associative laws, we can find this product by multiplying $4 \cdot 2$, to get 8, and $10^2 \cdot 10^3$, to get 10^5.

EXAMPLE 22 Multiply: $(1.8 \times 10^6) \cdot (2.3 \times 10^{-4})$.

We apply the commutative and associative laws to get

$$
\begin{aligned}
(1.8 \times 10^6) \cdot (2.3 \times 10^{-4}) &= (1.8 \cdot 2.3) \times (10^6 \cdot 10^{-4}) \\
&= 4.14 \times 10^{6+(-4)} \\
&= 4.14 \times 10^2.
\end{aligned}
$$

We get 4.14 by multiplying 1.8 and 2.3. We get 10^2 by adding the exponents 6 and -4.

EXAMPLE 23 Multiply: $(3.1 \times 10^5) \cdot (4.5 \times 10^{-3})$.

$$
\begin{aligned}
(3.1 \times 10^5) \cdot (4.5 \times 10^{-3}) &= (3.1 \times 4.5)(10^5 \cdot 10^{-3}) \\
&= 13.95 \times 10^2 & &\text{Not scientific notation; } \\
& & &\text{13.95 is greater than 10.} \\
&= (1.395 \times 10^1) \times 10^2 & &\text{Substituting } 1.395 \times 10^1 \\
& & &\text{for 13.95} \\
&= 1.395 \times (10^1 \times 10^2) & &\text{Associative law} \\
&= 1.395 \times 10^3 & &\text{Adding exponents.} \\
& & &\text{The answer is now in} \\
& & &\text{scientific notation.}
\end{aligned}
$$

Do Exercises 20 and 21.

Dividing

Consider the quotient $800{,}000 \div 400 = 2000$. In scientific notation, this is

$$(8 \times 10^5) \div (4 \times 10^2) = \frac{8 \times 10^5}{4 \times 10^2} = \frac{8}{4} \times \frac{10^5}{10^2} = 2 \times 10^3.$$

We found this product by dividing 8 by 4, to get 2, and 10^5 by 10^2, to get 10^3.

EXAMPLE 24 Divide: $(3.41 \times 10^5) \div (1.1 \times 10^{-3})$.

$$
\begin{aligned}
(3.41 \times 10^5) \div (1.1 \times 10^{-3}) &= \frac{3.41 \times 10^5}{1.1 \times 10^{-3}} = \frac{3.41}{1.1} \times \frac{10^5}{10^{-3}} \\
&= 3.1 \times 10^{5-(-3)} \\
&= 3.1 \times 10^8
\end{aligned}
$$

Calculator Corner

To find the product in Example 22 and express the result in scientific notation on a graphing calculator, we first set the calculator in Scientific mode by pressing **MODE**, positioning the cursor over Sci on the first line, and pressing **ENTER**. Then we go to the home screen and enter the computation by pressing $\boxed{1}\ \boxed{.}\ \boxed{8}\ \boxed{\text{2ND}}\ \boxed{\text{EE}}\ \boxed{6}\ \boxed{\times}\ \boxed{2}\ \boxed{.}$ $\boxed{3}\ \boxed{\text{2ND}}\ \boxed{\text{EE}}\ \boxed{(-)}\ \boxed{4}\ \boxed{\text{ENTER}}$. (EE is the second operation associated with the $\boxed{,}$ key.) The decimal portion of a number written in scientific notation appears before a small E and the exponent follows the E.

```
1.8E6*2.3E−4
              4.14E2
```

Exercises: Multiply or divide and express the answer in scientific notation.

1. $(3.15 \times 10^7)(4.3 \times 10^{-12})$

2. $(8 \times 10^9)(4 \times 10^{-5})$

3. $\dfrac{4.5 \times 10^6}{1.5 \times 10^{12}}$

4. $\dfrac{4 \times 10^{-9}}{5 \times 10^{16}}$

Answers

18. 689,300,000,000 **19.** 0.0000567
20. 5.6×10^{-15} **21.** 7.462×10^{-13}

EXAMPLE 25 Divide: $(6.4 \times 10^{-7}) \div (8.0 \times 10^6)$.

$$(6.4 \times 10^{-7}) \div (8.0 \times 10^6) = \frac{6.4 \times 10^{-7}}{8.0 \times 10^6}$$

$$= \frac{6.4}{8.0} \times \frac{10^{-7}}{10^6}$$

$$= 0.8 \times 10^{-7-6}$$

$$= 0.8 \times 10^{-13} \qquad \text{Not scientific notation;}$$
$$\qquad\qquad\qquad\qquad 0.8 \text{ is less than 1.}$$

$$= (8.0 \times 10^{-1}) \times 10^{-13} \qquad \text{Substituting}$$
$$\qquad\qquad\qquad\qquad\qquad 8.0 \times 10^{-1} \text{ for 0.8}$$

$$= 8.0 \times (10^{-1} \times 10^{-13}) \qquad \text{Associative law}$$

$$= 8.0 \times 10^{-14} \qquad\qquad \text{Adding exponents}$$

Do Exercises 22 and 23.

e Applications with Scientific Notation

EXAMPLE 26 *Distance from the Sun to Earth.* Light from the sun traveling at a rate of 300,000 kilometers per second (km/s) reaches Earth in 499 sec. Find the distance, expressed in scientific notation, from the sun to Earth.

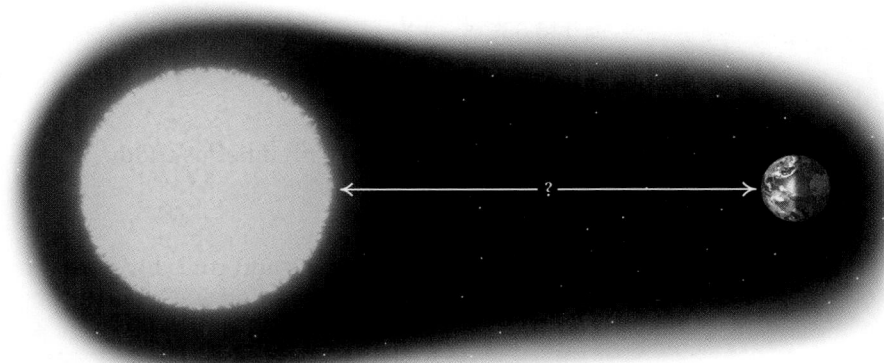

The time t that it takes for light to reach Earth from the sun is 4.99×10^2 sec (s). The speed is 3.0×10^5 km/s. Recall that distance can be expressed in terms of speed and time as

$$\text{Distance} = \text{Speed} \cdot \text{Time}$$
$$d = rt.$$

We substitute 3.0×10^5 for r and 4.99×10^2 for t:

$$d = rt$$

$$= (3.0 \times 10^5)(4.99 \times 10^2) \qquad \text{Substituting}$$

$$= 14.97 \times 10^7$$

$$= (1.497 \times 10^1) \times 10^7$$

$$= 1.497 \times (10^1 \times 10^7)$$

$$= 1.497 \times 10^8 \text{ km.} \qquad \text{Converting to scientific notation}$$

Thus the distance from the sun to Earth is 1.497×10^8 km.

Do Exercise 24.

EXAMPLE 27 *DNA.* A strand of DNA (deoxyribonucleic acid) is about 150 cm long and 1.3×10^{-10} cm wide. The length of a strand of DNA is how many times the width?

Source: Human Genome Project Information

25. Earth vs. Saturn. The mass of Earth is about 6×10^{21} metric tons. The mass of Saturn is about 5.7×10^{23} metric tons. About how many times the mass of Earth is the mass of Saturn? Express the answer in scientific notation.

To determine how many times longer DNA is than it is wide, we divide the length by the width:

$$\frac{150}{1.3 \times 10^{-10}} = \frac{150}{1.3} \times \frac{1}{10^{-10}}$$

$$\approx 115.385 \times 10^{10}$$

$$= (1.15385 \times 10^2) \times 10^{10}$$

$$= 1.15385 \times 10^{12}.$$

Thus the length of DNA is about 1.15385×10^{12} times its width.

Do Exercise 25.

The following is a summary of the definitions and rules for exponents that we have considered in this section and the preceding one.

DEFINITIONS AND RULES FOR EXPONENTS

Exponent of 1:	$a^1 = a$
Exponent of 0:	$a^0 = 1, a \neq 0$
Negative exponents:	$a^{-n} = \dfrac{1}{a^n}, \dfrac{1}{a^{-n}} = a^n, a \neq 0$
Product Rule:	$a^m \cdot a^n = a^{m+n}$
Quotient Rule:	$\dfrac{a^m}{a^n} = a^{m-n}, a \neq 0$
Power Rule:	$(a^m)^n = a^{mn}$
Raising a product to a power:	$(ab)^n = a^n b^n$
Raising a quotient to a power:	$\left(\dfrac{a}{b}\right)^n = \dfrac{a^n}{b^n}, b \neq 0;$
	$\left(\dfrac{a}{b}\right)^{-n} = \dfrac{b^n}{a^n}, b \neq 0, a \neq 0$
Scientific notation:	$M \times 10^n$, or 10^n, where $1 \leq M < 10$

Answer

25. The mass of Saturn is 9.5×10 times the mass of Earth.

a, **b** Simplify.

1. $(2^3)^2$

2. $(5^2)^4$

3. $(5^2)^{-3}$

4. $(7^{-3})^5$

5. $(x^{-3})^{-4}$

6. $(a^{-5})^{-6}$

7. $(a^{-2})^9$

8. $(x^{-5})^6$

9. $(t^{-3})^{-6}$

10. $(a^{-4})^{-7}$

11. $(t^4)^{-3}$

12. $(t^5)^{-2}$

13. $(x^{-2})^{-4}$

14. $(t^{-6})^{-5}$

15. $(ab)^3$

16. $(xy)^2$

17. $(ab)^{-3}$

18. $(xy)^{-6}$

19. $(mn^2)^{-3}$

20. $(x^3y)^{-2}$

21. $(4x^3)^2$

22. $4(x^3)^2$

23. $(3x^{-4})^2$

24. $(2a^{-5})^3$

25. $(x^4y^5)^{-3}$

26. $(t^5x^3)^{-4}$

27. $(x^{-6}y^{-2})^{-4}$

28. $(x^{-2}y^{-7})^{-5}$

29. $(a^{-2}b^7)^{-5}$

30. $(q^5r^{-1})^{-3}$

31. $(5r^{-4}t^3)^2$

32. $(4x^5y^{-6})^3$

33. $(a^{-5}b^7c^{-2})^3$

34. $(x^{-4}y^{-2}z^9)^2$

35. $(3x^3y^{-8}z^{-3})^2$

36. $(2a^2y^{-4}z^{-5})^3$

37. $(-4x^3y^{-2})^2$

38. $(-8x^3y^{-2})^3$

39. $(-a^{-3}b^{-2})^{-4}$

40. $(-p^{-4}q^{-3})^{-2}$

41. $\left(\dfrac{y^3}{2}\right)^2$

42. $\left(\dfrac{a^5}{3}\right)^3$

43. $\left(\dfrac{a^2}{b^3}\right)^4$

44. $\left(\dfrac{x^3}{y^4}\right)^5$

45. $\left(\dfrac{y^2}{2}\right)^{-3}$

46. $\left(\dfrac{a^4}{3}\right)^{-2}$

47. $\left(\dfrac{7}{x^{-3}}\right)^2$

48. $\left(\dfrac{3}{a^{-2}}\right)^3$

49. $\left(\dfrac{x^2y}{z}\right)^3$

50. $\left(\dfrac{m}{n^4p}\right)^3$

51. $\left(\dfrac{a^2b}{cd^3}\right)^{-2}$

52. $\left(\dfrac{2a^2}{3b^4}\right)^{-3}$

c Convert to scientific notation.

53. 28,000,000,000

54. 4,900,000,000,000

55. 907,000,000,000,000,000

56. 168,000,000,000,000

57. 0.00000304

58. 0.000000000865

59. 0.000000018

60. 0.00000000002

61. 100,000,000,000

62. 0.0000001

63. *Population of the United States.* It is estimated that the population of the United States will be 419,854,000 in 2050. Convert 419,854,000 to scientific notation.
Source: U.S. Census Bureau

64. *Young Voters.* About 22,750,000 young people, ages 18–29, voted in the 2008 presidential election. Convert 22,750,000 to scientific notation.
Source: Center for Information & Research on Civic Learning and Engagement, Tufts University

65. *Political Spending.* A record $2,400,000,000 was spent on campaigning, advertising, conventions, and other political activities in the 2008 presidential election. Convert $2,400,000,000 to scientific notation.
Source: Center for Responsive Politics

66. *Advertising Spending.* Coca-Cola spent $2,600,000,000 on advertising in a recent year. Convert $2,600,000,000 to scientific notation.
Source: Nielsen Media Research

Convert to decimal notation.

67. 8.74×10^7

68. 1.85×10^8

69. 5.704×10^{-8}

70. 8.043×10^{-4}

71. 10^7

72. 10^6

73. 10^{-5}

74. 10^{-8}

d Multiply or divide and write scientific notation for the result.

75. $(3 \times 10^4)(2 \times 10^5)$

76. $(3.9 \times 10^8)(8.4 \times 10^{-3})$

77. $(5.2 \times 10^5)(6.5 \times 10^{-2})$

78. $(7.1 \times 10^{-7})(8.6 \times 10^{-5})$

79. $(9.9 \times 10^{-6})(8.23 \times 10^{-8})$

80. $(1.123 \times 10^4) \times 10^{-9}$

81. $\dfrac{8.5 \times 10^8}{3.4 \times 10^{-5}}$

82. $\dfrac{5.6 \times 10^{-2}}{2.5 \times 10^5}$

83. $(3.0 \times 10^6) \div (6.0 \times 10^9)$

84. $(1.5 \times 10^{-3}) \div (1.6 \times 10^{-6})$

85. $\dfrac{7.5 \times 10^{-9}}{2.5 \times 10^{12}}$

86. $\dfrac{4.0 \times 10^{-3}}{8.0 \times 10^{20}}$

 Solve.

87. *River Discharge.* The average discharge at the mouths of the Amazon River is 4,200,000 cubic feet per second. How much water is discharged from the Amazon River in 1 yr? Express the answer in scientific notation.

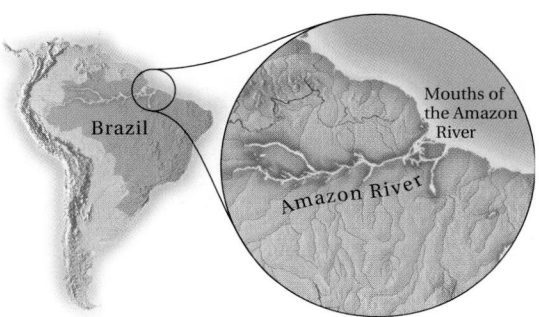

88. *Water Contamination.* Americans who change their own motor oil generate about 150 million gallons of used oil annually. If this oil is not disposed of properly, it can contaminate drinking water and soil. One gallon of used oil can contaminate one million gallons of drinking water. How many gallons of drinking water can 150 million gallons of oil contaminate? Express the answer in scientific notation. (1 million = 10^6).

Source: *New Car Buying Guide*

89. *Earth vs. Jupiter.* The mass of Earth is about 6×10^{21} metric tons. The mass of Jupiter is about 1.908×10^{24} metric tons. About how many times the mass of Earth is the mass of Jupiter? Express the answer in scientific notation.

90. *Computers.* A gigabyte is a measure of a computer's storage capacity. One gigabyte holds about one billion bytes of information. If a firm's computer network contains 2500 gigabytes of memory, how many bytes are in the network? Express the answer in scientific notation. (1 billion = 10^9)

91. *Stars.* It is estimated that there are 10 billion trillion stars in the known universe. Express the number of stars in scientific notation. (1 billion $= 10^9$; 1 trillion $= 10^{12}$)

92. *Closest Star.* Excluding the sun, the closest star to Earth is Proxima Centauri, which is 4.3 light-years away. (One light-year $= 5.88 \times 10^{12}$ mi.) How far, in miles, is Proxima Centauri from Earth? Express the answer in scientific notation.

93. *Red Light.* The wavelength of light is given by the velocity divided by the frequency. The velocity of red light is 300,000,000 m/sec, and its frequency is 400,000,000,000,000 cycles per second. What is the wavelength of red light? Express the answer in scientific notation.

94. *Earth vs. Sun.* The mass of Earth is about 6×10^{21} metric tons. The mass of the sun is about 1.998×10^{27} metric tons. About how many times the mass of Earth is the mass of the sun? Express the answer in scientific notation.

Space Travel. Use the following information for Exercises 95 and 96.

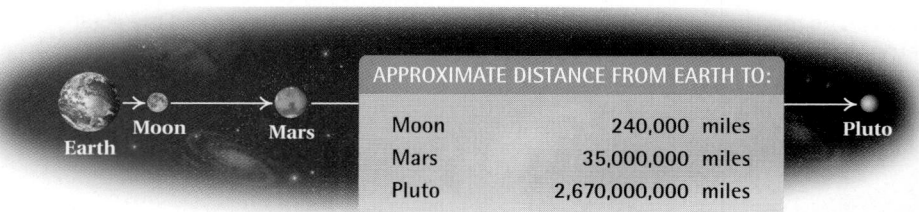

APPROXIMATE DISTANCE FROM EARTH TO:	
Moon	240,000 miles
Mars	35,000,000 miles
Pluto	2,670,000,000 miles

95. *Time to Reach Mars.* Suppose that it takes about 3 days for a space vehicle to travel from Earth to the moon. About how long would it take the same space vehicle traveling at the same speed to reach Mars? Express the answer in scientific notation.

96. *Time to Reach Pluto.* Suppose that it takes about 3 days for a space vehicle to travel from Earth to the moon. About how long would it take the same space vehicle traveling at the same speed to reach the dwarf planet Pluto? Express the answer in scientific notation.

Skill Maintenance

Factor.   [1.7d]

97. $9x - 36$

98. $4x - 2y + 16$

99. $3s + 3t + 24$

100. $-7x - 14$

Solve.   [2.3b]

101. $2x - 4 - 5x + 8 = x - 3$

102. $8x + 7 - 9x = 12 - 6x + 5$

Solve.   [2.3c]

103. $8(2x + 3) - 2(x - 5) = 10$

104. $4(x - 3) + 5 = 6(x + 2) - 8$

Graph.   [3.2a], [3.3a]

105. $y = x - 5$

106. $2x + y = 4$

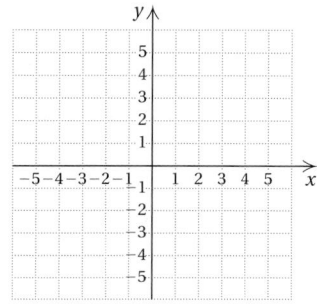

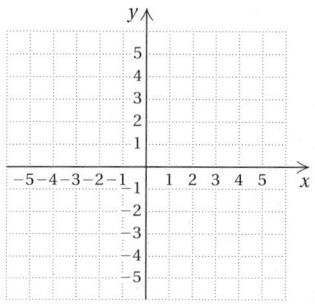

Synthesis

107. ▦ Carry out the indicated operations. Express the result in scientific notation.

$$\frac{(5.2 \times 10^6)(6.1 \times 10^{-11})}{1.28 \times 10^{-3}}$$

108. Find the reciprocal and express it in scientific notation.

$$6.25 \times 10^{-3}$$

Simplify.

109. $\dfrac{(5^{12})^2}{5^{25}}$

110. $\dfrac{a^{22}}{(a^2)^{11}}$

111. $\dfrac{(3^5)^4}{3^5 \cdot 3^4}$

112. $\left(\dfrac{5x^{-2}}{3y^{-2}z}\right)^0$

113. $\dfrac{49^{18}}{7^{35}}$

114. $\left(\dfrac{1}{a}\right)^{-n}$

115. $\dfrac{(0.4)^5}{[(0.4)^3]^2}$

116. $\left(\dfrac{4a^3b^{-2}}{5c^{-3}}\right)^1$

Determine whether each of the following is true for all pairs of integers m and n and all positive numbers x and y.

117. $x^m \cdot y^n = (xy)^{mn}$

118. $x^m \cdot y^m = (xy)^{2m}$

119. $(x - y)^m = x^m - y^m$

120. $-x^m = (-x)^m$

121. $(-x)^{2m} = x^{2m}$

122. $x^{-m} = \dfrac{-1}{x^m}$

4.3

Introduction to Polynomials

OBJECTIVES

a Evaluate a polynomial for a given value of the variable.

b Identify the terms of a polynomial.

c Identify the like terms of a polynomial.

d Identify the coefficients of a polynomial.

e Collect the like terms of a polynomial.

f Arrange a polynomial in descending order, or collect the like terms and then arrange in descending order.

g Identify the degree of each term of a polynomial and the degree of the polynomial.

h Identify the missing terms of a polynomial.

i Classify a polynomial as a monomial, a binomial, a trinomial, or none of these.

SKILL TO REVIEW
Objective 1.7e: Collect like terms.

Collect like terms.
1. $3x - 4y + 5x + y$
2. $2a - 7b + 6 - 3a + 4b - 1$

1. Write three polynomials.

We have already learned to evaluate and to manipulate certain kinds of algebraic expressions. We will now consider algebraic expressions called *polynomials*.

The following are examples of *monomials in one variable*:

$$3x^2, \quad 2x, \quad -5, \quad 37p^4, \quad 0.$$

Each expression is a constant or a constant times some variable to a nonnegative integer power.

> **MONOMIAL**
>
> A **monomial** is an expression of the type ax^n, where a is a real-number constant and n is a nonnegative integer.

Algebraic expressions like the following are **polynomials**:

$$\tfrac{3}{4}y^5, \quad -2, \quad 5y + 3, \quad 3x^2 + 2x - 5, \quad -7a^3 + \tfrac{1}{2}a, \quad 6x, \quad 37p^4, \quad x, \quad 0.$$

> **POLYNOMIAL**
>
> A **polynomial** is a monomial or a combination of sums and/or differences of monomials.

The following algebraic expressions are *not* polynomials:

$$\textbf{(1)} \ \frac{x+3}{x-4}, \quad \textbf{(2)} \ 5x^3 - 2x^2 + \frac{1}{x}, \quad \textbf{(3)} \ \frac{1}{x^3 - 2}.$$

Expressions (1) and (3) are not polynomials because they represent quotients, not sums or differences. Expression (2) is not a polynomial because

$$\frac{1}{x} = x^{-1},$$

and this is not a monomial because the exponent is negative.

Do Margin Exercise 1.

a Evaluating Polynomials and Applications

When we replace the variable in a polynomial with a number, the polynomial then represents a number called a **value** of the polynomial. Finding that number, or value, is called **evaluating the polynomial**. We evaluate a polynomial using the rules for order of operations (Section 1.8).

EXAMPLE 1 Evaluate the polynomial when $x = 2$.

a) $\begin{aligned}[t] 3x + 5 &= 3 \cdot 2 + 5 \\ &= 6 + 5 \\ &= 11 \end{aligned}$

b) $\begin{aligned}[t] 2x^2 - 7x + 3 &= 2 \cdot 2^2 - 7 \cdot 2 + 3 \\ &= 2 \cdot 4 - 7 \cdot 2 + 3 \\ &= 8 - 14 + 3 \\ &= -3 \end{aligned}$

Answers

Skill to Review:
1. $8x - 3y$ 2. $-a - 3b + 5$

Margin Exercise:
1. $4x^2 - 3x + \dfrac{5}{4}$; $15y^3$; $-7x^3 + 1.1$;

answers may vary

EXAMPLE 2 Evaluate the polynomial when $x = -4$.

a) $2 - x^3 = 2 - (-4)^3 = 2 - (-64)$
$$= 2 + 64 = 66$$

b) $-x^2 - 3x + 1 = -(-4)^2 - 3(-4) + 1$
$$= -16 + 12 + 1 = -3$$

Do Exercises 2–5.

Evaluate each polynomial when $x = 3$.

2. $-4x - 7$

3. $-5x^3 + 7x + 10$

Evaluate each polynomial when $x = -5$.

4. $5x + 7$

5. $2x^2 + 5x - 4$

※ Algebraic-Graphical Connection

Recall from Chapter 3 that in order to plot points before graphing an equation, we choose values for x and compute the corresponding y-values. An equation like $y = 2x - 2$, which has a polynomial on one side and only y on the other, is called a **polynomial equation**. For such an equation, determining y is the same as evaluating the polynomial. Once the graph of such an equation has been drawn, we can evaluate the polynomial for a given x-value by finding the y-value that is paired with it on the graph.

6. Use *only* the graph shown in Example 3 to evaluate the polynomial $2x - 2$ when $x = 4$ and when $x = -1$.

EXAMPLE 3 Use *only* the given graph of $y = 2x - 2$ to evaluate the polynomial $2x - 2$ when $x = 3$.

First, we locate 3 on the x-axis. From there we move vertically to the graph of the equation and then horizontally to the y-axis. There we locate the y-value that is paired with 3. Although our drawing may not be precise, it appears that the y-value 4 is paired with 3. Thus the value of $2x - 2$ is 4 when $x = 3$.

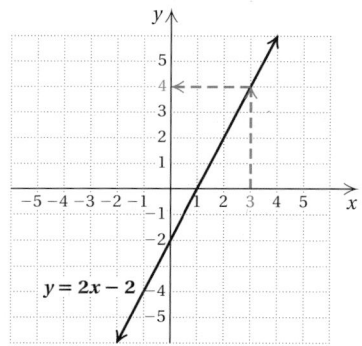

Do Exercise 6.

<hr>

※

Polynomial equations can be used to model many real-world situations.

7. Referring to Example 4, determine the total number of games to be played in a league of 12 teams.

8. Perimeter of a Baseball Diamond. The perimeter P of a square of side x is given by the polynomial equation $P = 4x$.

EXAMPLE 4 *Games in a Sports League.* In a sports league of x teams in which each team plays every other team twice, the total number of games N to be played is given by the polynomial equation

$$N = x^2 - x.$$

A women's slow-pitch softball league has 10 teams. What is the total number of games to be played?

We evaluate the polynomial when $x = 10$:

$$N = x^2 - x = 10^2 - 10 = 100 - 10 = 90.$$

The league plays 90 games.

A baseball diamond is a square 90 ft on a side. Find the perimeter of a baseball diamond.

Do Exercises 7 and 8.

Answers

2. -19 **3.** -104 **4.** -18 **5.** 21
6. 6; -4 **7.** 132 games   **8.** 360 ft

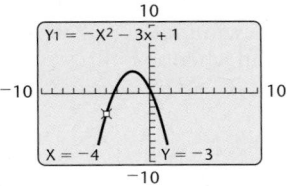
9. Medical Dosage.

   a) Referring to Example 5, determine the concentration after 3 hr by evaluating the polynomial when $t = 3$.

   b) Use *only* the graph showing medical dosage to check the value found in part (a).

10. Medical Dosage. Referring to Example 5, use *only* the graph showing medical dosage to estimate the value of the polynomial when $t = 26$.

Answers

9. (a) 7.55 parts per million; (b) When $t = 3$, $C \approx 7.5$ so the value found in part (a) appears to be correct. **10.** 20 parts per million

---

EXAMPLE 5 *Medical Dosage.* The concentration C, in parts per million, of a certain antibiotic in the bloodstream after t hours is given by the polynomial equation

$$C = -0.05t^2 + 2t + 2.$$

Find the concentration after 2 hr.

To find the concentration after 2 hr, we evaluate the polynomial when $t = 2$:

$$
\begin{aligned}
C &= -0.05t^2 + 2t + 2 \\
&= -0.05(2)^2 + 2(2) + 2 && \text{Substituting 2 for } t \\
&= -0.05(4) + 2(2) + 2 && \text{Carrying out the calculation using} \\
& && \text{the rules for order of operations} \\
&= -0.2 + 4 + 2 \\
&= 3.8 + 2 \\
&= 5.8.
\end{aligned}
$$

The concentration after 2 hr is 5.8 parts per million.

※ Algebraic–Graphical Connection

The polynomial equation in Example 5 can be graphed if we evaluate the polynomial for several values of t. We list the values in a table and show the graph below. Note that the concentration peaks at the 20-hr mark and after slightly more than 40 hr, the concentration is 0. Since neither time nor concentration can be negative, our graph uses only the first quadrant.

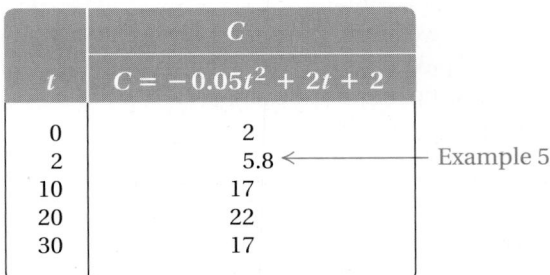

t	$C = -0.05t^2 + 2t + 2$	
0	2	
2	5.8	← Example 5
10	17	
20	22	
30	17	

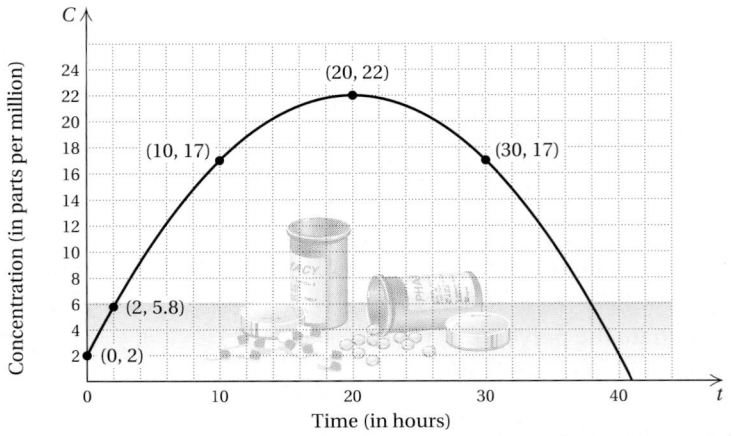

Do Exercises 9 and 10.

(b) Identifying Terms

As we saw in Section 1.4, subtractions can be rewritten as additions. For any polynomial that has some subtractions, we can find an equivalent polynomial using only additions.

EXAMPLES Find an equivalent polynomial using only additions.

6. $-5x^2 - x = -5x^2 + (-x)$

7. $4x^5 - 2x^6 + 4x - 7 = 4x^5 + (-2x^6) + 4x + (-7)$

> Do Exercises 11 and 12.

When a polynomial is written using only additions, the monomials being added are called **terms**. In Example 6, the terms are $-5x^2$ and $-x$. In Example 7, the terms are $4x^5$, $-2x^6$, $4x$, and -7.

EXAMPLE 8 Identify the terms of the polynomial

$$4x^7 + 3x + 12 + 8x^3 + 5x.$$

Terms: $4x^7$, $3x$, 12, $8x^3$, and $5x$.

If there are subtractions, you can *think* of them as additions without rewriting.

EXAMPLE 9 Identify the terms of the polynomial

$$3t^4 - 5t^6 - 4t + 2.$$

Terms: $3t^4$, $-5t^6$, $-4t$, and 2.

> Do Exercises 13 and 14.

(c) Like Terms

When terms have the same variable and the same exponent power, we say that they are **like terms**.

EXAMPLES Identify the like terms in the polynomials.

10. $4x^3 + 5x - 4x^2 + 2x^3 + x^2$

     Like terms: $4x^3$ and $2x^3$ Same variable and exponent

     Like terms: $-4x^2$ and x^2 Same variable and exponent

11. $6 - 3a^2 - 8 - a - 5a$

     Like terms: 6 and -8 Constant terms are like terms because $6 = 6x^0$ and $-8 = -8x^0$.

     Like terms: $-a$ and $-5a$

> Do Exercises 15–17.

(d) Coefficients

The coefficient of the term $5x^3$ is 5. In the following polynomial, the red numbers are the **coefficients**, 3, −2, 5, and 4:

$$3x^5 - 2x^3 + 5x + 4.$$

Find an equivalent polynomial using only additions.

11. $-9x^3 - 4x^5$

12. $-2y^3 + 3y^7 - 7y - 9$

Identify the terms of each polynomial.

13. $3x^2 + 6x + \dfrac{1}{2}$

14. $-4y^5 + 7y^2 - 3y - 2$

Identify the like terms in each polynomial.

15. $4x^3 - x^3 + 2$

16. $4t^4 - 9t^3 - 7t^4 + 10t^3$

17. $5x^2 + 3x - 10 + 7x^2 - 8x + 11$

Answers

11. $-9x^3 + (-4x^5)$
12. $-2y^3 + 3y^7 + (-7y) + (-9)$
13. $3x^2, 6x, \dfrac{1}{2}$ **14.** $-4y^5, 7y^2, -3y, -2$
15. $4x^3$ and $-x^3$ **16.** $4t^4$ and $-7t^4$; $-9t^3$ and $10t^3$ **17.** $5x^2$ and $7x^2$; $3x$ and $-8x$; -10 and 11

EXAMPLE 12 Identify the coefficient of each term in the polynomial

$$3x^4 - 4x^3 + \frac{1}{2}x^2 + x - 8.$$

The coefficient of the first term is 3.
The coefficient of the second term is -4.
The coefficient of the third term is $\frac{1}{2}$.
The coefficient of the fourth term is 1. $x = 1x$
The coefficient of the fifth term is -8.

Do Exercise 18.

18. Identify the coefficient of each term in the polynomial
$2x^4 - 7x^3 - 8.5x^2 - x - 4$.

(e) Collecting Like Terms

We can often simplify polynomials by **collecting like terms**, or **combining like terms**. To do this, we use the distributive laws. We factor out the variable expression and add or subtract the coefficients. We try to do this mentally as much as possible.

EXAMPLES Collect like terms.

13. $2x^3 - 6x^3 = (2 - 6)x^3$ Using a distributive law
$$= -4x^3$$

14. $5x^2 + 7 + 4x^4 + 2x^2 - 11 - 2x^4 = (5 + 2)x^2 + (4 - 2)x^4 + (7 - 11)$
$$= 7x^2 + 2x^4 - 4$$

Note that using the distributive laws in this manner allows us to collect like terms by adding or subtracting the coefficients. Often the middle step is omitted and we add or subtract mentally, writing just the answer. In collecting like terms, we may get 0.

EXAMPLE 15 Collect like terms: $3x^5 + 2x^2 - 3x^5 + 8$.

$$3x^5 + 2x^2 - 3x^5 + 8 = (3 - 3)x^5 + 2x^2 + 8$$
$$= 0x^5 + 2x^2 + 8$$
$$= 2x^2 + 8$$

Do Exercises 19–24.

Expressing a term like x^2 by showing 1 as a factor, $1 \cdot x^2$, may make it easier to understand how to factor or collect like terms.

EXAMPLES Collect like terms.

16. $5x^2 + x^2 = 5x^2 + 1x^2$ Replacing x^2 with $1x^2$
$$= (5 + 1)x^2$$ Using a distributive law
$$= 6x^2$$

Collect like terms.
19. $3x^2 + 5x^2$

20. $4x^3 - 2x^3 + 2 + 5$

21. $\frac{1}{2}x^5 - \frac{3}{4}x^5 + 4x^2 - 2x^2$

22. $24 - 4x^3 - 24$

23. $5x^3 - 8x^5 + 8x^5$

24. $-2x^4 + 16 + 2x^4 + 9 - 3x^5$

Answers
18. $2, -7, -8.5, -1, -4$ **19.** $8x^2$
20. $2x^3 + 7$ **21.** $-\frac{1}{4}x^5 + 2x^2$ **22.** $-4x^3$
23. $5x^3$ **24.** $25 - 3x^5$

17. $5x^8 - 6x^5 - x^8 = 5x^8 - 6x^5 - 1x^8 \qquad x^8 = 1x^8$

$$= (5-1)x^8 - 6x^5$$
$$= 4x^8 - 6x^5$$

18. $\frac{2}{3}x^4 - x^3 - \frac{1}{6}x^4 + \frac{2}{5}x^3 - \frac{3}{10}x^3$

$$= \left(\frac{2}{3} - \frac{1}{6}\right)x^4 + \left(-1 + \frac{2}{5} - \frac{3}{10}\right)x^3 \qquad -x^3 = -1 \cdot x^3$$
$$= \left(\frac{4}{6} - \frac{1}{6}\right)x^4 + \left(-\frac{10}{10} + \frac{4}{10} - \frac{3}{10}\right)x^3$$
$$= \frac{3}{6}x^4 - \frac{9}{10}x^3$$
$$= \frac{1}{2}x^4 - \frac{9}{10}x^3$$

> Do Exercises 25–28.

> Collect like terms.
>
> **25.** $7x - x$
>
> **26.** $5x^3 - x^3 + 4$
>
> **27.** $\dfrac{3}{4}x^3 + 4x^2 - x^3 + 7$
>
> **28.** $\dfrac{4}{5}x^4 - x^4 + x^5 - \dfrac{1}{5} - \dfrac{1}{4}x^4 + 10$

(f) Descending and Ascending Order

Note in the following polynomial that the exponents decrease from left to right. We say that the polynomial is arranged in **descending order**:

$$2x^4 - 8x^3 + 5x^2 - x + 3.$$

The term with the largest exponent is first. The term with the next largest exponent is second, and so on. The associative and commutative laws allow us to arrange the terms of a polynomial in descending order.

EXAMPLES Arrange the polynomial in descending order.

19. $6x^5 + 4x^7 + x^2 + 2x^3 = 4x^7 + 6x^5 + 2x^3 + x^2$

20. $\frac{2}{3} + 4x^5 - 8x^2 + 5x - 3x^3 = 4x^5 - 3x^3 - 8x^2 + 5x + \frac{2}{3}$

> Do Exercises 29–31.

> Arrange each polynomial in descending order.
>
> **29.** $x + 3x^5 + 4x^3 + 5x^2 + 6x^7 - 2x^4$
>
> **30.** $4x^2 - 3 + 7x^5 + 2x^3 - 5x^4$
>
> **31.** $-14 + 7t^2 - 10t^5 + 14t^7$

EXAMPLE 21 Collect like terms and then arrange in descending order:

$$2x^2 - 4x^3 + 3 - x^2 - 2x^3.$$

$$2x^2 - 4x^3 + 3 - x^2 - 2x^3 = x^2 - 6x^3 + 3 \qquad \text{Collecting like terms}$$
$$= -6x^3 + x^2 + 3 \qquad \text{Arranging in descending order}$$

> Do Exercises 32 and 33.

> Collect like terms and then arrange in descending order.
>
> **32.** $3x^2 - 2x + 3 - 5x^2 - 1 - x$
>
> **33.** $-x + \dfrac{1}{2} + 14x^4 - 7x - 1 - 4x^4$

We usually arrange polynomials in descending order, but not always. The opposite order is called **ascending order**. Generally, if an exercise is written in a certain order, we give the answer in that same order.

(g) Degrees

The **degree** of a term is the exponent of the variable. The degree of the term $-5x^3$ is 3.

EXAMPLE 22 Identify the degree of each term of $8x^4 - 3x + 7$.

The degree of $8x^4$ is 4.

The degree of $-3x$ is 1.   Recall that $x = x^1$.

The degree of 7 is 0.   Think of 7 as $7x^0$. Recall that $x^0 = 1$.

Answers

25. $6x$ **26.** $4x^3 + 4$ **27.** $-\dfrac{1}{4}x^3 + 4x^2 + 7$

28. $x^5 - \dfrac{9}{20}x^4 + \dfrac{49}{5}$

29. $6x^7 + 3x^5 - 2x^4 + 4x^3 + 5x^2 + x$
30. $7x^5 - 5x^4 + 2x^3 + 4x^2 - 3$
31. $14t^7 - 10t^5 + 7t^2 - 14$

32. $-2x^2 - 3x + 2$ **33.** $10x^4 - 8x - \dfrac{1}{2}$

The **degree of a polynomial** is the largest of the degrees of the terms, unless it is the polynomial 0. The polynomial 0 is a special case. We agree that it has *no* degree either as a term or as a polynomial. This is because we can express 0 as $0 = 0x^5 = 0x^7$, and so on, using any exponent we wish.

EXAMPLE 23 Identify the degree of the polynomial $5x^3 - 6x^4 + 7$.

$$5x^3 - 6x^4 + 7. \qquad \text{The largest exponent is 4.}$$

The degree of the polynomial is 4.

Do Exercises 34 and 35.

Let's summarize the terminology that we have learned, using the polynomial $3x^4 - 8x^3 + x^2 + 7x - 6$.

TERM	COEFFICIENT	DEGREE OF THE TERM	DEGREE OF THE POLYNOMIAL
$3x^4$	3	4	
$-8x^3$	-8	3	
x^2	1	2	4
$7x$	7	1	
-6	-6	0	

(h) Missing Terms

If a coefficient is 0, we generally do not write the term. We say that we have a **missing term**.

EXAMPLE 24 Identify the missing terms in the polynomial

$$8x^5 - 2x^3 + 5x^2 + 7x + 8.$$

There is no term with x^4. We say that the x^4-term is missing.

Do Exercises 36–39.

For certain skills or manipulations, we can write missing terms with zero coefficients or leave space.

EXAMPLE 25 Write the polynomial $x^4 - 6x^3 + 2x - 1$ in two ways: with its missing term and by leaving space for it.

a) $x^4 - 6x^3 + 2x - 1 = x^4 - 6x^3 + 0x^2 + 2x - 1$ Writing with the missing x^2-term

b) $x^4 - 6x^3 + 2x - 1 = x^4 - 6x^3 \qquad + 2x - 1$ Leaving space for the missing x^2-term

EXAMPLE 26 Write the polynomial $y^5 - 1$ in two ways: with its missing terms and by leaving space for them.

a) $y^5 - 1 = y^5 + 0y^4 + 0y^3 + 0y^2 + 0y - 1$

b) $y^5 - 1 = y^5 \qquad\qquad\qquad - 1$

Do Exercises 40 and 41.

Identify the degree of each term and the degree of the polynomial.

34. $-6x^4 + 8x^2 - 2x + 9$

35. $4 - x^3 + \dfrac{1}{2}x^6 - x^5$

Identify the missing terms in each polynomial.

36. $2x^3 + 4x^2 - 2$

37. $-3x^4$

38. $x^3 + 1$

39. $x^4 - x^2 + 3x + 0.25$

Write each polynomial in two ways: with its missing terms and by leaving space for them.

40. $2x^3 + 4x^2 - 2$

41. $a^4 + 10$

Answers

34. 4, 2, 1, 0; 4   **35.** 0, 3, 6, 5; 6   **36.** x
37. x^3, x^2, x, x^0 **38.** x^2, x **39.** x^3
40. $2x^3 + 4x^2 + 0x - 2$;
    $2x^3 + 4x^2 \qquad - 2$
41. $a^4 + 0a^3 + 0a^2 + 0a + 10$;
    $a^4 \qquad\qquad\qquad + 10$

(i) Classifying Polynomials

Polynomials with just one term are called **monomials**. Polynomials with just two terms are called **binomials**. Those with just three terms are called **trinomials**. Those with more than three terms are generally not specified with a name.

EXAMPLE 27

MONOMIALS	BINOMIALS	TRINOMIALS	NONE OF THESE
$4x^2$	$2x + 4$	$3x^3 + 4x + 7$	$4x^3 - 5x^2 + x - 8$
9	$3x^5 + 6x$	$6x^7 - 7x^2 + 4$	$z^5 + 2z^4 - z^3 + 7z + 3$
$-23x^{19}$	$-9x^7 - 6$	$4x^2 - 6x - \frac{1}{2}$	$4x^6 - 3x^5 + x^4 - x^3 + 2x - 1$

Do Exercises 42–45.

Classify each polynomial as a monomial, a binomial, a trinomial, or none of these.

42. $3x^2 + x$ **43.** $5x^4$

44. $4x^3 - 3x^2 + 4x + 2$

45. $3x^2 + 2x - 4$

Answers

42. Binomial   **43.** Monomial   **44.** None of these   **45.** Trinomial

---

4.3 Exercise Set

For Extra Help

MyMathLab | Math XL PRACTICE | WATCH | DOWNLOAD | READ | REVIEW

a Evaluate each polynomial when $x = 4$ and when $x = -1$.

1. $-5x + 2$

2. $-8x + 1$

3. $2x^2 - 5x + 7$

4. $3x^2 + x - 7$

5. $x^3 - 5x^2 + x$

6. $7 - x + 3x^2$

Evaluate each polynomial when $x = -2$ and when $x = 0$.

7. $\frac{1}{3}x + 5$

8. $8 - \frac{1}{4}x$

9. $x^2 - 2x + 1$

10. $5x + 6 - x^2$

11. $-3x^3 + 7x^2 - 3x - 2$

12. $-2x^3 + 5x^2 - 4x + 3$

13. *Skydiving.* During the first 13 sec of a jump, the distance S, in feet, that a skydiver falls in t seconds can be approximated by the polynomial equation

$$S = 11.12t^2.$$

Approximately how far has a skydiver fallen 10 sec after having jumped from a plane?

14. *Skydiving.* For jumps that exceed 13 sec, the polynomial equation

$$S = 173t - 369$$

can be used to approximate the distance S, in feet, that a skydiver has fallen in t seconds. Approximately how far has a skydiver fallen 20 sec after having jumped from a plane?

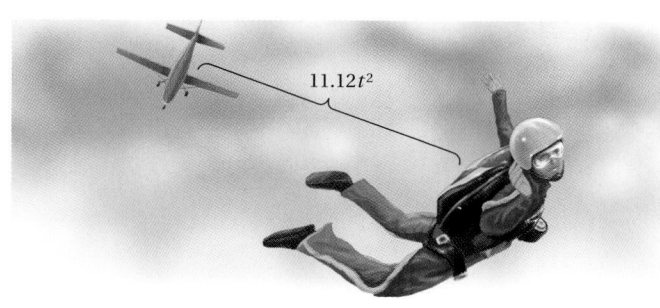

$11.12t^2$

15. *Total Revenue.* Hadley Electronics is marketing a new type of plasma TV. The firm determines that when it sells x TVs, its total revenue R (the total amount of money taken in) will be

$$R = 280x - 0.4x^2 \text{ dollars.}$$

What is the total revenue from the sale of 75 TVs? 100 TVs?

16. *Total Cost.* Hadley Electronics determines that the total cost C of producing x plasma TVs is given by

$$C = 5000 + 0.6x^2 \text{ dollars.}$$

What is the total cost of producing 500 TVs? 650 TVs?

17. The graph of the polynomial equation $y = 5 - x^2$ is shown below. Use *only* the graph to estimate the value of the polynomial when $x = -3$, $x = -1$, $x = 0$, $x = 1.5$, and $x = 2$.

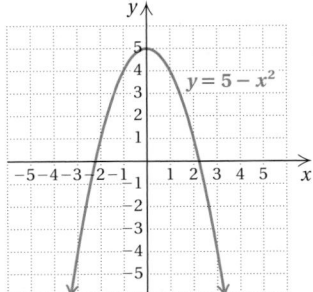

18. The graph of the polynomial equation $y = 6x^3 - 6x$ is shown below. Use *only* the graph to estimate the value of the polynomial when $x = -1$, $x = -0.5$, $x = 0.5$, $x = 1$, and $x = 1.1$.

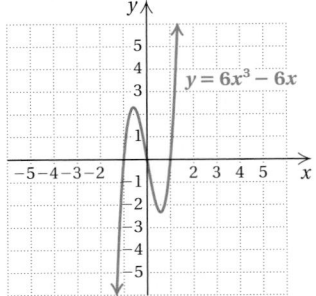

19. *Electricity Consumption.* The net consumption of electricity in China can be estimated by the polynomial equation

$$E = 158.68t + 2728.4,$$

where E is the consumption of electricity, in billions of kilowatt-hours, and t is the number of years after 2010. That is, $t = 0$ corresponds to 2010, $t = 5$ corresponds to 2015, and so on.

Source: Energy Information Administration

a) Use the equation to estimate the consumption of electricity, in billions of kilowatt-hours, in 2010, 2015, 2020, 2025, and 2030.

b) Check the results of part (a) using the graph below.

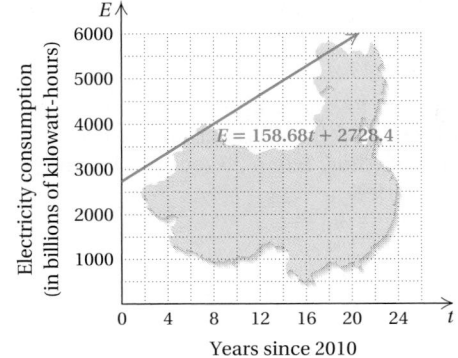

20. *Electricity Consumption.* The net consumption of electricity in the United States can be estimated by the polynomial equation

$$E = 72.9t + 4134.4,$$

where E is the consumption of electricity, in billions of kilowatt-hours, and t is the number of years after 2010. That is, $t = 0$ corresponds to 2010, $t = 5$ corresponds to 2015, and so on.

Source: Energy Information Administration

a) Use the equation to estimate the consumption of electricity, in billions of kilowatt-hours, in 2010, 2015, 2020, 2025, and 2030.

b) Check the results of part (a) using the graph below.

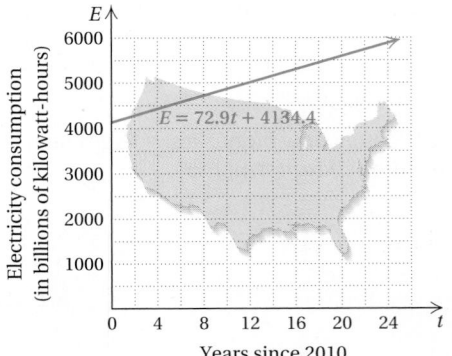

Memorizing Words.   Participants in a psychology experiment were able to memorize an average of M words in t minutes, where $M = -0.001t^3 + 0.1t^2$. Use the graph below for Exercises 21–26.

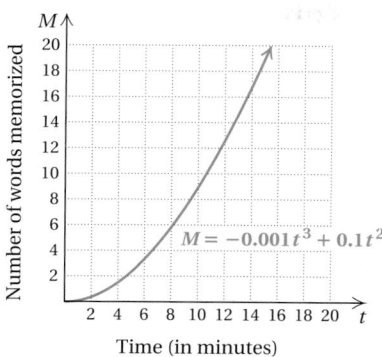

Time (in minutes)

21. Estimate the number of words memorized after 10 min.

22. Estimate the number of words memorized after 14 min.

23. Find the approximate value of M for $t = 8$.

24. Find the approximate value of M for $t = 12$.

25. Estimate the value of M when t is 13.

26. Estimate the value of M when t is 7.

b   Identify the terms of each polynomial.

27. $2 - 3x + x^2$

28. $2x^2 + 3x - 4$

29. $-2x^4 + \frac{1}{3}x^3 - x + 3$

30. $-\frac{2}{5}x^5 - x^3 + 6$

c   Identify the like terms in each polynomial.

31. $5x^3 + 6x^2 - 3x^2$

32. $3x^2 + 4x^3 - 2x^2$

33. $2x^4 + 5x - 7x - 3x^4$

34. $-3t + t^3 - 2t - 5t^3$

35. $3x^5 - 7x + 8 + 14x^5 - 2x - 9$

36. $8x^3 + 7x^2 - 11 - 4x^3 - 8x^2 - 29$

d   Identify the coefficient of each term of the polynomial.

37. $-3x + 6$

38. $2x - 4$

39. $5x^2 + \frac{3}{4}x + 3$

40. $\frac{2}{3}x^2 - 5x + 2$

41. $-5x^4 + 6x^3 - 2.7x^2 + x - 2$

42. $7x^3 - x^2 - 4.2x + 5$

e   Collect like terms.

43. $2x - 5x$

44. $2x^2 + 8x^2$

45. $x - 9x$

46. $x - 5x$

47. $5x^3 + 6x^3 + 4$

48. $6x^4 - 2x^4 + 5$

49. $5x^3 + 6x - 4x^3 - 7x$

50. $3a^4 - 2a + 2a + a^4$

51. $6b^5 + 3b^2 - 2b^5 - 3b^2$

52. $2x^2 - 6x + 3x + 4x^2$

53. $\dfrac{1}{4}x^5 - 5 + \dfrac{1}{2}x^5 - 2x - 37$

54. $\dfrac{1}{3}x^3 + 2x - \dfrac{1}{6}x^3 + 4 - 16$

55. $6x^2 + 2x^4 - 2x^2 - x^4 - 4x^2$

56. $8x^2 + 2x^3 - 3x^3 - 4x^2 - 4x^2$

57. $\dfrac{1}{4}x^3 - x^2 - \dfrac{1}{6}x^2 + \dfrac{3}{8}x^3 + \dfrac{5}{16}x^3$

58. $\dfrac{1}{5}x^4 + \dfrac{1}{5} - 2x^2 + \dfrac{1}{10} - \dfrac{3}{15}x^4 + 2x^2 - \dfrac{3}{10}$

(f) Arrange each polynomial in descending order.

59. $x^5 + x + 6x^3 + 1 + 2x^2$

60. $3 + 2x^2 - 5x^6 - 2x^3 + 3x$

61. $5y^3 + 15y^9 + y - y^2 + 7y^8$

62. $9p - 5 + 6p^3 - 5p^4 + p^5$

Collect like terms and then arrange in descending order.

63. $3x^4 - 5x^6 - 2x^4 + 6x^6$

64. $-1 + 5x^3 - 3 - 7x^3 + x^4 + 5$

65. $-2x + 4x^3 - 7x + 9x^3 + 8$

66. $-6x^2 + x - 5x + 7x^2 + 1$

67. $3x + 3x + 3x - x^2 - 4x^2$

68. $-2x - 2x - 2x + x^3 - 5x^3$

69. $-x + \dfrac{3}{4} + 15x^4 - x - \dfrac{1}{2} - 3x^4$

70. $2x - \dfrac{5}{6} + 4x^3 + x + \dfrac{1}{3} - 2x$

(g) Identify the degree of each term of the polynomial and the degree of the polynomial.

71. $2x - 4$

72. $6 - 3x$

73. $3x^2 - 5x + 2$

74. $5x^3 - 2x^2 + 3$

75. $-7x^3 + 6x^2 + \dfrac{3}{5}x + 7$

76. $5x^4 + \dfrac{1}{4}x^2 - x + 2$

77. $x^2 - 3x + x^6 - 9x^4$

78. $8x - 3x^2 + 9 - 8x^3$

79. Complete the following table for the polynomial $-7x^4 + 6x^3 - x^2 + 8x - 2$.

TERM	COEFFICIENT	DEGREE OF THE TERM	DEGREE OF THE POLYNOMIAL
$-7x^4$			
$6x^3$	6		
		2	
$8x$		1	
	-2		

80. Complete the following table for the polynomial $3x^2 + x^5 - 46x^3 + 6x - 2.4 - \frac{1}{2}x^4$.

TERM	COEFFICIENT	DEGREE OF THE TERM	DEGREE OF THE POLYNOMIAL
		5	
$-\frac{1}{2}x^4$		4	
	-46		
$3x^2$		2	
	6		
-2.4			

h   Identify the missing terms in each polynomial.

81. $x^3 - 27$

82. $x^5 + x$

83. $x^4 - x$

84. $5x^4 - 7x + 2$

85. $2x^3 - 5x^2 + x - 3$

86. $-6x^3$

Write each polynomial in two ways: with its missing terms and by leaving space for them.

87. $x^3 - 27$

88. $x^5 + x$

89. $x^4 - x$

90. $5x^4 - 7x + 2$

91. $2x^3 - 5x^2 + x - 3$

92. $-6x^3$

i   Classify each polynomial as a monomial, a binomial, a trinomial, or none of these.

93. $x^2 - 10x + 25$

94. $-6x^4$

95. $x^3 - 7x^2 + 2x - 4$

96. $x^2 - 9$

97. $4x^2 - 25$

98. $2x^4 - 7x^3 + x^2 + x - 6$

99. $40x$

100. $4x^2 + 12x + 9$

Skill Maintenance

101. Three tired hikers camped overnight. All they had to eat was a bag of apples. During the night, one awoke and ate one-third of the apples. Later, a second camper awoke and ate one-third of the apples that remained. Much later, the third camper awoke and ate one-third of those apples yet remaining after the other two had eaten. When they got up the next morning, 8 apples were left. How many apples did they begin with?   [2.6a]

Subtract.   [1.4a]

102. $1 - 20$

103. $\dfrac{1}{8} - \dfrac{5}{6}$

104. $\dfrac{3}{8} - \left(-\dfrac{1}{4}\right)$

105. $5.6 - 8.2$

106. Solve: $3(x + 2) = 5x - 9$.   [2.3c]

107. Solve $C = ab - r$ for b.   [2.4b]

108. A warehouse stores 1800 lb of peanuts, 1500 lb of cashews, and 700 lb of almonds. What percent of the total is peanuts? cashews? almonds?   [2.5a]

109. Factor: $3x - 15y + 63$.   [1.7d]

Synthesis

Collect like terms.

110. $6x^3 \cdot 7x^2 - (4x^3)^2 + (-3x^3)^2 - (-4x^2)(5x^3) - 10x^5 + 17x^6$

111. $(3x^2)^3 + 4x^2 \cdot 4x^4 - x^4(2x)^2 + ((2x)^2)^3 - 100x^2(x^2)^2$

112. Construct a polynomial in x (meaning that x is the variable) of degree 5 with four terms and coefficients that are integers.

113. What is the degree of $(5m^5)^2$?

114. A polynomial in x has degree 3. The coefficient of x^2 is 3 less than the coefficient of x^3. The coefficient of x is three times the coefficient of x^2. The remaining coefficient is 2 more than the coefficient of x^3. The sum of the coefficients is -4. Find the polynomial.

Use the CALC feature and choose VALUE on your graphing calculator to find the values in each of the following. (Refer to the Calculator Corner on p. 252.)

115. Exercise 17

116. Exercise 18

117. Exercise 21

118. Exercise 22

4.4 Addition and Subtraction of Polynomials

a) Addition of Polynomials

To add two polynomials, we can write a plus sign between them and then collect like terms. Depending on the situation, you may see polynomials written in descending order, ascending order, or neither. Generally, if an exercise is written in a particular order, we write the answer in that same order.

EXAMPLE 1 Add: $(-3x^3 + 2x - 4) + (4x^3 + 3x^2 + 2)$.

$$(-3x^3 + 2x - 4) + (4x^3 + 3x^2 + 2)$$
$$= (-3 + 4)x^3 + 3x^2 + 2x + (-4 + 2) \qquad \text{Collecting like terms}$$
$$= x^3 + 3x^2 + 2x - 2$$

EXAMPLE 2 Add:

$$\left(\tfrac{2}{3}x^4 + 3x^2 - 2x + \tfrac{1}{2}\right) + \left(-\tfrac{1}{3}x^4 + 5x^3 - 3x^2 + 3x - \tfrac{1}{2}\right).$$

We have

$$\left(\tfrac{2}{3}x^4 + 3x^2 - 2x + \tfrac{1}{2}\right) + \left(-\tfrac{1}{3}x^4 + 5x^3 - 3x^2 + 3x - \tfrac{1}{2}\right)$$
$$= \left(\tfrac{2}{3} - \tfrac{1}{3}\right)x^4 + 5x^3 + (3 - 3)x^2 + (-2 + 3)x + \left(\tfrac{1}{2} - \tfrac{1}{2}\right) \qquad \begin{array}{l}\text{Collecting}\\ \text{like terms}\end{array}$$
$$= \tfrac{1}{3}x^4 + 5x^3 + x.$$

We can add polynomials as we do because they represent numbers. After some practice, you will be able to add mentally.

> Do Margin Exercises 1–4.

EXAMPLE 3 Add: $(3x^2 - 2x + 2) + (5x^3 - 2x^2 + 3x - 4)$.

$$(3x^2 - 2x + 2) + (5x^3 - 2x^2 + 3x - 4)$$
$$= 5x^3 + (3 - 2)x^2 + (-2 + 3)x + (2 - 4) \qquad \begin{array}{l}\text{You might do this}\\ \text{step mentally.}\end{array}$$
$$= 5x^3 + x^2 + x - 2 \qquad \text{Then you would write only this.}$$

> Do Exercises 5 and 6 on the following page.

We can also add polynomials by writing like terms in columns.

EXAMPLE 4 Add: $9x^5 - 2x^3 + 6x^2 + 3$ and $5x^4 - 7x^2 + 6$ and $3x^6 - 5x^5 + x^2 + 5$.

We arrange the polynomials with the like terms in columns.

$$\begin{array}{lllll}
9x^5 & -2x^3 + 6x^2 + & 3 & \\
 5x^4 & -7x^2 + & 6 & \text{We leave spaces for missing terms.}\\
3x^6 - 5x^5 & +\; x^2 + & 5 & \\
\hline
3x^6 + 4x^5 + 5x^4 - 2x^3 & +14 & \text{Adding}
\end{array}$$

We write the answer as $3x^6 + 4x^5 + 5x^4 - 2x^3 + 14$ without the space.

OBJECTIVES

a Add polynomials.

b Simplify the opposite of a polynomial.

c Subtract polynomials.

d Use polynomials to represent perimeter and area.

SKILL TO REVIEW

Objective 1.4a: Subtract real numbers and simplify combinations of additions and subtractions.

Simplify.
1. $-4 - (-8)$
2. $-5 - 6 + 4$

Add.

1. $(3x^2 + 2x - 2) + (-2x^2 + 5x + 5)$

2. $(-4x^5 + x^3 + 4) + (7x^4 + 2x^2)$

3. $(31x^4 + x^2 + 2x - 1) + (-7x^4 + 5x^3 - 2x + 2)$

4. $(17x^3 - x^2 + 3x + 4) + \left(-15x^3 + x^2 - 3x - \tfrac{2}{3}\right)$

Answers

Skill to Review:
1. 4   2. -7

Margin Exercises:
1. $x^2 + 7x + 3$
2. $-4x^5 + 7x^4 + x^3 + 2x^2 + 4$
3. $24x^4 + 5x^3 + x^2 + 1$
4. $2x^3 + \dfrac{10}{3}$

Add mentally. Try to write just the answer.

5. $(4x^2 - 5x + 3) +$
$(-2x^2 + 2x - 4)$

6. $(3x^3 - 4x^2 - 5x + 3) +$
$\left(5x^3 + 2x^2 - 3x - \dfrac{1}{2}\right)$

Add.

7. $\quad\quad -2x^3 + 5x^2 - 2x + \;\;4$
$\quad x^4 \quad\quad\quad + 6x^2 + 7x - 10$
$\quad -9x^4 + 6x^3 + \;\;x^2 \quad\quad\quad - \;\;2$

8. $-3x^3 + 5x + 2$ and
$x^3 + x^2 + 5$ and
$x^3 - 2x - 4$

Simplify.

9. $-(4x^3 - 6x + 3)$

10. $-(5x^4 + 3x^2 + 7x - 5)$

11. $-\left(14x^{10} - \dfrac{1}{2}x^5 + 5x^3 - x^2 + 3x\right)$

Subtract.

12. $(7x^3 + 2x + 4) - (5x^3 - 4)$

13. $(-3x^2 + 5x - 4) -$
$(-4x^2 + 11x - 2)$

Answers

5. $2x^2 - 3x - 1$ **6.** $8x^3 - 2x^2 - 8x + \dfrac{5}{2}$

7. $-8x^4 + 4x^3 + 12x^2 + 5x - 8$
8. $-x^3 + x^2 + 3x + 3$ **9.** $-4x^3 + 6x - 3$
10. $-5x^4 - 3x^2 - 7x + 5$

11. $-14x^{10} + \dfrac{1}{2}x^5 - 5x^3 + x^2 - 3x$

12. $2x^3 + 2x + 8$ **13.** $x^2 - 6x - 2$

Do Exercises 7 and 8.

b Opposites of Polynomials

In Section 1.8, we used the property of -1 to show that we can find the opposite of an expression. For example, the opposite of $x - 2y + 5$ can be written as

$$-(x - 2y + 5).$$

We find an equivalent expression by changing the sign of every term:

$$-(x - 2y + 5) = -x + 2y - 5.$$

We use this concept when we subtract polynomials.

OPPOSITES OF POLYNOMIALS

To find an equivalent polynomial for the **opposite**, or **additive inverse**, of a polynomial, change the sign of every term. This is the same as multiplying by -1.

EXAMPLE 5 Simplify: $-(x^2 - 3x + 4)$.

$$-(x^2 - 3x + 4) = -x^2 + 3x - 4$$

EXAMPLE 6 Simplify: $-(-t^3 - 6t^2 - t + 4)$.

$$-(-t^3 - 6t^2 - t + 4) = t^3 + 6t^2 + t - 4$$

EXAMPLE 7 Simplify: $-\left(-7x^4 - \dfrac{5}{9}x^3 + 8x^2 - x + 67\right)$.

$$-\left(-7x^4 - \tfrac{5}{9}x^3 + 8x^2 - x + 67\right) = 7x^4 + \tfrac{5}{9}x^3 - 8x^2 + x - 67$$

Do Exercises 9–11.

c Subtraction of Polynomials

Recall that we can subtract a real number by adding its opposite, or additive inverse: $a - b = a + (-b)$. This allows us to subtract polynomials.

EXAMPLE 8 Subtract:

$$(9x^5 + x^3 - 2x^2 + 4) - (2x^5 + x^4 - 4x^3 - 3x^2).$$

We have

$(9x^5 + x^3 - 2x^2 + 4) - (2x^5 + x^4 - 4x^3 - 3x^2)$

$\quad = 9x^5 + x^3 - 2x^2 + 4 + [-(2x^5 + x^4 - 4x^3 - 3x^2)]$ Adding the opposite

$\quad = 9x^5 + x^3 - 2x^2 + 4 - 2x^5 - x^4 + 4x^3 + 3x^2$ Finding the opposite by changing the sign of *each* term

$\quad = 7x^5 - x^4 + 5x^3 + x^2 + 4.$ Adding (collecting like terms)

Do Exercises 12 and 13.

As with similar work in Section 1.8, we combine steps by changing the sign of each term of the polynomial being subtracted and collecting like terms. Try to do this mentally as much as possible.

EXAMPLE 9 Subtract: $(9x^5 + x^3 - 2x) - (-2x^5 + 5x^3 + 6)$.

$$(9x^5 + x^3 - 2x) - (-2x^5 + 5x^3 + 6)$$
$$= 9x^5 + x^3 - 2x + 2x^5 - 5x^3 - 6 \qquad \text{Finding the opposite by changing the sign of each term}$$
$$= 11x^5 - 4x^3 - 2x - 6 \qquad \text{Adding (collecting like terms)}$$

Do Exercises 14 and 15.

We can use columns to subtract. We replace coefficients with their opposites, as shown in Example 9.

EXAMPLE 10 Write in columns and subtract:

$$(5x^2 - 3x + 6) - (9x^2 - 5x - 3).$$

a)
$$\begin{array}{l} 5x^2 - 3x + 6 \\ -(9x^2 - 5x - 3) \end{array} \qquad \text{Writing like terms in columns}$$

b)
$$\begin{array}{l} 5x^2 - 3x + 6 \\ -9x^2 + 5x + 3 \end{array} \qquad \text{Changing signs}$$

c)
$$\begin{array}{l} 5x^2 - 3x + 6 \\ -9x^2 + 5x + 3 \\ \hline -4x^2 + 2x + 9 \end{array} \qquad \text{Adding}$$

If you can do so without error, you can arrange the polynomials in columns and write just the answer, remembering to change the signs and add.

EXAMPLE 11 Write in columns and subtract:

$$(x^3 + x^2 + 2x - 12) - (-2x^3 + x^2 - 3x).$$

$$\begin{array}{l} x^3 + x^2 + 2x - 12 \\ -(-2x^3 + x^2 - 3x \qquad) \\ \hline 3x^3 \qquad\quad + 5x - 12 \end{array}$$
Leaving space for the missing term
Changing the signs and adding

Do Exercises 16 and 17.

(d) Polynomials and Geometry

EXAMPLE 12 Find a polynomial for the sum of the areas of these rectangles.

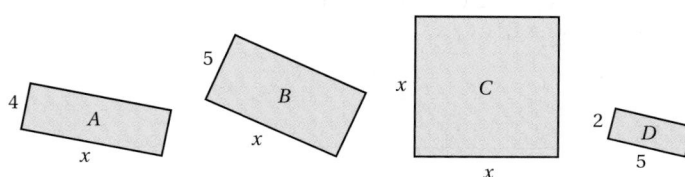

Recall that the area of a rectangle is the product of the length and the width. The sum of the areas is a sum of products. We find these products and then collect like terms.

Subtract.

14. $(-6x^4 + 3x^2 + 6) - (2x^4 + 5x^3 - 5x^2 + 7)$

15. $\left(\dfrac{3}{2}x^3 - \dfrac{1}{2}x^2 + 0.3\right) - \left(\dfrac{1}{2}x^3 + \dfrac{1}{2}x^2 + \dfrac{4}{3}x + 1.2\right)$

Write in columns and subtract.

16. $(4x^3 + 2x^2 - 2x - 3) - (2x^3 - 3x^2 + 2)$

17. $(2x^3 + x^2 - 6x + 2) - (x^5 + 4x^3 - 2x^2 - 4x)$

Answers

14. $-8x^4 - 5x^3 + 8x^2 - 1$

15. $x^3 - x^2 - \dfrac{4}{3}x - 0.9$

16. $2x^3 + 5x^2 - 2x - 5$

17. $-x^5 - 2x^3 + 3x^2 - 2x + 2$

18. Find a polynomial for the sums of the perimeters and of the areas of the rectangles.

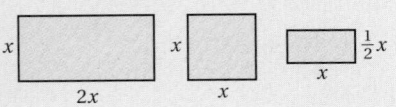

Area of A	plus	Area of B	plus	Area of C	plus	Area of D
$4 \cdot x$	$+$	$5 \cdot x$	$+$	$x \cdot x$	$+$	$2 \cdot 5$

We collect like terms:

$$4x + 5x + x^2 + 10 = x^2 + 9x + 10.$$

Do Exercise 18.

19. Lawn Area. An 8-ft by 8-ft shed is placed on a lawn x ft on a side. Find a polynomial for the remaining area.

EXAMPLE 13 *Lawn Area.* A water fountain with a 4-ft by 4-ft square base is placed in a park in a square grassy area that is x ft on a side. To determine the amount of grass seed needed for the lawn, find a polynomial for the grassy area.

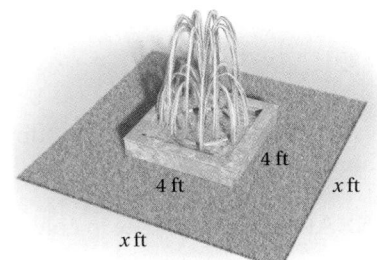

We make a drawing of the situation as shown here. We then reword the problem and write the polynomial as follows:

Area of grassy area	$-$	Area of base of fountain	$=$	Area left over
$x \cdot x$	$-$	$4 \cdot 4$	$=$	Area left over.

Then $(x^2 - 16)$ ft^2 = Area left over.

Do Exercise 19.

Calculator Corner

Checking Addition and Subtraction of Polynomials A table set in AUTO mode can be used to perform a partial check that polynomials have been added or subtracted correctly. To check Example 3, we enter $y_1 = (3x^2 - 2x + 2) + (5x^3 - 2x^2 + 3x - 4)$ and $y_2 = 5x^3 + x^2 + x - 2$. If the addition has been done correctly, the values of y_1 and y_2 will be the same regardless of the table settings used.

A graph can also be used to check addition and subtraction. See the Calculator Corner on p. 276 for the procedure.

X	Y1	Y2
-2	-40	-40
-1	-7	-7
0	-2	-2
1	5	5
2	44	44
3	145	145
4	338	338
X = -2		

Exercises: Use a table to determine whether the sum or the difference is correct.

1. $(-3x^3 + 2x - 4) + (4x^3 + 3x^2 + 2) = x^3 + 3x^2 + 2x - 2$

2. $(x^3 - 2x^2 + 3x - 7) + (3x^2 - 4x + 5) = x^3 + x^2 - x - 2$

3. $(5x^2 - 7x + 4) + (2x^2 + 3x - 6) = 7x^2 + 4x - 2$

4. $(9x^5 + x^3 - 2x) - (-2x^5 + 5x^3 + 6) = 11x^5 - 4x^3 - 2x - 6$

5. $(3x^4 - 2x^2 - 1) - (2x^4 - 3x^2 - 4) = x^4 + x^2 - 5$

6. $(-2x^3 + 3x^2 - 4x + 5) - (3x^2 + 2x + 8) = -2x^3 - 6x - 3$

Answers

18. Sum of perimeters: $13x$; sum of areas: $\frac{7}{2}x^2$

19. $(x^2 - 64)$ ft^2

a Add.

1. $(3x + 2) + (-4x + 3)$

2. $(6x + 1) + (-7x + 2)$

3. $(-6x + 2) + \left(x^2 + \frac{1}{2}x - 3\right)$

4. $\left(x^2 - \frac{5}{3}x + 4\right) + (8x - 9)$

5. $(x^2 - 9) + (x^2 + 9)$

6. $(x^3 + x^2) + (2x^3 - 5x^2)$

7. $(3x^2 - 5x + 10) + (2x^2 + 8x - 40)$

8. $(6x^4 + 3x^3 - 1) + (4x^2 - 3x + 3)$

9. $(1.2x^3 + 4.5x^2 - 3.8x) + (-3.4x^3 - 4.7x^2 + 23)$

10. $(0.5x^4 - 0.6x^2 + 0.7) + (2.3x^4 + 1.8x - 3.9)$

11. $(1 + 4x + 6x^2 + 7x^3) + (5 - 4x + 6x^2 - 7x^3)$

12. $(3x^4 - 6x - 5x^2 + 5) + (6x^2 - 4x^3 - 1 + 7x)$

13. $\left(\frac{1}{4}x^4 + \frac{2}{3}x^3 + \frac{5}{8}x^2 + 7\right) + \left(-\frac{3}{4}x^4 + \frac{3}{8}x^2 - 7\right)$

14. $\left(\frac{1}{3}x^9 + \frac{1}{5}x^5 - \frac{1}{2}x^2 + 7\right) +$ $\left(-\frac{1}{5}x^9 + \frac{1}{4}x^4 - \frac{3}{5}x^5 + \frac{3}{4}x^2 + \frac{1}{2}\right)$

15. $(0.02x^5 - 0.2x^3 + x + 0.08) +$ $(-0.01x^5 + x^4 - 0.8x - 0.02)$

16. $(0.03x^6 + 0.05x^3 + 0.22x + 0.05) +$ $\left(\frac{7}{100}x^6 - \frac{3}{100}x^3 + 0.5\right)$

17. $(9x^8 - 7x^4 + 2x^2 + 5) + (8x^7 + 4x^4 - 2x) +$ $(-3x^4 + 6x^2 + 2x - 1)$

18. $(4x^5 - 6x^3 - 9x + 1) + (6x^3 + 9x^2 + 9x) +$ $(-4x^3 + 8x^2 + 3x - 2)$

19. $\begin{aligned} & 0.15x^4 + 0.10x^3 - \ \ 0.9x^2 \\ & \qquad\quad -\ 0.01x^3 + 0.01x^2 + x \\ & 1.25x^4 \qquad\qquad +\ 0.11x^2 \qquad +\ 0.01 \\ & \qquad\ \ 0.27x^3 \qquad\qquad\qquad\quad +\ 0.99 \\ & \underline{-0.35x^4 \qquad\qquad +\ \ 15x^2 \qquad -\ 0.03} \end{aligned}$

20. $\begin{aligned} & 0.05x^4 + 0.12x^3 - \ \ 0.5x^2 \\ & \qquad\quad -\ 0.02x^3 + 0.02x^2 + 2x \\ & 1.5x^4 \qquad\qquad\ +\ 0.01x^2 \qquad +\ 0.15 \\ & \qquad\ \ 0.25x^3 \qquad\qquad\qquad\quad +\ 0.85 \\ & \underline{-0.25x^4 \qquad\qquad +\ \ 10x^2 \qquad -\ 0.04} \end{aligned}$

b Simplify.

21. $-(-5x)$

22. $-(x^2 - 3x)$

23. $-\left(-x^2 + \frac{3}{2}x - 2\right)$

24. $-\left(-4x^3 - x^2 - \frac{1}{4}x\right)$

25. $-(12x^4 - 3x^3 + 3)$

26. $-(4x^3 - 6x^2 - 8x + 1)$

27. $-(3x - 7)$

28. $-(-2x + 4)$

29. $-(4x^2 - 3x + 2)$

30. $-(-6a^3 + 2a^2 - 9a + 1)$

31. $-\left(-4x^4 + 6x^2 + \frac{3}{4}x - 8\right)$

32. $-(-5x^4 + 4x^3 - x^2 + 0.9)$

c Subtract.

33. $(3x + 2) - (-4x + 3)$

34. $(6x + 1) - (-7x + 2)$

35. $(-6x + 2) - (x^2 + x - 3)$

36. $(x^2 - 5x + 4) - (8x - 9)$

37. $(x^2 - 9) - (x^2 + 9)$

38. $(x^3 + x^2) - (2x^3 - 5x^2)$

39. $(6x^4 + 3x^3 - 1) - (4x^2 - 3x + 3)$

40. $(-4x^2 + 2x) - (3x^3 - 5x^2 + 3)$

41. $(1.2x^3 + 4.5x^2 - 3.8x) - (-3.4x^3 - 4.7x^2 + 23)$

42. $(0.5x^4 - 0.6x^2 + 0.7) - (2.3x^4 + 1.8x - 3.9)$

43. $\left(\frac{5}{8}x^3 - \frac{1}{4}x - \frac{1}{3}\right) - \left(-\frac{1}{8}x^3 + \frac{1}{4}x - \frac{1}{3}\right)$

44. $\left(\frac{1}{5}x^3 + 2x^2 - 0.1\right) - \left(-\frac{2}{5}x^3 + 2x^2 + 0.01\right)$

45. $(0.08x^3 - 0.02x^2 + 0.01x) - (0.02x^3 + 0.03x^2 - 1)$

46. $(0.8x^4 + 0.2x - 1) - \left(\frac{7}{10}x^4 + \frac{1}{5}x - 0.1\right)$

Subtract.

47.
$$x^2 + 5x + 6$$
$$\underline{-(x^2 + 2x)}$$

48.
$$x^3 + 1$$
$$\underline{-(x^3 + x^2)}$$

49.
$$5x^4 + 6x^3 - 9x^2$$
$$\underline{-(-6x^4 - 6x^3 + 8x + 9)}$$

50.
$$5x^4 + 6x^2 - 3x + 6$$
$$\underline{-(6x^3 + 7x^2 - 8x - 9)}$$

51.
$$x^5 - 1$$
$$\underline{-(x^5 - x^4 + x^3 - x^2 + x - 1)}$$

52.
$$x^5 + x^4 - x^3 + x^2 - x + 2$$
$$\underline{-(x^5 - x^4 + x^3 - x^2 - x + 2)}$$

 Solve.

Find a polynomial for the perimeter of each figure.

53.

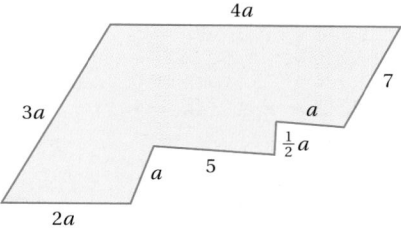

54.

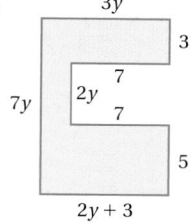

55. Find a polynomial for the sum of the areas of these rectangles.

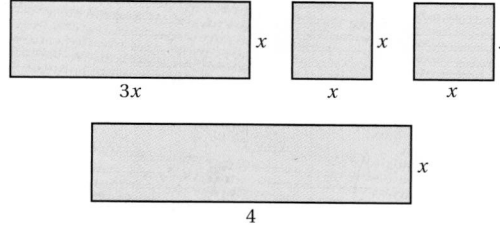

56. Find a polynomial for the sum of the areas of these circles.

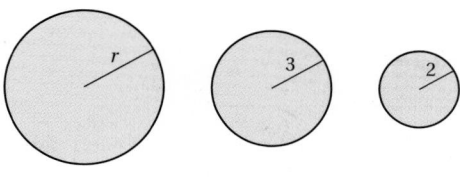

Find two algebraic expressions for the area of each figure. First, regard the figure as one large rectangle, and then regard the figure as a sum of four smaller rectangles.

57.

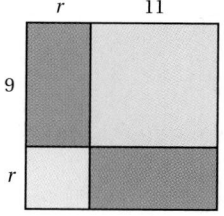

58.

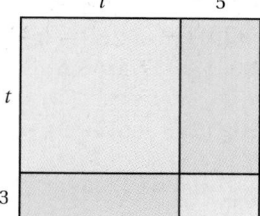

59.

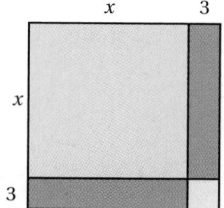

60.

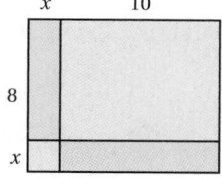

Find a polynomial for the shaded area of each figure.

61.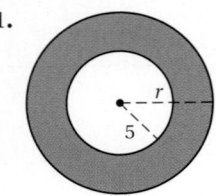

62.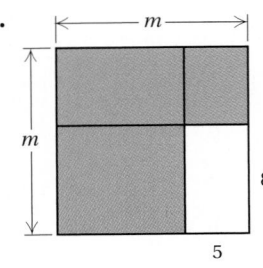

63.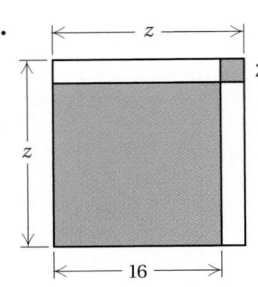

64.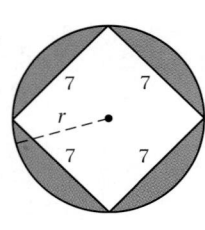

Skill Maintenance

Solve. [2.3b]

65. $8x + 3x = 66$

66. $5x - 7x = 38$

67. $\frac{3}{8}x + \frac{1}{4} - \frac{3}{4}x = \frac{11}{16} + x$

68. $5x - 4 = 26 - x$

69. $1.5x - 2.7x = 22 - 5.6x$

70. $3x - 3 = -4x + 4$

Solve. [2.3c]

71. $6(y - 3) - 8 = 4(y + 2) + 5$

72. $8(5x + 2) = 7(6x - 3)$

Solve. [2.7e]

73. $3x - 7 \le 5x + 13$

74. $2(x - 4) > 5(x - 3) + 7$

Synthesis

Find a polynomial for the surface area of each right rectangular solid.

75.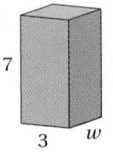

76.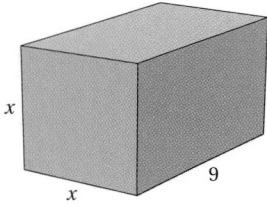

77.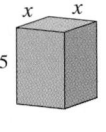

78.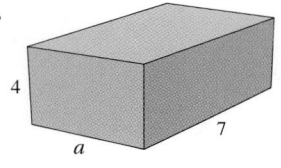

79. Find $(y - 2)^2$ using the four parts of this square.

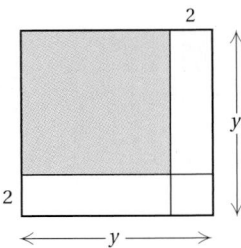

Simplify.

80. $(3x^2 - 4x + 6) - (-2x^2 + 4) + (-5x - 3)$

81. $(7y^2 - 5y + 6) - (3y^2 + 8y - 12) + (8y^2 - 10y + 3)$

82. $(-4 + x^2 + 2x^3) - (-6 - x + 3x^3) - (-x^2 - 5x^3)$

83. $(-y^4 - 7y^3 + y^2) + (-2y^4 + 5y - 2) - (-6y^3 + y^2)$

Mid-Chapter Review

Concept Reinforcement

Determine whether each statement is true or false.

_____ **1.** a^n and a^{-n} are reciprocals. [4.1f]

_____ **2.** $x^2 \cdot x^3 = x^6$ [4.1d]

_____ **3.** $-5y^4$ and $-5y^2$ are like terms. [4.3c]

_____ **4.** $4920^0 = 1$ [4.1b]

Guided Solutions

Fill in each blank with the number or variable that creates a correct statement or solution.

5. Collect like terms: $4w^3 + 6w - 8w^3 - 3w$. [4.3e]

$$4w^3 + 6w - 8w^3 - 3w = (4 - 8)\,\square + (6 - 3)\,\square$$
$$= \square\,w^3 + \square\,w$$

6. Subtract: $(3y^4 - y^2 + 11) - (y^4 - 4y^2 + 5)$. [4.4c]

$$(3y^4 - y^2 + 11) - (y^4 - 4y^2 + 5) = 3y^4 - y^2 + 11\,\square\,y^4\,\square\,4y^2\,\square\,5$$
$$= \square\,y^4 + \square\,y^2 + \square$$

Mixed Review

Evaluate. [4.1b, c]

7. z^1

8. 4.56^0

9. a^5, when $a = -2$

10. $-x^3$, when $x = -1$

Multiply and simplify. [4.1d, f]

11. $5^3 \cdot 5^4$

12. $(3a)^2 (3a)^7$

13. $x^{-8} \cdot x^5$

14. $t^4 \cdot t^{-4}$

Divide and simplify. [4.1e, f]

15. $\dfrac{7^8}{7^4}$

16. $\dfrac{x}{x^3}$

17. $\dfrac{w^5}{w^{-3}}$

18. $\dfrac{y^{-6}}{y^{-2}}$

Simplify. [4.2a, b]

19. $(3^5)^3$

20. $(x^{-3}y^2)^{-6}$

21. $\left(\dfrac{a^4}{5}\right)^6$

22. $\left(\dfrac{2y^3}{xz^2}\right)^{-2}$

Convert to scientific notation. [4.2c]

23. 25,430,000

24. 0.00012

Convert to decimal notation. [4.2c]

25. 3.6×10^{-5}

26. 1.44×10^8

Multiply or divide and write scientific notation for the result.   [4.2d]

27. $(3 \times 10^6)(2 \times 10^{-3})$ **28.** $\dfrac{1.2 \times 10^{-4}}{2.4 \times 10^2}$

Evaluate the polynomial when $x = -3$ and when $x = 2$.   [4.3a]

29. $-3x + 7$ **30.** $x^3 - 2x + 5$

Collect like terms and then arrange in descending order.   [4.3f]

31. $3x - 2x^5 + x - 5x^2 + 2$

32. $4x^3 - 9x^2 - 2x^3 + x^2 + 8x^6$

Identify the degree of each term of the polynomial and the degree of the polynomial.   [4.3g]

33. $5x^3 - x + 4$ **34.** $2x - x^4 + 3x^6$

Classify the polynomial as a monomial, a binomial, a trinomial, or none of these.   [4.3i]

35. $x - 9$ **36.** $x^5 - 2x^3 + 6x^2$

Add or subtract.   [4.4a, c]

37. $(3x^2 - 1) + (5x^2 + 6)$

38. $(x^3 + 2x - 5) + (4x^3 - 2x^2 - 6)$

39. $(5x - 8) - (9x + 2)$

40. $(0.1x^2 - 2.4x + 3.6) - (0.5x^2 + x - 5.4)$

41. Find a polynomial for the sum of the areas of these rectangles.   [4.4d]

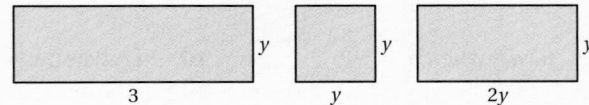

Understanding Through Discussion and Writing

42. Suppose that the length of a side of a square is three times the length of a side of a second square. How do the areas of the squares compare? Why?   [4.1d]

43. Suppose that the length of a side of a cube is twice the length of a side of a second cube. How do the volumes of the cubes compare? Why?   [4.1d]

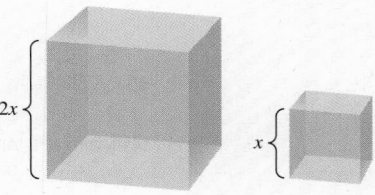

44. Explain in your own words when exponents should be added and when they should be multiplied.   [4.1d], [4.2a]

45. Without performing actual computations, explain why 3^{-29} is smaller than 2^{-29}.   [4.1f]

46. Is it better to evaluate a polynomial before or after like terms have been collected? Why?   [4.3a, e]

47. Is the sum of two binomials ever a trinomial? Why or why not?   [4.3i], [4.4a]

4.5 Multiplication of Polynomials

We now multiply polynomials using techniques based, for the most part, on the distributive laws, but also on the associative and commutative laws. As we proceed in this chapter, we will develop special ways to find certain products.

a Multiplying Monomials

Consider $(3x)(4x)$. We multiply as follows:

$$(3x)(4x) = 3 \cdot x \cdot 4 \cdot x \qquad \text{By the associative law of multiplication}$$
$$= 3 \cdot 4 \cdot x \cdot x \qquad \text{By the commutative law of multiplication}$$
$$= (3 \cdot 4)(x \cdot x) \qquad \text{By the associative law}$$
$$= 12x^2. \qquad \text{Using the product rule for exponents}$$

> **MULTIPLYING MONOMIALS**
>
> To find an equivalent expression for the product of two monomials, multiply the coefficients and then multiply the variables using the product rule for exponents.

EXAMPLES Multiply.

1. $5x \cdot 6x = (5 \cdot 6)(x \cdot x)$ By the associative and commutative laws

$\qquad = 30x^2$ Multiplying the coefficients and multiplying the variables

2. $(3x)(-x) = (3x)(-1x)$

$\qquad = (3)(-1)(x \cdot x) = -3x^2$

3. $(-7x^5)(4x^3) = (-7 \cdot 4)(x^5 \cdot x^3)$

$\qquad = -28x^{5+3}$ Adding the exponents

$\qquad = -28x^8$ Simplifying

After some practice, you will be able to multiply mentally. Multiply the coefficients and then the variables by keeping the base and adding the exponents. Write only the answer.

Do Margin Exercises 1–8.

b Multiplying a Monomial and Any Polynomial

To find an equivalent expression for the product of a monomial, such as $2x$, and a binomial, such as $5x + 3$, we use a distributive law and multiply each term of $5x + 3$ by $2x$.

EXAMPLE 4 Multiply: $2x(5x + 3)$.

$$2x(5x + 3) = (2x)(5x) + (2x)(3) \qquad \text{Using a distributive law}$$
$$= 10x^2 + 6x \qquad \text{Multiplying the monomials}$$

OBJECTIVES

a Multiply monomials.

b Multiply a monomial and any polynomial.

c Multiply two binomials.

d Multiply any two polynomials.

SKILL TO REVIEW

Objective 1.7c: Use the distributive laws to multiply expressions like 8 and $x - y$.

Multiply.

1. $3(x - 5)$

2. $2(3y + 4z - 1)$

Multiply.

1. $(3x)(-5)$ **2.** $(-x) \cdot x$

3. $(-x)(-x)$ **4.** $(-x^2)(x^3)$

5. $3x^5 \cdot 4x^2$ **6.** $(4y^5)(-2y^6)$

7. $(-7y^4)(-y)$ **8.** $7x^5 \cdot 0$

Answers

Skill to Review:
1. $3x - 15$ **2.** $6y + 8z - 2$

Margin Exercises:
1. $-15x$ **2.** $-x^2$ **3.** x^2 **4.** $-x^5$
5. $12x^7$ **6.** $-8y^{11}$ **7.** $7y^5$ **8.** 0

EXAMPLE 5 Multiply: $5x(2x^2 - 3x + 4)$.

$$5x(2x^2 - 3x + 4) = (5x)(2x^2) - (5x)(3x) + (5x)(4)$$
$$= 10x^3 - 15x^2 + 20x$$

> **MULTIPLYING A MONOMIAL AND A POLYNOMIAL**
>
> To multiply a monomial and a polynomial, multiply each term of the polynomial by the monomial.

EXAMPLE 6 Multiply: $-2x^2(x^3 - 7x^2 + 10x - 4)$.

$$-2x^2(x^3 - 7x^2 + 10x - 4)$$
$$= (-2x^2)(x^3) - (-2x^2)(7x^2) + (-2x^2)(10x) - (-2x^2)(4)$$
$$= -2x^5 + 14x^4 - 20x^3 + 8x^2$$

Do Exercises 9–11.

c) Multiplying Two Binomials

To find an equivalent expression for the product of two binomials, we use the distributive laws more than once. In Example 7, we use a distributive law three times.

EXAMPLE 7 Multiply: $(x + 5)(x + 4)$.

$$(x + 5)(x + 4) = x(x + 4) + 5(x + 4) \quad \text{Using a distributive law}$$
$$= x \cdot x + x \cdot 4 + 5 \cdot x + 5 \cdot 4 \quad \text{Using a distributive law on each part}$$
$$= x^2 + 4x + 5x + 20 \quad \text{Multiplying the monomials}$$
$$= x^2 + 9x + 20 \quad \text{Collecting like terms}$$

To visualize the product in Example 7, consider a rectangle of length $x + 5$ and width $x + 4$.

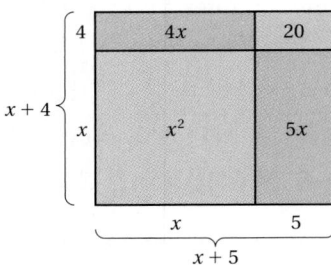

The total area can be expressed as $(x + 5)(x + 4)$ or, by adding the four smaller areas, $x^2 + 4x + 5x + 20$, or $x^2 + 9x + 20$.

Do Exercises 12–14.

Multiply.

9. $4x(2x + 4)$

10. $3t^2(-5t + 2)$

11. $-5x^3(x^3 + 5x^2 - 6x + 8)$

12. Multiply: $(y + 2)(y + 7)$.

a) Fill in the blanks in the steps of the solution below.

$(y + 2)(y + 7)$
$= y \cdot \underline{\quad} + 2 \cdot \underline{\quad}$
$= y \cdot \underline{\quad} + y \cdot \underline{\quad}$
$\quad + 2 \cdot \underline{\quad} + 2 \cdot \underline{\quad}$
$= \underline{\quad} + \underline{\quad}$
$\quad + \underline{\quad} + \underline{\quad}$
$= y^2 + \underline{\quad} + 14$

b) Write an algebraic expression that represents the total area of the four smaller rectangles in the figure shown here.

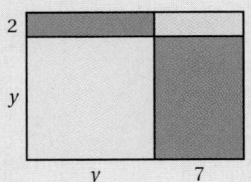

Multiply.

13. $(x + 8)(x + 5)$

14. $(x + 5)(x - 4)$

Answers

9. $8x^2 + 16x$ **10.** $-15t^3 + 6t^2$
11. $-5x^6 - 25x^5 + 30x^4 - 40x^3$
12. (a) $(y + 2)(y + 7)$
$\quad = y \cdot (y + 7) + 2 \cdot (y + 7)$
$\quad = y \cdot y + y \cdot 7$
$\quad \quad + 2 \cdot y + 2 \cdot 7$
$\quad = y^2 + 7y$
$\quad \quad + 2y + 14$
$\quad = y^2 + 9y + 14$
(b) $(y + 2)(y + 7)$, or $y^2 + 2y + 7y + 14$, or $y^2 + 9y + 14$ **13.** $x^2 + 13x + 40$
14. $x^2 + x - 20$

EXAMPLE 8 Multiply: $(4x + 3)(x - 2)$.

$$(4x + 3)(x - 2) = 4x(x - 2) + 3(x - 2)$$

Using a distributive law

$$= 4x \cdot x - 4x \cdot 2 + 3 \cdot x - 3 \cdot 2$$

Using a distributive law on each part

$$= 4x^2 - 8x + 3x - 6$$

Multiplying the monomials

$$= 4x^2 - 5x - 6$$

Collecting like terms

Do Exercises 15 and 16.

Multiply.

15. $(5x + 3)(x - 4)$

16. $(2x - 3)(3x - 5)$

(d) Multiplying Any Two Polynomials

Let's consider the product of a binomial and a trinomial. We use a distributive law four times. You may see ways to skip some steps and do the work mentally.

EXAMPLE 9 Multiply: $(x^2 + 2x - 3)(x^2 + 4)$.

$$(x^2 + 2x - 3)(x^2 + 4) = x^2(x^2 + 4) + 2x(x^2 + 4) - 3(x^2 + 4)$$

$$= x^2 \cdot x^2 + x^2 \cdot 4 + 2x \cdot x^2 + 2x \cdot 4 - 3 \cdot x^2 - 3 \cdot 4$$

$$= x^4 + 4x^2 + 2x^3 + 8x - 3x^2 - 12$$

$$= x^4 + 2x^3 + x^2 + 8x - 12$$

Do Exercises 17 and 18.

Multiply.

17. $(x^2 + 3x - 4)(x^2 + 5)$

18. $(3y^2 - 7)(2y^3 - 2y + 5)$

> **PRODUCT OF TWO POLYNOMIALS**
>
> To multiply two polynomials P and Q, select one of the polynomials—say, P. Then multiply each term of P by every term of Q and collect like terms.

To use columns for long multiplication, multiply each term in the top row by every term in the bottom row. We write like terms in columns, and then add the results. Such multiplication is like multiplying with whole numbers.

$$
\begin{array}{r}
3\ 2\ 1 \\
\times\ \ 1\ 2 \\
\hline
6\ 4\ 2 \\
3\ 2\ 1 \\
\hline
3\ 8\ 5\ 2
\end{array}
\qquad
\begin{array}{r}
300 + 20 + 1 \\
\times \qquad\ \ 10 + 2 \\
\hline
600 + 40 + 2 \\
3000 + 200 + 10 \\
\hline
3000 + 800 + 50 + 2
\end{array}
$$

Multiplying the top row by 2
Multiplying the top row by 10
Adding

EXAMPLE 10 Multiply: $(4x^3 - 2x^2 + 3x)(x^2 + 2x)$.

$$
\begin{array}{r}
4x^3 - 2x^2 + 3x \\
x^2 + 2x \\
\hline
8x^4 - 4x^3 + 6x^2 \\
4x^5 - 2x^4 + 3x^3 \\
\hline
4x^5 + 6x^4 - x^3 + 6x^2
\end{array}
$$

Multiplying the top row by $2x$
Multiplying the top row by x^2
Collecting like terms
Line up like terms in columns.

Answers

15. $5x^2 - 17x - 12$ **16.** $6x^2 - 19x + 15$
17. $x^4 + 3x^3 + x^2 + 15x - 20$
18. $6y^5 - 20y^3 + 15y^2 + 14y - 35$

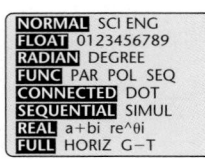

19. Multiply.

$$3x^2 - 2x - 5$$
$$2x^2 + x - 2$$

| **EXAMPLE 11** Multiply: $(2x^2 + 3x - 4)(2x^2 - x + 3)$.

$$
\begin{array}{r}
2x^2 + 3x - 4 \\
2x^2 - x + 3 \\
\hline
6x^2 + 9x - 12 \\
-2x^3 - 3x^2 + 4x \\
4x^4 + 6x^3 - 8x^2 \\
\hline
4x^4 + 4x^3 - 5x^2 + 13x - 12
\end{array}
$$

Multiplying by 3
Multiplying by $-x$
Multiplying by $2x^2$
Collecting like terms

Do Exercise 19.

| **EXAMPLE 12** Multiply: $(5x^3 - 3x + 4)(-2x^2 - 3)$.

When missing terms occur, it helps to leave spaces for them and align like terms as we multiply.

$$
\begin{array}{r}
5x^3 - 3x + 4 \\
-2x^2 - 3 \\
\hline
-15x^3 + 9x - 12 \\
-10x^5 + 6x^3 - 8x^2 \\
\hline
-10x^5 - 9x^3 - 8x^2 + 9x - 12
\end{array}
$$

Multiplying by -3
Multiplying by $-2x^2$
Collecting like terms

Do Exercises 20 and 21.

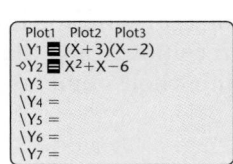

Multiply.

20. $3x^2 - 2x + 4$
$x + 5$

21. $-5x^2 + 4x + 2$
$-4x^2 - 8$

Calculator Corner

Checking Multiplication of Polynomials A partial check of multiplication of polynomials can be performed graphically. Consider the product $(x + 3)(x - 2) = x^2 + x - 6$. We will use two graph styles to determine whether this product is correct. First, we press **MODE** to determine whether SEQUENTIAL mode is selected. If it is not, we position the blinking cursor over SEQUENTIAL and then press **ENTER**. Next, on the Y= screen, we enter $y_1 = (x + 3)(x - 2)$ and $y_2 = x^2 + x - 6$. We will select the line-graph style for y_1 and the path style for y_2. To select these graph styles, we use ◁ to position the cursor over the icon to the left of the equation and press **ENTER** repeatedly until the desired style of icon appears, as shown below. Then we graph the equations.

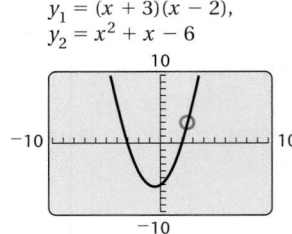

$y_1 = (x + 3)(x - 2)$,
$y_2 = x^2 + x - 6$

The graphing calculator will graph y_1 first as a solid line. Then it will graph y_2 as the circular cursor traces the leading edge of the graph, allowing us to determine visually whether the graphs coincide. In this case, the graphs appear to coincide, so the factorization is probably correct.

A table can also be used to perform a partial check of a product. See the Calculator Corner on p. 266 for the procedure.

Exercises Determine graphically whether each product is correct.

1. $(x + 5)(x + 4) = x^2 + 9x + 20$

2. $(4x + 3)(x - 2) = 4x^2 - 5x - 6$

3. $(5x + 3)(x - 4) = 5x^2 + 17x - 12$

4. $(2x - 3)(3x - 5) = 6x^2 - 19x - 15$

Answers

19. $6x^4 - x^3 - 18x^2 - x + 10$
20. $3x^3 + 13x^2 - 6x + 20$
21. $20x^4 - 16x^3 + 32x^2 - 32x - 16$

a Multiply.

1. $(8x^2)(5)$

2. $(4x^2)(-2)$

3. $(-x^2)(-x)$

4. $(-x^3)(x^2)$

5. $(8x^5)(4x^3)$

6. $(10a^2)(2a^2)$

7. $(0.1x^6)(0.3x^5)$

8. $(0.3x^4)(-0.8x^6)$

9. $\left(-\frac{1}{5}x^3\right)\left(-\frac{1}{3}x\right)$

10. $\left(-\frac{1}{4}x^4\right)\left(\frac{1}{5}x^8\right)$

11. $(-4x^2)(0)$

12. $(-4m^5)(-1)$

13. $(3x^2)(-4x^3)(2x^6)$

14. $(-2y^5)(10y^4)(-3y^3)$

b Multiply.

15. $2x(-x + 5)$

16. $3x(4x - 6)$

17. $-5x(x - 1)$

18. $-3x(-x - 1)$

19. $x^2(x^3 + 1)$

20. $-2x^3(x^2 - 1)$

21. $3x(2x^2 - 6x + 1)$

22. $-4x(2x^3 - 6x^2 - 5x + 1)$

23. $(-6x^2)(x^2 + x)$

24. $(-4x^2)(x^2 - x)$

25. $(3y^2)(6y^4 + 8y^3)$

26. $(4y^4)(y^3 - 6y^2)$

c Multiply.

27. $(x + 6)(x + 3)$

28. $(x + 5)(x + 2)$

29. $(x + 5)(x - 2)$

30. $(x + 6)(x - 2)$

31. $(x - 1)(x + 4)$

32. $(x - 8)(x + 7)$

33. $(x - 4)(x - 3)$

34. $(x - 7)(x - 3)$

35. $(x + 3)(x - 3)$

36. $(x + 6)(x - 6)$

37. $(x - 4)(x + 4)$

38. $(x - 9)(x + 9)$

39. $(3x + 5)(x + 2)$

40. $(2x + 6)(x + 3)$

41. $(5 - x)(5 - 2x)$

42. $(3 - 4x)(2 - x)$

43. $(2x + 5)(2x + 5)$

44. $(3x + 4)(3x + 4)$

45. $(x - 3)(x - 3)$

46. $(x - 6)(x - 6)$

47. $\left(x - \frac{5}{2}\right)\left(x + \frac{2}{5}\right)$

48. $\left(x + \frac{4}{3}\right)\left(x + \frac{3}{2}\right)$

49. $(x - 2.3)(x + 4.7)$

50. $(2x + 0.13)(2x - 0.13)$

Write an algebraic expression that represents the total area of the four smaller rectangles.

51.

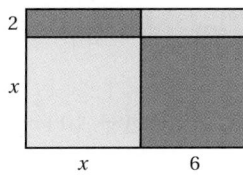

52.

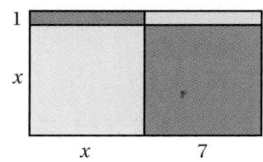

53.

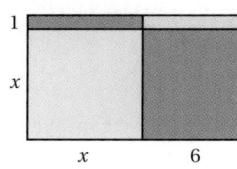

54.

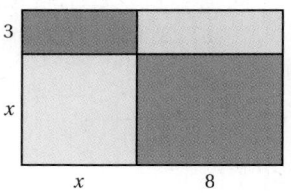

Draw and label rectangles similar to the one following Example 7 to illustrate each product.

55. $x(x + 5)$

56. $x(x + 2)$

57. $(x + 1)(x + 2)$

58. $(x + 3)(x + 1)$

59. $(x + 5)(x + 3)$

60. $(x + 4)(x + 6)$

d   Multiply.

61. $(x^2 + x + 1)(x - 1)$

62. $(x^2 + x - 2)(x + 2)$

63. $(2x + 1)(2x^2 + 6x + 1)$

64. $(3x - 1)(4x^2 - 2x - 1)$

65. $(y^2 - 3)(3y^2 - 6y + 2)$

66. $(3y^2 - 3)(y^2 + 6y + 1)$

67. $(x^3 + x^2)(x^3 + x^2 - x)$

68. $(x^3 - x^2)(x^3 - x^2 + x)$

69. $(-5x^3 - 7x^2 + 1)(2x^2 - x)$

70. $(-4x^3 + 5x^2 - 2)(5x^2 + 1)$

71. $(1 + x + x^2)(-1 - x + x^2)$

72. $(1 - x + x^2)(1 - x + x^2)$

73. $(2t^2 - t - 4)(3t^2 + 2t - 1)$

74. $(3a^2 - 5a + 2)(2a^2 - 3a + 4)$

75. $(x - x^3 + x^5)(x^2 - 1 + x^4)$

76. $(x - x^3 + x^5)(3x^2 + 3x^6 + 3x^4)$

77. $(x^3 + x^2 + x + 1)(x - 1)$

78. $(x + 2)(x^3 - x^2 + x - 2)$

79. $(x + 1)(x^3 + 7x^2 + 5x + 4)$

80. $(x + 2)(x^3 + 5x^2 + 9x + 3)$

81. $\left(x - \frac{1}{2}\right)\left(2x^3 - 4x^2 + 3x - \frac{2}{5}\right)$

82. $\left(x + \frac{1}{3}\right)\left(6x^3 - 12x^2 - 5x + \frac{1}{2}\right)$

Skill Maintenance

Simplify.

83. $-\dfrac{1}{4} - \dfrac{1}{2}$ [1.4a]

84. $-3.8 - (-10.2)$ [1.4a]

85. $(10 - 2)(10 + 2)$ [1.8d]

86. $10 - 2 + (-6)^2 \div 3 \cdot 2$ [1.8d]

Factor.   [1.7d]

87. $15x - 18y + 12$

88. $16x - 24y + 36$

89. $-9x - 45y + 15$

90. $100x - 100y + 1000a$

91. Graph: $y = \dfrac{1}{2}x - 3$.   [3.2a]

92. Solve: $4(x - 3) = 5(2 - 3x) + 1$.   [2.3c]

Synthesis

Find a polynomial for the shaded area of each figure.

93.

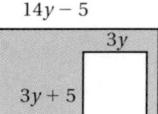

$14y - 5$

$3y$

$6y$

$3y + 5$

94.

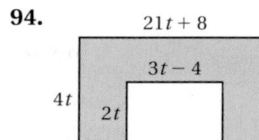

$21t + 8$

$3t - 4$

$4t$

$2t$

95. A box with a square bottom is to be made from a 12-in.-square piece of cardboard. Squares with side x are cut out of the corners and the sides are folded up. Find the polynomials for the volume and the outside surface area of the box.

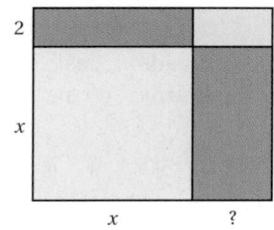

For each figure, determine what the missing number must be in order for the figure to have the given area.

96. Area $= x^2 + 7x + 10$

2

x

x $?$

97. Area $= x^2 + 8x + 15$

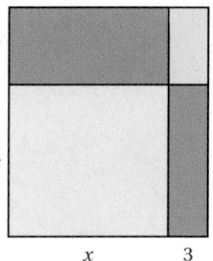

$?$

x

x 3

98. An open wooden box is a cube with side x cm. The box, including its bottom, is made of wood that is 1 cm thick. Find a polynomial for the interior volume of the cube.

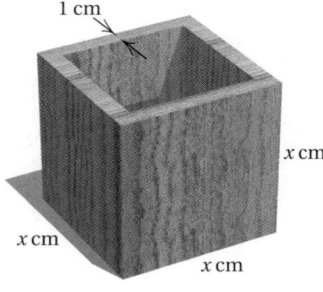

1 cm

x cm

x cm

x cm

99. Find a polynomial for the volume of the solid shown below.

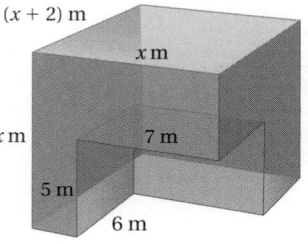

$(x + 2)$ m

x m

x m

7 m

5 m

6 m

Compute and simplify.

100. $(x + 3)(x + 6) + (x + 3)(x + 6)$

101. $(x - 2)(x - 7) - (x - 7)(x - 2)$

102. $(x + 5)^2 - (x - 3)^2$

103. Extend the pattern and simplify:

$$(x - a)(x - b)(x - c)(x - d) \cdots (x - z).$$

104. Use a graphing calculator to check your answers to Exercises 15, 29, and 61. Use graphs, tables, or both, as directed by your instructor.

4.6

Special Products

OBJECTIVES

a Multiply two binomials mentally using the FOIL method.

b Multiply the sum and the difference of two terms mentally.

c Square a binomial mentally.

d Find special products when polynomial products are mixed together.

We encounter certain products so often that it is helpful to have faster methods of computing. Such techniques are called *special products*. We now consider special ways of multiplying any two binomials.

a Products of Two Binomials Using FOIL

To multiply two binomials, we can select one binomial and multiply each term of that binomial by every term of the other. Then we collect like terms. Consider the product $(x + 3)(x + 7)$:

$$(x + 3)(x + 7) = x(x + 7) + 3(x + 7)$$
$$= x \cdot x + x \cdot 7 + 3 \cdot x + 3 \cdot 7$$
$$= x^2 + 7x + 3x + 21$$
$$= x^2 + 10x + 21.$$

This example illustrates a special technique for finding the product of two binomials:

First terms	Outside terms	Inside terms	Last terms

$$(x + 3)(x + 7) = x \cdot x + \quad 7 \cdot x \quad + 3 \cdot x \quad + 3 \cdot 7.$$

To remember this method of multiplying, we use the initials **FOIL**.

THE FOIL METHOD

To multiply two binomials, $A + B$ and $C + D$, multiply the First terms AC, the Outside terms AD, the Inside terms BC, and then the Last terms BD. Then collect like terms, if possible.

$$(A + B)(C + D) = AC + AD + BC + BD$$

1. Multiply First terms: AC.
2. Multiply Outside terms: AD.
3. Multiply Inside terms: BC.
4. Multiply Last terms: BD.

FOIL

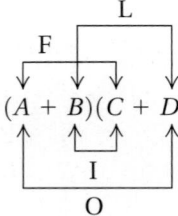

STUDY TIPS

MEMORIZING FORMULAS

Memorizing can be a very helpful tool in the study of mathematics. Don't underestimate its power as you consider the special products. Consider putting the rules, in words and in math symbols, on index cards and reviewing them many times.

EXAMPLE 1 Multiply: $(x + 8)(x^2 - 5)$.

We have

$$\overset{\text{F}}{} \quad \overset{\text{O}}{} \quad \overset{\text{I}}{} \quad \overset{\text{L}}{}$$
$$(x + 8)(x^2 - 5) = x \cdot x^2 + x \cdot (-5) + 8 \cdot x^2 + 8(-5)$$
$$= x^3 - 5x + 8x^2 - 40$$
$$= x^3 + 8x^2 - 5x - 40.$$

Since each of the original binomials is in descending order, we write the product in descending order, as is customary, but this is not a "must."

Often we can collect like terms after we have multiplied.

EXAMPLES Multiply.

2. $(x + 6)(x - 6) = x^2 - 6x + 6x - 36$ Using FOIL

$\qquad = x^2 - 36$ Collecting like terms

3. $(x + 7)(x + 4) = x^2 + 4x + 7x + 28$

$\qquad = x^2 + 11x + 28$

4. $(y - 3)(y - 2) = y^2 - 2y - 3y + 6$

$\qquad = y^2 - 5y + 6$

5. $(x^3 - 5)(x^3 + 5) = x^6 + 5x^3 - 5x^3 - 25$

$\qquad = x^6 - 25$

6. $(4t^3 + 5)(3t^2 - 2) = 12t^5 - 8t^3 + 15t^2 - 10$

> Do Exercises 1–8.

Multiply mentally, if possible. If you need extra steps, be sure to use them.

1. $(x + 3)(x + 4)$

2. $(x + 3)(x - 5)$

3. $(2x - 1)(x - 4)$

4. $(2x^2 - 3)(x - 2)$

5. $(6x^2 + 5)(2x^3 + 1)$

6. $(y^3 + 7)(y^3 - 7)$

7. $(t + 2)(t + 3)$

8. $(2x^4 + x^2)(-x^3 + x)$

EXAMPLES Multiply.

7. $\left(x - \frac{2}{3}\right)\left(x + \frac{2}{3}\right) = x^2 + \frac{2}{3}x - \frac{2}{3}x - \frac{4}{9}$

$\qquad = x^2 - \frac{4}{9}$

8. $(x^2 - 0.3)(x^2 - 0.3) = x^4 - 0.3x^2 - 0.3x^2 + 0.09$

$\qquad = x^4 - 0.6x^2 + 0.09$

9. $(3 - 4x)(7 - 5x^3) = 21 - 15x^3 - 28x + 20x^4$

$\qquad = 21 - 28x - 15x^3 + 20x^4$

(*Note*: If the original polynomials are in ascending order, it is natural to write the product in ascending order, but this is not a "must.")

10. $(5x^4 + 2x^3)(3x^2 - 7x) = 15x^6 - 35x^5 + 6x^5 - 14x^4$

$\qquad = 15x^6 - 29x^5 - 14x^4$

> Do Exercises 9–12.

Multiply.

9. $\left(x + \frac{4}{5}\right)\left(x - \frac{4}{5}\right)$

10. $(x^3 - 0.5)(x^2 + 0.5)$

11. $(2 + 3x^2)(4 - 5x^2)$

12. $(6x^3 - 3x^2)(5x^2 - 2x)$

We can show the FOIL method geometrically as follows.

The area of the large rectangle is $(A + B)(C + D)$.

The area of rectangle ① is AC.

The area of rectangle ② is AD.

The area of rectangle ③ is BC.

The area of rectangle ④ is BD.

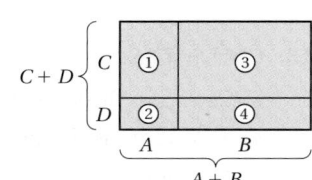

The area of the large rectangle is the sum of the areas of the smaller rectangles. Thus,

$$(A + B)(C + D) = AC + AD + BC + BD.$$

b Multiplying Sums and Differences of Two Terms

Consider the product of the sum and the difference of the same two terms, such as

$$(x + 2)(x - 2).$$

Answers

1. $x^2 + 7x + 12$ **2.** $x^2 - 2x - 15$
3. $2x^2 - 9x + 4$ **4.** $2x^3 - 4x^2 - 3x + 6$
5. $12x^5 + 10x^3 + 6x^2 + 5$ **6.** $y^6 - 49$
7. $t^2 + 5t + 6$ **8.** $-2x^7 + x^5 + x^3$
9. $x^2 - \dfrac{16}{25}$ **10.** $x^5 + 0.5x^3 - 0.5x^2 - 0.25$
11. $8 + 2x^2 - 15x^4$ **12.** $30x^5 - 27x^4 + 6x^3$

Since this is the product of two binomials, we can use FOIL. This type of product occurs so often, however, that it would be valuable if we could use an even faster method. To find a faster way to compute such a product, look for a pattern in the following:

a) $(x + 2)(x - 2) = x^2 - 2x + 2x - 4$ Using FOIL
$$= x^2 - 4;$$

b) $(3x - 5)(3x + 5) = 9x^2 + 15x - 15x - 25$
$$= 9x^2 - 25.$$

Do Exercises 13 and 14.

Do Exercises 13 and 14.

Perhaps you discovered in each case that when you multiply the two binomials, two terms are opposites, or additive inverses, which add to 0 and "drop out."

PRODUCT OF THE SUM AND THE DIFFERENCE OF TWO TERMS

The product of the sum and the difference of the same two terms is the square of the first term minus the square of the second term:

$$(A + B)(A - B) = A^2 - B^2.$$

It is helpful to memorize this rule in both words and symbols. (If you do forget it, you can, of course, use FOIL.)

EXAMPLES Multiply. (Carry out the rule and say the words as you go.)

$$(A + B)(A - B) = A^2 - B^2$$

11. $(x + 4)(x - 4) = x^2 - 4^2$ "The square of the first term, x^2, minus the square of the second, 4^2"

$$= x^2 - 16$$ Simplifying

12. $(5 + 2w)(5 - 2w) = 5^2 - (2w)^2$
$$= 25 - 4w^2$$

13. $(3x^2 - 7)(3x^2 + 7) = (3x^2)^2 - 7^2$
$$= 9x^4 - 49$$

14. $(-4x - 10)(-4x + 10) = (-4x)^2 - 10^2$
$$= 16x^2 - 100$$

15. $\left(x + \dfrac{3}{8}\right)\left(x - \dfrac{3}{8}\right) = x^2 - \left(\dfrac{3}{8}\right)^2 = x^2 - \dfrac{9}{64}$

Do Exercises 15–19.

Do Exercises 15–19.

c Squaring Binomials

Consider the square of a binomial, such as $(x + 3)^2$. This can be expressed as $(x + 3)(x + 3)$. Since this is the product of two binomials, we can use FOIL. But again, this type of product occurs so often that we would like to use an even faster method. Look for a pattern in the following.

Multiply.

13. $(x + 5)(x - 5)$

14. $(2x - 3)(2x + 3)$

Multiply.

15. $(x + 8)(x - 8)$

16. $(x - 7)(x + 7)$

17. $(6 - 4y)(6 + 4y)$

18. $(2x^3 - 1)(2x^3 + 1)$

19. $\left(x - \dfrac{2}{5}\right)\left(x + \dfrac{2}{5}\right)$

Answers

13. $x^2 - 25$ **14.** $4x^2 - 9$ **15.** $x^2 - 64$
16. $x^2 - 49$ **17.** $36 - 16y^2$ **18.** $4x^6 - 1$
19. $x^2 - \dfrac{4}{25}$

a) $(x + 3)^2 = (x + 3)(x + 3)$
$\quad\quad\quad\quad = x^2 + 3x + 3x + 9$
$\quad\quad\quad\quad = x^2 + 6x + 9;$

b) $(x - 3)^2 = (x - 3)(x - 3)$
$\quad\quad\quad\quad = x^2 - 3x - 3x + 9$
$\quad\quad\quad\quad = x^2 - 6x + 9;$

c) $(5 + 3p)^2 = (5 + 3p)(5 + 3p)$
$\quad\quad\quad\quad\quad = 25 + 15p + 15p + 9p^2$
$\quad\quad\quad\quad\quad = 25 + 30p + 9p^2;$

d) $(3x - 5)^2 = (3x - 5)(3x - 5)$
$\quad\quad\quad\quad\quad = 9x^2 - 15x - 15x + 25$
$\quad\quad\quad\quad\quad = 9x^2 - 30x + 25$

Do Exercises 20 and 21.

Multiply.

20. $(x + 8)(x + 8)$

21. $(x - 5)(x - 5)$

When squaring a binomial, we multiply a binomial by itself. Perhaps you noticed that two terms are the same and when added give twice the product of the terms in the binomial. The other two terms are squares.

SQUARE OF A BINOMIAL

The square of a sum or a difference of two terms is the square of the first term, plus twice the product of the two terms, plus the square of the last term:

$$(A + B)^2 = A^2 + 2AB + B^2; \quad (A - B)^2 = A^2 - 2AB + B^2.$$

It is helpful to memorize this rule in both words and symbols.

EXAMPLES Multiply. (Carry out the rule and say the words as you go.)

$(A + B)^2 = A^2 + 2 \cdot A \cdot B + B^2$

16. $(x + 3)^2 = x^2 + 2 \cdot x \cdot 3 + 3^2$ $\quad$ "x^2 plus 2 times x times 3 plus 3^2"
$\quad\quad\quad\quad = x^2 + 6x + 9$

$(A - B)^2 = A^2 - 2 \cdot A \cdot B + B^2$

17. $(t - 5)^2 = t^2 - 2 \cdot t \cdot 5 + 5^2$
$\quad\quad\quad\quad = t^2 - 10t + 25$

18. $(2x + 7)^2 = (2x)^2 + 2 \cdot 2x \cdot 7 + 7^2 = 4x^2 + 28x + 49$

19. $(5x - 3x^2)^2 = (5x)^2 - 2 \cdot 5x \cdot 3x^2 + (3x^2)^2 = 25x^2 - 30x^3 + 9x^4$

20. $(2.3 - 5.4m)^2 = 2.3^2 - 2(2.3)(5.4m) + (5.4m)^2$
$\quad\quad\quad\quad\quad\quad = 5.29 - 24.84m + 29.16m^2$

Do Exercises 22–27.

Multiply.

22. $(x + 2)^2$

23. $(a - 4)^2$

24. $(2x + 5)^2$

25. $(4x^2 - 3x)^2$

26. $(7.8 + 1.2y)(7.8 + 1.2y)$

27. $(3x^2 - 5)(3x^2 - 5)$

-------------------------------- *Caution!* --------------------------------

Although the square of a product is the product of the squares, the square of a sum is *not* the sum of the squares. That is, $(AB)^2 = A^2B^2$, but

$\quad\quad$ The term $2AB$ is missing.

$$(A + B)^2 \neq A^2 + B^2.$$

To illustrate this inequality, note, using the rules for order of operations, that

$$(7 + 5)^2 = 12^2 = 144,$$

whereas

$$7^2 + 5^2 = 49 + 25 = 74, \quad \text{and} \quad 74 \neq 144.$$

We can look at the rule for finding $(A + B)^2$ geometrically as follows. The area of the large square is

$$(A + B)(A + B) = (A + B)^2.$$

This is equal to the sum of the areas of the smaller rectangles:

$$A^2 + AB + AB + B^2 = A^2 + 2AB + B^2.$$

Thus, $(A + B)^2 = A^2 + 2AB + B^2$.

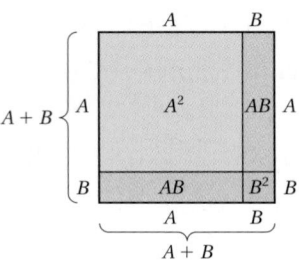

28. In the figure at right, describe in terms of area the sum $A^2 + B^2$. How can the figure be used to verify that $(A + B)^2 \neq A^2 + B^2$?

Do Exercise 28.

d Multiplication of Various Types

Let's now try several types of multiplications mixed together so that we can learn to sort them out. When you multiply, first see what kind of multiplication you have. Then use the best method.

MULTIPLYING TWO POLYNOMIALS

1. Is it the product of a monomial and a polynomial? If so, multiply each term of the polynomial by the monomial.

   Example: $5x(x + 7) = 5x \cdot x + 5x \cdot 7 = 5x^2 + 35x$

2. Is it the product of the sum and the difference of the *same* two terms? If so, use the following:

   $$(A + B)(A - B) = A^2 - B^2.$$

   The product of the sum and the difference of the same two terms is the difference of the squares. [The answer has 2 terms.]

   Example: $(x + 7)(x - 7) = x^2 - 7^2 = x^2 - 49$

3. Is the product the square of a binomial? If so, use the following:

   $$(A + B)(A + B) = (A + B)^2 = A^2 + 2AB + B^2,$$
   or $(A - B)(A - B) = (A - B)^2 = A^2 - 2AB + B^2.$

   The square of a binomial is the square of the first term, plus *twice* the product of the two terms, plus the square of the last term. [The answer has 3 terms.]

   Example: $(x + 7)(x + 7) = (x + 7)^2$
   $$= x^2 + 2 \cdot x \cdot 7 + 7^2 = x^2 + 14x + 49$$

4. Is it the product of two binomials other than those above? If so, use FOIL. [The answer will have 3 or 4 terms.]

   Example: $(x + 7)(x - 4) = x^2 - 4x + 7x - 28 = x^2 + 3x - 28$

5. Is it the product of two polynomials other than those above? If so, multiply each term of one by every term of the other. Use columns if you wish. [The answer will have 2 or more terms, usually more than 2 terms.]

   Example:

   $$(x^2 - 3x + 2)(x + 7) = x^2(x + 7) - 3x(x + 7) + 2(x + 7)$$
   $$= x^2 \cdot x + x^2 \cdot 7 - 3x \cdot x - 3x \cdot 7$$
   $$+ 2 \cdot x + 2 \cdot 7$$
   $$= x^3 + 7x^2 - 3x^2 - 21x + 2x + 14$$
   $$= x^3 + 4x^2 - 19x + 14$$

Answer

28. $(A + B)^2$ represents the area of the large square. This includes all four sections. $A^2 + B^2$ represents the area of only two of the sections.

Remember that FOIL will *always* work for two binomials. You can use it instead of either of rules 2 and 3, but those rules will make your work go faster.

EXAMPLE 21 Multiply: $(x + 3)(x - 3)$.

$(x + 3)(x - 3) = x^2 - 9$ Using method 2 (the product of the sum and the difference of two terms)

EXAMPLE 22 Multiply: $(t + 7)(t - 5)$.

$(t + 7)(t - 5) = t^2 + 2t - 35$ Using method 4, FOIL (the product of two binomials, but neither the square of a binomial nor the product of the sum and the difference of two terms)

EXAMPLE 23 Multiply: $(x + 6)(x + 6)$.

$(x + 6)(x + 6) = x^2 + 2(6)x + 36$ Using method 3 (the square of a binomial sum)

$\qquad\qquad\quad = x^2 + 12x + 36$

EXAMPLE 24 Multiply: $2x^3(9x^2 + x - 7)$.

$2x^3(9x^2 + x - 7) = 18x^5 + 2x^4 - 14x^3$ Using method 1 (the product of a monomial and a trinomial; multiplying each term of the trinomial by the monomial)

EXAMPLE 25 Multiply: $(5x^3 - 7x)^2$.

$(5x^3 - 7x)^2 = 25x^6 - 2(5x^3)(7x) + 49x^2$ Using method 3 (the square of a binomial)

$\qquad\qquad\quad = 25x^6 - 70x^4 + 49x^2$

EXAMPLE 26 Multiply: $\left(3x + \frac{1}{4}\right)^2$.

$\left(3x + \frac{1}{4}\right)^2 = 9x^2 + 2(3x)\left(\frac{1}{4}\right) + \frac{1}{16}$ Using method 3 (the square of a binomial. To get the middle term, we find twice the product of $3x$ and $\frac{1}{4}$.)

$\qquad\qquad\quad = 9x^2 + \frac{3}{2}x + \frac{1}{16}$

EXAMPLE 27 Multiply: $\left(4x - \frac{3}{4}\right)^2$.

$\left(4x - \frac{3}{4}\right)^2 = 16x^2 - 2(4x)\left(\frac{3}{4}\right) + \frac{9}{16}$ Using method 3 (the square of a binomial)

$\qquad\qquad\quad = 16x^2 - 6x + \frac{9}{16}$

EXAMPLE 28 Multiply: $(p + 3)(p^2 + 2p - 1)$.

$$
\begin{array}{r}
p^2 + 2p - 1 \\
p + 3 \\
\hline
3p^2 + 6p - 3 \\
p^3 + 2p^2 - p \\
\hline
p^3 + 5p^2 + 5p - 3
\end{array}
$$

Using method 5 (the product of two polynomials)

Multiplying by 3

Multiplying by p

Do Exercises 29–36.

Multiply.

29. $(x + 5)(x + 6)$

30. $(t - 4)(t + 4)$

31. $4x^2(-2x^3 + 5x^2 + 10)$

32. $(9x^2 + 1)^2$

33. $(2a - 5)(2a + 8)$

34. $\left(5x + \frac{1}{2}\right)^2$

35. $\left(2x - \frac{1}{2}\right)^2$

36. $(x^2 - x + 4)(x - 2)$

Answers

29. $x^2 + 11x + 30$ **30.** $t^2 - 16$

31. $-8x^5 + 20x^4 + 40x^2$ **32.** $81x^4 + 18x^2 + 1$

33. $4a^2 + 6a - 40$ **34.** $25x^2 + 5x + \frac{1}{4}$

35. $4x^2 - 2x + \frac{1}{4}$ **36.** $x^3 - 3x^2 + 6x - 8$

Visualizing for Success

1

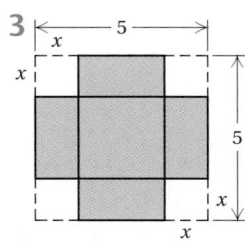

2

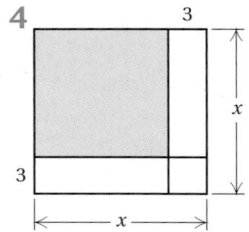

3

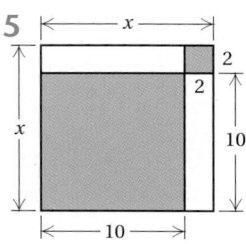

4

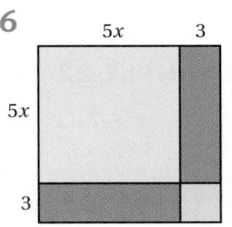

5

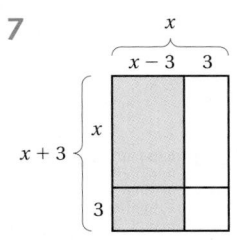

In each of Exercises 1–10, find two algebraic expressions for the shaded area of the figure from the list below.

A. $9 - 4x^2$

B. $x^2 - (x - 6)^2$

C. $(x + 3)(x - 3)$

D. $10^2 + 2^2$

E. $x^2 + 8x + 15$

F. $(x + 5)(x + 3)$

G. $x^2 - 6x + 9$

H. $(3 - 2x)^2 + 4x(3 - 2x)$

I. $(x + 3)^2$

J. $(5x + 3)^2$

K. $(5 - 2x)^2 + 4x(5 - 2x)$

L. $x^2 - 9$

M. 104

N. $x^2 - 15$

O. $12x - 36$

P. $25x^2 + 30x + 9$

Q. $(x - 5)(x - 3)$
　　$+ 3(x - 5) + 5(x - 3)$

R. $(x - 3)^2$

S. $25 - 4x^2$

T. $x^2 + 6x + 9$

Answers on page A-11

6

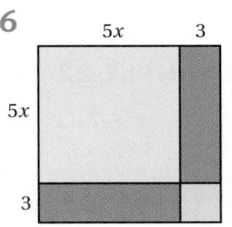

7

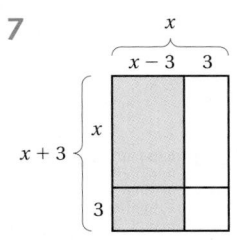

8

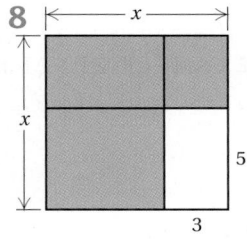

9

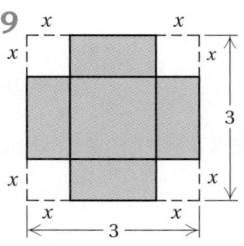

10

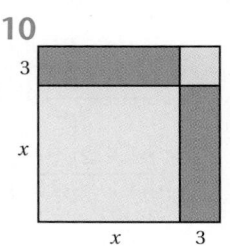

a Multiply. Try to write only the answer. If you need more steps, be sure to use them.

1. $(x + 1)(x^2 + 3)$

2. $(x^2 - 3)(x - 1)$

3. $(x^3 + 2)(x + 1)$

4. $(x^4 + 2)(x + 10)$

5. $(y + 2)(y - 3)$

6. $(a + 2)(a + 3)$

7. $(3x + 2)(3x + 2)$

8. $(4x + 1)(4x + 1)$

9. $(5x - 6)(x + 2)$

10. $(x - 8)(x + 8)$

11. $(3t - 1)(3t + 1)$

12. $(2m + 3)(2m + 3)$

13. $(4x - 2)(x - 1)$

14. $(2x - 1)(3x + 1)$

15. $\left(p - \frac{1}{4}\right)\left(p + \frac{1}{4}\right)$

16. $\left(q + \frac{3}{4}\right)\left(q + \frac{3}{4}\right)$

17. $(x - 0.1)(x + 0.1)$

18. $(x + 0.3)(x - 0.4)$

19. $(2x^2 + 6)(x + 1)$

20. $(2x^2 + 3)(2x - 1)$

21. $(-2x + 1)(x + 6)$

22. $(3x + 4)(2x - 4)$

23. $(a + 7)(a + 7)$

24. $(2y + 5)(2y + 5)$

25. $(1 + 2x)(1 - 3x)$

26. $(-3x - 2)(x + 1)$

27. $\left(\frac{3}{8}y - \frac{5}{6}\right)\left(\frac{3}{8}y - \frac{5}{6}\right)$

28. $\left(\frac{1}{5}x - \frac{2}{7}\right)\left(\frac{1}{5}x + \frac{2}{7}\right)$

29. $(x^2 + 3)(x^3 - 1)$

30. $(x^4 - 3)(2x + 1)$

31. $(3x^2 - 2)(x^4 - 2)$

32. $(x^{10} + 3)(x^{10} - 3)$

33. $(2.8x - 1.5)(4.7x + 9.3)$

34. $\left(x - \frac{3}{8}\right)\left(x + \frac{4}{7}\right)$

35. $(3x^5 + 2)(2x^2 + 6)$ **36.** $(1 - 2x)(1 + 3x^2)$ **37.** $(8x^3 + 1)(x^3 + 8)$ **38.** $(4 - 2x)(5 - 2x^2)$

39. $(4x^2 + 3)(x - 3)$ **40.** $(7x - 2)(2x - 7)$

41. $(4y^4 + y^2)(y^2 + y)$ **42.** $(5y^6 + 3y^3)(2y^6 + 2y^3)$

b Multiply mentally, if possible. If you need extra steps, be sure to use them.

43. $(x + 4)(x - 4)$ **44.** $(x + 1)(x - 1)$ **45.** $(2x + 1)(2x - 1)$ **46.** $(x^2 + 1)(x^2 - 1)$

47. $(5m - 2)(5m + 2)$ **48.** $(3x^4 + 2)(3x^4 - 2)$ **49.** $(2x^2 + 3)(2x^2 - 3)$ **50.** $(6x^5 - 5)(6x^5 + 5)$

51. $(3x^4 - 4)(3x^4 + 4)$ **52.** $(t^2 - 0.2)(t^2 + 0.2)$

53. $(x^6 - x^2)(x^6 + x^2)$ **54.** $(2x^3 - 0.3)(2x^3 + 0.3)$

55. $(x^4 + 3x)(x^4 - 3x)$ **56.** $\left(\frac{3}{4} + 2x^3\right)\left(\frac{3}{4} - 2x^3\right)$ **57.** $(x^{12} - 3)(x^{12} + 3)$ **58.** $(12 - 3x^2)(12 + 3x^2)$

59. $(2y^8 + 3)(2y^8 - 3)$ **60.** $\left(m - \frac{2}{3}\right)\left(m + \frac{2}{3}\right)$

61. $\left(\frac{5}{8}x - 4.3\right)\left(\frac{5}{8}x + 4.3\right)$ **62.** $(10.7 - x^3)(10.7 + x^3)$

c Multiply mentally, if possible. If you need extra steps, be sure to use them.

63. $(x + 2)^2$ **64.** $(2x - 1)^2$ **65.** $(3x^2 + 1)^2$ **66.** $\left(3x + \frac{3}{4}\right)^2$

67. $\left(a - \frac{1}{2}\right)^2$ **68.** $\left(2a - \frac{1}{5}\right)^2$ **69.** $(3 + x)^2$ **70.** $(x^3 - 1)^2$

71. $(x^2 + 1)^2$

72. $(8x - x^2)^2$

73. $(2 - 3x^4)^2$

74. $(6x^3 - 2)^2$

75. $(5 + 6t^2)^2$

76. $(3p^2 - p)^2$

77. $\left(x - \frac{5}{8}\right)^2$

78. $(0.3y + 2.4)^2$

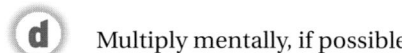 **Multiply mentally, if possible.**

79. $(3 - 2x^3)^2$

80. $(x - 4x^3)^2$

81. $4x(x^2 + 6x - 3)$

82. $8x(-x^5 + 6x^2 + 9)$

83. $\left(2x^2 - \frac{1}{2}\right)\left(2x^2 - \frac{1}{2}\right)$

84. $(-x^2 + 1)^2$

85. $(-1 + 3p)(1 + 3p)$

86. $(-3q + 2)(3q + 2)$

87. $3t^2(5t^3 - t^2 + t)$

88. $-6x^2(x^3 + 8x - 9)$

89. $(6x^4 + 4)^2$

90. $(8a + 5)^2$

91. $(3x + 2)(4x^2 + 5)$

92. $(2x^2 - 7)(3x^2 + 9)$

93. $(8 - 6x^4)^2$

94. $\left(\frac{1}{5}x^2 + 9\right)\left(\frac{3}{5}x^2 - 7\right)$

95. $(t - 1)(t^2 + t + 1)$

96. $(y + 5)(y^2 - 5y + 25)$

Compute each of the following and compare.

97. $3^2 + 4^2; (3 + 4)^2$

98. $6^2 + 7^2; (6 + 7)^2$

99. $9^2 - 5^2; (9 - 5)^2$

100. $11^2 - 4^2; (11 - 4)^2$

Find the total area of all the shaded rectangles.

101.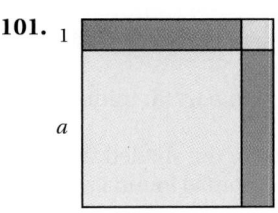

102.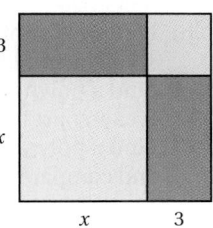

103.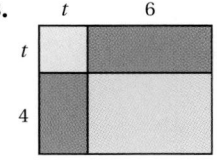

104.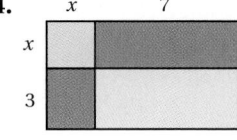

Skill Maintenance

105. *Electricity Usage.* In apartment 3B, lamps, an air conditioner, and a television set are all operating at the same time. The lamps use 10 times as many watts of electricity as the television set, and the air conditioner uses 40 times as many watts as the television set. The total wattage used in the apartment is 2550. How many watts are used by each appliance? [2.6a]

Solve. [2.3c]

106. $3x - 8x = 4(7 - 8x)$

107. $3(x - 2) = 5(2x + 7)$

108. $5(2x - 3) - 2(3x - 4) = 20$

Solve. [2.4b]

109. $3x - 2y = 12$, for y

110. $3a - 5d = 4$, for a

Synthesis

Multiply.

111. $5x(3x - 1)(2x + 3)$

112. $[(2x - 3)(2x + 3)](4x^2 + 9)$

113. $[(a - 5)(a + 5)]^2$

114. $(a - 3)^2(a + 3)^2$
(*Hint*: Examine Exercise 113.)

115. $(3t^4 - 2)^2(3t^4 + 2)^2$
(*Hint*: Examine Exercise 113.)

116. $[3a - (2a - 3)][3a + (2a - 3)]$

Solve.

117. $(x + 2)(x - 5) = (x + 1)(x - 3)$

118. $(2x + 5)(x - 4) = (x + 5)(2x - 4)$

119. *Factors and Sums.* To *factor* a number is to express it as a product. Since $12 = 4 \cdot 3$, we say that 12 is *factored* and that 4 and 3 are *factors* of 12. In the table below, the top number has been factored in such a way that the sum of the factors is the bottom number. For example, in the first column, 40 has been factored as $5 \cdot 8$, and $5 + 8 = 13$, the bottom number. Such thinking is important in algebra when we factor trinomials of the type $x^2 + bx + c$. Find the missing numbers in the table.

PRODUCT	40	63	36	72	−140	−96	48	168	110			
FACTOR	5									−9	−24	−3
FACTOR	8									−10	18	
SUM	13	16	−20	−38	−4	4	−14	−29	−21			18

120. Consider the rectangle below.

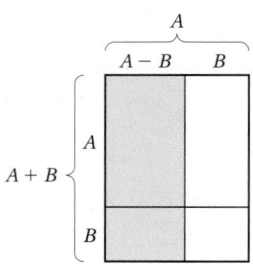

a) Find a polynomial for the area of the entire rectangle.

b) Find a polynomial for the sum of the areas of the two small unshaded rectangles.

c) Find a polynomial for the area in part (a) minus the area in part (b).

d) Find a polynomial for the area of the shaded region and compare this with the polynomial found in part (c).

Use the TABLE or GRAPH feature to check whether each of the following is correct.

121. $(x - 1)^2 = x^2 - 2x + 1$

122. $(x - 2)^2 = x^2 - 4x - 4$

123. $(x - 3)(x + 3) = x^2 - 6$

124. $(x - 3)(x + 2) = x^2 - x - 6$

4.7

Operations with Polynomials in Several Variables

The polynomials that we have been studying have only one variable. A **polynomial in several variables** is an expression like those you have already seen, but with more than one variable. Here are two examples:

$$3x + xy^2 + 5y + 4, \qquad 8xy^2z - 2x^3z - 13x^4y^2 + 15.$$

OBJECTIVES

a Evaluate a polynomial in several variables for given values of the variables.

b Identify the coefficients and the degrees of the terms of a polynomial and the degree of a polynomial.

c Collect like terms of a polynomial.

d Add polynomials.

e Subtract polynomials.

f Multiply polynomials.

a Evaluating Polynomials

EXAMPLE 1 Evaluate the polynomial $4 + 3x + xy^2 + 8x^3y^3$ when $x = -2$ and $y = 5$.

We replace x with -2 and y with 5:

$$\begin{aligned}
4 + 3x + xy^2 + 8x^3y^3 &= 4 + 3(-2) + (-2) \cdot 5^2 + 8(-2)^3 \cdot 5^3 \\
&= 4 + 3(-2) + (-2) \cdot 25 + 8(-8)(125) \\
&= 4 - 6 - 50 - 8000 \\
&= -8052.
\end{aligned}$$

EXAMPLE 2 *Male Caloric Needs.* The number of calories needed each day by a moderately active man who weighs w kilograms, is h centimeters tall, and is a years old can be estimated by the polynomial

$$19.18w + 7h - 9.52a + 92.4.$$

Steve is moderately active, weighs 82 kg, is 185 cm tall, and is 67 yr old. What are his daily caloric needs?

Source: Parker, M., *She Does Math.* Mathematical Association of America

Breakfast
Oatmeal with skim milk–231 calories
Cinnamon raisin bagel–350 calories
Orange juice–83 calories

Dinner
Chicken breast–142 calories
Wild rice–166 calories
Broccoli–42 calories
Cranberry sauce–209 calories

Lunch
Peanut butter and jelly sandwich–
1018 calories
Apple–81 calories

We evaluate the polynomial for $w = 82$, $h = 185$, and $a = 67$:

$$\begin{aligned}
19.18w &+ 7h - 9.52a + 92.4 \\
&= 19.18(82) + 7(185) - 9.52(67) + 92.4 \qquad \text{Substituting} \\
&= 2322.32.
\end{aligned}$$

Steve's daily caloric need is about 2322 calories.

Do Exercises 1–3.

1. Evaluate the polynomial
$$4 + 3x + xy^2 + 8x^3y^3$$
when $x = 2$ and $y = -5$.

2. Evaluate the polynomial
$$8xy^2 - 2x^3z - 13x^4y^2 + 5$$
when $x = -1$, $y = 3$, and $z = 4$.

3. Female Caloric Needs. The number of calories needed each day by a moderately active woman who weighs w pounds, is h inches tall, and is a years old can be estimated by the polynomial
$$917 + 6w + 6h - 6a.$$
Christine is moderately active, weighs 125 lb, is 64 in. tall, and is 27 yr old. What are her daily caloric needs?

Source: Parker, M., *She Does Math.* Mathematical Association of America

Answers

1. -7940 **2.** -176 **3.** 1889 calories

b Coefficients and Degrees

The **degree** of a term is the sum of the exponents of the variables. The **degree of a polynomial** is the degree of the term of highest degree.

EXAMPLE 3 Identify the coefficient and the degree of each term and the degree of the polynomial

$$9x^2y^3 - 14xy^2z^3 + xy + 4y + 5x^2 + 7.$$

TERM	COEFFICIENT	DEGREE	DEGREE OF THE POLYNOMIAL
$9x^2y^3$	9	5	
$-14xy^2z^3$	-14	6	6
xy	1	2	
$4y$	4	1	Think: $4y = 4y^1$.
$5x^2$	5	2	
7	7	0	Think: $7 = 7x^0$, or $7x^0y^0z^0$.

4. Identify the coefficient of each term:

$-3xy^2 + 3x^2y - 2y^3 + xy + 2.$

5. Identify the degree of each term and the degree of the polynomial

$4xy^2 + 7x^2y^3z^2 - 5x + 2y + 4.$

Do Exercises 4 and 5.

c Collecting Like Terms

Like terms have exactly the same variables with exactly the same exponents. For example,

$3x^2y^3$ and $-7x^2y^3$ are like terms;

$9x^4z^7$ and $12x^4z^7$ are like terms.

But

$13xy^5$ and $-2x^2y^5$ are *not* like terms, because the x-factors have different exponents;

and

$3xyz^2$ and $4xy$ are *not* like terms, because there is no factor of z^2 in the second expression.

Collecting like terms is based on the distributive laws.

EXAMPLES Collect like terms.

4. $5x^2y + 3xy^2 - 5x^2y - xy^2 = (5 - 5)x^2y + (3 - 1)xy^2 = 2xy^2$

5. $8a^2 - 2ab + 7b^2 + 4a^2 - 9ab - 17b^2 = 12a^2 - 11ab - 10b^2$

6. $7xy - 5xy^2 + 3xy^2 - 7 + 6x^3 + 9xy - 11x^3 + y - 1$
$\quad = 16xy - 2xy^2 - 5x^3 + y - 8$

Collect like terms.
6. $4x^2y + 3xy - 2x^2y$

7. $-3pq - 5pqr^3 - 12 + 8pq + 5pqr^3 + 4$

Do Exercises 6 and 7.

d Addition

We can find the sum of two polynomials in several variables by writing a plus sign between them and then collecting like terms.

EXAMPLE 7 Add: $(-5x^3 + 3y - 5y^2) + (8x^3 + 4x^2 + 7y^2)$.

$$(-5x^3 + 3y - 5y^2) + (8x^3 + 4x^2 + 7y^2)$$
$$= (-5 + 8)x^3 + 4x^2 + 3y + (-5 + 7)y^2$$
$$= 3x^3 + 4x^2 + 3y + 2y^2$$

EXAMPLE 8 Add:

$$(5xy^2 - 4x^2y + 5x^3 + 2) + (3xy^2 - 2x^2y + 3x^3y - 5).$$

We have

$$(5xy^2 - 4x^2y + 5x^3 + 2) + (3xy^2 - 2x^2y + 3x^3y - 5)$$
$$= (5 + 3)xy^2 + (-4 - 2)x^2y + 5x^3 + 3x^3y + (2 - 5)$$
$$= 8xy^2 - 6x^2y + 5x^3 + 3x^3y - 3.$$

Do Exercises 8–10.

e Subtraction

We subtract a polynomial by adding its opposite, or additive inverse. The opposite of the polynomial $4x^2y - 6x^3y^2 + x^2y^2 - 5y$ is

$$-(4x^2y - 6x^3y^2 + x^2y^2 - 5y) = -4x^2y + 6x^3y^2 - x^2y^2 + 5y.$$

EXAMPLE 9 Subtract:

$$(4x^2y + x^3y^2 + 3x^2y^3 + 6y + 10) - (4x^2y - 6x^3y^2 + x^2y^2 - 5y - 8).$$

We have

$$(4x^2y + x^3y^2 + 3x^2y^3 + 6y + 10) - (4x^2y - 6x^3y^2 + x^2y^2 - 5y - 8)$$
$$= 4x^2y + x^3y^2 + 3x^2y^3 + 6y + 10 - 4x^2y + 6x^3y^2 - x^2y^2 + 5y + 8$$

Finding the opposite by changing the sign of each term

$$= 7x^3y^2 + 3x^2y^3 - x^2y^2 + 11y + 18.$$ Collecting like terms. (Try to write just the answer!)

-------------------------- *Caution!* --------------------------

Do *not* add exponents when collecting like terms—that is,

$$7x^3 + 8x^3 \neq 15x^6; \leftarrow \text{Wrong}$$
$$7x^3 + 8x^3 = 15x^3. \leftarrow \text{Correct}$$

Do Exercises 11 and 12.

Add.

8. $(4x^3 + 4x^2 - 8y - 3) + (-8x^3 - 2x^2 + 4y + 5)$

9. $(13x^3y + 3x^2y - 5y) + (x^3y + 4x^2y - 3xy + 3y)$

10. $(-5p^2q^4 + 2p^2q^2 + 3q) + (6pq^2 + 3p^2q + 5)$

Subtract.

11. $(-4s^4t + s^3t^2 + 2s^2t^3) - (4s^4t - 5s^3t^2 + s^2t^2)$

12. $(-5p^4q + 5p^3q^2 - 3p^2q^3 - 7q^4 - 2) - (4p^4q - 4p^3q^2 + p^2q^3 + 2q^4 - 7)$

Answers

8. $-4x^3 + 2x^2 - 4y + 2$
9. $14x^3y + 7x^2y - 3xy - 2y$
10. $-5p^2q^4 + 2p^2q^2 + 3p^2q + 6pq^2 + 3q + 5$
11. $-8s^4t + 6s^3t^2 + 2s^2t^3 - s^2t^2$
12. $-9p^4q + 9p^3q^2 - 4p^2q^3 - 9q^4 + 5$

(f) Multiplication

To multiply polynomials in several variables, we can multiply each term of one by every term of the other. We can use columns for long multiplications as with polynomials in one variable. We multiply each term at the top by every term at the bottom. We write like terms in columns, and then we add.

EXAMPLE 10 Multiply: $(3x^2y - 2xy + 3y)(xy + 2y)$.

$$
\begin{array}{r}
3x^2y - 2xy + 3y \\
xy + 2y \\
\hline
6x^2y^2 - 4xy^2 + 6y^2 \\
3x^3y^2 - 2x^2y^2 + 3xy^2 \\
\hline
3x^3y^2 + 4x^2y^2 - xy^2 + 6y^2
\end{array}
$$

Multiplying by $2y$

Multiplying by xy

Adding

Do Exercises 13 and 14.

Where appropriate, we use the special products that we have learned.

EXAMPLES Multiply.

11. $(x^2y + 2x)(xy^2 + y^2) = x^3y^3 + x^2y^3 + 2x^2y^2 + 2xy^2$ Using FOIL

12. $(p + 5q)(2p - 3q) = 2p^2 - 3pq + 10pq - 15q^2$ Using FOIL
$$= 2p^2 + 7pq - 15q^2$$

$$
(A + B)^2 = A^2 + 2 \cdot A \cdot B + B^2
$$

13. $(3x + 2y)^2 = (3x)^2 + 2(3x)(2y) + (2y)^2 = 9x^2 + 12xy + 4y^2$

$$
(A - B)^2 = A^2 - 2 \cdot A \cdot B + B^2
$$

14. $(2y^2 - 5x^2y)^2 = (2y^2)^2 - 2(2y^2)(5x^2y) + (5x^2y)^2$
$$= 4y^4 - 20x^2y^3 + 25x^4y^2$$

$$
(A + B)(A - B) = A^2 - B^2
$$

15. $(3x^2y + 2y)(3x^2y - 2y) = (3x^2y)^2 - (2y)^2 = 9x^4y^2 - 4y^2$

16. $(-2x^3y^2 + 5t)(2x^3y^2 + 5t) = (5t - 2x^3y^2)(5t + 2x^3y^2)$

The sum and the difference of the same two terms

$$= (5t)^2 - (2x^3y^2)^2 = 25t^2 - 4x^6y^4$$

$$
(A - B)(A + B) = A^2 - B^2
$$

17. $(2x + 3 - 2y)(2x + 3 + 2y) = (2x + 3)^2 - (2y)^2$
$$= 4x^2 + 12x + 9 - 4y^2$$

Remember that FOIL will always work when you are multiplying binomials. You can use it instead of the rules for special products, but those rules will make your work go faster.

Do Exercises 15–22.

Multiply.
13. $(x^2y^3 + 2x)(x^3y^2 + 3x)$

14. $(p^4q - 2p^3q^2 + 3q^3)(p + 2q)$

Multiply.
15. $(3xy + 2x)(x^2 + 2xy^2)$

16. $(x - 3y)(2x - 5y)$

17. $(4x + 5y)^2$

18. $(3x^2 - 2xy^2)^2$

19. $(2xy^2 + 3x)(2xy^2 - 3x)$

20. $(3xy^2 + 4y)(-3xy^2 + 4y)$

21. $(3y + 4 - 3x)(3y + 4 + 3x)$

22. $(2a + 5b + c)(2a - 5b - c)$

Answers
13. $x^5y^5 + 2x^4y^2 + 3x^3y^3 + 6x^2$
14. $p^5q - 4p^3q^3 + 3pq^3 + 6q^4$
15. $3x^3y + 6x^2y^3 + 2x^3 + 4x^2y^2$
16. $2x^2 - 11xy + 15y^2$
17. $16x^2 + 40xy + 25y^2$
18. $9x^4 - 12x^3y^2 + 4x^2y^4$
19. $4x^2y^4 - 9x^2$ **20.** $16y^2 - 9x^2y^4$
21. $9y^2 + 24y + 16 - 9x^2$
22. $4a^2 - 25b^2 - 10bc - c^2$

a   Evaluate the polynomial when $x = 3$, $y = -2$, and $z = -5$.

1. $x^2 - y^2 + xy$ **2.** $x^2 + y^2 - xy$ **3.** $x^2 - 3y^2 + 2xy$ **4.** $x^2 - 4xy + 5y^2$

5. $8xyz$ **6.** $-3xyz^2$ **7.** $xyz^2 - z$ **8.** $xy - xz + yz$

Lung Capacity. The polynomial equation

$$C = 0.041h - 0.018A - 2.69$$

can be used to estimate the lung capacity C, in liters, of a person of height h, in centimeters, and age A, in years. Use this formula for Exercises 9 and 10.

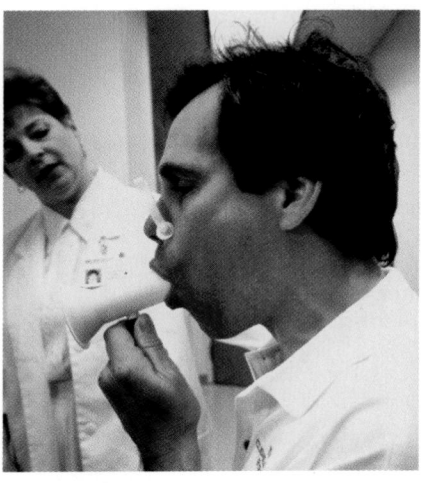

9. Find the lung capacity of a 20-year-old person who is 165 cm tall.

10. Find the lung capacity of a 50-year-old person who is 160 cm tall.

Altitude of a Launched Object. The altitude h, in meters, of a launched object is given by the polynomial equation

$$h = h_0 + vt - 4.9t^2,$$

where h_0 is the height, in meters, from which the launch occurs, v is the initial upward speed (or velocity), in meters per second (m/s), and t is the number of seconds for which the object is airborne. Use this formula for Exercises 11 and 12.

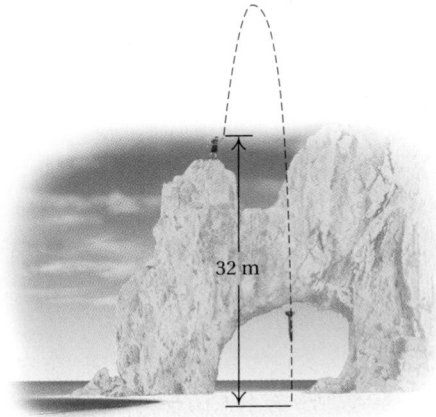

32 m

11. A golf ball is thrown upward with an initial speed of 30 m/s from the top of the Washington Monument, which is 160 m above the ground. How high above the ground will the ball be after 3 sec?

12. A model rocket is launched from the top of the Lands End Arch, near San Lucas, Baja, Mexico, 32 m above the ground. The upward speed is 40 m/s. How high will the rocket be 2 sec after the blastoff?

Surface Area of a Right Circular Cylinder. The surface area S of a right circular cylinder is given by the polynomial equation

$$S = 2\pi rh + 2\pi r^2,$$

where h is the height and r is the radius of the base. Use this formula for Exercises 13 and 14.

13. A 12-oz beverage can has a height of 4.7 in. and a radius of 1.2 in. Evaluate the polynomial when $h = 4.7$ and $r = 1.2$ to find the area of the can. Use 3.14 for π.

14. A 26-oz coffee can has a height of 6.5 in. and a radius of 2.5 in. Evaluate the polynomial when $h = 6.5$ and $r = 2.5$ to find the area of the can. Use 3.14 for π.

Surface Area of a Silo. A silo is a structure that is shaped like a right circular cylinder with a half sphere on top. The surface area S of a silo of height h and radius r (including the area of the base) is given by the polynomial equation $S = 2\pi rh + \pi r^2$. Note that h is the height of the entire silo.

15. A container of tennis balls is silo-shaped, with a height of $7\frac{1}{2}$ in. and a radius of $1\frac{1}{4}$ in. Find the surface area of the container. Use 3.14 for π.

16. A $1\frac{1}{2}$-oz bottle of roll-on deodorant has a height of 4 in. and a radius of $\frac{3}{4}$ in. Find the surface area of the bottle if the bottle is shaped like a silo. Use 3.14 for π.

b Identify the coefficient and the degree of each term of the polynomial. Then find the degree of the polynomial.

17. $x^3y - 2xy + 3x^2 - 5$

18. $5x^2y^2 - y^2 + 15xy + 1$

19. $17x^2y^3 - 3x^3yz - 7$

20. $6 - xy + 8x^2y^2 - y^5$

c Collect like terms.

21. $a + b - 2a - 3b$

22. $xy^2 - 1 + y - 6 - xy^2$

23. $3x^2y - 2xy^2 + x^2$

24. $m^3 + 2m^2n - 3m^2 + 3mn^2$

25. $6au + 3av + 14au + 7av$

26. $3x^2y - 2z^2y + 3xy^2 + 5z^2y$

27. $2u^2v - 3uv^2 + 6u^2v - 2uv^2$

28. $3x^2 + 6xy + 3y^2 - 5x^2 - 10xy - 5y^2$

d Add.

29. $(2x^2 - xy + y^2) + (-x^2 - 3xy + 2y^2)$

30. $(2zt - z^2 + 5t^2) + (z^2 - 3zt + t^2)$

31. $(r - 2s + 3) + (2r + s) + (s + 4)$

32. $(ab - 2a + 3b) + (5a - 4b) + (3a + 7ab - 8b)$

33. $(b^3a^2 - 2b^2a^3 + 3ba + 4) + (b^2a^3 - 4b^3a^2 + 2ba - 1)$

34. $(2x^2 - 3xy + y^2) + (-4x^2 - 6xy - y^2)$
$+ (x^2 + xy - y^2)$

e Subtract.

35. $(a^3 + b^3) - (a^2b - ab^2 + b^3 + a^3)$

36. $(x^3 - y^3) - (-2x^3 + x^2y - xy^2 + 2y^3)$

37. $(xy - ab - 8) - (xy - 3ab - 6)$

38. $(3y^4x^2 + 2y^3x - 3y - 7)$
$- (2y^4x^2 + 2y^3x - 4y - 2x + 5)$

39. $(-2a + 7b - c) - (-3b + 4c - 8d)$

40. Subtract $5a + 2b$ from the sum of $2a + b$ and $3a - b$.

f Multiply.

41. $(3z - u)(2z + 3u)$

42. $(a - b)(a^2 + b^2 + 2ab)$

43. $(a^2b - 2)(a^2b - 5)$

44. $(xy + 7)(xy - 4)$

45. $(a^3 + bc)(a^3 - bc)$

46. $(m^2 + n^2 - mn)(m^2 + mn + n^2)$

47. $(y^4x + y^2 + 1)(y^2 + 1)$

48. $(a - b)(a^2 + ab + b^2)$

49. $(3xy - 1)(4xy + 2)$

50. $(m^3n + 8)(m^3n - 6)$

51. $(3 - c^2d^2)(4 + c^2d^2)$

52. $(6x - 2y)(5x - 3y)$

53. $(m^2 - n^2)(m + n)$

54. $(pq + 0.2)(0.4pq - 0.1)$

55. $(xy + x^5y^5)(x^4y^4 - xy)$

56. $(x - y^3)(2y^3 + x)$

57. $(x + h)^2$

58. $(y - a)^2$

59. $(3a + 2b)^2$

60. $(2ab - cd)^2$

61. $(r^3t^2 - 4)^2$

62. $(3a^2b - b^2)^2$

63. $(p^4 + m^2n^2)^2$

64. $\left(2a^3 - \frac{1}{2}b^3\right)^2$

65. $3a(a - 2b)^2$

66. $-3x(x + 8y)^2$

67. $(m + n - 3)^2$

68. $(a^2 + b + 2)^2$

69. $(a + b)(a - b)$

70. $(x - y)(x + y)$

71. $(2a - b)(2a + b)$

72. $(w + 3z)(w - 3z)$

73. $(c^2 - d)(c^2 + d)$

74. $(p^3 - 5q)(p^3 + 5q)$

75. $(ab + cd^2)(ab - cd^2)$

76. $(xy + pq)(xy - pq)$

77. $(x + y - 3)(x + y + 3)$

78. $(p + q + 4)(p + q - 4)$

79. $[x + y + z][x - (y + z)]$

80. $[a + b + c][a - (b + c)]$

81. $(a + b + c)(a + b - c)$

82. $(3x + 2 - 5y)(3x + 2 + 5y)$

83. $(x^2 - 4y + 2)(3x^2 + 5y - 3)$

84. $(2x^2 - 7y + 4)(x^2 + y - 3)$

Skill Maintenance

In which quadrant is each point located?   [3.1a]

85. $(2, -5)$ **86.** $(-8, -9)$ **87.** $(16, 23)$ **88.** $(-3, 2)$

Graph.   [3.3b]

89. $2x = -10$ **90.** $y = -4$ **91.** $8y - 16 = 0$ **92.** $x = 4$

Synthesis

Find a polynomial for each shaded area. (Leave results in terms of π where appropriate.)

93.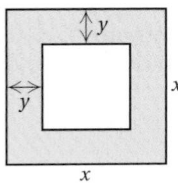

94.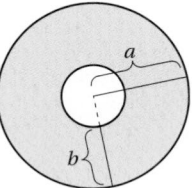

95.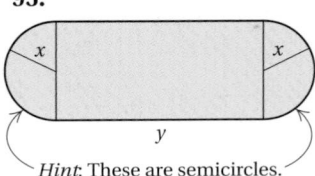

Hint: These are semicircles.

96.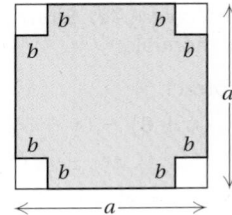

Find a formula for the surface area of each solid object. Leave results in terms of π.

97.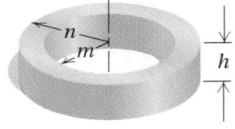

98.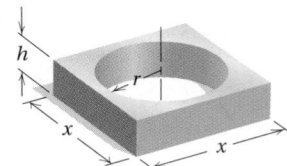

99. *Observatory Paint Costs.*   The observatory at Danville University is shaped like a silo that is 40 ft high and 30 ft wide (see Exercise 15). The Heavenly Bodies Astronomy Club is to paint the exterior of the observatory using paint that covers 250 ft^2 per gallon. How many gallons should they purchase?

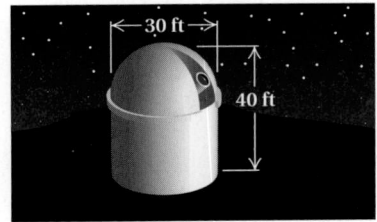

100. *Interest Compounded Annually.*   An amount of money P that is invested at the yearly interest rate r grows to the amount

$$P(1 + r)^t$$

after t years. Find a polynomial that can be used to determine the amount to which P will grow after 2 yr.

101. Suppose that $10,400 is invested at 3.5%, compounded annually. How much is in the account at the end of 5 yr? (See Exercise 100.)

102. Multiply: $(x + a)(x - b)(x - a)(x + b)$.

4.8

Division of Polynomials

OBJECTIVES

a Divide a polynomial by a monomial.

b Divide a polynomial by a divisor that is a binomial.

SKILL TO REVIEW

Objective 4.7e: Subtract polynomials.

Subtract.

1. $(x + 8) - (x + 3)$

2. $(5x^2 - 3x) - (5x^2 - 2x)$

Divide.

1. $\dfrac{20x^3}{5x}$

2. $\dfrac{-28x^{14}}{4x^3}$

3. $\dfrac{-56p^5q^7}{2p^2q^6}$

4. $\dfrac{x^5}{4x}$

In this section, we consider division of polynomials. You will see that such division is similar to what is done in arithmetic.

a Dividing by a Monomial

We first consider division by a monomial. When dividing a monomial by a monomial, we use the quotient rule of Section 4.1 to subtract exponents when the bases are the same. We also divide the coefficients.

EXAMPLES Divide.

1. $\dfrac{10x^2}{2x} = \dfrac{10}{2} \cdot \dfrac{x^2}{x} = 5x^{2-1} = 5x$

2. $\dfrac{x^9}{3x^2} = \dfrac{1x^9}{3x^2} = \dfrac{1}{3} \cdot \dfrac{x^9}{x^2} = \dfrac{1}{3}x^{9-2} = \dfrac{1}{3}x^7$

3. $\dfrac{-18x^{10}}{3x^3} = \dfrac{-18}{3} \cdot \dfrac{x^{10}}{x^3} = -6x^{10-3} = -6x^7$

4. $\dfrac{42a^2b^5}{-3ab^2} = \dfrac{42}{-3} \cdot \dfrac{a^2}{a} \cdot \dfrac{b^5}{b^2} = -14a^{2-1}b^{5-2} = -14ab^3$

> ---------- *Caution!* ----------
>
> The coefficients are divided but the exponents are subtracted.
> ---------------------------------

Do Margin Exercises 1–4.

To divide a polynomial by a monomial, we note that since

$$\frac{A}{C} + \frac{B}{C} = \frac{A + B}{C},$$

it follows that

$$\frac{A + B}{C} = \frac{A}{C} + \frac{B}{C}. \qquad \text{Switching the left and right sides of the equation}$$

This is actually the procedure we use when performing divisions like $86 \div 2$. Although we might write

$$\frac{86}{2} = 43,$$

we could also calculate as follows:

$$\frac{86}{2} = \frac{80 + 6}{2} = \frac{80}{2} + \frac{6}{2} = 40 + 3 = 43.$$

Similarly, to divide a polynomial by a monomial, we divide each term by the monomial.

EXAMPLE 5 Divide: $(9x^8 + 12x^6) \div (3x^2)$.

We have

$$(9x^8 + 12x^6) \div (3x^2) = \frac{9x^8 + 12x^6}{3x^2}$$

$$= \frac{9x^8}{3x^2} + \frac{12x^6}{3x^2}. \qquad \text{To see this, add and get the original expression.}$$

Answers

Skill to Review:

1. 5   **2.** $-x$

Margin Exercises:

1. $4x^2$ **2.** $-7x^{11}$ **3.** $-28p^3q$ **4.** $\dfrac{1}{4}x^4$

We now perform the separate divisions:

$$\frac{9x^8}{3x^2} + \frac{12x^6}{3x^2} = \frac{9}{3} \cdot \frac{x^8}{x^2} + \frac{12}{3} \cdot \frac{x^6}{x^2}$$

$$= 3x^{8-2} + 4x^{6-2}$$

$$= 3x^6 + 4x^4.$$

---------- **Caution!** ----------

The coefficients are *divided*, but the exponents are *subtracted*.

To check, we multiply the quotient, $3x^6 + 4x^4$, by the divisor, $3x^2$:

$$3x^2(3x^6 + 4x^4) = (3x^2)(3x^6) + (3x^2)(4x^4) = 9x^8 + 12x^6.$$

This is the polynomial that was being divided, so our answer is $3x^6 + 4x^4$.

Do Exercises 5–7.

EXAMPLE 6 Divide and check: $(10a^5b^4 - 2a^3b^2 + 6a^2b) \div (2a^2b)$.

$$\frac{10a^5b^4 - 2a^3b^2 + 6a^2b}{2a^2b} = \frac{10a^5b^4}{2a^2b} - \frac{2a^3b^2}{2a^2b} + \frac{6a^2b}{2a^2b}$$

$$= \frac{10}{2}a^{5-2}b^{4-1} - \frac{2}{2}a^{3-2}b^{2-1} + \frac{6}{2}$$

$$= 5a^3b^3 - ab + 3$$

Check: $2a^2b(5a^3b^3 - ab + 3) = 2a^2b \cdot 5a^3b^3 - 2a^2b \cdot ab + 2a^2b \cdot 3$

$$= 10a^5b^4 - 2a^3b^2 + 6a^2b$$

Our answer, $5a^3b^3 - ab + 3$, checks.

> To divide a polynomial by a monomial, divide each term of the polynomial by the monomial.

Do Exercises 8 and 9.

b Dividing by a Binomial

Let's first consider long division as it is performed in arithmetic. When we divide, we repeat the procedure at right.

We review this by considering the division $3711 \div 8$.

$$
\begin{array}{r}
4 \\
8 \overline{\smash{)}\ 3\ 7\ 1\ 1} \\
3\ 2 \\
\hline
5\ 1
\end{array}
$$

① Divide: $37 \div 8 \approx 4$.
② Multiply: $4 \times 8 = 32$.
③ Subtract: $37 - 32 = 5$.
④ Bring down the 1.

$$
\begin{array}{r}
4\ 6\ 3 \\
8\ \overline{\smash{)}\ 3\ 7\ 1\ 1} \\
3\ 2 \\
\hline
5\ 1 \\
4\ 8 \\
\hline
3\ 1 \\
2\ 4 \\
\hline
7
\end{array}
$$

Divide. Check the result.

5. $(28x^7 + 32x^5) \div (4x^3)$

6. $(2x^3 + 6x^2 + 4x) \div (2x)$

7. $(6x^2 + 3x - 2) \div 3$

Divide and check.

8. $(8x^2 - 3x + 1) \div 2$

9. $\dfrac{2x^4y^6 - 3x^3y^4 + 5x^2y^3}{x^2y^2}$

To carry out long division:

1. Divide,
2. Multiply,
3. Subtract, and
4. Bring down the next term.

Answers

5. $7x^4 + 8x^2$ **6.** $x^2 + 3x + 2$
7. $2x^2 + x - \dfrac{2}{3}$ **8.** $4x^2 - \dfrac{3}{2}x + \dfrac{1}{2}$
9. $2x^2y^4 - 3xy^2 + 5y$

Next, we repeat the process two more times. We obtain the complete division as shown on the right above. The quotient is 463. The remainder is 7, expressed as R = 7. We write the answer as

$$463 \text{ R } 7 \quad \text{or} \quad 463 + \frac{7}{8} = 463\frac{7}{8}.$$

We check by multiplying the quotient, 463, by the divisor, 8, and adding the remainder, 7:

$$8 \cdot 463 + 7 = 3704 + 7 = 3711.$$

Now let's look at long division with polynomials. We use this procedure when the divisor is not a monomial. We write polynomials in descending order and then write in missing terms.

EXAMPLE 7 Divide $x^2 + 5x + 6$ by $x + 2$.

$$
\begin{array}{r}
x \\
x + 2 \overline{)x^2 + 5x + 6} \\
x^2 + 2x \\
\hline
3x
\end{array}
$$

— Divide the first term by the first term: $x^2/x = x$.
  Ignore the term 2.
— Multiply x above by the divisor, $x + 2$.
— Subtract: $(x^2 + 5x) - (x^2 + 2x) = x^2 + 5x - x^2 - 2x$
  $= 3x$.

We now "bring down" the next term of the dividend—in this case, 6.

$$
\begin{array}{r}
x \;+\; 3 \\
x + 2 \overline{)x^2 + 5x + 6} \\
x^2 + 2x \\
\hline
3x + 6 \\
3x + 6 \\
\hline
0
\end{array}
$$

— Divide the first term by the first term: $3x/x = 3$.

— The 6 has been "brought down."
— Multiply 3 by the divisor, $x + 2$.
— Subtract: $(3x + 6) - (3x + 6) = 3x + 6 - 3x - 6 = 0$.

The quotient is $x + 3$. The remainder is 0, expressed as R = 0. A remainder of 0 is generally not included in an answer.

To check, we multiply the quotient by the divisor and add the remainder, if any, to see if we get the dividend:

Divisor	Quotient	Remainder	Dividend	
$(x + 2) \cdot$	$(x + 3) +$	0	$= x^2 + 5x + 6.$	The division checks.

10. Divide and check:

$(x^2 + x - 6) \div (x + 3)$.

Do Exercise 10.

EXAMPLE 8 Divide and check: $(x^2 + 2x - 12) \div (x - 3)$.

$$
\begin{array}{r}
x \\
x - 3 \overline{)x^2 + 2x - 12} \\
x^2 - 3x \\
\hline
5x
\end{array}
$$

— Divide the first term by the first term: $x^2/x = x$.

— Multiply x above by the divisor, $x - 3$.
— Subtract: $(x^2 + 2x) - (x^2 - 3x) = x^2 + 2x - x^2 + 3x$
  $= 5x$.

We now "bring down" the next term of the dividend—in this case, -12.

$$
\begin{array}{r}
x \;+\; 5 \\
x - 3 \overline{)x^2 + 2x - 12} \\
x^2 - 3x \\
\hline
5x - 12 \\
5x - 15 \\
\hline
3
\end{array}
$$

— Divide the first term by the first term: $5x/x = 5$.

— Bring down the -12.
— Multiply 5 above by the divisor, $x - 3$.
— Subtract:
  $(5x - 12) - (5x - 15) = 5x - 12 - 5x + 15$
  $= 3$.

Answer

10. $x - 2$

The answer is $x + 5$ with R $= 3$, or

$$\underset{\text{Quotient}}{\underbrace{x + 5}} + \underset{\text{Divisor}}{\underbrace{\dfrac{\overset{\text{Remainder}}{3}}{x - 3}}}.$$

(This is the way answers will be given at the back of the book.)

Check: We can check by multiplying the divisor by the quotient and adding the remainder, as follows:

$$(x - 3)(x + 5) + 3 = x^2 + 2x - 15 + 3$$
$$= x^2 + 2x - 12.$$

When dividing, an answer may "come out even" (that is, have a remainder of 0, as in Example 7), or it may not (as in Example 8). **If a remainder is not 0, we continue dividing until the degree of the remainder is less than the degree of the divisor**. Check this in each of Examples 7 and 8.

> Do Exercises 11 and 12.

EXAMPLE 9   Divide and check: $(x^3 + 1) \div (x + 1)$.

$$
\begin{array}{r}
x^2 - x + 1 \\
x + 1 \overline{)x^3 + 0x^2 + 0x + 1} \\
\end{array}
$$
$\leftarrow$ Fill in the missing terms. (See Section 4.3.)

$\dfrac{x^3 + x^2}{}$ ⟵ Subtract: $x^3 - (x^3 + x^2) = -x^2$.

$-x^2 + 0x$

$\dfrac{-x^2 - x}{}$ ⟵ Subtract: $-x^2 - (-x^2 - x) = x$.

$x + 1$

$\dfrac{x + 1}{0}$ ⟵ Subtract: $(x + 1) - (x + 1) = 0$.

The answer is $x^2 - x + 1$. The check is left to the student.

EXAMPLE 10   Divide and check: $(9x^4 - 7x^2 - 4x + 13) \div (3x - 1)$.

$$
\begin{array}{r}
3x^3 + x^2 - 2x - 2 \\
3x - 1 \overline{)9x^4 + 0x^3 - 7x^2 - 4x + 13} \\
\end{array}
$$
$\leftarrow$ Fill in the missing term.

$\dfrac{9x^4 - 3x^3}{}$ ⟵ Subtract: $9x^4 - (9x^4 - 3x^3) = 3x^3$.

$3x^3 - 7x^2$

$\dfrac{3x^3 - x^2}{}$ ⟵ Subtract: $(3x^3 - 7x^2) - (3x^3 - x^2) = -6x^2$.

$-6x^2 - 4x$

$\dfrac{-6x^2 + 2x}{}$ ⟵ Subtract: $(-6x^2 - 4x) - (-6x^2 + 2x) = -6x$.

$-6x + 13$

$\dfrac{-6x + 2}{11}$ ⟵ Subtract: $(-6x + 13) - (-6x + 2) = 11$.

The answer is $3x^3 + x^2 - 2x - 2$ with R $= 11$, or

$$3x^3 + x^2 - 2x - 2 + \dfrac{11}{3x - 1}.$$

Check: $(3x - 1)(3x^3 + x^2 - 2x - 2) + 11$
$$= 9x^4 + 3x^3 - 6x^2 - 6x - 3x^3 - x^2 + 2x + 2 + 11$$
$$= 9x^4 - 7x^2 - 4x + 13$$

> Do Exercises 13 and 14.

Divide and check.

11. $x - 2 \overline{)x^2 + 2x - 8}$

12. $x + 3 \overline{)x^2 + 7x + 10}$

Divide and check.

13. $(x^3 - 1) \div (x - 1)$

14. $(8x^4 + 10x^2 + 2x + 9) \div (4x + 2)$

Answers

11. $x + 4$ **12.** $x + 4$ with R $= -2$, or
$x + 4 + \dfrac{-2}{x + 3}$ **13.** $x^2 + x + 1$
14. $2x^3 - x^2 + 3x - 1$ with R $= 11$, or
$2x^3 - x^2 + 3x - 1 + \dfrac{11}{4x + 2}$

a   Divide and check.

1. $\dfrac{24x^4}{8}$

2. $\dfrac{-2u^2}{u}$

3. $\dfrac{25x^3}{5x^2}$

4. $\dfrac{16x^7}{-2x^2}$

5. $\dfrac{-54x^{11}}{-3x^8}$

6. $\dfrac{-75a^{10}}{3a^2}$

7. $\dfrac{64a^5b^4}{16a^2b^3}$

8. $\dfrac{-34p^{10}q^{11}}{-17pq^9}$

9. $\dfrac{24x^4 - 4x^3 + x^2 - 16}{8}$

10. $\dfrac{12a^4 - 3a^2 + a - 6}{6}$

11. $\dfrac{u - 2u^2 - u^5}{u}$

12. $\dfrac{50x^5 - 7x^4 + x^2}{x}$

13. $(15t^3 + 24t^2 - 6t) \div (3t)$

14. $(25t^3 + 15t^2 - 30t) \div (5t)$

15. $(20x^6 - 20x^4 - 5x^2) \div (-5x^2)$

16. $(24x^6 + 32x^5 - 8x^2) \div (-8x^2)$

17. $(24x^5 - 40x^4 + 6x^3) \div (4x^3)$

18. $(18x^6 - 27x^5 - 3x^3) \div (9x^3)$

19. $\dfrac{18x^2 - 5x + 2}{2}$

20. $\dfrac{15x^2 - 30x + 6}{3}$

21. $\dfrac{12x^3 + 26x^2 + 8x}{2x}$

22. $\dfrac{2x^4 - 3x^3 + 5x^2}{x^2}$

23. $\dfrac{9r^2s^2 + 3r^2s - 6rs^2}{3rs}$

24. $\dfrac{4x^4y - 8x^6y^2 + 12x^8y^6}{4x^4y}$

b Divide.

25. $(x^2 + 4x + 4) \div (x + 2)$

26. $(x^2 - 6x + 9) \div (x - 3)$

27. $(x^2 - 10x - 25) \div (x - 5)$

28. $(x^2 + 8x - 16) \div (x + 4)$

29. $(x^2 + 4x - 14) \div (x + 6)$

30. $(x^2 + 5x - 9) \div (x - 2)$

31. $\dfrac{x^2 - 9}{x + 3}$

32. $\dfrac{x^2 - 25}{x - 5}$

33. $\dfrac{x^5 + 1}{x + 1}$

34. $\dfrac{x^4 - 81}{x - 3}$

35. $\dfrac{8x^3 - 22x^2 - 5x + 12}{4x + 3}$

36. $\dfrac{2x^3 - 9x^2 + 11x - 3}{2x - 3}$

37. $(x^6 - 13x^3 + 42) \div (x^3 - 7)$

38. $(x^6 + 5x^3 - 24) \div (x^3 - 3)$

39. $(t^3 - t^2 + t - 1) \div (t - 1)$

40. $(y^3 + 3y^2 - 5y - 15) \div (y + 3)$

41. $(y^3 - y^2 - 5y - 3) \div (y + 2)$

42. $(t^3 - t^2 + t - 1) \div (t + 1)$

43. $(15x^3 + 8x^2 + 11x + 12) \div (5x + 1)$

44. $(20x^4 - 2x^3 + 5x + 3) \div (2x - 3)$

45. $(12y^3 + 42y^2 - 10y - 41) \div (2y + 7)$

46. $(15y^3 - 27y^2 - 35y + 60) \div (5y - 9)$

Skill Maintenance

In each of Exercises 47–54, fill in the blank with the correct term from the given list. Some of the choices may not be used.

47. The _____ rule asserts that when multiplying with exponential notation, if the bases are the same, we keep the base and add the exponents.   [4.1d]

48. A(n) _____ is an expression of the type ax^n, where a is a real-number constant and n is a nonnegative integer.   [4.3a, i]

49. The _____ principle asserts that when we multiply or divide by the same nonzero number on each side of an equation, we get _____ equations.   [2.2a]

50. Vertical lines are graphs of equations of the type _____.   [3.3b]

51. A(n) _____ is a polynomial with three terms, such as $5x^4 - 7x^2 + 4$.   [4.3i]

52. The _____ rule asserts that when dividing with exponential notation, if the bases are the same, we keep the base and subtract the exponent of the denominator from the exponent of the numerator.   [4.1e]

53. The _____ of a number is its distance from zero on the number line.   [1.2e]

54. The _____ of the line $y = mx + b$ is m.   [3.4b]

$x = a$

$y = b$

slope

y-intercept

opposite

absolute value

equivalent

inverse

quotient

product

monomial

binomial

trinomial

addition

multiplication

Synthesis

Divide.

55. $(x^4 + 9x^2 + 20) \div (x^2 + 4)$

56. $(y^4 + a^2) \div (y + a)$

57. $(5a^3 + 8a^2 - 23a - 1) \div (5a^2 - 7a - 2)$

58. $(15y^3 - 30y + 7 - 19y^2) \div (3y^2 - 2 - 5y)$

59. $(6x^5 - 13x^3 + 5x + 3 - 4x^2 + 3x^4) \div (3x^3 - 2x - 1)$

60. $(5x^7 - 3x^4 + 2x^2 - 10x + 2) \div (x^2 - x + 1)$

61. $(a^6 - b^6) \div (a - b)$

62. $(x^5 + y^5) \div (x + y)$

If the remainder is 0 when one polynomial is divided by another, the divisor is a *factor* of the dividend. Find the value(s) of c for which $x - 1$ is a factor of the polynomial.

63. $x^2 + 4x + c$

64. $2x^2 + 3cx - 8$

65. $c^2x^2 - 2cx + 1$

Summary and Review

Key Terms and Properties

exponent, p. 228
base, p. 228
scientific notation, p. 240
polynomial, p. 250
monomial, pp. 250, 257

binomial, p. 257
trinomial, p. 257
like terms, p. 253
coefficients, p. 254
collecting like terms, p. 254

descending/ascending order, p. 255
degree of a term/polynomial,
   pp. 255, 256
opposite of a polynomial, p. 264
polynomial in several variables, p. 291

Definitions and Rules for Exponents: See p. 244.

FOIL: $(A + B)(C + D) = AC + AD + BC + BD$

Square of a Sum: $(A + B)(A + B) = (A + B)^2 = A^2 + 2AB + B^2$

Square of a Difference: $(A - B)(A - B) = (A - B)^2 = A^2 - 2AB + B^2$

Product of a Sum and a Difference: $(A + B)(A - B) = A^2 - B^2$

Concept Reinforcement

Determine whether each statement is true or false.

_____ **1.** All trinomials are polynomials.   [4.3i]

_____ **2.** $(x + y)^2 = x^2 + y^2$ [4.6c]

_____ **3.** The square of the difference of two expressions is the difference of the squares of the two expressions.   [4.6c]

_____ **4.** The product of the sum and the difference of two expressions is the difference of the squares of the expressions.   [4.6b]

Important Concepts

Objective 4.1d   Use the product rule to multiply exponential expressions with like bases.

Example   Multiply and simplify: $x^3 \cdot x^4$.
   $x^3 \cdot x^4 = x^{3+4} = x^7$

Practice Exercise
   1. Multiply and simplify: $z^5 \cdot z^3$.

Objective 4.1e   Use the quotient rule to divide exponential expressions with like bases.

Example   Divide and simplify: $\dfrac{x^6 y^5}{xy^3}$.

$$\dfrac{x^6 y^5}{xy^3} = \dfrac{x^6}{x} \cdot \dfrac{y^5}{y^3}$$
$$= x^{6-1} y^{5-3}$$
$$= x^5 y^2$$

Practice Exercise
   2. Divide and simplify: $\dfrac{a^4 b^7}{a^2 b}$.

Objective 4.1f   Express an exponential expression involving negative exponents with positive exponents.

Objective 4.2a   Use the power rule to raise powers to powers.

Objective 4.2b   Raise a product to a power and a quotient to a power.

Example   Simplify: $\left(\dfrac{2a^3b^{-2}}{c^4}\right)^5$.

$$\left(\frac{2a^3b^{-2}}{c^4}\right)^5 = \frac{(2a^3b^{-2})^5}{(c^4)^5}$$

$$= \frac{2^5(a^3)^5(b^{-2})^5}{(c^4)^5}$$

$$= \frac{32a^{3\cdot5}b^{-2\cdot5}}{c^{4\cdot5}}$$

$$= \frac{32a^{15}b^{-10}}{c^{20}}$$

$$= \frac{32a^{15}}{b^{10}c^{20}}$$

Practice Exercise

3. Simplify: $\left(\dfrac{x^{-4}y^2}{3z^3}\right)^3$.

Objective 4.2c   Convert between scientific notation and decimal notation.

Example   Convert 0.00095 to scientific notation.

0.0009.5

4 places

The number is small, so the exponent is negative. (If the number were large, the exponent would be positive.)

$$0.00095 = 9.5 \times 10^{-4}$$

Example   Convert 3.409×10^6 to decimal notation.

3.409000.

6 places

The exponent is positive, so the number is large. (If the exponent were negative, the number would be small.)

$$3.409 \times 10^6 = 3,409,000$$

Practice Exercises

4. Convert to scientific notation: 763,000.

5. Convert to decimal notation: 3×10^{-4}.

Objective 4.2d   Multiply and divide using scientific notation.

Example   Multiply and express the result in scientific notation: $(5.3 \times 10^9) \cdot (2.4 \times 10^{-5})$.

$$(5.3 \times 10^9) \cdot (2.4 \times 10^{-5}) = (5.3 \cdot 2.4) \times (10^9 \cdot 10^{-5})$$

$$= 12.72 \times 10^4$$

The answer at this stage is not in scientific notation, because 12.72 is not a number between 1 and 10. We convert 12.72 to scientific notation and simplify:

$$12.72 \times 10^4 = (1.272 \times 10) \times 10^4$$

$$= 1.272 \times (10 \times 10^4)$$

$$= 1.272 \times 10^5.$$

Practice Exercise

6. Divide and express the result in scientific notation:

$$\frac{3.6 \times 10^3}{6.0 \times 10^{-2}}.$$

Objective 4.3e Collect the like terms of a polynomial.

Example Collect like terms: $4x^3 - 2x^2 + 5 + 3x^2 - 12$.

$4x^3 - 2x^2 + 5 + 3x^2 - 12$
$= 4x^3 + (-2 + 3)x^2 + (5 - 12)$
$= 4x^3 + x^2 - 7$

Practice Exercise

7. Collect like terms: $5x^4 - 6x^2 - 3x^4 + 2x^2 - 3$.

Objective 4.4a Add polynomials.

Example Add: $(4x^3 + x^2 - 8) + (2x^3 - 5x + 1)$.

$(4x^3 + x^2 - 8) + (2x^3 - 5x + 1)$
$= (4 + 2)x^3 + x^2 - 5x + (-8 + 1)$
$= 6x^3 + x^2 - 5x - 7$

Practice Exercise

8. Add: $(3x^4 - 5x^2 - 4) + (x^3 + 3x^2 + 6)$.

Objective 4.5d Multiply any two polynomials.

Example Multiply: $(z^2 - 2z + 3)(z - 1)$.

We use columns. First, we multiply the top row by -1 and then by z, placing like terms of the product in the same column. Finally, we collect like terms.

$$
\begin{array}{r}
z^2 - 2z + 3 \\
z - 1 \\
\hline
-z^2 + 2z - 3 \\
z^3 - 2z^2 + 3z \\
\hline
z^3 - 3z^2 + 5z - 3
\end{array}
$$

Practice Exercise

9. Multiply: $(x^4 - 3x^2 + 2)(x^2 - 3)$.

Objective 4.6a Multiply two binomials mentally using the FOIL method.

Example Multiply: $(3x + 5)(x - 1)$.

$$\qquad\qquad \text{F} \qquad\quad \text{O} \qquad\quad \text{I} \qquad\quad \text{L}$$
$(3x + 5)(x - 1) = 3x \cdot x + 3x \cdot (-1) + 5 \cdot x + 5 \cdot (-1)$
$\qquad\qquad\qquad = 3x^2 - 3x + 5x - 5$
$\qquad\qquad\qquad = 3x^2 + 2x - 5$

Practice Exercise

10. Multiply: $(y + 4)(2y + 3)$.

Objective 4.6b Multiply the sum and the difference of two terms mentally.

Example Multiply: $(3y + 2)(3y - 2)$.

$(3y + 2)(3y - 2) = (3y)^2 - 2^2$
$\qquad\qquad\qquad\quad = 9y^2 - 4$

Practice Exercise

11. Multiply: $(x + 5)(x - 5)$.

Objective 4.6c Square a binomial mentally.

Example Multiply: $(2x - 3)^2$.

$(2x - 3)^2 = (2x)^2 - 2 \cdot 2x \cdot 3 + 3^2$
$\qquad\qquad = 4x^2 - 12x + 9$

Practice Exercise

12. Multiply: $(3w + 4)^2$.

Objective 4.7e   Subtract polynomials.

Example   Subtract:
$(m^4n + 2m^3n^2 - m^2n^3) - (3m^4n + 2m^3n^2 - 4m^2n^2).$
$(m^4n + 2m^3n^2 - m^2n^3) - (3m^4n + 2m^3n^2 - 4m^2n^2)$
$\quad = m^4n + 2m^3n^2 - m^2n^3 - 3m^4n - 2m^3n^2 + 4m^2n^2$
$\quad = -2m^4n - m^2n^3 + 4m^2n^2$

Practice Exercise

13. Subtract:
$$(a^3b^2 - 5a^2b + 2ab) - (3a^3b^2 - ab^2 + 4ab).$$

Objective 4.8a   Divide a polynomial by a monomial.

Example   Divide: $(6x^3 - 8x^2 + 15x) \div (3x)$.
$$\frac{6x^3 - 8x^2 + 15x}{3x} = \frac{6x^3}{3x} - \frac{8x^2}{3x} + \frac{15x}{3x}$$
$$= \frac{6}{3}x^{3-1} - \frac{8}{3}x^{2-1} + \frac{15}{3}x^{1-1}$$
$$= 2x^2 - \frac{8}{3}x + 5$$

Practice Exercise

14. Divide: $(5y^2 - 20y + 8) \div 5$.

Objective 4.8b   Divide a polynomial by a divisor that is a binomial.

Example   Divide $x^2 - 3x + 7$ by $x + 1$.

$$
\begin{array}{r}
x - 4 \\
x + 1 \overline{)\, x^2 - 3x + 7} \\
\underline{x^2 + x} \\
-4x + 7 \\
\underline{-4x - 4} \\
11
\end{array}
$$

The answer is $x - 4 + \dfrac{11}{x + 1}$.

Practice Exercise

15. Divide: $(x^2 - 4x + 3) \div (x + 5)$.

Review Exercises

Multiply and simplify.   [4.1d, f]

1. $7^2 \cdot 7^{-4}$

2. $y^7 \cdot y^3 \cdot y$

3. $(3x)^5 \cdot (3x)^9$

4. $t^8 \cdot t^0$

Divide and simplify.   [4.1e, f]

5. $\dfrac{4^5}{4^2}$

6. $\dfrac{a^5}{a^8}$

7. $\dfrac{(7x)^4}{(7x)^4}$

Simplify.

8. $(3t^4)^2$ [4.2a, b]

9. $(2x^3)^2(-3x)^2$
  [4.1d], [4.2a, b]

10. $\left(\dfrac{2x}{y}\right)^{-3}$ [4.2b]

11. Express using a negative exponent: $\dfrac{1}{t^5}$.   [4.1f]

12. Express using a positive exponent: y^{-4}.   [4.1f]

13. Convert to scientific notation: 0.0000328.   [4.2c]

14. Convert to decimal notation: 8.3×10^6.   [4.2c]

Multiply or divide and write scientific notation for the result.   [4.2d]

15. $(3.8 \times 10^4)(5.5 \times 10^{-1})$

16. $\dfrac{1.28 \times 10^{-8}}{2.5 \times 10^{-4}}$

17. Pizza Consumption. Each man, woman, and child in the United States eats an average of 46 slices of pizza per year. The U.S. population is projected to be about 335.8 million in 2020. At this rate, how many slices of pizza would be consumed in 2020? Express the answer in scientific notation. [4.2e]

Sources: Packaged Facts; U.S. Census Bureau

18. Evaluate the polynomial $x^2 - 3x + 6$ when $x = -1$. [4.3a]

19. Identify the terms of the polynomial $-4y^5 + 7y^2 - 3y - 2$. [4.3b]

20. Identify the missing terms in $x^3 + x$. [4.3h]

21. Identify the degree of each term and the degree of the polynomial $4x^3 + 6x^2 - 5x + \frac{5}{3}$. [4.3g]

Classify the polynomial as a monomial, a binomial, a trinomial, or none of these. [4.3i]

22. $4x^3 - 1$

23. $4 - 9t^3 - 7t^4 + 10t^2$

24. $7y^2$

Collect like terms and then arrange in descending order. [4.3f]

25. $3x^2 - 2x + 3 - 5x^2 - 1 - x$

26. $-x + \frac{1}{2} + 14x^4 - 7x^2 - 1 - 4x^4$

Add. [4.4a]

27. $(3x^4 - x^3 + x - 4) + (x^5 + 7x^3 - 3x^2 - 5) + (-5x^4 + 6x^2 - x)$

28. $(3x^5 - 4x^4 + x^3 - 3) + (3x^4 - 5x^3 + 3x^2) + (-5x^5 - 5x^2) + (-5x^4 + 2x^3 + 5)$

Subtract. [4.4c]

29. $(5x^2 - 4x + 1) - (3x^2 + 1)$

30. $(3x^5 - 4x^4 + 3x^2 + 3) - (2x^5 - 4x^4 + 3x^3 + 4x^2 - 5)$

31. Find a polynomial for the perimeter and for the area. [4.4d], [4.5b]

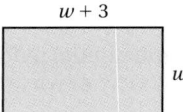

$w + 3$

w

32. Find two algebraic expressions for the area of this figure. First, regard the figure as one large rectangle, and then regard the figure as a sum of four smaller rectangles. [4.4d]

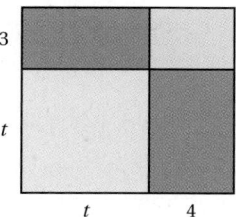

3

t

t ⠀⠀ 4

Multiply.

33. $\left(x + \frac{2}{3}\right)\left(x + \frac{1}{2}\right)$ [4.6a]　　**34.** $(7x + 1)^2$ [4.6c]

35. $(4x^2 - 5x + 1)(3x - 2)$ [4.5d]　　**36.** $(3x^2 + 4)(3x^2 - 4)$ [4.6b]

37. $5x^4(3x^3 - 8x^2 + 10x + 2)$ [4.5b]

38. $(x + 4)(x - 7)$ [4.6a]

39. $(3y^2 - 2y)^2$ [4.6c]　　**40.** $(2t^2 + 3)(t^2 - 7)$ [4.6a]

41. Evaluate the polynomial
$$2 - 5xy + y^2 - 4xy^3 + x^6$$
when $x = -1$ and $y = 2$. [4.7a]

42. Identify the coefficient and the degree of each term of the polynomial
$$x^5y - 7xy + 9x^2 - 8.$$
Then find the degree of the polynomial. [4.7b]

Collect like terms. [4.7c]

43. $y + w - 2y + 8w - 5$

44. $m^6 - 2m^2n + m^2n^2 + n^2m - 6m^3 + m^2n^2 + 7n^2m$

45. Add: [4.7d]
$$(5x^2 - 7xy + y^2) + (-6x^2 - 3xy - y^2) + (x^2 + xy - 2y^2).$$

46. Subtract: [4.7e]
$$(6x^3y^2 - 4x^2y - 6x) - (-5x^3y^2 + 4x^2y + 6x^2 - 6).$$

Multiply. [4.7f]

47. $(p - q)(p^2 + pq + q^2)$ **48.** $\left(3a^4 - \frac{1}{3}b^3\right)^2$

Divide.

49. $(10x^3 - x^2 + 6x) \div (2x)$ [4.8a]

50. $(6x^3 - 5x^2 - 13x + 13) \div (2x + 3)$ [4.8b]

51. The graph of the polynomial equation $y = 10x^3 - 10x$ is shown below. Use *only* the graph to estimate the value of the polynomial when $x = -1$, $x = -0.5$, $x = 0.5$, and $x = 1$.   [4.3a]

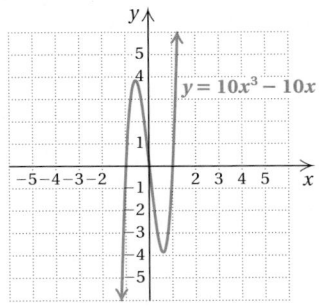

52. Subtract: $(2x^2 - 3x + 4) - (x^2 + 2x)$.   [4.4c]

   A. $x^2 - 3x - 2$ **B.** $x^2 - 5x + 4$

   C. $x^2 - x + 4$ **D.** $3x^2 - x + 4$

53. Multiply: $(x - 1)^2$.   [4.6c]

   A. $x^2 - 1$ **B.** $x^2 + 1$

   C. $x^2 - 2x - 1$ **D.** $x^2 - 2x + 1$

Synthesis

Find a polynomial for each shaded area.   [4.4d], [4.6b]

54.

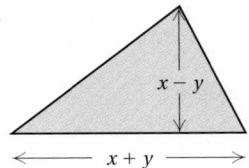

55.

56. Collect like terms: [4.1d], [4.2a], [4.3e]
$$-3x^5 \cdot 3x^3 - x^6(2x)^2 + (3x^4)^2 + (2x^2)^4 - 40x^2(x^3)^2.$$

57. Solve: [2.3b], [4.6a]
$$(x - 7)(x + 10) = (x - 4)(x - 6).$$

58. The product of two polynomials is $x^5 - 1$. One of the polynomials is $x - 1$. Find the other.   [4.8b]

59. A rectangular garden is twice as long as it is wide and is surrounded by a sidewalk that is 4 ft wide (see the figure below). The area of the sidewalk is 1024 ft². Find the dimensions of the garden.   [2.3b], [4.4d], [4.5a], [4.6a]

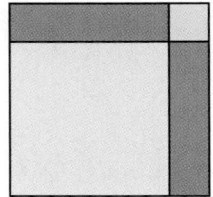

Understanding Through Discussion and Writing

1. Explain why the expression 578.6×10^{-7} is not in scientific notation.   [4.2c]

2. Explain why an understanding of the rules for order of operations is essential when evaluating polynomials. [4.3a]

3. How can the following figure be used to show that $(x + 3)^2 \neq x^2 + 9$?   [4.5c]

4. On an assignment, Emma *incorrectly* writes
$$\frac{12x^3 - 6x}{3x} = 4x^2 - 6x.$$
What mistake do you think she is making and how might you convince her that a mistake has been made?   [4.8a]

5. Can the sum of two trinomials in several variables be a trinomial in one variable? Why or why not?   [4.7d]

6. Is it possible for a polynomial in four variables to have a degree less than 4? Why or why not?   [4.7b]

CHAPTER

4

Test For Extra Help

CHAPTER
Test Prep
VIDEOS

Step-by-step test solutions are found on the Chapter Test Prep Videos available via the Video Resources on DVD, in **MyMathLab** , and on You Tube (search "BittingerIntroInter" and click on "Channels").

Multiply and simplify.

1. $6^{-2} \cdot 6^{-3}$

2. $x^6 \cdot x^2 \cdot x$

3. $(4a)^3 \cdot (4a)^8$

Divide and simplify.

4. $\dfrac{3^5}{3^2}$

5. $\dfrac{x^3}{x^8}$

6. $\dfrac{(2x)^5}{(2x)^5}$

Simplify.

7. $(x^3)^2$

8. $(-3y^2)^3$

9. $(2a^3b)^4$

10. $\left(\dfrac{ab}{c}\right)^3$

11. $(3x^2)^3(-2x^5)^3$

12. $3(x^2)^3(-2x^5)^3$

13. $2x^2(-3x^2)^4$

14. $(2x)^2(-3x^2)^4$

15. Express using a positive exponent: 5^{-3}.

16. Express using a negative exponent: $\dfrac{1}{y^8}$.

17. Convert to scientific notation: $3,900,000,000$.

18. Convert to decimal notation: 5×10^{-8}.

Multiply or divide and write scientific notation for the answer.

19. $\dfrac{5.6 \times 10^6}{3.2 \times 10^{-11}}$

20. $(2.4 \times 10^5)(5.4 \times 10^{16})$

21. *CD-ROM Memory.* A CD-ROM can contain about 600 million pieces of information (bytes). How many sound files, each containing 40,000 bytes, can a CD-ROM hold? Express the answer in scientific notation.

22. Evaluate the polynomial $x^5 + 5x - 1$ when $x = -2$.

23. Identify the coefficient of each term of the polynomial $\frac{1}{3}x^5 - x + 7$.

24. Identify the degree of each term and the degree of the polynomial $2x^3 - 4 + 5x + 3x^6$.

25. Classify the polynomial $7 - x$ as a monomial, a binomial, a trinomial, or none of these.

Collect like terms.

26. $4a^2 - 6 + a^2$

27. $y^2 - 3y - y + \dfrac{3}{4}y^2$

28. Collect like terms and then arrange in descending order:
$$3 - x^2 + 2x^3 + 5x^2 - 6x - 2x + x^5.$$

Add.

29. $(3x^5 + 5x^3 - 5x^2 - 3) +$
$(x^5 + x^4 - 3x^3 - 3x^2 + 2x - 4)$

30. $\left(x^4 + \dfrac{2}{3}x + 5\right) + \left(4x^4 + 5x^2 + \dfrac{1}{3}x\right)$

Subtract.

31. $(2x^4 + x^3 - 8x^2 - 6x - 3) - (6x^4 - 8x^2 + 2x)$

32. $(x^3 - 0.4x^2 - 12) - (x^5 + 0.3x^3 + 0.4x^2 + 9)$

Multiply.

33. $-3x^2(4x^2 - 3x - 5)$

34. $\left(x - \dfrac{1}{3}\right)^2$

35. $(3x + 10)(3x - 10)$

36. $(3b + 5)(b - 3)$

37. $(x^6 - 4)(x^8 + 4)$

38. $(8 - y)(6 + 5y)$

39. $(2x + 1)(3x^2 - 5x - 3)$

40. $(5t + 2)^2$

41. Collect like terms:
$$x^3y - y^3 + xy^3 + 8 - 6x^3y - x^2y^2 + 11.$$

42. Subtract:
$$(8a^2b^2 - ab + b^3) - (-6ab^2 - 7ab - ab^3 + 5b^3).$$

43. Multiply: $(3x^5 - 4y^5)(3x^5 + 4y^5)$.

Divide.

44. $(12x^4 + 9x^3 - 15x^2) \div (3x^2)$

45. $(6x^3 - 8x^2 - 14x + 13) \div (3x + 2)$

46. The graph of the polynomial equation $y = x^3 - 5x - 1$ is shown at right. Use *only* the graph to estimate the value of the polynomial when $x = -1$, $x = -0.5$, $x = 0.5$, $x = 1$, and $x = 1.1$.

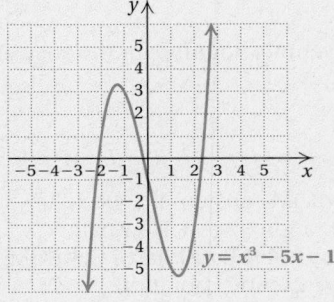

$y = x^3 - 5x - 1$

47. Find two algebraic expressions for the area of this figure. First, regard the figure as one large rectangle, and then regard the figure as a sum of four smaller rectangles.

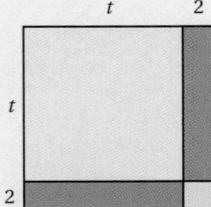

48. Find a polynomial for the surface area of this right rectangular solid.

A. $28a$

B. $28a + 90$

C. $14a + 45$

D. $45a$

Synthesis

49. The height of a box is 1 less than its length, and the length is 2 more than its width. Find the volume in terms of the length.

50. Solve: $(x - 5)(x + 5) = (x + 6)^2$.

1. Evaluate $\dfrac{x}{2y}$ when $x = 10$ and $y = 2$.

2. Evaluate $2x^3 + x^2 - 3$ when $x = -1$.

3. Evaluate $x^3y^2 + xy + 2xy^2$ when $x = -1$ and $y = 2$.

4. Find the absolute value: $|-4|$.

5. Find the reciprocal of 5.

Compute and simplify.

6. $-\dfrac{3}{5} + \dfrac{5}{12}$

7. $3.4 - (-0.8)$

8. $(-2)(-1.4)(2.6)$

9. $\dfrac{3}{8} \div \left(-\dfrac{9}{10}\right)$

10. $(1.1 \times 10^{10})(2 \times 10^{12})$

11. $(3.2 \times 10^{-10}) \div (8 \times 10^{-6})$

Simplify.

12. $\dfrac{-9x}{3x}$

13. $y - (3y + 7)$

14. $3(x - 1) - 2[x - (2x + 7)]$

15. $2 - [32 \div (4 + 2^2)]$

Add.

16. $(x^4 + 3x^3 - x + 7) + (2x^5 - 3x^4 + x - 5)$

17. $(x^2 + 2xy) + (y^2 - xy) + (2x^2 - 3y^2)$

Subtract.

18. $(x^3 + 3x^2 - 4) - (-2x^2 + x + 3)$

19. $\left(\dfrac{1}{3}x^2 - \dfrac{1}{4}x - \dfrac{1}{5}\right) - \left(\dfrac{2}{3}x^2 + \dfrac{1}{2}x - \dfrac{1}{5}\right)$

Multiply.

20. $3(4x - 5y + 7)$

21. $(-2x^3)(-3x^5)$

22. $2x^2(x^3 - 2x^2 + 4x - 5)$

23. $(y^2 - 2)(3y^2 + 5y + 6)$

24. $(2p^3 + p^2q + pq^2)(p - pq + q)$

25. $(2x + 3)(3x + 2)$

26. $(3x^2 + 1)^2$

27. $\left(t + \dfrac{1}{2}\right)\left(t - \dfrac{1}{2}\right)$

28. $(2y^2 + 5)(2y^2 - 5)$

29. $(2x^4 - 3)(2x^2 + 3)$

30. $(t - 2t^2)^2$

31. $(3p + q)(5p - 2q)$

Divide.

32. $(18x^3 + 6x^2 - 9x) \div (3x)$

33. $(3x^3 + 7x^2 - 13x - 21) \div (x + 3)$

Solve.

34. $1.5 = 2.7 + x$

35. $\dfrac{2}{7}x = -6$

36. $5x - 9 = 36$

37. $\dfrac{2}{3} = \dfrac{-m}{10}$

38. $5.4 - 1.9x = 0.8x$

39. $x - \dfrac{7}{8} = \dfrac{3}{4}$

40. $2(2 - 3x) = 3(5x + 7)$

41. $\dfrac{1}{4}x - \dfrac{2}{3} = \dfrac{3}{4} + \dfrac{1}{3}x$

42. $y + 5 - 3y = 5y - 9$

43. $\dfrac{1}{4}x - 7 < 5 - \dfrac{1}{2}x$

44. $2(x + 2) \geq 5(2x + 3)$

45. $A = Qx + P$, for x

Solve.

46. *Markup.* A bookstore sells books at a price that is 80% higher than the price the store pays for the books. A book is priced for sale at $6.30. How much did the store pay for the book?

47. A 6-ft by 3-ft raft is floating in a swimming pool of radius r. Find a polynomial for the area of the surface of the pool not covered by the raft.

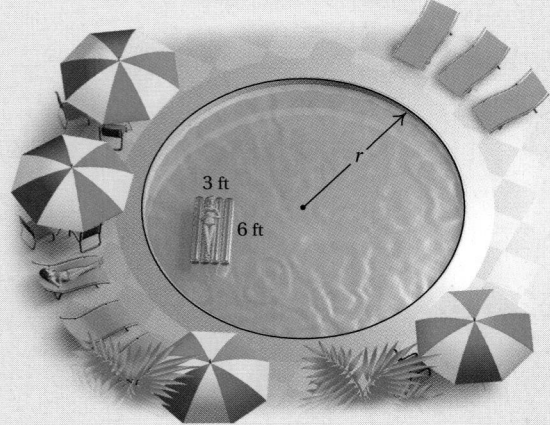

48. *Consecutive Page Numbers.* The sum of the page numbers on the facing pages of a book is 37. What are the page numbers?

49. *Room Perimeter.* The perimeter of a room is 88 ft. The width is 4 ft less than the length. Find the width and the length.

50. The second angle of a triangle is five times as large as the first. The third angle is twice the sum of the other two angles. Find the measure of the first angle.

Simplify.

51. $y^2 \cdot y^{-6} \cdot y^8$

52. $\dfrac{x^6}{x^7}$

53. $(-3x^3y^{-2})^3$

54. $\dfrac{x^3x^{-4}}{x^{-5}x}$

55. Find the intercepts of $4x - 5y = 20$ and then graph the equation using the intercepts.

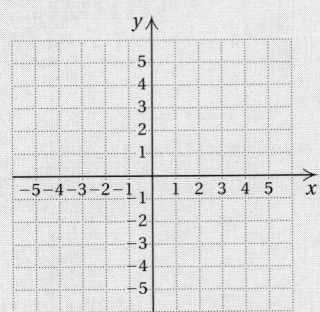

56. *Matching.* Match each item in the first column with the appropriate item in the second column by drawing connecting lines.

3^2	$\dfrac{1}{6}$
3^{-2}	$-\dfrac{1}{9}$
$\left(\dfrac{1}{3}\right)^2$	6
$\left(\dfrac{1}{3}\right)^{-2}$	9
-3^2	-9
$(-3)^2$	$\dfrac{1}{9}$
$\left(-\dfrac{1}{3}\right)^2$	-6
$\left(-\dfrac{1}{3}\right)^{-2}$	12

Synthesis

57. A picture frame is x inches square. The picture that it frames is 2 in. shorter than the frame in both length and width. Find a polynomial for the area of the frame.

Add.

58. $[(2x)^2 - (3x)^3 + 2x^2x^3 + (x^2)^2] + [5x^2(2x^3) - ((2x)^2)^2]$

59. $(x - 3)^2 + (2x + 1)^2$

Solve.

60. $(x + 3)(2x - 5) + (x - 1)^2 = (3x + 1)(x - 3)$

61. $(2x^2 + x - 6) \div (2x - 3) = (2x^2 - 9x - 5) \div (x - 5)$

62. $20 - 3|x| = 5$

63. $(x - 3)(x + 4) = (x^3 - 4x^2 - 17x + 60) \div (x - 5)$

Polynomials: Factoring

Real-World Application

Dr. Benton wants to investigate the potential spread of germs by contact. She knows that the number of possible handshakes within a group of x people, assuming each person shakes every other person's hand only once, is given by $N = \frac{1}{2}(x^2 - x)$. There are 40 people at a meeting. How many handshakes are possible?

This problem appears as Exercise 14 in Section 5.9.

5.1

Introduction to Factoring

OBJECTIVES

a Find the greatest common factor, the GCF, of monomials.

b Factor polynomials when the terms have a common factor, factoring out the greatest common factor.

c Factor certain expressions with four terms using factoring by grouping.

SKILL TO REVIEW

Objective 1.7d: Use the distributive laws to factor expressions like $4x - 12 + 24y$.

Factor.

1. $tz - tw$

2. $18a + 27b - 6$

We introduce factoring with a review of factoring natural numbers. Consider the product $15 = 3 \cdot 5$. We say that 3 and 5 are **factors** of 15 and that $3 \cdot 5$ is a **factorization** of 15. Since $15 = 15 \cdot 1$, we also know that 15 and 1 are factors of 15 and that $15 \cdot 1$ is a factorization of 15.

a Finding the Greatest Common Factor

The numbers 20 and 30 have several factors in common, among them 2 and 5. The greatest of the common factors is called the **greatest common factor**, **GCF**. One way to find the GCF is by making a list of factors of each number.

List all the factors of 20: $\underline{1}$, $\underline{2}$, 4, $\underline{5}$, $\underline{10}$, and 20.

List all the factors of 30: $\underline{1}$, $\underline{2}$, 3, $\underline{5}$, 6, $\underline{10}$, 15, and 30.

We now list the numbers common to both lists, the common factors:

  1, 2, 5, and 10.

The greatest common factor, the GCF, is 10, the largest number in the common list.

The preceding procedure gives meaning to the notion of a GCF, but the following method, using prime factorizations, is generally faster.

EXAMPLE 1 Find the GCF of 20 and 30.

We find the prime factorization of each number. Then we draw lines between the common factors.

$$20 = 2 \cdot 2 \cdot 5$$
$$30 = 2 \cdot 3 \cdot 5$$

The GCF $= 2 \cdot 5 = 10$.

EXAMPLE 2 Find the GCF of 180 and 420.

We find the prime factorization of each number. Then we draw lines between the common factors.

$$180 = 2 \cdot 2 \cdot 3 \cdot 3 \cdot 5 = 2^2 \cdot 3^2 \cdot 5^1$$
$$420 = 2 \cdot 2 \cdot 3 \cdot 5 \cdot 7 = 2^2 \cdot 3^1 \cdot 5^1 \cdot 7^1$$

The GCF $= 2 \cdot 2 \cdot 3 \cdot 5 = 2^2 \cdot 3^1 \cdot 5^1 = 60$. Note how we can use the exponents to determine the GCF. There are 2 lines for the 2's, 1 line for the 3, 1 line for the 5, and no line for the 7.

EXAMPLE 3 Find the GCF of 30 and 77.

We find the prime factorization of each number. Then we draw lines between the common factors, if any exist.

$$30 = 2 \cdot 3 \cdot 5 = 2^1 \cdot 3^1 \cdot 5^1$$

$$77 = 7 \cdot 11 = 7^1 \cdot 11^1$$

Since there is no common prime factor, the GCF is 1.

Answers

Skill to Review:

1. $t(z - w)$ **2.** $3(6a + 9b - 2)$

EXAMPLE 4   Find the GCF of 54, 90, and 252.

We find the prime factorization of each number. Then we draw lines between the common factors.

$$54 = 2 \cdot 3 \cdot 3 \cdot 3 = 2^1 \cdot 3^3,$$
$$90 = 2 \cdot 3 \cdot 3 \cdot 5 = 2^1 \cdot 3^2 \cdot 5^1,$$
$$252 = 2 \cdot 2 \cdot 3 \cdot 3 \cdot 7 = 2^2 \cdot 3^2 \cdot 7^1$$

The GCF $= 2^1 \cdot 3^2 = 18$.

Do Exercises 1–4.

Find the GCF.

1. 40, 100

2. 7, 21

3. 72, 360, 432

4. 3, 5, 22

Consider the product

$$12x^3(x^2 - 6x + 2) = 12x^5 - 72x^4 + 24x^3.$$

To factor the polynomial on the right, we reverse the process of multiplication:

$$12x^5 - 72x^4 + 24x^3 = \underline{12x^3(x^2 - 6x + 2)}.$$

This is a *factorization*. The *factors* are $(12x^3)$ and $(x^2 - 6x + 2)$.

FACTOR; FACTORIZATION

To **factor** a polynomial is to express it as a product.

A **factor** of a polynomial P is a polynomial that can be used to express P as a product.

A **factorization** of a polynomial is an expression that names that polynomial as a product.

In the factorization

$$12x^5 - 72x^4 + 24x^3 = 12x^3(x^2 - 6x + 2),$$

the monomial $12x^3$ is called the GCF of the terms, $12x^5$, $-72x^4$, and $24x^3$. The first step in factoring polynomials is to find the GCF of the terms.

Consider the monomials

$$x^3, \ x^4, \ x^6, \ \text{and} \ x^7.$$

The GCF of these monomials is x^3, found by noting that the smallest exponent of x is 3.

Consider

$$20x^2 \ \text{and} \ 30x^5.$$

The GCF of 20 and 30 is 10. The GCF of x^2 and x^5 is x^2. Then the GCF of $20x^2$ and $30x^5$ is the product of the individual GCFs, $10x^2$.

Answers

1. 20   **2.** 7   **3.** 72   **4.** 1

EXAMPLE 5 Find the GCF of $15x^5$, $-12x^4$, $27x^3$, and $-3x^2$.

First, we find a prime factorization of the coefficients, including a factor of -1 for the negative coefficients.

$$15x^5 = \quad\quad 3 \cdot 5 \cdot x^5,$$

$$-12x^4 = -1 \cdot 2 \cdot 2 \cdot 3 \cdot x^4,$$

$$27x^3 = \quad\quad 3 \cdot 3 \cdot 3 \cdot x^3,$$

$$-3x^2 = \quad\quad -1 \cdot 3 \cdot x^2$$

The greatest *positive* common factor of the coefficients is 3.

Next, we find the GCF of the powers of x. That GCF is x^2, because 2 is the smallest exponent of x. Thus the GCF of the set of monomials is $3x^2$.

What about the factors of -1 in Example 5? Strictly speaking, both 1 and -1 are factors of any number or expression. We see this as follows:

$$3x^2 = 1 \cdot 3x^2 = (-1)(-3x^2).$$

Because the coefficient -3 is less than the coefficient 3, we consider $3x^2$, and not $-3x^2$, the GCF.

EXAMPLE 6 Find the GCF of $14p^2y^3$, $-8py^2$, $2py$, and $4p^3$.

We have

$$14p^2y^3 = 2 \cdot 7 \cdot p^2 \cdot y^3,$$

$$-8py^2 = -1 \cdot 2 \cdot 2 \cdot 2 \cdot p \cdot y^2,$$

$$2py = 2 \cdot p \cdot y,$$

$$4p^3 = 2 \cdot 2 \cdot p^3.$$

The greatest positive common factor of the coefficients is 2, the GCF of the powers of p is p, and the GCF of the powers of y is 1 since there is no y-factor in the last monomial. Thus the GCF is $2p$.

> **TO FIND THE GCF OF TWO OR MORE MONOMIALS**
>
> 1. Find the prime factorization of the coefficients, including -1 as a factor if any coefficient is negative.
> 2. Determine any common prime factors of the coefficients. For each one that occurs, include it as a factor of the GCF. If none occurs, use 1 as a factor.
> 3. Examine each of the variables as factors. If any appear as a factor of all the monomials, include it as a factor, using the smallest exponent of the variable. If none occurs in all the monomials, use 1 as a factor.
> 4. The GCF is the product of the results of steps (2) and (3).

Find the GCF.

5. $12x^2$, $-16x^3$

6. $3y^6$, $-5y^3$, $2y^2$

7. $-24m^5n^6$, $12mn^3$, $-16m^2n^2$, $8m^4n^4$

8. $-35x^7$, $-49x^6$, $-14x^5$, $-63x^3$

Do Exercises 5–8.

Answers

5. $4x^2$ **6.** y^2 **7.** $4mn^2$ **8.** $7x^3$

(b) Factoring When Terms Have a Common Factor

The polynomials we consider most when factoring are those with more than one term. To multiply a monomial and a polynomial with more than one term, we multiply each term of the polynomial by the monomial using the distributive laws:

$$a(b + c) = ab + ac \quad \text{and} \quad a(b - c) = ab - ac.$$

To factor, we do the reverse. We express a polynomial as a product using the distributive laws in reverse:

$$ab + ac = a(b + c) \quad \text{and} \quad ab - ac = a(b - c).$$

Compare.

Multiply

$3x(x^2 + 2x - 4)$
$= 3x \cdot x^2 + 3x \cdot 2x - 3x \cdot 4$
$= 3x^3 + 6x^2 - 12x$

Factor

$3x^3 + 6x^2 - 12x$
$= 3x \cdot x^2 + 3x \cdot 2x - 3x \cdot 4$
$= 3x(x^2 + 2x - 4)$

Caution!

Consider the following:

$$3x^3 + 6x^2 - 12x = 3 \cdot x \cdot x \cdot x + 2 \cdot 3 \cdot x \cdot x - 2 \cdot 2 \cdot 3 \cdot x.$$

The terms of the polynomial, $3x^3$, $6x^2$, and $-12x$, have been factored but the polynomial itself has not been factored. This is not what we mean by a factorization of the polynomial. The *factorization* is

$$3x(x^2 + 2x - 4). \leftarrow \text{A product}$$

The expressions $3x$ and $x^2 + 2x - 4$ are *factors* of $3x^3 + 6x^2 - 12x$.

> Do Exercises 9 and 10.

To factor, we first find the GCF of all terms. It may be 1.

EXAMPLE 7 Factor: $7x^2 + 14$.

We have

$7x^2 + 14 = 7 \cdot x^2 + 7 \cdot 2$ Factoring each term
$\qquad\quad = 7(x^2 + 2).$ Factoring out the GCF, 7

Check: We multiply to check:

$$7(x^2 + 2) = 7 \cdot x^2 + 7 \cdot 2 = 7x^2 + 14.$$

9. a) Multiply: $3(x + 2)$.
    b) Factor: $3x + 6$.

10. a) Multiply: $2x(x^2 + 5x + 4)$.
    b) Factor: $2x^3 + 10x^2 + 8x$.

Answers

9. (a) $3x + 6$; (b) $3(x + 2)$
10. (a) $2x^3 + 10x^2 + 8x$; (b) $2x(x^2 + 5x + 4)$

EXAMPLE 8 Factor: $16x^3 + 20x^2$.

$$16x^3 + 20x^2 = (4x^2)(4x) + (4x^2)(5) \qquad \text{Factoring each term}$$
$$= 4x^2(4x + 5) \qquad \text{Factoring out the GCF, } 4x^2$$

Although it is always more efficient to begin by finding the GCF, suppose in Example 8 that you had not recognized the GCF and removed only part of it, as follows:

$$16x^3 + 20x^2 = (2x^2)(8x) + (2x^2)(10)$$
$$= 2x^2(8x + 10).$$

Note that $8x + 10$ still has a common factor of 2. You need not begin again. Just continue factoring out common factors, as follows, until finished:

$$= 2x^2(2 \cdot 4x + 2 \cdot 5)$$
$$= 2x^2[2(4x + 5)]$$
$$= (2x^2 \cdot 2)(4x + 5)$$
$$= 4x^2(4x + 5).$$

EXAMPLE 9 Factor: $15x^5 - 12x^4 + 27x^3 - 3x^2$.

$$15x^5 - 12x^4 + 27x^3 - 3x^2 = (3x^2)(5x^3) - (3x^2)(4x^2) + (3x^2)(9x) - (3x^2)(1)$$
$$= 3x^2(5x^3 - 4x^2 + 9x - 1) \qquad \text{Factoring out the GCF, } 3x^2$$

------------- *Caution!* -------------

Don't forget the term -1.

Check: We multiply to check:

$$3x^2(5x^3 - 4x^2 + 9x - 1)$$
$$= (3x^2)(5x^3) - (3x^2)(4x^2) + (3x^2)(9x) - (3x^2)(1)$$
$$= 15x^5 - 12x^4 + 27x^3 - 3x^2.$$

As you become more familiar with factoring, you will be able to spot the GCF without factoring each term. Then you can write just the answer.

EXAMPLES Factor.

10. $24x^2 + 12x - 36 = 12(2x^2 + x - 3)$

11. $8m^3 - 16m = 8m(m^2 - 2)$

12. $14p^2y^3 - 8py^2 + 2py = 2py(7py^2 - 4y + 1)$

13. $\dfrac{4}{5}x^2 + \dfrac{1}{5}x + \dfrac{2}{5} = \dfrac{1}{5}(4x^2 + x + 2)$

Do Exercises 11–16.

Factor. Check by multiplying.

11. $x^2 + 3x$

12. $3y^6 - 5y^3 + 2y^2$

13. $9x^4y^2 - 15x^3y + 3x^2y$

14. $\dfrac{3}{4}t^3 + \dfrac{5}{4}t^2 + \dfrac{7}{4}t + \dfrac{1}{4}$

15. $35x^7 - 49x^6 + 14x^5 - 63x^3$

16. $84x^2 - 56x + 28$

Answers

11. $x(x + 3)$ **12.** $y^2(3y^4 - 5y + 2)$

13. $3x^2y(3x^2y - 5x + 1)$

14. $\dfrac{1}{4}(3t^3 + 5t^2 + 7t + 1)$

15. $7x^3(5x^4 - 7x^3 + 2x^2 - 9)$

16. $28(3x^2 - 2x + 1)$

There are two important points to keep in mind as we study this chapter.

> **TIPS FOR FACTORING**
>
> • Before doing any other kind of factoring, first try to factor out the GCF.
> • Always check the result of factoring by multiplying.

(c) Factoring by Grouping: Four Terms

Certain polynomials with four terms can be factored using a method called *factoring by grouping*.

EXAMPLE 14 Factor: $x^2(x + 1) + 2(x + 1)$.

The binomial $x + 1$ is a common factor. We factor it out:

$$x^2(x + 1) + 2(x + 1) = (x + 1)(x^2 + 2).$$

The factorization is $(x + 1)(x^2 + 2)$.

Do Exercises 17 and 18.

Consider the four-term polynomial

$$x^3 + x^2 + 2x + 2.$$

There is no factor other than 1 that is common to all the terms. We can, however, factor $x^3 + x^2$ and $2x + 2$ separately:

$$x^3 + x^2 = x^2(x + 1); \qquad \text{Factoring } x^3 + x^2$$
$$2x + 2 = 2(x + 1). \qquad \text{Factoring } 2x + 2$$

When we group the terms as shown above and factor each polynomial separately, we see that $(x + 1)$ appears in *both* factorizations. Thus we can factor out the common binomial factor as in Example 14:

$$x^3 + x^2 + 2x + 2 = (x^3 + x^2) + (2x + 2)$$
$$= x^2(x + 1) + 2(x + 1)$$
$$= (x + 1)(x^2 + 2).$$

This method of factoring is called **factoring by grouping**. We began with a polynomial with four terms. After grouping and removing common factors, we obtained a polynomial with two parts, each having the common factor $x + 1$, which we then factored out. Not all polynomials with four terms can be factored by this procedure, but it does give us a method to try.

Factor.

17. $x^2(x + 7) + 3(x + 7)$

18. $x^3(a + b) - 5(a + b)$

Answers

17. $(x + 7)(x^2 + 3)$ **18.** $(a + b)(x^3 - 5)$

EXAMPLES Factor by grouping.

15. $6x^3 - 9x^2 + 4x - 6$

$= (6x^3 - 9x^2) + (4x - 6)$ Grouping the terms

$= 3x^2(2x - 3) + 2(2x - 3)$ Factoring each binomial

$= (2x - 3)(3x^2 + 2)$ Factoring out the common factor $2x - 3$

We think through this process as follows:

$$6x^3 - 9x^2 + 4x - 6 = \underbrace{3x^2(2x - 3)}\;\Box\;(2x - 3)$$

(1) Factor the first two terms.

(2) The factor $2x - 3$ gives us a hint to the factorization of the last two terms.

(3) Now we ask ourselves, "What times $2x - 3$ is $4x - 6$?" The answer is $+\,2$.

---- *Caution!* ----

16. $x^3 + x^2 + x + 1 = (x^3 + x^2) + (x + 1)$ Don't forget the 1.

$= x^2(x + 1) + 1(x + 1)$ Factoring each binomial

$= (x + 1)(x^2 + 1)$ Factoring out the common factor $x + 1$

17. $2x^3 - 6x^2 - x + 3$

$= (2x^3 - 6x^2) + (-x + 3)$ Grouping as two binomials

$= 2x^2(x - 3) - 1(x - 3)$ *Check*: $-1(x - 3) = -x + 3$.

$= (x - 3)(2x^2 - 1)$ Factoring out the common factor $x - 3$

We can think through this process as follows.

(1) Factor the first two terms: $2x^3 - 6x^2 = 2x^2(x - 3)$.

(2) The factor $x - 3$ gives us a hint for factoring the last two terms:

$$2x^3 - 6x^2 - x + 3 = 2x^2(x - 3)\;\Box\;(x - 3).$$

(3) Now we ask ourselves, "What times $x - 3$ is $-x + 3$?" The answer is -1.

18. $12x^5 + 20x^2 - 21x^3 - 35 = 4x^2(3x^3 + 5) - 7(3x^3 + 5)$

$$= (3x^3 + 5)(4x^2 - 7)$$

19. $x^3 + x^2 + 2x - 2 = x^2(x + 1) + 2(x - 1)$

This polynomial is not factorable using factoring by grouping. It may be factorable, but not by methods that we will consider in this text.

Do Exercises 19–24.

Factor by grouping.

19. $x^3 + 7x^2 + 3x + 21$

20. $8t^3 + 2t^2 + 12t + 3$

21. $3m^5 - 15m^3 + 2m^2 - 10$

22. $3x^3 - 6x^2 - x + 2$

23. $4x^3 - 6x^2 - 6x + 9$

24. $y^4 - 2y^3 - 2y - 10$

Answers

19. $(x + 7)(x^2 + 3)$ **20.** $(4t + 1)(2t^2 + 3)$
21. $(m^2 - 5)(3m^3 + 2)$
22. $(x - 2)(3x^2 - 1)$ **23.** $(2x - 3)(2x^2 - 3)$
24. Not factorable using factoring by grouping

a Find the GCF.

1. x^2, $-6x$

2. x^2, $5x$

3. $3x^4$, x^2

4. $8x^4$, $-24x^2$

5. $2x^2$, $2x$, -8

6. $8x^2$, $-4x$, -20

7. $-17x^5y^3$, $34x^3y^2$, $51xy$

8. $16p^6q^4$, $32p^3q^3$, $-48pq^2$

9. $-x^2$, $-5x$, $-20x^3$

10. $-x^2$, $-6x$, $-24x^5$

11. x^5y^5, x^4y^3, x^3y^3, $-x^2y^2$

12. $-x^9y^6$, $-x^7y^5$, x^4y^4, x^3y^3

b Factor. Check by multiplying.

13. $x^2 - 6x$

14. $x^2 + 5x$

15. $2x^2 + 6x$

16. $8y^2 - 8y$

17. $x^3 + 6x^2$

18. $3x^4 - x^2$

19. $8x^4 - 24x^2$

20. $5x^5 + 10x^3$

21. $2x^2 + 2x - 8$

22. $8x^2 - 4x - 20$

23. $17x^5y^3 + 34x^3y^2 + 51xy$

24. $16p^6q^4 + 32p^5q^3 - 48pq^2$

25. $6x^4 - 10x^3 + 3x^2$

26. $5x^5 + 10x^2 - 8x$

27. $x^5y^5 + x^4y^3 + x^3y^3 - x^2y^2$

28. $x^9y^6 - x^7y^5 + x^4y^4 + x^3y^3$

29. $2x^7 - 2x^6 - 64x^5 + 4x^3$

30. $8y^3 - 20y^2 + 12y - 16$

31. $1.6x^4 - 2.4x^3 + 3.2x^2 + 6.4x$

32. $2.5x^6 - 0.5x^4 + 5x^3 + 10x^2$

33. $\dfrac{5}{3}x^6 + \dfrac{4}{3}x^5 + \dfrac{1}{3}x^4 + \dfrac{1}{3}x^3$

34. $\dfrac{5}{9}x^7 + \dfrac{2}{9}x^5 - \dfrac{4}{9}x^3 - \dfrac{1}{9}x$

c Factor.

35. $x^2(x + 3) + 2(x + 3)$

36. $y^2(y + 4) + 6(y + 4)$

37. $4z^2(3z - 1) + 7(3z - 1)$

38. $2x^2(4x - 3) + 5(4x - 3)$

39. $2x^2(3x + 2) + (3x + 2)$

40. $3z^2(2z + 7) + (2z + 7)$

41. $5a^3(2a - 7) - (2a - 7)$

42. $m^4(8 - 3m) - 3(8 - 3m)$

Factor by grouping.

43. $x^3 + 3x^2 + 2x + 6$

44. $6z^3 + 3z^2 + 2z + 1$

45. $2x^3 + 6x^2 + x + 3$

46. $3x^3 + 2x^2 + 3x + 2$

47. $8x^3 - 12x^2 + 6x - 9$

48. $10x^3 - 25x^2 + 4x - 10$

49. $12p^3 - 16p^2 + 3p - 4$

50. $18x^3 - 21x^2 + 30x - 35$

51. $5x^3 - 5x^2 - x + 1$

52. $7x^3 - 14x^2 - x + 2$

53. $x^3 + 8x^2 - 3x - 24$

54. $2x^3 + 12x^2 - 5x - 30$

55. $2x^3 - 8x^2 - 9x + 36$

56. $20g^3 - 4g^2 - 25g + 5$

Skill Maintenance

Solve.

57. $-2x < 48$ [2.7d]

58. $4x - 8x + 16 \geq 6(x - 2)$ [2.7e]

59. Divide: $\dfrac{-108}{-4}$.   [1.6a]

60. Solve $A = \dfrac{p + q}{2}$ for p.   [2.4b]

Multiply.   [4.6d]

61. $(y + 5)(y + 7)$

62. $(y + 7)^2$

63. $(y + 7)(y - 7)$

64. $(y - 7)^2$

Find the intercepts of each equation. Then graph the equation.   [3.3a]

65. $x + y = 4$

66. $x - y = 3$

67. $5x - 3y = 15$

68. $y - 3x = 6$

Synthesis

Factor.

69. $4x^5 + 6x^3 + 6x^2 + 9$

70. $x^6 + x^4 + x^2 + 1$

71. $x^{12} + x^7 + x^5 + 1$

72. $x^3 - x^2 - 2x + 5$

73. $p^3 + p^2 - 3p + 10$

5.2

Factoring Trinomials of the Type $x^2 + bx + c$

(a) Factoring $x^2 + bx + c$

We now begin a study of the factoring of trinomials. We first factor trinomials like

$$x^2 + 5x + 6 \quad \text{and} \quad x^2 + 3x - 10$$

by a refined *trial-and-error process*. In this section, we restrict our attention to trinomials of the type $ax^2 + bx + c$, where $a = 1$. The coefficient a is called the **leading coefficient**.

To understand the factoring that follows, compare the following multiplications:

$$
\begin{array}{cccc}
\text{F} & \text{O} & \text{I} & \text{L} \\
\downarrow & \downarrow & \downarrow & \downarrow
\end{array}
$$

$$(x + 2)(x + 5) = x^2 + 5x + 2x + 2 \cdot 5$$
$$= x^2 + \quad 7x \quad + 10;$$

$$(x - 2)(x - 5) = x^2 - 5x - 2x + (-2)(-5)$$
$$= x^2 - \quad 7x \quad + 10;$$

$$(x + 3)(x - 7) = x^2 - 7x + 3x + 3(-7)$$
$$= x^2 - \quad 4x \quad - 21;$$

$$(x - 3)(x + 7) = x^2 + 7x - 3x + (-3)7$$
$$= x^2 + \quad 4x \quad - 21.$$

Note that for all four products:

- The product of the two binomials is a trinomial.
- The coefficient of x in the trinomial is the sum of the constant terms in the binomials.
- The constant term in the trinomial is the product of the constant terms in the binomials.

These observations lead to a method for factoring certain trinomials. The first type we consider has a positive constant term, just as in the first two multiplications above.

Constant Term Positive

To factor $x^2 + 7x + 10$, we think of FOIL in reverse. We multiplied x times x to get the first term of the trinomial, so we know that the first term of each binomial factor is x. Next, we look for numbers p and q such that

$$x^2 + 7x + 10 = (x + p)(x + q).$$

To get the middle term and the last term of the trinomial, we look for two numbers p and q whose product is 10 and whose sum is 7. Those numbers are 2 and 5. Thus the factorization is

$$(x + 2)(x + 5).$$

Check: $(x + 2)(x + 5) = x^2 + 5x + 2x + 10$
$$= x^2 + 7x + 10.$$

OBJECTIVE

 Factor trinomials of the type $x^2 + bx + c$ by examining the constant term c.

SKILL TO REVIEW
Objective 4.6a: Multiply two binomials mentally using the FOIL method.

Multiply.
1. $(x + 3)(x + 4)$
2. $(x - 1)(x + 2)$

Answers

Skill to Review:
1. $x^2 + 7x + 12$ 2. $x^2 + x - 2$

Do Exercises 1 and 2.

1. Consider the trinomial $x^2 + 7x + 12$.

a) Complete the following table.

PAIRS OF FACTORS	SUMS OF FACTORS
1, 12	13
−1, −12	
2, 6	
−2, −6	
3, 4	
−3, −4	

b) Explain why you need to consider only the positive factors in the table above.

c) Factor: $x^2 + 7x + 12$.

2. Factor: $x^2 + 13x + 36$.

3. Explain why you would *not* consider the pairs of factors listed below in factoring $y^2 − 8y + 12$.

PAIRS OF FACTORS	SUMS OF FACTORS
1, 12	
2, 6	
3, 4	

Factor.

4. $x^2 − 8x + 15$

5. $t^2 − 9t + 20$

Answers
1. (a) −13, 8, −8, 7, −7; **(b)** Both 7 and 12 are positive. **(c)** $(x + 3)(x + 4)$
2. $(x + 4)(x + 9)$ **3.** The coefficient of the middle term, −8, is negative.
4. $(x − 5)(x − 3)$
5. $(t − 5)(t − 4)$

EXAMPLE 1   Factor: $x^2 + 5x + 6$.

Think of FOIL in reverse. The first term of each factor is x: $(x + \square)(x + \square)$. Next, we look for two numbers whose product is 6 and whose sum is 5. All the pairs of factors of 6 are shown in the table on the left below. Since both the product, 6, and the sum, 5, of the pair of numbers must be positive, we need consider only the positive factors, listed in the table on the right.

PAIRS OF FACTORS	SUMS OF FACTORS
1, 6	7
−1, −6	−7
2, 3	5
−2, −3	−5

PAIRS OF FACTORS	SUMS OF FACTORS
1, 6	7
2, 3	5

↑
The numbers we need are 2 and 3.

The factorization is $(x + 2)(x + 3)$. We can check by multiplying to see whether we get the original trinomial.

Check: $(x + 2)(x + 3) = x^2 + 3x + 2x + 6 = x^2 + 5x + 6$.

Do Exercises 1 and 2.

Compare these multiplications:

$$(x − 2)(x − 5) = x^2 − 5x − 2x + 10 = x^2 − 7x + 10;$$
$$(x + 2)(x + 5) = x^2 + 5x + 2x + 10 = x^2 + 7x + 10.$$

TO FACTOR $x^2 + bx + c$ WHEN c IS POSITIVE

When the constant term of a trinomial is positive, look for two numbers with the same sign. The sign is that of the middle term:

$$x^2 − 7x + 10 = (x − 2)(x − 5);$$

$$x^2 + 7x + 10 = (x + 2)(x + 5).$$

EXAMPLE 2   Factor: $y^2 − 8y + 12$.

Since the constant term, 12, is positive and the coefficient of the middle term, −8, is negative, we look for a factorization of 12 in which both factors are negative. Their sum must be −8.

PAIRS OF FACTORS	SUMS OF FACTORS
−1, −12	−13
−2, −6	−8 ←
−3, −4	−7

The numbers we need are −2 and −6.

The factorization is $(y − 2)(y − 6)$. The student should check by multiplying.

Do Exercises 3-5.

Constant Term Negative

As we saw in two of the multiplications earlier in this section, the product of two binomials can have a negative constant term:

$$(x + 3)(x - 7) = x^2 - 4x - 21$$

and

$$(x - 3)(x + 7) = x^2 + 4x - 21.$$

Note that when the signs of the constants in the binomials are reversed, only the sign of the middle term in the product changes.

EXAMPLE 3 Factor: $x^2 - 8x - 20$.

The constant term, -20, must be expressed as the product of a negative number and a positive number. Since the sum of these two numbers must be negative (specifically, -8), the negative number must have the greater absolute value.

PAIRS OF FACTORS	SUMS OF FACTORS
1, −20	−19
2, −10	−8
4, −5	−1
5, −4	1
10, −2	8
20, −1	19

The numbers we need are 2 and −10.

Because these sums are all positive, for this problem all the corresponding pairs can be disregarded. Note that in all three pairs, the positive number has the greater absolute value.

The numbers that we are looking for are 2 and -10. The factorization is $(x + 2)(x - 10)$.

Check: $(x + 2)(x - 10) = x^2 - 10x + 2x - 20$
$$= x^2 - 8x - 20.$$

TO FACTOR $x^2 + bx + c$ WHEN c IS NEGATIVE

When the constant term of a trinomial is negative, look for two numbers whose product is negative. One must be positive and the other negative:

$$x^2 - 4x - 21 = (x + 3)(x - 7);$$

$$x^2 + 4x - 21 = (x - 3)(x + 7).$$

Consider pairs of numbers for which the number with the larger absolute value has the same sign as b, the coefficient of the middle term.

Do Exercises 6 and 7. (Exercise 7 is on the following page.)

6. Consider $x^2 - 5x - 24$.

   a) Explain why you would *not* consider the pairs of factors listed below in factoring $x^2 - 5x - 24$.

PAIRS OF FACTORS	SUMS OF FACTORS
−1, 24	
−2, 12	
−3, 8	
−4, 6	

   b) Explain why you *would* consider the pairs of factors listed below in factoring $x^2 - 5x - 24$.

PAIRS OF FACTORS	SUMS OF FACTORS
1, −24	
2, −12	
3, −8	
4, −6	

   c) Factor: $x^2 - 5x - 24$.

Answers

6. (a) The positive factor has the larger absolute value. (b) The negative factor has the larger absolute value. (c) $(x + 3)(x - 8)$

7. Consider $x^2 + 10x - 24$.

a) Explain why you would *not* consider the pairs of factors listed below in factoring $x^2 + 10x - 24$.

PAIRS OF FACTORS	SUMS OF FACTORS
1, −24	
2, −12	
3, −8	
4, −6	

b) Explain why you *would* consider the pairs of factors listed below in factoring $x^2 + 10x - 24$.

PAIRS OF FACTORS	SUMS OF FACTORS
−1, 24	
−2, 12	
−3, 8	
−4, 6	

c) Factor: $x^2 + 10x - 24$.

Factor.

8. $a^2 - 40 + 3a$

9. $-18 - 3t + t^2$

EXAMPLE 4 Factor: $t^2 - 24 + 5t$.

It helps to first write the trinomial in descending order: $t^2 + 5t - 24$. Since the constant term, -24, is negative, we look for a factorization of -24 in which one factor is positive and one factor is negative. Their sum must be 5, so we consider only pairs of factors in which the positive factor has the larger absolute value.

PAIRS OF FACTORS	SUMS OF FACTORS	
−1, 24	23	
−2, 12	10	
−3, 8	**5**	← — The numbers we need are −3 and 8.
−4, 6	2	

The factorization is $(t - 3)(t + 8)$. The check is left to the student.

Do Exercises 8 and 9.

EXAMPLE 5 Factor: $x^4 - x^2 - 110$.

Consider this trinomial as $(x^2)^2 - x^2 - 110$. We look for numbers p and q such that

$$x^4 - x^2 - 110 = (x^2 + p)(x^2 + q).$$

Since the constant term, -110, is negative, we look for a factorization of -110 in which one factor is positive and one factor is negative. Their sum must be -1. The middle-term coefficient, -1, is small compared to -110. This tells us that the desired factors are close to each other in absolute value. The numbers we want are 10 and -11. The factorization is

$$(x^2 + 10)(x^2 - 11).$$

EXAMPLE 6 Factor: $a^2 + 4ab - 21b^2$.

We consider the trinomial in the equivalent form

$$a^2 + 4ba - 21b^2.$$

This way we think of $-21b^2$ as the "constant" term and $4b$ as the "coefficient" of the middle term. Then we try to express $-21b^2$ as a product of two factors whose sum is $4b$. Those factors are $-3b$ and $7b$. The factorization is $(a - 3b)(a + 7b)$.

Check: $(a - 3b)(a + 7b) = a^2 + 7ab - 3ba - 21b^2$
$$= a^2 + 4ab - 21b^2.$$

There are polynomials that are not factorable.

EXAMPLE 7 Factor: $x^2 - x + 5$.

Since 5 has very few factors, we can easily check all possibilities.

PAIRS OF FACTORS	SUMS OF FACTORS
5, 1	6
−5, −1	−6

There are no factors whose sum is -1. Thus the polynomial is *not* factorable into factors that are polynomials with rational-number coefficients.

In this text, a polynomial like $x^2 - x + 5$ that cannot be factored further is said to be **prime**. In more advanced courses, polynomials like $x^2 - x + 5$ can be factored and are not considered prime.

Do Exercises 10-12.

Often factoring requires two or more steps. In general, when told to factor, we should *factor completely*. This means that the final factorization should not contain any factors that can be factored further.

EXAMPLE 8 Factor: $2x^3 - 20x^2 + 50x$.

Always look first for a common factor. This time there is one, $2x$, which we factor out first:

$$2x^3 - 20x^2 + 50x = 2x(x^2 - 10x + 25).$$

Now consider $x^2 - 10x + 25$. Since the constant term is positive and the coefficient of the middle term is negative, we look for a factorization of 25 in which both factors are negative. Their sum must be -10.

PAIRS OF FACTORS	SUMS OF FACTORS
$-25, -1$	-26
$-5, -5$	-10 ← — The numbers we need are -5 and -5.

The factorization of $x^2 - 10x + 25$ is $(x - 5)(x - 5)$, or $(x - 5)^2$. The final factorization is $2x(x - 5)^2$. We check by multiplying:

$$\begin{aligned}
2x(x - 5)^2 &= 2x(x^2 - 10x + 25) \\
&= (2x)(x^2) - (2x)(10x) + (2x)(25) \\
&= 2x^3 - 20x^2 + 50x.
\end{aligned}$$

Do Exercises 13-15.

Once any common factors have been factored out, the following summary can be used to factor $x^2 + bx + c$.

TO FACTOR $x^2 + bx + c$

1. First arrange in descending order.
2. Use a trial-and-error process that looks for factors of c whose sum is b.
3. If c is positive, the signs of the factors are the same as the sign of b.
4. If c is negative, one factor is positive and the other is negative. If the sum of two factors is the opposite of b, changing the sign of each factor will give the desired factors whose sum is b.
5. Check by multiplying.

Factor.

10. $y^2 - 12 - 4y$

11. $t^4 + 5t^2 - 14$

12. $x^2 + 2x + 7$

Factor.

13. $x^3 + 4x^2 - 12x$

14. $p^2 - pq - 3pq^2$

15. $3x^3 + 24x^2 + 48x$

Answers

10. $(y - 6)(y + 2)$ **11.** $(t^2 + 7)(t^2 - 2)$
12. Prime   **13.** $x(x + 6)(x - 2)$
14. $p(p - q - 3q^2)$ **15.** $3x(x + 4)^2$

Leading Coefficient −1

EXAMPLE 9 Factor: $10 - 3x - x^2$.

Note that the polynomial is written in ascending order. When we write it in descending order, we get

$$-x^2 - 3x + 10,$$

which has a leading coefficient of −1. Before factoring in such a case, we can factor out a −1, as follows:

$$-x^2 - 3x + 10 = -1 \cdot x^2 + (-1)(3x) + (-1)(-10)$$
$$= -1(x^2 + 3x - 10).$$

Then we proceed to factor $x^2 + 3x - 10$. We get

$$-x^2 - 3x + 10 = -1(x^2 + 3x - 10) = -1(x + 5)(x - 2).$$

We can also express this answer in two other ways by multiplying either binomial by −1. Thus each of the following is a correct answer:

$$-x^2 - 3x + 10 = -1(x + 5)(x - 2)$$
$$= (-x - 5)(x - 2) \qquad \text{Multiplying } x + 5 \text{ by } -1$$
$$= (x + 5)(-x + 2). \qquad \text{Multiplying } x - 2 \text{ by } -1$$

Factor.

16. $14 + 5x - x^2$

17. $-x^2 + 3x + 18$

Do Exercises 16 and 17.

STUDY TIPS

TIME MANAGEMENT

- **Are you a morning or an evening person?** If you are an evening person, it might be best to avoid scheduling early-morning classes. If you are a morning person, you will probably want to schedule morning classes if your work schedule and family obligations will allow it. Nothing can drain your study time and effectiveness like fatigue.

- **Keep on schedule.** Your course syllabus provides a plan for the semester's schedule. Read the entire syllabus at the beginning of the semester. Use a write-on calendar, daily planner, PDA, or laptop computer to outline your time for the semester. Be sure to note deadlines involving writing assignments and exams so you can begin a big task early, breaking it down into smaller segments that will not overwhelm you.

Answers

16. $-1(x + 2)(x - 7)$, or $(-x - 2)(x - 7)$, or $(x + 2)(-x + 7)$
17. $-1(x + 3)(x - 6)$, or $(-x - 3)(x - 6)$, or $(x + 3)(-x + 6)$

a Factor. Remember that you can check by multiplying.

1. $x^2 + 8x + 15$

PAIRS OF FACTORS	SUMS OF FACTORS

2. $x^2 + 5x + 6$

PAIRS OF FACTORS	SUMS OF FACTORS

3. $x^2 + 7x + 12$

PAIRS OF FACTORS	SUMS OF FACTORS

4. $x^2 + 9x + 8$

PAIRS OF FACTORS	SUMS OF FACTORS

5. $x^2 - 6x + 9$

PAIRS OF FACTORS	SUMS OF FACTORS

6. $y^2 - 11y + 28$

PAIRS OF FACTORS	SUMS OF FACTORS

7. $x^2 - 5x - 14$

PAIRS OF FACTORS	SUMS OF FACTORS

8. $a^2 + 7a - 30$

PAIRS OF FACTORS	SUMS OF FACTORS

9. $b^2 + 5b + 4$

PAIRS OF FACTORS	SUMS OF FACTORS

10. $z^2 - 8z + 7$

PAIRS OF FACTORS	SUMS OF FACTORS

11. $x^2 + \dfrac{2}{3}x + \dfrac{1}{9}$

PAIRS OF FACTORS	SUMS OF FACTORS

12. $x^2 - \dfrac{2}{5}x + \dfrac{1}{25}$

PAIRS OF FACTORS	SUMS OF FACTORS

13. $d^2 - 7d + 10$

14. $t^2 - 12t + 35$

15. $y^2 - 11y + 10$

16. $x^2 - 4x - 21$

17. $x^2 + x + 1$

18. $x^2 + 5x + 3$

19. $x^2 - 7x - 18$

20. $y^2 - 3y - 28$

21. $x^3 - 6x^2 - 16x$

22. $x^3 - x^2 - 42x$

23. $y^3 - 4y^2 - 45y$

24. $x^3 - 7x^2 - 60x$

25. $-2x - 99 + x^2$

26. $x^2 - 72 + 6x$

27. $c^4 + c^2 - 56$

28. $b^4 + 5b^2 - 24$

29. $a^4 + 2a^2 - 35$

30. $x^4 - x^2 - 6$

31. $x^2 + x - 42$

32. $x^2 + 2x - 15$

33. $7 - 2p + p^2$

34. $11 - 3w + w^2$

35. $x^2 + 20x + 100$

36. $a^2 + 19a + 88$

37. $2z^3 - 2z^2 - 24z$

38. $5w^4 - 20w^3 - 25w^2$

39. $3t^4 + 3t^3 + 3t^2$

40. $4y^5 - 4y^4 - 4y^3$

41. $x^4 - 21x^3 - 100x^2$

42. $x^4 - 20x^3 + 96x^2$

43. $x^2 - 21x - 72$

44. $4x^2 + 40x + 100$

45. $x^2 - 25x + 144$

46. $y^2 - 21y + 108$

47. $a^2 + a - 132$

48. $a^2 + 9a - 90$

49. $3t^2 + 6t + 3$

50. $2y^2 + 24y + 72$

51. $w^4 - 8w^3 + 16w^2$

52. $z^5 - 6z^4 + 9z^3$

53. $30 + 7x - x^2$

54. $45 + 4x - x^2$

55. $24 - a^2 - 10a$

56. $-z^2 + 36 - 9z$

57. $120 - 23x + x^2$

58. $96 + 22d + d^2$

59. $108 - 3x - x^2$

60. $112 + 9y - y^2$

61. $y^2 - 0.2y - 0.08$

62. $t^2 - 0.3t - 0.10$

63. $p^2 + 3pq - 10q^2$

64. $a^2 + 2ab - 3b^2$

65. $84 - 8t - t^2$

66. $72 - 6m - m^2$

67. $m^2 + 5mn + 4n^2$

68. $x^2 + 11xy + 24y^2$

69. $s^2 - 2st - 15t^2$

70. $p^2 + 5pq - 24q^2$

71. $6a^{10} - 30a^9 - 84a^8$

72. $7x^9 - 28x^8 - 35x^7$

Skill Maintenance

Multiply. [4.5b], [4.6d]

73. $8x(2x^2 - 6x + 1)$

74. $(7w + 6)(4w - 11)$

75. $(7w + 6)^2$

76. $(4w - 11)^2$

77. $(4w - 11)(4w + 11)$

78. $-y(-y^2 + 3y - 5)$

79. $(3x - 5y)(2x + 7y)$

80. Simplify: $(3x^4)^3$. [4.2a, b]

Solve. [2.3a]

81. $3x - 8 = 0$

82. $2x + 7 = 0$

Solve.

83. *Arrests for Counterfeiting.* In 2008, the U.S. Secret Service made 2231 arrests for counterfeiting. This was an increase of 28% over the number of arrests in 2007. How many arrests for counterfeiting were made in 2007? [2.5a]
Source: U.S. Secret Service

84. The first angle of a triangle is four times as large as the second. The measure of the third angle is 30° greater than that of the second. Find the angle measures. [2.6a]

Synthesis

85. Find all integers m for which $y^2 + my + 50$ can be factored.

86. Find all integers b for which $a^2 + ba - 50$ can be factored.

Factor completely.

87. $x^2 - \frac{1}{2}x - \frac{3}{16}$

88. $x^2 - \frac{1}{4}x - \frac{1}{8}$

89. $x^2 + \frac{30}{7}x - \frac{25}{7}$

90. $\frac{1}{3}x^3 + \frac{1}{3}x^2 - 2x$

91. $b^{2n} + 7b^n + 10$

92. $a^{2m} - 11a^m + 28$

Find a polynomial in factored form for the shaded area in each figure. (Leave answers in terms of π.)

93.

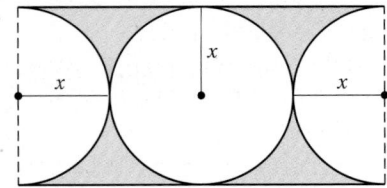

94.

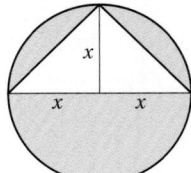

5.3

Factoring $ax^2 + bx + c, a \neq 1$: The FOIL Method

In Section 5.2, we learned a trial-and-error method to factor trinomials of the type $x^2 + bx + c$. In this section, we factor trinomials in which the coefficient of the leading term x^2 is not 1. Again, the procedure we use is a refined trial-and-error method.

a The FOIL Method

We want to factor trinomials of the type $ax^2 + bx + c$. Consider the following multiplication:

$$\begin{array}{cccc} & \text{F} \quad \text{O} \quad \text{I} \quad \text{L} & & \\ (2x + 5)(3x + 4) = & 6x^2 + 8x + 15x + 20 & & \\ = & 6x^2 + \quad 23x \quad + 20 & & \end{array}$$

F	O + I	L
$2 \cdot 3$	$2 \cdot 4 \quad 5 \cdot 3$	$5 \cdot 4$

To factor $6x^2 + 23x + 20$, we reverse the above multiplication, using what we might call an "unFOIL" process. We look for two binomials $rx + p$ and $sx + q$ whose product is $(rx + p)(sx + q) = 6x^2 + 23x + 20$. The product of the First terms must be $6x^2$. The product of the Outside terms plus the product of the Inside terms must be $23x$. The product of the Last terms must be 20. We know from the preceding discussion that the answer is $(2x + 5)(3x + 4)$. Generally, however, finding such an answer is a refined trial-and-error process. It turns out that $(-2x - 5)(-3x - 4)$ is also a correct answer, but we generally choose an answer in which the first coefficients are positive.

We will use the following trial-and-error method.

THE FOIL METHOD

To factor $ax^2 + bx + c, a \neq 1$, using the FOIL method:

1. Factor out the largest common factor, if one exists.

2. Find two First terms whose product is ax^2.

$$(\square x + \quad)(\square x + \quad) = ax^2 + bx + c.$$
$$\underline{\qquad\qquad\text{FOIL}}$$

3. Find two Last terms whose product is c:

$$(\quad x + \square)(\quad x + \square) = ax^2 + bx + c.$$
$$\underline{\qquad\qquad\text{FOIL}}$$

4. Look for Outer and Inner products resulting from steps (2) and (3) for which the sum is bx:

$$(\square x + \square)(\square x + \square) = ax^2 + bx + c.$$
$$\text{I} \qquad\qquad \text{FOIL}$$
$$\text{O}$$

5. Always check by multiplying.

OBJECTIVE

 a Factor trinomials of the type $ax^2 + bx + c, a \neq 1$, using the FOIL method.

SKILL TO REVIEW
Objective 4.6a: Multiply two binomials mentally using the FOIL method.

Multiply.
1. $(2x + 3)(x + 1)$
2. $(3x - 4)(2x - 1)$

The ac-method in Section 5.4

To the student: In Section 5.4, we will consider an alternative method for the same kind of factoring. It involves factoring by grouping and is called the *ac*-method.

To the instructor: We present two ways to factor general trinomials in Sections 5.3 and 5.4: the FOIL method in Section 5.3 and the *ac*-method in Section 5.4. You can teach both methods and let the student use the one that he or she prefers or you can select just one.

Answers

Skill to Review:
1. $2x^2 + 5x + 3$ 2. $6x^2 - 11x + 4$

EXAMPLE 1 Factor: $3x^2 - 10x - 8$.

1) First, we check for a common factor. Here there is none (other than 1 or -1).

2) Find two **First** terms whose product is $3x^2$.

The only possibilities for the **First** terms are $3x$ and x, so any factorization must be of the form

$$(3x + \square)(x + \square).$$

3) Find two **Last** terms whose product is -8.

Possible factorizations of -8 are

$$(-8) \cdot 1, \quad 8 \cdot (-1), \quad (-2) \cdot 4, \quad \text{and} \quad 2 \cdot (-4).$$

Since the First terms are not identical, we must also consider

$$1 \cdot (-8), \quad (-1) \cdot 8, \quad 4 \cdot (-2), \quad \text{and} \quad (-4) \cdot 2.$$

4) Inspect the **O**utside and **I**nside products resulting from steps (2) and (3). Look for a combination in which the sum of the products is the middle term, $-10x$:

Trial	Product	
$(3x - 8)(x + 1)$	$3x^2 + 3x - 8x - 8$ $= 3x^2 - 5x - 8$	← Wrong middle term
$(3x + 8)(x - 1)$	$3x^2 - 3x + 8x - 8$ $= 3x^2 + 5x - 8$	← Wrong middle term
$(3x - 2)(x + 4)$	$3x^2 + 12x - 2x - 8$ $= 3x^2 + 10x - 8$	← Wrong middle term
$(3x + 2)(x - 4)$	$3x^2 - 12x + 2x - 8$ $= 3x^2 - 10x - 8$	← **Correct middle term!**
$(3x + 1)(x - 8)$	$3x^2 - 24x + x - 8$ $= 3x^2 - 23x - 8$	← Wrong middle term
$(3x - 1)(x + 8)$	$3x^2 + 24x - x - 8$ $= 3x^2 + 23x - 8$	← Wrong middle term
$(3x + 4)(x - 2)$	$3x^2 - 6x + 4x - 8$ $= 3x^2 - 2x - 8$	← Wrong middle term
$(3x - 4)(x + 2)$	$3x^2 + 6x - 4x - 8$ $= 3x^2 + 2x - 8$	← Wrong middle term

The correct factorization is $(3x + 2)(x - 4)$.

5) **Check:** $(3x + 2)(x - 4) = 3x^2 - 10x - 8$.

Two observations can be made from Example 1. First, we listed all possible trials even though we could have stopped after having found the correct factorization. We did this to show that each trial differs only in the middle term of the product. **Second, note that as in Section 5.2, only the sign of the middle term changes when the signs in the binomials are reversed:**

Plus Minus
$$(3x + 4)(x - 2) = 3x^2 - 2x - 8$$
Minus Plus   — Middle term changes sign
$$(3x - 4)(x + 2) = 3x^2 + 2x - 8.$$

Factor.

1. $2x^2 - x - 15$

2. $12x^2 - 17x - 5$

EXAMPLE 2 Factor: $24x^2 - 76x + 40$.

1) First, we factor out the largest common factor, 4:

$$4(6x^2 - 19x + 10).$$

Now we factor the trinomial $6x^2 - 19x + 10$.

2) Because $6x^2$ can be factored as $3x \cdot 2x$ or $6x \cdot x$, we have these possibilities for factorizations:

$$(3x + \square)(2x + \square) \quad \text{or} \quad (6x + \square)(x + \square).$$

3) There are four pairs of factors of 10 and each pair can be listed in two ways:

$$10, 1 \quad -10, -1 \quad 5, 2 \quad -5, -2$$

and

$$1, 10 \quad -1, -10 \quad 2, 5 \quad -2, -5.$$

4) The two possibilities from step (2) and the eight possibilities from step (3) give $2 \cdot 8$, or 16 possibilities for factorizations. We look for **O**utside and **I**nside products resulting from steps (2) and (3) for which the sum is the middle term, $-19x$. Since the sign of the middle term is negative, but the sign of the last term, 10, is positive, both factors of 10 must be negative. This means only four pairings from step (3) need be considered. We first try these factors with

$$(3x + \square)(2x + \square).$$

If none gives the correct factorization, we will consider

$$(6x + \square)(x + \square).$$

Trial	Product
$(3x - 10)(2x - 1)$	$6x^2 - 3x - 20x + 10$
	$= 6x^2 - 23x + 10$ ← Wrong middle term
$(3x - 1)(2x - 10)$	$6x^2 - 30x - 2x + 10$
	$= 6x^2 - 32x + 10$ ← Wrong middle term
$(3x - 5)(2x - 2)$	$6x^2 - 6x - 10x + 10$
	$= 6x^2 - 16x + 10$ ← Wrong middle term
$(3x - 2)(2x - 5)$	$6x^2 - 15x - 4x + 10$
	$= 6x^2 - 19x + 10$ ← **Correct middle term!**

Since we have a correct factorization, we need not consider

$$(6x + \square)(x + \square).$$

The factorization of $6x^2 - 19x + 10$ is $(3x - 2)(2x - 5)$, but *do not forget the common factor*! We must include it in order to factor the original trinomial:

$$24x^2 - 76x + 40 = 4(6x^2 - 19x + 10)$$
$$= 4(3x - 2)(2x - 5).$$

5) Check: $4(3x - 2)(2x - 5) = 4(6x^2 - 19x + 10) = 24x^2 - 76x + 40$.

--------- *Caution!* ---------

When factoring any polynomial, always look for a common factor first. Failure to do so is such a common error that this caution bears repeating.

Answers

1. $(2x + 5)(x - 3)$ **2.** $(4x + 1)(3x - 5)$

In Example 2, look again at the possibility $(3x - 5)(2x - 2)$. Without multiplying, we can reject such a possibility. To see why, consider the following:

$$(3x - 5)(2x - 2) = (3x - 5)(2)(x - 1) = 2(3x - 5)(x - 1).$$

The expression $2x - 2$ has a common factor, 2. But we removed the *largest* common factor in the first step. If $2x - 2$ were one of the factors, then 2 would have to be a common factor in addition to the original 4. Thus, $(2x - 2)$ cannot be part of the factorization of the original trinomial.

> Given that the largest common factor is factored out at the outset, we need not consider factorizations that have a common factor.

Factor.

3. $3x^2 - 19x + 20$

4. $20x^2 - 46x + 24$

Do Exercises 3 and 4.

EXAMPLE 3   Factor: $10x^2 + 37x + 7$.

1) There is no common factor (other than 1 or -1).
2) Because $10x^2$ factors as $10x \cdot x$ or $5x \cdot 2x$, we have these possibilities for factorizations:

$$(10x + \square)(x + \square) \quad \text{or} \quad (5x + \square)(2x + \square).$$

3) There are two pairs of factors of 7 and each pair can be listed in two ways:

$$1, 7 \quad -1, -7 \qquad \text{and} \qquad 7, 1 \quad -7, -1.$$

4) From steps (2) and (3), we see that there are 8 possibilities for factorizations. Look for **O**uter and **I**nner products for which the sum is the middle term. Because all coefficients in $10x^2 + 37x + 7$ are positive, we need consider only positive factors of 7. The possibilities are

$$(10x + 1)(x + 7) = 10x^2 + 71x + 7,$$
$$(10x + 7)(x + 1) = 10x^2 + 17x + 7,$$
$$(5x + 7)(2x + 1) = 10x^2 + 19x + 7,$$
$$(5x + 1)(2x + 7) = 10x^2 + 37x + 7. \quad \leftarrow \textbf{Correct middle term}$$

The factorization is $(5x + 1)(2x + 7)$.

5) Check: $(5x + 1)(2x + 7) = 10x^2 + 37x + 7$.

5. Factor: $6x^2 + 7x + 2$.

Do Exercise 5.

> **TIPS FOR FACTORING** $ax^2 + bx + c$, $a \neq 1$
>
> - Always factor out the largest common factor first, if one exists.
> - Once the common factor has been factored out of the original trinomial, no binomial factor can contain a common factor (other than 1 or -1).
> - If c is positive, then the signs in both binomial factors must match the sign of b. (This assumes that $a > 0$.)
> - Reversing the signs in the binomials reverses the sign of the middle term of their product.
> - Organize your work so that you can keep track of which possibilities have or have not been checked.
> - Always check by multiplying.

Answers

3. $(3x - 4)(x - 5)$ **4.** $2(5x - 4)(2x - 3)$
5. $(2x + 1)(3x + 2)$

EXAMPLE 4 Factor: $10x + 8 - 3x^2$.

An important problem-solving strategy is to find a way to make new problems look like problems we already know how to solve. (See Example 9 in Section 5.2.) The factoring tips on the preceding page apply only to trinomials of the form $ax^2 + bx + c$, with $a > 0$. This leads us to rewrite $10x + 8 - 3x^2$ in descending order:

$$10x + 8 - 3x^2 = -3x^2 + 10x + 8. \qquad \text{Writing in descending order}$$

Although $-3x^2 + 10x + 8$ looks similar to the trinomials we have factored, the factoring tips require a positive leading coefficient. This can be attained by factoring out -1:

$$-3x^2 + 10x + 8 = -1(3x^2 - 10x - 8) \qquad \text{Factoring out } -1 \text{ changes the signs of the coefficients.}$$
$$= -1(3x + 2)(x - 4). \qquad \text{Using the result from Example 1}$$

The factorization of $10x + 8 - 3x^2$ is $-1(3x + 2)(x - 4)$. Other correct answers are

$$10x + 8 - 3x^2 = (3x + 2)(-x + 4) \qquad \text{Multiplying } x - 4 \text{ by } -1$$
$$= (-3x - 2)(x - 4). \qquad \text{Multiplying } 3x + 2 \text{ by } -1$$

Do Exercises 6 and 7.

EXAMPLE 5 Factor: $6p^2 - 13pq - 28q^2$.

1) Factor out a common factor, if any.

There is none (other than 1 or -1).

2) Factor the first term, $6p^2$.

Possibilities are $2p, 3p$ and $6p, p$. We have these as possibilities for factorizations:

$$(2p + \square)(3p + \square) \quad \text{or} \quad (6p + \square)(p + \square).$$

3) Factor the last term, $-28q^2$, which has a negative coefficient.

There are six pairs of factors and each can be listed in two ways:

$$-28q, q \qquad 28q, -q \qquad -14q, 2q \qquad 14q, -2q \qquad -7q, 4q \qquad 7q, -4q$$

and

$$q, -28q \qquad -q, 28q \qquad 2q, -14q \qquad -2q, 14q \qquad 4q, -7q \qquad -4q, 7q.$$

4) The coefficient of the middle term is negative, so we look for combinations of factors from steps (2) and (3) such that the sum of their products has a negative coefficient. We try some possibilities:

$$(2p + q)(3p - 28q) = 6p^2 - 53pq - 28q^2,$$
$$(2p - 7q)(3p + 4q) = 6p^2 - 13pq - 28q^2. \qquad \leftarrow \textbf{Correct middle term}$$

The factorization of $6p^2 - 13pq - 28q^2$ is $(2p - 7q)(3p + 4q)$.

5) The check is left to the student.

Do Exercises 8 and 9.

STUDY TIPS

READING EXAMPLES

A careful study of the examples in these sections on factoring is critical. *Read them carefully* to ensure success!

Factor.

6. $2 - x - 6x^2$

7. $2x + 8 - 6x^2$

Factor.

8. $6a^2 - 5ab + b^2$

9. $6x^2 + 15xy + 9y^2$

Answers

6. $-1(2x - 1)(3x + 2)$, or $(2x - 1)(-3x - 2)$, or $(-2x + 1)(3x + 2)$ **7.** $-2(3x - 4)(x + 1)$, or $2(3x - 4)(-x - 1)$, or $2(-3x + 4)(x + 1)$ **8.** $(2a - b)(3a - b)$ **9.** $3(2x + 3y)(x + y)$

5.3 **Exercise Set**

For Extra Help

MyMathLab

Math XL
PRACTICE

WATCH

DOWNLOAD

READ

REVIEW

a Factor.

1. $2x^2 - 7x - 4$

2. $3x^2 - x - 4$

3. $5x^2 - x - 18$

4. $4x^2 - 17x + 15$

5. $6x^2 + 23x + 7$

6. $6x^2 - 23x + 7$

7. $3x^2 + 4x + 1$

8. $7x^2 + 15x + 2$

9. $4x^2 + 4x - 15$

10. $9x^2 + 6x - 8$

11. $2x^2 - x - 1$

12. $15x^2 - 19x - 10$

13. $9x^2 + 18x - 16$

14. $2x^2 + 5x + 2$

15. $3x^2 - 5x - 2$

16. $18x^2 - 3x - 10$

17. $12x^2 + 31x + 20$

18. $15x^2 + 19x - 10$

19. $14x^2 + 19x - 3$

20. $35x^2 + 34x + 8$

21. $9x^2 + 18x + 8$

22. $6 - 13x + 6x^2$

23. $49 - 42x + 9x^2$

24. $16 + 36x^2 + 48x$

25. $24x^2 + 47x - 2$

26. $16p^2 - 78p + 27$

27. $35x^2 - 57x - 44$

28. $9a^2 + 12a - 5$

29. $20 + 6x - 2x^2$

30. $15 + x - 2x^2$

31. $12x^2 + 28x - 24$

32. $6x^2 + 33x + 15$

33. $30x^2 - 24x - 54$

34. $18t^2 - 24t + 6$

35. $4y + 6y^2 - 10$

36. $-9 + 18x^2 - 21x$

37. $3x^2 - 4x + 1$

38. $6t^2 + 13t + 6$

39. $12x^2 - 28x - 24$

40. $6x^2 - 33x + 15$

41. $-1 + 2x^2 - x$

42. $-19x + 15x^2 + 6$

43. $9x^2 - 18x - 16$

44. $14y^2 + 35y + 14$

45. $15x^2 - 25x - 10$

46. $18x^2 + 3x - 10$

47. $12p^3 + 31p^2 + 20p$

48. $15x^3 + 19x^2 - 10x$

49. $16 + 18x - 9x^2$

50. $33t - 15 - 6t^2$

51. $-15x^2 + 19x - 6$

52. $1 + p - 2p^2$

53. $14x^4 + 19x^3 - 3x^2$

54. $70x^4 + 68x^3 + 16x^2$

55. $168x^3 - 45x^2 + 3x$

56. $144x^5 + 168x^4 + 48x^3$

57. $15x^4 - 19x^2 + 6$

58. $9x^4 + 18x^2 + 8$

59. $25t^2 + 80t + 64$

60. $9x^2 - 42x + 49$

61. $6x^3 + 4x^2 - 10x$

62. $18x^3 - 21x^2 - 9x$

63. $25x^2 + 79x + 64$

64. $9y^2 + 42y + 47$

65. $6x^2 - 19x - 5$

66. $2x^2 + 11x - 9$

67. $12m^2 - mn - 20n^2$

68. $12a^2 - 17ab + 6b^2$

69. $6a^2 - ab - 15b^2$

70. $3p^2 - 16pq - 12q^2$

71. $9a^2 + 18ab + 8b^2$

72. $10s^2 + 4st - 6t^2$

73. $35p^2 + 34pq + 8q^2$

74. $30a^2 + 87ab + 30b^2$

75. $18x^2 - 6xy - 24y^2$

76. $15a^2 - 5ab - 20b^2$

Skill Maintenance

Solve. [2.4b]

77. $A = pq - 7$, for q

78. $y = mx + b$, for x

79. $3x + 2y = 6$, for y

80. $p - q + r = 2$, for q

Solve. [2.7e]

81. $5 - 4x < -11$

82. $2x - 4(x + 3x) \geq 6x - 8 - 9x$

83. Graph: $y = \dfrac{2}{5}x - 1$. [3.2a]

84. Divide: $\dfrac{y^{12}}{y^4}$. [4.1e]

Find the intercepts of each equation. [3.3a]

85. $4x - 16y = 64$

86. $4x + 16y = 64$

87. $x - 1.3y = 6.5$

88. $\frac{2}{3}x + \frac{5}{8}y = \frac{5}{12}$

89. $y = 4 - 5x$

90. $y = 2x - 5$

Synthesis

Factor.

91. $20x^{2n} + 16x^n + 3$

92. $-15x^{2m} + 26x^m - 8$

93. $3x^{6a} - 2x^{3a} - 1$

94. $x^{2n+1} - 2x^{n+1} + x$

95.–104. ⌁ Use the TABLE feature to check the factoring in Exercises 15–24. (See the Calculator Corner on p. 338.)

5.4

Factoring $ax^2 + bx + c$, $a \neq 1$: The *ac*-Method

(a) The *ac*-Method

Another method for factoring trinomials of the type $ax^2 + bx + c$, $a \neq 1$, involves the product, ac, of the leading coefficient a and the last term c. It is called the ***ac*-method**. Because it uses factoring by grouping, it is also referred to as the **grouping method**.

We know how to factor the trinomial $x^2 + 5x + 6$. We look for factors of the constant term, 6, whose sum is the coefficient of the middle term, 5. What happens when the leading coefficient is not 1? To factor a trinomial like $3x^2 - 10x - 8$, we can use a method similar to the one that we used for $x^2 + 5x + 6$. That method is outlined as follows.

> **OBJECTIVE**
>
> (a) Factor trinomials of the type $ax^2 + bx + c$, $a \neq 1$, using the *ac*-method.

> **THE *ac*-METHOD**
>
> To factor $ax^2 + bx + c$, $a \neq 1$, using the *ac*-method:
>
> 1. Factor out a common factor, if any.
> 2. Multiply the leading coefficient a and the constant c.
> 3. Try to factor the product ac so that the sum of the factors is b. That is, find integers p and q such that $pq = ac$ and $p + q = b$.
> 4. Split the middle term, writing it as a sum using the factors found in step (3).
> 5. Factor by grouping.
> 6. Check by multiplying.

EXAMPLE 1 Factor: $3x^2 - 10x - 8$.

1) First, we factor out a common factor, if any. There is none (other than 1 or −1).

2) We multiply the leading coefficient, 3, and the constant, −8:
$$3(-8) = -24.$$

3) Then we look for a factorization of −24 in which the sum of the factors is the coefficient of the middle term, −10.

PAIRS OF FACTORS	SUMS OF FACTORS	
−1, 24	23	
1, −24	−23	
−2, 12	10	
2, −12	−10	← $2 + (-12) = -10$
−3, 8	5	
3, −8	−5	
−4, 6	2	
4, −6	−2	

4) Next, we split the middle term as a sum or a difference using the factors found in step (3): $-10x = 2x - 12x$.

5) Finally, we factor by grouping, as follows:

$$3x^2 - 10x - 8 = 3x^2 + 2x - 12x - 8 \qquad \text{Substituting } 2x - 12x \\ \text{for } -10x$$

$$= (3x^2 + 2x) + (-12x - 8)$$

$$= x(3x + 2) - 4(3x + 2) \qquad \text{Factoring by grouping}$$

$$= (3x + 2)(x - 4).$$

We can also split the middle term as $-12x + 2x$. We still get the same factorization, although the factors may be in a different order. Note the following:

$$3x^2 - 10x - 8 = 3x^2 - 12x + 2x - 8 \qquad \text{Substituting } -12x + 2x \\ \text{for } -10x$$

$$= (3x^2 - 12x) + (2x - 8)$$

$$= 3x(x - 4) + 2(x - 4) \qquad \text{Factoring by grouping}$$

$$= (x - 4)(3x + 2).$$

6) **Check:** $(3x + 2)(x - 4) = 3x^2 - 10x - 8.$

Do Exercises 1 and 2.

Factor.

1. $6x^2 + 7x + 2$

2. $12x^2 - 17x - 5$

EXAMPLE 2 Factor: $8x^2 + 8x - 6.$

1) First, we factor out a common factor, if any. The number 2 is common to all three terms, so we factor it out: $2(4x^2 + 4x - 3).$

2) Next, we factor the trinomial $4x^2 + 4x - 3$. We multiply the leading coefficient and the constant, 4 and -3: $4(-3) = -12.$

3) We try to factor -12 so that the sum of the factors is 4.

PAIRS OF FACTORS	SUMS OF FACTORS
$-1, \quad 12$	11
$1, -12$	-11
$-2, \quad 6$	$4 \leftarrow \qquad -2 + 6 = 4$
$2, \quad -6$	-4
$-3, \quad 4$	1
$3, \quad -4$	-1

4) Then we split the middle term, $4x$, as follows: $4x = -2x + 6x.$

5) Finally, we factor by grouping:

$$4x^2 + 4x - 3 = 4x^2 - 2x + 6x - 3 \qquad \text{Substituting } -2x + 6x \\ \text{for } 4x$$

$$= (4x^2 - 2x) + (6x - 3)$$

$$= 2x(2x - 1) + 3(2x - 1) \qquad \text{Factoring by grouping}$$

$$= (2x - 1)(2x + 3).$$

The factorization of $4x^2 + 4x - 3$ is $(2x - 1)(2x + 3)$. But don't forget the common factor! We must include it to get a factorization of the original trinomial: $8x^2 + 8x - 6 = 2(2x - 1)(2x + 3).$

6) **Check:** $2(2x - 1)(2x + 3) = 2(4x^2 + 4x - 3) = 8x^2 + 8x - 6.$

Factor.

3. $6x^2 + 15x + 9$

4. $20x^2 - 46x + 24$

Answers

1. $(2x + 1)(3x + 2)$ **2.** $(4x + 1)(3x - 5)$
3. $3(2x + 3)(x + 1)$ **4.** $2(5x - 4)(2x - 3)$

Do Exercises 3 and 4.

a Factor. Note that the middle term has already been split.

1. $x^2 + 2x + 7x + 14$

2. $x^2 + 3x + x + 3$

3. $x^2 - 4x - x + 4$

4. $a^2 + 5a - 2a - 10$

5. $6x^2 + 4x + 9x + 6$

6. $3x^2 - 2x + 3x - 2$

7. $3x^2 - 4x - 12x + 16$

8. $24 - 18y - 20y + 15y^2$

9. $35x^2 - 40x + 21x - 24$

10. $8x^2 - 6x - 28x + 21$

11. $4x^2 + 6x - 6x - 9$

12. $2x^4 - 6x^2 - 5x^2 + 15$

13. $2x^4 + 6x^2 + 5x^2 + 15$

14. $9x^4 - 6x^2 - 6x^2 + 4$

Factor using the *ac*-method.

15. $2x^2 + 7x - 4$

16. $5x^2 + x - 18$

17. $3x^2 - 4x - 15$

18. $3x^2 + x - 4$

19. $6x^2 + 23x + 7$

20. $6x^2 + 13x + 6$

21. $3x^2 - 4x + 1$

22. $7x^2 - 15x + 2$

23. $4x^2 - 4x - 15$

24. $9x^2 - 6x - 8$

25. $2x^2 + x - 1$

26. $15x^2 + 19x - 10$

27. $9x^2 - 18x - 16$ **28.** $2x^2 - 5x + 2$ **29.** $3x^2 + 5x - 2$ **30.** $18x^2 + 3x - 10$

31. $12x^2 - 31x + 20$ **32.** $15x^2 - 19x - 10$ **33.** $14x^2 - 19x - 3$ **34.** $35x^2 - 34x + 8$

35. $9x^2 + 18x + 8$ **36.** $6 - 13x + 6x^2$ **37.** $49 - 42x + 9x^2$ **38.** $25x^2 + 40x + 16$

39. $24x^2 - 47x - 2$ **40.** $16a^2 + 78a + 27$ **41.** $5 - 9a^2 - 12a$ **42.** $17x - 4x^2 + 15$

43. $20 + 6x - 2x^2$ **44.** $15 + x - 2x^2$ **45.** $12x^2 + 28x - 24$ **46.** $6x^2 + 33x + 15$

47. $30x^2 - 24x - 54$ **48.** $18t^2 - 24t + 6$ **49.** $4y + 6y^2 - 10$ **50.** $-9 + 18x^2 - 21x$

51. $3x^2 - 4x + 1$ **52.** $6t^2 + t - 15$ **53.** $12x^2 - 28x - 24$ **54.** $6x^2 - 33x + 15$

55. $-1 + 2x^2 - x$ **56.** $-19x + 15x^2 + 6$ **57.** $9x^2 + 18x - 16$ **58.** $14y^2 + 35y + 14$

59. $15x^2 - 25x - 10$

60. $18x^2 + 3x - 10$

61. $12p^3 + 31p^2 + 20p$

62. $15x^3 + 19x^2 - 10x$

63. $4 - x - 5x^2$

64. $1 - p - 2p^2$

65. $33t - 15 - 6t^2$

66. $-15x^2 - 19x - 6$

67. $14x^4 + 19x^3 - 3x^2$

68. $70x^4 + 68x^3 + 16x^2$

69. $168x^3 - 45x^2 + 3x$

70. $144x^5 + 168x^4 + 48x^3$

71. $15x^4 - 19x^2 + 6$

72. $9x^4 + 18x^2 + 8$

73. $25t^2 + 80t + 64$

74. $9x^2 - 42x + 49$

75. $6x^3 + 4x^2 - 10x$

76. $18x^3 - 21x^2 - 9x$

77. $25x^2 + 79x + 64$

78. $9y^2 + 42y + 47$

79. $6x^2 - 19x - 5$

80. $2x^2 + 11x - 9$

81. $12m^2 - mn - 20n^2$

82. $12a^2 - 17ab + 6b^2$

83. $6a^2 - ab - 15b^2$

84. $3p^2 - 16pq - 12q^2$

85. $9a^2 - 18ab + 8b^2$

86. $10s^2 + 4st - 6t^2$

87. $35p^2 + 34pq + 8q^2$ **88.** $30a^2 + 87ab + 30b^2$ **89.** $18x^2 - 6xy - 24y^2$ **90.** $15a^2 - 5ab - 20b^2$

91. $60x + 18x^2 - 6x^3$ **92.** $60x + 4x^2 - 8x^3$ **93.** $35x^5 - 57x^4 - 44x^3$ **94.** $15x^3 + 33x^4 + 6x^5$

Skill Maintenance

Solve. [2.7d, e]

95. $-10x > 1000$ **96.** $-3.8x \leq -824.6$ **97.** $6 - 3x \geq -18$

98. $3 - 2x - 4x > -9$ **99.** $\frac{1}{2}x - 6x + 10 \leq x - 5x$ **100.** $-2(x + 7) > -4(x - 5)$

101. $3x - 6x + 2(x - 4) > 2(9 - 4x)$ **102.** $-6(x - 4) + 8(4 - x) \leq 3(x - 7)$

Solve. [2.6a]

103. The earth is a sphere (or ball) that is about 40,000 km in circumference. Find the radius of the earth, in kilometers and in miles. Use 3.14 for π. (*Hint*: 1 km ≈ 0.62 mi.)

104. The second angle of a triangle is 10° less than twice the first. The third angle is 15° more than four times the first. Find the measure of the second angle.

Synthesis

Factor.

105. $9x^{10} - 12x^5 + 4$ **106.** $24x^{2n} + 22x^n + 3$

107. $16x^{10} + 8x^5 + 1$ **108.** $(a + 4)^2 - 2(a + 4) + 1$

109.–118. 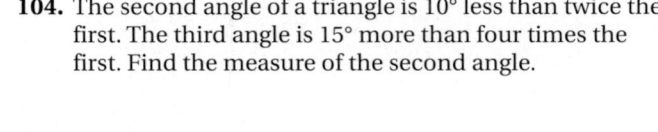 Use graphs to check the factoring in Exercises 15–24. (See the Calculator Corner on p. 338.)

Mid-Chapter Review

Concept Reinforcement

Determine whether each statement is true or false.

_____ **1.** The greatest common factor (GCF) of a set of natural numbers is at least 1 and always less than or equal to the smallest number in the set.   [5.1a]

_____ **2.** To factor $x^2 + bx + c$, we use a trial-and-error process that looks for factors of b whose sum is c.   [5.2a]

_____ **3.** A prime polynomial has no common factor other than 1 and -1.   [5.2a]

_____ **4.** When factoring $x^2 - 14x + 45$, we need consider only positive pairs of factors of 45.   [5.2a]

Guided Solutions

Fill in each blank with the number, variable, or expression that creates a correct statement or solution.

5. Factor: $10y^3 - 18y^2 + 12y$.   [5.1b]

$$10y^3 - 18y^2 + 12y = \square \cdot 5y^2 - \square \cdot 9y + \square \cdot 6$$
$$= 2y(\square)$$

6. Factor $2x^2 - x - 6$ using the ac-method.   [5.4a]

$a \cdot c = \square \cdot \square = -12;$ Multiplying the leading coefficient and the constant

$-x = \square + 3x;$ Splitting the middle term

$$2x^2 - x - 6 = 2x^2 - 4x + \square - 6$$
$$= \square(x - 2) + \square(x - 2)$$
$$= (x - 2)(\square)$$

Mixed Review

Find the GCF.   [5.1a]

7. $x^3,\ 3x$

8. $5x^4,\ x^2$

9. $6x^5,\ -12x^3$

10. $-8x,\ -12,\ 16x^2$

11. $15x^3y^2,\ 5x^2y,\ 40x^4y^3$

12. $x^2y^4,\ -x^3y^3,\ x^3y^2,\ x^5y^4$

Factor completely.   [5.1b, c], [5.2a], [5.3a], [5.4a]

13. $x^3 - 8x$

14. $3x^2 + 12x$

15. $2y^2 + 8y - 4$

16. $3t^6 - 5t^4 - 2t^3$

17. $x^2 + 4x + 3$

18. $z^2 - 4z + 4$

19. $x^3 + 4x^2 + 3x + 12$

20. $8y^5 - 48y^3$

21. $6x^3y + 24x^2y^2 - 42xy^3$

22. $6 - 11t - 4t^2$

23. $z^2 + 4z - 5$

24. $2z^3 + 8z^2 + 5z + 20$

25. $3p^3 - 2p^2 - 9p + 6$

26. $10x^8 - 25x^6 - 15x^5 + 35x^3$

27. $2w^3 + 3w^2 - 6w - 9$

28. $4x^4 - 5x^3 + 3x^2$

29. $6y^2 + 7y - 10$

30. $3x^2 - 3x - 18$

31. $6x^3 + 4x^2 + 3x + 2$

32. $15 - 8w + w^2$

33. $8x^3 + 20x^2 + 2x + 5$

34. $10z^2 - 21z - 10$

35. $6x^2 + 7x + 2$

36. $x^2 - 10xy + 24y^2$

37. $6z^3 + 3z^2 + 2z + 1$

38. $a^3b^7 + a^4b^5 - a^2b^3 + a^5b^6$

39. $4y^2 - 7yz - 15z^2$

40. $3x^3 + 21x^2 + 30x$

41. $x^3 - 3x^2 - 2x + 6$

42. $9y^2 + 6y + 1$

43. $y^2 + 6y + 8$

44. $6y^2 + 33y + 45$

45. $x^3 - 7x^2 + 4x - 28$

46. $4 + 3y - y^2$

47. $16x^2 - 16x - 60$

48. $10a^2 - 11ab + 3b^2$

49. $6w^3 - 15w^2 - 10w + 25$

50. $y^3 + 9y^2 + 18y$

51. $4x^2 + 11xy + 6y^2$

52. $6 - 5z - 6z^2$

53. $12t^3 + 8t^2 - 9t - 6$

54. $y^2 + yz - 20z^2$

55. $9x^2 - 6xy - 8y^2$

56. $-3 + 8z + 3z^2$

57. $m^2 - 6mn - 16n^2$

58. $2w^2 - 12w + 18$

59. $18t^3 - 18t^2 + 4t$

60. $5z^3 + 15z^2 + z + 3$

61. $-14 + 5t + t^2$

62. $4t^2 - 20t + 25$

63. $t^2 + 4t - 12$

64. $12 + 5z - 2z^2$

65. $12 + 4y - y^2$

Understanding Through Discussion and Writing

66. Explain how one could construct a polynomial with four terms that can be factored by grouping.   [5.1a], [5.4a]

67. When searching for a factorization, why do we list pairs of numbers with the correct *product* instead of pairs of numbers with the correct *sum*?   [5.2a]

68. Without multiplying $(x - 17)(x - 18)$, explain why it cannot possibly be a factorization of $x^2 + 35x + 306$. [5.2a]

69. A student presents the following work:
$$4x^2 + 28x + 48 = (2x + 6)(2x + 8)$$
$$= 2(x + 3)(x + 4).$$
Is it correct? Explain.   [5.3a], [5.4a]

5.5 Factoring Trinomial Squares and Differences of Squares

In this section, we first learn to factor trinomials that are squares of binomials. Then we factor binomials that are differences of squares.

a Recognizing Trinomial Squares

Some trinomials are squares of binomials. For example, the trinomial $x^2 + 10x + 25$ is the square of the binomial $x + 5$. To see this, we can calculate $(x + 5)^2$. It is $x^2 + 2 \cdot x \cdot 5 + 5^2$, or $x^2 + 10x + 25$. A trinomial that is the square of a binomial is called a **trinomial square**, or a **perfect-square trinomial**.

In Chapter 4, we considered squaring binomials as special-product rules:

$$(A + B)^2 = A^2 + 2AB + B^2;$$
$$(A - B)^2 = A^2 - 2AB + B^2.$$

We can use these equations in reverse to factor trinomial squares.

TRINOMIAL SQUARES

$$A^2 + 2AB + B^2 = (A + B)^2;$$
$$A^2 - 2AB + B^2 = (A - B)^2$$

How can we recognize when an expression to be factored is a trinomial square? Look at $A^2 + 2AB + B^2$ and $A^2 - 2AB + B^2$. In order for an expression to be a trinomial square:

a) The two expressions A^2 and B^2 must be squares, such as

$$4, \quad x^2, \quad 25x^4, \quad 16t^2.$$

When the coefficient is a perfect square and the power(s) of the variable(s) is (are) even, then the expression is a perfect square.

b) There must be no minus sign before A^2 or B^2.

c) If we multiply A and B and double the result, $2 \cdot AB$, we get either the remaining term or its opposite.

EXAMPLE 1 Determine whether $x^2 + 6x + 9$ is a trinomial square.

a) We know that x^2 and 9 are squares.

b) There is no minus sign before x^2 or 9.

c) If we multiply the square roots, x and 3, and double the product, we get the remaining term: $2 \cdot x \cdot 3 = 6x$.

Thus, $x^2 + 6x + 9$ is the square of a binomial. In fact, $x^2 + 6x + 9 = (x + 3)^2$.

EXAMPLE 2 Determine whether $x^2 + 6x + 11$ is a trinomial square.

The answer is no, because only one term, x^2, is a square.

OBJECTIVES

a Recognize trinomial squares.

b Factor trinomial squares.

c Recognize differences of squares.

d Factor differences of squares, being careful to factor completely.

It would be helpful to memorize this table of perfect squares.

NUMBER, N	PERFECT SQUARE, N^2
1	1
2	4
3	9
4	16
5	25
6	36
7	49
8	64
9	81
10	100
11	121
12	144
13	169
14	196
15	225
16	256
17	289
18	324
19	361
20	400
21	441
22	484
23	529
24	576
25	625

Determine whether each is a trinomial square. Write "yes" or "no."

1. $x^2 + 8x + 16$

2. $25 - x^2 + 10x$

3. $t^2 - 12t + 4$

4. $25 + 20y + 4y^2$

5. $5x^2 + 16 - 14x$

6. $16x^2 + 40x + 25$

7. $p^2 + 6p - 9$

8. $25a^2 + 9 - 30a$

EXAMPLE 3 Determine whether $16x^2 + 49 - 56x$ is a trinomial square.

It helps to first write the trinomial in descending order:

$$16x^2 - 56x + 49.$$

a) We know that $16x^2$ and 49 are squares.

b) There is no minus sign before $16x^2$ or 49.

c) If we multiply the square roots, $4x$ and 7, and double the product, we get the opposite of the remaining term: $2 \cdot 4x \cdot 7 = 56x$; $56x$ is the opposite of $-56x$.

Thus, $16x^2 + 49 - 56x$ is a trinomial square. In fact, $16x^2 - 56x + 49 = (4x - 7)^2$.

Do Exercises 1-8.

(b) Factoring Trinomial Squares

We can use the factoring methods from Sections 5.2–5.4 to factor trinomial squares, but there is a faster method using the following equations.

> **FACTORING TRINOMIAL SQUARES**
> $A^2 + 2AB + B^2 = (A + B)^2;$
> $A^2 - 2AB + B^2 = (A - B)^2$

We consider 3 to be a square root of 9 because $3^2 = 9$. Similarly, A is a square root of A^2. We use square roots of the squared terms and the sign of the remaining term to factor a trinomial square.

EXAMPLE 4 Factor: $x^2 + 6x + 9$.

$$x^2 + 6x + 9 = x^2 + 2 \cdot x \cdot 3 + 3^2 = (x + 3)^2$$
$$A^2 + 2 \; A \; B + B^2 = (A + B)^2$$

The sign of the middle term is positive.

EXAMPLE 5 Factor: $x^2 + 49 - 14x$.

$$x^2 + 49 - 14x = x^2 - 14x + 49 \qquad \text{Changing to descending order}$$
$$= x^2 - 2 \cdot x \cdot 7 + 7^2 \qquad \text{The sign of the middle term is negative.}$$
$$= (x - 7)^2$$

EXAMPLE 6 Factor: $16x^2 - 40x + 25$.

$$16x^2 - 40x + 25 = (4x)^2 - 2 \cdot 4x \cdot 5 + 5^2 = (4x - 5)^2$$
$$A^2 \; - 2 \; A \; B + B^2 = (A - B)^2$$

Do Exercises 9-13.

Factor.

9. $x^2 + 2x + 1$

10. $1 - 2x + x^2$

11. $4 + t^2 + 4t$

12. $25x^2 - 70x + 49$

13. $49 - 56y + 16y^2$

Answers

1. Yes 2. No 3. No 4. Yes
5. No 6. Yes 7. No 8. Yes
9. $(x + 1)^2$ 10. $(x - 1)^2$, or $(1 - x)^2$
11. $(t + 2)^2$ 12. $(5x - 7)^2$
13. $(4y - 7)^2$, or $(7 - 4y)^2$

EXAMPLE 7 Factor: $t^4 + 20t^2 + 100$.

$$t^4 + 20t^2 + 100 = (t^2)^2 + 2(t^2)(10) + 10^2$$
$$= (t^2 + 10)^2$$

EXAMPLE 8 Factor: $75m^3 + 210m^2 + 147m$.

Always look first for a common factor. This time there is one, $3m$:

$$75m^3 + 210m^2 + 147m = 3m(25m^2 + 70m + 49)$$
$$= 3m[(5m)^2 + 2(5m)(7) + 7^2]$$
$$= 3m(5m + 7)^2.$$

EXAMPLE 9 Factor: $4p^2 - 12pq + 9q^2$.

$$4p^2 - 12pq + 9q^2 = (2p)^2 - 2(2p)(3q) + (3q)^2$$
$$= (2p - 3q)^2$$

Do Exercises 14–17.

Factor.

14. $48m^2 + 75 + 120m$

15. $p^4 + 18p^2 + 81$

16. $4z^5 - 20z^4 + 25z^3$

17. $9a^2 + 30ab + 25b^2$

(c) Recognizing Differences of Squares

The following polynomials are *differences of squares:*

$$x^2 - 9, \quad 4t^2 - 49, \quad a^2 - 25b^2.$$

To factor a difference of squares such as $x^2 - 9$, think about the formula we used in Chapter 4:

$$(A + B)(A - B) = A^2 - B^2.$$

Equations are reversible, so we also know the following.

> **DIFFERENCE OF SQUARES**
>
> $A^2 - B^2 = (A + B)(A - B)$

Thus,

$$x^2 - 9 = (x + 3)(x - 3).$$

To use this formula, we must be able to recognize when it applies. A **difference of squares** is an expression like the following:

$$A^2 - B^2.$$

How can we recognize such expressions? Look at $A^2 - B^2$. In order for a binomial to be a difference of squares:

a) There must be two expressions, both squares, such as

$$4x^2, \quad 9, \quad 25t^4, \quad 1, \quad x^6, \quad 49y^8.$$

b) The terms must have different signs.

EXAMPLE 10 Is $9x^2 - 64$ a difference of squares?

a) The first expression is a square: $9x^2 = (3x)^2$.
The second expression is a square: $64 = 8^2$.

b) The terms have different signs, $+9x^2$ and -64.

Thus we have a difference of squares, $(3x)^2 - 8^2$.

EXAMPLE 11 Is $25 - t^3$ a difference of squares?

a) The expression t^3 is not a square.

The expression is not a difference of squares.

EXAMPLE 12 Is $-4x^2 + 16$ a difference of squares?

a) The expressions $4x^2$ and 16 are squares: $4x^2 = (2x)^2$ and $16 = 4^2$.

b) The terms have different signs, $-4x^2$ and $+16$.

Thus we have a difference of squares. We can also see this by rewriting in the equivalent form: $16 - 4x^2$.

> Do Exercises 18–24.

(d) Factoring Differences of Squares

To factor a difference of squares, we use the following equation.

> **FACTORING A DIFFERENCE OF SQUARES**
> $$A^2 - B^2 = (A + B)(A - B)$$

To factor a difference of squares $A^2 - B^2$, we find A and B, which are square roots of the expressions A^2 and B^2. We then use A and B to form two factors. One is the sum $A + B$, and the other is the difference $A - B$.

EXAMPLE 13 Factor: $x^2 - 4$.

$$x^2 - 4 = x^2 - 2^2 = (x + 2)(x - 2)$$
$$A^2 - B^2 = (A + B)(A - B)$$

EXAMPLE 14 Factor: $9 - 16t^4$.

$$9 - 16t^4 = 3^2 - (4t^2)^2 = (3 + 4t^2)(3 - 4t^2)$$
$$A^2 - B^2 = (A + B)(A - B)$$

Determine whether each is a difference of squares. Write "yes" or "no."

18. $x^2 - 25$

19. $t^2 - 24$

20. $y^2 + 36$

21. $4x^2 - 15$

22. $16x^4 - 49$

23. $9w^6 - 1$

24. $-49 + 25t^2$

Answers

18. Yes   **19.** No   **20.** No   **21.** No
22. Yes   **23.** Yes   **24.** Yes

EXAMPLE 15 Factor: $m^2 - 4p^2$.

$$m^2 - 4p^2 = m^2 - (2p)^2 = (m + 2p)(m - 2p)$$

EXAMPLE 16 Factor: $x^2 - \dfrac{1}{9}$.

$$x^2 - \frac{1}{9} = x^2 - \left(\frac{1}{3}\right)^2 = \left(x + \frac{1}{3}\right)\left(x - \frac{1}{3}\right)$$

EXAMPLE 17 Factor: $18x^2 - 50x^6$.

Always look first for a factor common to all terms. This time there is one, $2x^2$.

$$18x^2 - 50x^6 = 2x^2(9 - 25x^4)$$
$$= 2x^2[3^2 - (5x^2)^2]$$
$$= 2x^2(3 + 5x^2)(3 - 5x^2)$$

EXAMPLE 18 Factor: $49x^4 - 9x^6$.

$$49x^4 - 9x^6 = x^4(49 - 9x^2)$$
$$= x^4[7^2 - (3x)^2]$$
$$= x^4(7 + 3x)(7 - 3x)$$

Do Exercises 25–29.

Factor.

25. $x^2 - 9$

26. $4t^2 - 64$

27. $a^2 - 25b^2$

28. $64x^4 - 25x^6$

29. $5 - 20t^6$
[*Hint*: $1 = 1^2$, $t^6 = (t^3)^2$.]

--------------------------------------- *Caution!* ---------------------------------------

Note carefully in these examples that a difference of squares is *not* the square of the difference; that is,

$$A^2 - B^2 \neq (A - B)^2.$$

For example,

$$(45 - 5)^2 = 40^2 = 1600,$$

but

$$45^2 - 5^2 = 2025 - 25 = 2000.$$

Similarly,

$$A^2 - 2AB + B^2 \neq (A - B)(A + B).$$

For example,

$$(10 - 3)(10 + 3) = 7 \cdot 13 = 91,$$

but

$$10^2 - 2 \cdot 10 \cdot 3 + 3^2 = 100 - 2 \cdot 10 \cdot 3 + 9$$
$$= 100 - 60 + 9$$
$$= 49.$$

Factoring Completely

If a factor with more than one term can still be factored, you should do so. When no factor can be factored further, you have **factored completely**. Always factor completely whenever told to factor.

EXAMPLE 19 Factor: $p^4 - 16$.

$$p^4 - 16 = (p^2)^2 - 4^2$$
$$= (p^2 + 4)(p^2 - 4) \qquad \text{Factoring a difference of squares}$$
$$= (p^2 + 4)(p + 2)(p - 2) \qquad \text{Factoring further; } p^2 - 4 \text{ is a difference of squares.}$$

The polynomial $p^2 + 4$ cannot be factored further into polynomials with real coefficients.

------------------------------- *Caution!* -------------------------------

Apart from possibly removing a common factor, you cannot factor a sum of squares as a product of binomials. In particular,

$$A^2 + B^2 \neq (A + B)^2.$$

Consider $25x^2 + 100$. Here a sum of squares has a common factor, 25. Factoring, we get $25(x^2 + 4)$, where $x^2 + 4$ is prime. For example,

$$x^2 + 4 \neq (x + 2)^2.$$

EXAMPLE 20 Factor: $y^4 - 16x^{12}$.

$$y^4 - 16x^{12} = (y^2 + 4x^6)(y^2 - 4x^6) \qquad \text{Factoring a difference of squares}$$
$$= (y^2 + 4x^6)(y + 2x^3)(y - 2x^3) \qquad \text{Factoring further. The factor } y^2 - 4x^6 \text{ is a difference of squares.}$$

The polynomial $y^2 + 4x^6$ cannot be factored further into polynomials with real coefficients.

EXAMPLE 21 Factor: $\dfrac{1}{16}x^8 - 81$.

$$\frac{1}{16}x^8 - 81 = \left(\frac{1}{4}x^4 + 9\right)\left(\frac{1}{4}x^4 - 9\right) \qquad \text{Factoring a difference of squares}$$
$$= \left(\frac{1}{4}x^4 + 9\right)\left(\frac{1}{2}x^2 + 3\right)\left(\frac{1}{2}x^2 - 3\right) \qquad \text{Factoring further. The factor } \frac{1}{4}x^4 - 9 \text{ is a difference of squares.}$$

Factor completely.

30. $81x^4 - 1$

31. $16 - \dfrac{1}{81}y^8$

32. $49p^4 - 25q^6$

> **TIPS FOR FACTORING**
> - Always look first for a common factor. If there is one, factor it out.
> - Be alert for trinomial squares and differences of squares. Once recognized, they can be factored without trial and error.
> - Always factor completely.
> - Check by multiplying.

Answers

30. $(9x^2 + 1)(3x + 1)(3x - 1)$
31. $\left(4 + \dfrac{1}{9}y^4\right)\left(2 + \dfrac{1}{3}y^2\right)\left(2 - \dfrac{1}{3}y^2\right)$
32. $(7p^2 + 5q^3)(7p^2 - 5q^3)$

Do Exercises 30–32.

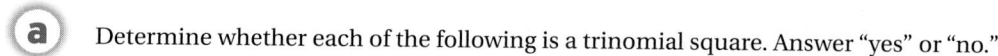
a Determine whether each of the following is a trinomial square. Answer "yes" or "no."

1. $x^2 - 14x + 49$

2. $x^2 - 16x + 64$

3. $x^2 + 16x - 64$

4. $x^2 - 14x - 49$

5. $x^2 - 2x + 4$

6. $x^2 + 3x + 9$

7. $9x^2 - 24x + 16$

8. $25x^2 + 30x + 9$

b Factor completely. Remember to look first for a common factor and to check by multiplying.

9. $x^2 - 14x + 49$

10. $x^2 - 20x + 100$

11. $x^2 + 16x + 64$

12. $x^2 + 20x + 100$

13. $x^2 - 2x + 1$

14. $x^2 + 2x + 1$

15. $4 + 4x + x^2$

16. $4 + x^2 - 4x$

17. $y^2 + 12y + 36$

18. $y^2 + 18y + 81$

19. $16 + t^2 - 8t$

20. $9 + t^2 - 6t$

21. $q^4 - 6q^2 + 9$

22. $64 + 16a^2 + a^4$

23. $49 + 56y + 16y^2$

24. $75 + 48a^2 - 120a$

25. $2x^2 - 4x + 2$

26. $2x^2 - 40x + 200$

27. $x^3 - 18x^2 + 81x$

28. $x^3 + 24x^2 + 144x$

29. $12q^2 - 36q + 27$

30. $20p^2 + 100p + 125$

31. $49 - 42x + 9x^2$

32. $64 - 112x + 49x^2$

33. $5y^4 + 10y^2 + 5$

34. $a^4 + 14a^2 + 49$

35. $1 + 4x^4 + 4x^2$

36. $1 - 2a^5 + a^{10}$

37. $4p^2 + 12pq + 9q^2$

38. $25m^2 + 20mn + 4n^2$

39. $a^2 - 6ab + 9b^2$

40. $x^2 - 14xy + 49y^2$

41. $81a^2 - 18ab + b^2$

42. $64p^2 + 16pq + q^2$

43. $36a^2 + 96ab + 64b^2$

44. $16m^2 - 40mn + 25n^2$

c Determine whether each of the following is a difference of squares. Answer "yes" or "no."

45. $x^2 - 4$

46. $x^2 - 36$

47. $x^2 + 25$

48. $x^2 + 9$

49. $x^2 - 45$

50. $x^2 - 80y^2$

51. $-25y^2 + 16x^2$

52. $-1 + 36x^2$

d Factor completely. Remember to look first for a common factor.

53. $y^2 - 4$

54. $q^2 - 1$

55. $p^2 - 9$

56. $x^2 - 36$

57. $-49 + t^2$

58. $-64 + m^2$

59. $a^2 - b^2$

60. $p^2 - q^2$

61. $25t^2 - m^2$

62. $w^2 - 49z^2$

63. $100 - k^2$

64. $81 - w^2$

65. $16a^2 - 9$

66. $25x^2 - 4$

67. $4x^2 - 25y^2$

68. $9a^2 - 16b^2$

69. $8x^2 - 98$

70. $24x^2 - 54$

71. $36x - 49x^3$

72. $16x - 81x^3$

73. $\dfrac{1}{16} - 49x^8$

74. $\dfrac{1}{625}x^8 - 49$

75. $0.09y^2 - 0.0004$

76. $0.16p^2 - 0.0025$

77. $49a^4 - 81$

78. $25a^4 - 9$

79. $a^4 - 16$

80. $y^4 - 1$

81. $5x^4 - 405$

82. $4x^4 - 64$

83. $1 - y^8$

84. $x^8 - 1$

85. $x^{12} - 16$

86. $x^8 - 81$

87. $y^2 - \dfrac{1}{16}$

88. $x^2 - \dfrac{1}{25}$

89. $25 - \dfrac{1}{49}x^2$

90. $\dfrac{1}{4} - 9q^2$

91. $16m^4 - t^4$

92. $p^4q^4 - 1$

Skill Maintenance

Divide. [1.6a, c]

93. $(-110) \div 10$

94. $-1000 \div (-2.5)$

95. $\left(-\dfrac{2}{3}\right) \div \dfrac{4}{5}$

96. $8.1 \div (-9)$

97. $-64 \div (-32)$

98. $-256 \div 1.6$

Find a polynomial for the shaded area in each figure. (Leave results in terms of π where appropriate.) [4.4d]

99.

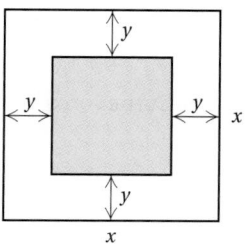

100.

Simplify.

101. $y^5 \cdot y^7$ [4.1d]

102. $(5a^2b^3)^2$ [4.2a, b]

Find the intercepts. Then graph each equation.   [3.3a]

103. $y - 6x = 6$

104. $3x - 5y = 15$

Synthesis

Factor completely, if possible.

105. $49x^2 - 216$

106. $27x^3 - 13x$

107. $x^2 + 22x + 121$

108. $x^2 - 5x + 25$

109. $18x^3 + 12x^2 + 2x$

110. $162x^2 - 82$

111. $x^8 - 2^8$

112. $4x^4 - 4x^2$

113. $3x^5 - 12x^3$

114. $3x^2 - \frac{1}{3}$

115. $18x^3 - \frac{8}{25}x$

116. $x^2 - 2.25$

117. $0.49p - p^3$

118. $3.24x^2 - 0.81$

119. $0.64x^2 - 1.21$

120. $1.28x^2 - 2$

121. $(x + 3)^2 - 9$

122. $(y - 5)^2 - 36q^2$

123. $x^2 - \left(\frac{1}{x}\right)^2$

124. $a^{2n} - 49b^{2n}$

125. $81 - b^{4k}$

126. $9x^{18} + 48x^9 + 64$

127. $9b^{2n} + 12b^n + 4$

128. $(x + 7)^2 - 4x - 24$

129. $(y + 3)^2 + 2(y + 3) + 1$

130. $49(x + 1)^2 - 42(x + 1) + 9$

Find c such that the polynomial is the square of a binomial.

131. $cy^2 + 6y + 1$

132. $cy^2 - 24y + 9$

⌐∿ Use the TABLE feature or graphs to determine whether each factorization is correct. (See the Calculator Corner on p. 338.)

133. $x^2 + 9 = (x + 3)(x + 3)$

134. $x^2 - 49 = (x - 7)(x + 7)$

135. $x^2 + 9 = (x + 3)^2$

136. $x^2 - 49 = (x - 7)^2$

5.6

Factoring Sums or Differences of Cubes

a Sums or Differences of Cubes

We can factor the sum or the difference of two expressions that are cubes. Consider the following products:

$$(A + B)(A^2 - AB + B^2) = A(A^2 - AB + B^2) + B(A^2 - AB + B^2)$$
$$= A^3 - A^2B + AB^2 + A^2B - AB^2 + B^3$$
$$= A^3 + B^3$$

and

$$(A - B)(A^2 + AB + B^2) = A(A^2 + AB + B^2) - B(A^2 + AB + B^2)$$
$$= A^3 + A^2B + AB^2 - A^2B - AB^2 - B^3$$
$$= A^3 - B^3.$$

The above equations (reversed) show how we can factor a sum or a difference of two cubes. Each factors as a product of a binomial and a trinomial.

> **SUM OR DIFFERENCE OF CUBES**
>
> $A^3 + B^3 = (A + B)(A^2 - AB + B^2);$
> $A^3 - B^3 = (A - B)(A^2 + AB + B^2)$

Note that what we are considering here is a sum or a difference of cubes. We are not cubing a binomial. For example, $(A + B)^3$ is *not* the same as $A^3 + B^3$. The table of cubes in the margin is helpful.

EXAMPLE 1   Factor: $x^3 - 27$.

We have

$$\overset{\displaystyle A^3 \;-\; B^3}{\underset{}{\downarrow \qquad \downarrow}}$$
$$x^3 - 27 = x^3 - 3^3.$$

In one set of parentheses, we write the cube root of the first term, x. Then we write the cube root of the second term, -3. This gives us the expression $x - 3$:

$$(x - 3)(\qquad\qquad).$$

To get the next factor, we think of $x - 3$ and do the following:

— Square the first term: $x \cdot x = x^2$.

— Multiply the terms, $x(-3) = -3x$, and then change the sign: $3x$.

— Square the second term: $(-3)^2 = 9$.

$$(x - 3)(x^2 + 3x + 9).$$
$$(A - B)(A^2 + AB + B^2)$$

Note that we cannot factor $x^2 + 3x + 9$. It is not a trinomial square nor can it be factored by trial and error. Check this on your own.

Do Exercises 1 and 2.

OBJECTIVE

a Factor sums or differences of cubes.

SKILL TO REVIEW
Objective 4.5d: Multiply any two polynomials.

Multiply.

1. $(x + 3)(x^2 - x + 5)$
2. $(y - 9)(2y^2 + y - 1)$

N	N^3
0.2	0.008
0.1	0.001
0	0
1	1
2	8
3	27
4	64
5	125
6	216
7	343
8	512
9	729
10	1000

Factor.

1. $x^3 - 8$
2. $64 - y^3$

Answers

Skill to Review:
1. $x^3 + 2x^2 + 2x + 15$
2. $2y^3 - 17y^2 - 10y + 9$

Margin Exercises:
1. $(x - 2)(x^2 + 2x + 4)$
2. $(4 - y)(16 + 4y + y^2)$

EXAMPLE 2 Factor: $125x^3 + y^3$.

We have

$$125x^3 + y^3 = (5x)^3 + y^3.$$

In one set of parentheses, we write the cube root of the first term, $5x$. Then we write a plus sign, and then the cube root of the second term, y. This gives us the expression $5x + y$:

$$(5x + y)(\qquad\qquad).$$

To get the next factor, we think of $5x + y$ and do the following:

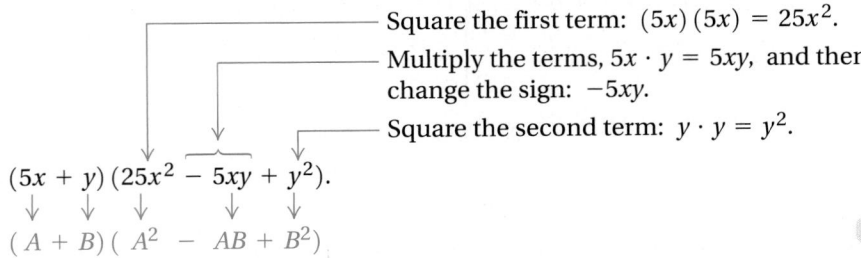

Square the first term: $(5x)(5x) = 25x^2$.

Multiply the terms, $5x \cdot y = 5xy$, and then change the sign: $-5xy$.

Square the second term: $y \cdot y = y^2$.

$$(5x + y)(25x^2 - 5xy + y^2).$$

$$(A + B)(A^2 - AB + B^2)$$

Factor.

3. $27x^3 + y^3$

4. $8y^3 + z^3$

Do Exercises 3 and 4.

EXAMPLE 3 Factor: $128y^7 - 250x^6y$.

We first look for the largest common factor:

$$\begin{aligned} 128y^7 - 250x^6y &= 2y(64y^6 - 125x^6) \\ &= 2y[(4y^2)^3 - (5x^2)^3] \\ &= 2y(4y^2 - 5x^2)(16y^4 + 20x^2y^2 + 25x^4). \end{aligned}$$

EXAMPLE 4 Factor: $a^6 - b^6$.

We can express this polynomial as a difference of squares:

$$a^6 - b^6 = (a^3)^2 - (b^3)^2.$$

We factor as follows:

$$a^6 - b^6 = (a^3 + b^3)(a^3 - b^3).$$

One factor is a sum of two cubes, and the other factor is a difference of two cubes. We factor them:

$$a^6 - b^6 = (a + b)(a^2 - ab + b^2)(a - b)(a^2 + ab + b^2).$$

We have now factored completely.

In Example 4, had we thought of factoring first as a difference of two cubes, we would have had

$$\begin{aligned} (a^2)^3 - (b^2)^3 &= (a^2 - b^2)(a^4 + a^2b^2 + b^4) \\ &= (a + b)(a - b)(a^4 + a^2b^2 + b^4). \end{aligned}$$

In this case, we might have missed some factors; $a^4 + a^2b^2 + b^4$ can be factored as $(a^2 - ab + b^2)(a^2 + ab + b^2)$, but we probably would not have known to do such factoring.

Answers

3. $(3x + y)(9x^2 - 3xy + y^2)$
4. $(2y + z)(4y^2 - 2yz + z^2)$

When you can factor as a difference of squares or a difference of cubes, factor as a difference of squares first.

EXAMPLE 5 Factor: $64a^6 - 729b^6$.

We have

$$64a^6 - 729b^6 = (8a^3)^2 - (27b^3)^2$$
$$= (8a^3 - 27b^3)(8a^3 + 27b^3) \quad \text{Factoring a difference of squares}$$
$$= [(2a)^3 - (3b)^3][(2a)^3 + (3b)^3].$$

Each factor is a sum or a difference of cubes. We factor each:

$$= (2a - 3b)(4a^2 + 6ab + 9b^2)(2a + 3b)(4a^2 - 6ab + 9b^2).$$

FACTORING SUMMARY

Sum of cubes: $A^3 + B^3 = (A + B)(A^2 - AB + B^2)$;

Difference of cubes: $A^3 - B^3 = (A - B)(A^2 + AB + B^2)$;

Difference of squares: $A^2 - B^2 = (A + B)(A - B)$;

Sum of squares: There is no formula for factoring $A^2 + B^2$.

Do Exercises 5-8.

Factor.

5. $m^6 - n^6$

6. $16x^7y + 54xy^7$

7. $729x^6 - 64y^6$

8. $x^3 - 0.027$

Answers

5. $(m + n)(m^2 - mn + n^2)(m - n)(m^2 + mn + n^2)$
6. $2xy(2x^2 + 3y^2)(4x^4 - 6x^2y^2 + 9y^4)$
7. $(3x + 2y)(9x^2 - 6xy + 4y^2)(3x - 2y)(9x^2 + 6xy + 4y^2)$
8. $(x - 0.3)(x^2 + 0.3x + 0.09)$

5.6   **Exercise Set**

a   Factor.

1. $z^3 + 27$

2. $a^3 + 8$

3. $x^3 - 1$

4. $c^3 - 64$

5. $y^3 + 125$

6. $x^3 + 1$

7. $8a^3 + 1$

8. $27x^3 + 1$

9. $y^3 - 8$

10. $p^3 - 27$

11. $8 - 27b^3$

12. $64 - 125x^3$

13. $64y^3 + 1$

14. $125x^3 + 1$

15. $8x^3 + 27$

16. $27y^3 + 64$

17. $a^3 - b^3$

18. $x^3 - y^3$

19. $a^3 + \frac{1}{8}$

20. $b^3 + \frac{1}{27}$

21. $2y^3 - 128$

22. $3z^3 - 3$

23. $24a^3 + 3$

24. $54x^3 + 2$

25. $rs^3 + 64r$

26. $ab^3 + 125a$

27. $5x^3 - 40z^3$

28. $2y^3 - 54z^3$

29. $x^3 + 0.001$

30. $y^3 + 0.125$

31. $64x^6 - 8t^6$

32. $125c^6 - 8d^6$

33. $2y^4 - 128y$

34. $3z^5 - 3z^2$

35. $z^6 - 1$

36. $t^6 + 1$

37. $t^6 + 64y^6$

38. $p^6 - q^6$

39. $8w^9 - z^9$

40. $a^9 + 64b^9$

41. $\frac{1}{8}c^3 + d^3$

42. $\frac{27}{125}x^3 - y^3$

43. $0.001x^3 - 0.008y^3$

44. $0.125r^3 - 0.216s^3$

Skill Maintenance

Simplify. [4.2b]

45. $(7y^{-5})^3$

46. $(a^{-4}b^{-9})^{-2}$

47. $\left(\dfrac{x^3}{4}\right)^{-2}$

Multiply.

48. $(2y^5 + 3)(2y^5 - 3)$ [4.6b]

49. $\left(w - \dfrac{1}{3}\right)^2$ [4.6c]

50. $(x - 0.1)(x + 0.5)$ [4.6a]

Synthesis

51. *Volume of Carpeting.* The volume of a carpet that is rolled up can be estimated by the polynomial $\pi R^2 h - \pi r^2 h$.

   a) Factor the polynomial.
   b) Use both the original form and the factored form to find the volume of a roll for which $R = 50$ cm, $r = 10$ cm, and $h = 4$ m. Use 3.14 for π.

52. Show how the geometric model below can be used to verify the formula for factoring $a^3 - b^3$.

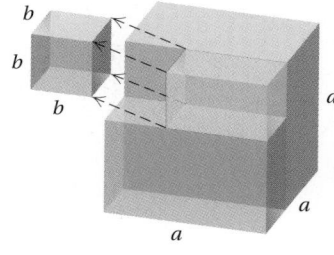

Factor. Assume that variables in exponents represent positive integers.

53. $3x^{3a} + 24y^{3b}$

54. $\dfrac{8}{27}x^3 + \dfrac{1}{64}y^3$

55. $\dfrac{1}{24}x^3y^3 + \dfrac{1}{3}z^3$

56. $7x^3 - \dfrac{7}{8}$

57. $(x + y)^3 - x^3$

58. $(1 - x)^3 + (x - 1)^6$

59. $(a + 2)^3 - (a - 2)^3$

60. $y^4 - 8y^3 - y + 8$

5.7

Factoring: A General Strategy

OBJECTIVE

a Factor polynomials completely using any of the methods considered in this chapter.

We now combine all of our factoring techniques and consider a general strategy for factoring polynomials. Here we will encounter polynomials of all the types we have considered, in random order, so you will have the opportunity to determine which method to use.

> **FACTORING STRATEGY**
>
> To factor a polynomial:
>
> **a)** Always look first for a common factor. If there is one, factor out the largest common factor.
>
> **b)** Then look at the number of terms.
>
> *Two terms*: Determine whether you have a difference of squares, $A^2 - B^2$, or a sum or difference of cubes, $A^3 + B^3$ or $A^3 - B^3$. Do not try to factor a sum of squares: $A^2 + B^2$.
>
> *Three terms*: Determine whether the trinomial is a square. If it is, you know how to factor. If not, try trial and error, using FOIL or the *ac*-method.
>
> *Four terms*: Try factoring by grouping.
>
> **c)** *Always factor completely.* If a factor with more than one term can still be factored, you should factor it. When no factor can be factored further, you have finished.
>
> **d)** Check by multiplying.

EXAMPLE 1   Factor: $5t^4 - 80$.

a) We look for a common factor. There is one, 5.

$$5t^4 - 80 = 5(t^4 - 16)$$

b) The factor $t^4 - 16$ has only two terms. It is a difference of squares: $(t^2)^2 - 4^2$. We factor $t^4 - 16$ and then include the common factor:

$$5(t^2 + 4)(t^2 - 4).$$

c) We see that one of the factors, $t^2 - 4$, is again a difference of squares. We factor it:

$$5(t^2 + 4)(t + 2)(t - 2).$$

This is a sum of squares. It cannot be factored.

We have factored completely because no factor with more than one term can be factored further.

d) Check:
$$5(t^2 + 4)(t + 2)(t - 2) = 5(t^2 + 4)(t^2 - 4)$$
$$= 5(t^4 - 16)$$
$$= 5t^4 - 80.$$

EXAMPLE 2 Factor: $2x^3 + 10x^2 + x + 5$.

a) We look for a common factor. There isn't one.

b) There are four terms. We try factoring by grouping:

$$2x^3 + 10x^2 + x + 5$$
$$= (2x^3 + 10x^2) + (x + 5) \qquad \text{Separating into two binomials}$$
$$= 2x^2(x + 5) + 1(x + 5) \qquad \text{Factoring each binomial}$$
$$= (x + 5)(2x^2 + 1). \qquad \text{Factoring out the common factor } x + 5$$

c) None of these factors can be factored further, so we have factored completely.

d) Check: $(x + 5)(2x^2 + 1) = x \cdot 2x^2 + x \cdot 1 + 5 \cdot 2x^2 + 5 \cdot 1$
$$= 2x^3 + x + 10x^2 + 5, \text{ or}$$
$$2x^3 + 10x^2 + x + 5.$$

EXAMPLE 3 Factor: $x^5 - 2x^4 - 35x^3$.

a) We look first for a common factor. This time there is one, x^3:

$$x^5 - 2x^4 - 35x^3 = x^3(x^2 - 2x - 35).$$

b) The factor $x^2 - 2x - 35$ has three terms, but it is not a trinomial square. We factor it using trial and error:

$$x^5 - 2x^4 - 35x^3 = x^3(x^2 - 2x - 35) = x^3(x - 7)(x + 5).$$

> Don't forget to include the common factor in the final answer!

c) No factor with more than one term can be factored further, so we have factored completely.

d) Check: $x^3(x - 7)(x + 5) = x^3(x^2 - 2x - 35) = x^5 - 2x^4 - 35x^3$.

EXAMPLE 4 Factor: $x^4 - 10x^2 + 25$.

a) We look first for a common factor. There isn't one.

b) There are three terms. We see that this polynomial is a trinomial square. We factor it:

$$x^4 - 10x^2 + 25 = (x^2)^2 - 2 \cdot x^2 \cdot 5 + 5^2 = (x^2 - 5)^2.$$

We could use trial and error if we have not recognized that we have a trinomial square.

c) Since $x^2 - 5$ cannot be factored further, we have factored completely.

d) Check: $(x^2 - 5)^2 = (x^2)^2 - 2(x^2)(5) + 5^2 = x^4 - 10x^2 + 25$.

Do Exercises 1–5.

Factor.

1. $3m^4 - 3$

2. $x^6 + 8x^3 + 16$

3. $2x^4 + 8x^3 + 6x^2$

4. $3x^3 + 12x^2 - 2x - 8$

5. $8x^3 - 200x$

Answers

1. $3(m^2 + 1)(m + 1)(m - 1)$
2. $(x^3 + 4)^2$ **3.** $2x^2(x + 1)(x + 3)$
4. $(x + 4)(3x^2 - 2)$
5. $8x(x - 5)(x + 5)$

EXAMPLE 5 Factor: $6x^2y^4 - 21x^3y^5 + 3x^2y^6$.

a) We look first for a common factor:

$$6x^2y^4 - 21x^3y^5 + 3x^2y^6 = 3x^2y^4(2 - 7xy + y^2).$$

b) There are three terms in $2 - 7xy + y^2$. We determine whether the trinomial is a square. Since only y^2 is a square, we do not have a trinomial square. Can the trinomial be factored by trial and error? A key to the answer is that x is only in the term $-7xy$. The polynomial might be in a form like $(1 - y)(2 + y)$, but there would be no x in the middle term. Thus, $2 - 7xy + y^2$ cannot be factored.

c) Have we factored completely? Yes, because no factor with more than one term can be factored further.

d) The check is left to the student.

EXAMPLE 6 Factor: $(p + q)(x + 2) + (p + q)(x + y)$.

a) We look for a common factor:

$$(p + q)(x + 2) + (p + q)(x + y) = (p + q)[(x + 2) + (x + y)]$$
$$= (p + q)(2x + y + 2).$$

b) There are three terms in $2x + y + 2$, but this trinomial cannot be factored further.

c) Neither factor can be factored further, so we have factored completely.

d) The check is left to the student.

EXAMPLE 7 Factor: $px + py + qx + qy$.

a) We look first for a common factor. There isn't one.

b) There are four terms. We try factoring by grouping:

$$px + py + qx + qy = p(x + y) + q(x + y)$$
$$= (x + y)(p + q).$$

c) Have we factored completely? Since neither factor can be factored further, we have factored completely.

d) Check: $(x + y)(p + q) = px + qx + py + qy$, or
$$px + py + qx + qy.$$

EXAMPLE 8 Factor: $25x^2 + 20xy + 4y^2$.

a) We look first for a common factor. There isn't one.

b) There are three terms. We determine whether the trinomial is a square. The first term and the last term are squares:

$$25x^2 = (5x)^2 \quad \text{and} \quad 4y^2 = (2y)^2.$$

Since twice the product of $5x$ and $2y$ is the other term,

$$2 \cdot 5x \cdot 2y = 20xy,$$

the trinomial is a perfect square.

We factor by writing the square roots of the square terms and the sign of the middle term:

$$25x^2 + 20xy + 4y^2 = (5x + 2y)^2.$$

c) Since $5x + 2y$ cannot be factored further, we have factored completely.

d) Check: $(5x + 2y)^2 = (5x)^2 + 2(5x)(2y) + (2y)^2$
$$= 25x^2 + 20xy + 4y^2.$$

EXAMPLE 9 Factor: $p^2q^2 + 7pq + 12$.

a) We look first for a common factor. There isn't one.

b) There are three terms. We determine whether the trinomial is a square. The first term is a square, but neither of the other terms is a square, so we do not have a trinomial square. We factor, thinking of the product pq as a single variable. We consider this possibility for factorization:

$$(pq + \square)(pq + \square).$$

We factor the last term, 12. All the signs are positive, so we consider only positive factors. Possibilities are 1, 12 and 2, 6 and 3, 4. The pair 3, 4 gives a sum of 7 for the coefficient of the middle term. Thus,

$$p^2q^2 + 7pq + 12 = (pq + 3)(pq + 4).$$

c) No factor with more than one term can be factored further, so we have factored completely.

d) Check: $(pq + 3)(pq + 4) = (pq)(pq) + 4 \cdot pq + 3 \cdot pq + 3 \cdot 4$
$$= p^2q^2 + 7pq + 12.$$

EXAMPLE 10 Factor: $8x^4 - 20x^2y - 12y^2$.

a) We look first for a common factor:

$$8x^4 - 20x^2y - 12y^2 = 4(2x^4 - 5x^2y - 3y^2).$$

b) There are three terms in $2x^4 - 5x^2y - 3y^2$. We determine whether the trinomial is a square. Since none of the terms is a square, we do not have a trinomial square. We factor $2x^4$. Possibilities are $2x^2$, x^2 and $2x$, x^3 and others. We also factor the last term, $-3y^2$. Possibilities are $3y$, $-y$ and $-3y$, y and others. We look for factors such that the sum of their products is the middle term. The x^2 in the middle term, $-5x^2y$, should lead us to try $(2x^2)(x^2)$. We try some possibilities:

$$(2x^2 - y)(x^2 + 3y) = 2x^4 + 5x^2y - 3y^2,$$
$$(2x^2 + y)(x^2 - 3y) = 2x^4 - 5x^2y - 3y^2.$$

c) No factor with more than one term can be factored further, so we have factored completely. The factorization, including the common factor, is

$$4(2x^2 + y)(x^2 - 3y).$$

d) Check: $4(2x^2 + y)(x^2 - 3y) = 4[(2x^2)(x^2) + 2x^2(-3y) + yx^2 + y(-3y)]$
$$= 4[2x^4 - 6x^2y + x^2y - 3y^2]$$
$$= 4(2x^4 - 5x^2y - 3y^2)$$
$$= 8x^4 - 20x^2y - 12y^2.$$

EXAMPLE 11 Factor: $a^4 - 16b^4$.

a) We look first for a common factor. There isn't one.

b) There are two terms. Since $a^4 = (a^2)^2$ and $16b^4 = (4b^2)^2$, we see that we do have a difference of squares. Thus,

$$a^4 - 16b^4 = (a^2 + 4b^2)(a^2 - 4b^2).$$

c) The last factor can be factored further. It is also a difference of squares.

$$a^4 - 16b^4 = (a^2 + 4b^2)(a + 2b)(a - 2b)$$

d) Check: $(a^2 + 4b^2)(a + 2b)(a - 2b) = (a^2 + 4b^2)(a^2 - 4b^2)$
$$= a^4 - 16b^4.$$

EXAMPLE 12 Factor: $40t^3 - 5s^3$.

a) We look first for a common factor:

$$40t^3 - 5s^3 = 5(8t^3 - s^3).$$

b) The factor $8t^3 - s^3$ has only two terms. It is a difference of two cubes. We factor as follows:

$$(2t - s)(4t^2 + 2ts + s^2).$$

c) No factor with more than one term can be factored further, so we have factored completely. The factorization, including the common factor, is

$$5(2t - s)(4t^2 + 2ts + s^2).$$

Do Exercises 6–13.

Factor.

6. $15x^4 + 5x^2y - 10y^2$

7. $10p^6q^2 + 4p^5q^3 + 2p^4q^4$

8. $(a - b)(x + 5) + (a - b)(x + y^2)$

9. $ax^2 + ay + bx^2 + by$

10. $x^4 + 2x^2y^2 + y^4$

11. $x^2y^2 + 5xy + 4$

12. $p^4 - 81q^4$

13. $15a^3 - 120b^3$

Answers

6. $5(3x^2 - 2y)(x^2 + y)$
7. $2p^4q^2(5p^2 + 2pq + q^2)$
8. $(a - b)(2x + 5 + y^2)$
9. $(x^2 + y)(a + b)$ **10.** $(x^2 + y^2)^2$
11. $(xy + 1)(xy + 4)$
12. $(p^2 + 9q^2)(p + 3q)(p - 3q)$
13. $15(a - 2b)(a^2 + 2ab + 4b^2)$

5.7

Exercise Set

For Extra Help

MyMathLab

Math XL
PRACTICE

WATCH

DOWNLOAD

READ

REVIEW

a   Factor completely.

1. $3x^2 - 192$

2. $2t^2 - 18$

3. $a^2 + 25 - 10a$

4. $y^2 + 49 + 14y$

5. $2x^2 - 11x + 12$

6. $8y^2 - 18y - 5$

7. $x^3 + 24x^2 + 144x$

8. $x^3 - 18x^2 + 81x$

9. $x^3 + 3x^2 - 4x - 12$

10. $x^3 - 5x^2 - 25x + 125$

11. $48x^2 - 3$

12. $50x^2 - 32$

13. $9x^3 + 12x^2 - 45x$

14. $20x^3 - 4x^2 - 72x$

15. $x^2 + 4$

16. $t^2 + 25$

17. $x^4 + 7x^2 - 3x^3 - 21x$

18. $m^4 + 8m^3 + 8m^2 + 64m$

19. $x^5 - 14x^4 + 49x^3$

20. $2x^6 + 8x^5 + 8x^4$

21. $20 - 6x - 2x^2$

22. $45 - 3x - 6x^2$

23. $x^2 - 6x + 1$

24. $x^2 + 8x + 5$

25. $4x^4 - 64$

26. $5x^5 - 80x$

27. $1 - y^8$

28. $t^8 - 1$

29. $x^5 - 4x^4 + 3x^3$

30. $x^6 - 2x^5 + 7x^4$

31. $\dfrac{1}{81}x^6 - \dfrac{8}{27}x^3 + \dfrac{16}{9}$

32. $36a^2 - 15a + \dfrac{25}{16}$

33. $\dfrac{1}{1000}m^3 - \dfrac{1}{27}n^3$

34. $125a^3 - 8b^3$

35. $9x^2y^2 - 36xy$

36. $x^2y - xy^2$

37. $2\pi rh + 2\pi r^2$

38. $10p^4q^4 + 35p^3q^3 + 10p^2q^2$

39. $(a + b)(x - 3) + (a + b)(x + 4)$

40. $5c(a^3 + b) - (a^3 + b)$

41. $(x - 1)(x + 1) - y(x + 1)$

42. $3(p - q) - q^2(p - q)$

43. $n^2 + 2n + np + 2p$

44. $a^2 - 3a + ay - 3y$

45. $6q^2 - 3q + 2pq - p$

46. $2x^2 - 4x + xy - 2y$

47. $4b^2 + a^2 - 4ab$

48. $x^2 + y^2 - 2xy$

49. $16x^2 + 24xy + 9y^2$

50. $9c^2 + 6cd + d^2$

51. $49m^4 - 112m^2n + 64n^2$

52. $4x^2y^2 + 12xyz + 9z^2$

53. $y^4 + 10y^2z^2 + 25z^4$

54. $0.01x^4 - 0.1x^2y^2 + 0.25y^4$

55. $\frac{1}{4}a^2 + \frac{1}{3}ab + \frac{1}{9}b^2$

56. $4p^2q + pq^2 + 4p^3$

57. $a^2 - ab - 2b^2$

58. $3b^2 - 17ab - 6a^2$

59. $2mn - 360n^2 + m^2$

60. $15 + x^2y^2 + 8xy$

61. $m^2n^2 - 4mn - 32$

62. $p^2q^2 + 7pq + 6$

63. $r^5s^2 - 10r^4s + 16r^3$

64. $p^5q^2 + 3p^4q - 10p^3$

65. $a^5 + 4a^4b - 5a^3b^2$

66. $2s^6t^2 + 10s^3t^3 + 12t^4$

67. $a^2 - \frac{1}{25}b^2$

68. $p^2 - \frac{1}{49}b^2$

69. $7x^6 - 7y^6$

70. $16p^3 + 54q^3$

71. $16 - p^4q^4$

72. $15a^4 - 15b^4$

73. $1 - 16x^{12}y^{12}$

74. $81a^4 - b^4$

75. $q^3 + 8q^2 - q - 8$

76. $m^3 - 7m^2 - 4m + 28$

77. $6a^3b^3 - a^2b^2 - 2ab$

78. $4ab^5 - 32b^4 + a^2b^6$

79. $m^4 - 5m^2 + 4$

80. $8x^3y^3 - 6x^2y^2 - 5xy$

81. $t^4 - 2t^2 + 1$

Skill Maintenance

Multiply [4.6d], [4.7f]

82. $(3x - 5y)(3x + 5y)$

83. $(3x - 5y)^2$

84. $(3x + 5y)^2$

85. $(3x - 5y)(2x + 7y)$

86. Multiply: $(5x - t)^2$. [4.6d]

87. Divide: $\dfrac{7}{5} \div \left(-\dfrac{11}{10}\right)$. [1.6c]

88. Solve: $4(x - 9) - 2(x + 7) < 14$. [2.7e]

89. Solve $A = aX + bX - 7$ for X. [2.4b]

Synthesis

Factor completely.

90. $x^4 + 9$

91. $x^3 + 20 - (5x^2 + 4x)$

92. $\dfrac{1}{5}x^2 - x + \dfrac{4}{5}$

93. $12.25x^2 - 7x + 1$

94. $x^3 - (x - 3x^2) - 3$

95. $5x^2 + 13x + 7.2$

96. $-(x^4 - 7x^2 - 18)$

97. $18 + y^3 - 9y - 2y^2$

98. $x^3 + x^2 - (4x + 4)$

99. $x^3 - x^2 - 4x + 4$

100. $3x^4 - 15x^2 + 12$

101. $a^3 - 4a^2 - a - 4$

102. $y^2(y + 1) - 4y(y + 1) - 21(y + 1)$

103. $y^2(y - 1) - 2y(y - 1) + (y - 1)$

104. $6(x - 1)^2 + 7y(x - 1) - 3y^2$

105. $(y + 4)^2 + 2x(y + 4) + x^2$

5.8

Solving Quadratic Equations by Factoring

Second-degree equations like $x^2 + x - 156 = 0$ and $9 - x^2 = 0$ are examples of *quadratic equations*.

> ### QUADRATIC EQUATION
>
> A **quadratic equation** is an equation equivalent to an equation of the type
>
> $$ax^2 + bx + c = 0, \ a \neq 0.$$

In order to solve quadratic equations, we need a new equation-solving principle.

(a) The Principle of Zero Products

The product of two numbers is 0 if one or both of the numbers is 0. Furthermore, *if any product is* 0, *then a factor must be* 0. For example:

If $7x = 0$, then we know that $x = 0$.

If $x(2x - 9) = 0$, then we know that $x = 0 \ or \ 2x - 9 = 0$.

If $(x + 3)(x - 2) = 0$, then we know that $x + 3 = 0 \ or \ x - 2 = 0$.

-------------------------------- *Caution!* --------------------------------

In a product such as $ab = 24$, we cannot conclude with certainty that a is 24 or that b is 24, but if $ab = 0$, we can conclude that $a = 0$ or $b = 0$.

EXAMPLE 1 Solve: $(x + 3)(x - 2) = 0$.

We have a product of 0. This equation will be true when either factor is 0. Thus it is true when

$$x + 3 = 0 \quad or \quad x - 2 = 0.$$

Here we have two simple equations that we know how to solve:

$$x = -3 \quad or \quad x = 2.$$

Each of the numbers -3 and 2 is a solution of the original equation, as we can see in the following checks.

Check: For -3:

$$\frac{(x + 3)(x - 2) = 0}{(-3 + 3)(-3 - 2) \ ? \ 0}$$
$$0(-5) \ \big| $$
$$0 \ \big| \quad \text{TRUE}$$

For 2:

$$\frac{(x + 3)(x - 2) = 0}{(2 + 3)(2 - 2) \ ? \ 0}$$
$$5(0) \ \big| $$
$$0 \ \big| \quad \text{TRUE}$$

OBJECTIVES

a Solve equations (already factored) using the principle of zero products.

b Solve quadratic equations by factoring and then using the principle of zero products.

SKILL TO REVIEW
Objective 2.3a: Solve equations using both the addition principle and the multiplication principle.

Solve.

1. $3x - 7 = 2$
2. $4y + 5 = 1$

STUDY TIPS

WORKING WITH A CLASSMATE

If you are finding it difficult to master a particular topic or concept, try talking about it with a classmate. Verbalizing your questions about the material might help clarify it. If your classmate is also finding the material difficult, it is possible that the majority of the people in your class are confused and you can ask your instructor to explain the concept again.

Answers

Skill to Review:
1. 3   2. -1

We now have a principle to help in solving quadratic equations.

> ### THE PRINCIPLE OF ZERO PRODUCTS
>
> An equation $ab = 0$ is true if and only if $a = 0$ is true or $b = 0$ is true, or both are true. (A product is 0 if and only if one or both of the factors is 0.)

EXAMPLE 2 Solve: $(5x + 1)(x - 7) = 0$.

We have

$$(5x + 1)(x - 7) = 0$$

$5x + 1 = 0$ *or* $x - 7 = 0$ Using the principle of zero products

$5x = -1$ *or* $x = 7$ Solving the two equations separately

$x = -\frac{1}{5}$ *or* $x = 7.$

Check: For $-\frac{1}{5}$:

$$\frac{(5x + 1)(x - 7) = 0}{\left(5\left(-\frac{1}{5}\right) + 1\right)\left(-\frac{1}{5} - 7\right) \overset{?}{=} 0}$$
$$(-1 + 1)\left(-7\frac{1}{5}\right)$$
$$0\left(-7\frac{1}{5}\right)$$
$$0 \quad | \quad \text{TRUE}$$

For 7:

$$\frac{(5x + 1)(x - 7) = 0}{(5(7) + 1)(7 - 7) \overset{?}{=} 0}$$
$$(35 + 1) \cdot 0$$
$$36 \cdot 0$$
$$0 \quad | \quad \text{TRUE}$$

The solutions are $-\frac{1}{5}$ and 7.

When you solve an equation using the principle of zero products, a check by substitution, as in Examples 1 and 2, will detect errors in solving.

> Do Exercises 1–3.

When some factors have only one term, you can still use the principle of zero products.

EXAMPLE 3 Solve: $x(2x - 9) = 0$.

We have

$$x(2x - 9) = 0$$

$x = 0$ *or* $2x - 9 = 0$ Using the principle of zero products

$x = 0$ *or* $2x = 9$

$x = 0$ *or* $x = \dfrac{9}{2}.$

Check: For 0:

$$\frac{x(2x - 9) = 0}{0 \cdot (2 \cdot 0 - 9) \overset{?}{=} 0}$$
$$0 \cdot (-9)$$
$$0 \quad | \quad \text{TRUE}$$

For $\frac{9}{2}$:

$$\frac{x(2x - 9) = 0}{\frac{9}{2} \cdot \left(2 \cdot \frac{9}{2} - 9\right) \overset{?}{=} 0}$$
$$\frac{9}{2} \cdot (9 - 9)$$
$$\frac{9}{2} \cdot 0$$
$$0 \quad | \quad \text{TRUE}$$

> Do Exercise 4.

Solve using the principle of zero products.

1. $(x - 3)(x + 4) = 0$

2. $(x - 7)(x - 3) = 0$

3. $(4t + 1)(3t - 2) = 0$

4. Solve: $y(3y - 17) = 0$.

Answers

1. $3, -4$ **2.** $7, 3$ **3.** $-\dfrac{1}{4}, \dfrac{2}{3}$ **4.** $0, \dfrac{17}{3}$

(b) Using Factoring to Solve Equations

Using factoring and the principle of zero products, we can solve some new kinds of equations. Thus we have extended our equation-solving abilities.

EXAMPLE 4 Solve: $x^2 + 5x + 6 = 0$.

There are no like terms to collect, and we have a squared term. We first factor the polynomial. Then we use the principle of zero products.

$$x^2 + 5x + 6 = 0$$
$$(x + 2)(x + 3) = 0 \qquad \text{Factoring}$$
$$x + 2 = 0 \quad or \quad x + 3 = 0 \qquad \text{Using the principle of zero products}$$
$$x = -2 \quad or \qquad x = -3$$

Check: For -2:

$$\begin{array}{c|c} x^2 + 5x + 6 = 0 \\ \hline (-2)^2 + 5(-2) + 6 \overset{?}{\,} 0 \\ 4 - 10 + 6 \\ -6 + 6 \\ 0 & \text{TRUE} \end{array}$$

For -3:

$$\begin{array}{c|c} x^2 + 5x + 6 = 0 \\ \hline (-3)^2 + 5(-3) + 6 \overset{?}{\,} 0 \\ 9 - 15 + 6 \\ -6 + 6 \\ 0 & \text{TRUE} \end{array}$$

The solutions are -2 and -3.

-- *Caution!* --

Keep in mind that you *must* have 0 on one side of the equation before you can use the principle of zero products. Get all nonzero terms on one side and 0 on the other.

| Do Exercise 5. |

5. Solve: $x^2 - x - 6 = 0$.

EXAMPLE 5 Solve: $x^2 - 8x = -16$.

We first add 16 to get a 0 on one side:

$$x^2 - 8x = -16$$
$$x^2 - 8x + 16 = 0 \qquad \text{Adding 16}$$
$$(x - 4)(x - 4) = 0 \qquad \text{Factoring}$$
$$x - 4 = 0 \quad or \quad x - 4 = 0 \qquad \text{Using the principle of zero products}$$
$$x = 4 \quad or \qquad x = 4. \qquad \text{Solving each equation}$$

There is only one solution, 4. The check is left to the student.

| Do Exercises 6 and 7. |

Solve.

6. $x^2 - 3x = 28$

7. $x^2 = 6x - 9$

EXAMPLE 6 Solve: $x^2 + 5x = 0$.

$$x^2 + 5x = 0$$
$$x(x + 5) = 0 \qquad \text{Factoring out a common factor}$$
$$x = 0 \quad or \quad x + 5 = 0 \qquad \text{Using the principle of zero products}$$
$$x = 0 \quad or \qquad x = -5$$

The solutions are 0 and -5. The check is left to the student.

Answers

5. $-2, 3$ **6.** $-4, 7$ **7.** 3

EXAMPLE 7 Solve: $4x^2 = 25$.

$$4x^2 = 25$$

$$4x^2 - 25 = 0 \qquad \text{Subtracting 25 on both sides to get 0 on one side}$$

$$(2x - 5)(2x + 5) = 0 \qquad \text{Factoring a difference of squares}$$

$$2x - 5 = 0 \quad or \quad 2x + 5 = 0 \qquad \text{Using the principle of zero products}$$

$$2x = 5 \quad or \qquad 2x = -5 \qquad \text{Solving each equation}$$

$$x = \frac{5}{2} \quad or \qquad x = -\frac{5}{2}$$

The solutions are $\frac{5}{2}$ and $-\frac{5}{2}$. The check is left to the student.

Do Exercises 8 and 9.

Solve.

8. $x^2 - 4x = 0$

9. $9x^2 = 16$

EXAMPLE 8 Solve: $-5x^2 + 2x + 3 = 0$.

In this case, the leading coefficient of the trinomial is negative. Thus we first multiply by -1 and then proceed as we have in Examples 4–7.

$$-5x^2 + 2x + 3 = 0$$

$$-1(-5x^2 + 2x + 3) = -1 \cdot 0 \qquad \text{Multiplying by } -1$$

$$5x^2 - 2x - 3 = 0 \qquad \text{Simplifying}$$

$$(5x + 3)(x - 1) = 0 \qquad \text{Factoring}$$

$$5x + 3 = 0 \quad or \quad x - 1 = 0 \qquad \text{Using the principle of zero products}$$

$$5x = -3 \quad or \qquad x = 1$$

$$x = -\frac{3}{5} \quad or \qquad x = 1$$

The solutions are $-\frac{3}{5}$ and 1. The check is left to the student.

Do Exercises 10 and 11.

Solve.

10. $-2x^2 + 13x - 21 = 0$

11. $10 - 3x - x^2 = 0$

EXAMPLE 9 Solve: $(x + 2)(x - 2) = 5$.

Be careful with an equation like this one! It might be tempting to set each factor equal to 5. **Remember: We must have a 0 on one side**. We first carry out the multiplication on the left. Next, we subtract 5 on both sides to get 0 on one side. Then we proceed with the principle of zero products.

$$(x + 2)(x - 2) = 5$$

$$x^2 - 4 = 5 \qquad \text{Multiplying on the left}$$

$$x^2 - 4 - 5 = 5 - 5 \qquad \text{Subtracting 5}$$

$$x^2 - 9 = 0 \qquad \text{Simplifying}$$

$$(x + 3)(x - 3) = 0 \qquad \text{Factoring}$$

$$x + 3 = 0 \quad or \quad x - 3 = 0 \qquad \text{Using the principle of zero products}$$

$$x = -3 \quad or \qquad x = 3$$

The solutions are -3 and 3. The check is left to the student.

Do Exercise 12.

12. Solve: $(x + 1)(x - 1) = 8$.

Answers

8. $0, 4$ **9.** $-\frac{4}{3}, \frac{4}{3}$ **10.** $-3, \frac{7}{2}$ **11.** $-5, 2$
12. $-3, 3$

✳ Algebraic-Graphical Connection

In Chapter 3, we graphed linear equations of the type $y = mx + b$ and $Ax + By = C$. Recall that to find the x-intercept, we replaced y with 0 and solved for x. This procedure can also be used to find the x-intercepts when an equation of the form $y = ax^2 + bx + c$, $a \neq 0$, is to be graphed. Although the details of creating such graphs will be left to Chapter 11, we consider them briefly here from the standpoint of finding the x-intercepts. The graph of $y = ax^2 + bx + c$, $a \neq 0$, is shaped like one of the following curves. Note that each x-intercept represents a solution of $ax^2 + bx + c = 0$.

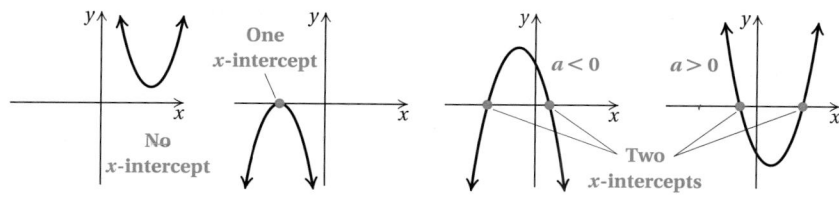

✳

EXAMPLE 10 Find the x-intercepts of the graph of $y = x^2 - 4x - 5$ shown at right. (The grid is intentionally not included.)

To find the x-intercepts, we let $y = 0$ and solve for x:

$$y = x^2 - 4x - 5$$
$$0 = x^2 - 4x - 5 \qquad \text{Substituting } 0 \text{ for } y$$
$$0 = (x - 5)(x + 1) \qquad \text{Factoring}$$
$$x - 5 = 0 \quad or \quad x + 1 = 0 \qquad \text{Using the principle of zero products}$$
$$x = 5 \quad or \qquad x = -1.$$

The solutions of the equation $0 = x^2 - 4x - 5$ are 5 and -1. Thus the x-intercepts of the graph of $y = x^2 - 4x - 5$ are $(5, 0)$ and $(-1, 0)$. We can now label them on the graph.

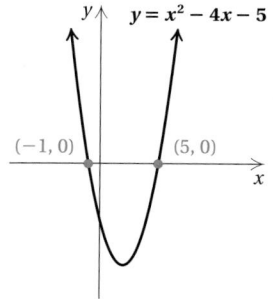

Do Exercises 13 and 14.

13. Find the x-intercepts of the graph shown below.

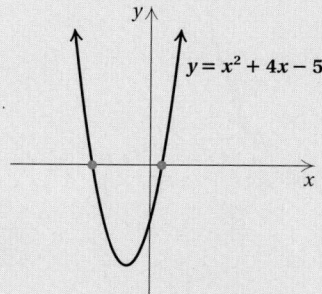

14. Use *only* the graph shown below to solve $3x - x^2 = 0$.

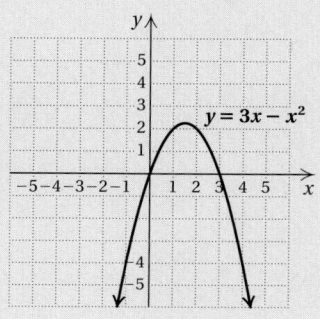

Answers

13. $(-5, 0), (1, 0)$ **14.** $0, 3$

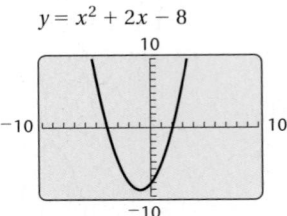

Calculator Corner

Solving Quadratic Equations   We can solve quadratic equations graphically. Consider the equation $x^2 + 2x = 8$. First, we must write the equation with 0 on one side. To do this, we subtract 8 on both sides of the equation; we get $x^2 + 2x - 8 = 0$. Next, we graph $y = x^2 + 2x - 8$ in a window that shows the x-intercepts. The standard window works well in this case.

$$y = x^2 + 2x - 8$$

The solutions of the equation are the values of x for which $x^2 + 2x - 8 = 0$. These are also the first coordinates of the x-intercepts of the graph. We use the ZERO feature from the CALC menu to find these numbers. To find the solution corresponding to the leftmost x-intercept, we first press **2ND** **CALC** **2** to select the ZERO feature. The prompt "Left Bound?" appears. Next, we use the **◁** or the **▷** key to move the cursor to the left of the intercept and press **ENTER**. Now the prompt "Right Bound?" appears. Then we move the cursor to the right of the intercept and press **ENTER**. The prompt "Guess?" appears. We move the cursor close to the intercept and press **ENTER** again. We now see the cursor positioned at the leftmost x-intercept and the coordinates of that point, $x = -4$, $y = 0$, are displayed. Thus, $x^2 + 2x - 8 = 0$ when $x = -4$. This is one solution of the equation.

We can repeat this procedure to find the first coordinate of the other x-intercept. We see that $x = 2$ at that point. Thus the solutions of the equation $x^2 + 2x - 8 = 0$ are -4 and 2. Note that the x-intercepts of the graph of $y = x^2 + 2x - 8$ are $(-4, 0)$ and $(2, 0)$.

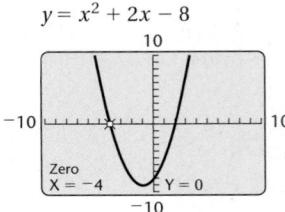
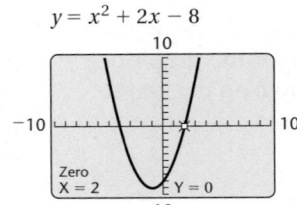

Exercises:

1. Solve each of the equations in Examples 4–8 graphically.

65. Use the following graph to solve $-x^2 + 2x + 3 = 0$.

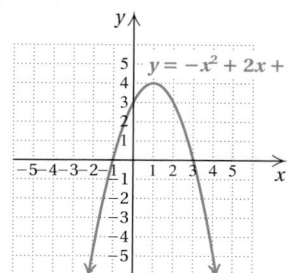

66. Use the following graph to solve $-x^2 - x + 6 = 0$.

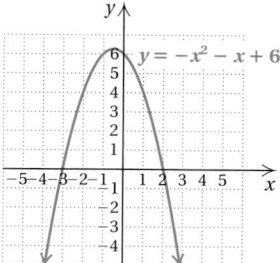

Skill Maintenance

Translate to an algebraic expression. [1.1b]

67. The square of the sum of a and b

68. The sum of the squares of a and b

Divide. [1.6a, c]

69. $144 \div (-9)$

70. $-24.3 \div 5.4$

71. $-\frac{5}{8} \div \frac{3}{16}$

72. $-\frac{3}{16} \div \left(-\frac{5}{8}\right)$

Synthesis

Solve.

73. $b(b + 9) = 4(5 + 2b)$

74. $y(y + 8) = 16(y - 1)$

75. $(t - 3)^2 = 36$

76. $(t - 5)^2 = 2(5 - t)$

77. $x^2 - \frac{1}{64} = 0$

78. $x^2 - \frac{25}{36} = 0$

79. $\frac{5}{16}x^2 = 5$

80. $\frac{27}{25}x^2 = \frac{1}{3}$

81. Find an equation that has the given numbers as solutions. For example, 3 and -2 are solutions of $x^2 - x - 6 = 0$.

a) $-3, 4$ b) $-3, -4$ c) $\frac{1}{2}, \frac{1}{2}$
d) $5, -5$ e) $0, 0.1, \frac{1}{4}$

82. *Matching.* Match each equation in the first column with the equivalent equation in the second column.

$x^2 + 10x - 2 = 0$ $4x^2 + 8x + 36 = 0$

$(x - 6)(x + 3) = 0$ $(2x + 8)(2x - 5) = 0$

$5x^2 - 5 = 0$ $9x^2 - 12x + 24 = 0$

$(2x - 5)(x + 4) = 0$ $(x + 1)(5x - 5) = 0$

$x^2 + 2x + 9 = 0$ $x^2 - 3x - 18 = 0$

$3x^2 - 4x + 8 = 0$ $2x^2 + 20x - 4 = 0$

Use a graphing calculator to find the solutions of each equation. Round solutions to the nearest hundredth.

83. $x^2 - 9.10x + 15.77 = 0$

84. $-x^2 + 0.63x + 0.22 = 0$

85. $0.84x^2 - 2.30x = 0$

86. $6.4x^2 - 8.45x - 94.06 = 0$

5.9

Applications of Quadratic Equations

OBJECTIVE

a) Solve applied problems involving quadratic equations that can be solved by factoring.

a) Applied Problems, Quadratic Equations, and Factoring

We can solve problems that translate to quadratic equations using the five steps for solving problems.

EXAMPLE 1 *Kitchen Island.* Lisa buys a kitchen island with a butcher-block top as part of a remodeling project. The top of the island is a rectangle that is twice as long as it is wide and that has an area of 800 in². What are the dimensions of the top of the island?

1. **Familiarize.** We first make a drawing. Recall that the area of a rectangle is Length · Width. We let $x =$ the width of the top, in inches. The length is then $2x$.

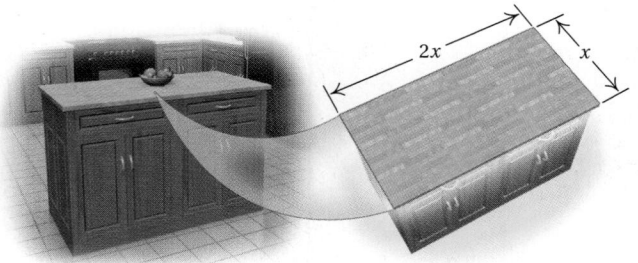

2. **Translate.** We reword and translate as follows:

   Rewording: The area of the rectangle is 800 in².

   Translating: $2x \cdot x = 800$

3. **Solve.** We solve the equation as follows:

$$2x \cdot x = 800$$
$$2x^2 = 800$$
$$2x^2 - 800 = 0 \qquad \text{Subtracting 800 to get 0 on one side}$$
$$2(x^2 - 400) = 0 \qquad \text{Removing a common factor of 2}$$
$$2(x - 20)(x + 20) = 0 \qquad \text{Factoring a difference of squares}$$
$$(x - 20)(x + 20) = 0 \qquad \text{Dividing by 2}$$
$$x - 20 = 0 \quad or \quad x + 20 = 0 \qquad \text{Using the principle of zero products}$$
$$x = 20 \quad or \quad x = -20. \qquad \text{Solving each equation}$$

4. **Check.** The solutions of the equation are 20 and -20. Since the width must be positive, -20 cannot be a solution. To check 20 in., we note that if the width is 20 in., then the length is $2 \cdot 20$ in., or 40 in., and the area is 20 in. · 40 in., or 800 in². Thus the solution 20 checks.

5. **State.** The top is 20 in. wide and 40 in. long.

Do Exercise 1.

---

1. **Dimensions of Picture.** A rectangular picture is twice as long as it is wide. If the area of the picture is 288 in², what are its dimensions?

$2w$

w

Answer

1. Length: 24 in.; width: 12 in.

EXAMPLE 2 *Racing Sailboat.* The height of a triangular sail on a racing sailboat is 9 ft more than the base. The area of the triangle is 110 ft². Find the height and the base of the sail.

Source: Whitney Gladstone, North Graphics, San Diego, CA

1. **Familiarize.** We first make a drawing. If you don't remember the formula for the area of a triangle, look it up in the list of formulas at the back of this book or in a geometry book. The area is $\frac{1}{2}$(base)(height).

   We let b = the base of the triangle, in feet. Then $b + 9$ = the height.

2. **Translate.** It helps to reword this problem before translating:

   Rewording: $\frac{1}{2}$ times Base times Height is 110

   Translating: $\frac{1}{2}$ $\cdot$ b $\cdot$ $(b + 9)$ = 110.

3. **Solve.** We solve the equation as follows:

$$\frac{1}{2} \cdot b \cdot (b + 9) = 110$$

$$\frac{1}{2}(b^2 + 9b) = 110 \qquad \text{Multiplying}$$

$$2 \cdot \frac{1}{2}(b^2 + 9b) = 2 \cdot 110 \qquad \text{Multiplying by 2}$$

$$b^2 + 9b = 220 \qquad \text{Simplifying}$$

$$b^2 + 9b - 220 = 220 - 220 \qquad \text{Subtracting 220 to get 0 on one side}$$

$$b^2 + 9b - 220 = 0$$

$$(b - 11)(b + 20) = 0 \qquad \text{Factoring}$$

$$b - 11 = 0 \quad or \quad b + 20 = 0 \qquad \text{Using the principle of zero products}$$

$$b = 11 \quad or \qquad b = -20.$$

4. **Check.** The base of a triangle cannot have a negative length, so -20 cannot be a solution. Suppose the base is 11 ft. The height is 9 ft more than the base, so the height is 11 ft + 9 ft, or 20 ft, and the area is $\frac{1}{2}(11)(20)$, or 110 ft². These numbers check in the original problem.

5. **State.** The height is 20 ft and the base is 11 ft.

Do Exercise 2.

2. Dimensions of a Sail. The triangular mainsail on Stacey's Lightning-styled sailboat has an area of 125 ft². The height of the sail is 15 ft more than the base. Find the height and the base of the sail.

Answer

2. Height: 25 ft; base: 10 ft

EXAMPLE 3 *Games in a Sports League.* In a sports league of x teams in which each team plays every other team twice, the total number N of games to be played is given by

$$x^2 - x = N.$$

Maggie's basketball league plays a total of 240 games. How many teams are in the league?

1., 2. Familiarize and **Translate.** We are given that x is the number of teams in a league and N is the number of games. To familiarize yourself with this problem, reread Example 4 in Section 4.3 where we first considered it. To find the number of teams x in a league in which 240 games are played, we substitute 240 for N in the equation:

$$x^2 - x = 240. \qquad \text{Substituting 240 for } N$$

3. Solve. We solve the equation as follows:

$$x^2 - x = 240$$
$$x^2 - x - 240 = 240 - 240 \qquad \text{Subtracting 240 to get 0 on one side}$$
$$x^2 - x - 240 = 0$$
$$(x - 16)(x + 15) = 0 \qquad \text{Factoring}$$
$$x - 16 = 0 \quad or \quad x + 15 = 0 \qquad \text{Using the principle of zero products}$$
$$x = 16 \quad or \qquad x = -15.$$

4. Check. The solutions of the equation are 16 and -15. Since the number of teams cannot be negative, -15 cannot be a solution. But 16 checks, since $16^2 - 16 = 256 - 16 = 240$.

5. State. There are 16 teams in the league.

Do Exercise 3.

---

3. Use $N = x^2 - x$ for each of the following.

a) **Volleyball League.** Amy's volleyball league has 19 teams. What is the total number of games to be played if each team plays every other team twice?

b) **Softball League.** Barry's slow-pitch softball league plays a total of 72 games. How many teams are in the league if each team plays every other team twice?

STUDY TIPS

FIVE STEPS FOR PROBLEM SOLVING

Recall the five steps for problem solving that were developed in Section 2.6.

1. **Familiarize** yourself with the situation.
   a) Carefully read and reread until you understand *what* you are being asked to find.
   b) Draw a diagram or see if there is a formula that applies.
   c) Assign a letter, or *variable*, to the unknown.
2. **Translate** the problem to an equation using the letter or variable.
3. **Solve** the equation.
4. **Check** the answer in the original wording of the problem.
5. **State** the answer to the problem clearly with appropriate units.

Answer

3. (a) 342 games; (b) 9 teams

EXAMPLE 4 *Marathoners' Numbers.* The product of the numbers of two consecutive entrants in a marathon race is 156. Find the numbers.

1. **Familiarize.** The numbers are consecutive integers. Recall that consecutive integers are one unit apart, like 49 and 50, or -6 and -5. Let $x =$ the smaller integer; then $x + 1 =$ the larger integer.

2. **Translate.** It helps to reword the problem before translating:

   Rewording: First integer times Second integer is 156

   Translating: x · $(x + 1)$ = 156.

3. **Solve.** We solve the equation as follows:

   $$x(x + 1) = 156$$
   $$x^2 + x = 156 \qquad \text{Multiplying}$$
   $$x^2 + x - 156 = 156 - 156 \qquad \text{Subtracting 156 to get 0 on one side}$$
   $$x^2 + x - 156 = 0 \qquad \text{Simplifying}$$
   $$(x - 12)(x + 13) = 0 \qquad \text{Factoring}$$
   $$x - 12 = 0 \quad or \quad x + 13 = 0 \qquad \text{Using the principle of zero products}$$
   $$x = 12 \quad or \qquad x = -13.$$

4. **Check.** The solutions of the equation are 12 and -13. When x is 12, then $x + 1$ is 13, and $12 \cdot 13 = 156$. The numbers 12 and 13 are consecutive integers that are solutions to the problem. When x is -13, then $x + 1$ is -12, and $(-13)(-12) = 156$. The numbers -13 and -12 are consecutive integers, but they are not solutions of the problem because negative numbers are not used as entry numbers.

5. **State.** The entry numbers are 12 and 13.

Do Exercise 4.

4. **Page Numbers.** The product of the page numbers on two facing pages of a book is 506. Find the page numbers.

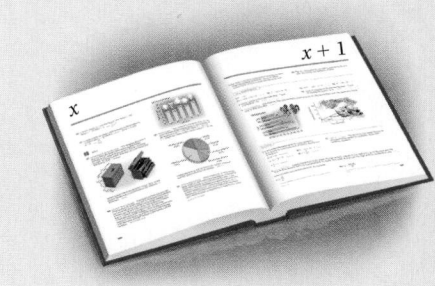

The Pythagorean Theorem

The problems that follow involve the Pythagorean theorem, which states a relationship involving the lengths of the sides of a *right* triangle. A triangle is a **right triangle** if it has a 90°, or *right*, angle. The side opposite the 90° angle is called the **hypotenuse**. The other sides are called **legs**.

Answer

4. 22 and 23

In any right triangle, if a and b are the lengths of the legs and c is the length of the hypotenuse, then

$$a^2 + b^2 = c^2.$$

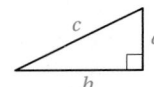

The symbol $\llcorner$ denotes a 90° angle.

EXAMPLE 5 *Wood Scaffold.* Jonah is building a wood scaffold to use for a home improvement project. He designs the scaffold with diagonal braces that are 5 ft long and that span a distance of 3 ft. How high does each brace reach vertically?

1. **Familiarize.** We make a drawing as shown above and let $h =$ the height, in feet, to which each brace rises vertically.

2. **Translate.** A right triangle is formed, so we can use the Pythagorean theorem:

$$a^2 + b^2 = c^2$$
$$3^2 + h^2 = 5^2. \qquad \text{Substituting 3 and } h \text{ for the lengths of the legs and} $$
$$\text{5 for the length of the hypotenuse}$$

3. **Solve.** We solve the equation as follows:

$$3^2 + h^2 = 5^2$$
$$9 + h^2 = 25 \qquad \text{Squaring 3 and 5}$$
$$9 + h^2 - 25 = 25 - 25 \qquad \text{Subtracting 25 to get 0 on one side}$$
$$h^2 - 16 = 0 \qquad \text{Simplifying}$$
$$(h - 4)(h + 4) = 0 \qquad \text{Factoring}$$
$$h - 4 = 0 \quad or \quad h + 4 = 0 \qquad \text{Using the principle of zero products}$$
$$h = 4 \quad or \qquad h = -4.$$

4. **Check.** Since height cannot be negative, -4 cannot be a solution. If the height is 4 ft, we have $3^2 + 4^2 = 9 + 16 = 25$, which is 5^2. Thus, 4 checks and is the solution.

5. **State.** Each brace reaches a height of 4 ft.

Do Exercise 5.

5. Reach of a Ladder. Twila has a 26-ft ladder leaning against her house. If the bottom of the ladder is 10 ft from the base of the house, how high does the ladder reach?

Answer

5. 24 ft

EXAMPLE 6 *Ladder Settings.* A ladder of length 13 ft is placed against a building in such a way that the distance from the top of the ladder to the ground is 7 ft more than the distance from the bottom of the ladder to the building. Find both distances.

1. **Familiarize.** We first make a drawing. The ladder and the missing distances form the hypotenuse and the legs of a right triangle. We let $x =$ the length of the side (leg) across the bottom, in feet. Then $x + 7 =$ the length of the other side (leg). The hypotenuse has length 13 ft.

2. **Translate.** Since a right triangle is formed, we can use the Pythagorean theorem:

$$a^2 + b^2 = c^2$$
$$x^2 + (x + 7)^2 = 13^2. \quad \text{Substituting}$$

3. **Solve.** We solve the equation as follows:

$x^2 + (x^2 + 14x + 49) = 169$	Squaring the binomial and 13
$2x^2 + 14x + 49 = 169$	Collecting like terms
$2x^2 + 14x + 49 - 169 = 169 - 169$	Subtracting 169 to get 0 on one side
$2x^2 + 14x - 120 = 0$	Simplifying
$2(x^2 + 7x - 60) = 0$	Factoring out a common factor
$x^2 + 7x - 60 = 0$	Dividing by 2
$(x + 12)(x - 5) = 0$	Factoring
$x + 12 = 0 \quad or \quad x - 5 = 0$	Using the principle of zero products
$x = -12 \quad or \quad x = 5.$	

4. **Check.** The negative integer -12 cannot be the length of a side. When $x = 5$, $x + 7 = 12$, and $5^2 + 12^2 = 13^2$. Thus, 5 and 12 check.

5. **State.** The distance from the top of the ladder to the ground is 12 ft. The distance from the bottom of the ladder to the building is 5 ft.

Do Exercise 6.

6. Right-Triangle Geometry.
The length of one leg of a right triangle is 1 m longer than the other. The length of the hypotenuse is 5 m. Find the lengths of the legs.

Answer

6. 3 m, 4 m

Translating for Success

1. *Angle Measures.* The measures of the angles of a triangle are three consecutive integers. Find the measures of the angles.

2. *Rectangle Dimensions.* The area of a rectangle is 3599 ft². The length is 2 ft longer than the width. Find the dimensions of the rectangle.

3. *Sales Tax.* Claire paid $40,704 for a new SUV. This included 6% for sales tax. How much did the SUV cost before tax?

4. *Wire Cutting.* A 180-m wire is cut into three pieces. The third piece is 2 m longer than the first. The second is two-thirds as long as the first. How long is each piece?

5. *Perimeter.* The perimeter of a rectangle is 240 ft. The length is 2 ft greater than the width. Find the length and the width.

The goal of these matching questions is to practice step (2), *Translate*, of the five-step problem-solving process. Translate each word problem to an equation and select a correct translation from equations A–O.

A. $2x \cdot x = 288$

B. $x(x + 60) = 7021$

C. $59 = x \cdot 60$

D. $x^2 + (x + 2)^2 = 3599$

E. $x^2 + (x + 70)^2 = 130^2$

F. $6\% \cdot x = 40{,}704$

G. $2(x + 2) + 2x = 240$

H. $\frac{1}{2}x(x - 1) = 1770$

I. $x + \frac{2}{3}x + (x + 2) = 180$

J. $59\% \cdot x = 60$

K. $x + 6\% \cdot x = 40{,}704$

L. $2x^2 + x = 288$

M. $x(x + 2) = 3599$

N. $x^2 + 60 = 7021$

O. $x + (x + 1) + (x + 2) = 180$

Answers on page A-15

6. *Cell-Phone Tower.* A guy wire on a cell-phone tower is 130 ft long and is attached to the top of the tower. The height of the tower is 70 ft longer than the distance from the point on the ground where the wire is attached to the bottom of the tower. Find the height of the tower.

7. *Sales Meeting Attendance.* PTQ Corporation holds a sales meeting in Tucson. Of the 60 employees, 59 of them attend the meeting. What percent attend the meeting?

8. *Dimensions of a Pool.* A rectangular swimming pool is twice as long as it is wide. The area of the surface is 288 ft². Find the dimensions of the pool.

9. *Dimensions of a Triangle.* The height of a triangle is 1 cm less than the length of the base. The area of the triangle is 1770 cm². Find the height and the length of the base.

10. *Width of a Rectangle.* The length of a rectangle is 60 ft longer than the width. Find the width if the area of the rectangle is 7021 ft².

Solve.

1. *Dimensions of a Painting.* A rectangular painting is three times as long as it is wide. The area of the picture is 588 in². Find the dimensions of the painting.

2. *Area of a Garden.* The length of a rectangular garden is 4 m greater than the width. The area of the garden is 96 m². Find the length and the width.

3. *Furnishings.* A rectangular table in Arlo's House of Tunes is six times as long as it is wide. The area of the table is 24 ft². Find the length and the width of the table.

4. *Design.* The screen of the TI-84 Plus graphing calculator is nearly rectangular. The length of the rectangle is 2 cm more than the width. If the area of the rectangle is 24 cm², find the length and the width.

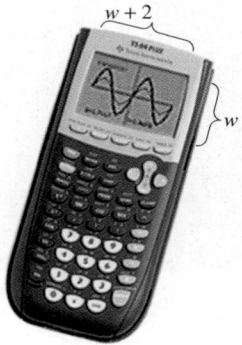

5. *Dimensions of a Triangle.* A triangle is 10 cm wider than it is tall. The area is 28 cm². Find the height and the base.

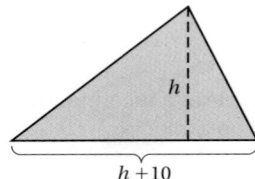

6. *Dimensions of a Triangle.* The height of a triangle is 3 cm less than the length of the base. The area of the triangle is 35 cm². Find the height and the length of the base.

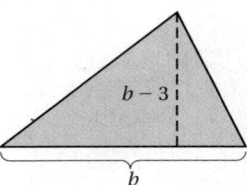

7. *Road Design.* A triangular traffic island has a base half as long as its height. The island has an area of 64 m². Find the base and the height.

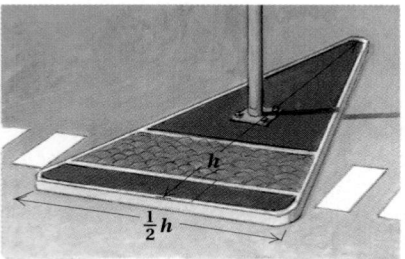

8. *Dimensions of a Sail.* The height of the jib sail on a Lightning sailboat is 5 ft greater than the length of its "foot." The area of the sail is 42 ft². Find the length of the foot and the height of the sail.

Games in a League. Use $x^2 - x = N$ for Exercises 9–12.

9. A chess league has 14 teams. What is the total number of games to be played if each team plays every other team twice?

10. A women's volleyball league has 23 teams. What is the total number of games to be played if each team plays every other team twice?

11. A slow-pitch softball league plays a total of 132 games. How many teams are in the league if each team plays every other team twice?

12. A basketball league plays a total of 90 games. How many teams are in the league if each team plays every other team twice?

Handshakes. Dr. Benton wants to investigate the potential spread of germs by contact. She knows that the number of possible handshakes within a group of x people, assuming each person shakes every other person's hand only once, is given by

$$N = \tfrac{1}{2}(x^2 - x).$$

Use this formula for Exercises 13–16.

13. There are 100 people at a party. How many handshakes are possible?

14. There are 40 people at a meeting. How many handshakes are possible?

15. Everyone at a meeting shook hands with each other. There were 300 handshakes in all. How many people were at the meeting?

16. Everyone at a party shook hands with each other. There were 153 handshakes in all. How many people were at the party?

17. *Toasting.* During a toast at the party celebrating James's fourth birthday, there were 190 "clicks" of paper cups. How many people took part in the toast?

18. *High-Fives.* After the Wildcats won the city baseball championship, all the teammates exchanged "high-fives." Altogether there were 66 high-fives. How many players were there?

19. *Consecutive Page Numbers.* The product of the page numbers on two facing pages of a book is 210. Find the page numbers.

20. *Consecutive Page Numbers.* The product of the page numbers on two facing pages of a book is 420. Find the page numbers.

21. The product of two consecutive even integers is 168. Find the integers. (See Section 2.6.)

22. The product of two consecutive even integers is 224. Find the integers. (See Section 2.6.)

23. The product of two consecutive odd integers is 255. Find the integers. (See Section 2.6.)

24. The product of two consecutive odd integers is 143. Find the integers. (See Section 2.6.)

25. *Right-Triangle Geometry.* The length of one leg of a right triangle is 8 ft. The length of the hypotenuse is 2 ft longer than the other leg. Find the length of the hypotenuse and the other leg.

26. *Right-Triangle Geometry.* The length of one leg of a right triangle is 24 ft. The length of the other leg is 16 ft shorter than the hypotenuse. Find the length of the hypotenuse and the other leg.

27. *Roadway Design.* Elliott Street is 24 ft wide when it ends at Main Street in Brattleboro, Vermont. A 40-ft long diagonal crosswalk allows pedestrians to cross Main Street to or from either corner of Elliott Street (see the figure). Determine the width of Main Street.

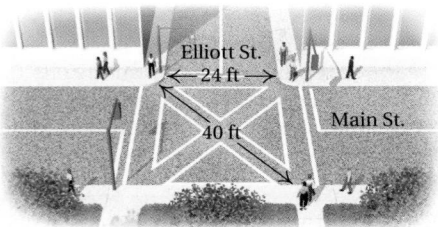

28. *Sailing.* The mainsail of a Lightning sailboat is a right triangle in which the hypotenuse is called the leech. If a 24-ft tall mainsail has a leech length of 26 ft and if Dacron® sailcloth costs $10 per square foot, find the cost of a new mainsail.

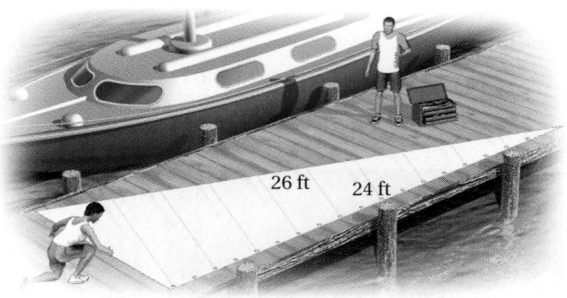

29. *Lookout Tower.* The diagonal braces in a lookout tower are 15 ft long and span a distance of 12 ft. How high does each brace reach vertically?

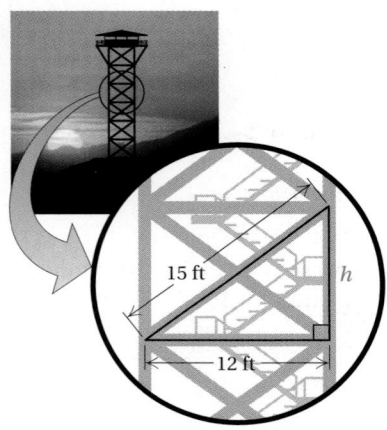

30. *Aviation.* Engine failure forced Geraldine to pilot her Cessna 150 to an emergency landing. To land, Geraldine's plane glided 17,000 ft over a 15,000-ft stretch of deserted highway. From what altitude did the descent begin?

31. *Architecture.* An architect has allocated a rectangular space of 264 ft² for a square dining room and a 10-ft wide kitchen, as shown in the figure. Find the dimensions of each room.

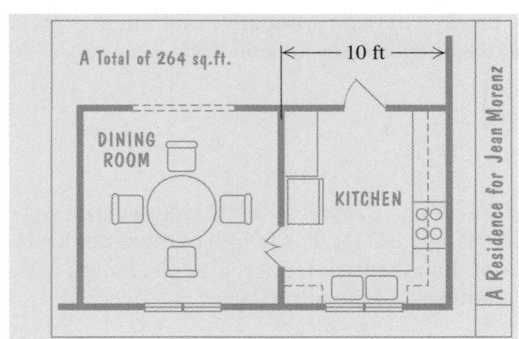

32. *Guy Wire.* The guy wire on a TV antenna is 1 m longer than the height of the antenna. If the guy wire is anchored 3 m from the foot of the antenna, how tall is the antenna?

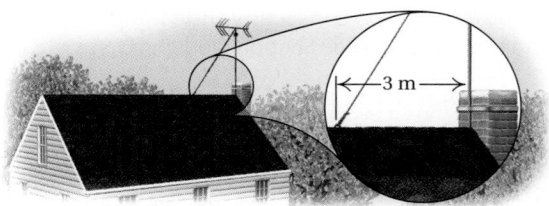

Rocket Launch. A model rocket is launched with an initial velocity of 180 ft/sec. Its height h, in feet, after t seconds is given by the formula

$$h = 180t - 16t^2.$$

Use this formula for Exercises 33 and 34.

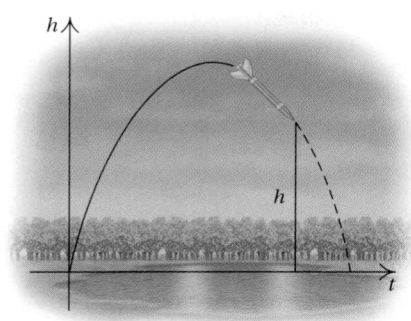

33. After how many seconds will the rocket first reach a height of 464 ft?

34. After how many seconds from launching will the rocket again be at that same height of 464 ft? (See Exercise 33.)

35. The sum of the squares of two consecutive odd positive integers is 74. Find the integers.

36. The sum of the squares of two consecutive odd positive integers is 130. Find the integers.

Skill Maintenance

In each of Exercises 37–44, fill in the blank with the correct term from the given list. Some of the choices may not be used and some may be used more than once.

37. To _____ a polynomial is to express it as a product. [5.1b]

38. A(n) _____ of a polynomial P is a polynomial that can be used to express P as a product. [5.1b]

39. A factorization of a polynomial is an expression that names that polynomial as a(n) _____ . [5.1b]

40. When factoring, always look first for a(n) _____ . [5.1b]

41. The expression $-5x^2 + 8x - 7$ is an example of a(n) _____ . [4.3i]

42. The _____ asserts that when dividing with exponential notation, if the bases are the same, keep the base and subtract the exponent of the denominator from the exponent of the numerator. [4.1e]

43. For the graph of the equation $4x - 3y = 12$, the pair $(0, -4)$ is known as the _____ . [3.3a]

44. For the graph of the equation $4x - 3y = 12$, the _____ is $\frac{4}{3}$. [3.4b]

quotient rule

product rule

slope

common factor

common multiple

factor

x-intercept

y-intercept

binomial

trinomial

quotient

product

Synthesis

45. *Telephone Service.* Use the information in the figure below to determine the height of the telephone pole.

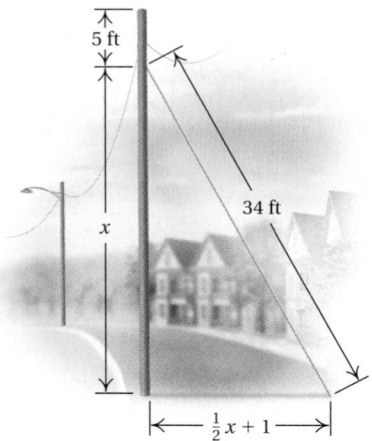

46. *Roofing.* A *square* of shingles covers 100 ft² of surface area. How many squares will be needed to reshingle the roof of the house shown?

47. *Pool Sidewalk.* A cement walk of constant width is built around a 20-ft by 40-ft rectangular pool. The total area of the pool and the walk is 1500 ft². Find the width of the walk.

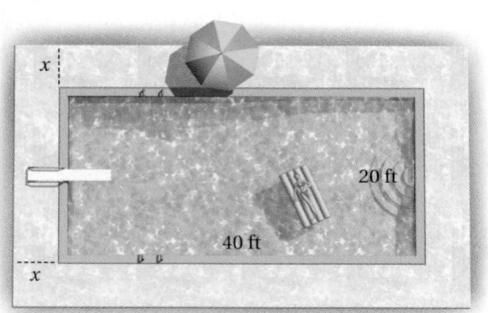

48. *Rain-Gutter Design.* An open rectangular gutter is made by turning up the sides of a piece of metal 20 in. wide. The area of the cross-section of the gutter is 50 in². Find the depth of the gutter.

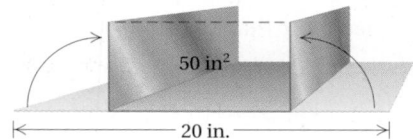

49. *Dimensions of an Open Box.* A rectangular piece of cardboard is twice as long as it is wide. A 4-cm square is cut out of each corner, and the sides are turned up to make a box with an open top. The volume of the box is 616 cm³. Find the original dimensions of the cardboard.

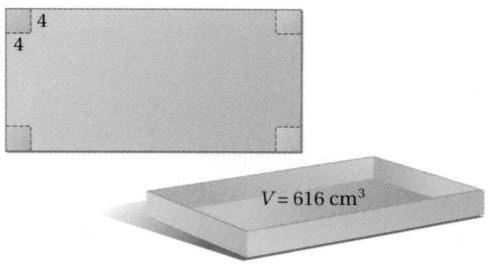

50. Solve for x.

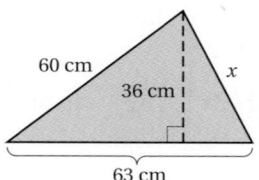

51. *Dimensions of a Closed Box.* The total surface area of a closed box is 350 in². The box is 9 in. high and has a square base and lid. Find the length of a side of the base.

52. The ones digit of a number less than 100 is 4 greater than the tens digit. The sum of the number and the product of the digits is 58. Find the number.

Summary and Review

Key Terms and Properties

greatest common factor (GCF),
p. 318
factor, p. 319
factorization, p. 319
factoring by grouping, p. 323
leading coefficient, p. 327

FOIL method, p. 337
ac-method, p. 345
trinomial square, or perfect-square
trinomial, p. 353
difference of squares, p. 355
sum of cubes, p. 363

difference of cubes, p. 363
quadratic equation, p. 377
right triangle, p. 389
hypotenuse, p. 389
legs of a right triangle, p. 389

Factoring Formulas:

$$A^2 - B^2 = (A + B)(A - B),$$
$$A^2 + 2AB + B^2 = (A + B)^2,$$
$$A^2 - 2AB + B^2 = (A - B)^2$$

The Principle of Zero Products: An equation $ab = 0$ is true if and only if $a = 0$ is true or $b = 0$ is true, or both are true.

The Pythagorean Theorem: $a^2 + b^2 = c^2$

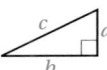

Concept Reinforcement

Determine whether each statement is true or false.

_____ **1.** Every polynomial with four terms can be factored by grouping. [5.1c]

_____ **2.** When factoring $x^2 + 5x + 6$, we need consider only positive pairs of factors of 6. [5.2a]

_____ **3.** A product is 0 if and only if all the factors are 0. [5.8a]

_____ **4.** If the principle of zero products is to be used, one side of the equation must be 0. [5.8b]

Important Concepts

Objective 5.1a Find the greatest common factor, the GCF, of monomials.

Example Find the GCF of $15x^4y^2$, $-18x$, and $12x^3y$.

$$15x^4y^2 = 3 \cdot 5 \cdot x^4 \cdot y^2;$$
$$-18x = -1 \cdot 2 \cdot 3 \cdot 3 \cdot x;$$
$$12x^3y = 2 \cdot 2 \cdot 3 \cdot x^3 \cdot y$$

Each coefficient has a factor of 3. There are no other common prime factors. The GCF of the powers of x is x because 1 is the smallest exponent of x. The GCF of the powers of y is 1 because $18x$ has no y-factor. Thus the GCF is $3 \cdot x \cdot 1$, or $3x$.

Practice Exercise

1. Find the GCF of $8x^3y^2$, $-20xy^3$, and $32x^2y$.

Objective 5.1b   Factor polynomials when the terms have a common factor, factoring out the greatest common factor.

Example   Factor: $16y^4 + 8y^3 - 24y^2$.

The *largest* common factor is $8y^2$.

$$16y^4 + 8y^3 - 24y^2 = (8y^2)(2y^2) + (8y^2)(y) - (8y^2)(3)$$
$$= 8y^2(2y^2 + y - 3)$$

Practice Exercise

2. Factor $27x^5 - 9x^3 + 18x^2$, factoring out the largest common factor.

Objective 5.1c   Factor certain expressions with four terms using factoring by grouping.

Example   Factor $6x^3 + 4x^2 - 15x - 10$ by grouping.

$$6x^3 + 4x^2 - 15x - 10 = (6x^3 + 4x^2) + (-15x - 10)$$
$$= 2x^2(3x + 2) - 5(3x + 2)$$
$$= (3x + 2)(2x^2 - 5)$$

Practice Exercise

3. Factor $z^3 - 3z^2 + 4z - 12$ by grouping.

Objective 5.2a   Factor trinomials of the type $x^2 + bx + c$ by examining the constant term c.

Example   Factor: $x^2 - x - 12$.

Since the constant term, -12, is negative, we look for a factorization of -12 in which one factor is positive and one factor is negative. The sum of the factors must be the coefficient of the middle term, -1, so the negative factor must have the larger absolute value. The possible pairs of factors that meet these criteria are $1, -12$ and $2, -6$ and $3, -4$. The numbers we need are 3 and -4:

$$x^2 - x - 12 = (x + 3)(x - 4).$$

Practice Exercise

4. Factor: $x^2 + 6x + 8$.

Objective 5.3a   Factor trinomials of the type $ax^2 + bx + c, a \neq 1$, using the FOIL method.

Example   Factor $2y^3 + 5y^2 - 3y$.

1) Factor out the largest common factor, y:

$y(2y^2 + 5y - 3).$

Now we factor $2y^2 + 5y - 3$.

2) Because $2y^2$ factors as $2y \cdot y$, we have this possibility for a factorization:

$(2y + \quad)(y + \quad).$

3) There are two pairs of factors of -3 and each can be written in two ways:

$$3, -1 \qquad -3, 1$$
$$\text{and} \quad -1, 3 \qquad 1, -3.$$

4) From steps (2) and (3), we see that there are 4 possibilities for factorizations. We look for **O**utside and **I**nside products for which the sum is the middle term, $5y$. We try some possibilities and find that the factorization of $2y^2 + 5y - 3$ is $(2y - 1)(y + 3)$.

We must include the common factor to get a factorization of the original trinomial:

$$2y^3 + 5y^2 - 3y = y(2y - 1)(y + 3).$$

Practice Exercise

5. Factor: $6z^2 - 21z - 12$.

Objective 5.4a Factor trinomials of the type $ax^2 + bx + c$, $a \neq 1$, using the *ac*-method.

Example Factor $5x^2 + 7x - 6$ using the *ac*-method.

1) There is no common factor (other than 1 or −1).

2) Multiply the leading coefficient 5 and the constant, −6:
$$5(-6) = -30.$$

3) Look for a factorization of −30 in which the sum of the factors is the coefficient of the middle term, 7. One number will be positive and the other will be negative. Since their sum, 7, is positive, the positive number will have the larger absolute value. The numbers we need are 10 and −3.

4) Split the middle term, writing it as a sum or a difference using the factors found in step (3):
$$7x = 10x - 3x.$$

5) Factor by grouping:
$$
\begin{aligned}
5x^2 + 7x - 6 &= 5x^2 + 10x - 3x - 6 \\
&= 5x(x + 2) - 3(x + 2) \\
&= (x + 2)(5x - 3).
\end{aligned}
$$

6) Check: $(x + 2)(5x - 3) = 5x^2 + 7x - 6.$

Practice Exercise

6. Factor $6y^2 + 7y - 3$ using the *ac*-method.

Objective 5.5b Factor trinomial squares.

Example Factor: $9x^2 - 12x + 4$.
$$
\begin{aligned}
9x^2 - 12x + 4 &= (3x)^2 - 2 \cdot 3x \cdot 2 + 2^2 \\
&= (3x - 2)^2
\end{aligned}
$$

Practice Exercise

7. Factor: $4x^2 + 4x + 1$.

Objective 5.5d Factor differences of squares, being careful to factor completely.

Example Factor: $b^6 - b^2$.
$$
\begin{aligned}
b^6 - b^2 &= b^2(b^4 - 1) = b^2(b^2 + 1)(b^2 - 1) \\
&= b^2(b^2 + 1)(b + 1)(b - 1)
\end{aligned}
$$

Practice Exercise

8. Factor $18x^2 - 8$ completely.

Objective 5.6a Factor sums or differences of cubes.

Examples Factor: **(a)** $w^3 - 512$ and **(b)** $125a^3 + b^3$.

a) $w^3 - 512 = w^3 - 8^3$
$$\qquad\qquad = (w - 8)(w^2 + 8w + 64)$$

b) $125a^3 + b^3 = (5a)^3 + b^3$
$$\qquad\qquad = (5a + b)(25a^2 - 5ab + b^2)$$

Practice Exercises

9. Factor: $27 - 125x^3$.

10. Factor: $\dfrac{1}{8}q^3 + 8a^3$.

Objective 5.8b Solve quadratic equations by factoring and then using the principle of zero products.

Example Solve: $x^2 - 3x = 28$.
$$
\begin{aligned}
x^2 - 3x - 28 &= 0 \quad \text{Subtracting 28} \\
(x + 4)(x - 7) &= 0 \\
x + 4 = 0 \quad &or \quad x - 7 = 0 \\
x = -4 \quad &or \qquad x = 7
\end{aligned}
$$
The solutions are −4 and 7.

Practice Exercise

11. Solve: $x^2 + 4x = 5$.

Review Exercises

Find the GCF.   [5.1a]

1. $-15y^2,\ 25y^6$

2. $12x^3,\ -60x^2y,\ 36xy$

Factor completely.   [5.7a]

3. $5 - 20x^6$

4. $x^2 - 3x$

5. $9x^2 - 4$

6. $x^2 + 4x - 12$

7. $x^2 + 14x + 49$

8. $6x^3 + 12x^2 + 3x$

9. $x^3 + x^2 + 3x + 3$

10. $6x^2 - 5x + 1$

11. $x^4 - 81$

12. $9x^3 + 12x^2 - 45x$

13. $2x^2 - 50$

14. $x^4 + 4x^3 - 2x - 8$

15. $16x^4 - 1$

16. $8x^6 - 32x^5 + 4x^4$

17. $75 + 12x^2 + 60x$

18. $x^2 + 9$

19. $x^3 - x^2 - 30x$

20. $8x^3 - 125$

21. $9x^2 + 25 - 30x$

22. $6x^2 - 28x - 48$

23. $x^2 - 6x + 9$

24. $2x^2 - 7x - 4$

25. $18x^2 - 12x + 2$

26. $3x^2 - 27$

27. $15 - 8x + x^2$

28. $25x^2 - 20x + 4$

29. $49b^{10} + 4a^8 - 28a^4b^5$

30. $x^2y^2 + xy - 12$

31. $12a^2 + 84ab + 147b^2$

32. $m^2 + 5m + mt + 5t$

33. $32x^4 - 128y^4z^4$

34. $5y^3 + 40t^3$

Solve.   [5.8a, b]

35. $(x - 1)(x + 3) = 0$

36. $x^2 + 2x - 35 = 0$

37. $x^2 + 4x = 0$

38. $3x^2 + 2 = 5x$

39. $x^2 = 64$

40. $16 = x(x - 6)$

Find the x-intercepts of the graph of each equation. [5.8b]

41. $y = x^2 + 9x + 20$

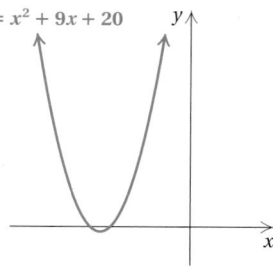

42. $y = 2x^2 - 7x - 15$

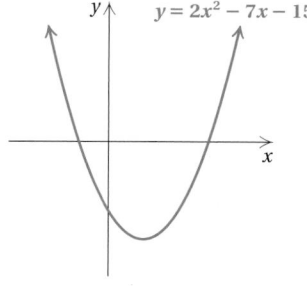

Solve. [5.9a]

43. *Sharks' Teeth.* Sharks' teeth are shaped like triangles. The height of a tooth of a great white shark is 1 cm longer than the base. The area is 15 cm^2. Find the height and the base.

44. The product of two consecutive even integers is 288. Find the integers.

45. The product of two consecutive odd integers is 323. Find the integers.

46. *Tree Supports.* A duckbill-anchor system is used to support a newly planted Bradford pear tree. Each cable is 5 ft long. The distance from the base of the tree to the point on the ground where each cable is anchored is 1 ft more than the distance from the base of the tree to the point where the cable is attached to the tree. Find both distances.

5 ft

47. If the sides of a square are lengthened by 3 km, the area increases to 81 km^2. Find the length of a side of the original square.

48. Factor: $x^2 - 9x + 8$. Which of the following is one factor? [5.2a], [5.7a]

A. $(x + 1)$ **B.** $(x - 1)$
C. $(x + 8)$ **D.** $(x - 4)$

49. Factor $15x^2 + 5x - 20$ completely. Which of the following is one factor? [5.3a], [5.4a], [5.7a]

A. $(3x + 4)$ **B.** $(3x - 4)$
C. $(5x - 5)$ **D.** $(15x + 20)$

Synthesis

Solve. [5.9a]

50. The pages of a book measure 15 cm by 20 cm. Margins of equal width surround the printing on each page and constitute one-half of the area of the page. Find the width of the margins.

15 cm

20 cm

51. The cube of a number is the same as twice the square of the number. Find all such numbers.

52. The length of a rectangle is two times its width. When the length is increased by 20 in. and the width is decreased by 1 in., the area is 160 in². Find the original length and width.

Solve. [5.8b]

53. $x^2 + 25 = 0$

54. $(x - 2)(x + 3)(2x - 5) = 0$

55. $(x - 3)4x^2 + 3x(x - 3) - (x - 3)10 = 0$

56. Find a polynomial in factored form for the shaded area in the figure below. Leave the answer in terms of π. [5.1b]

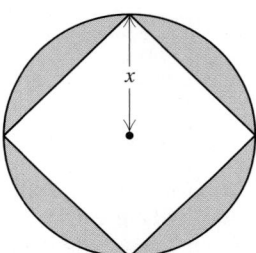

x

Understanding Through Discussion and Writing

1. Gwen factors $x^3 - 8x^2 + 15x$ as $(x^2 - 5x)(x - 3)$. Is she wrong? Why or why not? What advice would you offer? [5.2a]

2. After a test, Josh told a classmate that he was sure he had not written any incorrect factorizations. How could he be certain? [5.7a]

3. Kelly factored $16 - 8x + x^2$ as $(x - 4)^2$, while Tony factored it as $(4 - x)^2$. Evaluate each expression for several values of x. Then explain why both answers are correct. [5.5b]

4. What is wrong with the following? Explain the correct method of solution. [5.8b]

$$(x - 3)(x + 4) = 8$$
$$x - 3 = 8 \quad or \quad x + 4 = 8$$
$$x = 11 \quad or \qquad x = 4$$

5. What is incorrect about solving $x^2 = 3x$ by dividing by x on both sides? [5.8b]

6. An archaeologist has measuring sticks of 3 ft, 4 ft, and 5 ft. Explain how she could draw a 7-ft by 9-ft rectangle on a piece of land being excavated. [5.9a]

CHAPTER
5

Test

For Extra Help

CHAPTER
Test Prep
VIDEOS

Step-by-step test solutions are found on the Chapter Test Prep Videos available via the Video Resources on DVD, in *MyMathLab* , and on YouTube (search "BittingerIntroInter" and click on "Channels").

1. Find the GCF: $28x^3, 48x^7$.

Factor completely.

2. $x^2 - 7x + 10$

3. $x^2 + 25 - 10x$

4. $6y^2 - 8y^3 + 4y^4$

5. $x^3 + x^2 + 2x + 2$

6. $x^2 - 5x$

7. $x^3 + 2x^2 - 3x$

8. $28x - 48 + 10x^2$

9. $4x^2 - 9$

10. $x^2 - x - 12$

11. $6m^3 + 9m^2 + 3m$

12. $3w^2 - 75$

13. $60x + 45x^2 + 20$

14. $3x^4 - 48$

15. $49x^2 - 84x + 36$

16. $5x^2 - 26x + 5$

17. $x^4 + 2x^3 - 3x - 6$

18. $80 - 5x^4$

19. $6t^3 + 9t^2 - 15t$

20. $4x^2 - 4x - 15$

21. $3m^2 - 9mn - 30n^2$

22. $1000a^3 - 27b^3$

Solve.

23. $x^2 - 3x = 0$

24. $2x^2 = 32$

25. $x^2 - x - 20 = 0$

26. $2x^2 + 7x = 15$

27. $x(x - 3) = 28$

Find the *x*-intercepts of the graph of each equation.

28.

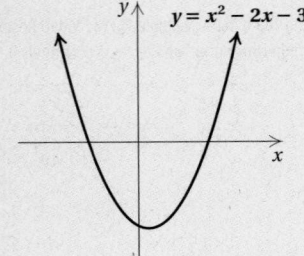

$y = x^2 - 2x - 35$

29.

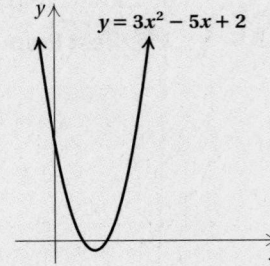

$y = 3x^2 - 5x + 2$

Solve.

30. The length of a rectangle is 2 m more than the width. The area of the rectangle is 48 m². Find the length and the width.

31. The base of a triangle is 6 cm greater than twice the height. The area is 28 cm². Find the height and the base.

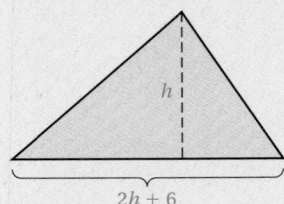

h

$2h + 6$

32. *Masonry Corner.*   A mason wants to be sure he has a right angle corner of a building's foundation. He marks a point 3 ft from the corner along one wall and another point 4 ft from the corner along the other wall. If the corner is a right angle, what should the distance be between the two marked points?

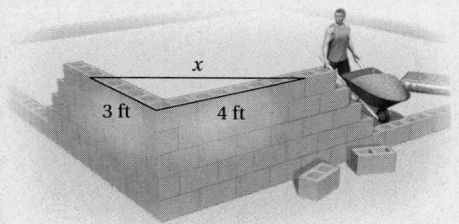

x

3 ft   4 ft

33. Factor $2y^4 - 32$ completely. Which of the following is one factor?

A. $(y + 2)$ **B.** $(y + 4)$
C. $(y^2 - 4)$ **D.** $(2y^2 + 8)$

Synthesis

34. The length of a rectangle is five times its width. When the length is decreased by 3 m and the width is increased by 2 m, the area of the new rectangle is 60 m². Find the original length and width.

35. Factor: $(a + 3)^2 - 2(a + 3) - 35$.

36. Solve: $20x(x + 2)(x - 1) = 5x^3 - 24x - 14x^2$.

37. If $x + y = 4$ and $x - y = 6$, then $x^2 - y^2 =$ which of the following?

A. 2   **B.** 10
C. 34   **D.** 24

Use either $<$ or $>$ for $\square$ to write a true sentence.

1. $\dfrac{2}{3} \square \dfrac{5}{7}$

2. $-\dfrac{4}{7} \square -\dfrac{8}{11}$

Compute and simplify.

3. $2.06 + (-4.79) - (-3.08)$

4. $5.652 \div (-3.6)$

5. $\left(\dfrac{2}{9}\right)\left(-\dfrac{3}{8}\right)\left(\dfrac{6}{7}\right)$

6. $\dfrac{21}{5} \div \left(-\dfrac{7}{2}\right)$

Simplify.

7. $[3x + 2(x - 1)] - [2x - (x + 3)]$

8. $1 - [14 + 28 \div 7 - (6 + 9 \div 3)]$

9. $(2x^2 y^{-1})^3$

10. $\dfrac{3x^5}{4x^3} \cdot \dfrac{-2x^{-3}}{9x^2}$

11. Add: $(2x^2 - 3x^3 + x - 4) + (x^4 - x - 5x^2)$.

12. Subtract: $(2x^2 y^2 + xy - 2xy^2) - (2xy - 2xy^2 + x^2 y)$.

13. Divide: $(x^3 + 2x^2 - x + 1) \div (x - 1)$.

Multiply.

14. $(2t - 3)^2$

15. $(x^2 - 3)(x^2 + 3)$

16. $(2x + 4)(3x - 4)$

17. $2x(x^3 + 3x^2 + 4x)$

18. $(2y - 1)(2y^2 + 3y + 4)$

19. $\left(x + \dfrac{2}{3}\right)\left(x - \dfrac{2}{3}\right)$

Factor.

20. $x^2 + 2x - 8$

21. $4x^2 - 25$

22. $3x^3 - 4x^2 + 3x - 4$

23. $x^2 - 26x + 169$

24. $75x^2 - 108y^2$

25. $6x^2 - 13x - 63$

26. $x^4 - 2x^2 - 3$

27. $4y^3 - 6y^2 - 4y + 6$

28. $6p^2 + pq - q^2$

29. $10x^3 + 52x^2 + 10x$

30. $49x^3 - 42x^2 + 9x$

31. $3x^2 + 5x - 4$

32. $75x^3 + 27x$

33. $3x^8 - 48y^8$

34. $14x^2 + 28 + 42x$

35. $2x^5 - 2x^3 + x^2 - 1$

Solve.

36. $3x - 5 = 2x + 10$

37. $3y + 4 > 5y - 8$

38. $(x - 15)\left(x + \dfrac{1}{4}\right) = 0$

39. $-98x(x + 37) = 0$

40. $x^3 + x^2 = 25x + 25$

41. $2x^2 = 72$

42. $9x^2 + 1 = 6x$

43. $x^2 + 17x + 70 = 0$

44. $14y^2 = 21y$

45. $1.6 - 3.5x = 0.9$

46. $(x + 3)(x - 4) = 8$

47. $1.5x - 3.6 \le 1.3x + 0.4$

48. $2x - [3x - (2x + 3)] = 3x + [4 - (2x + 1)]$

49. $y = mx + b$, for m

Solve.

50. The sum of two consecutive even integers is 102. Find the integers.

51. The product of two consecutive even integers is 360. Find the integers.

52. The length of a rectangular window is 3 ft longer than the height. The area of the window is 18 ft². Find the length and the height.

53. The length of a rectangular lot is 200 m longer than the width. The perimeter of the lot is 1000 m. Find the dimensions of the lot.

54. Money is borrowed at 12% simple interest. After 1 year, $7280 pays off the loan. How much was originally borrowed?

55. The length of one leg of a right triangle is 15 m. The length of the other leg is 9 m shorter than the length of the hypotenuse. Find the length of the hypotenuse.

56. A 100-m wire is cut into three pieces. The second piece is twice as long as the first piece. The third piece is one-third as long as the first piece. How long is each piece?

57. After a 25% price reduction, a pair of shoes is on sale for $21.75. What was the price before reduction?

58. The height of a triangle is 2 cm more than the base. The area of the triangle is 144 cm². Find the height and the base.

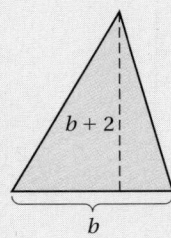

59. Find the intercepts. Then graph the equation.
$$3x + 4y = -12$$

Synthesis

Solve.

60. $(x + 3)(x - 5) \le (x + 2)(x - 1)$

61. $\dfrac{x - 3}{2} - \dfrac{2x + 5}{26} = \dfrac{4x + 11}{13}$

62. $(x + 1)^2 = 25$

Factor.

63. $x^2(x - 3) - x(x - 3) - 2(x - 3)$

64. $4a^2 - 4a + 1 - 9b^2 - 24b - 16$

Solve.

65. Find c such that the polynomial will be the square of a binomial: $cx^2 - 40x + 16$.

66. The length of the radius of a circle is increased by 2 cm to form a new circle. The area of the new circle is four times the area of the original circle. Find the length of the radius of the original circle.

Rational Expressions and Equations

Real-World Application

A company that prepares and sells gift boxes and baskets of fruit must order quantities of fruit larger than what they need to allow for selecting fruit that meets their quality standards. The packing-room supervisor keeps records and notes that approximately 87 pears from a shipment of 1000 do not meet the company standards. Over the holidays, a shipment of 3200 pears is ordered. How many pears can the company expect will not meet the quality required?

This problem appears as Example 5 in Section 6.8.

6.1

Multiplying and Simplifying Rational Expressions

OBJECTIVES

a Find all numbers for which a rational expression is not defined.

b Multiply a rational expression by 1, using an expression such as A/A.

c Simplify rational expressions by factoring the numerator and the denominator and removing factors of 1.

d Multiply rational expressions and simplify.

SKILL TO REVIEW
Objective 5.3a: Factor trinomials of the type $ax^2 + bx + c$, $a \neq 1$, using the FOIL method.

Factor.

1. $2x^2 - x - 21$

2. $40x^2 - 43x - 6$

Find all numbers for which the rational expression is not defined.

1. $\dfrac{16}{x - 3}$

2. $\dfrac{2x - 7}{x^2 + 5x - 24}$

3. $\dfrac{x + 5}{8}$

Answers

Skill to Review:
1. $(2x - 7)(x + 3)$ **2.** $(5x - 6)(8x + 1)$

Margin Exercises:
1. 3 **2.** $-8, 3$ **3.** None

a Rational Expressions and Replacements

Rational numbers are quotients of integers. Some examples are

$$\frac{2}{3}, \quad \frac{4}{-5}, \quad \frac{-8}{17}, \quad \frac{563}{1}.$$

The following are called **rational expressions** or **fraction expressions**. They are quotients, or ratios, of polynomials:

$$\frac{3}{4}, \quad \frac{z}{6}, \quad \frac{5}{x + 2}, \quad \frac{t^2 + 3t - 10}{7t^2 - 4}.$$

A rational expression is also a division. For example,

$$\frac{3}{4} \quad \text{means} \quad 3 \div 4 \quad \text{and} \quad \frac{x - 8}{x + 2} \quad \text{means} \quad (x - 8) \div (x + 2).$$

Because rational expressions indicate division, we must be careful to avoid denominators of zero. When a variable is replaced with a number that produces a denominator equal to zero, the rational expression is not defined. For example, in the expression

$$\frac{x - 8}{x + 2},$$

when x is replaced with -2, the denominator is 0, and the expression is *not* defined:

$$\frac{x - 8}{x + 2} = \frac{-2 - 8}{-2 + 2} = \frac{-10}{0}. \longleftarrow \text{Division by 0 is not defined.}$$

When x is replaced with a number other than -2, such as 3, the expression *is* defined because the denominator is nonzero:

$$\frac{x - 8}{x + 2} = \frac{3 - 8}{3 + 2} = \frac{-5}{5} = -1.$$

EXAMPLE 1 Find all numbers for which the rational expression

$$\frac{x + 4}{x^2 - 3x - 10}$$

is not defined.

The value of the numerator has no bearing on whether or not a rational expression is defined. To determine which numbers make the rational expression not defined, we set the *denominator* equal to 0 and solve:

$$
\begin{aligned}
x^2 - 3x - 10 &= 0 \\
(x - 5)(x + 2) &= 0 \qquad \text{Factoring} \\
x - 5 = 0 \quad &\text{or} \quad x + 2 = 0 \qquad \text{Using the principle of zero} \\
& \qquad\qquad\qquad\quad \text{products (See Section 5.8.)} \\
x = 5 \quad &\text{or} \qquad x = -2.
\end{aligned}
$$

The rational expression is not defined for the replacement numbers 5 and -2.

Do Margin Exercises 1–3.

b Multiplying by 1

We multiply rational expressions in the same way that we multiply fraction notation in arithmetic. For a review, see Appendix B. We know that

$$\frac{3}{7} \cdot \frac{2}{5} = \frac{3 \cdot 2}{7 \cdot 5} = \frac{6}{35}.$$

MULTIPLYING RATIONAL EXPRESSIONS

To multiply rational expressions, multiply numerators and multiply denominators:

$$\frac{A}{B} \cdot \frac{C}{D} = \frac{AC}{BD}.$$

For example,

$$\frac{x - 2}{3} \cdot \frac{x + 2}{x + 7} = \frac{(x - 2)(x + 2)}{3(x + 7)}.$$ Multiplying the numerators and the denominators

Note that we leave the numerator, $(x - 2)(x + 2)$, and the denominator, $3(x + 7)$, in factored form because it is easier to simplify if we do not multiply. In order to learn to simplify, we first need to consider multiplying the rational expression by 1.

Any rational expression with the same numerator and denominator is a symbol for 1:

$$\frac{19}{19} = 1, \qquad \frac{x + 8}{x + 8} = 1, \qquad \frac{3x^2 - 4}{3x^2 - 4} = 1, \qquad \frac{-1}{-1} = 1.$$

EQUIVALENT EXPRESSIONS

Expressions that have the same value for all allowable (or meaningful) replacements are called **equivalent expressions**.

We can multiply by 1 to obtain an *equivalent expression*. At this point, we select expressions for 1 arbitrarily. Later, we will have a system for our choices when we add and subtract.

EXAMPLES Multiply.

2. $\dfrac{3x + 2}{x + 1} \cdot 1 = \dfrac{3x + 2}{x + 1} \cdot \dfrac{2x}{2x} = \dfrac{(3x + 2)2x}{(x + 1)2x}$ Using the identity property of 1. We arbitrarily choose $2x/2x$ as a symbol for 1.

3. $\dfrac{x + 2}{x - 7} \cdot \dfrac{x + 3}{x + 3} = \dfrac{(x + 2)(x + 3)}{(x - 7)(x + 3)}$ We arbitrarily choose $(x + 3)/(x + 3)$ as a symbol for 1.

4. $\dfrac{2 + x}{2 - x} \cdot \dfrac{-1}{-1} = \dfrac{(2 + x)(-1)}{(2 - x)(-1)}$ Using $(-1)/(-1)$ as a symbol for 1

Do Exercises 4–6.

Multiply.

4. $\dfrac{2x + 1}{3x - 2} \cdot \dfrac{x}{x}$

5. $\dfrac{x + 1}{x - 2} \cdot \dfrac{x + 2}{x + 2}$

6. $\dfrac{x - 8}{x - y} \cdot \dfrac{-1}{-1}$

Answers

4. $\dfrac{(2x + 1)x}{(3x - 2)x}$ **5.** $\dfrac{(x + 1)(x + 2)}{(x - 2)(x + 2)}$

6. $\dfrac{(x - 8)(-1)}{(x - y)(-1)}$

(c) Simplifying Rational Expressions

Simplifying rational expressions is similar to simplifying fraction expressions in arithmetic. For a review, see Appendix B. We know, for example, that an expression like $\frac{15}{40}$ can be simplified as follows:

$$\frac{15}{40} = \frac{3 \cdot 5}{8 \cdot 5}$$ Factoring the numerator and the denominator. Note the common factor, 5.

$$= \frac{3}{8} \cdot \frac{5}{5}$$ Factoring the fraction expression

$$= \frac{3}{8} \cdot 1 \qquad \frac{5}{5} = 1$$

$$= \frac{3}{8}.$$ Using the identity property of 1, or "removing a factor of 1"

Similar steps are followed when simplifying rational expressions: We factor and remove a factor of 1, using the fact that

$$\frac{ab}{cb} = \frac{a}{c} \cdot \frac{b}{b} = \frac{a}{c} \cdot 1 = \frac{a}{c}.$$

In algebra, instead of simplifying a fraction like

$$\frac{15}{40},$$

we may need to simplify an expression like

$$\frac{x^2 - 16}{x + 4}.$$

Just as factoring is important in simplifying in arithmetic, so too is it important in simplifying rational expressions. The factoring we use most is the factoring of polynomials, which we studied in Chapter 5.

To simplify, we can do the reverse of multiplying. We factor the numerator and the denominator and "remove" a factor of 1.

EXAMPLE 5 Simplify: $\frac{8x^2}{24x}$.

$$\frac{8x^2}{24x} = \frac{8 \cdot x \cdot x}{3 \cdot 8 \cdot x}$$ Factoring the numerator and the denominator. Note the common factor, $8x$.

$$= \frac{8x}{8x} \cdot \frac{x}{3}$$ Factoring the rational expression

$$= 1 \cdot \frac{x}{3} \qquad \frac{8x}{8x} = 1$$

$$= \frac{x}{3}$$ We removed a factor of 1.

Do Exercises 7 and 8.

Simplify.

7. $\frac{5y}{y}$

8. $\frac{9x^2}{36x}$

Answers

7. 5   8. $\frac{x}{4}$

EXAMPLES Simplify.

6. $\dfrac{5a + 15}{10} = \dfrac{5(a + 3)}{5 \cdot 2}$ Factoring the numerator and the denominator

$\qquad = \dfrac{5}{5} \cdot \dfrac{a + 3}{2}$ Factoring the rational expression

$\qquad = 1 \cdot \dfrac{a + 3}{2} \qquad \dfrac{5}{5} = 1$

$\qquad = \dfrac{a + 3}{2}$ Removing a factor of 1

7. $\dfrac{6a + 12}{7a + 14} = \dfrac{6(a + 2)}{7(a + 2)}$ Factoring the numerator and the denominator

$\qquad = \dfrac{6}{7} \cdot \dfrac{a + 2}{a + 2}$ Factoring the rational expression

$\qquad = \dfrac{6}{7} \cdot 1 \qquad \dfrac{a + 2}{a + 2} = 1$

$\qquad = \dfrac{6}{7}$ Removing a factor of 1

8. $\dfrac{6x^2 + 4x}{2x^2 + 2x} = \dfrac{2x(3x + 2)}{2x(x + 1)}$ Factoring the numerator and the denominator

$\qquad = \dfrac{2x}{2x} \cdot \dfrac{3x + 2}{x + 1}$ Factoring the rational expression

$\qquad = 1 \cdot \dfrac{3x + 2}{x + 1} \qquad \dfrac{2x}{2x} = 1$

$\qquad = \dfrac{3x + 2}{x + 1}$ Removing a factor of 1

--------------------------------- *Caution!* ---------------------------------

Note that you *cannot* simplify further by removing the x's because x is not a *factor* of the entire numerator, $3x + 2$, and the entire denominator, $x + 1$.

----

9. $\dfrac{x^2 + 3x + 2}{x^2 - 1} = \dfrac{(x + 2)(x + 1)}{(x + 1)(x - 1)}$ Factoring the numerator and the denominator

$\qquad = \dfrac{x + 1}{x + 1} \cdot \dfrac{x + 2}{x - 1}$ Factoring the rational expression

$\qquad = 1 \cdot \dfrac{x + 2}{x - 1} \qquad \dfrac{x + 1}{x + 1} = 1$

$\qquad = \dfrac{x + 2}{x - 1}$ Removing a factor of 1

Canceling

You may have encountered canceling when working with rational expressions. With great concern, we mention it as a possible way to speed up your work. Our concern is that canceling be done with care and understanding. Example 9 might have been done faster as follows:

$$\frac{x^2 + 3x + 2}{x^2 - 1} = \frac{(x + 2)(x + 1)}{(x + 1)(x - 1)} \qquad \text{Factoring the numerator and the denominator}$$

$$= \frac{(x + 2)\cancel{(x + 1)}}{\cancel{(x + 1)}(x - 1)} \qquad \text{When a factor of 1 is noted, it is canceled, as shown: } \frac{x + 1}{x + 1} = 1.$$

$$= \frac{x + 2}{x - 1}. \qquad \text{Simplifying}$$

Caution!

The difficulty with canceling is that it is often applied incorrectly, as in the following situations:

$$\frac{\cancel{x} + 3}{\cancel{x}} = 3; \qquad \frac{\cancel{4} + 1}{\cancel{4} + 2} = \frac{1}{2}; \qquad \frac{1\cancel{5}}{\cancel{5}4} = \frac{1}{4}.$$

Wrong!   Wrong!   Wrong!

In each of these situations, the expressions canceled were *not* factors of 1. Factors are parts of products. For example, in $2 \cdot 3$, 2 and 3 are factors, but in $2 + 3$, 2 and 3 are *not* factors. If you can't factor, you can't cancel. If in doubt, don't cancel!

Simplify.

9. $\dfrac{2x^2 + x}{3x^2 + 2x}$

10. $\dfrac{x^2 - 1}{2x^2 - x - 1}$

11. $\dfrac{7x + 14}{7}$

12. $\dfrac{12y + 24}{48}$

Do Exercises 9–12.

Opposites in Rational Expressions

Expressions of the form $a - b$ and $b - a$ are **opposites** of each other. When either of these binomials is multiplied by -1, the result is the other binomial:

$$\left.\begin{array}{l} -1(a - b) = -a + b = b + (-a) = b - a; \\ -1(b - a) = -b + a = a + (-b) = a - b. \end{array}\right\} \begin{array}{l}\text{Multiplication by } -1 \\ \text{reverses the order in} \\ \text{which subtraction} \\ \text{occurs.}\end{array}$$

Consider, for example,

$$\frac{x - 4}{4 - x}.$$

At first glance, it appears as though the numerator and the denominator do not have any common factors other than 1. But $x - 4$ and $4 - x$ are opposites, or additive inverses, of each other. Thus we can rewrite one as the opposite of the other by factoring out a -1.

Simplify.

13. $\dfrac{x - 8}{8 - x}$

14. $\dfrac{c - d}{d - c}$

15. $\dfrac{-x - 7}{x + 7}$

EXAMPLE 10   Simplify: $\dfrac{x - 4}{4 - x}$.

$$\frac{x - 4}{4 - x} = \frac{x - 4}{-(x - 4)} = \frac{1(x - 4)}{-1(x - 4)} \qquad \begin{array}{l}4 - x = -(x - 4); 4 - x \text{ and} \\ x - 4 \text{ are opposites.}\end{array}$$

$$= -1 \cdot \frac{x - 4}{x - 4} \qquad 1/-1 = -1$$

$$= -1 \cdot 1$$

$$= -1$$

Do Exercises 13–15.

(d) Multiplying and Simplifying

We try to simplify after we multiply. That is why we leave the numerator and the denominator in factored form.

EXAMPLE 11 Multiply and simplify: $\dfrac{5a^3}{4} \cdot \dfrac{2}{5a}$.

$$\dfrac{5a^3}{4} \cdot \dfrac{2}{5a} = \dfrac{5a^3(2)}{4(5a)} \qquad \text{Multiplying the numerators and the denominators}$$

$$= \dfrac{5 \cdot a \cdot a \cdot a \cdot 2}{2 \cdot 2 \cdot 5 \cdot a} \qquad \text{Factoring the numerator and the denominator}$$

$$= \dfrac{\cancel{5} \cdot \cancel{a} \cdot a \cdot a \cdot 2}{2 \cdot 2 \cdot \cancel{5} \cdot \cancel{a}} \qquad \text{Removing a factor of 1: } \dfrac{2 \cdot 5 \cdot a}{2 \cdot 5 \cdot a} = 1$$

$$= \dfrac{a^2}{2} \qquad \text{Simplifying}$$

EXAMPLE 12 Multiply and simplify: $\dfrac{x^2 + 6x + 9}{x^2 - 4} \cdot \dfrac{x - 2}{x + 3}$.

$$\dfrac{x^2 + 6x + 9}{x^2 - 4} \cdot \dfrac{x - 2}{x + 3} = \dfrac{(x^2 + 6x + 9)(x - 2)}{(x^2 - 4)(x + 3)} \qquad \text{Multiplying the numerators and the denominators}$$

$$= \dfrac{(x + 3)(x + 3)(x - 2)}{(x + 2)(x - 2)(x + 3)} \qquad \text{Factoring the numerator and the denominator}$$

$$= \dfrac{\cancel{(x + 3)}(x + 3)\cancel{(x - 2)}}{(x + 2)\cancel{(x - 2)}\cancel{(x + 3)}} \qquad \begin{array}{l}\text{Removing a factor of 1:}\\ \dfrac{(x + 3)(x - 2)}{(x + 3)(x - 2)} = 1\end{array}$$

$$= \dfrac{x + 3}{x + 2} \qquad \text{Simplifying}$$

Do Exercise 16.

16. Multiply and simplify:
$$\dfrac{a^2 - 4a + 4}{a^2 - 9} \cdot \dfrac{a + 3}{a - 2}.$$

EXAMPLE 13 Multiply and simplify: $\dfrac{x^2 + x - 2}{15} \cdot \dfrac{5}{2x^2 - 3x + 1}$.

$$\dfrac{x^2 + x - 2}{15} \cdot \dfrac{5}{2x^2 - 3x + 1} = \dfrac{(x^2 + x - 2)5}{15(2x^2 - 3x + 1)} \qquad \begin{array}{l}\text{Multiplying the numerators and the denominators}\end{array}$$

$$= \dfrac{(x + 2)(x - 1)5}{5(3)(x - 1)(2x - 1)} \qquad \begin{array}{l}\text{Factoring the numerator and the denominator}\end{array}$$

$$= \dfrac{(x + 2)\cancel{(x - 1)}\cancel{5}}{\cancel{5}(3)\cancel{(x - 1)}(2x - 1)} \qquad \begin{array}{l}\text{Removing a factor of 1: } \dfrac{(x - 1)5}{(x - 1)5} = 1\end{array}$$

$$= \dfrac{x + 2}{3(2x - 1)} \qquad \text{Simplifying}$$

You need not carry out this multiplication.

Do Exercise 17.

17. Multiply and simplify:
$$\dfrac{x^2 - 25}{6} \cdot \dfrac{3}{x + 5}.$$

Answers

16. $\dfrac{a - 2}{a - 3}$ **17.** $\dfrac{x - 5}{2}$

a   Find all numbers for which the rational expression is not defined.

1. $\dfrac{-3}{2x}$

2. $\dfrac{24}{-8y}$

3. $\dfrac{5}{x-8}$

4. $\dfrac{y-4}{y+6}$

5. $\dfrac{3}{2y+5}$

6. $\dfrac{x^2-9}{4x-15}$

7. $\dfrac{x^2+11}{x^2-3x-28}$

8. $\dfrac{p^2-9}{p^2-7p+10}$

9. $\dfrac{m^3-2m}{m^2-25}$

10. $\dfrac{7-3x+x^2}{49-x^2}$

11. $\dfrac{x-4}{3}$

12. $\dfrac{x^2-25}{14}$

b   Multiply. Do not simplify. Note that in each case you are multiplying by 1.

13. $\dfrac{4x}{4x}\cdot\dfrac{3x^2}{5y}$

14. $\dfrac{5x^2}{5x^2}\cdot\dfrac{6y^3}{3z^4}$

15. $\dfrac{2x}{2x}\cdot\dfrac{x-1}{x+4}$

16. $\dfrac{2a-3}{5a+2}\cdot\dfrac{a}{a}$

17. $\dfrac{3-x}{4-x}\cdot\dfrac{-1}{-1}$

18. $\dfrac{x-5}{5-x}\cdot\dfrac{-1}{-1}$

19. $\dfrac{y+6}{y+6}\cdot\dfrac{y-7}{y+2}$

20. $\dfrac{x^2+1}{x^3-2}\cdot\dfrac{x-4}{x-4}$

c   Simplify.

21. $\dfrac{8x^3}{32x}$

22. $\dfrac{4x^2}{20x}$

23. $\dfrac{48p^7q^5}{18p^5q^4}$

24. $\dfrac{-76x^8y^3}{-24x^4y^3}$

25. $\dfrac{4x-12}{4x}$

26. $\dfrac{5a-40}{5}$

27. $\dfrac{3m^2 + 3m}{6m^2 + 9m}$

28. $\dfrac{4y^2 - 2y}{5y^2 - 5y}$

29. $\dfrac{a^2 - 9}{a^2 + 5a + 6}$

30. $\dfrac{t^2 - 25}{t^2 + t - 20}$

31. $\dfrac{a^2 - 10a + 21}{a^2 - 11a + 28}$

32. $\dfrac{x^2 - 2x - 8}{x^2 - x - 6}$

33. $\dfrac{x^2 - 25}{x^2 - 10x + 25}$

34. $\dfrac{x^2 + 8x + 16}{x^2 - 16}$

35. $\dfrac{a^2 - 1}{a - 1}$

36. $\dfrac{t^2 - 1}{t + 1}$

37. $\dfrac{x^2 + 1}{x + 1}$

38. $\dfrac{m^2 + 9}{m + 3}$

39. $\dfrac{6x^2 - 54}{4x^2 - 36}$

40. $\dfrac{8x^2 - 32}{4x^2 - 16}$

41. $\dfrac{6t + 12}{t^2 - t - 6}$

42. $\dfrac{4x + 32}{x^2 + 9x + 8}$

43. $\dfrac{2t^2 + 6t + 4}{4t^2 - 12t - 16}$

44. $\dfrac{3a^2 - 9a - 12}{6a^2 + 30a + 24}$

45. $\dfrac{t^2 - 4}{(t + 2)^2}$

46. $\dfrac{m^2 - 36}{(m - 6)^2}$

47. $\dfrac{6 - x}{x - 6}$

48. $\dfrac{t - 3}{3 - t}$

49. $\dfrac{a - b}{b - a}$

50. $\dfrac{y - x}{-x + y}$

51. $\dfrac{6t - 12}{2 - t}$

52. $\dfrac{5a - 15}{3 - a}$

53. $\dfrac{x^2 - 1}{1 - x}$

54. $\dfrac{a^2 - b^2}{b^2 - a^2}$

d   Multiply and simplify.

55. $\dfrac{4x^3}{3x} \cdot \dfrac{14}{x}$

56. $\dfrac{18}{x^3} \cdot \dfrac{5x^2}{6}$

57. $\dfrac{3c}{d^2} \cdot \dfrac{4d}{6c^3}$

58. $\dfrac{3x^2 y}{2} \cdot \dfrac{4}{xy^3}$

59. $\dfrac{x + 4}{x} \cdot \dfrac{x^2 - 3x}{x^2 + x - 12}$

60. $\dfrac{t^2}{t^2 - 4} \cdot \dfrac{t^2 - 5t + 6}{t^2 - 3t}$

61. $\dfrac{a^2 - 9}{a^2} \cdot \dfrac{a^2 - 3a}{a^2 + a - 12}$

62. $\dfrac{x^2 + 10x - 11}{x^2 - 1} \cdot \dfrac{x + 1}{x + 11}$

63. $\dfrac{4a^2}{3a^2 - 12a + 12} \cdot \dfrac{3a - 6}{2a}$

64. $\dfrac{5v + 5}{v - 2} \cdot \dfrac{v^2 - 4v + 4}{v^2 - 1}$

65. $\dfrac{t^4 - 16}{t^4 - 1} \cdot \dfrac{t^2 + 1}{t^2 + 4}$

66. $\dfrac{x^4 - 1}{x^4 - 81} \cdot \dfrac{x^2 + 9}{x^2 + 1}$

67. $\dfrac{(x+4)^3}{(x+2)^3} \cdot \dfrac{x^2+4x+4}{x^2+8x+16}$

68. $\dfrac{(t-2)^3}{(t-1)^3} \cdot \dfrac{t^2-2t+1}{t^2-4t+4}$

69. $\dfrac{5a^2-180}{10a^2-10} \cdot \dfrac{20a+20}{2a-12}$

70. $\dfrac{2t^2-98}{4t^2-4} \cdot \dfrac{8t+8}{16t-112}$

Skill Maintenance

Graph.

71. $x+y=-1$ [3.3a]

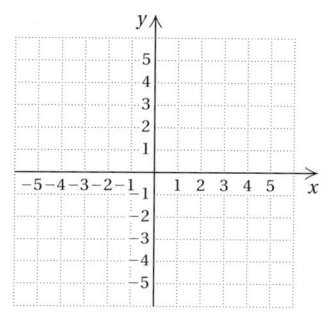

72. $y = -\dfrac{7}{2}$ [3.3b]

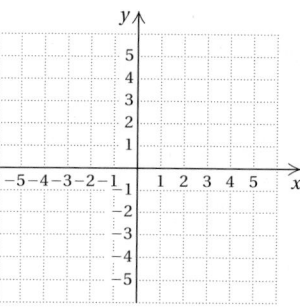

Solve.

73. *Consecutive Even Integers.* The product of two consecutive even integers is 360. Find the integers. [5.9a]

74. *Chemistry.* About 5 L of oxygen can be dissolved in 100 L of water at 0°C. This is 1.6 times the amount that can be dissolved in the same volume of water at 20°C. How much oxygen can be dissolved in 100 L at 20°C? [2.6a]

Factor. [5.7a]

75. x^2-x-56

76. $a^2-16a+64$

77. $x^5-2x^4-35x^3$

78. $2y^3-10y^2+y-5$

79. $16-t^4$

80. $10x^2+80x+70$

81. $x^2-9x+14$

82. x^2+x+7

83. $16x^2-40xy+25y^2$

84. $a^2-9ab+14b^2$

Synthesis

Simplify.

85. $\dfrac{x^4-16y^4}{(x^2+4y^2)(x-2y)}$

86. $\dfrac{(a-b)^2}{b^2-a^2}$

87. $\dfrac{t^4-1}{t^4-81} \cdot \dfrac{t^2-9}{t^2+1} \cdot \dfrac{(t-9)^2}{(t+1)^2}$

88. $\dfrac{(t+2)^3}{(t+1)^3} \cdot \dfrac{t^2+2t+1}{t^2+4t+4} \cdot \dfrac{t+1}{t+2}$

89. $\dfrac{x^2-y^2}{(x-y)^2} \cdot \dfrac{x^2-2xy+y^2}{x^2-4xy-5y^2}$

90. $\dfrac{x-1}{x^2+1} \cdot \dfrac{x^4-1}{(x-1)^2} \cdot \dfrac{x^2-1}{x^4-2x^2+1}$

91. Select any number x, multiply by 2, add 5, multiply by 5, subtract 25, and divide by 10. What do you get? Explain how this procedure can be used for a number trick.

6.2

Division and Reciprocals

OBJECTIVES

a Find the reciprocal of a rational expression.

b Divide rational expressions and simplify.

SKILL TO REVIEW
Objective 5.7a: Factor polynomials.

Factor.

1. $10 - 35x$
2. $3y^2 - 12w^2$

Find the reciprocal.

1. $\dfrac{7}{2}$

2. $\dfrac{x^2 + 5}{2x^3 - 1}$

3. $x - 5$

4. $\dfrac{1}{x^2 - 3}$

5. Divide: $\dfrac{3}{5} \div \dfrac{7}{10}$.

There is a similarity between what we do with rational expressions and what we do with rational numbers. In fact, after variables have been replaced with rational numbers, a rational expression represents a rational number.

a Finding Reciprocals

Two expressions are **reciprocals** of each other if their product is 1. The reciprocal of a rational expression is found by interchanging the numerator and the denominator.

EXAMPLES

1. The reciprocal of $\dfrac{2}{5}$ is $\dfrac{5}{2}$. $\left(\text{This is because } \dfrac{2}{5} \cdot \dfrac{5}{2} = \dfrac{10}{10} = 1.\right)$

2. The reciprocal of $\dfrac{2x^2 - 3}{x + 4}$ is $\dfrac{x + 4}{2x^2 - 3}$.

3. The reciprocal of $x + 2$ is $\dfrac{1}{x + 2}$. $\left(\text{Think of } x + 2 \text{ as } \dfrac{x + 2}{1}.\right)$

Do Margin Exercises 1–4.

b Division

We divide rational expressions in the same way that we divide fraction notation in arithmetic. For a review, see Appendix B.

> **DIVIDING RATIONAL EXPRESSIONS**
>
> To divide by a rational expression, multiply by its reciprocal:
>
> $$\frac{A}{B} \div \frac{C}{D} = \frac{A}{B} \cdot \frac{D}{C} = \frac{AD}{BC}.$$
>
> Then factor and, if possible, simplify.

EXAMPLE 4 Divide: $\dfrac{3}{4} \div \dfrac{9}{5}$.

$\dfrac{3}{4} \div \dfrac{9}{5} = \dfrac{3}{4} \cdot \dfrac{5}{9}$ Multiplying by the reciprocal of the divisor

$= \dfrac{3 \cdot 5}{4 \cdot 9} = \dfrac{3 \cdot 5}{2 \cdot 2 \cdot 3 \cdot 3}$ Factoring

$= \dfrac{3 \cdot 5}{2 \cdot 2 \cdot 3 \cdot 3}$ Removing a factor of 1: $\dfrac{3}{3} = 1$

$= \dfrac{5}{12}$ Simplifying

Do Exercise 5.

Answers

Skill to Review:
1. $5(2 - 7x)$ 2. $3(y - 2w)(y + 2w)$

Margin Exercises:
1. $\dfrac{2}{7}$ 2. $\dfrac{2x^3 - 1}{x^2 + 5}$ 3. $\dfrac{1}{x - 5}$
4. $x^2 - 3$ 5. $\dfrac{6}{7}$

EXAMPLE 5 Divide: $\dfrac{2}{x} \div \dfrac{3}{x}$.

$$\frac{2}{x} \div \frac{3}{x} = \frac{2}{x} \cdot \frac{x}{3} \qquad \text{Multiplying by the reciprocal of the divisor}$$

$$= \frac{2 \cdot x}{x \cdot 3} = \frac{2 \cdot \cancel{x}}{\cancel{x} \cdot 3} \qquad \text{Removing a factor of 1: } \frac{x}{x} = 1$$

$$= \frac{2}{3}$$

Do Exercise 6.

6. Divide: $\dfrac{x}{8} \div \dfrac{x}{5}$.

EXAMPLE 6 Divide: $\dfrac{x+1}{x+2} \div \dfrac{x-1}{x+3}$.

$$\frac{x+1}{x+2} \div \frac{x-1}{x+3} = \frac{x+1}{x+2} \cdot \frac{x+3}{x-1} \qquad \begin{array}{l}\text{Multiplying by the reciprocal}\\ \text{of the divisor}\end{array}$$

$$= \frac{(x+1)(x+3)}{(x+2)(x-1)} \left\}\right.$$

We usually do not carry out the multiplication in the numerator or the denominator. It is not wrong to do so, but the factored form is often more useful.

Do Exercise 7.

7. Divide:

$$\frac{x-3}{x+5} \div \frac{x+5}{x-2}.$$

EXAMPLE 7 Divide: $\dfrac{4}{x^2 - 7x} \div \dfrac{28x}{x^2 - 49}$.

$$\frac{4}{x^2 - 7x} \div \frac{28x}{x^2 - 49} = \frac{4}{x^2 - 7x} \cdot \frac{x^2 - 49}{28x} \qquad \begin{array}{l}\text{Multiplying by the}\\ \text{reciprocal}\end{array}$$

$$= \frac{4(x^2 - 49)}{(x^2 - 7x)(28x)}$$

$$= \frac{2 \cdot 2 \cdot (x - 7)(x + 7)}{x(x - 7) \cdot 2 \cdot 2 \cdot 7 \cdot x} \qquad \begin{array}{l}\text{Factoring the numerator}\\ \text{and the denominator}\end{array}$$

$$= \frac{2 \cdot 2 \cdot \cancel{(x - 7)}(x + 7)}{x\cancel{(x - 7)} \cdot 2 \cdot 2 \cdot 7 \cdot x} \qquad \begin{array}{l}\text{Removing a factor of 1:}\\ \dfrac{2 \cdot 2 \cdot (x - 7)}{2 \cdot 2 \cdot (x - 7)} = 1\end{array}$$

$$= \frac{x + 7}{7x^2}$$

Do Exercise 8.

8. Divide:

$$\frac{a^2 + 5a}{6} \div \frac{a^2 - 25}{18a}.$$

Answers

6. $\dfrac{5}{8}$ **7.** $\dfrac{(x-3)(x-2)}{(x+5)(x+5)}$ **8.** $\dfrac{3a^2}{a-5}$

EXAMPLE 8 Divide and simplify: $\dfrac{x + 1}{x^2 - 1} \div \dfrac{x + 1}{x^2 - 2x + 1}$.

$$\dfrac{x + 1}{x^2 - 1} \div \dfrac{x + 1}{x^2 - 2x + 1}$$

$$= \dfrac{x + 1}{x^2 - 1} \cdot \dfrac{x^2 - 2x + 1}{x + 1} \qquad \text{Multiplying by the reciprocal}$$

$$= \dfrac{(x + 1)(x^2 - 2x + 1)}{(x^2 - 1)(x + 1)}$$

$$= \dfrac{(x + 1)(x - 1)(x - 1)}{(x - 1)(x + 1)(x + 1)} \qquad \begin{array}{l}\text{Factoring the numerator}\\ \text{and the denominator}\end{array}$$

$$= \dfrac{\cancel{(x + 1)}\cancel{(x - 1)}(x - 1)}{\cancel{(x - 1)}\cancel{(x + 1)}(x + 1)} \qquad \text{Removing a factor of 1: } \dfrac{(x + 1)(x - 1)}{(x + 1)(x - 1)} = 1$$

$$= \dfrac{x - 1}{x + 1}$$

EXAMPLE 9 Divide and simplify: $\dfrac{x^2 - 2x - 3}{x^2 - 4} \div \dfrac{x + 1}{x + 5}$.

$$\dfrac{x^2 - 2x - 3}{x^2 - 4} \div \dfrac{x + 1}{x + 5}$$

$$= \dfrac{x^2 - 2x - 3}{x^2 - 4} \cdot \dfrac{x + 5}{x + 1} \qquad \text{Multiplying by the reciprocal}$$

$$= \dfrac{(x^2 - 2x - 3)(x + 5)}{(x^2 - 4)(x + 1)}$$

$$= \dfrac{(x - 3)(x + 1)(x + 5)}{(x - 2)(x + 2)(x + 1)} \qquad \begin{array}{l}\text{Factoring the numerator and}\\ \text{the denominator}\end{array}$$

$$= \dfrac{(x - 3)\cancel{(x + 1)}(x + 5)}{(x - 2)(x + 2)\cancel{(x + 1)}} \qquad \text{Removing a factor of 1: } \dfrac{x + 1}{x + 1} = 1$$

$$= \dfrac{(x - 3)(x + 5)}{(x - 2)(x + 2)} \quad \left\{\begin{array}{l}\text{You need not carry out the}\\ \text{multiplications in the numerator}\\ \text{and the denominator.}\end{array}\right.$$

Divide and simplify.

9. $\dfrac{x - 3}{x + 5} \div \dfrac{x + 2}{x + 5}$

10. $\dfrac{x^2 - 5x + 6}{x + 5} \div \dfrac{x + 2}{x + 5}$

11. $\dfrac{y^2 - 1}{y + 1} \div \dfrac{y^2 - 2y + 1}{y + 1}$

Do Exercises 9–11.

Answers

9. $\dfrac{x - 3}{x + 2}$ 10. $\dfrac{(x - 3)(x - 2)}{x + 2}$ 11. $\dfrac{y + 1}{y - 1}$

a Find the reciprocal.

1. $\dfrac{4}{x}$

2. $\dfrac{a+3}{a-1}$

3. $x^2 - y^2$

4. $x^2 - 5x + 7$

5. $\dfrac{1}{a+b}$

6. $\dfrac{x^2}{x^2-3}$

7. $\dfrac{x^2+2x-5}{x^2-4x+7}$

8. $\dfrac{(a-b)(a+b)}{(a+4)(a-5)}$

b Divide and simplify.

9. $\dfrac{2}{5} \div \dfrac{4}{3}$

10. $\dfrac{3}{10} \div \dfrac{3}{2}$

11. $\dfrac{2}{x} \div \dfrac{8}{x}$

12. $\dfrac{t}{3} \div \dfrac{t}{15}$

13. $\dfrac{a}{b^2} \div \dfrac{a^2}{b^3}$

14. $\dfrac{x^2}{y} \div \dfrac{x^3}{y^3}$

15. $\dfrac{a+2}{a-3} \div \dfrac{a-1}{a+3}$

16. $\dfrac{x-8}{x+9} \div \dfrac{x+2}{x-1}$

17. $\dfrac{x^2-1}{x} \div \dfrac{x+1}{x-1}$

18. $\dfrac{4y-8}{y+2} \div \dfrac{y-2}{y^2-4}$

19. $\dfrac{x+1}{6} \div \dfrac{x+1}{3}$

20. $\dfrac{a}{a-b} \div \dfrac{b}{a-b}$

21. $\dfrac{5x-5}{16} \div \dfrac{x-1}{6}$

22. $\dfrac{4y-12}{12} \div \dfrac{y-3}{3}$

23. $\dfrac{-6+3x}{5} \div \dfrac{4x-8}{25}$

24. $\dfrac{-12+4x}{4} \div \dfrac{-6+2x}{6}$

25. $\dfrac{a+2}{a-1} \div \dfrac{3a+6}{a-5}$

26. $\dfrac{t-3}{t+2} \div \dfrac{4t-12}{t+1}$

27. $\dfrac{x^2-4}{x} \div \dfrac{x-2}{x+2}$

28. $\dfrac{x+y}{x-y} \div \dfrac{x^2+y}{x^2-y^2}$

29. $\dfrac{x^2-9}{4x+12} \div \dfrac{x-3}{6}$

30. $\dfrac{a-b}{2a} \div \dfrac{a^2-b^2}{8a^3}$

31. $\dfrac{c^2+3c}{c^2+2c-3} \div \dfrac{c}{c+1}$

32. $\dfrac{y+5}{2y} \div \dfrac{y^2-25}{4y^2}$

33. $\dfrac{2y^2 - 7y + 3}{2y^2 + 3y - 2} \div \dfrac{6y^2 - 5y + 1}{3y^2 + 5y - 2}$

34. $\dfrac{x^2 + x - 20}{x^2 - 7x + 12} \div \dfrac{x^2 + 10x + 25}{x^2 - 6x + 9}$

35. $\dfrac{x^2 - 1}{4x + 4} \div \dfrac{2x^2 - 4x + 2}{8x + 8}$

36. $\dfrac{5t^2 + 5t - 30}{10t + 30} \div \dfrac{2t^2 - 8}{6t^2 + 36t + 54}$

Skill Maintenance

Solve.

37. Bonnie is taking an astronomy course. In order to receive an A, she must average at least 90 after four exams. Bonnie scored 96, 98, and 89 on the first three tests. Determine (in terms of an inequality) what scores on the last test will earn her an A. [2.8b]

38. *Triangle Dimensions.* The base of a triangle is 4 in. less than twice the height. The area is 35 in². Find the height and the base. [5.9a]

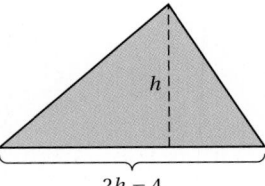

$2h - 4$

Subtract. [4.4c]

39. $(8x^3 - 3x^2 + 7) - (8x^2 + 3x - 5)$

40. $(3p^2 - 6pq + 7q^2) - (5p^2 - 10pq + 11q^2)$

Simplify. [4.2a, b]

41. $(2x^{-3}y^4)^2$

42. $(5x^6y^{-4})^3$

43. $\left(\dfrac{2x^3}{y^5}\right)^2$

44. $\left(\dfrac{a^{-3}}{b^4}\right)^5$

Synthesis

Simplify.

45. $\dfrac{3a^2 - 5ab - 12b^2}{3ab + 4b^2} \div (3b^2 - ab)$

46. $\dfrac{3x + 3y + 3}{9x} \div \dfrac{x^2 + 2xy + y^2 - 1}{x^4 + x^2}$

47. $\dfrac{a^2b^2 + 3ab^2 + 2b^2}{a^2b^4 + 4b^4} \div (5a^2 + 10a)$

48. The volume of this rectangular solid is $x - 3$. What is its height?

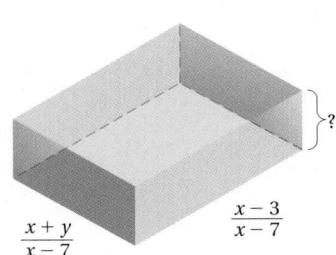

$\dfrac{x + y}{x - 7}$

$\dfrac{x - 3}{x - 7}$

6.3

Least Common Multiples and Denominators

a Least Common Multiples

To add when denominators are different, we first find a common denominator. For a review, see Appendixes A and B. We know, for example, that to add $\frac{5}{12}$ and $\frac{7}{30}$, we first look for the **least common multiple, LCM**, of 12 and 30. That number becomes the **least common denominator, LCD**. To find the LCM of 12 and 30, we factor:

$$12 = 2 \cdot 2 \cdot 3;$$
$$30 = 2 \cdot 3 \cdot 5.$$

The LCM is the number that has 2 as a factor twice, 3 as a factor once, and 5 as a factor once:

$$\text{LCM} = 2 \cdot 2 \cdot 3 \cdot 5 = 60.$$

12 is a factor of the LCM.

30 is a factor of the LCM.

> **FINDING LCMS**
>
> To find the LCM, use each factor the greatest number of times that it appears in any one factorization.

EXAMPLE 1   Find the LCM of 24 and 36.

$$\left.\begin{array}{l} 24 = 2 \cdot 2 \cdot 2 \cdot 3 \\ 36 = 2 \cdot 2 \cdot 3 \cdot 3 \end{array}\right\} \quad \text{LCM} = 2 \cdot 2 \cdot 2 \cdot 3 \cdot 3, \text{ or } 72$$

Do Margin Exercises 1–4.

b Adding Using the LCD

Let's finish adding $\frac{5}{12}$ and $\frac{7}{30}$:

$$\frac{5}{12} + \frac{7}{30} = \frac{5}{2 \cdot 2 \cdot 3} + \frac{7}{2 \cdot 3 \cdot 5}.$$

The least common denominator, LCD, is $2 \cdot 2 \cdot 3 \cdot 5$. To get the LCD in the first denominator, we need a 5. To get the LCD in the second denominator, we need another 2. We get these numbers by multiplying by forms of 1:

$$\frac{5}{12} + \frac{7}{30} = \frac{5}{2 \cdot 2 \cdot 3} \cdot \frac{5}{5} + \frac{7}{2 \cdot 3 \cdot 5} \cdot \frac{2}{2} \qquad \text{Multiplying by 1}$$

$$= \frac{25}{2 \cdot 2 \cdot 3 \cdot 5} + \frac{14}{2 \cdot 3 \cdot 5 \cdot 2} \qquad \begin{array}{l}\text{Each denominator is} \\ \text{now the LCD.}\end{array}$$

$$= \frac{39}{2 \cdot 2 \cdot 3 \cdot 5} \qquad \begin{array}{l}\text{Adding the numerators} \\ \text{and keeping the LCD}\end{array}$$

$$= \frac{3 \cdot 13}{2 \cdot 2 \cdot 3 \cdot 5} \qquad \begin{array}{l}\text{Factoring the numerator and} \\ \text{removing a factor of 1: } \frac{3}{3} = 1\end{array}$$

$$= \frac{13}{20}. \qquad \text{Simplifying}$$

OBJECTIVES

a Find the LCM of several numbers by factoring.

b Add fractions, first finding the LCD.

c Find the LCM of algebraic expressions by factoring.

SKILL TO REVIEW
Objective 5.5b: Factor trinomial squares.

Factor.
1. $x^2 - 16x + 64$
2. $4x^2 + 12x + 9$

Find the LCM by factoring.
1. 16, 18
2. 6, 12
3. 2, 5
4. 24, 30, 20

Answers

Skill to Review:
1. $(x - 8)^2$ **2.** $(2x + 3)^2$

Margin Exercises:
1. 144   **2.** 12   **3.** 10   **4.** 120

Add, first finding the LCD. Simplify if possible.

5. $\dfrac{3}{16} + \dfrac{1}{18}$

6. $\dfrac{1}{6} + \dfrac{1}{12}$

7. $\dfrac{1}{2} + \dfrac{3}{5}$

8. $\dfrac{1}{24} + \dfrac{1}{30} + \dfrac{3}{20}$

EXAMPLE 2 Add: $\dfrac{5}{12} + \dfrac{11}{18}$.

$$\left. \begin{array}{l} 12 = 2 \cdot 2 \cdot 3 \\ 18 = 2 \cdot 3 \cdot 3 \end{array} \right\} \quad \text{LCD} = 2 \cdot 2 \cdot 3 \cdot 3, \text{ or } 36$$

$$\frac{5}{12} + \frac{11}{18} = \frac{5}{2 \cdot 2 \cdot 3} \cdot \frac{3}{3} + \frac{11}{2 \cdot 3 \cdot 3} \cdot \frac{2}{2} = \frac{15 + 22}{2 \cdot 2 \cdot 3 \cdot 3} = \frac{37}{36}$$

Do Exercises 5–8.

c LCMs of Algebraic Expressions

To find the LCM of two or more algebraic expressions, we factor them. Then we use each factor the greatest number of times that it occurs in any one expression. In Section 6.4, each LCM will become an LCD used to add rational expressions.

EXAMPLE 3 Find the LCM of $12x$, $16y$, and $8xyz$.

$$\left. \begin{array}{l} 12x = 2 \cdot 2 \cdot 3 \cdot x \\ 16y = 2 \cdot 2 \cdot 2 \cdot 2 \cdot y \\ 8xyz = 2 \cdot 2 \cdot 2 \cdot x \cdot y \cdot z \end{array} \right\} \quad \begin{array}{l} \text{LCM} = 2 \cdot 2 \cdot 2 \cdot 2 \cdot 3 \cdot x \cdot y \cdot z \\ \qquad\quad = 48xyz \end{array}$$

EXAMPLE 4 Find the LCM of $x^2 + 5x - 6$ and $x^2 - 1$.

$$\left. \begin{array}{l} x^2 + 5x - 6 = (x + 6)(x - 1) \\ x^2 - 1 = (x + 1)(x - 1) \end{array} \right\} \quad \text{LCM} = (x + 6)(x - 1)(x + 1)$$

EXAMPLE 5 Find the LCM of $x^2 + 4$, $x + 1$, and 5.

These expressions do not share a common factor other than 1, so the LCM is their product:

$$5(x^2 + 4)(x + 1).$$

EXAMPLE 6 Find the LCM of $x^2 - 25$ and $2x - 10$.

$$\left. \begin{array}{l} x^2 - 25 = (x + 5)(x - 5) \\ 2x - 10 = 2(x - 5) \end{array} \right\} \quad \text{LCM} = 2(x + 5)(x - 5)$$

Find the LCM.

9. $12xy^2$, $15x^3y$

10. $y^2 + 5y + 4$, $y^2 + 2y + 1$

11. $t^2 + 16$, $t - 2$, 7

12. $x^2 + 2x + 1$, $3x^2 - 3x$, $x^2 - 1$

EXAMPLE 7 Find the LCM of $x^2 - 4y^2$, $x^2 - 4xy + 4y^2$, and $x - 2y$.

$$\left. \begin{array}{l} x^2 - 4y^2 = (x - 2y)(x + 2y) \\ x^2 - 4xy + 4y^2 = (x - 2y)(x - 2y) \\ x - 2y = x - 2y \end{array} \right\} \quad \begin{array}{l} \text{LCM} = (x + 2y)(x - 2y)(x - 2y) \\ \qquad\quad = (x + 2y)(x - 2y)^2 \end{array}$$

Do Exercises 9–12.

Answers

5. $\dfrac{35}{144}$ 6. $\dfrac{1}{4}$ 7. $\dfrac{11}{10}$ 8. $\dfrac{9}{40}$ 9. $60x^3y^2$
10. $(y + 1)^2(y + 4)$ 11. $7(t^2 + 16)(t - 2)$
12. $3x(x + 1)^2(x - 1)$

a Find the LCM.

1. 12, 27

2. 10, 15

3. 8, 9

4. 12, 18

5. 6, 9, 21

6. 8, 36, 40

7. 24, 36, 40

8. 4, 5, 20

9. 10, 100, 500

10. 28, 42, 60

b Add, first finding the LCD. Simplify if possible.

11. $\dfrac{7}{24} + \dfrac{11}{18}$

12. $\dfrac{7}{60} + \dfrac{2}{25}$

13. $\dfrac{1}{6} + \dfrac{3}{40}$

14. $\dfrac{5}{24} + \dfrac{3}{20}$

15. $\dfrac{1}{20} + \dfrac{1}{30} + \dfrac{2}{45}$

16. $\dfrac{2}{15} + \dfrac{5}{9} + \dfrac{3}{20}$

c Find the LCM.

17. $6x^2,\ 12x^3$

18. $2a^2b,\ 8ab^3$

19. $2x^2,\ 6xy,\ 18y^2$

20. $p^3q,\ p^2q,\ pq^2$

21. $2(y - 3),\ 6(y - 3)$

22. $5(m + 2),\ 15(m + 2)$

23. $t,\ t + 2,\ t - 2$

24. $y,\ y - 5,\ y + 5$

25. $x^2 - 4,\ x^2 + 5x + 6$

26. $x^2 - 4,\ x^2 - x - 2$

27. $t^3 + 4t^2 + 4t,\ t^2 - 4t$

28. $m^4 - m^2,\ m^3 - m^2$

29. $a + 1,\ (a - 1)^2,\ a^2 - 1$

30. $a^2 - 2ab + b^2,\ a^2 - b^2,\ 3a + 3b$

31. $m^2 - 5m + 6,\ m^2 - 4m + 4$

32. $2x^2 + 5x + 2,\ 2x^2 - x - 1$

33. $2 + 3x,\ 4 - 9x^2,\ 2 - 3x$

34. $9 - 4x^2,\ 3 + 2x,\ 3 - 2x$

35. $10v^2 + 30v,\ 5v^2 + 35v + 60$

36. $12a^2 + 24a,\ 4a^2 + 20a + 24$

37. $9x^3 - 9x^2 - 18x$, $6x^5 - 24x^4 + 24x^3$

38. $x^5 - 4x^3$, $x^3 + 4x^2 + 4x$

39. $x^5 + 4x^4 + 4x^3$, $3x^2 - 12$, $2x + 4$

40. $x^5 + 2x^4 + x^3$, $2x^3 - 2x$, $5x - 5$

41. $24w^4$, w^2, $10w^3$, w^6

42. t, $6t^4$, t^2, $15t^{15}$, $2t^3$

Skill Maintenance

Factor. [5.7a]

43. $x^2 - 6x + 9$

44. $6x^2 + 4x$

45. $x^2 - 9$

46. $x^2 + 4x - 21$

47. $x^2 + 6x + 9$

48. $x^2 - 4x - 21$

Complete the table below, finding the LCM, the GCF, and the product of each pair of expressions.   [4.5a], [5.1a], [6.3a]

	EXPRESSIONS	LCM	GCF	PRODUCT
	$12x^3$, $8x^2$	$24x^3$	$4x^2$	$96x^5$
49.	$40x^3$, $24x^4$			
50.	$16x^5$, $48x^6$			
51.	$20x^2$, $10x$			
52.	$12ab$, $16ab^3$			
53.	$10x^2$, $24x^3$			
54.	a^5, a^{15}			

Synthesis

55. Look for a pattern in Exercises 49–54. See if you can discover a formula connecting the LCM and the GCF.

56. *Running.*   Pedro and Maria leave the starting point of a fitness loop at the same time. Pedro jogs a lap in 6 min and Maria jogs one in 8 min. Assuming they continue to run at the same pace, when will they next meet at the starting place?

6.4

Adding Rational Expressions

a Adding Rational Expressions

We add rational expressions as we do rational numbers.

> **ADDING RATIONAL EXPRESSIONS WITH LIKE DENOMINATORS**
>
> To add when the denominators are the same, add the numerators and keep the same denominator. Then simplify if possible.

EXAMPLES Add.

1. $\dfrac{x}{x+1} + \dfrac{2}{x+1} = \dfrac{x+2}{x+1}$

2. $\dfrac{2x^2 + 3x - 7}{2x+1} + \dfrac{x^2 + x - 8}{2x+1} = \dfrac{(2x^2 + 3x - 7) + (x^2 + x - 8)}{2x+1}$

$= \dfrac{3x^2 + 4x - 15}{2x+1}$

$= \dfrac{(x+3)(3x-5)}{2x+1}$ Factoring the numerator to determine whether we can simplify

3. $\dfrac{x-5}{x^2-9} + \dfrac{2}{x^2-9} = \dfrac{(x-5)+2}{x^2-9} = \dfrac{x-3}{x^2-9}$

$= \dfrac{x-3}{(x-3)(x+3)}$ Factoring

$= \dfrac{1\cancel{(x-3)}}{\cancel{(x-3)}(x+3)}$ Removing a factor of 1: $\dfrac{x-3}{x-3} = 1$

$= \dfrac{1}{x+3}$ Simplifying

Do Margin Exercises 1–3.

When denominators are different, we find the least common denominator, LCD. The procedure we use follows.

> **ADDING RATIONAL EXPRESSIONS WITH DIFFERENT DENOMINATORS**
>
> To add rational expressions with different denominators:
>
> **1.** Find the LCM of the denominators. This is the least common denominator (LCD).
>
> **2.** For each rational expression, find an equivalent expression with the LCD. Multiply by 1 using an expression for 1 made up of factors of the LCD that are missing from the original denominator.
>
> **3.** Add the numerators. Write the sum over the LCD.
>
> **4.** Simplify if possible.

SKILL TO REVIEW
Objective 6.1c: Simplify rational expressions by factoring the numerator and the denominator and removing factors of 1.

Simplify.

1. $\dfrac{a^2 - b^2}{a+b}$

2. $\dfrac{x^2 - x - 6}{x^2 + 2x - 15}$

Add.

1. $\dfrac{5}{9} + \dfrac{2}{9}$

2. $\dfrac{3}{x-2} + \dfrac{x}{x-2}$

3. $\dfrac{4x+5}{x-1} + \dfrac{2x-1}{x-1}$

Answers

Skill to Review:

1. $a - b$ 2. $\dfrac{x+2}{x+5}$

Margin Exercises:

1. $\dfrac{7}{9}$ 2. $\dfrac{3+x}{x-2}$ 3. $\dfrac{6x+4}{x-1}$

EXAMPLE 4 Add: $\dfrac{5x^2}{8} + \dfrac{7x}{12}$.

First, we find the LCD:

$$\left.\begin{array}{l} 8 = 2 \cdot 2 \cdot 2 \\ 12 = 2 \cdot 2 \cdot 3 \end{array}\right\} \quad \text{LCD} = 2 \cdot 2 \cdot 2 \cdot 3, \text{ or } 24.$$

Compare the factorization $8 = 2 \cdot 2 \cdot 2$ with the factorization of the LCD, $24 = 2 \cdot 2 \cdot 2 \cdot 3$. The factor of 24 that is missing from 8 is 3. Compare $12 = 2 \cdot 2 \cdot 3$ and $24 = 2 \cdot 2 \cdot 2 \cdot 3$. The factor of 24 that is missing from 12 is 2.

We multiply each term by a symbol for 1 to get the LCD in each expression, and then add and, if possible, simplify:

$$\frac{5x^2}{8} + \frac{7x}{12} = \frac{5x^2}{2 \cdot 2 \cdot 2} + \frac{7x}{2 \cdot 2 \cdot 3}$$

$$= \frac{5x^2}{2 \cdot 2 \cdot 2} \cdot \frac{3}{3} + \frac{7x}{2 \cdot 2 \cdot 3} \cdot \frac{2}{2} \qquad \text{Multiplying by 1 to get the same denominators}$$

$$= \frac{15x^2}{24} + \frac{14x}{24} = \frac{15x^2 + 14x}{24} = \frac{x(15x + 14)}{24}.$$

EXAMPLE 5 Add: $\dfrac{3}{8x} + \dfrac{5}{12x^2}$.

First, we find the LCD:

$$\left.\begin{array}{l} 8x = 2 \cdot 2 \cdot 2 \cdot x \\ 12x^2 = 2 \cdot 2 \cdot 3 \cdot x \cdot x \end{array}\right\} \quad \text{LCD} = 2 \cdot 2 \cdot 2 \cdot 3 \cdot x \cdot x, \text{ or } 24x^2.$$

The factors of the LCD missing from $8x$ are 3 and x. The factor of the LCD missing from $12x^2$ is 2. We multiply each term by 1 to get the LCD in each expression, and then add and, if possible, simplify:

$$\frac{3}{8x} + \frac{5}{12x^2} = \frac{3}{8x} \cdot \frac{3 \cdot x}{3 \cdot x} + \frac{5}{12x^2} \cdot \frac{2}{2}$$

$$= \frac{9x}{24x^2} + \frac{10}{24x^2} = \frac{9x + 10}{24x^2}.$$

Do Exercises 4 and 5.

Add.

4. $\dfrac{3x}{16} + \dfrac{5x^2}{24}$

5. $\dfrac{3}{16x} + \dfrac{5}{24x^2}$

EXAMPLE 6 Add: $\dfrac{2a}{a^2 - 1} + \dfrac{1}{a^2 + a}$.

First, we find the LCD:

$$\left.\begin{array}{l} a^2 - 1 = (a - 1)(a + 1) \\ a^2 + a = a(a + 1) \end{array}\right\} \quad \text{LCD} = a(a - 1)(a + 1).$$

We multiply each term by 1 to get the LCD in each expression, and then add and simplify:

$$\frac{2a}{(a - 1)(a + 1)} \cdot \frac{a}{a} + \frac{1}{a(a + 1)} \cdot \frac{a - 1}{a - 1}$$

$$= \frac{2a^2}{a(a - 1)(a + 1)} + \frac{a - 1}{a(a - 1)(a + 1)}$$

$$= \frac{2a^2 + a - 1}{a(a - 1)(a + 1)}$$

$$= \frac{(a + 1)(2a - 1)}{a(a - 1)(a + 1)}. \qquad \text{Factoring the numerator in order to simplify}$$

Answers

4. $\dfrac{x(10x + 9)}{48}$ **5.** $\dfrac{9x + 10}{48x^2}$

Then

$$= \frac{(a+1)(2a-1)}{a(a-1)(a+1)}$$ Removing a factor of 1: $\frac{a+1}{a+1} = 1$

$$= \frac{2a-1}{a(a-1)}.$$

Do Exercise 6.

6. Add:

$$\frac{3}{x^3 - x} + \frac{4}{x^2 + 2x + 1}.$$

EXAMPLE 7   Add: $\dfrac{x+4}{x-2} + \dfrac{x-7}{x+5}$.

First, we find the LCD. It is just the product of the denominators:

$$\text{LCD} = (x-2)(x+5).$$

We multiply by 1 to get the LCD in each expression, and then add and simplify:

$$\frac{x+4}{x-2} \cdot \frac{x+5}{x+5} + \frac{x-7}{x+5} \cdot \frac{x-2}{x-2}$$

$$= \frac{(x+4)(x+5)}{(x-2)(x+5)} + \frac{(x-7)(x-2)}{(x-2)(x+5)}$$

$$= \frac{x^2 + 9x + 20}{(x-2)(x+5)} + \frac{x^2 - 9x + 14}{(x-2)(x+5)}$$

$$= \frac{x^2 + 9x + 20 + x^2 - 9x + 14}{(x-2)(x+5)} = \frac{2x^2 + 34}{(x-2)(x+5)} = \frac{2(x^2 + 17)}{(x-2)(x+5)}.$$

Do Exercise 7.

7. Add:

$$\frac{x-2}{x+3} + \frac{x+7}{x+8}.$$

EXAMPLE 8   Add: $\dfrac{x}{x^2 + 11x + 30} + \dfrac{-5}{x^2 + 9x + 20}$.

$$\frac{x}{x^2 + 11x + 30} + \frac{-5}{x^2 + 9x + 20}$$

$$= \frac{x}{(x+5)(x+6)} + \frac{-5}{(x+5)(x+4)}$$ Factoring the denominators in order to find the LCD. The LCD is $(x+4)(x+5)(x+6)$.

$$= \frac{x}{(x+5)(x+6)} \cdot \frac{x+4}{x+4} + \frac{-5}{(x+5)(x+4)} \cdot \frac{x+6}{x+6}$$ Multiplying by 1

$$= \frac{x(x+4) + (-5)(x+6)}{(x+4)(x+5)(x+6)} = \frac{x^2 + 4x - 5x - 30}{(x+4)(x+5)(x+6)}$$

$$= \frac{x^2 - x - 30}{(x+4)(x+5)(x+6)}$$

$$= \frac{(x-6)(x+5)}{(x+4)(x+5)(x+6)}$$ Always simplify at the end if possible: $\dfrac{x+5}{x+5} = 1$.

$$= \frac{x-6}{(x+4)(x+6)}$$

Do Exercise 8.

8. Add:

$$\frac{5}{x^2 + 17x + 16} + \frac{3}{x^2 + 9x + 8}.$$

Denominators That Are Opposites

When one denominator is the opposite of the other, we can first multiply either expression by 1 using $-1/-1$.

EXAMPLES

9. $\dfrac{x}{2} + \dfrac{3}{-2} = \dfrac{x}{2} + \dfrac{3}{-2} \cdot \dfrac{-1}{-1}$ Multiplying by 1 using $\dfrac{-1}{-1}$

$= \dfrac{x}{2} + \dfrac{-3}{2}$ The denominators are now the same.

$= \dfrac{x + (-3)}{2} = \dfrac{x - 3}{2}$

10. $\dfrac{3x + 4}{x - 2} + \dfrac{x - 7}{2 - x} = \dfrac{3x + 4}{x - 2} + \dfrac{x - 7}{2 - x} \cdot \dfrac{-1}{-1}$

> We could have chosen to multiply this expression by $-1/-1$. We multiply only one expression, *not* both.

$= \dfrac{3x + 4}{x - 2} + \dfrac{-x + 7}{x - 2}$ *Note:* $(2 - x)(-1) = -2 + x$
$= x - 2.$

$= \dfrac{(3x + 4) + (-x + 7)}{x - 2} = \dfrac{2x + 11}{x - 2}$

Do Exercises 9 and 10.

Factors That Are Opposites

Suppose that when we factor to find the LCD, we find factors that are opposites. The easiest way to handle this is to first go back and multiply by $-1/-1$ appropriately to change factors so that they are not opposites.

EXAMPLE 11   Add: $\dfrac{x}{x^2 - 25} + \dfrac{3}{10 - 2x}$.

First, we factor to find the LCD:

$x^2 - 25 = (x - 5)(x + 5);$
$10 - 2x = 2(5 - x).$

We note that $x - 5$ is one factor of $x^2 - 25$ and $5 - x$ is one factor of $10 - 2x$. If the denominator of the second expression were $2x - 10$, then $x - 5$ would be a factor of both denominators. To rewrite the second expression with a denominator of $2x - 10$, we multiply by 1 using $-1/-1$, and then continue as before:

$\dfrac{x}{x^2 - 25} + \dfrac{3}{10 - 2x} = \dfrac{x}{(x - 5)(x + 5)} + \dfrac{3}{10 - 2x} \cdot \dfrac{-1}{-1}$

$= \dfrac{x}{(x - 5)(x + 5)} + \dfrac{-3}{2x - 10}$

$= \dfrac{x}{(x - 5)(x + 5)} + \dfrac{-3}{2(x - 5)}$ LCD $= 2(x - 5)(x + 5)$

$= \dfrac{x}{(x - 5)(x + 5)} \cdot \dfrac{2}{2} + \dfrac{-3}{2(x - 5)} \cdot \dfrac{x + 5}{x + 5}$

$= \dfrac{2x}{2(x - 5)(x + 5)} + \dfrac{-3(x + 5)}{2(x - 5)(x + 5)}$

$= \dfrac{2x - 3(x + 5)}{2(x - 5)(x + 5)} = \dfrac{2x - 3x - 15}{2(x - 5)(x + 5)}$

$= \dfrac{-x - 15}{2(x - 5)(x + 5)}.$ Collecting like terms

Do Exercise 11.

Add.

9. $\dfrac{x}{4} + \dfrac{5}{-4}$

10. $\dfrac{2x + 1}{x - 3} + \dfrac{x + 2}{3 - x}$

11. Add:

$\dfrac{x + 3}{x^2 - 16} + \dfrac{5}{12 - 3x}.$

Answers

9. $\dfrac{x - 5}{4}$ **10.** $\dfrac{x - 1}{x - 3}$ **11.** $\dfrac{-2x - 11}{3(x + 4)(x - 4)}$

6.5

Subtracting Rational Expressions

a Subtracting Rational Expressions

We subtract rational expressions as we do rational numbers.

SUBTRACTING RATIONAL EXPRESSIONS WITH LIKE DENOMINATORS

To subtract when the denominators are the same, subtract the numerators and keep the same denominator. Then simplify if possible.

EXAMPLE 1 Subtract: $\dfrac{8}{x} - \dfrac{3}{x}$.

$$\frac{8}{x} - \frac{3}{x} = \frac{8 - 3}{x} = \frac{5}{x}$$

EXAMPLE 2 Subtract: $\dfrac{3x}{x + 2} - \dfrac{x - 2}{x + 2}$.

$$\frac{3x}{x + 2} - \frac{x - 2}{x + 2} = \frac{3x - (x - 2)}{x + 2}$$

----- Caution! -----

The parentheses are important to make sure that you subtract the entire numerator.

$$= \frac{3x - x + 2}{x + 2} \qquad \text{Removing parentheses}$$

$$= \frac{2x + 2}{x + 2} = \frac{2(x + 1)}{x + 2}$$

Do Margin Exercises 1–3.

To subtract rational expressions with different denominators, we use a procedure similar to what we used for addition, except that we subtract numerators and write the difference over the LCD.

SUBTRACTING RATIONAL EXPRESSIONS WITH DIFFERENT DENOMINATORS

To subtract rational expressions with different denominators:

1. Find the LCM of the denominators. This is the least common denominator (LCD).

2. For each rational expression, find an equivalent expression with the LCD. To do so, multiply by 1 using a symbol for 1 made up of factors of the LCD that are missing from the original denominator.

3. Subtract the numerators. Write the difference over the LCD.

4. Simplify if possible.

SKILL TO REVIEW
Objective 1.8a: Find an equivalent expression for an opposite without parentheses, where an expression has several terms.

Find an expression without parentheses.

1. $-(3x - 11)$
2. $-(-x + 8)$

Subtract.

1. $\dfrac{7}{11} - \dfrac{3}{11}$

2. $\dfrac{7}{y} - \dfrac{2}{y}$

3. $\dfrac{2x^2 + 3x - 7}{2x + 1} - \dfrac{x^2 + x - 8}{2x + 1}$

Answers

Skill to Review:
1. $-3x + 11$ 2. $x - 8$

Margin Exercises:
1. $\dfrac{4}{11}$ 2. $\dfrac{5}{y}$ 3. $\dfrac{(x + 1)^2}{2x + 1}$

EXAMPLE 3 Subtract: $\dfrac{x+2}{x-4} - \dfrac{x+1}{x+4}$.

The LCD $= (x-4)(x+4)$.

$$\frac{x+2}{x-4} \cdot \frac{x+4}{x+4} - \frac{x+1}{x+4} \cdot \frac{x-4}{x-4} \qquad \text{Multiplying by 1}$$

$$= \frac{(x+2)(x+4)}{(x-4)(x+4)} - \frac{(x+1)(x-4)}{(x-4)(x+4)}$$

$$= \frac{x^2+6x+8}{(x-4)(x+4)} - \frac{x^2-3x-4}{(x-4)(x+4)}$$

Subtracting this numerator.
Don't forget the parentheses.

$$= \frac{x^2+6x+8-(x^2-3x-4)}{(x-4)(x+4)}$$

$$= \frac{x^2+6x+8-x^2+3x+4}{(x-4)(x+4)} \qquad \text{Removing parentheses}$$

$$= \frac{9x+12}{(x-4)(x+4)} = \frac{3(3x+4)}{(x-4)(x+4)}$$

4. Subtract:

$$\frac{x-2}{3x} - \frac{2x-1}{5x}.$$

Do Exercise 4.

EXAMPLE 4 Subtract: $\dfrac{x}{x^2+5x+6} - \dfrac{2}{x^2+3x+2}$.

$$\frac{x}{x^2+5x+6} - \frac{2}{x^2+3x+2}$$

$$= \frac{x}{(x+2)(x+3)} - \frac{2}{(x+2)(x+1)} \qquad \text{LCD} = (x+1)(x+2)(x+3)$$

$$= \frac{x}{(x+2)(x+3)} \cdot \frac{x+1}{x+1} - \frac{2}{(x+2)(x+1)} \cdot \frac{x+3}{x+3}$$

$$= \frac{x^2+x}{(x+1)(x+2)(x+3)} - \frac{2x+6}{(x+1)(x+2)(x+3)}$$

Subtracting this numerator.
Don't forget the parentheses.

$$= \frac{x^2+x-(2x+6)}{(x+1)(x+2)(x+3)}$$

$$= \frac{x^2+x-2x-6}{(x+1)(x+2)(x+3)} = \frac{x^2-x-6}{(x+1)(x+2)(x+3)}$$

$$= \frac{(x+2)(x-3)}{(x+1)(x+2)(x+3)}$$

$$= \frac{(x+2)(x-3)}{(x+1)(x+2)(x+3)} \qquad \text{Simplifying by removing a factor of 1: } \frac{x+2}{x+2} = 1$$

5. Subtract:

$$\frac{x}{x^2+15x+56} - \frac{6}{x^2+13x+42}.$$

$$= \frac{x-3}{(x+1)(x+3)}$$

Do Exercise 5.

Denominators That Are Opposites

When one denominator is the opposite of the other, we can first multiply one expression by $-1/-1$ to obtain a common denominator.

EXAMPLE 5 Subtract: $\dfrac{x}{5} - \dfrac{3x-4}{-5}$.

$$\frac{x}{5} - \frac{3x-4}{-5} = \frac{x}{5} - \frac{3x-4}{-5} \cdot \frac{-1}{-1} \qquad \text{Multiplying by 1 using } \frac{-1}{-1} \leftarrow$$

> This is equal to 1 (not −1).

$$= \frac{x}{5} - \frac{(3x-4)(-1)}{(-5)(-1)}$$

$$= \frac{x}{5} - \frac{4-3x}{5}$$

$$= \frac{x - (4-3x)}{5} \qquad \text{Remember the parentheses!}$$

$$= \frac{x - 4 + 3x}{5} = \frac{4x-4}{5} = \frac{4(x-1)}{5}$$

EXAMPLE 6 Subtract: $\dfrac{5y}{y-5} - \dfrac{2y-3}{5-y}$.

$$\frac{5y}{y-5} - \frac{2y-3}{5-y} = \frac{5y}{y-5} - \frac{2y-3}{5-y} \cdot \frac{-1}{-1}$$

$$= \frac{5y}{y-5} - \frac{(2y-3)(-1)}{(5-y)(-1)}$$

$$= \frac{5y}{y-5} - \frac{3-2y}{y-5}$$

$$= \frac{5y - (3-2y)}{y-5} \qquad \text{Remember the parentheses!}$$

$$= \frac{5y - 3 + 2y}{y-5} = \frac{7y-3}{y-5}$$

Do Exercises 6 and 7.

> Subtract.
>
> **6.** $\dfrac{x}{3} - \dfrac{2x-1}{-3}$
>
> **7.** $\dfrac{3x}{x-2} - \dfrac{x-3}{2-x}$

Factors That Are Opposites

Suppose that when we factor to find the LCD, we find factors that are opposites. Then we multiply by $-1/-1$ appropriately to change factors so that they are not opposites.

EXAMPLE 7 Subtract: $\dfrac{p}{64 - p^2} - \dfrac{5}{p-8}$.

Factoring $64 - p^2$, we get $(8-p)(8+p)$. Note that the factors $8 - p$ in the first denominator and $p - 8$ in the second denominator are opposites. We multiply the first expression by $-1/-1$ to avoid this situation. Then we proceed as before.

$$\frac{p}{64-p^2} - \frac{5}{p-8} = \frac{p}{64-p^2} \cdot \frac{-1}{-1} - \frac{5}{p-8}$$

$$= \frac{-p}{p^2-64} - \frac{5}{p-8}$$

$$= \frac{-p}{(p-8)(p+8)} - \frac{5}{p-8} \qquad \text{LCD} = (p-8)(p+8)$$

$$= \frac{-p}{(p-8)(p+8)} - \frac{5}{p-8} \cdot \frac{p+8}{p+8}$$

Answers

6. $\dfrac{3x-1}{3}$ **7.** $\dfrac{4x-3}{x-2}$

Multiplying, we have

$$\frac{-p}{(p-8)(p+8)} - \frac{5p+40}{(p-8)(p+8)}$$

⎣── Subtracting this numerator.
   Don't forget the parentheses.

$$= \frac{-p - (5p+40)}{(p-8)(p+8)}$$

$$= \frac{-p-5p-40}{(p-8)(p+8)} = \frac{-6p-40}{(p-8)(p+8)} = \frac{-2(3p+20)}{(p-8)(p+8)}.$$

8. Subtract:

$$\frac{y}{16-y^2} - \frac{7}{y-4}.$$

Do Exercise 8.

b Combined Additions and Subtractions

Now let's look at some combined additions and subtractions.

EXAMPLE 8 Perform the indicated operations and simplify:

$$\frac{x+9}{x^2-4} + \frac{5-x}{4-x^2} - \frac{2+x}{x^2-4}.$$

$$\frac{x+9}{x^2-4} + \frac{5-x}{4-x^2} - \frac{2+x}{x^2-4}$$

$$= \frac{x+9}{x^2-4} + \frac{5-x}{4-x^2} \cdot \frac{-1}{-1} - \frac{2+x}{x^2-4}$$

$$= \frac{x+9}{x^2-4} + \frac{x-5}{x^2-4} - \frac{2+x}{x^2-4} = \frac{(x+9) + (x-5) - (2+x)}{x^2-4}$$

$$= \frac{x+9+x-5-2-x}{x^2-4} = \frac{x+2}{x^2-4} = \frac{(x+2) \cdot 1}{(x+2)(x-2)} = \frac{1}{x-2}$$

9. Perform the indicated operations and simplify:

$$\frac{x+2}{x^2-9} - \frac{x-7}{9-x^2} + \frac{-8-x}{x^2-9}.$$

Do Exercise 9.

EXAMPLE 9 Perform the indicated operations and simplify:

$$\frac{1}{x} - \frac{1}{x^2} + \frac{2}{x+1}.$$

The LCD $= x \cdot x(x+1)$, or $x^2(x+1)$.

$$\frac{1}{x} \cdot \frac{x(x+1)}{x(x+1)} - \frac{1}{x^2} \cdot \frac{(x+1)}{(x+1)} + \frac{2}{x+1} \cdot \frac{x^2}{x^2}$$

$$= \frac{x(x+1)}{x^2(x+1)} - \frac{x+1}{x^2(x+1)} + \frac{2x^2}{x^2(x+1)}$$

⎣── Subtracting this numerator.
   Don't forget the parentheses.

$$= \frac{x(x+1) - (x+1) + 2x^2}{x^2(x+1)}$$

$$= \frac{x^2 + x - x - 1 + 2x^2}{x^2(x+1)} \qquad \text{Removing parentheses}$$

$$= \frac{3x^2 - 1}{x^2(x+1)}$$

10. Perform the indicated operations and simplify:

$$\frac{1}{x} - \frac{5}{3x} + \frac{2x}{x+1}.$$

Do Exercise 10.

Answers

8. $\dfrac{-4(2y+7)}{(y+4)(y-4)}$ **9.** $\dfrac{x-13}{(x+3)(x-3)}$

10. $\dfrac{2(3x^2 - x - 1)}{3x(x+1)}$

a Subtract. Simplify if possible.

1. $\dfrac{7}{x} - \dfrac{3}{x}$

2. $\dfrac{5}{a} - \dfrac{8}{a}$

3. $\dfrac{y}{y-4} - \dfrac{4}{y-4}$

4. $\dfrac{t^2}{t+5} - \dfrac{25}{t+5}$

5. $\dfrac{2x-3}{x^2+3x-4} - \dfrac{x-7}{x^2+3x-4}$

6. $\dfrac{x+1}{x^2-2x+1} - \dfrac{5-3x}{x^2-2x+1}$

7. $\dfrac{a-2}{10} - \dfrac{a+1}{5}$

8. $\dfrac{y+3}{2} - \dfrac{y-4}{4}$

9. $\dfrac{4z-9}{3z} - \dfrac{3z-8}{4z}$

10. $\dfrac{a-1}{4a} - \dfrac{2a+3}{a}$

11. $\dfrac{4x+2t}{3xt^2} - \dfrac{5x-3t}{x^2t}$

12. $\dfrac{5x+3y}{2x^2y} - \dfrac{3x+4y}{xy^2}$

13. $\dfrac{5}{x+5} - \dfrac{3}{x-5}$

14. $\dfrac{3t}{t-1} - \dfrac{8t}{t+1}$

15. $\dfrac{3}{2t^2-2t} - \dfrac{5}{2t-2}$

16. $\dfrac{11}{x^2-4} - \dfrac{8}{x+2}$

17. $\dfrac{2s}{t^2-s^2} - \dfrac{s}{t-s}$

18. $\dfrac{3}{12+x-x^2} - \dfrac{2}{x^2-9}$

19. $\dfrac{y-5}{y} - \dfrac{3y-1}{4y}$

20. $\dfrac{3x-2}{4x} - \dfrac{3x+1}{6x}$

21. $\dfrac{a}{x+a} - \dfrac{a}{x-a}$

22. $\dfrac{a}{a-b} - \dfrac{a}{a+b}$

23. $\dfrac{11}{6} - \dfrac{5}{-6}$

24. $\dfrac{5}{9} - \dfrac{7}{-9}$

25. $\dfrac{5}{a} - \dfrac{8}{-a}$

26. $\dfrac{8}{x} - \dfrac{3}{-x}$

27. $\dfrac{4}{y-1} - \dfrac{4}{1-y}$

28. $\dfrac{5}{a-2} - \dfrac{3}{2-a}$

29. $\dfrac{3-x}{x-7} - \dfrac{2x-5}{7-x}$

30. $\dfrac{t^2}{t-2} - \dfrac{4}{2-t}$

31. $\dfrac{a-2}{a^2-25} - \dfrac{6-a}{25-a^2}$

32. $\dfrac{x-8}{x^2-16} - \dfrac{x-8}{16-x^2}$

33. $\dfrac{4-x}{x-9} - \dfrac{3x-8}{9-x}$

34. $\dfrac{4x-6}{x-5} - \dfrac{7-2x}{5-x}$

35. $\dfrac{5x}{x^2-9} - \dfrac{4}{3-x}$

36. $\dfrac{8x}{16-x^2} - \dfrac{5}{x-4}$

37. $\dfrac{t^2}{2t^2 - 2t} - \dfrac{1}{2t - 2}$

38. $\dfrac{4}{5a^2 - 5a} - \dfrac{2}{5a - 5}$

39. $\dfrac{x}{x^2 + 5x + 6} - \dfrac{2}{x^2 + 3x + 2}$

40. $\dfrac{a}{a^2 + 11a + 30} - \dfrac{5}{a^2 + 9a + 20}$

b   Perform the indicated operations and simplify.

41. $\dfrac{3(2x + 5)}{x - 1} - \dfrac{3(2x - 3)}{1 - x} + \dfrac{6x - 1}{x - 1}$

42. $\dfrac{a - 2b}{b - a} - \dfrac{3a - 3b}{a - b} + \dfrac{2a - b}{a - b}$

43. $\dfrac{x - y}{x^2 - y^2} + \dfrac{x + y}{x^2 - y^2} - \dfrac{2x}{x^2 - y^2}$

44. $\dfrac{x - 3y}{2(y - x)} + \dfrac{x + y}{2(x - y)} - \dfrac{2x - 2y}{2(x - y)}$

45. $\dfrac{2(x - 1)}{2x - 3} - \dfrac{3(x + 2)}{2x - 3} - \dfrac{x - 1}{3 - 2x}$

46. $\dfrac{5(2y + 1)}{2y - 3} - \dfrac{3(y - 1)}{3 - 2y} - \dfrac{3(y - 2)}{2y - 3}$

47. $\dfrac{10}{2y - 1} - \dfrac{6}{1 - 2y} + \dfrac{y}{2y - 1} + \dfrac{y - 4}{1 - 2y}$

48. $\dfrac{(x + 1)(2x - 1)}{(2x - 3)(x - 3)} - \dfrac{(x - 3)(x + 1)}{(3 - x)(3 - 2x)} + \dfrac{(2x + 1)(x + 3)}{(3 - 2x)(x - 3)}$

49. $\dfrac{a + 6}{4 - a^2} - \dfrac{a + 3}{a + 2} + \dfrac{a - 3}{2 - a}$

50. $\dfrac{4t}{t^2 - 1} - \dfrac{2}{t} - \dfrac{2}{t + 1}$

51. $\dfrac{2z}{1 - 2z} + \dfrac{3z}{2z + 1} - \dfrac{3}{4z^2 - 1}$

52. $\dfrac{1}{x - y} - \dfrac{2x}{x^2 - y^2} + \dfrac{1}{x + y}$

53. $\dfrac{1}{x + y} - \dfrac{1}{x - y} + \dfrac{2x}{x^2 - y^2}$

54. $\dfrac{2b}{a^2 - b^2} - \dfrac{1}{a + b} + \dfrac{1}{a - b}$

Skill Maintenance

Simplify.

55. $\dfrac{x^8}{x^3}$ [4.1e]

56. $3x^4 \cdot 10x^8$ [4.1d]

57. $(a^2 b^{-5})^{-4}$ [4.2a, b]

58. $\dfrac{54x^{10}}{3x^7}$ [4.1e]

59. $\dfrac{66x^2}{11x^5}$ [4.1e]

60. $5x^{-7} \cdot 2x^4$ [4.1d]

Solve. [2.3b]

61. $\dfrac{4}{7} + 3x = \dfrac{1}{2}x - \dfrac{3}{14}$

62. $2.5x + 15.5 = 0.5 + 4x$

Find a polynomial for the shaded area of each figure. [4.4d]

63.

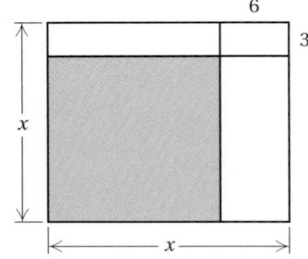

64.

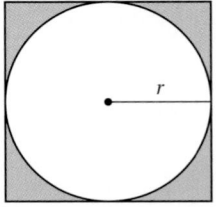

Synthesis

Perform the indicated operations and simplify.

65. $\dfrac{2x + 11}{x - 3} \cdot \dfrac{3}{x + 4} + \dfrac{2x + 1}{4 + x} \cdot \dfrac{3}{3 - x}$

66. $\dfrac{x^2}{3x^2 - 5x - 2} - \dfrac{2x}{3x + 1} \cdot \dfrac{1}{x - 2}$

67. $\dfrac{x}{x^4 - y^4} - \left(\dfrac{1}{x + y}\right)^2$

68. $\left(\dfrac{a}{a - b} + \dfrac{b}{a + b}\right)\left(\dfrac{1}{3a + b} + \dfrac{2a + 6b}{9a^2 - b^2}\right)$

69. The perimeter of the following right triangle is $2a + 5$. Find the missing length of the third side and the area.

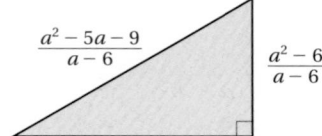

Concept Reinforcement

Determine whether each statement is true or false.

_____ **1.** The reciprocal of $\dfrac{3-w}{w+2}$ is $\dfrac{w-3}{w+2}$. [6.2a]

_____ **2.** The value of the numerator has no bearing on whether or not a rational expression is defined. [6.1a]

_____ **3.** To add or subtract rational expressions when the denominators are the same, add or subtract the numerators and keep the same denominator. [6.4a], [6.5a]

_____ **4.** For the rational expression $\dfrac{x(x-2)}{x+3}$, x is a factor of the numerator and a factor of the denominator. [6.1c]

_____ **5.** To find the LCM, use each factor the greatest number of times that it appears in any one factorization. [6.3a, c]

Guided Solutions

Fill in each blank with the number or expression that creates a correct solution.

6. Subtract: $\dfrac{x-1}{x-2} - \dfrac{x+1}{x+2} - \dfrac{x-6}{4-x^2}$. [6.5b]

$$\dfrac{x-1}{x-2} - \dfrac{x+1}{x+2} - \dfrac{x-6}{4-x^2} = \dfrac{x-1}{x-2} - \dfrac{x+1}{x+2} - \dfrac{x-6}{4-x^2} \cdot \dfrac{\square}{\square}$$

$$= \dfrac{x-1}{x-2} - \dfrac{x+1}{x+2} - \dfrac{6-\square}{\square-4}$$

$$= \dfrac{x-1}{x-2} - \dfrac{x+1}{x+2} - \dfrac{6-x}{(x-\square)(\square+2)}$$

$$= \dfrac{x-1}{x-2}\cdot\dfrac{\square}{\square} - \dfrac{x+1}{x+2}\cdot\dfrac{\square}{\square} - \dfrac{6-x}{(x-2)(x+2)}$$

$$= \dfrac{x^2+\square-2}{(x-2)(x+2)} - \dfrac{\square-x-2}{(x-2)(x+2)} - \dfrac{6-x}{(x-2)(x+2)}$$

$$= \dfrac{x^2+x-\square-x^2+\square+2-\square+x}{(x-2)(x+2)}$$

$$= \dfrac{\square-\square}{(x-2)(x+2)}$$

$$= \dfrac{3(\square-\square)}{(x-2)(x+2)} = \dfrac{\square}{\square}\cdot\dfrac{\square}{x+2} = \dfrac{3}{\square}$$

Mixed Review

Find all numbers for which the rational expression is not defined. [6.1a]

7. $\dfrac{t^2-16}{3}$

8. $\dfrac{x-8}{x^2-11x+24}$

9. $\dfrac{7}{2w-7}$

Simplify. [6.1c]

10. $\dfrac{x^2+2x-3}{x^2-9}$

11. $\dfrac{6y^2+12y-48}{3y^2-9y+6}$

12. $\dfrac{r-s}{s-r}$

13. Find the reciprocal of $-x + 3$. [6.2a]

14. Find the LCM of
$$x^2 - 100, 10x^3, \text{ and } x^2 - 20x + 100. \quad \text{[6.3c]}$$

Add, subtract, multiply, or divide and simplify if possible.

15. $\dfrac{a^2 - a - 2}{a^2 - a - 6} \div \dfrac{a^2 - 2a}{2a + a^2}$ [6.2b]

16. $\dfrac{3y}{y^2 - 7y + 10} - \dfrac{2y}{y^2 - 8y + 15}$ [6.5a]

17. $\dfrac{x^2}{x - 11} + \dfrac{121}{11 - x}$ [6.4a]

18. $\dfrac{x^2 - y^2}{(x - y)^2} \cdot \dfrac{1}{x + y}$ [6.1d]

19. $\dfrac{3a - b}{a^2 b} + \dfrac{a + 2b}{ab^2}$ [6.4a]

20. $\dfrac{5x}{x^2 - 4} - \dfrac{3}{x} + \dfrac{4}{x + 2}$ [6.5b]

Matching.   Perform the indicated operation and simplify. Then select the correct answer from selections A–G listed in the second column.   [6.1d], [6.2b], [6.4a], [6.5a]

21. $\dfrac{2}{x - 2} \div \dfrac{1}{x + 3}$

22. $\dfrac{1}{x + 3} - \dfrac{2}{x - 2}$

23. $\dfrac{2}{x - 2} - \dfrac{1}{x + 3}$

24. $\dfrac{1}{x + 3} \div \dfrac{2}{x - 2}$

25. $\dfrac{2}{x - 2} + \dfrac{1}{x + 3}$

26. $\dfrac{2}{x - 2} \cdot \dfrac{1}{x + 3}$

A. $\dfrac{-x - 8}{(x - 2)(x + 3)}$

B. $\dfrac{x - 2}{2(x + 3)}$

C. $\dfrac{2}{(x - 2)(x + 3)}$

D. $\dfrac{x + 8}{(x - 2)(x + 3)}$

E. $\dfrac{2(x + 3)}{x - 2}$

F. $\dfrac{3x + 4}{(x - 2)(x + 3)}$

G. $\dfrac{x + 3}{x - 2}$

Understanding Through Discussion and Writing

27. Explain why the product of two numbers is not always their least common multiple.   [6.3a]

28. Is the reciprocal of a product the product of the reciprocals? Why or why not? [6.2a]

29. A student insists on finding a common denominator by always multiplying the denominators of the expressions being added. How could this approach be improved?   [6.4a]

30. Explain why the expressions
$$\frac{1}{3 - x} \quad \text{and} \quad \frac{1}{x - 3}$$
are opposites.   [6.4a]

31. Explain why 5, -1, and 7 are *not* allowable replacements in the division
$$\frac{x + 3}{x - 5} \div \frac{x - 7}{x + 1}. \quad \text{[6.1a], [6.2a, b]}$$

32. If the LCM of a binomial and a trinomial is the trinomial, what relationship exists between the two expressions?   [6.3c]

6.6 Complex Rational Expressions

a Simplifying Complex Rational Expressions

A **complex rational expression**, or **complex fraction expression**, is a rational expression that has one or more rational expressions within its numerator or denominator. Here are some examples:

$$\frac{1 + \dfrac{2}{x}}{3}, \qquad \frac{x + y}{\dfrac{2x}{x + 1}}, \qquad \frac{\dfrac{1}{3} + \dfrac{1}{5}}{\dfrac{2}{x} - \dfrac{x}{y}}.$$

These are rational expressions within the complex rational expression.

There are two methods used to simplify complex rational expressions. We will consider both.

Method 1

> **MULTIPLYING BY THE LCM OF ALL THE DENOMINATORS**
>
> To simplify a complex rational expression:
>
> 1. First, find the LCM of all the denominators of all the rational expressions occurring *within* both the numerator and the denominator of the complex rational expression.
> 2. Then multiply by 1 using LCM/LCM.
> 3. If possible, simplify by removing a factor of 1.

> **EXAMPLE 1** Simplify: $\dfrac{\dfrac{1}{2} + \dfrac{3}{4}}{\dfrac{5}{6} - \dfrac{3}{8}}$.

We have

$$\frac{\dfrac{1}{2} + \dfrac{3}{4}}{\dfrac{5}{6} - \dfrac{3}{8}}$$

The denominators *within* the complex rational expression are 2, 4, 6, and 8. The LCM of these denominators is 24. We multiply by 1 using $\frac{24}{24}$. This amounts to multiplying both the numerator *and* the denominator by 24.

$$= \frac{\dfrac{1}{2} + \dfrac{3}{4}}{\dfrac{5}{6} - \dfrac{3}{8}} \cdot \frac{24}{24} \qquad \text{Multiplying by 1}$$

$$= \frac{\left(\dfrac{1}{2} + \dfrac{3}{4}\right)24}{\left(\dfrac{5}{6} - \dfrac{3}{8}\right)24} \quad \begin{array}{l} \leftarrow \text{Multiplying the numerator by 24} \\[1.5em] \leftarrow \text{Multiplying the denominator by 24} \end{array}$$

OBJECTIVE

a Simplify complex rational expressions.

SKILL TO REVIEW
Objective 6.3c: Find the LCM of algebraic expressions by factoring.

1. Find the LCM of 2, 4, 6, and 8.
2. Find the LCM of x, x^2, and $5x$.

To the instructor and the student: Students can be instructed to either try both methods and then choose the one that works best for them, or use the method chosen by the instructor.

Answers
Skill to Review:
1. 24 **2.** $5x^2$

Using the distributive laws, we carry out the multiplications:

$$= \frac{\dfrac{1}{2}(24) + \dfrac{3}{4}(24)}{\dfrac{5}{6}(24) - \dfrac{3}{8}(24)}$$

$$= \frac{12 + 18}{20 - 9} \quad \text{Simplifying}$$

$$= \frac{30}{11}.$$

Multiplying in this manner has the effect of clearing fractions in both the numerator and the denominator of the complex rational expression.

Do Exercise 1.

1. Simplify.

$$\frac{\dfrac{1}{3} + \dfrac{4}{5}}{\dfrac{7}{8} - \dfrac{5}{6}}$$

EXAMPLE 2 Simplify: $\dfrac{\dfrac{3}{x} + \dfrac{1}{2x}}{\dfrac{1}{3x} - \dfrac{3}{4x}}$.

The denominators within the complex expression are x, $2x$, $3x$, and $4x$. The LCM of these denominators is $12x$. We multiply by 1 using $12x/12x$.

$$\frac{\dfrac{3}{x} + \dfrac{1}{2x}}{\dfrac{1}{3x} - \dfrac{3}{4x}} \cdot \frac{12x}{12x} = \frac{\left(\dfrac{3}{x} + \dfrac{1}{2x}\right)12x}{\left(\dfrac{1}{3x} - \dfrac{3}{4x}\right)12x} = \frac{\dfrac{3}{x}(12x) + \dfrac{1}{2x}(12x)}{\dfrac{1}{3x}(12x) - \dfrac{3}{4x}(12x)}$$

$$= \frac{36 + 6}{4 - 9} = \frac{42}{-5} = -\frac{42}{5}$$

2. Simplify.

$$\frac{\dfrac{x}{2} + \dfrac{2x}{3}}{\dfrac{1}{x} - \dfrac{x}{2}}$$

Do Exercise 2.

EXAMPLE 3 Simplify: $\dfrac{1 - \dfrac{1}{x}}{1 - \dfrac{1}{x^2}}$.

The denominators within the complex expression are x and x^2. The LCM of these denominators is x^2. We multiply by 1 using x^2/x^2. Then, after obtaining a single rational expression, we simplify:

$$\frac{1 - \dfrac{1}{x}}{1 - \dfrac{1}{x^2}} \cdot \frac{x^2}{x^2} = \frac{\left(1 - \dfrac{1}{x}\right)x^2}{\left(1 - \dfrac{1}{x^2}\right)x^2} = \frac{1(x^2) - \dfrac{1}{x}(x^2)}{1(x^2) - \dfrac{1}{x^2}(x^2)} = \frac{x^2 - x}{x^2 - 1}$$

$$= \frac{x(x-1)}{(x+1)(x-1)} = \frac{x}{x+1}.$$

3. Simplify.

$$\frac{1 + \dfrac{1}{x}}{1 - \dfrac{1}{x^2}}$$

Do Exercise 3.

Answers

1. $\dfrac{136}{5}$ 2. $\dfrac{7x^2}{3(2 - x^2)}$ 3. $\dfrac{x}{x - 1}$

Method 2

ADDING IN THE NUMERATOR AND THE DENOMINATOR

To simplify a complex rational expression:

1. Add or subtract, as necessary, to get a single rational expression in the numerator.
2. Add or subtract, as necessary, to get a single rational expression in the denominator.
3. Divide the numerator by the denominator.
4. If possible, simplify by removing a factor of 1.

We will redo Examples 1–3 using this method.

EXAMPLE 4 Simplify: $\dfrac{\frac{1}{2} + \frac{3}{4}}{\frac{5}{6} - \frac{3}{8}}$.

The LCM of 2 and 4 in the numerator is 4. The LCM of 6 and 8 in the denominator is 24. We have

$$\frac{\frac{1}{2} + \frac{3}{4}}{\frac{5}{6} - \frac{3}{8}} = \frac{\frac{1}{2} \cdot \frac{2}{2} + \frac{3}{4}}{\frac{5}{6} \cdot \frac{4}{4} - \frac{3}{8} \cdot \frac{3}{3}}$$

$\left.\begin{array}{l}\end{array}\right\} \leftarrow$ Multiplying the $\frac{1}{2}$ by 1 to get the common denominator, 4

$\left.\begin{array}{l}\end{array}\right\} \leftarrow$ Multiplying the $\frac{5}{6}$ and the $\frac{3}{8}$ by 1 to get the common denominator, 24

$$= \frac{\frac{2}{4} + \frac{3}{4}}{\frac{20}{24} - \frac{9}{24}}$$

$$= \frac{\frac{5}{4}}{\frac{11}{24}} \qquad \text{Adding in the numerator;} \atop \text{subtracting in the denominator}$$

$$= \frac{5}{4} \div \frac{11}{24}$$

$$= \frac{5}{4} \cdot \frac{24}{11} \qquad \text{Multiplying by the reciprocal of the divisor}$$

$$= \frac{5 \cdot 3 \cdot 2 \cdot 2 \cdot 2}{2 \cdot 2 \cdot 11} \qquad \text{Factoring}$$

$$= \frac{5 \cdot 3 \cdot 2 \cdot \cancel{2} \cdot \cancel{2}}{\cancel{2} \cdot \cancel{2} \cdot 11} \qquad \text{Removing a factor of 1: } \frac{2 \cdot 2}{2 \cdot 2} = 1$$

$$= \frac{30}{11}.$$

Do Exercise 4.

4. Simplify. Use method 2.

$$\frac{\frac{1}{3} + \frac{4}{5}}{\frac{7}{8} - \frac{5}{6}}$$

Answer

4. $\frac{136}{5}$

EXAMPLE 5 Simplify: $\dfrac{\dfrac{3}{x} + \dfrac{1}{2x}}{\dfrac{1}{3x} - \dfrac{3}{4x}}$.

We have

$$\frac{\dfrac{3}{x} + \dfrac{1}{2x}}{\dfrac{1}{3x} - \dfrac{3}{4x}} = \frac{\left.\dfrac{3}{x} \cdot \dfrac{2}{2} + \dfrac{1}{2x}\right\}}{\left.\dfrac{1}{3x} \cdot \dfrac{4}{4} - \dfrac{3}{4x} \cdot \dfrac{3}{3}\right\}}$$

$\longleftarrow$ Finding the LCD, $2x$, and multiplying by 1 in the numerator

$\longleftarrow$ Finding the LCD, $12x$, and multiplying by 1 in the denominator

$$= \frac{\dfrac{6}{2x} + \dfrac{1}{2x}}{\dfrac{4}{12x} - \dfrac{9}{12x}} = \frac{\dfrac{7}{2x}}{\dfrac{-5}{12x}}$$

$\longleftarrow$ Adding in the numerator and subtracting in the denominator

$$= \frac{7}{2x} \div \frac{-5}{12x}$$

$$= \frac{7}{2x} \cdot \frac{12x}{-5}$$
Multiplying by the reciprocal of the divisor

$$= \frac{7 \cdot 6 \cdot (2x)}{(2x)(-5)}$$
Multiplying, factoring, and removing a factor of 1: $\dfrac{2x}{2x} = 1$

$$= \frac{42}{-5} = -\frac{42}{5}.$$

Do Exercise 5.

5. Simplify. Use method 2.

$$\frac{\dfrac{x}{2} + \dfrac{2x}{3}}{\dfrac{1}{x} - \dfrac{x}{2}}$$

EXAMPLE 6 Simplify: $\dfrac{1 - \dfrac{1}{x}}{1 - \dfrac{1}{x^2}}$.

We have

$$\frac{1 - \dfrac{1}{x}}{1 - \dfrac{1}{x^2}} = \frac{\left.1 \cdot \dfrac{x}{x} - \dfrac{1}{x}\right\}}{\left.1 \cdot \dfrac{x^2}{x^2} - \dfrac{1}{x^2}\right\}}$$

$\longleftarrow$ Finding the LCD, x, and multiplying by 1 in the numerator

$\longleftarrow$ Finding the LCD, x^2, and multiplying by 1 in the denominator

$$= \frac{\dfrac{x - 1}{x}}{\dfrac{x^2 - 1}{x^2}}$$

$\longleftarrow$ Subtracting in the numerator and subtracting in the denominator

$$= \frac{x - 1}{x} \div \frac{x^2 - 1}{x^2}$$

$$= \frac{x - 1}{x} \cdot \frac{x^2}{x^2 - 1}$$
Multiplying by the reciprocal of the divisor

$$= \frac{(x - 1)x \cdot x}{x(x - 1)(x + 1)}$$
Multiplying, factoring, and removing a factor of 1: $\dfrac{x(x - 1)}{x(x - 1)} = 1$

$$= \frac{x}{x + 1}.$$

Do Exercise 6.

6. Simplify. Use method 2.

$$\frac{1 + \dfrac{1}{x}}{1 - \dfrac{1}{x^2}}$$

a Simplify.

1. $\dfrac{1 + \dfrac{9}{16}}{1 - \dfrac{3}{4}}$

2. $\dfrac{6 - \dfrac{3}{8}}{4 + \dfrac{5}{6}}$

3. $\dfrac{1 - \dfrac{3}{5}}{1 + \dfrac{1}{5}}$

4. $\dfrac{2 + \dfrac{2}{3}}{2 - \dfrac{2}{3}}$

5. $\dfrac{\dfrac{1}{2} + \dfrac{3}{4}}{\dfrac{5}{8} - \dfrac{5}{6}}$

6. $\dfrac{\dfrac{3}{4} + \dfrac{7}{8}}{\dfrac{2}{3} - \dfrac{5}{6}}$

7. $\dfrac{\dfrac{1}{x} + 3}{\dfrac{1}{x} - 5}$

8. $\dfrac{2 - \dfrac{1}{a}}{4 + \dfrac{1}{a}}$

9. $\dfrac{4 - \dfrac{1}{x^2}}{2 - \dfrac{1}{x}}$

10. $\dfrac{\dfrac{2}{y} + \dfrac{1}{2y}}{y + \dfrac{y}{2}}$

11. $\dfrac{8 + \dfrac{8}{d}}{1 + \dfrac{1}{d}}$

12. $\dfrac{3 + \dfrac{2}{t}}{3 - \dfrac{2}{t}}$

13. $\dfrac{\dfrac{x}{8} - \dfrac{8}{x}}{\dfrac{1}{8} + \dfrac{1}{x}}$

14. $\dfrac{\dfrac{2}{m} + \dfrac{m}{2}}{\dfrac{m}{3} - \dfrac{3}{m}}$

15. $\dfrac{1 + \dfrac{1}{y}}{1 - \dfrac{1}{y^2}}$

16. $\dfrac{\dfrac{1}{q^2} - 1}{\dfrac{1}{q} + 1}$

17. $\dfrac{\dfrac{1}{5} - \dfrac{1}{a}}{\dfrac{5 - a}{5}}$

18. $\dfrac{\dfrac{4}{t}}{4 + \dfrac{1}{t}}$

19. $\dfrac{\dfrac{1}{a} + \dfrac{1}{b}}{\dfrac{1}{a^2} - \dfrac{1}{b^2}}$

20. $\dfrac{\dfrac{1}{x^2} - \dfrac{1}{y^2}}{\dfrac{2}{x} - \dfrac{2}{y}}$

21. $\dfrac{\dfrac{p}{q} + \dfrac{q}{p}}{\dfrac{1}{p} + \dfrac{1}{q}}$

22. $\dfrac{x - 3 + \dfrac{2}{x}}{x - 4 + \dfrac{3}{x}}$

23. $\dfrac{\dfrac{2}{a} + \dfrac{4}{a^2}}{\dfrac{5}{a^3} - \dfrac{3}{a}}$

24. $\dfrac{\dfrac{5}{x^3} - \dfrac{1}{x^2}}{\dfrac{2}{x} + \dfrac{3}{x^2}}$

25. $\dfrac{\dfrac{2}{7a^4} - \dfrac{1}{14a}}{\dfrac{3}{5a^2} + \dfrac{2}{15a}}$

26. $\dfrac{\dfrac{5}{4x^3} - \dfrac{3}{8x}}{\dfrac{3}{2x} + \dfrac{3}{4x^3}}$

27. $\dfrac{\dfrac{a}{b} + \dfrac{c}{d}}{\dfrac{b}{a} + \dfrac{d}{c}}$

28. $\dfrac{\dfrac{a}{b} - \dfrac{c}{d}}{\dfrac{b}{a} - \dfrac{d}{c}}$

29. $\dfrac{\dfrac{x}{5y^3} + \dfrac{3}{10y}}{\dfrac{3}{10y} + \dfrac{x}{5y^3}}$

30. $\dfrac{\dfrac{a}{6b^3} + \dfrac{4}{9b^2}}{\dfrac{5}{6b} - \dfrac{1}{9b^3}}$

31. $\dfrac{\dfrac{3}{x+1} + \dfrac{1}{x}}{\dfrac{2}{x+1} + \dfrac{3}{x}}$

32. $\dfrac{x - 7 + \dfrac{5}{x-1}}{x - 3 + \dfrac{1}{x-1}}$

Skill Maintenance

Solve. [2.7e]

33. $4 - \dfrac{1}{6}x \geq -12$

34. $3(b - 8) > -2(3b + 1)$

35. $1.5x + 19.2 < 4.2 - 3.5x$

36. $10(4 - 5m) - 6 \leq -10$

Add. [4.4a]

37. $(2x^3 - 4x^2 + x - 7) + (4x^4 + x^3 + 4x^2 + x)$

38. $(2x^3 - 4x^2 + x - 7) + (-2x^3 + 4x^2 - x + 7)$

Factor. [5.7a]

39. $p^2 - 10p + 25$

40. $p^2 + 10p + 25$

41. $50p^2 - 100$

42. $5p^2 - 40p - 100$

Solve. [5.9a]

43. *Perimeter of a Rectangle.* The length of a rectangle is 3 yd greater than the width. The area of the rectangle is 10 yd². Find the perimeter.

44. *Ladder Distances.* A ladder of length 13 ft is placed against a building in such a way that the distance from the top of the ladder to the ground is 7 ft more than the distance from the bottom of the ladder to the building. Find these distances.

Synthesis

45. Find the reciprocal of $\dfrac{2}{x-1} - \dfrac{1}{3x-2}$.

Simplify.

46. $\left[\dfrac{\dfrac{x+1}{x-1} + 1}{\dfrac{x+1}{x-1} - 1}\right]^5$

47. $1 + \dfrac{1}{1 + \dfrac{1}{1 + \dfrac{1}{1 + \dfrac{1}{x}}}}$

48. $\dfrac{\dfrac{z}{1 - \dfrac{z}{2 + 2z}} - 2z}{\dfrac{2z}{5z - 2} - 3}$

6.7

Solving Rational Equations

a Rational Equations

In Sections 6.1–6.6, we studied operations with *rational expressions*. These expressions have no equals signs. We can add, subtract, multiply, or divide and simplify expressions, but we cannot solve if there are no equals signs—as, for example, in

$$\frac{x^2 + 6x + 9}{x^2 - 4} \cdot \frac{x - 2}{x + 3}, \qquad \frac{x + y}{x - y} \div \frac{x^2 + y}{x^2 - y^2}, \quad \text{and} \quad \frac{a + 3}{a^2 - 16} + \frac{5}{12 - 3a}.$$

Operation signs occur. There are no equals signs!

Most often, the result of our calculation is another rational expression that has not been cleared of fractions.

Equations *do have* equals signs, and we can clear them of fractions as we did in Section 2.3. A **rational**, or **fraction**, **equation** is an equation containing one or more rational expressions. Here are some examples:

$$\frac{2}{3} + \frac{5}{6} = \frac{x}{9}, \qquad x + \frac{6}{x} = -5, \quad \text{and} \quad \frac{x^2}{x - 1} = \frac{1}{x - 1}.$$

There are equals signs as well as operation signs.

SOLVING RATIONAL EQUATIONS

To solve a rational equation, the first step is to clear the equation of fractions. To do this, multiply all terms on both sides of the equation by the LCM of all the denominators. Then carry out the equation-solving process as we learned it in Chapters 2 and 5.

When clearing an equation of fractions, we use the terminology LCM instead of LCD because we are *not* adding or subtracting rational expressions.

EXAMPLE 1 Solve: $\dfrac{2}{3} + \dfrac{5}{6} = \dfrac{x}{9}$.

The LCM of all denominators is $2 \cdot 3 \cdot 3$, or 18. We multiply all terms on both sides by 18:

$$18\left(\frac{2}{3} + \frac{5}{6}\right) = 18 \cdot \frac{x}{9} \qquad \text{Multiplying by the LCM on both sides}$$

$$18 \cdot \frac{2}{3} + 18 \cdot \frac{5}{6} = 18 \cdot \frac{x}{9} \qquad \text{Multiplying each term by the LCM to remove parentheses}$$

$$12 + 15 = 2x \qquad \text{Simplifying. Note that we have now cleared fractions.}$$

$$27 = 2x$$

$$\frac{27}{2} = x.$$

The solution is $\dfrac{27}{2}$.

Do Margin Exercise 1.

OBJECTIVE

a Solve rational equations.

SKILL TO REVIEW
Objective 2.3b: Solve equations in which like terms may need to be collected.

Solve. Clear fractions first.

1. $4 - \dfrac{5}{6}y = y + \dfrac{7}{12}$

2. $\dfrac{2}{5}x + \dfrac{1}{3} = \dfrac{7}{10}x - 2$

--------- *Caution!* ---------

We are introducing a new use of the LCM in this section. We previously used the LCM in adding or subtracting rational expressions. *Now* we have equations with equals signs. We clear fractions by multiplying by the LCM on both sides of the equation. This eliminates the denominators. Do *not* make the mistake of trying to clear fractions when you do not have an equation.

1. Solve: $\dfrac{3}{4} + \dfrac{5}{8} = \dfrac{x}{12}$.

Answers

Skill to Review:
1. $\dfrac{41}{22}$ **2.** $\dfrac{70}{9}$

Margin Exercise:
1. $\dfrac{33}{2}$

✖ Algebraic-Graphical Connection

We can obtain a visual check of the solutions of a rational equation by graphing. For example, consider the equation

$$\frac{x}{4} + \frac{x}{2} = 6.$$

We can examine the solution by graphing the equations

$$y = \frac{x}{4} + \frac{x}{2} \quad \text{and} \quad y = 6$$

using the same set of axes.

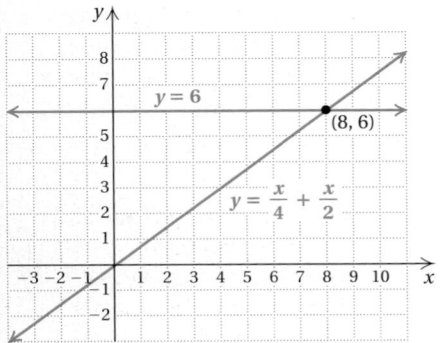

The first coordinate of the point of intersection of the graphs is the value of x for which $\frac{x}{4} + \frac{y}{2} = 6$, so it is the solution of the equation. It appears from the graph that when $x = 8$, the value of $x/4 + x/2$ is 6. We can check by substitution:

$$\frac{x}{4} + \frac{x}{2} = \frac{8}{4} + \frac{8}{2} = 2 + 4 = 6.$$

Thus the solution is 8.

✖

Solve.

2. $\dfrac{x}{4} - \dfrac{x}{6} = \dfrac{1}{8}$

3. $\dfrac{1}{x} = \dfrac{1}{6 - x}$

Answers

2. $\dfrac{3}{2}$ 3. 3

EXAMPLE 2 Solve: $\dfrac{x}{6} - \dfrac{x}{8} = \dfrac{1}{12}$.

The LCM is 24. We multiply all terms on both sides by 24:

$$\frac{x}{6} - \frac{x}{8} = \frac{1}{12}$$

$$24\left(\frac{x}{6} - \frac{x}{8}\right) = 24 \cdot \frac{1}{12} \qquad \text{Multiplying by the LCM on both sides}$$

$$24 \cdot \frac{x}{6} - 24 \cdot \frac{x}{8} = 24 \cdot \frac{1}{12} \qquad \text{Multiplying to remove parentheses}$$

> Be sure to multiply each term by the LCM.

$$4x - 3x = 2 \qquad \text{Simplifying}$$
$$x = 2.$$

Check:

$$\begin{array}{c|c} \dfrac{x}{6} - \dfrac{x}{8} & \dfrac{1}{12} \\[6pt] \dfrac{2}{6} - \dfrac{2}{8} & \dfrac{1}{12} \\[6pt] \dfrac{1}{3} - \dfrac{1}{4} & \\[6pt] \dfrac{4}{12} - \dfrac{3}{12} & \\[6pt] \dfrac{1}{12} & \text{TRUE} \end{array}$$

This checks, so the solution is 2.

EXAMPLE 3 Solve: $\dfrac{1}{x} = \dfrac{1}{4 - x}$.

The LCM is $x(4 - x)$. We multiply all terms on both sides by $x(4 - x)$:

$$\frac{1}{x} = \frac{1}{4 - x}$$

$$x(4 - x) \cdot \frac{1}{x} = x(4 - x) \cdot \frac{1}{4 - x} \qquad \text{Multiplying by the LCM on both sides}$$

$$4 - x = x \qquad \text{Simplifying}$$
$$4 = 2x$$
$$x = 2.$$

Check:

$$\begin{array}{c|c} \dfrac{1}{x} & \dfrac{1}{4 - x} \\[6pt] \dfrac{1}{2} & \dfrac{1}{4 - 2} \\[6pt] & \dfrac{1}{2} \qquad \text{TRUE} \end{array}$$

This checks, so the solution is 2.

> Do Exercises 2 and 3.

EXAMPLE 4 Solve: $\dfrac{2}{3x} + \dfrac{1}{x} = 10$.

The LCM is $3x$. We multiply all terms on both sides by $3x$:

$$\frac{2}{3x} + \frac{1}{x} = 10$$

$$3x\left(\frac{2}{3x} + \frac{1}{x}\right) = 3x \cdot 10 \qquad \text{Multiplying by the LCM on both sides}$$

$$3x \cdot \frac{2}{3x} + 3x \cdot \frac{1}{x} = 3x \cdot 10 \qquad \text{Multiplying to remove parentheses}$$

$$2 + 3 = 30x \qquad \text{Simplifying}$$

$$5 = 30x$$

$$\frac{5}{30} = x$$

$$\frac{1}{6} = x.$$

The check is left to the student. The solution is $\frac{1}{6}$.

> Do Exercise 4.

EXAMPLE 5 Solve: $x + \dfrac{6}{x} = -5$.

The LCM is x. We multiply all terms on both sides by x:

$$x + \frac{6}{x} = -5$$

$$x\left(x + \frac{6}{x}\right) = x \cdot (-5) \qquad \text{Multiplying by } x \text{ on both sides}$$

$$x \cdot x + x \cdot \frac{6}{x} = -5x \qquad \begin{array}{l}\text{Note that each rational expression}\\ \text{on the left is now multiplied by } x.\end{array}$$

$$x^2 + 6 = -5x \qquad \text{Simplifying}$$

$$x^2 + 5x + 6 = 0 \qquad \text{Adding } 5x \text{ to get a 0 on one side}$$

$$(x + 3)(x + 2) = 0 \qquad \text{Factoring}$$

$$x + 3 = 0 \quad or \quad x + 2 = 0 \qquad \text{Using the principle of zero products}$$

$$x = -3 \quad or \qquad x = -2.$$

Check: For -3:

$$\begin{array}{c|c} x + \dfrac{6}{x} = -5 & \\ \hline -3 + \dfrac{6}{-3} & -5 \\ -3 - 2 & \\ -5 & \text{TRUE} \end{array}$$

For -2:

$$\begin{array}{c|c} x + \dfrac{6}{x} = -5 & \\ \hline -2 + \dfrac{6}{-2} & -5 \\ -2 - 3 & \\ -5 & \text{TRUE} \end{array}$$

Both of these check, so there are two solutions, -3 and -2.

> Do Exercise 5.

---

4. Solve: $\dfrac{1}{2x} + \dfrac{1}{x} = -12$.

CHECKING POSSIBLE SOLUTIONS

When we multiply by the LCM on both sides of an equation, the resulting equation might have solutions that are *not* solutions of the original equation. Thus we must *always* check possible solutions in the original equation.

1. If you have carried out all algebraic procedures correctly, you need only check if a number makes a denominator 0 in the original equation. If it does make a denominator 0, it is *not* a solution.

2. To be sure that no computational errors have been made and that you indeed have a solution, a complete check is necessary, as we did in Chapter 2.

5. Solve: $x + \dfrac{1}{x} = 2$.

Answers

4. $-\dfrac{1}{8}$ **5.** 1

Example 6 illustrates the importance of checking all possible solutions.

EXAMPLE 6 Solve: $\dfrac{x^2}{x-1} = \dfrac{1}{x-1}$.

The LCM is $x-1$. We multiply all terms on both sides by $x-1$:

$$\frac{x^2}{x-1} = \frac{1}{x-1}$$

$$(x-1)\cdot\frac{x^2}{x-1} = (x-1)\cdot\frac{1}{x-1} \qquad \text{Multiplying by } x-1 \text{ on both sides}$$

$$x^2 = 1 \qquad \text{Simplifying}$$

$$x^2 - 1 = 0 \qquad \text{Subtracting 1 to get a 0 on one side}$$

$$(x-1)(x+1) = 0 \qquad \text{Factoring}$$

$$x - 1 = 0 \quad or \quad x + 1 = 0 \qquad \text{Using the principle of zero products}$$

$$x = 1 \quad or \qquad x = -1.$$

The numbers 1 and -1 are possible solutions.

Check: For 1:

$$\frac{x^2}{x-1} = \frac{1}{x-1}$$

$$\frac{1^2}{1-1} \overset{?}{\,} \frac{1}{1-1}$$

$$\frac{1}{0} \,\Big|\, \frac{1}{0} \qquad \text{NOT DEFINED}$$

For -1:

$$\frac{x^2}{x-1} = \frac{1}{x-1}$$

$$\frac{(-1)^2}{(-1)-1} \overset{?}{\,} \frac{1}{(-1)-1}$$

$$-\frac{1}{2} \,\Big|\, -\frac{1}{2} \qquad \text{TRUE}$$

We look at the original equation and see that 1 makes a denominator 0 and is therefore not a solution. The number -1 checks and is a solution.

EXAMPLE 7 Solve: $\dfrac{3}{x-5} + \dfrac{1}{x+5} = \dfrac{2}{x^2-25}$.

The LCM is $(x-5)(x+5)$. We multiply all terms on both sides by $(x-5)(x+5)$:

$$(x-5)(x+5)\left(\frac{3}{x-5} + \frac{1}{x+5}\right) = (x-5)(x+5)\left(\frac{2}{x^2-25}\right)$$

$$\text{Multiplying by the LCM on both sides}$$

$$(x-5)(x+5)\cdot\frac{3}{x-5} + (x-5)(x+5)\cdot\frac{1}{x+5} = (x-5)(x+5)\cdot\frac{2}{x^2-25}$$

$$3(x+5) + (x-5) = 2 \qquad \text{Simplifying}$$

$$3x + 15 + x - 5 = 2 \qquad \text{Removing parentheses}$$

$$4x + 10 = 2$$

$$4x = -8$$

$$x = -2.$$

The check is left to the student. The number -2 checks and is the solution.

Solve.

6. $\dfrac{x^2}{x+2} = \dfrac{4}{x+2}$

7. $\dfrac{4}{x-2} + \dfrac{1}{x+2} = \dfrac{26}{x^2-4}$

Do Exercises 6 and 7.

Answers

6. 2   **7.** 4

ARE YOU CALCULATING OR SOLVING?

One of the common difficulties with this chapter is knowing for sure the task at hand. Are you combining expressions using operations to get another *rational expression,* or are you solving equations for which the results are numbers that are *solutions* of an equation? To learn to make these decisions, complete the following list by writing in the blank the type of answer you should get: "Rational expression" or "Solutions." You need not complete the mathematical operations.

TASK	ANSWER (Just write "Rational expression" or "Solutions.")
1. Add: $\dfrac{4}{x-2} + \dfrac{1}{x+2}$.	
2. Solve: $\dfrac{4}{x-2} = \dfrac{1}{x+2}$.	
3. Subtract: $\dfrac{4}{x-2} - \dfrac{1}{x+2}$.	
4. Multiply: $\dfrac{4}{x-2} \cdot \dfrac{1}{x+2}$.	
5. Divide: $\dfrac{4}{x-2} \div \dfrac{1}{x+2}$.	
6. Solve: $\dfrac{4}{x-2} + \dfrac{1}{x+2} = \dfrac{26}{x^2-4}$.	
7. Perform the indicated operations and simplify: $\dfrac{4}{x-2} + \dfrac{1}{x+2} - \dfrac{26}{x^2-4}$.	
8. Solve: $\dfrac{x^2}{x-1} = \dfrac{1}{x-1}$.	
9. Solve: $\dfrac{2}{y^2-25} = \dfrac{3}{y-5} + \dfrac{1}{y-5}$.	
10. Solve: $\dfrac{x}{x+4} - \dfrac{4}{x-4} = \dfrac{x^2+16}{x^2-16}$.	
11. Perform the indicated operations and simplify: $\dfrac{x}{x+4} - \dfrac{4}{x-4} - \dfrac{x^2+16}{x^2-16}$.	
12. Solve: $\dfrac{5}{y-3} - \dfrac{30}{y^2-9} = 1$.	
13. Add: $\dfrac{5}{y-3} + \dfrac{30}{y^2-9} + 1$.	

6.7

Exercise Set

For Extra Help

MyMathLab

Math XL
PRACTICE   WATCH   DOWNLOAD   READ   REVIEW

a Solve. Don't forget to check!

1. $\dfrac{4}{5} - \dfrac{2}{3} = \dfrac{x}{9}$

2. $\dfrac{x}{20} = \dfrac{3}{8} - \dfrac{4}{5}$

3. $\dfrac{3}{5} + \dfrac{1}{8} = \dfrac{1}{x}$

4. $\dfrac{2}{3} + \dfrac{5}{6} = \dfrac{1}{x}$

5. $\dfrac{3}{8} + \dfrac{4}{5} = \dfrac{x}{20}$

6. $\dfrac{3}{5} + \dfrac{2}{3} = \dfrac{x}{9}$

7. $\dfrac{1}{x} = \dfrac{2}{3} - \dfrac{5}{6}$

8. $\dfrac{1}{x} = \dfrac{1}{8} - \dfrac{3}{5}$

9. $\dfrac{1}{6} + \dfrac{1}{8} = \dfrac{1}{t}$

10. $\dfrac{1}{8} + \dfrac{1}{12} = \dfrac{1}{t}$

11. $x + \dfrac{4}{x} = -5$

12. $\dfrac{10}{x} - x = 3$

13. $\dfrac{x}{4} - \dfrac{4}{x} = 0$

14. $\dfrac{x}{5} - \dfrac{5}{x} = 0$

15. $\dfrac{5}{x} = \dfrac{6}{x} - \dfrac{1}{3}$

16. $\dfrac{4}{x} = \dfrac{5}{x} - \dfrac{1}{2}$

17. $\dfrac{5}{3x} + \dfrac{3}{x} = 1$

18. $\dfrac{5}{2y} + \dfrac{8}{y} = 1$

19. $\dfrac{t-2}{t+3} = \dfrac{3}{8}$

20. $\dfrac{x-7}{x+2} = \dfrac{1}{4}$

21. $\dfrac{2}{x+1} = \dfrac{1}{x-2}$

22. $\dfrac{8}{y-3} = \dfrac{6}{y+4}$

23. $\dfrac{x}{6} - \dfrac{x}{10} = \dfrac{1}{6}$

24. $\dfrac{x}{8} - \dfrac{x}{12} = \dfrac{1}{8}$

25. $\dfrac{t+2}{5} - \dfrac{t-2}{4} = 1$

26. $\dfrac{x+1}{3} - \dfrac{x-1}{2} = 1$

27. $\dfrac{5}{x-1} = \dfrac{3}{x+2}$

28. $\dfrac{x-7}{x-9} = \dfrac{2}{x-9}$

29. $\dfrac{a-3}{3a+2} = \dfrac{1}{5}$

30. $\dfrac{x+7}{8x-5} = \dfrac{2}{3}$

31. $\dfrac{x-1}{x-5} = \dfrac{4}{x-5}$

32. $\dfrac{y+11}{y+8} = \dfrac{3}{y+8}$

33. $\dfrac{2}{x+3} = \dfrac{5}{x}$

34. $\dfrac{6}{y} = \dfrac{5}{y-8}$

35. $\dfrac{x-2}{x-3} = \dfrac{x-1}{x+1}$

36. $\dfrac{t+5}{t-2} = \dfrac{t-2}{t+4}$

37. $\dfrac{1}{x+3} + \dfrac{1}{x-3} = \dfrac{1}{x^2-9}$

38. $\dfrac{4}{x-3} + \dfrac{2x}{x^2-9} = \dfrac{1}{x+3}$

39. $\dfrac{x}{x+4} - \dfrac{4}{x-4} = \dfrac{x^2+16}{x^2-16}$

40. $\dfrac{5}{y-3} - \dfrac{30}{y^2-9} = 1$

41. $\dfrac{4-a}{8-a} = \dfrac{4}{a-8}$

42. $\dfrac{3}{x-7} = \dfrac{x+10}{x-7}$

43. $2 - \dfrac{a-2}{a+3} = \dfrac{a^2-4}{a+3}$

44. $\dfrac{5}{x-1} + x + 1 = \dfrac{5x+4}{x-1}$

45. $\dfrac{x+1}{x+2} = \dfrac{x+3}{x+4}$

46. $\dfrac{x^2}{x^2-4} = \dfrac{x}{x+2} - \dfrac{2x}{2-x}$

47. $4a - 3 = \dfrac{a+13}{a+1}$

48. $\dfrac{3x-9}{x-3} = \dfrac{5x-4}{2}$

49. $\dfrac{4}{y-2} - \dfrac{2y-3}{y^2-4} = \dfrac{5}{y+2}$

50. $\dfrac{y^2-4}{y+3} = 2 - \dfrac{y-2}{y+3}$

Skill Maintenance

In each of Exercises 51–58, fill in the blank with the correct term from the given list. Some of the choices may not be used.

51. A rational expression is a(n) _____ of two polynomials.   [6.1a]

52. A factor of a polynomial P is a polynomial that can be used to express P as a(n) _____ .   [5.1a]

53. Two expressions are _____ of each other if their product is 1.   [1.6b]

54. When _____ , always remember to look first for the greatest common factor.   [5.1b]

55. To find the LCM, use each factor the _____ number of times that it appears in any one factorization.   [6.3a]

56. When solving rational equations, always check a possible solution to see if it makes a denominator 0. If it does, it is _____ a solution.   [6.7a]

57. The quotient rule asserts that when dividing with exponential notation, if the bases are the same, keep the base and _____ the exponent of the denominator from the exponent of the numerator.   [4.1e]

58. Two expressions are _____ of each other if their sum is 0.   [1.3b]

not

always

factor

add

subtract

sum

product

smallest

greatest

factoring

quotient

reciprocals

additive inverses

exponents

Synthesis

Solve.

59. $\dfrac{x}{x^2+3x-4} + \dfrac{x+1}{x^2+6x+8} = \dfrac{2x}{x^2+x-2}$

60. $\dfrac{3a-5}{a^2+4a+3} + \dfrac{2a+2}{a+3} = \dfrac{a-3}{a+1}$

61. ⌁ Use a graphing calculator to check the solutions to Exercises 1–4.

62. ⌁ Use a graphing calculator to check the solutions to Exercises 13, 15, and 25.

6.8

Applications Using Rational Equations and Proportions

OBJECTIVES

a Solve applied problems using rational equations.

b Solve proportion problems.

In many areas of study, applications involving rates, proportions, or reciprocals translate to rational equations. By using the five steps for problem solving and the skills of Sections 6.1–6.7, we can now solve such problems.

a Solving Applied Problems

Problems Involving Work

EXAMPLE 1 *Sodding a Yard.* Charlie's Lawn Care has two three-person crews who lay sod. Crew A can lay 7 skids of sod in 4 hr, while crew B requires 6 hr to do the same job. How long would it take the two crews working together to lay 7 skids of sod?

1. **Familiarize.** We familiarize ourselves with the problem by considering two *incorrect* ways of translating the problem to mathematical language.

   a) A common *incorrect* way to translate the problem is to add the two times: 4 hr + 6 hr = 10 hr. Let's think about this. Crew A can do the job in 4 hr. If crew A and crew B work together, the time it takes them should be *less* than 4 hr. Thus we reject 10 hr as a solution, but we do have a partial check on any answer we get. The answer should be less than 4 hr.

   b) Another *incorrect* way to translate the problem is as follows. Suppose the two crews split up the sodding job in such a way that crew A does half the sodding and crew B does the other half. Then

   $$\text{crew A lays } \frac{1}{2} \text{ the skids of sod in } \frac{1}{2}(4 \text{ hr}), \text{ or 2 hr,}$$

   and $\quad$ crew B lays $\frac{1}{2}$ the skids of sod in $\frac{1}{2}(6 \text{ hr})$, or 3 hr.

   But time is wasted since crew A would finish 1 hr earlier than crew B. In effect, they have not worked together to get the job done as fast as possible. If crew A helps crew B after completing their half, the entire job could be done in a time somewhere between 2 hr and 3 hr.

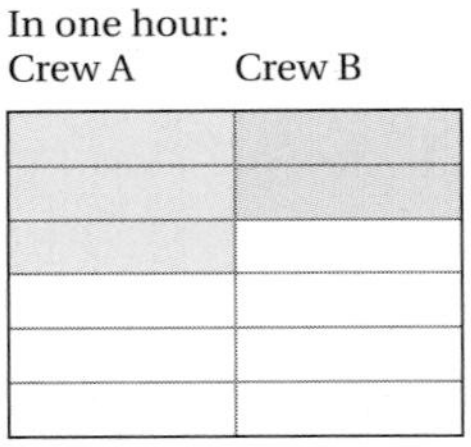

We proceed to a translation by considering how much of the job is finished in 1 hr, 2 hr, 3 hr, and so on. It takes crew A 4 hr to do the sodding job alone. Then, in 1 hr, crew A can do $\frac{1}{4}$ of the job. It takes crew B 6 hr to do the job alone. Then, in 1 hr, crew B can do $\frac{1}{6}$ of the job. Working together, the crews can do

$$\frac{1}{4} + \frac{1}{6}, \text{ or } \frac{3}{12} + \frac{2}{12}, \text{ or } \frac{5}{12} \text{ of the job in 1 hr.} \quad \text{(See Fig. 1.)}$$

In 2 hr, crew A can do $2\left(\frac{1}{4}\right)$ of the job and crew B can do $2\left(\frac{1}{6}\right)$ of the job. Working together, they can do

$$2\left(\frac{1}{4}\right) + 2\left(\frac{1}{6}\right), \text{ or } \frac{6}{12} + \frac{4}{12}, \text{ or } \frac{10}{12}, \text{ or } \frac{5}{6} \text{ of the job in 2 hr.}$$

(See Fig. 2.)

In one hour:
Crew A $\quad$ Crew B

FIGURE 1

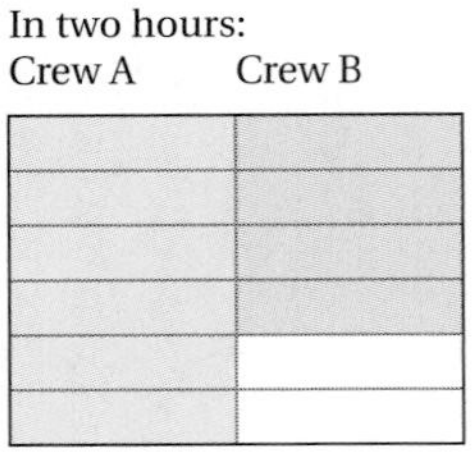

In two hours:
Crew A $\quad$ Crew B

FIGURE 2

Continuing this reasoning, we can create a table like the following one.

TIME	FRACTION OF THE JOB COMPLETED		
	CREW A	CREW B	TOGETHER
1 hr	$\frac{1}{4}$	$\frac{1}{6}$	$\frac{1}{4} + \frac{1}{6}$, or $\frac{5}{12}$
2 hr	$2\left(\frac{1}{4}\right)$	$2\left(\frac{1}{6}\right)$	$2\left(\frac{1}{4}\right) + 2\left(\frac{1}{6}\right)$, or $\frac{5}{6}$
3 hr	$3\left(\frac{1}{4}\right)$	$3\left(\frac{1}{6}\right)$	$3\left(\frac{1}{4}\right) + 3\left(\frac{1}{6}\right)$, or $1\frac{1}{4}$
t hr	$t\left(\frac{1}{4}\right)$	$t\left(\frac{1}{6}\right)$	$t\left(\frac{1}{4}\right) + t\left(\frac{1}{6}\right)$

From the table, we see that if the crews work together for 3 hr, the fraction of the job completed is $1\frac{1}{4}$, which is more of the job than needs to be done. We see again that the answer is somewhere between 2 hr and 3 hr. What we want is a number t such that the fraction of the job that gets completed is 1; that is, the job is just completed.

2. **Translate.** From the table, we see that the time we want is some number t for which

$$t\left(\frac{1}{4}\right) + t\left(\frac{1}{6}\right) = 1, \quad \text{or} \quad \frac{t}{4} + \frac{t}{6} = 1,$$

where 1 represents the idea that the entire job is completed in time t.

3. **Solve.** We solve the equation:

$$12\left(\frac{t}{4} + \frac{t}{6}\right) = 12 \cdot 1 \qquad \text{Multiplying by the LCM, which is } 2 \cdot 2 \cdot 3, \text{ or } 12$$

$$12 \cdot \frac{t}{4} + 12 \cdot \frac{t}{6} = 12$$

$$3t + 2t = 12$$

$$5t = 12$$

$$t = \frac{12}{5}, \text{ or } 2\frac{2}{5} \text{ hr.}$$

4. **Check.** In $\frac{12}{5}$ hr, crew A does $\frac{12}{5} \cdot \frac{1}{4}$, or $\frac{3}{5}$, of the job and crew B does $\frac{12}{5} \cdot \frac{1}{6}$, or $\frac{2}{5}$, of the job. Together, they do $\frac{3}{5} + \frac{2}{5}$, or 1 entire job.

   We also have another check in what we learned from the *Familiarize* step. The answer, $2\frac{2}{5}$ hr, is between 2 hr and 3 hr (see the table), and it is less than 4 hr, the time it takes crew A working alone.

5. **State.** It takes $2\frac{2}{5}$ hr for crew A and crew B working together to lay 7 skids of sod.

1. **Work Recycling.** Emma and Evan work as volunteers at a community recycling center. Emma can sort a morning's accumulation of recyclable objects in 3 hr, while Evan requires 5 hr to do the same job. How long would it take them, working together, to sort the recyclable material?

THE WORK PRINCIPLE

Suppose a = the time it takes A to do a job, b = the time it takes B to do the same job, and t = the time it takes them to do the job working together. Then

$$\frac{t}{a} + \frac{t}{b} = 1.$$

Do Exercise 1.

Problems Involving Motion

Problems that deal with distance, speed (or rate), and time are called **motion problems**. Translation of these problems involves the distance formula, $d = r \cdot t$, and/or the equivalent formulas $r = d/t$ and $t = d/r$.

MOTION FORMULAS

The following are the formulas for motion problems:

$d = rt$; Distance = Rate · Time (basic formula)

$r = \dfrac{d}{t}$; Rate = Distance/Time

$t = \dfrac{d}{r}$. Time = Distance/Rate

EXAMPLE 2 *Animal Speeds.* A zebra can run 15 mph faster than an elephant. A zebra can run 8 mi in the same time that an elephant can run 5 mi. Find the speed of each animal.

Source: *The World Almanac*, 2008, p. 279

1. **Familiarize.** We first make a drawing. We let r = the speed of the elephant. Then $r + 15$ = the speed of the zebra.

5 mi, r mph

8 mi, $r + 15$ mph

Recall that sometimes we need to find a formula in order to solve an application. As we see above, a formula that relates the notions of distance, speed, and time is $d = rt$, or

Distance = Speed · Time.

(Indeed, you may need to look up such a formula.)

Answer

1. $1\frac{7}{8}$ hr

Since each animal travels for the same length of time, we can use just t for time. We organize the information in a chart, as follows.

$$d = r \cdot t$$

	DISTANCE	SPEED	TIME	
Elephant	5	r	t	$\longrightarrow 5 = rt$
Zebra	8	$r + 15$	t	$\longrightarrow 8 = (r + 15)t$

2. **Translate.** We can apply the formula $d = rt$ along the rows of the table to obtain two equations:

$$5 = rt, \qquad \textbf{(1)}$$
$$8 = (r + 15)t. \qquad \textbf{(2)}$$

We know that the animals travel for the same length of time. Thus if we solve each equation for t and set the results equal to each other, we get an equation in terms of r.

Solving $5 = rt$ for t: $\qquad t = \dfrac{5}{r}$

Solving $8 = (r + 15)t$ for t: $\qquad t = \dfrac{8}{r + 15}$

Since the times are the same, we have the following equation:

$$\frac{5}{r} = \frac{8}{r + 15}.$$

3. **Solve.** To solve the equation, we first multiply on both sides by the LCM, which is $r(r + 15)$:

$$r(r + 15) \cdot \frac{5}{r} = r(r + 15) \cdot \frac{8}{r + 15} \qquad \text{Multiplying on both sides by the}$$
$$\text{LCM, which is } r(r + 15)$$
$$5(r + 15) = 8r \qquad \text{Simplifying}$$
$$5r + 75 = 8r \qquad \text{Removing parentheses}$$
$$75 = 3r$$
$$25 = r.$$

We now have a possible solution. The speed of the elephant is 25 mph, and the speed of the zebra is $r + 15 = 25 + 15$, or 40 mph.

4. **Check.** We first reread the problem to see what we were to find. We check the speeds of 25 for the elephant and 40 for the zebra. The zebra does travel 15 mph faster than the elephant and will travel farther than the elephant, which runs at a slower speed. If the zebra runs 8 mi at 40 mph, the time it has traveled is $\frac{8}{40}$, or $\frac{1}{5}$ hr. If the elephant runs 5 mi at 25 mph, the time it has traveled is $\frac{5}{25}$, or $\frac{1}{5}$ hr. Since the times are the same, the speeds check.

5. **State.** The speed of the elephant is 25 mph and the speed of the zebra is 40 mph.

Do Exercise 2.

2. **Driving Speed.** Nancy drives 20 mph faster than her father, Greg. In the same time that Nancy travels 180 mi, her father travels 120 mi. Find their speeds.

Nancy's car
180 mi, $r + 20$ mph

Greg's car
120 mi, r mph

3. Find the ratio of 145 km to 2.5 liters (L).

4. Batting Average. Recently, a baseball player got 7 hits in 25 times at bat. What was the rate, or batting average, in hits per times at bat?

5. Impulses in nerve fibers travel 310 km in 2.5 hr. What is the rate, or speed, in kilometers per hour?

6. A lake of area 550 yd² contains 1320 fish. What is the population density of the lake, in number of fish per square yard?

7. Mileage. In highway driving, a 2009 Toyota Venza can travel 261 mi on 9 gal of gas. How much gas will be required for an 820-mi trip?

Source: *Car and Driver,* January 2009

Answers

3. 58 km/L **4.** 0.28 hit per times at bat
5. 124 km/h **6.** 2.4 fish/yd²
7. About 28.3 gal

---

b Applications Involving Proportions

We now consider applications with proportions. A **proportion** involves ratios. A **ratio** of two quantities is their quotient. For example, 73% is the ratio of 73 to 100, $\frac{73}{100}$. The ratio of two different kinds of measure is called a **rate**. Suppose an animal travels 720 ft in 2.5 hr. Its **rate**, or **speed**, is then

$$\frac{720 \text{ ft}}{2.5 \text{ hr}} = 288 \ \frac{\text{ft}}{\text{hr}}.$$

Do Exercises 3–6.

PROPORTION

An equality of ratios, $A/B = C/D$, is called a **proportion**. The numbers within a proportion are said to be **proportional** to each other.

EXAMPLE 3 *Mileage.* A 2009 Chevrolet Cobalt SS can travel 176 mi in city driving on 8 gal of gas. Find the amount of gas required for 242 mi of city driving.

Source: *Road & Track,* November 2008

1. **Familiarize.** We know that the Chevrolet can travel 176 mi on 8 gal of gas. Thus we can set up a proportion, letting x = the amount of gas required to drive 242 mi.

2. **Translate.** We assume that the car uses gas at the same rate in all city driving. Thus the ratios are the same and we can write a proportion. Note that the units of *mileage* are in the numerators and the units of *gasoline* are in the denominators.

$$\begin{array}{ll} \text{Miles} \rightarrow \\ \text{Gas} \rightarrow \end{array} \frac{176}{8} = \frac{242}{x} \begin{array}{ll} \leftarrow \text{Miles} \\ \leftarrow \text{Gas} \end{array}$$

3. **Solve.** To solve for x, we multiply on both sides by the LCM, which is $8x$:

$$8x \cdot \frac{176}{8} = 8x \cdot \frac{242}{x} \qquad \text{Multiplying by } 8x$$

$$176x = 1936 \qquad \text{Simplifying}$$

$$\frac{176x}{176} = \frac{1936}{176} \qquad \text{Dividing by 176}$$

$$x = 11. \qquad \text{Simplifying}$$

We can also use cross products to solve the proportion:

$$\frac{176}{8} = \frac{242}{x} \qquad 176 \cdot x \text{ and } 8 \cdot 242 \text{ are cross products.}$$

$$176 \cdot x = 8 \cdot 242 \qquad \text{Equating cross products}$$

$$\frac{176 \cdot x}{176} = \frac{8 \cdot 242}{176} \qquad \text{Dividing by 176}$$

$$x = 11.$$

4. **Check.** The check is left to the student.

5. **State.** The Chevrolet Cobalt will require 11 gal of gas for 242 mi of city driving.

Do Exercise 7.

EXAMPLE 4 *Environmental Science.* The Fish and Wildlife Division of the Indiana Department of Natural Resources recently completed a study that determined the number of largemouth bass in Lake Monroe, near Bloomington, Indiana. For this project, anglers caught 300 largemouth bass, tagged them, and threw them back into the lake. Later, they caught 85 largemouth bass and found that 15 of them were tagged. Estimate how many largemouth bass are in the lake.

Source: Department of Natural Resources, Fish and Wildlife Division, Kevin Hoffman

Lake Monroe

1. **Familiarize.** The ratio of the number of largemouth bass tagged to the total number of fish in the lake, F, is $300/F$. Of the 85 largemouth bass caught later, 15 fish were tagged. The ratio of fish tagged to fish caught is $\frac{15}{85}$.

2. **Translate.** Assuming that the two ratios are the same, we can translate to a proportion.

$$\text{Fish tagged originally} \to \frac{300}{F} = \frac{15}{85} \leftarrow \text{Tagged fish caught later}$$
$$\text{Fish in lake} \to \qquad\qquad \leftarrow \text{Fish caught later}$$

3. **Solve.** We solve the proportion. We multiply by the LCM, which is $85F$.

$$85F \cdot \frac{300}{F} = 85F \cdot \frac{15}{85} \qquad \text{Multiplying by } 85F$$

$$85 \cdot 300 = F \cdot 15$$

$$\frac{85 \cdot 300}{15} = F \qquad\qquad \text{Dividing by 15}$$

$$1700 = F$$

4. **Check.** The check is left to the student.

5. **State.** We estimate that there are about 1700 largemouth bass in the lake.

Do Exercise 8.

8. Environmental Science.
To determine the number of humpback whales in a pod, a marine biologist, using tail markings, identifies 27 members of the pod. Several weeks later, 40 whales from the pod are randomly sighted. Of the 40 sighted, 12 are from the 27 originally identified. Estimate the number of whales in the pod.

EXAMPLE 5 *Fruit Quality.* A company that prepares and sells gift boxes and baskets of fruit must order quantities of fruit larger than what they need to allow for selecting fruit that meets their quality standards. The packing-room supervisor keeps records and notes that approximately 87 pears from a shipment of 1000 do not meet the company standards. Over the holidays, a shipment of 3200 pears is ordered. How many pears can the company expect will not meet the quality required?

Answer
8. 90 whales

1. **Familiarize.** The ratio of the number of pears *P* that do not meet the standards to the total order of 3200 is *P*/3200. The ratio of the average number of pears that do not meet the standard in an order of 1000 pears is $\frac{87}{1000}$.

2. **Translate.** Assuming that the two ratios are the same, we can translate to a proportion:

$$\frac{P}{3200} = \frac{87}{1000}.$$

3. **Solve.** We solve the proportion. We multiply by the LCM, which is 16,000.

$$16{,}000 \cdot \frac{P}{3200} = 16{,}000 \cdot \frac{87}{1000}$$

$$5 \cdot P = 16 \cdot 87$$

$$P = \frac{16 \cdot 87}{5}$$

$$P \approx 278.4$$

4. **Check.** The check is left to the student.

5. **State.** We estimate that there are about 278 pears in an order of 3200 that do not meet the quality standards.

Do Exercise 9.

> **9.** XYZ Pools and Spas, Inc., adds 2 gal of chlorine per 8000 gal of water in a newly constructed pool. How much chlorine is needed for a pool requiring 20,500 gal of water? Round the answer to the nearest tenth of a gallon.

Similar Triangles

Proportions arise in geometry when we are studying *similar triangles*. If two triangles are **similar**, then their corresponding angles have the same measure and their corresponding sides are proportional. To illustrate, if triangle *ABC* is similar to triangle *RST*, then angles *A* and *R* have the same measure, angles *B* and *S* have the same measure, angles *C* and *T* have the same measure, and

$$\frac{a}{r} = \frac{b}{s} = \frac{c}{t}.$$

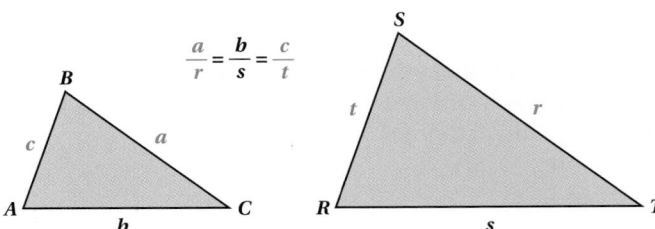

Answer

9. 5.1 gal

SIMILAR TRIANGLES

In **similar triangles**, corresponding angles have the same measure and the lengths of corresponding sides are proportional.

EXAMPLE 6 *Similar Triangles.* Triangles ABC and XYZ below are similar triangles. Solve for z if $a = 8$, $c = 5$, and $x = 10$.

We make a drawing, write a proportion, and then solve. Note that side a is always opposite angle A, side x is always opposite angle X, and so on.

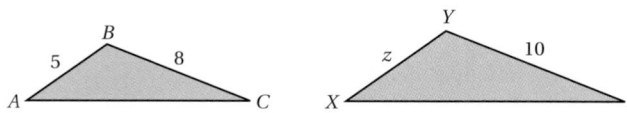

We have

$$\frac{z}{5} = \frac{10}{8}$$ The proportion $\frac{5}{z} = \frac{8}{10}$ could also be used.

$$40 \cdot \frac{z}{5} = 40 \cdot \frac{10}{8}$$ Multiplying by 40

$$8z = 50$$

$$z = \frac{50}{8}$$ Dividing by 8

$$z = \frac{25}{4}, \text{ or } 6.25.$$

Do Exercise 10.

EXAMPLE 7 *Rafters of a House.* Carpenters use similar triangles to determine the lengths of rafters for a house. They first choose the pitch of the roof, or the ratio of the rise over the run. Then using a triangle with that ratio, they calculate the length of the rafter needed for the house. Loren is constructing rafters for a roof with a 6/12 pitch on a house that is 30 ft wide. Using a rafter guide (see the figure at right), Loren knows that the rafter length corresponding to a 6-unit rise and a 12-unit run is 13.4. Find the length x of the rafter of the house.

We have the proportion

Length of rafter
in 6/12 triangle → $\dfrac{13.4}{x} = \dfrac{12}{15}$ ← Run in 6/12 triangle
Length of rafter →   ← Run in similar
on the house   triangle on the house

Solve: $13.4 \cdot 15 = x \cdot 12$ Equating cross products

$$\frac{13.4 \cdot 15}{12} = \frac{x \cdot 12}{12}$$ Dividing by 12 on both sides

$$\frac{13.4 \cdot 15}{12} = x$$

$$16.75 \text{ ft} = x$$

The length of the rafter x of the house is about 16.75 ft, or 16 ft 9 in.

Do Exercise 11.

10. Height of a Flagpole. How high is a flagpole that casts a 45-ft shadow at the same time that a 5.5-ft woman casts a 10-ft shadow?

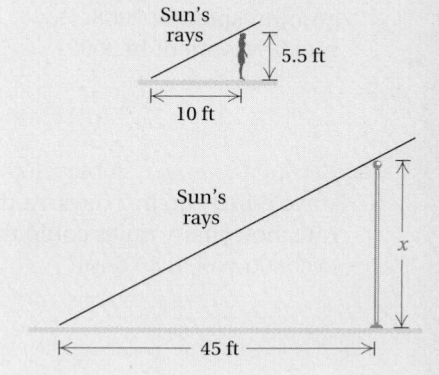

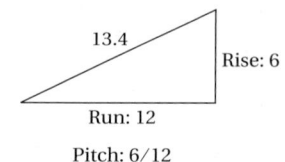

Pitch: 6/12

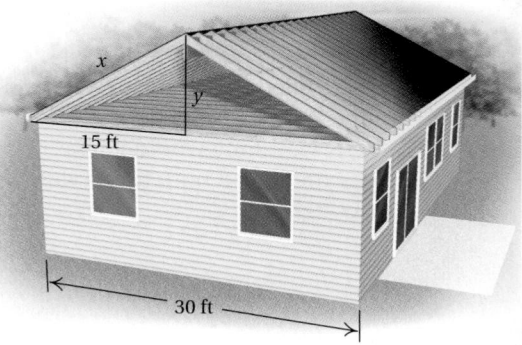

11. Rafters of a House. Referring to Example 7, find the length y in the rafter of the house.

Answers

10. 24.75 ft   **11.** 7.5 ft

Translating
for Success

1. *Search Engine Ads.* In 2009, it was estimated that $3.6 billion was spent in advertising on Internet search engines. This was a 25% increase over the amount spent in 2008. How much was spent in 2008?

2. *Cycling Distance.* A bicyclist traveled 197 mi in 7 days. At this rate, how many miles could the cyclist travel in 30 days?

3. *Bicycling.* The speed of one bicyclist is 2 km/h faster than the speed of another bicyclist. The first bicyclist travels 60 km in the same amount of time that it takes the second to travel 50 km. Find the speed of each bicyclist.

4. *Filling Time.* A swimming pool can be filled in 5 hr by hose A alone and in 6 hr by hose B alone. How long would it take to fill the tank if both hoses were working?

5. *Office Budget.* Emma has $36 budgeted for office stationery. Engraved stationery costs $20 for the first 25 sheets and $0.08 for each additional sheet. How many engraved sheets of stationery can Emma order and still stay within her budget?

The goal of these matching questions is to practice step (2), *Translate,* of the five-step problem-solving process. Translate each word problem to an equation and select a correct translation from equations A–O.

A. $2x + 2(x + 1) = 613$

B. $x^2 + (x + 1)^2 = 613$

C. $\dfrac{60}{x + 2} = \dfrac{50}{x}$

D. $20 + 0.08(x - 25) = 36$

E. $\dfrac{197}{7} = \dfrac{x}{30}$

F. $x + (x + 1) = 613$

G. $\dfrac{7}{197} = \dfrac{x}{30}$

H. $x^2 + (x + 2)^2 = 612$

I. $x^2 + (x + 1)^2 = 612$

J. $\dfrac{50}{x + 2} = \dfrac{60}{x}$

K. $x + 25\% \cdot x = 3.6$

L. $t + 5 = 7$

M. $x^2 + (x + 1)^2 = 452$

N. $\dfrac{1}{5} + \dfrac{1}{6} = \dfrac{1}{t}$

O. $x^2 + (x + 2)^2 = 452$

Answers on page A-18

6. *Sides of a Square.* If the sides of a square are increased by 2 ft, the area of the original square plus the area of the enlarged square is 452 ft^2. Find the length of a side of the original square.

7. *Consecutive Integers.* The sum of two consecutive integers is 613. Find the integers.

8. *Sums of Squares.* The sum of the squares of two consecutive odd integers is 612. Find the integers.

9. *Sums of Squares.* The sum of the squares of two consecutive integers is 613. Find the integers.

10. *Rectangle Dimensions.* The length of a rectangle is 1 ft longer than its width. Find the dimensions of the rectangle such that the perimeter of the rectangle is 613 ft.

a Solve.

1. *Construction.* It takes Mandy 4 hr to put up paneling in a room. Omar takes 5 hr to do the same job. How long would it take them, working together, to panel the room?

2. *Carpentry.* By checking work records, a carpenter finds that Juanita can build a small shed in 12 hr. Anton can do the same job in 16 hr. How long would it take if they worked together?

3. *Shoveling.* Vern can shovel the snow from his driveway in 45 min. Nina can do the same job in 60 min. How long would it take Nina and Vern to shovel the driveway if they worked together?

4. *Raking.* Zoë can rake her yard in 4 hr. Steffi does the same job in 3 hr. How long would it take the two of them, working together, to rake the yard?

5. *Wiring.* By checking work records, a contractor finds that Peggy Ann can wire a room addition in 9 hr. It takes Matthew 7 hr to wire the same room. How long would it take if they worked together?

6. *Plumbing.* By checking work records, a plumber finds that Raul can plumb a house in 48 hr. Mira can do the same job in 36 hr. How long would it take if they worked together?

7. *Gardening.* Nicole can weed her vegetable garden in 50 min. Glen can weed the same garden in 40 min. How long would it take if they worked together?

8. *Harvesting.* Bobbi can pick a quart of raspberries in 20 min. Blanche can pick a quart in 25 min. How long would it take if Bobbi and Blanche worked together?

9. *Office Printers.* The HP Officejet 4215 All-In-One printer, fax, scanner, and copier can print in black one copy of a company's year-end report in 10 min. The HP Officejet 7410 All-In-One can print the same report in 6 min. How long would it take the two printers, working together, to print one copy of the report?

HP Officejet 4215   HP Officejet 7410

10. *Office Copiers.* The HP Officejet 7410 All-In-One printer, fax, scanner, and copier can copy in color a staff training manual in 9 min. The HP Officejet 4215 All-In-One can copy the same report in 15 min. How long would it take the two copiers, working together, to make one copy of the manual?

11. *Car Speed.* Rick drives his four-wheel-drive truck 40 km/h faster than Sarah drives her Saturn. While Sarah travels 150 km, Rick travels 350 km. Find their speeds.

Complete this table and the equations as part of the *Familiarize* step.

d	=	r	$\cdot$	t

	DISTANCE	SPEED	TIME	
Car	150	r		$\longrightarrow 150 = r(\ \ )$
Truck	350		t	$\longrightarrow 350 = (\ \ )t$

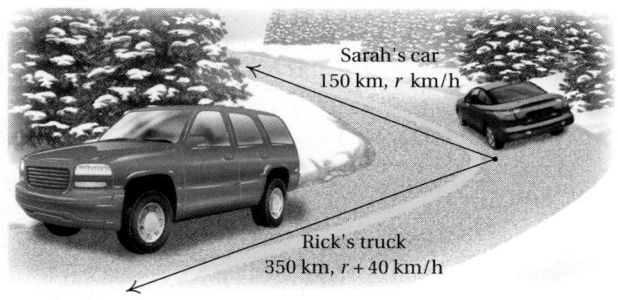

Sarah's car
150 km, r km/h

Rick's truck
350 km, $r + 40$ km/h

12. *Car Speed.* A passenger car travels 30 km/h faster than a delivery truck. While the car goes 400 km, the truck goes 250 km. Find their speeds.

13. *Train Speed.* The speed of a B & M freight train is 14 mph slower than the speed of an Amtrak passenger train. The freight train travels 330 mi in the same time that it takes the passenger train to travel 400 mi. Find the speed of each train.

Complete this table and the equations as part of the *Familiarize* step.

d	=	r	$\cdot$	t

	DISTANCE	SPEED	TIME	
B & M	330		t	$\longrightarrow 330 = (\ \ )t$
Amtrak	400	r		$\longrightarrow 400 = r(\ \ )$

14. *Train Speed.* The speed of a freight train is 15 mph slower than the speed of a passenger train. The freight train travels 390 mi in the same time that it takes the passenger train to travel 480 mi. Find the speed of each train.

15. *Trucking Speed.* A long-distance trucker traveled 120 mi in one direction during a snowstorm. The return trip in rainy weather was accomplished at double the speed and took 3 hr less time. Find the speed going.

120 mi, *r, t*

120 mi, 2*r, t* − 3

16. *Car Speed.* After driving 126 mi, Syd found that the drive would have taken 1 hr less time by increasing the speed by 8 mph. What was the actual speed?

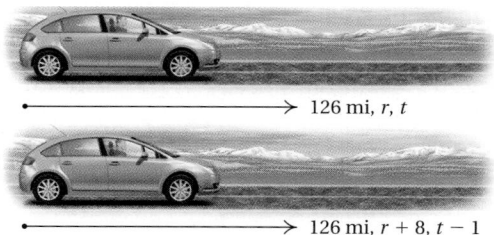

126 mi, *r, t*

126 mi, *r* + 8, *t* − 1

17. *Bicycle Speed.* Hank bicycles 5 km/h slower than Kelly. In the time that it takes Hank to bicycle 42 km, Kelly can bicycle 57 km. How fast does each bicyclist travel?

18. *Driving Speed.* Kaylee's Lexus travels 30 mph faster than Gavin's Harley. In the same time that Gavin travels 75 mi, Kaylee travels 120 mi. Find their speeds.

19. *Walking Speed.* Bonnie power walks 3 km/h faster than Ralph. In the time that it takes Ralph to walk 7.5 km, Bonnie walks 12 km. Find their speeds.

20. *Cross-Country Skiing.* Gerard cross-country skis 4 km/h faster than Sally. In the time that it takes Sally to ski 18 km, Gerard skis 24 km. Find their speeds.

21. *Tractor Speed.* Hobart's tractor is just as fast as Evan's. It takes Hobart 1 hr more than it takes Evan to drive to town. If Hobart is 20 mi from town and Evan is 15 mi from town, how long does it take Evan to drive to town?

22. *Boat Speed.* Tory and Emilio's motorboats travel at the same speed. Tory pilots her boat 40 km before docking. Emilio continues for another 2 hr, traveling a total of 100 km before docking. How long did it take Tory to navigate the 40 km?

Find the ratio of each of the following. Simplify if possible.

23. 60 students, 18 teachers

24. 800 mi, 50 gal

25. *Speed of Black Racer.* A black racer snake travels 4.6 km in 2 hr. What is the speed, in kilometers per hour?

26. *Speed of Light.* Light travels 558,000 mi in 3 sec. What is the speed, in miles per second?

Solve.

27. *Protein Needs.* A 120-lb person should eat a minimum of 44 g of protein each day. How much protein should a 180-lb person eat each day?

28. *Coffee Beans.* The coffee beans from 14 trees are required to produce 7.7 kg of coffee. (This is the average amount that each person in the United States drinks each year.) How many trees are required to produce 320 kg of coffee?

29. *Hemoglobin.* A normal 10-cc specimen of human blood contains 1.2 g of hemoglobin. How much hemoglobin would 16 cc of the same blood contain?

30. *Walking Speed.* Wanda walked 234 km in 14 days. At this rate, how far would she walk in 42 days?

31. *Honey Bees.* Making 1 lb of honey requires 20,000 trips by bees to flowers to gather nectar. How many pounds of honey would 35,000 trips produce?

Source: Tom Turpin, Professor of Entomology, Purdue University

32. *Cockroaches and Horses.* A cockroach can run about 2 mi/hr (mph). The average body length of a cockroach is 1 in. The average body length of a horse is 8 ft (96 in.). If a horse's speed-to-length ratio were the same as that of a cockroach, how fast would a horse run?

Source: Tom Turpin, Professor of Entomology, Purdue University

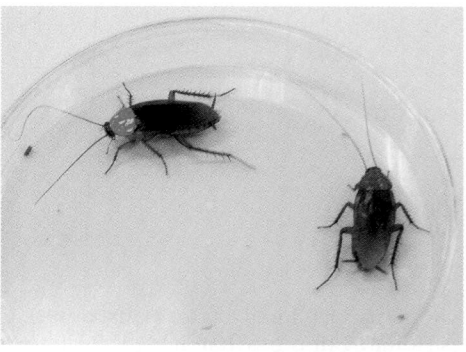

Professor Turpin founded the annual cockroach race at Purdue University.

33. *Money.* The ratio of the weight of copper to the weight of zinc in a U.S. penny is $\frac{1}{39}$. If 50 kg of zinc is being turned into pennies, how much copper is needed?

34. *Baking.* In a potato bread recipe, the ratio of milk to flour is $\frac{3}{13}$. If 5 cups of milk are used, how many cups of flour are used?

35. *Ryan Howard.* In the 2009 Major League Baseball season, Ryan Howard, playing for the Philadelphia Phillies of the National League, collected 118 hits in 439 at-bats in his first 114 games.

  a) The ratio of number of hits to number of at-bats, rounded to the nearest thousandth, is a player's *batting average.* What was Howard's batting average in his first 114 games?

  b) Based on the ratio of number of hits to number of games, how many hits would he get in the 162-game season?

  c) Based on the ratio of number of hits to number of at-bats and assuming he bats 700 times in 2009, how many hits would he get?

36. *Evan Longoria.* In the 2009 Major League Baseball season, Evan Longoria, playing for the Tampa Bay Rays of the American League, collected 116 hits in 416 at-bats in his first 112 games.

  a) The ratio of number of hits to number of at-bats, rounded to the nearest thousandth, is a player's *batting average.* What was Longoria's batting average in his first 112 games?

  b) Based on the ratio of number of hits to number of games, how many hits would he get in the 162-game season?

  c) Based on the ratio of number of hits to number of at-bats and assuming he bats 620 times in 2009, how many hits would he get?

Hat Sizes. Hat sizes are determined by measuring the circumference of one's head in either inches or centimeters. Use ratio and proportion to complete the missing parts of the following table.

	HAT SIZE	HEAD CIRCUMFERENCE (in inches)	HEAD CIRCUMFERENCE (in centimeters)
	$6\frac{3}{4}$	$21\frac{1}{5}$ in.	53.8 cm
37.	7		
38.			56.8 cm
39.		$22\frac{4}{5}$ in.	
40.	$7\frac{3}{8}$		
41.			59.8 cm
42.		24 in.	

43. *Estimating Trout Population.* To determine the number of trout in a lake, a conservationist catches 112 trout, tags them, and throws them back into the lake. Later, 82 trout are caught; 32 of them are tagged. Estimate the number of trout in the lake.

44. *Grass Seed.* It takes 60 oz of grass seed to seed 3000 ft^2 of lawn. At this rate, how much would be needed to seed 5000 ft^2 of lawn?

45. *Quality Control.* A sample of 144 firecrackers contained 9 "duds." How many duds would you expect in a sample of 3200 firecrackers?

46. *Frog Population.* To estimate how many frogs there are in a rain forest, a research team tags 600 frogs and then releases them. Later, the team catches 300 frogs and notes that 25 of them have been tagged. Estimate the total frog population in the rain forest.

47. *Weight on Mars.* The ratio of the weight of an object on Mars to the weight of the same object on Earth is 0.4 to 1.
   a) How much would a 12-ton rocket weigh on Mars?
   b) How much would a 120-lb astronaut weigh on Mars?

48. *Weight on Moon.* The ratio of the weight of an object on the moon to the weight of the same object on Earth is 0.16 to 1.
   a) How much would a 12-ton rocket weigh on the moon?
   b) How much would a 180-lb astronaut weigh on the moon?

Geometry. For each pair of similar triangles, find the length of the indicated side.

49. *b*:

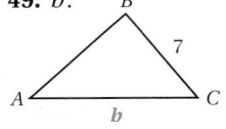

50. *a*:

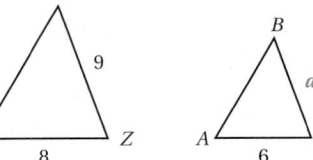

51. *f*:

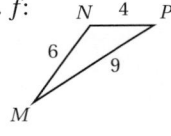

52. *r*:

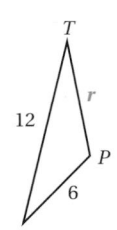

53. *h*:

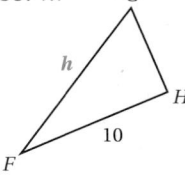

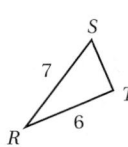

54. *n*:

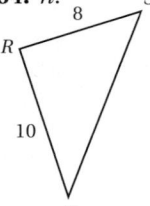

55. l:

56. h:

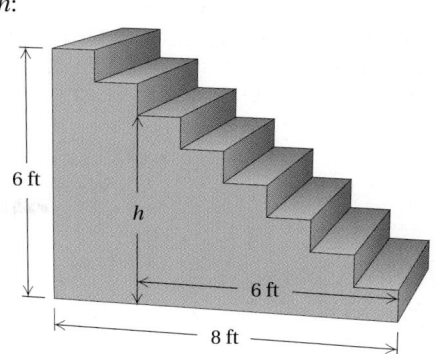

Skill Maintenance

Find the slope, if it exists, of the line containing the given pair of points.   [3.4a]

57. $(7, -6), (0, -6)$

58. $(3, -11), (-4, 3)$

Simplify.   [4.1d]

59. $x^5 \cdot x^6$

60. $x^{-5} \cdot x^6$

61. $x^{-5} \cdot x^{-6}$

62. $x^5 \cdot x^{-6}$

Graph.   [3.2a], [3.3a]

63. $y = 2x - 6$

64. $y = -2x + 6$

65. $3x + 2y = 12$

66. $x - 3y = 6$

67. $y = -\dfrac{3}{4}x + 2$

68. $y = \dfrac{2}{5}x - 4$

Synthesis

69. Ann and Betty work together and complete a sales report in 4 hr. It would take Betty 6 hr longer, working alone, to do the job than it would Ann. How long would it take each of them to do the job working alone?

70. Express 100 as the sum of two numbers for which the ratio of one number, increased by 5, to the other number, decreased by 5, is 4.

71. How soon, in minutes, after 5 o'clock will the hands on a clock first be together?

72. Rachel allows herself 1 hr to reach a sales appointment 50 mi away. After she has driven 30 mi, she realizes that she must increase her speed by 15 mph in order to arrive on time. What was her speed for the first 30 mi?

73. Solve $\dfrac{t}{a} + \dfrac{t}{b} = 1$ for t.

Variation and Applications

OBJECTIVES

a Find an equation of direct variation given a pair of values of the variables.

b Solve applied problems involving direct variation.

c Find an equation of inverse variation given a pair of values of the variables.

d Solve applied problems involving inverse variation.

e Find equations of other kinds of variation given values of the variables.

f Solve applied problems involving other kinds of variation.

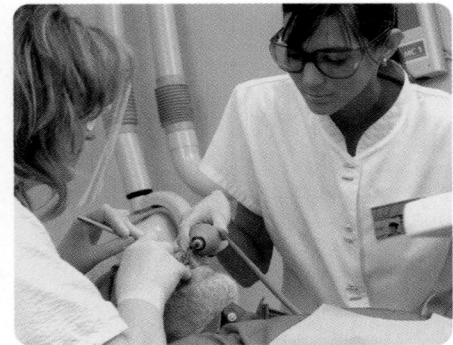

We now extend our study of formulas and functions by considering applications involving variation.

a Equations of Direct Variation

A dental hygienist earns $24 per hour. In 1 hr, $24 is earned; in 2 hr, $48 is earned; in 3 hr, $72 is earned; and so on. We plot this information on a graph, using the number of hours as the first coordinate and the amount earned as the second coordinate to form a set of ordered pairs:

$(1, 24)$, $(2, 48)$,
$(3, 72)$, $(4, 96)$,

and so on.

Note that the ratio of the second coordinate to the first is the same number for each point:

$$\frac{24}{1} = 24, \qquad \frac{48}{2} = 24, \qquad \frac{72}{3} = 24, \qquad \frac{96}{4} = 24, \quad \text{and so on.}$$

Whenever a situation produces pairs of numbers in which the *ratio is constant*, we say that there is **direct variation**. Here the amount earned varies directly as the time:

$$\frac{E}{t} = 24 \, (\text{a constant}), \quad \text{or} \quad E = 24t,$$

or, using function notation, $E(t) = 24t$. The equation is an **equation of direct variation**. The coefficient, 24 in the situation above, is called the **variation constant**. In this case, it is the rate of change of earnings with respect to time.

> ### DIRECT VARIATION
>
> If a situation gives rise to a linear function $f(x) = kx$, or $y = kx$, where k is a positive constant, we say that we have **direct variation**, or that **y varies directly as x**, or that **y is directly proportional to x**. The number k is called the **variation constant**, or **constant of proportionality**.

EXAMPLE 1 Find the variation constant and an equation of variation in which y varies directly as x, and $y = 32$ when $x = 2$.

We know that $(2, 32)$ is a solution of $y = kx$. Thus,

$$y = kx$$
$$32 = k \cdot 2 \qquad \text{Substituting}$$
$$\frac{32}{2} = k, \text{ or } k = 16. \qquad \text{Solving for } k$$

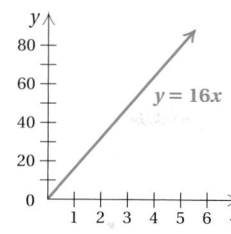

The variation constant, 16, is the rate of change of y with respect to x. The equation of variation is $y = 16x$.

> The graph of $y = kx$, $k > 0$, always goes through the origin and rises from left to right. Note that as x increases, y increases. The constant k is also the slope of the line.
>
>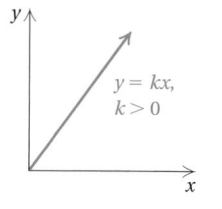

Do Exercises 1 and 2.

1. Find the variation constant and an equation of variation in which y varies directly as x, and $y = 8$ when $x = 20$.

2. Find the variation constant and an equation of variation in which y varies directly as x, and $y = 5.6$ when $x = 8$.

b Applications of Direct Variation

EXAMPLE 2 *Water from Melting Snow.* The number of centimeters W of water produced from melting snow varies directly as S, the number of centimeters of snow. Meteorologists have found that, under certain conditions, 150 cm of snow will melt to 16.8 cm of water. To how many centimeters of water will 200 cm of snow melt?

We first find the variation constant using the data and then find an equation of variation:

$$W = kS \qquad \text{W varies directly as S.}$$
$$16.8 = k \cdot 150 \qquad \text{Substituting}$$
$$\frac{16.8}{150} = k \qquad \text{Solving for } k$$
$$0.112 = k. \qquad \text{This is the variation constant.}$$

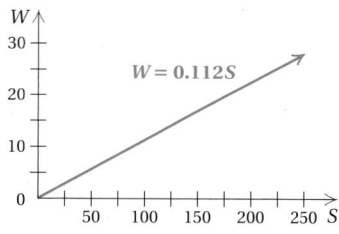

The equation of variation is $W = 0.112S$.

Next, we use the equation to find how many centimeters of water will result from melting 200 cm of snow:

$$W = 0.112S$$
$$W = 0.112(200) \qquad \text{Substituting}$$
$$W = 22.4.$$

Thus, 200 cm of snow will melt to 22.4 cm of water.

Do Exercises 3 and 4. (Exercise 4 is on the following page.)

3. Ohm's Law. Ohm's Law states that the voltage V in an electric circuit varies directly as the number of amperes I of electric current in the circuit. If the voltage is 10 volts when the current is 3 amperes, what is the voltage when the current is 15 amperes?

Answers

1. $\frac{2}{5}$; $y = \frac{2}{5}x$ 2. 0.7; $y = 0.7x$ 3. 50 volts

(c) Equations of Inverse Variation

A bus is traveling a distance of 20 mi. At a speed of 5 mph, the trip will take 4 hr; at 20 mph, it will take 1 hr; at 40 mph, it will take $\frac{1}{2}$ hr; and so on. We plot this information on a graph, using speed as the first coordinate and time as the second coordinate to determine a set of ordered pairs:

$(5, 4), \quad (10, 2),$
$(20, 2), \quad \left(40, \frac{1}{2}\right),$
and so on.

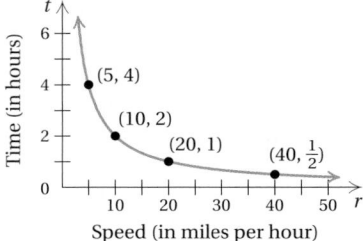

Note that the products of the coordinates are all the same number:

$$5 \cdot 4 = 20, \quad 20 \cdot 1 = 20, \quad 40 \cdot \tfrac{1}{2} = 20, \quad \text{and so on.}$$

Whenever a situation produces pairs of numbers in which the *product is constant*, we say that there is **inverse variation**. Here the time varies inversely as the speed:

$$rt = 20 \,(\text{a constant}), \quad \text{or} \quad t = \frac{20}{r}.$$

The equation is an **equation of inverse variation**. The coefficient, 20, in the situation above, is called the **variation constant**. Note that as the first number (speed) increases, the second number (time) decreases.

INVERSE VARIATION

If a situation gives rise to a function $f(x) = k/x$, or $y = k/x$, where k is a positive constant, we say that we have **inverse variation**, or that y **varies inversely as** x, or that y **is inversely proportional to** x. The number k is called the **variation constant**, or **constant of proportionality**.

EXAMPLE 3 Find the variation constant and an equation of variation in which y varies inversely as x, and $y = 32$ when $x = 0.2$.

We know that $(0.2, 32)$ is a solution of $y = k/x$. We substitute:

$$y = \frac{k}{x}$$

$$32 = \frac{k}{0.2} \qquad \text{Substituting}$$

$$(0.2)32 = k \qquad \text{Solving for } k$$

$$6.4 = k.$$

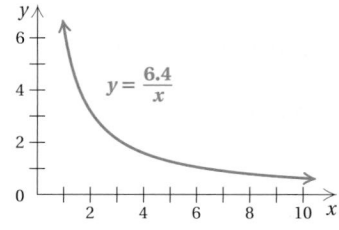

The variation constant is 6.4. The equation of variation is $y = \dfrac{6.4}{x}$.

Answer

4. $66\dfrac{2}{3}$ lb

It is helpful to look at the graph of $y = k/x$, $k > 0$. The graph is like the one shown at right for positive values of x. Note that as x increases, y decreases.

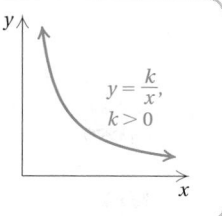

Do Exercise 5.

5. Find the variation constant and an equation of variation in which y varies inversely as x, and $y = 0.012$ when $x = 50$.

d Applications of Inverse Variation

EXAMPLE 4 *Musical Pitch.* The pitch P of a musical tone varies inversely as its wavelength W. One tone has a pitch of 550 vibrations per second and a wavelength of 1.92 ft. Find the pitch of another tone that has a wavelength of 3.2 ft.

We first find the variation constant using the data given and then find an equation of variation:

$$P = \frac{k}{W} \qquad P \text{ varies inversely as } W.$$

$$550 = \frac{k}{1.92} \qquad \text{Substituting}$$

$$1056 = k. \qquad \text{Solving for } k, \text{ the variation constant}$$

The equation of variation is $P = \dfrac{1056}{W}$.

Next, we use the equation to find the pitch of a tone that has a wavelength of 3.2 ft:

$$P = \frac{1056}{W} \qquad \text{Equation of variation}$$

$$P = \frac{1056}{3.2} \qquad \text{Substituting}$$

$$P = 330.$$

The pitch of a musical tone that has a wavelength of 3.2 ft is 330 vibrations per second.

Do Exercise 6.

6. Cleaning Bleachers. The time t to do a job varies inversely as the number of people P who work on the job (assuming that all work at the same rate). It takes 4.5 hr for 12 people to clean a section of bleachers after a NASCAR race. How long would it take 15 people to complete the same job?

Answers

5. $0.6; y = \dfrac{0.6}{x}$ **6.** 3.6 hr

e Other Kinds of Variation

We now look at other kinds of variation. Consider the equation for the area of a circle, in which A and r are variables and π is a constant:

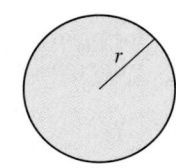

$$A = \pi r^2, \quad \text{or, as a function,} \quad A(r) = \pi r^2.$$

We say that the area *varies directly* as the square of the radius.

> y varies directly as the nth power of x if there is some positive constant k such that $y = kx^n$.

EXAMPLE 5 Find an equation of variation in which y varies directly as the square of x, and $y = 12$ when $x = 2$.

We write an equation of variation and find k:

$$y = kx^2$$
$$12 = k \cdot 2^2$$
$$12 = k \cdot 4$$
$$3 = k.$$

Thus, $y = 3x^2$.

Do Exercise 7.

7. Find an equation of variation in which y varies directly as the square of x, and $y = 175$ when $x = 5$.

From the law of gravity, we know that the weight W of an object *varies inversely* as the square of its distance d from the center of the earth:

$$W = \frac{k}{d^2}.$$

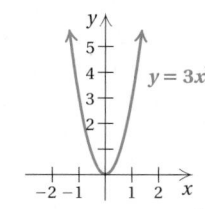

Earth

> y varies inversely as the nth power of x if there is some positive constant k such that
>
> $$y = \frac{k}{x^n}.$$

EXAMPLE 6 Find an equation of variation in which W varies inversely as the square of d, and $W = 3$ when $d = 5$.

$$W = \frac{k}{d^2}$$

$$3 = \frac{k}{5^2} \quad \text{Substituting}$$

$$3 = \frac{k}{25}$$

$$75 = k$$

Thus, $W = \frac{75}{d^2}$.

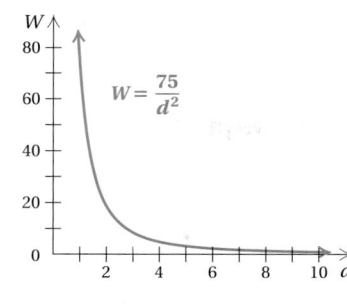

Do Exercise 8.

8. Find an equation of variation in which y varies inversely as the square of x, and $y = \frac{1}{4}$ when $x = 6$.

Consider the equation for the area A of a triangle with height h and base b: $A = \frac{1}{2}bh$. We say that the area **varies jointly** as the height and the base.

> y varies jointly as x and z if there is some positive constant k such that
>
> $$y = kxz.$$

EXAMPLE 7 Find an equation of variation in which y varies jointly as x and z, and $y = 42$ when $x = 2$ and $z = 3$.

$$y = kxz$$

$$42 = k \cdot 2 \cdot 3 \quad \text{Substituting}$$

$$42 = k \cdot 6$$

$$7 = k$$

Thus, $y = 7xz$.

Do Exercise 9.

9. Find an equation of variation in which y varies jointly as x and z, and $y = 65$ when $x = 10$ and $z = 13$.

Different types of variation can be combined. For example, the equation

$$y = k \cdot \frac{xz^2}{w}$$

asserts that y varies jointly as x and the square of z, and inversely as w.

EXAMPLE 8 Find an equation of variation in which y varies jointly as x and z and inversely as the square of w, and $y = 105$ when $x = 3$, $z = 20$, and $w = 2$.

$$y = k \cdot \frac{xz}{w^2}$$

$$105 = k \cdot \frac{3 \cdot 20}{2^2} \quad \text{Substituting}$$

$$105 = k \cdot 15$$

$$7 = k$$

Thus, $y = 7 \cdot \frac{xz}{w^2}$.

10. Find an equation of variation in which y varies jointly as x and the square of z and inversely as w, and $y = 80$ when $x = 4$, $z = 10$, and $w = 25$.

Answers

8. $y = \dfrac{9}{x^2}$ **9.** $y = \dfrac{1}{2}xz$ **10.** $y = \dfrac{5xz^2}{w}$

Do Exercise 10.

Other Applications of Variation

Many problem situations can be described with equations of variation.

EXAMPLE 9 *Volume of a Tree.* The volume of wood V in a tree varies jointly as the height h and the square of the girth g (girth is distance around). If the volume of a redwood tree is 216 m³ when the height is 30 m and the girth is 1.5 m, what is the height of a tree whose volume is 960 m³ and girth is 2 m?

We first find k using the first set of data. Then we solve for h using the second set of data.

$$V = khg^2$$
$$216 = k \cdot 30 \cdot 1.5^2$$
$$3.2 = k$$

Then the equation of variation is $V = 3.2hg^2$. We substitute the second set of data into the equation:

$$960 = 3.2 \cdot h \cdot 2^2$$
$$75 = h.$$

Therefore, the height of the tree is 75 m.

EXAMPLE 10 *TV Signal.* The intensity I of a TV signal varies inversely as the square of the distance d from the transmitter. If the intensity is 23 watts per square meter (W/m²) at a distance of 2 km, what is the intensity at a distance of 6 km?

We first find k using the first set of data. Then we solve for I using the second set of data.

$$I = \frac{k}{d^2}$$
$$23 = \frac{k}{2^2}$$
$$92 = k$$

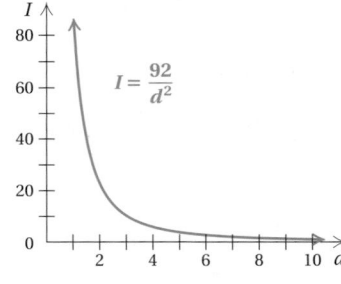

Then the equation of variation is $I = 92/d^2$. We substitute the second distance into the equation:

$$I = \frac{92}{d^2} = \frac{92}{6^2} \approx 2.56. \qquad \text{Rounded to the nearest hundredth}$$

Therefore, at 6 km, the intensity is about 2.56 W/m².

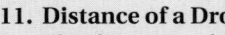

11. Distance of a Dropped Object. The distance s that an object falls when dropped from some point above the ground varies directly as the square of the time t that it falls. If the object falls 19.6 m in 2 sec, how far will the object fall in 10 sec?

12. Electrical Resistance. At a fixed temperature, the resistance R of a wire varies directly as the length l and inversely as the square of its diameter d. If the resistance is 0.1 ohm when the diameter is 1 mm and the length is 50 cm, what is the resistance when the length is 2000 cm and the diameter is 2 mm?

Do Exercises 11 and 12.

Answers

11. 490 m   **12.** 1 ohm

a   Find the variation constant and an equation of variation in which y varies directly as x and the following are true.

1. $y = 40$ when $x = 8$

2. $y = 54$ when $x = 12$

3. $y = 4$ when $x = 30$

4. $y = 3$ when $x = 33$

5. $y = 0.9$ when $x = 0.4$

6. $y = 0.8$ when $x = 0.2$

b   Solve.

7. *Shipping by Semi Truck.*   The number of semi trucks T needed to ship metal varies directly as the weight W of the metal. It takes 75 semi trucks to ship 1500 tons of metal. How many trucks are needed for 3500 tons of metal?

Source: www.scrappy.com/bargePage05.htm

8. *Shipping by Rail Cars.*   The number of rail cars R needed to ship metal varies directly as the weight W of the metal. It takes approximately 21 rail cars to ship 1500 tons of metal. How many rail cars are needed for 3500 tons of metal?

Source: www.scrappy.com/bargePage05.htm

9. *Aluminum Usage.*   The number N of aluminum cans used each year varies directly as the number of people using the cans. If 250 people use 60,000 cans in one year, how many cans are used each year in Seattle, Washington, which has a population of 563,374?

10. *Hooke's Law.*   Hooke's law states that the distance d that a spring is stretched by a hanging object varies directly as the weight w of the object. If a spring is stretched 40 cm by a 3-kg barbell, what is the distance stretched by a 5-kg barbell?

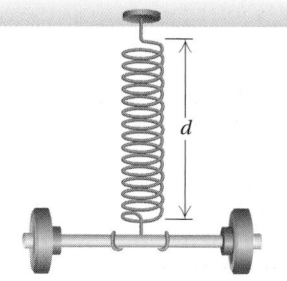

11. *Fat Intake.*   The maximum number of grams of fat that should be in a diet varies directly as a person's weight. A person weighing 120 lb should have no more than 60 g of fat per day. What is the maximum daily fat intake for a person weighing 180 lb?

12. *Relative Aperture.*   The relative aperture, or f-stop, of a 23.5-mm diameter lens is directly proportional to the focal length F of the lens. If a 150-mm focal length has an f-stop of 6.3, find the f-stop of a 23.5-mm diameter lens with a focal length of 80 mm.

13. *Mass of Water in Body.*   The number of kilograms W of water in a human body varies directly as the mass of the body. A 96-kg person contains 64 kg of water. How many kilograms of water are in a 60-kg person?

14. *Weight on Mars.*   The weight M of an object on Mars varies directly as its weight E on Earth. A person who weighs 95 lb on Earth weighs 38 lb on Mars. How much would a 100-lb person weigh on Mars?

Find the variation constant and an equation of variation in which *y* varies inversely as *x* and the following are true.

15. $y = 14$ when $x = 7$

16. $y = 1$ when $x = 8$

17. $y = 3$ when $x = 12$

18. $y = 12$ when $x = 5$

19. $y = 0.1$ when $x = 0.5$

20. $y = 1.8$ when $x = 0.3$

 Solve.

21. *Work Rate.* The time *T* required to do a job varies inversely as the number of people *P* working. It takes 5 hr for 7 bricklayers to build a park wall. How long will it take 10 bricklayers to complete the job?

22. *Pumping Rate.* The time *t* required to empty a tank varies inversely as the rate *r* of pumping. If a pump can empty a tank in 45 min at the rate of 600 kL/min, how long will it take the pump to empty the same tank at the rate of 1000 kL/min?

23. *Current and Resistance.* The current *I* in an electrical conductor varies inversely as the resistance *R* of the conductor. If the current is $\frac{1}{2}$ ampere when the resistance is 240 ohms, what is the current when the resistance is 540 ohms?

24. *Wavelength and Frequency.* The wavelength *W* of a radio wave varies inversely as its frequency *F*. A wave with a frequency of 1200 kilohertz has a length of 300 meters. What is the length of a wave with a frequency of 800 kilohertz?

25. *Beam Weight.* The weight *W* that a horizontal beam can support varies inversely as the length *L* of the beam. Suppose that a 12-ft beam can support 1200 lb. How many pounds can a 15-ft beam support?

26. *Musical Pitch.* The pitch *P* of a musical tone varies inversely as its wavelength *W*. One tone has a pitch of 440 vibrations per second and a wavelength of 2.4 ft. Find the wavelength of another tone that has a pitch of 275 vibrations per second.

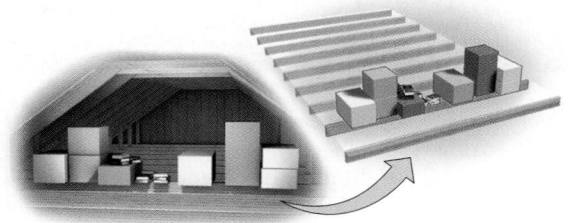

27. *Rate of Travel.* The time *t* required to drive a fixed distance varies inversely as the speed *r*. It takes 5 hr at a speed of 80 km/h to drive a fixed distance. How long will it take to drive the same distance at a speed of 70 km/h?

28. *Volume and Pressure.* The volume *V* of a gas varies inversely as the pressure *P* upon it. The volume of a gas is 200 cm³ under a pressure of 32 kg/cm². What will be its volume under a pressure of 40 kg/cm²?

 Find an equation of variation in which the following are true.

29. *y* varies directly as the square of *x*, and $y = 0.15$ when $x = 0.1$

30. *y* varies directly as the square of *x*, and $y = 6$ when $x = 3$

31. y varies inversely as the square of x, and $y = 0.15$ when $x = 0.1$

32. y varies inversely as the square of x, and $y = 6$ when $x = 3$

33. y varies jointly as x and z, and $y = 56$ when $x = 7$ and $z = 8$

34. y varies directly as x and inversely as z, and $y = 4$ when $x = 12$ and $z = 15$

35. y varies jointly as x and the square of z, and $y = 105$ when $x = 14$ and $z = 5$

36. y varies jointly as x and z and inversely as w, and $y = \frac{3}{2}$ when $x = 2$, $z = 3$, and $w = 4$

37. y varies jointly as x and z and inversely as the product of w and p, and $y = \frac{3}{28}$ when $x = 3$, $z = 10$, $w = 7$, and $p = 8$

38. y varies jointly as x and z and inversely as the square of w, and $y = \frac{12}{5}$ when $x = 16$, $z = 3$, and $w = 5$

 Solve.

39. *Intensity of Light.* The intensity I of light from a light bulb varies inversely as the square of the distance d from the bulb. Suppose that I is $90 \, \text{W/m}^2$ (watts per square meter) when the distance is 5 m. How much *further* would it be to a point where the intensity is $40 \, \text{W/m}^2$?

40. *Stopping Distance of a Car.* The stopping distance d of a car after the brakes have been applied varies directly as the square of the speed r. If a car traveling 60 mph can stop in 200 ft, how fast can a car travel and still stop in 72 ft?

41. *Weight of an Astronaut.* The weight W of an object varies inversely as the square of the distance d from the center of the earth. At sea level (3978 mi from the center of the earth), an astronaut weighs 220 lb. Find his weight when he is 200 mi above the surface of the earth and the spacecraft is not in motion.

42. *Combined Gas Law.* The volume V of a given mass of a gas varies directly as the temperature T and inversely as the pressure P. If $V = 231 \, \text{cm}^3$ when $T = 42°$ and $P = 20 \, \text{kg/cm}^2$, what is the volume when $T = 30°$ and $P = 15 \, \text{kg/cm}^2$?

43. *Earned-Run Average.* A pitcher's earned-run average E varies directly as the number R of earned runs allowed and inversely as the number I of innings pitched. In 2009, CC Sabathia of the New York Yankees had an earned-run average of 3.37. He gave up 86 earned runs in 230 innings. How many earned runs would he have given up had he pitched 255 innings with the same average? Round to the nearest whole number.

Source: Major League Baseball

44. *Atmospheric Drag.* Wind resistance, or atmospheric drag, tends to slow down moving objects. Atmospheric drag varies jointly as an object's surface area A and velocity v. If a car traveling at a speed of 40 mph with a surface area of $37.8 \, \text{ft}^2$ experiences a drag of 222 N (newtons), how fast must a car with $51 \, \text{ft}^2$ of surface area travel in order to experience a drag force of 430 N?

45. *Water Flow.* The amount Q of water emptied by a pipe varies directly as the square of the diameter d. A pipe 5 in. in diameter will empty 225 gal of water over a fixed time period. If we assume the same kind of flow, how many gallons of water are emptied in the same amount of time by a pipe that is 9 in. in diameter?

46. *Weight of a Sphere.* The weight W of a sphere of a given material varies directly as its volume V, and its volume V varies directly as the cube of its diameter.

a) Find an equation of variation relating the weight W to the diameter d.

b) An iron ball that is 5 in. in diameter is known to weigh 25 lb. Find the weight of an iron ball that is 8 in. in diameter.

Skill Maintenance

Factor completely.   [5.7a]

47. $x^2 - x - 56$

48. $a^2 - 16a + 64$

49. $x^5 - 2x^4 - 35x^3$

50. $2y^3 - 10y^2 + y - 5$

51. $16 - t^4$

52. $10x^2 + 80x + 70$

53. $x^2 - 9x + 14$

54. $x^2 + x + 7$

55. $16x^2 - 40xy + 25y^2$

56. $a^2 - 9ab + 14b^2$

57. $3x^3 - 3y^3$

58. $w^6 - t^6$

Synthesis

59. In each of the following equations, state whether y varies directly as x, inversely as x, or neither directly nor inversely as x.

a) $7xy = 14$

b) $x - 2y = 12$

c) $-2x + 3y = 0$

d) $x = \dfrac{3}{4}y$

60. *Area of a Circle.* The area of a circle varies directly as the square of the length of a diameter. What is the variation constant?

61. *Volume and Cost.* A jar of peanut butter in the shape of a right circular cylinder is 4 in. high and 3 in. in diameter and sells for $1.20. If we assume that cost is proportional to volume, how much should a jar 6 in. high and 6 in. in diameter cost?

Summary and Review

Key Terms

rational expression, p. 410
fraction expression, p. 410
equivalent expressions, p. 411
opposites, p. 414
reciprocals, p. 420
least common multiple, LCM, p. 425
least common denominator, LCD, p. 425

complex rational expression, p. 447
complex fraction expression, p. 447
rational equation, p. 453
fraction equation, p. 453
proportion, p. 466
rate, p. 466
speed, p. 466

proportional, p. 466
similar triangles, p. 468
direct variation, p. 478
equation of direct variation, p. 478
variation constant, p. 478
inverse variation, p. 480
equation of inverse variation, p. 480

Concept Reinforcement

Determine whether each statement is true or false.

_____ **1.** To determine the numbers for which a rational expression is not defined, we set the denominator equal to 0 and solve. [6.1a]

_____ **2.** When a situation translates to an equation described by $y = k/x$, with k a positive constant, y varies directly as x. [6.9a]

_____ **3.** The opposite of $2 - x$ is $x - 2$. [6.4a]

_____ **4.** When clearing an equation of fractions, we multiply by the LCM of all the denominators on both sides of the equation. [6.7a]

_____ **5.** The expressions $y + 5$ and $y - 5$ are opposites of each other. [6.1c]

Important Concepts

Objective 6.1c Simplify rational expressions by factoring the numerator and the denominator and removing factors of 1.

Example Simplify: $\dfrac{6y - 12}{2y^2 + y - 10}$.

$$\dfrac{6y - 12}{2y^2 + y - 10} = \dfrac{6(y - 2)}{(2y + 5)(y - 2)}$$

$$= \dfrac{y - 2}{y - 2} \cdot \dfrac{6}{2y + 5} = 1 \cdot \dfrac{6}{2y + 5} = \dfrac{6}{2y + 5}$$

Practice Exercise

1. Simplify:

$$\dfrac{2x^2 - 2}{4x^2 + 24x + 20}.$$

Objective 6.1d Multiply rational expressions and simplify.

Example Multiply and simplify: $\dfrac{x^2 + 14x + 49}{x^2 - 25} \cdot \dfrac{x + 5}{x + 7}$.

$$\dfrac{x^2 + 14x + 49}{x^2 - 25} \cdot \dfrac{x + 5}{x + 7} = \dfrac{(x^2 + 14x + 49)(x + 5)}{(x^2 - 25)(x + 7)}$$

$$= \dfrac{(x + 7)(x + 7)(x + 5)}{(x + 5)(x - 5)(x + 7)}$$

$$= \dfrac{x + 7}{x - 5}$$

Practice Exercise

2. Multiply and simplify:

$$\dfrac{2y^2 + 7y - 15}{5y^2 - 45} \cdot \dfrac{y - 3}{2y - 3}.$$

Objective 6.2b Divide rational expressions and simplify.

Example Divide and simplify: $\dfrac{a^2 - 9a}{a^2 - a - 6} \div \dfrac{a}{a + 2}$.

$$\dfrac{a^2 - 9a}{a^2 - a - 6} \div \dfrac{a}{a + 2} = \dfrac{a^2 - 9a}{a^2 - a - 6} \cdot \dfrac{a + 2}{a}$$

$$= \dfrac{(a^2 - 9a)(a + 2)}{(a^2 - a - 6)a}$$

$$= \dfrac{a(a - 9)(a + 2)}{(a + 2)(a - 3)a}$$

$$= \dfrac{a - 9}{a - 3}$$

Practice Exercise

3. Divide and simplify:

$$\dfrac{b^2 + 3b - 28}{b^2 + 5b - 24} \div \dfrac{b - 4}{b - 3}.$$

Objective 6.4a Add rational expressions.

Example Add and simplify: $\dfrac{6x - 5}{x - 1} + \dfrac{x}{1 - x}$.

$$\dfrac{6x - 5}{x - 1} + \dfrac{x}{1 - x} = \dfrac{6x - 5}{x - 1} + \dfrac{x}{1 - x} \cdot \dfrac{-1}{-1}$$

$$= \dfrac{6x - 5}{x - 1} + \dfrac{-x}{x - 1}$$

$$= \dfrac{6x - 5 - x}{x - 1}$$

$$= \dfrac{5x - 5}{x - 1}$$

$$= \dfrac{5(x - 1)}{x - 1} = 5$$

Practice Exercise

4. Add and simplify:

$$\dfrac{x}{x - 4} + \dfrac{2x - 4}{4 - x}.$$

Objective 6.5a Subtract rational expressions.

Example Subtract: $\dfrac{3}{x^2 - 1} - \dfrac{2x - 1}{x^2 + x - 2}$.

$$\dfrac{3}{x^2 - 1} - \dfrac{2x - 1}{x^2 + x - 2}$$

$$= \dfrac{3}{(x + 1)(x - 1)} - \dfrac{2x - 1}{(x + 2)(x - 1)}$$

The LCM is $(x + 1)(x - 1)(x + 2)$.

$$= \dfrac{3}{(x + 1)(x - 1)} \cdot \dfrac{x + 2}{x + 2} - \dfrac{2x - 1}{(x + 2)(x - 1)} \cdot \dfrac{x + 1}{x + 1}$$

$$= \dfrac{3(x + 2)}{(x + 1)(x - 1)(x + 2)} - \dfrac{(2x - 1)(x + 1)}{(x + 2)(x - 1)(x + 1)}$$

$$= \dfrac{3x + 6 - (2x^2 + x - 1)}{(x + 1)(x - 1)(x + 2)}$$

$$= \dfrac{3x + 6 - 2x^2 - x + 1}{(x + 1)(x - 1)(x + 2)}$$

$$= \dfrac{-2x^2 + 2x + 7}{(x + 1)(x - 1)(x + 2)}$$

Practice Exercise

5. Subtract:

$$\dfrac{x}{x^2 + x - 2} - \dfrac{5}{x^2 - 1}.$$

Objective 6.6a Simplify complex rational expressions.

Example Simplify $\dfrac{\dfrac{1}{3} - \dfrac{1}{x}}{\dfrac{1}{x} - \dfrac{1}{2}}$ using method 1.

The LCM of 3, x, and 2 is $6x$.

$$\frac{\dfrac{1}{3} - \dfrac{1}{x}}{\dfrac{1}{x} - \dfrac{1}{2}} = \frac{\dfrac{1}{3} - \dfrac{1}{x}}{\dfrac{1}{x} - \dfrac{1}{2}} \cdot \frac{6x}{6x} = \frac{\dfrac{1}{3} \cdot 6x - \dfrac{1}{x} \cdot 6x}{\dfrac{1}{x} \cdot 6x - \dfrac{1}{2} \cdot 6x}$$

$$= \frac{2x - 6}{6 - 3x} = \frac{2(x - 3)}{3(2 - x)}$$

Practice Exercise

6. Simplify: $\dfrac{\dfrac{2}{5} - \dfrac{1}{y}}{\dfrac{3}{y} - \dfrac{1}{3}}$.

Objective 6.7a Solve rational equations.

Example Solve: $12 = \dfrac{1}{5x} + \dfrac{4}{x}$.

The LCM of the denominators is $5x$. We multiply by $5x$ on both sides.

$$12 = \frac{1}{5x} + \frac{4}{x}$$

$$5x \cdot 12 = 5x\left(\frac{1}{5x} + \frac{4}{x}\right)$$

$$5x \cdot 12 = 5x \cdot \frac{1}{5x} + 5x \cdot \frac{4}{x}$$

$$60x = 1 + 20$$

$$60x = 21$$

$$x = \frac{21}{60} = \frac{7}{20}$$

This checks, so the solution is $\dfrac{7}{20}$.

Practice Exercise

7. Solve: $\dfrac{1}{x} = \dfrac{2}{3 - x}$.

Objective 6.9a Find an equation of direct variation given a pair of values of the variables.

Example Find an equation of variation in which y varies directly as x, and $y = 30$ when $x = 200$. Then find the value of y when $x = \frac{1}{2}$.

$$y = kx \qquad \text{Direct variation}$$

$$30 = k \cdot 200 \qquad \text{Substituting 30 for } y \text{ and 200 for } x$$

$$\frac{30}{200} = k, \text{ or } k = \frac{3}{20}$$

The equation of variation is $y = \frac{3}{20}x$.

Next, we substitute $\frac{1}{2}$ for x in $y = \frac{3}{20}x$ and solve for y:

$$y = \frac{3}{20}x = \frac{3}{20} \cdot \frac{1}{2} = \frac{3}{40}.$$

When $x = \frac{1}{2}$, $y = \frac{3}{40}$.

Practice Exercise

8. Find an equation of variation in which y varies directly as x, and $y = 60$ when $x = 0.4$. Then find the value of y when $x = 2$.

Objective 6.9c Find an equation of inverse variation given a pair of values of the variables.

Example Find an equation of variation in which y varies inversely as x, and $y = 0.5$ when $x = 20$. Then find the value of y when $x = 6$.

$$y = \frac{k}{x} \qquad \text{Inverse variation}$$

$$0.5 = \frac{k}{20} \qquad \text{Substituting 0.5 for } y \text{ and 20 for } x$$

$$10 = k$$

The equation of variation is $y = \dfrac{10}{x}$.

Next, we substitute 6 for x in $y = 10/x$ and solve for y:

$$y = \frac{10}{x} = \frac{10}{6} = \frac{5}{3}.$$

When $x = 6$, $y = \frac{5}{3}$.

Practice Exercise

9. Find an equation of variation in which y varies inversely as x, and $y = 150$ when $x = 1.5$. Then find the value of y when $x = 10$.

Review Exercises

Find all numbers for which the rational expression is not defined.   [6.1a]

1. $\dfrac{3}{x}$

2. $\dfrac{4}{x - 6}$

3. $\dfrac{x + 5}{x^2 - 36}$

4. $\dfrac{x^2 - 3x + 2}{x^2 + x - 30}$

5. $\dfrac{-4}{(x + 2)^2}$

6. $\dfrac{x - 5}{5}$

Simplify.   [6.1c]

7. $\dfrac{4x^2 - 8x}{4x^2 + 4x}$

8. $\dfrac{14x^2 - x - 3}{2x^2 - 7x + 3}$

9. $\dfrac{(y - 5)^2}{y^2 - 25}$

Multiply and simplify.   [6.1d]

10. $\dfrac{a^2 - 36}{10a} \cdot \dfrac{2a}{a + 6}$

11. $\dfrac{6t - 6}{2t^2 + t - 1} \cdot \dfrac{t^2 - 1}{t^2 - 2t + 1}$

Divide and simplify.   [6.2b]

12. $\dfrac{10 - 5t}{3} \div \dfrac{t - 2}{12t}$

13. $\dfrac{4x^4}{x^2 - 1} \div \dfrac{2x^3}{x^2 - 2x + 1}$

Find the LCM.   [6.3c]

14. $3x^2$, $10xy$, $15y^2$

15. $a - 2$, $4a - 8$

16. $y^2 - y - 2$, $y^2 - 4$

Add and simplify.   [6.4a]

17. $\dfrac{x + 8}{x + 7} + \dfrac{10 - 4x}{x + 7}$

18. $\dfrac{3}{3x - 9} + \dfrac{x - 2}{3 - x}$

19. $\dfrac{2a}{a + 1} + \dfrac{4a}{a^2 - 1}$

20. $\dfrac{d^2}{d - c} + \dfrac{c^2}{c - d}$

Subtract and simplify. [6.5a]

21. $\dfrac{6x - 3}{x^2 - x - 12} - \dfrac{2x - 15}{x^2 - x - 12}$

22. $\dfrac{3x - 1}{2x} - \dfrac{x - 3}{x}$

23. $\dfrac{x + 3}{x - 2} - \dfrac{x}{2 - x}$

24. $\dfrac{1}{x^2 - 25} - \dfrac{x - 5}{x^2 - 4x - 5}$

25. Perform the indicated operations and simplify: [6.5b]

$$\frac{3x}{x + 2} - \frac{x}{x - 2} + \frac{8}{x^2 - 4}.$$

Simplify. [6.6a]

26. $\dfrac{\dfrac{1}{z} + 1}{\dfrac{1}{z^2} - 1}$

27. $\dfrac{\dfrac{c}{d} - \dfrac{d}{c}}{\dfrac{1}{c} + \dfrac{1}{d}}$

Solve. [6.7a]

28. $\dfrac{3}{y} - \dfrac{1}{4} = \dfrac{1}{y}$

29. $\dfrac{15}{x} - \dfrac{15}{x + 2} = 2$

Solve. [6.8a]

30. *Highway Work.* In checking records, a contractor finds that crew A can pave a certain length of highway in 9 hr, while crew B can do the same job in 12 hr. How long would it take if they worked together?

31. *Airplane Speed.* One plane travels 80 mph faster than another. While one travels 1750 mi, the other travels 950 mi. Find the speed of each plane.

32. *Train Speed.* A manufacturer is testing two high-speed trains. One train travels 40 km/h faster than the other. While one train travels 70 km, the other travels 60 km. Find the speed of each train.

70 km, $r + 1$

60 km, r

Solve. [6.8b]

33. *Quality Control.* A sample of 250 calculators contained 8 defective calculators. How many defective calculators would you expect to find in a sample of 5000?

34. *Pizza Proportions.* At Finnelli's Pizzeria, the following ratios are used: 5 parts sausage to 7 parts cheese, 6 parts onion to 13 parts green pepper, and 9 parts pepperoni to 14 parts cheese.
   a) Finnelli's makes several pizzas with green pepper and onion. They use 2 cups of green pepper. How much onion would they use?
   b) Finnelli's makes several pizzas with sausage and cheese. They use 3 cups of sausage. How much cheese would they use?
   c) Finnelli's makes several pizzas with pepperoni and cheese. They use 6 cups of pepperoni. How much cheese would they use?

35. *Estimating Whale Population.* To determine the number of blue whales in the world's oceans, marine biologists tag 500 blue whales in various parts of the world. Later, 400 blue whales are checked, and it is found that 20 of them are tagged. Estimate the blue whale population.

36. Triangles ABC and XYZ below are similar. Find the value of x.

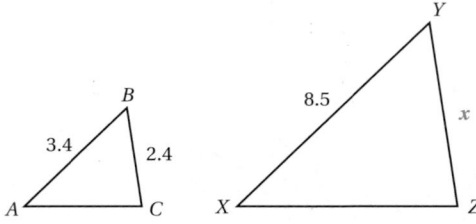

37. Find an equation of variation in which y varies directly as x, and $y = 100$ when $x = 25$. [6.9a]

38. Find an equation of variation in which y varies inversely as x, and $y = 100$ when $x = 25$. [6.9c]

39. Find an equation of variation in which y varies jointly as x and z, and $y = 210$ when $x = 5$ and $z = 7$. [6.9e]

40. Find an equation of variation in which y varies directly as x and inversely as z, and $y = 16$ when $x = 40$ and $z = 5$. [6.9e]

41. *Pumping Time.*   The time t required to empty a tank varies inversely as the rate r of pumping. If a pump can empty a tank in 35 min at the rate of 800 kL per minute, how long will it take the pump to empty the same tank at the rate of 1400 kL per minute? [6.9d]

42. *Test Score.*   The score N on a test varies directly as the number of correct responses a. Ellen answers 28 questions correctly and earns a score of 87. What would Ellen's score have been if she had answered 25 questions correctly? [6.9b]

43. *Power of Electric Current.*   The power P expended by heat in an electric circuit of fixed resistance varies directly as the square of the current C in the circuit. A circuit expends 180 watts when a current of 6 amperes is flowing. What is the amount of heat expended when the current is 10 amperes? [6.9f]

44. Find all numbers for which
$$\frac{3x^2 - 2x - 1}{3x^2 + x}$$
is not defined.   [6.1a]

A. $1, -\dfrac{1}{3}$ **B.** $-\dfrac{1}{3}$

C. $0, -\dfrac{1}{3}$ **D.** $0, \dfrac{1}{3}$

45. Subtract: $\dfrac{1}{x - 5} - \dfrac{1}{x + 5}$.   [6.5a]

A. $\dfrac{10}{(x - 5)(x + 5)}$ **B.** 0

C. $\dfrac{5}{x - 5}$ **D.** $\dfrac{10}{x + 5}$

Synthesis

46. Simplify: [6.1d], [6.2b]
$$\frac{2a^2 + 5a - 3}{a^2} \cdot \frac{5a^3 + 30a^2}{2a^2 + 7a - 4} \div \frac{a^2 + 6a}{a^2 + 7a + 12}.$$

47. Compare
$$\frac{A + B}{B} = \frac{C + D}{D}$$
with the proportion
$$\frac{A}{B} = \frac{C}{D}.$$
[6.8b]

Understanding Through Discussion and Writing

1. Are parentheses as important when adding rational expressions as they are when subtracting? Why or why not?   [6.4a], [6.5a]

2. How can a graph be used to determine how many solutions an equation has?   [6.7a]

3. How is the process of canceling related to the identity property of 1?   [6.1c]

4. Determine whether the situation represents direct variation, inverse variation, or neither. Give a reason for your answer.   [6.9a, c]

The number of plays that it takes to go 80 yd for a touchdown and the average gain per play

5. Explain how a rational expression can be formed for which -3 and 4 are not allowable replacements.   [6.1a]

6. Why is it especially important to check the possible solutions to a rational equation?   [6.7a]

CHAPTER
6

Test

For Extra Help

CHAPTER Test Prep VIDEOS

Step-by-step test solutions are found on the Chapter Test Prep Videos available via the Video Resources on DVD, in *MyMathLab*, and on You Tube (search "BittingerIntroInter" and click on "Channels").

Find all numbers for which the rational expression is not defined.

1. $\dfrac{8}{2x}$

2. $\dfrac{5}{x+8}$

3. $\dfrac{x-7}{x^2-49}$

4. $\dfrac{x^2+x-30}{x^2-3x+2}$

5. $\dfrac{11}{(x-1)^2}$

6. $\dfrac{x+2}{2}$

7. Simplify:

$$\frac{6x^2+17x+7}{2x^2+7x+3}.$$

8. Multiply and simplify:

$$\frac{a^2-25}{6a} \cdot \frac{3a}{a-5}.$$

9. Divide and simplify:

$$\frac{25x^2-1}{9x^2-6x} \div \frac{5x^2+9x-2}{3x^2+x-2}.$$

10. Find the LCM:

$$y^2-9,\ y^2+10y+21,\ y^2+4y-21.$$

Add or subtract. Simplify if possible.

11. $\dfrac{16+x}{x^3}+\dfrac{7-4x}{x^3}$

12. $\dfrac{5-t}{t^2+1}-\dfrac{t-3}{t^2+1}$

13. $\dfrac{x-4}{x-3}+\dfrac{x-1}{3-x}$

14. $\dfrac{x-4}{x-3}-\dfrac{x-1}{3-x}$

15. $\dfrac{5}{t-1}+\dfrac{3}{t}$

16. $\dfrac{1}{x^2-16}-\dfrac{x+4}{x^2-3x-4}$

17. $\dfrac{1}{x-1}+\dfrac{4}{x^2-1}-\dfrac{2}{x^2-2x+1}$

18. Simplify: $\dfrac{9-\dfrac{1}{y^2}}{3-\dfrac{1}{y}}.$

Solve.

19. $\dfrac{7}{y}-\dfrac{1}{3}=\dfrac{1}{4}$

20. $\dfrac{15}{x}-\dfrac{15}{x-2}=-2$

Find an equation of variation in which y varies directly as x and the following are true. Then find the value of y when $x=25$.

21. $y=6$ when $x=3$

22. $y=1.5$ when $x=3$

Find an equation of variation in which y varies inversely as x and the following are true. Then find the value of y when $x = 100$.

23. $y = 6$ when $x = 3$

24. $y = 11$ when $x = 2$

25. Find an equation of variation in which Q varies jointly as x and y, and $Q = 25$ when $x = 2$ and $y = 5$.

Solve.

26. *Train Travel.* The distance d traveled by a train varies directly as the time t that it travels. The train travels 60 km in $\frac{1}{2}$ hr. How far will it travel in 2 hr?

27. *Concrete Work.* It takes 3 hr for 2 concrete mixers to mix a fixed amount of concrete. The number of hours varies inversely as the number of concrete mixers used. How long would it take 5 concrete mixers to do the same job?

28. *Quality Control.* A sample of 125 spark plugs contained 4 defective spark plugs. How many defective spark plugs would you expect to find in a sample of 500?

29. *Zebra Population.* A game warden catches, tags, and then releases 15 zebras. A month later, a sample of 20 zebras is collected and 6 of them have tags. Use this information to estimate the size of the zebra population in that area.

30. *Copying Time.* Kopy Kwik has 2 copiers. One can copy a year-end report in 20 min. The other can copy the same document in 30 min. How long would it take both machines, working together, to copy the report?

31. *Driving Speed.* Craig drives 20 km/h faster than Marilyn. In the same time that Marilyn drives 225 km, Craig drives 325 km. Find the speed of each car.

32. This pair of triangles is similar. Find the missing length x.

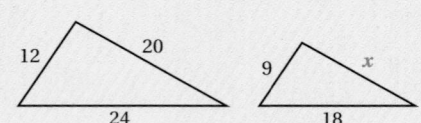

33. Solve: $\dfrac{2}{x - 4} + \dfrac{2x}{x^2 - 16} = \dfrac{1}{x + 4}$.

   A. -4 **B.** 4

   C. $4, -4$ **D.** No solution

Synthesis

34. Reggie and Rema work together to mulch the flower beds around an office complex in $2\frac{6}{7}$ hr. Working alone, it would take Reggie 6 hr more than it would take Rema. How long would it take each of them to complete the landscaping working alone?

35. Simplify: $1 + \dfrac{1}{1 + \dfrac{1}{1 + \dfrac{1}{a}}}$.

Cumulative Review

1. Find the absolute value: $|3.5|$.

2. Identify the degree of each term and the degree of the polynomial:
$$x^3 - 2x^2 + x - 1.$$

3. *Free-Range Eggs.* In an egg-testing project, it was found that eggs produced by hens raised on pasture contain approximately 35% less cholesterol than factory farm eggs. If a standard factory egg contains 423 mg of cholesterol, how much cholesterol does a free-range egg contain?
Source: "Meet Real Free-Range Eggs," by Cheryl Long and Tabitha Alterman, *Mother Earth News*, October/November 2007

4. *Square Footage.* In the third quarter of 2008, the size of new single-family homes averaged 2438 ft^2, down from 2629 ft^2 in the second quarter. What was the percent of decrease?
Source: Gopal Ahluwalia, Director of Research, National Association of Home Builders

5. *Principal Borrowed.* Money is borrowed at 6% simple interest. After 1 year, $2650 pays off the loan. How much was originally borrowed?

6. *Car Travel.* One car travels 105 mi in the same time that a car traveling 10 mph slower travels 75 mi. Find the speed of each car.

7. *Areas.* If the sides of a square are increased by 2 ft, the sum of the areas of the two squares is 452 ft^2. Find the length of a side of the original square.

8. *Muscle Weight.* The number of pounds of muscle M in the human body varies directly as body weight B. A person who weighs 175 lb has a muscle weight of 70 lb.
   a) Write an equation of variation that describes this situation.
   b) Mike weighs 192 lb. What is his muscle weight?

9. Collect like terms: $x^2 - 3x^3 - 4x^2 + 5x^3 - 2$.

Simplify.

10. $\dfrac{1}{2}x - \left[\dfrac{3}{8}x - \left(\dfrac{2}{3} + \dfrac{1}{4}x\right) - \dfrac{1}{3}\right]$

11. $\left(\dfrac{2x^3}{3x^{-1}}\right)^{-2}$

12. $\dfrac{\dfrac{4}{x} - \dfrac{6}{x^2}}{\dfrac{5}{x} + \dfrac{7}{2x}}$

Perform the indicated operations. Simplify if possible.

13. $(5xy^2 - 6x^2y^2 - 3xy^3) - (-4xy^3 + 7xy^2 - 2x^2y^2)$

14. $(4x^4 + 6x^3 - 6x^2 - 4) +$
$$(2x^5 + 2x^4 - 4x^3 - 4x^2 + 3x - 5)$$

15. $\dfrac{2y + 4}{21} \cdot \dfrac{7}{y^2 + 4y + 4}$

16. $\dfrac{x^2 - 9}{x^2 + 8x + 15} \div \dfrac{x - 3}{2x + 10}$

17. $\dfrac{x^2}{x - 4} + \dfrac{16}{4 - x}$

18. $\dfrac{5x}{x^2 - 4} - \dfrac{-3}{2 - x}$

Multiply.

19. $(2.5a + 7.5)(0.4a - 1.2)$ **20.** $(6x - 5)^2$

21. $(2x^3 + 1)(2x^3 - 1)$

Factor.

22. $9a^2 + 52a - 12$ **23.** $9x^2 - 30xy + 25y^2$

24. $49x^2 - 1$

Solve.

25. $x - [x - (x - 1)] = 2$ **26.** $2x^2 + 7x = 4$

27. $x^2 = 10x$ **28.** $3(x - 2) \leq 4(x + 5)$

29. $\dfrac{5x - 2}{4} - \dfrac{4x - 5}{3} = 1$ **30.** $t = ax + ay$, for a

Find the slope, if it exists, of the line containing the given pair of points.

31. $(-2, 6)$ and $(-2, -1)$ **32.** $(-4, 1)$ and $(3, -2)$

33. $\left(-\dfrac{1}{2}, 4\right)$ and $\left(3\dfrac{1}{2}, -5\right)$ **34.** $\left(-7, \dfrac{3}{4}\right)$ and $\left(-4, \dfrac{3}{4}\right)$

For each equation, find the coordinates of the y-intercept and the x-intercept. Do not graph.

35. $-8x - 24y = 48$ **36.** $15 - 40x = -120y$

37. $y = 25$ **38.** $x = -\dfrac{1}{4}$

Graph on a plane.

39. $x = -3$ **40.** $y = -3$

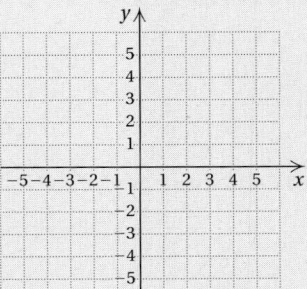   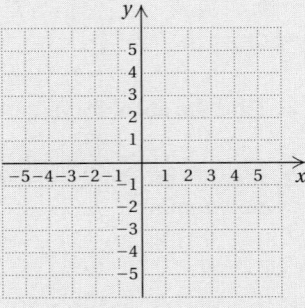

41. $3x - 5y = 15$ **42.** $2x - 6y = 12$

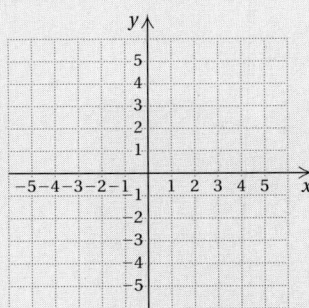   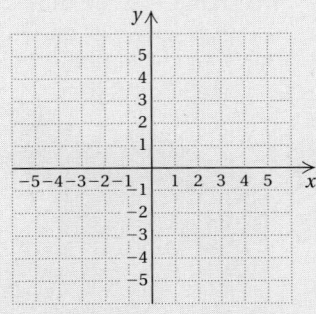

43. $y = -\dfrac{1}{3}x - 2$ **44.** $x - y = -5$

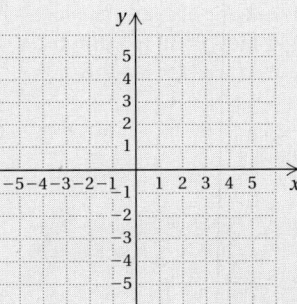   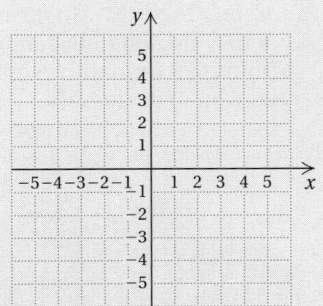

Synthesis

45. Find all numbers for which the following complex rational expression is not defined:

$$\dfrac{\dfrac{1}{x} + x}{2 + \dfrac{1}{x - 3}}.$$

Graphs, Functions, and Applications

Real-World Application

Amelia's Beads offers a class in designing necklaces. For a necklace made of 6-mm beads, 4.23 beads per inch are needed. The cost of a necklace made of 6-mm gemstone beads that sell for 40¢ each is $7 for the clasp and the crimps and approximately $1.70 per inch. Formulate a linear function that models the total cost of a necklace $C(n)$, where n is the length of the necklace, in inches. Then graph the model and use the model to determine the cost of a 30-in. necklace.

This problem appears as Example 6 in Section 7.5.

7.1

Functions and Graphs

OBJECTIVES

a Determine whether a correspondence is a function.

b Given a function described by an equation, find function values (outputs) for specified values (inputs).

c Draw the graph of a function.

d Determine whether a graph is that of a function using the vertical-line test.

e Solve applied problems involving functions and their graphs.

SKILL TO REVIEW
Objective 1.1a: Evaluate algebraic expressions by substitution.

Evaluate.

1. $-\dfrac{1}{4}x$, when $x = 40$

2. $y^2 - 2y + 6$, when $y = -1$

a Identifying Functions

Consider the equation $y = 2x - 3$. If we substitute a value for x—say, 5—we get a value for y, 7:

$$y = 2x - 3 = 2(5) - 3 = 10 - 3 = 7.$$

The equation $y = 2x - 3$ is an example of a *function*. We now develop the concept of a *function*, one of the most important concepts in mathematics.

In much the same way that ordered pairs form correspondences between first and second coordinates, a *function* is a correspondence from one set to another. For example:

To each student in a college, there corresponds his or her student ID.

To each item in a store, there corresponds its price.

To each real number, there corresponds the cube of that number.

In each case, the first set is called the **domain** and the second set is called the **range**. Each of these correspondences is a **function**, because given a member of the domain, there is *just one* member of the range to which it corresponds. Given a student, there is *just one* ID. Given an item, there is *just one* price. Given a real number, there is *just one* cube.

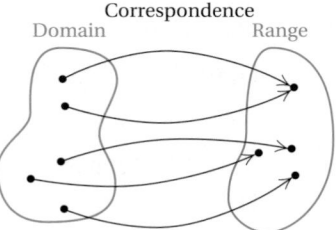

Correspondence
Domain   Range

EXAMPLE 1 Determine whether the correspondence is a function.

	Domain	Range
f:	1	$107.40
	2	$ 34.10
	3	$ 29.60
	4	$ 19.60

	Domain	Range
g:	3	5
	4	9
	5	-7
	6	

	Domain	Range
h:	Chicago	Cubs
		White Sox
	Baltimore	Orioles
	San Diego	Padres

	Domain	Range
p:	Cubs	Chicago
	White Sox	
	Orioles	Baltimore
	Padres	San Diego

The correspondence f is a function because each member of the domain is matched to *only one* member of the range.

The correspondence g is a function because each member of the domain is matched to *only one* member of the range. Note that a function allows two or more members of the domain to correspond to the same member of the range.

The correspondence h *is not* a function because one member of the domain, Chicago, is matched to *more than one* member of the range.

The correspondence p *is* a function because each member of the domain is matched to *only one* member of the range.

FUNCTION; DOMAIN; RANGE

A **function** is a correspondence between a first set, called the **domain**, and a second set, called the **range**, such that each member of the domain corresponds to **exactly one** member of the range.

> Do Exercises 1–4.

EXAMPLE 2 Determine whether each correspondence is a function.

	Domain	*Correspondence*	*Range*
a)	The integers	Each number's square	A set of nonnegative integers
b)	The set of all states	Each state's members of the U.S. Senate	The set of U.S. Senators
c)	The set of U.S. Senators	The state that a Senator represents	The set of all states

a) The correspondence *is* a function because each integer has *only one* square.

b) The correspondence *is not* a function because each state has two U.S. Senators.

Richard Burr   Kay R. Hagan

NORTH CAROLINA

c) The correspondence *is* a function because each Senator represents *only one* state.

> Do Exercises 5–7. (Exercise 7 is on the following page.)

When a correspondence between two sets is not a function, it is still an example of a **relation**.

RELATION

A **relation** is a correspondence between a first set, called the **domain**, and a second set, called the **range**, such that each member of the domain corresponds to **at least one** member of the range.

Determine whether each correspondence is a function.

1. *Domain*   *Range*

Cheetah ⟶ 70 mph
Human ⟶ 28 mph
Lion ⟶ 50 mph
Chicken ⟶ 9 mph

2. *Domain*   *Range*

A ⟶ a
B ⟶ b
C ⟶ c
D ⟶ d
⟶ e

3. *Domain*   *Range*

−2 ⟶ 4
2 ⟶ 4
−3 ⟶ 9
3 ⟶ 9
0 ⟶ 0

4. *Domain*   *Range*

4 ⟶ −2
4 ⟶ 2
9 ⟶ −3
9 ⟶ 3
0 ⟶ 0

Determine whether each correspondence is a function.

5. *Domain*
A set of numbers

Correspondence
Square each number and subtract 10.

Range
A set of numbers

6. *Domain*
A set of polygons

Correspondence
Find the perimeter of each polygon.

Range
A set of numbers

Answers

1. Yes	**2.** No	**3.** Yes
4. No	**5.** Yes	**6.** Yes

Thus, although the correspondences of Examples 1 and 2 are not all functions, they *are* all relations. A function is a special type of relation—one in which each member of the domain is paired with *exactly one* member of the range.

b Finding Function Values

Most functions considered in mathematics are described by equations like $y = 2x + 3$ or $y = 4 - x^2$. We graph the function $y = 2x + 3$ by first performing calculations like the following:

for $x = 4, y = 2x + 3 = 2 \cdot 4 + 3 = 8 + 3 = 11$;

for $x = -5, y = 2x + 3 = 2 \cdot (-5) + 3 = -10 + 3 = -7$;

for $x = 0, y = 2x + 3 = 2 \cdot 0 + 3 = 0 + 3 = 3$;   and so on.

For $y = 2x + 3$, the **inputs** (members of the domain) are values of x substituted into the equation. The **outputs** (members of the range) are the resulting values of y. If we call the function f, we can use x to represent an arbitrary *input* and $f(x)$—read "f of x," or "f at x," or "the value of f at x"—to represent the corresponding *output*. In this notation, the function given by $y = 2x + 3$ is written as $f(x) = 2x + 3$ and the calculations above can be written more concisely as follows:

$y = f(4) = 2 \cdot 4 + 3 = 8 + 3 = 11$;

$y = f(-5) = 2 \cdot (-5) + 3 = -10 + 3 = -7$;

$y = f(0) = 2 \cdot 0 + 3 = 0 + 3 = 3$;   and so on.

Thus instead of writing "when $x = 4$, the value of y is 11," we can simply write "$f(4) = 11$," which can also be read as "f of 4 is 11" or "for the input 4, the output of f is 11."

We can think of a function as a machine. Think of $f(4) = 11$ as putting 4, a member of the domain (an input), into the machine. The machine knows the correspondence $f(x) = 2x + 3$, multiplies 4 by 2 and adds 3, and produces 11, a member of the range (the output).

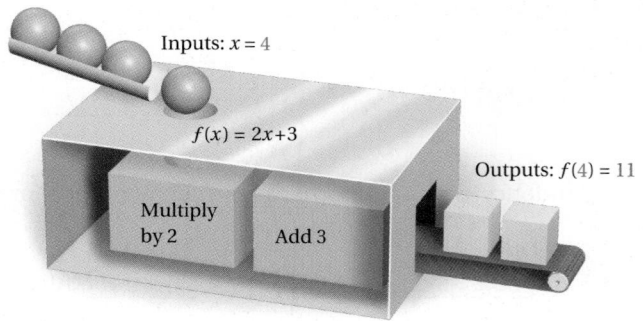

Inputs: $x = 4$

$f(x) = 2x + 3$

Multiply by 2

Add 3

Outputs: $f(4) = 11$

------ *Caution!* ------

The notation $f(x)$ *does not mean* "f times x" and should not be read that way.

EXAMPLE 3   A function f is given by $f(x) = 3x^2 - 2x + 8$. Find each of the indicated function values.

a) $f(0)$ **b)** $f(1)$ **c)** $f(-5)$ **d)** $f(7a)$

One way to find function values when a formula is given is to think of the formula with blanks, or placeholders, replacing the variable as follows:

$$f(\square) = 3\square^2 - 2\square + 8.$$

Answer

7. No

To find an output for a given input, we think: "Whatever goes in the blank on the left goes in the blank(s) on the right." With this in mind, let's complete the example.

a) $f(0) = 3 \cdot 0^2 - 2 \cdot 0 + 8 = 8$

b) $f(1) = 3 \cdot 1^2 - 2 \cdot 1 + 8 = 3 \cdot 1 - 2 + 8 = 3 - 2 + 8 = 9$

c) $f(-5) = 3(-5)^2 - 2 \cdot (-5) + 8 = 3 \cdot 25 + 10 + 8 = 75 + 10 + 8 = 93$

d) $f(7a) = 3(7a)^2 - 2(7a) + 8 = 3 \cdot 49a^2 - 14a + 8 = 147a^2 - 14a + 8$

Do Exercise 8.

8. Find the indicated function values for the following function:
$$f(x) = 2x^2 + 3x - 4.$$
a) $f(0)$ **b)** $f(8)$
c) $f(-5)$ **d)** $f(2a)$

EXAMPLE 4 Find the indicated function value.

a) $f(5)$, for $f(x) = 3x + 2$ **b)** $g(-2)$, for $g(x) = 7$
c) $F(a + 1)$, for $F(x) = 5x - 8$ **d)** $f(a + h)$, for $f(x) = -2x + 1$

a) $f(5) = 3 \cdot 5 + 2 = 15 + 2 = 17$

b) For the function given by $g(x) = 7$, all inputs share the same output, 7. Thus, $g(-2) = 7$. The function g is an example of a **constant function**.

c) $F(a + 1) = 5(a + 1) - 8 = 5a + 5 - 8 = 5a - 3$

d) $f(a + h) = -2(a + h) + 1 = -2a - 2h + 1$

Do Exercise 9.

9. Find the indicated function value.
a) $f(-6)$, for $f(x) = 5x - 3$
b) $g(55)$, for $g(x) = -3$
c) $F(a + 2)$, for $F(x) = -5x + 8$
d) $f(a - h)$, for $f(x) = 6x - 7$

Answers
8. (a) -4; (b) 148; (c) 31; (d) $8a^2 + 6a - 4$
9. (a) -33; (b) -3; (c) $-5a - 2$;
(d) $6a - 6h - 7$

Calculator Corner

Finding Function Values We can find function values on a graphing calculator. One method is to substitute inputs directly into the formula. Consider the function $f(x) = x^2 + 3x - 4$. To find $f(-5)$, we press ( ) ((-)) (5) ( )) (x^2) (+) (3) (( ) ((-)) (5) ( )) (−) (4) **ENTER**. We find that $f(-5) = 6$.

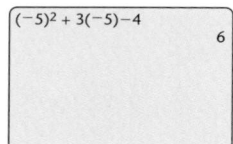

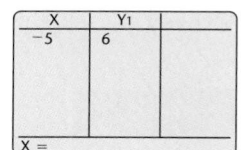

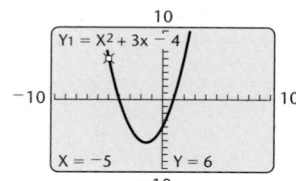

After we have entered the function as $y_1 = x^2 + 3x - 4$ on the equation-editor screen, there are several other methods that we can use to find function values. We can use a table set in ASK mode and enter $x = -5$. We see that the function value, y_1, is 6. We can also use the VALUE feature to evaluate the function. To do this, we first graph the function. Then we press **2ND** (CALC) (1) to access the VALUE feature. Next, we supply the desired x-value by pressing ((-)) (5). Finally, we press **ENTER** to see X $= -5$, Y $= 6$ at the bottom of the screen. Again we see that the function value is 6. Note that when the VALUE feature is used to find a function value, the x-value must be in the viewing window.

A fourth method for finding function values uses the TRACE feature. With the function graphed in a window that includes the x-value -5, we press (TRACE). The coordinates of the point where the blinking cursor is positioned on the graph are displayed at the bottom of the screen. To move the cursor to the point with x-coordinate -5, we press ((-)) (5) **ENTER**. Now we see X $= -5$, Y $= 6$ displayed at the bottom of the screen. This tells us that $f(-5) = 6$. The final calculator display for this method is the same as the one shown above for the VALUE feature. There are other ways to find function values, but we will not discuss them here.

Exercises: Find each function value.

1. $f(-5.1)$, for $f(x) = 3x + 2$ **2.** $f(4)$, for $f(x) = -3.6x$
3. $f(-3)$, for $f(x) = x^2 + 5$ **4.** $f(3)$, for $f(x) = 4x^2 + x - 5$

(c) Graphs of Functions

To graph a function, we find ordered pairs (x, y) or $(x, f(x))$, plot them, and connect the points. Note that y and $f(x)$ are used interchangeably—that is, $y = f(x)$—when we are working with functions and their graphs.

EXAMPLE 5 Graph: $f(x) = x + 2$.

A list of some function values is shown in this table. We plot the points and connect them. The graph is a straight line. The "y" on the vertical axis could also be labeled "$f(x)$."

x	$f(x)$
-4	-2
-3	-1
-2	0
-1	1
0	2
1	3
2	4
3	5
4	6

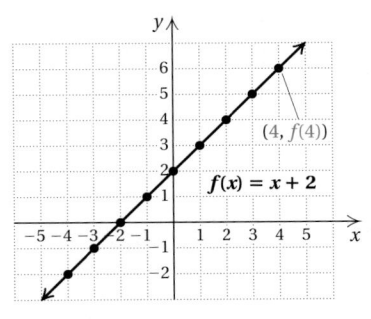

Do Exercise 10.

EXAMPLE 6 Graph: $g(x) = 4 - x^2$.

We calculate some function values, plot the corresponding points, and draw the curve.

$$g(0) = 4 - 0^2 = 4 - 0 = 4,$$
$$g(-1) = 4 - (-1)^2 = 4 - 1 = 3,$$
$$g(2) = 4 - 2^2 = 4 - 4 = 0,$$
$$g(-3) = 4 - (-3)^2 = 4 - 9 = -5$$

x	$g(x)$
-3	-5
-2	0
-1	3
0	4
1	3
2	0
3	-5

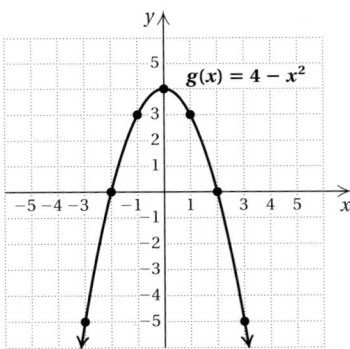

Do Exercise 11.

10. Graph: $f(x) = x - 4$.

x	$f(x)$

11. Graph: $g(x) = 5 - x^2$.

x	$g(x)$

Answers

10.

$f(x) = x - 4$

11.

$g(x) = 5 - x^2$

EXAMPLE 7 Graph: $h(x) = |x|$.

A list of some function values is shown in the following table. We plot the points and connect them. The graph is a V-shaped "curve" that rises on either side of the vertical axis.

x	h(x)
-3	3
-2	2
-1	1
0	0
1	1
2	2
3	3

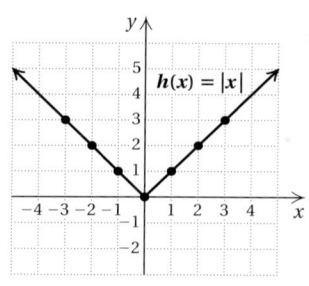

Do Exercise 12.

d The Vertical-Line Test

Consider the graph of the function f described by $f(x) = x^2 - 5$ shown at right. It is also the graph of the equation $y = x^2 - 5$.

To find a function value, like $f(3)$, from a graph, we locate the input on the horizontal axis, move directly up or down to the graph of the function, and then move left or right to find the output on the vertical axis. Thus, $f(3) = 4$. Keep in mind that members of the domain are found on the horizontal axis, members of the range are found on the vertical axis, and the y on the vertical axis could also be labeled $f(x)$.

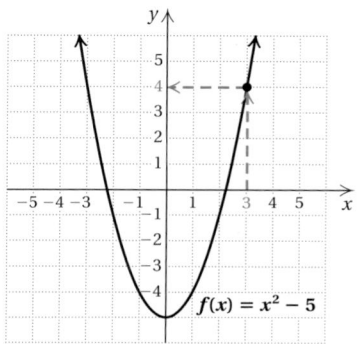

When one member of the domain is paired with two or more different members of the range, the correspondence is not a function. Thus, when a graph contains two or more different points with the same first coordinate, the graph cannot represent a function. Points sharing a common first coordinate are vertically above or below each other. (See the following graph.) This observation leads to the *vertical-line test*.

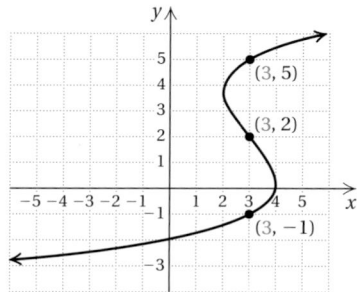

Since 3 is paired with more than one member of the range, the graph does not represent a function.

THE VERTICAL-LINE TEST

If it is possible for a vertical line to cross a graph more than once, then the graph is *not* the graph of a function.

---

12. Graph: $t(x) = 3 - |x|$.

x	t(x)

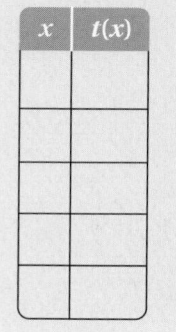

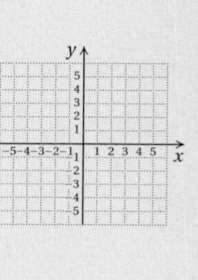

Calculator Corner

Graphing Functions

To graph a function using a graphing calculator, we replace the function notation with y and proceed as described in the Calculator Corner on p. 187. To graph $f(x) = 2x^2 + x$ in the standard window, for example, we replace $f(x)$ with y and enter $y_1 = 2x^2 + x$ on the Y= screen and then press (ZOOM) (6).

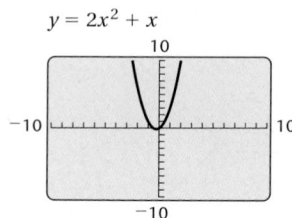

$y = 2x^2 + x$

Exercises: Graph each function.

1. $f(x) = x - 4$

2. $f(x) = -2x - 3$

3. $h(x) = 1 - x^2$

4. $f(x) = 3x^2 - 4x + 1$

5. $f(x) = x^3$

6. $f(x) = |x + 3|$

Answer

12.

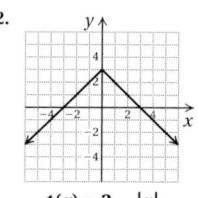

$t(x) = 3 - |x|$

EXAMPLE 8 Determine whether each of the following is the graph of a function.

a)
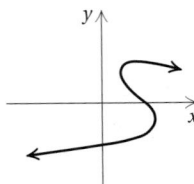

b)
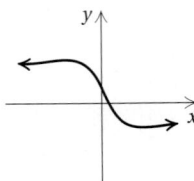

c)
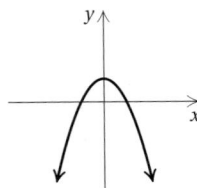

d)
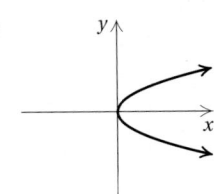

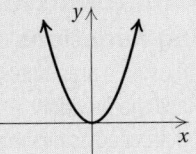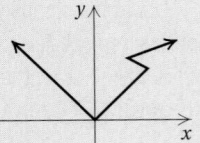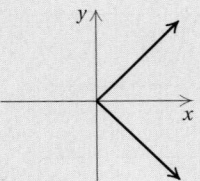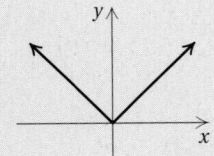
a) The graph *is not* that of a function because a vertical line can cross the graph at more than one point.

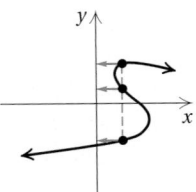

b) The graph *is* that of a function because no vertical line can cross the graph at more than one point. This can be confirmed with a ruler or straightedge.

c) The graph *is* that of a function because no vertical line can cross the graph more than once.

d) The graph *is not* that of a function because a vertical line can cross the graph more than once.

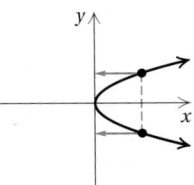

Do Exercises 13–16.

(e) Applications of Functions and Their Graphs

Functions are often described by graphs, whether or not an equation is given. To use a graph in an application, we note that each point on the graph represents a pair of values.

EXAMPLE 9 *World Population.* The following graph represents the world population, in billions. The population is a function of the year. Note that no equation is given for the function.

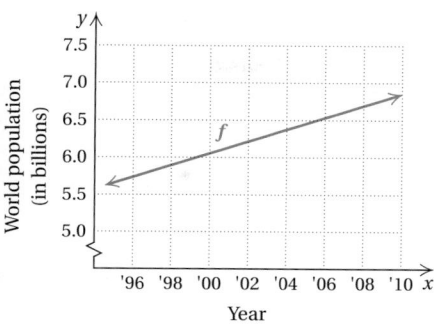

SOURCE: U.S. Census Bureau, Population Division/ International Programs Center

a) What was the world population in 1998? That is, find $f(1998)$.

b) What was the world population in 2008? That is, find $f(2008)$.

a) To estimate the world population in 1998, we locate 1998 on the horizontal axis and move directly up until we reach the graph. Then we move across to the vertical axis. We come to a point that is about 5.9, so we estimate that the population was about 5.9 billion in 1998.

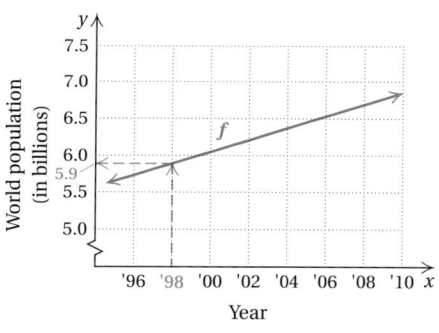

SOURCE: U.S. Census Bureau, Population Division/ International Programs Center

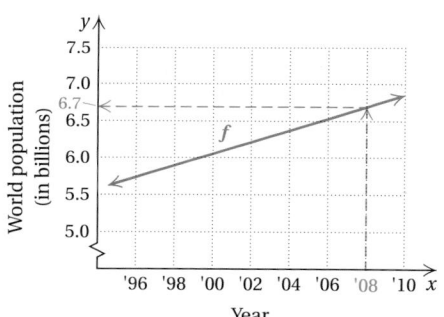

SOURCE: U.S. Census Bureau, Population Division/ International Programs Center

b) To estimate the world population in 2008, we locate 2008 on the horizontal axis and move directly up until we reach the graph. Then we move across to the vertical axis. We come to a point that is about 6.7, so we estimate that the population was about 6.7 billion in 2008.

Do Exercises 17 and 18.

Refer to the graph in Example 9 for Margin Exercises 17 and 18.

17. What was the world population in 2000?

18. What was the world population in 2010?

Answers

17. About 6.1 billion
18. About 6.8 billion

a Determine whether each correspondence is a function.

1. Domain Range
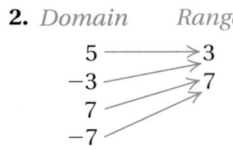

2 → 9
5 → 8
19

2. Domain Range
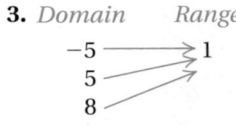

5 → 3
−3 → 7
7
−7

3. Domain Range
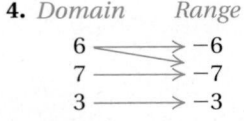

−5 → 1
5
8

4. Domain Range

6 → −6
7 → −7
3 → −3

5. Domain Range
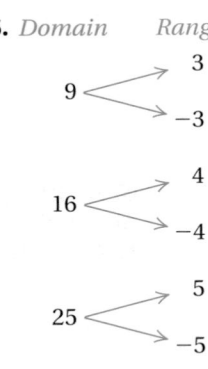

9 ⟨ 3
      −3

16 ⟨ 4
      −4

25 ⟨ 5
      −5

6. Domain Range

LIFE EXPECTANCY
COUNTRY **(in years)**

China → 64.7
India → 73.0
Japan → 78.2
Spain → 80.9
Sweden → 82.6
United States

Source: *CIA World Factbook*, 2008

7. Domain Range

Florida ⟨ Florida State University
          University of Florida
          University of Miami
Kansas ⟨ Baker University
          Kansas State University
          University of Kansas

8. Domain Range

Colorado State University
University of Colorado → Colorado
University of Denver
Gonzaga University
University of Washington → Washington
Washington State University

9. Domain Range
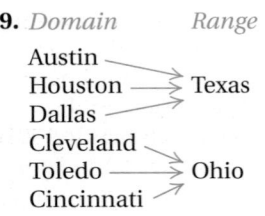

Austin
Houston → Texas
Dallas
Cleveland
Toledo → Ohio
Cincinnati

10. Domain Range
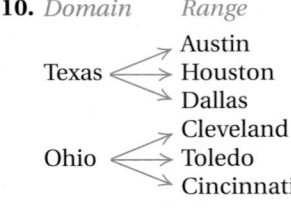

          Austin
Texas ⟨ Houston
          Dallas
          Cleveland
Ohio ⟨ Toledo
          Cincinnati

Domain	Correspondence	Range
11. A set of numbers	The area of a triangle	A set of triangles
12. A family	Each person's height, in inches	A set of positive numbers
13. A set of numbers	Square each number and then add 4.	A set of positive numbers
14. A set of years	A student's year of birth	A first-grade class

b Find the function values.

15. $f(x) = x + 5$
    a) $f(4)$ **b)** $f(7)$
    c) $f(-3)$ **d)** $f(0)$
    e) $f(2.4)$ **f)** $f\left(\frac{2}{3}\right)$

16. $g(t) = t - 6$
    a) $g(0)$ **b)** $g(6)$
    c) $g(13)$ **d)** $g(-1)$
    e) $g(-1.08)$ **f)** $g\left(\frac{7}{8}\right)$

17. $h(p) = 3p$
    a) $h(-7)$ **b)** $h(5)$
    c) $h\left(\frac{2}{3}\right)$ **d)** $h(0)$
    e) $h(6a)$ **f)** $h(a + 1)$

18. $f(x) = -4x$
    a) $f(6)$ **b)** $f\left(-\frac{1}{2}\right)$
    c) $f(0)$ **d)** $f(-1)$
    e) $f(3a)$ **f)** $f(a - 1)$

19. $g(s) = 3s + 4$
    a) $g(1)$ **b)** $g(-7)$
    c) $g\left(\frac{2}{3}\right)$ **d)** $g(0)$
    e) $g(a - 2)$ **f)** $g(a + h)$

20. $h(x) = 19$, a constant function
    a) $h(4)$ **b)** $h(-6)$
    c) $h(12.5)$ **d)** $h(0)$
    e) $h\left(\frac{2}{3}\right)$ **f)** $h(a + 3)$

21. $f(x) = 2x^2 - 3x$
    a) $f(0)$ **b)** $f(-1)$
    c) $f(2)$ **d)** $f(10)$
    e) $f(-5)$ **f)** $f(4a)$

22. $f(x) = 3x^2 - 2x + 1$
    a) $f(0)$ **b)** $f(1)$
    c) $f(-1)$ **d)** $f(10)$
    e) $f(-3)$ **f)** $f(2a)$

23. $f(x) = |x| + 1$
    a) $f(0)$ **b)** $f(-2)$
    c) $f(2)$ **d)** $f(-10)$
    e) $f(a - 1)$ **f)** $f(a + h)$

24. $g(t) = |t - 1|$
    a) $g(4)$ **b)** $g(-2)$
    c) $g(-1)$ **d)** $g(100)$
    e) $g(5a)$ **f)** $g(a + 1)$

25. $f(x) = x^3$
    a) $f(0)$ **b)** $f(-1)$
    c) $f(2)$ **d)** $f(10)$
    e) $f(-5)$ **f)** $f(-3a)$

26. $f(x) = x^4 - 3$
    a) $f(1)$ **b)** $f(-1)$
    c) $f(0)$ **d)** $f(2)$
    e) $f(-2)$ **f)** $f(-a)$

27. *Average Age of Senators.* The function $A(s)$ given by
$$A(s) = 0.321s + 54$$
can be used to estimate the average age of senators in the U.S. Senate in the years 1981 to 2009. Let $A(s) =$ the average age of the senators and $s =$ the number of years since 1981—that is, $s = 0$ for 1981 and $s = 9$ for 1990. What was the average age of the U.S. Senators in 2003? in 2009?

Source: House and Senate Historical Offices

28. *Average Age of House Members.* The function $A(h)$ given by
$$A(h) = 0.314h + 48$$
can be used to estimate the average age of house members in the U.S. House of Representatives in the years 1981 to 2009. Let $A(h) =$ the average age of the house members and $h =$ the number of years since 1981. What is the average age of U.S. House members in 1981? in 2005?

Source: House and Senate Historical Offices

29. *Pressure at Sea Depth.* The function $P(d) = 1 + (d/33)$ gives the pressure, in *atmospheres* (atm), at a depth of d feet in the sea. Note that $P(0) = 1$ atm, $P(33) = 2$ atm, and so on. Find the pressure at 20 ft, 30 ft, and 100 ft.

30. *Temperature as a Function of Depth.* The function $T(d) = 10d + 20$ gives the temperature, in degrees Celsius, inside the earth as a function of the depth d, in kilometers. Find the temperature at 5 km, 20 km, and 1000 km.

31. *Melting Snow.* The function $W(d) = 0.112d$ approximates the amount, in centimeters, of water that results from d centimeters of snow melting. Find the amount of water that results from snow melting from depths of 16 cm, 25 cm, and 100 cm.

32. *Temperature Conversions.* The function $C(F) = \frac{5}{9}(F - 32)$ determines the Celsius temperature that corresponds to F degrees Fahrenheit. Find the Celsius temperature that corresponds to 62°F, 77°F, and 23°F.

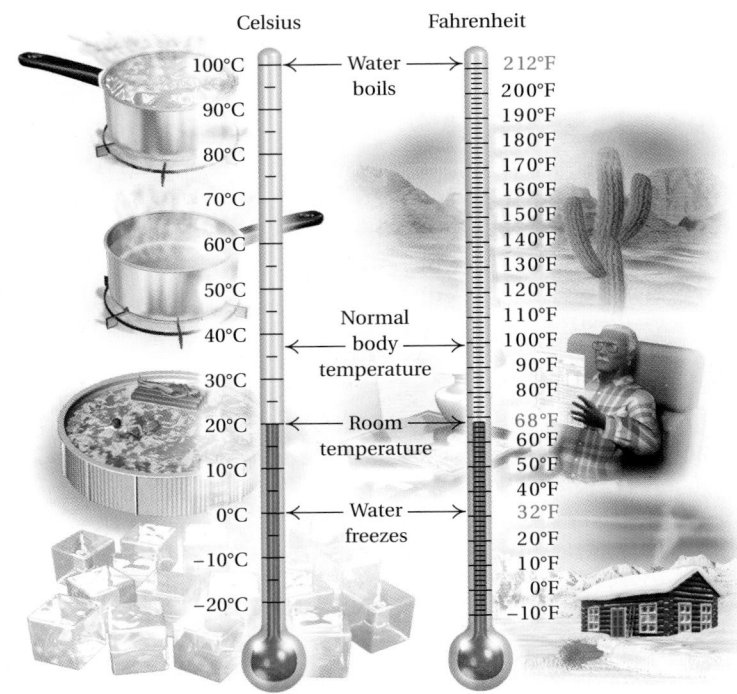

c Graph each function.

33. $f(x) = -2x$

x	y

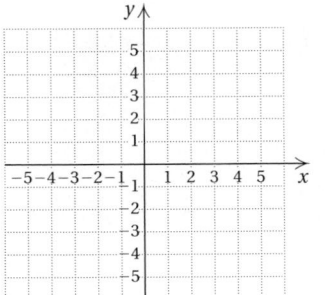

34. $g(x) = 3x$

x	y

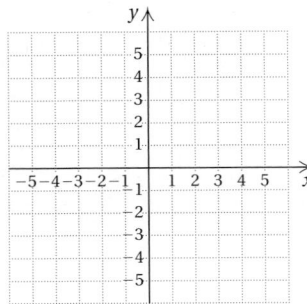

35. $f(x) = 3x - 1$

x	y

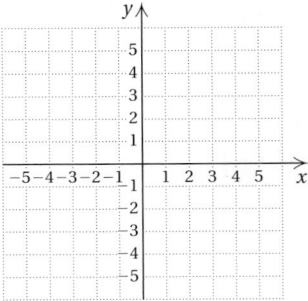

36. $g(x) = 2x + 5$

x	y

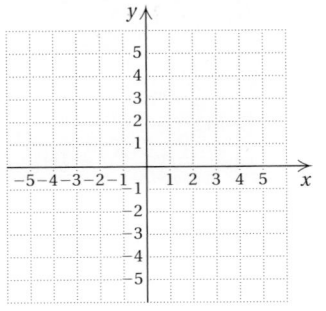

37. $g(x) = -2x + 3$

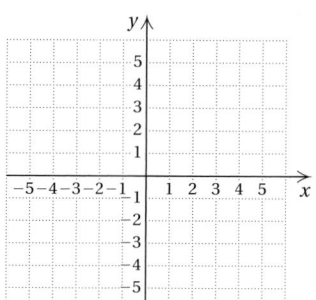

38. $f(x) = -\frac{1}{2}x + 2$

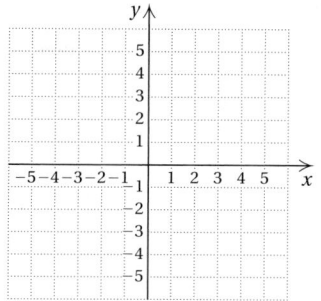

39. $f(x) = \frac{1}{2}x + 1$

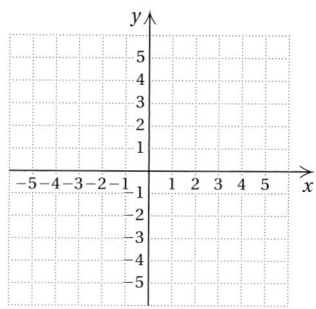

40. $f(x) = -\frac{3}{4}x - 2$

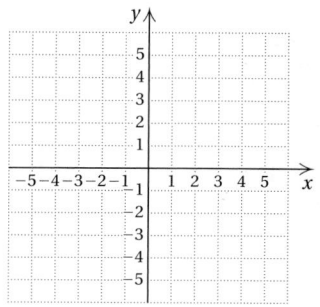

41. $f(x) = 2 - |x|$

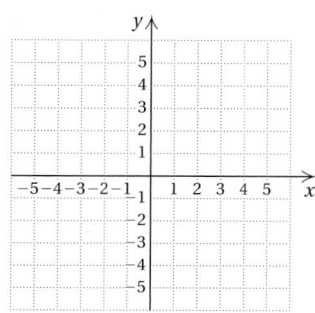

42. $f(x) = |x| - 4$

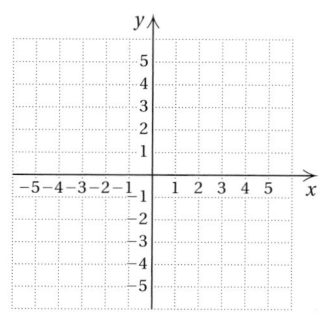

43. $g(x) = |x - 1|$

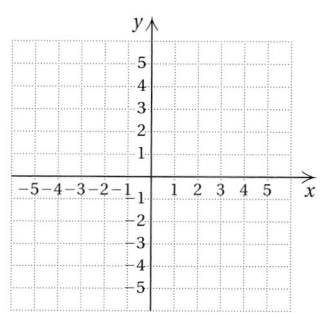

44. $g(x) = |x + 3|$

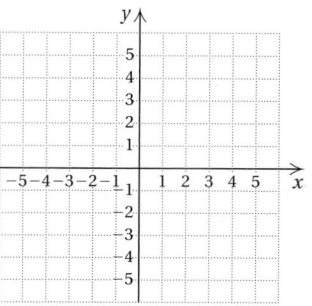

45. $f(x) = x^2$

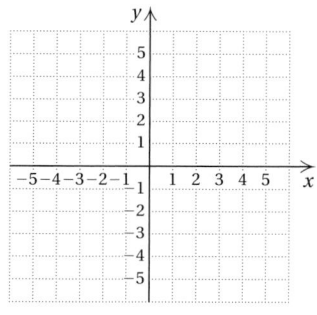

46. $f(x) = x^2 - 1$

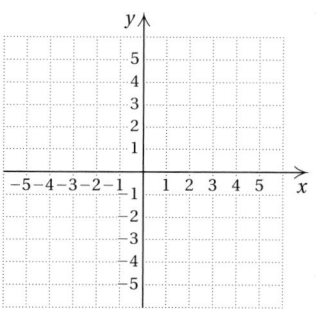

47. $f(x) = x^2 - x - 2$

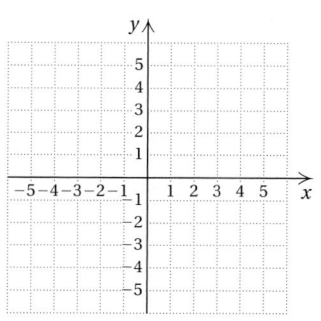

48. $f(x) = x^2 + 6x + 5$

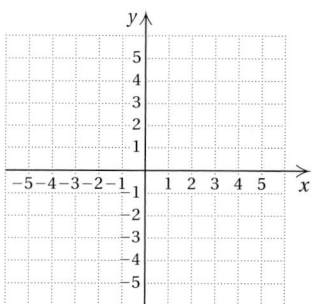

49. $f(x) = 2 - x^2$

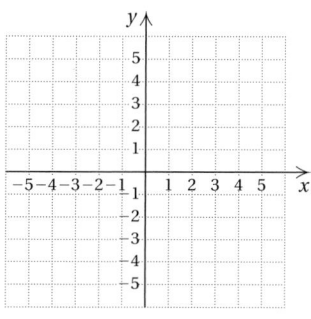

50. $f(x) = 1 - x^2$

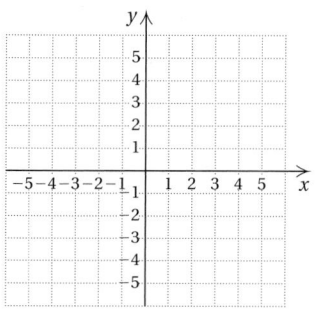

51. $f(x) = x^3 + 1$

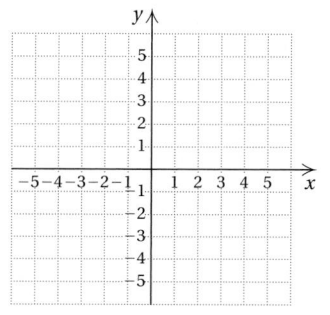

52. $f(x) = x^3 - 2$

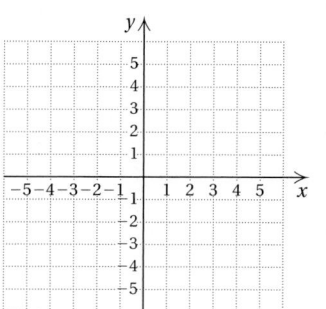

d   Determine whether each of the following is the graph of a function.

53.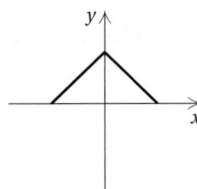

54.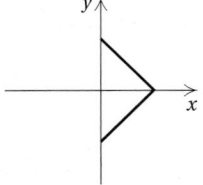

55.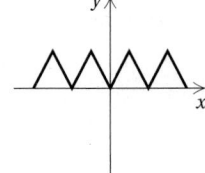

56.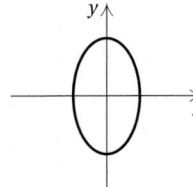

57.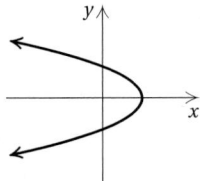

58.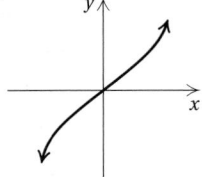

59.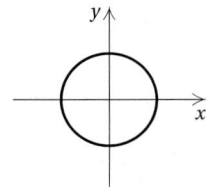

60.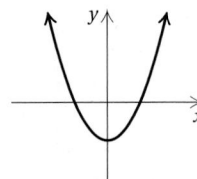

e   Solve.

News/Talk Radio Stations. The following graph approximates the number of U.S. commercial radio stations with a news/talk format. The number of stations is a function *f* of the year *x*.

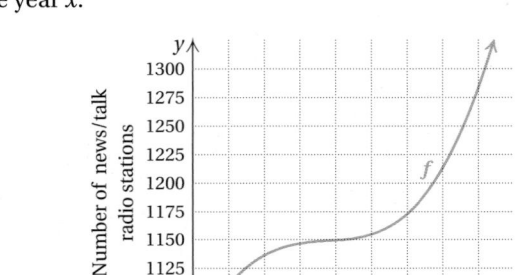

SOURCE: M Street Corporation

61. Approximate the number of news/talk radio stations in 2000. That is, find $f(2000)$.

62. Approximate the number of news/talk radio stations in 2003. That is, find $f(2003)$.

Digital Photographs. The following graph approximates the number of digital photos taken but not printed, in billions. The number of photos that are not printed is a function *f* of the year *x*.

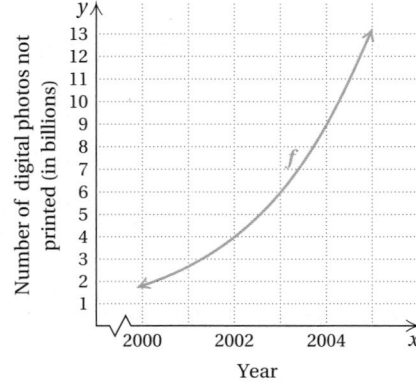

63. Approximate the number of digital photos taken but not printed in 2000. That is, find $f(2000)$.

64. Approximate the number of digital photos taken but not printed in 2002. That is, find $f(2002)$.

Skill Maintenance

In each of Exercises 65–72, fill in the blank with the correct term from the given list. Some of the choices may not be used and some may be used more than once.

65. The axes divide the plane into four regions called _____. [3.1a]

66. A(n) _____ is a correspondence between two sets such that each member of the first set corresponds to at least one member of the second set. [7.1a]

67. A(n) _____ is a correspondence between a first set, called the _____, and a second set, called the _____, such that each member of the _____ corresponds to exactly one member of the _____. [7.1a]

68. The _____ of an equation is a drawing that represents all of its solutions. [3.1c]

69. Members of the domain of a function are its _____. [7.1b]

70. The replacements for the variable that make an equation true are its _____. [2.1a]

71. The _____ states that for any real numbers a, b, and c, $a = b$ is equivalent to $a + c = b + c$. [2.1b]

72. The _____ can be used to determine whether a graph represents a function. [7.1d]

axes

coordinates

quadrants

addition principle

multiplication principle

vertical-line test

graph

domain

range

relation

function

inputs

outputs

solutions

values

Synthesis

73. Suppose that for some function g, $g(x - 6) = 10x - 1$. Find $g(-2)$.

74. Suppose that for some function h, $h(x + 5) = x^2 - 4$. Find $h(3)$.

For Exercises 75 and 76, let $f(x) = 3x^2 - 1$ and $g(x) = 2x + 5$.

75. Find $f(g(-4))$ and $g(f(-4))$.

76. Find $f(g(-1))$ and $g(f(-1))$.

77. Suppose that a function g is such that $g(-1) = -7$ and $g(3) = 8$. Find a formula for g if $g(x)$ is of the form $g(x) = mx + b$, where m and b are constants.

7.2

Finding Domain and Range

OBJECTIVE

a Find the domain and the range of a function.

SKILL TO REVIEW
Objective 2.3a: Solve equations using both the addition principle and the multiplication principle.

Solve.

1. $6x - 3 = 51$
2. $15 - 2x = 0$

1. Find the domain and the range of the function f whose graph is shown below.

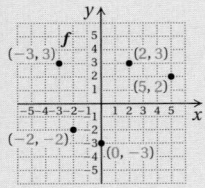

a Finding Domain and Range

The solutions of an equation in two variables consist of a set of ordered pairs. A set of ordered pairs is called a **relation**. When a set of ordered pairs is such that no two different pairs share a common first coordinate, we have a **function**. The **domain** is the set of all first coordinates, and the **range** is the set of all second coordinates.

EXAMPLE 1 Find the domain and the range of the function f whose graph is shown below.

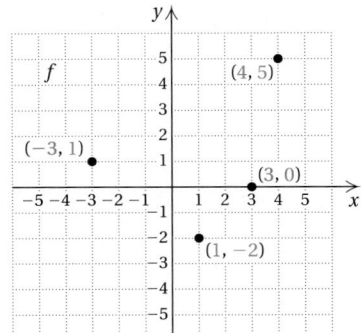

This function contains just four ordered pairs and it can be written as

$$\{(-3, 1), (1, -2), (3, 0), (4, 5)\}.$$

We can determine the domain and the range by reading the x- and y-values directly from the graph.

The domain is the set of all first coordinates, or x-values, $\{-3, 1, 3, 4\}$. The range is the set of all second coordinates, or y-values, $\{1, -2, 0, 5\}$.

Do Margin Exercise 1.

EXAMPLE 2 For the function f whose graph is shown below, determine each of the following.

a) The number in the range that is paired with 1 from the domain. That is, find $f(1)$.

b) The domain of f

c) The numbers in the domain that are paired with 1 from the range. That is, find all x such that $f(x) = 1$.

d) The range of f

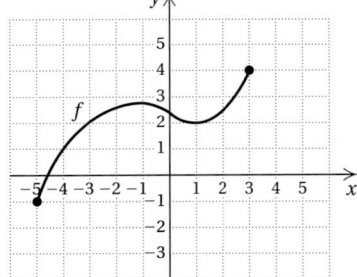

a) To determine which number in the range is paired with 1 in the domain, we locate 1 on the horizontal axis. Next, we find the point on the graph of f for which 1 is the first coordinate. From that point, we can look to the vertical axis to find the corresponding y-coordinate, 2. The input 1 has the output 2—that is, $f(1) = 2$.

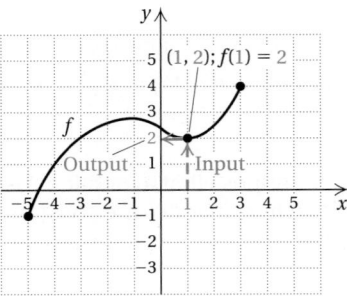

Answers

Skill to Review:
1. 9   2. $\dfrac{15}{2}$, or 7.5

Margin Exercise:
1. Domain = $\{-3, -2, 0, 2, 5\}$;
range = $\{-3, -2, 2, 3\}$

b) The domain of the function is the set of all *x*-values, or inputs, of the points on the graph. These extend from -5 to 3 and can be viewed as the curve's shadow, or projection, onto the *x*-axis. Thus the domain is the set $\{x \mid -5 \leq x \leq 3\}$.

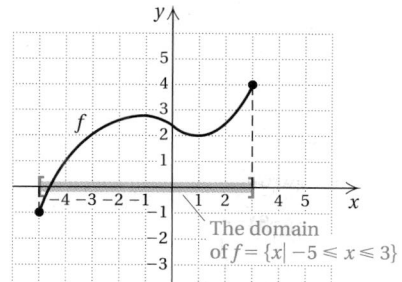

c) To determine which numbers in the domain are paired with 1 in the range, we locate 1 on the vertical axis. From there, we look left and right to the graph of *f* to find any points for which 1 is the second coordinate (output). One such point exists, $(-4, 1)$. For this function, we note that $x = -4$ is the only member of the domain paired with 1. For other functions, there might be more than one member of the domain paired with a member of the range.

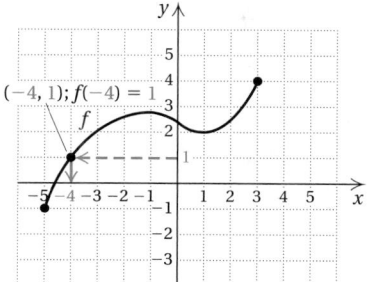

d) The range of the function is the set of all *y*-values, or outputs, of the points on the graph. These extend from -1 to 4 and can be viewed as the curve's shadow, or projection, onto the *y*-axis. Thus the range is the set $\{y \mid -1 \leq y \leq 4\}$.

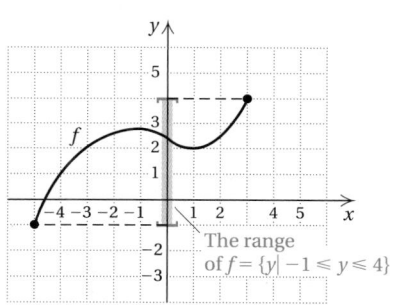

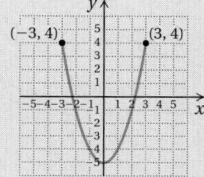

Do Exercise 2.

EXAMPLE 3 Find the domain and the range of the function *f* whose graph is shown below.

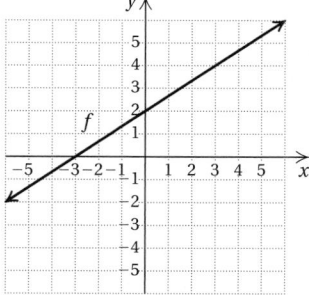

Since no endpoints are indicated, the graph extends indefinitely both horizontally and vertically. Thus the domain is the set of all real numbers. Likewise, the range is the set of all real numbers.

Do Exercise 3.

2. For the function *f* whose graph is shown below, determine each of the following.

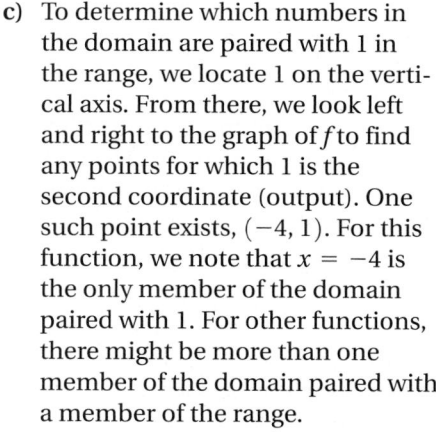

a) The number in the range that is paired with the input 1. That is, find $f(1)$.

b) The domain of *f*

c) The numbers in the domain that are paired with 4

d) The range of *f*

3. Find the domain and the range of the function *f* whose graph is shown below.

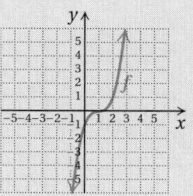

When a function is given by an equation or a formula, the domain is understood to be the largest set of real numbers (inputs) for which function values (outputs) can be calculated. That is, the domain is the set of all possible allowable inputs into the formula. To find the domain, think, "What can we substitute?"

EXAMPLE 4 Find the domain: $f(x) = |x|$.

We ask, "What can we substitute?" Is there any number x for which we cannot calculate $|x|$? The answer is no. Thus the domain of f is the set of all real numbers.

EXAMPLE 5 Find the domain: $f(x) = \dfrac{3}{2x - 5}$.

We ask, "What can we substitute?" Is there any number x for which we cannot calculate $3/(2x - 5)$? Since $3/(2x - 5)$ cannot be calculated when the denominator $2x - 5$ is 0, we solve the following equation to find those real numbers that must be excluded from the domain of f:

$$2x - 5 = 0 \qquad \text{Setting the denominator equal to 0}$$
$$2x = 5 \qquad \text{Adding 5}$$
$$x = \tfrac{5}{2}. \qquad \text{Dividing by 2}$$

Find the domain.

4. $f(x) = x^3 - |x|$

5. $f(x) = \dfrac{4}{3x + 2}$

Thus, $\tfrac{5}{2}$ is not in the domain, whereas all other real numbers are.

The domain of f is $\left\{ x \,\middle|\, x \text{ is a real number } \textit{and } x \neq \tfrac{5}{2} \right\}$.

Do Exercises 4 and 5.

The task of determining the domain and the range of a function is one that we will return to several times as we consider other types of functions in this book.

Functions: A Review

The following is a review of the function concepts considered in Sections 7.1 and 7.2.

Function Concepts

- Formula for f: $f(x) = x^2 - 7$
- For every input of f, there is exactly one output.
- When 1 is the input, -6 is the output.
- $f(1) = -6$
- $(1, -6)$ is on the graph.
- Domain = The set of all inputs
    = The set of all
        real numbers
- Range = The set of all outputs
    = $\{ y \,|\, y \geq -7 \}$

Graph

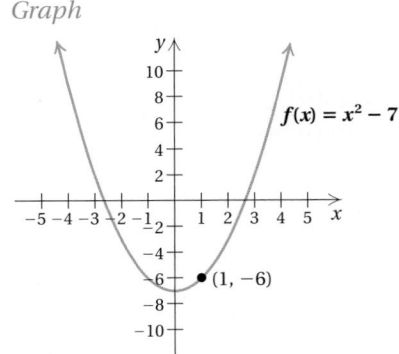

For Extra Help

MyMathLab | Math XL PRACTICE | WATCH | DOWNLOAD | READ | REVIEW

a In Exercises 1–12, the graph is that of a function. Determine for each one **(a)** $f(1)$; **(b)** the domain; **(c)** all x-values such that $f(x) = 2$; and **(d)** the range. An open dot indicates that the point does not belong to the graph.

1.

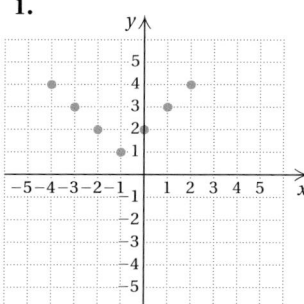

2.

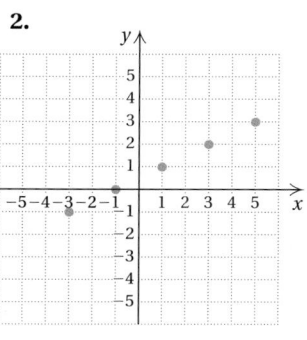

3.

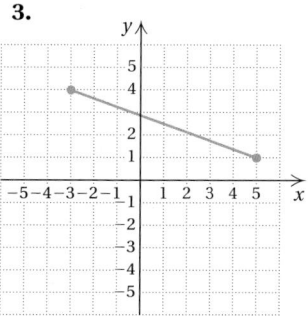

4.

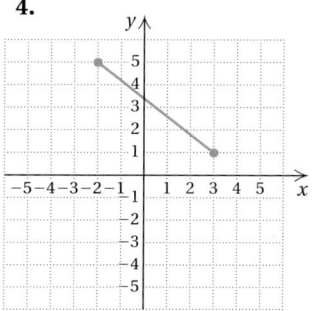

5.

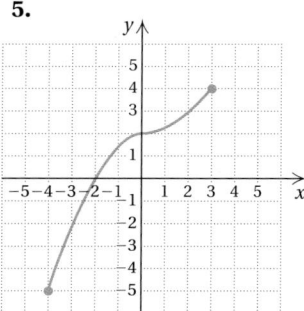

6.

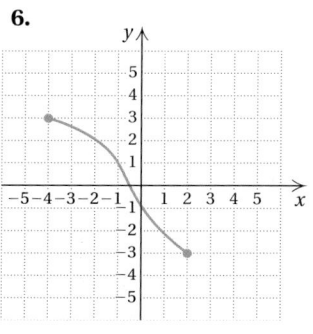

7.

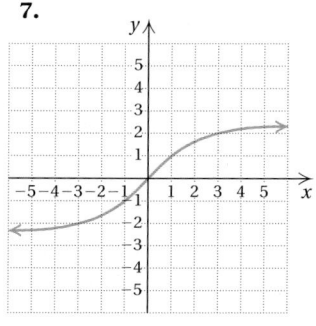

8.

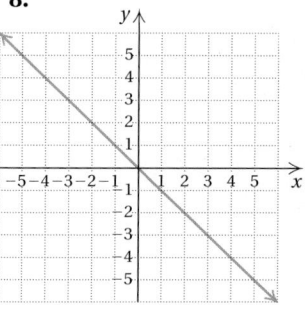

9.

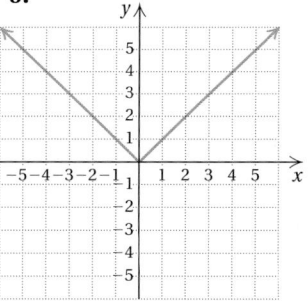

10.

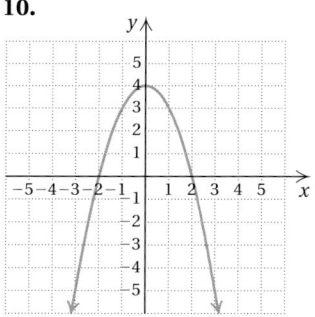

11.

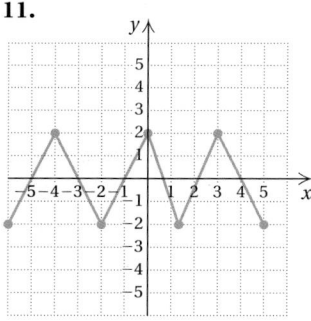

12.

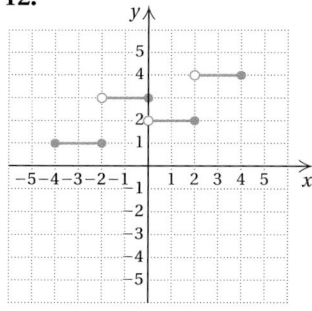

Find the domain.

13. $f(x) = \dfrac{2}{x + 3}$

14. $f(x) = \dfrac{7}{5 - x}$

15. $f(x) = 2x + 1$

16. $f(x) = 4 - 5x$

17. $f(x) = x^2 + 3$

18. $f(x) = x^2 - 2x + 3$

19. $f(x) = \dfrac{8}{5x - 14}$

20. $f(x) = \dfrac{x - 2}{3x + 4}$

21. $f(x) = |x| - 4$

22. $f(x) = |x - 4|$

23. $f(x) = \dfrac{x^2 - 3x}{|4x - 7|}$

24. $f(x) = \dfrac{4}{|2x - 3|}$

25. $g(x) = \dfrac{1}{x - 1}$

26. $g(x) = \dfrac{-11}{4 + x}$

27. $g(x) = x^2 - 2x + 1$

28. $g(x) = 8 - x^2$

29. $g(x) = x^3 - 1$

30. $g(x) = 4x^3 + 5x^2 - 2x$

31. $g(x) = \dfrac{7}{20 - 8x}$

32. $g(x) = \dfrac{2x - 3}{6x - 12}$

33. $g(x) = |x + 7|$

34. $g(x) = |x| + 1$

35. $g(x) = \dfrac{-2}{|4x + 5|}$

36. $g(x) = \dfrac{x^2 + 2x}{|10x - 20|}$

37. For the function f whose graph is shown below, find $f(-1), f(0),$ and $f(1)$.

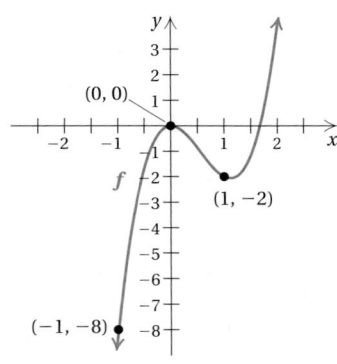

38. For the function g whose graph is shown below, find all the x-values for which $g(x) = 1$.

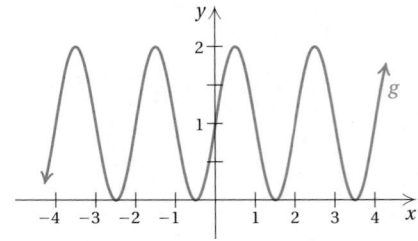

Skill Maintenance

Simplify.   [6.1c]

39. $\dfrac{a^2 - 1}{a + 1}$

40. $\dfrac{10y^2 + 10y - 20}{35y^2 + 210y - 245}$

41. $\dfrac{5x - 15}{x^2 - x - 6}$

Divide.   [4.8b]

42. $(t^2 + 3t - 28) \div (t + 7)$

43. $(w^2 + 4w + 5) \div (w + 3)$

44. $(x^6 + x^5 - 3x^4 + x + 5) \div (x^2 - 1)$

Multiply.   [4.6d]

45. $(7x - 3)(2x + 9)$

46. $(a - 1)(a + 1)$

47. $(9y + 10)^2$

48. $(2w - \frac{1}{2})(4w + \frac{1}{2})$

Synthesis

49. Determine the range of each of the functions in Exercises 13, 18, 21, and 22.

50. Determine the range of each of the functions in Exercises 26, 27, 28, and 34.

Find the domain of each function.

51. $f(x) = \sqrt[3]{x - 1}$

52. $g(x) = \sqrt{2 - x}$

Mid-Chapter Review

Concept Reinforcement

Determine whether each statement is true or false.

_____ **1.** Every function is a relation.   [7.1a]

_____ **2.** It is possible for one input of a function to have two or more outputs.   [7.1a]

_____ **3.** It is possible for all the inputs of a function to have the same output.   [7.1a]

_____ **4.** If it is possible for a vertical line to cross a graph more than once, the graph is not the graph of a function.   [7.1d]

_____ **5.** If the domain of a function is the set of real numbers, then the range is the set of real numbers.   [7.2a]

Guided Solutions

Use the graph to complete the table of ordered pairs that name points on the graph.

6. [7.1c]

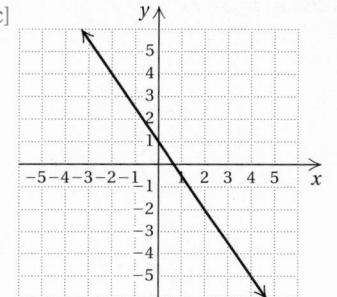

x	$f(x)$
0	☐
☐	−2
−2	☐
4	☐

7. [7.1c]

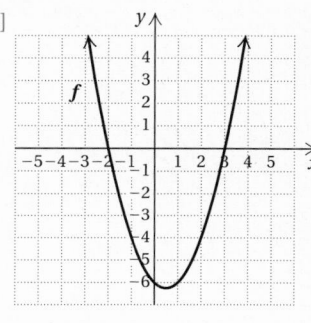

x	$f(x)$
−2	☐
☐	0
2	☐
☐	−4
1	☐

Mixed Review

Determine whether the correspondence is a function.   [7.1a]

8. Domain   Range

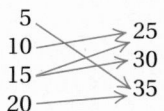

9. Domain   Range

10. Find the domain and the range.   [7.2a]

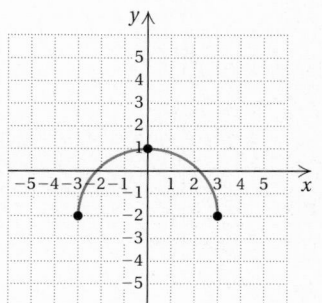

Find the function values.   [7.1b]

11. $g(x) = 2 + x$; $g(-5)$

12. $f(x) = x - 7$; $f(0)$

13. $h(x) = 8$; $h\left(\dfrac{1}{2}\right)$

14. $f(x) = 3x^2 - x + 5$; $f(-1)$

15. $g(p) = p^4 - p^3$; $g(10)$

16. $f(t) = \dfrac{1}{2}t + 3$; $f(-6)$

Determine whether each of the following is the graph of a function. [7.1d]

17.

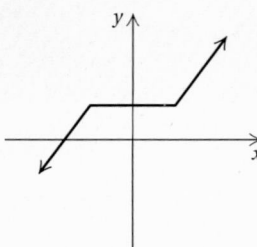

18.

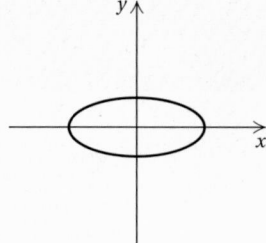

19.

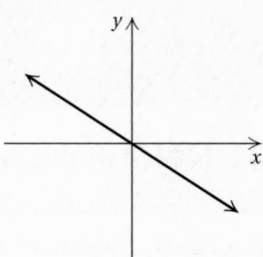

Find the domain. [7.2a]

20. $g(x) = \dfrac{3}{12 - 3x}$

21. $f(x) = x^2 - 10x + 3$

22. $h(x) = \dfrac{x - 2}{x + 2}$

23. $f(x) = |x - 4|$

Graph. [7.1c]

24. $g(x) = -\dfrac{2}{3}x - 2$

25. $f(x) = x - 1$

26. $h(x) = 2x + \dfrac{1}{2}$

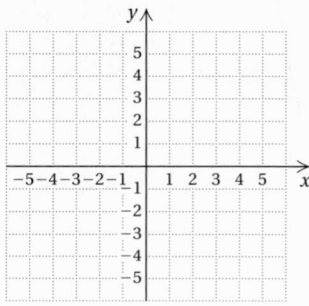

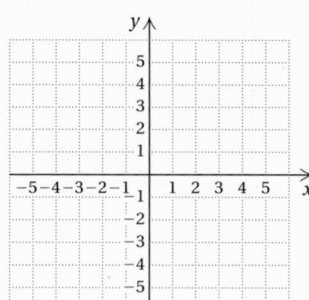

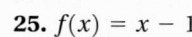

27. $g(x) = |x| - 3$

28. $f(x) = 1 + x^2$

29. $f(x) = -\dfrac{1}{4}x$

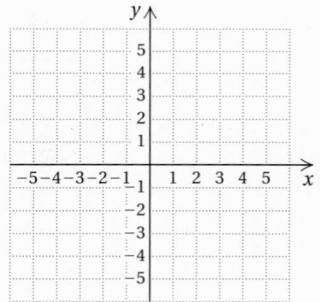

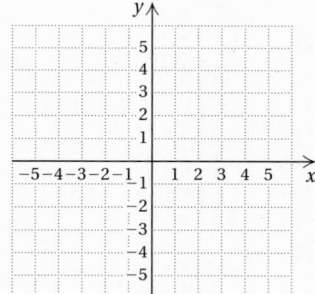

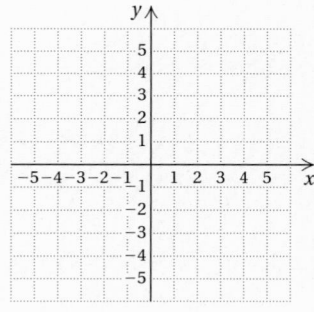

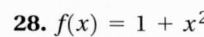

Understanding Through Discussion and Writing

30. Is it possible for a function to have more numbers as outputs than as inputs? Why or why not? [7.1a]

31. Without making a drawing, how can you tell that the graph of $f(x) = x - 30$ passes through three quadrants? [7.1c]

32. For a given function f, it is known that $f(2) = -3$. Give as many interpretations of this fact as you can. [7.1b], [7.2a]

33. Explain the difference between the domain and the range of a function. [7.2a]

Linear Functions: Graphs and Slope

We now turn our attention to functions whose graphs are straight lines. Such functions are called **linear** and can be written in the form $f(x) = mx + b$.

> ### LINEAR FUNCTION
>
> A **linear function** f is any function that can be described by $f(x) = mx + b$.

Compare the two equations $7y + 2x = 11$ and $y = 3x + 5$. Both are linear equations because their graphs are straight lines. Each can be expressed in an equivalent form that is a linear function.

The equation $y = 3x + 5$ can be expressed as $f(x) = mx + b$, where $m = 3$ and $b = 5$.

The equation $7y + 2x = 11$ also has an equivalent form $f(x) = mx + b$. To see this, we solve for y:

$$7y + 2x = 11$$
$$7y + 2x - 2x = -2x + 11 \qquad \text{Subtracting } 2x$$
$$7y = -2x + 11$$
$$\frac{7y}{7} = \frac{-2x + 11}{7} \qquad \text{Dividing by 7}$$
$$y = -\frac{2}{7}x + \frac{11}{7}. \qquad \text{Simplifying}$$

(It might be helpful to review the discussion on solving formulas in Section 2.4.) We now have an equivalent equation in the form

$$f(x) = -\frac{2}{7}x + \frac{11}{7}, \qquad \text{where} \quad m = -\frac{2}{7} \quad \text{and} \quad b = \frac{11}{7}.$$

In this section, we consider the effects of the constants m and b on the graphs of linear functions.

a The Constant b: The y-Intercept

Let's first explore the effect of the constant b.

EXAMPLE 1 Graph $y = 2x$ and $y = 2x + 3$ using the same set of axes. Compare the graphs.

We first make a table of solutions of both equations.

x	y $y = 2x$	y $y = 2x + 3$
0	0	3
1	2	5
−1	−2	1
2	4	7
−2	−4	−1

SKILL TO REVIEW
Objective 1.4a: Subtract real numbers.

Subtract.
  1. $11 - (-8)$ **2.** $-6 - (-6)$

Calculator Corner

Exploring b We can use a graphing calculator to explore the effect of the constant b on the graph of a function of the form $f(x) = mx + b$. Graph $y_1 = x$ in the standard $[-10, 10, -10, 10]$ viewing window. Then graph $y_2 = x + 4$, followed by $y_3 = x - 3$, in the same viewing window.

Exercises:

1. Compare the graph of y_2 with the graph of y_1.

2. Compare the graph of y_3 with the graph of y_1.

3. Visualize the graphs of $y = x + 8$ and $y = x - 5$. Compare each graph with the graph of y_1.

Answers

Skill to Review:
1. 19 **2.** 0

1. Graph $y = 3x$ and $y = 3x - 6$ using the same set of axes. Compare the graphs.

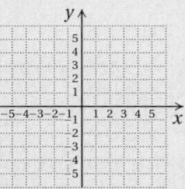

2. Graph $y = -2x$ and $y = -2x + 3$ using the same set of axes. Compare the graphs.

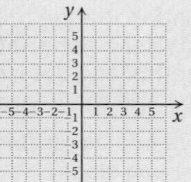

Next, we plot these points. Drawing a red line for $y = 2x$ and a blue line for $y = 2x + 3$, we note that the graph of $y = 2x + 3$ is simply the graph of $y = 2x$ shifted, or *translated*, up 3 units. The lines are parallel.

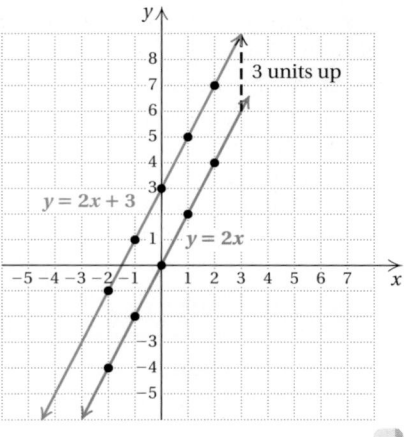

Do Exercises 1 and 2.

EXAMPLE 2 Graph $f(x) = \frac{1}{3}x$ and $g(x) = \frac{1}{3}x - 2$ using the same set of axes. Compare the graphs.

We first make a table of solutions of both equations. By choosing multiples of 3, we can avoid fractions.

x	$f(x)$ $f(x) = \frac{1}{3}x$	$g(x)$ $g(x) = \frac{1}{3}x - 2$
0	0	−2
3	1	−1
−3	−1	−3
6	2	0

We then plot these points. Drawing a red line for $f(x) = \frac{1}{3}x$ and a blue line for $g(x) = \frac{1}{3}x - 2$, we see that the graph of $g(x) = \frac{1}{3}x - 2$ is simply the graph of $f(x) = \frac{1}{3}x$ shifted, or translated, down 2 units. The lines are parallel.

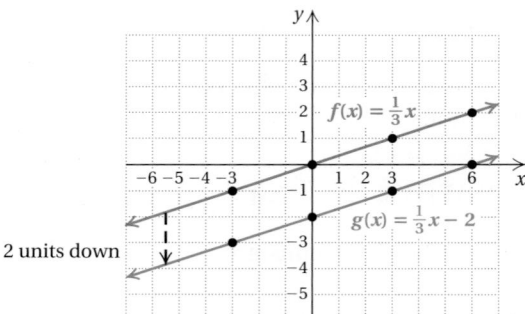

Answers

1.
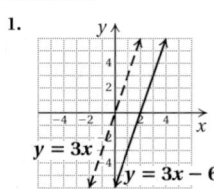
The graph of $y = 3x - 6$ looks just like the graph of $y = 3x$, but it is shifted down 6 units.

2.
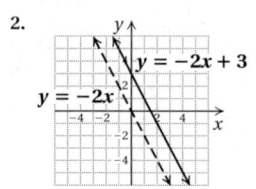
The graph of $y = -2x + 3$ looks just like the graph of $y = -2x$, but it is shifted up 3 units.

In Example 1, we saw that the graph of $y = 2x + 3$ is parallel to the graph of $y = 2x$ and that it passes through the point $(0, 3)$. Similarly, in Example 2, we saw that the graph of $y = \frac{1}{3}x - 2$ is parallel to the graph of $y = \frac{1}{3}x$ and that it passes through the point $(0, -2)$. In general, the graph of $y = mx + b$ is a line parallel to $y = mx$, passing through the point $(0, b)$. The point $(0, b)$ is called the **y-intercept** because it is the point at which the graph crosses the y-axis. Often it is convenient to refer to the number b as the y-intercept. The constant b has the effect of moving the graph of $y = mx$ up or down $|b|$ units to obtain the graph of $y = mx + b$.

Do Exercise 3.

3. Graph $f(x) = \frac{1}{3}x$ and $g(x) = \frac{1}{3}x + 2$ using the same set of axes. Compare the graphs.

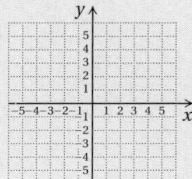

y-INTERCEPT

The y-intercept of the graph of $f(x) = mx + b$ is the point $(0, b)$ or, simply, b.

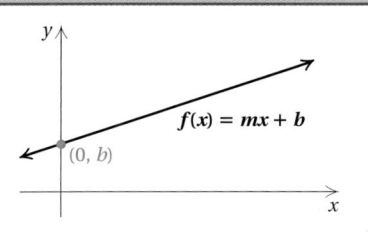

EXAMPLE 3 Find the y-intercept: $y = -5x + 4$.

$$y = -5x + 4 \qquad (0, 4) \text{ is the } y\text{-intercept.}$$

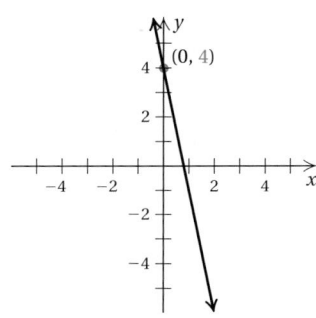

EXAMPLE 4 Find the y-intercept: $f(x) = 6.3x - 7.8$.

$$f(x) = 6.3x - 7.8 \qquad (0, -7.8) \text{ is the } y\text{-intercept.}$$

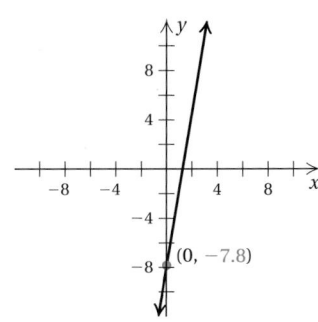

Do Exercises 4 and 5.

Find the y-intercept.

4. $y = 7x + 8$

5. $f(x) = -6x - \frac{2}{3}$

Answers

3.
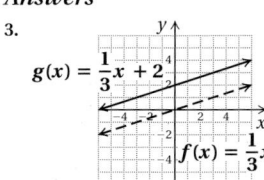
The graph of $g(x)$ looks just like the graph of $f(x)$, but it is shifted up 2 units.

4. $(0, 8)$ **5.** $\left(0, -\frac{2}{3}\right)$

ⓑ The Constant *m*: Slope

Look again at the graphs in Examples 1 and 2. Note that the slant of each red line seems to match the slant of each blue line. This leads us to believe that the number *m* in the equation $y = mx + b$ is related to the slant of the line. Let's consider some examples.

Graphs with $m < 0$:

$m = -1$

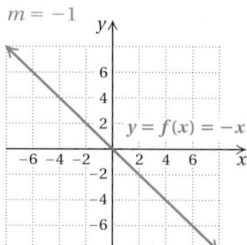

$m = -5$

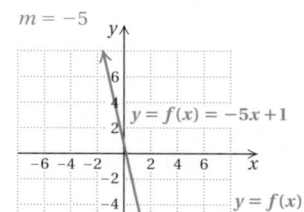

$m = -\frac{1}{2}$

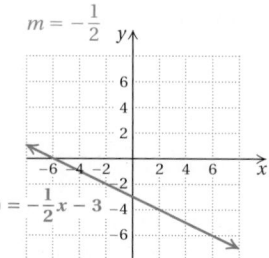

Graphs with $m = 0$:

$m = 0$

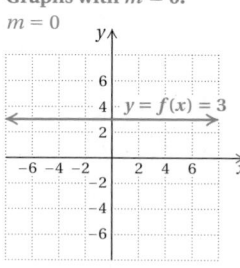

$m = 0$

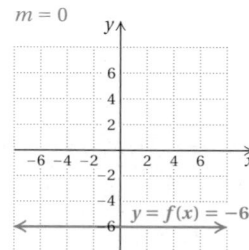

$m = 0$

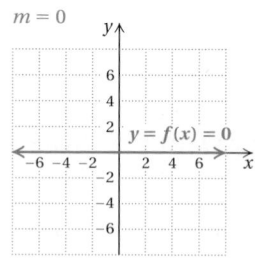

Graphs with $m > 0$:

$m = 1$

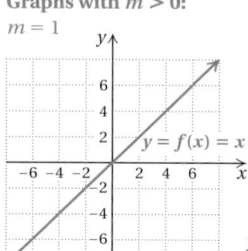

$m = 6$

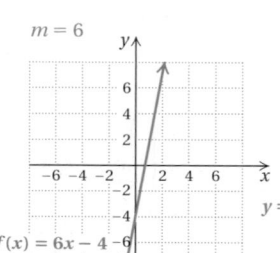

$m = \frac{1}{3}$

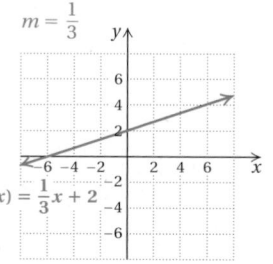

Note that

$m < 0 \rightarrow$ The graph slants down from left to right;

$m = 0 \rightarrow$ the graph is horizontal; and

$m > 0 \rightarrow$ the graph slants up from left to right.

The following definition enables us to visualize the slant and attach a number, a geometric ratio, or *slope*, to the line.

SLOPE

The **slope** of a line containing points (x_1, y_1) and (x_2, y_2) is given by

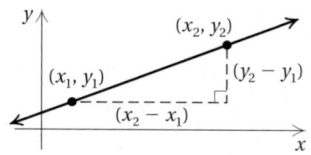

$$m = \frac{\text{rise}}{\text{run}}$$

$$= \frac{\text{change in } y}{\text{change in } x} = \frac{y_2 - y_1}{x_2 - x_1} = \frac{y_1 - y_2}{x_1 - x_2}.$$

Consider a line with two points marked P_1 and P_2, as follows. As we move from P_1 to P_2, the y-coordinate changes from 1 to 3 and the x-coordinate changes from 2 to 7. The change in y is $3 - 1$, or 2. The change in x is $7 - 2$, or 5.

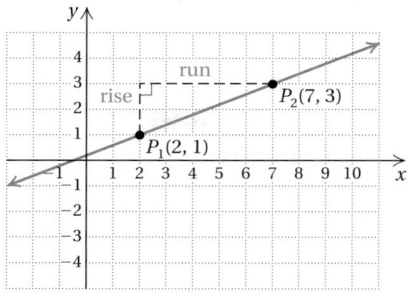

We call the change in y the **rise** and the change in x the **run**. The ratio rise/run is the same for any two points on a line. We call this ratio the **slope**. Slope describes the slant of a line. The slope of the line in the graph above is given by

$$\frac{\text{rise}}{\text{run}}, \quad \text{or} \quad \frac{\text{change in } y}{\text{change in } x}, \quad \text{or} \quad \frac{2}{5}.$$

Whenever x increases by 5 units, y increases by 2 units. Equivalently, whenever x increases by 1 unit, y increases by $\frac{2}{5}$ unit.

EXAMPLE 5 Graph the line containing the points $(-4, 3)$ and $(2, -5)$ and find the slope.

The graph is shown below. Going from $(-4, 3)$ to $(2, -5)$, we see that the change in y, or the rise, is $-5 - 3$, or -8. The change in x, or the run, is $2 - (-4)$, or 6.

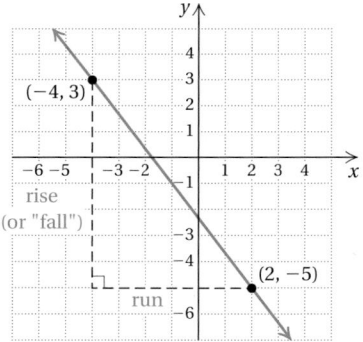

$$\text{Slope} = \frac{\text{rise}}{\text{run}} = \frac{\text{change in } y}{\text{change in } x}$$

$$= \frac{-5 - 3}{2 - (-4)}$$

$$= \frac{-8}{6} = -\frac{8}{6}, \text{ or } -\frac{4}{3}$$

Visualizing Slope

Exercises: Use the window settings $[-6, 6, -4, 4]$, with Xscl $= 1$ and Yscl $= 1$.

1. Graph $y = x$, $y = 2x$, and $y = 5x$ in the same window. What do you think the graph of $y = 10x$ will look like?

2. Graph $y = x$, $y = \frac{1}{2}x$, and $y = 0.1x$ in the same window. What do you think the graph of $y = 0.005x$ will look like?

3. Graph $y = -x$, $y = -2x$, and $y = -5x$ in the same window. What do you think the graph of $y = -10x$ will look like?

4. Graph $y = -x$, $y = -\frac{1}{2}x$, and $y = -0.1x$ in the same window. What do you think the graph of $y = -0.005x$ will look like?

Graph the line through the given points and find its slope.

6. $(-1, -1)$ and $(2, -4)$

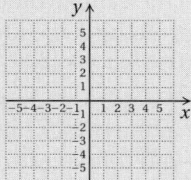

7. $(0, 2)$ and $(3, 1)$

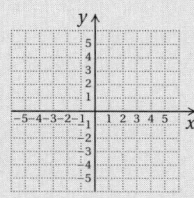

8. Find the slope of the line $f(x) = -\frac{2}{3}x + 1$. Use the points $(9, -5)$ and $(3, -1)$.

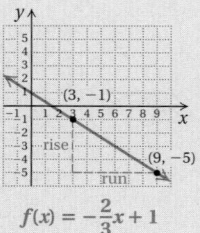

$$f(x) = -\frac{2}{3}x + 1$$

The formula

$$m = \frac{y_2 - y_1}{x_2 - x_1} = \frac{y_1 - y_2}{x_1 - x_2}$$

tells us that we can subtract in two ways. We must remember, however, to subtract the x-coordinates in the same order that we subtract the y-coordinates.

Let's do Example 5 again:

$$\text{Slope} = \frac{\text{change in } y}{\text{change in } x} = \frac{3 - (-5)}{-4 - 2} = \frac{8}{-6} = -\frac{8}{6} = -\frac{4}{3}.$$

We see that both ways give the same value for the slope.

The slope of a line tells how it slants. A line with positive slope slants up from left to right. The larger the positive number, the steeper the slant. A line with negative slope slants downward from left to right. The smaller the negative number, the steeper the line.

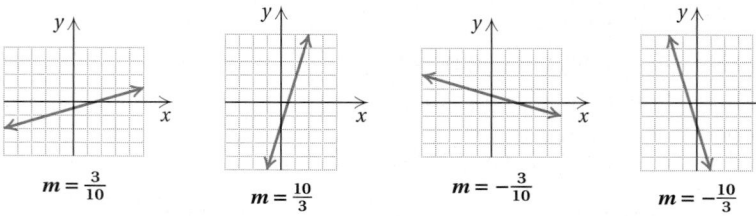

$$m = \frac{3}{10} \qquad m = \frac{10}{3} \qquad m = -\frac{3}{10} \qquad m = -\frac{10}{3}$$

Do Exercises 6 and 7.

How can we find the slope from a given equation? Let's consider the equation $y = 2x + 3$, which is in the form $y = mx + b$. We can find two points by choosing convenient values for x—say, 0 and 1—and substituting to find the corresponding y-values.

If $x = 0$, $y = 2 \cdot 0 + 3 = 3$.

If $x = 1$, $y = 2 \cdot 1 + 3 = 5$.

We find two points on the line to be

$$(0, 3) \quad \text{and} \quad (1, 5).$$

The slope of the line is found as follows, using the definition of slope:

$$m = \frac{\text{change in } y}{\text{change in } x}$$
$$= \frac{5 - 3}{1 - 0} = \frac{2}{1} = 2.$$

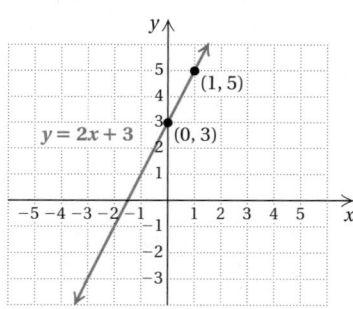

$$y = 2x + 3$$

The slope is 2. Note that this is the coefficient of the x-term in the equation $y = 2x + 3$.

If we had chosen different points on the line—say, $(-2, -1)$ and $(4, 11)$—the slope would still be 2, as we see in the following calculation:

$$m = \frac{11 - (-1)}{4 - (-2)} = \frac{11 + 1}{4 + 2} = \frac{12}{6} = 2.$$

Do Exercise 8.

Answers

6. 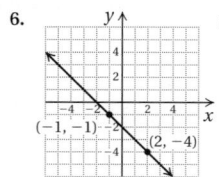 ; $m = -1$

7. 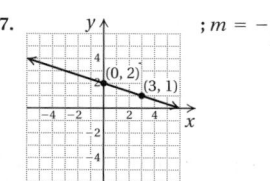 ; $m = -\frac{1}{3}$

8. $m = -\frac{2}{3}$

We see that the slope of the line $y = mx + b$ is indeed the constant m, the coefficient of x.

> ### SLOPE
>
> The **slope** of the line $y = mx + b$ is m.

From a linear equation in the form $y = mx + b$, we can read the slope and the y-intercept of the graph directly.

> ### SLOPE–INTERCEPT EQUATION
>
> The equation $y = mx + b$ is called the **slope–intercept equation**. The slope is m and the y-intercept is $(0, b)$.

Note that any graph of an equation $y = mx + b$ passes the vertical-line test and thus represents a function.

EXAMPLE 6 Find the slope and the y-intercept of $y = 5x - 4$.

Since the equation is already in the form $y = mx + b$, we simply read the slope and the y-intercept from the equation:

$$y = 5x - 4.$$

The slope is 5. The y-intercept is $(0, -4)$.

EXAMPLE 7 Find the slope and the y-intercept of $2x + 3y = 8$.

We first solve for y so we can easily read the slope and the y-intercept:

$$2x + 3y = 8$$
$$3y = -2x + 8 \qquad \text{Subtracting } 2x$$
$$\frac{3y}{3} = \frac{-2x + 8}{3} \qquad \text{Dividing by } 3$$
$$y = -\frac{2}{3}x + \frac{8}{3}. \qquad \text{Finding the form } y = mx + b$$

The slope is $-\frac{2}{3}$. The y-intercept is $\left(0, \frac{8}{3}\right)$.

Do Exercises 9 and 10.

Find the slope and the y-intercept.

9. $f(x) = -8x + 23$

10. $5x - 10y = 25$

(c) Applications

Slope has many real-world applications. For example, numbers like 2%, 3%, and 6% are often used to represent the *grade* of a road, a measure of how steep a road on a hill or mountain is. A 3% grade $\left(3\% = \frac{3}{100}\right)$ means that for every horizontal distance of 100 ft that the road runs, the road rises 3 ft, and a -3% grade means that for every horizontal distance of 100 ft, the road drops 3 ft. (Normally, the road signs do not include negative signs, since it is obvious

Answers

9. Slope: -8; y-intercept: $(0, 23)$

10. Slope: $\frac{1}{2}$; y-intercept: $\left(0, -\frac{5}{2}\right)$

whether you are climbing or descending.) An athlete might change the grade of a treadmill during a workout. An escape ramp on an airliner might have a slope of about −0.6.

Architects and carpenters use slope when designing and building stairs, ramps, or roof pitches. Another application occurs in hydrology. The strength or force of a river depends on how far the river falls vertically compared to how far it flows horizontally. Slope can also be considered as a **rate of change**.

EXAMPLE 8 *Health Insurance.* Premiums for family health insurance plans have increased steadily in recent years. In 2000, the average premium was $6438 per year. By 2008, this amount had increased to $12,680 per year. Find the rate of change of the average yearly family health insurance premium with respect to time, in years.

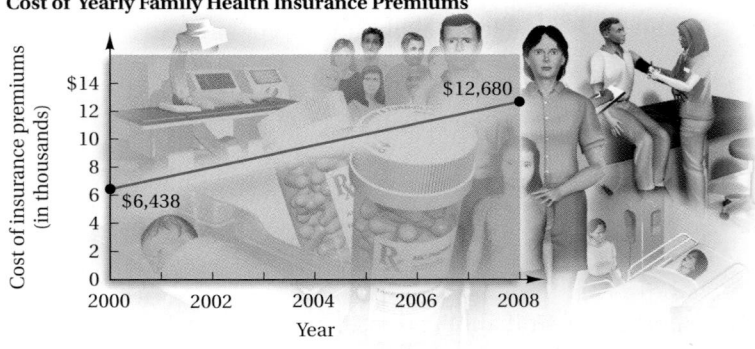

Cost of Yearly Family Health Insurance Premiums

SOURCE: The Kaiser Family Foundation

The rate of change with respect to time, in years, is given by

$$\text{Rate of change} = \frac{\$12,680 - \$6438}{2008 - 2000}$$

$$= \frac{\$6242}{8 \text{ yr}}$$

$$= \$780.25.$$

The average yearly family health insurance premium is increasing at a rate of about $780.25 per year.

Do Exercise 11.

EXAMPLE 9 *Volume of Mail.* The volume of mail through the U.S. Postal Service has been dropping since 2006, as shown in the graph below.

Volume of Mail Through U.S. Post Office

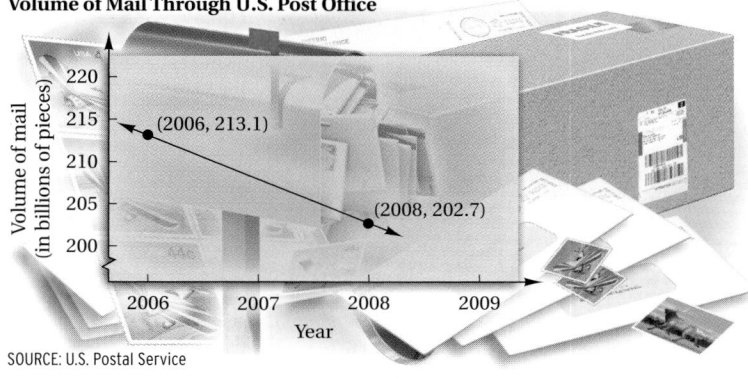

SOURCE: U.S. Postal Service

Since the graph is linear, we can use any pair of points to determine the rate of change:

$$\text{Rate of change} = \frac{202.7 \text{ billion} - 213.1 \text{ billion}}{2008 - 2006}$$

$$= \frac{-10.4 \text{ billion}}{2 \text{ yr}} = -5.2 \text{ billion per year.}$$

The volume of mail through the U.S. Postal Service is decreasing at a rate of about 5.2 billion pieces per year.

Do Exercise 12.

11. Haircutting. The graph below displays data from a day's work at Lee's Barbershop. At 2:00, 8 haircuts had been completed. At 4:00, 14 haircuts had been done. Use the graph to determine the rate of change of the number of haircuts with respect to time.

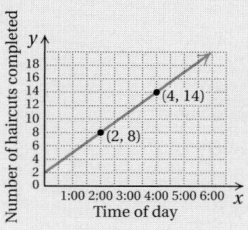

12. Newspaper Circulation. Daily newspaper circulation has decreased in recent years. The graph below shows the circulation of daily newspapers, in millions, for three years. Find the rate of change of the circulation of daily newspapers per year.

Circulation of Daily Newspapers

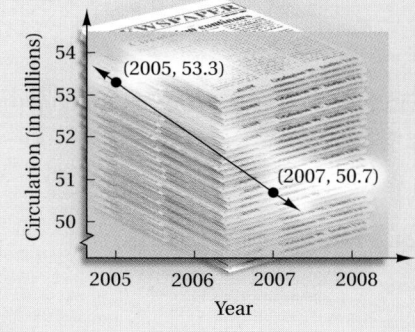

a , **b** Find the slope and the *y*-intercept.

1. $y = 4x + 5$

2. $y = -5x + 10$

3. $f(x) = -2x - 6$

4. $g(x) = -5x + 7$

5. $y = -\frac{3}{8}x - \frac{1}{5}$

6. $y = \frac{15}{7}x + \frac{16}{5}$

7. $g(x) = 0.5x - 9$

8. $f(x) = -3.1x + 5$

9. $2x - 3y = 8$

10. $-8x - 7y = 24$

11. $9x = 3y + 6$

12. $9y + 36 - 4x = 0$

13. $3 - \frac{1}{4}y = 2x$

14. $5x = \frac{2}{3}y - 10$

15. $17y + 4x + 3 = 7 + 4x$

16. $3y - 2x = 5 + 9y - 2x$

b Find the slope of each line.

17.

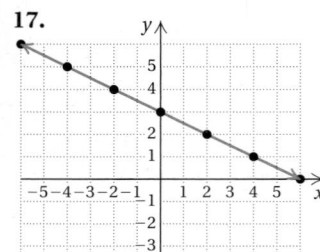

18.

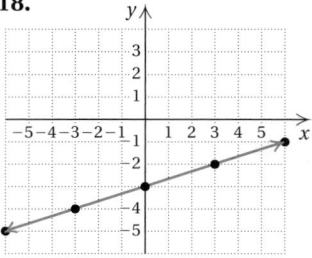

19.

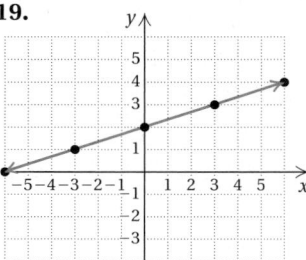

20.

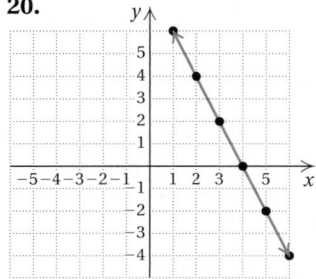

Find the slope of the line containing the given pair of points.

21. $(6, 9)$ and $(4, 5)$

22. $(8, 7)$ and $(2, -1)$

23. $(9, -4)$ and $(3, -8)$

24. $(17, -12)$ and $(-9, -15)$

25. $(-16.3, 12.4)$ and $(-5.2, 8.7)$

26. $(14.4, -7.8)$ and $(-12.5, -17.6)$

c Find the slope (or rate of change).

27. Find the slope (or grade) of the treadmill.

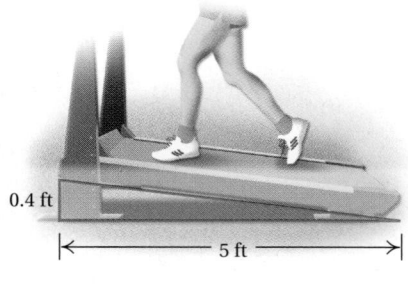

0.4 ft

5 ft

28. Find the slope (or head) of the river.

43.33 ft

1238 ft

29. Find the slope (or pitch) of the roof.

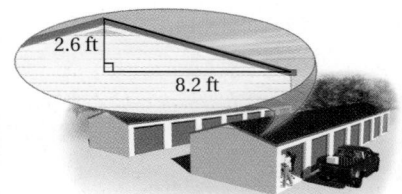

2.6 ft

8.2 ft

30. Public buildings regularly include steps with 7-in. risers and 11-in. treads. Find the grade of such a stairway.

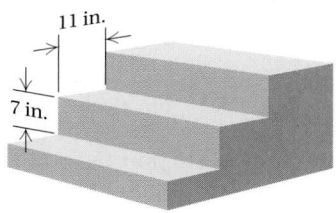

31. *Mine Deaths.* Find the rate of change of the number of mine deaths in the United States with respect to time, in years.

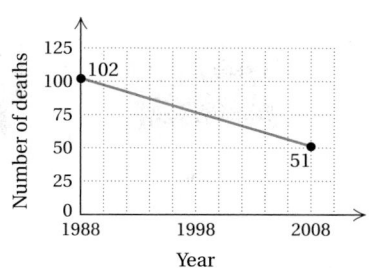

SOURCE: U.S. Mine Safety and Health Administration

32. Find the rate of change of the cost of a formal wedding.

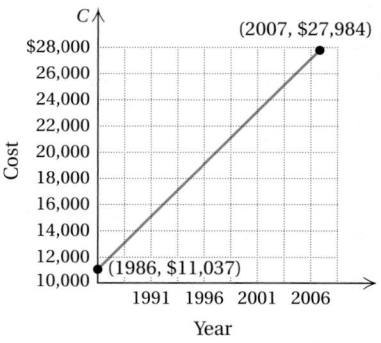

SOURCE: Modern Bride Magazine

Find the rate of change.

33.

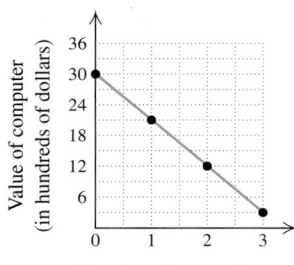

34.

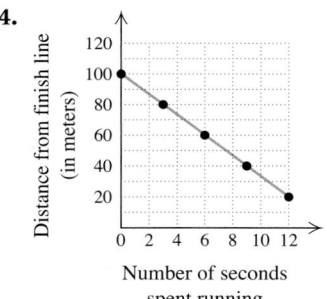

35.

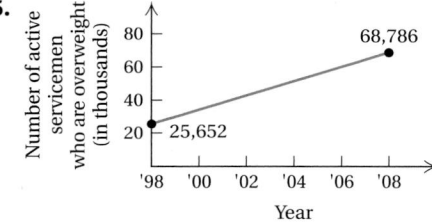

SOURCE: U.S. Department of Defense

36.

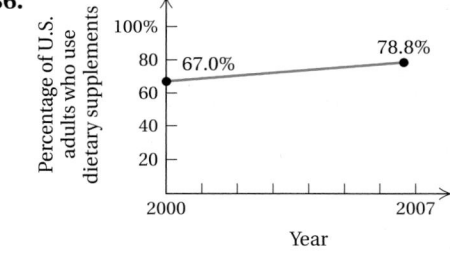

Skill Maintenance

Simplify.   [1.8c, d], [4.1b]

37. $3^2 - 24 \cdot 56 + 144 \div 12$

38. $9\{2x - 3[5x + 2(-3x + y^0 - 2)]\}$

39. $10\{2x + 3[5x - 2(-3x + y^1 - 2)]\}$

40. $5^4 \div 625 \div 5^2 \cdot 5^7 \div 5^3$

Solve.   [2.6a]

41. One side of a square is 5 yd less than a side of an equilateral triangle. If the perimeter of the square is the same as the perimeter of the triangle, what is the length of a side of the square? of the triangle?

Factor.   [5.6a]

42. $8 - 125x^3$

43. $c^6 - d^6$

44. $56x^3 - 7$

45. Divide: $(a^2 - 11a + 6) \div (a - 1)$.   [4.8b]

7.4

More on Graphing Linear Equations

OBJECTIVES

a Graph linear equations using intercepts.

b Given a linear equation in slope-intercept form, use the slope and the *y*-intercept to graph the line.

c Graph linear equations of the form $x = a$ or $y = b$.

d Given the equations of two lines, determine whether their graphs are parallel or whether they are perpendicular.

a Graphing Using Intercepts

The **x-intercept** of the graph of a linear equation or function is the point at which the graph crosses the *x*-axis. The **y-intercept** is the point at which the graph crosses the *y*-axis. We know from geometry that only one line can be drawn through two given points. Thus, if we know the intercepts, we can graph the line. To ensure that a computation error has not been made, it is a good idea to calculate a third point as a check.

Many equations of the type $Ax + By = C$ can be graphed conveniently using intercepts.

> ### x- AND y-INTERCEPTS
>
> A **y-intercept** is a point $(0, b)$. To find b, let $x = 0$ and solve for y.
> An **x-intercept** is a point $(a, 0)$. To find a, let $y = 0$ and solve for x.

EXAMPLE 1 Find the intercepts of $3x + 2y = 12$ and then graph the line.

y-intercept: To find the *y*-intercept, we let $x = 0$ and solve for y:

$$3x + 2y = 12$$
$$3 \cdot 0 + 2y = 12 \qquad \text{Substituting 0 for } x$$
$$2y = 12$$
$$y = 6.$$

The *y*-intercept is $(0, 6)$.

x-intercept: To find the *x*-intercept, we let $y = 0$ and solve for x:

$$3x + 2y = 12$$
$$3x + 2 \cdot 0 = 12 \qquad \text{Substituting 0 for } y$$
$$3x = 12$$
$$x = 4.$$

The *x*-intercept is $(4, 0)$.

We plot these points and draw the line, using a third point as a check. We choose $x = 6$ and solve for y:

$$3(6) + 2y = 12$$
$$18 + 2y = 12$$
$$2y = -6$$
$$y = -3.$$

We plot $(6, -3)$ and note that it is on the line.

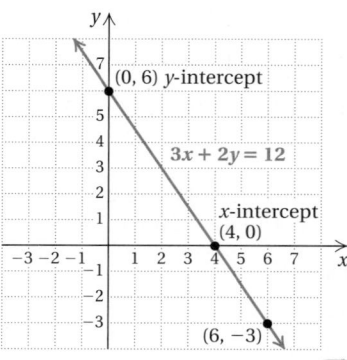

SKILL TO REVIEW

Objective 3.1a: Plot points associated with ordered pairs of numbers.

1. Plot the following points:
$A(0, 4)$, $B(0, -1)$, $C(0, 0)$, $D(3, 0)$, and $E\left(-\frac{7}{2}, 0\right)$.

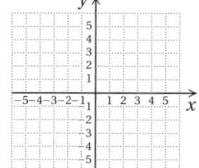

Answer

Skill to Review:

1.

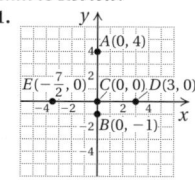

When both the x-intercept and the y-intercept are $(0, 0)$, as is the case with an equation such as $y = 2x$, whose graph passes through the origin, another point would have to be calculated and a third point used as a check.

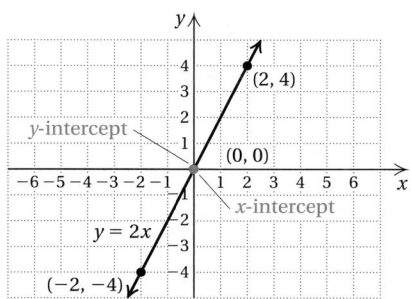

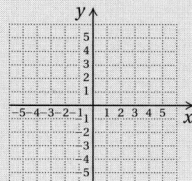

1. Find the intercepts of $4y - 12 = -6x$ and then graph the line.

Do Exercise 1.

Calculator Corner

Viewing the Intercepts   Knowing the intercepts of a linear equation helps us determine a good viewing window for the graph of the equation. For example, when we graph the equation $y = -x + 15$ in the standard window, we see only a small portion of the graph in the upper right-hand corner of the screen, as shown on the left below.

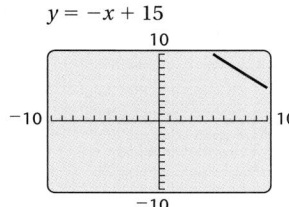

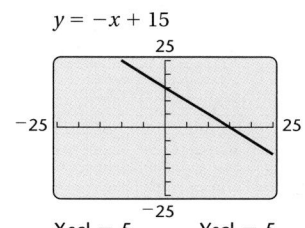

Using algebra, as we did in Example 1, we can find that the intercepts of the graph of this equation are $(0, 15)$ and $(15, 0)$. This tells us that, if we are to see a portion of the graph that includes the intercepts, both Xmax and Ymax should be greater than 15. We can try different window settings until we find one that suits us. One good choice, shown on the right above, is $[-25, 25, -25, 25]$, with Xscl $= 5$ and Yscl $= 5$.

Exercises: Find the intercepts of the equation algebraically. Then graph the equation on a graphing calculator, choosing window settings that allow the intercepts to be seen clearly. (Settings may vary.)

1. $y = -3.2x - 16$

2. $y - 4.25x = 85$

3. $6x + 5y = 90$

4. $5x - 6y = 30$

5. $8x + 3y = 9$

6. $y = 0.4x - 5$

7. $y = 1.2x - 12$

8. $4x - 5y = 2$

Answer

1.
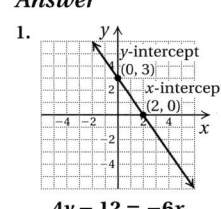

$4y - 12 = -6x$

Graph using the slope and the y-intercept.

2. $y = \dfrac{3}{2}x + 1$

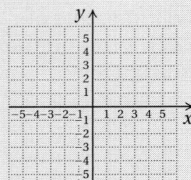

3. $f(x) = \dfrac{3}{4}x - 2$

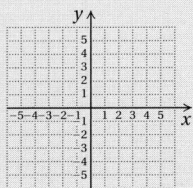

4. $g(x) = -\dfrac{3}{5}x + 5$

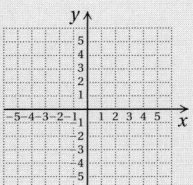

5. $y = -\dfrac{5}{3}x - 4$

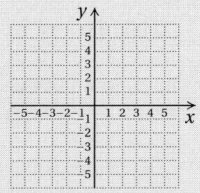

b Graphing Using the Slope and the y-Intercept

We can also graph a line using its slope and y-intercept.

EXAMPLE 2 Graph: $y = -\frac{2}{3}x + 1$.

This equation is in slope–intercept form, $y = mx + b$. The y-intercept is $(0, 1)$. We plot $(0, 1)$. We can think of the slope $\left(m = -\frac{2}{3}\right)$ as $\frac{-2}{3}$.

$$m = \frac{\text{Rise}}{\text{Run}} = \frac{-2}{3} \quad \begin{array}{l}\text{Move 2 units down.}\\ \text{Move 3 units right.}\end{array}$$

Starting at the y-intercept and using the slope, we find another point by moving down 2 units (since the numerator is *negative* and corresponds to the change in y) and to the right 3 units (since the denominator is *positive* and corresponds to the change in x). We get to a new point, $(3, -1)$. In a similar manner, we can move from the point $(3, -1)$ to find another point, $(6, -3)$.

We could also think of the slope $\left(m = -\frac{2}{3}\right)$ as $\frac{2}{-3}$.

$$m = \frac{\text{Rise}}{\text{Run}} = \frac{2}{-3} \quad \begin{array}{l}\text{Move 2 units up.}\\ \text{Move 3 units left.}\end{array}$$

Then we can start again at $(0, 1)$, but this time we move up 2 units (since the numerator is *positive* and corresponds to the change in y) and to the left 3 units (since the denominator is *negative* and corresponds to the change in x). We get another point on the graph, $(-3, 3)$, and from it we can obtain $(-6, 5)$ and others in a similar manner. We plot the points and draw the line.

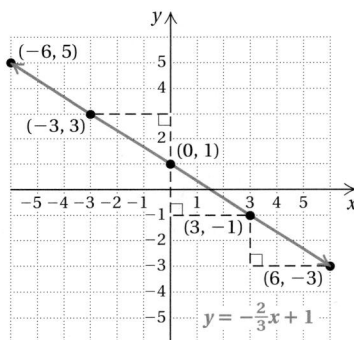

EXAMPLE 3 Graph: $f(x) = \frac{2}{5}x + 4$.

First, we plot the y-intercept, $(0, 4)$. We then consider the slope $\frac{2}{5}$. A slope of $\frac{2}{5}$ tells us that, for every 2 units that the graph rises, it runs 5 units horizontally in the positive direction, or to the right. Thus, starting at the y-intercept and using the slope, we find another point by moving up 2 units (since the numerator is *positive* and corresponds to the change in y) and to the right 5 units (since the denominator is *positive* and corresponds to the change in x). We get to a new point, $(5, 6)$.

Answers

2.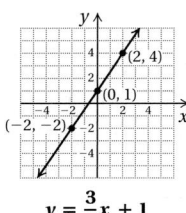

$y = \dfrac{3}{2}x + 1$

3.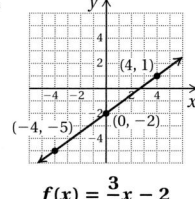

$f(x) = \dfrac{3}{4}x - 2$

4.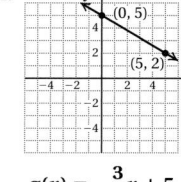

$g(x) = -\dfrac{3}{5}x + 5$

5.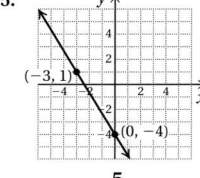

$y = -\dfrac{5}{3}x - 4$

We can also think of the slope $\frac{2}{5}$ as $\frac{-2}{-5}$. A slope of $\frac{-2}{-5}$ tells us that, for every 2 units that the graph drops, it runs 5 units horizontally in the negative direction, or to the left. We again start at the y-intercept, $(0, 4)$. We move down 2 units (since the numerator is *negative* and corresponds to the change in y) and to the left 5 units (since the denominator is *negative* and corresponds to the change in x). We get to another new point, $(-5, 2)$. We plot the points and draw the line.

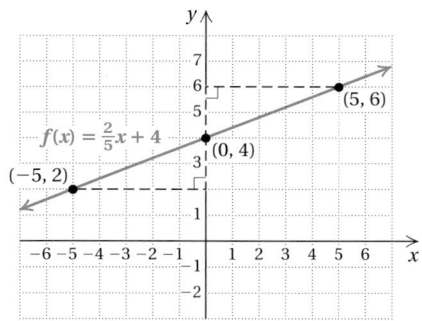

Do Exercises 2–5 on the preceding page.

c Horizontal and Vertical Lines

Some equations have graphs that are parallel to one of the axes. This happens when either A or B is 0 in $Ax + By = C$. These equations have a missing variable; that is, there is only one variable in the equation. In the following example, x is missing.

EXAMPLE 4   Graph: $y = 3$.

Since x is missing, any number for x will do. Thus all ordered pairs $(x, 3)$ are solutions. The graph is a **horizontal line** parallel to the x-axis.

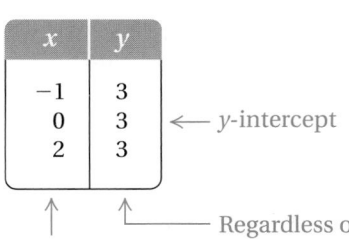

x	y
-1	3
0	3	← y-intercept
2	3

↑ Choose *any* number for x.   ↑ Regardless of x, y must be 3.

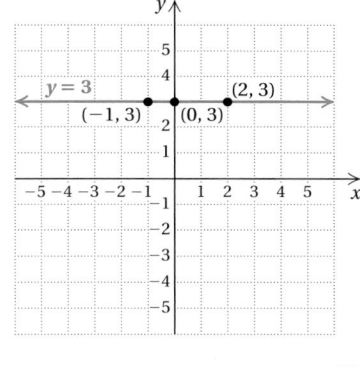

What about the slope of a horizontal line? In Example 4, consider the points $(-1, 3)$ and $(2, 3)$, which are on the line $y = 3$. The change in y is $3 - 3$, or 0. The change in x is $-1 - 2$, or -3. Thus,

$$m = \frac{3 - 3}{-1 - 2} = \frac{0}{-3} = 0.$$

Any two points on a horizontal line have the same y-coordinate. Thus the change in y is always 0, so the slope is 0.

Do Exercises 6 and 7.

Graph and determine the slope.

6. $f(x) = -4$

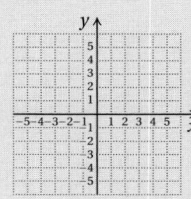

7. $y = 3.6$

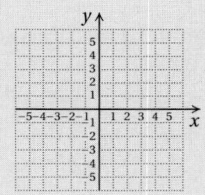

Answers

6. $m = 0$

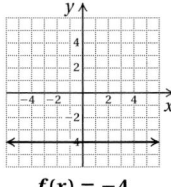

$f(x) = -4$

7. $m = 0$

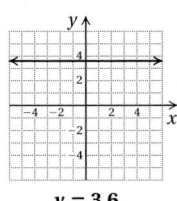

$y = 3.6$

We can also determine the slope by noting that $y = 3$ can be written in slope–intercept form as $y = 0x + 3$, or $f(x) = 0x + 3$. From this equation, we read that the slope is 0. A function of this type is called a **constant function**. We can express it in the form $y = b$, or $f(x) = b$. Its graph is a horizontal line that crosses the y-axis at $(0, b)$.

In the following example, y is missing and the graph is parallel to the y-axis.

EXAMPLE 5 Graph: $x = -2$.

Since y is missing, any number for y will do. Thus all ordered pairs $(-2, y)$ are solutions. The graph is a **vertical line** parallel to the y-axis.

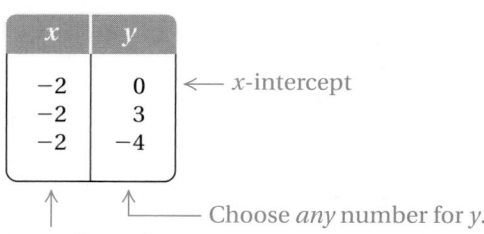

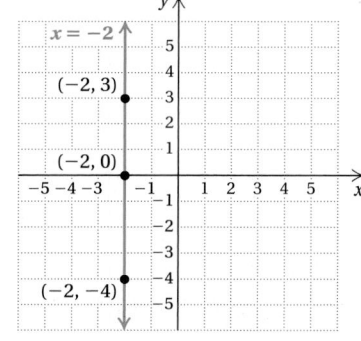

This graph is not the graph of a function because it fails the vertical-line test. The vertical line itself crosses the graph more than once.

Do Exercises 8 and 9.

What about the slope of a vertical line? In Example 5, consider the points $(-2, 3)$ and $(-2, -4)$, which are on the line $x = -2$. The change in y is $3 - (-4)$, or 7. The change in x is $-2 - (-2)$, or 0. Thus,

$$m = \frac{3 - (-4)}{-2 - (-2)} = \frac{7}{0}. \quad \text{Not defined}$$

Since division by 0 is not defined, the slope of this line is not defined. Any two points on a vertical line have the same x-coordinate. Thus the change in x is always 0, so the slope of any vertical line is not defined.

The following summarizes horizontal and vertical lines and their equations.

HORIZONTAL LINE; VERTICAL LINE

The graph of $y = b$, or $f(x) = b$, is a **horizontal line** with y-intercept $(0, b)$. It is the graph of a constant function with slope 0.

The graph of $x = a$ is a **vertical line** through the point $(a, 0)$. The slope is not defined. It is not the graph of a function.

Do Exercise 10.

Graph.

8. $x = -5$

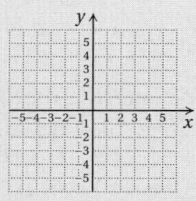

9. $8x - 5 = 19$ (*Hint*: Solve for x.)

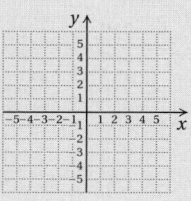

10. Determine, if possible, the slope of each line.

a) $x = -12$ **b)** $y = 6$

c) $2y + 7 = 11$ **d)** $x = 0$

e) $y = -\frac{3}{4}$ **f)** $10 - 5x = 15$

Answers

8.

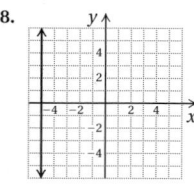

$x = -5$

9.

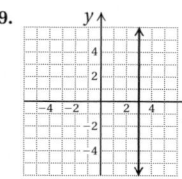

$8x - 5 = 19$

10. (a) Not defined; (b) $m = 0$; (c) $m = 0$; (d) not defined; (e) $m = 0$; (f) not defined

We have graphed linear equations in several ways in this chapter. Although, in general, you can use any method that works best for you, we list some guidelines in the margin at right.

 d **Parallel and Perpendicular Lines**

Parallel Lines

Parallel lines extend indefinitely without intersecting. If two lines are vertical, they are parallel. How can we tell whether nonvertical lines are parallel? We examine their slopes and y-intercepts.

PARALLEL LINES

Two nonvertical lines are **parallel** if they have the *same* slope and *different* y-intercepts.

EXAMPLE 6 Determine whether the graphs of
$$y - 3x = 1 \quad \text{and} \quad 3x + 2y = -2$$
are parallel.

To determine whether lines are parallel, we first find their slopes. To do this, we find the slope–intercept form of each equation by solving for y:

$$
\begin{aligned}
y - 3x &= 1 & 3x + 2y &= -2 \\
y &= 3x + 1; & 2y &= -3x - 2 \\
& & y &= \tfrac{1}{2}(-3x - 2) \\
& & y &= -\tfrac{3}{2}x - 1.
\end{aligned}
$$

The slopes, 3 and $-\frac{3}{2}$, are different. Thus the lines are not parallel, as the graphs at right confirm.

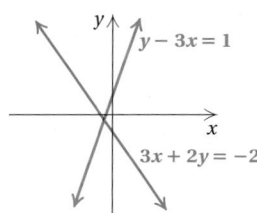

EXAMPLE 7 Determine whether the graphs of
$$3x - y = -5 \quad \text{and} \quad y - 3x = -2$$
are parallel.

We first find the slope–intercept form of each equation by solving for y:

$$
\begin{aligned}
3x - y &= -5 & y - 3x &= -2 \\
-y &= -3x - 5 & y &= 3x - 2. \\
-1(-y) &= -1(-3x - 5) \\
y &= 3x + 5;
\end{aligned}
$$

The slopes, 3, are the same. The y-intercepts, $(0, 5)$ and $(0, -2)$, are different. Thus the lines are parallel, as the graphs appear to confirm.

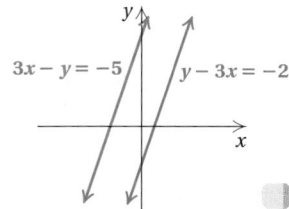

Do Exercises 11–13.

To graph a linear equation:

1. Is the equation of the type $x = a$ or $y = b$? If so, the graph will be a line parallel to an axis; $x = a$ is vertical and $y = b$ is horizontal.

2. If the line is of the type $y = mx$, both intercepts are the origin, $(0, 0)$. Plot $(0, 0)$ and one other point.

3. If the line is of the type $y = mx + b$, plot the y-intercept and one other point.

4. If the equation is of the form $Ax + By = C$, graph using intercepts. If the intercepts are too close together, choose another point farther from the origin.

5. In all cases, use a third point as a check.

Determine whether the graphs of the given pair of lines are parallel.

11. $x + 4 = y$,
$\quad y - x = -3$

12. $y + 4 = 3x$,
$\quad 4x - y = -7$

13. $y = 4x + 5$,
$\quad 2y = 8x + 10$

Answers

11. Yes **12.** No **13.** No; they are the same line.

Perpendicular Lines

If one line is vertical and another is horizontal, they are perpendicular. For example, the lines $x = 5$ and $y = -3$ are perpendicular. Otherwise, how can we tell whether two lines are perpendicular?

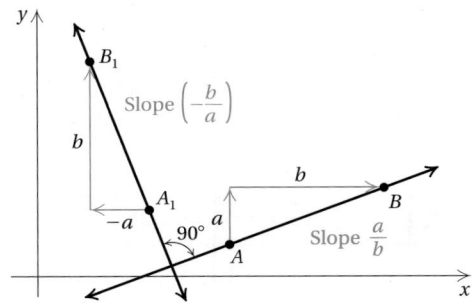

Consider a line $\overleftrightarrow{AB}$, as shown in the figure above, with slope a/b. Then think of rotating the line 90° to get a line $\overleftrightarrow{A_1 B_1}$ perpendicular to $\overleftrightarrow{AB}$. For the new line, the rise and the run are interchanged, but the run is now negative. Thus the slope of the new line is $-b/a$, which is the opposite of the reciprocal of the slope of the first line. Also note that when we multiply the slopes, we get

$$\frac{a}{b}\left(-\frac{b}{a}\right) = -1.$$

This is the condition under which lines will be perpendicular.

PERPENDICULAR LINES

Two nonvertical lines are **perpendicular** if the product of their slopes is -1. (If one line has slope m, the slope of a line perpendicular to it is $-1/m$. That is, to find the slope of a line perpendicular to a given line, we take the reciprocal of the given slope and change the sign.)

Lines are also perpendicular if one of them is vertical ($x = a$) and one of them is horizontal ($y = b$).

EXAMPLE 8 Determine whether the graphs of $5y = 4x + 10$ and $4y = -5x + 4$ are perpendicular.

To determine whether the lines are perpendicular, we determine whether the product of their slopes is -1. We first find the slope–intercept form of each equation by solving for y.

We have

$$5y = 4x + 10 \qquad\qquad 4y = -5x + 4$$
$$y = \tfrac{1}{5}(4x + 10) \qquad\quad y = \tfrac{1}{4}(-5x + 4)$$
$$= \tfrac{1}{5}(4x) + \tfrac{1}{5}(10) \qquad\quad = \tfrac{1}{4}(-5x) + \tfrac{1}{4}(4)$$
$$= \tfrac{4}{5}x + 2; \qquad\qquad\quad = -\tfrac{5}{4}x + 1.$$

The slope of the first line is $\frac{4}{5}$, and the slope of the second line is $-\frac{5}{4}$. The product of the slopes is $\frac{4}{5} \cdot \left(-\frac{5}{4}\right) = -1$. Thus the lines are perpendicular.

Do Exercises 14 and 15.

Determine whether the graphs of the given pair of lines are perpendicular.

14. $2y - x = 2,$
$\quad y + 2x = 4$

15. $3y = 2x + 15,$
$\quad 2y = 3x + 10$

Answers

14. Yes   **15.** No

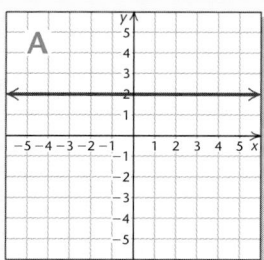

A

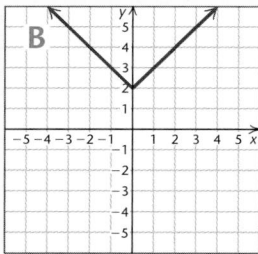

B

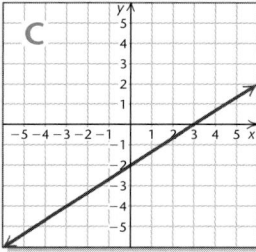

C

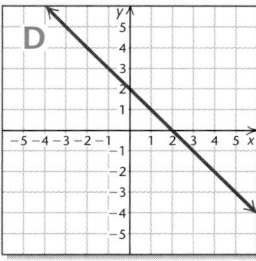

D

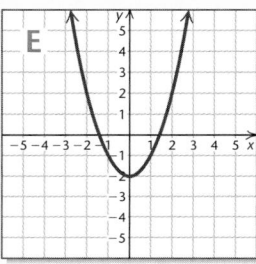

E

Visualizing
for Success

Match each equation with its graph.

1. $y = 2 - x$

2. $x - y = 2$

3. $x + 2y = 2$

4. $2x - 3y = 6$

5. $x = 2$

6. $y = 2$

7. $y = |x + 2|$

8. $y = |x| + 2$

9. $y = x^2 - 2$

10. $y = 2 - x^2$

Answers on page A-22

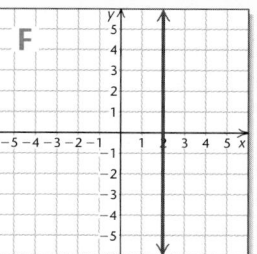

F

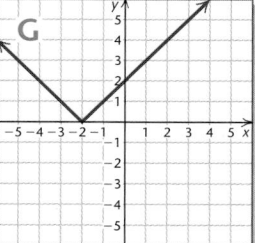

G

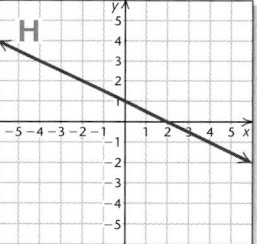

H

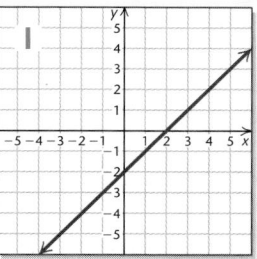

I

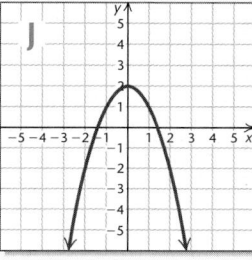

J

7.4 **Exercise Set**

For Extra Help

MyMathLab

Math XL
PRACTICE

WATCH

DOWNLOAD

READ

REVIEW

a Find the intercepts and then graph the line.

1. $x - 2 = y$

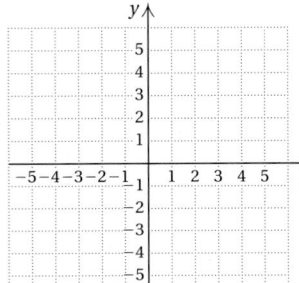

2. $x + 3 = y$

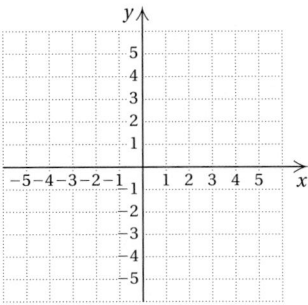

3. $x + 3y = 6$

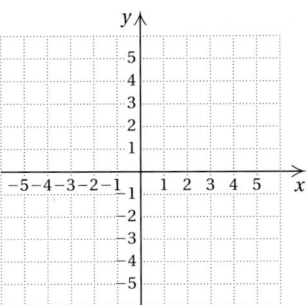

4. $x - 2y = 4$

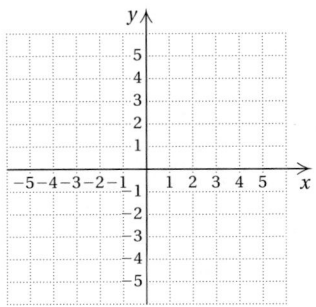

5. $2x + 3y = 6$

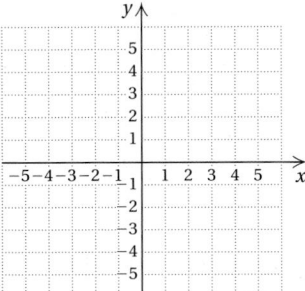

6. $5x - 2y = 10$

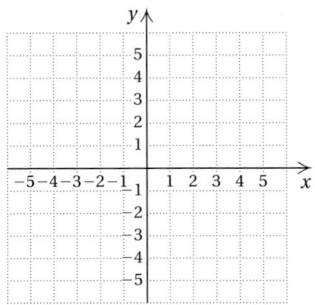

7. $f(x) = -2 - 2x$

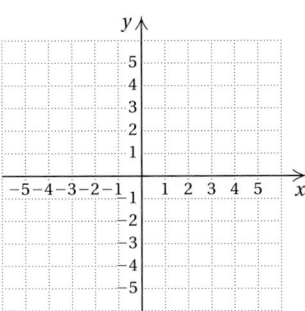

8. $g(x) = 5x - 5$

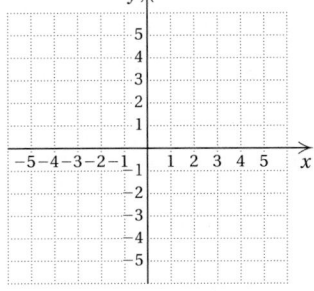

9. $5y = -15 + 3x$

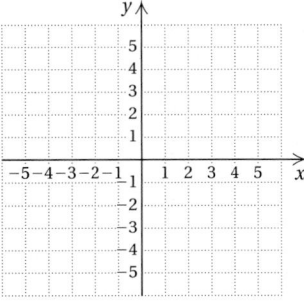

10. $5x - 10 = 5y$

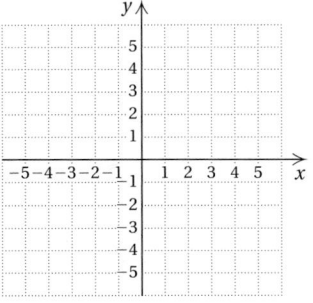

11. $2x - 3y = 6$

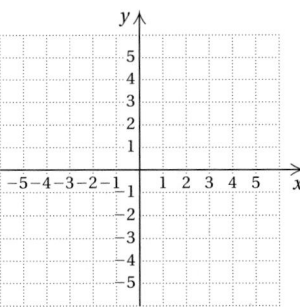

12. $4x + 5y = 20$

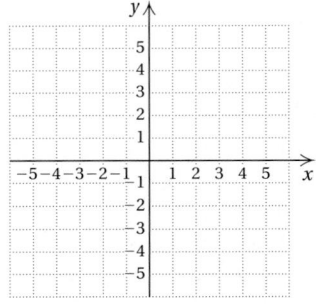

13. $2.8y - 3.5x = -9.8$

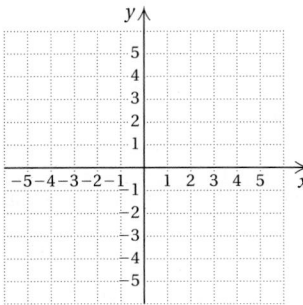

14. $10.8x - 22.68 = 4.2y$

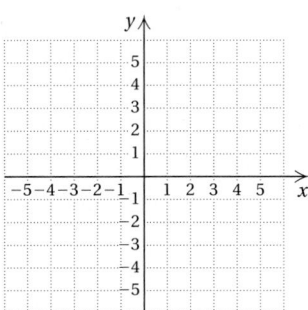

15. $5x + 2y = 7$

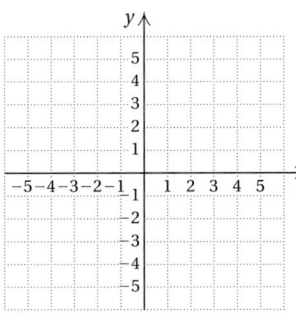

16. $3x - 4y = 10$

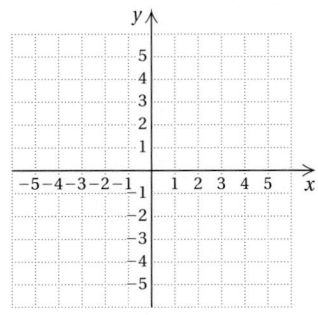

17. $y = \dfrac{5}{2}x + 1$

18. $y = \dfrac{2}{5}x - 4$

19. $f(x) = -\dfrac{5}{2}x - 4$

20. $f(x) = \dfrac{2}{5}x + 3$

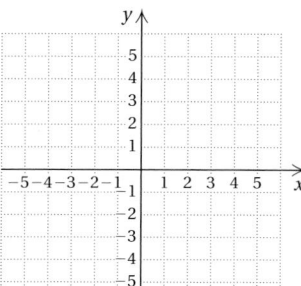

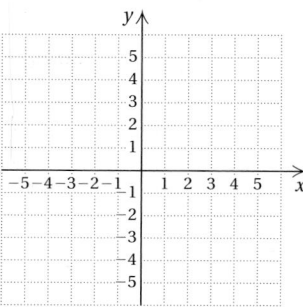

21. $x + 2y = 4$

22. $x - 3y = 6$

23. $4x - 3y = 12$

24. $2x + 6y = 12$

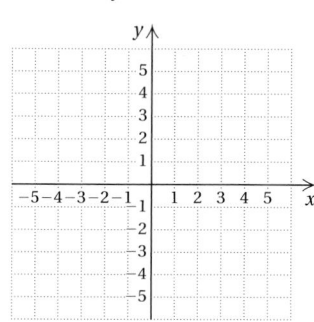

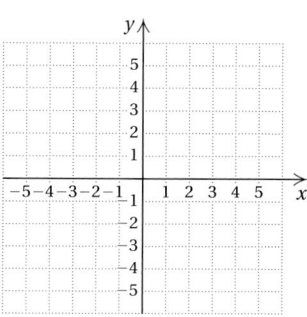

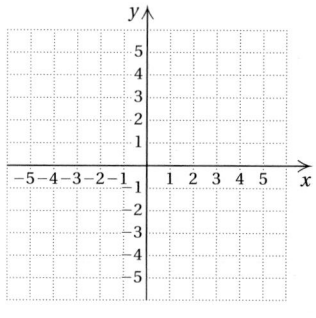

25. $f(x) = \dfrac{1}{3}x - 4$

26. $g(x) = -0.25x + 2$

27. $5x + 4 \cdot f(x) = 4$
(*Hint*: Solve for $f(x)$.)

28. $3 \cdot f(x) = 4x + 6$

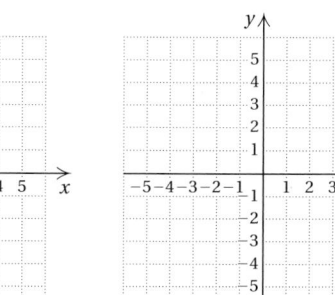

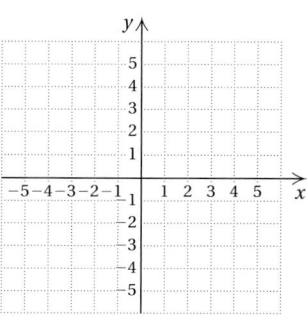

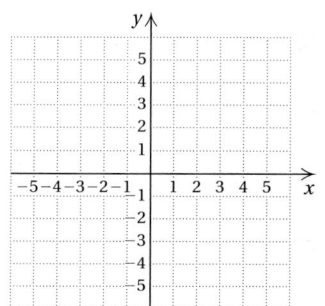

Graph and, if possible, determine the slope.

29. $x = 1$

30. $x = -4$

31. $y = -1$

32. $y = \dfrac{3}{2}$

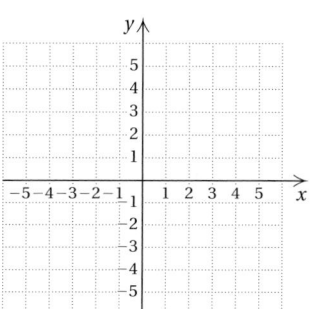

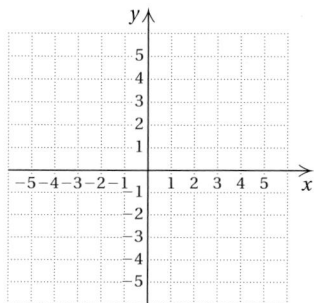

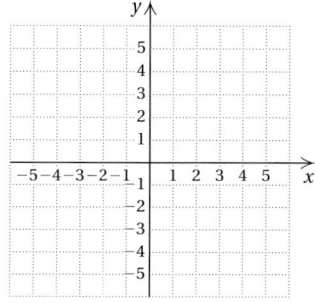

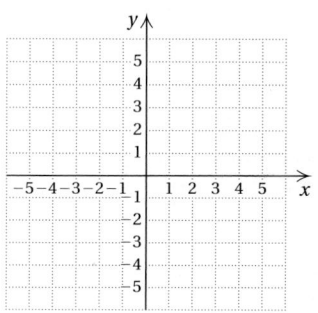

33. $f(x) = -6$

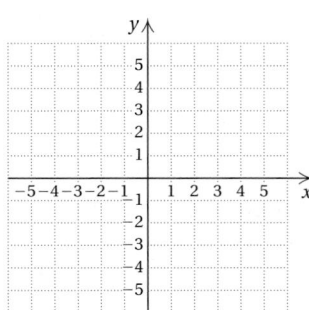

34. $f(x) = 2$

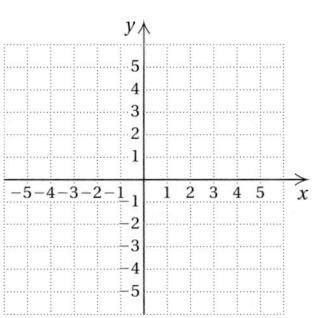

35. $y = 0$

36. $x = 0$

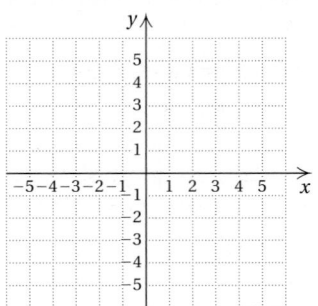

37. $2 \cdot f(x) + 5 = 0$

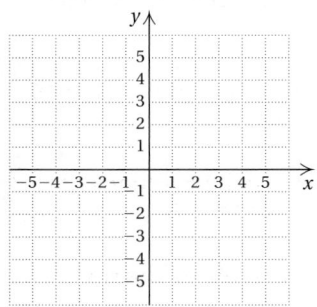

38. $4 \cdot g(x) + 3x = 12 + 3x$

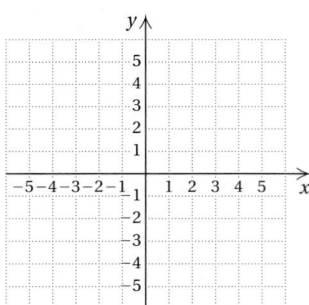

39. $7 - 3x = 4 + 2x$

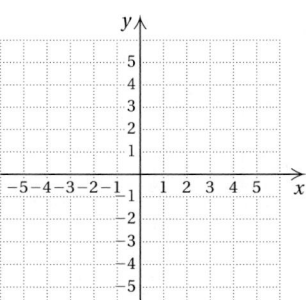

40. $3 - f(x) = 2$

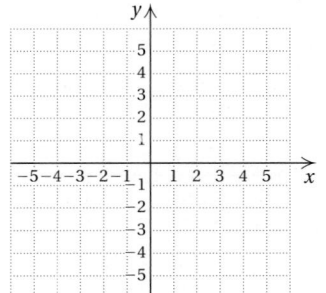

d Determine whether the graphs of the given pair of lines are parallel.

41. $x + 6 = y$,
    $y - x = -2$

42. $2x - 7 = y$,
    $y - 2x = 8$

43. $y + 3 = 5x$,
    $3x - y = -2$

44. $y + 8 = -6x$,
    $-2x + y = 5$

45. $y = 3x + 9$,
    $2y = 6x - 2$

46. $y + 7x = -9$,
    $-3y = 21x + 7$

47. $12x = 3$,
    $-7x = 10$

48. $5y = -2$,
    $\frac{3}{4}x = 16$

Determine whether the graphs of the given pair of lines are perpendicular.

49. $y = 4x - 5$,
    $4y = 8 - x$

50. $2x - 5y = -3$,
    $2x + 5y = 4$

51. $x + 2y = 5$,
    $2x + 4y = 8$

52. $y = -x + 7$,
    $y = x + 3$

53. $2x - 3y = 7$,
    $2y - 3x = 10$

54. $x = y$,
    $y = -x$

55. $2x = 3$,
    $-3y = 6$

56. $-5y = 10$,
    $y = -\frac{4}{9}$

Skill Maintenance

Write in scientific notation.   [4.2c]

57. 53,000,000,000

58. 0.000047

59. 0.018

60. 99,902,000

Write in decimal notation.   [4.2c]

61. 2.13×10^{-5}

62. 9.01×10^{8}

63. 2×10^{4}

64. 8.5677×10^{-2}

Factor.   [1.7d]

65. $9x - 15y$

66. $12a + 21ab$

67. $21p - 7pq + 14p$

68. $64x - 128y + 256$

Synthesis

69. Find an equation of a horizontal line that passes through the point $(-2, 3)$.

70. Find an equation of a vertical line that passes through the point $(-2, 3)$.

71. Find the value of a such that the graphs of $5y = ax + 5$ and $\frac{1}{4}y = \frac{1}{10}x - 1$ are parallel.

72. Find the value of k such that the graphs of $x + 7y = 70$ and $y + 3 = kx$ are perpendicular.

73. Write an equation of the line that has x-intercept $(-3, 0)$ and y-intercept $\left(0, \frac{2}{5}\right)$.

74. Find the coordinates of the point of intersection of the graphs of the equations $x = -4$ and $y = 5$.

75. Write an equation for the x-axis. Is this equation a function?

76. Write an equation for the y-axis. Is this equation a function?

77. Find the value of m in $y = mx + 3$ so that the x-intercept of its graph will be $(4, 0)$.

78. Find the value of b in $2y = -7x + 3b$ so that the y-intercept of its graph will be $(0, -13)$.

79. Match each sentence with the most appropriate graph from those at right.

    a) The rate at which fluids were given intravenously was doubled after 3 hr.

    b) The rate at which fluids were given intravenously was gradually reduced to 0.

    c) The rate at which fluids were given intravenously remained constant for 5 hr.

    d) The rate at which fluids were given intravenously was gradually increased.

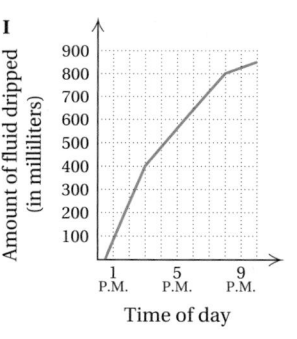

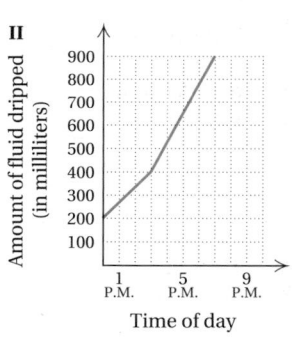

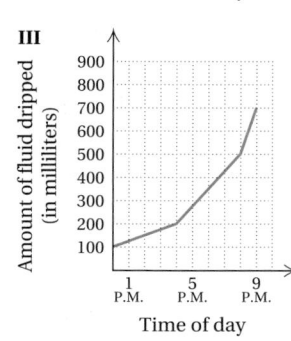

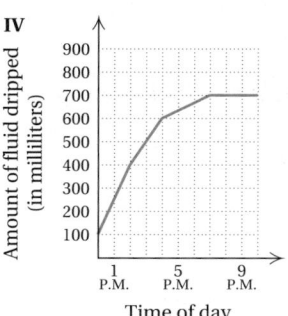

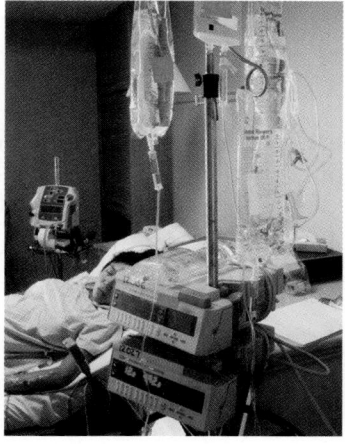

7.5

Finding Equations of Lines; Applications

OBJECTIVES

a Find an equation of a line when the slope and the *y*-intercept are given.

b Find an equation of a line when the slope and a point are given.

c Find an equation of a line when two points are given.

d Given a line and a point not on the given line, find an equation of the line parallel to the line and containing the point, and find an equation of the line perpendicular to the line and containing the point.

e Solve applied problems involving linear functions.

SKILL TO REVIEW

Objective 1.6b: Find the reciprocal of a real number.

Find the reciprocal of the number.

1. 3

2. $-\dfrac{4}{9}$

---

1. A line has slope 3.4 and *y*-intercept $(0, -8)$. Find an equation of the line.

In this section, we will learn to find an equation of a line for which we have been given two pieces of information.

a Finding an Equation of a Line When the Slope and the *y*-Intercept Are Given

If we know the slope and the *y*-intercept of a line, we can find an equation of the line using the slope–intercept equation $y = mx + b$.

EXAMPLE 1 A line has slope -0.7 and *y*-intercept $(0, 13)$. Find an equation of the line.

We use the slope–intercept equation and substitute -0.7 for m and 13 for b:

$$y = mx + b$$
$$y = -0.7x + 13.$$

Do Margin Exercise 1.

b Finding an Equation of a Line When the Slope and a Point Are Given

Suppose we know the slope of a line and the coordinates of one point on the line. We can use the slope–intercept equation to find an equation of the line. Or, we can use the **point–slope equation**. We first develop a formula for such a line.

Suppose that a line of slope m passes through the point (x_1, y_1). For any other point (x, y) on this line, we must have

$$\frac{y - y_1}{x - x_1} = m.$$

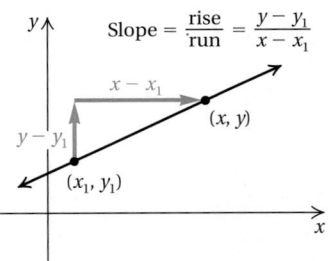

$$\text{Slope} = \frac{\text{rise}}{\text{run}} = \frac{y - y_1}{x - x_1}$$

It is tempting to use this last equation as an equation of the line of slope m that passes through (x_1, y_1). The only problem with this form is that when x and y are replaced with x_1 and y_1, we have $\frac{0}{0} = m$, a false equation. To avoid this difficulty, we multiply by $x - x_1$ on both sides and simplify:

$$\frac{y - y_1}{x - x_1}(x - x_1) = m(x - x_1) \qquad \text{Multiplying by } x - x_1 \text{ on both sides}$$

$$y - y_1 = m(x - x_1). \qquad \text{Removing a factor of 1: } \frac{x - x_1}{x - x_1} = 1$$

This is the *point–slope* form of a linear equation.

Answers

Skill to Review:

1. $\dfrac{1}{3}$ **2.** $-\dfrac{9}{4}$

Margin Exercise:

1. $y = 3.4x - 8$

The **point–slope equation** of a line with slope m, passing through (x_1, y_1), is

$$y - y_1 = m(x - x_1).$$

If we know the slope of a line and a certain point on the line, we can find an equation of the line using either the point–slope equation,

$$y - y_1 = m(x - x_1),$$

or the slope–intercept equation,

$$y = mx + b.$$

EXAMPLE 2 Find an equation of the line with slope 5 and containing the point $\left(\frac{1}{2}, -1\right)$.

Using the Point–Slope Equation: We consider $\left(\frac{1}{2}, -1\right)$ to be (x_1, y_1) and 5 to be the slope m, and substitute:

$$y - y_1 = m(x - x_1) \qquad \text{Point–slope equation}$$
$$y - (-1) = 5\left(x - \tfrac{1}{2}\right) \qquad \text{Substituting}$$
$$y + 1 = 5x - \tfrac{5}{2} \qquad \text{Simplifying}$$
$$y = 5x - \tfrac{5}{2} - 1$$
$$y = 5x - \tfrac{5}{2} - \tfrac{2}{2}$$
$$y = 5x - \tfrac{7}{2}.$$

Using the Slope–Intercept Equation: The point $\left(\frac{1}{2}, -1\right)$ is on the line, so it is a solution. Thus we can substitute $\frac{1}{2}$ for x and -1 for y in $y = mx + b$. We also substitute 5 for m, the slope. Then we solve for b:

$$y = mx + b \qquad \text{Slope–intercept equation}$$
$$-1 = 5 \cdot \left(\tfrac{1}{2}\right) + b \qquad \text{Substituting}$$
$$-1 = \tfrac{5}{2} + b$$
$$-1 - \tfrac{5}{2} = b$$
$$-\tfrac{2}{2} - \tfrac{5}{2} = b$$
$$-\tfrac{7}{2} = b. \qquad \text{Solving for } b$$

We then use the slope–intercept equation $y = mx + b$ again and substitute 5 for m and $-\frac{7}{2}$ for b:

$$y = 5x - \tfrac{7}{2}.$$

Do Exercises 2–5.

Find an equation of the line with the given slope and containing the given point.

2. $m = -5, \ (-4, 2)$

3. $m = 3, \ (1, -2)$

4. $m = 8, \ (3, 5)$

5. $m = -\dfrac{2}{3}, \ (1, 4)$

Answers

2. $y = -5x - 18$ **3.** $y = 3x - 5$
4. $y = 8x - 19$ **5.** $y = -\dfrac{2}{3}x + \dfrac{14}{3}$

(c) Finding an Equation of a Line When Two Points Are Given

We can also use the slope–intercept equation or the point–slope equation to find an equation of a line when two points are given.

EXAMPLE 3 Find an equation of the line containing the points $(2, 3)$ and $(-6, 1)$.

First, we find the slope:

$$m = \frac{3 - 1}{2 - (-6)} = \frac{2}{8}, \text{ or } \frac{1}{4}.$$

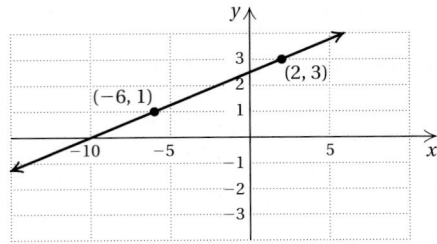

Now we have the slope and two points. We then proceed as we did in Example 2, using either point, and either the point–slope equation or the slope–intercept equation.

Using the Point–Slope Equation: We choose $(2, 3)$ and substitute 2 for x_1, 3 for y_1, and $\frac{1}{4}$ for m:

$$y - y_1 = m(x - x_1) \qquad \text{Point–slope equation}$$
$$y - 3 = \tfrac{1}{4}(x - 2) \qquad \text{Substituting}$$
$$y - 3 = \tfrac{1}{4}x - \tfrac{1}{2}$$
$$y = \tfrac{1}{4}x - \tfrac{1}{2} + 3$$
$$y = \tfrac{1}{4}x - \tfrac{1}{2} + \tfrac{6}{2}$$
$$y = \tfrac{1}{4}x + \tfrac{5}{2}.$$

Using the Slope–Intercept Equation: We choose $(2, 3)$ and substitute 2 for x, 3 for y, and $\frac{1}{4}$ for m:

$$y = mx + b \qquad \text{Slope–intercept equation}$$
$$3 = \tfrac{1}{4} \cdot 2 + b \qquad \text{Substituting}$$
$$3 = \tfrac{1}{2} + b$$
$$3 - \tfrac{1}{2} = \tfrac{1}{2} + b - \tfrac{1}{2}$$
$$\tfrac{6}{2} - \tfrac{1}{2} = b$$
$$\tfrac{5}{2} = b. \qquad \text{Solving for } b$$

6. Find an equation of the line containing the points $(4, -3)$ and $(1, 2)$.

7. Find an equation of the line containing the points $(-3, -5)$ and $(-4, 12)$.

Finally, we use the slope–intercept equation $y = mx + b$ again and substitute $\frac{1}{4}$ for m and $\frac{5}{2}$ for b:

$$y = \tfrac{1}{4}x + \tfrac{5}{2}.$$

Do Exercises 6 and 7.

(d) Finding an Equation of a Line Parallel or Perpendicular to a Given Line Through a Point Off the Line

We can also use the methods of Example 2 to find an equation of a line through a point off the line parallel or perpendicular to a given line.

EXAMPLE 4 Find an equation of the line containing the point $(-1, 3)$ and parallel to the line $2x + y = 10$.

A line parallel to the given line $2x + y = 10$ must have the same slope as the given line. To find that slope, we first find the slope–intercept equation by solving for y:

$$2x + y = 10$$
$$y = -2x + 10.$$

Thus the line we want to find through $(-1, 3)$ must also have slope -2.

Using the Point–Slope Equation: We use the point $(-1, 3)$ and the slope -2, substituting -1 for x_1, 3 for y_1, and -2 for m:

$$y - y_1 = m(x - x_1)$$
$$y - 3 = -2(x - (-1)) \quad \text{Substituting}$$
$$y - 3 = -2(x + 1) \quad \text{Simplifying}$$
$$y - 3 = -2x - 2$$
$$y = -2x - 2 + 3$$
$$y = -2x + 1.$$

Using the Slope–Intercept Equation: We substitute -1 for x and 3 for y in $y = mx + b$, and -2 for m, the slope. Then we solve for b:

$$y = mx + b$$
$$3 = -2(-1) + b \quad \text{Substituting}$$
$$3 = 2 + b$$
$$1 = b. \quad \text{Solving for } b$$

We then use the equation $y = mx + b$ again and substitute -2 for m and 1 for b:

$$y = -2x + 1.$$

The given line $2x + y = 10$, or $y = -2x + 10$, and the line $y = -2x + 1$ have the same slope but different y-intercepts. Thus their graphs are parallel.

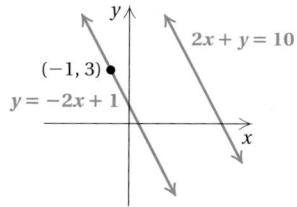

Do Exercise 8.

8. Find an equation of the line containing the point $(2, -1)$ and parallel to the line $9x - 3y = 5$.

Answer

8. $y = 3x - 7$

EXAMPLE 5 Find an equation of the line containing the point $(2, -3)$ and perpendicular to the line $4y - x = 20$.

To find the slope of the given line, we first find its slope–intercept form by solving for y:

$$4y - x = 20$$
$$4y = x + 20$$
$$\frac{4y}{4} = \frac{x + 20}{4} \qquad \text{Dividing by 4}$$
$$y = \tfrac{1}{4}x + 5.$$

We know that the slope of the perpendicular line must be the opposite of the reciprocal of $\frac{1}{4}$. Thus the new line through $(2, -3)$ must have slope -4.

Using the Point–Slope Equation: We use the point $(2, -3)$ and the slope -4, substituting 2 for x_1, -3 for y_1, and -4 for m:

$$y - y_1 = m(x - x_1)$$
$$y - (-3) = -4(x - 2) \qquad \text{Substituting}$$
$$y + 3 = -4(x - 2) \qquad \text{Simplifying}$$
$$y + 3 = -4x + 8$$
$$y = -4x + 8 - 3$$
$$y = -4x + 5.$$

Using the Slope–Intercept Equation: We now substitute 2 for x and -3 for y in $y = mx + b$. We also substitute -4 for m, the slope. Then we solve for b:

$$y = mx + b$$
$$-3 = -4(2) + b \qquad \text{Substituting}$$
$$-3 = -8 + b$$
$$5 = b. \qquad \text{Solving for } b$$

Finally, we use the equation $y = mx + b$ again and substitute -4 for m and 5 for b:

$$y = -4x + 5.$$

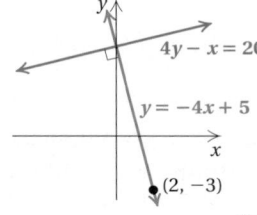

The product of the slopes of the lines $4y - x = 20$ and $y = -4x + 5$ is $\frac{1}{4} \cdot (-4) = -1$. Thus their graphs are perpendicular.

9. Find an equation of the line containing the point $(5, 4)$ and perpendicular to the line $2x - 4y = 9$.

Do Exercise 9.

(e) Applications of Linear Functions

When the essential parts of a problem are described in mathematical language, we say that we have a **mathematical model**. We have already studied many kinds of mathematical models in this text—for example, the formulas in Section 2.4 and the functions in Section 7.1. Here we study linear functions as models.

EXAMPLE 6 *Cost of a Necklace.* Amelia's Beads offers a class in designing necklaces. For a necklace made of 6-mm beads, 4.23 beads per inch are needed. The cost of a necklace of 6-mm gemstone beads that sell for 40¢ each is $7 for the clasp and the crimps and approximately $1.70 per inch.

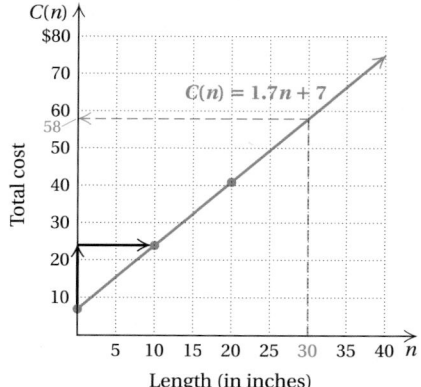

a) Formulate a linear function that models the total cost of a necklace $C(n)$, where n is the length of the necklace, in inches.

b) Graph the model.

c) Use the model to determine the cost of a 30-in. necklace.

a) The problem describes a situation in which cost per inch is charged in addition to the fixed cost of the clasp and the crimps. The total cost of a 16-in. necklace is

$$\$7 + \$1.70 \cdot 16 = \$34.20.$$

For a 17-in. necklace, the total cost is

$$\$7 + \$1.70 \cdot 17 = \$35.90.$$

These calculations lead us to generalize that for a necklace that is n inches long, the total cost is given by $C(n) = 7 + 1.7n$, where $n \geq 0$ since the length of the necklace cannot be negative. (Actually most necklaces are at least 14 in. long.) The notation $C(n)$ indicates that the cost C is a function of the length n.

b) Before we draw the graph, we rewrite the model in slope–intercept form:

$$C(n) = 1.7n + 7.$$

The y-intercept is $(0, 7)$ and the slope, or rate of change, is $1.70 per inch, or $\frac{17}{10}$. We first plot $(0, 7)$; from that point, we move up 17 units and to the right 10 units to the point $(10, 24)$. We then draw a line through these points. We also calculate a third value as a check:

$$C(20) = 1.7 \cdot 20 + 7 = 41.$$

The point $(20, 41)$ lines up with the other two points so the graph is correct.

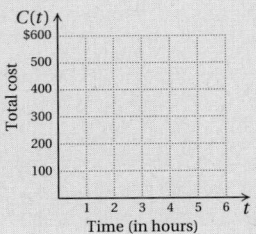
c) To determine the total cost of a 30-in. necklace, we find $C(30)$:

$$C(30) = 1.7 \cdot 30 + 7 = 58.$$

From the graph, we see that the input 30 corresponds to the output 58. Thus we see that a 30-in. necklace costs $58.

Do Exercise 10.

In the following example, we use two points and find an equation for the linear function through these points. Then we use the equation to make a prediction.

EXAMPLE 7 *Retail Trade.* Sales at warehouse clubs and superstores have increased steadily in recent years. The table below lists data regarding the correspondence between the year and the total sales at warehouses and superstores, in billions of dollars.

YEAR, x (in years since 2000)	TOTAL SALES (in billions)
2000, 0	$139.6
2007, 7	323.3

SOURCE: U.S. Census Bureau

a) Assuming a constant rate of change, use the two data points to find a linear function that fits the data.

b) Use the function to determine the total sales in 2005.

c) In which year will the total sales reach $400 billion?

a) We let $x =$ the number of years since 2000 and $S =$ total sales. The table gives us two ordered pairs, $(0, 139.6)$ and $(7, 323.3)$. We use them to find a linear function that fits the data. First, we find the slope:

$$m = \frac{323.3 - 139.6}{7 - 0} = \frac{183.7}{7} \approx 26.24.$$

Next, we find an equation $S = mx + b$ that fits the data. We can use either the point–slope equation or the slope–intercept equation to do this.

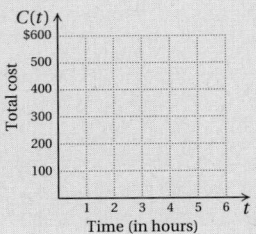

Using the Point–Slope Equation: We substitute one of the points—say, $(0, 139.6)$—and the slope, 26.24, in the point–slope equation:

$$S - S_1 = m(x - x_1) \qquad \text{Point–slope equation}$$
$$S - 139.6 = 26.24(x - 0) \qquad \text{Substituting}$$
$$S - 139.6 = 26.24x - 0$$
$$S = 26.24x + 139.6.$$

Using the Slope–Intercept Equation: One of the data points $(0, 139.6)$ is the y-intercept. Thus we know b in the slope–intercept equation, $y = mx + b$. We use the equation $S = mx + b$ and substitute 26.24 for m and 139.6 for b:

$$S = 26.24x + 139.6.$$

Using function notation, we have

$$S(x) = 26.24x + 139.6.$$

b) To determine the total sales in 2005, we substitute 5 for x (2005 is 5 yr since 2000) in the function $S(x) = 26.24x + 139.6$:

$$S(x) = 26.24x + 139.6$$
$$S(5) = 26.24 \cdot 5 + 139.6 \qquad \text{Substituting}$$
$$= 131.2 + 139.6$$
$$= 270.8.$$

The total sales in 2005 were $270.8 billion.

c) To find the year in which total sales will reach $400 billion, we substitute 400 for $S(x)$ and solve for x:

$$S(x) = 26.24x + 139.6$$
$$400 = 26.24x + 139.6 \qquad \text{Substituting}$$
$$260.4 = 26.24x \qquad \text{Subtracting 139.6}$$
$$10 \approx x. \qquad \text{Dividing by 26.24}$$

Total sales will reach $400 billion about 10 yr after 2000, or in 2010.

> Do Exercise 11.

11. Hat Size as a Function of Head Circumference. The table below lists data relating hat size to head circumference.

HEAD CIRCUMFERENCE, C (in inches)	HAT SIZE, H
21.2	$6\frac{3}{4}$
22	7

SOURCE: Shushan's New Orleans

a) Assuming a constant rate of change, use the two data points to find a linear function that fits the data.

b) Use the function to determine the hat size of a person whose head has a circumference of 24.8 in.

c) Jerome's hat size is 8. What is the circumference of his head?

a Find an equation of the line having the given slope and y-intercept.

1. Slope: -8; y-intercept: $(0, 4)$

2. Slope: 5; y-intercept: $(0, -3)$

3. Slope: 2.3; y-intercept: $(0, -1)$

4. Slope: -9.1; y-intercept: $(0, 2)$

Find a linear function $f(x) = mx + b$ whose graph has the given slope and y-intercept.

5. Slope: $-\frac{7}{3}$; y-intercept: $(0, -5)$

6. Slope: $\frac{4}{5}$; y-intercept: $(0, 28)$

7. Slope: $\frac{2}{3}$; y-intercept: $\left(0, \frac{5}{8}\right)$

8. Slope: $-\frac{7}{8}$; y-intercept: $\left(0, -\frac{7}{11}\right)$

b Find an equation of the line having the given slope and containing the given point.

9. $m = 5$, $(4, 3)$

10. $m = 4$, $(5, 2)$

11. $m = -3$, $(9, 6)$

12. $m = -2$, $(2, 8)$

13. $m = 1$, $(-1, -7)$

14. $m = 3$, $(-2, -2)$

15. $m = -2$, $(8, 0)$

16. $m = -3$, $(-2, 0)$

17. $m = 0$, $(0, -7)$

18. $m = 0$, $(0, 4)$

19. $m = \frac{2}{3}$, $(1, -2)$

20. $m = -\frac{4}{5}$, $(2, 3)$

c Find an equation of the line containing the given pair of points.

21. $(1, 4)$ and $(5, 6)$

22. $(2, 5)$ and $(4, 7)$

23. $(-3, -3)$ and $(2, 2)$

24. $(-1, -1)$ and $(9, 9)$

25. $(-4, 0)$ and $(0, 7)$

26. $(0, -5)$ and $(3, 0)$

27. $(-2, -3)$ and $(-4, -6)$

28. $(-4, -7)$ and $(-2, -1)$

29. $(0, 0)$ and $(6, 1)$

30. $(0, 0)$ and $(-4, 7)$

31. $\left(\frac{1}{4}, -\frac{1}{2}\right)$ and $\left(\frac{3}{4}, 6\right)$

32. $\left(\frac{2}{3}, \frac{3}{2}\right)$ and $\left(-3, \frac{5}{6}\right)$

d Write an equation of the line containing the given point and parallel to the given line.

33. $(3, 7)$; $x + 2y = 6$

34. $(0, 3)$; $2x - y = 7$

35. $(2, -1)$; $5x - 7y = 8$

36. $(-4, -5)$; $2x + y = -3$

37. $(-6, 2)$; $3x = 9y + 2$

38. $(-7, 0)$; $2y + 5x = 6$

Write an equation of the line containing the given point and perpendicular to the given line.

39. $(2, 5)$; $2x + y = 3$

40. $(4, 1)$; $x - 3y = 9$

41. $(3, -2)$; $3x + 4y = 5$

42. $(-3, -5)$; $5x - 2y = 4$

43. $(0, 9)$; $2x + 5y = 7$

44. $(-3, -4)$; $6y - 3x = 2$

e Solve.

45. *Moving Costs.* Metro Movers charges $85 plus $40 an hour to move households across town.

  a) Formulate a linear function for the total cost $C(t)$ of t hours of moving. $C = 40(t) + 85$

  b) Graph the model.

  c) Use the model to determine the cost of $6\frac{1}{2}$ hr of moving service.

46. *Deluxe Cable TV Service.* Vista Cable TV Service charges a $35 installation fee and $20 per month for basic service.

  a) Formulate a linear function for the total cost $C(t)$ of t months of cable TV service.

  b) Graph the model.

  c) Use the model to determine the cost of 9 months of service.

47. *Value of a Fax Machine.* Melton Corporation bought a multifunction fax machine for $750. The value $V(t)$ of the machine depreciates (declines) at a rate of $25 per month.

  a) Formulate a linear function for the value $V(t)$ of the machine after t months.

  b) Graph the model.

  c) Use the model to determine the value of the machine after 13 months.

48. *Value of a Computer.* True Tone Graphics bought a computer for $3800. The value $V(t)$ of the computer depreciates at a rate of $50 per month.

  a) Formulate a linear function for the value $V(t)$ of the computer after t months.

  b) Graph the model.

  c) Use the model to determine the value of the computer after $10\frac{1}{2}$ months.

In Exercises 49–54, assume that a constant rate of change exists for each model formed.

49. *Whooping Cough.* The table below lists data regarding the number of cases of whooping cough in 1987 and in 2007.

YEAR, x (in years since 1987)	NUMBER OF CASES
1987, 0	2,862
2007, 20	10,454

SOURCE: Centers for Disease Control and Prevention

a) Use the two data points to find a linear function that fits the data. Let x = the number of years since 1987 and $W(x)$ = the number of cases of whooping cough.
b) Use the function of part (a) to estimate and predict the number of cases of whooping cough in 1990 and in 2012.

50. *Lobbying Expenses.* The table below lists data regarding spending, in billions of dollars, on lobbying Congress and the federal government in 2004 and in 2008.

YEAR, x (in years since 2004)	AMOUNT SPENT ON LOBBYING (in billions)
2004, 0	$1.5
2008, 4	$3.3

SOURCE: CQ MoneyLine

a) Use the two data points to find a linear function that fits the data. Let x = the number of years since 2004 and $L(x)$ = the amount spent, in billions of dollars, on lobbying Congress and the federal government.
b) Use the function of part (a) to estimate the amount spent on lobbying in 2005 and in 2010.

51. *Auto Dealers.* At the close of 1991, there were 24,026 auto dealers in the United States. By the end of 2008, this number had dropped to 20,084. Let $D(x)$ = the number of auto dealerships and x = the number of years since 1991.

Source: Urban Science Automotive Dealer Census

a) Find a linear function that fits the data.
b) Use the function of part (a) to estimate the number of auto dealerships in 2000.
c) At this rate of decrease, when will the number of auto dealerships be 18,000?

52. *Records in the 400-Meter Run.* In 1930, the record for the 400-m run was 46.8 sec. In 1970, it was 43.8 sec. Let $R(t)$ = the record in the 400-m run and t = the number of years since 1930.

a) Find a linear function that fits the data.
b) Use the function of part (a) to estimate the record in 2003; in 2006.
c) When will the record be 40 sec?

53. *Life Expectancy of Males in the United States.* In 1990, the life expectancy of males was 71.8 yr. In 2001, it was 74.4 yr. Let $M(t)$ = life expectancy and t = the number of years since 1990.

Source: U.S. National Center for Health Statistics

a) Find a linear function that fits the data.
b) Use the function of part (a) to estimate the life expectancy of males in 2007.

54. *Life Expectancy of Females in the United States.* In 1990, the life expectancy of females was 78.8 yr. In 2001, it was 79.8 yr. Let $F(t)$ = life expectancy and t = the number of years since 1990.

Source: U.S. National Center for Health Statistics

a) Find a linear function that fits the data.
b) Use the function of part (a) to estimate the life expectancy of females in 2010.

Skill Maintenance

Simplify. [6.1c]

55. $\dfrac{w - t}{t - w}$

56. $\dfrac{b^2 - 1}{b - 1}$

57. $\dfrac{3x^2 + 15x - 72}{6x^2 + 18x - 240}$

58. $\dfrac{4y + 32}{y^2 - y - 72}$

Find the slope, if it exists, of the line. [7.3b]

59. $2x - 7y = -10$

60. $y = -1$

61. $x = 42$

62. $4 - 6y = 12x$

Summary and Review

Key Terms, Properties, and Formulas

function, p. 500
domain, p. 500
range, p. 500
relation, p. 501
inputs, p. 502
outputs, p. 502
constant function, p. 503
vertical-line test, p. 505

linear function, p. 521
slope, p. 524
rise, p. 525
run, p. 525
slope–intercept equation, p. 527
grade, p. 527
rate of change, p. 528
y-intercept, pp. 523, 532

x-intercept, p. 532
horizontal line, p. 535
vertical line, p. 536
parallel lines, p. 537
perpendicular lines, p. 538
point–slope equation, p. 544
mathematical model, p. 549

$$Slope = m = \frac{y_2 - y_1}{x_2 - x_1}, \text{ or } \frac{y_1 - y_2}{x_1 - x_2}$$

Slope–Intercept Equation: $f(x) = mx + b$, or $y = mx + b$

Point–Slope Equation: $y - y_1 = m(x - x_1)$

Horizontal Line: $f(x) = b$, or $y = b$; slope $= 0$

Vertical Line: $x = a$, slope is not defined.

Parallel Lines: $m_1 = m_2, b_1 \neq b_2$

Perpendicular Lines: $m_1 = -\dfrac{1}{m_2}, m_1, m_2 \neq 0$

Concept Reinforcement

Determine whether each statement is true or false.

_____ **1.** The slope of a vertical line is 0.   [7.4c]

_____ **2.** A line with slope 1 slants less steeply than a line with slope -5.   [7.3b]

_____ **3.** Parallel lines have the same slope and y-intercept.   [7.4d]

Important Concepts

Objective 7.1a   Determine whether a correspondence is a function.

Example   Determine whether each correspondence is a function.

Domain   Range   Domain   Range

$f:$
   4, -4, 6, -6 → 2, 0, -2

$g:$
   Q, R → A, C, E

The correspondence f *is* a function because each member of the domain is matched to *only one* member of the range. The correspondence g *is not* a function because one member of the domain, Q, is matched to more than one member of the range.

Practice Exercise

1. Determine whether the correspondence is a function.

Domain   Range

$h:$
   11, 15, 19 → 20, 31

Objective 7.1b Given a function described by an equation, find function values (outputs) for specified values (inputs).

Example Find the indicated function value.

a) $f(0)$, for $f(x) = -x + 6$ **b)** $g(5)$, for $g(x) = -10$

c) $h(-1)$, for $h(x) = 4x^2 + x$

a) $f(x) = -x + 6$: $f(0) = -0 + 6 = 6$

b) $g(x) = -10$: $g(5) = -10$

c) $h(x) = 4x^2 + x$: $h(-1) = 4(-1)^2 + (-1) = 4 \cdot 1 - 1 = 3$

Practice Exercise

2. Find $g(0), g(-2)$, and $g(6)$ for $g(x) = \frac{1}{2}x - 2$.

Objective 7.1c Draw the graph of a function.

Example Graph: $f(x) = -\dfrac{2}{3}x + 2$.

By choosing multiples of 3 for x, we can avoid fraction values for $f(x)$. If $x = -3$, then $f(-3) = -\frac{2}{3} \cdot (-3) + 2 = 2 + 2 = 4$. We list three ordered pairs in a table, plot the points, draw the line, and label the graph.

x	$f(x)$
3	0
0	2
-3	4

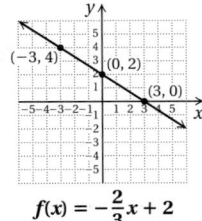

$$f(x) = -\frac{2}{3}x + 2$$

Practice Exercise

3. Graph: $f(x) = \dfrac{2}{5}x - 3$.

x	$f(x)$

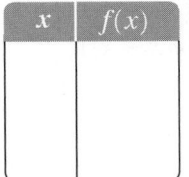 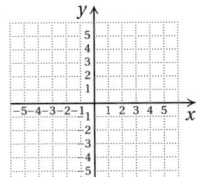

Objective 7.1d Determine whether a graph is that of a function using the vertical-line test.

Example Determine whether each of the following is the graph of a function.

a) 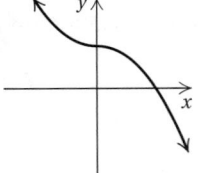   **b)**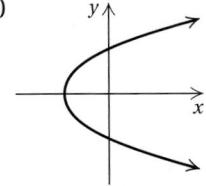

a) The graph is that of a function because no vertical line can cross the graph at more than one point.

b) The graph is not that of a function because a vertical line can cross the graph more than once.

Practice Exercise

4. Determine whether the graph is the graph of a function.

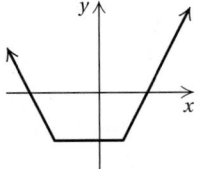

Objective 7.2a Find the domain and the range of a function.

Example For the function f whose graph is shown below, determine the domain and the range.

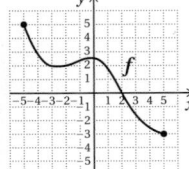 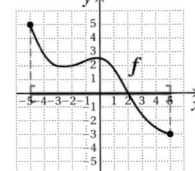 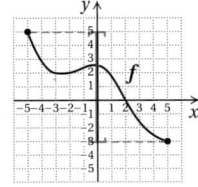

Domain: $\{x | -5 \le x \le 5\}$; range: $\{y | -3 \le y \le 5\}$

Practice Exercises

5. For the function g whose graph is shown below, determine the domain and the range.

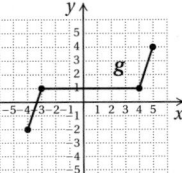

Example   Find the domain of $g(x) = \dfrac{x + 1}{2x - 6}$.

Since $(x + 1)/(2x - 6)$ cannot be calculated when the denominator $2x - 6$ is 0, we solve $2x - 6 = 0$ to find the real numbers that must be excluded from the domain of g:

$$2x - 6 = 0$$
$$2x = 6$$
$$x = 3.$$

Thus, 3 is not in the domain. The domain of g is $\{x | x$ is a real number *and* $x \neq 3\}$.

6. Find the domain of
$$h(x) = \dfrac{x - 3}{3x + 9}.$$

Objective 7.3b   Given two points on a line, find the slope. Given a linear equation, derive the equivalent slope–intercept equation and determine the slope and the y-intercept.

Example   Find the slope of the line containing $(-5, 6)$ and $(-1, -4)$.

$$m = \frac{\text{change in } y}{\text{change in } x} = \frac{6 - (-4)}{-5 - (-1)} = \frac{6 + 4}{-5 + 1} = \frac{10}{-4} = -\frac{5}{2}$$

Example   Find the slope and the y-intercept of
$$4x - 2y = 20.$$

We first solve for y:

$$4x - 2y = 20$$
$$-2y = -4x + 20 \qquad \text{Subtracting } 4x$$
$$y = 2x - 10. \qquad \text{Dividing by } -2$$

The slope is 2, and the y-intercept is $(0, -10)$.

Practice Exercises

7. Find the slope of the line containing $(2, -8)$ and $(-3, 2)$.

8. Find the slope and the y-intercept of
$$3x = -6y + 12.$$

Objective 7.4a   Graph linear equations using intercepts.

Example   Find the intercepts of $x - 2y = 6$ and then graph the line.

To find the y-intercept, we let $x = 0$ and solve for y:

$$0 - 2y = 6 \qquad \text{Substituting 0 for } x$$
$$-2y = 6$$
$$y = -3.$$

The y-intercept is $(0, -3)$.

To find the x-intercept, we let $y = 0$ and solve for x:

$$x - 2 \cdot 0 = 6 \qquad \text{Substituting 0 for } y$$
$$x - 0 = 6$$
$$x = 6.$$

The x-intercept is $(6, 0)$.

We plot these points and draw the line, using a third point as a check. We let $x = -2$ and solve for y:

$$-2 - 2y = 6$$
$$-2y = 8$$
$$y = -4.$$

We plot $(-2, -4)$ and note that it is on the line.

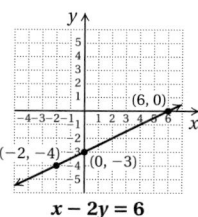

Practice Exercise

9. Find the intercepts of $3y - 3 = x$ and then graph the line.

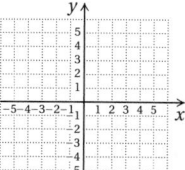

Objective 7.4b Given a linear equation in slope–intercept form, use the slope and the y-intercept to graph the line.

Example Graph using the slope and the y-intercept:
$$y = -\frac{3}{2}x + 5.$$

This equation is in slope–intercept form, $y = mx + b$. The y-intercept is $(0, 5)$. We plot $(0, 5)$. We can think of the slope $\left(m = -\frac{3}{2}\right)$ as $\frac{-3}{2}$.

Starting at the y-intercept, we use the slope to find another point on the graph. We move down 3 units and to the right 2 units. We get a new point: $(2, 2)$.

To get a third point for a check, we start at $(2, 2)$ and move down 3 units and to the right 2 units to the point $(4, -1)$. We plot the points and draw the line.

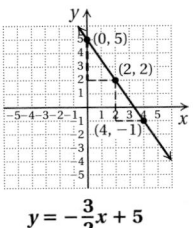

$$y = -\frac{3}{2}x + 5$$

Practice Exercise

10. Graph using the slope and the y-intercept: $y = \frac{1}{4}x - 3$.

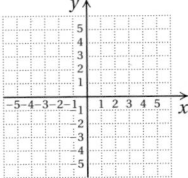

Objective 7.4c Graph linear equations of the form $x = a$ or $y = b$.

Example Graph: $y = -1$.

All ordered pairs $(x, -1)$ are solutions; y is -1 at each point. The graph is a horizontal line that intersects the y-axis at $(0, -1)$.

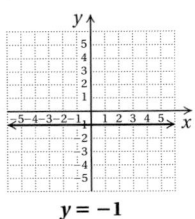

$$y = -1$$

Example Graph: $x = 2$.

All ordered pairs $(2, y)$ are solutions; x is 2 at each point. The graph is a vertical line that intersects the x-axis at $(2, 0)$.

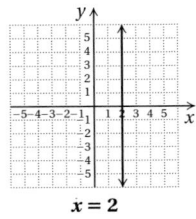

$$x = 2$$

Practice Exercises

11. Graph: $y = 3$.

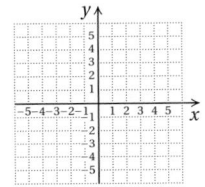

12. Graph: $x = -\dfrac{5}{2}$.

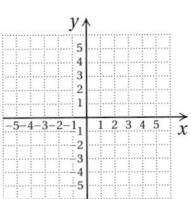

Objective 7.4d Given the equations of two lines, determine whether their graphs are parallel or whether they are perpendicular.

Example Determine whether the graphs of the given pair of lines are parallel, perpendicular, or neither.

a) $2y - x = 16$,
$x + \frac{1}{2}y = 4$

b) $5x - 3 = 2y$,
$2y + 12 = 5x$

a) Writing each equation in slope–intercept form, we have $y = \frac{1}{2}x + 8$ and $y = -2x + 8$. The slopes are $\frac{1}{2}$ and -2. The product of the slopes is -1: $\frac{1}{2} \cdot (-2) = -1$. The graphs are perpendicular.

b) Writing each equation in slope–intercept form, we have $y = \frac{5}{2}x - \frac{3}{2}$ and $y = \frac{5}{2}x - 6$. The slopes are the same, $\frac{5}{2}$, and the y-intercepts are different. The graphs are parallel.

Practice Exercises

Determine whether the graphs of the given pair of lines are parallel, perpendicular, or neither.

13. $-3x + 8y = -8$,
$8y = 3x + 40$

14. $5x - 2y = -8$,
$2x + 5y = 15$

Objective 7.5a Find an equation of a line when the slope and the y-intercept are given.

Example A line has slope 0.8 and y-intercept $(0, -17)$. Find an equation of the line.

We use the slope–intercept equation and substitute 0.8 for m and -17 for b:

$y = mx + b$ Slope–intercept equation

$y = 0.8x - 17.$

Practice Exercise

15. A line has slope -8 and y-intercept $(0, 0.3)$. Find an equation of the line.

Objective 7.5b Find an equation of a line when the slope and a point are given.

Example Find an equation of the line with slope -2 and containing the point $\left(\frac{1}{3}, -1\right)$.

Using the *point–slope equation*, we substitute -2 for m, $\frac{1}{3}$ for x_1, and -1 for y_1:

$y - (-1) = -2\left(x - \frac{1}{3}\right)$ Using $y - y_1 = m(x - x_1)$

$y + 1 = -2x + \frac{2}{3}$

$y = -2x - \frac{1}{3}.$

Using the *slope–intercept equation*, we substitute -2 for m, $\frac{1}{3}$ for x, and -1 for y, and then solve for b:

$-1 = -2 \cdot \frac{1}{3} + b$ Using $y = mx + b$

$-1 = -\frac{2}{3} + b$

$-\frac{1}{3} = b.$

Then, substituting -2 for m and $-\frac{1}{3}$ for b in the slope–intercept equation $y = mx + b$, we have $y = -2x - \frac{1}{3}$.

Practice Exercise

16. Find an equation of the line with slope -4 and containing the point $\left(\frac{1}{2}, -3\right)$.

Objective 7.5c   Find an equation of a line when two points are given.

Example   Find an equation of the line containing the points $(-3, 9)$ and $(1, -2)$.

We first find the slope:

$$\frac{9 - (-2)}{-3 - 1} = \frac{11}{-4} = -\frac{11}{4}.$$

Using the slope–intercept equation and point $(1, -2)$, we substitute $-\frac{11}{4}$ for m, 1 for x, and -2 for y, and then solve for b. We could also have used the point $(-3, 9)$.

$$y = mx + b$$
$$-2 = -\tfrac{11}{4} \cdot 1 + b$$
$$-\tfrac{8}{4} = -\tfrac{11}{4} + b$$
$$\tfrac{3}{4} = b$$

Then substituting $-\frac{11}{4}$ for m and $\frac{3}{4}$ for b in $y = mx + b$, we have $y = -\frac{11}{4}x + \frac{3}{4}$.

Practice Exercise

17. Find an equation of the line containing the points $(-2, 7)$ and $(4, -3)$.

Objective 7.5d   Given a line and a point not on the given line, find an equation of the line parallel to the line and containing the point, and find an equation of the line perpendicular to the line and containing the point.

Example   Write an equation of the line containing $(-1, 1)$ and parallel to $3y - 6x = 5$.

Solving $3y - 6x = 5$ for y, we get $y = 2x + \frac{5}{3}$. The slope of the given line is 2.

A line parallel to the given line must have the same slope, 2. We substitute 2 for m, -1 for x_1, and 1 for y_1 in the point–slope equation:

$$y - 1 = 2[x - (-1)] \qquad \text{Using } y - y_1 = m(x - x_1)$$
$$y - 1 = 2(x + 1)$$
$$y - 1 = 2x + 2$$
$$y = 2x + 3. \qquad \text{Line parallel to the given line and passing through } (-1, 1)$$

Example   Write an equation of the line containing the point $(2, -4)$ and perpendicular to $6x + 2y = 13$.

Solving $6x + 2y = 13$ for y, we get $y = -3x + \frac{13}{2}$. The slope of the given line is -3.

The slope of a line perpendicular to the given line is the opposite of the reciprocal of -3, or $\frac{1}{3}$. We substitute $\frac{1}{3}$ for m, 2 for x_1, and -4 for y_1 in the point–slope equation:

$$y - (-4) = \tfrac{1}{3}(x - 2) \qquad \text{Using } y - y_1 = m(x - x_1)$$
$$y + 4 = \tfrac{1}{3}x - \tfrac{2}{3}$$
$$y = \tfrac{1}{3}x - \tfrac{14}{3}. \qquad \text{Line perpendicular to the given line and passing through } (2, -4)$$

Practice Exercises

18. Write an equation of the line containing the point $(2, -5)$ and parallel to $4x - 3y = 6$.

19. Write an equation of the line containing $(2, -5)$ and perpendicular to $4x - 3y = 6$.

Review Exercises

Determine whether each correspondence is a function. [7.1a]

1.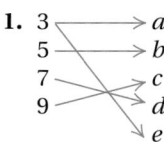

2.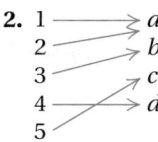

Find the function values. [7.1b]

3. $g(x) = -2x + 5$; $g(0)$ and $g(-1)$

4. $f(x) = 3x^2 - 2x + 7$; $f(0)$ and $f(-1)$

5. *Tuition Cost.* The function $C(t) = 309.2t + 3717.7$ can be used to approximate the average cost of tuition and fees for in-state students at public four-year colleges, where t is the number of years after 2000. Estimate the average cost of tuition and fees in 2010. That is, find $C(10)$. [7.1b]

Source: U.S. National Center for Education Statistics

Graph. [7.1c]

6. $f(x) = -3x + 2$

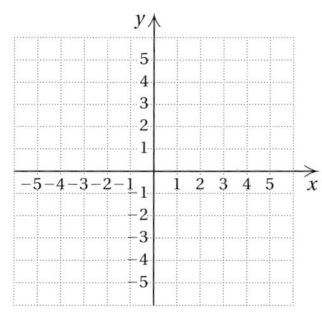

7. $g(x) = \frac{5}{2}x - 3$

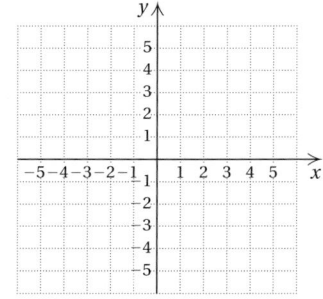

8. $f(x) = |x - 3|$

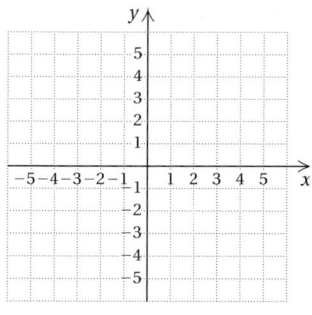

9. $h(x) = 3 - x^2$

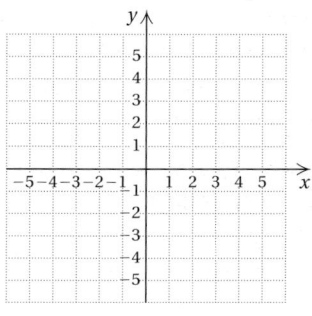

Determine whether each of the following is the graph of a function. [7.1d]

10.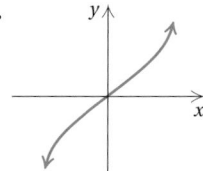

11.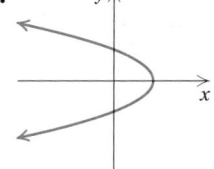

12. For the following graph of a function f, determine **(a)** $f(2)$; **(b)** the domain; **(c)** all x-values such that $f(x) = 2$; and **(d)** the range.   [7.2a]

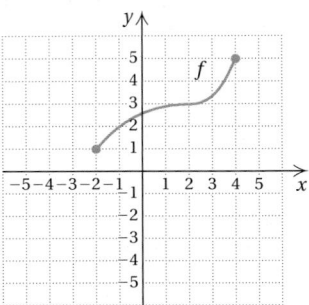

Find the domain.   [7.2a]

13. $f(x) = \dfrac{5}{x - 4}$ **14.** $g(x) = x - x^2$

Find the slope and the y-intercept.   [7.3a, b]

15. $y = -3x + 2$ **16.** $4y + 2x = 8$

17. Find the slope, if it exists, of the line containing the points $(13, 7)$ and $(10, -4)$.   [7.3b]

Find the intercepts. Then graph the equation.   [7.4a]

18. $2y + x = 4$

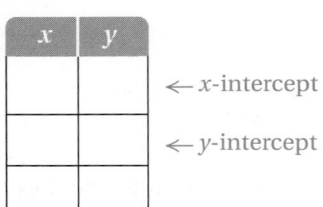
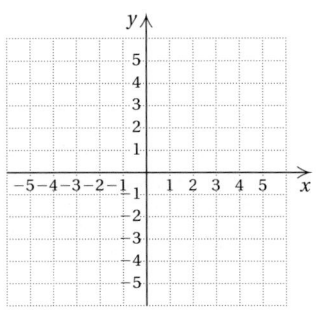

19. $2y = 6 - 3x$

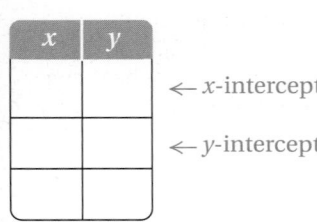
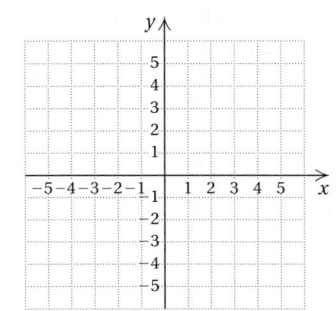

Graph using the slope and the y-intercept.   [7.4b]

20. $g(x) = -\dfrac{2}{3}x - 4$ **21.** $f(x) = \dfrac{5}{2}x + 3$

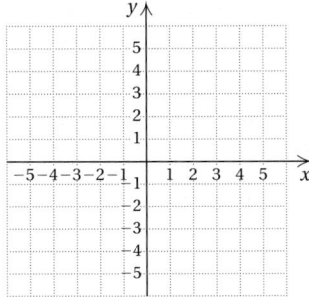
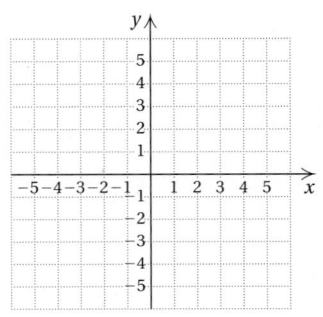

Graph.   [7.4c]

22. $x = -3$ **23.** $f(x) = 4$

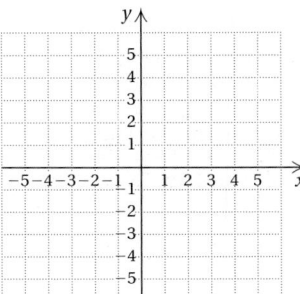
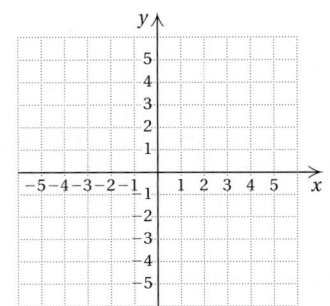

Determine whether the graphs of the given pair of lines are parallel or perpendicular.   [7.4d]

24. $y + 5 = -x,$
$\quad x - y = 2$

25. $3x - 5 = 7y,$
$\quad 7y - 3x = 7$

26. $4y + x = 3,$
$\quad 2x + 8y = 5$

27. $x = 4,$
$\quad y = -3$

28. Find a linear function $f(x) = mx + b$ whose graph has the given slope and y-intercept: [7.5a]

slope: 4.7; y-intercept: $(0, -23)$.

29. Find an equation of the line having the given slope and containing the given point: [7.5b]

$m = -3;$ $(3, -5)$.

30. Find an equation of the line containing the given pair of points: [7.5c]
$$(-2,3) \quad \text{and} \quad (-4,6).$$

31. Find an equation of the line containing the given point and parallel to the given line: [7.5d]
$$(14,-1); \quad 5x + 7y = 8.$$

32. Find an equation of the line containing the given point and perpendicular to the given line: [7.5d]
$$(5,2); \quad 3x + y = 5.$$

33. *Records in the 400-Meter Run.*   The table below shows data regarding the world indoor records in the men's 400-m run. [7.5e]

YEAR	RECORDS IN THE 400-M RUN (in seconds)
1970	46.8
2004	44.63

a) Use the two data points to find a linear function that fits the data. Let $x =$ the number of years since 1970 and $R(x) =$ the world record x years from 1970.

b) Use the function to estimate the world record in the men's 400-m run in 2008 and in 2010.

34. What is the domain of $f(x) = \dfrac{x+3}{x-2}$? [7.2a]

A. $\{x | x \geq -3\}$
B. $\{x | x \text{ is a real number } and\ x \neq 0 \ and\ x \neq 2\}$
C. $\{x | x \text{ is a real number } and\ x \neq 2\}$
D. $\{x | x > -3\}$

35. Find an equation of the line containing the point $(-2,1)$ and perpendicular to $3y - \frac{1}{2}x = 0$. [7.5d]

A. $6x + y = -11$ **B.** $y = -\dfrac{1}{6}x - 11$

C. $y = -2x - 3$ **D.** $2x + \dfrac{1}{3} = 0$

Synthesis

36. Homespun Jellies charges $2.49 for each jar of preserves. Shipping charges are $3.75 for handling, plus $0.60 per jar. Find a linear function for determining the cost of buying and shipping x jars of preserves. [7.5e]

Understanding Through Discussion and Writing

1. Under what conditions will the x-intercept and the y-intercept of a line be the same? What would the equation for such a line look like? [7.4a]

2. Explain the usefulness of the concept of slope when describing a line. [7.3b, c], [7.4b], [7.5a, b, c, d]

3. A student makes a mistake when using a graphing calculator to draw $4x + 5y = 12$ and the following screen appears. Use algebra to show that a mistake has been made. What do you think the mistake was? [7.3b]

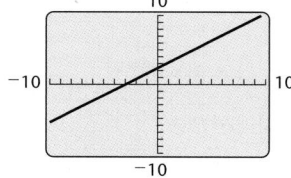

4. *Computer Repair.*   The cost $R(t)$, in dollars, of computer repair at PC Pros is given by
$$R(t) = 50t + 35,$$
where t is the number of hours that the repair requires. Determine m and b in this application and explain their meaning. [7.5e]

5. Explain why the slope of a vertical line is not defined but the slope of a horizontal line is 0. [7.4c]

6. A student makes a mistake when using a graphing calculator to draw $5x - 2y = 3$ and the following screen appears. Use algebra to show that a mistake has been made. What do you think the mistake was? [7.3b]

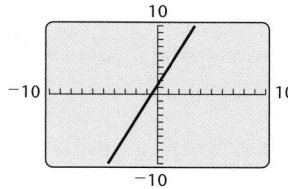

Test

For Extra Help

CHAPTER
Test Prep
VIDEOS

Step-by-step test solutions are found on the Chapter Test Prep Videos available via the Video Resources on DVD, in *MyMathLab*, and on You Tube (search "BittingerIntroInter" and click on "Channels").

Determine whether each correspondence is a function.

1.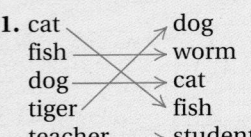
cat   dog
fish   worm
dog   cat
tiger   fish
teacher student

2.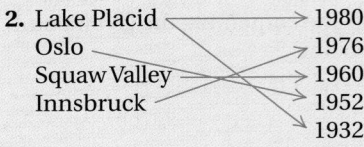
Lake Placid   1980
Oslo   1976
Squaw Valley   1960
Innsbruck   1952
              1932

Find the function values.

3. $f(x) = -3x - 4$; $f(0)$ and $f(-2)$

4. $g(x) = x^2 + 7$; $g(0)$ and $g(-1)$

5. $h(x) = -6$; $h(-4)$ and $h(-6)$

6. $f(x) = |x + 7|$; $f(-10)$ and $f(-7)$

Graph.

7. $h(x) = -2x - 5$

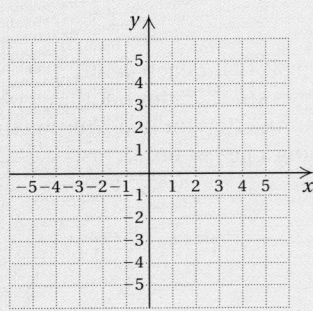

8. $f(x) = -\dfrac{3}{5}x$

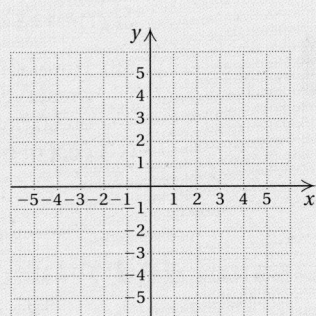

9. $g(x) = 2 - |x|$

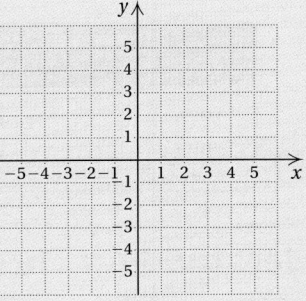

10. $f(x) = x^2 + 2x - 3$

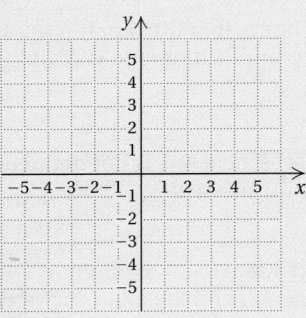

11. $y = f(x) = -3$

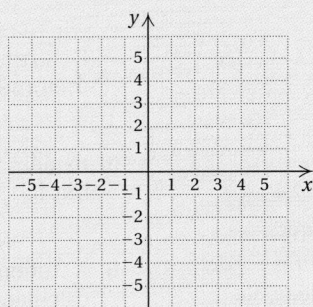

12. $2x = -4$

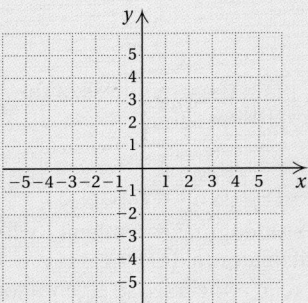

13. *Median Age of Cars.* The function

$$A(t) = 0.233t + 5.87$$

can be used to estimate the median age of cars in the United States t years after 1990. (In this context, we mean that if the median age of cars is 3 yr, then half the cars are older than 3 yr and half are younger.)
Source: The Polk Co.

a) Find the median age of cars in 2002.
b) In what year was the median age of cars 7.734 yr?

Determine whether each of the following is the graph of a function.

14.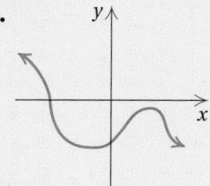

15.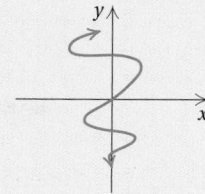

Find the domain.

16. $f(x) = \dfrac{8}{2x + 3}$

17. $g(x) = 5 - x^2$

18. For the following graph of function f, determine (a) $f(1)$; (b) the domain; (c) all x-values such that $f(x) = 2$; and (d) the range.

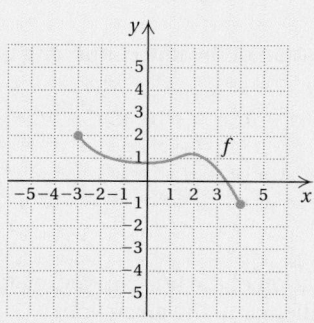

Find the slope and the y-intercept.

19. $f(x) = -\dfrac{3}{5}x + 12$

20. $-5y - 2x = 7$

Find the slope, if it exists, of the line containing the following points.

21. $(-2, -2)$ and $(6, 3)$

22. $(-3.1, 5.2)$ and $(-4.4, 5.2)$

23. Find the slope, or rate of change, of the graph at right.

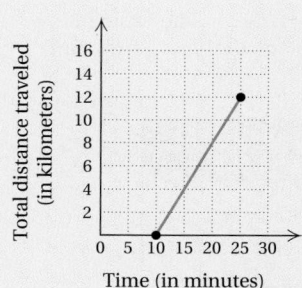

24. Find the intercepts. Then graph the equation.

$$2x + 3y = 6$$

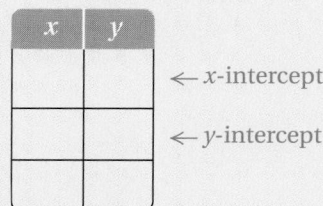

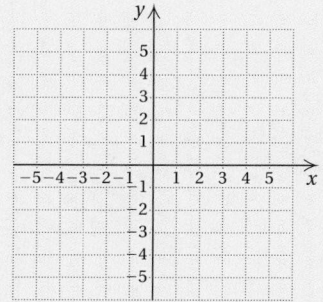

25. Graph using the slope and the y-intercept:

$$f(x) = -\dfrac{2}{3}x - 1.$$

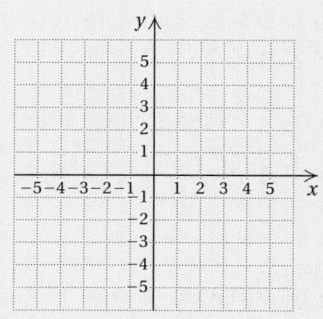

Determine whether the graphs of the given pair of lines are parallel or perpendicular.

26. $4y + 2 = 3x,$
$-3x + 4y = -12$

27. $y = -2x + 5,$
$2y - x = 6$

28. Find an equation of the line that has the given characteristics:

slope: -3; y-intercept: $(0, 4.8)$.

29. Find a linear function $f(x) = mx + b$ whose graph has the given slope and y-intercept:

slope: 5.2; y-intercept: $\left(0, -\frac{5}{8}\right)$.

30. Find an equation of the line having the given slope and containing the given point:

$m = -4$; $(1, -2)$.

31. Find an equation of the line containing the given pair of points:

$(4, -6)$ and $(-10, 15)$.

32. Find an equation of the line containing the given point and parallel to the given line:

$(4, -1)$; $x - 2y = 5$.

33. Find an equation of the line containing the given point and perpendicular to the given line:

$(2, 5)$; $x + 3y = 2$.

34. *Median Age of Men at First Marriage.* The table below lists data regarding the median age of men at first marriage in 1970 and in 2007.

YEAR	MEDIAN AGE OF MEN AT FIRST MARRIAGE
1970	23.2
2007	27.7

SOURCE: U.S. Census Bureau

a) Use the two data points to find a linear function that fits the data. Let $x =$ the number of years since 1970 and $A =$ the median age at first marriage x years from 1970.

b) Use the function to estimate the median age of men at first marriage in 2008 and in 2015.

35. Find an equation of the line having slope -2 and containing the point $(3, 1)$.

A. $y - 1 = 2(x - 3)$ **B.** $y - 1 = -2(x - 3)$
C. $x - 1 = -2(y - 3)$ **D.** $x - 1 = 2(y - 3)$

Synthesis

36. Find k such that the line $3x + ky = 17$ is perpendicular to the line $8x - 5y = 26$.

37. Find a formula for a function f for which $f(-2) = 3$.

Cumulative Review

Compute and simplify.

1. $-2[1.4 - (-0.8 - 1.2)]$

2. $(1.3 \times 10^8)(2.4 \times 10^{-10})$

3. $\left(-\dfrac{1}{6}\right) \div \left(\dfrac{2}{9}\right)$

4. $\dfrac{2^{12}2^{-7}}{2^8}$

Simplify.

5. $\dfrac{x^2 - 9}{2x^2 - 7x + 3}$

6. $\dfrac{t^2 - 16}{(t + 4)^2}$

7. $\dfrac{x - \dfrac{x}{x + 2}}{\dfrac{2}{x} - \dfrac{1}{x + 2}}$

Perform the indicated operations and simplify.

8. $(1 - 3x^2)(2 - 4x^2)$

9. $(2a^2b - 5ab^2)^2$

10. $(3x^2 + 4y)(3x^2 - 4y)$

11. $-2x^2(x - 2x^2 + 3x^3)$

12. $(1 + 2x)(4x^2 - 2x + 1)$

13. $\left(8 - \dfrac{1}{3}x\right)\left(8 + \dfrac{1}{3}x\right)$

14. $(-8y^2 - y + 2) - (y^3 - 6y^2 + y - 5)$

15. $(2x^3 - 3x^2 - x - 1) \div (2x - 1)$

16. $\dfrac{7}{5x - 25} + \dfrac{x + 7}{5 - x}$

17. $\dfrac{2x - 1}{x - 2} - \dfrac{2x}{2 - x}$

18. $\dfrac{y^2 + y}{y^2 + y - 2} \cdot \dfrac{y + 2}{y^2 - 1}$

19. $\dfrac{7x + 7}{x^2 - 2x} \div \dfrac{14}{3x - 6}$

Factor completely.

20. $6x^5 - 36x^3 + 9x^2$

21. $16y^4 - 81$

22. $3x^2 + 10x - 8$

23. $4x^4 - 12x^2y + 9y^2$

24. $3m^3 + 6m^2 - 45m$

25. $x^3 + x^2 - x - 1$

Solve.

26. $3x - 4(x + 1) = 5$

27. $x(2x - 5) = 0$

28. $5x + 3 \geq 6(x - 4) + 7$

29. $1.5x - 2.3x = 0.4(x - 0.9)$

30. $2x^2 = 338$

31. $3x^2 + 15 = 14x$

32. $\dfrac{2}{x} - \dfrac{3}{x - 2} = \dfrac{1}{x}$

33. $1 + \dfrac{3}{x} + \dfrac{x}{x + 1} = \dfrac{1}{x^2 + x}$

34. $w - \dfrac{9}{10} = -\dfrac{9}{10} + w$

35. $20 - 3y - 2y \leq 45 + 5y$

36. $9x - 4 - 2x = 5x - 5 + 2x$

37. $N = rx - t$, for x

Solve.

38. *Digital Photo Frame.* Joel paid $37.10, including 6% sales tax, for a digital photo frame. What was the price of the frame itself?

39. *Roofing Time.* It takes David 15 hr to put a roof on a house. It takes Loren 12 hr to put a roof on the same type of house. How long would it take to complete the job if they worked together?

40. *Triangle Dimensions.* The length of one leg of a right triangle is 12 in. The length of the hypotenuse is 8 in. longer than the length of the other leg. Find the lengths of the hypotenuse and the other leg.

41. *Quality Control.* A sample of 120 computer chips contained 5 defective chips. How many defective chips would you expect to find in a batch of 1800 chips?

42. *Triangle Dimensions.* The height of a triangle is 5 ft more than the base. The area is 18 ft^2. Find the height and the base.

43. *Height of a Parallelogram.* The height h of a parallelogram of fixed area varies inversely as the base b. Suppose that the height is 24 ft when the base is 15 ft. Find the height when the base is 5 ft. What is the variation constant?

44. Find an equation of variation in which y varies directly as x and $y = 2.4$ when $x = 12$.

Find the function values.

45. $g(x) = -\frac{1}{2}x + 6$; $g(0)$ and $g(-6)$

46. $f(x) = |2x - 3|$; $f(-4)$ and $f(0)$

47. Find the slope of the line containing the points $(2, 3)$ and $(-1, 3)$.

48. Find the slope and the y-intercept of the line $2x + 3y = 6$.

49. Find an equation of the line that contains the points $(-5, 6)$ and $(2, -4)$.

50. Find an equation of the line containing the point $(0, -3)$ and having the slope $m = 6$.

Graph on a plane.

51. $y = -2$

52. $2x + 5y = 10$

53. $f(x) = -\frac{3}{4}x + 2$

54. $2x - y = 3$

55. $x - 4 = 0$

56. $3x - y = -3$

Synthesis

57. Solve: $x^2 + 2 < 0$.

58. Simplify:
$$\frac{x - 5}{x + 3} - \frac{x^2 - 6x + 5}{x^2 + x - 2} \div \frac{x^2 + 4x + 3}{x^2 + 3x + 2}.$$

59. Find the value of k such that $y - kx = 4$ and $10x - 3y = -12$ are perpendicular.

Systems of Equations

Real-World Application

To stimulate the economy in his town of Brewton, Alabama, in 2009, Danny Cottrell, co-owner of The Medical Center Pharmacy, gave each of his full-time employees $700 and each part-time employee $300. He asked that each person donate 15% to a charity of his or her choice and spend the rest locally. The money was paid in $2 bills, a rarely used currency, so that the business community could easily see how the money circulated. Cottrell gave away a total of $16,000 to his 24 employees. How many full-time employees and how many part-time employees were there?

This problem appears as Example 7 in Section 8.3.

8.1

Systems of Equations in Two Variables

OBJECTIVE

a Solve a system of two linear equations or two functions by graphing and determine whether a system is consistent or inconsistent and whether the equations in a system are dependent or independent.

SKILL TO REVIEW

Objective 3.2a: Graph linear equations of the type $y = mx + b$ and $Ax + By = C$.

Graph.

1. $x + y = 3$ **2.** $y = x - 2$

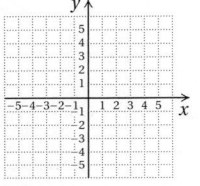 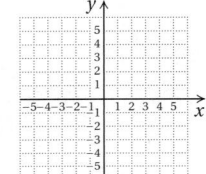

We can solve many applied problems more easily by translating them to two or more equations in two or more variables than by translating to a single equation. Let's look at such a problem.

Mother's Day Spending. Mother's Day ranks fourth in spending in the United States behind the winter holidays, back-to-school buying, and Valentine's Day. About $15.8 billion was spent to celebrate Mother's Day in 2008. Of this amount, a total of $5 billion was spent on meals in restaurants and flowers. The amount spent on restaurant meals was $1 billion more than the amount spent on flowers. How much was spent on each?

Source: National Retail Association

To solve, we first let

$x =$ the amount spent on restaurant meals,   and

$y =$ the amount spent on flowers,

where x and y are in billions of dollars. The problem gives us two statements that can be translated to equations.

First, we consider the total amount spent on meals and flowers:

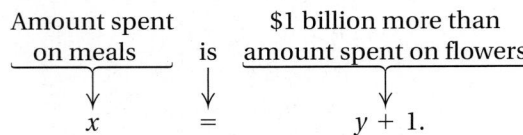

Amount spent on meals	plus	Amount spent on flowers	is	Total amount spent
↓	↓	↓	↓	↓
x	$+$	y	$=$	$5.$

The second statement compares the two different amounts spent:

Amount spent on meals	is	$1 billion more than amount spent on flowers
↓	↓	↓
x	$=$	$y + 1.$

We have now translated the problem to a pair, or **system, of equations**:

$$x + y = 5,$$
$$x = y + 1.$$

A **solution** of a system of two equations in two variables is an ordered pair that makes *both* equations true. If we graph a system of equations, the point at which the graphs intersect will be a solution of *both* equations.

Answers

Skill to Review:

1.

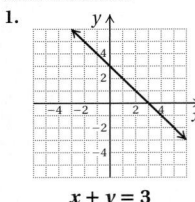

$x + y = 3$

2.

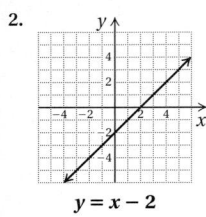

$y = x - 2$

We graph the equations listed on the preceding page.

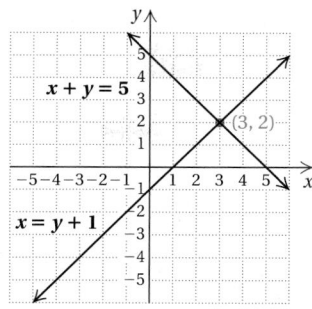

We see that the graphs intersect at the point $(3, 2)$—that is, $x = 3$ and $y = 2$. These numbers check in the statement of the original problem. This tells us that \$3 billion was spent on restaurant meals and \$2 billion was spent on flowers.

(a) Solving Systems of Equations Graphically

As we have just seen, we can solve systems of equations graphically.

One Solution

EXAMPLE 1 Solve this system graphically:

$$y - x = 1,$$
$$y + x = 3.$$

We draw the graph of each equation using any method studied in Chapter 3 and find the coordinates of the point of intersection.

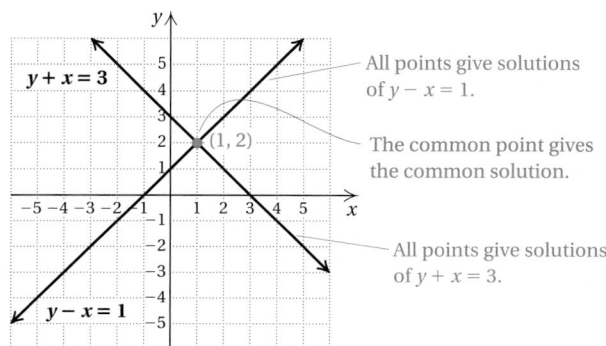

The point of intersection has coordinates that make *both* equations true. The solution seems to be the point $(1, 2)$. However, solving by graphing may give only approximate answers. Thus we check the pair $(1, 2)$ in both equations.

Check:

$y - x = 1$	$y + x = 3$
$2 - 1$? 1	$2 + 1$? 3
1 | TRUE	3 | TRUE

The solution is $(1, 2)$.

Do Exercises 1 and 2.

Solve each system graphically.

1. $-2x + y = 1,$
    $3x + y = 1$

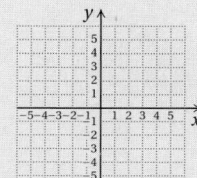

2. $y = \frac{1}{2}x,$
    $y = -\frac{1}{4}x + \frac{3}{2}$

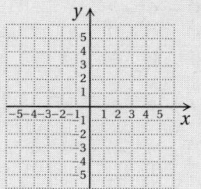

Answers

1. $(0, 1)$ **2.** $(2, 1)$

Solving Systems of Equations We can solve a system of two equations in two variables using a graphing calculator. Consider the system of equations in Example 1:

$$y - x = 1,$$
$$y + x = 3.$$

First, we solve the equations for y, obtaining $y = x + 1$ and $y = -x + 3$. Next, we enter $y_1 = x + 1$ and $y_2 = -x + 3$ on the equation-editor screen and graph the equations. We can use the standard viewing window, $[-10, 10, -10, 10]$.

We will use the INTERSECT feature to find the coordinates of the point of intersection of the lines. To access this feature, we press **2ND** **CALC** **5**. (CALC is the second operation associated with the **TRACE** key.) The query "First curve?" appears on the graph screen. The blinking cursor is positioned on the graph of y_1. We press **ENTER** to indicate that this is the first curve involved in the intersection. Next, the query "Second curve?" appears and the blinking cursor is positioned on the graph of y_2. We press **ENTER** to indicate that this is the second curve. Now the query "Guess?" appears. We use the ▷ and ◁ keys to move the cursor close to the point of intersection or we enter an x-value close to the first coordinate of the point of intersection. Then we press **ENTER**. The coordinates of the point of intersection of the graphs, $x = 1$, $y = 2$, appear at the bottom of the screen. Thus the solution of the system of equations is $(1, 2)$.

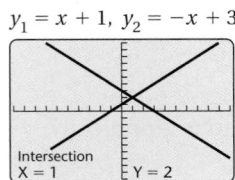

$$y_1 = x + 1, \; y_2 = -x + 3$$

Intersection
X = 1   Y = 2

Exercises: Use a graphing calculator to solve each system of equations.

1. $x + y = 5,$
$\quad y = x + 1$

2. $y = x + 3,$
$\quad 2x - y = -7$

3. $x - y = -6,$
$\quad y = 2x + 7$

4. $x + 4y = -1,$
$\quad x - y = 4$

No Solution

Sometimes the equations in a system have graphs that are parallel lines.

EXAMPLE 2 Solve graphically:

$$f(x) = -3x + 5,$$
$$g(x) = -3x - 2.$$

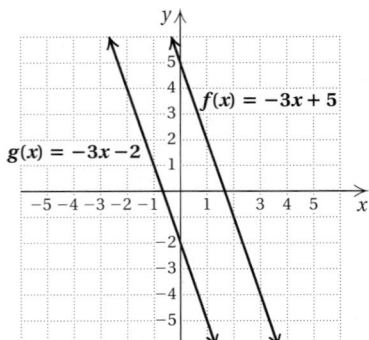

Note that this system is written using function notation. We graph the functions. The graphs have the same slope, -3, and different y-intercepts, so they are parallel. There is no point at which they cross, so the system has no solution. No matter what point we try, it will *not* check in *both* equations. The solution set is thus the empty set, denoted $\varnothing$ or $\{\ \}$.

CONSISTENT SYSTEMS AND INCONSISTENT SYSTEMS

If a system of equations has at least one solution, it is **consistent**.

If a system of equations has no solution, it is **inconsistent**.

The system in Example 1 is consistent. The system in Example 2 is inconsistent.

Do Exercises 3 and 4.

Infinitely Many Solutions

Sometimes the equations in a system have the same graph. In such a case, the equations have an *infinite* number of solutions in common.

EXAMPLE 3   Solve graphically:

$$3y - 2x = 6,$$
$$-12y + 8x = -24.$$

We graph the equations and see that the graphs are the same. Thus any solution of one of the equations is a solution of the other. Each equation has an infinite number of solutions, two of which are shown on the graph.

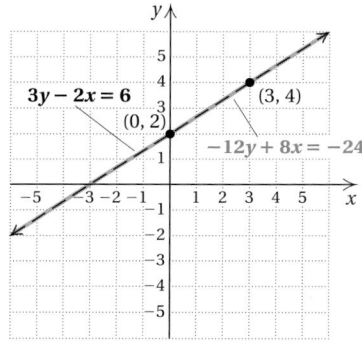

We check one such solution, $(0, 2)$, which is the y-intercept of each equation.

Check:

$$\begin{array}{c|c}
3y - 2x = 6 & -12y + 8x = -24 \\
\hline
3(2) - 2(0) \;?\; 6 & -12(2) + 8(0) \;?\; -24 \\
6 - 0 & -24 + 0 \\
6 \quad \text{TRUE} & -24 \quad \text{TRUE}
\end{array}$$

On your own, check that $(3, 4)$ is a solution of both equations. If $(0, 2)$ and $(3, 4)$ are solutions, then all points on the line containing them will be solutions. The system has an infinite number of solutions.

3. Solve graphically:

$$y + 2x = 3,$$
$$y + 2x = -4.$$

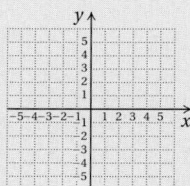

4. Classify each of the systems in Margin Exercises 1–3 as consistent or inconsistent.

When we graph a system of two equations, one of the following three things can happen.

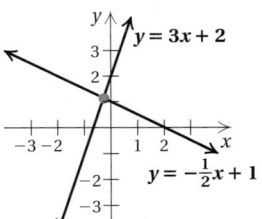

One solution.
Graphs intersect.
The system is consistent
and *the equations are
independent.*

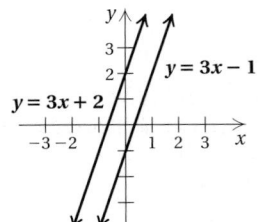

No solution.
Graphs are parallel.
The system is inconsistent
and *the equations are
independent.*

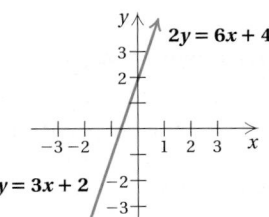

Infinitely many solutions.
Equations have the same
graph. *The system is consistent*
and *the equations are
dependent.*

Let's summarize what we know about the systems of equations in Examples 1–3. The system in Example 1 has exactly one solution, and the system in Example 3 has an infinite number of solutions. Since each system has at least one solution, both systems are *consistent*. The system of equations in Example 2 has no solution, so it is *inconsistent*.

The system of equations in Example 1 has exactly one solution, and the system in Example 2 has no solutions. Thus the equations in each of these systems are *independent*. In a system of equations with infinitely many solutions, the equations are dependent. This tells us that the equations in Example 3 are *dependent*. In a system with dependent equations, one equation can be obtained by multiplying the other equation by a constant.

Do Exercises 5 and 6.

5. Solve graphically:

$$2x - 5y = 10,$$
$$-6x + 15y = -30.$$

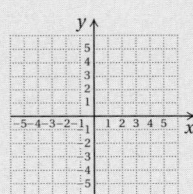

6. Classify the equations in Margin Exercises 1, 2, 3, and 5 as dependent or independent.

※ Algebraic-Graphical Connection

Consider the equation $-2x + 13 = 4x - 17$. Let's solve it algebraically as we did in Chapter 2:

$$-2x + 13 = 4x - 17$$
$$13 = 6x - 17 \qquad \text{Adding } 2x$$
$$30 = 6x \qquad \text{Adding } 17$$
$$5 = x. \qquad \text{Dividing by } 6$$

Answers

5. Infinitely many solutions
6. Independent: Margin Exercises 1, 2, and 3;
dependent: Margin Exercise 5

Could we also solve the equation graphically? The answer is yes, as we see in the following two methods.

METHOD 1: Solve $-2x + 13 = 4x - 17$ graphically.

We let $f(x) = -2x + 13$ and $g(x) = 4x - 17$. Graphing the system of equations, we get the graph shown below.

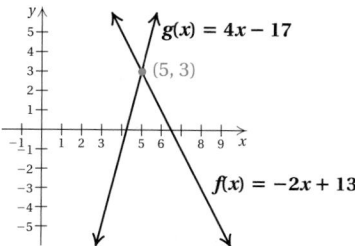

The point of intersection of the two graphs is $(5, 3)$. Note that the x-coordinate of this point is 5. This is the value of x for which $-2x + 13 = 4x - 17$, so it is the solution of the equation.

Do Exercises 7 and 8.

METHOD 2: Solve $-2x + 13 = 4x - 17$ graphically.

Adding $-4x$ and 17 on both sides, we obtain an equation with 0 on one side: $-6x + 30 = 0$. This time we let $f(x) = -6x + 30$ and $g(x) = 0$. Since the graph of $g(x) = 0$, or $y = 0$, is the x-axis, we need only graph $f(x) = -6x + 30$ and see where it crosses the x-axis.

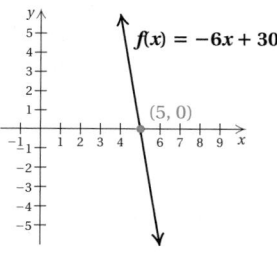

Note that the x-intercept of $f(x) = -6x + 30$ is $(5, 0)$, or just 5. This x-value is the solution of the equation $-2x + 13 = 4x - 17$.

Do Exercise 9.

Let's compare the two methods. Using Method 1, we graph two functions. The solution of the original equation is the x-coordinate of the point of intersection. Using Method 2, we graph one function. The solution of the original equation is the x-coordinate of the x-intercept of the graph.

※

Do Exercise 10.

7. a) Solve $x + 1 = \frac{2}{3}x$ algebraically.
   b) Solve $x + 1 = \frac{2}{3}x$ graphically using Method 1.

8. Solve $\frac{1}{2}x + 3 = 2$ graphically using Method 1.

9. a) Solve $x + 1 = \frac{2}{3}x$ graphically using Method 2.
   b) Compare your answers to Margin Exercises 7(a), 7(b), and 9(a).

10. Solve $\frac{1}{2}x + 3 = 2$ graphically using Method 2.

Answers

7. (a) -3; (b) the same: -3 8. -2
9. (a) -3; (b) All are -3.   10. -2

8.1 **Exercise Set**

For Extra Help

MyMathLab

Math XL
PRACTICE

WATCH

DOWNLOAD

READ

REVIEW

a Solve each system of equations graphically. Then classify the system as consistent or inconsistent and the equations as dependent or independent. Complete the check for Exercises 1–4.

1. $x + y = 4,$
$x - y = 2$

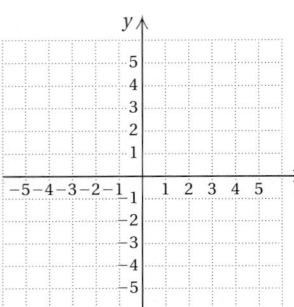

Check:
$$\frac{x + y = 4}{\ ?\ }$$

$$\frac{x - y = 2}{\ ?\ }$$

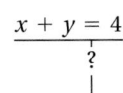

2. $x - y = 3,$
$x + y = 5$

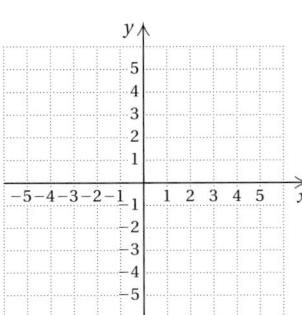

Check:
$$\frac{x - y = 3}{\ ?\ }$$

$$\frac{x + y = 5}{\ ?\ }$$

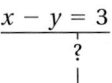

3. $2x - y = 4,$
$2x + 3y = -4$

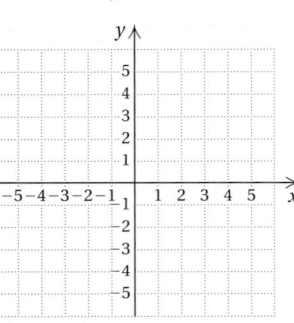

Check:
$$\frac{2x - y = 4}{\ ?\ }$$

$$\frac{2x + 3y = -4}{\ ?\ }$$

4. $3x + y = 5,$
$x - 2y = 4$

Check:
$$\frac{3x + y = 5}{\ ?\ }$$

$$\frac{x - 2y = 4}{\ ?\ }$$

5. $2x + y = 6,$
$3x + 4y = 4$

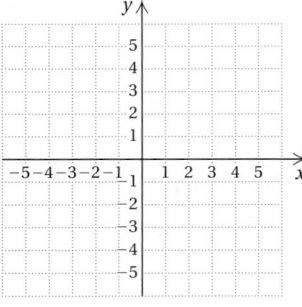

6. $2y = 6 - x,$
$3x - 2y = 6$

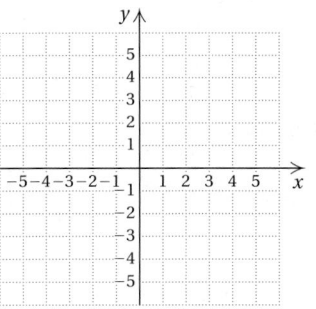

7. $f(x) = x - 1,$
$g(x) = -2x + 5$

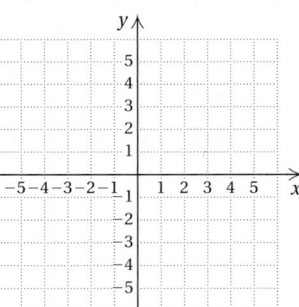

8. $f(x) = x + 1,$
$g(x) = \frac{2}{3}x$

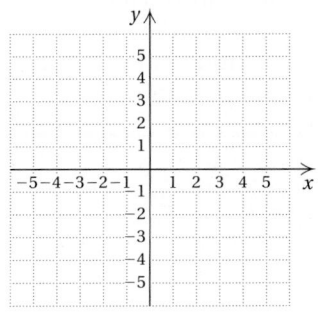

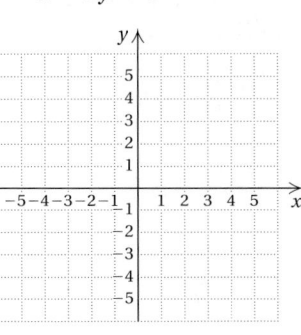

9. $2u + v = 3,$
     $2u = v + 7$

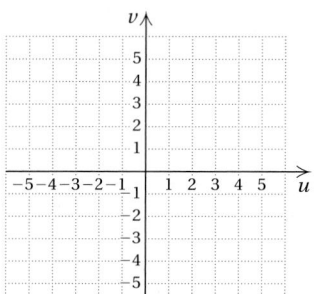

10. $2b + a = 11,$
     $a - b = 5$

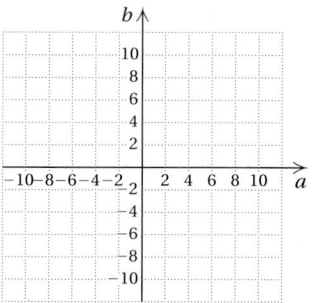

11. $f(x) = -\frac{1}{3}x - 1,$
     $g(x) = \frac{4}{3}x - 6$

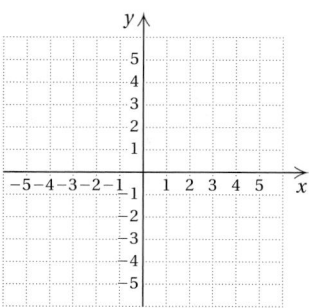

12. $f(x) = -\frac{1}{4}x + 1,$
     $g(x) = \frac{1}{2}x - 2$

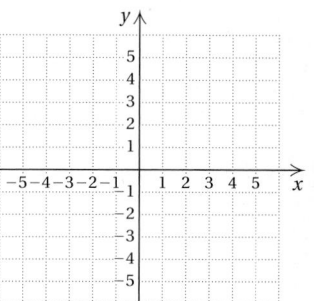

13. $6x - 2y = 2,$
     $9x - 3y = 1$

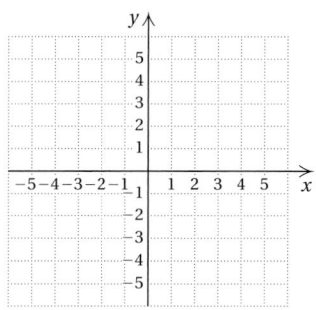

14. $y - x = 5,$
     $2x - 2y = 10$

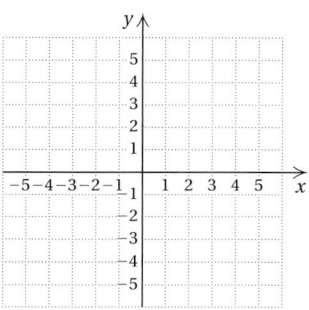

15. $2x - 3y = 6,$
     $3y - 2x = -6$

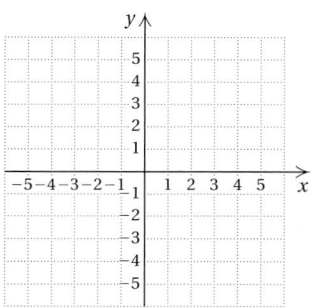

16. $y = 3 - x,$
     $2x + 2y = 6$

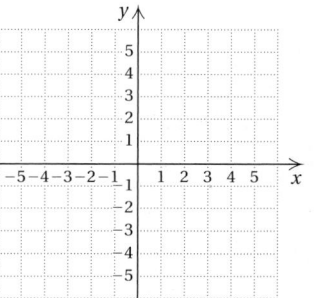

17. $x = 4,$
     $y = -5$

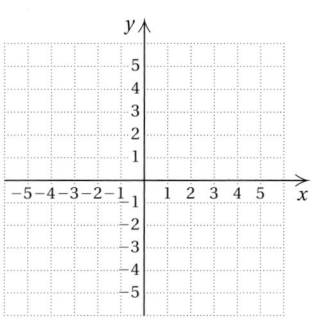

18. $x = -3,$
     $y = 2$

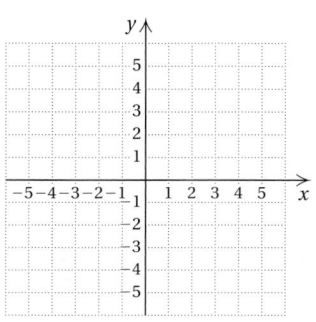

19. $y = -x - 1,$
     $4x - 3y = 17$

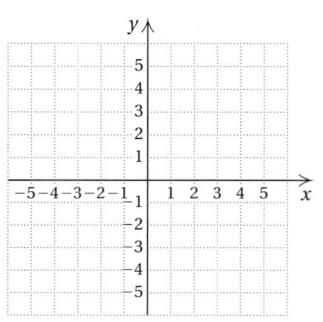

20. $a + 2b = -3,$
     $b - a = 6$

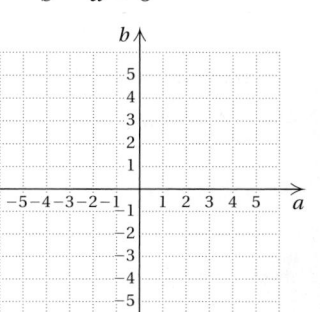

Matching. Each of Exercises 21–26 shows the graph of a system of equations and its solution. First, classify the system as consistent or inconsistent and the equations as dependent or independent. Then match it with one of the appropriate systems of equations (A)–(F), which follow.

21. Solution: $(3, 3)$

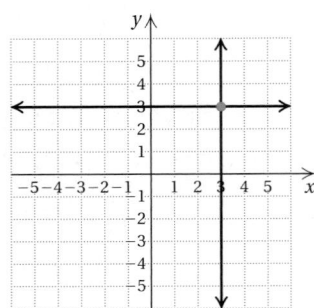

22. Solution: $(1, 1)$

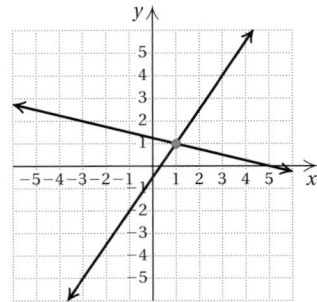

23. Solutions: Infinitely many

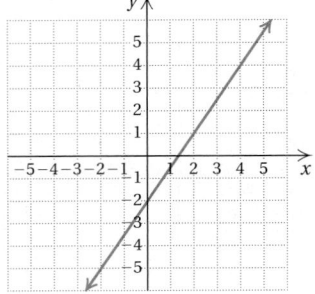

24. Solution: $(4, -3)$

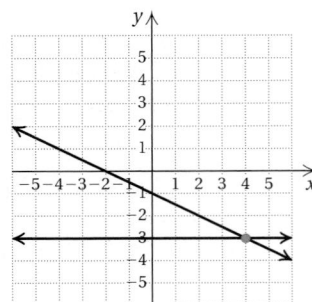

25. Solution: No solution

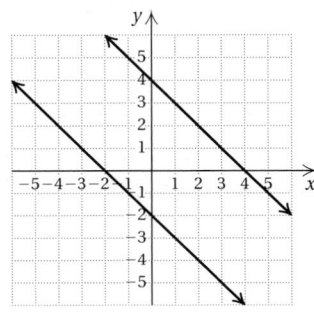

26. Solution: $(-1, 3)$

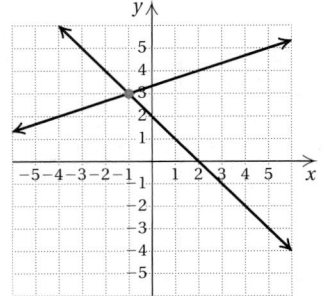

A. $3y - x = 10,$
$\quad x = -y + 2$

B. $9x - 6y = 12,$
$\quad y = \frac{3}{2}x - 2$

C. $2y - 3x = -1,$
$\quad x + 4y = 5$

D. $x + y = 4,$
$\quad y = -x - 2$

E. $\frac{1}{2}x + y = -1,$
$\quad y = -3$

F. $x = 3,$
$\quad y = 3$

Skill Maintenance

Write an equation of the line containing the given point and parallel to the given line. [7.5d]

27. $(-4, 2); 3x = 5y - 4$

28. $(-6, 0); 8y - 3x = 2$

Write an equation of the line containing the given point and perpendicular to the given line. [7.5d]

29. $(-4, 6); 2x = 3y - 12$

30. $(3, -10); 8y - 4 = -6x$

Synthesis

Use a graphing calculator to solve each system of equations. Round all answers to the nearest hundredth. You may need to solve for y first.

31. $2.18x + 7.81y = 13.78,$
$\quad 5.79x - 3.45y = 8.94$

32. $f(x) = 123.52x + 89.32,$
$\quad g(x) = -89.22x + 33.76$

Solve graphically.

33. $y = |x|,$
$\quad x + 4y = 15$

34. $x - y = 0,$
$\quad y = x^2$

8.2 Solving by Substitution

Consider this system of equations:

$$5x + 9y = 2,$$
$$4x - 9y = 10.$$

What is the solution? It is rather difficult to tell exactly by graphing. It would appear that fractions are involved. It turns out that the solution is

$$\left(\frac{4}{3}, -\frac{14}{27}\right).$$

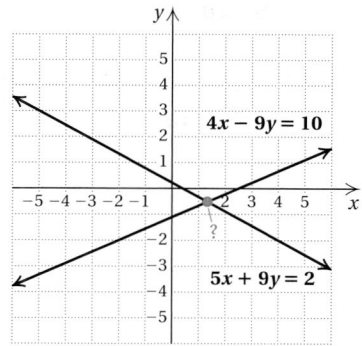

Solving by graphing, though useful in many applied situations, is not always fast or accurate in cases where solutions are not integers. We need techniques involving algebra to determine the solution exactly. Because they use algebra, they are called **algebraic methods**.

a The Substitution Method

One nongraphical method for solving systems is known as the **substitution method**.

EXAMPLE 1 Solve this system:

$$x + y = 4, \quad \textbf{(1)}$$
$$x = y + 1. \quad \textbf{(2)}$$

Equation (2) says that x and $y + 1$ name the same number. Thus we can substitute $y + 1$ for x in equation (1):

$$x + y = 4 \qquad \text{Equation (1)}$$
$$(y + 1) + y = 4. \qquad \text{Substituting } y + 1 \text{ for } x$$

Since this equation has only one variable, we can solve for y using methods learned earlier:

$$(y + 1) + y = 4$$
$$2y + 1 = 4 \qquad \text{Removing parentheses and collecting like terms}$$
$$2y = 3 \qquad \text{Subtracting 1}$$
$$y = \tfrac{3}{2}. \qquad \text{Dividing by 2}$$

We return to the original pair of equations and substitute $\frac{3}{2}$ for y in *either* equation so that we can solve for x. Calculation will be easier if we choose equation (2) since it is already solved for x:

$$x = y + 1 \qquad \text{Equation (2)}$$
$$x = \tfrac{3}{2} + 1 \qquad \text{Substituting } \tfrac{3}{2} \text{ for } y$$
$$x = \tfrac{3}{2} + \tfrac{2}{2} = \tfrac{5}{2}.$$

We obtain the ordered pair $\left(\frac{5}{2}, \frac{3}{2}\right)$. Even though we solved for y *first*, it is still the *second* coordinate since x is before y alphabetically. We check to be sure that the ordered pair is a solution.

Answers

Skill to Review:

1. 1   **2.** −3

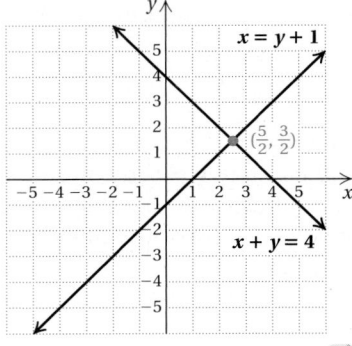

Check:

$$x + y = 4$$
$$\frac{5}{2} + \frac{3}{2} \overset{?}{} 4$$
$$\frac{8}{2}$$
$$4 \quad \text{TRUE}$$

$$x = y + 1$$
$$\frac{5}{2} \overset{?}{} \frac{3}{2} + 1$$
$$\frac{3}{2} + \frac{2}{2}$$
$$\frac{5}{2} \quad \text{TRUE}$$

Since $\left(\frac{5}{2}, \frac{3}{2}\right)$ checks, it is the solution. Even though exact fraction solutions are difficult to determine graphically, a graph can help us to visualize whether the solution is reasonable.

Solve by the substitution method.

1. $x + y = 6,$
$\quad y = x + 2$

2. $y = 7 - x,$
$\quad 2x - y = 8$

(*Caution*: Use parentheses when you substitute, being careful about removing them. Remember to solve for both variables.)

Do Exercises 1 and 2.

Suppose neither equation of a pair has a variable alone on one side. We then solve one equation for one of the variables.

EXAMPLE 2 Solve this system:

$$2x + y = 6, \qquad \textbf{(1)}$$
$$3x + 4y = 4. \qquad \textbf{(2)}$$

First, we solve one equation for one variable. Since the coefficient of y is 1 in equation (1), it is the easier one to solve for y:

$$y = 6 - 2x. \qquad \textbf{(3)}$$

Next, we substitute $6 - 2x$ for y in equation (2) and solve for x:

Solve by the substitution method.

3. $2y + x = 1,$
$\quad y - 2x = 8$

4. $8x - 5y = 12,$
$\quad x - y = 3$

$3x + 4(6 - 2x) = 4$	Substituting $6 - 2x$ for y
$3x + 24 - 8x = 4$	Multiplying to remove parentheses
$24 - 5x = 4$	Collecting like terms
$-5x = -20$	Subtracting 24
$x = 4.$	Dividing by -5

-------- *Caution!* --------

Remember to use parentheses when you substitute. Then remove them properly.

In order to find y, we return to either of the original equations, (1) or (2), or equation (3), which we solved for y. It is generally easier to use an equation like (3), where we have solved for the specific variable. We substitute 4 for x in equation (3) and solve for y:

$$y = 6 - 2x = 6 - 2(4) = 6 - 8 = -2.$$

We obtain the ordered pair $(4, -2)$.

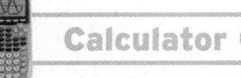

Calculator Corner

Solving Systems of Equations Use the INTERSECT feature to solve the systems of equations in Margin Exercises 1-4. (See the Calculator Corner on p. 572 for the procedure.)

Check:

$$2x + y = 6$$
$$2(4) + (-2) \overset{?}{} 6$$
$$8 - 2$$
$$6 \quad \text{TRUE}$$

$$3x + 4y = 4$$
$$3(4) + 4(-2) \overset{?}{} 4$$
$$12 - 8$$
$$4 \quad \text{TRUE}$$

Since $(4, -2)$ checks, it is the solution.

Do Exercises 3 and 4.

Answers

1. $(2, 4)$ **2.** $(5, 2)$ **3.** $(-3, 2)$
4. $(-1, -4)$

EXAMPLE 3   Solve this system of equations:

$$y = -3x + 5, \quad \textbf{(1)}$$
$$y = -3x - 2. \quad \textbf{(2)}$$

We solved this system graphically in Example 2 of Section 8.1. We found that the graphs are parallel and the system has no solution. Let's try to solve this system algebraically using substitution.

We substitute $-3x - 2$ for y in equation (1):

$$-3x - 2 = -3x + 5 \qquad \text{Substituting } -3x - 2 \text{ for } y$$
$$-2 = 5. \qquad \text{Adding } 3x$$

We have a false equation. The equation has **no solution**. (See also Example 13 of Section 2.3.)

> Do Exercise 5.

5. a) Solve this system of equations algebraically using substitution:

$$y + 2x = 3,$$
$$y + 2x = -4.$$

b) Check your answer in part (a) with the one you found graphically in Margin Exercise 3 of Section 8.1.

b Solving Applied Problems Involving Two Equations

Many applied problems are easier to solve if we first translate to a system of two equations rather than to a single equation. Here we will solve a few problems that can be solved using substitution. Section 8.4 is devoted entirely to applied problems.

EXAMPLE 4   *Architecture.*   The architects who designed the John Hancock Building in Chicago created a visually appealing building that slants on the sides. The ground floor is a rectangle that is larger than the rectangle formed by the top floor. The ground floor has a perimeter of 860 ft. The length is 100 ft more than the width. Find the length and the width.

1. **Familiarize.**   We first make a drawing and label it, using l for length and w for width. We recall or look up the formula for perimeter: $P = 2l + 2w$. (This formula can be found inside the back cover of this book.)

$l = w + 100$ w

2. **Translate.**   We translate as follows:

The perimeter is 860 ft.
$$2l + 2w = 860.$$

We can also write a second equation:

The length is 100 ft more than the width.
$$l = w + 100.$$

We now have a system of equations:

$$2l + 2w = 860, \quad \textbf{(1)}$$
$$l = w + 100. \quad \textbf{(2)}$$

Answers

5. (a) No solution; **(b)** the same—no solution

6. Architecture. The top floor of the John Hancock Building is also a rectangle, but its perimeter is 520 ft. The width is 60 ft less than the length. Find the length and the width.

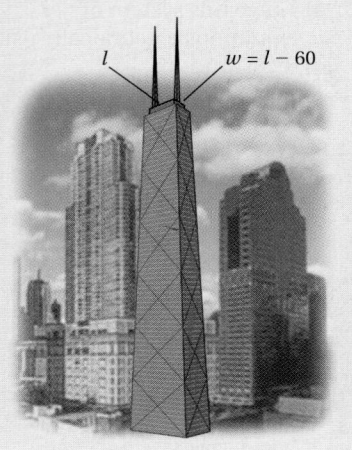

l $w = l - 60$

3. Solve. We substitute $w + 100$ for l in equation (1):

$$2(w + 100) + 2w = 860 \qquad \text{Substituting in equation (1)}$$

$$2w + 200 + 2w = 860 \qquad \text{Multiplying to remove parentheses on the left}$$

$$4w + 200 = 860 \qquad \text{Collecting like terms}$$

$$\left. \begin{array}{l} 4w = 660 \\ w = 165. \end{array} \right\} \text{Solving for } w$$

Next, we substitute 165 for w in equation (2) and solve for l:

$$l = 165 + 100 = 265.$$

4. Check. Consider the dimensions 265 ft and 165 ft. The length is 100 ft more than the width. The perimeter is 2(265 ft) + 2(165 ft), or 860 ft. The dimensions 265 ft and 165 ft check in the original problem.

5. State. The length is 265 ft, and the width is 165 ft.

Do Exercise 6.

STUDY TIPS

USING THIS TEXTBOOK

One of the most important ways to improve your math study skills is to learn the proper use of the textbook. Here we highlight a few points that we consider most helpful.

- **Be sure to note the special symbols** (a), (b), (c), **and so on, that correspond to the objectives you are to be able to master.** The first time you see them is in the margin at the beginning of each section; the second time is in the subheadings of each section; and the third time is in the exercise set for the section. You will also find them referred to in the skill maintenance exercises in each exercise set, in the mid-chapter review, and in the review exercises at the end of the chapter, as well as in the answers to the chapter tests and the cumulative reviews. These objective symbols allow you to refer to the appropriate place in the text whenever you need to review a topic.

- **Read and study each step of each example.** The examples include important side comments that explain each step. These carefully chosen examples and notes prepare you for success in the exercise set.

- **Stop and do the margin exercises as you study a section.** Doing the margin exercises is one of the most effective ways to enhance your ability to learn mathematics from this text. Don't deprive yourself of its benefits!

- **Note the icons listed at the top of each exercise set.** These refer to the many distinctive multimedia study aids that accompany the book.

- **Odd-numbered exercises.** Usually an instructor assigns some odd-numbered exercises. When you complete these, you can check your answers at the back of the book. If you miss any, check your work in the *Student's Solutions Manual* or ask your instructor for guidance.

- **Even-numbered exercises.** Whether or not your instructor assigns the even-numbered exercises, always do some on your own. Remember, there are no answers given for the class tests, so you need to practice doing exercises without answers. Check your answers later with a friend or your instructor.

Answer

6. Length: 160 ft; width: 100 ft

8.2

Exercise Set

For Extra Help

MyMathLab

Math XL
PRACTICE   WATCH   DOWNLOAD   READ   REVIEW

a   Solve each system of equations by the substitution method.

1. $y = 5 - 4x,$
     $2x - 3y = 13$

2. $x = 8 - 4y,$
     $3x + 5y = 3$

3. $2y + x = 9,$
     $x = 3y - 3$

4. $9x - 2y = 3,$
     $3x - 6 = y$

5. $3s - 4t = 14,$
     $5s + t = 8$

6. $m - 2n = 3,$
     $4m + n = 1$

7. $9x - 2y = -6,$
     $7x + 8 = y$

8. $t = 4 - 2s,$
     $t + 2s = 6$

9. $-5s + t = 11,$
     $4s + 12t = 4$

10. $5x + 6y = 14,$
     $-3y + x = 7$

11. $2x + 2y = 2,$
     $3x - y = 1$

12. $4p - 2q = 16,$
     $5p + 7q = 1$

13. $3a - b = 7,$
     $2a + 2b = 5$

14. $5x + 3y = 4,$
     $x - 4y = 3$

15. $2x - 3 = y,$
     $y - 2x = 1$

16. $4x + 13y = 5,$
     $-6x + y = 13$

b   Solve.

17. *Racquetball Court.*   A regulation racquetball court has a perimeter of 120 ft, with a length that is twice the width. Find the length and the width of such a court.

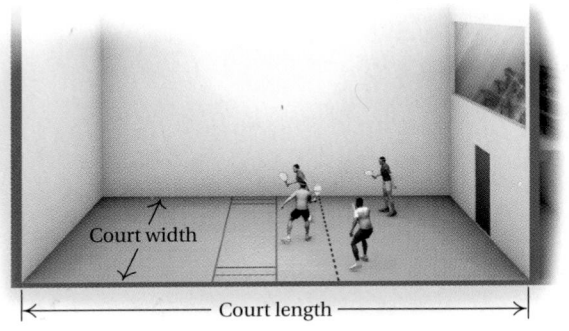

18. *Soccer Field.*   The perimeter of a soccer field is 340 m. The length exceeds the width by 50 m. Find the length and the width.

19. *Supplementary Angles.* **Supplementary angles** are angles whose sum is 180°. Two supplementary angles are such that one angle is 12° less than three times the other. Find the measures of the angles.

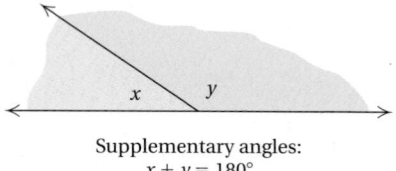

Supplementary angles:
$x + y = 180°$

20. *Complementary Angles.* **Complementary angles** are angles whose sum is 90°. Two complementary angles are such that one angle is 6° more than five times the other. Find the measures of the angles.

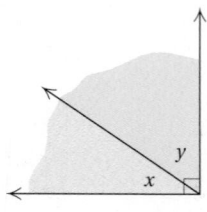

Complementary angles:
$x + y = 90°$

21. *Hockey Points.* At one time, hockey teams received two points when they won a game and one point when they tied. One season, a team won a championship with 60 points. They won 9 more games than they tied. How many wins and how many ties did the team have?

22. *Airplane Seating.* An airplane has a total of 152 seats. The number of coach-class seats is 5 more than six times the number of first-class seats. How many of each type of seat are there on the plane?

Skill Maintenance

23. Find the slope of the line $y = 1.3x - 7$.   [7.3b]

24. Simplify: $-9(y + 7) - 6(y - 4)$.   [1.8b]

25. Solve $A = \dfrac{pq}{7}$ for p.   [2.4b]

26. Find the slope of the line containing the points $(-2, 3)$ and $(-5, -4)$.   [7.3b]

Solve.   [2.3c]

27. $-4x + 5(x - 7) = 8x - 6(x + 2)$

28. $-12(2x - 3) = 16(4x - 5)$

Synthesis

29. Two solutions of $y = mx + b$ are $(1, 2)$ and $(-3, 4)$. Find m and b.

30. Solve for x and y in terms of a and b:
$$5x + 2y = a,$$
$$x - y = b.$$

31. *Design.* A piece of posterboard has a perimeter of 156 in. If you cut 6 in. off the width, the length becomes four times the width. What were the dimensions of the original piece of posterboard?

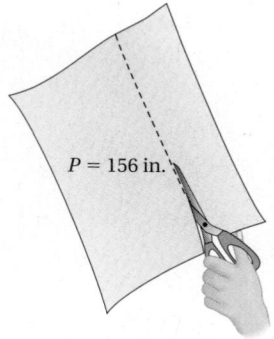

$P = 156$ in.

32. *Nontoxic Scouring Powder.* A nontoxic scouring powder is made up of 4 parts baking soda and 1 part vinegar. How much of each ingredient is needed for a 16-oz mixture?

8.3

Solving by Elimination

a) The Elimination Method

OBJECTIVES

a) Solve systems of equations in two variables by the elimination method.

b) Solve applied problems by solving systems of two equations using elimination.

The **elimination method** for solving systems of equations makes use of the *addition principle* for equations. Some systems are much easier to solve using the elimination method rather than the substitution method.

EXAMPLE 1 Solve this system:

$$2x - 3y = 0, \quad (1)$$
$$-4x + 3y = -1. \quad (2)$$

The key to the advantage of the elimination method in this case is the $-3y$ in one equation and the $3y$ in the other. These terms are opposites. If we add them, these terms will add to 0, and in effect, the variable y will have been "eliminated."

We will use the addition principle for equations, adding the same number on both sides of the equation. According to equation (2), $-4x + 3y$ and -1 are the same number. Thus we can use a vertical form and add $-4x + 3y$ to the left side of equation (1) and -1 to the right side:

$$
\begin{array}{ll}
2x - 3y = 0 & (1) \\
\underline{-4x + 3y = -1} & (2) \\
-2x + 0y = -1 & \text{Adding} \\
-2x + 0 = -1 & \\
-2x = -1.
\end{array}
$$

We have eliminated the variable y, which is why we call this the *elimination method*.* We now have an equation with just one variable, which we solve for x:

$$-2x = -1$$
$$x = \tfrac{1}{2}.$$

Next, we substitute $\tfrac{1}{2}$ for x in either equation and solve for y:

$$
\begin{array}{ll}
2 \cdot \tfrac{1}{2} - 3y = 0 & \text{Substituting in equation (1)} \\
1 - 3y = 0 & \\
-3y = -1 & \text{Subtracting 1} \\
y = \tfrac{1}{3}. & \text{Dividing by } -3
\end{array}
$$

We obtain the ordered pair $\left(\tfrac{1}{2}, \tfrac{1}{3}\right)$.

Check:

$$
\begin{array}{c|c}
2x - 3y = 0 \\
\hline
2\left(\tfrac{1}{2}\right) - 3\left(\tfrac{1}{3}\right) \;?\; 0 \\
1 - 1 \\
0 \;\bigm|\; \text{TRUE}
\end{array}
\qquad
\begin{array}{c|c}
-4x + 3y = -1 \\
\hline
-4\left(\tfrac{1}{2}\right) + 3\left(\tfrac{1}{3}\right) \;?\; -1 \\
-2 + 1 \\
-1 \;\bigm|\; \text{TRUE}
\end{array}
$$

SKILL TO REVIEW

Objective 2.3b: Solve equations by first clearing the equations of fractions or decimals.

Solve. Clear fractions or decimals first.

1. $4.2x - 10.4 = 45.4 - 5.1x$

2. $\tfrac{1}{4}x - \tfrac{2}{5} + \tfrac{1}{2}x = \tfrac{3}{5} + x$

---

*This method is also called the *addition method*.

Answers

Skill to Review:
1. 6　**2.** −4

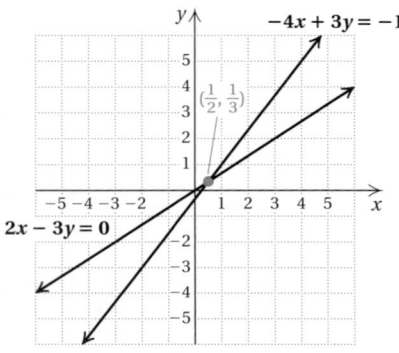

$-4x + 3y = -1$

$2x - 3y = 0$

Since $\left(\frac{1}{2}, \frac{1}{3}\right)$ checks, it is the solution. We can also see this in the graph shown at left.

Do Exercises 1 and 2.

In order to eliminate a variable, we sometimes use the multiplication principle to multiply one or both of the equations by a particular number before adding.

EXAMPLE 2 Solve this system:

$$3x + 3y = 15, \quad \textbf{(1)}$$
$$2x + 6y = 22. \quad \textbf{(2)}$$

If we add directly, we get $5x + 9y = 37$, and we have not eliminated a variable. However, note that if the $3y$ in equation (1) were $-6y$, we could eliminate y. Thus we multiply by -2 on both sides of equation (1) and add:

$$
\begin{array}{ll}
-6x - 6y = -30 & \text{Multiplying by } -2 \text{ on both sides of equation (1)} \\
\underline{2x + 6y = 22} & \text{Equation (2)} \\
-4x + 0 = -8 & \text{Adding} \\
-4x = -8 & \\
x = 2. & \text{Solving for } x
\end{array}
$$

Then

$$
\begin{array}{ll}
2 \cdot 2 + 6y = 22 & \text{Substituting 2 for } x \text{ in equation (2)} \\
\left.\begin{array}{l}
4 + 6y = 22 \\
6y = 18 \\
y = 3.
\end{array}\right\} & \text{Solving for } y
\end{array}
$$

We obtain $(2, 3)$, or $x = 2$, $y = 3$. This checks, so it is the solution. We can also see this in the graph at left.

Do Exercise 3.

Sometimes we must multiply twice in order to make two terms opposites.

EXAMPLE 3 Solve this system:

$$2x + 3y = 17, \quad \textbf{(1)}$$
$$5x + 7y = 29. \quad \textbf{(2)}$$

We must first multiply in order to make one pair of terms with the same variable opposites. We decide to do this with the x-terms in each equation. We multiply equation (1) by 5 and equation (2) by -2. Then we get $10x$ and $-10x$, which are opposites.

$$
\begin{array}{lll}
\textit{From equation (1):} & 10x + 15y = 85 & \text{Multiplying by 5} \\
\textit{From equation (2):} & \underline{-10x - 14y = -58} & \text{Multiplying by } -2 \\
& 0 + y = 27 & \text{Adding} \\
& y = 27 & \text{Solving for } y
\end{array}
$$

Solve by the elimination method.

1. $5x + 3y = 17,$
    $-5x + 2y = 3$

2. $-3a + 2b = 0,$
    $3a - 4b = -1$

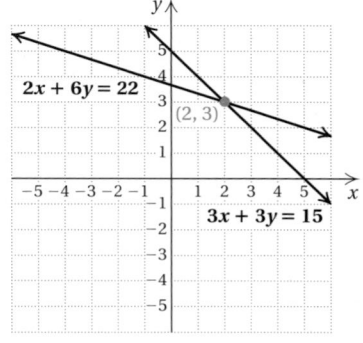

$2x + 6y = 22$

$(2, 3)$

$3x + 3y = 15$

3. Solve by the elimination method:

    $2y + 3x = 12,$
    $-4y + 5x = -2.$

Answers

1. $(1, 4)$ 2. $\left(\frac{1}{3}, \frac{1}{2}\right)$ 3. $(2, 3)$

Then

$$2x + 3 \cdot 27 = 17 \qquad \text{Substituting 27 for } y \text{ in equation (1)}$$

$$\left.\begin{array}{r} 2x + 81 = 17 \\ 2x = -64 \\ x = -32. \end{array}\right\} \quad \text{Solving for } x$$

We check the ordered pair $(-32, 27)$.

Check:

$$\begin{array}{c} 2x + 3y = 17 \\ \hline 2(-32) + 3(27) \; \overset{?}{\vert} \; 17 \\ -64 + 81 \; \Big\vert \\ 17 \; \Big\vert \quad \text{TRUE} \end{array} \qquad \begin{array}{c} 5x + 7y = 29 \\ \hline 5(-32) + 7(27) \; \overset{?}{\vert} \; 29 \\ -160 + 189 \; \Big\vert \\ 29 \; \Big\vert \quad \text{TRUE} \end{array}$$

We obtain $(-32, 27)$, or $x = -32$, $y = 27$, as the solution.

> Do Exercises 4 and 5.

Solve by the elimination method.

4. $4x + 5y = -8,$
$7x + 9y = 11$

5. $4x - 5y = 38,$
$7x - 8y = -22$

Some systems have no solution, as we saw graphically in Section 8.1 and algebraically in Example 3 of Section 8.2. How do we recognize such systems if we are solving using elimination?

EXAMPLE 4 Solve this system:

$$y + 3x = 5, \qquad \textbf{(1)}$$
$$y + 3x = -2. \qquad \textbf{(2)}$$

If we find the slope–intercept equations for this system, we get

$$y = -3x + 5,$$
$$y = -3x - 2.$$

The graphs are parallel lines.
The system has no solution.

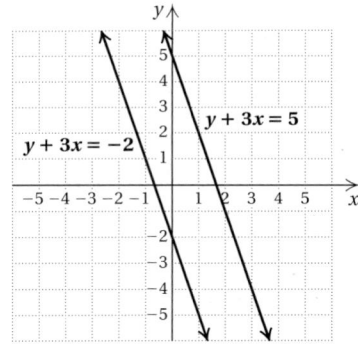

Let's see what happens if we attempt to solve the system by the elimination method. We multiply by -1 on both sides of equation (2) and add:

$$\begin{array}{ll} y + 3x = 5 & \text{Equation (1)} \\ \underline{-y - 3x = 2} & \text{Multiplying equation (2) by } -1 \\ \quad\quad 0 = 7. & \text{Adding, we obtain a false equation.} \end{array}$$

The x-terms and the y-terms are eliminated and we have a *false* equation. Thus, if we obtain a false equation, such as $0 = 7$, when solving algebraically, we know that the system has **no solution**. The system is inconsistent, and the equations are independent.

> Do Exercise 6.

6. Solve by the elimination method:

$$y + 2x = 3,$$
$$y + 2x = -1.$$

Answers

4. $(-127, 100)$ **5.** $(-138, -118)$
6. No solution

Some systems have infinitely many solutions. How can we recognize such a situation when we are solving systems using an algebraic method?

EXAMPLE 5 Solve this system:

$$3y - 2x = 6, \qquad (1)$$
$$-12y + 8x = -24. \qquad (2)$$

We see from the figure at left that the graphs are the same line. The system has an infinite number of solutions.

Suppose we try to solve this system by the elimination method:

$12y - 8x = 24$	Multiplying equation (1) by 4
$-12y + 8x = -24$	Equation (2)
$0 = 0.$	Adding, we obtain a true equation.

We have eliminated both variables, and what remains is a true equation, $0 = 0$. It can be expressed as $0 \cdot x + 0 \cdot y = 0$, and is true for all numbers x and y. If an ordered pair is a solution of one of the original equations, then it will be a solution of the other. The system has an **infinite number of solutions**. The system is consistent, and the equations are dependent.

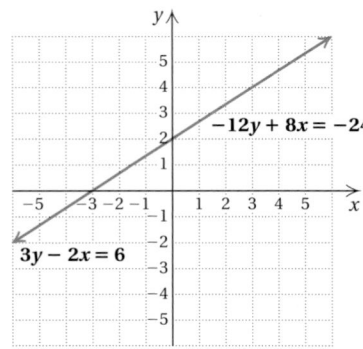

$-12y + 8x = -24$

$3y - 2x = 6$

SPECIAL CASES

When solving a system of two linear equations in two variables:

1. If a false equation is obtained, such as $0 = 7$, then the system has no solution. The system is *inconsistent*, and the equations are *independent*.

2. If a true equation is obtained, such as $0 = 0$, then the system has an infinite number of solutions. The system is *consistent*, and the equations are *dependent*.

---

7. Solve by the elimination method:

$$2x - 5y = 10,$$
$$-6x + 15y = -30.$$

Do Exercise 7.

When solving a system of equations using the elimination method, it helps to first write the equations in the form $Ax + By = C$. When decimals or fractions occur, it also helps to *clear* before solving.

EXAMPLE 6 Solve this system:

$$0.2x + 0.3y = 1.7,$$
$$\tfrac{1}{7}x + \tfrac{1}{5}y = \tfrac{29}{35}.$$

We have

$$0.2x + 0.3y = 1.7, \xrightarrow[\text{to clear decimals}]{\text{Multiplying by 10}} 2x + 3y = 17,$$
$$\tfrac{1}{7}x + \tfrac{1}{5}y = \tfrac{29}{35} \xrightarrow[\text{to clear fractions}]{\text{Multiplying by 35}} 5x + 7y = 29.$$

We multiplied by 10 to clear the decimals. Multiplication by 35, the least common denominator, clears the fractions. The problem is now identical to Example 3. The solution is $(-32, 27)$, or $x = -32$, $y = 27$.

Do Exercises 8 and 9.

8. Clear the decimals. Then solve.

$$0.02x + 0.03y = 0.01,$$
$$0.3x - 0.1y = 0.7$$

(*Hint*: Multiply the first equation by 100 and the second one by 10.)

9. Clear the fractions. Then solve.

$$\frac{3}{5}x + \frac{2}{3}y = \frac{1}{3},$$
$$\frac{3}{4}x - \frac{1}{3}y = \frac{1}{4}$$

Answers

7. Infinitely many solutions
8. $2x + 3y = 1,$
$\quad 3x - y = 7; (2, -1)$
9. $9x + 10y = 5,$
$\quad 9x - 4y = 3; \left(\dfrac{25}{63}, \dfrac{1}{7}\right)$

To use the elimination method to solve systems of two equations:

1. Write both equations in the form $Ax + By = C$.
2. Clear any decimals or fractions.
3. Choose a variable to eliminate.
4. Make the chosen variable's terms opposites by multiplying one or both equations by appropriate numbers if necessary.
5. Eliminate a variable by adding the respective sides of the equations and then solve for the remaining variable.
6. Substitute in either of the original equations to find the value of the other variable.

Comparing Methods

When deciding which method to use, consider this table and directions from your instructor. The situation is analogous to having a piece of wood to cut and three different types of saws available. Although all three saws can cut the wood, the "best" choice depends on the particular piece of wood, the type of cut being made, and your level of skill with each saw.

METHOD	STRENGTHS	WEAKNESSES
Graphical	Can "see" solutions.	Inexact when solutions involve numbers that are not integers. Solutions may not appear on the part of the graph drawn.
Substitution	Yields exact solutions. Convenient to use when a variable has a coefficient of 1.	Can introduce extensive computations with fractions. Cannot "see" solutions quickly.
Elimination	Yields exact solutions. Convenient to use when no variable has a coefficient of 1. The preferred method for systems of 3 or more equations in 3 or more variables. (See Section 8.5.)	Cannot "see" solutions quickly.

b) Solving Applied Problems Using Elimination

Let's now solve an applied problem using the elimination method. (We will solve many more problems in Section 8.4, which is devoted entirely to applied problems.)

EXAMPLE 7 *Stimulating the Hometown Economy.* To stimulate the economy in his town of Brewton, Alabama, in 2009, Danny Cottrell, co-owner of The Medical Center Pharmacy, gave each of his full-time employees $700 and each part-time employee $300. He asked that each person donate 15% to a charity of his or her choice and spend the rest locally. The money was paid in $2 bills, a rarely used currency, so that the business community could easily see how the money circulated. Cottrell gave away a total of $16,000 to his 24 employees. How many full-time employees and how many part-time employees were there?

Source: *The Press-Register*, March 4, 2009

1. **Familiarize.** We let f = the number of full-time employees and p = the number of part-time employees. Each full-time employee received $700, so a total of $700f$ was paid to them. Similarly, the part-time employees received a total of $300p$. Thus a total of $700f + 300p$ was given away.

2. **Translate.** We translate to two equations.

$$\underbrace{\text{Total amount given away}} \quad \text{is} \quad \$16{,}000.$$
$$700f + 300p \quad = \quad 16{,}000$$

$$\underbrace{\text{Total number of employees}} \quad \text{is} \quad 24.$$
$$f + p \quad = \quad 24$$

We now have a system of equations:

$$700f + 300p = 16{,}000, \quad \textbf{(1)}$$
$$f + p = 24. \quad \textbf{(2)}$$

3. **Solve.** First, we multiply by -300 on both sides of equation (2) and add:

$700f + 300p = 16{,}000$	Equation (1)
$\underline{-300f - 300p = -7200}$	Multiplying by -300 on both sides of equation (2)
$400f = 8800$	Adding
$f = 22.$	Solving for f

Next, we substitute 22 for f in equation (2) and solve for p:

$$22 + p = 24$$
$$p = 2.$$

4. **Check.** If there are 22 full-time employees and 2 part-time employees, there is a total of $22 + 2$, or 24, employees. The 22 full-time employees received a total of $\$700 \cdot 22$, or $15{,}400, and the 2 part-time employees received a total of $\$300 \cdot 2$, or $600. Then a total of $15{,}400 + $600, or $16{,}000, was given away. The numbers check in the original problem.

5. **State.** There were 22 full-time employees and 2 part-time employees.

10. **Bonuses.** Monica gave each of the full-time employees in her small business a year-end bonus of $500 while each part-time employee received $250. She gave a total of $4000 in bonuses to her 10 employees. How many full-time employees and how many part-time employees did Monica have?

Answer

10. Full-time: 6; part-time: 4

Do Exercise 10.

a Solve each system of equations by the elimination method.

1. $x + 3y = 7,$
$-x + 4y = 7$

2. $x + y = 9,$
$2x - y = -3$

3. $9x + 5y = 6,$
$2x - 5y = -17$

4. $2x - 3y = 18,$
$2x + 3y = -6$

5. $5x + 3y = -11,$
$3x - y = -1$

6. $2x + 3y = -9,$
$5x - 6y = -9$

7. $5r - 3s = 19,$
$2r - 6s = -2$

8. $2a + 3b = 11,$
$4a - 5b = -11$

9. $2x + 3y = 1,$
$4x + 6y = 2$

10. $3x - 2y = 1,$
$-6x + 4y = -2$

11. $5x - 9y = 7,$
$7y - 3x = -5$

12. $5x + 4y = 2,$
$2x - 8y = 4$

13. $3x + 2y = 24,$
$2x + 3y = 26$

14. $5x + 3y = 25,$
$3x + 4y = 26$

15. $2x - 4y = 5,$
$2x - 4y = 6$

16. $3x - 5y = -2,$
$5y - 3x = 7$

17. $2a + b = 12,$
$a + 2b = -6$

18. $10x + y = 306,$
$10y + x = 90$

19. $\frac{1}{3}x + \frac{1}{5}y = 7,$
$\frac{1}{6}x - \frac{2}{5}y = -4$

20. $\frac{2}{3}x + \frac{1}{7}y = -11,$
$\frac{1}{7}x - \frac{1}{3}y = -10$

21. $\frac{1}{5}x + \frac{1}{2}y = 6,$
$\frac{2}{5}x - \frac{3}{2}y = -8$

22. $\frac{2}{3}x + \frac{3}{5}y = -17,$
$\frac{1}{2}x - \frac{1}{3}y = -1$

23. $\frac{1}{2}x - \frac{1}{3}y = -4,$
$\frac{1}{4}x + \frac{5}{6}y = 4$

24. $\frac{4}{3}x + \frac{3}{2}y = 4,$
$\frac{5}{6}x - \frac{1}{8}y = -6$

25. $0.3x - 0.2y = 4,$
$0.2x + 0.3y = 0.5$

26. $0.7x - 0.3y = 0.5,$
$-0.4x + 0.7y = 1.3$

27. $0.05x + 0.25y = 22,$
$0.15x + 0.05y = 24$

28. $1.3x - 0.2y = 12,$
$0.4x + 17y = 89$

 Solve. Use the elimination method when solving the translated system.

29. *Finding Numbers.* The sum of two numbers is 63. The larger number minus the smaller number is 9. Find the numbers.

30. *Finding Numbers.* The sum of two numbers is 2. The larger number minus the smaller number is 20. Find the numbers.

31. *Finding Numbers.* The sum of two numbers is 3. Three times the larger number plus two times the smaller number is 24. Find the numbers.

32. *Finding Numbers.* The sum of two numbers is 9. Two times the larger number plus three times the smaller number is 2. Find the numbers.

33. *Complementary Angles.* Two angles are complementary. (**Complementary angles** are angles whose sum is 90°.) Their difference is 6°. Find the angles.

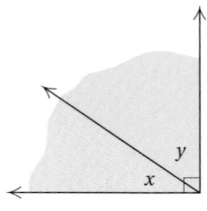

Complementary angles:
$x + y = 90°$

34. *Supplementary Angles.* Two angles are supplementary. (**Supplementary angles** are angles whose sum is 180°.) Their difference is 22°. Find the angles.

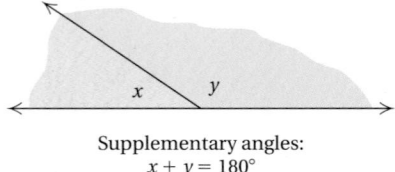

Supplementary angles:
$x + y = 180°$

35. *Basketball Scoring.* Jared's Youth League basketball team scored on 27 shots, some two-point field goals and the rest one-point free throws. The team scored a total of 48 points in the game. How many of each kind of shot was made?

36. *Hockey Scoring.* At one time, hockey teams received two points when they won a game and one point when they tied. One season, a team won a championship with 65 points. They played 35 games. How many wins and how many ties did the team have?

37. *Sales Promotion.* Rick's Sporting Goods ran a promotion offering either a free rechargeable lantern or a free portable propane grill to each customer who bought a deluxe family tent. The store's cost for each lantern was $20, and its cost for each grill was $25. At the end of the promotion, 12 tents had been sold. The store's total cost for the items given away was $280. How many of each type of free item did the customers choose?

38. *Sales Promotion.* The Serenity Yoga Center offered patrons who bought a 24-class pass either a free eye pillow or a free yoga DVD. The center's cost for each eye pillow was $10, and its cost for each DVD was $8. A total of 15 people took advantage of the offer. The center's total cost for the promotional items was $136. How many of each item did the patrons choose?

Skill Maintenance

Given the function $f(x) = 3x^2 - x + 1$, find each of the following function values. [7.1b]

39. $f(0)$

40. $f(-1)$

41. $f(1)$

42. $f(10)$

43. $f(-2)$

44. $f(2a)$

45. $f(-4)$

46. $f(1.8)$

47. Find the domain of the function

$$f(x) = \frac{x - 5}{x + 7}. \quad [7.2a]$$

48. Find the domain and the range of the function

$$g(x) = 5 - x^2. \quad [7.2a]$$

49. Find an equation of the line with slope $-\frac{3}{5}$ and y-intercept $(0, -7)$. [7.5a]

50. Find an equation of the line containing the points $(-10, 2)$ and $(-2, 10)$. [7.5c]

Synthesis

51. Use the INTERSECT feature to solve the following system of equations. You may need to first solve for y. Round answers to the nearest hundredth.

$$3.5x - 2.1y = 106.2,$$
$$4.1x + 16.7y = -106.28$$

52. Solve:

$$\frac{x + y}{2} - \frac{x - y}{5} = 1,$$
$$\frac{x - y}{2} + \frac{x + y}{6} = -2.$$

53. The solution of this system is $(-5, -1)$. Find A and B.

$$Ax - 7y = -3,$$
$$x - By = -1$$

54. Find an equation to pair with $6x + 7y = -4$ such that $(-3, 2)$ is a solution of the system.

55. The points $(0, -3)$ and $\left(-\frac{3}{2}, 6\right)$ are two of the solutions of the equation $px - qy = -1$. Find p and q.

56. Determine a and b for which $(-4, -3)$ will be a solution of the system

$$ax + by = -26,$$
$$bx - ay = 7.$$

8.4

Solving Applied Problems: Two Equations

OBJECTIVES

a) Solve applied problems involving total value and mixture using systems of two equations.

b) Solve applied problems involving motion using systems of two equations.

a Total-Value Problems and Mixture Problems

Systems of equations can be a useful tool in solving applied problems. Using systems often makes the *Translate* step easier than using a single equation. The first kind of problem we consider involves quantities of items purchased and the total value, or cost, of the items. We refer to this type of problem as a **total-value problem**.

EXAMPLE 1 *School Lunches.* To serve lunch to students, school cafeterias receive up to $2.47 per lunch from the U.S. government. After expenses such as labor, transportation, utilities, and equipment, schools are left with a little more than $1 to spend on food. Of this amount, about 25 cents is spent for a carton of milk, another 25 cents for fruit and/or vegetables, and the remaining 50 cents for a main dish. In buying food for a week's lunches, one school purchased 580 servings of two menu items: the ingredients for turkey/cheese wraps at $0.56 per serving and mixed vegetables at $0.22 per serving. The total cost of these two menu items was $246.60. How many servings of each type of item were purchased?

Source: *USA Today*, May 1, 2008

1. **Familiarize.** Let's begin by making a guess that the ingredients for 300 servings of turkey/cheese wraps were purchased along with 280 servings of mixed vegetables. This is a total of 580 servings. Now let's find the total cost of this order. Since the turkey/cheese wraps cost $0.56 per serving and the vegetables cost $0.22 per serving, the total cost would be

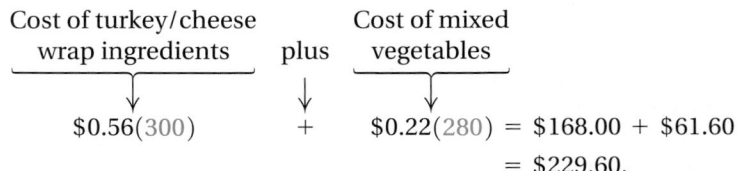

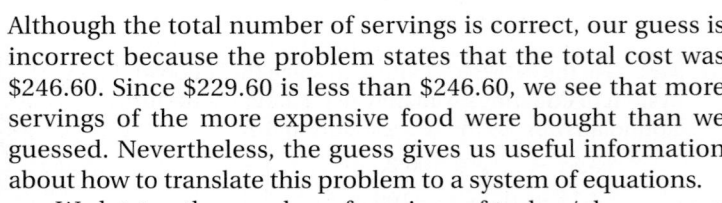

Although the total number of servings is correct, our guess is incorrect because the problem states that the total cost was $246.60. Since $229.60 is less than $246.60, we see that more servings of the more expensive food were bought than we guessed. Nevertheless, the guess gives us useful information about how to translate this problem to a system of equations.

We let $t =$ the number of servings of turkey/cheese wrap ingredients and $v =$ the number of servings of mixed vegetables that were purchased. The ingredients for each serving of the turkey/cheese wraps cost $0.56, so the cost of t servings is $0.56t$. Similarly, the cost for each serving of mixed vegetables is $0.22, so the cost of v servings of mixed vegetables is $0.22v$.

It is helpful to organize the information we have in a table, as follows.

	TURKEY/CHEESE WRAPS	MIXED VEGETABLES	TOTAL	
NUMBER OF SERVINGS	t	v	580	→ $t + v = 580$
COST PER SERVING	$0.56	$0.22		
TOTAL COST	$0.56t	$0.22v	$246.60	→ $0.56t + 0.22v = 246.60$

2. Translate. The first row of the table gives us one equation:

$t + v = 580.$

The last row of the table gives us a second equation:

$0.56t + 0.22v = 246.60.$

We can multiply by 100 on both sides of the second equation to clear the decimals. This gives us the following system of equations:

$t + v = 580,$ $\qquad$ **(1)**

$56t + 22v = 24{,}660.$ $\qquad$ **(2)**

3. Solve. We use the elimination method to solve the system of equations. We eliminate v by multiplying by -22 on both sides of equation (1) and then adding the result to equation (2):

$$\begin{array}{ll} -22t - 22v = -12{,}760 & \text{Multiplying equation (1) by } -22 \\ \underline{56t + 22v = 24{,}660} & \text{Equation (2)} \\ 34t \qquad\quad = 11{,}900 & \text{Adding} \\ \qquad\quad t = 350. & \text{Dividing by 34} \end{array}$$

Next, we substitute 350 for t in equation (1) and solve for v:

$$\begin{array}{ll} t + v = 580 & \text{Equation (1)} \\ 350 + v = 580 & \text{Substituting 350 for } t \\ v = 230. & \text{Solving for } v \end{array}$$

We obtain $(350, 230)$, or $t = 350$, $v = 230$.

4. Check. We check in the original problem.

$$\begin{array}{ll} \textit{Total number of servings}: & t + v = 350 + 230 = 580 \\ \textit{Cost of turkey/cheese wraps}: & \$0.56t = \$0.56(350) = \$196.00 \\ \textit{Cost of mixed vegetables}: & \$0.22v = \$0.22(230) = \underline{\$\ \ 50.60} \\ & \qquad\qquad\qquad\quad \text{Total} = \$246.60 \end{array}$$

The numbers check.

5. State. The school bought the ingredients for 350 servings of turkey/cheese wraps and 230 servings of mixed vegetables.

Do Exercise 1.

1. Retail Sales of Sweatshirts.
A campus bookstore sells college sweatshirts. White sweatshirts sell for $18.95 each and red ones sell for $19.50 each. If receipts for the sale of 30 sweatshirts total $572.90, how many of each color did the shop sell?

Complete the following table, letting w = the number of white sweatshirts and r = the number of red sweatshirts.

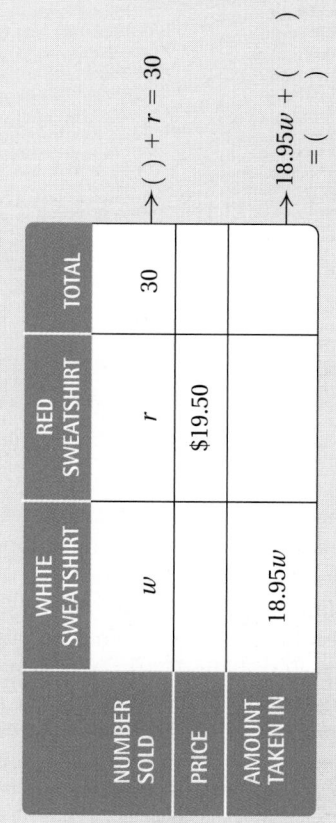

Answer

1. White: 22; red: 8

White	Red	Total	
w	r	30	→ $w + r = 30$
$18.95	$19.50		
$18.95w$	$19.50r$	572.90	→ $18.95w + 19.50r = 572.90$

The following problem, similar to Example 1, is called a **mixture problem**.

EXAMPLE 2 *Blending Flower Seeds.* Tara's Web site, Garden Edibles, specializes in the sale of herbs and flowers for colorful meals and garnishes. Tara sells packets of nasturtium seeds for $0.95 each and packets of Johnny-jump-up seeds for $1.43 each. She decides to offer a 16-packet spring-garden mixture, combining packets of both types of seeds at $1.10 per packet. How many packets of each type of seed should be put in her garden mix?

1. **Familiarize.** To familiarize ourselves with the problem situation, we make a guess and do some calculations. The total number of packets of seeds is 16. Let's try 12 packets of nasturtiums and 4 packets of Johnny-jump-ups.

   The sum of the number of packets is $12 + 4$, or 16.

   The value of these seed packets is found by multiplying the cost per packet by the number of packets and adding:

   $0.95(12) + $1.43(4), or $17.12.

   The desired cost is $1.10 per packet. If we multiply $1.10 by 16, we get $16($1.10)$, or $17.60. This shows us that the guess is incorrect, but these calculations give us a basis for understanding how to translate.

   We let a = the number of packets of nasturtium seeds and b = the number of packets of Johnny-jump-up seeds. Next, we organize the information in a table, as follows.

	NASTURTIUM	JOHNNY-JUMP-UP	SPRING	
NUMBER OF PACKETS	a	b	16	$\rightarrow a + b = 16$
PRICE PER PACKET	$0.95	$1.43	$1.10	
VALUE OF PACKETS	$0.95a$	$1.43b$	$16 \cdot 1.10$, or 17.60	$\rightarrow 0.95a + 1.43b = 17.60$

2. Translate. The total number of packets is 16, so we have one equation:

$a + b = 16$.

The value of the nasturtium seeds is $0.95a$ and the value of the Johnny-jump-up seeds is $1.43b$. These amounts are in dollars. Since the total value is to be 16($1.10), or $17.60, we have

$0.95a + 1.43b = 17.60$.

We can multiply by 100 on both sides of this equation in order to clear the decimals. Thus we have translated to a system of equations:

$a + b = 16,$ **(1)**
$95a + 143b = 1760.$ **(2)**

3. Solve. We decide to use substitution, although elimination could be used as we did in Example 1. When equation (1) is solved for b, we get $b = 16 - a$. Substituting $16 - a$ for b in equation (2) and solving gives us

$95a + 143(16 - a) = 1760$	Substituting
$95a + 2288 - 143a = 1760$	Using the distributive law
$-48a = -528$	Subtracting 2288 and collecting like terms
$a = 11.$	

We have $a = 11$. Substituting this value in the equation $b = 16 - a$, we obtain $b = 16 - 11$, or 5.

4. Check. We check in a manner similar to our guess in the *Familiarize* step. The total number of packets is $11 + 5$, or 16. The value of the packet mixture is

$0.95(11) + $1.43(5)$, or $17.60.

Thus the numbers of packets check.

5. State. The spring-garden mixture can be made by combining 11 packets of nasturtium seeds with 5 packets of Johnny-jump-up seeds.

> Do Exercise 2.

2. Blending Coffees. The Coffee Counter charges $9.00 per pound for Kenyan French Roast coffee and $8.00 per pound for Sumatran coffee. How much of each type should be used to make a 20-lb blend that sells for $8.40 per pound?

EXAMPLE 3 *Student Loans.* Jed's student loans totaled $16,200. Part was a Perkins loan made at 5% interest and the rest was a Stafford loan made at 4% interest. After one year, Jed's loans accumulated $715 in interest. What was the amount of each loan?

1. Familiarize. Listing the given information in a table will help. The columns in the table come from the formula for simple interest: $I = Prt$. We let $x =$ the number of dollars in the Perkins loan and $y =$ the number of dollars in the Stafford loan.

	PERKINS LOAN	STAFFORD LOAN	TOTAL	
PRINCIPAL	x	y	$16,200	→ $x + y = 16{,}200$
RATE OF INTEREST	5%	4%		
TIME	1 year	1 year		
INTEREST	$0.05x$	$0.04y$	$715	→ $0.05x + 0.04y = 715$

Answer

2. Kenyan: 8 lb; Sumatran: 12 lb

3. Client Investments. Kaufman Financial Corporation makes investments for corporate clients. It makes an investment of $3700 for one year at simple interest, yielding $297. Part of the money is invested at 7% and the rest at 9%. How much was invested at each rate?

Do the *Familiarize* and *Translate* steps by completing the following table. Let x = the number of dollars invested at 7% and y = the number of dollars invested at 9%.

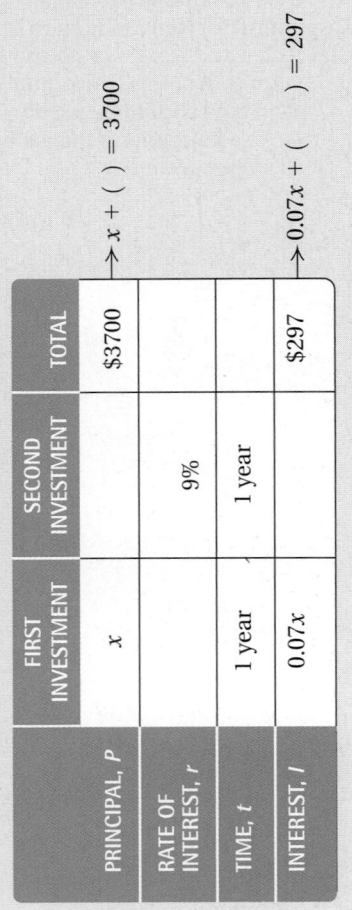

	FIRST INVESTMENT	SECOND INVESTMENT	TOTAL	
PRINCIPAL, P	x		$3700	$\rightarrow x + (\ ) = 3700$
RATE OF INTEREST, r		9%		
TIME, t	1 year	1 year		
INTEREST, I	$0.07x$		$297	$\rightarrow 0.07x + (\quad) = 297$

Answer

3. $1800 at 7%; $1900 at 9%

First Investment	Second Investment	Total	
x	y	$3700	$\rightarrow x + y = 3700$
7%	9%		
1 year	1 year		
$0.07x$	$0.09y$	$297	$\rightarrow 0.07x + 0.09y = 297$

2. Translate. The total of the amounts of the loans is found in the first row of the table. This gives us one equation:

$$x + y = 16{,}200.$$

Look at the last row of the table. The interest totals $715. This gives us a second equation:

$$5\%x + 4\%y = 715, \quad \text{or} \quad 0.05x + 0.04y = 715.$$

After we multiply on both sides to clear the decimals, we have

$$5x + 4y = 71{,}500.$$

3. Solve. Using either elimination or substitution, we solve the resulting system:

$$x + y = 16{,}200,$$
$$5x + 4y = 71{,}500.$$

We find that $x = 6700$ and $y = 9500$.

4. Check. The sum is $6700 + $9500, or $16,200. The interest from $6700 at 5% for one year is 5%($6700), or $335. The interest from $9500 at 4% for one year is 4%($9500), or $380. The total interest is $335 + $380, or $715. The numbers check in the problem.

5. State. The Perkins loan was for $6700 and the Stafford loan was for $9500.

Do Exercise 3.

EXAMPLE 4 *Mixing Fertilizers.* Yardbird Gardening carries two kinds of fertilizer containing nitrogen and water. "Gently Green" is 5% nitrogen and "Sun Saver" is 15% nitrogen. Yardbird Gardening needs to combine the two types of solution to make 90 L of a solution that is 12% nitrogen. How much of each brand should be used?

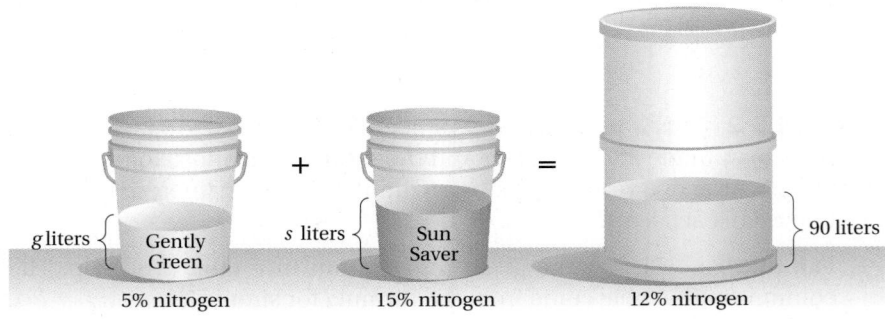

1. Familiarize. We first make a drawing and a guess to become familiar with the problem.

We choose two numbers that total 90 L—say, 40 L of Gently Green and 50 L of Sun Saver—for the amounts of each fertilizer. Will the resulting mixture have the correct percentage of nitrogen?

To find out, we multiply as follows:

$5\%(40\,\text{L}) = 2\,\text{L of nitrogen}$ and $15\%(50\,\text{L}) = 7.5\,\text{L of nitrogen}.$

Thus the total amount of nitrogen in the mixture is 2 L + 7.5 L, or 9.5 L. The final mixture of 90 L is supposed to be 12% nitrogen. Now

$12\%(90\,\text{L}) = 10.8\,\text{L}.$

Since 9.5 L and 10.8 L are not the same, our guess is incorrect. But these calculations help us to become familiar with the problem and to make the translation.

We let $g = $ the number of liters of Gently Green and $s = $ the number of liters of Sun Saver in the mixture.

The information can be organized in a table, as follows.

	GENTLY GREEN	SUN SAVER	MIXTURE	
NUMBER OF LITERS	g	s	90	$\rightarrow g + s = 90$
PERCENT OF NITROGEN	5%	15%	12%	
AMOUNT OF NITROGEN	$0.05g$	$0.15s$	$0.12 \times 90,$ or 10.8 liters	$\rightarrow 0.05g + 0.15s = 10.8$

2. **Translate.**   If we add g and s in the first row, we get 90, and this gives us one equation:

$g + s = 90.$

If we add the amounts of nitrogen listed in the third row, we get 10.8, and this gives us another equation:

$5\%g + 15\%s = 10.8,$ or $0.05g + 0.15s = 10.8.$

After clearing the decimals, we have the following system:

$$g + s = 90, \qquad \textbf{(1)}$$
$$5g + 15s = 1080. \qquad \textbf{(2)}$$

3. **Solve.**   We solve the system using elimination. We multiply equation (1) by -5 and add the result to equation (2):

$$
\begin{array}{ll}
-5g - 5s = -450 & \text{Multiplying equation (1) by } -5 \\
\underline{5g + 15s = 1080} & \text{Equation (2)} \\
10s = 630 & \text{Adding} \\
s = 63. & \text{Dividing by 10}
\end{array}
$$

Next, we substitute 63 for s in equation (1) and solve for g:

$$
\begin{array}{ll}
g + 63 = 90 & \text{Substituting in equation (1)} \\
g = 27. & \text{Solving for } g
\end{array}
$$

We obtain $(27, 63)$, or $g = 27, s = 63$.

4. Mixing Cleaning Solutions.
King's Service Station uses two kinds of cleaning solution containing acid and water. "Attack" is 2% acid and "Blast" is 6% acid. They want to mix the two to get 60 qt of a solution that is 5% acid. How many quarts of each should they use?

Do the *Familiarize* and *Translate* steps by completing the following table. Let a = the number of quarts of Attack and b = the number of quarts of Blast.

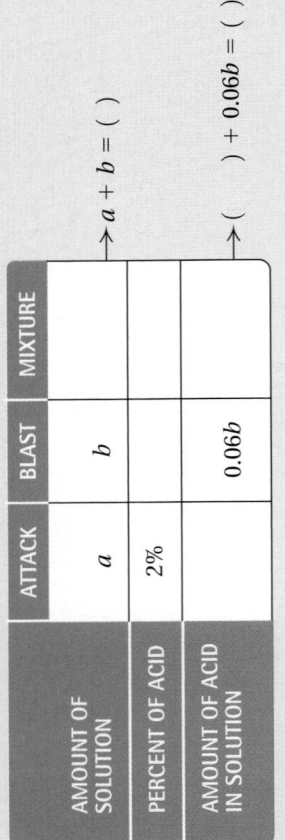

	ATTACK	BLAST	MIXTURE
AMOUNT OF SOLUTION	a	b	$\rightarrow a + b = (\ )$
PERCENT OF ACID	2%		
AMOUNT OF ACID IN SOLUTION		$0.06b$	$\rightarrow (\ ) + 0.06b = (\ )$

4. Check. Remember that g is the number of liters of Gently Green, with 5% nitrogen, and s is the number of liters of Sun Saver, with 15% nitrogen.

Total number of liters of mixture: $g + s = 27 + 63 = 90$ L

Amount of nitrogen: $5\%(27) + 15\%(63) = 1.35 + 9.45 = 10.8$ L

Percentage of nitrogen in mixture: $\dfrac{10.8}{90} = 0.12 = 12\%$

The numbers check in the original problem.

5. State. Yardbird Gardening should mix 27 L of Gently Green and 63 L of Sun Saver.

Do Exercise 4.

b Motion Problems

When a problem deals with speed, distance, and time, we can expect to use the following *motion formula*.

> **THE MOTION FORMULA**
>
> Distance = Rate (or speed) · Time
>
> $$d = rt$$

> **TIPS FOR SOLVING MOTION PROBLEMS**
>
> 1. Make a drawing using an arrow or arrows to represent distance and the direction of each object in motion.
> 2. Organize the information in a table or a chart.
> 3. Look for as many things as you can that are the same, so you can write equations.

Answer

4. Attack: 15 qt; Blast: 45 qt

Attack	Blast	Mixture	
a	b	60	$\rightarrow a + b = 60$
2%	6%	5%	
$0.02a$	$0.06b$	0.05×60, or 3	$\rightarrow 0.02a + 0.06b = 3$

EXAMPLE 5 *Auto Travel.* Your brother leaves your home on a trip, forgetting his suitcase. You know that he normally drives at a speed of 55 mph. You do not discover the suitcase until 1 hr after he has left. If you follow him at a speed of 65 mph, how long will it take you to catch up with him?

1. **Familiarize.** We first make a drawing. From the drawing, we see that when you catch up with your brother, the distances from home are the same. We let d = the distance, in miles. If we let t = the time, in hours, for you to catch your brother, then $t + 1$ = the time traveled by your brother at a slower speed.

We organize the information in a table as follows.

	DISTANCE	RATE	TIME	
BROTHER	d	55	$t + 1$	$\rightarrow d = 55(t + 1)$
YOU	d	65	t	$\rightarrow d = 65t$

$d = r \cdot t$

2. **Translate.** Using $d = rt$ in each row of the table, we get an equation. Thus we have a system of equations:

$$d = 55(t + 1), \quad \textbf{(1)}$$
$$d = 65t. \quad \textbf{(2)}$$

3. **Solve.** We solve the system using the substitution method:

$65t = 55(t + 1)$ Substituting $65t$ for d in equation (1)

$65t = 55t + 55$ Multiplying to remove parentheses on the right

$\left.\begin{array}{l} 10t = 55 \\ t = 5.5. \end{array}\right\}$ Solving for t

Your time is 5.5 hr, which means that your brother's time is $5.5 + 1$, or 6.5 hr.

4. **Check.** At 65 mph, you will travel $65 \cdot 5.5$, or 357.5 mi, in 5.5 hr. At 55 mph, your brother will travel $55 \cdot 6.5$, or the same 357.5 mi, in 6.5 hr. The numbers check.

5. **State.** You will catch up with your brother in 5.5 hr.

> Do Exercise 5.

---

5. **Train Travel.** A train leaves Barstow traveling east at 35 km/h. One hour later, a faster train leaves Barstow, also traveling east on a parallel track at 40 km/h. How far from Barstow will the faster train catch up with the slower one?

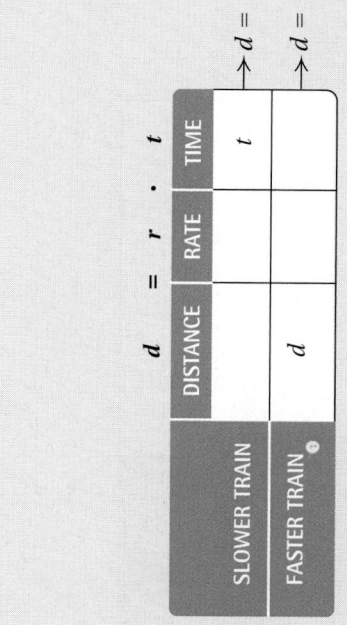

Answer

5. 280 km

Distance	Rate	Time	
d	35 km/h	t	$\rightarrow d = 35t$
d	40 km/h	$t - 1$	$\rightarrow d = 40(t - 1)$

EXAMPLE 6 *Marine Travel.* A Coast-Guard patrol boat travels 4 hr on a trip downstream with a 6-mph current. The return trip against the same current takes 5 hr. Find the speed of the boat in still water.

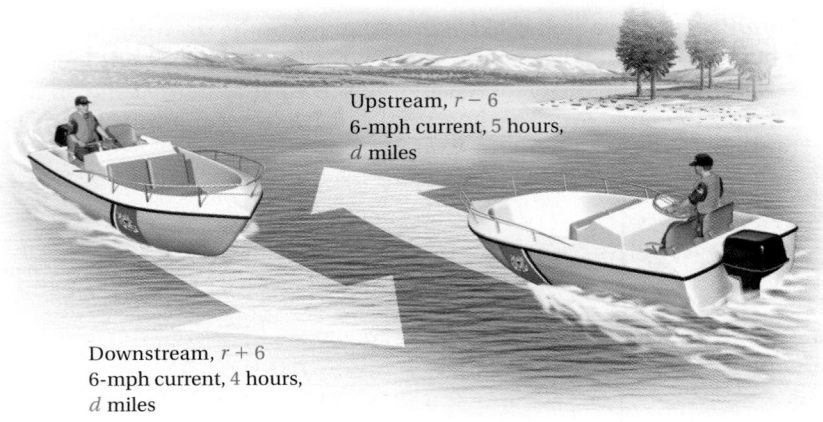

Upstream, $r - 6$
6-mph current, 5 hours,
d miles

Downstream, $r + 6$
6-mph current, 4 hours,
d miles

1. **Familiarize.** We first make a drawing. From the drawing, we see that the distances are the same. We let d = the distance, in miles, and r = the speed of the boat in still water, in miles per hour. Then, when the boat is traveling downstream, its speed is $r + 6$. (The current helps the boat along.) When it is traveling upstream, its speed is $r - 6$. (The current holds the boat back.) We can organize the information in a table. We use the formula $d = rt$.

	d	=	r	·	t
	DISTANCE		RATE		TIME
DOWNSTREAM	d		$r + 6$		4	→ $d = (r + 6)4$
UPSTREAM	d		$r - 6$		5	→ $d = (r - 6)5$

2. **Translate.** From each row of the table, we get an equation, $d = rt$:

$$d = 4r + 24, \quad \textbf{(1)}$$
$$d = 5r - 30. \quad \textbf{(2)}$$

3. **Solve.** We solve the system by the substitution method:

$4r + 24 = 5r - 30$ Substituting $4r + 24$ for d in equation (2)

$\left. \begin{array}{l} 24 = r - 30 \\ 54 = r. \end{array} \right\}$ Solving for r

4. **Check.** If $r = 54$, then $r + 6 = 60$; and $60 \cdot 4 = 240$, the distance traveled downstream. If $r = 54$, then $r - 6 = 48$; and $48 \cdot 5 = 240$, the distance traveled upstream. The distances are the same. In this type of problem, a problem-solving tip to keep in mind is "Have I found what the problem asked for?" We could solve for a certain variable but still have not answered the question of the original problem. For example, we might have found speed when the problem wanted distance. In this problem, we want the speed of the boat in still water, and that is r.

5. **State.** The speed in still water is 54 mph.

Do Exercise 6.

---

6. Air Travel. An airplane flew for 4 hr with a 20-mph tailwind. The return flight against the same wind took 5 hr. Find the speed of the plane in still air.

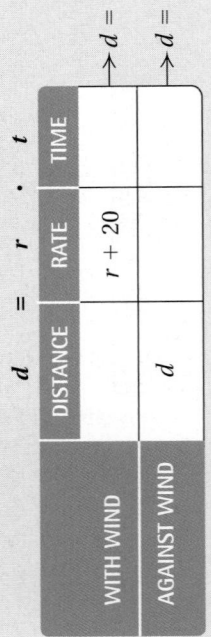

$d = r \cdot t$

	DISTANCE	RATE	TIME	
WITH WIND		$r + 20$		$d =$
AGAINST WIND	d			$d =$

Answer

6. 180 mph

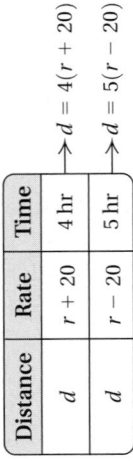

Distance	Rate	Time	
d	$r + 20$	4 hr	$d = 4(r + 20)$
d	$r - 20$	5 hr	$d = 5(r - 20)$

Translating for Success

1. *Office Expense.* The monthly telephone expense for an office is $1094 less than the janitorial expense. Three times the janitorial expense minus four times the telephone expense is $248. What is the total of the two expenses?

2. *Dimensions of a Triangle.* The sum of the base and the height of a triangle is 192 in. The height is twice the base. Find the base and the height.

3. *Supplementary Angles.* Two supplementary angles are such that twice one angle is 7° more than the other. Find the measures of the angles.

4. *SAT Scores.* The total of Megan's writing and math scores on the SAT was 1094. Her math score was 248 points higher than her writing score. What were her math and writing SAT scores?

5. *Sightseeing Boat.* A sightseeing boat travels 3 hr on a trip downstream with a 2.5-mph current. The return trip against the same current takes 3.5 hr. Find the speed of the boat in still water.

The goal of these matching questions is to practice step (2), *Translate*, of the five-step problem-solving process. Translate each word problem to a system of equations and select a correct translation from systems A–J.

A. $x = y + 248,$
$x + y = 1094$

B. $5x = 2y - 3,$
$y = \frac{2}{3}x + 5$

C. $y = \frac{1}{2}x,$
$2x + 2y = 192$

D. $2x = 7 + y,$
$x + y = 180$

E. $x + y = 192,$
$x = 2y$

F. $x + y = 180,$
$x = 2y + 7$

G. $x - 1094 = y,$
$3x - 4y = 248$

H. $3\%x + 2.5\%y = 97.50,$
$x + y = 2500$

I. $2x = 5 + \frac{2}{3}y,$
$3y = 15x - 4$

J. $x = (y + 2.5) \cdot 3,$
$3.5(y - 2.5) = x$

Answers on page A-24

---

6. *Running Distances.* Each day Tricia runs 5 mi more than two-thirds the distance that Chris runs. Five times the distance that Chris runs is 3 mi less than twice the distance that Tricia runs. How far does Tricia run daily?

7. *Dimensions of a Rectangle.* The perimeter of a rectangle is 192 in. The width is half the length. Find the length and the width.

8. *Mystery Numbers.* Teka asked her students to determine the two numbers that she placed in a sealed envelope. Twice the smaller number is 5 more than two-thirds the larger number. Three times the larger number is 4 less than fifteen times the smaller. Find the numbers.

9. *Supplementary Angles.* Two supplementary angles are such that one angle is 7° more than twice the other. Find the measures of the angles.

10. *Student Loans.* Brandt's student loans totaled $2500. Part was borrowed at 3% interest and the rest at 2.5%. After one year, Brandt had accumulated $97.50 in interest. What was the amount of each loan?

8.4 **Exercise Set**

For Extra Help

MyMathLab

Math XL
PRACTICE

WATCH

DOWNLOAD

READ

REVIEW

a Solve.

1. *Retail Sales.* Paint Town sold 45 paintbrushes, one kind at $8.50 each and another at $9.75 each. In all, $398.75 was taken in for the brushes. How many of each kind were sold?

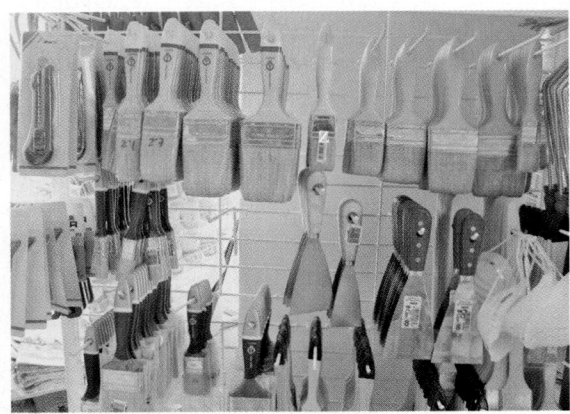

2. *Retail Sales.* Mountainside Fleece sold 40 neckwarmers. Solid-color neckwarmers sold for $9.90 each and print ones sold for $12.75 each. In all, $421.65 was taken in for the neckwarmers. How many of each type were sold?

Mountainside Fleece

3. *Sales of Pharmaceuticals.* In 2009, the Diabetic Express charged $39.95 for a vial of Humulin insulin and $30.49 for a vial of Novolin insulin. If a total of $1723.16 was collected for 50 vials of insulin, how many vials of each type were sold?

4. *Fundraising.* The St. Mark's Community Barbecue served 250 dinners. A child's plate cost $3.50 and an adult's plate cost $7.00. A total of $1347.50 was collected. How many of each type of plate was served?

5. *Radio Airplay.* Rudy must play 12 commercials during his 1-hr radio show. Each commercial is either 30 sec or 60 sec long. If the total commercial time during the hour is 10 min, how many commercials of each type does Rudy play?

6. *Nontoxic Floor Wax.* A nontoxic floor wax can be made by combining lemon juice and food-grade linseed oil. The amount of oil should be twice the amount of lemon juice. How much of each ingredient is needed in order to make 32 oz of floor wax? (The mix should be spread with a rag and buffed when dry.)

7. *Catering.* Stella's Catering is planning a wedding reception. The bride and groom would like to serve a nut mixture containing 25% peanuts. Stella has available mixtures that are either 40% or 10% peanuts. How much of each type should be mixed to get a 10-lb mixture that is 25% peanuts?

8. *Blending Granola.* Deep Thought Granola is 25% nuts and dried fruit. Oat Dream Granola is 10% nuts and dried fruit. How much of Deep Thought and how much of Oat Dream should be mixed to form a 20-lb batch of granola that is 19% nuts and dried fruit?

9. *Ink Remover.* Etch Clean Graphics uses one cleanser that is 25% acid and a second that is 50% acid. How many liters of each should be mixed to get 10 L of a solution that is 40% acid?

10. *Livestock Feed.* Soybean meal is 16% protein and corn meal is 9% protein. How many pounds of each should be mixed to get a 350-lb mixture that is 12% protein?

11. *Dry Cleaners.* Claudio, a banking vice-president, took 17 neckties to Milto Cleaners. The rate for non-silk ties is $3.25 per tie and for silk ties is $3.60 per tie. His total bill was $58.75. How many silk ties did he have dry-cleaned?

12. *Laundry.* While on a four-week hiking trip in the mountains, the Tryon family washed 11 loads of clothes at The Mountain View Laundry. The 20-lb capacity washing machine costs $1.50 per load while the 30-lb costs $2.50. Their total laundry expense was $20.50. How many loads were laundered in each size washing machine?

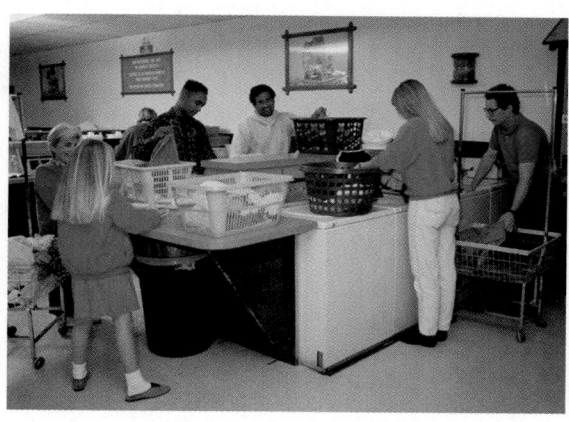

13. *Student Loans.* Sarah's two student loans totaled $12,000. One of her loans was at 6% simple interest and the other at 9%. After one year, Sarah owed $855 in interest. What was the amount of each loan?

14. *Investments.* An executive nearing retirement made two investments totaling $45,000. In one year, these investments yielded $2430 in simple interest. Part of the money was invested at 4% and the rest at 6%. How much was invested at each rate?

15. *Food Science.* The following bar graph shows the milk fat percentages in three dairy products. How many pounds each of whole milk and cream should be mixed in order to form 200 lb of milk for cream cheese?

Dairy Product Milk Fat

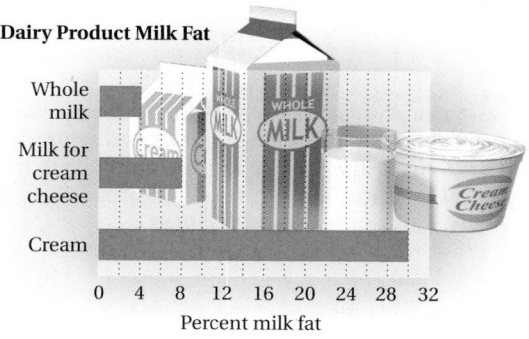

Percent milk fat

16. *Automotive Maintenance.* Arctic Antifreeze is 18% alcohol and Frost No-More is 10% alcohol. How many liters of Arctic Antifreeze should be mixed with 7.5 L of Frost No-More in order to get a mixture that is 15% alcohol?

17. *Teller Work.* Juan goes to a bank and gets change for a $50 bill consisting of all $5 bills and $1 bills. There are 22 bills in all. How many of each kind are there?

18. *Making Change.* Christina makes a $9.25 purchase at a bookstore in Reno with a $20 bill. The store has no bills and gives her the change in quarters and dollar coins. There are 19 coins in all. How many of each kind are there?

19. Investments. William opened two investment accounts for his grandson's college fund. The first year, these investments, which totaled $18,000, yielded $831 in simple interest. Part of the money was invested at 5.5% and the rest at 4%. How much was invested at each rate?

20. Student Loans. Cole's two student loans totaled $31,000. One of his loans was at 2.8% simple interest and the other at 4.5%. After one year, Cole owed $1024.40 in interest. What was the amount of each loan?

 Solve.

21. Train Travel. A train leaves Danville Junction and travels north at a speed of 75 mph. Two hours later, a second train leaves on a parallel track and travels north at 125 mph. How far from the station will they meet?

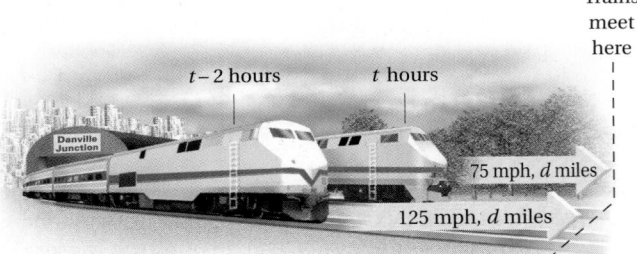

Trains meet here

t − 2 hours   t hours

75 mph, d miles

125 mph, d miles

22. Car Travel. Two cars leave Denver traveling in opposite directions. One car travels at a speed of 80 km/h and the other at 96 km/h. In how many hours will they be 528 km apart?

23. Canoeing. Darren paddled for 4 hr with a 6-km/h current to reach a campsite. The return trip against the same current took 10 hr. Find the speed of Darren's canoe in still water.

24. Boating. Mia's motorboat took 3 hr to make a trip downstream with a 6-mph current. The return trip against the same current took 5 hr. Find the speed of the boat in still water.

25. Car Travel. Donna is late for a sales meeting after traveling from one town to another at a speed of 32 mph. If she had traveled 4 mph faster, she could have made the trip in $\frac{1}{2}$ hr less time. How far apart are the towns?

26. Air Travel. Rod is a pilot for Crossland Airways. He computes his flight time against a headwind for a trip of 2900 mi at 5 hr. The flight would take 4 hr and 50 min if the headwind were half as great. Find the headwind and the plane's air speed.

27. Air Travel. Two planes travel toward each other from cities that are 780 km apart at rates of 190 km/h and 200 km/h. They started at the same time. In how many hours will they meet?

28. Motorcycle Travel. Sally and Rocky travel on motorcycles toward each other from Chicago and Indianapolis, which are about 350 km apart, and they are biking at rates of 110 km/h and 90 km/h. They started at the same time. In how many hours will they meet?

29. Air Travel. Two airplanes start at the same time and fly toward each other from points 1000 km apart at rates of 420 km/h and 330 km/h. After how many hours will they meet?

30. Truck and Car Travel. A truck and a car leave a service station at the same time and travel in the same direction. The truck travels at 55 mph and the car at 40 mph. They can maintain CB radio contact within a range of 10 mi. When will they lose contact?

31. 🖩 *Point of No Return.* A plane flying the 3458-mi trip from New York City to London has a 50-mph tailwind. The flight's *point of no return* is the point at which the flight time required to return to New York is the same as the time required to continue to London. If the speed of the plane in still air is 360 mph, how far is New York from the point of no return?

32. 🖩 *Point of No Return.* A plane is flying the 2553-mi trip from Los Angeles to Honolulu into a 60-mph headwind. If the speed of the plane in still air is 310 mph, how far from Los Angeles is the plane's point of no return? (See Exercise 31.)

Skill Maintenance

Given the function $f(x) = 4x - 7$, find each of the following function values. [7.1b]

33. $f(0)$

34. $f(-1)$

35. $f(1)$

36. $f(10)$

37. $f(-2)$

38. $f(2a)$

39. $f(-4)$

40. $f(1.8)$

41. $f\left(\frac{3}{4}\right)$

42. $f(-2.5)$

43. $f(-3h)$

44. $f(1000)$

Synthesis

45. Automotive Maintenance. The radiator in Michelle's car contains 16 L of antifreeze and water. This mixture is 30% antifreeze. How much of this mixture should she drain and replace with pure antifreeze so that there will be a mixture of 50% antifreeze?

46. Physical Exercise. Natalie jogs and walks to school each day. She averages 4 km/h walking and 8 km/h jogging. The distance from home to school is 6 km and Natalie makes the trip in 1 hr. How far does she jog in a trip?

47. Fuel Economy. Sally Cline's SUV gets 18 miles per gallon (mpg) in city driving and 24 mpg in highway driving. The SUV is driven 465 mi on 23 gal of gasoline. How many miles were driven in the city and how many were driven on the highway?

48. Siblings. Phil and Phyllis are siblings. Phyllis has twice as many brothers as she has sisters. Phil has the same number of brothers as sisters. How many girls and how many boys are in the family?

49. Wood Stains. Bennet Custom Flooring has 0.5 gal of stain that is 20% brown and 80% neutral. A customer orders 1.5 gal of a stain that is 60% brown and 40% neutral. How much pure brown stain and how much neutral stain should be added to the original 0.5 gal in order to make up the order?

50. 📈 See Exercise 49. Let $x =$ the amount of pure brown stain added to the original 0.5 gal. Find a function $P(x)$ that can be used to determine the percentage of brown stain in the 1.5-gal mixture. On a graphing calculator, draw the graph of P and use ZOOM and TRACE or the TABLE feature to confirm the answer to Exercise 49.

Mid-Chapter Review

Concept Reinforcement

Determine whether each statement is true or false.

_____ **1.** If, when solving a system of two linear equations in two variables, a false equation is obtained, the system has infinitely many solutions.   [8.2a], [8.3a]

_____ **2.** Every system of equations has at least one solution.   [8.1a]

_____ **3.** If the graphs of two linear equations intersect, then the system is consistent.   [8.1a]

_____ **4.** The intersection of the graphs of the lines $x = a$ and $y = b$ is (a, b).   [8.1a]

Guided Solutions

Fill in each box with the number, variable, or expression that creates a correct statement or solution.

Solve.   [8.2a], [8.3a]

5. $x + 2y = 3,$ **(1)**
$\quad y = x - 6$ **(2)**

$\quad x + 2(\square) = 3$ Substituting for y in equation (1)
$\quad x + \square x - \square = 3$ Removing parentheses
$\quad \square x - 12 = 3$ Collecting like terms
$\quad\quad\quad 3x = \square$
$\quad\quad\quad\ x = \square$

$\quad y = \square - 6$ Substituting in equation (2)
$\quad y = \square$ Subtracting

The solution is $(\square, \square)$.

6. $3x - 2y = 5,$ **(1)**
$\quad 2x + 4y = 14$ **(2)**

$\quad \square x - \square y = \square$ Multiplying equation (1) by 2
$\quad \underline{2x + \ 4y = 14}$ Equation (2)
$\quad \square x \quad\quad\ = \square$ Adding
$\quad\quad\ x = \square$

$\quad 2 \cdot \square + 4y = 14$ Substituting for x in equation (2)
$\quad \square + 4y = 14$ Multiplying
$\quad\quad\quad 4y = \square$
$\quad\quad\quad\ y = \square$

The solution is $(\square, \square)$.

Mixed Review

Solve each system of equations graphically. Then classify the system as consistent or inconsistent and the equations as dependent or independent.   [8.1a]

7. $y = x - 6,$
$\quad y = 4 - x$

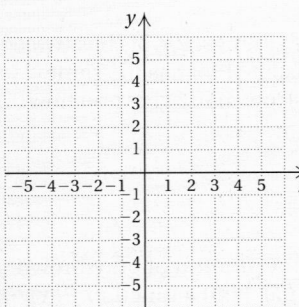

8. $x + y = 3,$
$\quad 3x + y = 3$

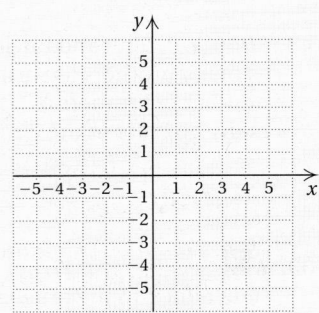

9. $y = 2x - 3,$
$\quad 4x - 2y = 6$

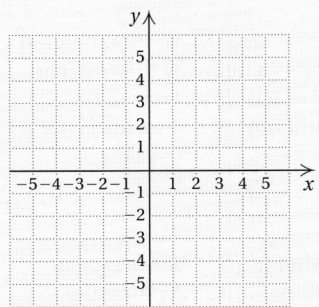

10. $x - y = 3,$
$\quad 2y - 2x = 6$

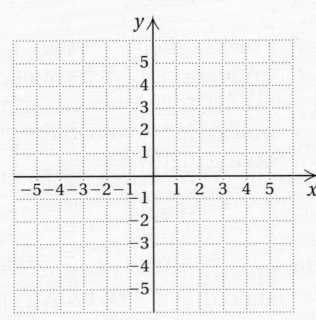

Solve using the substitution method.   [8.2a]

11. $x = y + 2$,
    $2x - 3y = -2$

12. $y = x - 5$,
    $x - 2y = 8$

13. $4x + 3y = 3$,
    $y = x + 8$

14. $3x - 2y = 1$,
    $x = y + 1$

Solve using the elimination method.   [8.3a]

15. $2x + y = 2$,
    $x - y = 4$

16. $x - 2y = 13$,
    $x + 2y = -3$

17. $3x - 4y = 5$,
    $5x - 2y = -1$

18. $3x + 2y = 11$,
    $2x + 3y = 9$

19. $x - 2y = 5$,
    $3x - 6y = 10$

20. $4x - 6y = 2$,
    $-2x + 3y = -1$

21. $\frac{1}{2}x + \frac{1}{3}y = 1$,
    $\frac{1}{5}x - \frac{3}{4}y = 11$

22. $0.2x + 0.3y = 0.6$,
    $0.1x - 0.2y = -2.5$

Solve.

23. *Garden Dimensions.*   A landscape architect designs a garden with a perimeter of 44 ft. The width is 2 ft less than the length. Find the length and the width.   [8.2b]

24. *Investments.*   Sandy made two investments totaling $5000. Part of the money was invested at 2% and the rest at 3%. In one year, these investments earned $129 in simple interest. How much was invested at each rate?   [8.4a]

25. *Mixing Solutions.*   A lab technician wants to mix a solution that is 20% acid with a second solution that is 50% acid in order to get 84 L of a solution that is 30% acid. How many liters of each solution should be used?   [8.4a]

26. *Boating.*   Monica's motorboat took 5 hr to make a trip downstream with a 6-mph current. The return trip against the same current took 8 hr. Find the speed of the boat in still water.   [8.4b]

Understanding Through Discussion and Writing

27. Explain how to find the solution of $\frac{3}{4}x + 2 = \frac{2}{5}x - 5$ in two ways graphically and in two ways algebraically.
[8.1a], [8.2a], [8.3a]

28. Write a system of equations with the given solution. Answers may vary.   [8.1a], [8.2a], [8.3a]
a) $(4, -3)$ b) No solution
c) Infinitely many solutions

29. Describe a method that could be used to create an inconsistent system of equations.   [8.1a], [8.2a], [8.3a]

30. Describe a method that could be used to create a system of equations with dependent equations.   [8.1a], [8.2a], [8.3a]

8.5

Systems of Equations in Three Variables

OBJECTIVE

a Solve systems of three equations in three variables.

SKILL TO REVIEW

Objective 8.3a: Solve systems of equations in two variables by the elimination method.

Solve.

1. $3x + y = 1,$
   $5x - y = 7$
2. $2x + 3y = 9,$
   $3x + 2y = 1$

a Solving Systems in Three Variables

A **linear equation in three variables** is an equation equivalent to one of the type $Ax + By + Cz = D$. A **solution** of a system of three equations in three variables is an ordered triple (x, y, z) that makes *all three* equations true.

The substitution method can be used to solve systems of three equations, but it is not efficient unless a variable has already been eliminated from one or more of the equations. Therefore, we will use only the elimination method—essentially the same procedure for systems of three equations as for systems of two equations.* The first step is to eliminate a variable and obtain a system of two equations in two variables.

EXAMPLE 1 Solve the following system of equations:

$$x + y + z = 4, \quad \text{(1)}$$
$$x - 2y - z = 1, \quad \text{(2)}$$
$$2x - y - 2z = -1. \quad \text{(3)}$$

a) We first use *any* two of the three equations to get an equation in two variables. In this case, let's use equations (1) and (2) and add to eliminate z:

$$\begin{array}{ll} x + y + z = 4 & \text{(1)} \\ \underline{x - 2y - z = 1} & \text{(2)} \\ 2x - y \phantom{{}+ z} = 5. & \text{(4)} \quad \text{Adding to eliminate } z \end{array}$$

b) We use a *different* pair of equations and eliminate the **same variable** that we did in part (a). Let's use equations (1) and (3) and again eliminate z.

-- *Caution!* --

A common error is to eliminate a different variable the second time.

$$\begin{array}{ll} x + y + z = 4, & \text{(1)} \\ 2x - y - 2z = -1; & \text{(3)} \end{array}$$

$$\begin{array}{ll} 2x + 2y + 2z = 8 & \text{Multiplying equation (1) by 2} \\ \underline{2x - y - 2z = -1} & \text{(3)} \\ 4x + y \phantom{{}- 2z} = 7 & \text{(5)} \quad \text{Adding to eliminate } z \end{array}$$

c) Now we solve the resulting system of equations, (4) and (5). That solution will give us two of the numbers. Note that we now have two equations in two variables. Had we eliminated two *different* variables in parts (a) and (b), this would not be the case.

$$\begin{array}{ll} 2x - y = 5 & \text{(4)} \\ \underline{4x + y = 7} & \text{(5)} \\ 6x \phantom{{}- y} = 12 & \text{Adding} \\ x = 2 \end{array}$$

STUDY TIPS

WORKED-OUT-SOLUTIONS

The *Student's Solutions Manual* is an excellent resource if you need additional help with an exercise in the exercise sets. It contains step-by-step solutions to each odd-numbered exercise.

Answers

Skill to Review:
1. $(1, -2)$ 2. $(-3, 5)$

*Other methods for solving systems of equations are considered in Appendixes I and J.

We can use either equation (4) or (5) to find y. We choose equation (5):

$$4x + y = 7 \qquad \textbf{(5)}$$
$$4(2) + y = 7 \qquad \text{Substituting 2 for } x$$
$$8 + y = 7$$
$$y = -1.$$

d) We now have $x = 2$ and $y = -1$. To find the value for z, we use any of the original three equations and substitute to find the third number, z. Let's use equation (1) and substitute our two numbers in it:

$$x + y + z = 4 \qquad \textbf{(1)}$$
$$2 + (-1) + z = 4 \qquad \text{Substituting 2 for } x \text{ and } -1 \text{ for } y$$
$$\left.\begin{array}{l} 1 + z = 4 \\ z = 3. \end{array}\right\} \quad \text{Solving for } z$$

We have obtained the ordered triple $(2, -1, 3)$. We check as follows, substituting $(2, -1, 3)$ into each of the three equations using alphabetical order.

Check:

$$\frac{x + y + z = 4}{2 + (-1) + 3 \;?\; 4}$$
$$\qquad\qquad\quad 4 \;\big|\; \quad \text{TRUE}$$

$$\frac{x - 2y - z = 1}{2 - 2(-1) - 3 \;?\; 1}$$
$$\qquad 2 + 2 - 3 \;\big|$$
$$\qquad\qquad\quad 1 \;\big|\; \quad \text{TRUE}$$

$$\frac{2x - y - 2z = -1}{2(2) - (-1) - 2 \cdot 3 \;?\; -1}$$
$$\qquad\quad 4 + 1 - 6 \;\big|$$
$$\qquad\qquad\qquad -1 \;\big|\; \quad \text{TRUE}$$

The triple $(2, -1, 3)$ checks and is the solution.

To use the elimination method to solve systems of three equations:

1. Write all equations in the standard form, $Ax + By + Cz = D$.
2. Clear any decimals or fractions.
3. Choose a variable to eliminate. Then use *any* two of the three equations to eliminate that variable, getting an equation in two variables.
4. Next, use a different pair of equations and get another equation in *the same two variables*. That is, eliminate the same variable that you did in step (3).
5. Solve the resulting system (pair) of equations. That will give two of the numbers.
6. Then use any of the original three equations to find the third number.

Do Exercise 1.

1. Solve. Don't forget to check.
$$4x - y + z = 6,$$
$$-3x + 2y - z = -3,$$
$$2x + y + 2z = 3$$

Answer

1. $(2, 1, -1)$

EXAMPLE 2 Solve this system:

$$4x - 2y - 3z = 5, \qquad (1)$$
$$-8x - y + z = -5, \qquad (2)$$
$$2x + y + 2z = 5. \qquad (3)$$

a) The equations are in standard form and do not contain decimals or fractions.

b) We decide to eliminate the variable y since the y-terms are opposites in equations (2) and (3). We add:

$$\begin{array}{ll} -8x - y + z = -5 & (2) \\ \underline{2x + y + 2z = 5} & (3) \\ -6x + 3z = 0. & (4) \quad \text{Adding} \end{array}$$

c) We use another pair of equations to get an equation in the same two variables, x and z. That is, we eliminate the same variable, y, that we did in step (b). We use equations (1) and (3) and eliminate y:

$$4x - 2y - 3z = 5, \qquad (1)$$
$$2x + y + 2z = 5; \qquad (3)$$

$$\begin{array}{ll} 4x - 2y - 3z = 5 & (1) \\ \underline{4x + 2y + 4z = 10} & \text{Multiplying equation (3) by 2} \\ 8x + z = 15. & (5) \quad \text{Adding} \end{array}$$

d) Now we solve the resulting system of equations (4) and (5). That will give us two of the numbers:

$$-6x + 3z = 0, \qquad (4)$$
$$8x + z = 15. \qquad (5)$$

We multiply equation (5) by -3. $\left(\text{We could also have multiplied equation (4) by } -\frac{1}{3}.\right)$

$$\begin{array}{ll} -6x + 3z = 0 & (4) \\ \underline{-24x - 3z = -45} & \text{Multiplying equation (5) by } -3 \\ -30x = -45 & \text{Adding} \\ x = \frac{-45}{-30} = \frac{3}{2} \end{array}$$

We now use equation (5) to find z:

$$\begin{array}{ll} 8x + z = 15 & (5) \\ 8\left(\frac{3}{2}\right) + z = 15 & \text{Substituting } \frac{3}{2} \text{ for } x \\ \left.\begin{array}{l} 12 + z = 15 \\ z = 3. \end{array}\right\} & \text{Solving for } z \end{array}$$

e) Next, we use any of the original equations and substitute to find the third number, y. We choose equation (3) since the coefficient of y there is 1:

$$\begin{array}{ll} 2x + y + 2z = 5 & (3) \\ 2\left(\frac{3}{2}\right) + y + 2(3) = 5 & \text{Substituting } \frac{3}{2} \text{ for } x \text{ and 3 for } z \\ \left.\begin{array}{l} 3 + y + 6 = 5 \\ y + 9 = 5 \\ y = -4. \end{array}\right\} & \text{Solving for } y \end{array}$$

The solution is $\left(\frac{3}{2}, -4, 3\right)$. The check is as follows.

Check:

$$\frac{4x - 2y - 3z = 5}{4 \cdot \frac{3}{2} - 2(-4) - 3(3) \ ? \ 5}$$
$$6 + 8 - 9$$
$$5 \ \big| \quad \text{TRUE}$$

$$\frac{-8x - y + z = -5}{-8 \cdot \frac{3}{2} - (-4) + 3 \ ? \ -5}$$
$$-12 + 4 + 3$$
$$-5 \ \big| \quad \text{TRUE}$$

$$\frac{2x + y + 2z = 5}{2 \cdot \frac{3}{2} + (-4) + 2(3) \ ? \ 5}$$
$$3 - 4 + 6$$
$$5 \ \big| \quad \text{TRUE}$$

Do Exercise 2.

2. Solve. Don't forget to check.
$$2x + \ y - 4z = 0,$$
$$x - \ y + 2z = 5,$$
$$3x + 2y + 2z = 3$$

In Example 3, two of the equations have a missing variable.

EXAMPLE 3 Solve this system:

$$x + y + z = 180, \quad \textbf{(1)}$$
$$x \quad - z = -70, \quad \textbf{(2)}$$
$$2y - z = 0. \quad \textbf{(3)}$$

We note that there is no y in equation (2). In order to have a system of two equations in the variables x and z, we need to find another equation without a y. We use equations (1) and (3) to eliminate y:

$$x + y + z = 180, \quad \textbf{(1)}$$
$$2y - z = 0; \quad \textbf{(3)}$$

$$-2x - 2y - 2z = -360 \qquad \text{Multiplying equation (1) by } -2$$
$$\underline{ 2y - \ z = 0} \qquad \textbf{(3)}$$
$$-2x - 3z = -360. \qquad \textbf{(4)} \qquad \text{Adding}$$

Now we solve the resulting system of equations (2) and (4):

$$x - \ z = -70, \quad \textbf{(2)}$$
$$-2x - 3z = -360; \quad \textbf{(4)}$$

$$2x - 2z = -140 \qquad \text{Multiplying equation (2) by 2}$$
$$\underline{-2x - 3z = -360} \qquad \textbf{(4)}$$
$$-5z = -500 \qquad \text{Adding}$$
$$z = 100.$$

To find x, we substitute 100 for z in equation (2) and solve for x:

$$x - z = -70$$
$$x - 100 = -70$$
$$x = 30.$$

Answer

2. $\left(2, -2, \dfrac{1}{2}\right)$

To find y, we substitute 100 for z in equation (3) and solve for y:

$$2y - z = 0$$
$$2y - 100 = 0$$
$$2y = 100$$
$$y = 50.$$

The triple $(30, 50, 100)$ is the solution. The check is left to the student.

Do Exercise 3.

> **3.** Solve. Don't forget to check.
> $$x + y + z = 100,$$
> $$x - y = -10,$$
> $$x - z = -30$$

It is possible for a system of three equations to have no solution, that is, to be inconsistent. An example is the system

$$x + y + z = 14,$$
$$x + y + z = 11,$$
$$2x - 3y + 4z = -3.$$

Note the first two equations. It is not possible for a sum of three numbers to be both 14 and 11. Thus the system has no solution. We will not consider such systems here, nor will we consider systems with infinitely many solutions, which also exist.

STUDY TIPS

LEARNING RESOURCES

Please see the preface for more information on these resources and others. To order any of our products, call (800) 824-7799 in the United States or (201) 767-5021 outside the United States, or visit your campus bookstore.

- The *Student's Solutions Manual* contains fully worked-out solutions to the odd-numbered exercises in the exercise sets, as well as solutions to all exercises in the Mid-Chapter Reviews, end-of-chapter Review Exercises, Chapter Tests, and Cumulative Reviews. (ISBN: 978-0-321-61362-2)
- *Worksheets for Classroom or Lab Practice* provide a list of learning objectives, vocabulary and practice problems, and extra practice problems with ample work space. (ISBN: 978-0-321-61368-4)

- As described on p. 56 and in the Preface, Video Resources on DVD Featuring Chapter Test Prep Videos provide section-level lectures for every objective and step-by-step solutions to all the Chapter Test exercises in this textbook. The Chapter Test videos are also available on YouTube (search using BittingerIntroInter) and in MyMathLab.
- InterAct Math Tutorial Website (www.interactmath.com) provides algorithmically generated practice exercises that correlate directly to the exercises in the textbook.
- MathXL® Tutorials on CD provide practice exercises correlated at the objective level to the exercises in the textbook. Every practice exercise is accompanied by an example and a guided solution, and selected exercises may also include a video clip to help illustrate a concept.

Answer

3. $(20, 30, 50)$

a) Solve.

1. $x + y + z = 2,$
$2x - y + 5z = -5,$
$-x + 2y + 2z = 1$

2. $2x - y - 4z = -12,$
$2x + y + z = 1,$
$x + 2y + 4z = 10$

3. $2x - y + z = 5,$
$6x + 3y - 2z = 10,$
$x - 2y + 3z = 5$

4. $x - y + z = 4,$
$3x + 2y + 3z = 7,$
$2x + 9y + 6z = 5$

5. $2x - 3y + z = 5,$
$x + 3y + 8z = 22,$
$3x - y + 2z = 12$

6. $6x - 4y + 5z = 31,$
$5x + 2y + 2z = 13,$
$x + y + z = 2$

7. $3a - 2b + 7c = 13,$
$a + 8b - 6c = -47,$
$7a - 9b - 9c = -3$

8. $x + y + z = 0,$
$2x + 3y + 2z = -3,$
$-x + 2y - 3z = -1$

9. $2x + 3y + z = 17,$
$x - 3y + 2z = -8,$
$5x - 2y + 3z = 5$

10. $2x + y - 3z = -4,$
$4x - 2y + z = 9,$
$3x + 5y - 2z = 5$

11. $2x + y + z = -2,$
$2x - y + 3z = 6,$
$3x - 5y + 4z = 7$

12. $2x + y + 2z = 11,$
$3x + 2y + 2z = 8,$
$x + 4y + 3z = 0$

13. $x - y + z = 4,$
$5x + 2y - 3z = 2,$
$3x - 7y + 4z = 8$

14. $2x + y + 2z = 3,$
$x + 6y + 3z = 4,$
$3x - 2y + z = 0$

15. $4x - y - z = 4,$
$2x + y + z = -1,$
$6x - 3y - 2z = 3$

16. $2r + s + t = 6,$
$3r - 2s - 5t = 7,$
$r + s - 3t = -10$

17. $a - 2b - 5c = -3,$
$3a + b - 2c = -1,$
$2a + 3b + c = 4$

18. $x + 4y - z = 5,$
$2x - y + 3z = -5,$
$4x + 3y + z = 5$

19. $2r + 3s + 12t = 4,$
$4r - 6s + 6t = 1,$
$r + s + t = 1$

20. $10x + 6y + z = 7,$
$5x - 9y - 2z = 3,$
$15x - 12y + 2z = -5$

21. $a + 2b + c = 1,$
$7a + 3b - c = -2,$
$a + 5b + 3c = 2$

22. $3p + 2r = 11,$
$q - 7r = 4,$
$p - 6q = 1$

23. $x + y + z = 57,$
$-2x + y = 3,$
$x - z = 6$

24. $4a + 9b = 8,$
$8a + 6c = -1,$
$6b + 6c = -1$

25. $r + s = 5,$
$3s + 2t = -1,$
$4r + t = 14$

26. $a - 5c = 17,$
$b + 2c = -1,$
$4a - b - 3c = 12$

27. $x + y + z = 105,$
$10y - z = 11,$
$2x - 3y = 7$

Skill Maintenance

Solve for the indicated letter.   [2.4b]

28. $F = 3ab,$ for a

29. $Q = 4(a + b),$ for a

30. $F = \frac{1}{2}t(c - d),$ for d

31. $F = \frac{1}{2}t(c - d),$ for c

32. $Ax + By = c,$ for y

33. $Ax - By = c,$ for y

Find the slope and the y-intercept.   [7.3b]

34. $y = -\frac{2}{3}x - \frac{5}{4}$

35. $y = 5 - 4x$

36. $2x - 5y = 10$

37. $7x - 6.4y = 20$

Synthesis

Solve.

38. $w + x - y + z = 0,$
$w - 2x - 2y - z = -5,$
$w - 3x - y + z = 4,$
$2w - x - y + 3z = 7$

39. $w + x + y + z = 2,$
$w + 2x + 2y + 4z = 1,$
$w - x + y + z = 6,$
$w - 3x - y + z = 2$

8.6

Solving Applied Problems: Three Equations

a Using Systems of Three Equations

OBJECTIVE

a Solve applied problems using systems of three equations.

Solving systems of three or more equations is important in many applications occurring in the natural and social sciences, business, and engineering.

EXAMPLE 1 *Jewelry Design.* Kim is designing a triangular-shaped pendant for a client of her custom jewelry business. The largest angle of the triangle is 70° greater than the smallest angle. The largest angle is twice as large as the remaining angle. Find the measure of each angle.

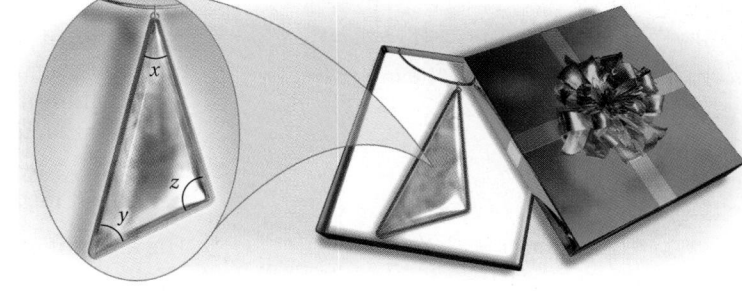

1. **Familiarize.** We first make a drawing. Since we do not know the size of any angle, we use x, y, and z for the measures of the angles. We let $x =$ the smallest angle, $z =$ the largest angle, and $y =$ the remaining angle.

2. **Translate.** In order to translate the problem, we need to make use of a geometric fact—that is, the sum of the measures of the angles of a triangle is 180°. This fact about triangles gives us one equation:

    $$x + y + z = 180.$$

    There are two statements in the problem that we can translate directly.

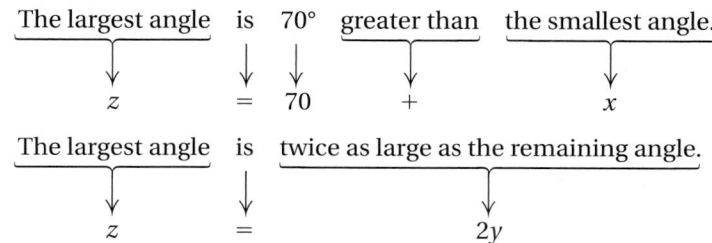

We now have a system of three equations:

$$\begin{aligned} x + y + z &= 180, & & & x + y + z &= 180, \\ x + 70 &= z, & & \text{or} \quad x \qquad -z &= -70, \\ 2y &= z; & & & 2y - z &= 0. \end{aligned}$$

3. **Solve.** The system was solved in Example 3 of Section 8.5. The solution is $(30, 50, 100)$.

4. **Check.** The sum of the numbers is 180. The largest angle measures 100° and the smallest measures 30°, so the largest angle is 70° greater than the smallest. The largest angle is twice as large as 50°, the remaining angle. We have an answer to the problem.

5. **State.** The measures of the angles of the triangle are 30°, 50°, and 100°.

Do Exercise 1.

1. **Triangle Measures.** One angle of a triangle is twice as large as a second angle. The remaining angle is 20° greater than the first angle. Find the measure of each angle.

Answer

1. 64°, 32°, 84°

EXAMPLE 2 *Cholesterol Levels.* Americans have become very conscious of their cholesterol levels. Recent studies indicate that a child's intake of cholesterol should be no more than 300 mg per day. By eating 1 egg, 1 cupcake, and 1 slice of pizza, a child consumes 302 mg of cholesterol. If the child eats 2 cupcakes and 3 slices of pizza, he or she takes in 65 mg of cholesterol. By eating 2 eggs and 1 cupcake, a child consumes 567 mg of cholesterol. How much cholesterol is in each item?

1. **Familiarize.** After we have read the problem a few times, it becomes clear that an egg contains considerably more cholesterol than the other foods. Let's guess that one egg contains 200 mg of cholesterol and one cupcake contains 50 mg. Because of the third sentence in the problem, it would follow that a slice of pizza contains 52 mg of cholesterol since $200 + 50 + 52 = 302$.

   To see if our guess satisfies the other statements in the problem, we find the amount of cholesterol that 2 cupcakes and 3 slices of pizza would contain: $2 \cdot 50 + 3 \cdot 52 = 256$. Since this does not match the 65 mg listed in the fourth sentence of the problem, our guess was incorrect. Rather than guess again, we examine how we checked our guess and let g, c, and s = the number of milligrams of cholesterol in an egg, a cupcake, and a slice of pizza, respectively.

2. **Translate.** By rewording some of the sentences in the problem, we can translate it into three equations.

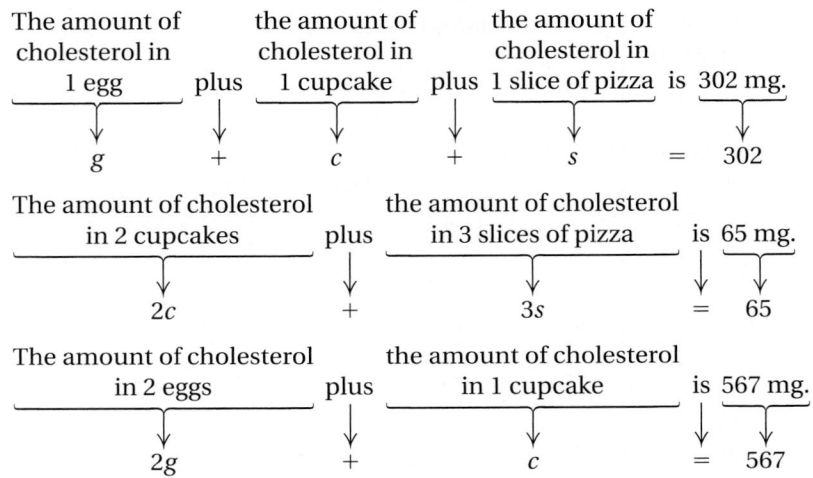

We now have a system of three equations:

$$
\begin{aligned}
g + c + s &= 302, &\textbf{(1)} \\
2c + 3s &= 65, &\textbf{(2)} \\
2g + c\phantom{{}+3s} &= 567. &\textbf{(3)}
\end{aligned}
$$

3. Solve. To solve, we first note that the variable g does not appear in equation (2). In order to have a system of two equations in the variables c and s, we need to find another equation without the variable g. We use equations (1) and (3) to eliminate g:

$$g + c + s = 302, \quad \textbf{(1)}$$
$$2g + c \quad\quad = 567; \quad \textbf{(3)}$$

$$
\begin{array}{ll}
-2g - 2c - 2s = -604 & \text{Multiplying equation (1) by } -2 \\
\underline{2g + \quad c \quad\quad = \quad 567} & \textbf{(3)} \\
-c - 2s = -37. & \textbf{(4)} \quad \text{Adding}
\end{array}
$$

Next, we solve the resulting system of equations (2) and (4):

$$2c + 3s = 65, \quad \textbf{(2)}$$
$$-c - 2s = -37; \quad \textbf{(4)}$$

$$
\begin{array}{ll}
2c + 3s = \quad 65 & \textbf{(2)} \\
\underline{-2c - 4s = -74} & \text{Multiplying equation (4) by 2} \\
-s = -9 & \text{Adding} \\
s = 9.
\end{array}
$$

To find c, we substitute 9 for s in equation (4) and solve for c:

$$
\begin{array}{ll}
-c - 2s = -37 & \textbf{(4)} \\
-c - 2(9) = -37 & \text{Substituting} \\
-c - 18 = -37 \\
-c = -19 \\
c = 19.
\end{array}
$$

To find g, we substitute 19 for c in equation (3) and solve for g:

$$
\begin{array}{ll}
2g + c = 567 & \textbf{(3)} \\
2g + 19 = 567 & \text{Substituting} \\
2g = 548 \\
g = 274.
\end{array}
$$

The solution is $c = 19, g = 274, s = 9$, or $(19, 274, 9)$.

4. Check. The sum of 19, 274, and 9 is 302 so the total cholesterol in 1 cupcake, 1 egg, and 1 slice of pizza checks. Two cupcakes and three slices of pizza would contain $2 \cdot 19 + 3 \cdot 9$, or 65 mg, while two eggs and one cupcake would contain $2 \cdot 274 + 19$, or 567 mg of cholesterol. The answer checks.

5. State. A cupcake contains 19 mg of cholesterol, an egg contains 274 mg of cholesterol, and a slice of pizza contains 9 mg of cholesterol.

Do Exercise 2.

2. Client Investments. Kaufman Financial Corporation makes investments for corporate clients. One year, a client receives $1620 in simple interest from three investments that total $25,000. Part is invested at 5%, part at 6%, and part at 7%. There is $11,000 more invested at 7% than at 6%. How much was invested at each rate?

a Solve.

1. *Scholastic Aptitude Test.* More than two million high-school students take the Scholastic Aptitude Test each year as part of the college admission process. Students receive a critical reading score, a mathematics score, and a writing score. The average total score of students who graduated from high school in 2008 was 1511. The average math score exceeded the average reading score by 13 points. The average math score was 481 points less than the sum of the average reading and writing scores. Find the average score on each part of the test.

Source: College Board

2. *Fat Content of Fast Food.* A meal at McDonald's consisting of a Big Mac, a medium order of fries, and a 21-oz vanilla milkshake contains 66 g of fat. The Big Mac has 11 more grams of fat than the milkshake. The total fat content of the fries and the shake exceeds that of the Big Mac by 8 g. Find the fat content of each food item.

Source: McDonald's

3. *Triangle Measures.* In triangle *ABC*, the measure of angle *B* is three times that of angle *A*. The measure of angle *C* is 20° more than that of angle *A*. Find the measure of each angle.

4. *Triangle Measures.* In triangle *ABC*, the measure of angle *B* is twice the measure of angle *A*. The measure of angle *C* is 80° more than that of angle *A*. Find the measure of each angle.

5. The sum of three numbers is 55. The difference of the largest and the smallest is 49, and the sum of the two smaller is 13. Find the numbers.

6. The sum of three numbers is −30. The largest minus twice the smallest is 45, and the largest is 20 more than the middle number. Find the numbers.

7. *Automobile Pricing.*   A recent basic model of a particular automobile had a price of $12,685. The basic model with the added features of automatic transmission and power door locks was $14,070. The basic model with air conditioning (AC) and power door locks was $13,580. The basic model with AC and automatic transmission was $13,925. What was the individual cost of each of the three options?

8. *Telemarketing.*   Steve, Teri, and Isaiah can process 740 telephone orders per day. Steve and Teri together can process 470 orders, while Teri and Isaiah together can process 520 orders per day. How many orders can each person process alone?

9. *Low-Fat Fruit Drinks.*   A Smoothie King® on a large college campus recently sold small low-fat fruit Smoothies for $4.30 each, medium Smoothies for $6.50 each, and large Smoothies for $8.00 each. One hot summer afternoon, Jake sold 34 Smoothies for a total of $211. The number of small and large Smoothies, combined, was 2 more than the number of medium Smoothies. How many of each size were sold?

Source: campusfood.com

10. *Cappuccinos.*   A Starbucks® on campus sells cappuccinos in three sizes: tall for $2.65, grande for $3.20, and venti® for $3.50. One morning, Brianna served 50 cappuccinos. The number of tall and venti® cappuccinos, combined, was 2 fewer than the number of grande cappuccinos. If she collected a total of $157, how many cappuccinos of each size were sold?

Source: Starbucks® Corporation

11. *Investments.*   A business class divided an imaginary investment of $80,000 among three mutual funds. The first fund grew by 2%, the second by 6%, and the third by 3%. Total earnings were $2250. The earnings from the first fund were $150 more than the earnings from the third. How much was invested in each fund?

12. *Crying Rate.*   The sum of the average number of times that a man, a woman, and a one-year-old child cry each month is 71.7. A one-year-old cries 46.4 more times than a man. The average number of times that a one-year-old cries per month is 28.3 more than the average number of times combined that a man and a woman cry. What is the average number of times per month that each cries?

13. *Veterinary Expenditure.* The sum of the average amounts Americans spent, per animal, for veterinary expenses for dogs, cats, and birds in a recent year was $290. The average expenditure per dog exceeded the sum of the averages for cats and birds by $110. The amount spent per cat was 9 times the amount spent per bird. Find the average amount spent on each type of animal.

Source: American Veterinary Medical Association

14. *Welding Rates.* Eldon, Dana, and Casey can weld 74 linear feet per hour when working together. Eldon and Dana together can weld 44 linear feet per hour, while Eldon and Casey can weld 50 linear feet per hour. How many linear feet per hour can each weld alone?

15. *Nutrition.* A dietician in a hospital prepares meals under the guidance of a physician. Suppose that for a particular patient a physician prescribes a meal to have 800 calories, 55 g of protein, and 220 mg of vitamin C. The dietician prepares a meal of roast beef, baked potato, and broccoli according to the data in the following table.

16. *Nutrition.* Repeat Exercise 15 but replace the broccoli with asparagus, for which one 180-g serving contains 50 calories, 5 g of protein, and 44 mg of vitamin C. Which meal would you prefer eating?

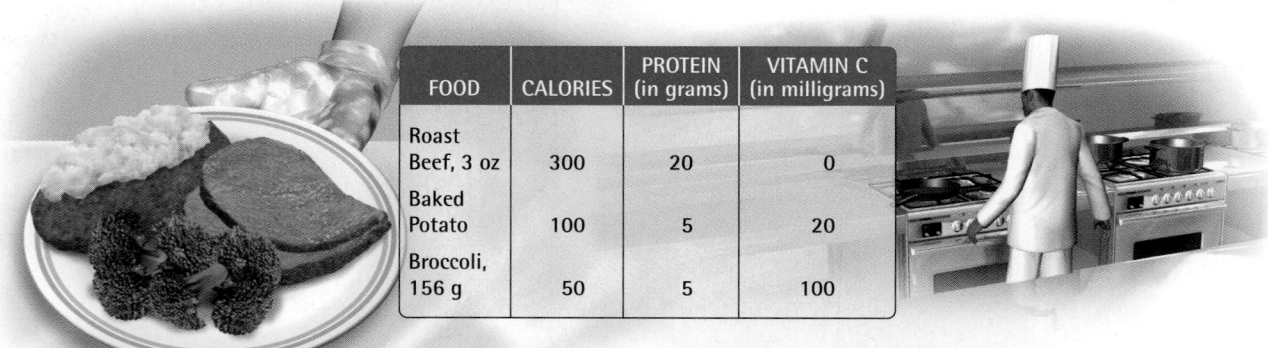

FOOD	CALORIES	PROTEIN (in grams)	VITAMIN C (in milligrams)
Roast Beef, 3 oz	300	20	0
Baked Potato	100	5	20
Broccoli, 156 g	50	5	100

How many servings of each food are needed in order to satisfy the doctor's orders?

17. *Lens Production.* When Sight-Rite's three polishing machines, A, B, and C, are all working, 5700 lenses can be polished in one week. When only A and B are working, 3400 lenses can be polished in one week. When only B and C are working, 4200 lenses can be polished in one week. How many lenses can be polished in a week by each machine alone?

18. *Nutrition Facts.* A meal at Subway consisting of a 6-in. turkey breast sandwich, a bowl of minestrone soup, and a chocolate chip cookie contains 580 calories. The number of calories in the sandwich is 20 less than in the soup and the cookie together. The cookie has 120 calories more than the soup. Find the number of calories in each item.
Source: Subway

19. *Golf.* On an 18-hole golf course, there are par-3 holes, par-4 holes, and par-5 holes. A golfer who shoots par on every hole has a total of 70. There are twice as many par-4 holes as there are par-5 holes. How many of each type of hole are there on the golf course?

20. *Golf.* On an 18-hole golf course, there are par-3 holes, par-4 holes, and par-5 holes. A golfer who shoots par on every hole has a total of 72. The sum of the number of par-3 holes and the number of par-5 holes is 8. How many of each type of hole are there on the golf course?

21. *Basketball Scoring.* The New York Knicks once scored a total of 92 points on a combination of 2-point field goals, 3-point field goals, and 1-point foul shots. Altogether, the Knicks made 50 baskets and 19 more 2-pointers than foul shots. How many shots of each kind were made?

22. *History.* Find the year in which the first U.S. transcontinental railroad was completed. The following are some facts about the number. The sum of the digits in the year is 24. The ones digit is 1 more than the hundreds digit. Both the tens and the ones digits are multiples of 3.

Skill Maintenance

In each of Exercises 23–30, fill in the blank with the correct term from the given list. Some of the choices may not be used.

23. The expression $x \leq q$ means x is _____ q.
[2.8a]

24. The expression $x \geq q$ means x is _____ q.
[2.8a]

25. The graph of a(n) _____ equation is a line.
[3.2a]

26. When the slope of a line is _____ , the graph of the line slants down from left to right.   [3.4a]

27. A(n) _____ system of equations has at least one solution.   [8.1a]

28. Two lines are _____ if the product of their slopes is -1.   [7.4d]

29. The _____ of the graph of $f(x) = mx + b$ is the point $(0, b)$.   [7.3a]

30. When the slope of a line is zero, the graph of the line is _____ .   [7.3b]

parallel

perpendicular

consistent

inconsistent

linear

x-intercept

y-intercept

positive

zero

negative

vertical

horizontal

at least

at most

Synthesis

31. Find the sum of the angle measures at the tips of the star in this figure.

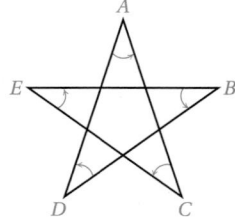

32. *Sharing Raffle Tickets.*   Hal gives Tom as many raffle tickets as Tom has and Gary as many as Gary has. In like manner, Tom then gives Hal and Gary as many tickets as each then has. Similarly, Gary gives Hal and Tom as many tickets as each then has. If each finally has 40 tickets, with how many tickets does Tom begin?

33. *Digits.*   Find a three-digit positive integer such that the sum of all three digits is 14, the tens digit is 2 more than the ones digit, and if the digits are reversed, the number is unchanged.

34. *Ages.*   Tammy's age is the sum of the ages of Carmen and Dennis. Carmen's age is 2 more than the sum of the ages of Dennis and Mark. Dennis's age is four times Mark's age. The sum of all four ages is 42. How old is Tammy?

Summary and Review

Key Terms and Formulas

system of equations, p. 570
solution of a system of equations, p. 570
consistent system of equations, p. 573

inconsistent system of equations, p. 573
dependent equations, p. 574
independent equations, p. 574

substitution method, p. 579
elimination method, p. 585
linear equation in three variables, p. 610

Motion formula: $d = rt$

Concept Reinforcement

Determine whether each statement is true or false.

_____ **1.** A system of equations with infinitely many solutions is inconsistent. [8.1a]

_____ **2.** It is not possible for the equations in an inconsistent system of two equations to be dependent. [8.1a]

_____ **3.** When $(0, b)$ is a solution of each equation in a system of two equations, the graphs of the two equations have the same y-intercept. [8.1a]

_____ **4.** The system of equations $x = 4$ and $y = -4$ is inconsistent. [8.1a]

Important Concepts

Objective 8.1a Solve a system of two linear equations or two functions by graphing and determine whether a system is consistent or inconsistent and whether the equations in a system are dependent or independent.

Example Solve this system of equations graphically. Then classify the system as consistent or inconsistent and the equations as dependent or independent.

$$x - y = 3,$$
$$y = 2x - 4$$

We graph the equations.

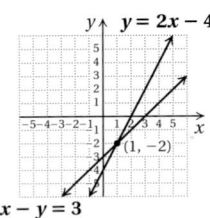

The point of intersection appears to be $(1, -2)$. This checks in both equations, so it is the solution. The system has one solution, so it is consistent and the equations are independent.

Practice Exercise

1. Solve this system of equations graphically. Then classify the system as consistent or inconsistent and the equations as dependent or independent.

$$x + 3y = 1,$$
$$x + y = 3$$

Objective 8.2a   Solve systems of equations in two variables by the substitution method.

Example   Solve the system

$$x - 2y = 1, \quad \textbf{(1)}$$
$$2x - 3y = 3. \quad \textbf{(2)}$$

We solve equation (1) for x, since the coefficient of x is 1 in that equation:

$$x - 2y = 1$$
$$x = 2y + 1. \quad \textbf{(3)}$$

Next, we substitute for x in equation (2) and solve for y:

$$2x - 3y = 3$$
$$2(2y + 1) - 3y = 3$$
$$4y + 2 - 3y = 3$$
$$y + 2 = 3$$
$$y = 1.$$

Then we substitute 1 for y in equation (1), (2), or (3) and find x. We choose equation (3) since it is already solved for x:

$$x = 2y + 1 = 2 \cdot 1 + 1 = 2 + 1 = 3.$$

Check:

$$\begin{array}{c|c} x - 2y = 1 & 2x - 3y = 3 \\ \hline 3 - 2 \cdot 1 \,?\, 1 & 2 \cdot 3 - 3 \cdot 1 \,?\, 3 \\ 3 - 2 \;\big|\; & 6 - 3 \;\big|\; \\ 1 \;\big|\; \text{TRUE} & 3 \;\big|\; \text{TRUE} \end{array}$$

The ordered pair (3, 1) checks in both equations, so it is the solution of the system of equations.

Practice Exercise

2. Solve the system

$$2x + y = 2,$$
$$3x + 2y = 5.$$

Objective 8.3a   Solve systems of equations in two variables by the elimination method.

Example   Solve the system

$$2a + 3b = -1, \quad \textbf{(1)}$$
$$3a + 2b = 6. \quad \textbf{(2)}$$

We could eliminate either a or b. In this case, we decide to eliminate the a-terms. We multiply equation (1) by 3 and equation (2) by -2 and then add and solve for b:

$$\begin{array}{r} 6a + 9b = -3 \\ -6a - 4b = -12 \\ \hline 5b = -15 \\ b = -3. \end{array}$$

Next, we substitute -3 for b in either of the original equations:

$$2a + 3b = -1 \quad \textbf{(1)}$$
$$2a + 3(-3) = -1$$
$$2a - 9 = -1$$
$$2a = 8$$
$$a = 4.$$

The ordered pair (4, −3) checks in both equations, so it is a solution of the system of equations.

Practice Exercise

3. Solve the system

$$2x + 3y = 5,$$
$$3x + 4y = 6.$$

Objective 8.4a Solve applied problems involving total value and mixture using system of two equations.

Example To start a small business, Michael took two loans totaling $18,000. One of the loans was at 7% and the other at 8%. After one year, Michael owed $1365 in interest. What was the amount of each loan?

1. **Familiarize.** We let x and y represent the two loans. Next we organize the information in a table and use the simple interest formula, $I = Prt$.

	LOAN 1	LOAN 2	TOTAL
PRINCIPAL	x	y	$18,000
RATE OF INTEREST	7%	8%	
TIME	1 year	1 year	
INTEREST	7%x, or 0.07x	8%y, or 0.08y	$1365

2. **Translate** The total of the loans is found in the first row of the table. This gives us one equation:

$$x + y = 18,000.$$

From the last row of the table, we see that the interest totals $1365. This gives us a second equation:

$$0.07x + 0.08y = 1365.$$

3. **Solve.** We solve the resulting system of equations:

$$x + y = 18,000, \quad \textbf{(1)}$$
$$0.07x + 0.08y = 1365. \quad \textbf{(2)}$$

We multiply by -0.07 on both sides of equation (1) and add:

$$-0.07x - 0.07y = -1260$$
$$\underline{0.07x + 0.08y = \quad 1365} \quad (2)$$
$$0.01y = 105 \qquad \text{Adding}$$
$$y = 10,500. \qquad \text{Solving for } y$$

Then

$$x + 10,500 = 18,000 \qquad \text{Substituting 10,500 for } y \text{ in equation (1)}$$
$$x = 7500. \qquad \text{Solving for } x$$

We find that $x = 7500$ and $y = 10,500$.

4. **Check.** The sum is $7500 + $10,500, or $18,000. The interest from $7500 at 7% for one year is 7%($7500), or $525. The interest from $10,500 at 8% for one year is 8%(10,500), or $840. The total interest is $525 + $840, or $1365. The numbers check in the problem.

5. **State.** Michael took loans of $7500 at 7% and $10,500 at 8%.

Practice Exercise

4. Jaretta made two investments totaling $23,000. In one year, these investments yielded $1237 in simple interest. Part of the money was invested at 6% and the rest at 5%. How much was invested at each rate?

Objective 8.5a Solve systems of three equations in three variables.

Example Solve:
$$x - y - z = -2, \quad \textbf{(1)}$$
$$2x + 3y + z = 3, \quad \textbf{(2)}$$
$$5x - 2y - 2z = -1. \quad \textbf{(3)}$$

The equations are in standard form and do not contain decimals or fractions. We choose to eliminate z since the z-terms in equations (1) and (2) are opposites. First, we add these two equations:

$$\begin{array}{r} x - y - z = -2 \\ 2x + 3y + z = 3 \\ \hline 3x + 2y = 1. \quad \textbf{(4)} \end{array}$$

Next, we multiply equation (2) by 2 and add it to equation (3) to eliminate z from another pair of equations:

$$\begin{array}{r} 4x + 6y + 2z = 6 \\ 5x - 2y - 2z = -1 \\ \hline 9x + 4y = 5. \quad \textbf{(5)} \end{array}$$

Now we solve the system consisting of equations (4) and (5). We multiply equation (4) by -2 and add:

$$\begin{array}{r} -6x - 4y = -2 \\ 9x + 4y = 5 \\ \hline 3x = 3 \\ x = 1. \end{array}$$

Then we use either equation (4) or (5) to find y:

$$3x + 2y = 1 \quad \textbf{(4)}$$
$$3 \cdot 1 + 2y = 1$$
$$3 + 2y = 1$$
$$2y = -2$$
$$y = -1.$$

Finally, we use one of the original equations to find z:

$$2x + 3y + z = 3 \quad \textbf{(2)}$$
$$2 \cdot 1 + 3(-1) + z = 3$$
$$-1 + z = 3$$
$$z = 4.$$

Check:

$$\begin{array}{c|c} x - y - z = -2 \\ \hline 1 - (-1) - 4 \;?\; -2 \\ 1 + 1 - 4 \\ -2 \;\bigm|\; \text{TRUE} \end{array}$$

$$\begin{array}{c|c} 2x + 3y + z = 3 \\ \hline 2 \cdot 1 + 3(-1) + 4 \;?\; 3 \\ 2 - 3 + 4 \\ 3 \;\bigm|\; \text{TRUE} \end{array}$$

$$\begin{array}{c|c} 5x - 2y - 2z = -1 \\ \hline 5 \cdot 1 - 2(-1) - 2 \cdot 4 \;?\; -1 \\ 5 + 2 - 8 \\ -1 \;\bigm|\; \text{TRUE} \end{array}$$

The ordered triple $(1, -1, 4)$ checks in all three equations, so it is the solution of the system of equations.

Practice Exercise

5. Solve:
$$x - y + z = 9,$$
$$2x + y + 2z = 3,$$
$$4x + 2y - 3z = -1.$$

Review Exercises

Solve graphically. Then classify the system as consistent or inconsistent and the equations as dependent or independent.   [8.1a]

1. $4x - y = -9,$
$\quad x - y = -3$

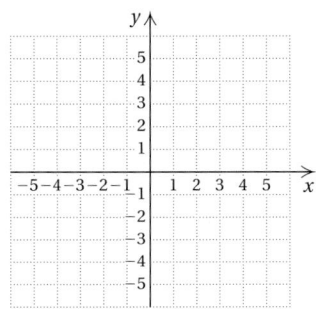

2. $15x + 10y = -20,$
$\quad 3x + \;\; 2y = -4$

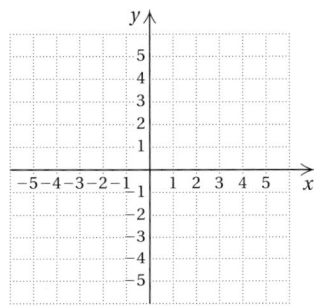

3. $y - 2x = 4,$
$\quad y - 2x = 5$

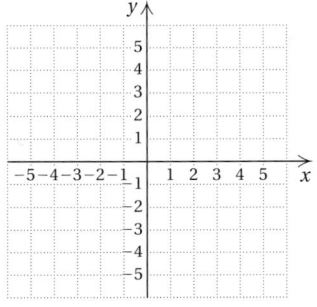

Solve by the substitution method.   [8.2a]

4. $2x - 3y = 5,$
$\quad x = 4y + 5$

5. $y = x + 2,$
$\quad y - x = 8$

6. $7x - 4y = 6,$
$\quad y - 3x = -2$

Solve by the elimination method.   [8.3a]

7. $x + 3y = -3,$
$\quad 2x - 3y = 21$

8. $3x - 5y = -4,$
$\quad 5x - 3y = 4$

9. $\dfrac{1}{3}x + \dfrac{2}{9}y = 1,$

$\quad \dfrac{3}{2}x + \dfrac{1}{2}y = 6$

10. $1.5x - 3 = -2y,$
$\quad 3x + 4y = 6$

11. *Air Travel.*   An airplane flew for 3 hr with a 30-mph tailwind. The return flight against the same wind took 4.5 hr. Find the speed of the plane in still air.   [8.4b]

12. *Spending Choices.*   Sean has \$86 to spend. He can spend all of it on one CD and two DVDs, or he can buy two CDs and one DVD and have \$16 left over. What is the price of a CD? of a DVD?   [8.4a]

13. *Orange Drink Mixtures.*   "Orange Thirst" is 15% orange juice and "Quencho" is 5% orange juice. How many liters of each should be combined in order to get 10 L of a mixture that is 10% orange juice?   [8.4a]

14. *Train Travel.*   A train leaves Watsonville at noon traveling north at 44 mph. One hour later, another train, going 52 mph, travels north on a parallel track. How many hours will the second train travel before it overtakes the first train?   [8.4b]

Solve. [8.5a]

15. $x + 2y + z = 10,$
$2x - y + z = 8,$
$3x + y + 4z = 2$

16. $3x + 2y + z = 1,$
$2x - y - 3z = 1,$
$-x + 3y + 2z = 6$

17. $2x - 5y - 2z = -4,$
$7x + 2y - 5z = -6,$
$-2x + 3y + 2z = 4$

18. $x + y + 2z = 1,$
$x - y + z = 1,$
$x + 2y + z = 2$

19. *Triangle Measure.* In triangle ABC, the measure of angle A is four times the measure of angle C, and the measure of angle B is 45° more than the measure of angle C. What are the measures of the angles of the triangle? [8.6a]

20. *Money Mixtures.* Elaine has $194, consisting of $20, $5, and $1 bills. The number of $1 bills is 1 less than the total number of $20 and $5 bills. If she has 39 bills in her purse, how many of each denomination does she have? [8.6a]

21. Solve using the elimination method:
$x - y = -9,$
$y - 2x = 9.$
The first coordinate of the solution is which of the following? [8.3a]

A. 9
B. −9
C. 0
D. $\frac{9}{2}$

22. The sum of two numbers is −2. The sum of twice one number and the other is 4. One number is which of the following? [8.3b]

A. −6
B. 2
C. 6
D. 8

23. *Distance Traveled.* Two cars leave Martinsville traveling in opposite directions. One car travels at a speed of 50 mph and the other at 60 mph. In how many hours will they be 275 mi apart? [8.4b]

A. 2.5 hr
B. 3 hr
C. 3.5 hr
D. 4 hr

Synthesis

24. Solve graphically: [7.1c], [8.1a]
$y = x + 2,$
$y = x^2 + 2.$

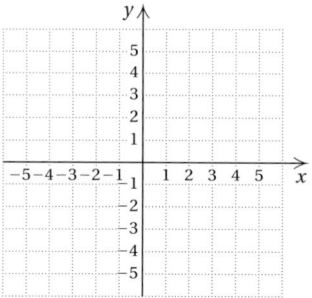

Understanding Through Discussion and Writing

1. Write a problem for a classmate to solve. Design the problem so the answer is "The florist sold 14 hanging baskets and 9 flats of petunias." [8.4a]

2. Exercise 14 in Exercise Set 8.6 can be solved mentally after a careful reading of the problem. Explain how this can be done. [8.6a]

3. *Ticket Revenue.* A pops-concert audience of 100 people consists of adults, senior citizens, and children. The ticket prices are $10 each for adults, $3 each for senior citizens, and $0.50 each for children. The total amount of money taken in is $100. How many adults, senior citizens, and children are in attendance? Does there seem to be some information missing? Do some careful reasoning and explain. [8.6a]

CHAPTER

8

Test

For Extra Help

CHAPTER
Test Prep
VIDEOS

Step-by-step test solutions are found on the Chapter Test Prep Videos available via the Video Resources on DVD, in *MyMathLab* , and on YouTube (search "BittingerIntroInter" and click on "Channels").

Solve graphically. Then classify the system as consistent or inconsistent and the equations as dependent or independent.

1. $y = 3x + 7,$
$\quad 3x + 2y = -4$

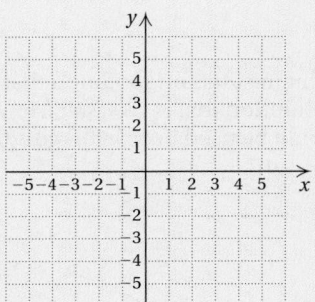

2. $y = 3x + 4,$
$\quad y = 3x - 2$

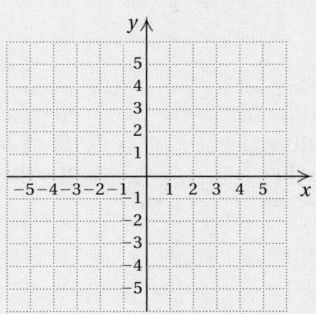

3. $y - 3x = 6,$
$\quad 6x - 2y = -12$

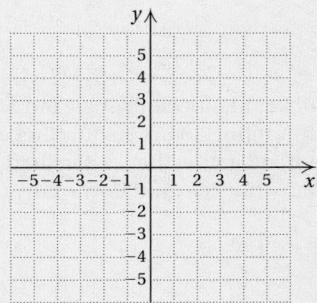

Solve by the substitution method.

4. $4x + 3y = -1,$
$\quad y = 2x - 7$

5. $x = 3y + 2,$
$\quad 2x - 6y = 4$

6. $x + 2y = 6,$
$\quad 2x + 3y = 7$

7. $t = 2 - r,$
$\quad 3r - 2t = 36$

Solve by the elimination method.

8. $2x + 5y = 3,$
$\quad -2x + 3y = 5$

9. $x + y = -2,$
$\quad 4x - 6y = -3$

10. $\dfrac{2}{3}x - \dfrac{4}{5}y = 1,$

$\quad \dfrac{1}{3}x - \dfrac{2}{5}y = 2$

11. $0.3a - 0.4b = 11,$
$\quad 0.7a + 1.2b = -17$

Solve.

12. *Tennis Court.* The perimeter of a standard tennis court used for playing doubles is 288 ft. The width of the court is 42 ft less than the length. Find the length and the width.

13. *Air Travel.* An airplane flew for 5 hr with a 20-km/h tailwind and returned in 7 hr against the same wind. Find the speed of the plane in still air.

14. *Chicken Dinners.* High Flyin' Wings charges $12 for a bucket of chicken wings and $7 for a chicken dinner. After filling 28 orders for buckets and dinners during a football game, the waiters had collected $281. How many buckets and how many dinners did they sell?

15. *Mixing Solutions.* A chemist has one solution that is 20% salt and a second solution that is 45% salt. How many liters of each should be used in order to get 20 L of a solution that is 30% salt?

16. Solve:

$$6x + 2y - 4z = 15,$$
$$-3x - 4y + 2z = -6,$$
$$4x - 6y + 3z = 8.$$

17. *Repair Rates.* An electrician, a carpenter, and a plumber are hired to work on a house. The electrician earns $21 per hour, the carpenter $19.50 per hour, and the plumber $24 per hour. The first day on the job, they worked a total of 21.5 hr and earned a total of $469.50. If the plumber worked 2 hr more than the carpenter did, how many hours did the electrician work?

18. A business class divided an imaginary $30,000 investment among three funds. The first fund grew 2%, the second grew 3%, and the third grew 5%. Total earnings were $990. The earnings from the third fund were $280 more than the earnings from the first. How much was invested at 5%?

A. $9000   **B.** $10,000   **C.** $11,000   **D.** $12,000

Synthesis

19. The graph of the function $f(x) = mx + b$ contains the points $(-1, 3)$ and $(-2, -4)$. Find m and b.

Cumulative Review

Perform the indicated operations and simplify.

1. $(3x^4 - 2y^5)(3x^4 + 2y^5)$ **2.** $(x^2 + 4)^2$

3. $\left(2x + \dfrac{1}{4}\right)\left(4x - \dfrac{1}{2}\right)$ **4.** $\dfrac{x}{2x - 1} - \dfrac{3x + 2}{1 - 2x}$

5. $(3x^2 - 2x^3) - (x^3 - 2x^2 + 5) + (3x^2 - 5x + 5)$

6. $\dfrac{2x + 2}{3x - 9} \cdot \dfrac{x^2 - 8x + 15}{x^2 - 1}$

7. $\dfrac{2x^2 - 2}{2x^2 + 7x + 3} \div \dfrac{4x - 4}{2x^2 - 5x - 3}$

8. $(3x^3 - 2x^2 + x - 5) \div (x - 2)$

Factor completely.

9. $3 - 12x^8$ **10.** $12t - 4t^2 - 48t^4$

11. $6x^2 - 28x + 16$ **12.** $4x^3 + 4x^2 - x - 1$

13. $16x^4 - 56x^2 + 49$ **14.** $x^2 + 3x - 180$

15. Find the slope and the y-intercept of $5y - 4x = 20$.

16. Find an equation of the line with slope -3 and containing the point $(5, 2)$.

17. Find an equation of the line parallel to $3x - 9y = 2$ and containing the point $(-6, 2)$.

18. Determine whether the graphs of the given lines are parallel, perpendicular, or neither.

$$x - 2y = 4,$$
$$4x + 2y = 1$$

Solve.

19. $x^2 = -17x$

20. $\dfrac{1}{4}x + \dfrac{2}{3}x = \dfrac{2}{3} - \dfrac{3}{4}x$

21. $\dfrac{1}{x} + \dfrac{2}{3} = \dfrac{1}{4}$

22. $x^2 - 30 = x$

23. $-4(x + 5) \geq 2(x + 5) - 3$

24. $\dfrac{x}{x - 1} - \dfrac{x}{x + 1} = \dfrac{1}{2x - 2}$

25. Solve $4A = pr + pq$ for p.

Solve.

26. $3x + 4y = 4,$
$\quad\;\; x = 2y + 2$

27. $3x + y = 2,$
$\quad\;\; 6x - y = 7$

28. $4x + 3y = 5,$
$\quad\;\; 3x + 2y = 3$

29. $\;\;x - y + z = 1,$
$\;\;2x + y + z = 3,$
$\quad x + y - 2z = 4$

Graph on a plane.

30. $3y = 9$

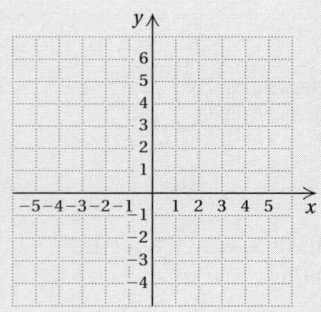

31. $f(x) = -\dfrac{1}{2}x - 3$

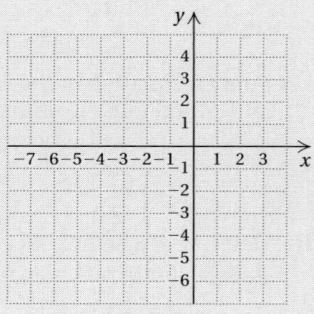

32. $3x - 1 = y$

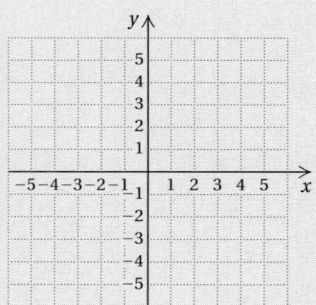

33. $3x + 5y = 15$

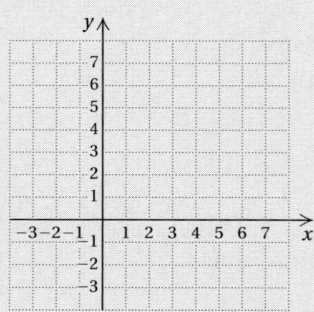

34. For the function f whose graph is shown below, determine **(a)** the domain, **(b)** the range, **(c)** $f(-3)$, and **(d)** any input for which $f(x) = 5$.

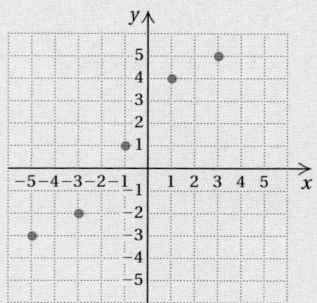

35. Find the domain of the function given by
$$f(x) = \dfrac{7}{2x - 1}.$$

36. Given $g(x) = 1 - 2x^2$, find $g(-1)$, $g(0)$, and $g(3)$.

37. *Mixing Solutions.* A technician wants to mix one solution that is 15% alcohol with another solution that is 25% alcohol in order to get 30 L of a solution that is 18% alcohol. How much of each solution should be used?

38. *Utility Cost.* One month Ladi and Bo spent $680 for electricity, rent, and telephone. The electric bill was one-fourth of the rent and the rent was $400 more than the phone bill. How much was the electric bill?

39. *Quality Control.* A sample of 150 resistors contained 12 defective resistors. How many defective resistors would you expect to find in a sample of 250 resistors?

40. *Rectangle Dimensions.* The length of a rectangle is 3 m greater than the width. The area of the rectangle is 180 m^2. Find the length and the width.

41. *Apparent Size.* The apparent size A of an object varies inversely as the distance d of the object from the eye. You are sitting at a concert 100 ft from the stage. The musicians appear to be 4 ft tall. How tall would they appear to be if you were sitting 1000 ft away in the lawn seats?

42. *Angles of a Triangle.* The second angle of a triangle is twice as large as the first. The third angle is 48° less than the sum of the other two angles. Find the measures of the angles.

Synthesis

43. *Radio Advertising.* An automotive dealer discovers that when $1000 is spent on radio advertising, weekly sales increase by $101,000. When $1250 is spent on radio advertising, weekly sales increase by $126,000. Assuming that sales increase according to a linear function, by what amount would sales increase when $1500 is spent on radio advertising?

44. Given that $f(x) = mx + b$ and that $f(5) = -3$ when $f(-4) = 2$, find m and b.

More on Inequalities

Real-World Application

The formula $I = 2(s + 10)$ can be used to convert dress sizes s in the United States to dress sizes I in Italy. For what dress sizes in the United States will dress sizes in Italy be larger than 36?

This problem appears as Exercise 83 in Section 9.1.

9.1

Sets, Inequalities, and Interval Notation

OBJECTIVES

a Determine whether a given number is a solution of an inequality.

b Write interval notation for the solution set or the graph of an inequality.

c Solve an inequality using the addition principle and the multiplication principle and then graph the inequality.

d Solve applied problems by translating to inequalities.

SKILL TO REVIEW

Objective 2.7b: Graph an inequality on the number line.

Graph each inequality.

1. $x > -2$

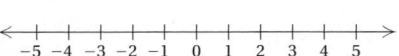

2. $x \leq 1$

Determine whether the given number is a solution of the inequality.

1. $3 - x < 2$; 8

2. $3x + 2 > -1$; -2

3. $3x + 2 \leq 4x - 3$; 5

Answers

Skill to Review:

1.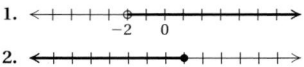

2.

Margin Exercises:

1. Yes **2.** No **3.** Yes

We begin this chapter with a review of solving inequalities. In Chapter 2, we wrote solution sets of inequalities using *set-builder notation*. Here we will also write solution sets using *interval notation*.

a Inequalities

> **INEQUALITY**
>
> An **inequality** is a sentence containing $<, >, \leq, \geq,$ or $\neq$.

Some examples of inequalities are

$$-2 < a, \qquad x > 4, \qquad x + 3 \leq 6, \qquad 6 - 7y \geq 10y - 4, \quad \text{and} \quad 5x \neq 10.$$

> **SOLUTION OF AN INEQUALITY**
>
> Any replacement or value for the variable that makes an inequality true is called a **solution** of the inequality. The set of all solutions is called the **solution set**. When all the solutions of an inequality have been found, we say that we have **solved** the inequality.

EXAMPLES Determine whether the given number is a solution of the inequality.

1. $x + 3 < 6$; 5

We substitute 5 for x and get $5 + 3 < 6$, or $8 < 6$, a *false* sentence. Therefore, 5 is not a solution.

2. $2x - 3 > -3$; 1

We substitute 1 for x and get $2(1) - 3 > -3$, or $-1 > -3$, a *true* sentence. Therefore, 1 is a solution.

3. $4x - 1 \leq 3x + 2$; -3

We substitute -3 for x and get $4(-3) - 1 \leq 3(-3) + 2$, or $-13 \leq -7$, a *true* sentence. Therefore, -3 is a solution.

Do Margin Exercises 1–3.

b Inequalities and Interval Notation

The **graph** of an inequality is a drawing that represents its solutions. An inequality in one variable can be graphed on the number line.

EXAMPLE 4 Graph $x < 4$ on the number line.

The solutions are all real numbers less than 4, so we shade all numbers less than 4 on the number line. To indicate that 4 is not a solution, we use a right parenthesis ")" at 4.

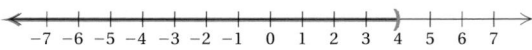

We can write the solution set for $x < 4$ using **set-builder notation**: $\{x | x < 4\}$. This is read "The set of all x such that x is less than 4."

In Chapter 2, we used an open circle to indicate that a number *is not* part of a graph and a closed circle to indicate that a number *is* part of a graph. Beginning in this chapter, we use a parenthesis "(" or ")" to indicate that a number *is not* part of a graph and a bracket "[" or "]" to indicate that a number *is* part of a graph.

Another way to write solutions of an inequality in one variable is to use **interval notation**. Interval notation uses parentheses ( ) and brackets [ ].

If a and b are real numbers such that $a < b$, we define the interval (a, b) as the set of all numbers between but not including a and b—that is, the set of all x for which $a < x < b$. Thus,

$$(a, b) = \{x | a < x < b\}.$$

The points a and b are the **endpoints** of the interval. The parentheses indicate that the endpoints are *not* included in the graph.

The interval $[a, b]$ is defined as the set of all numbers x for which $a \le x \le b$. Thus,

$$[a, b] = \{x | a \le x \le b\}.$$

The brackets indicate that the endpoints *are* included in the graph.*

The following intervals include one endpoint and exclude the other:

$$(a, b] = \{x | a < x \le b\}. \quad \text{The graph excludes } a \text{ and includes } b.$$

$$[a, b) = \{x | a \le x < b\}. \quad \text{The graph includes } a \text{ and excludes } b.$$

Some intervals extend without bound in one or both directions. We use the symbols ∞, read "infinity," and $-\infty$, read "negative infinity," to name these intervals. The notation (a, ∞) represents the set of all numbers greater than a—that is,

$$(a, \infty) = \{x | x > a\}.$$

Similarly, the notation $(-\infty, a)$ represents the set of all numbers less than a—that is,

$$(-\infty, a) = \{x | x < a\}.$$

---

*Some books use the representations ⊸⊸⊸⊸ and ⊢⊸⊸⊣ instead of, respectively, ⟮⟶⟯ and ⟦⟶⟧.

> **Caution!**
>
> Do not confuse the *interval* (a, b) with the *ordered pair* (a, b), which denotes a point in the plane, as we saw in Chapter 3. The context in which the notation appears usually makes the meaning clear.

The notations $[a, \infty)$ and $(-\infty, a]$ are used when we want to include the endpoint a. The interval $(-\infty, \infty)$ names the set of all real numbers.

$$(-\infty, \infty) = \{x \mid x \text{ is a real number}\}$$

Interval notation is summarized in the following table.

INTERVALS: NOTATION AND GRAPHS

INTERVAL NOTATION	SET NOTATION	GRAPH
(a, b)	$\{x \mid a < x < b\}$	(————) a b
$[a, b]$	$\{x \mid a \le x \le b\}$	[————] a b
$[a, b)$	$\{x \mid a \le x < b\}$	[————) a b
$(a, b]$	$\{x \mid a < x \le b\}$	(————] a b
(a, ∞)	$\{x \mid x > a\}$	(———→ a
$[a, \infty)$	$\{x \mid x \ge a\}$	[———→ a
$(-\infty, b)$	$\{x \mid x < b\}$	←———) b
$(-\infty, b]$	$\{x \mid x \le b\}$	←———] b
$(-\infty, \infty)$	$\{x \mid x \text{ is a real number}\}$	←————→

-- *Caution!* --

Whenever the symbol ∞ is included in interval notation, a right parenthesis ")" is used. Similarly, when $-\infty$ is included, a left parenthesis "(" is used.

EXAMPLES Write interval notation for the given set or graph.

5. $\{x \mid -4 < x < 5\} = (-4, 5)$

6. $\{x \mid x \ge -2\} = [-2, \infty)$

7. $\{x \mid 7 > x \ge 1\} = \{x \mid 1 \le x < 7\} = [1, 7)$

8.
←——+——+——+——(——+——+——+——+——]——+——+——→
   −6 −5 −4 −3 −2 −1 0 1 2 3 4 5 6
                    $(-2, 4]$

9.
←——+——+——+——+——+——)——+——+——+——+——+——+——→
   −6 −5 −4 −3 −2 −1 0 1 2 3 4 5 6
                $(-\infty, -1)$

Do Exercises 4–8.

(c) Solving Inequalities

Two inequalities are **equivalent** if they have the same solution set. For example, the inequalities $x > 4$ and $4 < x$ are equivalent. Just as the addition principle for equations gives us equivalent equations, the addition principle for inequalities gives us equivalent inequalities.

Write interval notation for the given set or graph.

4. $\{x \mid -4 \le x < 5\}$

5. $\{x \mid x \le -2\}$

6. $\{x \mid 6 \ge x > 2\}$

7.
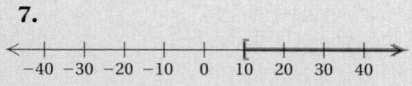
←——+——+——+——[——+——+——+——+——→
−40 −30 −20 −10 0 10 20 30 40

8.
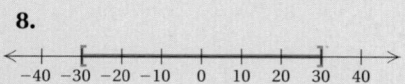
←——+——[——+——+——+——+——+——]——+——→
−40 −30 −20 −10 0 10 20 30 40

Answers

4. $[-4, 5)$ **5.** $(-\infty, -2]$ **6.** $(2, 6]$
7. $[10, \infty)$ **8.** $[-30, 30]$

THE ADDITION PRINCIPLE FOR INEQUALITIES

For any real numbers a, b, and c:

$a < b$ is equivalent to $a + c < b + c$;

$a > b$ is equivalent to $a + c > b + c$.

Similar statements hold for $\leq$ and $\geq$.

Since subtracting c is the same as adding $-c$, there is no need for a separate subtraction principle.

EXAMPLE 10 Solve and graph: $x + 5 > 1$.

We have

$$x + 5 > 1$$
$$x + 5 - 5 > 1 - 5 \qquad \text{Using the addition principle:}$$
$$\qquad\qquad\qquad\qquad \text{adding } -5 \text{ or subtracting } 5$$
$$x > -4.$$

We used the addition principle to show that the inequalities $x + 5 > 1$ and $x > -4$ are equivalent. The solution set is $\{x \mid x > -4\}$ and consists of an infinite number of solutions. We cannot possibly check them all. Instead, we can perform a partial check by substituting one member of the solution set (here we use -1) into the original inequality:

$$\dfrac{x + 5 > 1}{-1 + 5 \; ? \; 1}$$
$$\qquad\; 4 \mid \quad \text{TRUE}$$

Since $4 > 1$ is true, we have a partial check. The solution set is $\{x \mid x > -4\}$, or $(-4, \infty)$. The graph is as follows:

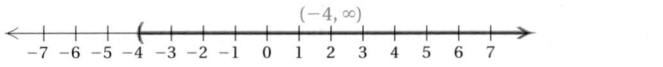

Do Exercises 9 and 10.

EXAMPLE 11 Solve and graph: $4x - 1 \geq 5x - 2$.

We have

$$4x - 1 \geq 5x - 2$$
$$4x - 1 + 2 \geq 5x - 2 + 2 \qquad \text{Adding 2}$$
$$4x + 1 \geq 5x \qquad\qquad\qquad \text{Simplifying}$$
$$4x + 1 - 4x \geq 5x - 4x \qquad \text{Subtracting } 4x$$
$$1 \geq x. \qquad\qquad\qquad\quad \text{Simplifying}$$

The inequalities $1 \geq x$ and $x \leq 1$ have the same meaning and the same solutions. The solution set is $\{x \mid 1 \geq x\}$ or, more commonly, $\{x \mid x \leq 1\}$. Using interval notation, we write that the solution set is $(-\infty, 1]$. The graph is as follows:

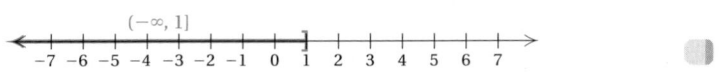

Do Exercise 11.

Solve and graph.

9. $x + 6 > 9$

10. $x + 4 \leq 7$

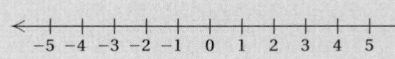

11. Solve and graph:
$$2x - 3 \geq 3x - 1.$$

Answers

9. $\{x \mid x > 3\}$, or $(3, \infty)$;

10. $\{x \mid x \leq 3\}$, or $(-\infty, 3]$;

11. $\{x \mid x \leq -2\}$, or $(-\infty, -2]$;

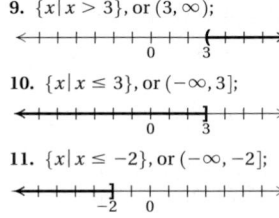

The multiplication principle for inequalities differs from the multiplication principle for equations. Consider the true inequality

$$-4 < 9.$$

If we multiply both numbers by 2, we get another true inequality:

$$-4(2) < 9(2), \quad \text{or} \quad -8 < 18. \qquad \text{True}$$

If we multiply both numbers by -3, we get a false inequality:

$$-4(-3) < 9(-3), \quad \text{or} \quad 12 < -27. \qquad \text{False}$$

However, if we now *reverse* the inequality symbol above, we get a true inequality:

$$12 > -27. \qquad \text{True}$$

THE MULTIPLICATION PRINCIPLE FOR INEQUALITIES

For any real numbers a and b, and any *positive* number c:

    $a < b$ is equivalent to $ac < bc$;

    $a > b$ is equivalent to $ac > bc$.

For any real numbers a and b, and any *negative* number c:

    $a < b$ is equivalent to $ac > bc$;

    $a > b$ is equivalent to $ac < bc$.

Similar statements hold for $\leq$ and $\geq$.

Since division by c is the same as multiplication by $1/c$, there is no need for a separate division principle.

The multiplication principle tells us that when we multiply or divide on both sides of an inequality by a negative number, we must reverse the inequality symbol to obtain an equivalent inequality.

EXAMPLE 12 Solve and graph: $3y < \frac{3}{4}$.

We have

$$3y < \frac{3}{4}$$

$$\frac{1}{3} \cdot 3y < \frac{1}{3} \cdot \frac{3}{4} \qquad \text{Multiplying by } \tfrac{1}{3}. \text{ Since } \tfrac{1}{3} > 0, \text{ the symbol stays the same.}$$

$$y < \frac{1}{4}. \qquad \text{Simplifying}$$

Any number less than $\frac{1}{4}$ is a solution. The solution set is $\left\{ y \,\middle|\, y < \frac{1}{4} \right\}$, or $\left(-\infty, \frac{1}{4} \right)$. The graph is as follows:

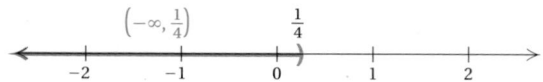

EXAMPLE 13   Solve and graph: $-5x \geq -80$.

We have

$$-5x \geq -80$$

$$\frac{-5x}{-5} \leq \frac{-80}{-5} \quad\boxed{\begin{array}{l}\text{Dividing by } -5. \text{ Since } -5 < 0, \text{ the}\\ \text{inequality symbol must be reversed.}\end{array}}$$

$$x \leq 16.$$

The solution set is $\{x \mid x \leq 16\}$, or $(-\infty, 16]$. The graph is as follows:

$(-\infty, 16]$

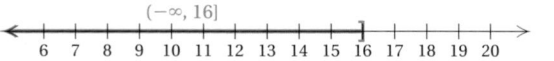

6 7 8 9 10 11 12 13 14 15 16 17 18 19 20

Do Exercises 12–14.

We use the addition and multiplication principles together in solving inequalities in much the same way as in solving equations.

EXAMPLE 14   Solve: $16 - 7y \geq 10y - 4$.

We have

$$16 - 7y \geq 10y - 4$$
$$-16 + 16 - 7y \geq -16 + 10y - 4 \qquad \text{Adding } -16$$
$$-7y \geq 10y - 20 \qquad \text{Collecting like terms}$$
$$-10y + (-7y) \geq -10y + 10y - 20 \qquad \text{Adding } -10y$$
$$-17y \geq -20 \qquad \text{Collecting like terms}$$
$$\frac{-17y}{-17} \leq \frac{-20}{-17} \qquad\boxed{\begin{array}{l}\text{Dividing by } -17. \text{ The symbol}\\ \text{must be reversed.}\end{array}}$$
$$y \leq \frac{20}{17}. \qquad \text{Simplifying}$$

The solution set is $\left\{y \mid y \leq \frac{20}{17}\right\}$, or $\left(-\infty, \frac{20}{17}\right]$.

In some cases, we can avoid multiplying or dividing by a negative number by using the addition principle in a different way. Let's rework Example 14 by adding $7y$ instead of $-10y$:

$$16 - 7y \geq 10y - 4$$
$$16 - 7y + 7y \geq 10y - 4 + 7y \qquad\begin{array}{l}\text{Adding } 7y. \text{ This makes the coefficient}\\ \text{of the } y\text{-term positive.}\end{array}$$
$$16 \geq 17y - 4 \qquad \text{Collecting like terms}$$
$$16 + 4 \geq 17y - 4 + 4 \qquad \text{Adding } 4$$
$$20 \geq 17y \qquad \text{Collecting like terms}$$
$$\frac{20}{17} \geq \frac{17y}{17} \qquad\boxed{\begin{array}{l}\text{Dividing by } 17. \text{ The symbol}\\ \text{stays the same.}\end{array}}$$
$$\frac{20}{17} \geq y.$$

Note that $\frac{20}{17} \geq y$ is equivalent to $y \leq \frac{20}{17}$.

Solve and graph.

12. $5y \leq \dfrac{3}{2}$

$\xleftarrow{\hspace{0.3cm}}\!\!\underset{\substack{-5\ -4\ -3\ -2\ -1\ \ 0\ \ 1\ \ 2\ \ 3\ \ 4\ \ 5}}{+\!+\!+\!+\!+\!+\!+\!+\!+\!+}\!\!\xrightarrow{\hspace{0.3cm}}$

13. $-2y > 10$

$\xleftarrow{\hspace{0.3cm}}\!\!\underset{\substack{-5\ -4\ -3\ -2\ -1\ \ 0\ \ 1\ \ 2\ \ 3\ \ 4\ \ 5}}{+\!+\!+\!+\!+\!+\!+\!+\!+\!+}\!\!\xrightarrow{\hspace{0.3cm}}$

14. $-\dfrac{1}{3}x \leq -4$

$\xleftarrow{\hspace{0.3cm}}\!\!\underset{\substack{-40\ -30\ -20\ -10\ \ 0\ \ 10\ \ 20\ \ 30\ \ 40}}{+\!+\!+\!+\!+\!+\!+\!+}\!\!\xrightarrow{\hspace{0.3cm}}$

Answers

12. $\left\{y \mid y \leq \dfrac{3}{10}\right\}$, or $\left(-\infty, \dfrac{3}{10}\right]$

$\overset{\dfrac{3}{10}}{\xleftarrow{\hspace{0.3cm}}\!\!\underset{0}{+\!+\!+\!\rlap{[}+\!+\!+\!+\!+\!+}\!\!\xrightarrow{\hspace{0.3cm}}}$

13. $\{y \mid y < -5\}$, or $(-\infty, -5)$

$\xleftarrow{\hspace{0.3cm}}\!\!\underset{\substack{-5\hspace{2.2cm}0}}{\rlap{)}+\!+\!+\!+\!+\!+\!+\!+\!+\!+}\!\!\xrightarrow{\hspace{0.3cm}}$

14. $\{x \mid x \geq 12\}$, or $[12, \infty)$

$\xleftarrow{\hspace{0.3cm}}\!\!\underset{\substack{0\hspace{0.6cm}4\hspace{1.2cm}12}}{+\!+\!+\!+\!+\!+\!\rlap{[}+\!+\!+}\!\!\xrightarrow{\hspace{0.3cm}}$

EXAMPLE 15 Solve: $-3(x + 8) - 5x > 4x - 9$.

We have

$$-3(x + 8) - 5x > 4x - 9$$

$\quad -3x - 24 - 5x > 4x - 9 \qquad$ Using the distributive law

$\quad\quad\quad -24 - 8x > 4x - 9 \qquad$ Collecting like terms

$\quad -24 - 8x + 8x > 4x - 9 + 8x \qquad$ Adding $8x$

$\quad\quad\quad\quad\quad -24 > 12x - 9 \qquad$ Collecting like terms

$\quad\quad\quad -24 + 9 > 12x - 9 + 9 \qquad$ Adding 9

$\quad\quad\quad\quad\quad\quad -15 > 12x$

Dividing by 12. The symbol stays the same.

$$\frac{-15}{12} > \frac{12x}{12}$$

$$-\frac{5}{4} > x.$$

The solution set is $\left\{x \mid -\frac{5}{4} > x\right\}$, or $\left\{x \mid x < -\frac{5}{4}\right\}$, or $\left(-\infty, -\frac{5}{4}\right)$.

Do Exercises 15–17.

(d) Applications and Problem Solving

Many problem-solving and applied situations translate to inequalities. In addition to "is less than" and "is more than," other phrases are commonly used.

IMPORTANT WORDS	SAMPLE SENTENCE	TRANSLATION
is at least	Max is at least 5 years old.	$m \geq 5$
is at most	At most 6 people could fit in the elevator.	$n \leq 6$
cannot exceed	Total weight in the elevator cannot exceed 2000 pounds.	$w \leq 2000$
must exceed	The speed must exceed 15 mph.	$s > 15$
is between	Heather's income is between $23,000 and $35,000.	$23{,}000 < h < 35{,}000$
no more than	Bing weighs no more than 90 pounds.	$w \leq 90$
no less than	Saul would accept no less than $4000 for the piano.	$t \geq 4000$

The following phrases deserve special attention.

TRANSLATING "AT LEAST" AND "AT MOST"

A quantity x is **at least** some amount q: $x \geq q$.
(If x is at least q, it cannot be less than q.)

A quantity x is **at most** some amount q: $x \leq q$.
(If x is at most q, it cannot be more than q.)

Do Exercises 18–24.

Solve.

15. $6 - 5y \geq 7$

16. $3x + 5x < 4$

17. $17 - 5(y - 2) \leq 45y + 8(2y - 3) - 39y$

Translate.

18. Russell will pay at most $250 for that plane ticket.

19. Emma scored at least an 88 on her Spanish test.

20. The time of the test was between 50 and 60 min.

21. The University of Northern Kentucky is more than 25 mi away.

22. Sarah's weight is less than 110 lb.

23. That number is greater than -8.

24. The costs of production of that DVD player cannot exceed $135,000.

Answers

15. $\left\{y \mid y \leq -\frac{1}{5}\right\}$, or $\left(-\infty, -\frac{1}{5}\right]$

16. $\left\{x \mid x < \frac{1}{2}\right\}$, or $\left(-\infty, \frac{1}{2}\right)$

17. $\left\{y \mid y \geq \frac{17}{9}\right\}$, or $\left[\frac{17}{9}, \infty\right)$

18. $t \leq 250$ **19.** $s \geq 88$ **20.** $50 < t < 60$
21. $d > 25$ **22.** $w < 110$ **23.** $n > -8$
24. $c \leq 135{,}000$

EXAMPLE 16 *Cost of Higher Education.* The equation

$$C = 126t + 1293$$

can be used to estimate the average cost of tuition and fees at two-year public institutions of higher education, where t is the number of years after 2000. Determine, in terms of an inequality, the years for which the cost will be more than $3000.

Source: National Center for Education Statistics

1. **Familiarize.** We already have a formula. To become more familiar with it, we might make a substitution for t. Suppose we want to know the cost 15 yr after 2000, or in 2015. We substitute 15 for t:

   $$C = 126(15) + 1293 = \$3183.$$

   We see that in 2015, the cost of tuition and fees at two-year public institutions will be more than $3000. To find all the years in which the cost exceeds $3000, we could make other guesses less than 15, but it is more efficient to proceed to the next step.

2. **Translate.** The cost C is to be *more than* $3000. Thus we have

   $$C > 3000.$$

   We replace C with $126t + 1293$ to find the values of t that are solutions of the inequality:

   $$126t + 1293 > 3000.$$

3. **Solve.** We solve the inequality:

   $$126t + 1293 > 3000$$
   $$126t > 1707 \qquad \text{Subtracting 1293}$$
   $$t > 13.55. \qquad \text{Dividing by 126 and rounding}$$

4. **Check.** A partial check is to substitute a value for t greater than 13.55. We did that in the *Familiarize* step and found that the cost was more than $3000.

5. **State.** The average cost of tuition and fees at two-year public institutions of higher education will be more than $3000 for more than 13.55 yr after 2000, so we have $\{t \mid t > 13.55\}$.

Do Exercise 25.

25. **Cost of Higher Education.** Refer to Example 16. Determine, in terms of an inequality, the years for which the average cost of tuition and fees is more than $2500.

Answer

25. More than 9.58 yr after 2000, or $\{t \mid t > 9.58\}$

EXAMPLE 17 *Salary Plans.* On her new job, Rose can be paid in one of two ways: *Plan A* is a salary of $600 per month, plus a commission of 4% of sales; and *Plan B* is a salary of $800 per month, plus a commission of 6% of sales in excess of $10,000. For what amount of monthly sales is plan A better than plan B, if we assume that sales are always more than $10,000?

1. **Familiarize.** Listing the given information in a table will be helpful.

PLAN A: MONTHLY INCOME	PLAN B: MONTHLY INCOME
$600 salary	$800 salary
4% of sales	6% of sales over $10,000
Total: $600 + 4% of sales	*Total*: $800 + 6% of sales over $10,000

Next, suppose that Rose had sales of $12,000 in one month. Which plan would be better? Under plan A, she would earn $600 plus 4% of $12,000, or

$$600 + 0.04(12,000) = \$1080.$$

Since with plan B commissions are paid only on sales in excess of $10,000, Rose would earn $800 plus 6% of ($12,000 − $10,000), or

$$800 + 0.06(12,000 - 10,000) = 800 + 0.06(2000) = \$920.$$

This shows that for monthly sales of $12,000, plan A is better. Similar calculations will show that for sales of $30,000 a month, plan B is better. To determine *all* values for which plan A pays more money, we must solve an inequality that is based on the calculations above.

2. **Translate.** We let S = the amount of monthly sales. If we examine the calculations in the *Familiarize* step, we see that the monthly income from plan A is $600 + 0.04S$ and from plan B is $800 + 0.06(S - 10,000)$. Thus we want to find all values of S for which

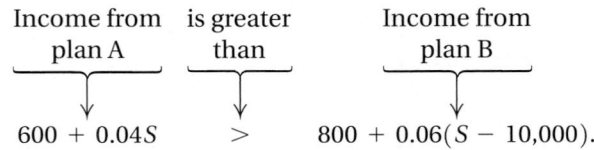

Income from plan A / is greater than / Income from plan B

$$600 + 0.04S \quad > \quad 800 + 0.06(S - 10,000).$$

3. **Solve.** We solve the inequality:

$$600 + 0.04S > 800 + 0.06(S - 10,000)$$

$600 + 0.04S > 800 + 0.06S - 600$	Using the distributive law
$600 + 0.04S > 200 + 0.06S$	Collecting like terms
$400 > 0.02S$	Subtracting 200 and $0.04S$
$20,000 > S$, or $S < 20,000$.	Dividing by 0.02

4. **Check.** For $S = 20,000$, the income from plan A is

$$600 + 4\% \cdot 20,000, \text{ or } \$1400.$$

The income from plan B is

$$800 + 6\% \cdot (20,000 - 10,000), \text{ or } \$1400.$$

This confirms that for sales of $20,000, Rose's pay is the same under either plan.

In the *Familiarize* step, we saw that for sales of $12,000, plan A pays more. Since $12,000 < 20,000$, this is a partial check. Since we cannot check all possible values of S, we will stop here.

5. **State.** For monthly sales of less than $20,000, plan A is better.

26. Salary Plans. A painter can be paid in one of two ways:

Plan A: $500 plus $4 per hour;

Plan B: Straight $9 per hour.

Suppose that the job takes n hours. For what values of n is plan A better for the painter?

Answer

26. For $\{n | n < 100\}$, plan A is better.

Do Exercise 26.

Translating for Success

1. **Consecutive Integers.** The sum of two consecutive even integers is 102. Find the integers.

2. **Salary Increase.** After Susanna earned a 5% raise, her new salary was $25,750. What was her former salary?

3. **Dimensions of a Rectangle.** The length of a rectangle is 6 in. more than the width. The perimeter of the rectangle is 102 in. Find the length and the width.

4. **Population.** The population of Doddville is decreasing at a rate of 5% per year. The current population is 25,750. What was the population the previous year?

5. **Reading Assignment.** Quinn has 6 days to complete a 150-page reading assignment. How many pages must he read the first day so that he has no more than 102 pages left to read on the 5 remaining days?

The goal of these matching questions is to practice step (2), *Translate*, of the five-step problem-solving process. Translate each word problem to an equation or an inequality and select a correct translation from A–O.

A. $0.05(25{,}750) = x$

B. $x + 2x = 102$

C. $2x + 2(x + 6) = 102$

D. $150 - x \leq 102$

E. $x - 0.05x = 25{,}750$

F. $x + (x + 2) = 102$

G. $x + (x + 6) > 102$

H. $x + 5x = 150$

I. $x + 0.05x = 25{,}750$

J. $x + (2x + 6) = 102$

K. $x + (x + 1) = 102$

L. $102 + x > 150$

M. $0.05x = 25{,}750$

N. $102 + 5x > 150$

O. $x + (x + 6) = 102$

Answer on page A-26

---

6. **Numerical Relationship.** One number is 6 more than twice another. The sum of the numbers is 102. Find the numbers.

7. **DVD Collections.** Together Ella and Ken have 102 DVDs. If Ken has 6 more DVDs than Ella, how many does each have?

8. **Sales Commissions.** Will earns a commission of 5% on his sales. One year he earned commissions totaling $25,750. What were his total sales for the year?

9. **Fencing.** Jess has 102 ft of fencing that he plans to use to enclose two dog runs. The perimeter of one run is to be twice the perimeter of the other. Into what lengths should the fencing be cut?

10. **Quiz Scores.** Lupe has a total of 102 points on the first 6 quizzes in her sociology class. How many total points must she earn on the 5 remaining quizzes in order to have more than 150 points for the semester?

a Determine whether the given numbers are solutions of the inequality.

1. $x - 2 \geq 6$; $-4, 0, 4, 8$

2. $3x + 5 \leq -10$; $-5, -10, 0, 27$

3. $t - 8 > 2t - 3$; $0, -8, -9, -3, -\frac{7}{8}$

4. $5y - 7 < 8 - y$; $2, -3, 0, 3, \frac{2}{3}$

b Write interval notation for the given set or graph.

5. $\{x | x < 5\}$

6. $\{t | t \geq -5\}$

7. $\{x | -3 \leq x \leq 3\}$

8. $\{t | -10 < t \leq 10\}$

9. $\{x | -4 > x > -8\}$

10. $\{x | 13 > x \geq 5\}$

11.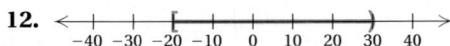

12.

13.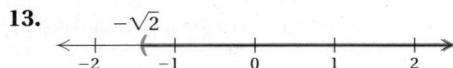

14.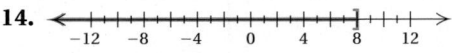

c Solve and graph.

15. $x + 2 > 1$

16. $x + 8 > 4$

17. $y + 3 < 9$

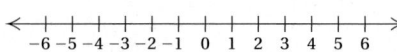 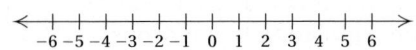

18. $y + 4 < 10$

19. $a - 9 \leq -31$

20. $a + 6 \leq -14$

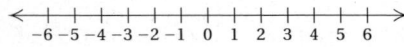 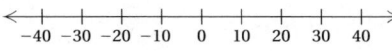 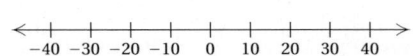

21. $t + 13 \geq 9$

22. $x - 8 \leq 17$

23. $y - 8 > -14$

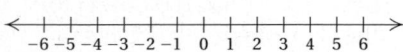 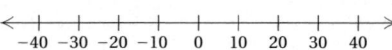 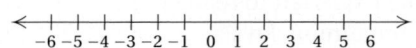

24. $y - 9 > -18$

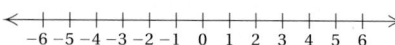

25. $x - 11 \leq -2$

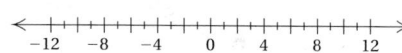

26. $y - 18 \leq -4$

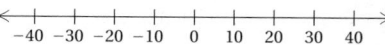

27. $8x \geq 24$

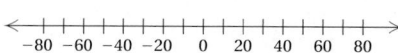

28. $8t < -56$

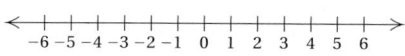

29. $0.3x < -18$

30. $0.6x < 30$

31. $\frac{2}{3}x > 2$

32. $\frac{3}{5}x > -3$

Solve.

33. $-9x \geq -8.1$

34. $-5y \leq 3.5$

35. $-\frac{3}{4}x \geq -\frac{5}{8}$

36. $-\frac{1}{8}y \leq -\frac{9}{8}$

37. $2x + 7 < 19$

38. $5y + 13 > 28$

39. $5y + 2y \leq -21$

40. $-9x + 3x \geq -24$

41. $2y - 7 < 5y - 9$

42. $8x - 9 < 3x - 11$

43. $0.4x + 5 \leq 1.2x - 4$

44. $0.2y + 1 > 2.4y - 10$

45. $5x - \frac{1}{12} \leq \frac{5}{12} + 4x$

46. $2x - 3 < \frac{13}{4}x + 10 - 1.25x$

47. $4(4y - 3) \geq 9(2y + 7)$

48. $2m + 5 \geq 16(m - 4)$

49. $3(2 - 5x) + 2x < 2(4 + 2x)$

50. $2(0.5 - 3y) + y > (4y - 0.2)8$

51. $5[3m - (m + 4)] > -2(m - 4)$

52. $[8x - 3(3x + 2)] - 5 \geq 3(x + 4) - 2x$

53. $3(r - 6) + 2 > 4(r + 2) - 21$

54. $5(t + 3) + 9 < 3(t - 2) + 6$

55. $19 - (2x + 3) \leq 2(x + 3) + x$

56. $13 - (2c + 2) \geq 2(c + 2) + 3c$

57. $\frac{1}{4}(8y + 4) - 17 < -\frac{1}{2}(4y - 8)$

58. $\frac{1}{3}(6x + 24) - 20 > -\frac{1}{4}(12x - 72)$

59. $2[4 - 2(3 - x)] - 1 \geq 4[2(4x - 3) + 7] - 25$

60. $5[3(7 - t) - 4(8 + 2t)] - 20 \leq -6[2(6 + 3t) - 4]$

61. $\frac{4}{5}(7x - 6) < 40$

62. $\frac{2}{3}(4x - 3) > 30$

63. $\frac{3}{4}(3 + 2x) + 1 \geq 13$

64. $\frac{7}{8}(5 - 4x) - 17 \geq 38$

65. $\frac{3}{4}\left(3x - \frac{1}{2}\right) - \frac{2}{3} < \frac{1}{3}$

66. $\frac{2}{3}\left(\frac{7}{8} - 4x\right) - \frac{5}{8} < \frac{3}{8}$

67. $0.7(3x + 6) \geq 1.1 - (x + 2)$

68. $0.9(2x + 8) < 20 - (x + 5)$

69. $a + (a - 3) \le (a + 2) - (a + 1)$

70. $0.8 - 4(b - 1) > 0.2 + 3(4 - b)$

d Solve.

Body Mass Index. *Body mass index I* can be used to determine whether an individual has a healthy weight for his or her height. An index in the range 18.5–24.9 indicates a normal weight. Body mass index is given by the formula, or model,

$$I = \frac{703W}{H^2},$$

where W is weight, in pounds, and H is height, in inches. Use this formula for Exercises 71 and 72.

Source: Centers for Disease Control and Prevention

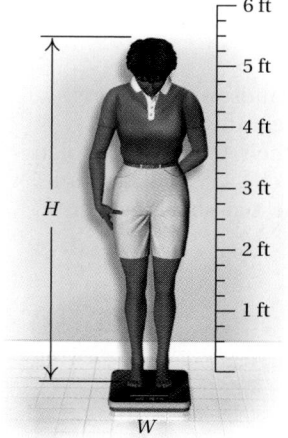

71. *Body Mass Index.* Marv's height is 73 in. Determine, in terms of an inequality, those weights W that will keep his body mass index below 25.

72. *Body Mass Index.* Elaine's height is 67 in. Determine, in terms of an inequality, those weights W that will keep her body mass index below 25.

73. *Grades.* Morris is taking a European history course in which there will be 4 tests, each worth 100 points. He has scores of 89, 92, and 95 on the first three tests. He must make a total of at least 360 in order to get an A. What scores on the last test will give Morris an A?

74. *Grades.* Eve is taking a literature course in which there will be 5 tests, each worth 100 points. She has scores of 94, 90, and 89 on the first three tests. She must make a total of at least 450 in order to get an A. What scores on the fourth test will keep Eve eligible for an A?

75. *Insurance Claims.* After a serious automobile accident, most insurance companies will replace the damaged car with a new one if repair costs exceed 80% of the N.A.D.A., or "blue-book," value of the car. Miguel's car recently sustained $9200 worth of damage but was not replaced. What was the blue-book value of his car?

76. *Delivery Service.* Jay's Express prices cross-town deliveries at $15 for the first 10 miles plus $1.25 for each additional mile. PDQ, Inc., prices its cross-town deliveries at $25 for the first 10 miles plus $0.75 for each additional mile. For what number of miles is PDQ less expensive?

77. *Salary Plans.* Toni can be paid in one of two ways:

Plan A: A salary of $400 per month plus a commission of 8% of gross sales;

Plan B: A salary of $610 per month, plus a commission of 5% of gross sales.

For what amount of gross sales should Toni select plan A?

78. *Salary Plans.* Branford can be paid for his masonry work in one of two ways:

Plan A: $300 plus $9.00 per hour;

Plan B: Straight $12.50 per hour.

Suppose that the job takes n hours. For what values of n is plan B better for Branford?

79. *Checking-Account Rates.* The Hudson Bank offers two checking-account plans. Their Anywhere plan charges 20¢ per check whereas their Acu-checking plan costs $2 per month plus 12¢ per check. For what numbers of checks per month will the Acu-checking plan cost less?

80. *Insurance Benefits.* Bayside Insurance offers two plans. Under plan A, Giselle would pay the first $50 of her medical bills and 20% of all bills after that. Under plan B, Giselle would pay the first $250 of bills, but only 10% of the rest. For what amount of medical bills will plan B save Giselle money? (Assume that her bills will exceed $250.)

81. *Wedding Costs.* The Arnold Inn offers two plans for wedding parties. Under plan A, the inn charges $30 for each person in attendance. Under plan B, the inn charges $1300 plus $20 for each person in excess of the first 25 who attend. For what size parties will plan B cost less? (Assume that more than 25 guests will attend.)

82. *Investing.* Lillian is about to invest $20,000, part at 3% and the rest at 4%. What is the most that she can invest at 3% and still be guaranteed at least $650 in interest per year?

83. *Converting Dress Sizes.* The formula

$$I = 2(s + 10)$$

can be used to convert dress sizes s in the United States to dress sizes I in Italy. For what dress sizes in the United States will dress sizes in Italy be larger than 36?

84. *Temperatures of Solids.* The formula

$$C = \tfrac{5}{9}(F - 32)$$

can be used to convert Fahrenheit temperatures F to Celsius temperatures C.

a) Gold is a solid at Celsius temperatures less than 1063°C. Find the Fahrenheit temperatures for which gold is a solid.

b) Silver is a solid at Celsius temperatures less than 960.8°C. Find the Fahrenheit temperatures for which silver is a solid.

85. *Bottled Water Consumption.* Bottled water consumption has increased steadily in recent years. The number N of gallons, in millions, consumed in the United States t years after 2006 is approximated by the equation

$$N = 0.6t + 8.2.$$

Source: Beverage Marketing Corporation

a) How many gallons of bottled water were consumed in the United States in 2006 ($t = 0$)? in 2008 ($t = 2$)? in 2010 ($t = 4$)?

b) For what years will the amount of bottled water consumed in the United States exceed 12 million gal?

86. *Dewpoint Spread.* Pilots use the *dewpoint spread*, or the difference between the current temperature and the dewpoint (the temperature at which dew occurs), to estimate the height of the cloud cover. Each 3° of dewpoint spread corresponds to an increased height of cloud cover of 1000 ft. A plane, flying with limited instruments, must have a cloud cover higher than 3500 ft. What dewpoint spreads will allow the plane to fly?

Decrease of 3° per 1000 ft

3500 ft

Skill Maintenance

Multiply.   [4.6a]

87. $(3x - 4)(x + 8)$

88. $(r - 4s)(6r + s)$

89. $(2a - 5)(3a + 11)$

90. $(t + 2s)(t - 9s)$

Factor.   [5.7a]

91. $4x^2 - 36x + 81$

92. $400y^2 - 16$

93. $27w^3 - 8$

94. $80 - 14x - 6x^2$

Find the domain.   [7.2a]

95. $f(x) = \dfrac{-3}{x + 8}$

96. $f(x) = 3x - 5$

97. $f(x) = |x| - 4$

98. $f(x) = \dfrac{x + 7}{3x - 2}$

Synthesis

99. *Supply and Demand.* The supply S and demand D for a certain product are given by

$$S = 460 + 94p \quad \text{and} \quad D = 2000 - 60p.$$

a) Find those values of p for which supply exceeds demand.

b) Find those values of p for which supply is less than demand.

Determine whether each statement is true or false. If false, give a counterexample.

100. For any real numbers x and y, if $x < y$, then $x^2 < y^2$.

101. For any real numbers a, b, c, and d, if $a < b$ and $c < d$, then $a + c < b + d$.

102. Determine whether the inequalities

$$x < 3 \quad \text{and} \quad 0 \cdot x < 0 \cdot 3$$

are equivalent. Give reasons to support your answer.

Solve.

103. $x + 5 \le 5 + x$

104. $x + 8 < 3 + x$

105. $x^2 + 1 > 0$

9.2

Intersections, Unions, and Compound Inequalities

Cholesterol is a substance that is found in every cell of the human body. High levels of cholesterol can cause fatty deposits in the blood vessels that increase the risk of heart attack or stroke. A blood test can be used to measure *total cholesterol*. The following table shows the health risks associated with various cholesterol levels.

TOTAL CHOLESTEROL	RISK LEVEL
Less than 200	Normal
From 200 to 239	Borderline high
240 or higher	High

A total-cholesterol level T from 200 to 239 is considered borderline high. We can express this by the sentence

$$200 \leq T \quad and \quad T \leq 239$$

or more simply by

$$200 \leq T \leq 239.$$

This is an example of a *compound inequality*. **Compound inequalities** consist of two or more inequalities joined by the word *and* or the word *or*. We now "solve" such sentences—that is, we find the set of all solutions.

a Intersections of Sets and Conjunctions of Inequalities

> ### INTERSECTION
>
> The **intersection** of two sets A and B is the set of all members that are common to A and B. We denote the intersection of sets A and B as
>
> $$A \cap B.$$

The intersection of two sets is often illustrated as shown at right.

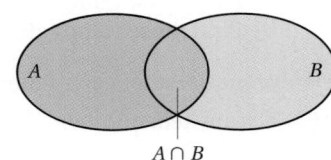

$A \cap B$

EXAMPLE 1 Find the intersection: $\{1, 2, 3, 4, 5\} \cap \{-2, -1, 0, 1, 2, 3\}$.

The numbers 1, 2, and 3 are common to the two sets, so the intersection is $\{1, 2, 3\}$.

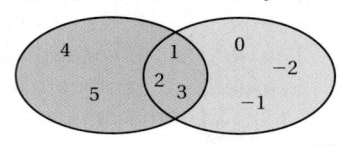

Do Exercises 1 and 2.

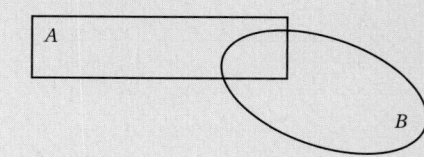

1. Find the intersection:
$$\{0, 3, 5, 7\} \cap \{0, 1, 3, 11\}.$$

2. Shade the intersection of sets A and B.

CONJUNCTION

When two or more sentences are joined by the word *and* to make a compound sentence, the new sentence is called a **conjunction** of the sentences.

The following is a conjunction of inequalities:

$$-2 < x \quad and \quad x < 1.$$

A number is a solution of a conjunction if it is a solution of *both* inequalities. For example, 0 is a solution of $-2 < x$ *and* $x < 1$ because $-2 < 0$ *and* $0 < 1$. Shown below is the graph of $-2 < x$, followed by the graph of $x < 1$, and then by the graph of the conjunction $-2 < x$ *and* $x < 1$. As the graphs demonstrate, *the solution set of a conjunction is the intersection of the solution sets of the individual inequalities.*

$\{x \mid -2 < x\}$ ⟶ $(-2, \infty)$

$\{x \mid x < 1\}$ ⟶ $(-\infty, 1)$

$\{x \mid -2 < x\} \cap \{x \mid x < 1\}$
$= \{x \mid -2 < x \text{ and } x < 1\}$ ⟶ $(-2, 1)$

Because there are numbers that are both greater than -2 and less than 1, the conjunction $-2 < x$ *and* $x < 1$ can be abbreviated by $-2 < x < 1$. Thus the interval $(-2, 1)$ can be represented as $\{x \mid -2 < x < 1\}$, the set of all numbers that are *simultaneously* greater than -2 *and* less than 1. Note that, in general, for $a < b$,

$$a < x \quad and \quad x < b \quad \text{can be abbreviated} \quad a < x < b;$$
$$and \quad b > x \quad and \quad x > a \quad \text{can be abbreviated} \quad b > x > a.$$

Caution!

"$a > x$ *and* $x < b$" cannot be abbreviated as "$a > x < b$".

Answers

1. $\{0, 3\}$

2.

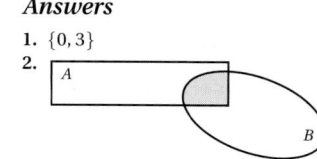

3. Graph and write interval notation:

$$-1 < x \text{ and } x < 4.$$

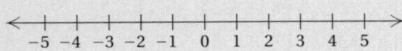

> ## "AND"; "INTERSECTION"
>
> The word **"and"** corresponds to **"intersection"** and to the symbol "∩". In order for a number to be a solution of a conjunction, it must make each part of the conjunction true.

> Do Exercise 3.

EXAMPLE 2   Solve and graph: $-1 \le 2x + 5 < 13$.

This inequality is an abbreviation for the conjunction

$$-1 \le 2x + 5 \quad \text{and} \quad 2x + 5 < 13.$$

The word *and* corresponds to set *intersection*, ∩. To solve the conjunction, we solve each of the two inequalities separately and then find the intersection of the solution sets:

$-1 \le 2x + 5$	*and*	$2x + 5 < 13$	
$-6 \le 2x$	*and*	$2x < 8$	Subtracting 5
$-3 \le x$	*and*	$x < 4$.	Dividing by 2

We now abbreviate the result:

$$-3 \le x < 4.$$

The solution set is $\{x \mid -3 \le x < 4\}$, or, in interval notation, $[-3, 4)$. The graph is the intersection of the two separate solution sets.

$\{x \mid -3 \le x\}$ 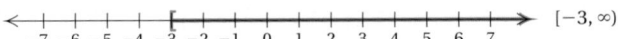 $[-3, \infty)$

$\{x \mid x < 4\}$ 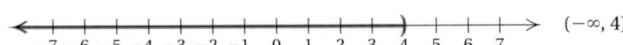 $(-\infty, 4)$

$\{x \mid -3 \le x\} \cap \{x \mid x < 4\}$
$= \{x \mid -3 \le x < 4\}$ 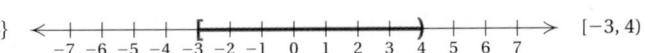 $[-3, 4)$

The steps above are generally combined as follows:

$-1 \le 2x + 5 < 13$	$2x + 5$ appears in both inequalities.
$-6 \le 2x < 8$	Subtracting 5
$-3 \le x < 4$.	Dividing by 2

Such an approach saves some writing and will prove useful in Section 9.3.

4. Solve and graph:

$$-22 < 3x - 7 \le 23.$$

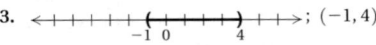

> Do Exercise 4.

EXAMPLE 3   Solve and graph: $2x - 5 \ge -3 \text{ and } 5x + 2 \ge 17$.

We first solve each inequality separately:

$2x - 5 \ge -3$	*and*	$5x + 2 \ge 17$
$2x \ge 2$	*and*	$5x \ge 15$
$x \ge 1$	*and*	$x \ge 3$.

Answers

3. ; $(-1, 4)$

4. $\{x \mid -5 < x \le 10\}$, or $(-5, 10]$;

Next, we find the intersection of the two separate solution sets:

$\{x \mid x \geq 1\}$ 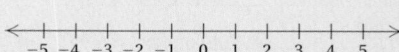 $[1, \infty)$

$\{x \mid x \geq 3\}$ $[3, \infty)$

$\{x \mid x \geq 1\} \cap \{x \mid x \geq 3\}$ $[3, \infty)$
$= \{x \mid x \geq 3\}$

The numbers common to both sets are those that are greater than or equal to 3. Thus the solution set is $\{x \mid x \geq 3\}$, or, in interval notation, $[3, \infty)$. You should check that any number in $[3, \infty)$ satisfies the conjunction whereas numbers outside $[3, \infty)$ do not.

Do Exercise 5.

5. Solve and graph:
$$3x + 4 < 10 \text{ } and \text{ } 2x - 7 < -13.$$

EMPTY SET; DISJOINT SETS

Sometimes two sets have no elements in common. In such a case, we say that the intersection of the two sets is the **empty set**, denoted $\{ \}$ or $\varnothing$. Two sets with an empty intersection are said to be **disjoint**.

$A \cap B = \varnothing$

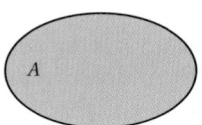 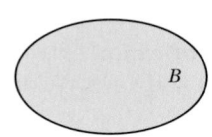

EXAMPLE 4 Solve and graph: $2x - 3 > 1 \text{ } and \text{ } 3x - 1 < 2.$

We solve each inequality separately:

$$2x - 3 > 1 \quad and \quad 3x - 1 < 2$$
$$2x > 4 \quad and \quad 3x < 3$$
$$x > 2 \quad and \quad x < 1.$$

The solution set is the intersection of the solution sets of the individual inequalities.

$\{x \mid x > 2\}$ $(2, \infty)$

$\{x \mid x < 1\}$ $(-\infty, 1)$

$\{x \mid x > 2\} \cap \{x \mid x < 1\}$ $\varnothing$
$= \{x \mid x > 2 \text{ } and \text{ } x < 1\}$
$= \varnothing$

Since no number is both greater than 2 and less than 1, the solution set is the empty set, $\varnothing$.

Do Exercise 6.

6. Solve and graph:
$$3x - 7 \leq -13 \text{ } and \text{ } 4x + 3 > 8.$$

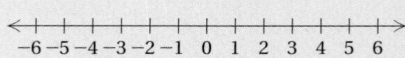

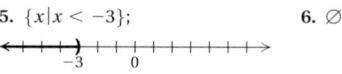

EXAMPLE 5 Solve: $3 \leq 5 - 2x < 7$.

We have

$$3 \leq 5 - 2x < 7$$
$$3 - 5 \leq 5 - 2x - 5 < 7 - 5 \qquad \text{Subtracting 5}$$
$$-2 \leq \qquad -2x \qquad < 2 \qquad \text{Simplifying}$$
$$\frac{-2}{-2} \geq \frac{-2x}{-2} \qquad > \frac{2}{-2} \qquad \begin{array}{l}\text{Dividing by } -2. \text{ The symbols must} \\ \text{be reversed.}\end{array}$$
$$1 \geq x > -1. \qquad \text{Simplifying}$$

The solution set is $\{x | 1 \geq x > -1\}$, or $\{x | -1 < x \leq 1\}$, since the inequalities $1 \geq x > -1$ and $-1 < x \leq 1$ are equivalent. The solution, in interval notation, is $(-1, 1]$.

7. Solve: $-4 \leq 8 - 2x \leq 4$.

Do Exercise 7.

b Unions of Sets and Disjunctions of Inequalities

UNION

The **union** of two sets A and B is the collection of elements belonging to A and/or B. We denote the union of A and B by

$$A \cup B.$$

The union of two sets is often pictured as shown below.

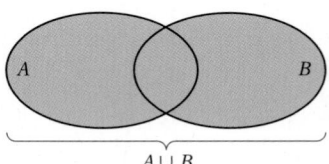

$A \cup B$

8. Find the union:

$\{0, 1, 3, 4\} \cup \{0, 1, 7, 9\}$.

9. Shade the union of sets A and B.

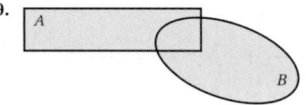

EXAMPLE 6 Find the union: $\{2, 3, 4\} \cup \{3, 5, 7\}$.

The numbers in either or both sets are 2, 3, 4, 5, and 7, so the union is $\{2, 3, 4, 5, 7\}$. We don't list the number 3 twice.

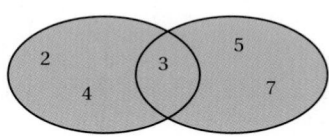

Do Exercises 8 and 9.

DISJUNCTION

When two or more sentences are joined by the word *or* to make a compound sentence, the new sentence is called a **disjunction** of the sentences.

Answers

7. $\{x | 2 \leq x \leq 6\}$, or $[2, 6]$ **8.** $\{0, 1, 3, 4, 7, 9\}$

9.

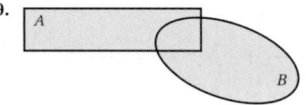

The following is an example of a disjunction:

$$x < -3 \quad or \quad x > 3.$$

A number is a solution of a disjunction if it is a solution of at least one of the individual inequalities. For example, -7 is a solution of $x < -3$ *or* $x > 3$ because $-7 < -3$. Similarly, 5 is also a solution because $5 > 3$.

Shown below is the graph of $x < -3$, followed by the graph of $x > 3$, and then by the graph of the disjunction $x < -3$ *or* $x > 3$. As the graphs demonstrate, *the solution set of a disjunction is the union of the solution sets of the individual sentences.*

$\{x \mid x < -3\}$ $(-\infty, -3)$

$\{x \mid x > 3\}$ $(3, \infty)$

$\{x \mid x < -3\} \cup \{x \mid x > 3\}$
$= \{x \mid x < -3 \ or \ x > 3\}$ $(-\infty, -3)$ $\cup (3, \infty)$

The solution set of

$$x < -3 \quad or \quad x > 3$$

is written $\{x \mid x < -3 \ or \ x > 3\}$, or, in interval notation, $(-\infty, -3) \cup (3, \infty)$. This cannot be written in a more condensed form.

"OR"; "UNION"

The word **"or"** corresponds to **"union"** and the symbol "$\cup$". In order for a number to be in the solution set of a disjunction, it must be in *at least one* of the solution sets of the individual sentences.

Do Exercise 10.

10. Graph and write interval notation:

$$x \leq -2 \ or \ x > 4.$$

EXAMPLE 7 Solve and graph: $7 + 2x < -1$ *or* $13 - 5x \leq 3$.

We solve each inequality separately, retaining the word *or*:

$$7 + 2x < -1 \quad or \quad 13 - 5x \leq 3$$
$$2x < -8 \quad or \quad -5x \leq -10$$

Dividing by -5. The symbol must be reversed.

$$x < -4 \quad or \quad x \geq 2.$$

To find the solution set of the disjunction, we consider the individual graphs. We graph $x < -4$ and then $x \geq 2$. Then we take the union of the graphs.

$\{x \mid x < -4\}$ $(-\infty, -4)$

$\{x \mid x \geq 2\}$ $[2, \infty)$

$\{x \mid x < -4 \ or \ x \geq 2\}$ $(-\infty, -4)$ $\cup [2, \infty)$

The solution set is written $\{x \mid x < -4 \ or \ x \geq 2\}$, or, in interval notation, $(-\infty, -4) \cup [2, \infty)$.

Answer

10.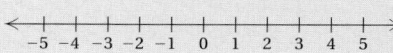
$(-\infty, -2] \cup (4, \infty)$

Solve and graph.

11. $x - 4 < -3 \ or \ x - 3 \geq 3$

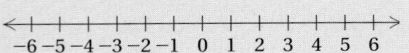

12. $-2x + 4 \leq -3 \ or \ x + 5 < 3$

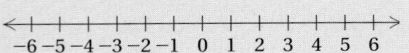

13. Solve:

$$-3x - 7 < -1 \ or \ x + 4 < -1.$$

14. Solve and graph:

$$5x - 7 \leq 13 \ or \ 2x - 1 \geq -7.$$

---------- *Caution!* ----------

A compound inequality like

$$x < -4 \quad or \quad x \geq 2,$$

as in Example 7, *cannot* be expressed as $2 \leq x < -4$ because to do so would be to say that x is *simultaneously* less than -4 and greater than or equal to 2. No number is both less than -4 *and* greater than or equal to 2, but many are less than -4 *or* greater than or equal to 2.

Do Exercises 11 and 12.

EXAMPLE 8 Solve: $-2x - 5 < -2 \ or \ x - 3 < -10$.

We solve the individual inequalities separately, retaining the word *or*:

$$-2x - 5 < -2 \quad or \quad x - 3 < -10$$
$$-2x < 3 \quad or \quad x < -7$$

Reversing the symbol

$$x > -\tfrac{3}{2} \quad or \quad x < -7.$$

Keep the word "or."

The solution set is written $\{x | x < -7 \ or \ x > -\tfrac{3}{2}\}$, or, in interval notation, $(-\infty, -7) \cup \left(-\tfrac{3}{2}, \infty\right)$.

Do Exercise 13.

EXAMPLE 9 Solve: $3x - 11 < 4 \ or \ 4x + 9 \geq 1$.

We solve the individual inequalities separately, retaining the word *or*:

$$3x - 11 < 4 \quad or \quad 4x + 9 \geq 1$$
$$3x < 15 \quad or \quad 4x \geq -8$$
$$x < 5 \quad or \quad x \geq -2.$$

To find the solution set, we first look at the individual graphs.

$\{x | x < 5\}$ $(-\infty, 5)$

$\{x | x \geq -2\}$ $[-2, \infty)$

$\{x | x < 5\} \cup \{x | x \geq -2\}$
$= \{x | x < 5 \ or \ x \geq -2\}$
$= \{x | x \text{ is a real number}\}$
$(-\infty, \infty)$
= The set of all real numbers

Since any number is either less than 5 or greater than or equal to -2, the two sets fill the entire number line. Thus the solution set is the set of all real numbers, $(-\infty, \infty)$.

Do Exercise 14.

Answers

11. $\{x | x < 1 \ or \ x \geq 6\}$, or $(-\infty, 1) \cup [6, \infty)$;

12. $\left\{x | x \geq \tfrac{7}{2} \ or \ x < -2\right\}$, or

$(-\infty, -2) \cup \left[\tfrac{7}{2}, \infty\right)$;

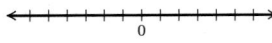

13. $\{x | x < -5 \ or \ x > -2\}$, or $(-\infty, -5) \cup (-2, \infty)$

14. All real numbers;

(c) Applications and Problem Solving

EXAMPLE 10 *Converting Dress Sizes.* The equation

$$I = 2(s + 10)$$

can be used to convert dress sizes s in the United States to dress sizes I in Italy. Which dress sizes in the United States correspond to dress sizes between 32 and 46 in Italy?

1. **Familiarize.** We have a formula for converting the dress sizes. Thus we can substitute a value into the formula. For a dress of size 6 in the United States, we get the corresponding dress size in Italy as follows:

$$I = 2(6 + 10) = 2 \cdot 16 = 32.$$

This familiarizes us with the formula and also tells us that the United States sizes that we are looking for must be larger than size 6.

2. **Translate.** We want the Italian sizes *between* 32 and 46, so we want to find those values of s for which

$$32 < I < 46 \qquad I \text{ is between 32 and 46}$$

or

$$32 < 2(s + 10) < 46. \qquad \text{Substituting } 2(s + 10) \text{ for } I$$

Thus we have translated the problem to an inequality.

3. **Solve.** We solve the inequality:

$$32 < 2(s + 10) < 46$$
$$\frac{32}{2} < \frac{2(s + 10)}{2} < \frac{46}{2} \qquad \text{Dividing by 2}$$
$$16 < s + 10 < 23$$
$$6 < s < 13. \qquad \text{Subtracting 10}$$

4. **Check.** We substitute some values as we did in the *Familiarize* step.

5. **State.** Dress sizes between 6 and 13 in the United States correspond to dress sizes between 32 and 46 in Italy.

Do Exercise 15.

15. Converting Dress Sizes. Refer to Example 10. Which dress sizes in the United States correspond to dress sizes between 36 and 58 in Italy?

Answer

15. $\{s \mid 8 < s < 19\}$

a , **b**   Find the intersection or union.

1. $\{9, 10, 11\} \cap \{9, 11, 13\}$

2. $\{1, 5, 10, 15\} \cap \{5, 15, 20\}$

3. $\{a, b, c, d\} \cap \{b, f, g\}$

4. $\{m, n, o, p\} \cap \{m, o, p\}$

5. $\{9, 10, 11\} \cup \{9, 11, 13\}$

6. $\{1, 5, 10, 15\} \cup \{5, 15, 20\}$

7. $\{a, b, c, d\} \cup \{b, f, g\}$

8. $\{m, n, o, p\} \cup \{m, o, p\}$

9. $\{2, 5, 7, 9\} \cap \{1, 3, 4\}$

10. $\{a, e, i, o, u\} \cap \{m, q, w, s, t\}$

11. $\{3, 5, 7\} \cup \varnothing$

12. $\{3, 5, 7\} \cap \varnothing$

a   Graph and write interval notation.

13. $-4 < a$ *and* $a \leq 1$

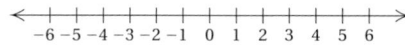

14. $-\frac{5}{2} \leq m$ *and* $m < \frac{3}{2}$

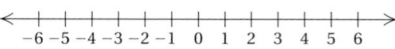

15. $1 < x < 6$

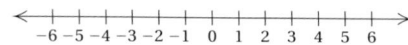

16. $-3 \leq y \leq 4$

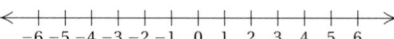

Solve and graph.

17. $-10 \leq 3x + 2$ *and* $3x + 2 < 17$

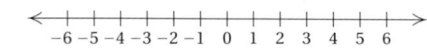

18. $-11 < 4x - 3$ *and* $4x - 3 \leq 13$

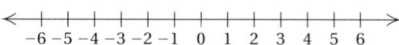

19. $3x + 7 \geq 4$ *and* $2x - 5 \geq -1$

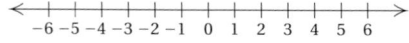

20. $4x - 7 < 1$ *and* $7 - 3x > -8$

21. $4 - 3x \geq 10$ *and* $5x - 2 > 13$

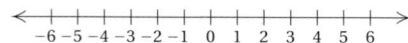

22. $5 - 7x > 19$ *and* $2 - 3x < -4$

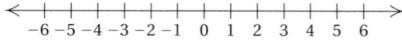

Solve.

23. $-4 < x + 4 < 10$

24. $-6 < x + 6 \leq 8$

25. $6 > -x \geq -2$

26. $3 > -x \geq -5$

27. $2 < x + 3 \le 9$

28. $-6 \le x + 1 < 9$

29. $1 < 3y + 4 \le 19$

30. $5 \le 8x + 5 \le 21$

31. $-10 \le 3x - 5 \le -1$

32. $-6 \le 2x - 3 < 6$

33. $-18 \le -2x - 7 < 0$

34. $4 > -3m - 7 \ge 2$

35. $-\dfrac{1}{2} < \dfrac{1}{4}x - 3 \le \dfrac{1}{2}$

36. $-\dfrac{2}{3} \le 4 - \dfrac{1}{4}x < \dfrac{2}{3}$

37. $-4 \le \dfrac{7 - 3x}{5} \le 4$

38. $-3 < \dfrac{2x - 5}{4} < 8$

b Graph and write interval notation.

39. $x < -2 \ or \ x > 1$

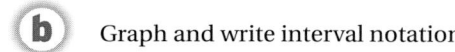

40. $x < -4 \ or \ x > 0$

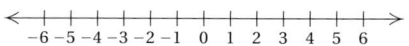

41. $x \le -3 \ or \ x > 1$

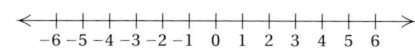

42. $x \le -1 \ or \ x > 3$

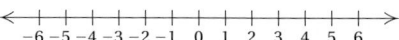

Solve and graph.

43. $x + 3 < -2 \ or \ x + 3 > 2$

44. $x - 2 < -1 \ or \ x - 2 > 3$

45. $2x - 8 \le -3 \ or \ x - 1 \ge 3$

46. $x - 5 \le -4 \ or \ 2x - 7 \ge 3$

47. $7x + 4 \ge -17 \ or \ 6x + 5 \ge -7$

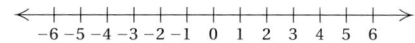

48. $4x - 4 < -8 \ or \ 4x - 4 < 12$

Solve.

49. $7 > -4x + 5 \ or \ 10 \le -4x + 5$

50. $6 > 2x - 1 \ or \ -4 \le 2x - 1$

51. $3x - 7 > -10 \text{ or } 5x + 2 \le 22$

52. $3x + 2 < 2 \text{ or } 4 - 2x < 14$

53. $-2x - 2 < -6 \text{ or } -2x - 2 > 6$

54. $-3m - 7 < -5 \text{ or } -3m - 7 > 5$

55. $\frac{2}{3}x - 14 < -\frac{5}{6} \text{ or } \frac{2}{3}x - 14 > \frac{5}{6}$

56. $\frac{1}{4} - 3x \le -3.7 \text{ or } \frac{1}{4} - 5x \ge 4.8$

57. $\frac{2x - 5}{6} \le -3 \text{ or } \frac{2x - 5}{6} \ge 4$

58. $\frac{7 - 3x}{5} < -4 \text{ or } \frac{7 - 3x}{5} > 4$

C   Solve.

59. *Pressure at Sea Depth.* The equation
$$P = 1 + \frac{d}{33}$$
gives the pressure P, in atmospheres (atm), at a depth of d feet in the sea. For what depths d is the pressure at least 1 atm and at most 7 atm?

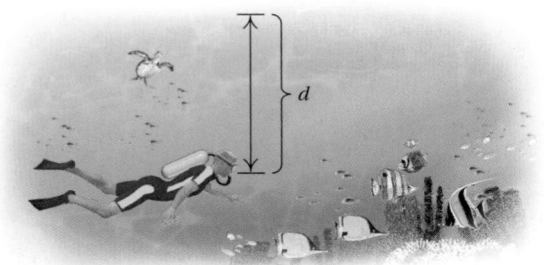

60. *Temperatures of Liquids.* The formula
$$C = \tfrac{5}{9}(F - 32)$$
can be used to convert Fahrenheit temperatures F to Celsius temperatures C.

a) Gold is a liquid for Celsius temperatures C such that $1063° \le C < 2660°$. Find such an inequality for the corresponding Fahrenheit temperatures.

b) Silver is a liquid for Celsius temperatures C such that $960.8° \le C < 2180°$. Find such an inequality for the corresponding Fahrenheit temperatures.

61. *Aerobic Exercise.* In order to achieve maximum results from aerobic exercise, one should maintain one's heart rate at a certain level. A 30-year-old woman with a resting heart rate of 60 beats per minute should keep her heart rate between 138 and 162 beats per minute while exercising. She checks her pulse for 10 sec while exercising. What should the number of beats be?

62. *Minimizing Tolls.* A $6.00 toll is charged to cross the bridge from mainland Florida to Sanibel Island. A six-month pass, costing $50.00, reduces the toll to $2.00. A one-year pass, costing $400, allows for free crossings. How many crossings per year does it take, on average, for the two six-month passes to be the most economical choice? Assume a constant number of trips per month.

Source: leewayinfo.com

63. *Body Mass Index.* Refer to Exercises 71 and 72 in Exercise Set 9.1. Marv's height is 73 in. What weights W will allow Marv to keep his body mass index I in the 18.5–24.9 range?

64. *Body Mass Index.* Refer to Exercises 71 and 72 in Exercise Set 9.1. Elaine's height is 67 in. What weight W will allow Elaine to keep her body mass index in the 18.5–24.9 range?

65. *Young's Rule in Medicine.* Young's rule for determining the amount of a medicine dosage for a child is given by

$$c = \frac{ad}{a + 12},$$

where a is the child's age and d is the usual adult dosage, in milligrams. (*Warning!* Do not apply this formula without checking with a physician!) An 8-year-old child needs medication. What adult dosage can be used if a child's dosage must stay between 100 mg and 200 mg?

Source: Olsen, June L., et al., *Medical Dosage Calculations*, 6th ed. Reading, MA: Addison Wesley Longman, p. A-31.

66. *Young's Rule in Medicine.* Refer to Exercise 65. The dosage of a medication for a 5-year-old child must stay between 50 mg and 100 mg. Find the equivalent adult dosage.

Skill Maintenance

Solve. [8.2a], [8.3a]

67. $3x - 2y = -7,$
$\quad 2x + 5y = 8$

68. $4x - 7y = 23,$
$\quad x + 6y = -33$

69. $x + y = 0,$
$\quad x - y = 8$

Find an equation of the line containing the given pair of points. [7.5c]

70. $(2, 7), (3, -4)$

71. $(0, 7), (2, -1)$

72. $(4, -2), (-2, 4)$

Multiply. [4.6a]

73. $(2a - b)(3a + 5b)$

74. $(5y + 6)(5y + 1)$

75. $(7x - 8)(3x - 5)$

76. $(13x - 2y)(x + 3y)$

Synthesis

Solve.

77. $x - 10 < 5x + 6 \leq x + 10$

78. $4m - 8 > 6m + 5 \ or \ 5m - 8 < -2$

79. $-\frac{2}{15} \leq \frac{2}{3}x - \frac{2}{5} \leq \frac{2}{15}$

80. $2[5(3 - y) - 2(y - 2)] > y + 4$

81. $3x < 4 - 5x < 5 + 3x$

82. $2x - \frac{3}{4} < -\frac{1}{10} \ or \ 2x - \frac{3}{4} > \frac{1}{10}$

83. $x + 4 < 2x - 6 \leq x + 12$

84. $2x + 3 \leq x - 6 \ or \ 3x - 2 \leq 4x + 5$

Determine whether each sentence is true or false for all real numbers a, b, and c.

85. If $-b < -a$, then $a < b$.

86. If $a \leq c$ and $c \leq b$, then $b \geq a$.

87. If $a < c$ and $b < c$, then $a < b$.

88. If $-a < c$ and $-c > b$, then $a > b$.

89. What is the union of the set of all rational numbers with the set of all irrational numbers? the intersection?

Concept Reinforcement

Determine whether each statement is true or false.

_____ **1.** The inequalities $x - 5 > 2$ and $x > 7$ are equivalent.   [9.1c]

_____ **2.** If a is at most c, then it cannot be less than c.   [9.1d]

_____ **3.** Sets A and B where $A = \{x | x < 2\}$ and $B = \{x | x \geq 2\}$ are disjoint sets.   [9.2a]

_____ **4.** The union of two sets A and B is the collection of elements belonging to A and/or B [9.2b]

Guided Solutions

Fill in each blank with the number, variable, or symbol that creates a correct solution.

5. Solve: $8 - 5x \leq x + 20$.   [9.1c]

$$8 - 5x \leq x + 20$$
$$-5x \leq x + \square$$
$$\square x \leq 12$$
$$x \,\square\, -2$$

6. Solve: $-17 < 3 - x < 36$.   [9.2a]

$$-17 < 3 - x < 36$$
$$\square < -x < 33$$
$$20 \,\square\, x \,\square\, -33$$

Mixed Review

Match each graph of a solution with the correct set-builder notation or interval notation in selections A–H.   [9.1b], [9.2a, b]

7.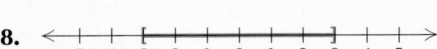

8.

9.

10.

11.

12.

A. $(-\infty, 3]$

B. $\{x | -3 \leq x \leq 3\}$

C. $\{x | x \leq -3 \text{ or } x > 3\}$

D. $[-3, 3)$

E. $(-\infty, -3) \cup (3, \infty)$

F. $\{x | -3 \leq x < 3\}$

G. $\{x | x > -3\}$

H. $(3, \infty)$

Find the intersection or union.   [9.2a, b]

13. $\{-1, 0, 10, 21, 40\} \cap \{-10, 0, 10\}$

14. $\{e, f, g, h\} \cup \{b, d, e\}$

15. $\left\{\dfrac{1}{4}, \dfrac{3}{8}\right\} \cup \varnothing$

16. $\{3, 6, 9, 12, 15\} \cap \{-12, -6, 7, 8\}$

Solve. Express the answer in both set-builder notation and interval notation.

17. $y - 8 \leq -10$ [9.1c]

18. $-\dfrac{5}{11}x \geq -\dfrac{20}{11}$ [9.1c]

19. $x - 6 < -15$ *or* $x + 2 > 3$ [9.2b]

20. $-6 \leq x - 9 < 15$ [9.2a]

21. $x + 6 < -4$ *or* $x + 8 > 9$ [9.2b]

22. $4(3t - 4) > 2(6 - t)$ [9.1c]

23. $3x - 2 \geq -11$ *or* $5x + 3 \geq -7$
[9.2b]

24. $0.1y + 3 < 5.6y - 2$ [9.1c]

25. $-6 < \dfrac{2x - 1}{3} < 8$ [9.2a]

26. $20 - (2x - 9) \leq 3(x - 2) + x$
[9.1c]

27. $-\dfrac{1}{2} < 8 - \dfrac{1}{2}x < 6$ [9.2a]

28. $2x - 7 > -18$ *or* $3x - 7 \leq 40$
[9.2b]

Solve.   [9.1d]

29. James is taking a chemistry course in which there will be 5 exams, each worth 100 points. He has scores of 85, 96, 88, and 95 on the first four exams. He must make a total of at least 450 points in order to get an A. What scores on the last exam will give James an A?

30. Lauren is about to invest $12,500, part at 4.5% and the rest at 5%. What is the most she can invest at 4.5% and still be guaranteed at least $610 in interest per year?

Understanding Through Discussion and Writing

31. Explain in your own words why the inequality symbol must be reversed when both sides of an inequality are multiplied or divided by a negative number.   [9.1c]

32. Find the error or errors in each of the following steps:
[9.1c]
$$7 - 9x + 6x < -9(x + 2) + 10x$$
$$7 - 9x + 6x < -9x + 2 + 10x \qquad \textbf{(1)}$$
$$7 + 6x > 2 + 10x \qquad \textbf{(2)}$$
$$-4x > 8 \qquad \textbf{(3)}$$
$$x > -2. \qquad \textbf{(4)}$$

33. Explain why the conjunction $3 < x$ *and* $x < 5$ is equivalent to $3 < x < 5$, but the disjunction $3 < x$ *or* $x < 5$ is not.   [9.2a, b]

9.3

Absolute-Value Equations and Inequalities

OBJECTIVES

a Simplify expressions containing absolute-value symbols.

b Find the distance between two points on the number line.

c Solve equations with absolute-value expressions.

d Solve equations with two absolute-value expressions.

e Solve inequalities with absolute-value expressions.

SKILL TO REVIEW
Objective 1.2e: Find the absolute value of a real number.

Find each absolute value.

1. $|-4|$ **2.** $|3.5|$

Simplify, leaving as little as possible inside the absolute-value signs.

1. $|7x|$ **2.** $|x^8|$

3. $|5a^2b|$ **4.** $\left|\dfrac{7a}{b^2}\right|$

5. $|-9x|$

Answers

Skill to Review:
1. 4   **2.** 3.5

Margin Exercises:
1. $7|x|$ **2.** x^8 **3.** $5a^2|b|$

4. $\dfrac{7|a|}{b^2}$ **5.** $9|x|$

---

a Properties of Absolute Value

We can think of the **absolute value** of a number as its distance from zero on the number line.

> **ABSOLUTE VALUE**
>
> The **absolute value** of x, denoted $|x|$, is defined as follows:
>
> $$x \geq 0 \;\longrightarrow\; |x| = x; \qquad x < 0 \;\longrightarrow\; |x| = -x.$$

This definition tells us that, when x is nonnegative, the absolute value of x is x and, when x is negative, the absolute value of x is the opposite of x. For example, $|3| = 3$ and $|-3| = -(-3) = 3$.

We see that absolute value is never negative. We can also think of a number's absolute value as its distance from zero on the number line.

Some simple properties of absolute value allow us to manipulate or simplify algebraic expressions.

> **PROPERTIES OF ABSOLUTE VALUE**
>
> a) $|ab| = |a| \cdot |b|$, for any real numbers a and b.
>
> (The absolute value of a product is the product of the absolute values.)
>
> b) $\left|\dfrac{a}{b}\right| = \dfrac{|a|}{|b|}$, for any real numbers a and $b \neq 0$.
>
> (The absolute value of a quotient is the quotient of the absolute values.)
>
> c) $|-a| = |a|$, for any real number a.
>
> (The absolute value of the opposite of a number is the same as the absolute value of the number.)

EXAMPLES Simplify, leaving as little as possible inside the absolute-value signs.

1. $|5x| = |5| \cdot |x| = 5|x|$

2. $|-3y| = |-3| \cdot |y| = 3|y|$

3. $|7x^2| = |7| \cdot |x^2| = 7|x^2| = 7x^2$ Since x^2 is never negative for any number x

4. $\left|\dfrac{6x}{-3x^2}\right| = \left|\dfrac{2}{-x}\right| = \dfrac{|2|}{|-x|} = \dfrac{2}{|x|}$

Do Margin Exercises 1–5.

b) Distance on the Number Line

The number line below shows that the distance between -3 and 2 is 5.

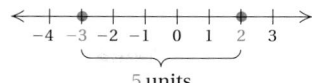

Another way to find the distance between two numbers on the number line is to determine the absolute value of the difference, as follows:

$$|-3 - 2| = |-5| = 5, \quad \text{or} \quad |2 - (-3)| = |5| = 5.$$

Note that the order in which we subtract does not matter because we are taking the absolute value after we have subtracted.

DISTANCE AND ABSOLUTE VALUE

For any real numbers a and b, the **distance** between them is $|a - b|$.

We should note that the distance is also $|b - a|$, because $a - b$ and $b - a$ are opposites and hence have the same absolute value.

EXAMPLE 5 Find the distance between -8 and -92 on the number line.

$$|-8 - (-92)| = |84| = 84, \quad \text{or} \quad |-92 - (-8)| = |-84| = 84$$

EXAMPLE 6 Find the distance between x and 0 on the number line.

$$|x - 0| = |x|$$

Do Exercises 6–8.

Find the distance between the points.

6. $-6, \ -35$

7. $19, \ 14$

8. $0, \ p$

c) Equations with Absolute Value

EXAMPLE 7 Solve: $|x| = 4$. Then graph on the number line.

Note that $|x| = |x - 0|$, so that $|x - 0|$ is the distance from x to 0. Thus solutions of the equation $|x| = 4$, or $|x - 0| = 4$, are those numbers x whose distance from 0 is 4. Those numbers are -4 and 4. The solution set is $\{-4, 4\}$. The graph consists of just two points, as shown.

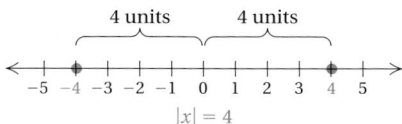

EXAMPLE 8 Solve: $|x| = 0$.

The only number whose absolute value is 0 is 0 itself. Thus the solution is 0. The solution set is $\{0\}$.

EXAMPLE 9 Solve: $|x| = -7$.

The absolute value of a number is always nonnegative. Thus there is no number whose absolute value is -7; consequently, the equation has no solution. The solution set is $\varnothing$.

Answers

6. 29   **7.** 5   **8.** $|p|$

Examples 7–9 lead us to the following principle for solving linear equations with absolute value.

> ### THE ABSOLUTE-VALUE PRINCIPLE
>
> For any positive number p and any algebraic expression X:
>
> **a)** The solution of $|X| = p$ is those numbers that satisfy $X = -p$ or $X = p$.
>
> **b)** The equation $|X| = 0$ is equivalent to the equation $X = 0$.
>
> **c)** The equation $|X| = -p$ has no solution.

Do Exercises 9–11.

We can use the absolute-value principle with the addition and multiplication principles to solve equations with absolute value.

EXAMPLE 10   Solve: $2|x| + 5 = 9$.

We first use the addition and multiplication principles to get $|x|$ by itself. Then we use the absolute-value principle.

$$2|x| + 5 = 9$$
$$2|x| = 4 \qquad \text{Subtracting 5}$$
$$|x| = 2 \qquad \text{Dividing by 2}$$
$$x = -2 \quad or \quad x = 2 \qquad \text{Using the absolute-value principle}$$

The solutions are -2 and 2. The solution set is $\{-2, 2\}$.

Do Exercises 12–14.

EXAMPLE 11   Solve: $|x - 2| = 3$.

We can consider solving this equation in two different ways.

METHOD 1:   This allows us to see the meaning of the solutions graphically. The solution set consists of those numbers that are 3 units from 2 on the number line.

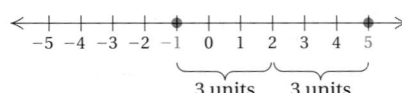

The solutions of $|x - 2| = 3$ are -1 and 5. The solution set is $\{-1, 5\}$.

METHOD 2:   This method is more efficient. We use the absolute-value principle, replacing X with $x - 2$ and p with 3. Then we solve each equation separately.

$$|X| = p$$
$$|x - 2| = 3$$
$$x - 2 = -3 \quad or \quad x - 2 = 3 \qquad \text{Absolute-value principle}$$
$$x = -1 \quad or \quad x = 5$$

The solutions are -1 and 5. The solution set is $\{-1, 5\}$.

Do Exercise 15.

---

9. Solve: $|x| = 6$. Then graph on the number line.

$$\overset{\longleftarrow \;+\!+\!+\!+\!+\!+\!+\!+\!+\!+\!+\!+\!+\!+\!+\!+\!+\; \longrightarrow}{\underset{-8 \;\; -6 \;\; -4 \;\; -2 \;\;\; 0 \;\;\; 2 \;\;\; 4 \;\;\; 6 \;\;\; 8}{}}$$

10. Solve: $|x| = -6$.

11. Solve: $|p| = 0$.

Solve.

12. $|3x| = 6$

13. $4|x| + 10 = 27$

14. $3|x| - 2 = 10$

15. Solve: $|x - 4| = 1$. Use two methods as in Example 11.

Answers

9. $\{6, -6\}$;

$$\overset{\longleftarrow \;+\!+\!+\!+\!+\!+\!+\!+\!+\!+\!+\!+\; \longrightarrow}{\underset{-6 \qquad\quad 0 \qquad\quad 6}{\bullet \qquad\qquad\qquad \bullet}}$$

10. $\varnothing$ **11.** $\{0\}$ **12.** $\{-2, 2\}$

13. $\left\{-\dfrac{17}{4}, \dfrac{17}{4}\right\}$ **14.** $\{-4, 4\}$ **15.** $\{3, 5\}$

EXAMPLE 12 Solve: $|2x + 5| = 13$.

We use the absolute-value principle, replacing X with $2x + 5$ and p with 13:

$$|X| = p$$
$$|2x + 5| = 13$$

$$2x + 5 = -13 \quad or \quad 2x + 5 = 13 \qquad \text{Absolute-value principle}$$
$$2x = -18 \quad or \qquad 2x = 8$$
$$x = -9 \quad or \qquad x = 4.$$

The solutions are -9 and 4. The solution set is $\{-9, 4\}$.

Do Exercise 16.

16. Solve: $|3x - 4| = 17$.

EXAMPLE 13 Solve: $|4 - 7x| = -8$.

Since absolute value is always nonnegative, this equation has no solution. The solution set is $\varnothing$.

Do Exercise 17.

17. Solve: $|6 + 2x| = -3$.

(d) Equations with Two Absolute-Value Expressions

Sometimes equations have two absolute-value expressions. Consider $|a| = |b|$. This means that a and b are the same distance from 0. If a and b are the same distance from 0, then either they are the same number or they are opposites.

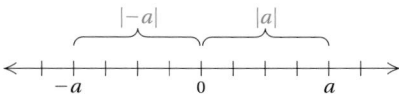

EXAMPLE 14 Solve: $|2x - 3| = |x + 5|$.

Either $2x - 3 = x + 5$ or $2x - 3 = -(x + 5)$. We solve each equation:

$$2x - 3 = x + 5 \quad or \quad 2x - 3 = -(x + 5)$$
$$x - 3 = 5 \qquad or \quad 2x - 3 = -x - 5$$
$$x = 8 \qquad or \quad 3x - 3 = -5$$
$$x = 8 \qquad or \qquad 3x = -2$$
$$x = 8 \qquad or \qquad x = -\tfrac{2}{3}.$$

The solutions are 8 and $-\tfrac{2}{3}$. The solution set is $\left\{8, -\tfrac{2}{3}\right\}$.

EXAMPLE 15 Solve: $|x + 8| = |x - 5|$.

$$x + 8 = x - 5 \quad or \quad x + 8 = -(x - 5)$$
$$8 = -5 \qquad or \quad x + 8 = -x + 5$$
$$8 = -5 \qquad or \qquad 2x = -3$$
$$8 = -5 \qquad or \qquad x = -\tfrac{3}{2}$$

The first equation has no solution. The second equation has $-\tfrac{3}{2}$ as a solution. The solution set is $\left\{-\tfrac{3}{2}\right\}$.

Do Exercises 18 and 19.

Solve:

18. $|5x - 3| = |x + 4|$

19. $|x - 3| = |x + 10|$

Answers

16. $\left\{-\dfrac{13}{3}, 7\right\}$ **17.** $\varnothing$

18. $\left\{\dfrac{7}{4}, -\dfrac{1}{6}\right\}$ **19.** $\left\{-\dfrac{7}{2}\right\}$

(e) Inequalities with Absolute Value

We can extend our methods for solving equations with absolute value to those for solving inequalities with absolute value.

EXAMPLE 16 Solve: $|x| = 4$. Then graph on the number line.

From Example 7, we know that the solutions are -4 and 4. The solution set is $\{-4, 4\}$. The graph consists of just two points, as shown here.

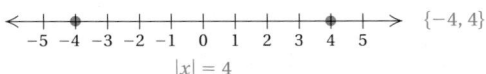

$|x| = 4$

Do Exercise 20.

EXAMPLE 17 Solve: $|x| < 4$. Then graph.

The solutions of $|x| < 4$ are the solutions of $|x - 0| < 4$ and are those numbers x whose distance from 0 is less than 4. We can check by substituting or by looking at the number line that numbers like $-3, -2, -1, -\frac{1}{2}, -\frac{1}{4}, 0, \frac{1}{4}, \frac{1}{2}$, 1, 2, and 3 are all solutions. In fact, the solutions are all the real numbers x between -4 and 4. The solution set is $\{x \mid -4 < x < 4\}$ or, in interval notation, $(-4, 4)$. The graph is as follows.

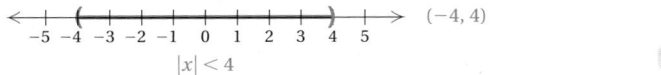

$|x| < 4$

Do Exercise 21.

EXAMPLE 18 Solve: $|x| \geq 4$. Then graph.

The solutions of $|x| \geq 4$ are solutions of $|x - 0| \geq 4$ and are those numbers whose distance from 0 is greater than or equal to 4—in other words, those numbers x such that $x \leq -4$ *or* $x \geq 4$. The solution set is $\{x \mid x \leq -4$ *or* $x \geq 4\}$, or $(-\infty, -4] \cup [4, \infty)$. The graph is as follows.

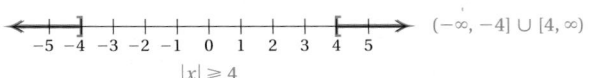

$|x| \geq 4$

Do Exercise 22.

Examples 16–18 illustrate three cases of solving equations and inequalities with absolute value. The following is a general principle for solving equations and inequalities with absolute value.

---

20. Solve: $|x| = 5$. Then graph on the number line.

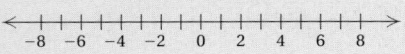

21. Solve: $|x| < 5$. Then graph.

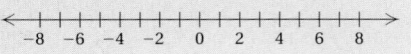

22. Solve: $|x| \geq 5$. Then graph.

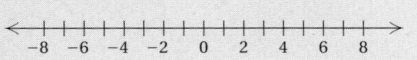

Answers

20. $\{-5, 5\}$;

$\xleftarrow{\hspace{1cm}} \overset{\bullet}{\underset{-5}{\vert}} \hspace{0.5cm} \overset{}{\underset{0}{\vert}} \hspace{0.5cm} \overset{\bullet}{\underset{5}{\vert}} \xrightarrow{\hspace{1cm}}$

21. $\{x \mid -5 < x < 5\}$, or $(-5, 5)$;

$\xleftarrow{\hspace{1cm}} \overset{(}{\underset{-5}{\vert}} \hspace{0.5cm} \overset{}{\underset{0}{\vert}} \hspace{0.5cm} \overset{)}{\underset{5}{\vert}} \xrightarrow{\hspace{1cm}}$

22. $\{x \mid x \leq -5$ *or* $x \geq 5\}$, or $(-\infty, -5] \cup [5, \infty)$;

$\xleftarrow{\hspace{1cm}} \overset{]}{\underset{-5}{\vert}} \hspace{0.5cm} \overset{}{\underset{0}{\vert}} \hspace{0.5cm} \overset{[}{\underset{5}{\vert}} \xrightarrow{\hspace{1cm}}$

For any positive number p and any algebraic expression X:

a) The solutions of $|X| = p$ are those numbers that satisfy $X = -p$ or $X = p$.

As an example, replacing X with $5x - 1$ and p with 8, we see that the solutions of $|5x - 1| = 8$ are those numbers x for which

$$5x - 1 = -8 \quad \text{or} \quad 5x - 1 = 8$$
$$5x = -7 \quad \text{or} \quad 5x = 9$$
$$x = -\tfrac{7}{5} \quad \text{or} \quad x = \tfrac{9}{5}.$$

The solution set is $\left\{-\tfrac{7}{5}, \tfrac{9}{5}\right\}$.

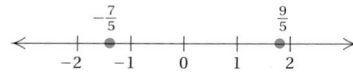

b) The solutions of $|X| < p$ are those numbers that satisfy $-p < X < p$.

As an example, replacing X with $6x + 7$ and p with 5, we see that the solutions of $|6x + 7| < 5$ are those numbers x for which

$$-5 < 6x + 7 < 5$$
$$-12 < 6x < -2$$
$$-2 < x < -\tfrac{1}{3}.$$

The solution set is $\left\{x | -2 < x < -\tfrac{1}{3}\right\}$, or $\left(-2, -\tfrac{1}{3}\right)$.

c) The solutions of $|X| > p$ are those numbers that satisfy $X < -p$ or $X > p$.

As an example, replacing X with $2x - 9$ and p with 4, we see that the solutions of $|2x - 9| > 4$ are those numbers x for which

$$2x - 9 < -4 \quad \text{or} \quad 2x - 9 > 4$$
$$2x < 5 \quad \text{or} \quad 2x > 13$$
$$x < \tfrac{5}{2} \quad \text{or} \quad x > \tfrac{13}{2}.$$

The solution set is $\left\{x | x < \tfrac{5}{2} \, or \, x > \tfrac{13}{2}\right\}$, or $\left(-\infty, \tfrac{5}{2}\right) \cup \left(\tfrac{13}{2}, \infty\right)$.

EXAMPLE 19 Solve: $|3x - 2| < 4$. Then graph.

We use part (b). In this case, X is $3x - 2$ and p is 4:

$$|X| < p$$
$$|3x - 2| < 4 \qquad \text{Replacing } X \text{ with } 3x - 2 \text{ and } p \text{ with } 4$$
$$-4 < 3x - 2 < 4$$
$$-2 < 3x < 6$$
$$-\tfrac{2}{3} < x < 2.$$

The solution set is $\{x | -\tfrac{2}{3} < x < 2\}$, or $\left(-\tfrac{2}{3}, 2\right)$. The graph is as follows.

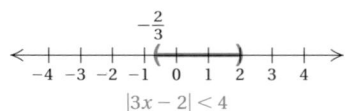

$$|3x - 2| < 4$$

EXAMPLE 20 Solve: $|8 - 4x| \leq 5$. Then graph.

We use part (b). In this case, X is $8 - 4x$ and p is 5:

$$|X| \leq p$$
$$|8 - 4x| \leq 5 \qquad \text{Replacing } X \text{ with } 8 - 4x \text{ and } p \text{ with } 5$$
$$-5 \leq 8 - 4x \leq 5$$
$$-13 \leq -4x \leq -3$$
$$\tfrac{13}{4} \geq x \geq \tfrac{3}{4}. \qquad \text{Dividing by } -4 \text{ and reversing the inequality symbols}$$

The solution set is $\{x | \tfrac{13}{4} \geq x \geq \tfrac{3}{4}\}$, or $\{x | \tfrac{3}{4} \leq x \leq \tfrac{13}{4}\}$, or $\left[\tfrac{3}{4}, \tfrac{13}{4}\right]$.

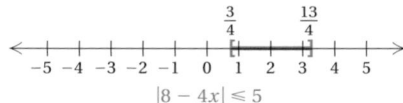

$$|8 - 4x| \leq 5$$

EXAMPLE 21 Solve: $|4x + 2| \geq 6$. Then graph.

We use part (c). In this case, X is $4x + 2$ and p is 6:

$$|X| \geq p$$
$$|4x + 2| \geq 6 \qquad \text{Replacing } X \text{ with } 4x + 2 \text{ and } p \text{ with } 6$$
$$4x + 2 \leq -6 \quad or \quad 4x + 2 \geq 6$$
$$4x \leq -8 \quad or \qquad 4x \geq 4$$
$$x \leq -2 \quad or \qquad x \geq 1.$$

The solution set is $\{x | x \leq -2 \text{ or } x \geq 1\}$, or $(-\infty, -2] \cup [1, \infty)$.

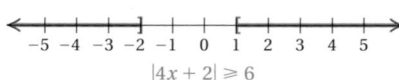

$$|4x + 2| \geq 6$$

Do Exercises 23–25.

Solve. Then graph.

23. $|2x - 3| < 7$

```
<-+++++++++++++++++++->
 -8  -6  -4  -2   0   2   4   6   8
```

24. $|7 - 3x| \leq 4$

```
<-+++++++++++++++++++->
 -8  -6  -4  -2   0   2   4   6   8
```

25. $|3x + 2| \geq 5$

```
<-+++++++++++++++++++->
 -8  -6  -4  -2   0   2   4   6   8
```

Answers

23. $\{x | -2 < x < 5\}$, or $(-2, 5)$;

```
<-+++++(+++++++)+-+->
      -2    0        5
```

24. $\{x | 1 \leq x \leq \tfrac{11}{3}\}$, or $\left[1, \tfrac{11}{3}\right]$;

```
<-+++++++++[+++]++++->
              0  1    11
                      --
                       3
```

25. $\{x | x \leq -\tfrac{7}{3} \text{ or } x \geq 1\}$, or $\left(-\infty, -\tfrac{7}{3}\right] \cup [1, \infty)$;

```
<-+++++]+++[+++++++->
       -7      0  1
       --
        3
```

a Simplify, leaving as little as possible inside absolute-value signs.

1. $|9x|$

2. $|26x|$

3. $|2x^2|$

4. $|8x^2|$

5. $|-2x^2|$

6. $|-20x^2|$

7. $|-6y|$

8. $|-17y|$

9. $\left|\dfrac{-2}{x}\right|$

10. $\left|\dfrac{y}{3}\right|$

11. $\left|\dfrac{x^2}{-y}\right|$

12. $\left|\dfrac{x^4}{-y}\right|$

13. $\left|\dfrac{-8x^2}{2x}\right|$

14. $\left|\dfrac{-9y^2}{3y}\right|$

15. $\left|\dfrac{4y^3}{-12y}\right|$

16. $\left|\dfrac{5x^3}{-25x}\right|$

b Find the distance between the points on the number line.

17. $-8,\ -46$

18. $-7,\ -32$

19. $36,\ 17$

20. $52,\ 18$

21. $-3.9,\ 2.4$

22. $-1.8,\ -3.7$

23. $-5,\ 0$

24. $\dfrac{2}{3},\ -\dfrac{5}{6}$

c Solve.

25. $|x| = 3$

26. $|x| = 5$

27. $|x| = -3$

28. $|x| = -9$

29. $|q| = 0$

30. $|y| = 7.4$

31. $|x - 3| = 12$

32. $|3x - 2| = 6$

33. $|2x - 3| = 4$

34. $|5x + 2| = 3$

35. $|4x - 9| = 14$

36. $|9y - 2| = 17$

37. $|x| + 7 = 18$

38. $|x| - 2 = 6.3$

39. $574 = 283 + |t|$

40. $-562 = -2000 + |x|$

41. $|5x| = 40$

42. $|2y| = 18$

43. $|3x| - 4 = 17$

44. $|6x| + 8 = 32$

45. $7|w| - 3 = 11$

46. $5|x| + 10 = 26$

47. $\left|\dfrac{2x - 1}{3}\right| = 5$

48. $\left|\dfrac{4 - 5x}{6}\right| = 7$

49. $|m + 5| + 9 = 16$

50. $|t - 7| - 5 = 4$

51. $10 - |2x - 1| = 4$

52. $2|2x - 7| + 11 = 25$

53. $|3x - 4| = -2$

54. $|x - 6| = -8$

55. $\left|\dfrac{5}{9} + 3x\right| = \dfrac{1}{6}$

56. $\left|\dfrac{2}{3} - 4x\right| = \dfrac{4}{5}$

d   Solve.

57. $|3x + 4| = |x - 7|$

58. $|2x - 8| = |x + 3|$

59. $|x + 3| = |x - 6|$

60. $|x - 15| = |x + 8|$

61. $|2a + 4| = |3a - 1|$

62. $|5p + 7| = |4p + 3|$

63. $|y - 3| = |3 - y|$

64. $|m - 7| = |7 - m|$

65. $|5 - p| = |p + 8|$

66. $|8 - q| = |q + 19|$

67. $\left|\dfrac{2x - 3}{6}\right| = \left|\dfrac{4 - 5x}{8}\right|$

68. $\left|\dfrac{6 - 8x}{5}\right| = \left|\dfrac{7 + 3x}{2}\right|$

69. $\left|\dfrac{1}{2}x - 5\right| = \left|\dfrac{1}{4}x + 3\right|$

70. $\left|2 - \dfrac{2}{3}x\right| = \left|4 + \dfrac{7}{8}x\right|$

e Solve.

71. $|x| < 3$

72. $|x| \le 5$

73. $|x| \ge 2$

74. $|y| > 12$

75. $|x - 1| < 1$

76. $|x + 4| \le 9$

77. $5|x + 4| \le 10$

78. $2|x - 2| > 6$

79. $|2x - 3| \le 4$

80. $|5x + 2| \le 3$

81. $|2y - 7| > 10$

82. $|3y - 4| > 8$

83. $|4x - 9| \ge 14$

84. $|9y - 2| \ge 17$

85. $|y - 3| < 12$

86. $|p - 2| < 6$

87. $|2x + 3| \le 4$

88. $|5x + 2| \le 13$

89. $|4 - 3y| > 8$

90. $|7 - 2y| > 5$

91. $|9 - 4x| \ge 14$

92. $|2 - 9p| \ge 17$

93. $|3 - 4x| < 21$

94. $|-5 - 7x| \le 30$

95. $\left| \dfrac{1}{2} + 3x \right| \ge 12$

96. $\left| \dfrac{1}{4}y - 6 \right| > 24$

97. $\left| \dfrac{x - 7}{3} \right| < 4$

98. $\left| \dfrac{x + 5}{4} \right| \le 2$

99. $\left| \dfrac{2 - 5x}{4} \right| \ge \dfrac{2}{3}$

100. $\left| \dfrac{1 + 3x}{5} \right| > \dfrac{7}{8}$

101. $|m + 5| + 9 \le 16$

102. $|t - 7| + 3 \ge 4$

103. $7 - |3 - 2x| \ge 5$

104. $16 \le |2x - 3| + 9$

105. $\left| \dfrac{2x - 1}{3} \right| \le 1$

106. $\left| \dfrac{3x - 2}{5} \right| \ge 1$

Skill Maintenance

In each of Exercises 107–114, fill in the blank with the correct term from the given list. Some of the choices may not be used.

107. The _____ of two sets A and B is the collection of elements belonging to A and/or B.   [9.2b]

108. Two sets with an empty intersection are said to be _____.   [9.2a]

109. The expression $x \geq q$ means x is _____ q.   [9.1d]

110. Interval notation for $\{x \mid a \leq x \leq b\}$ is _____.   [9.1b]

111. The _____ of a number is its distance from zero on the number line.   [9.3a]

112. A(n) _____ is a number sentence that says that the expression on either side of the equals sign represents the same number.   [2.1a]

113. Equations with the same solutions are called _____ equations.   [2.1b]

114. A(n) _____ is any sentence containing $<, >, \leq$, $\geq$, or $\neq$.   [9.1a]

$[a, b]$

$[a, b)$

(a, b)

disjoint

union

intersection

equivalent

absolute value

equation

inequality

at least

at most

Synthesis

115. *Motion of a Spring.* A weighted spring is bouncing up and down so that its distance d above the ground satisfies the inequality $|d - 6\,\text{ft}| \leq \frac{1}{2}$ ft. Find all possible distances d.

116. *Container Sizes.* A container company is manufacturing rectangular boxes of various sizes. The length of any box must exceed the width by at least 3 in., but the perimeter cannot exceed 24 in. What widths are possible?

$$l \geq w + 3,$$
$$2l + 2w \leq 24$$

Solve.

117. $|x + 5| = x + 5$

118. $1 - \left|\frac{1}{4}x + 8\right| = \frac{3}{4}$

119. $|7x - 2| = x + 4$

120. $|x - 1| = x - 1$

121. $|x - 6| \leq -8$

122. $|3x - 4| > -2$

123. $|x + 5| > x$

124. $\left|\frac{5}{9} + 3x\right| < -\frac{1}{6}$

Find an equivalent inequality with absolute value.

125. $-3 < x < 3$

126. $-5 \leq y \leq 5$

127. $x \leq -6 \, or \, x \geq 6$

128. $-5 < x < 1$

129. $x < -8 \, or \, x > 2$

9.4 Systems of Inequalities in Two Variables

A **graph** of an inequality is a drawing that represents its solutions. An inequality in one variable can be graphed on the number line. (See Section 9.1.) An inequality in two variables can be graphed on a coordinate plane.

A **linear inequality** is one that we can get from a related linear equation by changing the equals symbol to an inequality symbol. The graph of a linear inequality is a region on one side of a line. This region is called a **half-plane**. The graph sometimes includes the graph of the related line at the boundary of the half-plane.

OBJECTIVES

(a) Determine whether an ordered pair of numbers is a solution of an inequality in two variables.

(b) Graph linear inequalities in two variables.

(c) Graph systems of linear inequalities and find coordinates of any vertices.

(a) Solutions of Inequalities in Two Variables

The solutions of an inequality in two variables are ordered pairs.

EXAMPLES Determine whether the ordered pair is a solution of the inequality $5x - 4y > 13$.

1. $(-3, 2)$

We have

$$\frac{5x - 4y > 13}{5(-3) - 4 \cdot 2 \overset{?}{} 13}$$
$$\begin{array}{c|c} -15 - 8 & \text{We use alphabetical order to replace } x \\ -23 & \text{with } -3 \text{ and } y \text{ with } 2. \\ & \text{FALSE} \end{array}$$

Since $-23 > 13$ is false, $(-3, 2)$ is not a solution.

2. $(4, -3)$

We have

$$\frac{5x - 4y > 13}{5(4) - 4(-3) \overset{?}{} 13}$$
$$\begin{array}{c|c} 20 + 12 & \text{Replacing } x \text{ with 4 and } y \text{ with } -3 \\ 32 & \text{TRUE} \end{array}$$

Since $32 > 13$ is true, $(4, -3)$ is a solution.

> Do Margin Exercises 1 and 2.

(b) Graphing Inequalities in Two Variables

Let's visualize the results of Examples 1 and 2. The equation $5x - 4y = 13$ is represented by the dashed line in the graphs on the following page. The solutions of the inequality $5x - 4y > 13$ are shaded below that dashed line. As shown in the graph on the left on the following page, the pair $(-3, 2)$ is not a solution of the inequality $5x - 4y > 13$ and is not in the shaded region.

SKILL TO REVIEW
Objective 7.4a: Graph linear equations using intercepts.

Find the intercepts. Then graph the equation.

1. $3x - 2y = 6$ **2.** $2x + y = 4$

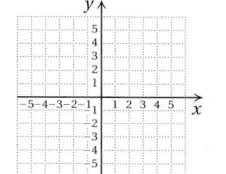   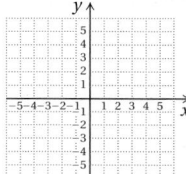

1. Determine whether $(1, -4)$ is a solution of $4x - 5y < 12$.

$$\frac{4x - 5y < 12}{?}$$

2. Determine whether $(4, -3)$ is a solution of $3y - 2x \leq 6$.

$$\frac{3y - 2x \leq 6}{?}$$

Answers

Answers are on p. 678.

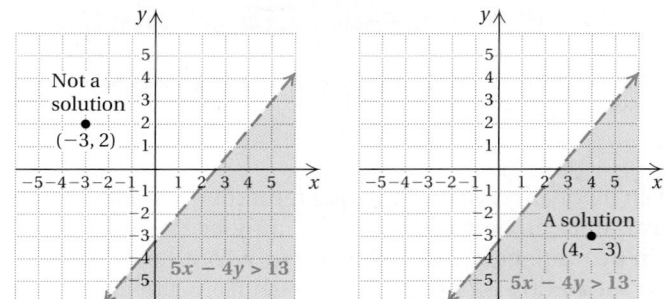

The pair $(4, -3)$ is a solution of the inequality $5x - 4y > 13$ and is in the shaded region. See the graph on the right above.

We now consider how to graph inequalities.

EXAMPLE 3 Graph: $y < x$.

We first graph the line $y = x$. Every solution of $y = x$ is an ordered pair like $(3, 3)$, where the first and second coordinates are the same. The graph of $y = x$ is shown on the left below. We draw it dashed because these points are *not* solutions of $y < x$.

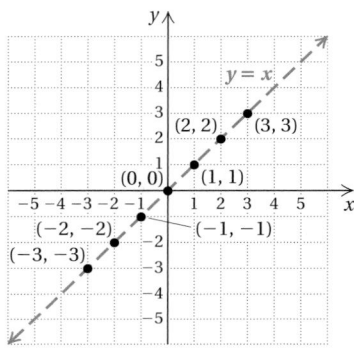
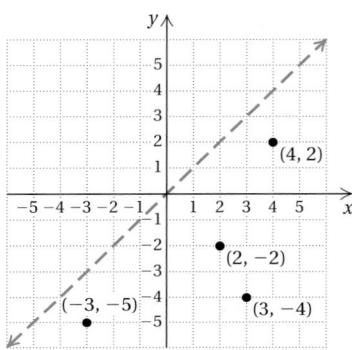

Now look at the graph on the right above. Several ordered pairs are plotted on the half-plane below $y = x$. Each is a solution of $y < x$. We can check the pair $(4, 2)$ as follows:

$$\frac{y < x}{2 \ ? \ 4} \quad \text{TRUE}$$

It turns out that any point on the same side of $y = x$ as $(4, 2)$ is also a solution. Thus, *if you know that one point in a half-plane is a solution of an inequality, then all points in that half-plane are solutions.* In this text, we will usually indicate this by color shading. We shade the half-plane below $y = x$.

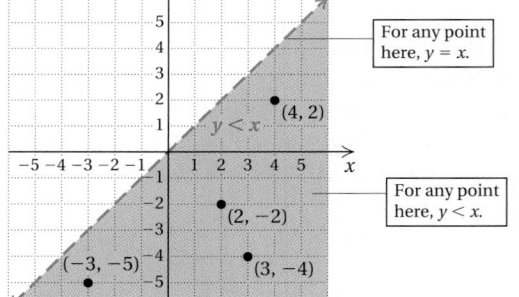

Answers

Skill to Review:

1.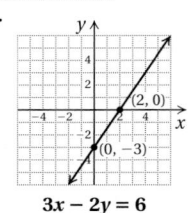

   $3x - 2y = 6$

2.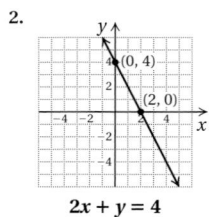

   $2x + y = 4$

Margin Exercises:

1. No   2. Yes

EXAMPLE 4 Graph: $8x + 3y \geq 24$.

First, we sketch the line $8x + 3y = 24$. Points on the line $8x + 3y = 24$ are also in the graph of $8x + 3y \geq 24$, so we draw the line solid. This indicates that all points on the line are solutions. The rest of the solutions are in the half-plane either to the left or to the right of the line. To determine which, we select a point that is not on the line and determine whether it is a solution of $8x + 3y \geq 24$. We try $(-3, 4)$ as a test point:

$$\frac{8x + 3y \geq 24}{\begin{array}{l} 8(-3) + 3(4) \text{ ? } 24 \\ -24 + 12 \\ -12 \end{array}} \quad \text{FALSE}$$

We see that $-12 \geq 24$ is *false*. Since $(-3, 4)$ is not a solution, none of the points in the half-plane containing $(-3, 4)$ is a solution. Thus the points in the opposite half-plane are solutions. We shade that half-plane and obtain the graph shown below.

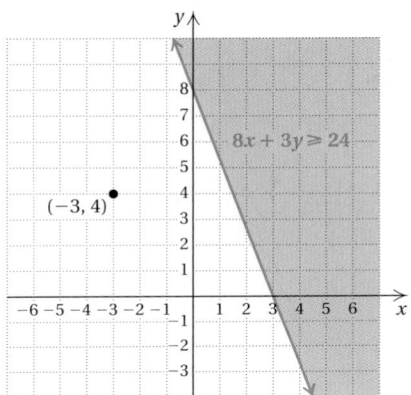

To graph an inequality in two variables:

1. Replace the inequality symbol with an equals sign and graph this related equation. This separates points that represent solutions from those that do not.

2. If the inequality symbol is $<$ or $>$, draw the line dashed. If the inequality symbol is $\leq$ or $\geq$, draw the line solid.

3. The graph consists of a half-plane that is either above or below or to the left or to the right of the line and, if the line is solid, the line as well. To determine which half-plane to shade, choose a point not on the line as a test point. If the line does not go through the origin, $(0, 0)$ is an easy point to use. Substitute to determine whether that point is a solution. If so, shade the half-plane containing that point. If not, shade the opposite half-plane.

EXAMPLE 5 Graph: $6x - 2y < 12$.

1. We first graph the related equation $6x - 2y = 12$.

2. Since the inequality uses the symbol $<$, points on the line are not solutions of the inequality, so we draw a dashed line.

3. To determine which half-plane to shade, we consider a test point *not* on the line. We try $(0, 0)$ and substitute:

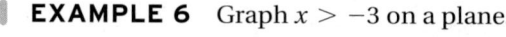

$$\frac{6x - 2y < 12}{\begin{array}{c|c} 6(0) - 2(0) \ ? \ 12 \\ 0 - 0 & \\ 0 & \text{TRUE} \end{array}}$$

Since the inequality $0 < 12$ is *true*, the point $(0, 0)$ is a solution; each point in the half-plane containing $(0, 0)$ is a solution. Thus each point in the opposite half-plane is *not* a solution. The graph is shown below.

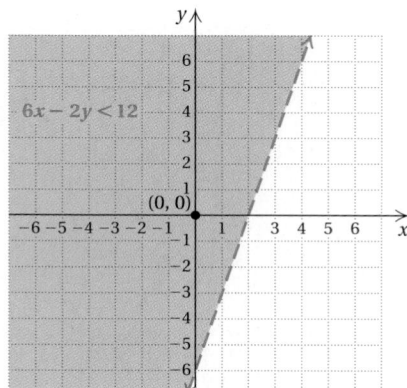

Do Exercises 3 and 4.

EXAMPLE 6 Graph $x > -3$ on a plane.

There is a missing variable in this inequality. If we graph the inequality on the number line, its graph is as follows:

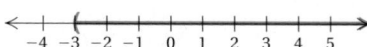

However, we can also write this inequality as $x + 0y > -3$ and consider graphing it in the plane. We are, in effect, determining which ordered pairs have x-values greater than -3. We use the same technique that we have used with the other examples. We first graph the related equation $x = -3$ in the plane. We draw the boundary with a dashed line. The rest of the graph is a half-plane to the right or to the left of the line $x = -3$. To determine which, we consider a test point, $(2, 5)$:

$$\frac{x + 0y > -3}{\begin{array}{c|c} 2 + 0(5) \ ? \ -3 \\ 2 & \text{TRUE} \end{array}}$$

Graph.

3. $6x - 3y < 18$

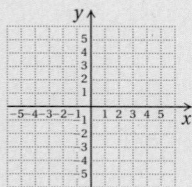

4. $4x + 3y \geq 12$

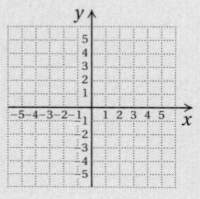

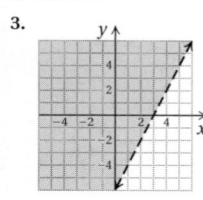
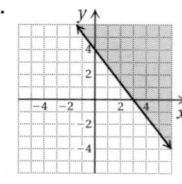

Since $(2, 5)$ is a solution, all the points in the half-plane containing $(2, 5)$ are solutions. We shade that half-plane.

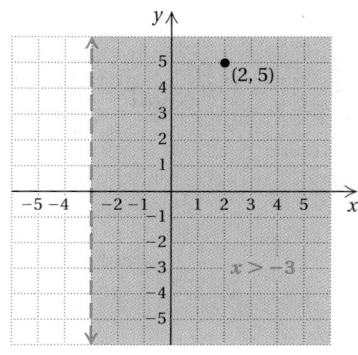

EXAMPLE 7 Graph $y \leq 4$ on a plane.

We first graph $y = 4$ using a solid line. We then use $(-2, 5)$ as a test point and substitute in $0x + y \leq 4$:

$$\frac{0x + y \leq 4}{\begin{array}{c|c} 0(-2) + 5 \;?\; 4 \\ 0 + 5 \\ 5 \end{array}} \quad \text{FALSE}$$

We see that $(-2, 5)$ is *not* a solution, so all the points in the half-plane containing $(-2, 5)$ are not solutions. Thus each point in the opposite half-plane is a solution.

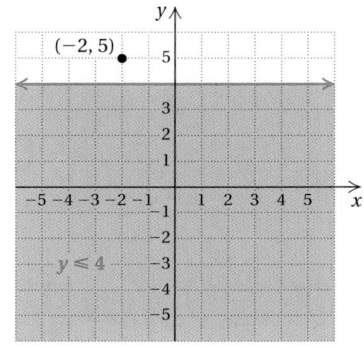

Do Exercises 5 and 6.

(c) Systems of Linear Inequalities

The following is an example of a system of two linear inequalities in two variables:

$$x + y \leq 4,$$
$$x - y < 4.$$

A **solution** of a system of linear inequalities is an ordered pair that is a solution of *both* inequalities. We now graph solutions of systems of linear inequalities. To do so, we graph each inequality and determine where the graphs overlap, or intersect. That will be a region in which the ordered pairs are solutions of both inequalities.

Graph on a plane.

5. $x < 3$

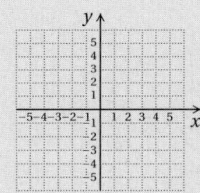

6. $y \geq -4$

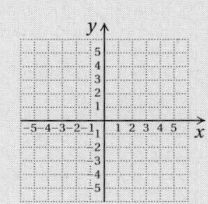

Answers

5.

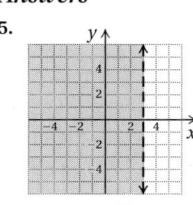

$x < 3$

6.

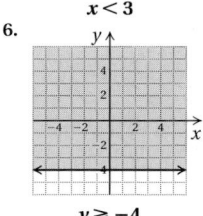

$y \geq -4$

EXAMPLE 8 Graph the solutions of the system

$$x + y \le 4,$$
$$x - y < 4.$$

We graph $x + y \le 4$ by first graphing the equation $x + y = 4$ using a solid red line. We consider $(0, 0)$ as a test point and find that it is a solution, so we shade all points on that side of the line using red shading. (See the graph on the left below.) The arrows near the ends of the line also indicate the half-plane, or region, that contains the solutions.

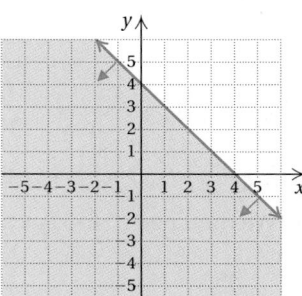 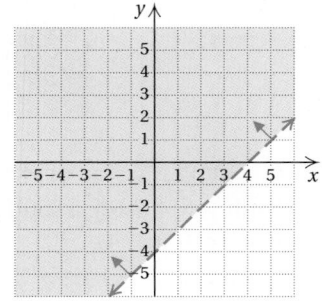

Next, we graph $x - y < 4$. We begin by graphing the equation $x - y = 4$ using a dashed blue line and consider $(0, 0)$ as a test point. Again, $(0, 0)$ is a solution so we shade that side of the line using blue shading. (See the graph on the right above.) The solution set of the system is the region that is shaded both red and blue and part of the line $x + y = 4$. (See the graph below.)

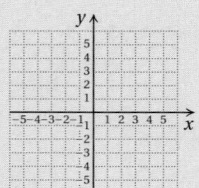

7. Graph:

$$x + y \ge 1,$$
$$y - x \ge 2.$$

Do Exercise 7.

Answer

7.
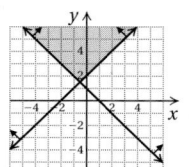

EXAMPLE 9 Graph: $-2 < x \leq 5$.

This is actually a system of inequalities:

$$-2 < x,$$
$$x \leq 5.$$

We graph the equation $-2 = x$ and see that the graph of the first inequality is the half-plane to the right of the line $-2 = x$. (See the graph on the left below.)

Next, we graph the second inequality, starting with the line $x = 5$, and find that its graph is the line and also the half-plane to the left of it. (See the graph on the right below.)

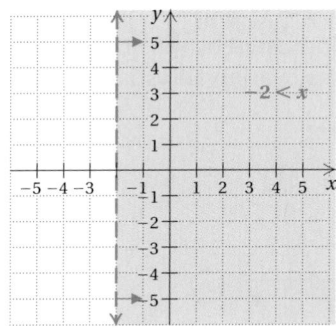   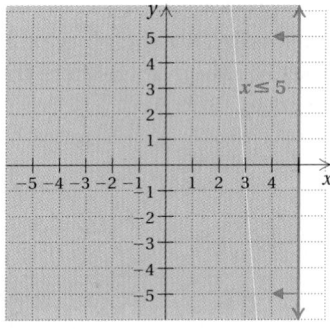

We shade the intersection of these graphs.

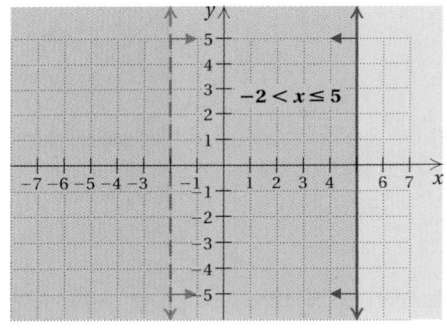

Do Exercise 8.

8. Graph: $-3 \leq y < 4$.

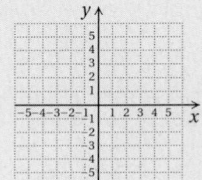

Answer

8.

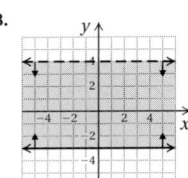

A system of inequalities may have a graph that consists of a polygon and its interior. In *linear programming*, which is a topic rich in application that you may study in a later course, it is important to be able to find the vertices of such a polygon.

EXAMPLE 10 Graph the following system of inequalities. Find the coordinates of any vertices formed.

$$6x - 2y \leq 12, \quad \textbf{(1)}$$
$$y - 3 \leq 0, \quad \textbf{(2)}$$
$$x + y \geq 0 \quad \textbf{(3)}$$

We graph the lines $6x - 2y = 12$, $y - 3 = 0$, and $x + y = 0$ using solid lines. The regions for each inequality are indicated by the arrows at the ends of the lines. We then note where the regions overlap and shade the region of solutions using one color.

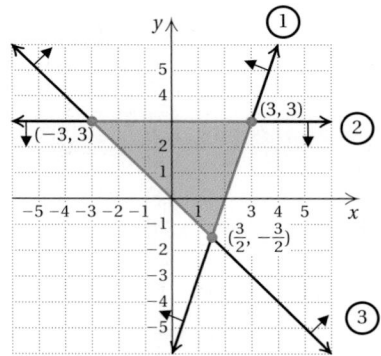

To find the vertices, we solve three different systems of equations. The system of equations from inequalities (1) and (2) is

$$6x - 2y = 12, \quad \textbf{(1)}$$
$$y - 3 = 0. \quad \textbf{(2)}$$

Solving, we obtain the vertex $(3, 3)$.
  The system of equations from inequalities (1) and (3) is

$$6x - 2y = 12, \quad \textbf{(1)}$$
$$x + y = 0. \quad \textbf{(3)}$$

Solving, we obtain the vertex $\left(\frac{3}{2}, -\frac{3}{2}\right)$.
  The system of equations from inequalities (2) and (3) is

$$y - 3 = 0, \quad \textbf{(2)}$$
$$x + y = 0. \quad \textbf{(3)}$$

Solving, we obtain the vertex $(-3, 3)$.

Do Exercise 9.

9. Graph the system of inequalities. Find the coordinates of any vertices formed.

$$5x + 6y \leq 30,$$
$$0 \leq y \leq 3,$$
$$0 \leq x \leq 4$$

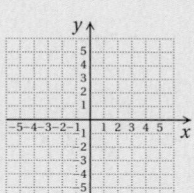

Answer

9.

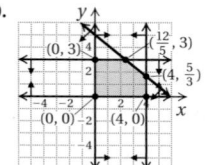

EXAMPLE 11 Graph the following system of inequalities. Find the coordinates of any vertices formed.

$$x + y \leq 16, \quad \text{(1)}$$
$$3x + 6y \leq 60, \quad \text{(2)}$$
$$x \geq 0, \quad \text{(3)}$$
$$y \geq 0 \quad \text{(4)}$$

We graph each inequality using solid lines. The regions for each inequality are indicated by the arrows at the ends of the lines. We then note where the regions overlap and shade the region of solutions using one color.

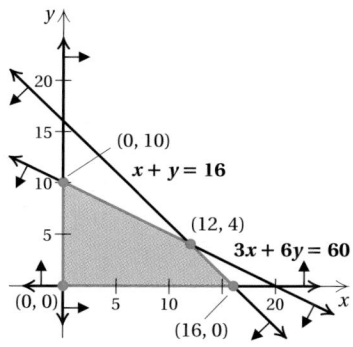

To find the vertices, we solve four different systems of equations. The system of equations from inequalities (1) and (2) is

$$x + y = 16, \quad \text{(1)}$$
$$3x + 6y = 60. \quad \text{(2)}$$

Solving, we obtain the vertex $(12, 4)$.

The system of equations from inequalities (1) and (4) is

$$x + y = 16, \quad \text{(1)}$$
$$y = 0. \quad \text{(4)}$$

Solving, we obtain the vertex $(16, 0)$.

The system of equations from inequalities (3) and (4) is

$$x = 0, \quad \text{(3)}$$
$$y = 0. \quad \text{(4)}$$

The vertex is $(0, 0)$.

The system of equations from inequalities (2) and (3) is

$$3x + 6y = 60, \quad \text{(2)}$$
$$x = 0. \quad \text{(3)}$$

Solving, we obtain the vertex $(0, 10)$.

Do Exercise 10.

10. Graph the system of inequalities. Find the coordinates of any vertices formed.

$$2x + 4y \leq 8,$$
$$x + y \leq 3,$$
$$x \geq 0,$$
$$y \geq 0$$

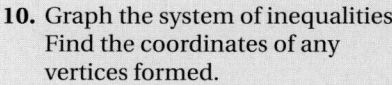

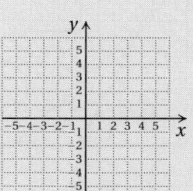

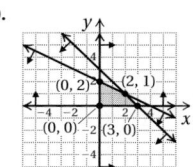

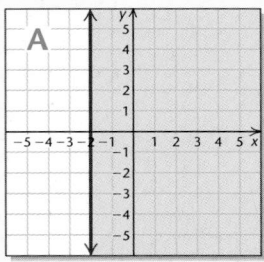

A

Visualizing for Success

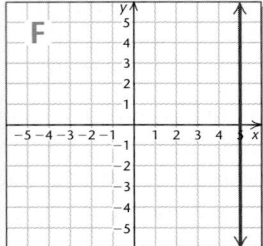

F

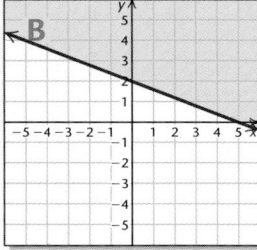

B

Match the equation, inequality, system of equations, or system of inequalities with its graph.

1. $x + y = -4,$
   $2x + y = -8$

2. $2x + 5y \geq 10$

3. $2x - 2y = 5$

4. $2x - 5y = 10$

5. $-2y < 8$

6. $5x - 2y = 10$

7. $2x = 10$

8. $5x + 2y < 10,$
   $2x - 5y > 10$

9. $5x \geq -10$

10. $y - 2x < 8$

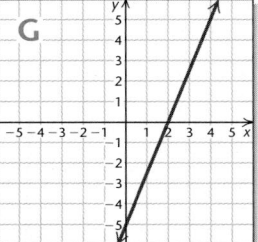

G

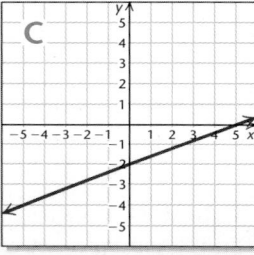

C

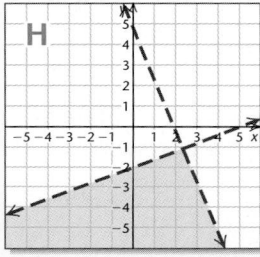

H

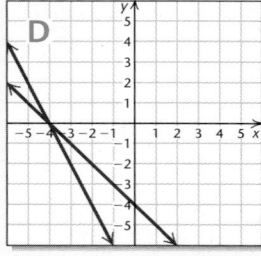

D

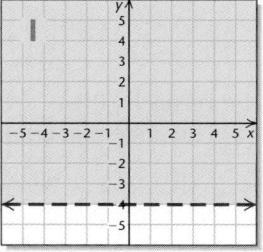

I

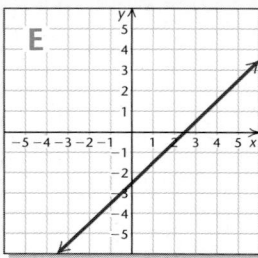

E

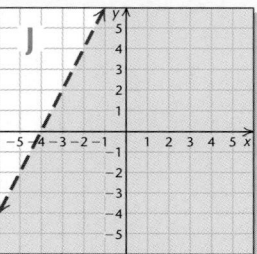

J

Answers on page A-27

Exercise Set

a Determine whether the given ordered pair is a solution of the given inequality.

1. $(-3, 3)$; $3x + y < -5$

2. $(6, -8)$; $4x + 3y \geq 0$

3. $(5, 9)$; $2x - y > -1$

4. $(5, -2)$; $6y - x > 2$

b Graph each inequality on a plane.

5. $y > 2x$

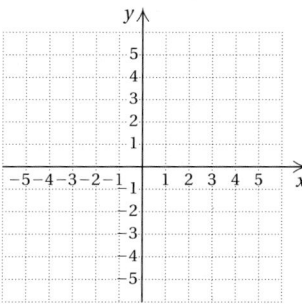

6. $y < 3x$

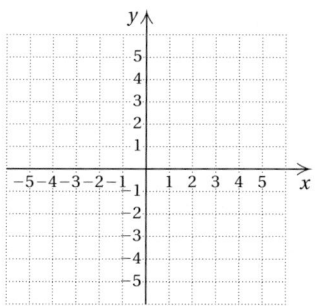

7. $y < x + 1$

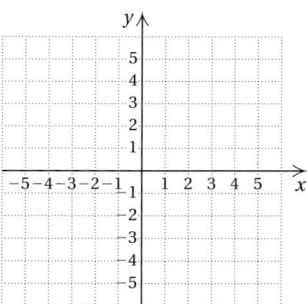

8. $y \leq x - 3$

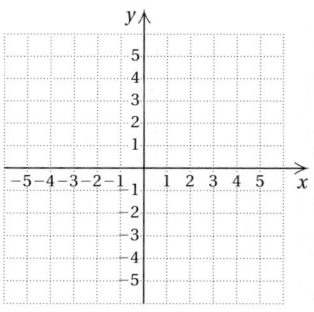

9. $y > x - 2$

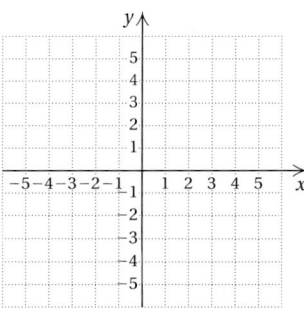

10. $y \geq x + 4$

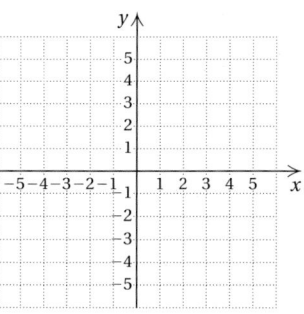

11. $x + y < 4$

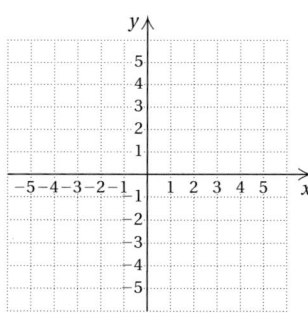

12. $x - y \geq 3$

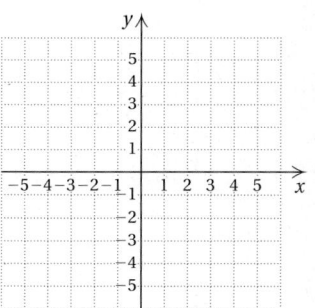

13. $3x + 4y \leq 12$

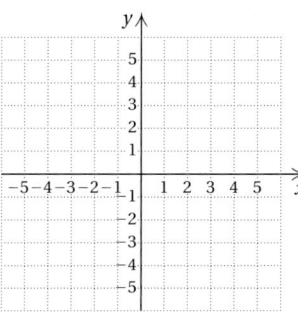

14. $2x + 3y < 6$

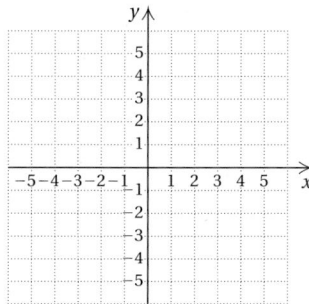

15. $2y - 3x > 6$

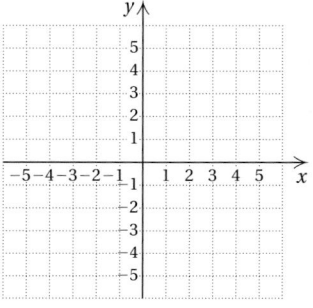

16. $2y - x \leq 4$

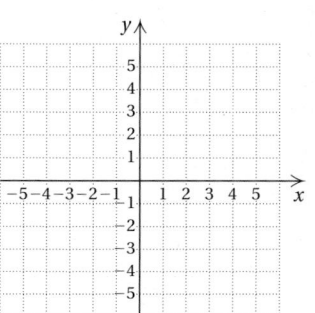

17. $3x - 2 \le 5x + y$

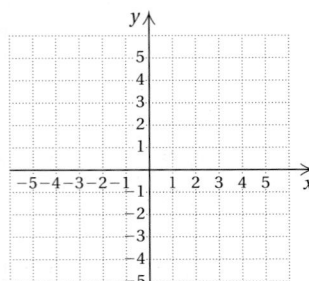

18. $2x - 2y \ge 8 + 2y$

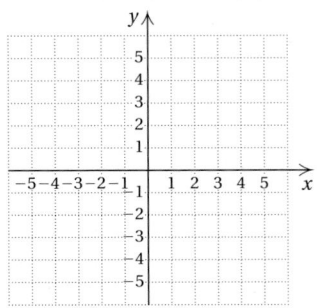

19. $x < 5$

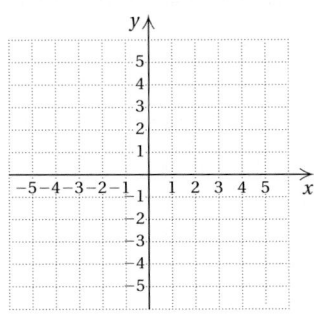

20. $y \ge -2$

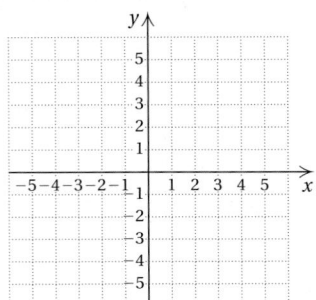

21. $y > 2$

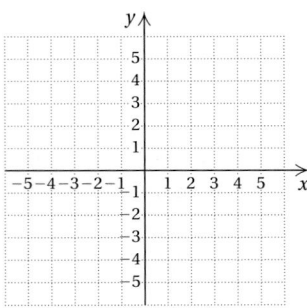

22. $x \le -4$

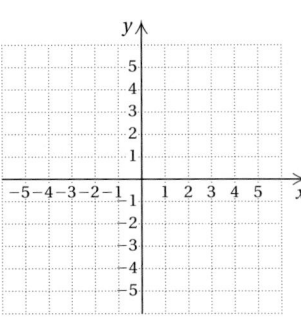

23. $2x + 3y \le 6$

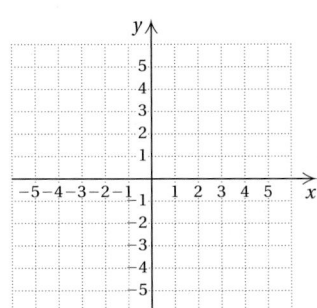

24. $7x + 2y \ge 21$

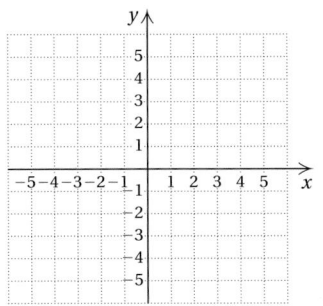

Matching. Each of Exercises 25–30 shows the graph of an inequality. Match the graph with one of the appropriate inequalities (A)–(F) that follow.

25.

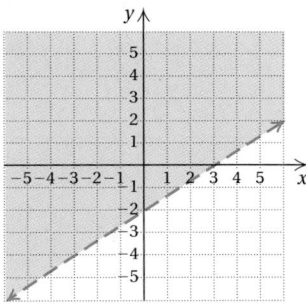

26.

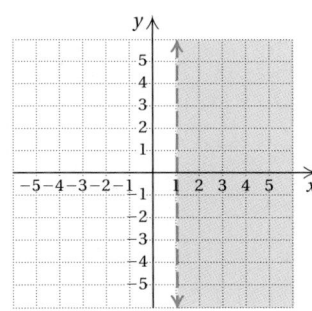

27.

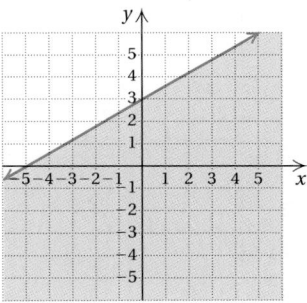

28.

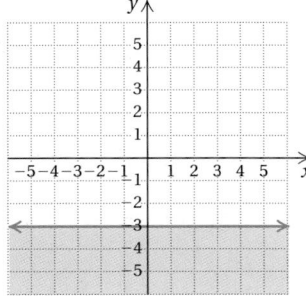

29.

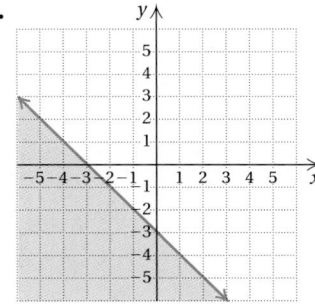

30.

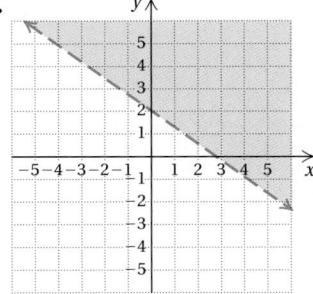

A. $4y > 8 - 3x$ **B.** $3x \ge 5y - 15$ **C.** $y + x \le -3$ **D.** $x > 1$ **E.** $y \le -3$ **F.** $2x - 3y < 6$

Graph each system of inequalities. Find the coordinates of any vertices formed.

31. $y \geq x,$
 $\quad y \leq -x + 2$

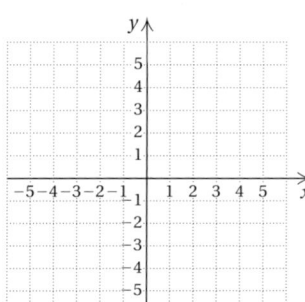

32. $y \geq x,$
 $\quad y \leq -x + 4$

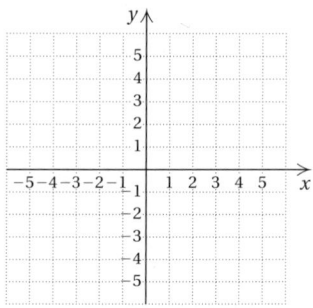

33. $y > x,$
 $\quad y < -x + 1$

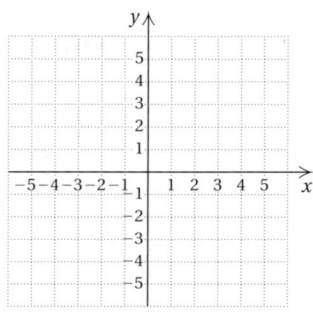

34. $y < x,$
 $\quad y > -x + 3$

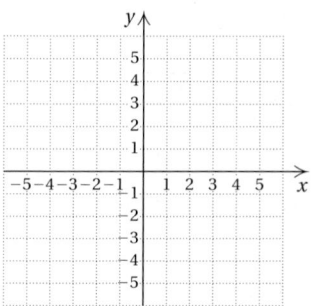

35. $x \leq 3,$
 $\quad y \geq -3x + 2$

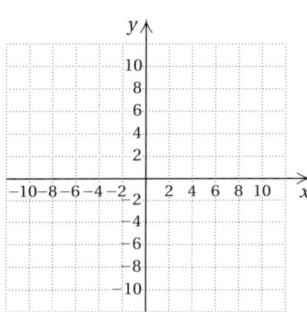

36. $x \geq -2,$
 $\quad y \leq -2x + 3$

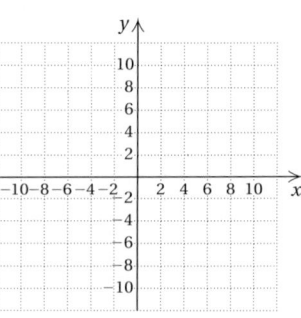

37. $x + y \leq 1,$
 $\quad x - y \leq 2$

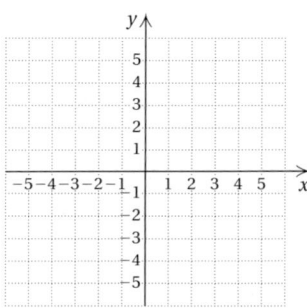

38. $x + y \leq 3,$
 $\quad x - y \leq 4$

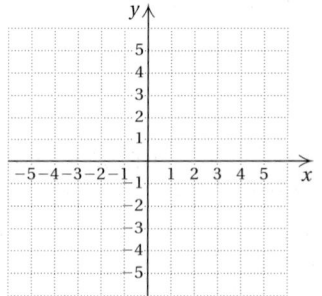

39. $y \leq 2x + 1,$
 $\quad y \geq -2x + 1,$
 $\quad x \leq 2$

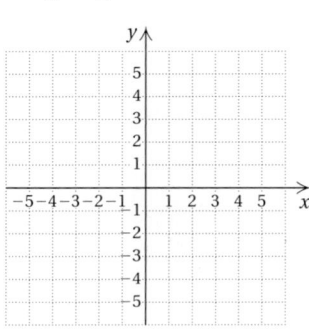

40. $x - y \leq 2,$
 $\quad x + 2y \geq 8,$
 $\quad y \leq 4$

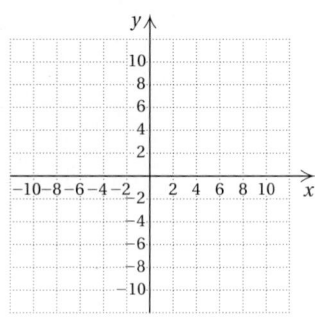

41. $x + 2y \leq 12,$
 $\quad 2x + y \leq 12,$
 $\quad x \geq 0,$
 $\quad y \geq 0$

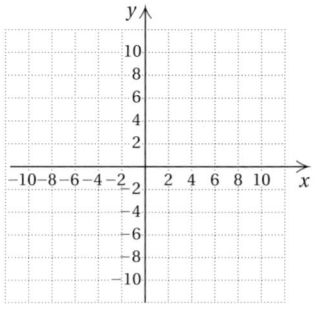

42. $y - x \geq 1,$
 $\quad y - x \leq 3,$
 $\quad 2 \leq x \leq 5$

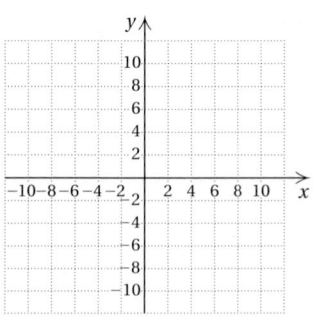

Skill Maintenance

Solve.   [2.3b, c]

43. $5(3x - 4) = -2(x + 5)$

44. $4(3x + 4) = 2 - x$

45. $2(x - 1) + 3(x - 2) - 4(x - 5) = 10$

46. $10x - 8(3x - 7) = 2(4x - 1)$

47. $5x + 7x = -144$

48. $0.5x - 2.34 + 2.4x = 7.8x - 9$

Given the function $f(x) = |2 - x|$, find each of the following function values.   [7.1b]

49. $f(0)$

50. $f(-1)$

51. $f(1)$

52. $f(10)$

53. $f(-2)$

54. $f(2a)$

55. $f(-4)$

56. $f(1.8)$

Synthesis

57. *Luggage Size.*   Unless an additional fee is paid, most major airlines will not check any luggage for which the sum of the item's length, width, and height exceeds 62 in. The U.S. Postal Service will ship a package only if the sum of the package's length and girth (distance around its midsection) does not exceed 130 in. Video Promotions is ordering several 30-in. long cases that will be both mailed and checked as luggage. Using w and h for width and height (in inches), respectively, write and graph an inequality that represents all acceptable combinations of width and height.

Source: U.S. Postal Service

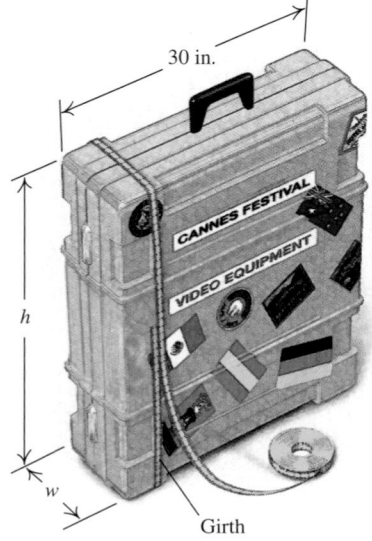

58. *Exercise Danger Zone.*   It is dangerous to exercise when the weather is hot and humid. The solutions of the following system of inequalities give a "danger zone" for which it is dangerous to exercise intensely:

$$4H - 3F < 70,$$
$$F + H > 160,$$
$$2F + 3H > 390,$$

where F is the temperature, in degrees Fahrenheit, and H is the humidity.

a) Draw the danger zone by graphing the system of inequalities.

b) Is it dangerous to exercise when $F = 80°$ and $H = 80\%$?

Summary and Review

Key Terms and Properties

inequality, p. 636
graph of an inequality, p. 636
set-builder notation, p. 637
interval notation, p. 637
compound inequality, p. 652
intersection of sets, p. 652

conjunction, p. 653
empty set, p. 655
disjoint sets, p. 655
union of sets, p. 656
disjunction, p. 656
absolute value, p. 666

linear inequality, p. 677
half-plane, p. 677
system of linear inequalities, p. 681
solution of a system of linear
   inequalities, p. 681

The Addition Principle for Inequalities: For any real numbers a, b, and c: $a < b$ is equivalent to $a + c < b + c$; $a > b$ is equivalent to $a + c > b + c$.

The Multiplication Principle for Inequalities: For any real numbers a and b, and any *positive* number c: $a < b$ is equivalent to $ac < bc$; $a > b$ is equivalent to $ac > bc$.

For any real numbers a and b, and any *negative* number c: $a < b$ is equivalent to $ac > bc$; $a > b$ is equivalent to $ac < bc$.

Similar statements hold for $\le$ and $\ge$.

Set Intersection: $A \cap B = \{x \mid x$ is in A and x is in $B\}$

Set Union: $A \cup B = \{x \mid x$ is in A or x is in B, or x is in both A and $B\}$

"$a < x$ and $x < b$" is equivalent to "$a < x < b$."

Properties of Absolute Value

$$|ab| = |a| \cdot |b|, \qquad \left|\frac{a}{b}\right| = \frac{|a|}{|b|}, \qquad |-a| = |a|, \qquad \text{The distance between } a \text{ and } b \text{ is } |a - b|.$$

Principles for Solving Equations and Inequalities Involving Absolute Value:

For any positive number p and any algebraic expression X:

a) The solutions of $|X| = p$ are those numbers that satisfy $X = -p$ or $X = p$.
b) The solutions of $|X| < p$ are those numbers that satisfy $-p < X < p$.
c) The solutions of $|X| > p$ are those numbers that satisfy $X < -p$ or $X > p$.

Concept Reinforcement

Determine whether each statement is true or false.

_____ **1.** If one point in a half-plane is a solution of a linear inequality, then all points in that half-plane are solutions.   [9.4b]

_____ **2.** Every system of linear inequalities has at least one solution.   [9.4b]

_____ **3.** For any real numbers a, b, and c, $c \ne 0$, $a \le b$ is equivalent to $ac \le bc$.   [9.1c]

_____ **4.** The inequalities $x < 2$ and $x \le 1$ are equivalent.   [9.1c]

_____ **5.** If x is negative, $|x| = -x$.   [9.3a]

_____ **6.** $|x|$ is always positive.   [9.3a]

_____ **7.** $|a - b| = |b - a|$.   [9.3b]

Important Concepts

Objective 9.1a   Determine whether a given number is a solution of an inequalitiy.

Example   Determine whether -3 and 1 are solutions of the inequality $4 - x \geq 2 - 5x$.

We substitute -3 for x and get $4 - (-3) \geq 2 - 5(-3)$, or $7 \geq 17$, a *false* sentence. Therefore, -3 is not a solution.

We substitute 1 for x and get $4 - 1 \geq 2 - 5 \cdot 1$, or $3 \geq -3$, a *true* sentence. Therefore, 1 is a solution.

Practice Exercise

1. Determine whether -2 and 5 are solutions of the inequality $8 - 3x \leq 3x + 6$.

Objective 9.1b   Write interval notation for the solution set of an inequality.

Example   Write interval notation for the solution set.

a) $\{x \mid x \leq -12\} = (-\infty, -12]$

b) $\{r \mid r > -1\} = (-1, \infty)$

c) $\{y \mid -8 \leq y < 9\} = [-8, 9)$

d) $\{x \mid 0 \geq x \geq -6\} = [-6, 0]$

e) $\{c \mid -25 < c \leq 25\} = (-25, 25]$

Practice Exercise

2. Write interval notation for the solution set.

a) $\{t \mid t < -8\}$

b) $\{x \mid -7 \leq x < 10\}$

c) $\{b \mid b \geq 3\}$

Objective 9.1c   Solve an inequality using the addition principle and the multiplication principle and then graph the inequality.

Example   Solve and graph: $6x - 7 \leq 3x + 2$.

$$6x - 7 \leq 3x + 2$$

$3x - 7 \leq 2$ Subtracting $3x$

$3x \leq 9$ Adding 7

$x \leq 3$ Dividing by 3

The solution set is $\{x \mid x \leq 3\}$, or $(-\infty, 3]$. We graph the solution set.

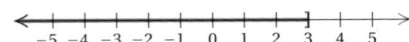

Practice Exercise

3. Solve and graph: $5y + 5 < 2y - 1$.

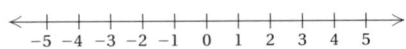

Objective 9.2a   Find the intersection of two sets. Solve and graph conjunctions of inequalities.

Example   Solve and graph: $-5 < 2x - 3 \leq 3$.

$-5 < 2x - 3 \leq 3$

$-2 < 2x \leq 6$ Adding 3

$-1 < x \leq 3$ Dividing by 2

The solution set is $\{x \mid -1 < x \leq 3\}$, or $(-1, 3]$. We graph the solution set.

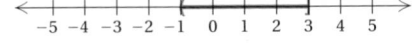

Practice Exercise

4. Solve and graph: $-4 \leq 5z + 6 < 11$.

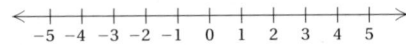

Objective 9.2b Find the union of two sets. Solve and graph disjunctions of inequalities.

Example Solve and graph: $2x + 1 \leq -5$ *or* $3x + 1 > 7$.

$$2x + 1 \leq -5 \quad or \quad 3x + 1 > 7$$
$$2x \leq -6 \quad or \quad 3x > 6$$
$$x \leq -3 \quad or \quad x > 2$$

The solution set is $\{x | x \leq -3 \ or \ x > 2\}$, or $(-\infty, -3] \cup (2, \infty)$. We graph the solution set.

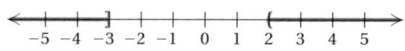

Practice Exercise

5. Solve and graph: $z + 4 < 3 \ or \ 4z + 1 \geq 5$.

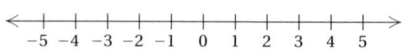

Objective 9.3c Solve equations with absolute-value expressions.

Example Solve: $|y - 2| = 1$.

$$y - 2 = -1 \quad or \quad y - 2 = 1$$
$$y = 1 \quad or \quad y = 3$$

The solution set is $\{1, 3\}$.

Practice Exercise

6. Solve: $|5x - 1| = 9$.

Objective 9.3d Solve equations with two absolute-value expressions.

Example Solve: $|4x - 4| = |2x + 8|$.

$$4x - 4 = 2x + 8 \quad or \quad 4x - 4 = -(2x + 8)$$
$$2x - 4 = 8 \quad or \quad 4x - 4 = -2x - 8$$
$$2x = 12 \quad or \quad 6x - 4 = -8$$
$$x = 6 \quad or \quad 6x = -4$$
$$x = 6 \quad or \quad x = -\frac{2}{3}$$

The solution set is $\left\{6, -\frac{2}{3}\right\}$.

Practice Exercise

7. Solve: $|z + 4| = |3z - 2|$.

Objective 9.3e Solve inequalities with absolute-value expressions.

Example Solve: **(a)** $|5x + 3| < 2$; **(b)** $|x + 3| \geq 1$.

a) $|5x + 3| < 2$

$$-2 < 5x + 3 < 2$$
$$-5 < 5x < -1$$
$$-1 < x < -\frac{1}{5}$$

The solution set is $\left\{x | -1 < x < -\frac{1}{5}\right\}$, or $\left(-1, -\frac{1}{5}\right)$.

b) $|x + 3| \geq 1$

$$x + 3 \leq -1 \quad or \quad x + 3 \geq 1$$
$$x \leq -4 \quad or \quad x \geq -2$$

The solution set is $\{x | x \leq -4 \ or \ x \geq -2\}$, or $(-\infty, -4] \cup [-2, \infty)$.

Practice Exercise

8. Solve: **(a)** $|2x + 3| < 5$; **(b)** $|3x + 2| \geq 8$.

Objective 9.4b Graph linear inequalities in two variables.

Example Graph: $2x + y \leq 4$.

First, we graph the line $2x + y = 4$. The intercepts are $(0, 4)$ and $(2, 0)$. We draw the line solid since the inequality symbol is $\leq$. Next, we choose a test point not on the line and determine whether it is a solution of the inequality. We choose $(0, 0)$, since it is usually an easy point to use.

$$\frac{2x + y \leq 4}{2 \cdot 0 + 0 \; ? \; 4}$$
$$0 \quad | \quad \text{TRUE}$$

Since $(0, 0)$ is a solution, we shade the half-plane that contains $(0, 0)$.

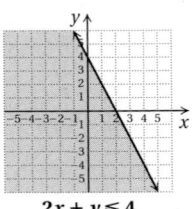

$$2x + y \leq 4$$

Practice Exercise

9. Graph: $3x - 2y > 6$.

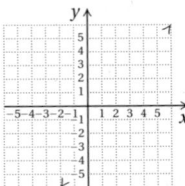

Objective 9.4c Graph systems of linear inequalities and find coordinates of any vertices.

Example Graph this system of inequalities and find the coordinates of any vertices formed:

$$x - 2y \geq -2, \quad \textbf{(1)}$$
$$3x - y \leq 4, \quad \textbf{(2)}$$
$$y \geq -1. \quad \textbf{(3)}$$

We graph the related equations using solid lines. Then we indicate the region for each inequality by arrows at the ends of the line. Next, we shade the region of overlap.

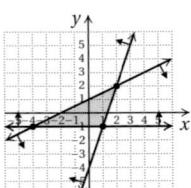

To find the vertices, we solve three different systems of related equations. From (1) and (2), we solve

$$x - 2y = -2,$$
$$3x - y = 4$$

to find the vertex $(2, 2)$. From (1) and (3), we solve

$$x - 2y = -2,$$
$$y = -1$$

to find the vertex $(-4, -1)$. From (2) and (3), we solve

$$3x - y = 4,$$
$$y = -1$$

to find the vertex $(1, -1)$.

Practice Exercise

10. Graph this system of inequalities and find the coordinates of any vertices found:

$$x - 2y \leq 4,$$
$$x + y \leq 4,$$
$$x - 1 \geq 0.$$

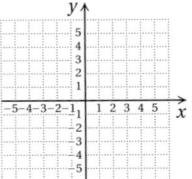

Review Exercises

1. Determine whether -3 and 7 are solutions of the inequality $2(1 - x) \leq 3x + 15$.   [9.1a]

Write interval notation for the given set or graph.   [9.1b]

2. $\{x \mid -8 \leq x < 9\}$

3.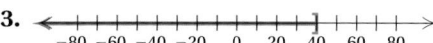

Solve and graph. Write interval notation for the solution set. [9.1c]

4. $x - 2 \leq -4$

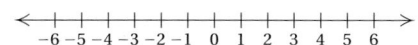

5. $x + 5 > 6$

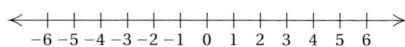

Solve.   [9.1c]

6. $a + 7 \leq -14$

7. $y - 5 \geq -12$

8. $4y > -16$

9. $-0.3y < 9$

10. $-6x - 5 < 13$

11. $4y + 3 \leq -6y - 9$

12. $-\frac{1}{2}x - \frac{1}{4} > \frac{1}{2} - \frac{1}{4}x$

13. $0.3y - 8 < 2.6y + 15$

14. $-2(x - 5) \geq 6(x + 7) - 12$

15. *Moving Costs.*   Metro Movers charges $85 plus $40 an hour to move households across town. Champion Moving charges $60 an hour for cross-town moves. For what lengths of time is Champion more expensive?   [9.1d]

16. *Investments.*   Joe plans to invest $30,000, part at 3% and part at 4%, for one year. What is the most that can be invested at 3% in order to make at least $1100 interest in one year?   [9.1d]

Graph and write interval notation.   [9.2a, b]

17. $-2 \leq x < 5$

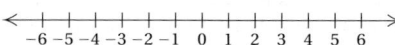

18. $x \leq -2 \ or \ x > 5$

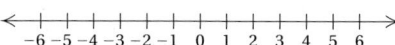

19. Find the intersection: [9.2a]
$$\{1, 2, 5, 6, 9\} \cap \{1, 3, 5, 9\}.$$

20. Find the union: [9.2b]
$$\{1, 2, 5, 6, 9\} \cup \{1, 3, 5, 9\}.$$

Solve.   [9.2a, b]

21. $2x - 5 < -7 \ and \ 3x + 8 \geq 14$

22. $-4 < x + 3 \leq 5$

23. $-15 < -4x - 5 < 0$

24. $3x < -9 \ or \ -5x < -5$

25. $2x + 5 < -17 \ or \ -4x + 10 \leq 34$

26. $2x + 7 \leq -5 \ or \ x + 7 \geq 15$

Simplify.   [9.3a]

27. $\left| -\dfrac{3}{x} \right|$

28. $\left| \dfrac{2x}{y^2} \right|$

29. $\left| \dfrac{12y}{-3y^2} \right|$

30. Find the distance between -23 and 39.   [9.3b]

Solve.   [9.3c, d]

31. $|x| = 6$

32. $|x - 2| = 7$

33. $|2x + 5| = |x - 9|$

34. $|5x + 6| = -8$

Solve.   [9.3e]

35. $|2x + 5| < 12$

36. $|x| \geq 3.5$

37. $|3x - 4| \geq 15$

38. $|x| < 0$

Graph. [9.4b]

39. $2x + 3y < 12$

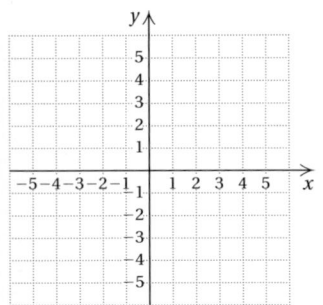

40. $y \le 0$

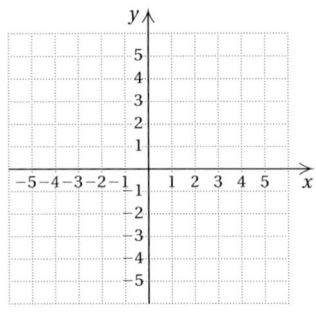

41. $x + y \ge 1$

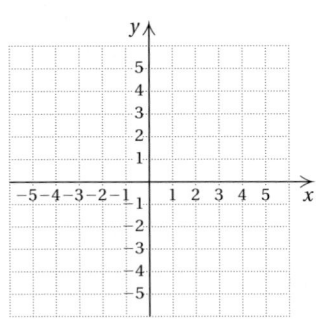

Graph. Find the coordinates of any vertices formed. [9.4c]

42. $y \ge -3,$
$\quad x \ge 2$

43. $x + 3y \ge -1,$
$\quad x + 3y \le 4$

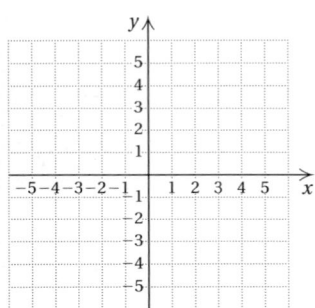

44. $x - y \le 3,$
$\quad x + y \ge -1,$
$\quad y \le 2$

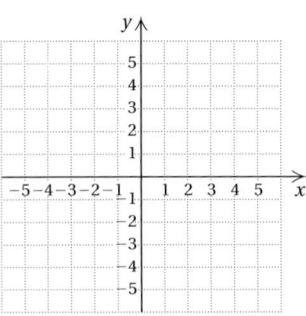

45. Solve: $-2(x + 3) - 1 < -2(2 - x)$. [9.1c]

  A. $\left(-\infty, -\dfrac{3}{4}\right)$ **B.** $(-6, \infty)$

  C. $(6, \infty)$ **D.** $\left(-\dfrac{3}{4}, \infty\right)$

46. Solve: $\left|-\dfrac{1}{3}x - 10\right| \ge 30$. [9.3e]

  A. $(-\infty, -100] \cup [80, \infty)$
  B. $(-\infty, -60] \cup [120, \infty)$
  C. $(-\infty, -120] \cup [60, \infty)$
  D. $(-\infty, -80] \cup [100, \infty)$

Synthesis

47. Solve: $|2x + 5| \le |x + 3|$. [9.3d, e]

Understanding Through Discussion and Writing

1. Describe the circumstances under which, for intervals, $[a, b] \cup [c, d] = [a, d]$. [9.2b]

2. Explain in your own words why the solutions of the inequality $|x + 5| \le 2$ can be interpreted as "all those numbers x whose distance from -5 is at most 2 units." [9.3e]

3. When graphing linear inequalities, Ron always shades above the line when he sees a $\ge$ symbol. Is this wise? Why or why not? [9.4a]

4. Explain in your own words why the interval $[6, \infty)$ is only part of the solution set of $|x| \ge 6$. [9.3e]

Test

For Extra Help

CHAPTER
Test Prep
VIDEOS

Step-by-step test solutions are found on the Chapter Test Prep Videos available via the Video Resour
on DVD, in *MyMathLab* , and on You Tube (search "BittingerIntroInter" and click on "Channels").

Write interval notation for the given set or graph.

1. $\{x | -3 < x \leq 2\}$

2.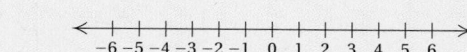

Solve and graph. Write interval notation for the solution set.

3. $x - 2 \leq 4$

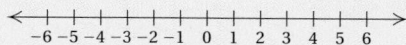

4. $-4y - 3 \geq 5$

Solve.

5. $x - 4 \geq 6$

6. $-0.6y < 30$

7. $3a - 5 \leq -2a + 6$

8. $-5y - 1 > -9y + 3$

9. $4(5 - x) < 2x + 5$

10. $-8(2x + 3) + 6(4 - 5x) \geq 2(1 - 7x) - 4(4 + 6x)$

Solve.

11. *Moving Costs.* Mitchell Moving Company charges $105 plus $30 an hour to move households across town. Quick-Pak Moving charges $80 an hour for cross-town moves. For what lengths of time is Quick-Pak more expensive?

12. *Pressure at Sea Depth.* The equation

$$P = 1 + \frac{d}{33}$$

gives the pressure P, in atmospheres (atm), at a depth of d feet in the sea. For what depths d is the pressure at least 2 atm and at most 8 atm?

Graph and write interval notation.

13. $-3 \leq x \leq 4$

14. $x < -3 \, or \, x > 4$

Solve.

15. $5 - 2x \leq 1 \, and \, 3x + 2 \geq 14$

16. $-3 < x - 2 < 4$

17. $-11 \leq -5x - 2 < 0$

18. $-3x > 12 \, or \, 4x > -10$

19. $x - 7 \leq -5 \, or \, x - 7 \geq -10$

20. $3x - 2 < 7 \, or \, x - 2 > 4$

Simplify.

21. $\left|\dfrac{7}{x}\right|$

22. $\left|\dfrac{-6x^2}{3x}\right|$

23. Find the distance between 4.8 and -3.6.

24. Find the intersection:

$$\{1, 3, 5, 7, 9\} \cap \{3, 5, 11, 13\}.$$

25. Find the union:

$$\{1, 3, 5, 7, 9\} \cup \{3, 5, 11, 13\}.$$

Solve.

26. $|x| = 9$

27. $|x - 3| = 9$

28. $|x + 10| = |x - 12|$

29. $|2 - 5x| = -10$

30. $|4x - 1| < 4.5$

31. $|x| > 3$

32. $\left|\dfrac{6 - x}{7}\right| \leq 15$

33. $|-5x - 3| \geq 10$

Graph. Find the coordinates of any vertices formed.

34. $x - 6y < -6$

35. $x + y \geq 3,$
$\quad\ \ x - y \geq 5$

36. $2y - x \geq -4,$
$\quad\ \ 2y + 3x \leq -6,$
$\quad\ \ \ \ \ \ \ \ y \leq 0,$
$\quad\ \ \ \ \ \ \ \ x \leq 0$

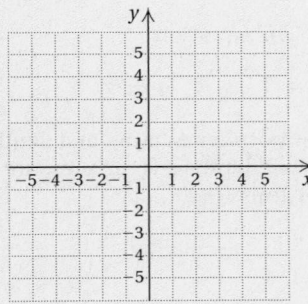

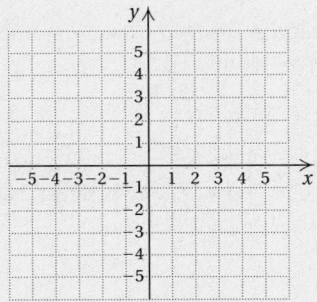

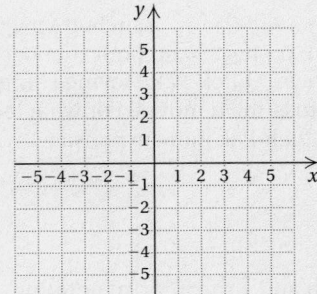

37. Solve: $\left|\dfrac{1}{2}x - 2\right| \geq 2.2.$

    A. $[-0.4, 8.4]$

    B. $(-\infty, -0.4] \cup [8.4, \infty)$

    C. $(-\infty, 8.4]$

    D. $[-0.4, \infty)$

Synthesis

Solve.

38. $|3x - 4| \leq -3$

39. $7x < 8 - 3x < 6 + 7x$

Cumulative Review

Graph.

1. $y = -5x + 4$

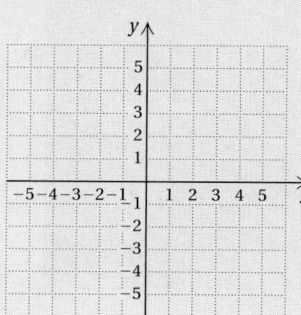

2. $3x - 18 = 0$

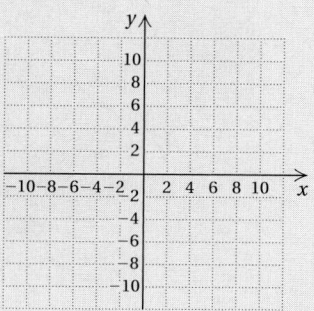

3. $x + 3y < 4$

4. $x + y \geq 4,$
$x - y > 1$

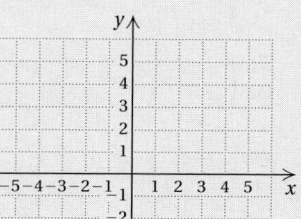

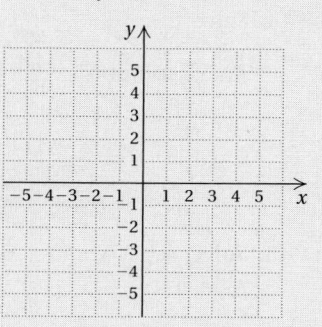

5. Given that $g(x) = |x - 4| + 5$, find $g(-2)$.

6. Given that

$$f(x) = \frac{x - 2}{x^2 - 25},$$

find the domain.

7. Find the domain and the range of the function graphed below.

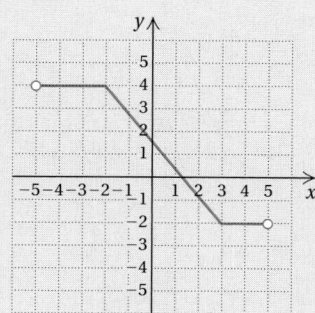

Simplify.

8. $(6m - n)^2$

9. $(3a - 4b)(5a + 2b)$

10. $\dfrac{y^2 - 4}{3y + 33} \cdot \dfrac{y + 11}{y + 2}$

11. $\dfrac{9x^2 - 25}{x^2 - 16} \div \dfrac{3x + 5}{x - 4}$

12. $\dfrac{2x + 1}{4x - 12} - \dfrac{x - 2}{5x - 15}$

13. $\dfrac{1 - \dfrac{2}{y^2}}{1 - \dfrac{1}{y^3}}$

14. $(6p^2 - 2p + 5) - (-10p^2 + 6p + 5)$

15. $\dfrac{2}{x + 2} + \dfrac{3}{x - 2} - \dfrac{x + 1}{x^2 - 4}$

16. $(2x^3 - 7x^2 + x - 3) \div (x + 2)$

Solve.

17. $9y - (5y - 3) = 33$

18. $-3 < -2x - 6 < 0$

19. $\dfrac{3x}{x - 2} - \dfrac{6}{x + 2} = \dfrac{24}{x^2 - 4}$

20. $P = \dfrac{3a}{a + b}$, for a

21. $F = \dfrac{9}{5}C + 32$, for C

22. $|x| \geq 2.1$

23. $\dfrac{6}{x - 5} = \dfrac{2}{2x}$

24. $8x = 1 + 16x^2$

25. $14 + 3x = 2x^2$

Solve.

26. $4x - 2y = 6,$
$\quad 6x - 3y = 9$

27. $4x + 5y = -3,$
$\quad x = 1 - 3y$

28. $x + 2y - 2z = 9,$
$\quad 2x - 3y + 4z = -4,$
$\quad 5x - 4y + 2z = 5$

29. $x + 6y + 4z = -2,$
$\quad 4x + 4y + \ z = 2,$
$\quad 3x + 2y - 4z = 5$

Factor.

30. $4x^3 + 18x^2$

31. $8a^3 - 4a^2 - 6a + 3$

32. $x^2 + 8x - 84$

33. $6x^2 + 11x - 10$

34. $16y^2 - 81$

35. $t^2 - 16t + 64$

36. $64x^3 + 8$

37. $0.027b^3 - 0.008c^3$

38. $x^6 - x^2$

39. $20x^2 + 7x - 3$

40. Find an equation of the line with slope $-\frac{1}{2}$ passing through the point $(2, -2)$.

41. Find an equation of the line that is perpendicular to the line $2x + y = 5$ and passes through the point $(3, -1)$.

42. *Hockey Results.* A hockey team played 81 games in a season. They won 1 fewer game than three times the number of ties and lost 8 fewer games than they won. How many games did they win? lose? tie?

43. *Waste Generation.* The amount of waste generated by a fast-food restaurant varies directly as the number of customers served. A typical restaurant that serves 2000 customers per day generates 238 lb of waste daily. How many pounds of waste would be generated daily by a restaurant that serves 1700 customers a day?

44. Solve: $\dfrac{x}{x - 4} - \dfrac{4}{x + 3} = \dfrac{28}{x^2 - x - 12}.$
  A. No solution   **B.** 0
  C. $-4, 3$ **D.** $4, -3$

45. Solve: $x^2 - x - 6 = 6.$
  A. $4, 9$ **B.** $3, 8$
  C. $4, -3$ **D.** $0, 1$

46. *Tank Filling.* An oil storage tank can be filled in 10 hr by ship A working alone and in 15 hr by ship B working alone. How many hours would it take to fill the oil storage tank if both ships A and B are working?
  A. 8 hr   **B.** 6 hr
  C. $12\frac{1}{2}$ hr   **D.** 25 hr

Synthesis

47. The graph of $y = ax^2 + bx + c$ contains the three points $(4, 2)$, $(2, 0)$, and $(1, 2)$. Find a, b, and c.

Solve.

48. $16x^3 = x$

49. $\dfrac{18}{x - 9} + \dfrac{10}{x + 5} = \dfrac{28x}{x^2 - 4x - 45}$

Radical Expressions, Equations, and Functions

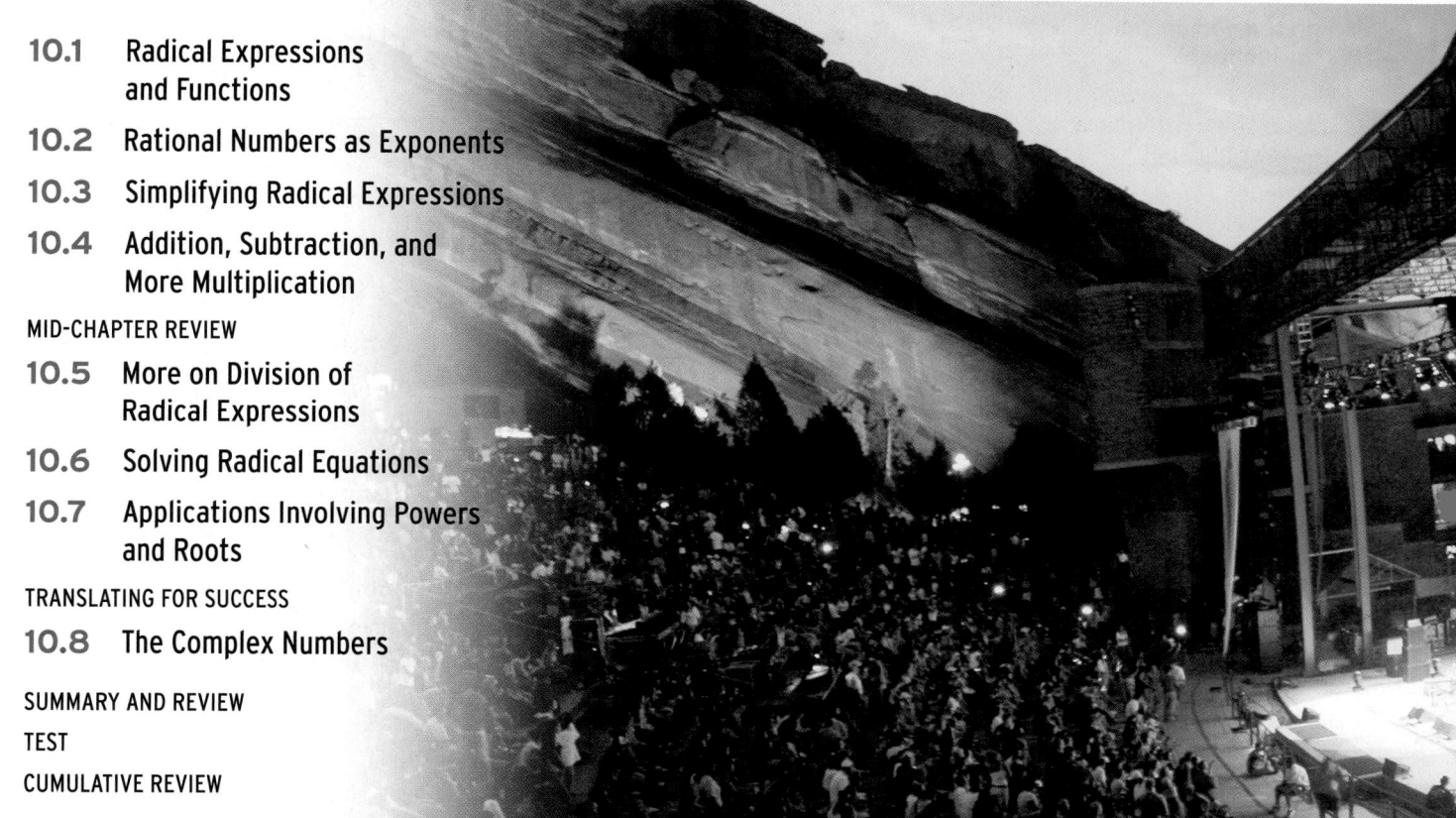

Real-World Application

The geologically formed, open-air Red Rocks Amphitheatre near Denver, Colorado, hosts a series of concerts. A scientific instrument at one of these concerts determined that the sound of the music was traveling at a rate of 1170 ft/sec. What was the air temperature at the concert?

This problem appears as Example 9 in Section 10.6.

10.1

Radical Expressions and Functions

OBJECTIVES

a Find principal square roots and their opposites, approximate square roots, identify radicands, find outputs of square-root functions, graph square-root functions, and find the domains of square-root functions.

b Simplify radical expressions with perfect-square radicands.

c Find cube roots, simplifying certain expressions, and find outputs of cube-root functions.

d Simplify expressions involving odd roots and even roots.

SKILL TO REVIEW
Objective 7.2a: Find the domain of a function.

Find the domain of the function.

1. $g(x) = \dfrac{2x + 1}{x - 3}$

2. $f(x) = 4 - x^2$

Find the square roots.
1. 9   **2.** 36   **3.** 121

Simplify.
4. $\sqrt{1}$ **5.** $\sqrt{36}$

6. $\sqrt{\dfrac{81}{100}}$ **7.** $\sqrt{0.0064}$

Answers

Skill to Review:
1. $\{x \mid x$ is a real number *and* $x \neq 3\}$, or $(-\infty, 3) \cup (3, \infty)$ **2.** All real numbers

Margin Exercises:
1. $3, -3$ **2.** $6, -6$ **3.** $11, -11$
4. 1   **5.** 6   **6.** $\dfrac{9}{10}$ **7.** 0.08

In this section, we consider roots, such as square roots and cube roots. We define the symbolism and consider methods of manipulating symbols to get equivalent expressions.

a Square Roots and Square-Root Functions

When we raise a number to the second power, we say that we have **squared** the number. Sometimes we may need to find the number that was squared. We call this process **finding a square root** of a number.

> ### SQUARE ROOT
>
> The number c is a **square root** of a if $c^2 = a$.

For example:

> 5 is a *square root* of 25 because $5^2 = 5 \cdot 5 = 25$;
>
> -5 is a *square root* of 25 because $(-5)^2 = (-5)(-5) = 25$.

The number -4 does not have a real-number square root because there is no real number c such that $c^2 = -4$.

> ### PROPERTIES OF SQUARE ROOTS
>
> Every positive real number has two real-number square roots.
>
> The number 0 has just one square root, 0 itself.
>
> Negative numbers do not have real-number square roots.*

EXAMPLE 1 Find the two square roots of 64.

The square roots of 64 are 8 and -8 because $8^2 = 64$ and $(-8)^2 = 64$.

Do Exercises 1–3.

> ### PRINCIPAL SQUARE ROOT
>
> The **principal square root** of a nonnegative number is its nonnegative square root. The symbol $\sqrt{a}$ represents the principal square root of a. To name the negative square root of a, we can write $-\sqrt{a}$.

---

*In Section 10.8, we will consider a number system in which negative numbers do have square roots.

EXAMPLES Simplify.

2. $\sqrt{25} = 5$

> Remember: $\sqrt{}$ indicates the principal (nonnegative) square root.

3. $-\sqrt{25} = -5$

4. $\sqrt{\dfrac{81}{64}} = \dfrac{9}{8}$ because $\left(\dfrac{9}{8}\right)^2 = \dfrac{9}{8} \cdot \dfrac{9}{8} = \dfrac{81}{64}$.

5. $\sqrt{0.0049} = 0.07$ because $(0.07)^2 = (0.07)(0.07) = 0.0049$.

6. $-\sqrt{0.000001} = -0.001$

7. $\sqrt{0} = 0$

8. $\sqrt{-25}$ Does not exist as a real number. Negative numbers do not have real-number square roots.

Do Exercises 4–13. (Exercises 4–7 are on the preceding page.)

We found exact square roots in Examples 1–8. We often need to use rational numbers to *approximate* square roots that are irrational. Such expressions can be found using a calculator with a square-root key.

EXAMPLES Use a calculator to approximate each of the following.

Number	Using a calculator with a 10-digit readout	Rounded to three decimal places
9. $\sqrt{11}$	3.316624790	3.317
10. $\sqrt{487}$	22.06807649	22.068
11. $-\sqrt{7297.8}$	−85.42716196	−85.427
12. $\sqrt{\dfrac{463}{557}}$	.9117229728	0.912

Do Exercises 14–19.

RADICAL; RADICAL EXPRESSION; RADICAND

The symbol $\sqrt{}$ is called a **radical**.

An expression written with a radical is called a **radical expression**.

The expression written under the radical is called the **radicand**.

These are radical expressions:

$$\sqrt{5}, \quad \sqrt{a}, \quad -\sqrt{5x}, \quad \sqrt{y^2 + 7}.$$

The radicands in these expressions are 5, a, $5x$, and $y^2 + 7$, respectively.

EXAMPLE 13 Identify the radicand in $x\sqrt{x^2 - 9}$.

The radicand is the expression under the radical, $x^2 - 9$.

Do Exercises 20 and 21 on the following page.

Find each of the following.

8. a) $\sqrt{16}$
   b) $-\sqrt{16}$
   c) $\sqrt{-16}$

9. a) $\sqrt{49}$
   b) $-\sqrt{49}$
   c) $\sqrt{-49}$

10. $\sqrt{\dfrac{25}{64}}$

11. $\sqrt{\dfrac{16}{9}}$

12. $-\sqrt{0.81}$

13. $\sqrt{1.44}$

It would be helpful to memorize the following table of exact square roots.

TABLE OF COMMON SQUARE ROOTS	
$\sqrt{1} = 1$	$\sqrt{196} = 14$
$\sqrt{4} = 2$	$\sqrt{225} = 15$
$\sqrt{9} = 3$	$\sqrt{256} = 16$
$\sqrt{16} = 4$	$\sqrt{289} = 17$
$\sqrt{25} = 5$	$\sqrt{324} = 18$
$\sqrt{36} = 6$	$\sqrt{361} = 19$
$\sqrt{49} = 7$	$\sqrt{400} = 20$
$\sqrt{64} = 8$	$\sqrt{441} = 21$
$\sqrt{81} = 9$	$\sqrt{484} = 22$
$\sqrt{100} = 10$	$\sqrt{529} = 23$
$\sqrt{121} = 11$	$\sqrt{576} = 24$
$\sqrt{144} = 12$	$\sqrt{625} = 25$
$\sqrt{169} = 13$	

Use a calculator to approximate each square root to three decimal places.

14. $\sqrt{17}$

15. $\sqrt{40}$

16. $\sqrt{1138}$

17. $-\sqrt{867.6}$

18. $\sqrt{\dfrac{22}{35}}$

19. $-\sqrt{\dfrac{2103.4}{67.82}}$

Answers

8. (a) 4; (b) −4; (c) does not exist as a real number   9. (a) 7; (b) −7; (c) does not exist as a real number   10. $\dfrac{5}{8}$ 11. $\dfrac{4}{3}$ 12. −0.9
13. 1.2   14. 4.123   15. 6.325   16. 33.734
17. −29.455   18. 0.793   19. −5.569

Since each nonnegative real number x has exactly one principal square root, the symbol $\sqrt{x}$ represents exactly one real number and thus can be used to define a square-root function:

$$f(x) = \sqrt{x}.$$

The domain of this function is the set of nonnegative real numbers. In interval notation, the domain is $[0, \infty)$. This function will be discussed further in Example 16.

EXAMPLE 14 For the given function, find the indicated function values:

$$f(x) = \sqrt{3x - 2}; \quad f(1), f(5), \text{ and } f(0).$$

We have

$$f(1) = \sqrt{3 \cdot 1 - 2} \qquad \text{Substituting 1 for } x$$
$$= \sqrt{3 - 2} = \sqrt{1} = 1; \qquad \text{Simplifying and taking the square root}$$

$$f(5) = \sqrt{3 \cdot 5 - 2} \qquad \text{Substituting 5 for } x$$
$$= \sqrt{13} \approx 3.606; \qquad \text{Simplifying and approximating}$$

$$f(0) = \sqrt{3 \cdot 0 - 2} \qquad \text{Substituting 0 for } x$$
$$= \sqrt{-2}. \qquad \text{Negative radicand. No real-number function value exists; 0 is not in the domain of } f.$$

Do Exercises 22 and 23.

EXAMPLE 15 Find the domain of $g(x) = \sqrt{x + 2}$.

The expression $\sqrt{x + 2}$ is a real number only when $x + 2$ is nonnegative. Thus the domain of $g(x) = \sqrt{x + 2}$ is the set of all x-values for which $x + 2 \geq 0$. We solve as follows:

$$x + 2 \geq 0$$
$$x \geq -2. \qquad \text{Adding } -2$$

The domain of $g = \{x | x \geq -2\} = [-2, \infty)$.

EXAMPLE 16 Graph: **(a)** $f(x) = \sqrt{x}$; **(b)** $g(x) = \sqrt{x + 2}$.

We first find outputs as we did in Example 14. We can either select inputs that have exact outputs or use a calculator to make approximations. Once ordered pairs have been calculated, a smooth curve can be drawn.

a)

x	$f(x) = \sqrt{x}$	$(x, f(x))$
0	0	$(0, 0)$
1	1	$(1, 1)$
3	1.7	$(3, 1.7)$
4	2	$(4, 2)$
7	2.6	$(7, 2.6)$
9	3	$(9, 3)$

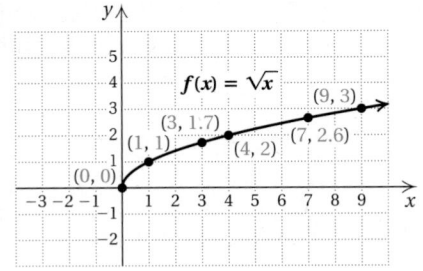

We can see from the table and the graph that the domain of f is $[0, \infty)$. The range is also the set of nonnegative real numbers $[0, \infty)$.

b)

x	$g(x) = \sqrt{x + 2}$	$(x, g(x))$
-2	0	$(-2, 0)$
-1	1	$(-1, 1)$
0	1.4	$(0, 1.4)$
3	2.2	$(3, 2.2)$
5	2.6	$(5, 2.6)$
10	3.5	$(10, 3.5)$

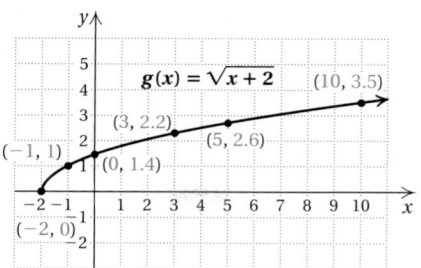

We can see from the table, the graph, and Example 15 that the domain of g is $[-2, \infty)$. The range is the set of nonnegative real numbers $[0, \infty)$.

Do Exercises 24–27.

b Finding $\sqrt{a^2}$

In the expression $\sqrt{a^2}$, the radicand is a perfect square. It is tempting to think that $\sqrt{a^2} = a$, but we see below that this is not the case.

Suppose $a = 5$. Then we have $\sqrt{5^2}$, which is $\sqrt{25}$, or 5.

Suppose $a = -5$. Then we have $\sqrt{(-5)^2}$, which is $\sqrt{25}$, or 5.

Suppose $a = 0$. Then we have $\sqrt{0^2}$, which is $\sqrt{0}$, or 0.

The symbol $\sqrt{a^2}$ never represents a negative number. It represents the principal square root of a^2. Note the following.

SIMPLIFYING $\sqrt{a^2}$

$a \geq 0 \longrightarrow \sqrt{a^2} = a$

If a is positive or 0, the principal square root of a^2 is a.

$a < 0 \longrightarrow \sqrt{a^2} = -a$

If a is negative, the principal square root of a^2 is the opposite of a.

In all cases, the radical expression represents the absolute value of a.

PRINCIPAL SQUARE ROOT OF a^2

For any real number a, $\sqrt{a^2} = |a|$. The principal (nonnegative) square root of a^2 is the absolute value of a.

The absolute value is used to ensure that the principal square root is nonnegative, which is as it is defined.

Find the domain of each function.

24. $f(x) = \sqrt{x - 5}$

25. $g(x) = \sqrt{2x + 3}$

Graph.

26. $g(x) = -\sqrt{x}$

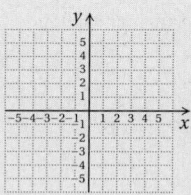

27. $f(x) = 2\sqrt{x} + 3$

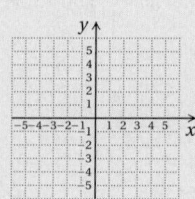

Answers

24. $\{x | x \geq 5\}$, or $[5, \infty)$
25. $\{x | x \geq -\frac{3}{2}\}$, or $\left[-\frac{3}{2}, \infty\right)$
26.

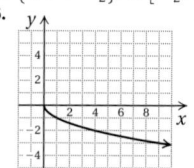

$g(x) = -\sqrt{x}$

27.

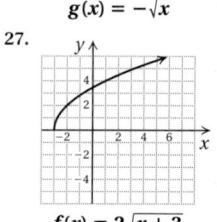

$f(x) = 2\sqrt{x} + 3$

Find each of the following. Assume that letters can represent *any* real number.

28. $\sqrt{y^2}$ **29.** $\sqrt{(-24)^2}$

30. $\sqrt{(5y)^2}$ **31.** $\sqrt{16y^2}$

32. $\sqrt{(x+7)^2}$

33. $\sqrt{4(x-2)^2}$

34. $\sqrt{49(y+5)^2}$

35. $\sqrt{x^2 - 6x + 9}$

EXAMPLES Find each of the following. Assume that letters can represent any real number.

17. $\sqrt{(-16)^2} = |-16|$, or 16

18. $\sqrt{(3b)^2} = |3b| = |3| \cdot |b| = 3|b|$

$|3b|$ can be simplified to $3|b|$ because the absolute value of any product is the product of the absolute values. That is, $|a \cdot b| = |a| \cdot |b|$.

19. $\sqrt{(x-1)^2} = |x-1|$ ---- *Caution!* ----

20. $\sqrt{x^2 + 8x + 16} = \sqrt{(x+4)^2}$ $|x+4|$ is *not* the

$= |x+4| \longleftarrow$ same as $|x| + 4$.

Do Exercises 28–35.

(c) Cube Roots

CUBE ROOT

The number c is the **cube root** of a, written $\sqrt[3]{a}$, if the third power of c is a—that is, if $c^3 = a$, then $\sqrt[3]{a} = c$.

For example:

2 is the *cube root* of 8 because $2^3 = 2 \cdot 2 \cdot 2 = 8$;

-4 is the *cube root* of -64 because $(-4)^3 = (-4)(-4)(-4) = -64$.

We talk about *the* cube root of a number rather than *a* cube root because of the following.

Every real number has exactly one cube root in the system of real numbers. The symbol $\sqrt[3]{a}$ represents *the* cube root of a.

EXAMPLES Find each of the following.

21. $\sqrt[3]{8} = 2$ because $2^3 = 8$.   **22.** $\sqrt[3]{-27} = -3$

23. $\sqrt[3]{-\dfrac{216}{125}} = -\dfrac{6}{5}$ **24.** $\sqrt[3]{0.001} = 0.1$

25. $\sqrt[3]{x^3} = x$ **26.** $\sqrt[3]{-8} = -2$

27. $\sqrt[3]{0} = 0$ **28.** $\sqrt[3]{-8y^3} = \sqrt[3]{(-2y)^3} = -2y$

Find each of the following.

36. $\sqrt[3]{-64}$ **37.** $\sqrt[3]{27y^3}$

38. $\sqrt[3]{8(x+2)^3}$ **39.** $\sqrt[3]{-\dfrac{343}{64}}$

When we are determining a cube root, no absolute-value signs are needed because a real number has just one cube root. The real-number cube root of a positive number is positive. The real-number cube root of a negative number is negative. The cube root of 0 is 0. That is, $\sqrt[3]{a^3} = a$ whether $a > 0$, $a < 0$, or $a = 0$.

Do Exercises 36–39.

Answers

28. $|y|$ **29.** 24   **30.** $5|y|$ **31.** $4|y|$
32. $|x+7|$ **33.** $2|x-2|$ **34.** $7|y+5|$
35. $|x-3|$ **36.** -4 **37.** $3y$
38. $2(x+2)$ **39.** $-\dfrac{7}{4}$

Since the symbol $\sqrt[3]{x}$ represents exactly one real number, it can be used to define a cube-root function: $f(x) = \sqrt[3]{x}$.

EXAMPLE 29 For the given function, find the indicated function values:

$$f(x) = \sqrt[3]{x}; \quad f(125), f(0), f(-8), \text{ and } f(-10).$$

We have

$$f(125) = \sqrt[3]{125} = 5;$$
$$f(0) = \sqrt[3]{0} = 0;$$
$$f(-8) = \sqrt[3]{-8} = -2;$$
$$f(-10) = \sqrt[3]{-10} \approx -2.154.$$

Do Exercise 40.

The graph of $f(x) = \sqrt[3]{x}$ is shown below for reference. Note that both the domain and the range consist of the entire set of real numbers, $(-\infty, \infty)$.

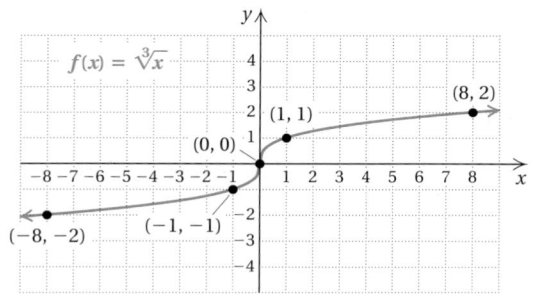

d Odd and Even *k*th Roots

In the expression $\sqrt[k]{a}$, we call k the **index** and assume $k \geq 2$.

Odd Roots

The 5th root of a number a is the number c for which $c^5 = a$. There are also 7th roots, 9th roots, and so on. Whenever the number k in $\sqrt[k]{}$ is an odd number, we say that we are taking an **odd root**.

Every number has just one real-number odd root. For example, $\sqrt[3]{8} = 2$, $\sqrt[3]{-8} = -2$, and $\sqrt[3]{0} = 0$. If the number is positive, then the root is positive. If the number is negative, then the root is negative. If the number is 0, then the root is 0. Absolute-value signs are *not* needed when we are finding odd roots.

> If k is an *odd* natural number, then for any real number a,
> $$\sqrt[k]{a^k} = a.$$

---

40. For the given function, find the indicated function values:
$$g(x) = \sqrt[3]{x} - 4; \quad g(-23),$$
$$g(4), g(-1), \text{ and } g(11).$$

Answer

40. -3; 0; $\sqrt[3]{-5} \approx -1.710$; $\sqrt[3]{7} \approx 1.913$

Find each of the following.

41. $\sqrt[5]{243}$ **42.** $\sqrt[5]{-243}$

43. $\sqrt[5]{x^5}$ **44.** $\sqrt[7]{y^7}$

45. $\sqrt[5]{0}$ **46.** $\sqrt[5]{-32x^5}$

47. $\sqrt[7]{(3x + 2)^7}$

EXAMPLES Find each of the following.

30. $\sqrt[5]{32} = 2$ **31.** $\sqrt[5]{-32} = -2$

32. $-\sqrt[5]{32} = -2$ **33.** $-\sqrt[5]{-32} = -(-2) = 2$

34. $\sqrt[7]{x^7} = x$ **35.** $\sqrt[7]{128} = 2$

36. $\sqrt[7]{-128} = -2$ **37.** $\sqrt[7]{0} = 0$

38. $\sqrt[5]{a^5} = a$ **39.** $\sqrt[9]{(x - 1)^9} = x - 1$

Do Exercises 41–47.

Even Roots

When the index k in $\sqrt[k]{}$ is an even number, we say that we are taking an **even root**. When the index is 2, we do not write it. Every positive real number has two real-number kth roots when k is even. One of those roots is positive and one is negative. Negative real numbers do not have real-number kth roots when k is even. When we are finding even kth roots, absolute-value signs are sometimes necessary, as we have seen with square roots. For example,

$$\sqrt{64} = 8, \qquad \sqrt[6]{64} = 2, \qquad -\sqrt[6]{64} = -2, \qquad \sqrt[6]{64x^6} = \sqrt[6]{(2x)^6} = |2x| = 2|x|.$$

Note that in $\sqrt[6]{64x^6}$, we need absolute-value signs because a variable is involved.

EXAMPLES Find each of the following. Assume that variables can represent any real number.

40. $\sqrt[4]{16} = 2$

41. $-\sqrt[4]{16} = -2$

42. $\sqrt[4]{-16}$ Does not exist as a real number.

43. $\sqrt[4]{81x^4} = \sqrt[4]{(3x)^4} = |3x| = 3|x|$

44. $\sqrt[6]{(y + 7)^6} = |y + 7|$

45. $\sqrt{81y^2} = \sqrt{(9y)^2} = |9y| = 9|y|$

The following is a summary of how absolute value is used when we are taking even roots or odd roots.

Find each of the following. Assume that letters can represent any real number.

48. $\sqrt[4]{81}$ **49.** $-\sqrt[4]{81}$

50. $\sqrt[4]{-81}$ **51.** $\sqrt[4]{0}$

52. $\sqrt[4]{16(x - 2)^4}$ **53.** $\sqrt[6]{x^6}$

54. $\sqrt[8]{(x + 3)^8}$ **55.** $\sqrt[7]{(x + 3)^7}$

56. $\sqrt[5]{243x^5}$

> **SIMPLIFYING**
>
> For any real number a:
>
> **a)** $\sqrt[k]{a^k} = |a|$ when k is an *even* natural number. We use absolute value when k is even unless a is nonnegative.
>
> **b)** $\sqrt[k]{a^k} = a$ when k is an *odd* natural number greater than 1. We do not use absolute value when k is odd.

Do Exercises 48–56.

Answers

41. 3   **42.** −3   **43.** x **44.** y **45.** 0
46. $-2x$ **47.** $3x + 2$ **48.** 3   **49.** −3
50. Does not exist as a real number   **51.** 0
52. $2|x - 2|$ **53.** $|x|$ **54.** $|x + 3|$
55. $x + 3$ **56.** $3x$

10.1 **Exercise Set**

For Extra Help

MyMathLab

Math XL
PRACTICE

WATCH

DOWNLOAD

READ

REVIEW

a Find the square roots.

1. 16

2. 225

3. 144

4. 9

5. 400

6. 81

Simplify.

7. $-\sqrt{\dfrac{49}{36}}$

8. $-\sqrt{\dfrac{361}{9}}$

9. $\sqrt{196}$

10. $\sqrt{441}$

11. $\sqrt{0.0036}$

12. $\sqrt{0.04}$

13. $\sqrt{-225}$

14. $\sqrt{-64}$

Use a calculator to approximate to three decimal places.

15. $\sqrt{347}$

16. $-\sqrt{1839.2}$

17. $\sqrt{\dfrac{285}{74}}$

18. $\sqrt{\dfrac{839.4}{19.7}}$

Identify the radicand.

19. $9\sqrt{y^2 + 16}$

20. $-3\sqrt{p^2 - 10}$

21. $x^4y^5\sqrt{\dfrac{x}{y - 1}}$

22. $a^2b^2\sqrt{\dfrac{a^2 - b}{b}}$

For the given function, find the indicated function values.

23. $f(x) = \sqrt{5x - 10}$; $f(6), f(2), f(1)$, and $f(-1)$

24. $t(x) = -\sqrt{2x + 1}$; $t(4), t(0), t(-1)$, and $t\left(-\tfrac{1}{2}\right)$

25. $g(x) = \sqrt{x^2 - 25}$; $g(-6), g(3), g(6)$, and $g(13)$

26. $F(x) = \sqrt{x^2 + 1}$; $F(0), F(-1)$, and $F(-10)$

27. Find the domain of the function f in Exercise 23.

28. Find the domain of the function t in Exercise 24.

29. *Parking-Lot Arrival Spaces.* The attendants at a parking lot park cars in temporary spaces before the cars are taken to long-term parking stalls. The number N of such spaces needed is approximated by the function

$$N(a) = 2.5\sqrt{a},$$

where a is the average number of arrivals in peak hours. What is the number of spaces needed when the average number of arrivals is 66? 100?

30. *Body Surface Area.* Body surface area B can be estimated using the Mosteller formula

$$B = \sqrt{\dfrac{h \times w}{3600}},$$

where B is in square meters, h is height, in centimeters, and w is weight, in kilograms. Estimate the body surface area of a woman whose height is 165 cm and whose weight is 63 kg; of a man whose height is 183 cm and whose weight is 100 kg. Round to the nearest tenth.

Graph.

31. $f(x) = 2\sqrt{x}$

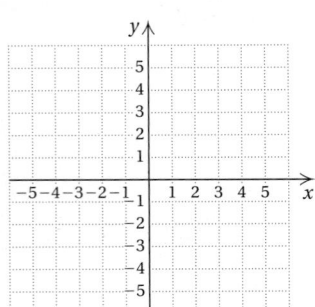

32. $g(x) = 3 - \sqrt{x}$

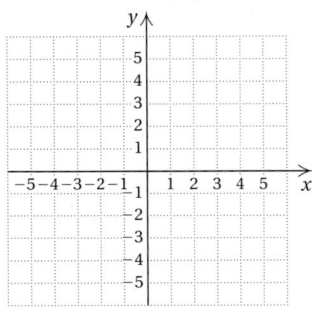

33. $F(x) = -3\sqrt{x}$

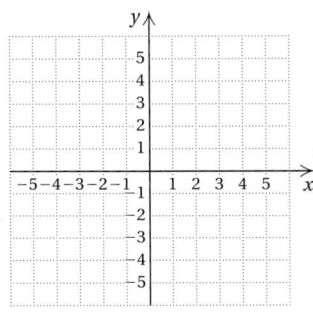

34. $f(x) = 2 + \sqrt{x - 1}$

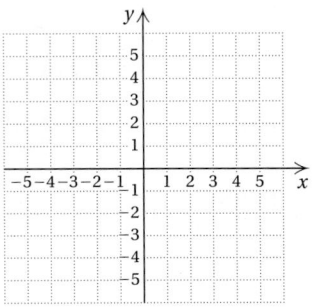

35. $f(x) = \sqrt{x}$

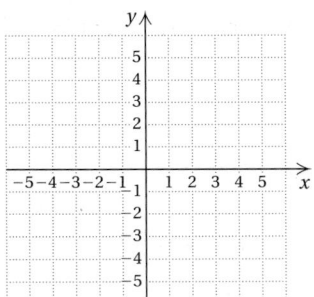

36. $g(x) = -\sqrt{x}$

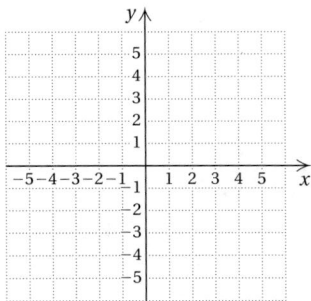

37. $f(x) = \sqrt{x - 2}$

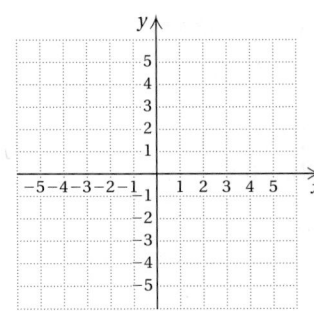

38. $g(x) = \sqrt{x + 3}$

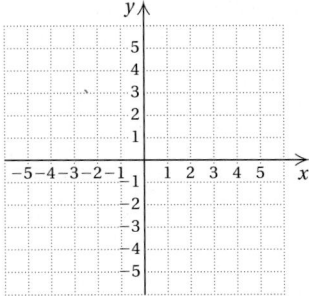

39. $f(x) = \sqrt{12 - 3x}$

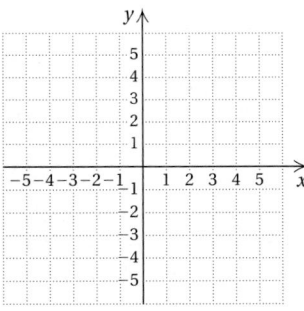

40. $g(x) = \sqrt{8 - 4x}$

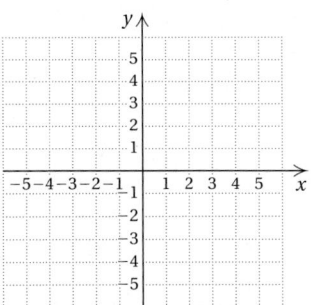

41. $g(x) = \sqrt{3x + 9}$

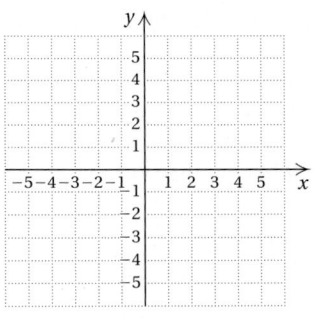

42. $f(x) = \sqrt{3x - 6}$

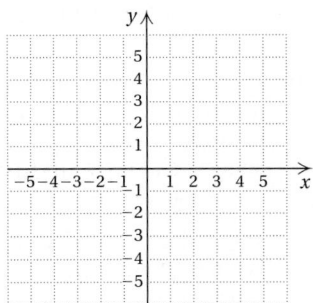

b Find each of the following. Assume that letters can represent *any* real number.

43. $\sqrt{16x^2}$

44. $\sqrt{25t^2}$

45. $\sqrt{(-12c)^2}$

46. $\sqrt{(-9d)^2}$

47. $\sqrt{(p + 3)^2}$

48. $\sqrt{(2 - x)^2}$

49. $\sqrt{x^2 - 4x + 4}$

50. $\sqrt{9t^2 - 30t + 25}$

Simplify.

51. $\sqrt[3]{27}$

52. $-\sqrt[3]{64}$

53. $\sqrt[3]{-64x^3}$

54. $\sqrt[3]{-125y^3}$

55. $\sqrt[3]{-216}$

56. $-\sqrt[3]{-1000}$

57. $\sqrt[3]{0.343(x+1)^3}$

58. $\sqrt[3]{0.000008(y-2)^3}$

For the given function, find the indicated function values.

59. $f(x) = \sqrt[3]{x+1}$; $f(7), f(26), f(-9),$ and $f(-65)$

60. $g(x) = -\sqrt[3]{2x-1}$; $g(-62), g(0), g(-13),$ and $g(63)$

61. $f(x) = -\sqrt[3]{3x+1}$; $f(0), f(-7), f(21),$ and $f(333)$

62. $g(t) = \sqrt[3]{t-3}$; $g(30), g(-5), g(1),$ and $g(67)$

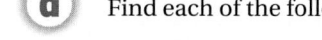 Find each of the following. Assume that letters can represent *any* real number.

63. $-\sqrt[4]{625}$

64. $-\sqrt[4]{256}$

65. $\sqrt[5]{-1}$

66. $\sqrt[5]{-32}$

67. $\sqrt[5]{-\dfrac{32}{243}}$

68. $\sqrt[5]{-\dfrac{1}{32}}$

69. $\sqrt[6]{x^6}$

70. $\sqrt[8]{y^8}$

71. $\sqrt[4]{(5a)^4}$

72. $\sqrt[4]{(7b)^4}$

73. $\sqrt[10]{(-6)^{10}}$

74. $\sqrt[12]{(-10)^{12}}$

75. $\sqrt[414]{(a+b)^{414}}$

76. $\sqrt[1999]{(2a+b)^{1999}}$

77. $\sqrt{y^7}$

78. $\sqrt[3]{(-6)^3}$

79. $\sqrt[5]{(x-2)^5}$

80. $\sqrt[9]{(2xy)^9}$

Skill Maintenance

Solve. [5.8b]

81. $x^2 + x - 2 = 0$

82. $x^2 + x = 0$

83. $4x^2 - 49 = 0$

84. $2x^2 - 26x + 72 = 0$

85. $3x^2 + x = 10$

86. $4x^2 - 20x + 25 = 0$

87. $4x^3 - 20x^2 + 25x = 0$

88. $x^3 - x^2 = 0$

Simplify.

89. $(a^3b^2c^5)^3$ [4.2a]

90. $(5a^7b^8)(2a^3b)$ [4.1d]

Synthesis

91. Find the domain of
$$f(x) = \frac{\sqrt{x+3}}{\sqrt{2-x}}.$$

92. ⬜ Use a graphing calculator to check your answers to Exercises 35, 39, and 41.

93. Use only the graph of $f(x) = \sqrt{x}$, shown below, to approximate $\sqrt{3}$, $\sqrt{5}$, and $\sqrt{10}$. Answers may vary.

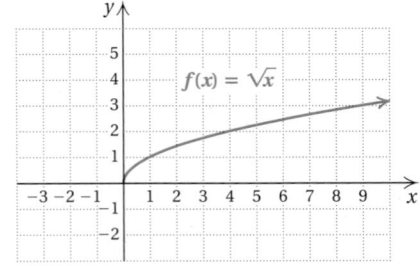

94. Use only the graph of $f(x) = \sqrt[3]{x}$, shown below, to approximate $\sqrt[3]{4}$, $\sqrt[3]{6}$, and $\sqrt[3]{-5}$. Answers may vary.

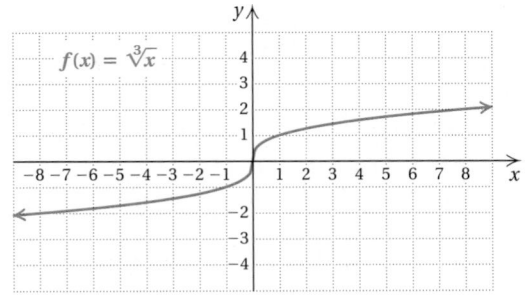

95. ⬜ Use the TABLE, TRACE, and GRAPH features of a graphing calculator to find the domain and the range of each of the following functions.

a) $f(x) = \sqrt[3]{x}$

b) $g(x) = \sqrt[3]{4x-5}$

c) $q(x) = 2 - \sqrt{x+3}$

d) $h(x) = \sqrt[4]{x}$

e) $t(x) = \sqrt[4]{x-3}$

10.2 Rational Numbers as Exponents

In this section, we give meaning to expressions such as $a^{1/3}, 7^{-1/2}$, and $(3x)^{0.84}$, which have rational numbers as exponents. We will see that using such notation can help simplify certain radical expressions.

a Rational Exponents

Expressions like $a^{1/2}$, $5^{-1/4}$, and $(2y)^{4/5}$ have not yet been defined. We will define such expressions so that the general properties of exponents hold.

Consider $a^{1/2} \cdot a^{1/2}$. If we want to multiply by adding exponents, it must follow that $a^{1/2} \cdot a^{1/2} = a^{1/2+1/2}$, or a^1. Thus we should define $a^{1/2}$ to be a square root of a. Similarly, $a^{1/3} \cdot a^{1/3} \cdot a^{1/3} = a^{1/3+1/3+1/3}$, or a^1, so $a^{1/3}$ should be defined to mean $\sqrt[3]{a}$.

$a^{1/n}$

For any *nonnegative* real number a and any natural number index n ($n \neq 1$),

$$a^{1/n} \quad \text{means} \quad \sqrt[n]{a} \quad \text{(the nonnegative nth root of a)}.$$

Whenever we use rational exponents, we assume that the bases are nonnegative.

EXAMPLES Rewrite without rational exponents, and simplify, if possible.

1. $27^{1/3} = \sqrt[3]{27} = 3$

2. $(abc)^{1/5} = \sqrt[5]{abc}$

3. $x^{1/2} = \sqrt{x}$ An index of 2 is not written.

> Do Exercises 1–5.

EXAMPLES Rewrite with rational exponents.

4. $\sqrt[5]{7xy} = (7xy)^{1/5}$ We need parentheses around the radicand here.

5. $8\sqrt[3]{xy} = 8(xy)^{1/3}$

6. $\sqrt[7]{\dfrac{x^3y}{9}} = \left(\dfrac{x^3y}{9}\right)^{1/7}$

> Do Exercises 6–9.

How should we define $a^{2/3}$? If the general properties of exponents are to hold, we have $a^{2/3} = (a^{1/3})^2$, or $(a^2)^{1/3}$, or $\left(\sqrt[3]{a}\right)^2$, or $\sqrt[3]{a^2}$. We define this accordingly.

$a^{m/n}$

For any natural numbers m and n ($n \neq 1$) and any nonnegative real number a,

$$a^{m/n} \quad \text{means} \quad \sqrt[n]{a^m}, \quad \text{or} \quad \left(\sqrt[n]{a}\right)^m.$$

OBJECTIVES

a Write expressions with or without rational exponents, and simplify, if possible.

b Write expressions without negative exponents, and simplify, if possible.

c Use the laws of exponents with rational exponents.

d Use rational exponents to simplify radical expressions.

Rewrite without rational exponents, and simplify, if possible.

1. $y^{1/4}$

2. $(3a)^{1/2}$

3. $16^{1/4}$

4. $(125)^{1/3}$

5. $(a^3b^2c)^{1/5}$

Rewrite with rational exponents.

6. $\sqrt[3]{19ab}$

7. $19\sqrt[3]{ab}$

8. $\sqrt[5]{\dfrac{x^2y}{16}}$

9. $7\sqrt[4]{2ab}$

Answers

1. $\sqrt[4]{y}$ 2. $\sqrt{3a}$ 3. 2   4. 5
5. $\sqrt[5]{a^3b^2c}$ 6. $(19ab)^{1/3}$ 7. $19(ab)^{1/3}$
8. $\left(\dfrac{x^2y}{16}\right)^{1/5}$ 9. $7(2ab)^{1/4}$

Rewrite without rational exponents, and simplify, if possible.

10. $x^{3/5}$ **11.** $8^{2/3}$

12. $4^{5/2}$

EXAMPLES Rewrite without rational exponents, and simplify, if possible.

7. $(27)^{2/3} = \sqrt[3]{27^2}$
   $= \left(\sqrt[3]{27}\right)^2$
   $= 3^2$
   $= 9$

8. $4^{3/2} = \sqrt[2]{4^3}$
   $= \left(\sqrt[2]{4}\right)^3$
   $= 2^3$
   $= 8$

Do Exercises 10–12.

EXAMPLES Rewrite with rational exponents.

The index becomes the denominator of the rational exponent.

9. $\sqrt[3]{9^4} = 9^{4/3}$ **10.** $\left(\sqrt[4]{7xy}\right)^5 = (7xy)^{5/4}$

Do Exercises 13 and 14.

Rewrite with rational exponents.

13. $\left(\sqrt[3]{7abc}\right)^4$ **14.** $\sqrt[5]{6^7}$

b Negative Rational Exponents

Negative rational exponents have a meaning similar to that of negative integer exponents.

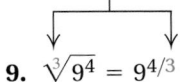

> **$a^{-m/n}$**
>
> For any rational number m/n and any positive real number a,
> $$a^{-m/n} \quad \text{means} \quad \frac{1}{a^{m/n}};$$
> that is, $a^{m/n}$ and $a^{-m/n}$ are reciprocals.

EXAMPLES Rewrite with positive exponents, and simplify, if possible.

11. $9^{-1/2} = \dfrac{1}{9^{1/2}} = \dfrac{1}{\sqrt{9}} = \dfrac{1}{3}$

12. $(5xy)^{-4/5} = \dfrac{1}{(5xy)^{4/5}}$

13. $64^{-2/3} = \dfrac{1}{64^{2/3}} = \dfrac{1}{\left(\sqrt[3]{64}\right)^2} = \dfrac{1}{4^2} = \dfrac{1}{16}$

14. $4x^{-2/3}y^{1/5} = 4 \cdot \dfrac{1}{x^{2/3}} \cdot y^{1/5} = \dfrac{4y^{1/5}}{x^{2/3}}$

15. $\left(\dfrac{3r}{7s}\right)^{-5/2} = \left(\dfrac{7s}{3r}\right)^{5/2}$ Since $\left(\dfrac{a}{b}\right)^{-n} = \left(\dfrac{b}{a}\right)^{n}$

Do Exercises 15–19.

Rewrite with positive exponents, and simplify, if possible.

15. $16^{-1/4}$ **16.** $(3xy)^{-7/8}$

17. $81^{-3/4}$ **18.** $7p^{3/4}q^{-6/5}$

19. $\left(\dfrac{11m}{7n}\right)^{-2/3}$

Answers

10. $\sqrt[5]{x^3}$ **11.** 4   **12.** 32   **13.** $(7abc)^{4/3}$
14. $6^{7/5}$ **15.** $\dfrac{1}{2}$ **16.** $\dfrac{1}{(3xy)^{7/8}}$ **17.** $\dfrac{1}{27}$
18. $\dfrac{7p^{3/4}}{q^{6/5}}$ **19.** $\left(\dfrac{7n}{11m}\right)^{2/3}$

c   Laws of Exponents

The same laws hold for rational-number exponents as for integer exponents. We list them for review.

> For any real number a and any rational exponents m and n:
>
> **1.** $a^m \cdot a^n = a^{m+n}$ In multiplying, we can add exponents if the bases are the same.
>
> **2.** $\dfrac{a^m}{a^n} = a^{m-n}$ In dividing, we can subtract exponents if the bases are the same.
>
> **3.** $(a^m)^n = a^{m \cdot n}$ To raise a power to a power, we can multiply the exponents.
>
> **4.** $(ab)^m = a^m b^m$ To raise a product to a power, we can raise each factor to the power.
>
> **5.** $\left(\dfrac{a}{b}\right)^n = \dfrac{a^n}{b^n}$ To raise a quotient to a power, we can raise both the numerator and the denominator to the power.

EXAMPLES Use the laws of exponents to simplify.

16. $3^{1/5} \cdot 3^{3/5} = 3^{1/5+3/5} = 3^{4/5}$ Adding exponents

17. $\dfrac{7^{1/4}}{7^{1/2}} = 7^{1/4-1/2} = 7^{1/4-2/4} = 7^{-1/4} = \dfrac{1}{7^{1/4}}$ Subtracting exponents

18. $(7.2^{2/3})^{3/4} = 7.2^{2/3 \cdot 3/4} = 7.2^{6/12} = 7.2^{1/2}$ Multiplying exponents

19. $(a^{-1/3}b^{2/5})^{1/2} = a^{-1/3 \cdot 1/2} \cdot b^{2/5 \cdot 1/2}$ Raising a product to a power and multiplying exponents

$$= a^{-1/6}b^{1/5} = \dfrac{b^{1/5}}{a^{1/6}}$$

Do Exercises 20–23.

Use the laws of exponents to simplify.

20. $7^{1/3} \cdot 7^{3/5}$

21. $\dfrac{5^{7/6}}{5^{5/6}}$

22. $(9^{3/5})^{2/3}$

23. $(p^{-2/3}q^{1/4})^{1/2}$

Answers

20. $7^{14/15}$ **21.** $5^{1/3}$ **22.** $9^{2/5}$ **23.** $\dfrac{q^{1/8}}{p^{1/3}}$

(d) Simplifying Radical Expressions

Rational exponents can be used to simplify some radical expressions. The procedure is as follows.

SIMPLIFYING RADICAL EXPRESSIONS

1. Convert radical expressions to exponential expressions.
2. Use arithmetic and the laws of exponents to simplify.
3. Convert back to radical notation when appropriate.

Important: This procedure works only when we assume that a negative number has not been raised to an even power in the radicand. With this assumption, no absolute-value signs will be needed.

EXAMPLES Use rational exponents to simplify.

20. $\sqrt[6]{x^3} = x^{3/6}$ Converting to an exponential expression

$\quad\quad = x^{1/2}$ Simplifying the exponent

$\quad\quad = \sqrt{x}$ Converting back to radical notation

21. $\sqrt[6]{4} = 4^{1/6}$ Converting to exponential notation

$\quad\quad = (2^2)^{1/6}$ Renaming 4 as 2^2

$\quad\quad = 2^{2/6}$ Using $(a^m)^n = a^{mn}$; multiplying exponents

$\quad\quad = 2^{1/3}$ Simplifying the exponent

$\quad\quad = \sqrt[3]{2}$ Converting back to radical notation

22. $\sqrt[8]{a^2b^4} = (a^2b^4)^{1/8}$ Converting to exponential notation

$\quad\quad = a^{2/8} \cdot b^{4/8}$ Using $(ab)^n = a^nb^n$

$\quad\quad = a^{1/4} \cdot b^{1/2}$ Simplifying the exponents

$\quad\quad = a^{1/4} \cdot b^{2/4}$ Rewriting $\frac{1}{2}$ with a denominator of 4

$\quad\quad = (ab^2)^{1/4}$ Using $a^nb^n = (ab)^n$

$\quad\quad = \sqrt[4]{ab^2}$ Converting back to radical notation

Use rational exponents to simplify.

24. $\sqrt[4]{a^2}$ **25.** $\sqrt[4]{x^4}$

26. $\sqrt[6]{8}$ **27.** $\sqrt[12]{x^3y^6}$

28. $\sqrt[6]{a^{12}b^3}$ **29.** $\sqrt[5]{a^5b^{10}}$

Do Exercises 24–29.

We can use properties of rational exponents to write a single radical expression for a product or a quotient.

EXAMPLE 23 Use rational exponents to write a single radical expression for $\sqrt[3]{5} \cdot \sqrt{2}$.

$\sqrt[3]{5} \cdot \sqrt{2} = 5^{1/3} \cdot 2^{1/2}$ Converting to exponential notation

$\quad\quad = 5^{2/6} \cdot 2^{3/6}$ Rewriting so that exponents have a common denominator

$\quad\quad = (5^2 \cdot 2^3)^{1/6}$ Using $a^nb^n = (ab)^n$

$\quad\quad = \sqrt[6]{5^2 \cdot 2^3}$ Converting back to radical notation

$\quad\quad = \sqrt[6]{200}$ Multiplying under the radical

30. Use rational exponents to write a single radical expression for $\sqrt[4]{7} \cdot \sqrt{3}$.

Do Exercise 30.

Answers

24. $\sqrt{a}$ **25.** x **26.** $\sqrt{2}$ **27.** $\sqrt[4]{xy^2}$
28. $a^2\sqrt{b}$ **29.** ab^2 **30.** $\sqrt[4]{63}$

EXAMPLE 24 Write a single radical expression for $a^{1/2}b^{-1/2}c^{5/6}$.

$$a^{1/2}b^{-1/2}c^{5/6} = a^{3/6}b^{-3/6}c^{5/6}$$ Rewriting so that exponents have a common denominator

$$= (a^3b^{-3}c^5)^{1/6}$$ Using $a^n b^n = (ab)^n$

$$= \sqrt[6]{a^3b^{-3}c^5}$$ Converting to radical notation

EXAMPLE 25 Write a single radical expression for $\dfrac{x^{5/6} \cdot y^{3/8}}{x^{4/9} \cdot y^{1/4}}$.

$$\dfrac{x^{5/6} \cdot y^{3/8}}{x^{4/9} \cdot y^{1/4}} = x^{5/6-4/9} \cdot y^{3/8-1/4}$$ Subtracting exponents

$$= x^{15/18-8/18} \cdot y^{3/8-2/8}$$ Finding common denominators so that exponents can be subtracted

$$= x^{7/18} \cdot y^{1/8}$$ Carrying out the subtraction of exponents

$$= x^{28/72} \cdot y^{9/72}$$ Rewriting so that all exponents have a common denominator

$$= (x^{28}y^9)^{1/72}$$ Using $a^n b^n = (ab)^n$

$$= \sqrt[72]{x^{28}y^9}$$ Converting to radical notation

Do Exercises 31 and 32.

Write a single radical expression.

31. $x^{2/3}y^{1/2}z^{5/6}$

32. $\dfrac{a^{1/2}b^{3/8}}{a^{1/4}b^{1/8}}$

EXAMPLES Use rational exponents to simplify.

26. $\sqrt[6]{(5x)^3} = (5x)^{3/6}$ Converting to exponential notation

$= (5x)^{1/2}$ Simplifying the exponent

$= \sqrt{5x}$ Converting back to radical notation

27. $\sqrt[5]{t^{20}} = t^{20/5}$ Converting to exponential notation

$= t^4$ Simplifying the exponent

28. $\left(\sqrt[3]{pq^2c}\right)^{12} = (pq^2c)^{12/3}$ Converting to exponential notation

$= (pq^2c)^4$ Simplifying the exponent

$= p^4q^8c^4$ Using $(ab)^n = a^n b^n$

29. $\sqrt{\sqrt[3]{x}} = \sqrt{x^{1/3}}$ Converting the radicand to exponential notation

$= (x^{1/3})^{1/2}$ Try to go directly to this step.

$= x^{1/6}$ Multiplying exponents

$= \sqrt[6]{x}$ Converting back to radical notation

Do Exercises 33–36.

Use rational exponents to simplify.

33. $\sqrt[14]{(5m)^2}$ **34.** $\sqrt[18]{m^3}$

35. $\left(\sqrt[6]{a^5b^3c}\right)^{24}$ **36.** $\sqrt[5]{\sqrt{x}}$

Answers

31. $\sqrt[6]{x^4y^3z^5}$ **32.** $\sqrt[4]{ab}$ **33.** $\sqrt[7]{5m}$
34. $\sqrt[6]{m}$ **35.** $a^{20}b^{12}c^4$ **36.** $\sqrt[10]{x}$

a Rewrite without rational exponents, and simplify, if possible.

1. $y^{1/7}$

2. $x^{1/6}$

3. $8^{1/3}$

4. $16^{1/2}$

5. $(a^3b^3)^{1/5}$

6. $(x^2y^2)^{1/3}$

7. $16^{3/4}$

8. $4^{7/2}$

9. $49^{3/2}$

10. $27^{4/3}$

Rewrite with rational exponents.

11. $\sqrt{17}$

12. $\sqrt{x^3}$

13. $\sqrt[3]{18}$

14. $\sqrt[3]{23}$

15. $\sqrt[5]{xy^2z}$

16. $\sqrt[7]{x^3y^2z^2}$

17. $\left(\sqrt{3mn}\right)^3$

18. $\left(\sqrt[3]{7xy}\right)^4$

19. $\left(\sqrt[7]{8x^2y}\right)^5$

20. $\left(\sqrt[6]{2a^5b}\right)^7$

b Rewrite with positive exponents, and simplify, if possible.

21. $27^{-1/3}$

22. $100^{-1/2}$

23. $100^{-3/2}$

24. $16^{-3/4}$

25. $3x^{-1/4}$

26. $8y^{-1/7}$

27. $(2rs)^{-3/4}$

28. $(5xy)^{-5/6}$

29. $2a^{3/4}b^{-1/2}c^{2/3}$

30. $5x^{-2/3}y^{4/5}z$

31. $\left(\dfrac{7x}{8yz}\right)^{-3/5}$

32. $\left(\dfrac{2ab}{3c}\right)^{-5/6}$

33. $\dfrac{1}{x^{-2/3}}$

34. $\dfrac{1}{a^{-7/8}}$

35. $2^{-1/3}x^4y^{-2/7}$

36. $3^{-5/2}a^3b^{-7/3}$

37. $\dfrac{7x}{\sqrt[3]{z}}$

38. $\dfrac{6a}{\sqrt[4]{b}}$

39. $\dfrac{5a}{3c^{-1/2}}$

40. $\dfrac{2z}{5x^{-1/3}}$

c Use the laws of exponents to simplify. Write the answers with positive exponents.

41. $5^{3/4} \cdot 5^{1/8}$

42. $11^{2/3} \cdot 11^{1/2}$

43. $\dfrac{7^{5/8}}{7^{3/8}}$

44. $\dfrac{3^{5/8}}{3^{-1/8}}$

45. $\dfrac{4.9^{-1/6}}{4.9^{-2/3}}$

46. $\dfrac{2.3^{-3/10}}{2.3^{-1/5}}$

47. $(6^{3/8})^{2/7}$

48. $(3^{2/9})^{3/5}$

49. $a^{2/3} \cdot a^{5/4}$

50. $x^{3/4} \cdot x^{2/3}$

51. $(a^{2/3} \cdot b^{5/8})^4$ **52.** $(x^{-1/3} \cdot y^{-2/5})^{-15}$ **53.** $(x^{2/3})^{-3/7}$ **54.** $(a^{-3/2})^{2/9}$

55. $\left(\dfrac{x^{3/4}}{y^{1/2}}\right)^{-2/3}$ **56.** $\left(\dfrac{a^{-3/2}}{b^{-5/3}}\right)^{1/3}$ **57.** $(m^{-1/4} \cdot n^{-5/6})^{-12/5}$ **58.** $(x^{3/8} \cdot y^{5/2})^{4/3}$

(d) Use rational exponents to simplify. Write the answer in radical notation if appropriate.

59. $\sqrt[6]{a^2}$ **60.** $\sqrt[6]{t^4}$ **61.** $\sqrt[3]{x^{15}}$ **62.** $\sqrt[4]{a^{12}}$ **63.** $\sqrt[6]{x^{-18}}$

64. $\sqrt[5]{a^{-10}}$ **65.** $\left(\sqrt[3]{ab}\right)^{15}$ **66.** $\left(\sqrt[7]{cd}\right)^{14}$ **67.** $\sqrt[14]{128}$ **68.** $\sqrt[6]{81}$

69. $\sqrt[6]{4x^2}$ **70.** $\sqrt[3]{8y^6}$ **71.** $\sqrt{x^4y^6}$ **72.** $\sqrt[4]{16x^4y^2}$ **73.** $\sqrt[5]{32c^{10}d^{15}}$

Use rational exponents to write a single radical expression.

74. $\sqrt[3]{3}\sqrt{3}$ **75.** $\sqrt[3]{7} \cdot \sqrt[4]{5}$ **76.** $\sqrt[7]{11} \cdot \sqrt[6]{13}$ **77.** $\sqrt[4]{5} \cdot \sqrt[5]{7}$ **78.** $\sqrt[3]{y}\sqrt[5]{3y}$

79. $\sqrt{x}\,\sqrt[3]{2x}$ **80.** $\left(\sqrt[3]{x^2y^5}\right)^{12}$ **81.** $\left(\sqrt[5]{a^2b^4}\right)^{15}$ **82.** $\sqrt[4]{\sqrt{x}}$ **83.** $\sqrt[3]{\sqrt[6]{m}}$

84. $a^{2/3} \cdot b^{3/4}$ **85.** $x^{1/3} \cdot y^{1/4} \cdot z^{1/6}$ **86.** $\dfrac{x^{8/15} \cdot y^{7/5}}{x^{1/3} \cdot y^{-1/5}}$ **87.** $\left(\dfrac{c^{-4/5}d^{5/9}}{c^{3/10}d^{1/6}}\right)^3$ **88.** $\sqrt[3]{\sqrt[4]{xy}}$

Skill Maintenance

Solve. [9.3c]

89. $|7x - 5| = 9$ **90.** $|3x| = 120$ **91.** $8 - |2x + 5| = -2$ **92.** $\left|\dfrac{1}{2} + x\right| = \dfrac{7}{8}$

Synthesis

93. Use the SIMULTANEOUS mode to graph
$$y_1 = x^{1/2}, \quad y_2 = 3x^{2/5}, \quad y_3 = x^{4/7}, \quad y_4 = \tfrac{1}{5}x^{3/4}.$$
Then, looking only at coordinates, match each graph with its equation.

94. Simplify:
$$\left(\sqrt[10]{\sqrt[5]{x^{15}}}\right)^5 \left(\sqrt[5]{\sqrt[10]{x^{15}}}\right)^5.$$

10.3

Simplifying Radical Expressions

OBJECTIVES

a Multiply and simplify radical expressions.

b Divide and simplify radical expressions.

SKILL TO REVIEW
Objective 4.1e: Use the quotient rule to divide exponential expressions with like bases.

Divide and simplify.

1. $\dfrac{x^{-6}}{x^3}$

2. $\dfrac{a^{10}}{a^{-2}}$

Multiply.

1. $\sqrt{19}\,\sqrt{7}$ 2. $\sqrt{3p}\,\sqrt{7q}$

3. $\sqrt[4]{403}\,\sqrt[4]{7}$ 4. $\sqrt[3]{\dfrac{5}{p}}\cdot\sqrt[3]{\dfrac{2}{q}}$

Multiply.

5. $\sqrt{5}\,\sqrt[3]{2}$ 6. $\sqrt[4]{x}\,\sqrt[3]{2y}$

Answers

Skill to Review:

1. $\dfrac{1}{x^9}$ 2. a^{12}

Margin Exercises:

1. $\sqrt{133}$ 2. $\sqrt{21pq}$ 3. $\sqrt[4]{2821}$
4. $\sqrt[3]{\dfrac{10}{pq}}$ 5. $\sqrt[6]{500}$ 6. $\sqrt[12]{16x^3y^4}$

a Multiplying and Simplifying Radical Expressions

Note that $\sqrt{4}\sqrt{25} = 2 \cdot 5 = 10$. Also $\sqrt{4 \cdot 25} = \sqrt{100} = 10$. Likewise,

$$\sqrt[3]{27}\sqrt[3]{8} = 3 \cdot 2 = 6 \quad \text{and} \quad \sqrt[3]{27 \cdot 8} = \sqrt[3]{216} = 6.$$

These examples suggest the following.

THE PRODUCT RULE FOR RADICALS

For any nonnegative real numbers a and b and any index k,

$$\sqrt[k]{a} \cdot \sqrt[k]{b} = \sqrt[k]{a \cdot b}, \quad \text{or} \quad a^{1/k} \cdot b^{1/k} = (ab)^{1/k}. \qquad \text{The index must be the same throughout.}$$

(To multiply, multiply the radicands.)

EXAMPLES Multiply.

1. $\sqrt{3} \cdot \sqrt{5} = \sqrt{3 \cdot 5} = \sqrt{15}$

2. $\sqrt{5a}\sqrt{2b} = \sqrt{5a \cdot 2b} = \sqrt{10ab}$

3. $\sqrt[3]{4}\sqrt[3]{5} = \sqrt[3]{4 \cdot 5} = \sqrt[3]{20}$

4. $\sqrt[4]{\dfrac{y}{5}}\,\sqrt[4]{\dfrac{7}{x}} = \sqrt[4]{\dfrac{y}{5} \cdot \dfrac{7}{x}} = \sqrt[4]{\dfrac{7y}{5x}}$

Caution!
A common error is to omit the index in the answer.

Do Exercises 1–4.

Keep in mind that the product rule can be used only when the indexes are the same. When indexes differ, we can use rational exponents as we did in Examples 23 and 24 of Section 10.2.

EXAMPLE 5 Multiply: $\sqrt{5x} \cdot \sqrt[4]{3y}$.

$$
\begin{aligned}
\sqrt{5x} \cdot \sqrt[4]{3y} &= (5x)^{1/2}(3y)^{1/4} && \text{Converting to exponential notation}\\
&= (5x)^{2/4}(3y)^{1/4} && \text{Rewriting so that exponents have a common denominator}\\
&= [(5x)^2(3y)]^{1/4} && \text{Using } a^n b^n = (ab)^n\\
&= [(25x^2)(3y)]^{1/4} && \text{Squaring } 5x\\
&= \sqrt[4]{(25x^2)(3y)} && \text{Converting back to radical notation}\\
&= \sqrt[4]{75x^2y} && \text{Multiplying under the radical}
\end{aligned}
$$

Do Exercises 5 and 6.

We can reverse the product rule to simplify a product. We simplify the root of a product by taking the root of each factor separately.

FACTORING RADICAL EXPRESSIONS

For any nonnegative real numbers a and b and any index k,
$$\sqrt[k]{ab} = \sqrt[k]{a} \cdot \sqrt[k]{b}, \quad \text{or} \quad (ab)^{1/k} = a^{1/k} \cdot b^{1/k}.$$
(Take the kth root of each factor separately.)

Compare the following:
$$\sqrt{50} = \sqrt{10 \cdot 5} = \sqrt{10}\,\sqrt{5};$$
$$\sqrt{50} = \sqrt{25 \cdot 2} = \sqrt{25}\,\sqrt{2} = 5\sqrt{2}.$$

In the second case, the radicand is written with the perfect-square factor 25. If you do not recognize perfect-square factors, try factoring the radicand into its prime factors. For example,
$$\sqrt{50} = \sqrt{2 \cdot \underline{5 \cdot 5}} = 5\sqrt{2}.$$

Perfect square (a pair of the same numbers)

Square-root radical expressions in which the radicand has no perfect-square factors, such as $5\sqrt{2}$, are considered to be in simplest form. A procedure for simplifying kth roots follows.

SIMPLIFYING kth ROOTS

To simplify a radical expression by factoring:

1. Look for the largest factors of the radicand that are perfect kth powers (where k is the index).
2. Then take the kth root of the resulting factors.
3. A radical expression, with index k, is *simplified* when its radicand has no factors that are perfect kth powers.

EXAMPLES Simplify by factoring.

6. $\sqrt{50} = \sqrt{25 \cdot 2} = \sqrt{25} \cdot \sqrt{2} = \sqrt{5 \cdot 5} \cdot \sqrt{2} = 5\sqrt{2}$

This factor is a perfect square.

7. $\sqrt[3]{32} = \sqrt[3]{8 \cdot 4} = \sqrt[3]{8} \cdot \sqrt[3]{4} = \sqrt[3]{2 \cdot 2 \cdot 2} \cdot \sqrt[3]{2 \cdot 2} = 2\sqrt[3]{4}$

This factor is a perfect cube (third power).

8. $\sqrt[4]{48} = \sqrt[4]{16 \cdot 3} = \sqrt[4]{16} \cdot \sqrt[4]{3} = \sqrt[4]{2 \cdot 2 \cdot 2 \cdot 2} \cdot \sqrt[4]{3} = 2\sqrt[4]{3}$

This factor is a perfect fourth power.

Do Exercises 7 and 8.

Simplify by factoring.
7. $\sqrt{32}$ **8.** $\sqrt[3]{80}$

Answers
7. $4\sqrt{2}$ **8.** $2\sqrt[3]{10}$

Frequently, expressions under radicals do not contain negative numbers raised to even powers. In such cases, absolute-value notation is not necessary. **For this reason, we will no longer use absolute-value notation.**

EXAMPLES Simplify by factoring. Assume that no radicands were formed by raising negative numbers to even powers.

9. $\sqrt{5x^2} = \sqrt{5 \cdot x^2}$ Factoring the radicand

$= \sqrt{5} \cdot \sqrt{x^2}$ Factoring into two radicals

$= \sqrt{5} \cdot x$ Taking the square root of x^2

Absolute-value notation is not needed because we assume that x is not negative.

10. $\sqrt{18x^2y} = \sqrt{9 \cdot 2 \cdot x^2 \cdot y}$ Factoring the radicand and looking for perfect-square factors

$= \sqrt{9 \cdot x^2 \cdot 2 \cdot y}$

$= \sqrt{9} \cdot \sqrt{x^2} \cdot \sqrt{2} \cdot \sqrt{y}$ Factoring into several radicals

$= 3x\sqrt{2y}$ Taking square roots

11. $\sqrt{216x^5y^3} = \sqrt{36 \cdot 6 \cdot x^4 \cdot x \cdot y^2 \cdot y}$ Factoring the radicand and looking for perfect-square factors

$= \sqrt{36 \cdot x^4 \cdot y^2 \cdot 6 \cdot x \cdot y}$

$= \sqrt{36}\,\sqrt{x^4}\,\sqrt{y^2}\,\sqrt{6xy}$ Factoring into several radicals

$= 6x^2y\sqrt{6xy}$ Taking square roots

Let's look at this example another way. We do a complete factorization and look for pairs of factors. Each pair of factors makes a square:

$\sqrt{216x^5y^3} = \sqrt{2 \cdot 2 \cdot 2 \cdot 3 \cdot 3 \cdot 3 \cdot x \cdot x \cdot x \cdot x \cdot x \cdot y \cdot y \cdot y}$ Each pair of factors makes a perfect square.

$= 2 \cdot 3 \cdot x \cdot x \cdot y \cdot \sqrt{2 \cdot 3 \cdot x \cdot y}$

$= 6x^2y\sqrt{6xy}.$

12. $\sqrt[3]{16a^7b^{11}} = \sqrt[3]{8 \cdot 2 \cdot a^6 \cdot a \cdot b^9 \cdot b^2}$ Factoring the radicand. The index is 3, so we look for the largest powers that are multiples of 3 because these are perfect cubes.

$= \sqrt[3]{8} \cdot \sqrt[3]{a^6} \cdot \sqrt[3]{b^9} \cdot \sqrt[3]{2ab^2}$ Factoring into radicals

$= 2a^2b^3\sqrt[3]{2ab^2}$ Taking cube roots

Let's look at this example another way. We do a complete factorization and look for triples of factors. Each triple of factors makes a cube:

$\sqrt[3]{16a^7b^{11}}$

$= \sqrt[3]{2 \cdot 2 \cdot 2 \cdot 2 \cdot a \cdot a \cdot a \cdot a \cdot a \cdot a \cdot a \cdot b \cdot b \cdot b \cdot b \cdot b \cdot b \cdot b \cdot b \cdot b \cdot b \cdot b}$

Each triple of factors makes a cube.

$= 2 \cdot a \cdot a \cdot b \cdot b \cdot b \cdot \sqrt[3]{2 \cdot a \cdot b \cdot b}$

$= 2a^2b^3\sqrt[3]{2ab^2}.$

Do Exercises 9–14.

Simplify by factoring. Assume that no radicands were formed by raising negative numbers to even powers.

9. $\sqrt{300}$ **10.** $\sqrt{36y^2}$

11. $\sqrt{12a^2b}$ **12.** $\sqrt{12ab^3c^2}$

13. $\sqrt[3]{16}$ **14.** $\sqrt[3]{81x^4y^8}$

Answers

9. $10\sqrt{3}$ **10.** $6y$ **11.** $2a\sqrt{3b}$
12. $2bc\sqrt{3ab}$ **13.** $2\sqrt[3]{2}$ **14.** $3xy^2\sqrt[3]{3xy^2}$

Sometimes after we have multiplied, we can simplify by factoring.

EXAMPLES Multiply and simplify. Assume that no radicands were formed by raising negative numbers to even powers.

13. $\sqrt{20}\sqrt{8} = \sqrt{20 \cdot 8} = \sqrt{\underline{4} \cdot 5 \cdot \underline{4} \cdot 2} = 4\sqrt{10}$

14. $3\sqrt[3]{25} \cdot 2\sqrt[3]{5} = 3 \cdot 2 \cdot \sqrt[3]{25} \cdot \sqrt[3]{5} = 6 \cdot \sqrt[3]{25 \cdot 5}$
$$= 6 \cdot \sqrt[3]{5 \cdot 5 \cdot 5}$$
$$= 6 \cdot 5 = 30$$

15. $\sqrt[3]{18y^3}\sqrt[3]{4x^2} = \sqrt[3]{18y^3 \cdot 4x^2}$ Multiplying radicands
$$= \sqrt[3]{2 \cdot 3 \cdot 3 \cdot \underline{y \cdot y \cdot y} \cdot \underline{2 \cdot 2} \cdot x \cdot x}$$
$$= 2 \cdot y \cdot \sqrt[3]{3 \cdot 3 \cdot x \cdot x}$$
$$= 2y\sqrt[3]{9x^2}$$

Do Exercises 15–18.

Multiply and simplify. Assume that no radicands were formed by raising negative numbers to even powers.

15. $\sqrt{3}\,\sqrt{6}$

16. $\sqrt{18y}\,\sqrt{14y}$

17. $\sqrt[3]{3x^2y}\,\sqrt[3]{36x}$

18. $\sqrt{7a}\,\sqrt{21b}$

b Dividing and Simplifying Radical Expressions

Note that $\dfrac{\sqrt[3]{27}}{\sqrt[3]{8}} = \dfrac{3}{2}$ and that $\sqrt[3]{\dfrac{27}{8}} = \dfrac{3}{2}$. This example suggests the following.

THE QUOTIENT RULE FOR RADICALS

For any nonnegative number a, any positive number b, and any index k,

$$\frac{\sqrt[k]{a}}{\sqrt[k]{b}} = \sqrt[k]{\frac{a}{b}}, \quad \text{or} \quad \frac{a^{1/k}}{b^{1/k}} = \left(\frac{a}{b}\right)^{1/k}.$$

(To divide, divide the radicands. After doing this, you can sometimes simplify by taking roots.)

EXAMPLES Divide and simplify. Assume that no radicands were formed by raising negative numbers to even powers.

16. $\dfrac{\sqrt{80}}{\sqrt{5}} = \sqrt{\dfrac{80}{5}} = \sqrt{16} = 4$ We divide the radicands.

17. $\dfrac{5\sqrt[3]{32}}{\sqrt[3]{2}} = 5\sqrt[3]{\dfrac{32}{2}} = 5\sqrt[3]{16} = 5\sqrt[3]{8 \cdot 2} = 5\sqrt[3]{8}\,\sqrt[3]{2} = 5 \cdot 2\sqrt[3]{2} = 10\sqrt[3]{2}$

18. $\dfrac{\sqrt{72xy}}{2\sqrt{2}} = \dfrac{1}{2}\dfrac{\sqrt{72xy}}{\sqrt{2}} = \dfrac{1}{2}\sqrt{\dfrac{72xy}{2}} = \dfrac{1}{2}\sqrt{36xy} = \dfrac{1}{2}\sqrt{36}\sqrt{xy}$
$$= \dfrac{1}{2} \cdot 6\sqrt{xy} = 3\sqrt{xy}$$

Do Exercises 19–22.

Divide and simplify. Assume that no radicands were formed by raising negative numbers to even powers.

19. $\dfrac{\sqrt{75}}{\sqrt{3}}$

20. $\dfrac{14\sqrt{128xy}}{2\sqrt{2}}$

21. $\dfrac{\sqrt{50a^3}}{\sqrt{2a}}$

22. $\dfrac{4\sqrt[3]{250}}{7\sqrt[3]{2}}$

Answers

15. $3\sqrt{2}$ **16.** $6y\sqrt{7}$ **17.** $3x\sqrt[3]{4y}$
18. $7\sqrt{3ab}$ **19.** 5 **20.** $56\sqrt{xy}$ **21.** $5a$
22. $\dfrac{20}{7}$

We can reverse the quotient rule to simplify a quotient. We simplify the root of a quotient by taking the roots of the numerator and of the denominator separately.

> ### *k*th ROOTS OF QUOTIENTS
>
> For any nonnegative number a, any positive number b, and any index k,
>
> $$\sqrt[k]{\frac{a}{b}} = \frac{\sqrt[k]{a}}{\sqrt[k]{b}}, \quad \text{or} \quad \left(\frac{a}{b}\right)^{1/k} = \frac{a^{1/k}}{b^{1/k}}.$$
>
> (Take the kth roots of the numerator and of the denominator separately.)

EXAMPLES Simplify by taking the roots of the numerator and the denominator. Assume that no radicands were formed by raising negative numbers to even powers.

19. $\sqrt[3]{\dfrac{27}{125}} = \dfrac{\sqrt[3]{27}}{\sqrt[3]{125}} = \dfrac{3}{5}$ We take the cube root of the numerator and of the denominator.

20. $\sqrt{\dfrac{25}{y^2}} = \dfrac{\sqrt{25}}{\sqrt{y^2}} = \dfrac{5}{y}$ We take the square root of the numerator and of the denominator.

21. $\sqrt{\dfrac{16x^3}{y^4}} = \dfrac{\sqrt{16x^3}}{\sqrt{y^4}} = \dfrac{\sqrt{16x^2 \cdot x}}{\sqrt{y^4}} = \dfrac{\sqrt{16x^2} \cdot \sqrt{x}}{\sqrt{y^4}} = \dfrac{4x\sqrt{x}}{y^2}$

22. $\sqrt[3]{\dfrac{27y^5}{343x^3}} = \dfrac{\sqrt[3]{27y^5}}{\sqrt[3]{343x^3}} = \dfrac{\sqrt[3]{27y^3 \cdot y^2}}{\sqrt[3]{343x^3}} = \dfrac{\sqrt[3]{27y^3} \cdot \sqrt[3]{y^2}}{\sqrt[3]{343x^3}} = \dfrac{3y\sqrt[3]{y^2}}{7x}$

We are assuming here that no variable represents 0 or a negative number. Thus we need not be concerned about zero denominators.

Do Exercises 23–25.

When indexes differ, we can use rational exponents.

EXAMPLE 23 Divide and simplify: $\dfrac{\sqrt[3]{a^2b^4}}{\sqrt{ab}}$.

$$\frac{\sqrt[3]{a^2b^4}}{\sqrt{ab}} = \frac{(a^2b^4)^{1/3}}{(ab)^{1/2}}$$ Converting to exponential notation

$$= \frac{a^{2/3}b^{4/3}}{a^{1/2}b^{1/2}}$$ Using the product and power rules

$$= a^{2/3-1/2}b^{4/3-1/2}$$ Subtracting exponents

$$= a^{4/6-3/6}b^{8/6-3/6}$$ Finding common denominators so exponents can be subtracted

$$= a^{1/6}b^{5/6}$$

$$= (ab^5)^{1/6}$$ Using $a^nb^n = (ab)^n$

$$= \sqrt[6]{ab^5}$$ Converting back to radical notation

Do Exercise 26.

Simplify by taking the roots of the numerator and the denominator. Assume that no radicands were formed by raising negative numbers to even powers.

23. $\sqrt{\dfrac{25}{36}}$ **24.** $\sqrt{\dfrac{x^2}{100}}$

25. $\sqrt[3]{\dfrac{54x^5}{125}}$

26. Divide and simplify:
$$\frac{\sqrt[4]{x^3y^2}}{\sqrt[3]{x^2y}}.$$

a Simplify by factoring. Assume that no radicands were formed by raising negative numbers to even powers.

1. $\sqrt{24}$

2. $\sqrt{20}$

3. $\sqrt{90}$

4. $\sqrt{18}$

5. $\sqrt[3]{250}$

6. $\sqrt[3]{108}$

7. $\sqrt{180x^4}$

8. $\sqrt{175y^6}$

9. $\sqrt[3]{54x^8}$

10. $\sqrt[3]{40y^3}$

11. $\sqrt[3]{80t^8}$

12. $\sqrt[3]{108x^5}$

13. $\sqrt[4]{80}$

14. $\sqrt[4]{32}$

15. $\sqrt{32a^2b}$

16. $\sqrt{75p^3q^4}$

17. $\sqrt[4]{243x^8y^{10}}$

18. $\sqrt[4]{162c^4d^6}$

19. $\sqrt[5]{96x^7y^{15}}$

20. $\sqrt[5]{p^{14}q^9r^{23}}$

Multiply and simplify. Assume that no radicands were formed by raising negative numbers to even powers.

21. $\sqrt{10}\ \sqrt{5}$

22. $\sqrt{6}\ \sqrt{3}$

23. $\sqrt{15}\ \sqrt{6}$

24. $\sqrt{2}\ \sqrt{32}$

25. $\sqrt[3]{2}\ \sqrt[3]{4}$

26. $\sqrt[3]{9}\ \sqrt[3]{3}$

27. $\sqrt{45}\ \sqrt{60}$

28. $\sqrt{24}\ \sqrt{75}$

29. $\sqrt{3x^3}\ \sqrt{6x^5}$

30. $\sqrt{5a^7}\ \sqrt{15a^3}$

31. $\sqrt{5b^3}\ \sqrt{10c^4}$

32. $\sqrt{2x^3y}\ \sqrt{12xy}$

33. $\sqrt[3]{5a^2}\ \sqrt[3]{2a}$

34. $\sqrt[3]{7x}\ \sqrt[3]{3x^2}$

35. $\sqrt[3]{y^4}\ \sqrt[3]{16y^5}$

36. $\sqrt[3]{s^2t^4}\ \sqrt[3]{s^4t^6}$

37. $\sqrt[4]{16}\ \sqrt[4]{64}$

38. $\sqrt[5]{64}\ \sqrt[5]{16}$

39. $\sqrt{12a^3b}\ \sqrt{8a^4b^2}$

40. $\sqrt{30x^3y^4}\ \sqrt{18x^2y^5}$

41. $\sqrt{2}\ \sqrt[3]{5}$

42. $\sqrt{6}\ \sqrt[3]{5}$

43. $\sqrt[4]{3}\ \sqrt{2}$

44. $\sqrt[3]{5}\ \sqrt[4]{2}$

45. $\sqrt{a}\ \sqrt[4]{a^3}$

46. $\sqrt[3]{x^2}\ \sqrt[6]{x^5}$

47. $\sqrt[5]{b^2}\ \sqrt{b^3}$

48. $\sqrt[4]{a^3}\ \sqrt[3]{a^2}$

49. $\sqrt{xy^3}\ \sqrt[3]{x^2y}$

50. $\sqrt{y^5z}\ \sqrt[3]{yz^4}$

51. $\sqrt{2a^3b}\ \sqrt[4]{8ab^2}$

52. $\sqrt[4]{9ab^3}\ \sqrt{3a^4b}$

b   Divide and simplify. Assume that all expressions under radicals represent positive numbers.

53. $\dfrac{\sqrt{90}}{\sqrt{5}}$

54. $\dfrac{\sqrt{98}}{\sqrt{2}}$

55. $\dfrac{\sqrt{35q}}{\sqrt{7q}}$

56. $\dfrac{\sqrt{30x}}{\sqrt{10x}}$

57. $\dfrac{\sqrt[3]{54}}{\sqrt[3]{2}}$

58. $\dfrac{\sqrt[3]{40}}{\sqrt[3]{5}}$

59. $\dfrac{\sqrt{56xy^3}}{\sqrt{8x}}$

60. $\dfrac{\sqrt{52ab^3}}{\sqrt{13a}}$

61. $\dfrac{\sqrt[3]{96a^4b^2}}{\sqrt[3]{12a^2b}}$

62. $\dfrac{\sqrt[3]{189x^5y^7}}{\sqrt[3]{7x^2y^2}}$

63. $\dfrac{\sqrt{128xy}}{2\sqrt{2}}$

64. $\dfrac{\sqrt{48ab}}{2\sqrt{3}}$

65. $\dfrac{\sqrt[4]{48x^9y^{13}}}{\sqrt[4]{3xy^5}}$

66. $\dfrac{\sqrt[5]{64a^{11}b^{28}}}{\sqrt[5]{2ab^2}}$

67. $\dfrac{\sqrt[3]{a}}{\sqrt{a}}$

68. $\dfrac{\sqrt{x}}{\sqrt[4]{x}}$

69. $\dfrac{\sqrt[3]{a^2}}{\sqrt[4]{a}}$

70. $\dfrac{\sqrt[3]{x^2}}{\sqrt[5]{x}}$

71. $\dfrac{\sqrt[4]{x^2y^3}}{\sqrt[3]{xy}}$

72. $\dfrac{\sqrt[5]{a^4b^2}}{\sqrt[3]{ab^2}}$

Simplify.

73. $\sqrt{\dfrac{25}{36}}$

74. $\sqrt{\dfrac{49}{64}}$

75. $\sqrt{\dfrac{16}{49}}$

76. $\sqrt{\dfrac{100}{81}}$

77. $\sqrt[3]{\dfrac{125}{27}}$

78. $\sqrt[3]{\dfrac{343}{1000}}$

79. $\sqrt{\dfrac{49}{y^2}}$

80. $\sqrt{\dfrac{121}{x^2}}$

81. $\sqrt{\dfrac{25y^3}{x^4}}$

82. $\sqrt{\dfrac{36a^5}{b^6}}$

83. $\sqrt[3]{\dfrac{81y^5}{64}}$

84. $\sqrt[3]{\dfrac{8z^7}{125}}$

85. $\sqrt[3]{\dfrac{27a^4}{8b^3}}$

86. $\sqrt[3]{\dfrac{64x^7}{216y^6}}$

87. $\sqrt[4]{\dfrac{81x^4}{16}}$

88. $\sqrt[4]{\dfrac{256}{81x^8}}$

89. $\sqrt[4]{\dfrac{16a^{12}}{b^4c^{16}}}$

90. $\sqrt[4]{\dfrac{81x^4}{y^8z^4}}$

91. $\sqrt[5]{\dfrac{32x^8}{y^{10}}}$

92. $\sqrt[5]{\dfrac{32b^{10}}{243a^{20}}}$

93. $\sqrt[5]{\dfrac{w^7}{z^{10}}}$

94. $\sqrt[5]{\dfrac{z^{11}}{w^{20}}}$

95. $\sqrt[6]{\dfrac{x^{13}}{y^6z^{12}}}$

96. $\sqrt[6]{\dfrac{p^9q^{24}}{r^{18}}}$

Skill Maintenance

Solve. [5.9a]

97. The sum of a number and its square is 90. Find the number.

98. *Triangle Dimensions.* The base of a triangle is 2 in. longer than the height. The area is 12 in². Find the height and the base.

Solve. [6.7a]

99. $\dfrac{12x}{x-4} - \dfrac{3x^2}{x+4} = \dfrac{384}{x^2-16}$

100. $\dfrac{2}{3} + \dfrac{1}{t} = \dfrac{4}{5}$

101. $\dfrac{18}{x^2-3x} = \dfrac{2x}{x-3} - \dfrac{6}{x}$

102. $\dfrac{4x}{x+5} + \dfrac{20}{x} = \dfrac{100}{x^2+5x}$

Synthesis

103. *Pendulums.* The **period** of a pendulum is the time it takes to complete one cycle, swinging to and fro. For a pendulum that is L centimeters long, the period T is given by the function

$$T(L) = 2\pi\sqrt{\dfrac{L}{980}},$$

where T is in seconds. Find, to the nearest hundredth of a second, the period of a pendulum of length **(a)** 65 cm; **(b)** 98 cm; **(c)** 120 cm. Use a calculator's $\boxed{\pi}$ key if possible.

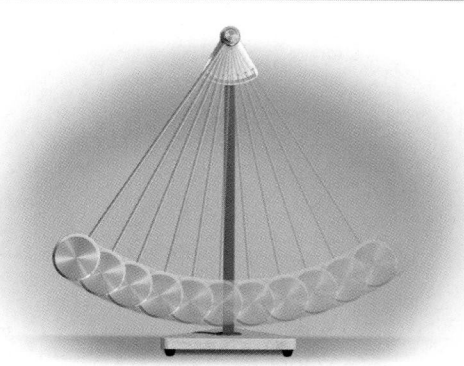

Simplify.

104. $\dfrac{\sqrt[3]{x^3-y^3}}{\sqrt[3]{x-y}}$

105. $\dfrac{\sqrt{44x^2y^9z}\ \sqrt{22y^9z^6}}{\left(\sqrt{11xy^8z^2}\right)^2}$

106. 📈 Use a graphing calculator to check your answers to Exercises 7, 12, 30, and 56.

10.4 Addition, Subtraction, and More Multiplication

a Addition and Subtraction

Any two real numbers can be added. For example, the sum of 7 and $\sqrt{3}$ can be expressed as $7 + \sqrt{3}$. We cannot simplify this sum. However, when we have **like radicals** (radicals having the same index and radicand), we can use the distributive laws to simplify by collecting like radical terms. For example,

$$7\sqrt{3} + \sqrt{3} = 7\sqrt{3} + 1 \cdot \sqrt{3} = (7 + 1)\sqrt{3} = 8\sqrt{3}.$$

EXAMPLES Add or subtract. Simplify by collecting like radical terms, if possible.

1. $6\sqrt{7} + 4\sqrt{7} = (6 + 4)\sqrt{7}$ Using a distributive law $\left(\text{factoring out } \sqrt{7}\right)$
$$= 10\sqrt{7}$$

2. $8\sqrt[3]{2} - 7x\sqrt[3]{2} + 5\sqrt[3]{2} = (8 - 7x + 5)\sqrt[3]{2}$ Factoring out $\sqrt[3]{2}$
$$= (13 - 7x)\sqrt[3]{2}$$

> These parentheses *are* necessary!

3. $6\sqrt[5]{4x} + 4\sqrt[5]{4x} - \sqrt[3]{4x} = (6 + 4)\sqrt[5]{4x} - \sqrt[3]{4x}$
$$= 10\sqrt[5]{4x} - \sqrt[3]{4x}$$

> Note that these expressions have the same *radicand,* but they are not like radicals because they do not have the same *index.*

> Do Margin Exercises 1 and 2.

Sometimes we need to simplify radicals by factoring in order to obtain terms with like radicals.

EXAMPLES Add or subtract. Simplify by collecting like radical terms, if possible.

4. $3\sqrt{8} - 5\sqrt{2} = 3\sqrt{4 \cdot 2} - 5\sqrt{2}$ Factoring 8
$$= 3\sqrt{4} \cdot \sqrt{2} - 5\sqrt{2} \quad \text{Factoring } \sqrt{4 \cdot 2} \text{ into two radicals}$$
$$= 3 \cdot 2\sqrt{2} - 5\sqrt{2} \quad \text{Taking the square root of 4}$$
$$= 6\sqrt{2} - 5\sqrt{2}$$
$$= (6 - 5)\sqrt{2} \quad \text{Collecting like radical terms}$$
$$= \sqrt{2}$$

5. $5\sqrt{2} - 4\sqrt{3}$ No simplification possible

6. $5\sqrt[3]{16y^4} + 7\sqrt[3]{2y} = 5\sqrt[3]{8y^3 \cdot 2y} + 7\sqrt[3]{2y}$ } Factoring the first radical
$$= 5\sqrt[3]{8y^3} \cdot \sqrt[3]{2y} + 7\sqrt[3]{2y}$$
$$= 5 \cdot 2y \cdot \sqrt[3]{2y} + 7\sqrt[3]{2y} \quad \text{Taking the cube root of } 8y^3$$
$$= 10y\sqrt[3]{2y} + 7\sqrt[3]{2y}$$
$$= (10y + 7)\sqrt[3]{2y} \quad \text{Collecting like radical terms}$$

> Do Exercises 3–5.

OBJECTIVES

a Add or subtract with radical notation and simplify.

b Multiply expressions involving radicals in which some factors contain more than one term.

SKILL TO REVIEW
Objective 1.7e: Collect like terms.

Collect like terms.
1. $2x + 5x$
2. $y + 3 - 4y + 1$

Add or subtract. Simplify by collecting like radical terms, if possible.
1. $5\sqrt{2} + 8\sqrt{2}$

2. $7\sqrt[4]{5x} + 3\sqrt[4]{5x} - \sqrt{7}$

Add or subtract. Simplify by collecting like radical terms, if possible.
3. $7\sqrt{45} - 2\sqrt{5}$

4. $3\sqrt[3]{y^5} + 4\sqrt[3]{y^2} + \sqrt[3]{8y^6}$

5. $\sqrt{25x - 25} - \sqrt{9x - 9}$

Answers

Skill to Review:
1. $7x$ **2.** $-3y + 4$

Margin Exercises:
1. $13\sqrt{2}$ **2.** $10\sqrt[4]{5x} - \sqrt{7}$ **3.** $19\sqrt{5}$
4. $(3y + 4)\sqrt[3]{y^2} + 2y^2$ **5.** $2\sqrt{x - 1}$

b More Multiplication

To multiply expressions in which some factors contain more than one term, we use the procedures for multiplying polynomials.

EXAMPLES Multiply.

7. $\sqrt{3}(x - \sqrt{5}) = \sqrt{3} \cdot x - \sqrt{3} \cdot \sqrt{5}$ Using a distributive law

           $= x\sqrt{3} - \sqrt{15}$ Multiplying radicals

8. $\sqrt[3]{y}(\sqrt[3]{y^2} + \sqrt[3]{2}) = \sqrt[3]{y} \cdot \sqrt[3]{y^2} + \sqrt[3]{y} \cdot \sqrt[3]{2}$ Using a distributive law

           $= \sqrt[3]{y^3} + \sqrt[3]{2y}$ Multiplying radicals

           $= y + \sqrt[3]{2y}$ Simplifying $\sqrt[3]{y^3}$

Do Exercises 6 and 7.

EXAMPLE 9 Multiply: $(4\sqrt{3} + \sqrt{2})(\sqrt{3} - 5\sqrt{2})$.

                                        F   O   I   L

$(4\sqrt{3} + \sqrt{2})(\sqrt{3} - 5\sqrt{2}) = 4(\sqrt{3})^2 - 20\sqrt{3} \cdot \sqrt{2} + \sqrt{2} \cdot \sqrt{3} - 5(\sqrt{2})^2$

                              $= 4 \cdot 3 - 20\sqrt{6} + \sqrt{6} - 5 \cdot 2$

                              $= 12 - 20\sqrt{6} + \sqrt{6} - 10$

                              $= 2 - 19\sqrt{6}$ Collecting like terms

EXAMPLE 10 Multiply: $(\sqrt{a} + \sqrt{3})(\sqrt{b} + \sqrt{3})$. Assume that all expressions under radicals represent nonnegative numbers.

$(\sqrt{a} + \sqrt{3})(\sqrt{b} + \sqrt{3}) = \sqrt{a}\sqrt{b} + \sqrt{a}\sqrt{3} + \sqrt{3}\sqrt{b} + \sqrt{3}\sqrt{3}$

                             $= \sqrt{ab} + \sqrt{3a} + \sqrt{3b} + 3$

EXAMPLE 11 Multiply: $(\sqrt{5} + \sqrt{7})(\sqrt{5} - \sqrt{7})$.

$(\sqrt{5} + \sqrt{7})(\sqrt{5} - \sqrt{7}) = (\sqrt{5})^2 - (\sqrt{7})^2$ This is now a difference of two squares: $(A - B)(A + B) = A^2 - B^2$.

                           $= 5 - 7 = -2$

EXAMPLE 12 Multiply: $(\sqrt{a} + \sqrt{b})(\sqrt{a} - \sqrt{b})$. Assume that no radicands were formed by raising negative numbers to even powers.

$(\sqrt{a} + \sqrt{b})(\sqrt{a} - \sqrt{b}) = (\sqrt{a})^2 - (\sqrt{b})^2$

                     $= a - b$ No radicals

Expressions of the form $\sqrt{a} + \sqrt{b}$ and $\sqrt{a} - \sqrt{b}$ are called **conjugates**. Their product is always an expression that has no radicals.

Do Exercises 8–11.

EXAMPLE 13 Multiply: $(\sqrt{3} + x)^2$.

$(\sqrt{3} + x)^2 = (\sqrt{3})^2 + 2x\sqrt{3} + x^2$ Squaring a binomial

           $= 3 + 2x\sqrt{3} + x^2$

Do Exercises 12 and 13.

Multiply. Assume that no radicands were formed by raising negative numbers to even powers.

6. $\sqrt{2}(5\sqrt{3} + 3\sqrt{7})$

7. $\sqrt[3]{a^2}(\sqrt[3]{3a} - \sqrt[3]{2})$

Multiply. Assume that no radicands were formed by raising negative numbers to even powers.

8. $(\sqrt{3} - 5\sqrt{2})(2\sqrt{3} + \sqrt{2})$

9. $(\sqrt{a} + 2\sqrt{3})(3\sqrt{b} - 4\sqrt{3})$

10. $(\sqrt{2} + \sqrt{5})(\sqrt{2} - \sqrt{5})$

11. $(\sqrt{p} - \sqrt{q})(\sqrt{p} + \sqrt{q})$

Multiply.

12. $(2\sqrt{5} - y)^2$ **13.** $(3\sqrt{6} + 2)^2$

Answers

6. $5\sqrt{6} + 3\sqrt{14}$ **7.** $a\sqrt[3]{3} - \sqrt[3]{2a^2}$
8. $-4 - 9\sqrt{6}$
9. $3\sqrt{ab} - 4\sqrt{3a} + 6\sqrt{3b} - 24$ **10.** -3
11. $p - q$ **12.** $20 - 4y\sqrt{5} + y^2$
13. $58 + 12\sqrt{6}$

a Add or subtract. Then simplify by collecting like radical terms, if possible. Assume that no radicands were formed by raising negative numbers to even powers.

1. $7\sqrt{5} + 4\sqrt{5}$

2. $2\sqrt{3} + 9\sqrt{3}$

3. $6\sqrt[3]{7} - 5\sqrt[3]{7}$

4. $13\sqrt[5]{3} - 8\sqrt[5]{3}$

5. $4\sqrt[3]{y} + 9\sqrt[3]{y}$

6. $6\sqrt[4]{t} - 3\sqrt[4]{t}$

7. $5\sqrt{6} - 9\sqrt{6} - 4\sqrt{6}$

8. $3\sqrt{10} - 8\sqrt{10} + 7\sqrt{10}$

9. $4\sqrt[3]{3} - \sqrt{5} + 2\sqrt[3]{3} + \sqrt{5}$

10. $5\sqrt{7} - 8\sqrt[4]{11} + \sqrt{7} + 9\sqrt[4]{11}$

11. $8\sqrt{27} - 3\sqrt{3}$

12. $9\sqrt{50} - 4\sqrt{2}$

13. $8\sqrt{45} + 7\sqrt{20}$

14. $9\sqrt{12} + 16\sqrt{27}$

15. $18\sqrt{72} + 2\sqrt{98}$

16. $12\sqrt{45} - 8\sqrt{80}$

17. $3\sqrt[3]{16} + \sqrt[3]{54}$

18. $\sqrt[3]{27} - 5\sqrt[3]{8}$

19. $2\sqrt{128} - \sqrt{18} + 4\sqrt{32}$

20. $5\sqrt{50} - 2\sqrt{18} + 9\sqrt{32}$

21. $\sqrt{5a} + 2\sqrt{45a^3}$

22. $4\sqrt{3x^3} - \sqrt{12x}$

23. $\sqrt[3]{24x} - \sqrt[3]{3x^4}$

24. $\sqrt[3]{54x} - \sqrt[3]{2x^4}$

25. $7\sqrt{27x^3} + \sqrt{3x}$

26. $2\sqrt{45x^3} - \sqrt{5x}$

27. $\sqrt{4} + \sqrt{18}$

28. $\sqrt[3]{8} - \sqrt[3]{24}$

29. $5\sqrt[3]{32} - \sqrt[3]{108} + 2\sqrt[3]{256}$

30. $3\sqrt[3]{8x} - 4\sqrt[3]{27x} + 2\sqrt[3]{64x}$

31. $\sqrt[3]{6x^4} + \sqrt[3]{48x} - \sqrt[3]{6x}$

32. $\sqrt[4]{80x^5} - \sqrt[4]{405x^9} + \sqrt[4]{5x}$

33. $\sqrt{4a - 4} + \sqrt{a - 1}$

34. $\sqrt{9y + 27} + \sqrt{y + 3}$

35. $\sqrt{x^3 - x^2} + \sqrt{9x - 9}$

36. $\sqrt{4x - 4} + \sqrt{x^3 - x^2}$

b   Multiply. Assume that no radicands were formed by raising negative numbers to even powers.

37. $\sqrt{5}(4 - 2\sqrt{5})$

38. $\sqrt{6}(2 + \sqrt{6})$

39. $\sqrt{3}(\sqrt{2} - \sqrt{7})$

40. $\sqrt{2}(\sqrt{5} - \sqrt{2})$

41. $\sqrt{3}(-4\sqrt{3} + 6)$

42. $\sqrt{2}(-5\sqrt{2} - 7)$

43. $\sqrt{3}(2\sqrt{5} - 3\sqrt{4})$

44. $\sqrt{2}(3\sqrt{10} - 2\sqrt{2})$

45. $\sqrt[3]{2}(\sqrt[3]{4} - 2\sqrt[3]{32})$

46. $\sqrt[3]{3}(\sqrt[3]{9} - 4\sqrt[3]{21})$

47. $3\sqrt[3]{y}(2\sqrt[3]{y^2} - 4\sqrt[3]{y})$

48. $2\sqrt[3]{y^2}(5\sqrt[3]{y} + 4\sqrt[3]{y^2})$

49. $\sqrt[3]{a}(\sqrt[3]{2a^2} + \sqrt[3]{16a^2})$

50. $\sqrt[3]{x}(\sqrt[3]{3x^2} - \sqrt[3]{81x^2})$

51. $(\sqrt{3} - \sqrt{2})(\sqrt{3} + \sqrt{2})$

52. $(\sqrt{5} + \sqrt{6})(\sqrt{5} - \sqrt{6})$

53. $\left(\sqrt{8} + 2\sqrt{5}\right)\left(\sqrt{8} - 2\sqrt{5}\right)$

54. $\left(\sqrt{18} + 3\sqrt{7}\right)\left(\sqrt{18} - 3\sqrt{7}\right)$

55. $\left(7 + \sqrt{5}\right)\left(7 - \sqrt{5}\right)$

56. $\left(4 - \sqrt{3}\right)\left(4 + \sqrt{3}\right)$

57. $\left(2 - \sqrt{3}\right)\left(2 + \sqrt{3}\right)$

58. $\left(11 - \sqrt{2}\right)\left(11 + \sqrt{2}\right)$

59. $\left(\sqrt{8} + \sqrt{5}\right)\left(\sqrt{8} - \sqrt{5}\right)$

60. $\left(\sqrt{6} - \sqrt{7}\right)\left(\sqrt{6} + \sqrt{7}\right)$

61. $\left(3 + 2\sqrt{7}\right)\left(3 - 2\sqrt{7}\right)$

62. $\left(6 - 3\sqrt{2}\right)\left(6 + 3\sqrt{2}\right)$

63. $\left(\sqrt{a} + \sqrt{b}\right)\left(\sqrt{a} - \sqrt{b}\right)$

64. $\left(\sqrt{x} - \sqrt{y}\right)\left(\sqrt{x} + \sqrt{y}\right)$

65. $\left(3 - \sqrt{5}\right)\left(2 + \sqrt{5}\right)$

66. $\left(2 + \sqrt{6}\right)\left(4 - \sqrt{6}\right)$

67. $\left(\sqrt{3} + 1\right)\left(2\sqrt{3} + 1\right)$

68. $\left(4\sqrt{3} + 5\right)\left(\sqrt{3} - 2\right)$

69. $\left(2\sqrt{7} - 4\sqrt{2}\right)\left(3\sqrt{7} + 6\sqrt{2}\right)$

70. $\left(4\sqrt{5} + 3\sqrt{3}\right)\left(3\sqrt{5} - 4\sqrt{3}\right)$

71. $\left(\sqrt{a} + \sqrt{2}\right)\left(\sqrt{a} + \sqrt{3}\right)$

72. $\left(2 - \sqrt{x}\right)\left(1 - \sqrt{x}\right)$

73. $\left(2\sqrt[3]{3} + \sqrt[3]{2}\right)\left(\sqrt[3]{3} - 2\sqrt[3]{2}\right)$

74. $\left(3\sqrt[3]{7} + \sqrt[3]{6}\right)\left(2\sqrt[3]{7} - 3\sqrt[3]{6}\right)$

75. $\left(2 + \sqrt{3}\right)^2$

76. $\left(\sqrt{5} + 1\right)^2$

77. $\left(\sqrt[5]{9} - \sqrt[5]{3}\right)\left(\sqrt[5]{8} + \sqrt[5]{27}\right)$

78. $\left(\sqrt[3]{8x} - \sqrt[3]{5y}\right)^2$

Skill Maintenance

Multiply or divide and simplify.

79. $\dfrac{x^3 + 4x}{x^2 - 16} \div \dfrac{x^2 + 8x + 15}{x^2 + x - 20}$ [6.2b]

80. $\dfrac{a^2 - 4}{a} \div \dfrac{a - 2}{a + 4}$ [6.2b]

81. $\dfrac{a^3 + 8}{a^2 - 4} \cdot \dfrac{a^2 - 4a + 4}{a^2 - 2a + 4}$ [6.1d]

82. $\dfrac{y^3 - 27}{y^2 - 9} \cdot \dfrac{y^2 - 6y + 9}{y^2 + 3y + 9}$ [6.1d]

Simplify.   [6.6a]

83. $\dfrac{x - \dfrac{1}{3}}{x + \dfrac{1}{4}}$

84. $\dfrac{1 - \dfrac{1}{x}}{1 - \dfrac{1}{x^2}}$

85. $\dfrac{\dfrac{1}{p} - \dfrac{1}{q}}{\dfrac{1}{p^2} - \dfrac{1}{q^2}}$

86. $\dfrac{\dfrac{1}{a} + \dfrac{1}{b}}{\dfrac{1}{a^3} + \dfrac{1}{b^3}}$

Solve.   [9.3c, d, e]

87. $|3x + 7| = 22$

88. $|3x + 7| < 22$

89. $|3x + 7| \geq 22$

90. $|3x + 7| = |2x - 5|$

Synthesis

91. ⌨ Graph the function $f(x) = \sqrt{(x - 2)^2}$. What is the domain?

92. ⌨ Use a graphing calculator to check your answers to Exercises 5, 22, and 72.

Multiply and simplify.

93. $\sqrt{9 + 3\sqrt{5}}\ \sqrt{9 - 3\sqrt{5}}$

94. $\left(\sqrt{x + 2} - \sqrt{x - 2}\right)^2$

95. $\left(\sqrt{3} + \sqrt{5} - \sqrt{6}\right)^2$

96. $\sqrt[3]{y}\left(1 - \sqrt[3]{y}\right)\left(1 + \sqrt[3]{y}\right)$

97. $\left(\sqrt[3]{9} - 2\right)\left(\sqrt[3]{9} + 4\right)$

98. $\left[\sqrt{3 + \sqrt{2 + \sqrt{1}}}\right]^4$

Mid-Chapter Review

Concept Reinforcement

Determine whether each statement is true or false.

_____ 1. Every real number has two real-number square roots. [10.1a]

_____ 2. If $\sqrt[3]{q}$ is negative, then q is negative. [10.1c]

_____ 3. $a^{m/n}$ and $a^{n/m}$ are reciprocals. [10.2b]

_____ 4. To multiply radicals with the same index, we multiply the radicands. [10.3a]

Guided Solutions

Fill in each blank with the number that creates a correct statement or solution.

Perform the indicated operations and simplify. [10.3a], [10.4a]

5. $\sqrt{6}\sqrt{10} = \sqrt{6 \cdot \square} = \sqrt{2 \cdot \square \cdot 2 \cdot \square} = \square\sqrt{\square}$

6. $5\sqrt{32} - 3\sqrt{18} = 5\sqrt{\square \cdot 2} - 3\sqrt{\square \cdot 2}$

$= 5 \cdot \square\sqrt{2} - 3 \cdot \square\sqrt{2}$

$= \square\sqrt{2} - \square\sqrt{2}$

$= \square\sqrt{2}$

Mixed Review

Simplify. [10.1a]

7. $\sqrt{81}$

8. $-\sqrt{144}$

9. $\sqrt{\dfrac{16}{25}}$

10. $\sqrt{-9}$

11. For $f(x) = \sqrt{2x + 3}$, find $f(3)$ and $f(-2)$. [10.1a]

12. Find the domain of $f(x) = \sqrt{4 - x}$. [10.1a]

Graph. [10.1a]

13. $f(x) = -2\sqrt{x}$

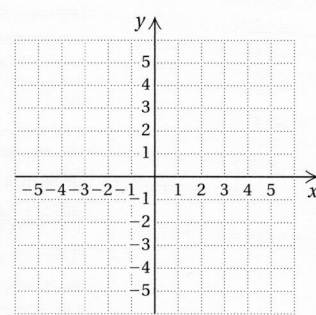

14. $g(x) = \sqrt{x + 1}$

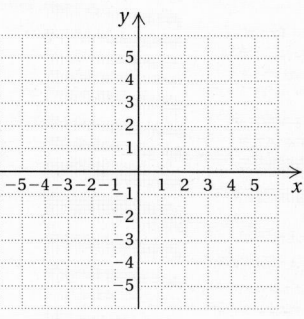

Find each of the following. Assume that letters can represent *any* real number. [10.1b, c, d]

15. $\sqrt{36z^2}$

16. $\sqrt{x^2 - 8x + 16}$

17. $\sqrt[3]{-64}$

18. $-\sqrt[3]{27a^3}$

19. $\sqrt[5]{32}$

20. $\sqrt[10]{y^{10}}$

Rewrite without rational exponents and simplify, if possible. [10.2a]

21. $125^{1/3}$

22. $(a^3b)^{1/4}$

Rewrite with rational exponents. [10.2a]

23. $\sqrt[5]{16}$

24. $\sqrt[3]{6m^2n}$

Simplify. Write the answer with positive exponents. [10.2c]

25. $3^{1/4} \cdot 3^{-5/8}$

26. $\dfrac{7^{6/5}}{7^{2/5}}$

27. $\left(x^{3/4}y^{-2/3}\right)^2$

28. $\left(n^{-3/5}\right)^{5/4}$

Use rational exponents to simplify. Write the answer in radical notation. [10.2d]

29. $\sqrt[6]{16}$

30. $\left(\sqrt[10]{ab}\right)^5$

Use rational exponents to write a single radical expression. [10.2d]

31. $\sqrt{y}\,\sqrt[3]{y}$

32. $a^{2/3}b^{3/5}$

Perform the indicated operation and simplify. Assume that no radicands were formed by raising negative numbers to even powers. [10.3a, b], [10.4a, b]

33. $\sqrt{5}\sqrt{15}$

34. $\sqrt[3]{4x^2y}\,\sqrt[3]{6xy^4}$

35. $\dfrac{\sqrt[3]{80}}{\sqrt[3]{2}}$

36. $\sqrt{\dfrac{49a^5}{b^8}}$

37. $5\sqrt{7} + 6\sqrt{7}$

38. $3\sqrt{18x^3} - 6\sqrt{32x}$

39. $\sqrt{3}\left(2 - 5\sqrt{3}\right)$

40. $\left(1 - \sqrt{x}\right)\left(3 - \sqrt{x}\right)$

41. $\left(\sqrt{m} - \sqrt{n}\right)\left(\sqrt{m} + \sqrt{n}\right)$

42. $\left(\sqrt{7} + 2\right)^2$

43. $\left(2\sqrt{3} + 3\sqrt{5}\right)\left(3\sqrt{3} - 4\sqrt{5}\right)$

Understanding Through Discussion and Writing

44. Does the nth root of x^2 always exist? Why or why not? [10.1a]

45. Explain how to formulate a radical expression that can be used to define a function f with a domain of $\{x | x \leq 5\}$. [10.1a]

46. Explain why $\sqrt[3]{x^6} = x^2$ for any value of x, but $\sqrt{x^6} = x^3$ only when $x \geq 0$. [10.2d]

47. Is the quotient of two irrational numbers always an irrational number? Why or why not? [10.3b]

10.5 More on Division of Radical Expressions

a Rationalizing Denominators

Sometimes in mathematics it is useful to find an equivalent expression without a radical in the denominator. This provides a standard notation for expressing results. The procedure for finding such an expression is called **rationalizing the denominator**. We carry this out by multiplying by 1.

EXAMPLE 1 Rationalize the denominator: $\sqrt{\dfrac{7}{3}}$.

We multiply by 1, using $\sqrt{3}/\sqrt{3}$. We do this so that the denominator of the radicand will be a perfect square.

$$\sqrt{\frac{7}{3}} = \frac{\sqrt{7}}{\sqrt{3}} \cdot \frac{\sqrt{3}}{\sqrt{3}}$$

$$= \frac{\sqrt{7} \cdot \sqrt{3}}{\sqrt{3} \cdot \sqrt{3}}$$

$$= \frac{\sqrt{21}}{\sqrt{3^2}} = \frac{\sqrt{21}}{3}$$

The radicand is a perfect square.

Do Margin Exercise 1.

EXAMPLE 2 Rationalize the denominator: $\sqrt[3]{\dfrac{7}{25}}$.

We first factor the denominator:

$$\sqrt[3]{\frac{7}{25}} = \sqrt[3]{\frac{7}{5 \cdot 5}}.$$

To get a perfect cube in the denominator, we consider the index 3 and the factors. We have 2 factors of 5, and we need 3 factors of 5. We achieve this by multiplying by 1, using $\sqrt[3]{5}/\sqrt[3]{5}$.

$$\sqrt[3]{\frac{7}{25}} = \sqrt[3]{\frac{7}{5 \cdot 5}} \cdot \frac{\sqrt[3]{5}}{\sqrt[3]{5}} \qquad \text{Multiplying by } \frac{\sqrt[3]{5}}{\sqrt[3]{5}} \text{ to make the denominator}$$

of the radicand a perfect cube

$$= \frac{\sqrt[3]{7} \cdot \sqrt[3]{5}}{\sqrt[3]{5 \cdot 5} \cdot \sqrt[3]{5}}$$

$$= \frac{\sqrt[3]{35}}{\sqrt[3]{5^3}} \qquad \text{The radicand is a perfect cube.}$$

$$= \frac{\sqrt[3]{35}}{5}$$

Do Exercise 2.

OBJECTIVES

a Rationalize the denominator of a radical expression having one term in the denominator.

b Rationalize the denominator of a radical expression having two terms in the denominator.

SKILL TO REVIEW
Objective 4.6b: Multiply the sum and the difference of two terms mentally.

Multiply.
1. $(x + 3)(x - 3)$
2. $(2y + 5)(2y - 5)$

1. Rationalize the denominator:
$$\sqrt{\frac{2}{5}}.$$

2. Rationalize the denominator:
$$\sqrt[3]{\frac{5}{4}}.$$

Answers

Skill to Review:
1. $x^2 - 9$ **2.** $4y^2 - 25$

Margin Exercises:
1. $\dfrac{\sqrt{10}}{5}$ **2.** $\dfrac{\sqrt[3]{10}}{2}$

EXAMPLE 3 Rationalize the denominator: $\sqrt{\dfrac{2a}{5b}}$. Assume that no radicands were formed by raising negative numbers to even powers.

$$\sqrt{\dfrac{2a}{5b}} = \dfrac{\sqrt{2a}}{\sqrt{5b}} \qquad \text{Converting to a quotient of radicals}$$

$$= \dfrac{\sqrt{2a}}{\sqrt{5b}} \cdot \dfrac{\sqrt{5b}}{\sqrt{5b}} \qquad \text{Multiplying by 1}$$

$$= \dfrac{\sqrt{10ab}}{\sqrt{5^2 b^2}} \qquad \begin{array}{l}\text{The radicand in the denominator}\\\text{is a perfect square.}\end{array}$$

$$= \dfrac{\sqrt{10ab}}{5b}$$

Do Exercise 3.

> **3.** Rationalize the denominator:
> $$\sqrt{\dfrac{4a}{3b}}.$$

EXAMPLE 4 Rationalize the denominator: $\dfrac{\sqrt[3]{a}}{\sqrt[3]{9x}}$.

We factor the denominator:

$$\dfrac{\sqrt[3]{a}}{\sqrt[3]{9x}} = \dfrac{\sqrt[3]{a}}{\sqrt[3]{3 \cdot 3 \cdot x}}.$$

To choose the symbol for 1, we look at $3 \cdot 3 \cdot x$. To make it a cube, we need another 3 and two more x's. Thus we multiply by 1, using $\sqrt[3]{3x^2}/\sqrt[3]{3x^2}$:

$$\dfrac{\sqrt[3]{a}}{\sqrt[3]{9x}} = \dfrac{\sqrt[3]{a}}{\sqrt[3]{3 \cdot 3 \cdot x}} \cdot \dfrac{\sqrt[3]{3x^2}}{\sqrt[3]{3x^2}} \qquad \text{Multiplying by 1}$$

$$= \dfrac{\sqrt[3]{3ax^2}}{\sqrt[3]{3^3 x^3}} \qquad \begin{array}{l}\text{The radicand in the denominator}\\\text{is a perfect cube.}\end{array}$$

$$= \dfrac{\sqrt[3]{3ax^2}}{3x}.$$

Do Exercises 4 and 5.

> Rationalize the denominator.
> **4.** $\dfrac{\sqrt[4]{7}}{\sqrt[4]{2}}$ **5.** $\sqrt[3]{\dfrac{3x^5}{2y}}$

EXAMPLE 5 Rationalize the denominator: $\dfrac{3x}{\sqrt[5]{2x^2 y^3}}$.

$$\dfrac{3x}{\sqrt[5]{2x^2 y^3}} = \dfrac{3x}{\sqrt[5]{2 \cdot x \cdot x \cdot y \cdot y \cdot y}}$$

$$= \dfrac{3x}{\sqrt[5]{2x^2 y^3}} \cdot \dfrac{\sqrt[5]{2^4 x^3 y^2}}{\sqrt[5]{2^4 x^3 y^2}}$$

$$= \dfrac{3x\sqrt[5]{16x^3 y^2}}{\sqrt[5]{2^5 x^5 y^5}} \qquad \begin{array}{l}\text{The radicand in the denominator}\\\text{is a perfect fifth power.}\end{array}$$

$$= \dfrac{3x\sqrt[5]{16x^3 y^2}}{2xy}$$

$$= \dfrac{x}{x} \cdot \dfrac{3\sqrt[5]{16x^3 y^2}}{2y}$$

$$= \dfrac{3\sqrt[5]{16x^3 y^2}}{2y}$$

Do Exercise 6.

> **6.** Rationalize the denominator:
> $$\dfrac{7x}{\sqrt[3]{4xy^5}}.$$

Answers

3. $\dfrac{2\sqrt{3ab}}{3b}$ **4.** $\dfrac{\sqrt[4]{56}}{2}$ **5.** $\dfrac{x\sqrt[3]{12x^2 y^2}}{2y}$

6. $\dfrac{7\sqrt[3]{2x^2 y}}{2y^2}$

b Rationalizing When There Are Two Terms

Do Exercises 7 and 8.

Certain pairs of expressions containing square roots, such as $c - \sqrt{b}$, $c + \sqrt{b}$ and $\sqrt{a} - \sqrt{b}$, $\sqrt{a} + \sqrt{b}$, are called **conjugates**. The product of such a pair of conjugates has no radicals in it. (See Example 12 of Section 10.4.) Thus when we wish to rationalize a denominator that has two terms and one or more of them involves a square-root radical, we multiply by 1 using the conjugate of the denominator to write a symbol for 1.

EXAMPLES In each of the following, what symbol for 1 would you use to rationalize the denominator?

Expression	Symbol for 1

6. $\dfrac{3}{x + \sqrt{7}}$ $\dfrac{x - \sqrt{7}}{x - \sqrt{7}}$

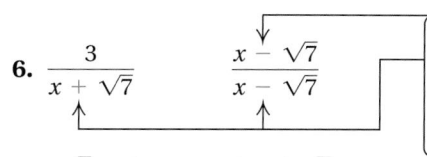

> Change the operation sign in the denominator to obtain the conjugate. Use the conjugate for the numerator and denominator of the symbol for 1.

7. $\dfrac{\sqrt{7} + 4}{3 - 2\sqrt{5}}$ $\dfrac{3 + 2\sqrt{5}}{3 + 2\sqrt{5}}$

Do Exercises 9 and 10.

EXAMPLE 8 Rationalize the denominator: $\dfrac{4}{\sqrt{3} + x}$.

$$\frac{4}{\sqrt{3} + x} = \frac{4}{\sqrt{3} + x} \cdot \frac{\sqrt{3} - x}{\sqrt{3} - x}$$

$$= \frac{4(\sqrt{3} - x)}{(\sqrt{3} + x)(\sqrt{3} - x)}$$

$$= \frac{4\sqrt{3} - 4x}{3 - x^2}$$

EXAMPLE 9 Rationalize the denominator: $\dfrac{4 + \sqrt{2}}{\sqrt{5} - \sqrt{2}}$.

$$\frac{4 + \sqrt{2}}{\sqrt{5} - \sqrt{2}} = \frac{4 + \sqrt{2}}{\sqrt{5} - \sqrt{2}} \cdot \frac{\sqrt{5} + \sqrt{2}}{\sqrt{5} + \sqrt{2}}$$ Multiplying by 1, using the conjugate of $\sqrt{5} - \sqrt{2}$, which is $\sqrt{5} + \sqrt{2}$

$$= \frac{(4 + \sqrt{2})(\sqrt{5} + \sqrt{2})}{(\sqrt{5} - \sqrt{2})(\sqrt{5} + \sqrt{2})}$$ Multiplying numerators and denominators

$$= \frac{4\sqrt{5} + 4\sqrt{2} + \sqrt{2}\sqrt{5} + (\sqrt{2})^2}{(\sqrt{5})^2 - (\sqrt{2})^2}$$ Using $(A - B)(A + B) = A^2 - B^2$ in the denominator

$$= \frac{4\sqrt{5} + 4\sqrt{2} + \sqrt{10} + 2}{5 - 2}$$

$$= \frac{4\sqrt{5} + 4\sqrt{2} + \sqrt{10} + 2}{3}$$

Do Exercises 11 and 12.

Multiply.

7. $(c - \sqrt{b})(c + \sqrt{b})$

8. $(\sqrt{a} + \sqrt{b})(\sqrt{a} - \sqrt{b})$

What symbol for 1 would you use to rationalize the denominator?

9. $\dfrac{\sqrt{5} + 1}{\sqrt{3} - y}$ **10.** $\dfrac{1}{\sqrt{2} + \sqrt{3}}$

Rationalize the denominator.

11. $\dfrac{14}{3 + \sqrt{2}}$ **12.** $\dfrac{5 + \sqrt{2}}{1 - \sqrt{2}}$

Answers

7. $c^2 - b$ **8.** $a - b$ **9.** $\dfrac{\sqrt{3} + y}{\sqrt{3} + y}$

10. $\dfrac{\sqrt{2} - \sqrt{3}}{\sqrt{2} - \sqrt{3}}$ **11.** $6 - 2\sqrt{2}$

12. $-7 - 6\sqrt{2}$

a Rationalize the denominator. Assume that no radicands were formed by raising negative numbers to even powers.

1. $\sqrt{\dfrac{5}{3}}$

2. $\sqrt{\dfrac{8}{7}}$

3. $\sqrt{\dfrac{11}{2}}$

4. $\sqrt{\dfrac{17}{6}}$

5. $\dfrac{2\sqrt{3}}{7\sqrt{5}}$

6. $\dfrac{3\sqrt{5}}{8\sqrt{2}}$

7. $\sqrt[3]{\dfrac{16}{9}}$

8. $\sqrt[3]{\dfrac{1}{3}}$

9. $\dfrac{\sqrt[3]{3a}}{\sqrt[3]{5c}}$

10. $\dfrac{\sqrt[3]{7x}}{\sqrt[3]{3y}}$

11. $\dfrac{\sqrt[3]{2y^4}}{\sqrt[3]{6x^4}}$

12. $\dfrac{\sqrt[3]{3a^4}}{\sqrt[3]{7b^2}}$

13. $\dfrac{1}{\sqrt[4]{st}}$

14. $\dfrac{1}{\sqrt[3]{yz}}$

15. $\sqrt{\dfrac{3x}{20}}$

16. $\sqrt{\dfrac{7a}{32}}$

17. $\sqrt[3]{\dfrac{4}{5x^5y^2}}$

18. $\sqrt[3]{\dfrac{7c}{100ab^5}}$

19. $\sqrt[4]{\dfrac{1}{8x^7y^3}}$

20. $\dfrac{2x}{\sqrt[5]{18x^8y^6}}$

Rationalize the denominator. Assume that no radicands were formed by raising negative numbers to even powers.

21. $\dfrac{9}{6 - \sqrt{10}}$

22. $\dfrac{3}{8 + \sqrt{5}}$

23. $\dfrac{-4\sqrt{7}}{\sqrt{5} + \sqrt{3}}$

24. $\dfrac{-5\sqrt{2}}{\sqrt{7} - \sqrt{5}}$

25. $\dfrac{6\sqrt{3}}{3\sqrt{2} - \sqrt{5}}$

26. $\dfrac{34\sqrt{5}}{2\sqrt{5} - \sqrt{3}}$

27. $\dfrac{3 + \sqrt{5}}{\sqrt{2} + \sqrt{5}}$

28. $\dfrac{2 + \sqrt{3}}{\sqrt{3} + \sqrt{5}}$

29. $\dfrac{\sqrt{3} - \sqrt{2}}{\sqrt{3} - \sqrt{7}}$

30. $\dfrac{\sqrt{5} - \sqrt{3}}{\sqrt{5} - \sqrt{2}}$

31. $\dfrac{\sqrt{5} - 2\sqrt{6}}{\sqrt{3} - 4\sqrt{5}}$

32. $\dfrac{\sqrt{6} - 3\sqrt{5}}{\sqrt{3} - 2\sqrt{7}}$

33. $\dfrac{2 - \sqrt{a}}{3 + \sqrt{a}}$

34. $\dfrac{5 + \sqrt{x}}{8 - \sqrt{x}}$

35. $\dfrac{2 + 3\sqrt{x}}{3 + 2\sqrt{x}}$

36. $\dfrac{5 + 2\sqrt{y}}{4 + 3\sqrt{y}}$

37. $\dfrac{5\sqrt{3} - 3\sqrt{2}}{3\sqrt{2} - 2\sqrt{3}}$

38. $\dfrac{7\sqrt{2} + 4\sqrt{3}}{4\sqrt{3} - 3\sqrt{2}}$

39. $\dfrac{\sqrt{x} - \sqrt{y}}{\sqrt{x} + \sqrt{y}}$

40. $\dfrac{\sqrt{a} + \sqrt{b}}{\sqrt{a} - \sqrt{b}}$

Skill Maintenance

Solve.   [6.7a]

41. $\dfrac{1}{2} - \dfrac{1}{3} = \dfrac{5}{t}$

42. $\dfrac{5}{x - 1} + \dfrac{9}{x^2 + x + 1} = \dfrac{15}{x^3 - 1}$

Divide and simplify.   [6.2b]

43. $\dfrac{1}{x^3 - y^3} \div \dfrac{1}{(x - y)(x^2 + xy + y^2)}$

44. $\dfrac{2x^2 - x - 6}{x^2 + 4x + 3} \div \dfrac{2x^2 + x - 3}{x^2 - 1}$

Synthesis

45. 〰 Use a graphing calculator to check your answers to Exercises 15 and 16.

46. Express each of the following as the product of two radical expressions.
  a) $x - 5$
  b) $x - a$

Simplify. (*Hint*: Rationalize the denominator.)

47. $\sqrt{a^2 - 3} - \dfrac{a^2}{\sqrt{a^2 - 3}}$

48. $\dfrac{1}{4 + \sqrt{3}} + \dfrac{1}{\sqrt{3}} + \dfrac{1}{\sqrt{3} - 4}$

10.6

Solving Radical Equations

OBJECTIVES

a Solve radical equations with one radical term.

b Solve radical equations with two radical terms.

c Solve applied problems involving radical equations.

SKILL TO REVIEW
Objective 5.8b: Solve quadratic equations by factoring and then using the principle of zero products.

Solve.

1. $x^2 - x = 6$

2. $x^2 - x = 2x + 4$

a The Principle of Powers

A **radical equation** has variables in one or more radicands—for example,

$$\sqrt[3]{2x} + 1 = 5, \qquad \sqrt{x} + \sqrt{4x - 2} = 7.$$

To solve such an equation, we need a new equation-solving principle. Suppose that an equation $a = b$ is true. If we square both sides, we get another true equation: $a^2 = b^2$. This can be generalized.

THE PRINCIPLE OF POWERS

For any natural number n, if an equation $a = b$ is true, then $a^n = b^n$ is true.

However, if an equation $a^n = b^n$ is true, it *may not* be true that $a = b$, if n is even. For example, $3^2 = (-3)^2$ is true, but $3 = -3$ is not true. Thus we *must check* the possible solutions when we solve an equation using the principle of powers.

To solve an equation with a radical term, we first isolate the radical term on one side of the equation. Then we use the principle of powers.

EXAMPLE 1 Solve: $\sqrt{x} - 3 = 4$.

We have

$$\sqrt{x} - 3 = 4$$
$$\sqrt{x} = 7 \qquad \text{Adding to isolate the radical}$$
$$\left(\sqrt{x}\right)^2 = 7^2 \qquad \text{Using the principle of powers (squaring)}$$
$$x = 49. \qquad \sqrt{x} \cdot \sqrt{x} = x$$

The number 49 is a possible solution. But we *must* check in order to be sure!

Check:
$$\begin{array}{c|c} \sqrt{x} - 3 = 4 \\ \hline \sqrt{49} - 3 \ ? \ 4 \\ 7 - 3 \\ 4 & \text{TRUE} \end{array}$$

The solution is 49.

-------------------------------- *Caution!* --------------------------------

The principle of powers does not always give equivalent equations. For this reason, a check is a must!

EXAMPLE 2 Solve: $\sqrt{x} = -3$.

We might observe at the outset that this equation has no solution because the principal square root of a number is never negative. Let's continue as above for comparison.

$$\sqrt{x} = -3$$
$$(\sqrt{x})^2 = (-3)^2$$
$$x = 9$$

Check:
$$\frac{\sqrt{x} = -3}{\sqrt{9} \overset{?}{} -3}$$
$$3 \mid \quad \text{FALSE}$$

The number 9 does *not* check. Thus the equation $\sqrt{x} = -3$ has no real-number solution. Note that the solution of the equation $x = 9$ is 9, but the equation $\sqrt{x} = -3$ has *no* solution. Thus the equations $x = 9$ and $\sqrt{x} = -3$ are *not* equivalent equations.

Do Exercises 1 and 2.

EXAMPLE 3 Solve: $x - 7 = 2\sqrt{x + 1}$.

The radical term is already isolated. We proceed with the principle of powers:

$$x - 7 = 2\sqrt{x + 1}$$
$$(x - 7)^2 = \left(2\sqrt{x + 1}\right)^2 \qquad \text{Using the principle of powers (squaring)}$$
$$(x - 7) \cdot (x - 7) = \left(2\sqrt{x + 1}\right)\left(2\sqrt{x + 1}\right)$$
$$x^2 - 14x + 49 = 2^2\left(\sqrt{x + 1}\right)^2$$
$$x^2 - 14x + 49 = 4(x + 1)$$
$$x^2 - 14x + 49 = 4x + 4$$
$$x^2 - 18x + 45 = 0$$
$$(x - 3)(x - 15) = 0 \qquad \text{Factoring}$$
$$x - 3 = 0 \quad or \quad x - 15 = 0 \qquad \text{Using the principle of zero products}$$
$$x = 3 \quad or \qquad x = 15.$$

The possible solutions are 3 and 15. We check.

For 3:
$$\frac{x - 7 = 2\sqrt{x + 1}}{3 - 7 \overset{?}{} 2\sqrt{3 + 1}}$$
$$-4 \mid 2\sqrt{4}$$
$$\mid 2(2)$$
$$\mid 4 \qquad \text{FALSE}$$

For 15:
$$\frac{x - 7 = 2\sqrt{x + 1}}{15 - 7 \overset{?}{} 2\sqrt{15 + 1}}$$
$$8 \mid 2\sqrt{16}$$
$$\mid 2(4)$$
$$\mid 8 \qquad \text{TRUE}$$

The number 3 does *not* check, but the number 15 does check. The solution is 15.

The number 3 in Example 3 is what is sometimes called an *extraneous solution,* but such terminology is at best risky to use because the number 3 is in *no way* a *solution* of the original equation.

Do Exercises 3 and 4.

Solve.

1. $\sqrt{x} - 7 = 3$

2. $\sqrt{x} = -2$

Solve.

3. $x + 2 = \sqrt{2x + 7}$

4. $x + 1 = 3\sqrt{x - 1}$

Answers

1. 100   **2.** No solution   **3.** 1   **4.** 2, 5

We can visualize or check the solutions of a radical equation graphically. Consider the equation of Example 3: $x - 7 = 2\sqrt{x + 1}$. We can examine the solutions by graphing the equations

$$y = x - 7 \quad \text{and} \quad y = 2\sqrt{x + 1}$$

using the same set of axes. A hand-drawn graph of $y = 2\sqrt{x + 1}$ would involve approximating square roots on a calculator.

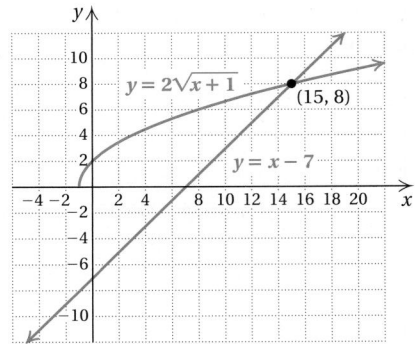

It appears from the graph that when $x = 15$, the values of $y = x - 7$ and $y = 2\sqrt{x + 1}$ are the same, 8. We can check this as we did in Example 3. Note too that the graphs *do not* intersect at $x = 3$, the extraneous solution.

Calculator Corner

Solving Radical Equations   We can solve radical equations graphically. Consider the equation in Example 3,

$x - 7 = 2\sqrt{x + 1}$.

We first graph each side of the equation. We enter $y_1 = x - 7$ and $y_2 = 2\sqrt{x + 1}$ on the equation-editor screen and graph the equations using the window $[-5, 20, -10, 10]$. Note that there is one point of intersection. We use the INTERSECT feature to find its coordinates. (See the Calculator Corner on p. 572 for the procedure.) The first coordinate, 15, is the value of x for which $y_1 = y_2$, or $x - 7 = 2\sqrt{x + 1}$. It is the solution of the equation. Note that the graph shows a single solution whereas the algebraic solution in Example 3 yields two possible solutions, 3 and 15, that must be checked. The algebraic check shows that 15 is the only solution.

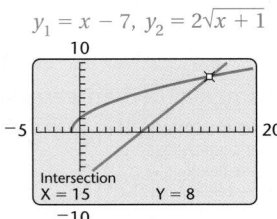

Exercises:

1. Solve the equations in Examples 1 and 4 graphically.

2. Solve the equations in Margin Exercises 1, 3, and 4 graphically.

EXAMPLE 4 Solve: $x = \sqrt{x + 7} + 5$.

We have

$$x = \sqrt{x + 7} + 5$$

$$x - 5 = \sqrt{x + 7} \qquad \text{Subtracting 5 to isolate the radical term}$$

$$(x - 5)^2 = \left(\sqrt{x + 7}\right)^2 \qquad \text{Using the principle of powers (squaring both sides)}$$

$$x^2 - 10x + 25 = x + 7$$

$$x^2 - 11x + 18 = 0$$

$$(x - 9)(x - 2) = 0 \qquad \text{Factoring}$$

$$x = 9 \quad or \quad x = 2. \qquad \text{Using the principle of zero products}$$

The possible solutions are 9 and 2. Let's check.

For 9:

$$\begin{array}{c|c} x = \sqrt{x + 7} + 5 \\ \hline 9 \;?\; \sqrt{9 + 7} + 5 \\ \sqrt{16} + 5 \\ 4 + 5 \\ 9 & \text{TRUE} \end{array}$$

For 2:

$$\begin{array}{c|c} x = \sqrt{x + 7} + 5 \\ \hline 2 \;?\; \sqrt{2 + 7} + 5 \\ \sqrt{9} + 5 \\ 3 + 5 \\ 8 & \text{FALSE} \end{array}$$

Since 9 checks but 2 does not, the solution is 9.

EXAMPLE 5 Solve: $\sqrt[3]{2x + 1} + 5 = 0$.

We have

$$\sqrt[3]{2x + 1} + 5 = 0$$

$$\sqrt[3]{2x + 1} = -5 \qquad \text{Subtracting 5. This isolates the radical term.}$$

$$\left(\sqrt[3]{2x + 1}\right)^3 = (-5)^3 \qquad \text{Using the principle of powers (raising to the third power)}$$

$$2x + 1 = -125$$

$$2x = -126 \qquad \text{Subtracting 1}$$

$$x = -63.$$

Check:

$$\begin{array}{c|c} \sqrt[3]{2x + 1} + 5 = 0 \\ \hline \sqrt[3]{2 \cdot (-63) + 1} + 5 \;?\; 0 \\ \sqrt[3]{-125} + 5 \\ -5 + 5 \\ 0 & \text{TRUE} \end{array}$$

The solution is -63.

Do Exercises 5 and 6.

Solve.

5. $x = \sqrt{x + 5} + 1$

6. $\sqrt[4]{x - 1} - 2 = 0$

b Equations with Two Radical Terms

A general strategy for solving radical equations, including those with two radical terms, is as follows.

> **SOLVING RADICAL EQUATIONS**
>
> To solve radical equations:
>
> 1. Isolate one of the radical terms.
> 2. Use the principle of powers.
> 3. If a radical remains, perform steps (1) and (2) again.
> 4. Check possible solutions.

EXAMPLE 6 Solve: $\sqrt{x-3} + \sqrt{x+5} = 4$.

$$\sqrt{x-3} + \sqrt{x+5} = 4$$

$$\sqrt{x-3} = 4 - \sqrt{x+5}$$
Subtracting $\sqrt{x+5}$. This isolates one of the radical terms.

$$\left(\sqrt{x-3}\right)^2 = \left(4 - \sqrt{x+5}\right)^2$$
Using the principle of powers (squaring both sides)

$$x - 3 = 16 - 8\sqrt{x+5} + (x+5)$$
Using $(A - B)^2 = A^2 - 2AB + B^2$. See this rule in Section 4.6.

$$-3 = 21 - 8\sqrt{x+5}$$
Subtracting x and collecting like terms

$$-24 = -8\sqrt{x+5}$$
Isolating the remaining radical term

$$3 = \sqrt{x+5}$$
Dividing by -8

$$3^2 = \left(\sqrt{x+5}\right)^2$$
Squaring

$$9 = x + 5$$

$$4 = x$$

The number 4 checks and is the solution.

EXAMPLE 7 Solve: $\sqrt{2x-5} = 1 + \sqrt{x-3}$.

$$\sqrt{2x-5} = 1 + \sqrt{x-3}$$

$$\left(\sqrt{2x-5}\right)^2 = \left(1 + \sqrt{x-3}\right)^2$$
One radical is already isolated. We square both sides.

$$2x - 5 = 1 + 2\sqrt{x-3} + \left(\sqrt{x-3}\right)^2$$

$$2x - 5 = 1 + 2\sqrt{x-3} + (x-3)$$

$$x - 3 = 2\sqrt{x-3}$$
Isolating the remaining radical term

$$(x-3)^2 = \left(2\sqrt{x-3}\right)^2$$
Squaring both sides

$$x^2 - 6x + 9 = 4(x-3)$$

$$x^2 - 6x + 9 = 4x - 12$$

$$x^2 - 10x + 21 = 0$$

$$(x-7)(x-3) = 0$$
Factoring

$$x = 7 \quad or \quad x = 3$$
Using the principle of zero products

The possible solutions are 7 and 3. We check.

For 7:

$$\frac{\sqrt{2x - 5} = 1 + \sqrt{x - 3}}{\sqrt{2(7) - 5} \ ? \ 1 + \sqrt{7 - 3}}$$
$$\sqrt{14 - 5} \ \bigg| \ 1 + \sqrt{4}$$
$$\sqrt{9} \ \bigg| \ 1 + 2$$
$$3 \ \bigg| \ 3 \qquad \text{TRUE}$$

For 3:

$$\frac{\sqrt{2x - 5} = 1 + \sqrt{x - 3}}{\sqrt{2(3) - 5} \ ? \ 1 + \sqrt{3 - 3}}$$
$$\sqrt{6 - 5} \ \bigg| \ 1 + \sqrt{0}$$
$$\sqrt{1} \ \bigg| \ 1 + 0$$
$$1 \ \bigg| \ 1 \qquad \text{TRUE}$$

The numbers 7 and 3 check and are the solutions.

Do Exercises 7 and 8.

Solve.

7. $\sqrt{x} - \sqrt{x - 5} = 1$

8. $\sqrt{2x - 5} - 2 = \sqrt{x - 2}$

EXAMPLE 8 Solve: $\sqrt{x + 2} - \sqrt{2x + 2} + 1 = 0$.

We first isolate one radical.

$$\sqrt{x + 2} - \sqrt{2x + 2} + 1 = 0$$
$$\sqrt{x + 2} + 1 = \sqrt{2x + 2} \qquad \text{Adding } \sqrt{2x + 2} \text{ to isolate a radical term}$$
$$\left(\sqrt{x + 2} + 1\right)^2 = \left(\sqrt{2x + 2}\right)^2 \qquad \text{Squaring both sides}$$
$$x + 2 + 2\sqrt{x + 2} + 1 = 2x + 2$$
$$2\sqrt{x + 2} = x - 1$$
$$\left(2\sqrt{x + 2}\right)^2 = (x - 1)^2$$
$$4(x + 2) = x^2 - 2x + 1$$
$$4x + 8 = x^2 - 2x + 1$$
$$0 = x^2 - 6x - 7$$
$$0 = (x - 7)(x + 1) \qquad \text{Factoring}$$
$$x - 7 = 0 \quad or \quad x + 1 = 0 \qquad \text{Using the principle of zero products}$$
$$x = 7 \quad or \quad x = -1$$

The possible solutions are 7 and −1. We check.

For 7:

$$\frac{\sqrt{x + 2} - \sqrt{2x + 2} + 1 = 0}{\sqrt{7 + 2} - \sqrt{2 \cdot 7 + 2} + 1 \ ? \ 0}$$
$$\sqrt{9} - \sqrt{16} + 1 \ \bigg|$$
$$3 - 4 + 1 \ \bigg|$$
$$0 \ \bigg| \qquad \text{TRUE}$$

For −1:

$$\frac{\sqrt{x + 2} - \sqrt{2x + 2} + 1 = 0}{\sqrt{-1 + 2} - \sqrt{2 \cdot (-1) + 2} + 1 \ ? \ 0}$$
$$\sqrt{1} - \sqrt{0} + 1 \ \bigg|$$
$$1 - 0 + 1 \ \bigg|$$
$$2 \ \bigg| \qquad \text{FALSE}$$

The number 7 checks, but −1 does not. The solution is 7.

Do Exercise 9.

9. Solve:
$$\sqrt{3x + 1} - 1 - \sqrt{x + 4} = 0.$$

(c) Applications

Speed of Sound. Many applications translate to radical equations. For example, at a temperature of *t* degrees Fahrenheit, sound travels at a rate of *S* feet per second, where

$$S = 21.9\sqrt{5t + 2457}.$$

EXAMPLE 9 *Outdoor Concert.* The geologically formed, open-air Red Rocks Amphitheatre near Denver, Colorado, hosts a series of concerts. A scientific instrument at one of these concerts determined that the sound of the music was traveling at a rate of 1170 ft/sec. What was the air temperature at the concert?

We substitute 1170 for *S* in the formula $S = 21.9\sqrt{5t + 2457}$:

$$1170 = 21.9\sqrt{5t + 2457}.$$

Then we solve the equation for *t*:

$$1170 = 21.9\sqrt{5t + 2457}$$

$$\frac{1170}{21.9} = \sqrt{5t + 2457} \qquad \text{Dividing by 21.9}$$

$$\left(\frac{1170}{21.9}\right)^2 = \left(\sqrt{5t + 2457}\right)^2 \qquad \text{Squaring both sides}$$

$$2854.2 \approx 5t + 2457 \qquad \text{Simplifying}$$

$$397.2 \approx 5t \qquad \text{Subtracting 2457}$$

$$79 \approx t. \qquad \text{Dividing by 5}$$

The temperature at the concert was about 79°F.

Do Exercise 10.

10. Marching Band Performance. When the Fulton High School marching band performed at half-time of a football game, the speed of sound from the music was measured by a scientific instrument to be 1162 ft/sec. What was the air temperature?

Answer

10. About 72°F

10.6 **Exercise Set**

For Extra Help

MyMathLab   PRACTICE   WATCH   DOWNLOAD   READ   REVIEW

(a) Solve.

1. $\sqrt{2x - 3} = 4$

2. $\sqrt{5x + 2} = 7$

3. $\sqrt{6x + 1} = 8$

4. $\sqrt{3x} - 4 = 6$

5. $\sqrt{y + 7} - 4 = 4$

6. $\sqrt{x - 1} - 3 = 9$

7. $\sqrt{5y + 8} = 10$

8. $\sqrt{2y + 9} = 5$

9. $\sqrt[3]{x} = -1$

10. $\sqrt[3]{y} = -2$

11. $\sqrt{x + 2} = -4$

12. $\sqrt{y - 3} = -2$

13. $\sqrt[3]{x + 5} = 2$

14. $\sqrt[3]{x - 2} = 3$

15. $\sqrt[4]{y - 3} = 2$

16. $\sqrt[4]{x + 3} = 3$

17. $\sqrt[3]{6x + 9} + 8 = 5$

18. $\sqrt[3]{3y + 6} + 2 = 3$

19. $8 = \dfrac{1}{\sqrt{x}}$

20. $\dfrac{1}{\sqrt{y}} = 3$

21. $x - 7 = \sqrt{x - 5}$

22. $x - 5 = \sqrt{x + 7}$

23. $2\sqrt{x + 1} + 7 = x$

24. $\sqrt{2x + 7} - 2 = x$

25. $3\sqrt{x - 1} - 1 = x$

26. $x - 1 = \sqrt{x + 5}$

27. $x - 3 = \sqrt{27 - 3x}$

28. $x - 1 = \sqrt{1 - x}$

 Solve.

29. $\sqrt{3y + 1} = \sqrt{2y + 6}$

30. $\sqrt{5x - 3} = \sqrt{2x + 3}$

31. $\sqrt{y - 5} + \sqrt{y} = 5$

32. $\sqrt{x - 9} + \sqrt{x} = 1$

33. $3 + \sqrt{z - 6} = \sqrt{z + 9}$

34. $\sqrt{4x - 3} = 2 + \sqrt{2x - 5}$

35. $\sqrt{20 - x} + 8 = \sqrt{9 - x} + 11$

36. $4 + \sqrt{10 - x} = 6 + \sqrt{4 - x}$

37. $\sqrt{4y + 1} - \sqrt{y - 2} = 3$

38. $\sqrt{y + 15} - \sqrt{2y + 7} = 1$

39. $\sqrt{x + 2} + \sqrt{3x + 4} = 2$

40. $\sqrt{6x + 7} - \sqrt{3x + 3} = 1$

41. $\sqrt{3x - 5} + \sqrt{2x + 3} + 1 = 0$

42. $\sqrt{2m - 3} + 2 - \sqrt{m + 7} = 0$

43. $2\sqrt{t-1} - \sqrt{3t-1} = 0$

44. $3\sqrt{2y+3} - \sqrt{y+10} = 0$

c Solve.

Sighting to the Horizon. How far can you see to the horizon from a given height? The function

$$D = 1.2\sqrt{h}$$

can be used to approximate the distance D, in miles, that a person can see to the horizon from a height h, in feet.

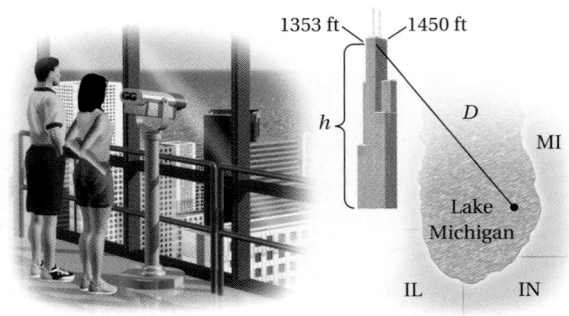

45. An observation deck near the top of the Willis Tower (formerly known as the Sears Tower) in Chicago is 1353 ft high. How far can a tourist see to the horizon from this deck?

46. The roof of the Willis Tower is 1450 ft high. How far can a worker see to the horizon from the top of the Willis Tower?

47. Sarah can see 31.3 mi to the horizon from the top of a cliff. What is the height of Sarah's eyes?

48. A technician can see 30.4 mi to the horizon from the top of a radio tower. How high is the tower?

49. A steeplejack can see 13 mi to the horizon from the top of a building. What is the height of the steeplejack's eyes?

50. A person can see 230 mi to the horizon from an airplane window. How high is the airplane?

Speed of a Skidding Car.   After an accident, how do police determine the speed at which the car had been traveling? The formula

$$r = 2\sqrt{5L}$$

can be used to approximate the speed r, in miles per hour, of a car that has left a skid mark of length L, in feet. Use this formula for Exercises 51 and 52.

51. How far will a car skid at 55 mph? at 75 mph?

52. How far will a car skid at 65 mph? at 100 mph?

Temperature and the Speed of Sound.   Solve Exercises 53 and 54 using the formula $S = 21.9\sqrt{5t + 2457}$ from Example 9.

53. During blasting for avalanche control in Utah's Wasatch Mountains, sound traveled at a rate of 1113 ft/sec. What was the temperature at the time?

54. At a recent concert by the Dave Matthews Band, sound traveled at a rate of 1176 ft/sec. What was the temperature at the time?

Period of a Swinging Pendulum.   The formula $T = 2\pi\sqrt{L/32}$ can be used to find the period T, in seconds, of a pendulum of length L, in feet.

55. What is the length of a pendulum that has a period of 1.0 sec? Use 3.14 for π.

56. What is the length of a pendulum that has a period of 2.0 sec? Use 3.14 for π.

57. The pendulum in Jean's grandfather clock has a period of 2.2 sec. Find the length of the pendulum. Use 3.14 for π.

58. A playground swing has a period of 3.1 sec. Find the length of the swing's chain. Use 3.14 for π.

Skill Maintenance

Solve. [6.8a]

59. *Painting a Room.* Julia can paint a room in 8 hr. George can paint the same room in 10 hr. How long will it take them, working together, to paint the same room?

60. *Delivering Leaflets.* Jeff can drop leaflets in mailboxes three times as fast as Grace can. If they work together, it takes them 1 hr to complete the job. How long would it take each to deliver the leaflets alone?

Solve. [6.8b]

61. *Bicycle Travel.* A cyclist traveled 702 mi in 14 days. At this same ratio, how far would the cyclist have traveled in 56 days?

62. *Earnings.* Dharma earned $696.64 working for 56 hr at a fruit stand. How many hours must she work in order to earn $1044.96?

Solve. [5.8b]

63. $x^2 + 2.8x = 0$

64. $3x^2 - 5x = 0$

65. $x^2 - 64 = 0$

66. $2x^2 = x + 21$

Find the function value. [7.1b]

67. $f(x) = 13 - 2x^2; f(-1)$

68. $h(x) = -7; h(-3)$

69. $g(x) = |5 - x|; g(8)$

70. $f(x) = \frac{1}{2}x - 9; f\left(-\frac{2}{9}\right)$

Synthesis

71. ⌧ Use a graphing calculator to check your answers to Exercises 4, 9, 33, and 38.

72. ⌧ Consider the equation
$$\sqrt{2x + 1} + \sqrt{5x - 4} = \sqrt{10x + 9}.$$
a) Use a graphing calculator to solve the equation.
b) Solve the equation algebraically.
c) Explain the advantages and disadvantages of using each method. Which do you prefer?

Solve.

73. $\sqrt[3]{\dfrac{z}{4}} - 10 = 2$

74. $\sqrt[4]{z^2 + 17} = 3$

75. $\sqrt{\sqrt{y + 49} - \sqrt{y}} = \sqrt{7}$

76. $\sqrt[3]{x^2 + x + 15} - 3 = 0$

77. $\sqrt{\sqrt{x^2 + 9x + 34}} = 2$

78. $\sqrt{8 - b} = b\sqrt{8 - b}$

79. $\sqrt{x - 2} - \sqrt{x + 2} + 2 = 0$

80. $6\sqrt{y} + 6y^{-1/2} = 37$

81. $\sqrt{a^2 + 30a} = a + \sqrt{5a}$

82. $\sqrt{\sqrt{x + 4}} = \sqrt{x} - 2$

83. $\dfrac{x - 1}{\sqrt{x^2 + 3x + 6}} = \dfrac{1}{4}$

84. $\sqrt{x + 1} - \dfrac{2}{\sqrt{x + 1}} = 1$

85. $\sqrt{y^2 + 6} + y - 3 = 0$

86. $2\sqrt{x - 1} - \sqrt{3x - 5} = \sqrt{x - 9}$

87. $\sqrt{y + 1} - \sqrt{2y - 5} = \sqrt{y - 2}$

88. Evaluate: $\sqrt{7 + 4\sqrt{3}} - \sqrt{7 - 4\sqrt{3}}$.

10.7 Applications Involving Powers and Roots

a Applications

There are many kinds of applied problems that involve powers and roots. Many also make use of right triangles and the Pythagorean theorem: $a^2 + b^2 = c^2$.

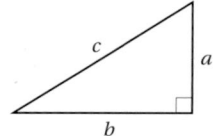

OBJECTIVE

a Solve applied problems involving the Pythagorean theorem and powers and roots.

EXAMPLE 1 *HDTV Dimensions.* An HDTV whose screen measures 46 in. diagonally has a width of 40 in. What is its height?

Using the Pythagorean theorem, $a^2 + b^2 = c^2$, we substitute 40 for b and 46 for c and then solve for a:

$$a^2 + b^2 = c^2$$
$$a^2 + 40^2 = 46^2 \quad \text{Substituting}$$
$$a^2 + 1600 = 2116$$
$$a^2 = 516$$
$$a = \sqrt{516}$$
$$a \approx 22.7.$$

> We consider only the positive root since length cannot be negative.

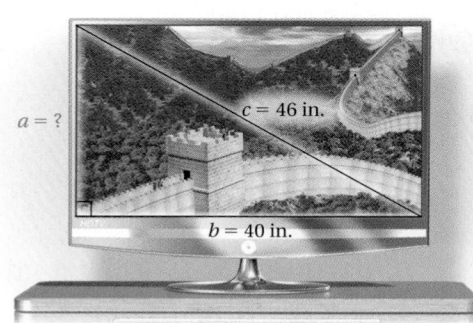

The exact answer is $\sqrt{516}$. This is approximately equal to 22.7. Thus the height of the HDTV screen is about 22.7 in.

EXAMPLE 2 Find the length of the hypotenuse of this right triangle. Give an exact answer and an approximation to three decimal places.

$$7^2 + 4^2 = c^2 \quad \text{Substituting}$$
$$49 + 16 = c^2$$
$$65 = c^2$$

Exact answer: $c = \sqrt{65}$

Approximation: $c \approx 8.062$ Using a calculator

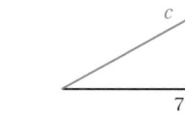

EXAMPLE 3 Find the missing length b in this right triangle. Give an exact answer and an approximation to three decimal places.

$$1^2 + b^2 = \left(\sqrt{11}\right)^2 \quad \text{Substituting}$$
$$1 + b^2 = 11$$
$$b^2 = 10$$

Exact answer: $b = \sqrt{10}$

Approximation: $b \approx 3.162$ Using a calculator

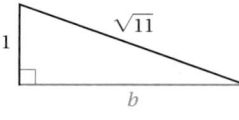

Do Exercises 1 and 2.

1. Find the length of the hypotenuse of this right triangle. Give an exact answer and an approximation to three decimal places.

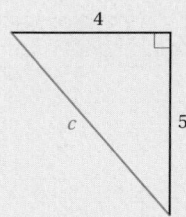

2. Find the length of the leg of this right triangle. Give an exact answer and an approximation to three decimal places.

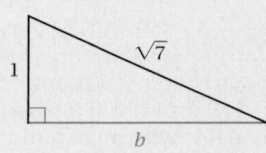

Answers

1. $\sqrt{41}$; 6.403   2. $\sqrt{6}$; 2.449

EXAMPLE 4 *Construction.* Darla is laying out the footer of a house. To see if the corner is square, she measures 16 ft from the corner along one wall and 12 ft from the corner along the other wall. How long should the diagonal be between those two points if the corner is a right angle?

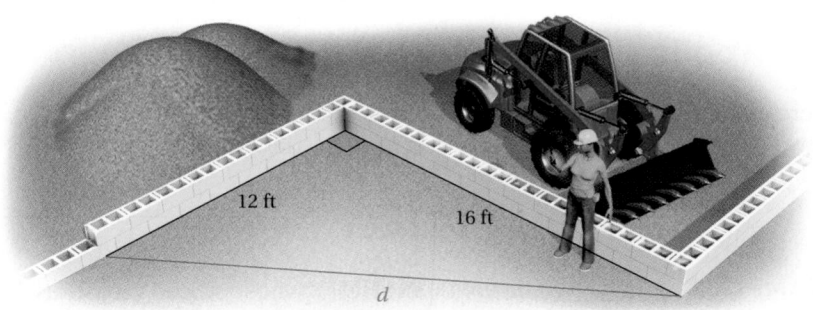

12 ft   16 ft

d

We make a drawing and let d = the length of the diagonal. It is the length of the hypotenuse of a right triangle whose legs are 12 ft and 16 ft. We substitute these values in the Pythagorean theorem to find d:

$$d^2 = 12^2 + 16^2$$
$$d^2 = 144 + 256$$
$$d^2 = 400$$
$$d = \sqrt{400}$$
$$d = 20.$$

The length of the diagonal should be 20 ft.

Do Exercise 3.

3. Baseball Diamond. A baseball diamond is actually a square 90 ft on a side. Suppose a catcher fields a bunt along the third-base line 10 ft from home plate. How far would the catcher have to throw the ball to first base? Give an exact answer and an approximation to three decimal places.

d

10 ft   90 ft

EXAMPLE 5 *Road-Pavement Messages.* In a psychological study, it was determined that the ideal length L of the letters of a word painted on pavement is given by

$$L = \frac{0.000169d^{2.27}}{h},$$

where d is the distance of a car from the lettering and h is the height of the eye above the road. All units are in feet. For a person h feet above the road, a message d feet away will be the most readable if the length of the letters is L. Find L, given that $h = 4$ ft and $d = 180$ ft.

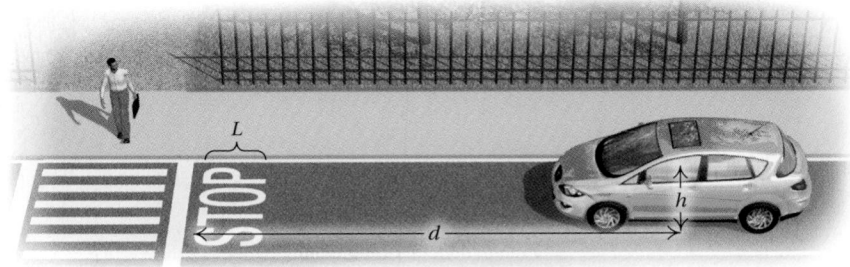

L

STOP

h

d

We substitute 4 for h and 180 for d and calculate L using a calculator with an exponentiation key $\boxed{y^x}$, or ⌃:

$$L = \frac{0.000169(180)^{2.27}}{4} \approx 5.6 \text{ ft.}$$

4. Referring to Example 5, find L given that $h = 3$ ft and $d = 180$ ft. You will need a calculator with an exponentiation key $\boxed{y^x}$, or ⌃.

Answers

3. $\sqrt{8200}$ ft; 90.554 ft   **4.** 7.4 ft

Do Exercise 4.

Translating for Success

1. *Angles of a Triangle.* The second angle of a triangle is four times as large as the first. The third is 27° less than the sum of the other angles. Find the measures of the angles.

2. *Lengths of a Rectangle.* The area of a rectangle is 180 ft². The length is 26 ft greater than the width. Find the length and the width.

3. *Boat Travel.* The speed of a river is 3 mph. A boat can go 72 mi upstream and 24 mi downstream in a total time of 16 hr. Find the speed of the boat in still water.

4. *Coin Mixture.* A collection of nickels and quarters is worth $13.85. There are 85 coins in all. How many of each coin are there?

5. *Perimeter.* The perimeter of a rectangle is 180 ft. The length is 26 ft greater than the width. Find the length and the width.

Translate each word problem to an equation or a system of equations and select a correct translation from equations A–O.

A. $12^2 + 12^2 = x^2$

B. $x(x + 26) = 180$

C. $10,311 + 5\%x = x$

D. $x + y = 85,$
$5x + 25y = 13.85$

E. $x^2 + 4^2 = 12^2$

F. $\dfrac{240}{x - 18} = \dfrac{384}{x}$

G. $x + 5\%x = 10,311$

H. $\dfrac{x}{65} + 1 = \dfrac{x}{85}$

I. $\dfrac{x}{65} + \dfrac{x}{85} = 1$

J. $x + y + z = 180,$
$y = 4x,$
$z = x + y - 27$

K. $2x + 2(x + 26) = 180$

L. $\dfrac{384}{x - 18} = \dfrac{240}{x}$

M. $x + y = 85,$
$0.05x + 0.25y = 13.85$

N. $2x + 2(x + 24) = 240$

O. $\dfrac{72}{x - 3} + \dfrac{24}{x + 3} = 16$

Answers on page A-31

6. *Shoveling Time.* It takes Marv 65 min to shovel 4 in. of snow from his driveway. It takes Elaine 85 min to do the same job. How long would it take if they worked together?

7. *Money Borrowed.* Claire borrows some money at 5% simple interest. After 1 year, $10,311 pays off her loan. How much did she originally borrow?

8. *Plank Height.* A 12-ft plank is leaning against a shed. The bottom of the plank is 4 ft from the building. How high up the side of the shed is the top of the plank?

9. *Train Speeds.* The speed of train A is 18 mph slower than the speed of train B. Train A travels 240 mi in the same time that it takes train B to travel 384 mi. Find the speed of train A.

10. *Diagonal of a Square.* Find the length of a diagonal of a square swimming pool whose sides are 12 ft long.

10.7

Exercise Set

For Extra Help

MyMathLab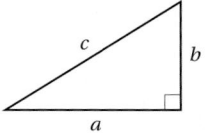

Math XL
PRACTICE

WATCH

DOWNLOAD

READ

REVIEW

a In a right triangle, find the length of the side not given. Give an exact answer and, where appropriate, an approximation to three decimal places.

1. $a = 3, \ b = 5$

2. $a = 8, \ b = 10$

3. $a = 15, \ b = 15$

4. $a = 8, \ b = 8$

5. $b = 12, \ c = 13$

6. $a = 5, \ c = 12$

7. $c = 7, \ a = \sqrt{6}$

8. $c = 10, \ a = 4\sqrt{5}$

9. $b = 1, \ c = \sqrt{13}$

10. $a = 1, \ c = \sqrt{12}$

11. $a = 1, \ c = \sqrt{n}$

12. $c = 2, \ a = \sqrt{n}$

In the following problems, give an exact answer and, where appropriate, an approximation to three decimal places.

13. *Guy Wire.* How long is a guy wire reaching from the top of a 10-ft pole to a point on the ground 4 ft from the pole?

10 ft   g

4 ft

14. *Softball Diamond.* A slow-pitch softball diamond is actually a square 65 ft on a side. How far is it from home to second base?

15. *Road-Pavement Messages.* Using the formula of Example 5, find the length L of a road-pavement message when $h = 4$ ft and $d = 200$ ft.

16. *Road-Pavement Messages.* Using the formula of Example 5, find the length L of a road-pavement message when $h = 8$ ft and $d = 300$ ft.

17. *Vegetable Garden.* Benito and Dominique are planting a vegetable garden in the backyard. They decide that it will be a 30-ft by 40-ft rectangle and begin to lay it out using string. They soon realize that it is difficult to form the right angles and that it would be helpful to know the length of a diagonal. Find the length of a diagonal.

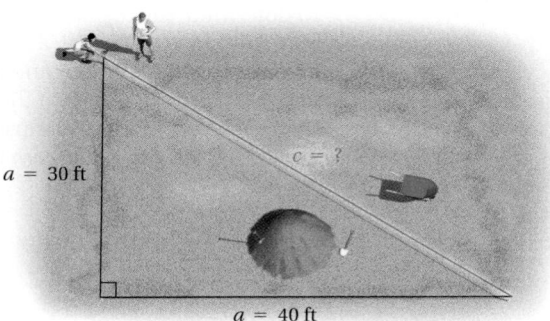

$a = 30$ ft

$c = ?$

$a = 40$ ft

18. *Screen Dimensions.* A television whose screen has a 25-in. diagonal has a height of 15 in. What is its width?

19. *Bridge Expansion.* During the summer heat, a 2-mi bridge expands 2 ft in length. If we assume that the bulge occurs straight up the middle, how high is the bulge? (The answer may surprise you. In reality, bridges are built with expansion spaces to avoid such buckling.)

20. *Triangle Areas.* Triangle *ABC* has sides of lengths 25 ft, 25 ft, and 30 ft. Triangle *PQR* has sides of lengths 25 ft, 25 ft, and 40 ft. Which triangle has the greater area and by how much?

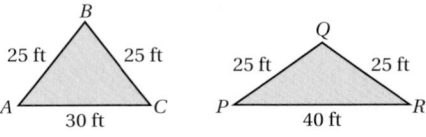

B
25 ft   25 ft
A   30 ft   *C*

Q
25 ft   25 ft
P   40 ft   *R*

21. Each side of a regular octagon has length *s*. Find a formula for the distance *d* between the parallel sides of the octagon,

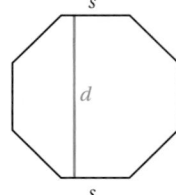

s

d

s

22. The two equal sides of an isosceles right triangle are of length *s*. Find a formula for the length of the hypotenuse.

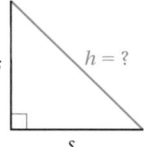

s $h = ?$

s

23. The length and the width of a rectangle are given by consecutive integers. The area of the rectangle is 90 cm². Find the length of a diagonal of the rectangle.

24. The diagonal of a square has length $8\sqrt{2}$ ft. Find the length of a side of the square.

25. Find all ordered pairs on the *x*-axis of a Cartesian coordinate system that are 5 units from the point (0, 4).

26. Find all ordered pairs on the *y*-axis of a Cartesian coordinate system that are 5 units from the point (3, 0).

27. *Speaker Placement.* A stereo receiver is in a corner of a 12-ft by 14-ft room. Speaker wire will run under a rug, diagonally, to a speaker in the far corner. If 4 ft of slack is required on each end, how long should the piece of wire be?

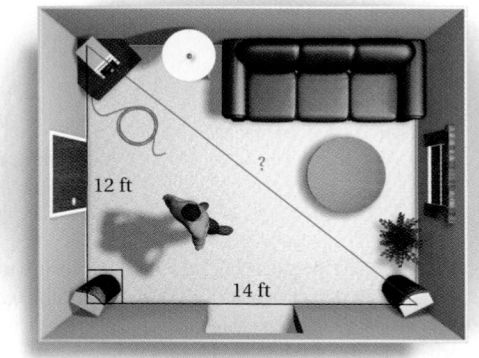

12 ft

14 ft

28. *Distance Over Water.* To determine the distance between two points on opposite sides of a pond, a surveyor locates two stakes at either end of the pond and uses instrumentation to place a third stake so that the distance across the pond is the length of a hypotenuse. If the third stake is 90 m from one stake and 70 m from the other, how wide is the pond?

90 m

70 m

?

29. *Plumbing.* Plumbers use the Pythagorean theorem to calculate pipe length. If a pipe is to be offset, as shown in the figure, the *travel*, or length, of the pipe, is calculated using the lengths of the *advance* and *offset*. Find the travel if the offset is 17.75 in. and the advance is 10.25 in.

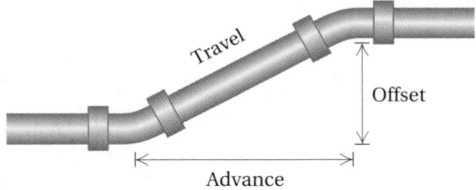

Travel

Offset

Advance

30. *Ramps for the Disabled.* Laws regarding access ramps for the disabled state that a ramp must be in the form of a right triangle, where every vertical length (leg) of 1 ft has a horizontal length (leg) of 12 ft. What is the length of a ramp with a 12-ft horizontal leg and a 1-ft vertical leg?

l

1 ft

12 ft

Skill Maintenance

Solve.   [6.8a]

31. *Commuter Travel.* The speed of the Zionsville Flash commuter train is 14 mph faster than that of the Carmel Crawler. The Flash travels 290 mi in the same time that it takes the Crawler to travel 230 mi. Find the speed of each train.

32. *Marine Travel.* A motor boat travels three times as fast as the current in the Saskatee River. A trip up the river and back takes 10 hr, and the total distance of the trip is 100 mi. Find the speed of the current.

Solve.

33. $2x^2 + 11x - 21 = 0$ [5.8b]

34. $x^2 + 24 = 11x$ [5.8b]

35. $\dfrac{x+2}{x+3} = \dfrac{x-4}{x-5}$ [6.7a]

36. $3x^2 - 12 = 0$ [5.8b]

37. $\dfrac{x-5}{x-7} = \dfrac{4}{3}$ [6.7a]

38. $\dfrac{x-1}{x-3} = \dfrac{6}{x-3}$ [6.7a]

Synthesis

39. *Roofing.* Kit's cottage, which is 24 ft wide and 32 ft long, needs a new roof. By counting clapboards that are 4 in. apart, Kit determines that the peak of the roof is 6 ft higher than the sides. If one packet of shingles covers $33\frac{1}{3}$ sq ft, how many packets will the job require?

40. *Painting.* (Refer to Exercise 39.) A gallon of paint covers about 275 sq ft. If Kit's first floor is 10 ft high, how many gallons of paint should be bought to paint the house? What assumption(s) is made in your answer?

41. *Cube Diagonal.* A cube measures 5 cm on each side. How long is the diagonal that connects two opposite corners of the cube? Give an exact answer.

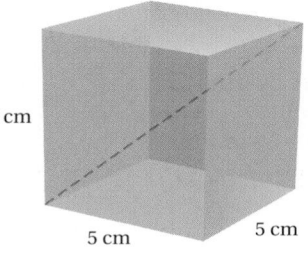

5 cm

5 cm   5 cm

42. *Wind Chill Temperature.* Because wind enhances the loss of heat from the skin, we feel colder when there is wind than when there is not. The *wind chill temperature* is what the temperature would have to be with no wind in order to give the same chilling effect as with the wind. A formula for finding the wind chill temperature, T_W, is

$$T_W = 35.74 + 0.6215T - 35.75V^{0.16} + 0.4275TV^{0.16},$$

where T is the actual temperature given by a thermometer, in degrees Fahrenheit, and V is the wind speed, in miles per hour.* Use a calculator to find the wind chill temperature in each case. Round to the nearest degree.

Source: National Weather Service

a) $T = 40°F$,
    $V = 25$ mph
b) $T = 20°F$,
    $V = 25$ mph
c) $T = 10°F$,
    $V = 20$ mph
d) $T = 10°F$,
    $V = 40$ mph
e) $T = -5°F$,
    $V = 35$ mph
f) $T = -15°F$,
    $V = 35$ mph

*This formula can be used only when the wind speed is *above* 3 mph.

10.8

The Complex Numbers

OBJECTIVES

a Express imaginary numbers as bi, where b is a nonzero real number, and complex numbers as $a + bi$, where a and b are real numbers.

b Add and subtract complex numbers.

c Multiply complex numbers.

d Write expressions involving powers of i in the form $a + bi$.

e Find conjugates of complex numbers and divide complex numbers.

f Determine whether a given complex number is a solution of an equation.

SKILL TO REVIEW
Objective 4.6a: Multiply two binomials mentally using the FOIL method.

Multiply.
1. $(w + 4)(w - 6)$
2. $(2x + 3y)(3x - 5y)$

Express in terms of i.
1. $\sqrt{-5}$ **2.** $\sqrt{-25}$

3. $-\sqrt{-11}$ **4.** $-\sqrt{-36}$

5. $\sqrt{-54}$

a Imaginary and Complex Numbers

Negative numbers do not have square roots in the real-number system. However, mathematicians have described a larger number system that contains the real-number system, such that negative numbers have square roots. That system is called the **complex-number system**. We begin by defining a number that is a square root of -1. We call this new number i.

THE COMPLEX NUMBER i

We define the number i to be $\sqrt{-1}$. That is,

$$i = \sqrt{-1} \quad \text{and} \quad i^2 = -1.$$

To express roots of negative numbers in terms of i, we can use the fact that in the complex numbers, $\sqrt{-p} = \sqrt{-1 \cdot p} = \sqrt{-1}\sqrt{p}$ when p is a positive real number.

EXAMPLES Express in terms of i.

1. $\sqrt{-7} = \sqrt{-1 \cdot 7} = \sqrt{-1} \cdot \sqrt{7} = i\sqrt{7}$, or $\sqrt{7}i$ *i is not under the radical.*

2. $\sqrt{-16} = \sqrt{-1 \cdot 16} = \sqrt{-1} \cdot \sqrt{16} = i \cdot 4 = 4i$

3. $-\sqrt{-13} = -\sqrt{-1 \cdot 13} = -\sqrt{-1} \cdot \sqrt{13} = -i\sqrt{13}$, or $-\sqrt{13}i$

4. $-\sqrt{-64} = -\sqrt{-1 \cdot 64} = -\sqrt{-1} \cdot \sqrt{64} = -i \cdot 8 = -8i$

5. $\sqrt{-48} = \sqrt{-1 \cdot 48} = \sqrt{-1} \cdot \sqrt{48} = i\sqrt{48} = i \cdot 4\sqrt{3} = 4\sqrt{3}i$, or $4i\sqrt{3}$

Do Margin Exercises 1–5.

IMAGINARY NUMBER

An **imaginary* number** is a number that can be named

    $bi,$

where b is some real number and $b \neq 0$.

To form the system of **complex numbers**, we take the imaginary numbers and the real numbers and all possible sums of real and imaginary numbers. These are complex numbers:

$$7 - 4i, \quad -\pi + 19i, \quad 37, \quad i\sqrt{8}.$$

Answers

Skill to Review:
1. $w^2 - 2w - 24$ 2. $6x^2 - xy - 15y^2$

Margin Exercises:
1. $i\sqrt{5}$, or $\sqrt{5}i$ 2. $5i$ 3. $-i\sqrt{11}$, or $-\sqrt{11}i$
4. $-6i$ 5. $3i\sqrt{6}$, or $3\sqrt{6}i$

---

*Don't let the name "imaginary" fool you. The imaginary numbers are very important in such fields as engineering and the physical sciences.

COMPLEX NUMBER

A **complex number** is any number that can be named

$$a + bi,$$

where a and b are any real numbers. (Note that either a or b or both can be 0.)

Since $0 + bi = bi$, every imaginary number is a complex number. Simi-larly, $a + 0i = a$, so every real number is a complex number. The relation-ships among various real and complex numbers are shown in the following diagram.

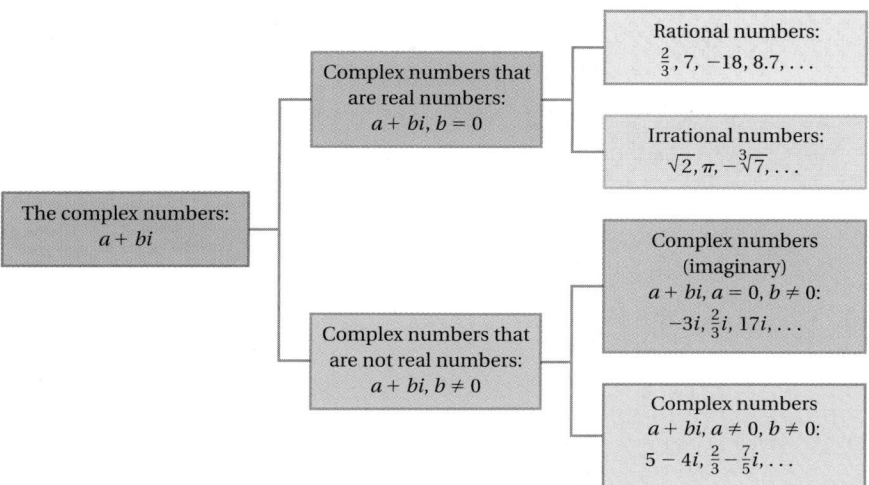

It is important to keep in mind some comparisons between numbers that have real-number roots and those that have complex-number roots that are not real. For example, $\sqrt{-48}$ is a complex number that is not a real number because we are taking the square root of a negative number. *But*, $\sqrt[3]{-125}$ is a real number because we are taking the cube root of a negative number and *any* real number has a cube root that is a real number.

b Addition and Subtraction

The complex numbers follow the commutative and associative laws of addition. Thus we can add and subtract them as we do binomials with real-number coefficients, that is, we collect like terms.

EXAMPLES Add or subtract.

6. $(8 + 6i) + (3 + 2i) = (8 + 3) + (6 + 2)i = 11 + 8i$

7. $(3 + 2i) - (5 - 2i) = (3 - 5) + [2 - (-2)]i = -2 + 4i$

> Do Exercises 6–9.

c Multiplication

The complex numbers obey the commutative, associative, and distributive laws. But although the property $\sqrt{a}\sqrt{b} = \sqrt{ab}$ does *not* hold for complex numbers in general, it does hold when $a = -1$ and b is a positive real number.

Add or subtract.

6. $(7 + 4i) + (8 - 7i)$

7. $(-5 - 6i) + (-7 + 12i)$

8. $(8 + 3i) - (5 + 8i)$

9. $(5 - 4i) - (-7 + 3i)$

Answers

6. $15 - 3i$ **7.** $-12 + 6i$ **8.** $3 - 5i$
9. $12 - 7i$

To multiply square roots of negative real numbers, we first express them in terms of i. For example,

$$\sqrt{-2} \cdot \sqrt{-5} = \sqrt{-1} \cdot \sqrt{2} \cdot \sqrt{-1} \cdot \sqrt{5} = i\sqrt{2} \cdot i\sqrt{5}$$
$$= i^2\sqrt{10} = -\sqrt{10} \quad \text{is correct!}$$

---------------- *Caution!* ----------------

The rule $\sqrt{a}\sqrt{b} = \sqrt{ab}$ holds only for nonnegative real numbers.

But $\sqrt{-2} \cdot \sqrt{-5} = \sqrt{(-2)(-5)} = \sqrt{10} \quad \text{is wrong!}$

Keeping this and the fact that $i^2 = -1$ in mind, we multiply in much the same way that we do with real numbers.

EXAMPLES Multiply.

8. $\sqrt{-49} \cdot \sqrt{-16} = \sqrt{-1} \cdot \sqrt{49} \cdot \sqrt{-1} \cdot \sqrt{16}$
$$= i \cdot 7 \cdot i \cdot 4$$
$$= i^2(28)$$
$$= (-1)(28) \quad i^2 = -1$$
$$= -28$$

9. $\sqrt{-3} \cdot \sqrt{-7} = \sqrt{-1} \cdot \sqrt{3} \cdot \sqrt{-1} \cdot \sqrt{7}$
$$= i \cdot \sqrt{3} \cdot i \cdot \sqrt{7}$$
$$= i^2(\sqrt{21})$$
$$= (-1)\sqrt{21} \quad i^2 = -1$$
$$= -\sqrt{21}$$

10. $-2i \cdot 5i = -10 \cdot i^2$
$$= (-10)(-1) \quad i^2 = -1$$
$$= 10$$

11. $(-4i)(3 - 5i) = (-4i) \cdot 3 - (-4i)(5i) \qquad$ Using a distributive law
$$= -12i + 20i^2$$
$$= -12i + 20(-1) \qquad i^2 = -1$$
$$= -12i - 20$$
$$= -20 - 12i$$

12. $(1 + 2i)(1 + 3i) = 1 + 3i + 2i + 6i^2 \qquad$ Multiplying each term of one number by every term of the other (FOIL)
$$= 1 + 3i + 2i + 6(-1) \qquad i^2 = -1$$
$$= 1 + 3i + 2i - 6$$
$$= -5 + 5i \qquad \text{Collecting like terms}$$

13. $(3 - 2i)^2 = 3^2 - 2(3)(2i) + (2i)^2 \qquad$ Squaring the binomial
$$= 9 - 12i + 4i^2$$
$$= 9 - 12i + 4(-1) \qquad i^2 = -1$$
$$= 9 - 12i - 4$$
$$= 5 - 12i$$

Do Exercises 10–17.

Multiply.

10. $\sqrt{-25} \cdot \sqrt{-4}$

11. $\sqrt{-2} \cdot \sqrt{-17}$

12. $-6i \cdot 7i$

13. $-3i(4 - 3i)$

14. $5i(-5 + 7i)$

15. $(1 + 3i)(1 + 5i)$

16. $(3 - 2i)(1 + 4i)$

17. $(3 + 2i)^2$

Answers

10. -10 **11.** $-\sqrt{34}$ **12.** 42
13. $-9 - 12i$ **14.** $-35 - 25i$
15. $-14 + 8i$ **16.** $11 + 10i$ **17.** $5 + 12i$

(d) Powers of *i*

We now want to simplify certain expressions involving powers of *i*. To do so, we first see how to simplify powers of *i*. Simplifying powers of *i* can be done by using the fact that $i^2 = -1$ and expressing the given power of *i* in terms of even powers, and then in terms of powers of i^2. Consider the following:

$$i,$$
$$i^2 = -1,$$
$$i^3 = i^2 \cdot i = (-1)i = -i,$$
$$i^4 = (i^2)^2 = (-1)^2 = 1,$$
$$i^5 = i^4 \cdot i = (i^2)^2 \cdot i = (-1)^2 \cdot i = i,$$
$$i^6 = (i^2)^3 = (-1)^3 = -1.$$

Note that the powers of *i* cycle through the values i, -1, $-i$, and 1.

EXAMPLES Simplify.

14. $i^{37} = i^{36} \cdot i = (i^2)^{18} \cdot i = (-1)^{18} \cdot i = 1 \cdot i = i$

15. $i^{58} = (i^2)^{29} = (-1)^{29} = -1$

16. $i^{75} = i^{74} \cdot i = (i^2)^{37} \cdot i = (-1)^{37} \cdot i = -1 \cdot i = -i$

17. $i^{80} = (i^2)^{40} = (-1)^{40} = 1$

Do Exercises 18–21.

Simplify.

18. i^{47}

19. i^{68}

20. i^{85}

21. i^{90}

Now let's simplify other expressions.

EXAMPLES Simplify to the form $a + bi$.

18. $8 - i^2 = 8 - (-1) = 8 + 1 = 9$

19. $17 + 6i^3 = 17 + 6 \cdot i^2 \cdot i = 17 + 6(-1)i = 17 - 6i$

20. $i^{22} - 67i^2 = (i^2)^{11} - 67(-1) = (-1)^{11} + 67 = -1 + 67 = 66$

21. $i^{23} + i^{48} = (i^{22}) \cdot i + (i^2)^{24} = (i^2)^{11} \cdot i + (-1)^{24} = (-1)^{11} \cdot i + (-1)^{24}$
$$= -i + 1 = 1 - i$$

Do Exercises 22–25.

Simplify.

22. $8 - i^5$

23. $7 + 4i^2$

24. $6i^{11} + 7i^{14}$

25. $i^{34} - i^{55}$

(e) Conjugates and Division

Conjugates of complex numbers are defined as follows.

> **CONJUGATE**
>
> The **conjugate** of a complex number $a + bi$ is $a - bi$, and the **conjugate** of $a - bi$ is $a + bi$.

Find the conjugate.
26. $6 + 3i$ **27.** $-9 - 5i$

28. $-\dfrac{1}{4}i$

EXAMPLES   Find the conjugate.

22. $5 + 7i$ The conjugate is $5 - 7i$.

23. $14 - 3i$ The conjugate is $14 + 3i$.

24. $-3 - 9i$ The conjugate is $-3 + 9i$.

25. $4i$ The conjugate is $-4i$.

Do Exercises 26–28.

> When we multiply a complex number by its conjugate, we get a real number.

EXAMPLES   Multiply.

26.
$$
\begin{aligned}
(5 + 7i)(5 - 7i) &= 5^2 - (7i)^2 &&\text{Using } (A + B)(A - B) = A^2 - B^2 \\
&= 25 - 49i^2 \\
&= 25 - 49(-1) &&i^2 = -1 \\
&= 25 + 49 \\
&= 74
\end{aligned}
$$

27.
$$
\begin{aligned}
(2 - 3i)(2 + 3i) &= 2^2 - (3i)^2 \\
&= 4 - 9i^2 \\
&= 4 - 9(-1) &&i^2 = -1 \\
&= 4 + 9 \\
&= 13
\end{aligned}
$$

Do Exercises 29 and 30.

We use conjugates when dividing complex numbers.

EXAMPLE 28   Divide and simplify to the form $a + bi$: $\dfrac{-5 + 9i}{1 - 2i}$.

$$
\begin{aligned}
\frac{-5 + 9i}{1 - 2i} \cdot \frac{1 + 2i}{1 + 2i} &= \frac{(-5 + 9i)(1 + 2i)}{(1 - 2i)(1 + 2i)} &&\text{Multiplying by 1 using the conjugate of the denominator in the symbol for 1} \\
&= \frac{-5 - 10i + 9i + 18i^2}{1^2 - 4i^2} \\
&= \frac{-5 - i + 18(-1)}{1 - 4(-1)} &&i^2 = -1 \\
&= \frac{-5 - i - 18}{1 + 4} \\
&= \frac{-23 - i}{5} \\
&= -\frac{23}{5} - \frac{1}{5}i
\end{aligned}
$$

Note the similarity between the preceding example and rationalizing denominators. In both cases, we used the conjugate of the denominator to write another name for 1. In Example 28, the symbol for the number 1 was chosen using the conjugate of the divisor, $1 - 2i$.

Multiply.
29. $(7 - 2i)(7 + 2i)$

30. $(-3 - i)(-3 + i)$

Answers

26. $6 - 3i$ **27.** $-9 + 5i$ **28.** $\dfrac{1}{4}i$ **29.** 53

30. 10

EXAMPLE 29 What symbol for 1 would you use to divide?

Division to be done

$$\frac{3 + 5i}{4 + 3i}$$

Symbol for 1

$$\frac{4 - 3i}{4 - 3i}$$

EXAMPLE 30 Divide and simplify to the form $a + bi$: $\frac{3 + 5i}{4 + 3i}$.

$$\frac{3 + 5i}{4 + 3i} \cdot \frac{4 - 3i}{4 - 3i} = \frac{(3 + 5i)(4 - 3i)}{(4 + 3i)(4 - 3i)} \qquad \text{Multiplying by 1}$$

$$= \frac{12 - 9i + 20i - 15i^2}{4^2 - 9i^2}$$

$$= \frac{12 + 11i - 15(-1)}{16 - 9(-1)} \qquad i^2 = -1$$

$$= \frac{27 + 11i}{25} = \frac{27}{25} + \frac{11}{25}i$$

Do Exercises 31 and 32.

Divide and simplify to the form $a + bi$.

31. $\frac{6 + 2i}{1 - 3i}$

32. $\frac{2 + 3i}{-1 + 4i}$

Calculator Corner

Complex Numbers   We can perform operations on complex numbers on a graphing calculator. To do so, we first set the calculator in complex, or $a + bi$, mode by pressing **MODE**, using the ⌄ and ▷ keys to position the blinking cursor over $a + bi$, and then pressing **ENTER**. We press **2ND** **QUIT** to go to the home screen. Now we can add, subtract, multiply, and divide complex numbers.

To find $(3 + 4i) - (7 - i)$, for example, we press **(** **3** **+** **4** **2ND** **i** **)** **−** **(** **7** **−** **2ND** **i** **)** **ENTER**. (i is the second operation associated with the ⊙ key.) Note that although the parentheses around $3 + 4i$ are optional, those around $7 - i$ are necessary to ensure that both parts of the second complex number are subtracted from the first number.

To find $\frac{5 - 2i}{-1 + 3i}$ and display the result using fraction notation, we press **(** **5** **−** **2** **2ND** **i** **)** **÷** **(** **(-)** **1** **+**
3 **2ND** **i** **)** **MATH** **1** **ENTER**. Since the fraction bar acts as a grouping symbol in the original expression, the parentheses must be used to group the numerator and the denominator when the expression is entered in the calculator. To find $\sqrt{-4} \cdot \sqrt{-9}$, we press **2ND**
√ **(-)** **4** **)** **×** **2ND** **√** **(-)** **9** **)** **ENTER**. Note that the calculator supplies the left parenthesis in each radicand and we supply the right parenthesis. The results of these operations are shown below.

```
(3+4i)−(7−i)
                  −4+5i
(5−2i)/(−1+3i)►Frac
            −11/10−13/10i
√(−4)*√(−9)
                     −6
■
```

Exercises: Carry out each operation.

1. $(9 + 4i) + (-11 - 13i)$

2. $(9 + 4i) - (-11 - 13i)$

3. $(9 + 4i) \cdot (-11 - 13i)$

4. $(9 + 4i) \div (-11 - 13i)$

5. $\sqrt{-16} \cdot \sqrt{-25}$

6. $\sqrt{-23} \cdot \sqrt{-35}$

7. $\frac{4 - 5i}{-6 + 8i}$

8. $(-3i)^4$

9. $(1 - i)^3 - (2 + 3i)^4$

10. $\frac{(1 - i)^3}{(2 + 3i)^2}$

Answers

31. $2i$ **32.** $\frac{10}{17} - \frac{11}{17}i$

f) Solutions of Equations

The equation $x^2 + 1 = 0$ has no real-number solution, but it has *two* nonreal complex solutions.

EXAMPLE 31 Determine whether i is a solution of the equation $x^2 + 1 = 0$.

We substitute i for x in the equation.

$$\frac{x^2 + 1 = 0}{i^2 + 1 \; ? \; 0}$$
$$\begin{array}{c|c} -1 + 1 & \\ 0 & \text{TRUE} \end{array}$$

The number i is a solution.

Do Exercise 33.

Any equation consisting of a polynomial in one variable on one side and 0 on the other has complex-number solutions. (Some may be real.) It is not always easy to find the solutions, but they always exist.

EXAMPLE 32 Determine whether $1 + i$ is a solution of the equation $x^2 - 2x + 2 = 0$.

We substitute $1 + i$ for x in the equation.

$$\frac{x^2 - 2x + 2 = 0}{(1 + i)^2 - 2(1 + i) + 2 \; ? \; 0}$$
$$\begin{array}{c|c} 1 + 2i + i^2 - 2 - 2i + 2 & \\ 1 + 2i - 1 - 2 - 2i + 2 & \\ (1 - 1 - 2 + 2) + (2 - 2)i & \\ 0 + 0i & \\ 0 & \text{TRUE} \end{array}$$

The number $1 + i$ is a solution.

EXAMPLE 33 Determine whether $2i$ is a solution of $x^2 + 3x - 4 = 0$.

$$\frac{x^2 + 3x - 4 = 0}{(2i)^2 + 3(2i) - 4 \; ? \; 0}$$
$$\begin{array}{c|c} 4i^2 + 6i - 4 & \\ -4 + 6i - 4 & \\ -8 + 6i & \text{FALSE} \end{array}$$

The number $2i$ is not a solution.

Do Exercise 34.

---

33. Determine whether $-i$ is a solution of $x^2 + 1 = 0$.

$$\frac{x^2 + 1 = 0}{?}$$

34. Determine whether $1 - i$ is a solution of $x^2 - 2x + 2 = 0$.

$$\frac{x^2 - 2x + 2 = 0}{?}$$

Answers

33. Yes **34.** Yes

a   Express in terms of i.

1. $\sqrt{-35}$

2. $\sqrt{-21}$

3. $\sqrt{-16}$

4. $\sqrt{-36}$

5. $-\sqrt{-12}$

6. $-\sqrt{-20}$

7. $\sqrt{-3}$

8. $\sqrt{-4}$

9. $\sqrt{-81}$

10. $\sqrt{-27}$

11. $\sqrt{-98}$

12. $-\sqrt{-18}$

13. $-\sqrt{-49}$

14. $-\sqrt{-125}$

15. $4 - \sqrt{-60}$

16. $6 - \sqrt{-84}$

17. $\sqrt{-4} + \sqrt{-12}$

18. $-\sqrt{-76} + \sqrt{-125}$

b   Add or subtract and simplify.

19. $(7 + 2i) + (5 - 6i)$

20. $(-4 + 5i) + (7 + 3i)$

21. $(4 - 3i) + (5 - 2i)$

22. $(-2 - 5i) + (1 - 3i)$

23. $(9 - i) + (-2 + 5i)$

24. $(6 + 4i) + (2 - 3i)$

25. $(6 - i) - (10 + 3i)$

26. $(-4 + 3i) - (7 + 4i)$

27. $(4 - 2i) - (5 - 3i)$

28. $(-2 - 3i) - (1 - 5i)$

29. $(9 + 5i) - (-2 - i)$

30. $(6 - 3i) - (2 + 4i)$

c Multiply.

31. $\sqrt{-36} \cdot \sqrt{-9}$

32. $\sqrt{-16} \cdot \sqrt{-64}$

33. $\sqrt{-7} \cdot \sqrt{-2}$

34. $\sqrt{-11} \cdot \sqrt{-3}$

35. $-3i \cdot 7i$

36. $8i \cdot 5i$

37. $-3i(-8 - 2i)$

38. $4i(5 - 7i)$

39. $(3 + 2i)(1 + i)$

40. $(4 + 3i)(2 + 5i)$

41. $(2 + 3i)(6 - 2i)$

42. $(5 + 6i)(2 - i)$

43. $(6 - 5i)(3 + 4i)$

44. $(5 - 6i)(2 + 5i)$

45. $(7 - 2i)(2 - 6i)$

46. $(-4 + 5i)(3 - 4i)$

47. $(3 - 2i)^2$

48. $(5 - 2i)^2$

49. $(1 + 5i)^2$

50. $(6 + 2i)^2$

51. $(-2 + 3i)^2$

52. $(-5 - 2i)^2$

d Simplify.

53. i^7

54. i^{11}

55. i^{24}

56. i^{35}

57. i^{42}

58. i^{64}

59. i^9

60. $(-i)^{71}$

61. i^6

62. $(-i)^4$

63. $(5i)^3$

64. $(-3i)^5$

Simplify to the form $a + bi$.

65. $7 + i^4$

66. $-18 + i^3$

67. $i^{28} - 23i$

68. $i^{29} + 33i$

69. $i^2 + i^4$

70. $5i^5 + 4i^3$

71. $i^5 + i^7$

72. $i^{84} - i^{100}$

73. $1 + i + i^2 + i^3 + i^4$

74. $i - i^2 + i^3 - i^4 + i^5$

75. $5 - \sqrt{-64}$

76. $\sqrt{-12} + 36i$

77. $\dfrac{8 - \sqrt{-24}}{4}$

78. $\dfrac{9 + \sqrt{-9}}{3}$

e   Divide and simplify to the form $a + bi$.

79. $\dfrac{4 + 3i}{3 - i}$

80. $\dfrac{5 + 2i}{2 + i}$

81. $\dfrac{3 - 2i}{2 + 3i}$

82. $\dfrac{6 - 2i}{7 + 3i}$

83. $\dfrac{8 - 3i}{7i}$

84. $\dfrac{3 + 8i}{5i}$

85. $\dfrac{4}{3 + i}$

86. $\dfrac{6}{2 - i}$

87. $\dfrac{2i}{5 - 4i}$

88. $\dfrac{8i}{6 + 3i}$

89. $\dfrac{4}{3i}$

90. $\dfrac{5}{6i}$

91. $\dfrac{2 - 4i}{8i}$

92. $\dfrac{5 + 3i}{i}$

93. $\dfrac{6 + 3i}{6 - 3i}$

94. $\dfrac{4 - 5i}{4 + 5i}$

f   Determine whether the complex number is a solution of the equation.

95. $1 - 2i$;
$$x^2 - 2x + 5 = 0$$
$$\underset{?}{\underline{\hspace{2cm}}}$$

96. $1 + 2i$;
$$x^2 - 2x + 5 = 0$$
$$\underset{?}{\underline{\hspace{2cm}}}$$

97. $2 + i$;
$$x^2 - 4x - 5 = 0$$
$$\underset{?}{\underline{\hspace{2cm}}}$$

98. $1 - i$;
$$x^2 + 2x + 2 = 0$$
$$\underset{?}{\underline{\hspace{2cm}}}$$

Skill Maintenance

In each of Exercises 99–106, fill in the blank with the correct term from the given list. Some of the choices may not be used.

99. An expression that consists of the quotient of two polynomials, where the polynomial in the denominator is nonzero, is called a(n) _____ expression.   [6.1a]

100. In the equation $(A + B)(A - B) = A^2 - B^2$, the expression $A^2 - B^2$ is called a(n) _____ .   [5.5c]

101. When being graphed, the numbers in an ordered pair are called _____ .   [3.1b]

102. Every _____ real number has two real-number square roots.   [10.1a]

103. An equality of ratios, $A/B = C/D$, read "A is to B as C is to D," is called a(n) _____ .   [6.8b]

104. A(n) _____ is a polynomial that can be expressed as a binomial squared.   [5.5a]

105. _____ numbers do not have real-number square roots.   [10.1a]

106. The principle of _____ states that if $ab = 0$, then $a = 0$ or $b = 0$ (or both).   [5.8a]

coordinates

intercepts

trinomial square

positive

negative

rational

irrational

proportion

zero products

difference of squares

cross product

Synthesis

107. A complex function g is given by
$$g(z) = \frac{z^4 - z^2}{z - 1}.$$
Find $g(2i)$, $g(1 + i)$, and $g(-1 + 2i)$.

108. Evaluate $\dfrac{1}{w - w^2}$ when $w = \dfrac{1 - i}{10}$.

Express in terms of i.

109. $\dfrac{1}{8}\left(-24 - \sqrt{-1024}\right)$

110. $12\sqrt{-\dfrac{1}{32}}$

111. $7\sqrt{-64} - 9\sqrt{-256}$

Simplify.

112. $\dfrac{i^5 + i^6 + i^7 + i^8}{(1 - i)^4}$

113. $(1 - i)^3(1 + i)^3$

114. $\dfrac{5 - \sqrt{5}\,i}{\sqrt{5}\,i}$

115. $\dfrac{6}{1 + \dfrac{3}{i}}$

116. $\left(\dfrac{1}{2} - \dfrac{1}{3}i\right)^2 - \left(\dfrac{1}{2} + \dfrac{1}{3}i\right)^2$

117. $\dfrac{i - i^{38}}{1 + i}$

118. Find all numbers a for which the opposite of a is the same as the reciprocal of a.

Summary and Review

Key Terms and Properties

square of a number, p. 702
square root, p. 702
principal square root, p. 702
radical symbol, p. 703
radical expression, p. 703
radicand, p. 703

cube root, p. 706
index, p. 707
odd root, p. 707
even root, p. 708
rationalizing the denominator, p. 737
conjugates, p. 739

radical equation, p. 742
complex-number system, p. 760
complex number i, p. 760
imaginary number, p. 760

$\sqrt{a^2} = |a|$; $\sqrt[k]{a^k} = |a|$, when k is even; $\sqrt[k]{a^k} = a$, when k is odd;

$\sqrt[k]{ab} = \sqrt[k]{a} \cdot \sqrt[k]{b}$; $\sqrt[k]{\dfrac{a}{b}} = \dfrac{\sqrt[k]{a}}{\sqrt[k]{b}}$; $a^{1/n} = \sqrt[n]{a}$;

$a^{m/n} = \sqrt[n]{a^m} = \left(\sqrt[n]{a}\right)^m$; $a^{-m/n} = \dfrac{1}{a^{m/n}}$

Principle of Powers: If $a = b$ is true, then $a^n = b^n$ is true.
Pythagorean Theorem: $a^2 + b^2 = c^2$, in a right triangle.
$i = \sqrt{-1}$, $i^2 = -1$, $i^3 = -i$, $i^4 = 1$

Imaginary Numbers: $bi, i^2 = -1, b \neq 0$
Complex Numbers: $a + bi, i^2 = -1$
Conjugates: $a + bi, a - bi$

Concept Reinforcement

Determine whether each statement is true or false.

_____ **1.** For any negative number a, we have $\sqrt{a^2} = -a$. [10.1a]

_____ **2.** For any real numbers $\sqrt[m]{a}$ and $\sqrt[n]{b}$, $\sqrt[m]{a} \cdot \sqrt[n]{b} = \sqrt[mn]{ab}$. [10.3a]

_____ **3.** For any real numbers $\sqrt[n]{a}$ and $\sqrt[n]{b}$, $\sqrt[n]{a} + \sqrt[n]{b} = \sqrt[n]{a + b}$. [10.4a]

_____ **4.** If $x^2 = 4$, then $x = 2$. [10.6a]

_____ **5.** All real numbers are complex numbers, but not every complex number is a real number. [10.8a]

_____ **6.** The product of a complex number and its conjugate is always a real number. [10.8e]

Important Concepts

Objective 10.1b   Simplify radical expressions with perfect-square radicands.

Example Simplify: $\sqrt{16x^2}$.
$\sqrt{16x^2} = \sqrt{(4x)^2} = |4x| = |4| \cdot |x| = 4|x|$

Example Simplify: $\sqrt{x^2 - 6x + 9}$.
$\sqrt{x^2 - 6x + 9} = \sqrt{(x - 3)^2} = |x - 3|$

Practice Exercises

1. Simplify: $\sqrt{36y^2}$.

2. Simplify: $\sqrt{a^2 + 4a + 4}$.

Objective 10.2a Write expressions with or without rational exponents, and simplify, if possible.

Example Rewrite $x^{1/4}$ without a rational exponent.
  Recall that $a^{1/n}$ means $\sqrt[n]{a}$. Then
  $$x^{1/4} = \sqrt[4]{x}.$$

Example Rewrite $\left(\sqrt[3]{4xy^2}\right)^4$ with a rational exponent.
  Recall that $\left(\sqrt[n]{a}\right)^m$ means $a^{m/n}$. Then
  $$\left(\sqrt[3]{4xy^2}\right)^4 = (4xy^2)^{4/3}.$$

Practice Exercises

3. Rewrite $z^{3/5}$ without a rational exponent.

4. Rewrite $\left(\sqrt{6ab}\right)^5$ with a rational exponent.

Objective 10.2b Write expressions without negative exponents, and simplify, if possible.

Example Rewrite $8^{-2/3}$ with a positive exponent, and simplify, if possible.

  Recall that $a^{-m/n}$ means $\dfrac{1}{a^{m/n}}$. Then

  $$8^{-2/3} = \frac{1}{8^{2/3}} = \frac{1}{\left(\sqrt[3]{8}\right)^2} = \frac{1}{2^2} = \frac{1}{4}.$$

Practice Exercise

5. Rewrite $9^{-3/2}$ with a positive exponent, and simplify, if possible.

Objective 10.2d Use rational exponents to simplify radical expressions.

Example Use rational exponents to simplify: $\sqrt[6]{x^2y^4}$.
  $$\begin{aligned}
  \sqrt[6]{x^2y^4} &= (x^2y^4)^{1/6} \\
  &= x^{2/6}y^{4/6} \\
  &= x^{1/3}y^{2/3} \\
  &= (xy^2)^{1/3} \\
  &= \sqrt[3]{xy^2}
  \end{aligned}$$

Practice Exercise

6. Use rational exponents to simplify: $\sqrt[8]{a^6b^2}$.

Objective 10.3a Multiply and simplify radical expressions.

Example Multiply and simplify: $\sqrt[3]{6xy^2}\,\sqrt[3]{9y}$.
  $$\begin{aligned}
  \sqrt[3]{6xy^2}\,\sqrt[3]{9y} &= \sqrt[3]{6xy^2 \cdot 9y} \\
  &= \sqrt[3]{54xy^3} \\
  &= \sqrt[3]{27y^3 \cdot 2x} \\
  &= \sqrt[3]{27y^3}\,\sqrt[3]{2x} \\
  &= 3y\sqrt[3]{2x}
  \end{aligned}$$

Practice Exercise

7. Multiply and simplify. Assume that all expressions under radicals represent nonnegative numbers.
  $$\sqrt{5y}\sqrt{30y}$$

Objective 10.3b Divide and simplify radical expressions.

Example Divide and simplify: $\dfrac{\sqrt{24x^5}}{\sqrt{6x}}$.

  $$\frac{\sqrt{24x^5}}{\sqrt{6x}} = \sqrt{\frac{24x^5}{6x}} = \sqrt{4x^4} = 2x^2$$

Practice Exercise

8. Divide and simplify: $\dfrac{\sqrt{20a}}{\sqrt{5}}$.

Objective 10.4a   Add or subtract with radical notation and simplify.

Example   Subtract: $5\sqrt{2} - 4\sqrt{8}$.
$$
\begin{aligned}
5\sqrt{2} - 4\sqrt{8} &= 5\sqrt{2} - 4\sqrt{4 \cdot 2} \\
&= 5\sqrt{2} - 4\sqrt{4}\sqrt{2} \\
&= 5\sqrt{2} - 4 \cdot 2\sqrt{2} = 5\sqrt{2} - 8\sqrt{2} \\
&= (5 - 8)\sqrt{2} = -3\sqrt{2}
\end{aligned}
$$

Practice Exercise

9. Subtract: $\sqrt{48} - 2\sqrt{3}$.

Objective 10.4b   Multiply expressions involving radicals in which some factors contain more than one term.

Example   Multiply: $\left(3 - \sqrt{6}\right)\left(2 + 4\sqrt{6}\right)$.

We use FOIL:
$$
\begin{aligned}
\left(3 - \sqrt{6}\right)&\left(2 + 4\sqrt{6}\right) \\
&= 3 \cdot 2 + 3 \cdot 4\sqrt{6} - \sqrt{6} \cdot 2 - \sqrt{6} \cdot 4\sqrt{6} \\
&= 6 + 12\sqrt{6} - 2\sqrt{6} - 4 \cdot 6 \\
&= 6 + 12\sqrt{6} - 2\sqrt{6} - 24 \\
&= -18 + 10\sqrt{6}.
\end{aligned}
$$

Practice Exercise

10. Multiply: $\left(5 - \sqrt{x}\right)^2$.

Objective 10.6a   Solve radical equations with one radical term.

Example   Solve: $x = \sqrt{x - 2} + 4$.

First, we subtract 4 on both sides to isolate the radical. Then we square both sides of the equation.
$$
\begin{aligned}
x &= \sqrt{x - 2} + 4 \\
x - 4 &= \sqrt{x - 2} \\
(x - 4)^2 &= \left(\sqrt{x - 2}\right)^2 \\
x^2 - 8x + 16 &= x - 2 \\
x^2 - 9x + 18 &= 0 \\
(x - 3)(x - 6) &= 0 \\
x - 3 = 0 \quad &or \quad x - 6 = 0 \\
x = 3 \quad &or \qquad x = 6
\end{aligned}
$$
We must check both possible solutions. When we do, we find that 6 checks, but 3 does not. Thus the solution is 6.

Practice Exercise

11. Solve: $3 + \sqrt{x - 1} = x$.

Objective 10.6b   Solve radical equations with two radical terms.

Example   Solve: $1 = \sqrt{x + 9} - \sqrt{x}$.
$$
\begin{aligned}
1 &= \sqrt{x + 9} - \sqrt{x} \\
\sqrt{x} + 1 &= \sqrt{x + 9} \qquad \text{Isolating one radical} \\
\left(\sqrt{x} + 1\right)^2 &= \left(\sqrt{x + 9}\right)^2 \qquad \text{Squaring both sides} \\
x + 2\sqrt{x} + 1 &= x + 9 \\
2\sqrt{x} &= 8 \qquad \text{Isolating the remaining radical} \\
\sqrt{x} &= 4 \\
\left(\sqrt{x}\right)^2 &= 4^2 \\
x &= 16
\end{aligned}
$$
The number 16 checks. It is the solution.

Practice Exercise

12. Solve: $\sqrt{x + 3} - \sqrt{x - 2} = 1$.

Objective 10.8c   Multiply complex numbers.

Example   Multiply: $(3 - 2i)(4 + i)$.

$(3 - 2i)(4 + i) = 12 + 3i - 8i - 2i^2$ Using FOIL

$= 12 + 3i - 8i - 2(-1)$

$= 12 + 3i - 8i + 2$

$= 14 - 5i$

Practice Exercise

13. Multiply: $(2 - 5i)^2$.

Objective 10.8e   Find conjugates of complex numbers and divide complex numbers.

Example   Divide and simplify to the form $a + bi$:

$\dfrac{5 - i}{4 + 3i}.$

The conjugate of the denominator is $4 - 3i$, so we multiply by 1 using $\dfrac{4 - 3i}{4 - 3i}$:

$\dfrac{5 - i}{4 + 3i} = \dfrac{5 - i}{4 + 3i} \cdot \dfrac{4 - 3i}{4 - 3i}$

$= \dfrac{20 - 15i - 4i + 3i^2}{16 - 9i^2}$

$= \dfrac{20 - 19i + 3(-1)}{16 - 9(-1)}$

$= \dfrac{20 - 19i - 3}{16 + 9}$

$= \dfrac{17 - 19i}{25} = \dfrac{17}{25} - \dfrac{19}{25}i.$

Practice Exercise

14. Divide and simplify to the form $a + bi$: $\dfrac{3 - 2i}{2 + i}.$

Review Exercises

Use a calculator to approximate to three decimal places. [10.1a]

1. $\sqrt{778}$

2. $\sqrt{\dfrac{963.2}{23.68}}$

3. For the given function, find the indicated function values.   [10.1a]
$f(x) = \sqrt{3x - 16}; \quad f(0), f(-1), f(1), \text{ and } f\left(\tfrac{41}{3}\right)$

4. Find the domain of the function f in Exercise 3.   [10.1a]

Simplify. Assume that letters represent *any* real number. [10.1b]

5. $\sqrt{81a^2}$

6. $\sqrt{(-7z)^2}$

7. $\sqrt{(6 - b)^2}$

8. $\sqrt{x^2 + 6x + 9}$

Simplify.   [10.1c]

9. $\sqrt[3]{-1000}$

10. $\sqrt[3]{-\dfrac{1}{27}}$

11. For the given function, find the indicated function values.   [10.1c]
$f(x) = \sqrt[3]{x + 2}; \quad f(6), f(-10), \text{ and } f(25)$

Simplify. Assume that letters represent *any* real number. [10.1d]

12. $\sqrt[10]{x^{10}}$

13. $-\sqrt[13]{(-3)^{13}}$

Rewrite without rational exponents, and simplify, if possible. [10.2a]

14. $a^{1/5}$

15. $64^{3/2}$

Rewrite with rational exponents. [10.2a]

16. $\sqrt{31}$ **17.** $\sqrt[5]{a^2b^3}$

Rewrite with positive exponents, and simplify, if possible. [10.2b]

18. $49^{-1/2}$ **19.** $(8xy)^{-2/3}$

20. $5a^{-3/4}b^{1/2}c^{-2/3}$ **21.** $\dfrac{3a}{\sqrt[4]{t}}$

Use the laws of exponents to simplify. Write answers with positive exponents. [10.2c]

22. $(x^{-2/3})^{3/5}$ **23.** $\dfrac{7^{-1/3}}{7^{-1/2}}$

Use rational exponents to simplify. Write the answer in radical notation if appropriate. [10.2d]

24. $\sqrt[3]{x^{21}}$ **25.** $\sqrt[3]{27x^6}$

Use rational exponents to write a single radical expression. [10.2d]

26. $x^{1/3}y^{1/4}$ **27.** $\sqrt[4]{x}\sqrt[3]{x}$

Simplify by factoring. Assume that all expressions under radicals represent nonnegative numbers. [10.3a]

28. $\sqrt{245}$ **29.** $\sqrt[3]{-108}$

30. $\sqrt[3]{250a^2b^6}$

Simplify. Assume that no radicands were formed by raising negative numbers to even powers. [10.3b]

31. $\sqrt{\dfrac{49}{36}}$ **32.** $\sqrt[3]{\dfrac{64x^6}{27}}$

33. $\sqrt[4]{\dfrac{16x^8}{81y^{12}}}$

Perform the indicated operations and simplify. Assume that no radicands were formed by raising negative numbers to even powers. [10.3a, b], [10.4a]

34. $\sqrt{5x}\sqrt{3y}$ **35.** $\sqrt[3]{a^5b}\sqrt[3]{27b}$

36. $\sqrt[3]{a}\sqrt[5]{b^3}$ **37.** $\dfrac{\sqrt[3]{60xy^3}}{\sqrt[3]{10x}}$

38. $\dfrac{\sqrt{75x}}{2\sqrt{3}}$ **39.** $\dfrac{\sqrt[3]{x^2}}{\sqrt[4]{x}}$

40. $5\sqrt[3]{x} + 2\sqrt[3]{x}$ **41.** $2\sqrt{75} - 7\sqrt{3}$

42. $\sqrt{50} + 2\sqrt{18} + \sqrt{32}$ **43.** $\sqrt[3]{8x^4} + \sqrt[3]{xy^6}$

Multiply. [10.4b]

44. $\left(\sqrt{5} - 3\sqrt{8}\right)\left(\sqrt{5} + 2\sqrt{8}\right)$

45. $\left(1 - \sqrt{7}\right)^2$

46. $\left(\sqrt[3]{27} - \sqrt[3]{2}\right)\left(\sqrt[3]{27} + \sqrt[3]{2}\right)$

Rationalize the denominator. [10.5a, b]

47. $\sqrt{\dfrac{8}{3}}$ **48.** $\dfrac{2}{\sqrt{a} + \sqrt{b}}$

Solve. [10.6a, b]

49. $x - 3 = \sqrt{5 - x}$ **50.** $\sqrt[4]{x + 3} = 2$

51. $\sqrt{x + 8} - \sqrt{3x + 1} = 1$

Automotive Repair. For an engine with a displacement of 2.8 L, the function given by

$$d(n) = 0.75\sqrt{2.8n}$$

can be used to determine the diameter of the carburetor's opening, $d(n)$, in millimeters, where n is the number of rpm's at which the engine achieves peak performance. [10.6c]

Source: macdizzy.com

52. 🖩 If a carburetor's opening is 81 mm, for what number of rpm's will the engine produce peak power?

53. 🖩 If a carburetor's opening is 84 mm, for what number of rpm's will the engine produce peak power?

54. *Length of a Side of a Square.* The diagonal of a square has length $9\sqrt{2}$ cm. Find the length of a side of the square. [10.7a]

55. *Bookcase Width.* A bookcase is 5 ft tall and has a 7-ft diagonal brace, as shown. How wide is the bookcase? [10.7a]

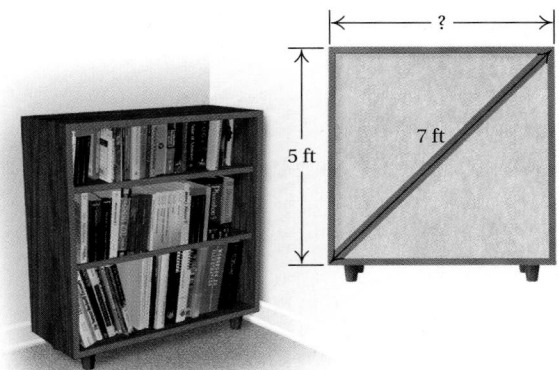

In a right triangle, find the length of the side not given. Give an exact answer and an answer to three decimal places. [10.7a]

56. $a = 7,\ b = 24$

57. $a = 2,\ c = 5\sqrt{2}$

58. Express in terms of i: $\sqrt{-25} + \sqrt{-8}$. [10.8a]

Add or subtract. [10.8b]

59. $(-4 + 3i) + (2 - 12i)$

60. $(4 - 7i) - (3 - 8i)$

Multiply. [10.8c, d]

61. $(2 + 5i)(2 - 5i)$

62. i^{13}

63. $(6 - 3i)(2 - i)$

Divide. [10.8e]

64. $\dfrac{-3 + 2i}{5i}$

65. $\dfrac{1 - 2i}{3 + i}$

66. Graph: $f(x) = \sqrt{x}$. [10.1a]

67. Which of the following is a solution of $x^2 + 4x + 5 = 0$? [10.8f]

A. $1 - i$ **B.** $1 + i$

C. $2 + i$ **D.** $-2 + i$

Synthesis

68. Simplify: $i \cdot i^2 \cdot i^3 \cdots i^{99} \cdot i^{100}$. [10.8c, d]

69. Solve: $\sqrt{11x + \sqrt{6 + x}} = 6$. [10.6a]

Understanding Through Discussion and Writing

1. Find the domain of
$$f(x) = (x + 5)^{1/2}(x + 7)^{-1/2}$$
and explain how you found your answer. [10.1a], [10.2b]

2. 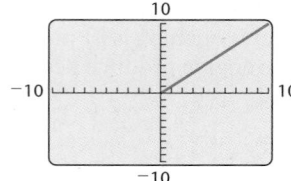 Ron is puzzled. When he uses a graphing calculator to graph $y = \sqrt{x} \cdot \sqrt{x}$, he gets the following screen. Explain why Ron did not get the complete line $y = x$. [10.1a], [10.3a]

3. In what way(s) is collecting like radical terms the same as collecting like monomial terms? [10.4a]

4. Is checking solutions of equations necessary when the principle of powers is used with an odd power n? Why or why not? [10.1d], [10.6a, b]

5. A student *incorrectly* claims that
$$\frac{5 + \sqrt{2}}{\sqrt{18}} = \frac{5 + \sqrt{1}}{\sqrt{9}} = \frac{5 + 1}{3} = 2.$$
How could you convince the student that a mistake has been made? How would you explain the correct way of rationalizing the denominator? [10.5a]

6. How are conjugates of complex numbers similar to the conjugates used in Section 10.5? [10.8e]

Test

For Extra Help

CHAPTER
Test Prep
VIDEOS

Step-by-step test solutions are found on the Chapter Test Prep Videos available via the Video Resources on DVD, in *MyMathLab* , and on YouTube (search "BittingerIntroInter" and click on "Channels").

1. Use a calculator to approximate $\sqrt{148}$ to three decimal places.

2. For the given function, find the indicated function values.
$$f(x) = \sqrt{8 - 4x}; \quad f(1) \text{ and } f(3)$$

3. Find the domain of the function f in Exercise 2.

Simplify. Assume that letters represent *any* real number.

4. $\sqrt{(-3q)^2}$

5. $\sqrt{x^2 + 10x + 25}$

6. $\sqrt[3]{-\dfrac{1}{1000}}$

7. $\sqrt[5]{x^5}$

8. $\sqrt[10]{(-4)^{10}}$

Rewrite without rational exponents, and simplify, if possible.

9. $a^{2/3}$

10. $32^{3/5}$

Rewrite with rational exponents.

11. $\sqrt{37}$

12. $\left(\sqrt{5xy^2}\right)^5$

Rewrite with positive exponents, and simplify, if possible.

13. $1000^{-1/3}$

14. $8a^{3/4}b^{-3/2}c^{-2/5}$

Use the laws of exponents to simplify. Write answers with positive exponents.

15. $(x^{2/3}y^{-3/4})^{12/5}$

16. $\dfrac{2.9^{-5/8}}{2.9^{2/3}}$

Use rational exponents to simplify. Write the answer in radical notation if appropriate. Assume that no radicands were formed by raising negative numbers to even powers.

17. $\sqrt[8]{x^2}$

18. $\sqrt[4]{16x^6}$

Use rational exponents to write a single radical expression.

19. $a^{2/5}b^{1/3}$

20. $\sqrt[4]{2y}\sqrt[3]{y}$

Simplify by factoring. Assume that no radicands were formed by raising negative numbers to even powers.

21. $\sqrt{148}$

22. $\sqrt[4]{80}$

23. $\sqrt[3]{24a^{11}b^{13}}$

Simplify. Assume that no radicands were formed by raising negative numbers to even powers.

24. $\sqrt[3]{\dfrac{16x^5}{y^6}}$

25. $\sqrt{\dfrac{25x^2}{36y^4}}$

Perform the indicated operations and simplify. Assume that no radicands were formed by raising negative numbers to even powers.

26. $\sqrt[3]{2x}\sqrt[3]{5y^2}$

27. $\sqrt[4]{x^3y^2}\sqrt[4]{xy}$

28. $\dfrac{\sqrt[5]{x^3y^4}}{\sqrt[5]{xy^2}}$

29. $\dfrac{\sqrt{300a}}{5\sqrt{3}}$

30. Add: $3\sqrt{128} + 2\sqrt{18} + 2\sqrt{32}$.

Multiply.

31. $\left(\sqrt{20} + 2\sqrt{5}\right)\left(\sqrt{20} - 3\sqrt{5}\right)$

32. $\left(3 + \sqrt{x}\right)^2$

33. Rationalize the denominator: $\dfrac{1 + \sqrt{2}}{3 - 5\sqrt{2}}$.

Solve.

34. $\sqrt[5]{x - 3} = 2$

35. $\sqrt{x - 6} = \sqrt{x + 9} - 3$

36. $\sqrt{x - 1} + 3 = x$

37. *Length of a Side of a Square.* The diagonal of a square has length $7\sqrt{2}$ ft. Find the length of a side of the square.

38. *Sighting to the Horizon.* A person can see 72 mi to the horizon from an airplane window. How high is the airplane? Use the formula $D = 1.2\sqrt{h}$, where D is in miles and h is in feet.

In a right triangle, find the length of the side not given. Give an exact answer and an answer to three decimal places.

39. $a = 7, \; b = 7$

40. $a = 1, \; c = \sqrt{5}$

41. Express in terms of i: $\sqrt{-9} + \sqrt{-64}$.

42. Subtract: $(5 + 8i) - (-2 + 3i)$.

Multiply.

43. $(3 - 4i)(3 + 7i)$

44. i^{95}

45. Divide: $\dfrac{-7 + 14i}{6 - 8i}$.

46. Determine whether $1 + 2i$ is a solution of $x^2 + 2x + 5 = 0$.

47. Which of the following describes the solution(s) of the equation $x - 4 = \sqrt{x - 2}$?
A. There is exactly one solution, and it is positive.
B. There are one positive solution and one negative solution.
C. There are two positive solutions.
D. There is no solution.

Synthesis

48. Simplify: $\dfrac{1 - 4i}{4i(1 + 4i)^{-1}}$.

49. Solve: $\sqrt{2x - 2} + \sqrt{7x + 4} = \sqrt{13x + 10}$.

Cumulative Review

Simplify. Assume that no radicands were formed by raising negative numbers to even powers.

1. $(2x^2 - 3x + 1) + (6x - 3x^3 + 7x^2 - 4)$

2. $(2x^2 - y)^2$

3. $(5x^2 - 2x + 1)(3x^2 + x - 2)$

4. $\dfrac{x^3 + 64}{x^2 - 49} \cdot \dfrac{x^2 - 14x + 49}{x^2 - 4x + 16}$

5. $\dfrac{\dfrac{y^2 - 5y - 6}{y^2 - 7y - 18}}{\dfrac{y^2 + 3y + 2}{y^2 + 4y + 4}}$

6. $\dfrac{x}{x + 2} + \dfrac{1}{x - 3} - \dfrac{x^2 - 2}{x^2 - x - 6}$

7. $(y^3 + 3y^2 - 5) \div (y + 2)$

8. $\sqrt[3]{-8x^3}$

9. $\sqrt{16x^2 - 32x + 16}$

10. $9\sqrt{75} + 6\sqrt{12}$

11. $\sqrt{2xy^2} \cdot \sqrt{8xy^3}$

12. $\dfrac{3\sqrt{5}}{\sqrt{6} - \sqrt{3}}$

13. $\sqrt[6]{\dfrac{m^{12}n^{24}}{64}}$

14. $6^{2/9} \cdot 6^{2/3}$

15. $(6 + i) - (3 - 4i)$

16. $\dfrac{2 - i}{6 + 5i}$

Solve.

17. $\dfrac{1}{5} + \dfrac{3}{10}x = \dfrac{4}{5}$

18. $M = \dfrac{1}{8}(c - 3)$, for c

19. $3a - 4 < 10 + 5a$

20. $-8 < x + 2 < 15$

21. $|3x - 6| = 2$

22. $625 = 49y^2$

23. $3x + 5y = 30,$
$5x + 3y = 34$

24. $3x + 2y - z = -7,$
$-x + y + 2z = 9,$
$5x + 5y + z = -1$

25. $\dfrac{6x}{x - 5} - \dfrac{300}{x^2 + 5x + 25} = \dfrac{2250}{x^3 - 125}$

26. $\dfrac{3x^2}{x + 2} + \dfrac{5x - 22}{x - 2} = \dfrac{-48}{x^2 - 4}$

27. $I = \dfrac{nE}{R + nr}$, for R

28. $\sqrt{4x + 1} - 2 = 3$

29. $2\sqrt{1 - x} = \sqrt{5}$

30. $13 - x = 5 + \sqrt{x + 4}$

Graph.

31. $f(x) = -\dfrac{2}{3}x + 2$

32. $4x - 2y = 8$

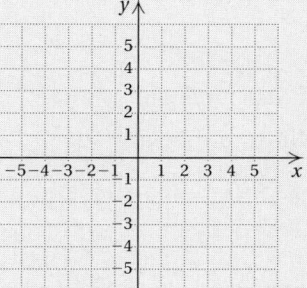

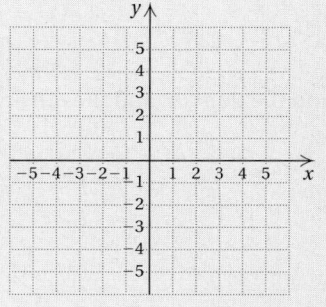

33. $4x \geq 5y + 20$

34. $y \geq -3,$
$y \leq 2x + 3$

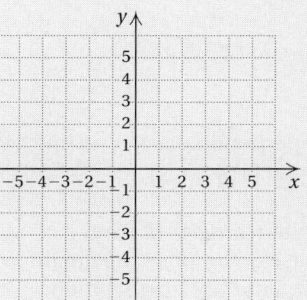

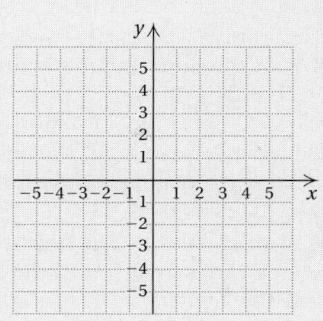

35. $g(x) = x^2 - x - 2$

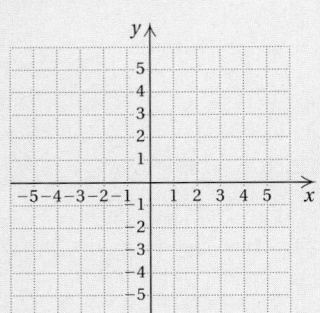

36. $f(x) = |x + 4|$

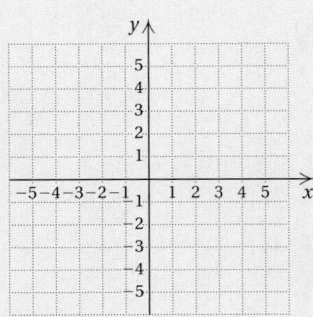

37. $g(x) = \dfrac{4}{x - 3}$

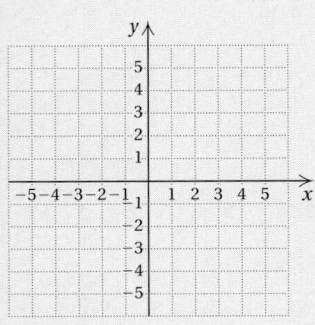

38. $f(x) = 2 - \sqrt{x}$

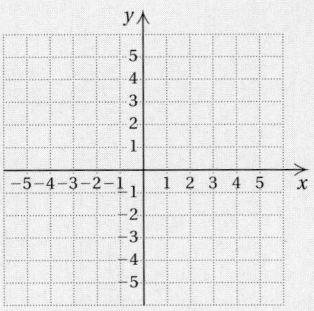

Factor.

39. $12x^2y^2 - 30xy^3$

40. $3x^2 - 17x - 28$

41. $y^2 - y - 132$

42. $27y^3 + 8$

43. $4x^2 - 625$

Find the domain and the range of each function.

44.

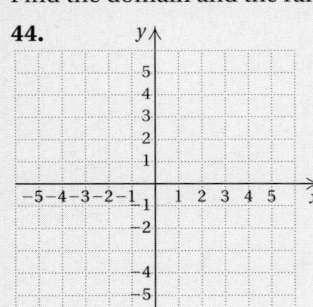

45.

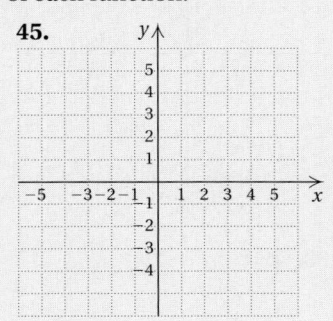

46. Find the slope and the y-intercept of the line
$3x - 2y = 8$.

47. Find an equation for the line perpendicular to the line
$3x - y = 5$ and passing through $(1, 4)$.

48. *Triangle Area.* The height h of triangles of fixed
area varies inversely as the base b. Suppose the height
is 100 ft when the base is 20 ft. Find the height when
the base is 16 ft. What is the fixed area?

Solve.

49. *Harvesting Time.* One combine can harvest a field in
3 hr. Another combine can harvest the same field in
1.5 hr. How long should it take them to harvest the
field together?

50. *Warning Dye.* A warning dye is used by people in
lifeboats to aid search planes. The volume V of the dye
used varies directly as the square of the diameter d of
the circular area formed by the dye in the water. If 4 L
of dye is required for a 10-m wide circle, how much dye
is needed for a 40-m wide circle?

51. Rewrite with rational exponents: $\sqrt[5]{xy^4}$.

A. $\dfrac{1}{(xy^4)^5}$ **B.** $(xy^4)^5$

C. $(xy)^{4/5}$ **D.** $(xy^4)^{1/5}$

52. A grain bin can be filled in 3 hr if the grain enters
through spout A alone or in 15 hr if the grain enters
through spout B alone. If grain is entering through
both spouts at the same time, how many hours will it
take to fill the bin?

A. $\frac{5}{2}$ hr   **B.** 9 hr

C. $22\frac{1}{2}$ hr   **D.** $10\frac{1}{2}$ hr

53. Divide: $(x^3 - x^2 + 2x + 4) \div (x - 3)$.

A. $x^2 + 2x + 8$, R -28 **B.** $x^2 + 2x - 4$, R -8

C. $x^2 - 4x - 10$, R -26 **D.** $x^2 - 4x + 14$, R 46

54. Solve: $2x + 6 = 8 + \sqrt{5x + 1}$.

A. $\frac{1}{4}$ **B.** 3

C. $3, \frac{1}{4}$ **D.** $4, 3$

Synthesis

55. Solve: $\dfrac{x + \sqrt{x + 1}}{x - \sqrt{x + 1}} = \dfrac{5}{11}$.

Quadratic Equations and Functions

Real-World Application

Canoes are deepest at the middle of the center line, with the depth decreasing to zero at the edges. Lou and Jen own a company that specializes in producing custom canoes. A customer provided suggested guidelines for measures of the depths D, in inches, along the center line of the canoe at distances x, in inches, from the edge. The measures are listed in a table on p. 849. Make a scatterplot of the data and decide whether the data seem to fit a quadratic function. Use data points to find a quadratic function that fits the data and use the function to estimate the depth of the canoe 10 in. from the edge along the center line.

This problem appears as Example 7 in Section 11.7.

11.1

The Basics of Solving Quadratic Equations

SKILL TO REVIEW

Objective 5.8b: Solve quadratic equations by factoring and then using the principle of zero products.

Solve.

1. $x^2 + 6x - 16 = 0$

2. $6x^2 - 13x - 5 = 0$

---

1. Consider solving the equation
$$x^2 - 6x + 8 = 0.$$
Below is the graph of
$$f(x) = x^2 - 6x + 8.$$

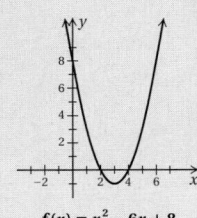

$f(x) = x^2 - 6x + 8$

a) What are the x-intercepts of the graph?

b) What are the solutions of $x^2 - 6x + 8 = 0$?

c) What relationship exists between the answers to parts (a) and (b)?

Answers

Answers to Skill to Review Exercises 1 and 2 and Margin Exercise 1 are on p. 783.

※ Algebraic–Graphical Connection

Let's reexamine the graphical connections to the algebraic equation-solving concepts we have studied before.

In Chapter 7, we introduced the graph of a quadratic function:

$$f(x) = ax^2 + bx + c, \quad a \neq 0.$$

For example, the graph of the function $f(x) = x^2 + 6x + 8$ and its x-intercepts are shown below.

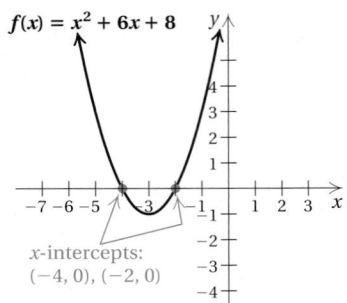

$f(x) = x^2 + 6x + 8$

x-intercepts: $(-4, 0), (-2, 0)$

The x-intercepts are $(-4, 0)$ and $(-2, 0)$. These pairs are also the points of intersection of the graphs of $f(x) = x^2 + 6x + 8$ and $g(x) = 0$ (the x-axis). We will analyze the graphs of quadratic functions in greater detail in Sections 11.5–11.7.

In Chapter 5, we solved quadratic equations like $x^2 + 6x + 8 = 0$ using factoring, as here:

$$x^2 + 6x + 8 = 0$$
$$(x + 4)(x + 2) = 0 \qquad \text{Factoring}$$
$$x + 4 = 0 \quad or \quad x + 2 = 0 \qquad \text{Using the principle of zero products}$$
$$x = -4 \quad or \quad x = -2.$$

We see that the solutions of $x^2 + 6x + 8 = 0$, -4 and -2, are the first coordinates of the x-intercepts, $(-4, 0)$ and $(-2, 0)$, of the graph of $f(x) = x^2 + 6x + 8$.

※

Do Margin Exercise 1.

We now extend our ability to solve quadratic equations.

a The Principle of Square Roots

The quadratic equation

$$5x^2 + 8x - 2 = 0$$

is said to be in **standard form**. The quadratic equation

$$5x^2 = 2 - 8x$$

is equivalent to the preceding equation, but it is *not* in standard form.

QUADRATIC EQUATION

An equation of the type $ax^2 + bx + c = 0$, where a, b, and c are real-number constants and $a > 0$, is called the **standard form of a quadratic equation**.

To find the standard form of the quadratic equation $-5x^2 + 4x - 7 = 0$, we find an equivalent equation by multiplying by -1 on both sides:

$$-1(-5x^2 + 4x - 7) = -1(0)$$
$$5x^2 - 4x + 7 = 0. \quad \text{Writing in standard form}$$

In Section 5.8, we studied the use of factoring and the principle of zero products to solve certain quadratic equations. Let's review that procedure and introduce a new one.

EXAMPLE 1

a) Solve: $x^2 = 25$.

b) Find the x-intercepts of $f(x) = x^2 - 25$.

a) We first find standard form and then factor:

$$x^2 - 25 = 0 \qquad \text{Subtracting 25}$$
$$(x - 5)(x + 5) = 0 \qquad \text{Factoring}$$
$$x - 5 = 0 \quad or \quad x + 5 = 0 \qquad \text{Using the principle of zero products}$$
$$x = 5 \quad or \qquad x = -5.$$

The solutions are 5 and -5.

b) The x-intercepts of $f(x) = x^2 - 25$ are $(-5, 0)$ and $(5, 0)$. The solutions of the equation $x^2 = 25$ are the first coordinates of the x-intercepts of the graph of $f(x) = x^2 - 25$.

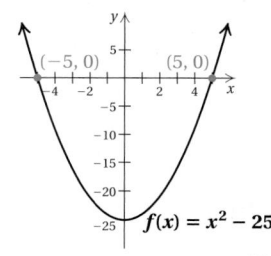

EXAMPLE 2 Solve: $6x^2 - 15x = 0$.

We factor and use the principle of zero products:

$$6x^2 - 15x = 0$$
$$3x(2x - 5) = 0$$
$$3x = 0 \quad or \quad 2x - 5 = 0$$
$$x = 0 \quad or \qquad 2x = 5$$
$$x = 0 \quad or \qquad x = \tfrac{5}{2}.$$

The solutions are 0 and $\frac{5}{2}$. The check is left to the student.

Do Exercises 2 and 3.

2. a) Solve: $x^2 = 16$.

b) Find the x-intercepts of $f(x) = x^2 - 16$.

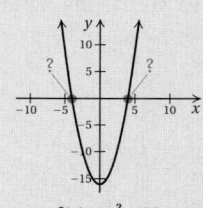

$f(x) = x^2 - 16$

3. a) Solve: $4x^2 + 14x = 0$.

b) Find the x-intercepts of $f(x) = 4x^2 + 14x$.

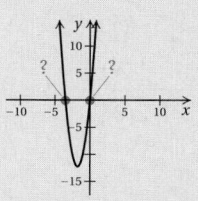

$f(x) = 4x^2 + 14x$

Answers

Skill to Review:

1. $-8, 2$ 2. $-\dfrac{1}{3}, \dfrac{5}{2}$

Margin Exercises:
1. **(a)** $(2, 0), (4, 0)$; **(b)** $2, 4$; **(c)** The solutions of $x^2 - 6x + 8 = 0$, 2 and 4, are the first coordinates of the x-intercepts, $(2, 0)$ and $(4, 0)$, of the graph of $f(x) = x^2 - 6x + 8$.
2. **(a)** 4 and -4; **(b)** $(-4, 0), (4, 0)$
3. **(a)** $0, -\frac{7}{2}$; **(b)** $\left(-\frac{7}{2}, 0\right), (0, 0)$

EXAMPLE 3

a) Solve: $3x^2 = 2 - x$.

b) Find the x-intercepts of $f(x) = 3x^2 + x - 2$.

a) We first find standard form. Then we factor and use the principle of zero products.

$$3x^2 = 2 - x$$

$$3x^2 + x - 2 = 0 \qquad \text{Adding } x \text{ and subtracting 2 to get the standard form}$$

$$(x + 1)(3x - 2) = 0 \qquad \text{Factoring}$$

$$x + 1 = 0 \quad or \quad 3x - 2 = 0 \qquad \text{Using the principle of zero products}$$

$$x = -1 \quad or \qquad 3x = 2$$

$$x = -1 \quad or \qquad x = \tfrac{2}{3}$$

Check: For -1:

$$\begin{array}{c|c} \multicolumn{2}{c}{3x^2 = 2 - x} \\ \hline 3(-1)^2 \ ? \ 2 - (-1) & \\ 3 \cdot 1 & 2 + 1 \\ 3 & 3 \qquad \text{TRUE} \end{array}$$

For $\tfrac{2}{3}$:

$$\begin{array}{c|c} \multicolumn{2}{c}{3x^2 = 2 - x} \\ \hline 3\left(\tfrac{2}{3}\right)^2 \ ? \ 2 - \left(\tfrac{2}{3}\right) & \\ 3 \cdot \tfrac{4}{9} & \tfrac{6}{3} - \tfrac{2}{3} \\ \tfrac{4}{3} & \tfrac{4}{3} \qquad \text{TRUE} \end{array}$$

The solutions are -1 and $\tfrac{2}{3}$.

b) The x-intercepts of $f(x) = 3x^2 + x - 2$ are $(-1, 0)$ and $\left(\tfrac{2}{3}, 0\right)$. The solutions of the equation $3x^2 = 2 - x$ are the first coordinates of the x-intercepts of the graph of $f(x) = 3x^2 + x - 2$.

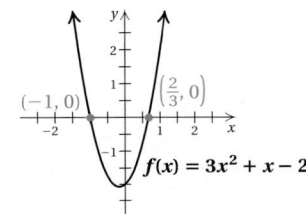

$$f(x) = 3x^2 + x - 2$$

Do Exercise 4.

---

4. a) Solve: $5x^2 = 8x - 3$.

b) Find the x-intercepts of $f(x) = 5x^2 - 8x + 3$.

$$f(x) = 5x^2 - 8x + 3$$

Solving Equations of the Type $x^2 = d$

Consider the equation $x^2 = 25$ again. We know from Chapter 6 that the number 25 has two real-number square roots, namely, 5 and -5. Note that these are the solutions of the equation in Example 1. This exemplifies the principle of square roots, which provides a quick method for solving equations of the type $x^2 = d$.

> ### THE PRINCIPLE OF SQUARE ROOTS
>
> The solutions of the equation $x^2 = d$ are $\sqrt{d}$ and $-\sqrt{d}$.
>
> When $d > 0$, the solutions are two real numbers.
> When $d = 0$, the only solution is 0.
> When $d < 0$, the solutions are two imaginary numbers.

EXAMPLE 4 Solve: $3x^2 = 6$. Give the exact solutions and approximate the solutions to three decimal places.

We have

$$3x^2 = 6$$
$$x^2 = 2$$
$$x = \sqrt{2} \quad or \quad x = -\sqrt{2}.$$

We often use the symbol $\pm\sqrt{2}$ to represent both of the solutions.

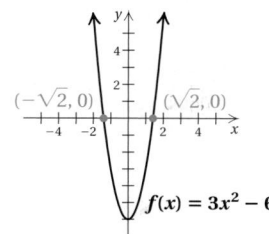

$f(x) = 3x^2 - 6$

Check: For $\sqrt{2}$:

$$\begin{array}{c|c} 3x^2 = 6 \\ \hline 3(\sqrt{2})^2 \stackrel{?}{\,} 6 \\ 3 \cdot 2 \\ 6 & \text{TRUE} \end{array}$$

For $-\sqrt{2}$:

$$\begin{array}{c|c} 3x^2 = 6 \\ \hline 3(-\sqrt{2})^2 \stackrel{?}{\,} 6 \\ 3 \cdot 2 \\ 6 & \text{TRUE} \end{array}$$

The solutions are $\sqrt{2}$ and $-\sqrt{2}$, or $\pm\sqrt{2}$, which are about 1.414 and -1.414, or ±1.414, when rounded to three decimal places.

Do Exercise 5.

5. Solve: $5x^2 = 15$. Give the exact solution and approximate the solutions to three decimal places.

Sometimes we rationalize denominators to simplify answers.

EXAMPLE 5 Solve: $-5x^2 + 2 = 0$. Give the exact solutions and approximate the solutions to three decimal places.

$$-5x^2 + 2 = 0$$
$$x^2 = \frac{2}{5} \qquad \text{Subtracting 2 and dividing by } -5$$
$$x = \sqrt{\frac{2}{5}} \quad or \quad x = -\sqrt{\frac{2}{5}} \qquad \text{Using the principle of square roots}$$
$$x = \sqrt{\frac{2}{5} \cdot \frac{5}{5}} \quad or \quad x = -\sqrt{\frac{2}{5} \cdot \frac{5}{5}} \qquad \text{Rationalizing the denominators}$$
$$x = \frac{\sqrt{10}}{5} \quad or \quad x = -\frac{\sqrt{10}}{5}$$

Check: We check both numbers at once, since there is no x-term in the equation. We could have checked both numbers at once in Example 4 as well.

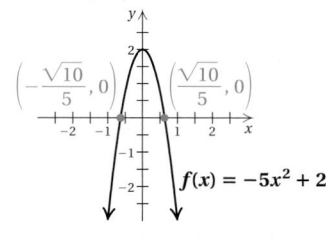

$f(x) = -5x^2 + 2$

$$\begin{array}{c|c} -5x^2 + 2 = 0 \\ \hline -5\left(\pm\dfrac{\sqrt{10}}{5}\right)^2 + 2 \stackrel{?}{\,} 0 \\ -5\left(\dfrac{10}{25}\right) + 2 \\ -2 + 2 \\ 0 & \text{TRUE} \end{array}$$

The solutions are $\dfrac{\sqrt{10}}{5}$ and $-\dfrac{\sqrt{10}}{5}$, or $\pm\dfrac{\sqrt{10}}{5}$. We can use a calculator for approximations:

$$\pm\frac{\sqrt{10}}{5} \approx \pm0.632.$$

Answer

5. $\sqrt{3}$ and $-\sqrt{3}$, or $\pm\sqrt{3}$; 1.732 and -1.732, or ±1.732

6. Solve: $-3x^2 + 8 = 0$. Give the exact solution and approximate the solutions to three decimal places.

Do Exercise 6.

Sometimes we get solutions that are imaginary numbers.

EXAMPLE 6   Solve: $4x^2 + 9 = 0$.

$$4x^2 + 9 = 0$$

$$x^2 = -\frac{9}{4} \qquad \text{Subtracting 9 and dividing by 4}$$

$$x = \sqrt{-\frac{9}{4}} \quad or \quad x = -\sqrt{-\frac{9}{4}} \qquad \text{Using the principle of square roots}$$

$$x = \frac{3}{2}i \quad or \quad x = -\frac{3}{2}i \qquad \text{Simplifying}$$

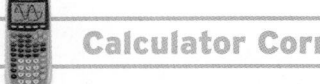

Calculator Corner

Imaginary Solutions of Quadratic Equations

What happens when you use the ZERO feature to solve the equation in Example 6? Explain why this happens.

Check:

$$\begin{array}{c|c} 4x^2 + 9 = 0 \\ \hline 4\left(\pm\frac{3}{2}i\right)^2 + 9 \; ? \; 0 \\ 4\left(-\frac{9}{4}\right) + 9 \\ -9 + 9 \\ 0 \; \bigg| \; \text{TRUE} \end{array}$$

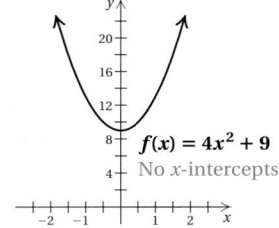

$f(x) = 4x^2 + 9$
No x-intercepts

The solutions are $\frac{3}{2}i$ and $-\frac{3}{2}i$, or $\pm\frac{3}{2}i$.

We see that the graph of $f(x) = 4x^2 + 9$ does not cross the x-axis. This is true because the equation $4x^2 + 9 = 0$ has *imaginary* complex-number solutions. Only real-number solutions correspond to x-intercepts.

7. Solve: $2x^2 + 1 = 0$.

Do Exercise 7.

Solving Equations of the Type $(x + c)^2 = d$

The equation $(x - 2)^2 = 7$ can also be solved using the principle of square roots.

EXAMPLE 7

a) Solve: $(x - 2)^2 = 7$.

b) Find the x-intercepts of $f(x) = (x - 2)^2 - 7$.

8. a) Solve: $(x - 1)^2 = 5$.

b) Find the x-intercepts of $f(x) = (x - 1)^2 - 5$.

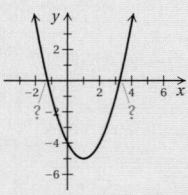

$f(x) = (x - 1)^2 - 5$

a) We have

$$(x - 2)^2 = 7$$

$$x - 2 = \sqrt{7} \qquad or \qquad x - 2 = -\sqrt{7} \qquad \text{Using the principle of square roots}$$

$$x = 2 + \sqrt{7} \quad or \qquad x = 2 - \sqrt{7}.$$

The solutions are $2 + \sqrt{7}$ and $2 - \sqrt{7}$, or $2 \pm \sqrt{7}$.

b) The x-intercepts of $f(x) = (x - 2)^2 - 7$ are $\left(2 - \sqrt{7}, 0\right)$ and $\left(2 + \sqrt{7}, 0\right)$.

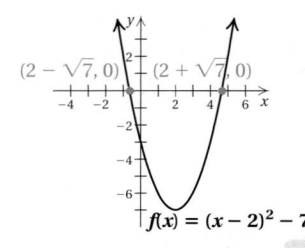

$(2 - \sqrt{7}, 0)$ $(2 + \sqrt{7}, 0)$

$f(x) = (x - 2)^2 - 7$

Answers

6. $\dfrac{2\sqrt{6}}{3}$ and $-\dfrac{2\sqrt{6}}{3}$, or $\pm\dfrac{2\sqrt{6}}{3}$; 1.633 and -1.633, or ± 1.633 **7.** $\dfrac{\sqrt{2}}{2}i$ and $-\dfrac{\sqrt{2}}{2}i$, or $\pm\dfrac{\sqrt{2}}{2}i$ **8. (a)** $1 \pm \sqrt{5}$; **(b)** $\left(1 - \sqrt{5}, 0\right)$, $\left(1 + \sqrt{5}, 0\right)$

Do Exercise 8.

If we can express the left side of an equation as the square of a binomial, we can proceed as we did in Example 7.

EXAMPLE 8 Solve: $x^2 + 6x + 9 = 2$.

We have

$$x^2 + 6x + 9 = 2 \qquad \text{The left side is the square of a binomial.}$$
$$(x + 3)^2 = 2$$
$$x + 3 = \sqrt{2} \qquad or \quad x + 3 = -\sqrt{2} \qquad \text{Using the principle of square roots}$$
$$x = -3 + \sqrt{2} \quad or \qquad x = -3 - \sqrt{2}.$$

The solutions are $-3 + \sqrt{2}$ and $-3 - \sqrt{2}$, or $-3 \pm \sqrt{2}$.

Do Exercise 9.

9. Solve: $x^2 + 16x + 64 = 11$.

b Completing the Square

We can solve quadratic equations like $3x^2 = 6$ and $(x - 2)^2 = 7$ by using the principle of square roots. We can also solve an equation such as $x^2 + 6x + 9 = 2$ in like manner because the expression on the left side is the square of a binomial, $(x + 3)^2$. This second procedure is the basis for a method called **completing the square**. *It can be used to solve any quadratic equation.*

Suppose we have the following quadratic equation:

$$x^2 + 14x = 4.$$

If we could add on both sides of the equation a constant that would make the expression on the left the square of a binomial, we could then solve the equation using the principle of square roots.

How can we determine what to add to $x^2 + 14x$ to construct the square of a binomial? We want to find a number a such that the following equation is satisfied:

$$x^2 + 14x + a^2 = (x + a)(x + a) = x^2 + 2ax + a^2.$$

Thus, a is such that $2a = 14$. Solving, we get $a = 7$. That is, a is half of the coefficient of x in $x^2 + 14x$. Since $a^2 = \left(\frac{14}{2}\right)^2 = 7^2 = 49$, we add 49 to our original expression:

$$x^2 + 14x + 49 \text{ is the square of } x + 7;$$

that is,

$$x^2 + 14x + 49 = (x + 7)^2.$$

> **COMPLETING THE SQUARE**
>
> When solving an equation, to **complete the square** of an expression like $x^2 + bx$, we take half the x-coefficient, which is $b/2$, and square it. Then we add that number, $(b/2)^2$, on both sides of the equation.

Answer

9. $-8 \pm \sqrt{11}$

Returning to solving our original equation, we first add 49 on *both* sides to *complete the square* on the left. Then we solve:

$$x^2 + 14x \qquad = 4 \qquad \text{Original equation}$$

$$x^2 + 14x + 49 = 4 + 49 \qquad \text{Adding 49: } \left(\tfrac{14}{2}\right)^2 = 7^2 = 49$$

$$(x + 7)^2 = 53$$

$$x + 7 = \sqrt{53} \qquad or \quad x + 7 = -\sqrt{53} \qquad \text{Using the principle of square roots}$$

$$x = -7 + \sqrt{53} \quad or \qquad x = -7 - \sqrt{53}.$$

The solutions are $-7 \pm \sqrt{53}$.

We have seen that a quadratic equation $(x + c)^2 = d$ can be solved using the principle of square roots. Any equation, such as $x^2 - 6x + 8 = 0$, can be put in this form by completing the square. Then we can solve as before.

EXAMPLE 9 Solve: $x^2 - 6x + 8 = 0$.

We have

$$x^2 - 6x + 8 = 0$$

$$x^2 - 6x \qquad = -8. \qquad \text{Subtracting 8}$$

We take half of -6 and square it, to get 9. Then we add 9 on *both* sides of the equation. This makes the left side the square of a binomial, $x - 3$. We have now *completed the square*.

$$x^2 - 6x + 9 = -8 + 9 \qquad \text{Adding 9: } \left(\tfrac{-6}{2}\right)^2 = (-3)^2 = 9$$

$$(x - 3)^2 = 1$$

$$x - 3 = 1 \quad or \quad x - 3 = -1 \qquad \text{Using the principle of square roots}$$

$$x = 4 \quad or \qquad x = 2$$

The solutions are 2 and 4.

Solve.
10. $x^2 + 6x + 8 = 0$

11. $x^2 - 8x - 20 = 0$

Do Exercises 10 and 11.

EXAMPLE 10 Solve $x^2 + 4x - 7 = 0$ by completing the square.

We have

$$x^2 + 4x - 7 = 0$$

$$x^2 + 4x \qquad = 7 \qquad \text{Adding 7}$$

$$x^2 + 4x + 4 = 7 + 4 \qquad \text{Adding 4: } \left(\tfrac{4}{2}\right)^2 = (2)^2 = 4$$

$$(x + 2)^2 = 11$$

$$x + 2 = \sqrt{11} \qquad or \quad x + 2 = -\sqrt{11} \qquad \text{Using the principle of square roots}$$

$$x = -2 + \sqrt{11} \quad or \qquad x = -2 - \sqrt{11}.$$

The solutions are $-2 \pm \sqrt{11}$.

12. Solve by completing the square:
$$x^2 + 6x - 1 = 0.$$

Do Exercise 12.

When the coefficient of x^2 is not 1, we can make it 1, as shown in the following example.

EXAMPLE 11 Solve $3x^2 + 7x = 2$ by completing the square.

We have

$$3x^2 + 7x = 2$$

$$\frac{1}{3}(3x^2 + 7x) = \frac{1}{3} \cdot 2 \qquad \text{Multiplying by } \tfrac{1}{3} \text{ to make the } x^2\text{-coefficient 1}$$

$$x^2 + \frac{7}{3}x = \frac{2}{3} \qquad \text{Multiplying and simplifying}$$

$$x^2 + \frac{7}{3}x + \frac{49}{36} = \frac{2}{3} + \frac{49}{36} \qquad \text{Adding } \frac{49}{36}: \left[\frac{1}{2} \cdot \frac{7}{3}\right]^2 = \frac{49}{36}$$

$$\left(x + \frac{7}{6}\right)^2 = \frac{24}{36} + \frac{49}{36} \qquad \text{Finding a common denominator}$$

$$\left(x + \frac{7}{6}\right)^2 = \frac{73}{36}$$

$$x + \frac{7}{6} = \sqrt{\frac{73}{36}} \qquad or \quad x + \frac{7}{6} = -\sqrt{\frac{73}{36}} \qquad \text{Using the principle of square roots}$$

$$x + \frac{7}{6} = \frac{\sqrt{73}}{6} \qquad or \quad x + \frac{7}{6} = -\frac{\sqrt{73}}{6}$$

$$x = -\frac{7}{6} + \frac{\sqrt{73}}{6} \quad or \qquad x = -\frac{7}{6} - \frac{\sqrt{73}}{6}.$$

The solutions are $-\dfrac{7}{6} \pm \dfrac{\sqrt{73}}{6}$.

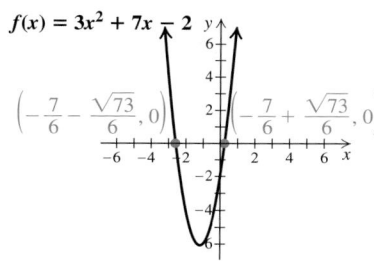

$$f(x) = 3x^2 + 7x - 2$$

$$\left(-\frac{7}{6} - \frac{\sqrt{73}}{6}, 0\right) \qquad \left(-\frac{7}{6} + \frac{\sqrt{73}}{6}, 0\right)$$

Do Exercises 13 and 14.

Solve by completing the square.

13. $2x^2 + 6x = 5$

14. $3x^2 - 2x = 7$

EXAMPLE 12 Solve $2x^2 = 3x - 7$ by completing the square.

$$2x^2 = 3x - 7$$

$$2x^2 - 3x = -7 \qquad \text{Subtracting } 3x$$

$$\frac{1}{2}(2x^2 - 3x) = \frac{1}{2} \cdot (-7) \qquad \text{Multiplying by } \tfrac{1}{2} \text{ to make the } x^2\text{-coefficient 1}$$

$$x^2 - \frac{3}{2}x = -\frac{7}{2} \qquad \text{Multiplying and simplifying}$$

$$x^2 - \frac{3}{2}x + \frac{9}{16} = -\frac{7}{2} + \frac{9}{16} \qquad \text{Adding } \tfrac{9}{16}: \left[\frac{1}{2}\left(-\frac{3}{2}\right)\right]^2 = \left[-\frac{3}{4}\right]^2 = \frac{9}{16}$$

$$\left(x - \frac{3}{4}\right)^2 = -\frac{56}{16} + \frac{9}{16} \qquad \text{Finding a common denominator}$$

$$\left(x - \frac{3}{4}\right)^2 = -\frac{47}{16}$$

$$x - \frac{3}{4} = \sqrt{-\frac{47}{16}} \quad or \quad x - \frac{3}{4} = -\sqrt{-\frac{47}{16}} \qquad \text{Using the principle of square roots}$$

$$x - \frac{3}{4} = \frac{\sqrt{47}}{4}i \quad or \quad x - \frac{3}{4} = -\frac{\sqrt{47}}{4}i \qquad \sqrt{-1} = i$$

$$x = \frac{3}{4} + \frac{\sqrt{47}}{4}i \quad or \quad x = \frac{3}{4} - \frac{\sqrt{47}}{4}i$$

The solutions are $\dfrac{3}{4} \pm \dfrac{\sqrt{47}}{4}i.$

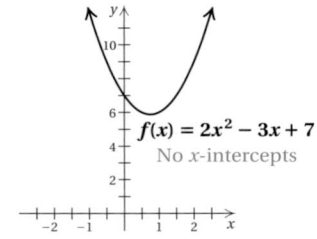

$f(x) = 2x^2 - 3x + 7$
No x-intercepts

We see that the graph of $f(x) = 2x^2 - 3x + 7$ does not cross the x-axis. This is true because the equation $2x^2 = 3x - 7$ has nonreal complex-number solutions.

Do Exercise 15.

15. Solve by completing the square:
$$3x^2 = 2x - 1.$$

Answer

15. $\dfrac{1}{3} \pm \dfrac{\sqrt{2}}{3}i$

SOLVING BY COMPLETING THE SQUARE

To solve an equation $ax^2 + bx + c = 0$ by completing the square:

1. If $a \neq 1$, multiply by $1/a$ so that the x^2-coefficient is 1.

2. If the x^2-coefficient is 1, add or subtract so that the equation is in the form

$$x^2 + bx = -c, \quad or \quad x^2 + \frac{b}{a}x = -\frac{c}{a} \text{ if step (1) has been applied.}$$

3. Take half of the x-coefficient and square it. Add the result on both sides of the equation.

4. Express the side with the variables as the square of a binomial.

5. Use the principle of square roots and complete the solution.

Completing the square provides a base for proving the quadratic formula in Section 11.2.

(c) Applications and Problem Solving

EXAMPLE 13 *Hang Time.* One of the most exciting plays in basketball is the dunk shot. The amount of time T that passes from the moment a player leaves the ground, goes up, makes the shot, and arrives back on the ground is called *hang time*. A function relating an athlete's vertical leap V, in inches, to hang time T, in seconds, is given by

$$V(T) = 48T^2.$$

a) Hall-of-Famer Michael Jordan had a hang time of about 0.889 sec. What was his vertical leap?

b) Although his height is only 5 ft 7 in., Spud Webb, formerly of the Sacramento Kings, had a vertical leap of about 44 in. What was his hang time?
Source: www.vertcoach.com/highest-vertical-leap.html

a) To find Jordan's vertical leap, we substitute 0.889 for T in the function and compute V:

$$V(0.889) = 48(0.889)^2 \approx 37.9 \text{ in.}$$

Jordan's vertical leap was about 37.9 in. Surprisingly, Jordan did not have the vertical leap most fans would expect.

b) To find Webb's hang time, we substitute 44 for V and solve for T:

$$44 = 48T^2 \qquad \text{Substituting 44 for } V$$

$$\frac{44}{48} = T^2 \qquad \text{Solving for } T^2$$

$$0.91\overline{6} = T^2$$

$$\sqrt{0.91\overline{6}} = T \qquad \text{Hang time is positive.}$$

$$0.957 \approx T. \qquad \text{Using a calculator}$$

Webb's hang time was 0.957 sec. Note that his hang time was greater than Jordan's.

Do Exercises 16 and 17.

16. Vertical Leap. Larry Bird, currently President of Basketball Operations for the Indiana Pacers, played for the Boston Celtics from 1979 though 1992. He had a hang time of about 0.764 sec. What was his vertical leap?

17. Hang Time. Vince Carter of the Orlando Magic has a vertical leap of 43 in. What is his hang time?

Answers
16. About 28 in. **17.** About 0.946 sec

a

1. a) Solve:
$6x^2 = 30$.
b) Find the x-intercepts of $f(x) = 6x^2 - 30$.

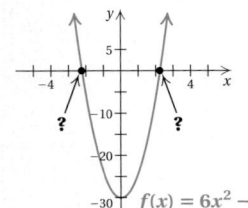

2. a) Solve:
$5x^2 = 35$.
b) Find the x-intercepts of $f(x) = 5x^2 - 35$.

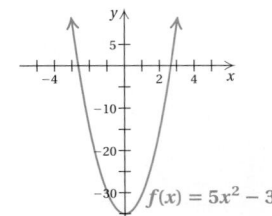

3. a) Solve:
$9x^2 + 25 = 0$.
b) Find the x-intercepts of $f(x) = 9x^2 + 25$.

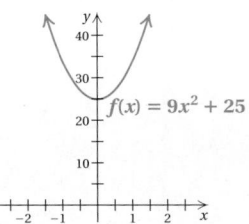

4. a) Solve:
$36x^2 + 49 = 0$.
b) Find the x-intercepts of $f(x) = 36x^2 + 49$.

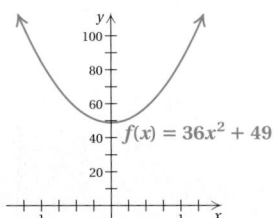

Solve. Give the exact solution and approximate solutions to three decimal places, when appropriate.

5. $2x^2 - 3 = 0$

6. $3x^2 - 7 = 0$

7. $(x + 2)^2 = 49$

8. $(x - 1)^2 = 6$

9. $(x - 4)^2 = 16$

10. $(x + 3)^2 = 9$

11. $(x - 11)^2 = 7$

12. $(x - 9)^2 = 34$

13. $(x - 7)^2 = -4$

14. $(x + 1)^2 = -9$

15. $(x - 9)^2 = 81$

16. $(t - 2)^2 = 25$

17. $\left(x - \frac{3}{2}\right)^2 = \frac{7}{2}$

18. $\left(y + \frac{3}{4}\right)^2 = \frac{17}{16}$

19. $x^2 + 6x + 9 = 64$

20. $x^2 + 10x + 25 = 100$

21. $y^2 - 14y + 49 = 4$

22. $p^2 - 8p + 16 = 1$

b Solve by completing the square. Show your work.

23. $x^2 + 4x = 2$

24. $x^2 + 2x = 5$

25. $x^2 - 22x = 11$

26. $x^2 - 18x = 10$

27. $x^2 + x = 1$

28. $x^2 - x = 3$

29. $t^2 - 5t = 7$

30. $y^2 + 9y = 8$

31. $x^2 + \frac{3}{2}x = 3$

32. $x^2 - \frac{4}{3}x = \frac{2}{3}$

33. $m^2 - \frac{9}{2}m = \frac{3}{2}$

34. $r^2 + \frac{2}{5}r = \frac{4}{5}$

35. $x^2 + 6x - 16 = 0$

36. $x^2 - 8x + 15 = 0$

37. $x^2 + 22x + 102 = 0$ **38.** $x^2 + 18x + 74 = 0$ **39.** $x^2 - 10x - 4 = 0$ **40.** $x^2 + 10x - 4 = 0$

41. a) Solve:
$x^2 + 7x - 2 = 0$.
b) Find the
x-intercepts of
$f(x) = x^2 + 7x - 2$.

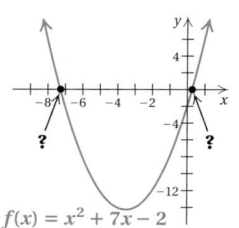

$f(x) = x^2 + 7x - 2$

42. a) Solve:
$x^2 - 7x - 2 = 0$.
b) Find the
x-intercepts of
$f(x) = x^2 - 7x - 2$.

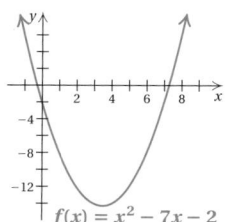

$f(x) = x^2 - 7x - 2$

43. a) Solve:
$2x^2 - 5x + 8 = 0$.
b) Find the
x-intercepts of
$f(x) = 2x^2 - 5x + 8$.

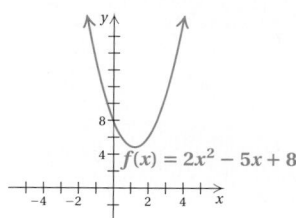

$f(x) = 2x^2 - 5x + 8$

44. a) Solve:
$2x^2 - 3x + 9 = 0$.
b) Find the
x-intercepts of
$f(x) = 2x^2 - 3x + 9$.

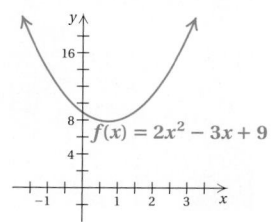

$f(x) = 2x^2 - 3x + 9$

Solve by completing the square. Show your work.

45. $x^2 - \frac{3}{2}x - \frac{1}{2} = 0$ **46.** $x^2 + \frac{3}{2}x - 2 = 0$ **47.** $2x^2 - 3x - 17 = 0$ **48.** $2x^2 + 3x - 1 = 0$

49. $3x^2 - 4x - 1 = 0$ **50.** $3x^2 + 4x - 3 = 0$ **51.** $x^2 + x + 2 = 0$ **52.** $x^2 - x + 1 = 0$

53. $x^2 - 4x + 13 = 0$ **54.** $x^2 - 6x + 13 = 0$

C *Hang Time.* For Exercises 55 and 56, use the hang-time function $V(T) = 48T^2$, relating vertical leap to hang time.

55. The NBA's Shaquille O'Neal, of the Cleveland Cavaliers, has a vertical leap of about 32 in. What is his hang time?

56. The NBA's Antonio McDyess, of the San Antonio Spurs, has a vertical leap of 42 in. What is his hang time?

Free-Falling Objects.   The function $s(t) = 16t^2$ is used to approximate the distance s, in feet, that an object falls freely from rest in t seconds. Use the formula for Exercises 57–62.

57. Reaching 745 ft above the water, the towers of California's Golden Gate Bridge are the world's tallest bridge towers. How long would it take an object to fall freely from the top?

58. Suspended 1053 ft above the water, the bridge over Colorado's Royal Gorge is the world's highest bridge. How long would it take an object to fall freely from the bridge?

59. The Washington Monument, near the west end of the National Mall in Washington, D.C., is the world's tallest stone structure and the world's tallest obelisk. It is 555.427 ft tall. How long would it take an object to fall freely from the top of the monument?

60. The Gateway Arch in St. Louis is 640 ft high. How long would it take an object to fall freely from the top?

61. The Millau viaduct is part of the E11 expressway connecting Paris and Barcelona. The viaduct has the tallest piers ever constructed. The tallest pier is 804 ft high. How long would it take an object to fall freely from the viaduct?

62. Completed in 2009, the Burj Dubai, in downtown Dubai, is the tallest free-standing structure in the world. It is 2684 ft tall. How long would it take an object to fall freely from the top?

Skill Maintenance

63. *Record Births.*   The following table lists data regarding the number of births in the United States in 1930 and in 2007.   [7.5e]

NUMBER OF YEARS SINCE 1930	NUMBER OF BIRTHS IN THE UNITED STATES (in millions)
0	2.6
77	4.3

Source: National Center for Health Statistics

a) Use the two data points to find a linear function $B(t) = mt + b$ that fits the data.
b) Use the function to estimate the number of births in 2012.
c) In what year will there be 4.5 million births?

Graph. [7.1c], [7.4a]

64. $f(x) = 5 - 2x^2$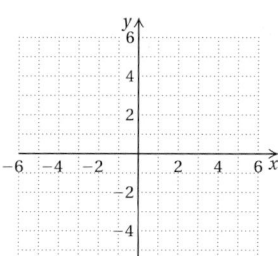

65. $f(x) = 5 - 2x$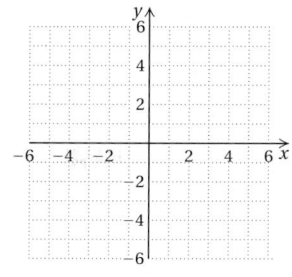

66. $2x - 5y = 10$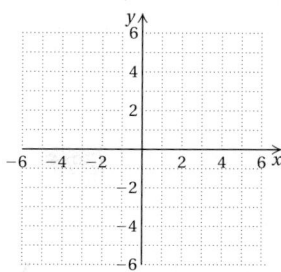

67. $f(x) = |5 - 2x|$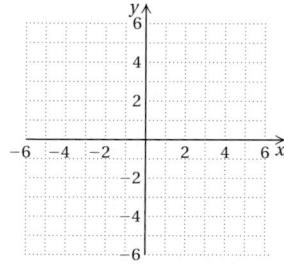

68. Simplify: $\sqrt{88}$. [10.3a]

69. Rationalize the denominator: $\sqrt{\dfrac{2}{5}}$. [10.5a]

Solve. [10.6a, b]

70. $\sqrt{5x - 4} + \sqrt{13 - x} = 7$

71. $\sqrt{4x - 4} = \sqrt{x + 4} + 1$

72. $\sqrt{7x - 5} = \sqrt{4x + 7}$

73. $-35 = \sqrt{2x + 5}$

Synthesis

74.  Use a graphing calculator to solve each of the following equations.
   a) $25.55x^2 - 1635.2 = 0$
   b) $-0.0644x^2 + 0.0936x + 4.56 = 0$
   c) $2.101x + 3.121 = 0.97x^2$

75. Problems such as those in Exercises 17, 21, and 25 can be solved without first finding standard form by using the INTERSECT feature on a graphing calculator. We let y_1 = the left side of the equation and y_2 = the right side. Use a graphing calculator to solve Exercises 17, 21, and 25 in this manner.

Find b such that the trinomial is a square.

76. $x^2 + bx + 75$

77. $x^2 + bx + 64$

Solve.

78. $\left(x - \frac{1}{3}\right)\left(x - \frac{1}{3}\right) + \left(x - \frac{1}{3}\right)\left(x + \frac{2}{9}\right) = 0$

79. $x(2x^2 + 9x - 56)(3x + 10) = 0$

80. *Boating.* A barge and a fishing boat leave a dock at the same time, traveling at right angles to each other. The barge travels 7 km/h slower than the fishing boat. After 4 hr, the boats are 68 km apart. Find the speed of each vessel.

68 km

11.2

The Quadratic Formula

OBJECTIVE

a Solve quadratic equations using the quadratic formula, and approximate solutions using a calculator.

SKILL TO REVIEW

Objective 10.8a: Express imaginary numbers as *bi*, where *b* is a nonzero real number, and complex numbers as *a* + *bi*, where *a* and *b* are real numbers.

Express in terms of *i*.

1. $\sqrt{-100}$
2. $10 - \sqrt{-68}$

There are at least two reasons for learning to complete the square. One is to enhance your ability to graph certain equations that are needed to solve problems in Section 11.7. The other is to prove a general formula for solving quadratic equations.

a Solving Using the Quadratic Formula

Each time you solve by completing the square, the procedure is the same. When we do the same kind of procedure many times, we look for a formula to speed up our work. Consider

$$ax^2 + bx + c = 0, \quad a > 0.$$

Note that if $a < 0$, we can get an equivalent form with $a > 0$ by first multiplying by -1.

Let's solve by *completing the square*. As we carry out the steps, compare them with Example 12 in the preceding section.

$$x^2 + \frac{b}{a}x + \frac{c}{a} = 0 \qquad \text{Multiplying by } \frac{1}{a}$$

$$x^2 + \frac{b}{a}x = -\frac{c}{a} \qquad \text{Subtracting } \frac{c}{a}$$

Half of $\frac{b}{a}$ is $\frac{b}{2a}$. The square is $\frac{b^2}{4a^2}$. We add $\frac{b^2}{4a^2}$ on both sides:

$$x^2 + \frac{b}{a}x + \frac{b^2}{4a^2} = -\frac{c}{a} + \frac{b^2}{4a^2} \qquad \text{Adding } \frac{b^2}{4a^2}$$

$$\left(x + \frac{b}{2a}\right)^2 = -\frac{4ac}{4a^2} + \frac{b^2}{4a^2} \qquad \begin{array}{l}\text{Factoring the left side and finding a}\\ \text{common denominator on the right}\end{array}$$

$$\left(x + \frac{b}{2a}\right)^2 = \frac{b^2 - 4ac}{4a^2}$$

$$x + \frac{b}{2a} = \sqrt{\frac{b^2 - 4ac}{4a^2}} \quad \text{or} \quad x + \frac{b}{2a} = -\sqrt{\frac{b^2 - 4ac}{4a^2}}. \qquad \begin{array}{l}\text{Using the principle}\\ \text{of square roots}\end{array}$$

Since $a > 0$, $\sqrt{4a^2} = 2a$, so we can simplify as follows:

$$x + \frac{b}{2a} = \frac{\sqrt{b^2 - 4ac}}{2a} \quad \text{or} \quad x + \frac{b}{2a} = -\frac{\sqrt{b^2 - 4ac}}{2a}.$$

Thus,

$$x = -\frac{b}{2a} \pm \frac{\sqrt{b^2 - 4ac}}{2a}, \quad \text{or} \quad x = \frac{-b \pm \sqrt{b^2 - 4ac}}{2a}.$$

We now have the following.

THE QUADRATIC FORMULA

The solutions of $ax^2 + bx + c = 0$ are given by

$$x = \frac{-b \pm \sqrt{b^2 - 4ac}}{2a}.$$

Answers

Skill to Review:
1. $10i$ **2.** $10 - 2\sqrt{17}i$

The formula also holds when $a < 0$. A similar proof would show this, but we will not consider it here.

※ Algebraic-Graphical Connection

The Quadratic Formula (Algebraic). The solutions of $ax^2 + bx + c = 0$, $a \neq 0$, are given by

$$x = \frac{-b \pm \sqrt{b^2 - 4ac}}{2a}.$$

The Quadratic Formula (Graphical). The x-intercepts of the graph of the function $f(x) = ax^2 + bx + c$, $a \neq 0$, if they exist, are given by

$$\left(\frac{-b \pm \sqrt{b^2 - 4ac}}{2a}, 0 \right).$$

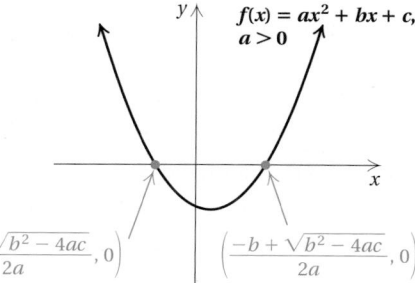

$f(x) = ax^2 + bx + c,$
$a > 0$

$$\left(\frac{-b - \sqrt{b^2 - 4ac}}{2a}, 0 \right) \qquad \left(\frac{-b + \sqrt{b^2 - 4ac}}{2a}, 0 \right)$$

EXAMPLE 1 Solve $5x^2 + 8x = -3$ using the quadratic formula.

We first find standard form and determine a, b, and c:

$$5x^2 + 8x + 3 = 0;$$
$$a = 5, \quad b = 8, \quad c = 3.$$

We then use the quadratic formula:

$$x = \frac{-b \pm \sqrt{b^2 - 4ac}}{2a}$$

$$x = \frac{-8 \pm \sqrt{8^2 - 4 \cdot 5 \cdot 3}}{2 \cdot 5} \qquad \text{Substituting}$$

$$x = \frac{-8 \pm \sqrt{64 - 60}}{10}$$

> Be sure to write the fraction bar all the way across.

$$x = \frac{-8 \pm \sqrt{4}}{10}$$

$$x = \frac{-8 \pm 2}{10}$$

$$x = \frac{-8 + 2}{10} \quad or \quad x = \frac{-8 - 2}{10}$$

$$x = \frac{-6}{10} \quad\quad or \quad x = \frac{-10}{10}$$

$$x = -\frac{3}{5} \quad\quad or \quad x = -1.$$

The solutions are $-\frac{3}{5}$ and -1.

$f(x) = 5x^2 + 8x + 3$

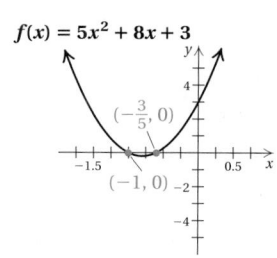

$\left(-\frac{3}{5}, 0\right)$

$(-1, 0)$

1. Consider the equation
$$2x^2 = 4 + 7x.$$

a) Solve using the quadratic formula.

b) Solve by factoring.

It turns out that we could have solved the equation in Example 1 more easily by factoring, as follows:

$$5x^2 + 8x + 3 = 0$$
$$(5x + 3)(x + 1) = 0$$
$$5x + 3 = 0 \quad or \quad x + 1 = 0$$
$$5x = -3 \quad or \quad x = -1$$
$$x = -\tfrac{3}{5} \quad or \quad x = -1.$$

To solve a quadratic equation:

1. Check for the form $x^2 = d$ or $(x + c)^2 = d$. If it is in this form, use the principle of square roots as in Section 11.1.

2. If it is not in the form of step (1), write it in standard form $ax^2 + bx + c = 0$ with a and b nonzero.

3. Then try factoring.

4. If it is not possible to factor or if factoring seems difficult, use the quadratic formula.

The solutions of a quadratic equation cannot always be found by factoring. They can *always* be found using the quadratic formula.

The solutions to all the exercises in this section could also be found by completing the square. However, the quadratic formula is the preferred method because it is faster.

> Do Exercise 1.

We will see in Example 2 that we cannot always rely on factoring.

EXAMPLE 2 Solve: $5x^2 - 8x = 3$. Give the exact solutions and approximate the solutions to three decimal places.

We first find standard form and determine a, b, and c:

$$5x^2 - 8x - 3 = 0;$$
$$a = 5, \quad b = -8, \quad c = -3.$$

We then use the quadratic formula, $x = \dfrac{-b \pm \sqrt{b^2 - 4ac}}{2a}$:

$$x = \frac{-(-8) \pm \sqrt{(-8)^2 - 4 \cdot 5 \cdot (-3)}}{2 \cdot 5} \qquad \text{Substituting}$$

$$x = \frac{8 \pm \sqrt{64 + 60}}{10} = \frac{8 \pm \sqrt{124}}{10} = \frac{8 \pm \sqrt{4 \cdot 31}}{10}$$

$$x = \frac{8 \pm 2\sqrt{31}}{10} = \frac{2(4 \pm \sqrt{31})}{2 \cdot 5} = \frac{2}{2} \cdot \frac{4 \pm \sqrt{31}}{5} = \frac{4 \pm \sqrt{31}}{5}.$$

- - - - - - - - - - - - - - - - - *Caution!* - - - - - - - - - - - - - - - - -

To avoid a common error in simplifying, remember to *factor the numerator and the denominator* and then remove a factor of 1.

- -

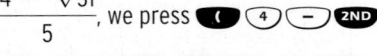

Answer

1. (a) $-\tfrac{1}{2}, 4$; (b) $-\tfrac{1}{2}, 4$

We can use a calculator to approximate the solutions:

$$\frac{4 + \sqrt{31}}{5} \approx 1.914; \qquad \frac{4 - \sqrt{31}}{5} \approx -0.314.$$

Check: Checking the exact solutions $\left(4 \pm \sqrt{31}\right)/5$ can be quite cumbersome. It could be done on a calculator or by using the approximations. Here we check 1.914; the check for -0.314 is left to the student.

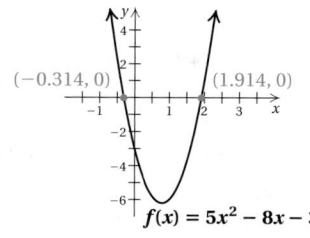

For 1.914:

$$\underline{ 5x^2 - 8x = 3}$$
$$5(1.914)^2 - 8(1.914) \;?\; 3$$
$$5(3.663396) - 15.312 \;\Big|$$
$$3.00498 \;\Big|$$

We do not have a perfect check due to the rounding error. But our check seems to confirm the solutions.

Do Exercise 2.

Some quadratic equations have solutions that are nonreal complex numbers.

EXAMPLE 3 Solve: $x^2 + x + 1 = 0$.

We have $a = 1$, $b = 1$, $c = 1$. We use the quadratic formula:

$$x = \frac{-1 \pm \sqrt{1^2 - 4 \cdot 1 \cdot 1}}{2 \cdot 1}$$

$$x = \frac{-1 \pm \sqrt{1 - 4}}{2}$$

$$x = \frac{-1 \pm \sqrt{-3}}{2}$$

$$x = \frac{-1 \pm \sqrt{3}i}{2}.$$

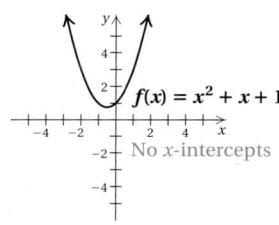

The solutions are

$$\frac{-1 + i\sqrt{3}}{2} \quad \text{and} \quad \frac{-1 - \sqrt{3}i}{2}.$$

The solutions can also be expressed in the form

$$-\frac{1}{2} + \frac{\sqrt{3}}{2}i \quad \text{and} \quad -\frac{1}{2} - \frac{\sqrt{3}}{2}i.$$

Do Exercise 3.

EXAMPLE 4 Solve: $2 + \dfrac{7}{x} = \dfrac{5}{x^2}$. Give the exact solutions and approximate solutions to three decimal places.

We first find an equivalent quadratic equation in standard form:

$$x^2\left(2 + \frac{7}{x}\right) = x^2 \cdot \frac{5}{x^2} \qquad \text{Multiplying by } x^2 \text{ to clear fractions, noting that } x \neq 0$$

$$2x^2 + 7x = 5$$

$$2x^2 + 7x - 5 = 0. \qquad \text{Subtracting 5}$$

2. Solve using the quadratic formula:
$$3x^2 + 2x = 7.$$
Give the exact solutions and approximate solutions to three decimal places.

3. Solve: $x^2 - x + 2 = 0$.

Answers

2. $\dfrac{-1 \pm \sqrt{22}}{3}$; $1.230, -1.897$

3. $\dfrac{1 \pm \sqrt{7}i}{2}$, or $\dfrac{1}{2} \pm \dfrac{\sqrt{7}}{2}i$

Then

$$a = 2, \quad b = 7, \quad c = -5$$

$$x = \frac{-7 \pm \sqrt{7^2 - 4 \cdot 2 \cdot (-5)}}{2 \cdot 2} \qquad \text{Substituting}$$

$$x = \frac{-7 \pm \sqrt{49 + 40}}{4} = \frac{-7 \pm \sqrt{89}}{4}$$

$$x = \frac{-7 + \sqrt{89}}{4} \quad \text{or} \quad x = \frac{-7 - \sqrt{89}}{4}.$$

Since we began with a rational equation, we need to check. We cleared the fractions before obtaining a quadratic equation in standard form, and this step could introduce numbers that do not check in the original rational equation. We need to show that neither of the numbers makes a denominator 0. Since neither of them does, the solutions are

$$\frac{-7 + \sqrt{89}}{4} \quad \text{and} \quad \frac{-7 - \sqrt{89}}{4}.$$

We can use a calculator to approximate the solutions:

$$\frac{-7 + \sqrt{89}}{4} \approx 0.608;$$

$$\frac{-7 - \sqrt{89}}{4} \approx -4.108.$$

Do Exercise 4.

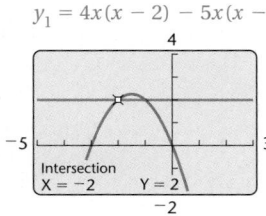
$f(x) = 2 + \dfrac{7}{x} - \dfrac{5}{x^2}$

$(-4.108, 0)$

$(0.608, 0)$

4. Solve:

$$3 = \frac{5}{x} + \frac{4}{x^2}.$$

Give the exact solutions and approximate solutions to three decimal places.

Calculator Corner

Solving Quadratic Equations A quadratic equation written with 0 on one side of the equals sign can be solved using the ZERO feature of a graphing calculator. See the Calculator Corner on p. 382 for the procedure.

We can also use the INTERSECT feature to solve a quadratic equation. Consider the equation in Exercise 19 in Exercise Set 11.2: $4x(x - 2) - 5x(x - 1) = 2$. First, we enter $y_1 = 4x(x - 2) - 5x(x - 1)$ and $y_2 = 2$ on the equation-editor screen and graph the equations in a window that shows the point(s) of intersection of the graphs.

We use the INTERSECT feature to find the coordinates of the left-hand point of intersection. (See the Calculator Corner on p. 572 for the procedure.) The first coordinate of this point, -2, is one solution of the equation. We use the INTERSECT feature again to find the other solution, -1.

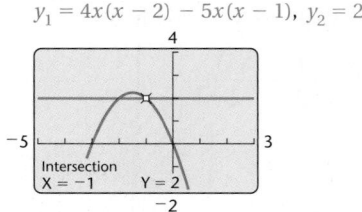

Exercises: Solve.

1. $5x^2 = -11x + 12$

2. $2x^2 - 15 = 7x$

3. $6(x - 3) = (x - 3)(x - 2)$

4. $(x + 1)(x - 4) = 3(x - 4)$

Answer

4. $\dfrac{5 \pm \sqrt{73}}{6}$; $2.257, -0.591$

11.2 Exercise Set

For Extra Help

MyMathLab

Math XL
PRACTICE

WATCH

DOWNLOAD

READ

REVIEW

a Solve.

1. $x^2 + 8x + 2 = 0$

2. $x^2 - 6x - 4 = 0$

3. $3p^2 = -8p - 1$

4. $3u^2 = 18u - 6$

5. $x^2 - x + 1 = 0$

6. $x^2 + x + 2 = 0$

7. $x^2 + 13 = 4x$

8. $x^2 + 13 = 6x$

9. $r^2 + 3r = 8$

10. $h^2 + 4 = 6h$

11. $1 + \dfrac{2}{x} + \dfrac{5}{x^2} = 0$

12. $1 + \dfrac{5}{x^2} = \dfrac{2}{x}$

13. a) Solve: $3x + x(x - 2) = 0$.
  b) Find the x-intercepts of
  $f(x) = 3x + x(x - 2)$.

14. a) Solve: $4x + x(x - 3) = 0$.
  b) Find the x-intercepts of
  $f(x) = 4x + x(x - 3)$.

15. a) Solve: $11x^2 - 3x - 5 = 0$.
  b) Find the x-intercepts of
  $f(x) = 11x^2 - 3x - 5$.

16. a) Solve: $7x^2 + 8x = -2$.
  b) Find the x-intercepts of
  $f(x) = 7x^2 + 8x + 2$.

17. a) Solve: $25x^2 = 20x - 4$.
  b) Find the x-intercepts of
  $f(x) = 25x^2 - 20x + 4$.

18. a) Solve: $49x^2 - 14x + 1 = 0$.
  b) Find the x-intercepts of
  $f(x) = 49x^2 - 14x + 1$.

Solve.

19. $4x(x - 2) - 5x(x - 1) = 2$

20. $3x(x + 1) - 7x(x + 2) = 6$

21. $14(x - 4) - (x + 2) = (x + 2)(x - 4)$

22. $11(x - 2) + (x - 5) = (x + 2)(x - 6)$

23. $5x^2 = 17x - 2$

24. $15x = 2x^2 + 16$

25. $x^2 + 5 = 4x$

26. $x^2 + 5 = 2x$

27. $x + \dfrac{1}{x} = \dfrac{13}{6}$

28. $\dfrac{3}{x} + \dfrac{x}{3} = \dfrac{5}{2}$

29. $\dfrac{1}{y} + \dfrac{1}{y + 2} = \dfrac{1}{3}$

30. $\dfrac{1}{x} + \dfrac{1}{x + 4} = \dfrac{1}{7}$

31. $(2t - 3)^2 + 17t = 15$

32. $2y^2 - (y + 2)(y - 3) = 12$

33. $(x - 2)^2 + (x + 1)^2 = 0$

34. $(x + 3)^2 + (x - 1)^2 = 0$

35. $x^3 - 1 = 0$
(*Hint*: Factor the difference of cubes. Then use the quadratic formula.)

36. $x^3 + 27 = 0$

Solve. Give the exact solutions and approximate solutions to three decimal places.

37. $x^2 + 6x + 4 = 0$

38. $x^2 + 4x - 7 = 0$

39. $x^2 - 6x + 4 = 0$

40. $x^2 - 4x + 1 = 0$

41. $2x^2 - 3x - 7 = 0$

42. $3x^2 - 3x - 2 = 0$

43. $5x^2 = 3 + 8x$

44. $2y^2 + 2y - 3 = 0$

Skill Maintenance

Solve. [10.6a, b]

45. $x = \sqrt{x + 2}$

46. $x = \sqrt{15 - 2x}$

47. $\sqrt{x + 2} = \sqrt{2x - 8}$

48. $\sqrt{x + 1} + 2 = \sqrt{3x + 1}$

49. $\sqrt{x + 5} = -7$

50. $\sqrt{2x - 6} + 11 = 2$

51. $\sqrt[3]{4x - 7} = 2$

52. $\sqrt[4]{3x - 1} = 2$

Synthesis

53. Use a graphing calculator to solve the equations in Exercises 3, 16, 17, and 43 using the INTERSECT feature, letting $y_1 =$ the left side and $y_2 =$ the right side. Then solve $2.2x^2 + 0.5x - 1 = 0$.

54. Use a graphing calculator to solve the equations in Exercises 9, 27, and 30. Then solve $5.33x^2 = 8.23x + 3.24$.

Solve.

55. $2x^2 - x - \sqrt{5} = 0$

56. $\dfrac{5}{x} + \dfrac{x}{4} = \dfrac{11}{7}$

57. $ix^2 - x - 1 = 0$

58. $\sqrt{3}x^2 + 6x + \sqrt{3} = 0$

59. $\dfrac{x}{x + 1} = 4 + \dfrac{1}{3x^2 - 3}$

60. $\left(1 + \sqrt{3}\right)x^2 - \left(3 + 2\sqrt{3}\right)x + 3 = 0$

61. Let $f(x) = (x - 3)^2$. Find all inputs x such that $f(x) = 13$.

62. Let $f(x) = x^2 + 14x + 49$. Find all inputs x such that $f(x) = 36$.

11.3 Applications Involving Quadratic Equations

(a) Applications and Problem Solving

OBJECTIVES

(a) Solve applied problems involving quadratic equations.

(b) Solve a formula for a given letter.

Sometimes when we translate a problem to mathematical language, the result is a quadratic equation.

EXAMPLE 1 *Quilt Dimensions.* Michelle is making a quilt for a wall hanging at the entrance of a state museum. The finished quilt will measure 8 ft by 6 ft. The quilt has a border of uniform width around it. The area of the interior rectangular section is one-half the area of the entire quilt. How wide is the border?

1. **Familiarize.** We first make a drawing and label it with the known information. We don't know how wide the border is, so we have called its width x.

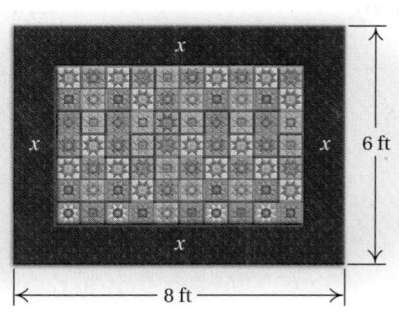

2. **Translate.** Remember, the area of a rectangle is lw (length times width). Then:

   Area of entire quilt $= 8 \cdot 6$;
   Area of interior section $= (8 - 2x)(6 - 2x)$.

   Since the area of the interior section is one-half the area of the entire quilt, we have

   $$(8 - 2x)(6 - 2x) = \frac{1}{2} \cdot 8 \cdot 6.$$

3. **Solve.** We solve the equation:

$48 - 28x + 4x^2 = 24$	Using FOIL on the left
$4x^2 - 28x + 24 = 0$	Finding standard form
$x^2 - 7x + 6 = 0$	Dividing by 4
$(x - 6)(x - 1) = 0$	Factoring
$x = 6 \quad or \quad x = 1.$	Using the principle of zero products

4. **Check.** We check in the original problem. We see that 6 is not a solution because an 8-ft by 6-ft quilt cannot have a 6-ft border. When $x = 6$, then $8 - 2x = -4$ and $6 - 2x = -6$ and the dimensions of the interior section of the quilt cannot be negative.

   If the border is 1 ft wide, then the interior will have length $8 - 2 \cdot 1$, or 6 ft. The width will be $6 - 2 \cdot 1$, or 4 ft. The area of the interior is thus $6 \cdot 4$, or 24 ft^2. The area of the entire quilt is $8 \cdot 6$, or 48 ft^2. The area of the interior is one-half of 48 ft^2, so the number 1 checks.

5. **State.** The border of the quilt is 1 ft wide.

Do Exercise 1.

1. **Landscaping.** A rectangular garden is 60 ft by 80 ft. Part of the garden is torn up to install a sidewalk of uniform width around it. The area of the new garden is one-half of the old area. How wide is the sidewalk?

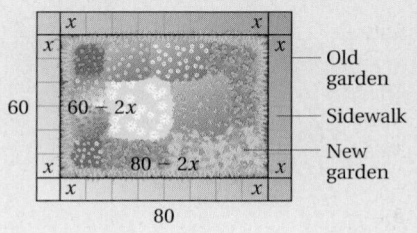

Answer

1. 10 ft

EXAMPLE 2 *Town Planning.* Three towns A, B, and C are situated as shown. The roads at A form a right angle. The distance from A to B is 2 mi less than the distance from A to C. The distance from B to C is 10 mi. Find the distance from A to B and the distance from A to C.

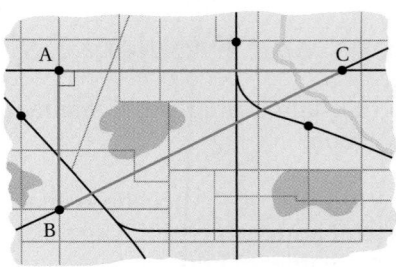

1. **Familiarize.** We first make a drawing and label it. We let d = the distance from A to C. Then the distance from A to B is $d - 2$.

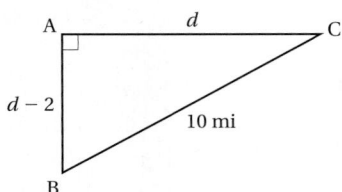

2. **Translate.** We see that a right triangle is formed. We can use the Pythagorean theorem, which we studied in Chapter 10: $c^2 = a^2 + b^2$. In this problem, we have

$$10^2 = d^2 + (d - 2)^2.$$

3. **Solve.** We solve the equation:

$$10^2 = d^2 + (d - 2)^2$$
$$100 = d^2 + d^2 - 4d + 4 \quad \text{Squaring}$$
$$2d^2 - 4d - 96 = 0 \quad \text{Finding standard form}$$
$$d^2 - 2d - 48 = 0 \quad \text{Multiplying by } \tfrac{1}{2}, \text{ or dividing by 2}$$
$$(d - 8)(d + 6) = 0 \quad \text{Factoring}$$
$$d - 8 = 0 \quad \text{or} \quad d + 6 = 0 \quad \text{Using the principle of zero products}$$
$$d = 8 \quad \text{or} \qquad d = -6.$$

4. **Check.** We know that -6 cannot be a solution because distances are not negative. If $d = 8$, then $d - 2 = 6$, and

$$d^2 + (d - 2)^2 = 8^2 + 6^2 = 64 + 36 = 100.$$

Since $10^2 = 100$, the distance 8 mi checks.

5. **State.** The distance from A to C is 8 mi, and the distance from A to B is 6 mi.

> Do Exercise 2.

EXAMPLE 3 *Town Planning.* Three towns A, B, and C are situated as shown in Example 2. The roads at A form a right angle. The distance from A to B is 2 mi less than the distance from A to C. The distance from B to C is 8 mi. Find the distance from A to B and the distance from A to C. Find exact and approximate answers to the nearest hundredth of a mile.

---

2. Ladder Location. A ladder leans against a building, as shown below. The ladder is 20 ft long. The distance to the top of the ladder is 4 ft greater than the distance d from the building. Find the distance d and the distance to the top of the ladder.

Answer

2. The distance d is 12 ft; the distance to the top of the ladder is 16 ft.

Using the same reasoning that we did in Example 2, we translate the problem to the equation

$$8^2 = d^2 + (d - 2)^2.$$

We solve as follows. Note that the quadratic equation we get is not easily factored, so we use the quadratic formula:

$$64 = d^2 + d^2 - 4d + 4 \qquad \text{Squaring}$$
$$2d^2 - 4d - 60 = 0 \qquad \text{Finding standard form}$$
$$d^2 - 2d - 30 = 0. \qquad \text{Multiplying by } \tfrac{1}{2}, \text{ or dividing by 2}$$

Then

$$d = \frac{-b \pm \sqrt{b^2 - 4ac}}{2a}$$

$$d = \frac{-(-2) \pm \sqrt{(-2)^2 - 4(1)(-30)}}{2(1)} \qquad \begin{array}{l}\text{Substituting 1 for } a, -2 \text{ for } b,\\ \text{and } -30 \text{ for } c\end{array}$$

$$d = \frac{2 \pm \sqrt{124}}{2} = \frac{2 \pm \sqrt{4(31)}}{2} = \frac{2 \pm 2\sqrt{31}}{2} = 1 \pm \sqrt{31}.$$

Since $1 - \sqrt{31} < 0$ and $1 + \sqrt{31} > 0$, it follows that $d = 1 + \sqrt{31}$. Using a calculator, we find that $d = 1 + \sqrt{31} \approx 6.57$ mi, and that $d - 2 \approx 4.57$ mi. Thus the distance from A to C is about 6.57 mi, and the distance from A to B is about 4.57 mi.

> **Do Exercise 3.**

3. Ladder Location. Refer to Margin Exercise 2. Suppose that the ladder has length 10 ft. Find the distance d and the distance $d + 4$.

EXAMPLE 4 *Motorcycle Travel.* Karin's motorcycle traveled 300 mi at a certain speed. Had she gone 10 mph faster, she could have made the trip in 1 hr less time. Find her speed.

1. **Familiarize.** We first make a drawing, labeling it with known and unknown information. We can also organize the information in a table as we did in Section 6.8. We let $r =$ the speed, in miles per hour, and $t =$ the time, in hours.

r mph   300 miles   t hours

r + 10 mph   300 miles   t − 1 hours

DISTANCE	SPEED	TIME
300	r	t
300	$r + 10$	$t - 1$

$\longrightarrow r = \dfrac{300}{t}$

$\longrightarrow r + 10 = \dfrac{300}{t - 1}$

Recalling the motion formula $d = rt$ and solving for r, we get $r = d/t$. From the rows of the table, we obtain

$$r = \frac{300}{t} \quad \text{and} \quad r + 10 = \frac{300}{t - 1}.$$

Answer

3. The distance d is about 4.782 ft; the distance to the top of the ladder is about 8.782 ft.

2. Translate.

We substitute for r from the first equation into the second and get a translation:

$$\frac{300}{t} + 10 = \frac{300}{t-1}.$$

3. Solve.

We solve as follows:

$$\frac{300}{t} + 10 = \frac{300}{t-1}$$

$$t(t-1)\left[\frac{300}{t} + 10\right] = t(t-1) \cdot \frac{300}{t-1} \qquad \text{Multiplying by the LCM}$$

$$t(t-1) \cdot \frac{300}{t} + t(t-1) \cdot 10 = t(t-1) \cdot \frac{300}{t-1}$$

$$300(t-1) + 10(t^2 - t) = 300t$$

$$300t - 300 + 10t^2 - 10t = 300t$$

$$10t^2 - 10t - 300 = 0 \qquad \text{Standard form}$$

$$t^2 - t - 30 = 0 \qquad \text{Dividing by 10}$$

$$(t-6)(t+5) = 0 \qquad \text{Factoring}$$

$$t = 6 \quad or \quad t = -5. \qquad \text{Using the principle of zero products}$$

4. Check.

Since negative time has no meaning in this problem, we try 6 hr. Remembering that $r = d/t$, we get $r = 300/6 = 50$ mph.

To check, we take the speed 10 mph faster, which is 60 mph, and see how long the trip would have taken at that speed:

$$t = \frac{d}{r} = \frac{300}{60} = 5 \text{ hr.}$$

This is 1 hr less than the trip actually took, so we have an answer.

5. State.

Karin's speed was 50 mph.

> Do Exercise 4.

---

4. Marine Travel. Two ships make the same voyage of 3000 nautical miles. The faster ship travels 10 knots faster than the slower one. (A *knot* is 1 nautical mile per hour.) The faster ship makes the voyage in 50 hr less time than the slower one. Find the speeds of the two ships.

Complete this table to help with the familiarization.

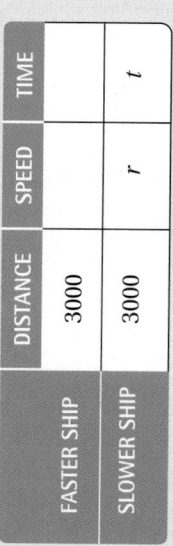

	DISTANCE	SPEED	TIME
FASTER SHIP	3000	r	t
SLOWER SHIP	3000		

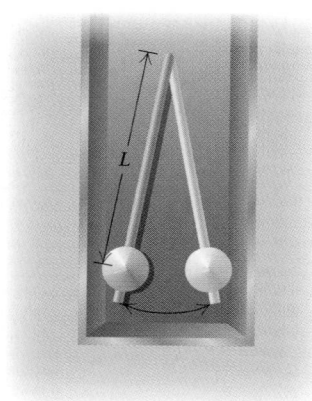

Answer

4.

	Distance	Speed	Time
Faster Ship	3000	$r + 10$	$t - 50$
Slower Ship	3000	r	t

20 knots, 30 knots

---

b Solving Formulas

Recall that to solve a formula for a certain letter, we use the principles for solving equations to get that letter alone on one side.

EXAMPLE 5 *Period of a Pendulum.* The time T required for a pendulum of length L to swing back and forth (complete one period) is given by the formula $T = 2\pi\sqrt{L/g}$, where g is the gravitational constant. Solve for L.

$$T = 2\pi\sqrt{\frac{L}{g}} \qquad \text{This is a radical equation. (See Section 10.6.)}$$

$$T^2 = \left(2\pi\sqrt{\frac{L}{g}}\right)^2 \qquad \text{Principle of powers (squaring)}$$

$$T^2 = 2^2\pi^2\frac{L}{g}$$

$$gT^2 = 4\pi^2 L \qquad \text{Clearing fractions}$$

$$\frac{gT^2}{4\pi^2} = L \qquad \text{Multiplying by } \frac{1}{4\pi^2}$$

We now have L alone on one side and L does not appear on the other side, so the formula is solved for L.

Do Exercise 5.

5. Solve $A = \sqrt{\dfrac{w_1}{w_2}}$ for w_2.

In most formulas, variables represent nonnegative numbers, so we need only the positive root when taking square roots.

EXAMPLE 6 *Hang Time.* An athlete's *hang time* is the amount of time that the athlete can remain airborne when jumping. A formula relating an athlete's vertical leap V, in inches, to hang time T, in seconds, is $V = 48T^2$. (See Example 13 in Section 11.1.) Solve for T.

We have

$$48T^2 = V$$

$$T^2 = \frac{V}{48} \qquad \text{Multiplying by } \tfrac{1}{48} \text{ to get } T^2 \text{ alone}$$

$$T = \sqrt{\frac{V}{48}} \qquad \text{Using the principle of square roots; note that } T \geq 0.$$

$$T = \sqrt{\frac{V}{2 \cdot 2 \cdot 2 \cdot 2 \cdot 3} \cdot \frac{3}{3}} = \frac{\sqrt{3V}}{2 \cdot 2 \cdot 3} = \frac{\sqrt{3V}}{12}.$$

Do Exercise 6.

6. Solve $V = \pi r^2 h$ for r.
(Volume of a right circular cylinder)

EXAMPLE 7 *Falling Distance.* An object that is tossed downward with an initial speed (velocity) of v_0 will travel a distance of s meters, where $s = 4.9t^2 + v_0t$ and t is measured in seconds. Solve for t.

Answers

5. $w_2 = \dfrac{w_1}{A^2}$ **6.** $r = \sqrt{\dfrac{V}{\pi h}}$

To solve a formula for a letter, say, t:

1. **Clear the fractions and use the principle of powers, as needed, until t does not appear in any radicand or denominator. (In some cases, you may clear the fractions first, and in some cases, you may use the principle of powers first.)**

2. **Collect all terms with t^2 in them. Also collect all terms with t in them.**

3. **If t^2 does not appear, you can finish by using just the addition and multiplication principles.**

4. **If t^2 appears but t does not, solve the equation for t^2. Then take square roots on both sides.**

5. **If there are terms containing both t and t^2, write the equation in standard form and use the quadratic formula.**

7. Solve $s = gt + 16t^2$ for t.

8. Solve $\dfrac{b}{\sqrt{a^2 - b^2}} = t$ for b.

Answers

7. $t = \dfrac{-g + \sqrt{g^2 + 64s}}{32}$

8. $b = \dfrac{ta}{\sqrt{1 + t^2}}$

Since t is squared in one term and raised to the first power in the other term, the equation is quadratic in t. The variable is t; v_0 and s are treated as constants.

We have

$$4.9t^2 + v_0 t = s$$

$$4.9t^2 + v_0 t - s = 0 \qquad \text{Writing standard form}$$

$$a = 4.9, \quad b = v_0, \quad c = -s$$

$$t = \frac{-v_0 \pm \sqrt{(v_0)^2 - 4(4.9)(-s)}}{2(4.9)} \qquad \begin{array}{l} \text{Using the quadratic formula:} \\ t = \dfrac{-b \pm \sqrt{b^2 - 4ac}}{2a} \end{array}$$

$$t = \frac{-v_0 \pm \sqrt{(v_0)^2 + 19.6s}}{9.8}.$$

Since the negative square root would yield a negative value for t, we use only the positive root:

$$t = \frac{-v_0 + \sqrt{(v_0)^2 + 19.6s}}{9.8}.$$

The steps listed in the margin should help you when solving formulas for a given letter. Try to remember that, when solving a formula, you do the same things you would do to solve any equation.

Do Exercise 7.

EXAMPLE 8 Solve $t = \dfrac{a}{\sqrt{a^2 + b^2}}$ for a.

In this case, we could either clear the fractions first or use the principle of powers first. Let's clear the fractions. Multiplying by $\sqrt{a^2 + b^2}$, we have

$$t\sqrt{a^2 + b^2} = a.$$

Now we square both sides and then continue:

$$\left(t\sqrt{a^2 + b^2}\right)^2 = a^2 \qquad \text{Squaring}$$

------- *Caution!* -------

Don't forget to square both t and $\sqrt{a^2 + b^2}$.

$$t^2\left(\sqrt{a^2 + b^2}\right)^2 = a^2$$

$$t^2(a^2 + b^2) = a^2$$

$$t^2 a^2 + t^2 b^2 = a^2$$

$$t^2 b^2 = a^2 - t^2 a^2 \qquad \text{Getting all } a^2\text{-terms together}$$

$$t^2 b^2 = a^2(1 - t^2) \qquad \text{Factoring out } a^2$$

$$\frac{t^2 b^2}{1 - t^2} = a^2 \qquad \text{Dividing by } 1 - t^2$$

$$\sqrt{\frac{t^2 b^2}{1 - t^2}} = a \qquad \text{Taking the square root}$$

$$\frac{tb}{\sqrt{1 - t^2}} = a. \qquad \text{Simplifying}$$

You need not rationalize denominators in situations such as this.

Do Exercise 8.

Translating for Success

1. *Car Travel.* Sarah drove her car 800 mi to see her friend. The return trip was 2 hr faster at a speed that was 10 mph more. Find her return speed.

2. *Coin Mixture.* A collection of dimes and quarters is worth $26.95. There are 117 coins in all. How many of each coin are there?

3. *Wire Cutting.* A 537-in. wire is cut into three pieces. The second piece is 7 in. shorter than the first. The third is half as long as the first. How long is each piece?

4. *Marine Travel.* The Columbia River flows at a rate of 2 mph for the length of a popular boating route. In order for a motorized dinghy to travel 3 mi upriver and return in a total of 4 hr, how fast must the boat be able to travel in still water?

5. *Locker Numbers.* The numbers on three adjoining lockers are consecutive integers whose sum is 537. Find the integers.

Translate each word problem to an equation or a system of equations and select a correct translation from equations A–O.

A. $(80 - 2x)(100 - 2x) = \frac{1}{3} \cdot 80 \cdot 100$

B. $\dfrac{800}{x} + 10 = \dfrac{800}{x - 2}$

C. $x + 18\% \cdot x = 3.24$

D. $x + 25y = 26.95,$
$x + y = 117$

E. $2x + 2(x - 7) = 537$

F. $x + (x - 7) + \frac{1}{2}x = 537$

G. $0.10x + 0.25y = 26.95,$
$x + y = 117$

H. $3.24 - 18\% \cdot 3.24 = x$

I. $\dfrac{4}{x + 2} + \dfrac{4}{x - 2} = 3$

J. $x^2 + (x + 1)^2 = 7^2$

K. $75^2 + x^2 = 78^2$

L. $\dfrac{3}{x + 2} + \dfrac{3}{x - 2} = 4$

M. $75^2 + 78^2 = x^2$

N. $x + (x + 1) + (x + 2) = 537$

O. $\dfrac{800}{x} + \dfrac{800}{x - 2} = 10$

Answers on page A-33

---

6. *Gasoline Prices.* One day the price of gasoline was increased 18% to a new price of $3.24 per gallon. What was the original price?

7. *Triangle Dimensions.* The hypotenuse of a right triangle is 7 ft. The length of one leg is 1 ft longer than the other. Find the lengths of the legs.

8. *Rectangle Dimensions.* The perimeter of a rectangle is 537 ft. The width of the rectangle is 7 ft shorter than the length. Find the length and the width.

9. *Guy Wire.* A guy wire is 78 ft long. It is attached to the top of a 75-ft cell-phone tower. How far is it from the base of the pole to the point where the wire is attached to the ground?

10. *Landscaping.* A rectangular garden is 80 ft by 100 ft. Part of the garden is torn up to install a sidewalk of uniform width around it. The area of the new garden is $\frac{1}{3}$ of the old area. How wide is the sidewalk?

a   Solve.

1. *Flower Bed.*   The width of a rectangular flower bed is 7 ft less than the length. The area is 18 ft². Find the length and the width.

2. *Feed Lot.*   The width of a rectangular feed lot is 8 m less than the length. The area is 20 m². Find the length and the width.

3. *Parking Lot.*   The length of a rectangular parking lot is twice the width. The area is 162 yd². Find the length and the width.

4. *Flag Dimensions.*   The length of an American flag that is displayed at a government office is 3 in. less than twice its width. The area is 1710 in². Find the length and the width of the flag.

5. *Sailing.*   The base of a triangular sail is 9 m less than its height. The area is 56 m². Find the base and the height of the sail.

6. *Parking Lot.*   The width of a rectangular parking lot is 50 ft less than its length. Determine the dimensions of the parking lot if it measures 250 ft diagonally.

Area = 56 m²

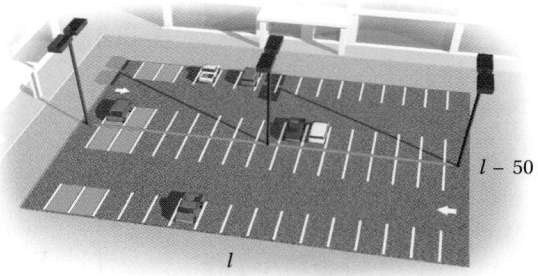

$l - 50$

l

7. *Parking Lot.*   The width of a rectangular parking lot is 51 ft less than its length. Determine the dimensions of the parking lot if it measures 250 ft diagonally.

8. *Sailing.*   The base of a triangular sail is 8 ft less than its height. The area is 56 ft². Find the base and the height of the sail.

9. *Mirror Framing.*   The outside of a mosaic mirror frame measures 14 in. by 20 in., and 160 in² of mirror shows. Find the width of the frame.

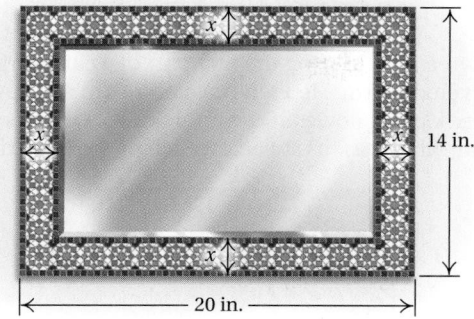

10. *Box Construction.*   An open box is to be made from a 10-ft by 20-ft rectangular piece of cardboard by cutting a square from each corner. The area of the bottom of the box is to be 96 ft². What is the length of the sides of the squares that are cut from the corners?

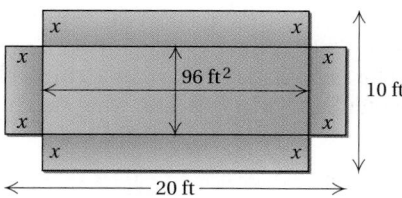

11. *Landscaping.*   A landscaper is designing a flower garden in the shape of a right triangle. She wants 10 ft of a perennial border to form the hypotenuse of the triangle, and one leg is to be 2 ft longer than the other. Find the lengths of the legs.

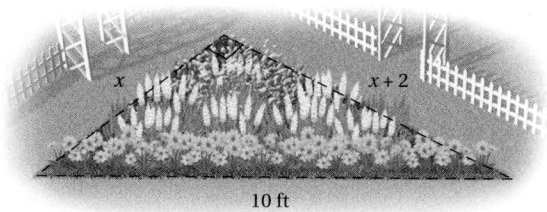

12. The hypotenuse of a right triangle is 25 m long. The length of one leg is 17 m less than the other. Find the lengths of the legs.

13. *Page Numbers.*   A student opens a literature book to two facing pages. The product of the page numbers is 812. Find the page numbers.

14. *Page Numbers.*   A student opens a mathematics book to two facing pages. The product of the page numbers is 1980. Find the page numbers.

Solve. Find exact answers and approximate answers rounded to three decimal places.

15. The width of a rectangle is 4 ft less than the length. The area is 10 ft². Find the length and the width.

16. The length of a rectangle is twice the width. The area is 328 cm². Find the length and the width.

17. *Page Dimensions.*   The outside of an oversized book page measures 14 in. by 20 in.; 100 in² of printed text shows. Find the width of the margin.

18. *Picture Framing.*   The outside of a picture frame measures 13 cm by 20 cm, and 80 cm² of picture shows. Find the width of the frame.

19. The hypotenuse of a right triangle is 24 ft long. The length of one leg is 14 ft more than the other. Find the lengths of the legs.

20. The hypotenuse of a right triangle is 22 m long. The length of one leg is 10 m less than the other. Find the lengths of the legs.

21. *Car Trips.* During the first part of a trip, Sam's Chevrolet Cobalt SS traveled 120 mi. Sam then drove another 100 mi at a speed that was 10 mph slower. If the total time for Sam's trip was 4 hr, what was his speed on each part of the trip?

DISTANCE	SPEED	TIME

22. *Canoeing.* During the first part of a canoe trip, Doug covered 60 km. He then traveled 24 km at a speed that was 4 km/h slower. If the total time for Doug's trip was 8 hr, what was his speed on each part of the trip?

DISTANCE	SPEED	TIME

23. *Car Trips.* Colleen's Hyundai Sonata travels 200 mi. If the car had gone 10 mph faster, the trip would have taken 1 hr less. Find Colleen's speed.

24. *Car Trips.* Katie's Nissan Altima travels 280 mi. If the car had gone 5 mph faster, the trip would have taken 1 hr less. Find Katie's speed.

25. *Air Travel.* A Cessna flies 600 mi. A Beechcraft flies 1000 mi at a speed that is 50 mph faster, but takes 1 hr longer. Find the speed of each plane.

26. *Air Travel.* A turbo-jet flies 50 mph faster than a super-prop plane. If a turbo-jet goes 2000 mi in 3 hr less time than it takes the super-prop to go 2800 mi, find the speed of each plane.

27. *Bicycling.* Naoki bikes 40 mi to Hillsboro. The return trip is made at a speed that is 6 mph slower. Total travel time for the round trip is 14 hr. Find Naoki's speed on each part of the trip.

28. *Car Speed.* On a sales trip, Gail drives the 600 mi to Richmond. The return trip is made at a speed that is 10 mph slower. Total travel time for the round trip is 22 hr. How fast did Gail travel on each part of the trip?

29. *Navigation.* The current in a typical Mississippi River shipping route flows at a rate of 4 mph. In order for a barge to travel 24 mi upriver and then return in a total of 5 hr, approximately how fast must the barge be able to travel in still water?

30. *Navigation.* The Hudson River flows at a rate of 3 mph. A patrol boat travels 60 mi upriver and returns in a total time of 9 hr. What is the speed of the boat in still water?

Solve each formula for the given letter. Assume that all variables represent nonnegative numbers.

31. $A = 6s^2$, for s
(Surface area of a cube)

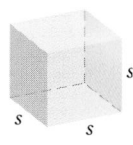

32. $A = 4\pi r^2$, for r
(Surface area of a sphere)

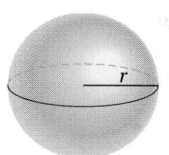

33. $F = \dfrac{Gm_1m_2}{r^2}$, for r

34. $N = \dfrac{kQ_1Q_2}{s^2}$, for s
(Number of phone calls between two cities)

35. $E = mc^2$, for c
(Einstein's energy–mass relationship)

36. $V = \frac{1}{3}s^2h$, for s
(Volume of a pyramid)

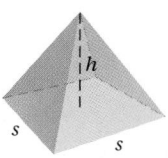

37. $a^2 + b^2 = c^2$, for b
(Pythagorean formula in two dimensions)

38. $a^2 + b^2 + c^2 = d^2$, for c
(Pythagorean formula in three dimensions)

39. $N = \dfrac{k^2 - 3k}{2}$, for k
(Number of diagonals of a polygon of k sides)

40. $s = v_0t + \dfrac{gt^2}{2}$, for t
(A motion formula)

41. $A = 2\pi r^2 + 2\pi rh$, for r
(Surface area of a cylinder)

42. $A = \pi r^2 + \pi rs$, for r
(Surface area of a cone)

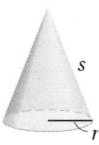

43. $T = 2\pi\sqrt{\dfrac{L}{g}}$, for g
(A pendulum formula)

44. $W = \sqrt{\dfrac{1}{LC}}$, for L
(An electricity formula)

45. $I = \dfrac{703W}{H^2}$, for H
(Body mass index; see Exercises 71 and 72 of Exercise Set 9.1)

46. $N + p = \dfrac{6.2A^2}{pR^2}$, for R

47. $m = \dfrac{m_0}{\sqrt{1 - \dfrac{v^2}{c^2}}}$, for v
(A relativity formula)

48. Solve the formula given in Exercise 47 for c.

Skill Maintenance

Add or subtract.

49. $\dfrac{1}{x-1} + \dfrac{1}{x^2 - 3x + 2}$ [6.4a]

50. $\dfrac{x+1}{x-1} - \dfrac{x+1}{x^2 + x + 1}$ [6.5a]

51. $\dfrac{2}{x+3} - \dfrac{x}{x-1} + \dfrac{x^2 + 2}{x^2 + 2x - 3}$
[6.5b]

52. Multiply and simplify: $\sqrt{3x^2}\sqrt{3x^3}$. [10.3a]

53. Express in terms of i: $\sqrt{-20}$. [10.8a]

Simplify. [6.6a]

54. $\dfrac{\dfrac{3}{x-1}}{\dfrac{1}{x+1} + \dfrac{2}{x-1}}$

55. $\dfrac{\dfrac{4}{a^2 b}}{\dfrac{3}{a} - \dfrac{4}{b^2}}$

Synthesis

56. Solve: $\dfrac{4}{2x+i} - \dfrac{1}{x-i} = \dfrac{2}{x+i}$.

57. Find a when the reciprocal of $a - 1$ is $a + 1$.

58. *Bungee Jumping.* Jesse is tied to one end of a 40-m elasticized (bungee) cord. The other end of the cord is tied to the middle of a train trestle. If Jesse jumps off the bridge, for how long will he fall before the cord begins to stretch? (See Example 7 and let $v_0 = 0$.)

40 m

59. *Surface Area.* A sphere is inscribed in a cube as shown in the figure below. Express the surface area of the sphere as a function of the surface area S of the cube.

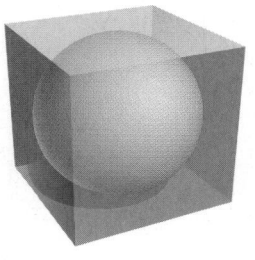

60. *Pizza Crusts.* At Pizza Perfect, Ron can make 100 large pizza crusts in 1.2 hr less than Chad. Together they can do the job in 1.8 hr. How long does it take each to do the job alone?

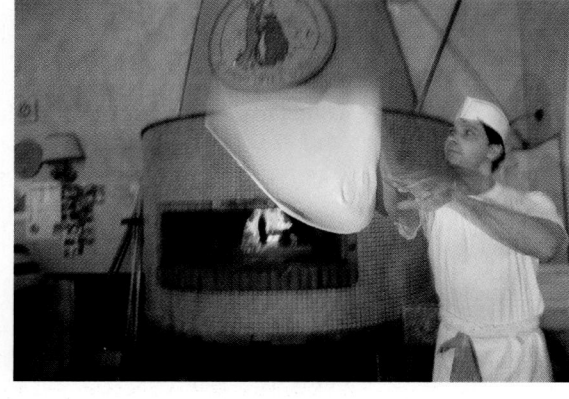

61. *The Golden Rectangle.* For over 2000 yr, the proportions of a "golden" rectangle have been considered visually appealing. A rectangle of width w and length l is considered "golden" if

$$\frac{w}{l} = \frac{l}{w+l}.$$

Solve for l.

l

w

11.4 More on Quadratic Equations

a The Discriminant

From the quadratic formula, we know that the solutions x_1 and x_2 of a quadratic equation are given by

$$x_1 = \frac{-b + \sqrt{b^2 - 4ac}}{2a} \quad \text{and} \quad x_2 = \frac{-b - \sqrt{b^2 - 4ac}}{2a}.$$

The expression $b^2 - 4ac$ is called the **discriminant**. When we are using the quadratic formula, it is helpful to compute the discriminant first. If it is 0, there will be just one real solution. If it is positive, there will be two real solutions. If it is negative, we will be taking the square root of a negative number; hence there will be two nonreal complex-number solutions, and they will be complex conjugates.

DISCRIMINANT $b^2 - 4ac$	NATURE OF SOLUTIONS	x-INTERCEPTS
0	Only one solution; it is a real number	Only one
Positive	Two different real-number solutions	Two different
Negative	Two different nonreal complex-number solutions (complex conjugates)	None

If the discriminant is a perfect square, we can solve the equation by factoring, not needing the quadratic formula.

EXAMPLE 1 Determine the nature of the solutions of $9x^2 - 12x + 4 = 0$.

We have

$$a = 9, \quad b = -12, \quad c = 4.$$

We compute the discriminant:

$$b^2 - 4ac = (-12)^2 - 4 \cdot 9 \cdot 4$$
$$= 144 - 144$$
$$= 0.$$

There is just one solution, and it is a real number. Since 0 is a perfect square, the equation can be solved by factoring.

$f(x) = 9x^2 - 12x + 4$

One x-intercept

EXAMPLE 2 Determine the nature of the solutions of $x^2 + 5x + 8 = 0$.

We have

$$a = 1, \quad b = 5, \quad c = 8.$$

We compute the discriminant:

$$b^2 - 4ac = 5^2 - 4 \cdot 1 \cdot 8$$
$$= 25 - 32$$
$$= -7.$$

Since the discriminant is negative, there are two nonreal complex-number solutions.

$f(x) = x^2 + 5x + 8$

No x-intercepts

OBJECTIVES

a Determine the nature of the solutions of a quadratic equation.

b Write a quadratic equation having two given numbers as solutions.

c Solve equations that are quadratic in form.

SKILL TO REVIEW
Objective 3.3a: Find the intercepts of a linear equation.

Find the coordinates of the y-intercept and the x-intercept.

1. $3x - 12y = -36$
2. $5 - 2y = -10x$

STUDY TIPS

TAKE THE TIME!

The foundation of all your study skills is making time to study! If you invest your time, you increase your likelihood of succeeding.

Answers

Skill to Review:
1. y-intercept: $(0, 3)$; x-intercept: $(-12, 0)$
2. y-intercept: $\left(0, \frac{5}{2}\right)$; x-intercept: $\left(-\frac{1}{2}, 0\right)$

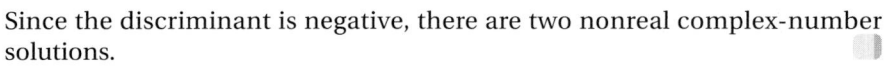

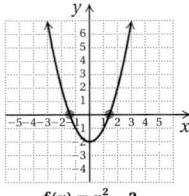

$f(x) = x^2 - 2$

$b^2 - 4ac = 8 > 0$

Two real solutions
Two x-intercepts

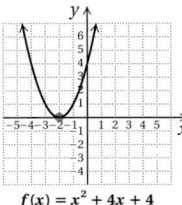

$f(x) = x^2 + 4x + 4$

$b^2 - 4ac = 0$

One real solution
One x-intercept

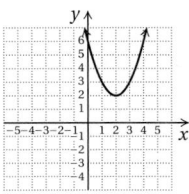

$f(x) = x^2 - 4x + 6$

$b^2 - 4ac = -8 < 0$

No real solutions
No x-intercept

Determine the nature of the solutions without solving.

1. $x^2 + 5x - 3 = 0$

2. $9x^2 - 6x + 1 = 0$

3. $3x^2 - 2x + 1 = 0$

EXAMPLE 3   Determine the nature of the solutions of $x^2 + 5x + 6 = 0$.

We have

$$a = 1, \quad b = 5, \quad c = 6;$$
$$b^2 - 4ac = 5^2 - 4 \cdot 1 \cdot 6 = 1.$$

Since the discriminant is positive, there are two solutions, and they are real numbers. The equation can be solved by factoring since the discriminant is a perfect square.

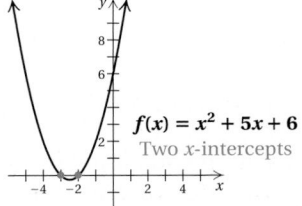

$f(x) = x^2 + 5x + 6$

Two x-intercepts

EXAMPLE 4   Determine the nature of the solutions of $5x^2 + x - 3 = 0$.

We have

$$a = 5, \quad b = 1, \quad c = -3;$$
$$b^2 - 4ac = 1^2 - 4 \cdot 5 \cdot (-3) = 1 + 60 = 61.$$

Since the discriminant is positive, there are two solutions, and they are real numbers. The equation cannot be solved by factoring because 61 is not a prefect square.

The discriminant, $b^2 - 4ac$, tells us how many real-number solutions the equation $ax^2 + bx + c = 0$ has, so it also indicates how many x-intercepts the graph of $f(x) = ax^2 + bx + c$ has. Compare the graphs at left.

Do Exercises 1–3.

b   Writing Equations from Solutions

We know by the principle of zero products that $(x - 2)(x + 3) = 0$ has solutions 2 and -3. If we know the solutions of an equation, we can write the equation, using this principle in reverse.

EXAMPLE 5   Find a quadratic equation whose solutions are 3 and $-\frac{2}{5}$.

We have

$$x = 3 \quad or \qquad x = -\frac{2}{5}$$
$$x - 3 = 0 \quad or \quad x + \frac{2}{5} = 0 \qquad \text{Getting the 0's on one side}$$
$$x - 3 = 0 \quad or \quad 5x + 2 = 0 \qquad \text{Clearing the fraction}$$
$$(x - 3)(5x + 2) = 0 \qquad \text{Using the principle of zero products in reverse}$$
$$5x^2 - 13x - 6 = 0. \qquad \text{Using FOIL}$$

EXAMPLE 6   Write a quadratic equation whose solutions are $2i$ and $-2i$.

We have

$$x = 2i \quad or \qquad x = -2i$$
$$x - 2i = 0 \quad or \quad x + 2i = 0 \qquad \text{Getting the 0's on one side}$$
$$(x - 2i)(x + 2i) = 0 \qquad \text{Using the principle of zero products in reverse}$$
$$x^2 - (2i)^2 = 0 \qquad \text{Using } (A - B)(A + B) = A^2 - B^2$$
$$x^2 - 4i^2 = 0$$
$$x^2 - 4(-1) = 0$$
$$x^2 + 4 = 0.$$

Answers

1. Two real   2. One real   3. Two nonreal

EXAMPLE 7 Write a quadratic equation whose solutions are $\sqrt{3}$ and $-2\sqrt{3}$.

We have

$$x = \sqrt{3} \quad \textit{or} \quad x = -2\sqrt{3}$$

$$x - \sqrt{3} = 0 \quad \textit{or} \quad x + 2\sqrt{3} = 0 \qquad \text{Getting the 0's}$$
$$\text{on one side}$$

$$\left(x - \sqrt{3}\right)\left(x + 2\sqrt{3}\right) = 0 \qquad \begin{array}{l}\text{Using the principle of}\\\text{zero products}\end{array}$$

$$x^2 + 2\sqrt{3}x - \sqrt{3}x - 2\left(\sqrt{3}\right)^2 = 0 \qquad \text{Using FOIL}$$

$$x^2 + \sqrt{3}x - 6 = 0. \qquad \text{Collecting like terms}$$

EXAMPLE 8 Write a quadratic equation whose solutions are $-12i$ and $12i$.

We have

$$x = -12i \quad \textit{or} \quad x = 12i$$

$$x + 12i = 0 \quad \textit{or} \quad x - 12i = 0 \qquad \text{Getting the 0's}$$
$$\text{on one side}$$

$$(x + 12i)(x - 12i) = 0 \qquad \text{Using the principle of zero products}$$

$$x^2 - 12ix + 12ix - 144i^2 = 0 \qquad \text{Using FOIL}$$

$$x^2 - 144(-1) = 0 \qquad \begin{array}{l}\text{Collecting like terms;}\\\text{substituting } -1 \text{ for } i^2\end{array}$$

$$x^2 + 144 = 0.$$

> Do Exercises 4–8.

> Find a quadratic equation having the following solutions.
>
> **4.** 7 and -2
>
> **5.** -4 and $\dfrac{5}{3}$
>
> **6.** $5i$ and $-5i$
>
> **7.** $-2\sqrt{2}$ and $\sqrt{2}$
>
> **8.** $-7i$ and $7i$

(c) Equations Quadratic in Form

Certain equations that are not really quadratic can still be solved as quadratic. Consider this fourth-degree equation.

$$\begin{array}{cccc} x^4 & - 9x^2 & + 8 & = 0 \\ \downarrow & \downarrow & \downarrow & \downarrow \\ (x^2)^2 & - 9(x^2) & + 8 & = 0 \qquad \text{Thinking of } x^4 \text{ as } (x^2)^2 \\ \downarrow & \downarrow & \downarrow & \downarrow \\ u^2 & - 9u & + 8 & = 0 \qquad \begin{array}{l}\text{To make this clearer,}\\\text{write } u \text{ instead of } x^2.\end{array}\end{array}$$

The equation $u^2 - 9u + 8 = 0$ can be solved by factoring or by the quadratic formula. After that, we can find x by remembering that $x^2 = u$. Equations that can be solved like this are said to be **quadratic in form**, or **reducible to quadratic**.

EXAMPLE 9 Solve: $x^4 - 9x^2 + 8 = 0$.

Let $u = x^2$. Then we solve the equation found by substituting u for x^2:

$$u^2 - 9u + 8 = 0$$

$$(u - 8)(u - 1) = 0 \qquad \text{Factoring}$$

$$u - 8 = 0 \quad \textit{or} \quad u - 1 = 0 \qquad \text{Using the principle of zero products}$$

$$u = 8 \quad \textit{or} \qquad u = 1.$$

Next, we substitute x^2 for u and solve these equations:

$$x^2 = 8 \quad or \quad x^2 = 1$$
$$x = \pm\sqrt{8} \quad or \quad x = \pm 1$$
$$x = \pm 2\sqrt{2} \quad or \quad x = \pm 1.$$

Note that when a number and its opposite are raised to an even power, the results are the same. Thus we can make one check for $\pm 2\sqrt{2}$ and one for ± 1.

Check:

For $\pm 2\sqrt{2}$:

$$\begin{array}{c} x^4 - 9x^2 + 8 = 0 \\ \hline (\pm 2\sqrt{2})^4 - 9(\pm 2\sqrt{2})^2 + 8 \; ? \; 0 \\ 64 - 9 \cdot 8 + 8 \quad \Big| \\ 0 \quad \Big| \quad \text{TRUE} \end{array}$$

For ± 1:

$$\begin{array}{c} x^4 - 9x^2 + 8 = 0 \\ \hline (\pm 1)^4 - 9(\pm 1)^2 + 8 \; ? \; 0 \\ 1 - 9 + 8 \quad \Big| \\ 0 \quad \Big| \quad \text{TRUE} \end{array}$$

The solutions are $1, -1, 2\sqrt{2},$ and $-2\sqrt{2}$.

--------- *Caution!* ---------

A common error is to solve for u and then forget to solve for x. Remember that you *must* find values for the *original* variable!

Do Exercise 9.

9. Solve: $x^4 - 10x^2 + 9 = 0$.

Solving equations quadratic in form can sometimes introduce numbers that are not solutions of the original equation. Thus a check by substitution in the original equation is necessary.

EXAMPLE 10 Solve: $x - 3\sqrt{x} - 4 = 0$.

Let $u = \sqrt{x}$. Then we solve the equation found by substituting u for $\sqrt{x}$ and u^2 for x:

$$u^2 - 3u - 4 = 0$$
$$(u - 4)(u + 1) = 0$$
$$u = 4 \quad or \quad u = -1.$$

Next, we substitute $\sqrt{x}$ for u and solve these equations:

$$\sqrt{x} = 4 \quad or \quad \sqrt{x} = -1.$$

Squaring the first equation, we get $x = 16$. Squaring the second equation, we get $x = 1$. We check both solutions.

Check:

For 16:

$$\begin{array}{c} x - 3\sqrt{x} - 4 = 0 \\ \hline 16 - 3\sqrt{16} - 4 \; ? \; 0 \\ 16 - 3 \cdot 4 - 4 \quad \Big| \\ 16 - 12 - 4 \quad \Big| \\ 0 \quad \Big| \quad \text{TRUE} \end{array}$$

For 1:

$$\begin{array}{c} x - 3\sqrt{x} - 4 = 0 \\ \hline 1 - 3\sqrt{1} - 4 \; ? \; 0 \\ 1 - 3 \cdot 1 - 4 \quad \Big| \\ -6 \quad \Big| \quad \text{FALSE} \end{array}$$

10. Solve: $x + 3\sqrt{x} - 10 = 0$.
Be sure to check.

Since 16 checks but 1 does not, the solution is 16.

Do Exercise 10.

Answers

9. $\pm 3, \pm 1$ **10.** 4

EXAMPLE 11 Solve: $y^{-2} - y^{-1} - 2 = 0$.

Let $u = y^{-1}$. Then we solve the equation found by substituting u for y^{-1} and u^2 for y^{-2}:

$$u^2 - u - 2 = 0$$
$$(u - 2)(u + 1) = 0$$
$$u = 2 \quad or \quad u = -1.$$

Next, we substitute y^{-1} or $1/y$ for u and solve these equations:

$$\frac{1}{y} = 2 \quad or \quad \frac{1}{y} = -1.$$

Solving, we get

$$y = \frac{1}{2} \quad or \quad y = \frac{1}{(-1)} = -1.$$

The numbers $\frac{1}{2}$ and -1 both check. They are the solutions.

Do Exercise 11.

11. Solve: $x^{-2} + x^{-1} - 6 = 0$.

EXAMPLE 12 Find the x-intercepts of the graph of

$$f(x) = (x^2 - 1)^2 - (x^2 - 1) - 2.$$

The x-intercepts occur where $f(x) = 0$, so we must have

$$(x^2 - 1)^2 - (x^2 - 1) - 2 = 0.$$

Let $u = x^2 - 1$. Then we solve the equation found by substituting u for $x^2 - 1$:

$$u^2 - u - 2 = 0$$
$$(u - 2)(u + 1) = 0$$
$$u = 2 \quad or \quad u = -1.$$

Next, we substitute $x^2 - 1$ for u and solve these equations:

$$x^2 - 1 = 2 \quad or \quad x^2 - 1 = -1$$
$$x^2 = 3 \quad or \quad x^2 = 0$$
$$x = \pm\sqrt{3} \quad or \quad x = 0.$$

The numbers $\sqrt{3}$, $-\sqrt{3}$, and 0 check. They are the solutions of $(x^2 - 1)^2 - (x^2 - 1) - 2 = 0$. Thus the x-intercepts of the graph of $f(x)$ are $\left(-\sqrt{3}, 0\right)$, $(0, 0)$, and $\left(\sqrt{3}, 0\right)$.

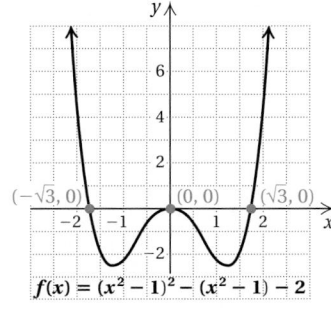

Do Exercise 12.

12. Find the x-intercepts of
$$f(x) = (x^2 - x)^2 - 14(x^2 - x) + 24.$$

Answers

11. $-\dfrac{1}{3}, \dfrac{1}{2}$ **12.** $(-3, 0), (-1, 0), (2, 0), (4, 0)$

a Determine the nature of the solutions of each equation.

1. $x^2 - 8x + 16 = 0$

2. $x^2 + 12x + 36 = 0$

3. $x^2 + 1 = 0$

4. $x^2 + 6 = 0$

5. $x^2 - 6 = 0$

6. $x^2 - 3 = 0$

7. $4x^2 - 12x + 9 = 0$

8. $4x^2 + 8x - 5 = 0$

9. $x^2 - 2x + 4 = 0$

10. $x^2 + 3x + 4 = 0$

11. $9t^2 - 3t = 0$

12. $4m^2 + 7m = 0$

13. $y^2 = \frac{1}{2}y + \frac{3}{5}$

14. $y^2 + \frac{9}{4} = 4y$

15. $4x^2 - 4\sqrt{3}x + 3 = 0$

16. $6y^2 - 2\sqrt{3}y - 1 = 0$

b Write a quadratic equation having the given numbers as solutions.

17. -4 and 4

18. -11 and 9

19. $-4i$ and $4i$

20. $-i$ and i

21. 8, only solution

[*Hint*: It must be a double solution, that is, $(x - 8)(x - 8) = 0$.]

22. -3, only solution

23. $-\frac{2}{5}$ and $\frac{6}{5}$

24. $-\frac{1}{4}$ and $-\frac{1}{2}$

25. $\frac{k}{3}$ and $\frac{m}{4}$

26. $\dfrac{c}{2}$ and $\dfrac{d}{2}$

27. $-\sqrt{3}$ and $2\sqrt{3}$

28. $\sqrt{2}$ and $3\sqrt{2}$

29. $6i$ and $-6i$

30. $8i$ and $-8i$

C   Solve.

31. $x^4 - 6x^2 + 9 = 0$

32. $x^4 - 7x^2 + 12 = 0$

33. $x - 10\sqrt{x} + 9 = 0$

34. $2x - 9\sqrt{x} + 4 = 0$

35. $(x^2 - 6x)^2 - 2(x^2 - 6x) - 35 = 0$

36. $(x^2 + 5x)^2 + 2(x^2 + 5x) - 24 = 0$

37. $x^{-2} - 5x^{-1} - 36 = 0$

38. $3x^{-2} - x^{-1} - 14 = 0$

39. $\left(1 + \sqrt{x}\right)^2 + \left(1 + \sqrt{x}\right) - 6 = 0$

40. $\left(2 + \sqrt{x}\right)^2 - 3\left(2 + \sqrt{x}\right) - 10 = 0$

41. $(y^2 - 5y)^2 - 2(y^2 - 5y) - 24 = 0$

42. $(2t^2 + t)^2 - 4(2t^2 + t) + 3 = 0$

43. $w^4 - 29w^2 + 100 = 0$

44. $t^4 - 10t^2 + 9 = 0$

45. $2x^{-2} + x^{-1} - 1 = 0$

46. $m^{-2} + 9m^{-1} - 10 = 0$

47. $6x^4 - 19x^2 + 15 = 0$

48. $6x^4 - 17x^2 + 5 = 0$

49. $x^{2/3} - 4x^{1/3} - 5 = 0$

50. $x^{2/3} + 2x^{1/3} - 8 = 0$

51. $\left(\dfrac{x-4}{x+1}\right)^2 - 2\left(\dfrac{x-4}{x+1}\right) - 35 = 0$

52. $\left(\dfrac{x+3}{x-3}\right)^2 - \left(\dfrac{x+3}{x-3}\right) - 6 = 0$

53. $9\left(\dfrac{x+2}{x+3}\right)^2 - 6\left(\dfrac{x+2}{x+3}\right) + 1 = 0$

54. $16\left(\dfrac{x-1}{x-8}\right)^2 + 8\left(\dfrac{x-1}{x-8}\right) + 1 = 0$

55. $\left(\dfrac{x^2-2}{x}\right)^2 - 7\left(\dfrac{x^2-2}{x}\right) - 18 = 0$

56. $\left(\dfrac{y^2-1}{y}\right)^2 - 4\left(\dfrac{y^2-1}{y}\right) - 12 = 0$

Find the x-intercepts of the graph of each function.

57. $f(x) = 5x + 13\sqrt{x} - 6$

58. $f(x) = 3x + 10\sqrt{x} - 8$

59. $f(x) = (x^2 - 3x)^2 - 10(x^2 - 3x) + 24$

60. $f(x) = (x^2 - x)^2 - 8(x^2 - x) + 12$

61. $f(x) = x^{2/3} + x^{1/3} - 2$

62. $f(x) = x^{2/5} + x^{1/5} - 6$

Skill Maintenance

Solve. [8.4a]

63. *Coffee Beans.* Twin Cities Roasters sells Kenyan coffee worth $6.75 per pound and Peruvian coffee worth $11.25 per pound. How many pounds of each kind should be mixed in order to obtain a 50-lb mixture that is worth $8.55 per pound?

64. *Solution Mixtures.* Solution A is 18% alcohol and solution B is 45% alcohol. How many liters of each should be mixed in order to get 12 L of a solution that is 36% alcohol?

Multiply and simplify. Assume that no radicands were formed by raising negative numbers to even powers. [10.3a]

65. $\sqrt{8x}\sqrt{2x}$

66. $\sqrt[3]{x^2}\sqrt[3]{27x^4}$

67. $\sqrt[4]{9a^2}\sqrt[4]{18a^3}$

68. $\sqrt[5]{16}\sqrt[5]{64}$

Graph. [7.1c], [7.4a, c]

69. $f(x) = -\frac{3}{5}x + 4$

70. $5x - 2y = 8$

71. $y = 4$

72. $f(x) = -x - 3$

Synthesis

73. 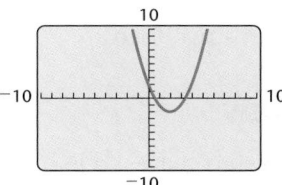 Use a graphing calculator to check your answers to Exercises 32, 34, 36, and 39.

74. 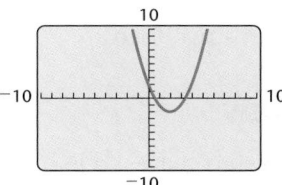 Use a graphing calculator to solve each of the following equations.
 a) $6.75x - 35\sqrt{x} - 5.26 = 0$
 b) $\pi x^4 - \pi^2 x^2 = \sqrt{99.3}$
 c) $x^4 - x^3 - 13x^2 + x + 12 = 0$

For each equation under the given condition, **(a)** find k and **(b)** find the other solution.

75. $kx^2 - 2x + k = 0$; one solution is -3.

76. $kx^2 - 17x + 33 = 0$; one solution is 3.

77. Find a quadratic equation for which the sum of the solutions is $\sqrt{3}$ and the product is 8.

78. Find k given that $kx^2 - 4x + (2k - 1) = 0$ and the product of the solutions is 3.

79. The graph of a function of the form
$$f(x) = ax^2 + bx + c$$
is a curve similar to the one shown below. Determine a, b, and c from the information given.

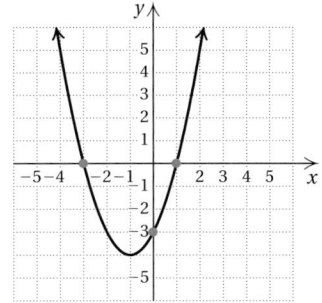

80. 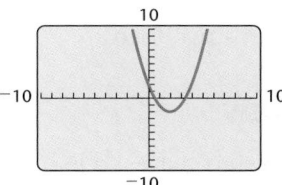 While solving a quadratic equation of the form $ax^2 + bx + c = 0$ with a graphing calculator, Shawn-Marie gets the following screen.

How could the discriminant help her check the graph?

Solve.

81. $\dfrac{x}{x - 1} - 6\sqrt{\dfrac{x}{x - 1}} - 40 = 0$

82. $\dfrac{x}{x - 3} - 24 = 10\sqrt{\dfrac{x}{x - 3}}$

83. $\sqrt{x - 3} - \sqrt[4]{x - 3} = 12$

84. $a^3 - 26a^{3/2} - 27 = 0$

85. $x^6 - 28x^3 + 27 = 0$

86. $x^6 + 7x^3 - 8 = 0$

Mid-Chapter Review

Concept Reinforcement

Determine whether each statement is true or false.

_____ **1.** Every quadratic equation has exactly two real-number solutions. [11.4a]

_____ **2.** The quadratic formula can be used to find all the solutions of any quadratic equation. [11.2a]

_____ **3.** If the graph of a quadratic equation crosses the x-axis, then it has exactly two real-number solutions. [11.4a]

_____ **4.** The x-intercepts of $f(x) = x^2 - t$ are $\left(0, \sqrt{t}\right)$ and $\left(0, -\sqrt{t}\right)$. [11.1a]

Guided Solutions

Fill in each blank with the number that creates a correct solution.

5. Solve $5x^2 + 3x = 4$ by completing the square. [11.1b]

$$5x^2 + 3x = 4$$

$$\Box(5x^2 + 3x) = \Box \cdot 4$$

$$x^2 + \frac{3}{\Box}x = \frac{4}{\Box}$$

$$x^2 + \frac{3}{5}x + \Box = \frac{4}{5} + \Box$$

$$(x + \Box)^2 = \frac{\Box}{100}$$

$$x + \frac{3}{10} = \sqrt{\Box} \qquad or \qquad x + \frac{3}{10} = -\sqrt{\Box}$$

$$x + \frac{3}{10} = \frac{\sqrt{\Box}}{\Box} \qquad or \qquad x + \frac{3}{10} = -\frac{\sqrt{\Box}}{\Box}$$

$$x = -\frac{\Box}{10} + \frac{\sqrt{\Box}}{10} \qquad or \qquad x = -\frac{\Box}{10} - \frac{\sqrt{\Box}}{10}.$$

The solutions are $-\dfrac{\Box}{10} \pm \dfrac{\sqrt{\Box}}{10}$.

6. Use the quadratic formula to solve $5x^2 + 3x = 4$. [11.2a]

$$5x^2 + 3x = 4$$

$$5x^2 + 3x - \Box = 0$$

$$5x^2 + 3x + \Box = 0$$

$$a = \Box, \quad b = \Box, \quad c = \Box$$

$$x = \frac{-b \pm \sqrt{b^2 - 4ac}}{2a}$$

$$x = \frac{-\Box \pm \sqrt{\Box^2 - 4 \cdot \Box \cdot \Box}}{2 \cdot \Box}$$

$$x = \frac{-3 \pm \sqrt{\Box + \Box}}{\Box}$$

$$x = \frac{-3 \pm \sqrt{\Box}}{\Box}$$

$$x = -\frac{3}{10} \pm \frac{\sqrt{\Box}}{\Box}$$

Mixed Review

Solve by completing the square. [11.1b]

7. $x^2 + 1 = -4x$

8. $2x^2 + 5x - 3 = 0$

9. $x^2 + 10x - 6 = 0$

10. $x^2 - x = 5$

Determine the nature of the solutions of each equation $ax^2 + bx + c = 0$ and the number of x-intercepts of the graph of the function $f(x) = ax^2 + bx + c$. [11.4a]

11. $x^2 - 10x + 25 = 0$

12. $x^2 - 11 = 0$

13. $y^2 = \frac{1}{3}y - \frac{4}{7}$

14. $x^2 + 5x + 9 = 0$

15. $x^2 - 4 = 2x$

16. $x^2 - 8x = 0$

Write a quadratic equation having the given numbers as solutions. [11.4b]

17. -1 and 10

18. -13 and 13

19. $-\sqrt{5}$ and $3\sqrt{5}$

20. $-4i$ and $4i$

21. -6, only solution

22. $-\dfrac{4}{3}$ and $\dfrac{2}{7}$

Solve.

23. Jacob traveled 780 mi by car. Had he gone 5 mph faster, he could have made the trip in 1 hr less time. Find his speed. [11.3a]

24. $R = as^2$, for s [11.3b]

Solve. [11.1a], [11.2a], [11.4c]

25. $3x^2 + x = 4$

26. $x^4 - 8x^2 + 15 = 0$

27. $4x^2 = 15x - 5$

28. $7x^2 + 2 = -9x$

29. $2x + x(x - 1) = 0$

30. $(x + 3)^2 = 64$

31. $49x^2 + 16 = 0$

32. $(x^2 - 2)^2 + 2(x^2 - 2) - 24 = 0$

33. $r^2 + 5r = 12$

34. $s^2 + 12s + 37 = 0$

35. $\left(x - \dfrac{5}{2}\right)^2 = \dfrac{11}{4}$

36. $x + \dfrac{1}{x} = \dfrac{7}{3}$

37. $4x + 1 = 4x^2$

38. $(x - 3)^2 + (x + 5)^2 = 0$

39. $b^2 - 16b + 64 = 3$

40. $(x - 3)^2 = -10$

41. $\dfrac{1}{x} + \dfrac{1}{x + 2} = \dfrac{1}{5}$

42. $x - \sqrt{x} - 6 = 0$

Understanding Through Discussion and Writing

43. Given the solutions of a quadratic equation, is it possible to reconstruct the original equation? Why or why not? [11.4b]

44. Explain how the quadratic formula can be used to factor a quadratic polynomial into two binomials. Use it to factor $5x^2 + 8x - 3$. [11.2a]

45. Describe a procedure that could be used to write an equation having the first seven natural numbers as solutions. [11.4b]

46. Describe a procedure that could be used to write an equation that is quadratic in $3x^2 + 1$ and has real-number solutions. [11.4c]

OBJECTIVES

a Graph quadratic functions of the type $f(x) = ax^2$ and then label the vertex and the line of symmetry.

b Graph quadratic functions of the type $f(x) = a(x - h)^2$ and then label the vertex and the line of symmetry.

c Graph quadratic functions of the type $f(x) = a(x - h)^2 + k$, finding the vertex, the line of symmetry, and the maximum or minimum function value, or y-value.

SKILL TO REVIEW
Objective 7.1c: Draw the graph of a function.

Graph the function.

1. $f(x) = -\dfrac{1}{2}x - 3$

2. $f(x) = x^2 + 1$

In this section and the next, we develop techniques for graphing quadratic functions.

a **Graphs of** $f(x) = ax^2$

The most basic quadratic function is $f(x) = x^2$.

EXAMPLE 1 Graph: $f(x) = x^2$.

We choose some values for x and compute $f(x)$ for each. Then we plot the ordered pairs and connect them with a smooth curve.

x	$f(x) = x^2$	$(x, f(x))$
-3	9	$(-3, 9)$
-2	4	$(-2, 4)$
-1	1	$(-1, 1)$
0	0	$(0, 0)$
1	1	$(1, 1)$
2	4	$(2, 4)$
3	9	$(3, 9)$

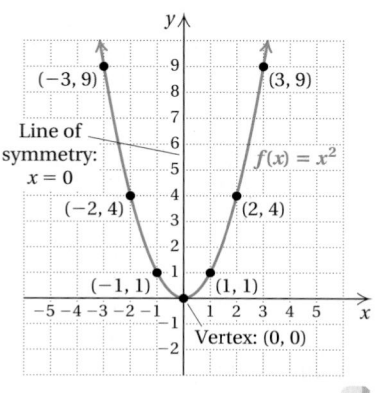

All quadratic functions have graphs similar to the one in Example 1. Such curves are called **parabolas**. They are cup-shaped curves that are symmetric with respect to a vertical line known as the parabola's **line of symmetry**, or **axis of symmetry**. In the graph of $f(x) = x^2$, shown above, the y-axis (or the line $x = 0$) is the line of symmetry. If the paper were to be folded on this line, the two halves of the curve would coincide. The point $(0, 0)$ is the **vertex** of this parabola.

Let's compare the graphs of $g(x) = \frac{1}{2}x^2$ and $h(x) = 2x^2$ with the graph of $f(x) = x^2$. We choose x-values and plot points for both functions.

x	$g(x) = \frac{1}{2}x^2$
-3	$\frac{9}{2}$
-2	2
-1	$\frac{1}{2}$
0	0
1	$\frac{1}{2}$
2	2
3	$\frac{9}{2}$

x	$h(x) = 2x^2$
-3	18
-2	8
-1	2
0	0
1	2
2	8
3	18

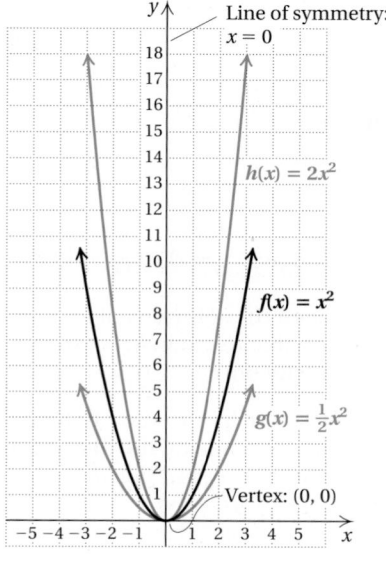

Note the symmetry: For equal increments to the left and right of the vertex, the y-values are the same.

Answers

Skill to Review:
1.

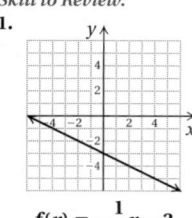

$f(x) = -\dfrac{1}{2}x - 3$

2.

$f(x) = x^2 + 1$

Note that the graph of $g(x) = \frac{1}{2}x^2$ is a wider parabola than the graph of $f(x) = x^2$, and the graph of $h(x) = 2x^2$ is narrower. The vertex and the line of symmetry, however, remain $(0, 0)$ and $x = 0$, respectively.

When we consider the graph of $k(x) = -\frac{1}{2}x^2$, we see that the parabola opens down and is the same shape as the graph of $g(x) = \frac{1}{2}x^2$.

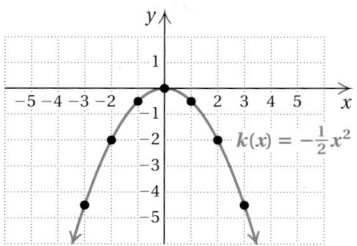

GRAPHS OF $f(x) = ax^2$

The graph of $f(x) = ax^2$, or $y = ax^2$, is a parabola with $x = 0$ as its line of symmetry; its vertex is the origin.

For $a > 0$, the parabola opens up; for $a < 0$, the parabola opens down.

If $|a|$ is greater than 1, the parabola is narrower than $y = x^2$.

If $|a|$ is between 0 and 1, the parabola is wider than $y = x^2$.

Do Exercises 1–3.

(b) Graphs of $f(x) = a(x - h)^2$

It would seem logical now to consider functions of the type

$$f(x) = ax^2 + bx + c.$$

We are heading in that direction, but it is convenient to first consider graphs of $f(x) = a(x - h)^2$ and then $f(x) = a(x - h)^2 + k$, where a, h, and k are constants.

EXAMPLE 2 Graph: $g(x) = (x - 3)^2$.

We choose some values for x and compute $g(x)$. Then we plot the points and draw the curve.

x	$g(x) = (x - 3)^2$
3	0
4	1
5	4
6	9
2	1
1	4
0	9

← Vertex

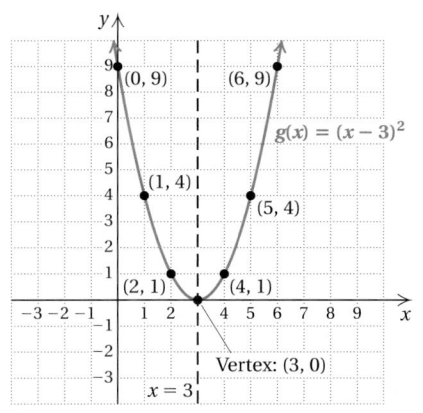

Graph.

1. $f(x) = -\dfrac{1}{3}x^2$

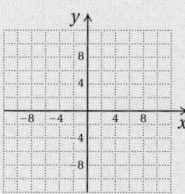

2. $f(x) = 3x^2$

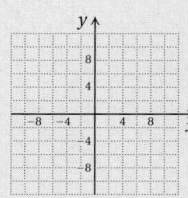

3. $f(x) = -2x^2$

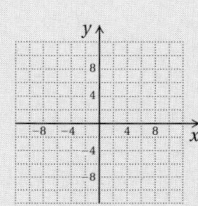

Answers

1.

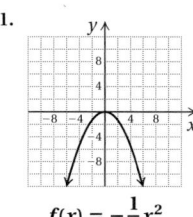

$f(x) = -\dfrac{1}{3}x^2$

2.

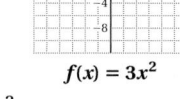

$f(x) = 3x^2$

3.

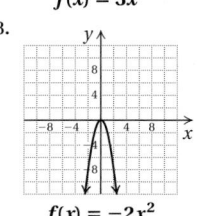

$f(x) = -2x^2$

First, note that for an x-value of 3, $g(3) = (3 - 3)^2 = 0$. As we increase x-values from 3, note that the corresponding y-values increase. Then as we decrease x-values from 3, note that the corresponding y-values increase again. The line $x = 3$ is the line of symmetry. Equal distances of x-values to the left and right of the vertex produce the same y-values.

EXAMPLE 3 Graph: $t(x) = (x + 3)^2$.

We choose some values for x and compute $t(x)$. Then we plot the points and draw the curve.

x	$t(x) = (x + 3)^2$
-3	0
-2	1
-1	4
0	9
-4	1
-5	4
-6	9

← Vertex (for row $x = -3$, $t(x) = 0$)

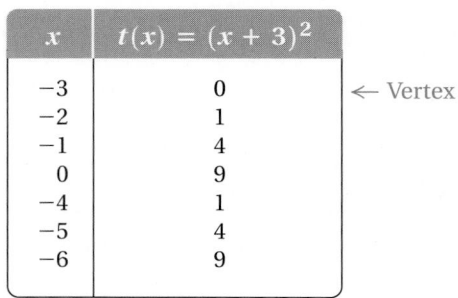

First, note that for an x-value of -3, $t(-3) = (-3 + 3)^2 = 0$. As we increase x-values from -3, note that the corresponding y-values increase. Then as we decrease x-values from -3, note that the y-values increase again. The line $x = -3$ is the line of symmetry.

The graph of $g(x) = (x - 3)^2$ in Example 2 looks just like the graph of $f(x) = x^2$ in Example 1, except that it is moved, or translated, 3 units to the right. Comparing the pairs for $g(x)$ with those for $f(x)$, we see that when an input for $g(x)$ is 3 more than an input for $f(x)$, the outputs are the same.

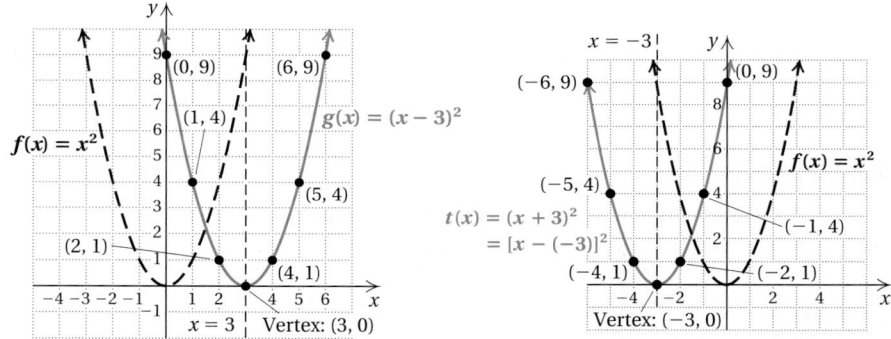

The graph of $t(x) = (x + 3)^2 = [x - (-3)]^2$ in Example 3 looks just like the graph of $f(x) = x^2$ in Example 1, except that it is moved, or translated, 3 units to the left. Comparing the pairs for $t(x)$ with those for $f(x)$, we see that when an input for $t(x)$ is 3 less than an input for $f(x)$, the outputs are the same.

GRAPHS OF $f(x) = a(x - h)^2$

The graph of $f(x) = a(x - h)^2$ has the same shape as the graph of $y = ax^2$.

If h is positive, the graph of $y = ax^2$ is shifted h units to the right.

If h is negative, the graph of $y = ax^2$ is shifted $|h|$ units to the left.

The vertex is $(h, 0)$, and the line of symmetry is $x = h$.

EXAMPLE 4 Graph: $f(x) = -2(x + 3)^2$.

We first rewrite the equation as $f(x) = -2[x - (-3)]^2$. In this case, $a = -2$ and $h = -3$, so the graph looks like that of $g(x) = 2x^2$ translated 3 units to the left and, since $-2 < 0$, the graph opens down. The vertex is $(-3, 0)$, and the line of symmetry is $x = -3$. Plotting points as needed, we obtain the graph shown below.

x	$f(x) = -2(x + 3)^2$
-3	0
-2	-2
-1	-8
-4	-2
-5	-8

← Vertex

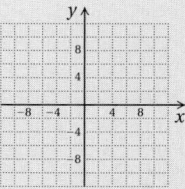

Graph. Find and label the vertex and the line of symmetry.

4. $f(x) = \dfrac{1}{2}(x - 4)^2$

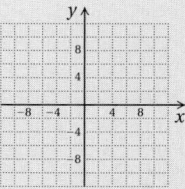

5. $f(x) = -\dfrac{1}{2}(x - 4)^2$

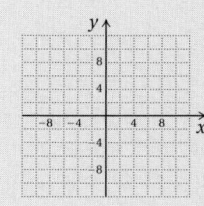

Do Exercises 4 and 5.

(c) Graphs of $f(x) = a(x - h)^2 + k$

Given a graph of $f(x) = a(x - h)^2$, what happens if we add a constant k? Suppose that we add 2. This increases each function value $f(x)$ by 2, so the curve is moved up. If k is negative, the curve is moved down. The line of symmetry for the parabola remains $x = h$, but the vertex will be at (h, k), or equivalently, $(h, f(h))$.

Note that if a parabola opens up ($a > 0$), the function value, or y-value, at the vertex is a least, or **minimum**, value. That is, it is less than the y-value at any other point on the graph. If the parabola opens down ($a < 0$), the function value at the vertex is a greatest, or **maximum**, value.

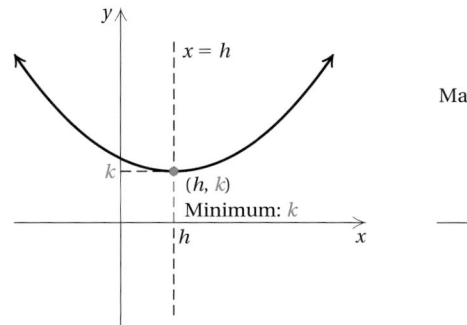

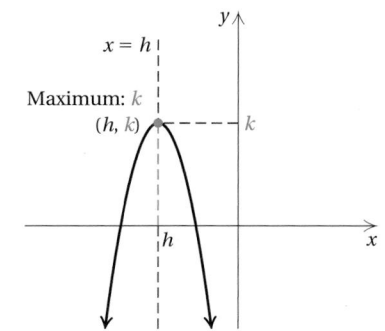

Answers

4.

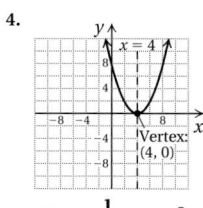

$f(x) = \dfrac{1}{2}(x - 4)^2$

5.

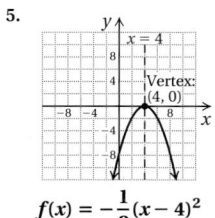

$f(x) = -\dfrac{1}{2}(x - 4)^2$

> ### GRAPHS OF $f(x) = a(x - h)^2 + k$
>
> The graph of $f(x) = a(x - h)^2 + k$ has the same shape as the graph of $y = a(x - h)^2$.
>
> If k is positive, the graph of $y = a(x - h)^2$ is shifted k units up.
>
> If k is negative, the graph of $y = a(x - h)^2$ is shifted $|k|$ down.
>
> The vertex is (h, k), and the line of symmetry is $x = h$.
>
> For $a > 0$, k is the minimum function value. For $a < 0$, k is the maximum function value.

EXAMPLE 5   Graph $f(x) = (x - 3)^2 - 5$, and find the minimum function value.

The graph will look like that of $g(x) = (x - 3)^2$ (see Example 2) but translated 5 units down. You can confirm this by plotting some points. For instance,

$$f(4) = (4 - 3)^2 - 5 = -4,$$

whereas in Example 2,

$$g(4) = (4 - 3)^2 = 1.$$

Note that the vertex is $(h, k) = (3, -5)$, so we begin calculating points on both sides of $x = 3$. The line of symmetry is $x = 3$, and the minimum function value is -5.

x	$f(x) = (x - 3)^2 - 5$	
3	-5	← Vertex
4	-4	
5	-1	
6	4	
2	-4	
1	-1	
0	4	

EXAMPLE 6 Graph $t(x) = \frac{1}{2}(x - 3)^2 + 5$, and find the minimum function value.

The graph looks just like that of $f(x) = \frac{1}{2}x^2$ but moved 3 units to the right and 5 units up. The vertex is $(3, 5)$, and the line of symmetry is $x = 3$. We draw $f(x) = \frac{1}{2}x^2$ and then shift the curve over and up. The minimum function value is 5. By plotting some points, we have a check.

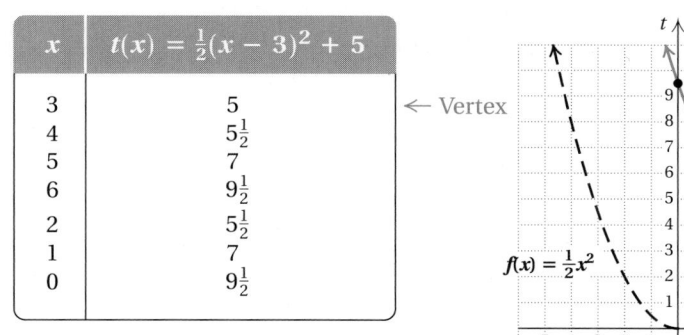

x	$t(x) = \frac{1}{2}(x - 3)^2 + 5$
3	5
4	$5\frac{1}{2}$
5	7
6	$9\frac{1}{2}$
2	$5\frac{1}{2}$
1	7
0	$9\frac{1}{2}$

← Vertex

EXAMPLE 7 Graph $f(x) = -2(x + 3)^2 + 5$. Find the vertex, the line of symmetry, and the maximum or minimum value.

We first express the equation in the equivalent form

$$f(x) = -2[x - (-3)]^2 + 5.$$

The graph looks like that of $g(x) = -2x^2$ translated 3 units to the left and 5 units up. The vertex is $(-3, 5)$, and the line of symmetry is $x = -3$. Since $-2 < 0$, we know that the graph opens down so 5, the second coordinate of the vertex, is the maximum y-value.

We compute a few points as needed and draw the graph.

x	$f(x) = -2(x + 3)^2 + 5$
-3	5
-2	3
-1	-3
-4	3
-5	-3

← Vertex

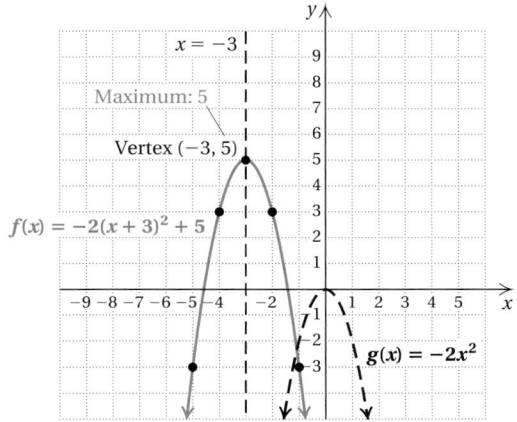

Do Exercises 6 and 7.

Graph. Find the vertex, the line of symmetry, and the maximum or minimum y-value.

6. $f(x) = \frac{1}{2}(x + 2)^2 - 4$

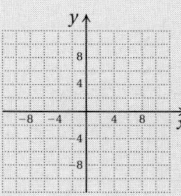

7. $f(x) = -2(x - 5)^2 + 3$

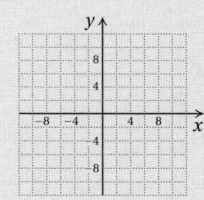

Answers

6.

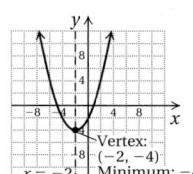

Vertex: $(-2, -4)$
$x = -2$ Minimum: -4

$f(x) = \frac{1}{2}(x + 2)^2 - 4$

7.

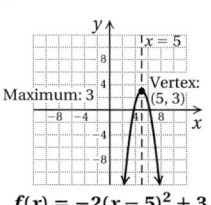

Maximum: 3   Vertex: $(5, 3)$ $x = 5$

$f(x) = -2(x - 5)^2 + 3$

11.5 **Exercise Set**

For Extra Help

MyMathLab

Math XL
PRACTICE

WATCH

DOWNLOAD

READ

REVIEW

a , **b**   Graph. Find and label the vertex and the line of symmetry.

1. $f(x) = 4x^2$

x	$f(x)$
0	
1	
2	
−1	
−2	

Vertex: (____ , ____)
Line of symmetry: $x =$ ____

2. $f(x) = 5x^2$

x	$f(x)$
0	
1	
2	
−1	
−2	

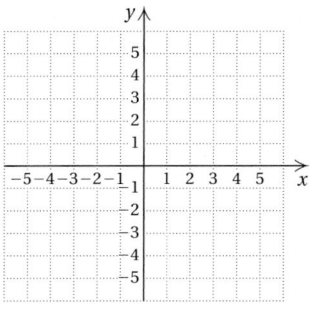

Vertex: (____ , ____)
Line of symmetry: $x =$ ____

3. $f(x) = \frac{1}{3}x^2$

x	$f(x)$
0	
1	
2	
−1	
−2	

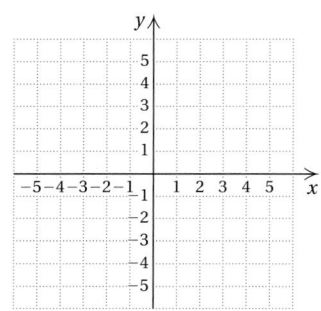

Vertex: (____ , ____)
Line of symmetry: $x =$ ____

4. $f(x) = \frac{1}{4}x^2$

x	$f(x)$
0	
1	
2	
−1	
−2	

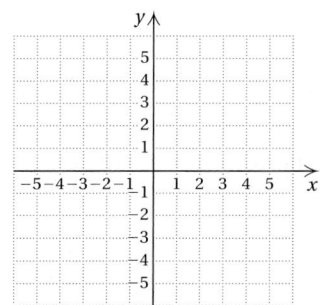

Vertex: (____ , ____)
Line of symmetry: $x =$ ____

5. $f(x) = (x + 3)^2$

x	$f(x)$
−3	
−2	
−1	
−4	
−5	

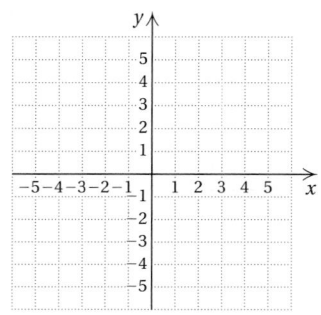

Vertex: (____ , ____)
Line of symmetry: $x =$ ____

6. $f(x) = (x + 1)^2$

x	$f(x)$
−1	
0	
1	
−2	
−3	

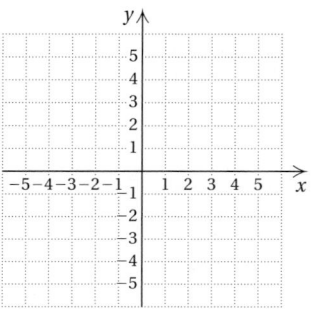

Vertex: (____ , ____)
Line of symmetry: $x =$ ____

7. $f(x) = -4x^2$

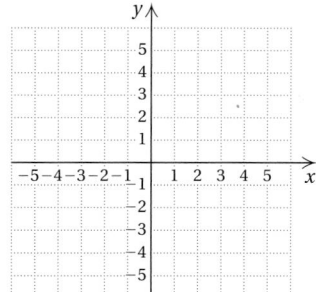

Vertex: (____, ____)
Line of symmetry: $x =$ ____

8. $f(x) = -3x^2$

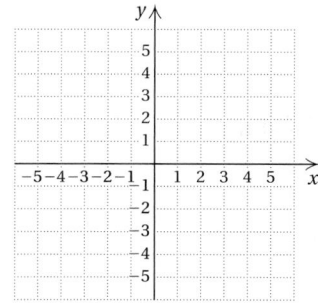

Vertex: (____, ____)
Line of symmetry: $x =$ ____

9. $f(x) = -\frac{1}{2}x^2$

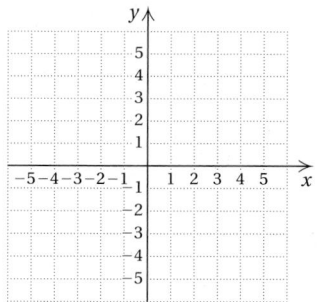

Vertex: (____, ____)
Line of symmetry: $x =$ ____

10. $f(x) = -\frac{1}{4}x^2$

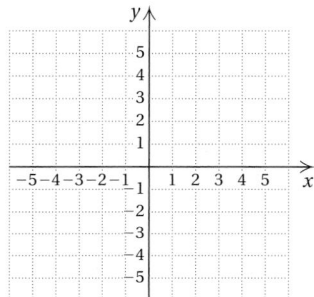

Vertex: (____, ____)
Line of symmetry: $x =$ ____

11. $f(x) = 2(x - 4)^2$

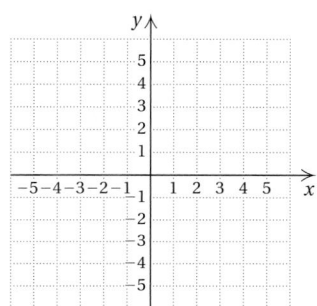

Vertex: (____, ____)
Line of symmetry: $x =$ ____

12. $f(x) = 4(x - 1)^2$

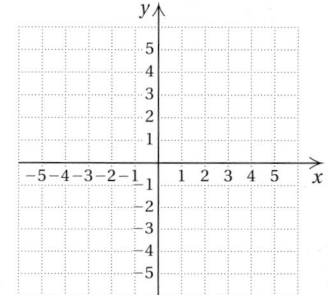

Vertex: (____, ____)
Line of symmetry: $x =$ ____

13. $f(x) = -2(x + 2)^2$

x	$f(x)$
-2	
-3	
-1	
-4	
0	

Vertex: (____, ____)
Line of symmetry: $x =$ ____

14. $f(x) = -2(x + 4)^2$

x	$f(x)$
-4	
-5	
-3	
-6	
-2	

Vertex: (____, ____)
Line of symmetry: $x =$ ____

15. $f(x) = 3(x - 1)^2$

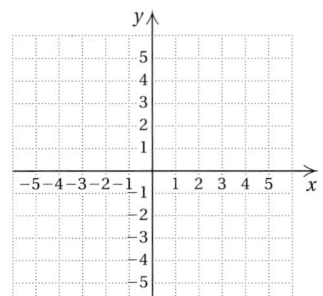

16. $f(x) = 4(x - 2)^2$

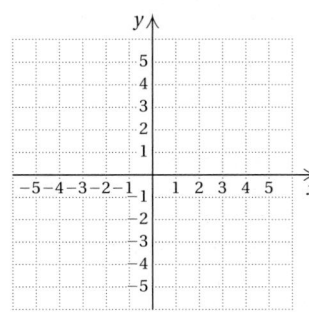

17. $f(x) = -\frac{3}{2}(x + 2)^2$

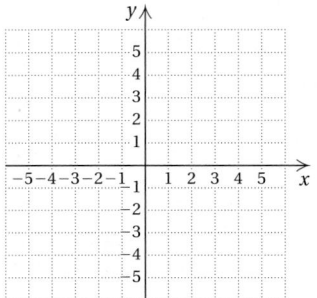

18. $f(x) = -\frac{5}{2}(x + 3)^2$

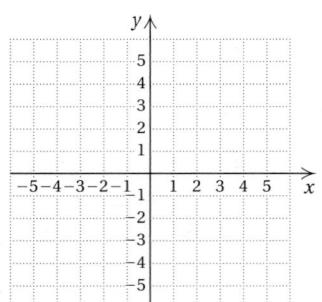

c   Graph. Find and label the vertex and the line of symmetry. Find the maximum or minimum value.

19. $f(x) = (x - 3)^2 + 1$

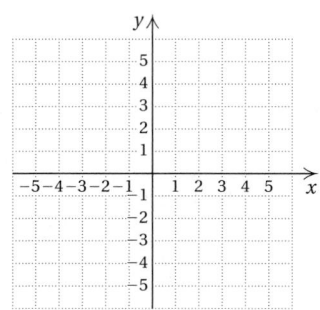

Vertex: (____, ____)
Line of symmetry: $x =$ ____
Minimum value: ____

20. $f(x) = (x + 2)^2 - 3$

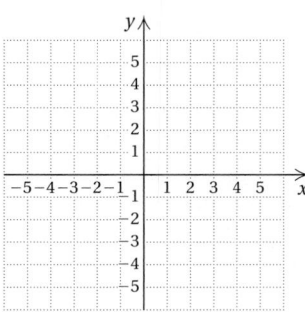

Vertex: (____, ____)
Line of symmetry: $x =$ ____
Minimum value: ____

21. $f(x) = -3(x + 4)^2 + 1$

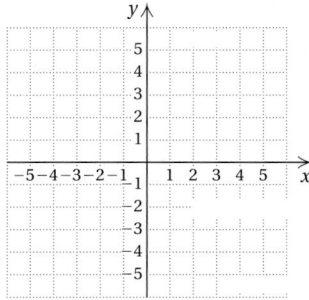

Vertex: (____, ____)
Line of symmetry: $x =$ ____
Maximum value: ____

22. $f(x) = -\frac{1}{2}(x - 1)^2 - 3$

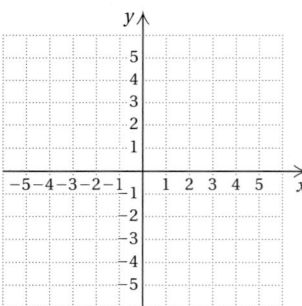

Vertex: (____, ____)
Line of symmetry: $x =$ ____
Maximum value: ____

23. $f(x) = \frac{1}{2}(x + 1)^2 + 4$

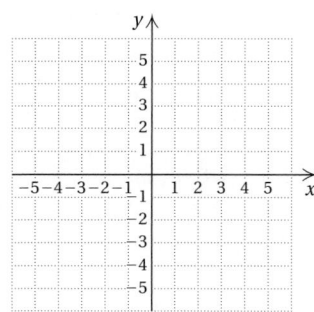

Vertex: (____, ____)
Line of symmetry: $x =$ ____
_____ value: ____

24. $f(x) = -2(x - 5)^2 - 3$

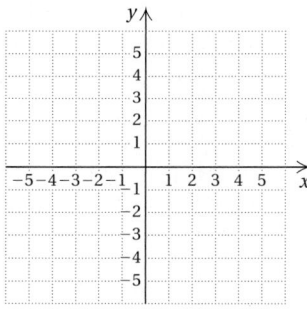

Vertex: (____, ____)
Line of symmetry: $x =$ ____
_____ value: ____

25. $f(x) = -(x + 1)^2 - 2$

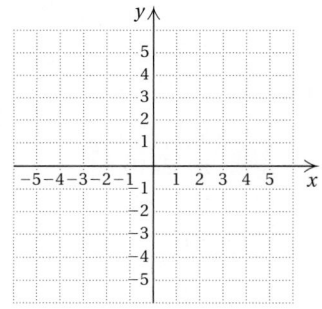

Vertex: (____, ____)
Line of symmetry: $x =$ ____
_____ value: ____

26. $f(x) = 3(x - 4)^2 + 2$

Vertex: (____, ____)
Line of symmetry: $x =$ ____
_____ value: ____

Skill Maintenance

Multiply and simplify. Assume that no radicands were formed by raising negative numbers to even powers.   [10.3a]

27. $\sqrt[4]{5x^3y^5}\sqrt[4]{125x^2y^3}$

28. $\sqrt{9a^3}\sqrt{16ab^4}$

11.6 Graphing $f(x) = ax^2 + bx + c$

(a) Analyzing and Graphing $f(x) = ax^2 + bx + c$

By *completing the square*, we can begin with any quadratic polynomial $ax^2 + bx + c$ and find an equivalent expression $a(x - h)^2 + k$. This allows us to combine the skills of Sections 11.1 and 11.5 to analyze and graph any quadratic function $f(x) = ax^2 + bx + c$.

EXAMPLE 1 For $f(x) = x^2 - 6x + 4$, find the vertex, the line of symmetry, and the maximum or the minimum value. Then graph.

We first find the vertex and the line of symmetry. To do so, we find the equivalent form $a(x - h)^2 + k$ by completing the square, beginning as follows:

$$f(x) = x^2 - 6x + 4 = (x^2 - 6x \qquad) + 4.$$

We complete the square inside the parentheses, but in a different manner than we did before. We take half the x-coefficient, $-6/2 = -3$, and square it: $(-3)^2 = 9$. Then we add 0, or $9 - 9$, inside the parentheses. (Because we are using function notation, instead of adding $(b/2)^2$ on both sides of an equation, we add and subtract it on the same side, effectively adding 0 and not changing the value of the expression.)

$$
\begin{aligned}
f(x) &= (x^2 - 6x + 0) + 4 && \text{Adding 0} \\
&= (x^2 - 6x + 9 - 9) + 4 && \text{Substituting } 9 - 9 \text{ for } 0 \\
&= (x^2 - 6x + 9) + (-9 + 4) && \text{Using the associative law} \\
& && \text{of addition to regroup} \\
&= (x - 3)^2 - 5 && \text{Factoring and simplifying}
\end{aligned}
$$

(This equation was graphed in Example 5 of Section 11.5.) The vertex is $(3, -5)$, and the line of symmetry is $x = 3$. The coefficient of x^2 is 1, which is positive, so the graph opens up. This tells us that -5 is the minimum value. We plot the vertex and draw the line of symmetry. We choose some x-values on both sides of the vertex and graph the parabola. Suppose we compute the pair $(5, -1)$:

$$f(5) = 5^2 - 6(5) + 4 = 25 - 30 + 4 = -1.$$

We note that it is 2 units to the right of the line of symmetry. There will also be a pair with the same y-coordinate on the graph 2 units to the *left* of the line of symmetry. Thus we get a second point, $(1, -1)$, without making another calculation.

x	$f(x)$	
3	-5	← Vertex
4	-4	
5	-1	
6	4	
2	-4	
1	-1	
0	4	

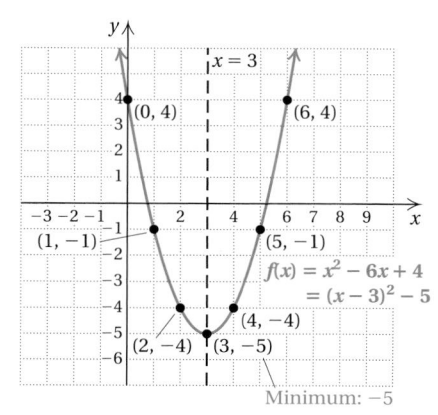

SKILL TO REVIEW
Objective 7.4a: Graph linear equations using intercepts.

Find the intercepts and then graph the line.
1. $3x - y = 3$
2. $2x + 4y = -8$

Answers

Skill to Review:
1. y-intercept: $(0, -3)$; x-intercept: $(1, 0)$
2. y-intercept: $(0, -2)$; x-intercept: $(-4, 0)$

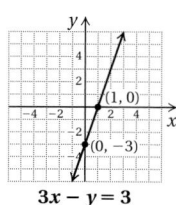

$3x - y = 3$

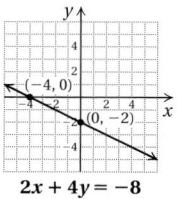

$2x + 4y = -8$

1. For $f(x) = x^2 - 4x + 7$, find the vertex, the line of symmetry, and the maximum or the minimum value. Then graph.

x	$f(x)$

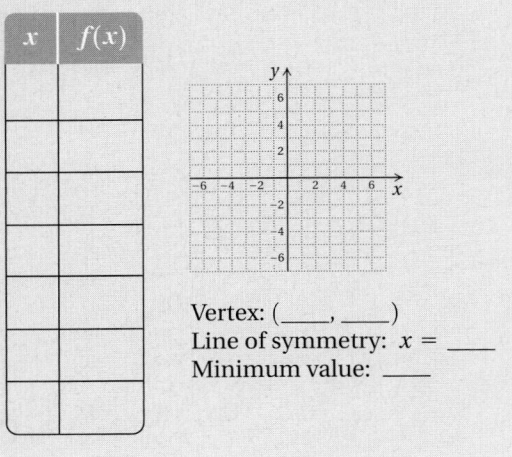

Vertex: (____ , ____)
Line of symmetry: $x =$ ____
Minimum value: ____

2. For $f(x) = 3x^2 - 24x + 43$, find the vertex, the line of symmetry, and the maximum or the minimum value. Then graph.

x	$f(x)$

Vertex: (____ , ____)
Line of symmetry: $x =$ ____
Minimum value: ____

Do Exercise 1.

EXAMPLE 2 For $f(x) = 3x^2 + 12x + 13$, find the vertex, the line of symmetry, and the maximum or the minimum value. Then graph.

Since the coefficient of x^2 is not 1, we factor out 3 from only the *first two* terms of the expression. Remember that we want to get to the form $f(x) = a(x - h)^2 + k$:

$$f(x) = 3x^2 + 12x + 13$$
$$= 3(x^2 + 4x) + 13. \qquad \text{Factoring 3 out of the first two terms}$$

Next, we complete the square inside the parentheses:

$$f(x) = 3(x^2 + 4x \qquad) + 13.$$

We take half the x-coefficient, $\frac{1}{2} \cdot 4 = 2$, and square it: $2^2 = 4$. Then we add 0, or $4 - 4$, inside the parentheses:

$$f(x) = 3(x^2 + 4x + 0) + 13 \qquad \text{Adding 0}$$
$$= 3(x^2 + 4x + 4 - 4) + 13 \qquad \text{Substituting } 4 - 4 \text{ for 0}$$
$$= 3(x^2 + 4x + 4 - 4) + 13 \qquad \left.\begin{array}{l}\text{Using the distributive}\\ \text{law to separate } -4 \\ \text{from the trinomial}\end{array}\right.$$
$$= 3(x^2 + 4x + 4) + 3(-4) + 13$$
$$= 3(x^2 + 4x + 4) - 12 + 13$$
$$= 3(x + 2)^2 + 1 \qquad \text{Factoring and simplifying}$$
$$= 3[x - (-2)]^2 + 1.$$

The vertex is $(-2, 1)$, and the line of symmetry is $x = -2$. The coefficient of x^2 is 3, so the graph is narrow and opens up. This tells us that 1 is the minimum value of the function. We choose a few x-values on one side of the line of symmetry, compute y-values, and use the resulting coordinates to find more points on the other side of the line of symmetry. We plot points and graph the parabola.

x	$f(x)$
-2	1	← Vertex
-1	4
-3	4
0	13
-4	13

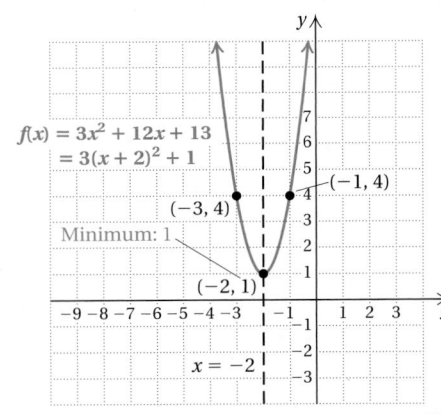

$f(x) = 3x^2 + 12x + 13$
$= 3(x + 2)^2 + 1$

$(-1, 4)$
$(-3, 4)$
Minimum: 1
$(-2, 1)$
$x = -2$

Do Exercise 2.

Answers

1.

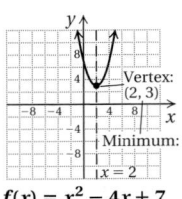

Vertex: (2, 3)
Minimum: 3
$x = 2$
$f(x) = x^2 - 4x + 7$
$= (x - 2)^2 + 3$

2.

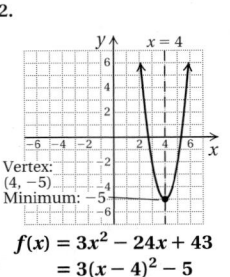

$x = 4$
Vertex: (4, −5)
Minimum: −5
$f(x) = 3x^2 - 24x + 43$
$= 3(x - 4)^2 - 5$

EXAMPLE 3 For $f(x) = -2x^2 + 10x - 7$, find the vertex, the line of symmetry, and the maximum or the minimum value. Then graph.

Again, the coefficient of x^2 is not 1. We factor out -2 from only the *first two* terms of the expression. This makes the coefficient of x^2 inside the parentheses 1:

$$f(x) = -2x^2 + 10x - 7$$
$$= -2(x^2 - 5x) - 7.$$

Next, we complete the square as before:

$$f(x) = -2(x^2 - 5x \qquad) - 7.$$

We take half the x-coefficient, $\frac{1}{2}(-5) = -\frac{5}{2}$, and square it: $\left(-\frac{5}{2}\right)^2 = \frac{25}{4}$. Then we add 0, or $\frac{25}{4} - \frac{25}{4}$, inside the parentheses:

$$f(x) = -2\left(x^2 - 5x + \frac{25}{4} - \frac{25}{4}\right) - 7 \qquad \text{Adding 0, or } \frac{25}{4} - \frac{25}{4}$$
$$= -2\left(x^2 - 5x + \frac{25}{4} - \frac{25}{4}\right) - 7$$

Using the distributive law to separate the $-\frac{25}{4}$ from the trinomial

$$= -2\left(x^2 - 5x + \frac{25}{4}\right) + (-2)\left(-\frac{25}{4}\right) - 7$$
$$= -2\left(x^2 - 5x + \frac{25}{4}\right) + \frac{25}{2} - 7$$
$$= -2\left(x - \frac{5}{2}\right)^2 + \frac{11}{2}.$$

Factoring and simplifying

The vertex is $\left(\frac{5}{2}, \frac{11}{2}\right)$, and the line of symmetry is $x = \frac{5}{2}$. The coefficient of x^2 is -2, so the graph is narrow and opens down. This tells us that $\frac{11}{2}$ is the maximum value of the function. We choose a few x-values on one side of the line of symmetry, compute y-values, and use the resulting coordinates to find more points on the other side of the line of symmetry. We plot points and graph the parabola.

x	$f(x)$	
$\frac{5}{2}$	$\frac{11}{2}$, or $5\frac{1}{2}$	← Vertex
3	5	
4	1	
5	-7	
2	5	
1	1	
0	-7	

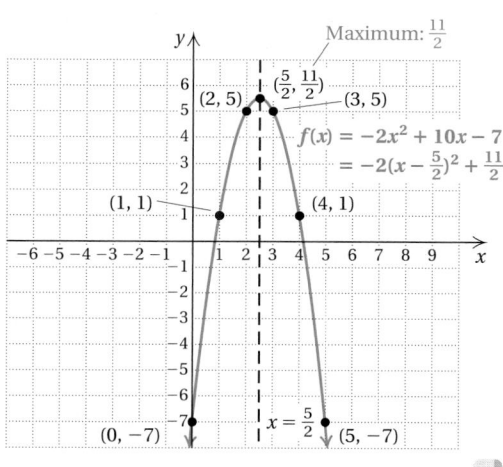

3. For $f(x) = -4x^2 + 12x - 5$, find the vertex, the line of symmetry, and the maximum or the minimum value. Then graph.

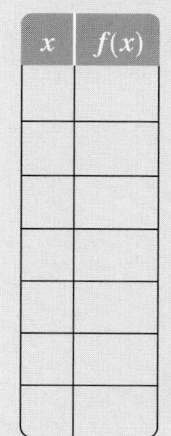

x	$f(x)$

Vertex: (____, ____)
Line of symmetry:
$x =$ ____
Maximum
value: ____

Do Exercise 3.

Answer

3.

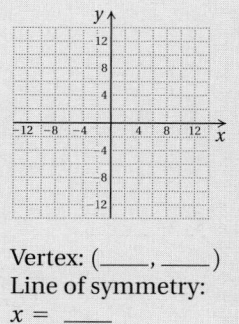

$$f(x) = -4x^2 + 12x - 5$$
$$= -4\left(x - \frac{3}{2}\right)^2 + 4$$

The method used in Examples 1–3 can be generalized to find a formula for locating the vertex. We complete the square as follows:

$$f(x) = ax^2 + bx + c$$

$$= a\left(x^2 + \frac{b}{a}x\right) + c. \qquad \text{Factoring } a \text{ out of the first two terms. Check by multiplying.}$$

Half of the x-coefficient, $\frac{b}{a}$, is $\frac{b}{2a}$. We square it to get $\frac{b^2}{4a^2}$ and add $\frac{b^2}{4a^2} - \frac{b^2}{4a^2}$ inside the parentheses. Then we distribute the a:

$$f(x) = a\left(x^2 + \frac{b}{a}x + \frac{b^2}{4a^2} - \frac{b^2}{4a^2}\right) + c$$

$$\left.\begin{array}{l} = a\left(x^2 + \frac{b}{a}x + \frac{b^2}{4a^2}\right) + a\left(-\frac{b^2}{4a^2}\right) + c \\[2mm] = a\left(x + \frac{b}{2a}\right)^2 + \frac{-b^2}{4a} + \frac{4ac}{4a} \\[2mm] = a\left[x - \left(-\frac{b}{2a}\right)\right]^2 + \frac{4ac - b^2}{4a}. \end{array}\right\}$$

Using the distributive law

Factoring and finding a common denominator

Thus we have the following.

> ## VERTEX; LINE OF SYMMETRY
>
> The **vertex** of a parabola given by $f(x) = ax^2 + bx + c$ is
>
> $$\left(-\frac{b}{2a}, \frac{4ac - b^2}{4a}\right), \quad \text{or} \quad \left(-\frac{b}{2a}, f\left(-\frac{b}{2a}\right)\right).$$
>
> The x-coordinate of the vertex is $-b/(2a)$. The **line of symmetry** is $x = -b/(2a)$. The second coordinate of the vertex is easiest to find by computing $f\left(-\frac{b}{2a}\right)$.

Let's reexamine Example 3 to see how we could have found the vertex directly. From the formula above,

$$\text{the } x\text{-coordinate of the vertex is } -\frac{b}{2a} = -\frac{10}{2(-2)} = \frac{5}{2}.$$

Substituting $\frac{5}{2}$ into $f(x) = -2x^2 + 10x - 7$, we find the second coordinate of the vertex:

$$f\left(\tfrac{5}{2}\right) = -2\left(\tfrac{5}{2}\right)^2 + 10\left(\tfrac{5}{2}\right) - 7$$

$$= -2\left(\tfrac{25}{4}\right) + 25 - 7$$

$$= -\tfrac{25}{2} + 18 = -\tfrac{25}{2} + \tfrac{36}{2} = \tfrac{11}{2}.$$

The vertex is $\left(\frac{5}{2}, \frac{11}{2}\right)$. The line of symmetry is $x = \frac{5}{2}$.

We have developed two methods for finding the vertex. One is by completing the square and the other is by using a formula. You should check with your instructor about which method to use.

Do Exercises 4–6.

Find the vertex of each parabola using the formula.

4. $f(x) = x^2 - 6x + 4$

5. $f(x) = 3x^2 - 24x + 43$

6. $f(x) = -4x^2 + 12x - 5$

Answers

4. $(3, -5)$ **5.** $(4, -5)$ **6.** $\left(\frac{3}{2}, 4\right)$

STUDY TIPS

BEGINNING TO STUDY FOR THE FINAL EXAM

It is never too soon to begin to study for the final examination. Take a few minutes each week to review the highlighted information, such as formulas, properties, and procedures. Make special use of the Mid-Chapter Reviews, Summary and Reviews, Chapter Tests, and Cumulative Reviews. The Cumulative Review for Chapters 1–12 is a sample final exam.

(b) Finding the Intercepts of a Quadratic Function

The points at which a graph crosses an axis are called **intercepts**. We determine the y-intercept by finding $f(0)$. For $f(x) = ax^2 + bx + c$, $f(0) = a \cdot 0^2 + b \cdot 0 + c = c$, so the y-intercept is $(0, c)$.

To find the x-intercepts, we look for values of x for which $f(x) = 0$. For $f(x) = ax^2 + bx + c$, we solve

$$0 = ax^2 + bx + c.$$

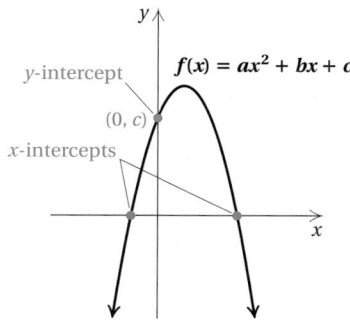

EXAMPLE 4 Find the intercepts of $f(x) = x^2 - 2x - 2$.

The y-intercept is $(0, f(0))$. Since $f(0) = 0^2 - 2 \cdot 0 - 2 = -2$, the y-intercept is $(0, -2)$. To find the x-intercepts, we solve

$$0 = x^2 - 2x - 2.$$

Using the quadratic formula, we have $x = 1 \pm \sqrt{3}$. Thus the x-intercepts are $\left(1 - \sqrt{3}, 0\right)$ and $\left(1 + \sqrt{3}, 0\right)$, or, approximately, $(-0.732, 0)$ and $(2.732, 0)$.

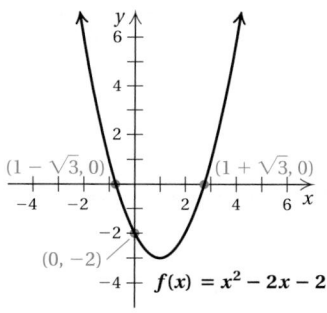

Do Exercises 7–9.

Find the intercepts.

7. $f(x) = x^2 + 2x - 3$

8. $f(x) = x^2 + 8x + 16$

9. $f(x) = x^2 - 4x + 1$

Answers

7. y-intercept: $(0, -3)$; x-intercepts: $(-3, 0)$, $(1, 0)$
8. y-intercept: $(0, 16)$; x-intercept: $(-4, 0)$
9. y-intercept: $(0, 1)$; x-intercepts: $\left(2 - \sqrt{3}, 0\right)$, $\left(2 + \sqrt{3}, 0\right)$, or $(0.268, 0)$, $(3.732, 0)$

Visualizing for Success

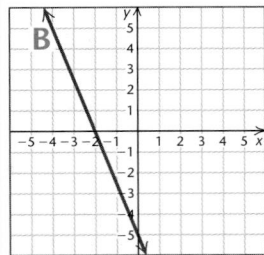

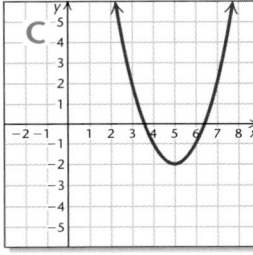

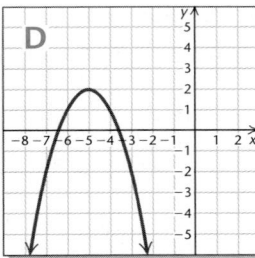

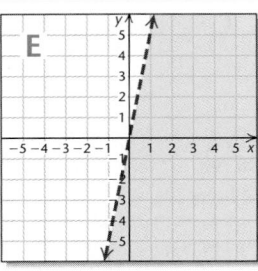

Match each equation or inequality with its graph.

1. $y = -(x - 5)^2 + 2$

2. $2x + 5y = 10$

3. $5x - 2y = 10$

4. $2x - 5y = 10$

5. $y = (x - 5)^2 - 2$

6. $y = x^2 - 5$

7. $5x + 2y \geq 10$

8. $5x + 2y = -10$

9. $y < 5x$

10. $y = -(x + 5)^2 + 2$

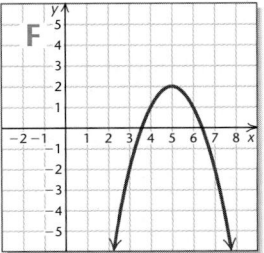

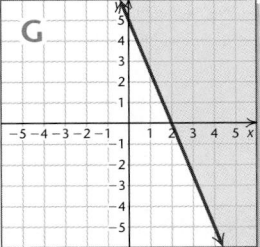

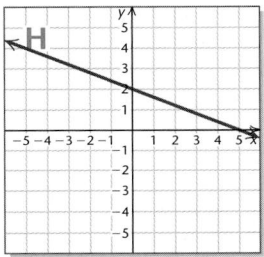

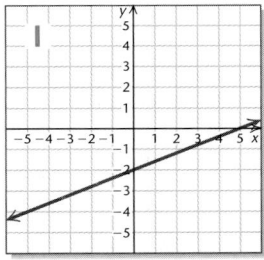

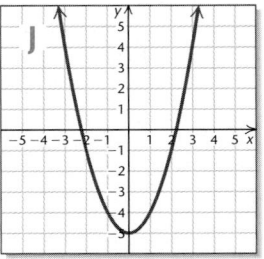

Answers on page A-35

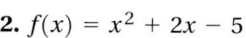
a For each quadratic function, find **(a)** the vertex, **(b)** the line of symmetry, and **(c)** the maximum or minimum value. Then **(d)** graph the function.

1. $f(x) = x^2 - 2x - 3$

x	$f(x)$

Vertex: (____, ____)
Line of symmetry: $x =$ ____
_____ value: ____

2. $f(x) = x^2 + 2x - 5$

x	$f(x)$

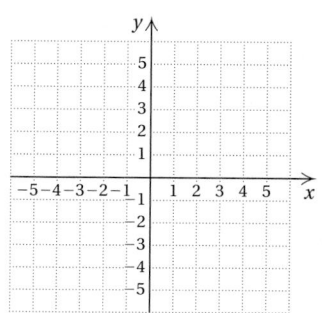

Vertex: (____, ____)
Line of symmetry: $x =$ ____
_____ value: ____

3. $f(x) = -x^2 - 4x - 2$

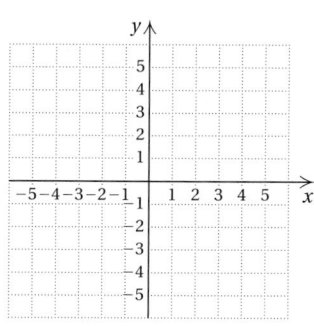

x	$f(x)$

Vertex: (____, ____)
Line of symmetry: $x =$ ____
_____ value: ____

4. $f(x) = -x^2 + 4x + 1$

x	$f(x)$

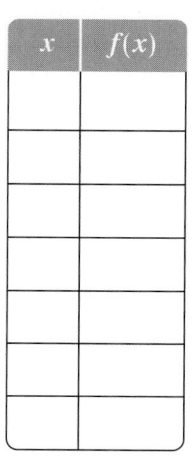
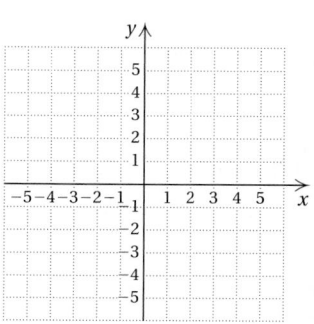

Vertex: (____, ____)
Line of symmetry: $x =$ ____
_____ value: ____

5. $f(x) = 3x^2 - 24x + 50$

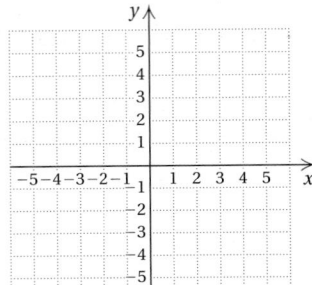

Vertex: (____, ____)
Line of symmetry: $x =$ ____
_____ value: ____

6. $f(x) = 4x^2 + 8x + 1$

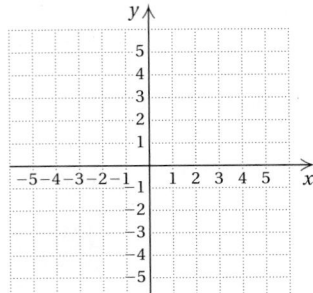

Vertex: (____, ____)
Line of symmetry: $x =$ ____
_____ value: ____

7. $f(x) = -2x^2 - 2x + 3$

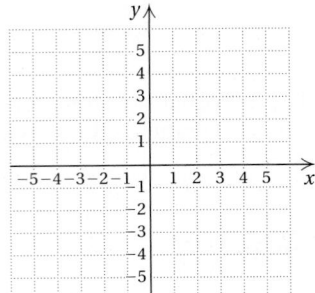

Vertex: (____, ____)
Line of symmetry: $x =$ ____
_____ value: ____

8. $f(x) = -2x^2 + 2x + 1$

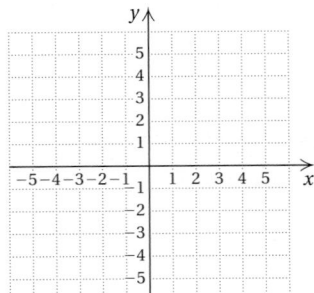

Vertex: (____, ____)
Line of symmetry: $x =$ ____
_____ value: ____

9. $f(x) = 5 - x^2$

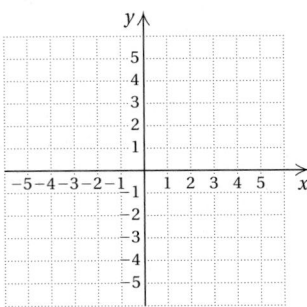

Vertex: (____, ____)
Line of symmetry: $x =$ ____
_____ value: ____

10. $f(x) = x^2 - 3x$

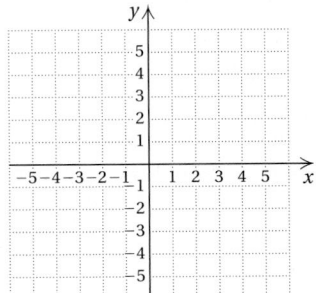

Vertex: (____, ____)
Line of symmetry: $x =$ ____
_____ value: ____

11. $f(x) = 2x^2 + 5x - 2$

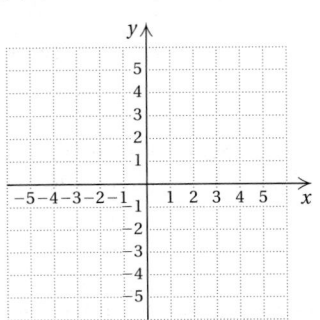

Vertex: (____, ____)
Line of symmetry: $x =$ ____
_____ value: ____

12. $f(x) = -4x^2 - 7x + 2$

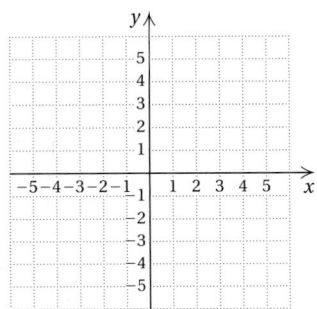

Vertex: (____, ____)
Line of symmetry: $x =$ ____
_____ value: ____

b Find the x- and y-intercepts.

13. $f(x) = x^2 - 6x + 1$ **14.** $f(x) = x^2 + 2x + 12$ **15.** $f(x) = -x^2 + x + 20$ **16.** $f(x) = -x^2 + 5x + 24$

17. $f(x) = 4x^2 + 12x + 9$ **18.** $f(x) = 3x^2 - 6x + 1$ **19.** $f(x) = 4x^2 - x + 8$ **20.** $f(x) = 2x^2 + 4x - 1$

Skill Maintenance

Solve.   [6.9a, b]

21. *Determining Medication Dosage.* A child's dosage D, in milligrams, of a medication varies directly as the child's weight w, in kilograms. To control a fever, a doctor suggests that a child who weighs 28 kg be given 420 mg of analgesic medication. Find an equation of variation.

22. *Calories Burned.* The number C of calories burned while exercising varies directly as the time t, in minutes, spent exercising. Harold exercises for 24 min on a StairMaster and burns 356 calories. Find an equation of variation.

Find the variation constant and an equation of variation in which y varies inversely as x and the following are true.   [6.9c]

23. $y = 125$ when $x = 2$

24. $y = 2$ when $x = 125$

Find the variation constant and an equation of variation in which y varies directly as x and the following are true.   [6.9a]

25. $y = 125$ when $x = 2$

26. $y = 2$ when $x = 125$

Synthesis

27. Use the TRACE and/or TABLE features of a graphing calculator to estimate the maximum or minimum values of the following functions.
    a) $f(x) = 2.31x^2 - 3.135x - 5.89$
    b) $f(x) = -18.8x^2 + 7.92x + 6.18$

28. Use the INTERSECT feature of a graphing calculator to find the points of intersection of the graphs of the functions.
$$f(x) = x^2 + 2x + 1, \quad g(x) = -2x^2 - 4x + 1$$

Graph.

29. $f(x) = |x^2 - 1|$ **30.** $f(x) = |x^2 + 6x + 4|$ **31.** $f(x) = |x^2 - 3x - 4|$ **32.** $f(x) = |2(x - 3)^2 - 5|$

33. A quadratic function has $(-1, 0)$ as one of its intercepts and $(3, -5)$ as its vertex. Find an equation for the function.

34. A quadratic function has $(4, 0)$ as one of its intercepts and $(-1, 7)$ as its vertex. Find an equation for the function.

35. Consider
$$f(x) = \frac{x^2}{8} + \frac{x}{4} - \frac{3}{8}.$$
Find the vertex, the line of symmetry, and the maximum or minimum value. Then draw the graph.

36. Use only the graph in Exercise 35 to approximate the solutions of each of the following equations.
    a) $\dfrac{x^2}{8} + \dfrac{x}{4} - \dfrac{3}{8} = 0$ **b)** $\dfrac{x^2}{8} + \dfrac{x}{4} - \dfrac{3}{8} = 1$
    c) $\dfrac{x^2}{8} + \dfrac{x}{4} - \dfrac{3}{8} = 2$

11.7

Mathematical Modeling with Quadratic Functions

We now consider some of the many situations in which quadratic functions can serve as mathematical models.

 ## Maximum-Minimum Problems

We have seen that for any quadratic function $f(x) = ax^2 + bx + c$, the value of $f(x)$ at the vertex is either a maximum or a minimum, meaning that either all outputs are smaller than that value for a maximum or larger than that value for a minimum.

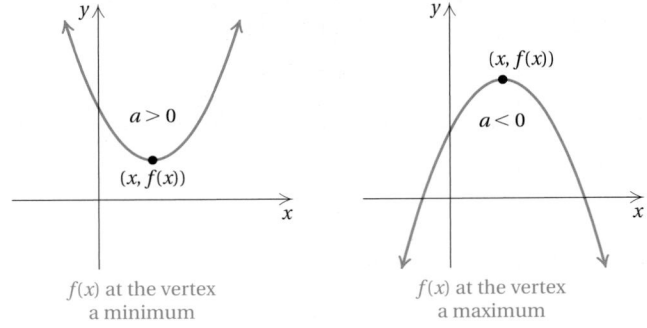

$f(x)$ at the vertex
a minimum

$f(x)$ at the vertex
a maximum

There are many applied problems in which we want to find a maximum or minimum value. If a quadratic function can be used as a model, we can find such maximums or minimums by finding coordinates of the vertex.

EXAMPLE 1 *Bordered Garden.* Millie is planting a garden to produce vegetables and fruit for the local food bank. She has enough raspberry plants to edge a 64-yd perimeter and wants to maximize the area within to plant the most vegetables possible. What are the dimensions of the largest rectangular garden that Millie can enclose with the raspberry plants?

1. **Familiarize.** We first make a drawing and label it. We let $l = $ the length of the garden and $w = $ the width. Recall the following formulas:

Perimeter: $2l + 2w$;

Area: $l \cdot w$.

l	*w*	*A*
22	10	220
20	12	240
18	14	252
18.5	13.5	249.75
12.4	19.6	243.04
15	17	255

To become familiar with the problem, let's choose some dimensions (shown at left) for which $2l + 2w = 64$ and then calculate the corresponding areas. What choice of l and w will maximize A?

2. **Translate.** We have two equations, one for perimeter and one for area:

$$2l + 2w = 64,$$
$$A = l \cdot w.$$

Let's use them to express A as a function of l or w, but not both. To express A in terms of w, for example, we solve for l in the first equation:

$$2l + 2w = 64$$
$$2l = 64 - 2w$$
$$l = \frac{64 - 2w}{2}$$
$$= 32 - w.$$

Substituting $32 - w$ for l, we get a quadratic function $A(w)$, or just A:

$$A = lw = (32 - w)w = 32w - w^2 = -w^2 + 32w.$$

3. **Carry out.** Note here that we are altering the third step of our five-step problem-solving strategy to "carry out" some kind of mathematical manipulation, because we are going to find the vertex rather than solve an equation. To do so, we complete the square as in Section 11.6:

$$A = -w^2 + 32w \qquad \text{This is a parabola opening down,}$$
$$\text{so a maximum exists.}$$
$$= -1(w^2 - 32w) \qquad \text{Factoring out } -1$$
$$= -1(\underbrace{w^2 - 32w + 256}\ - 256) \qquad \tfrac{1}{2}(-32) = -16; (-16)^2 = 256.$$
$$\text{We add 0, or } 256 - 256.$$

$$= -1(w^2 - 32w + 256) + (-1)(-256) \qquad \text{Using the distributive law}$$
$$= -(w - 16)^2 + 256.$$

The vertex is $(16, 256)$. Thus the maximum value is 256. It occurs when $w = 16$ and $l = 32 - w = 32 - 16 = 16$.

4. **Check.** We note that 256 is larger than any of the values found in the *Familiarize* step. To be more certain, we could make more calculations. We leave this to the student. We can also use the graph of the function to check the maximum value.

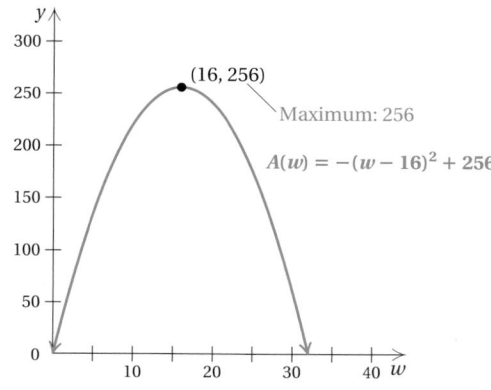

5. **State.** The largest rectangular garden that can be enclosed is 16 yd by 16 yd; that is, it is a square with sides of 16 ft.

Do Exercise 1.

1. **Fenced-In Land.** A farmer has 100 yd of fencing. What are the dimensions of the largest rectangular pen that the farmer can enclose?

To familiarize yourself with the problem, complete the following table.

l	w	A
12	38	456
15	35	
24	26	
25	25	
26.2	23.8	

Answer

1. 25 yd by 25 yd

Maximum and Minimum Values   We can use a graphing calculator to find the maximum or minimum value of a quadratic function. Consider the quadratic function in Example 1, $A = -w^2 + 32w$. First, we replace w with x and A with y and graph the function in a window that displays the vertex of the graph. We choose $[0, 40, 0, 300]$, with Xscl $= 5$ and Yscl $= 20$. Now, we press **2ND** **CALC** **4** or **2ND** **CALC** ⌄⌄⌄ **ENTER** to select the MAXIMUM feature from the CALC menu. We are prompted to select a left bound for the maximum point. This means that we must choose an x-value that is to the left of the x-value of the point where the maximum occurs. This can be done by using the left- and right-arrow keys to move the cursor to a point to the left of the maximum point or by keying in an appropriate value. Once this is done, we press **ENTER**. Now, we are prompted to select a right bound. We move the cursor to a point to the right of the maximum point or key in an appropriate value.

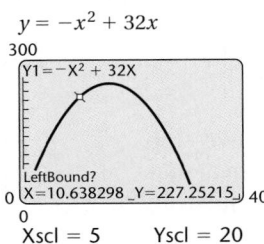

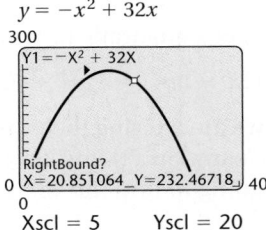

We press **ENTER** again. Finally, we are prompted to guess the x-value at which the maximum occurs. We move the cursor close to the maximum or key in an x-value. We press **ENTER** a third time and see that the maximum function value of 256 occurs when $x = 16$. (One or both coordinates of the maximum point might be approximations of the actual values, as shown with the x-value below, because of the method the calculator uses to find these values.)

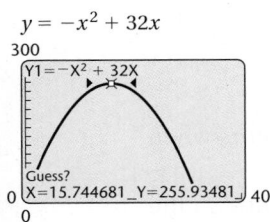

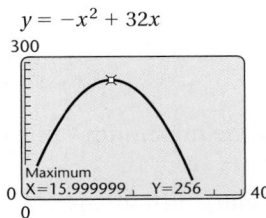

To find a minimum value, we select item 3, "minimum," from the CALC menu by pressing **2ND** **CALC** **3** or **2ND** **CALC** ⌄⌄ **ENTER**.

Exercises: Use the maximum or minimum feature on a graphing calculator to find the maximum or minimum value of each function.

1. $y = 3x^2 - 6x + 4$

2. $y = 2x^2 + x + 5$

3. $y = -x^2 + 4x + 2$

4. $y = -4x^2 + 5x - 1$

ⓑ Fitting Quadratic Functions to Data

As we move through our study of mathematics, we develop a library of functions. These functions can serve as models for many applications. Some of them are graphed below. We have not considered the cubic or quartic functions in detail other than in the Calculator Corners (we leave that discussion to a later course), but we show them here for reference.

Linear function:
$f(x) = mx + b$

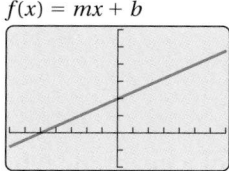

Quadratic function:
$f(x) = ax^2 + bx + c, \; a > 0$

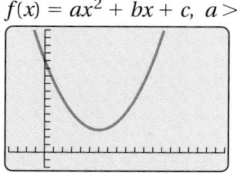

Quadratic function:
$f(x) = ax^2 + bx + c, \; a < 0$

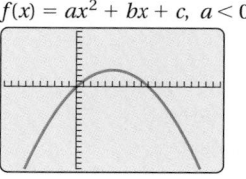

Absolute-value function:
$f(x) = |x|$

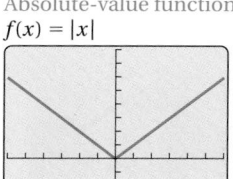

Cubic function:
$f(x) = ax^3 + bx^2 + cx + d, \; a > 0$

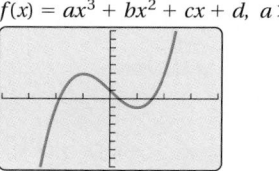

Quartic function:
$f(x) = ax^4 + bx^3 + cx^2 + dx + e, \; a > 0$

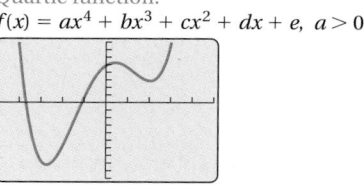

Now let's consider some real-world data. How can we decide which type of function might fit the data of a particular application? One simple way is to graph the data and look for a pattern resembling one of the graphs above. For example, data might be modeled by a linear function if the graph resembles a straight line. The data might be modeled by a quadratic function if the graph rises and then falls, or falls and then rises, in a curved manner resembling a parabola. For a quadratic, it might also just rise or fall in a curved manner as if following only one part of the parabola.

Let's now use our library of functions to see which, if any, might fit certain data situations.

EXAMPLES *Choosing Models.* For the scatterplots and graphs below, determine which, if any, of the following functions might be used as a model for the data.

Linear, $f(x) = mx + b$;

Quadratic, $f(x) = ax^2 + bx + c, a > 0$;

Quadratic, $f(x) = ax^2 + bx + c, a < 0$;

Polynomial, neither quadratic nor linear

2.

The data rise and then fall in a curved manner fitting a quadratic function $f(x) = ax^2 + bx + c, a < 0$.

3.

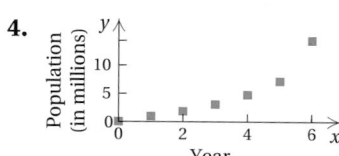

The data seem to fit a linear function $f(x) = mx + b$.

4.

The data rise in a manner fitting the right side of a quadratic function $f(x) = ax^2 + bx + c, a > 0$.

5. **Household Gas Bill from February 2008 through January 2009**

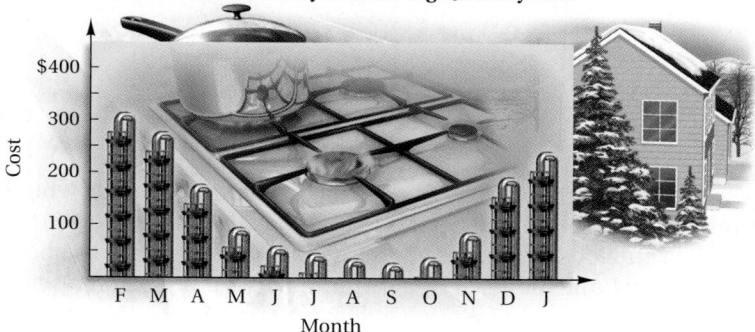

The data fall and then rise in a curved manner fitting a quadratic function $f(x) = ax^2 + bx + c, a > 0$.

6. **Average Number of Motorcyclists Killed per Hour on the Weekend**

SOURCE: Motor Vehicle Crash Data from FARS and GES

The data fall, then rise, then fall again. They do not appear to fit a linear or quadratic function but might fit a polynomial function that is neither quadratic nor linear.

Do Exercises 2–5.

Choosing Models. For the scatterplots in Margin Exercises 2–5, determine which, if any, of the following functions might be used as a model for the data.

Linear, $f(x) = mx + b$;

Quadratic, $f(x) = ax^2 + bx + c$, $a > 0$;

Quadratic, $f(x) = ax^2 + bx + c$, $a < 0$;

Polynomial, neither quadratic nor linear

2.

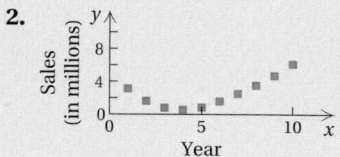

3.

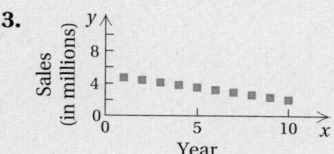

4.

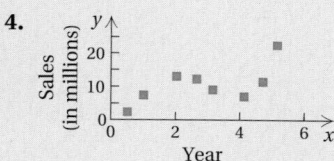

5.

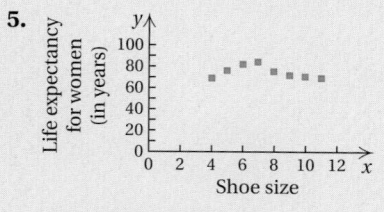

SOURCE: Orthopedic Quarterly

Answers

2. $f(x) = ax^2 + bx + c, a > 0$
3. $f(x) = mx + b$ **4.** Polynomial, neither quadratic nor linear **5.** Polynomial, neither quadratic nor linear

Whenever a quadratic function seems to fit a data situation, that function can be determined if at least three inputs and their outputs are known.

EXAMPLE 7 *Canoe Depth.* The drawing below shows the cross section of a canoe. Canoes are deepest at the middle of the center line, with the depth decreasing to zero at the edges. Lou and Jen own a company that specializes in producing custom canoes. A customer provided suggested guidelines for measures of the depths D, in inches, along the center line of the canoe at distances x, in inches, from the edge. The measures are listed in the table at right.

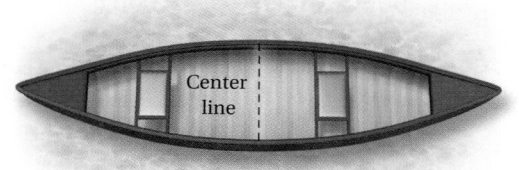

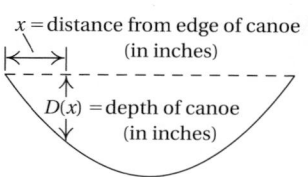

x = distance from edge of canoe (in inches)

$D(x)$ = depth of canoe (in inches)

DISTANCE x FROM THE EDGE OF THE CANOE ALONG THE CENTER LINE (in inches)	DEPTH D OF THE CANOE (in inches)
0	0
9	10.5
18	14
30	7.75
36	0

a) Make a scatterplot of the data.

b) Decide whether the data seem to fit a quadratic function.

c) Use the data points $(0, 0)$, $(18, 14)$, and $(36, 0)$ to find a quadratic function that fits the data.

d) Use the function to estimate the depth of the canoe 10 in. from the edge along the center line.

a) The red squares shown below comprise the scatterplot.

Canoe Depth

Depth (in inches) / Distance from the edge of the canoe along the center line (in inches)

b) The data seem to rise and fall in a manner similar to a quadratic function. The dashed black line in the graph represents a sample quadratic function of fit. Note that it may not necessarily go through each point.

c) We are looking for a quadratic function

$$D(x) = ax^2 + bx + c.$$

We need to determine the constants a, b, and c. We use the three data points $(0, 0)$, $(18, 14)$, and $(36, 0)$ and substitute as follows:

$$0 = a \cdot 0^2 + b \cdot 0 + c,$$
$$14 = a \cdot 18^2 + b \cdot 18 + c,$$
$$0 = a \cdot 36^2 + b \cdot 36 + c.$$

6. Ticket Profits. Valley Community College is presenting a play. The profit P, in dollars, after x days is given in the following table.

DAYS x	PROFIT P
0	$-100
90	560
180	872
270	870
360	548
450	-100

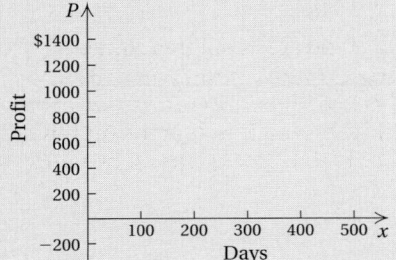

a) Make a scatterplot of the data.

b) Decide whether the data can be modeled by a quadratic function.

c) Use the data points $(0, -100)$, $(180, 872)$, and $(360, 548)$ to find a quadratic function that fits the data.

d) Use the function to estimate the profits after 225 days.

Answer

6. (a)

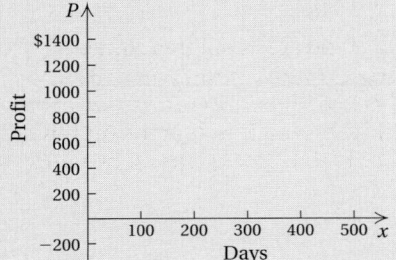

(b) yes; **(c)** $f(x) = -0.02x^2 + 9x - 100$;
(d) $912.50

After simplifying, we see that we need to solve the system

$$0 = c,$$
$$14 = 324a + 18b + c,$$
$$0 = 1296a + 36b + c.$$

Since $c = 0$, the system reduces to a system of two equations in two variables:

$$14 = 324a + 18b, \quad \textbf{(1)}$$
$$0 = 1296a + 36b. \quad \textbf{(2)}$$

We multiply equation (1) by -2, add, and solve for a (see Section 8.3):

$$\begin{aligned}
-28 &= -648a - 36b \\
0 &= 1296a + 36b \\
\hline
-28 &= 648a \qquad \text{Adding} \\
\frac{-28}{648} &= a \qquad \text{Solving for } a \\
-\frac{7}{162} &= a.
\end{aligned}$$

Next, we substitute $-\frac{7}{162}$ for a in equation (2) and solve for b:

$$0 = 1296\left(-\frac{7}{162}\right) + 36b$$
$$0 = -56 + 36b$$
$$56 = 36b$$
$$\frac{56}{36} = b$$
$$\frac{14}{9} = b.$$

This gives us the quadratic function:

$$D(x) = -\frac{7}{162}x^2 + \frac{14}{9}x, \text{ or}$$
$$D(x) \approx -0.043x^2 + 1.556x.$$

d) To find the depth 10 in. from the edge of the canoe, we substitute:

$$D(10) = -0.043(10)^2 + 1.556(10) = 11.26.$$

At a distance of 10 in. from the edge of the canoe, the depth of the canoe is about 11.26 in.

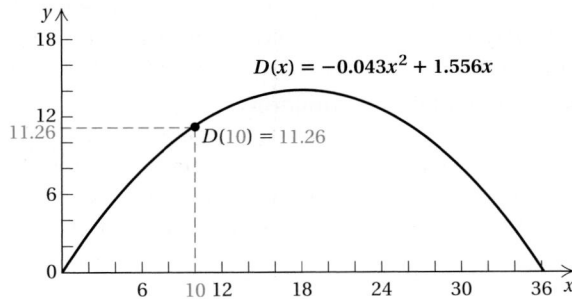

Do Exercise 6.

a Solve.

1. *Architecture.* An architect is designing a hotel with a central atrium. Each floor is to be rectangular and is allotted 720 ft of security piping around walls outside the rooms. What dimensions will allow the atrium to have maximum area?

3. *Molding Plastics.* Economite Plastics plans to produce a one-compartment vertical file by bending the long side of an 8-in. by 14-in. sheet of plastic along two lines to form a U shape. How tall should the file be in order to maximize the volume that the file can hold?

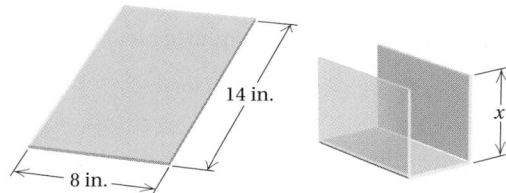

5. *Minimizing Cost.* Aki's Bicycle Designs has determined that when x hundred bicycles are built, the average cost per bicycle is given by

$$C(x) = 0.1x^2 - 0.7x + 2.425,$$

where $C(x)$ is in hundreds of dollars. How many bicycles should the shop build in order to minimize the average cost per bicycle?

2. *Stained-Glass Window Design.* An artist is designing a rectangular stained-glass window with a perimeter of 84 in. What dimensions will yield the maximum area?

4. *Patio Design.* A stone mason has enough stones to enclose a rectangular patio with a perimeter of 60 ft, assuming that the attached house forms one side of the rectangle. What is the maximum area that the mason can enclose? What should the dimensions of the patio be in order to yield this area?

6. *Corral Design.* A rancher needs to enclose two adjacent rectangular corrals, one for sheep and one for cattle. If a river forms one side of the corrals and 180 yd of fencing is available, what is the largest total area that can be enclosed?

7. *Garden Design.* A farmer decides to enclose a rectangular garden, using the side of a barn as one side of the rectangle. What is the maximum area that the farmer can enclose with 40 ft of fence? What should the dimensions of the garden be in order to yield this area?

8. *Composting.* A rectangular compost container is to be formed in a corner of a fenced yard, with 8 ft of chicken wire completing the other two sides of the rectangle. If the chicken wire is 3 ft high, what dimensions of the base will maximize the volume of the container?

9. *Ticket Sales.* The number of tickets sold each day for an upcoming performance of Handel's *Messiah* is given by

$$N(x) = -0.4x^2 + 9x + 11,$$

where x is the number of days since the concert was first announced. When will daily ticket sales peak and how many tickets will be sold that day?

10. *Stock Prices.* The value of a share of a particular stock, in dollars, can be represented by $V(x) = x^2 - 6x + 13$, where x is the number of months after January 2009. What is the lowest value $V(x)$ will reach, and when did that occur?

Maximizing Profit. Total profit P is the difference between total revenue R and total cost C. Given the following total-revenue and total-cost functions, find the total profit, the maximum value of the total profit, and the value of x at which it occurs.

11. $R(x) = 1000x - x^2,$
$C(x) = 3000 + 20x$

12. $R(x) = 200x - x^2,$
$C(x) = 5000 + 8x$

13. What is the maximum product of two numbers whose sum is 22? What numbers yield this product?

14. What is the maximum product of two numbers whose sum is 45? What numbers yield this product?

15. What is the minimum product of two numbers whose difference is 4? What are the numbers?

16. What is the minimum product of two numbers whose difference is 6? What are the numbers?

17. What is the maximum product of two numbers that add to -12? What numbers yield this product?

18. What is the minimum product of two numbers that differ by 9? What are the numbers?

Choosing Models. For the scatterplots and graphs in Exercises 19–26, determine which, if any, of the following functions might be used as a model for the data: Linear, $f(x) = mx + b$; quadratic, $f(x) = ax^2 + bx + c, a > 0$; quadratic, $f(x) = ax^2 + bx + c, a < 0$; polynomial, neither quadratic nor linear.

19.

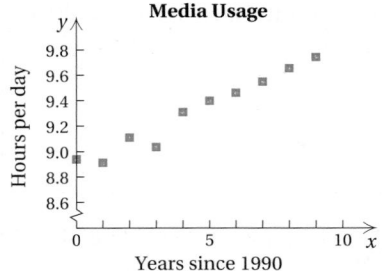

Media Usage

20.

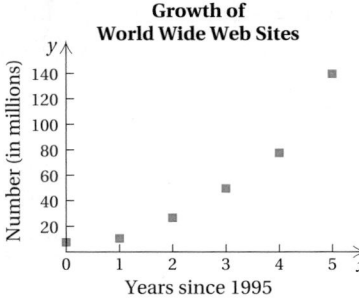

Growth of World Wide Web Sites

21.

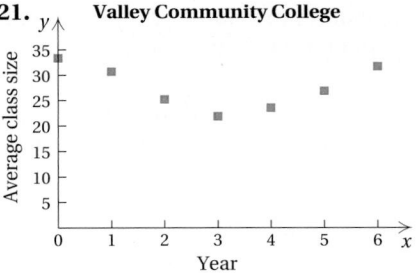

Valley Community College

Copyright © 2011 Pearson Education, Inc.

22.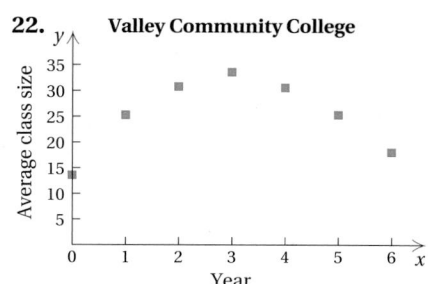
Valley Community College

23.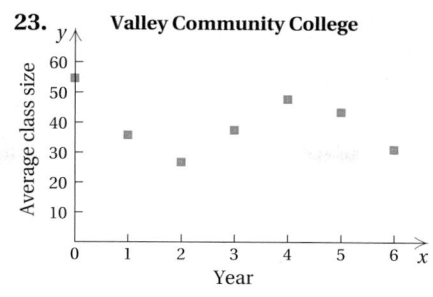
Valley Community College

24.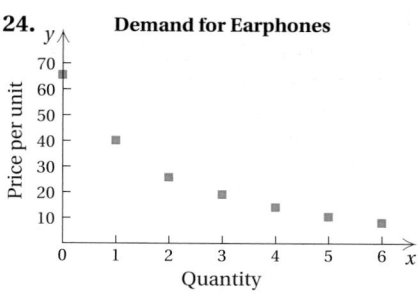
Demand for Earphones

25. Foreign Adoptions to the United States

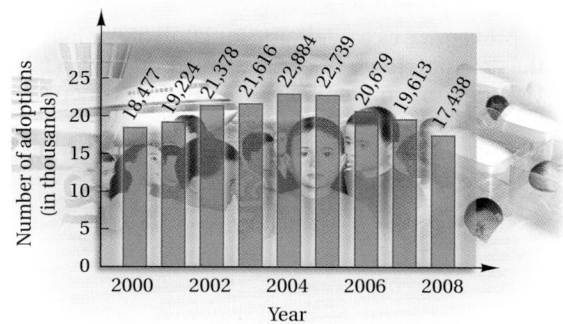

SOURCE: Intercountry Adoption Office of Children's Issues, U.S. Department of State

26. Hurricanes in the Atlantic Basin

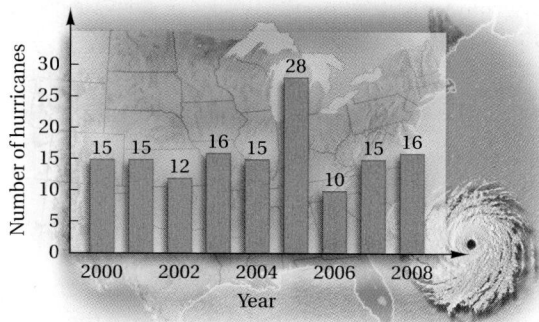

SOURCE: National Oceanic and Atmospheric Administration/ Hurricane Research Division

Find a quadratic function that fits the set of data points.

27. $(1, 4), (-1, -2), (2, 13)$

28. $(1, 4), (-1, 6), (-2, 16)$

29. $(2, 0), (4, 3), (12, -5)$

30. $(-3, -30), (3, 0), (6, 6)$

31. *Nighttime Accidents.*

  a) Find a quadratic function that fits the following data.

TRAVEL SPEED (in kilometers per hour)	NUMBER OF NIGHTTIME ACCIDENTS (for every 200 million kilometers driven)
60	400
80	250
100	250

  b) Use the function to estimate the number of nighttime accidents that occur at 50 km/h.

32. *Daytime Accidents.*

  a) Find a quadratic function that fits the following data.

TRAVEL SPEED (in kilometers per hour)	NUMBER OF DAYTIME ACCIDENTS (for every 200 million kilometers driven)
60	100
80	130
100	200

  b) Use the function to estimate the number of daytime accidents that occur at 50 km/h.

33. *River Depth.* Typically, rivers are deepest in the middle, with the depth decreasing to zero at the edges. A hydrologist measures the depths D, in feet, of a river at distances x, in feet, from one bank. The results are listed in the table below. Use the data points $(0, 0)$, $(50, 20)$, and $(100, 0)$ to find a quadratic function that fits the data. Then use the function to estimate the depth of the river at 75 ft from the bank.

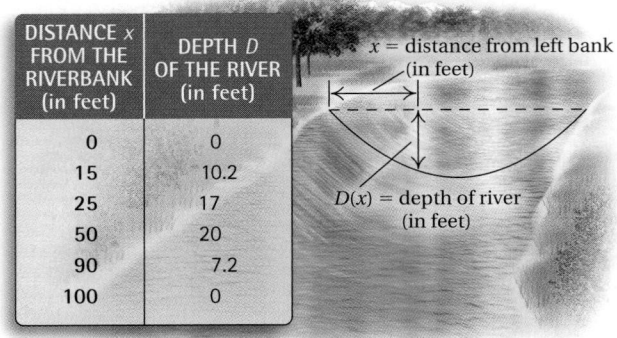

DISTANCE x FROM THE RIVERBANK (in feet)	DEPTH D OF THE RIVER (in feet)
0	0
15	10.2
25	17
50	20
90	7.2
100	0

x = distance from left bank (in feet)

$D(x)$ = depth of river (in feet)

34. *Pizza Prices.* Pizza Unlimited has the following prices for pizzas.

DIAMETER	PRICE
8 in.	$10.00
12 in.	$12.50
16 in.	$15.50

Is price a quadratic function of diameter? It probably should be, because the price should be proportional to the area, and the area is a quadratic function of the diameter. (The area of a circular region is given by $A = \pi r^2$ or $(\pi/4) \cdot d^2$.)

a) Express price as a quadratic function of diameter using the data points $(8, 10)$, $(12, 12.50)$, and $(16, 15.50)$.
b) Use the function to find the price of a 14-in. pizza.

Skill Maintenance

In each of Exercises 35–42, fill in the blank with the correct word(s) from the given list. Some of the choices may not be used.

35. In the expression $5\sqrt{2x - 9} + 3$, the symbol $\sqrt{}$ is called a(n) _____ and $2x - 9$ is called the _____ .
[10.1a]

36. When a system of two equations in two variables has infinitely many solutions, the equations are _____ . [8.1a]

37. The degree of a term of a polynomial is the _____ of the exponents of the variables. [4.3g]

38. A consistent system of equations has _____ solution. [8.1a]

39. The equation $y = k/x$, where k is a positive constant, is an equation of _____ variation. [6.9c]

40. When a system of two equations in two variables has one solution or no solutions, the equations are _____ . [8.1a]

41. If the exponents in a polynomial decrease from left to right, the polynomial is written in _____ order. [4.3f]

42. A(n) _____ is a point $(a, 0)$. [7.4a]

at least one
no
dependent
independent
ascending
descending
direct
inverse
sum
product
x-intercept
y-intercept
radical
radicand

Synthesis

43. The sum of the base and the height of a triangle is 38 cm. Find the dimensions for which the area is a maximum, and find the maximum area.

11.8

Polynomial Inequalities and Rational Inequalities

a) Quadratic and Other Polynomial Inequalities

Inequalities like the following are called **quadratic inequalities**:

$$x^2 + 3x - 10 < 0, \qquad 5x^2 - 3x + 2 \geq 0.$$

In each case, we have a polynomial of degree 2 on the left. We will solve such inequalities in two ways. The first method provides understanding and the second yields the more efficient method.

The first method for solving a quadratic inequality, such as $ax^2 + bx + c > 0$, is by considering the graph of a related function, $f(x) = ax^2 + bx + c$.

EXAMPLE 1 Solve: $x^2 + 3x - 10 > 0$.

Consider the function $f(x) = x^2 + 3x - 10$ and its graph. The graph opens up since the leading coefficient ($a = 1$) is positive. We find the x-intercepts by setting the polynomial equal to 0 and solving:

$$x^2 + 3x - 10 = 0$$
$$(x + 5)(x - 2) = 0$$
$$x + 5 = 0 \quad or \quad x - 2 = 0$$
$$x = -5 \quad or \qquad x = 2.$$

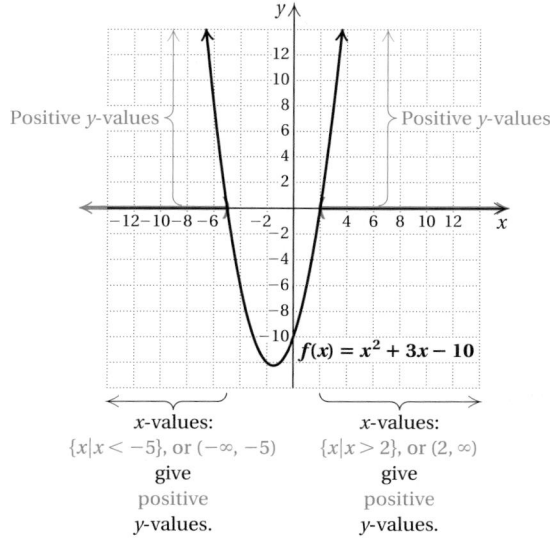

x-values:
$\{x \mid x < -5\}$, or $(-\infty, -5)$
give
positive
y-values.

x-values:
$\{x \mid x > 2\}$, or $(2, \infty)$
give
positive
y-values.

Values of y will be positive to the left and right of the intercepts, as shown. Thus the solution set of the inequality is

$$\{x \mid x < -5 \ or \ x > 2\}, \quad or \quad (-\infty, -5) \cup (2, \infty).$$

Do Margin Exercise 1.

We can solve any inequality by considering the graph of a related function and finding x-intercepts, as in Example 1. In some cases, we may need to use the quadratic formula to find the intercepts.

SKILL TO REVIEW
Objective 9.1b: Write interval notation for the solution set or graph of an inequality.

Write interval notation for the given set.

1. $\{x \mid -3 < x \leq 10\}$
2. $\{y \mid y > -\frac{1}{2}\}$

1. Solve by graphing:
$$x^2 + 2x - 3 > 0.$$

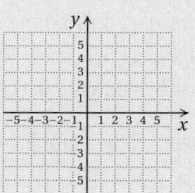

Answers

Skill to Review:
1. $(-3, 10]$　2. $\left(-\frac{1}{2}, \infty\right)$

Margin Exercise:
1. $\{x \mid x < -3 \ or \ x > 1\}$, or $(-\infty, -3) \cup (1, \infty)$

EXAMPLE 2 Solve: $x^2 + 3x - 10 < 0$.

Looking again at the graph of $f(x) = x^2 + 3x - 10$ or at least visualizing it tells us that y-values are negative for those x-values between -5 and 2.

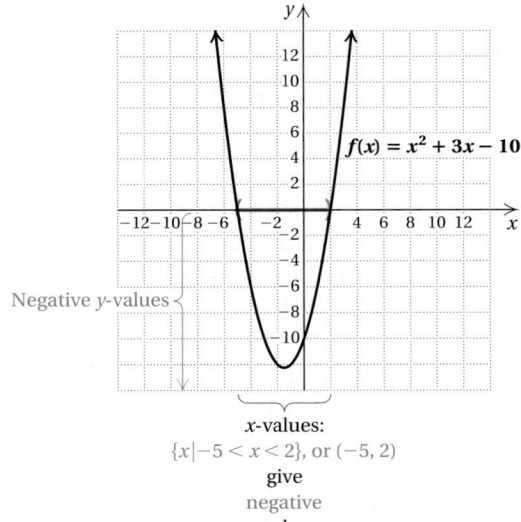

x-values:
$\{x \mid -5 < x < 2\}$, or $(-5, 2)$
give
negative
y-values.

The solution set is $\{x \mid -5 < x < 2\}$, or $(-5, 2)$.

When an inequality contains $\leq$ or $\geq$, the x-values of the x-intercepts must be included. Thus the solution set of the inequality $x^2 + 3x - 10 \leq 0$ is $\{x \mid -5 \leq x \leq 2\}$, or $[-5, 2]$.

Do Exercises 2 and 3.

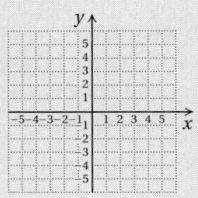

Solve by graphing.

2. $x^2 + 2x - 3 < 0$

3. $x^2 + 2x - 3 \leq 0$

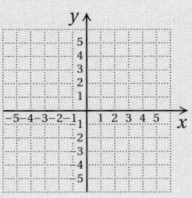

Calculator Corner

Solving Polynomial Inequalities We can solve polynomial inequalities graphically. Consider the inequality in Example 2, $x^2 + 3x - 10 < 0$. We first graph the function $f(x) = x^2 + 3x - 10$. Then we use the ZERO feature to find the solutions of the equation $f(x) = 0$, or $x^2 + 3x - 10 = 0$. The solutions, -5 and 2, divide the number line into three intervals, $(-\infty, -5)$, $(-5, 2)$, and $(2, \infty)$.

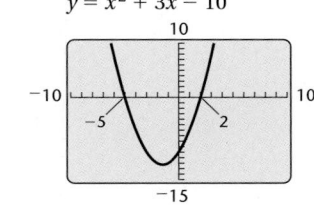

Since we want to find the values of x for which $f(x) < 0$, we look for the interval(s) on which the function values are negative. That is, we note where the graph lies below the x-axis. This occurs in the interval $(-5, 2)$, so the solution set is $\{x \mid -5 < x < 2\}$, or $(-5, 2)$.

If we were solving the inequality $x^2 + 3x - 10 > 0$, we would look for the intervals on which the graph lies above the x-axis. We can see that $x^2 + 3x - 10 > 0$ for $\{x \mid x < -5 \text{ or } x > 2\}$, or $(-\infty, -5) \cup (2, \infty)$. If the inequality symbol were $\leq$ or $\geq$, we would include the endpoints of the intervals as well.

Exercises: Solve graphically.

1. $x^2 + 3x - 4 > 0$ **2.** $x^2 - x - 6 < 0$

3. $6x^3 + 9x^2 - 6x \leq 0$ **4.** $x^3 - 16x \geq 0$

Answers

2. $\{x \mid -3 < x < 1\}$, or $(-3, 1)$
3. $\{x \mid -3 \leq x \leq 1\}$, or $[-3, 1]$

We now consider a more efficient method for solving polynomial inequalities. The preceding discussion provides the understanding for this method. In Examples 1 and 2, we see that the x-intercepts divide the number line into intervals.

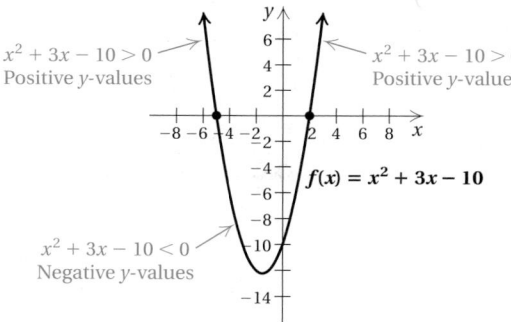

$x^2 + 3x - 10 > 0$
Positive y-values

$x^2 + 3x - 10 > 0$
Positive y-values

$f(x) = x^2 + 3x - 10$

$x^2 + 3x - 10 < 0$
Negative y-values

If a function has a positive output for one number in an interval, it will be positive for all the numbers in the interval. The same is true for negative outputs. Thus we can merely make a test substitution in each interval to solve the inequality. This is very similar to our method of using test points to graph a linear inequality in a plane.

EXAMPLE 3 Solve: $x^2 + 3x - 10 < 0$.

We set the polynomial equal to 0 and solve. The solutions of $x^2 + 3x - 10 = 0$, or $(x + 5)(x - 2) = 0$, are -5 and 2. We locate the solutions on the number line as follows. Note that the numbers divide the number line into three intervals, which we will call A, B, and C. Within each interval, the values of the function $f(x) = x^2 + 3x - 10$ will be all positive or will be all negative.

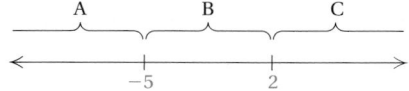

We choose a test number in interval A, say -7, and substitute -7 for x in the function $f(x) = x^2 + 3x - 10$:

$$f(-7) = (-7)^2 + 3(-7) - 10$$
$$= 49 - 21 - 10 = 18. \qquad \text{Thus}, f(-7) > 0.$$

Note that $18 > 0$, so the function values will be positive for any number in interval A.

Next, we try a test number in interval B, say 1, and find the corresponding function value:

$$f(1) = 1^2 + 3(1) - 10$$
$$= 1 + 3 - 10 = -6. \qquad \text{Thus}, f(1) < 0.$$

Note that $-6 < 0$, so the function values will be negative for any number in interval B.

Next, we try a test number in interval C, say 4, and find the corresponding function value:

$$f(4) = 4^2 + 3(4) - 10$$
$$= 16 + 12 - 10 = 18. \qquad \text{Thus,} f(4) > 0.$$

Note that $18 > 0$, so the function values will be positive for any number in interval C.

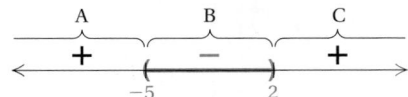

We are looking for numbers x for which $f(x) = x^2 + 3x - 10 < 0$. Thus any number x in interval B is a solution. If the inequality had been $\leq$, it would have been necessary to include the endpoints -5 and 2 in the solution set as well. The solution set is $\{x | -5 < x < 2\}$, or the interval $(-5, 2)$.

To solve a polynomial inequality:

1. Get 0 on one side, set the expression on the other side equal to 0, and solve to find the x-intercepts.
2. Use the numbers found in step (1) to divide the number line into intervals.
3. Substitute a number from each interval into the related function. If the function value is positive, then the expression will be positive for all numbers in the interval. If the function value is negative, then the expression will be negative for all numbers in the interval.
4. Select the intervals for which the inequality is satisfied and write set-builder notation or interval notation for the solution set.

Solve using the method of Example 3.

4. $x^2 + 3x > 4$

5. $x^2 + 3x \leq 4$

Do Exercises 4 and 5.

EXAMPLE 4 Solve: $5x(x + 3)(x - 2) \geq 0$.

The solutions of $f(x) = 0$, or $5x(x + 3)(x - 2) = 0$, are 0, -3, and 2. They divide the real-number line into four intervals, as shown below.

We try test numbers in each interval:

A: Test -5, $f(-5) = 5(-5)(-5 + 3)(-5 - 2) = -350 < 0$.

B: Test -2, $f(-2) = 5(-2)(-2 + 3)(-2 - 2) = 40 > 0$.

C: Test 1, $f(1) = 5(1)(1 + 3)(1 - 2) = -20 < 0$.

D: Test 3, $f(3) = 5(3)(3 + 3)(3 - 2) = 90 > 0$.

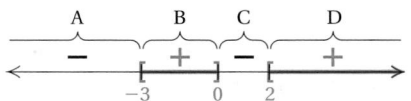

The expression is positive for values of x in intervals B and D. Since the inequality symbol is $\geq$, we need to include the x-intercepts. The solution set of the inequality is

$$\{x\,|\,{-3} \leq x \leq 0 \ or \ x \geq 2\}, \quad or \quad [-3, 0] \cup [2, \infty).$$

We visualize this with the graph below.

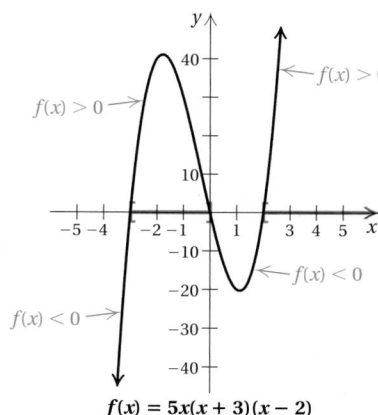

$$f(x) = 5x(x + 3)(x - 2)$$

Do Exercise 6.

6. Solve: $6x(x + 1)(x - 1) < 0$.

(b) Rational Inequalities

We adapt the preceding method for inequalities that involve rational expressions. We call these **rational inequalities**.

 EXAMPLE 5 Solve: $\dfrac{x - 3}{x + 4} \geq 2$.

We write a related equation by changing the $\geq$ symbol to $=$:

$$\frac{x - 3}{x + 4} = 2.$$

Then we solve this related equation. First, we multiply on both sides of the equation by the LCM, which is $x + 4$:

$$(x + 4) \cdot \frac{x - 3}{x + 4} = (x + 4) \cdot 2$$
$$x - 3 = 2x + 8$$
$$-11 = x.$$

With rational inequalities, we also need to determine those numbers for which the rational expression is not defined—that is, those numbers that make the denominator 0. We set the denominator equal to 0 and solve: $x + 4 = 0$, or $x = -4$. Next, we use the numbers -11 and -4 to divide the number line into intervals, as shown below.

We try test numbers in each interval to see if each satisfies the original inequality.

Answer

6. $\{x\,|\,x < -1 \ or \ 0 < x < 1\}$, or $(-\infty, -1) \cup (0, 1)$

A: Test -15, $\dfrac{x - 3}{x + 4} \geq 2$

$$\dfrac{-15 - 3}{-15 + 4} \;?\; 2$$

$$\dfrac{18}{11} \quad \text{FALSE}$$

Since the inequality is false for $x = -15$, the number -15 is not a solution of the inequality. Interval A is *not* part of the solution set.

B: Test -8, $\dfrac{x - 3}{x + 4} \geq 2$

$$\dfrac{-8 - 3}{-8 + 4} \;?\; 2$$

$$\dfrac{11}{4} \quad \text{TRUE}$$

Since the inequality is true for $x = -8$, the number -8 is a solution of the inequality. Interval B *is* part of the solution set.

C: Test 1, $\dfrac{x - 3}{x + 4} \geq 2$

$$\dfrac{1 - 3}{1 + 4} \;?\; 2$$

$$-\dfrac{2}{5} \quad \text{FALSE}$$

Since the inequality is false for $x = 1$, the number 1 is not a solution of the inequality. Interval C is *not* part of the solution set.

The solution set includes the interval B. The number -11 is also included since the inequality symbol is $\geq$ and -11 is a solution of the related equation. The number -4 is not included; it is not an allowable replacement because it results in division by 0. Thus the solution set of the original inequality is

$$\{x | -11 \leq x < -4\}, \quad \text{or} \quad [-11, -4).$$

To solve a rational inequality:

1. Change the inequality symbol to an equals sign and solve the related equation.

2. Find the numbers for which any denominator in the inequality is not defined.

3. Use the numbers found in steps (1) and (2) to divide the number line into intervals.

4. Substitute a number from each interval into the inequality. If the number is a solution, then the interval to which it belongs is part of the solution set.

5. Select the intervals for which the inequality is satisfied and write set-builder notation or interval notation for the solution set.

Solve.

7. $\dfrac{x + 1}{x - 2} \geq 3$

8. $\dfrac{x}{x - 5} < 2$

Answers

7. $\left\{ x | 2 < x \leq \dfrac{7}{2} \right\}$, or $\left(2, \dfrac{7}{2} \right]$

8. $\{x | x < 5 \; or \; x > 10\}$, or $(-\infty, 5) \cup (10, \infty)$

Do Exercises 7 and 8.

a Solve algebraically and verify results from the graph.

1. $(x - 6)(x + 2) > 0$

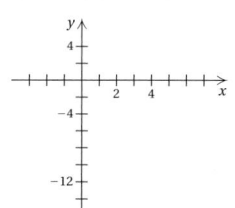

2. $(x - 5)(x + 1) > 0$

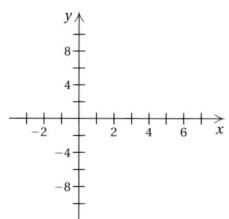

3. $4 - x^2 \geq 0$

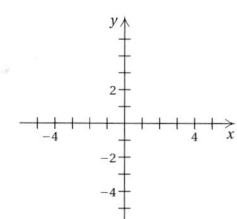

4. $9 - x^2 \leq 0$

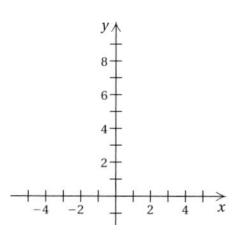

Solve.

5. $3(x + 1)(x - 4) \leq 0$

6. $(x - 7)(x + 3) \leq 0$

7. $x^2 - x - 2 < 0$

8. $x^2 + x - 2 < 0$

9. $x^2 - 2x + 1 \geq 0$

10. $x^2 + 6x + 9 < 0$

11. $x^2 + 8 < 6x$

12. $x^2 - 12 > 4x$

13. $3x(x + 2)(x - 2) < 0$

14. $5x(x + 1)(x - 1) > 0$

15. $(x + 9)(x - 4)(x + 1) > 0$

16. $(x - 1)(x + 8)(x - 2) < 0$

17. $(x + 3)(x + 2)(x - 1) < 0$

18. $(x - 2)(x - 3)(x + 1) < 0$

b  Solve.

19. $\dfrac{1}{x - 6} < 0$

20. $\dfrac{1}{x + 4} > 0$

21. $\dfrac{x + 1}{x - 3} > 0$

22. $\dfrac{x - 2}{x + 5} < 0$

23. $\dfrac{3x + 2}{x - 3} \leq 0$

24. $\dfrac{5 - 2x}{4x + 3} \leq 0$

25. $\dfrac{x - 1}{x - 2} > 3$

26. $\dfrac{x + 1}{2x - 3} < 1$

27. $\dfrac{(x - 2)(x + 1)}{x - 5} < 0$

28. $\dfrac{(x + 4)(x - 1)}{x + 3} > 0$

29. $\dfrac{x + 3}{x} \le 0$

30. $\dfrac{x}{x - 2} \ge 0$

31. $\dfrac{x}{x - 1} > 2$

32. $\dfrac{x - 5}{x} < 1$

33. $\dfrac{x - 1}{(x - 3)(x + 4)} < 0$

34. $\dfrac{x + 2}{(x - 2)(x + 7)} > 0$

35. $3 < \dfrac{1}{x}$

36. $\dfrac{1}{x} \le 2$

37. $\dfrac{x^2 + x - 2}{x^2 - x - 12}$

38. $\dfrac{x^2 - 11x + 30}{x^2 - 8x - 9} \ge 0$

Skill Maintenance

Simplify.   [10.3b]

39. $\sqrt[3]{\dfrac{125}{27}}$

40. $\sqrt{\dfrac{25}{4a^2}}$

41. $\sqrt{\dfrac{16a^3}{b^4}}$

42. $\sqrt[3]{\dfrac{27c^5}{343d^3}}$

Add or subtract.   [10.4a]

43. $3\sqrt{8} - 5\sqrt{2}$

44. $7\sqrt{45} - 2\sqrt{20}$

45. $5\sqrt[3]{16a^4} + 7\sqrt[3]{2a}$

46. $3\sqrt{10} + 8\sqrt{20} - 5\sqrt{80}$

Synthesis

47. Use a graphing calculator to solve Exercises 11, 22, and 25 by graphing two curves, one for each side of the inequality.

48. Use a graphing calculator to solve each of the following.

    a) $x + \dfrac{1}{x} < 0$ **b)** $x - \sqrt{x} \ge 0$

    c) $\frac{1}{3}x^3 - x + \frac{2}{3} \le 0$

Solve.

49. $x^2 - 2x \le 2$

50. $x^2 + 2x > 4$

51. $x^4 + 2x^2 > 0$

52. $x^4 + 3x^2 \le 0$

53. $\left| \dfrac{x + 2}{x - 1} \right| < 3$

54. *Total Profit.*   A company determines that its total profit from the production and sale of x units of a product is given by

$$P(x) = -x^2 + 812x - 9600.$$

    a) A company makes a profit for those nonnegative values of x for which $P(x) > 0$. Find the values of x for which the company makes a profit.

    b) A company loses money for those nonnegative values of x for which $P(x) < 0$. Find the values of x for which the company loses money.

55. *Height of a Thrown Object.*   The function

$$H(t) = -16t^2 + 32t + 1920$$

gives the height H of an object thrown from a cliff 1920 ft high, after time t seconds.

    a) For what times is the height greater than 1920 ft?

    b) For what times is the height less than 640 ft?

CHAPTER 11

Summary and Review

Key Terms, Properties, and Formulas

standard form, p. 782
completing the square, p. 787
quadratic formula, p. 796
discriminant, p. 815

parabola, p. 826
line of symmetry, p. 826
axis of symmetry, p. 826
vertex, p. 826

minimum, p. 829
maximum, p. 829
quadratic inequality, p. 855
rational inequality, p. 859

Principle of Square Roots: $x^2 = d$ has solutions $\sqrt{d}$ and $-\sqrt{d}$.

Quadratic Formula: $x = \dfrac{-b \pm \sqrt{b^2 - 4ac}}{2a}$ *Discriminant*: $b^2 - 4ac$

The *vertex* of the graph of $f(x) = ax^2 + bx + c$ is $\left(-\dfrac{b}{2a}, \dfrac{4ac - b^2}{4a}\right)$, or $\left(-\dfrac{b}{2a}, f\left(-\dfrac{b}{2a}\right)\right)$.

The *line of symmetry* of the graph of $f(x) = ax^2 + bx + c$ is $x = -\dfrac{b}{2a}$.

Concept Reinforcement

Determine whether each statement is true or false.

_____ **1.** The graph of $f(x) = -(-x^2 - 8x - 3)$ opens downward. [11.5a]

_____ **2.** If $(-5, 7)$ is the vertex of a parabola, then $x = -5$ is the line of symmetry. [11.6a]

_____ **3.** The graph of $f(x) = -3(x + 2)^2 - 5$ is a translation to the right of the graph of $f(x) = -3x^2 - 5$. [11.5b]

Important Concepts

Objective 11.1a Solve quadratic equations using the principle of square roots.

Example Solve: $(x - 3)^2 = -36$.

$\quad x - 3 = \sqrt{-36}$ *or* $x - 3 = -\sqrt{-36}$

$\quad\quad x = 3 + 6i$ *or* $x = 3 - 6i$

The solutions are $3 \pm 6i$.

Practice Exercise

  1. Solve: $(x - 2)^2 = -9$.

Objective 11.1b Solve quadratic equations by completing the square.

Example Solve by completing the square:

$\quad x^2 - 8x + 13 = 0$.

$\quad x^2 - 8x \quad\quad = -13$

$\quad x^2 - 8x + 16 = -13 + 16$

$\quad\quad (x - 4)^2 = 3$

$\quad x - 4 = \sqrt{3} \quad$ *or* $x - 4 = -\sqrt{3}$

$\quad\quad x = 4 + \sqrt{3}$ *or* $x = 4 - \sqrt{3}$

The solutions are $4 \pm \sqrt{3}$.

Practice Exercise

  2. Solve by completing the square:

$\quad x^2 - 12x + 31 = 0$.

Objective 11.2a Solve quadratic equations using the quadratic formula, and approximate solutions using a calculator.

Example Solve: $x^2 - 2x = 2$. Give the exact solutions and approximate solutions to three decimal places.

$x^2 - 2x - 2 = 0$ Standard form

$\quad a = 1, \quad b = -2, \quad c = -2$

$x = \dfrac{-(-2) \pm \sqrt{(-2)^2 - 4 \cdot 1 \cdot (-2)}}{2 \cdot 1}$ Using the quadratic formula

$x = \dfrac{2 \pm \sqrt{4 + 8}}{2} = \dfrac{2 \pm \sqrt{12}}{2} = \dfrac{2 \pm 2\sqrt{3}}{2}$

$\quad\quad\quad = 1 \pm \sqrt{3}$, or 2.732 and -0.732

Practice Exercise

3. Solve: $x^2 - 10x = -23$. Give the exact solutions and approximate solutions to three decimal places.

Objective 11.4a Determine the nature of the solutions of a quadratic equation.

Example Determine the nature of the solutions of the quadratic equation $x^2 - 7x = 1$.

In standard form, we have $x^2 - 7x - 1 = 0$. Thus, $a = 1, b = -7$, and $c = -1$. The discriminant, $b^2 - 4ac$, is $(-7)^2 - 4 \cdot 1 \cdot (-1)$, or 53. Since the discriminant is positive, there are two real solutions.

Practice Exercise

4. Determine the nature of the solutions of each quadratic equation.
a) $x^2 - 3x = 7$
b) $2x^2 - 5x + 5 = 0$

Objective 11.4b Write a quadratic equation having two given numbers as solutions.

Example Write a quadratic equation whose solutions are 7 and $-\frac{1}{4}$.

$\quad\quad x = 7 \quad or \quad\quad x = -\frac{1}{4}$

$\quad x - 7 = 0 \quad or \quad x + \frac{1}{4} = 0$

$\quad x - 7 = 0 \quad or \quad 4x + 1 = 0$ Clearing the fraction

$(x - 7)(4x + 1) = 0$ Using the principle of zero products in reverse

$\quad 4x^2 - 27x - 7 = 0$ Using FOIL

Practice Exercise

5. Write a quadratic equation whose solutions are $-\frac{2}{5}$ and 3.

Objective 11.4c Solve equations that are quadratic in form.

Example Solve: $x - 8\sqrt{x} - 9 = 0$.

Let $u = \sqrt{x}$. Then we substitute u for $\sqrt{x}$ and u^2 for x and solve for u:

$\quad\quad u^2 - 8u - 9 = 0$

$\quad (u - 9)(u + 1) = 0$

$\quad\quad\quad u = 9 \quad or \quad u = -1.$

Next, we substitute $\sqrt{x}$ for u and solve for x:

$\quad \sqrt{x} = 9 \quad or \quad \sqrt{x} = -1.$

Squaring each equation, we get

$\quad\quad x = 81 \quad or \quad x = 1.$

Checking both 81 and 1 in $x - 8\sqrt{x} - 9 = 0$, we find that 81 checks but 1 does not. The solution is 81.

Practice Exercise

6. Solve:
$$(x^2 - 3)^2 - 5(x^2 - 3) - 6 = 0.$$

Objective 11.6a For a quadratic function, find the vertex, the line of symmetry, and the maximum or minimum value, and then graph the function.

Example For $f(x) = -2x^2 + 4x + 1$, find the vertex, the line of symmetry, and the maximum or minimum value. Then graph.

We factor out -2 from only the first two terms:

$$f(x) = -2(x^2 - 2x) + 1.$$

Next, we complete the square, factor, and simplify:

$$\begin{aligned} f(x) &= -2(x^2 - 2x \quad\ ) + 1 \\ &= -2(x^2 - 2x + 1 - 1) + 1 \\ &= -2(x^2 - 2x + 1) + (-2)(-1) + 1 \\ &= -2(x - 1)^2 + 3. \end{aligned}$$

The vertex is $(1, 3)$. The line of symmetry is $x = 1$. The coefficient of x^2 is negative, so the graph opens down. Thus, 3 is the maximum value of the function.

We plot points and graph the parabola.

x	y
1	3
2	1
0	1
3	-5
-1	-5

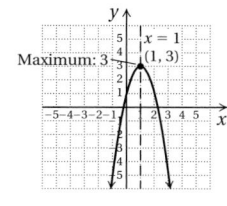

Practice Exercise

7. For $f(x) = -x^2 - 2x - 3$, find the vertex, the line of symmetry, and the maximum or minimum value. Then graph.

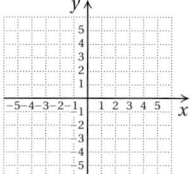

Objective 11.6b Find the intercepts of a quadratic function.

Example Find the intercepts of $f(x) = x^2 - 8x + 14$.

Since $f(0) = 0^2 - 8 \cdot 0 + 14$, the y-intercept is $(0, 14)$. To find the x-intercepts, we solve $0 = x^2 - 8x + 14$. Using the quadratic formula, we have $x = 4 \pm \sqrt{2}$. Thus the x-intercepts are $\left(4 - \sqrt{2}, 0\right)$ and $\left(4 + \sqrt{2}, 0\right)$.

Practice Exercise

8. Find the intercepts of $f(x) = x^2 - 6x + 4$.

Objective 11.8a Solve quadratic inequalities and other polynomial inequalities.

Example Solve: $x^2 - 15 > 2x$.

$$x^2 - 2x - 15 > 0 \qquad \text{Adding 15}$$

We set the polynomial equal to 0 and solve. The solutions of $x^2 - 2x - 15 = 0$, or $(x + 3)(x - 5) = 0$, are -3 and 5. They divide the number line into three intervals.

$$\xleftarrow[\ -6\,-5\,-4\,-3\,-2\,-1\ \ 0\ \ 1\ \ 2\ \ 3\ \ 4\ \ 5\ \ 6\ ]{}\rightarrow$$

We try a test point in each interval:

Test -5: $(-5)^2 - 2(-5) - 15 = 20 > 0$;

Test 0: $0^2 - 2 \cdot 0 - 15 = -15 < 0$;

Test 6: $6^2 - 2 \cdot 6 - 15 = 9 > 0$.

The expression $x^2 - 2x - 15$ is positive for values of x in the intervals $(-\infty, -3)$ and $(5, \infty)$. The inequality symbol is $>$, so -3 and 5 are not solutions. The solution set is $\{x \mid x < -3 \ or \ x > 5\}$, or $(-\infty, -3) \cup (5, \infty)$.

Practice Exercise

9. Solve: $x^2 + 40 > 14x$.

Objective 11.8b   Solve rational inequalities.

Example   Solve: $\dfrac{x + 3}{x - 6} \geq 2$.

    We first solve the related equation $\dfrac{x + 3}{x - 6} = 2$. The solution is 15. We also need to determine those numbers for which the rational expression is not defined. We set the denominator equal to 0 and solve: $x - 6 = 0$, or $x = 6$. The numbers 6 and 15 divide the number line into three intervals. We test a point in each interval.

$$\xleftarrow{\quad}\overset{\displaystyle 4\ \ 5\ \ 6\ \ 7\ \ 8\ \ 9\ \ 10\ 11\ 12\ 13\ 14\ 15\ 16}{\longmapsto\!\!\!\!+\!\!\!\!+\!\!\!\!+\!\!\!\!+\!\!\!\!+\!\!\!\!+\!\!\!\!+\!\!\!\!+\!\!\!\!+\!\!\!\!+\!\!\!\!+\!\!\!\!+\!\!\!\!+}\xrightarrow{\quad}$$

Test 5: $\dfrac{5 + 3}{5 - 6} \geq 2$, or $-8 \geq 2$, which is false.

Test 9: $\dfrac{9 + 3}{9 - 6} \geq 2$, or $4 \geq 2$, which is true.

Test 17: $\dfrac{17 + 3}{17 - 6} \geq 2$, or $\dfrac{20}{11} \geq 2$, which is false.

The solution set includes the interval $(6, 15)$ and the number 15, the solution of the related equation. The number 6 is not included. It is not an allowable replacement because it results in division by 0. The solution set is $\{x \mid 6 < x \leq 15\}$, or $(6, 15]$.

Practice Exercise

10.   Solve: $\dfrac{x + 7}{x - 5} \geq 3$.

Review Exercises

1. a) Solve: $2x^2 - 7 = 0$.   [11.1a]
   b) Find the x-intercepts of $f(x) = 2x^2 - 7$.

Solve.   [11.2a]

2. $14x^2 + 5x = 0$

3. $x^2 - 12x + 27 = 0$

4. $4x^2 + 3x + 1 = 0$

5. $x^2 - 7x + 13 = 0$

6. $4x(x - 1) + 15 = x(3x + 4)$

7. $x^2 + 4x + 1 = 0$. Give exact solutions and approximate solutions to three decimal places.

8. $\dfrac{x}{x - 2} + \dfrac{4}{x - 6} = 0$

9. $\dfrac{x}{4} - \dfrac{4}{x} = 2$ **10.** $15 = \dfrac{8}{x + 2} - \dfrac{6}{x - 2}$

11. Solve $x^2 + 6x + 2 = 0$ by completing the square. Show your work.   [11.1b]

12. *Hang Time.*   Use the function $V(T) = 48T^2$. A basketball player has a vertical leap of 39 in. What is his hang time?   [11.1c]

13. *DVD Player Screen.*   The width of a rectangular screen on a portable DVD player is 5 cm less than the length. The area is 126 cm^2. Find the length and the width.   [11.3a]

14. *Picture Matting.*   A picture mat measures 12 in. by 16 in., and 140 in^2 of picture shows. Find the width of the mat.   [11.3a]

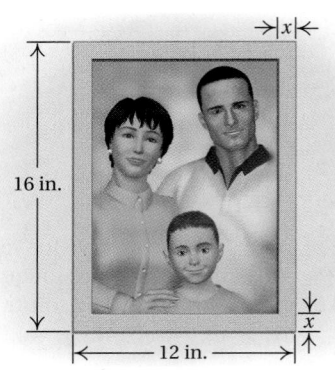

16 in.

$\rightarrow |x| \leftarrow$

12 in.

15. *Motorcycle Travel.* During the first part of a trip, a motorcyclist travels 50 mi. The rider travels 80 mi on the second part of the trip at a speed that is 10 mph slower. The total time for the trip is 3 hr. What is the speed on each part of the trip? [11.3a]

Determine the nature of the solutions of each equation. [11.4a]

16. $x^2 + 3x - 6 = 0$

17. $x^2 + 2x + 5 = 0$

Write a quadratic equation having the given solutions. [11.4b]

18. $\frac{1}{5}, -\frac{3}{5}$

19. -4, only solution

Solve for the indicated letter. [11.3b]

20. $N = 3\pi\sqrt{\dfrac{1}{p}}$, for p

21. $2A = \dfrac{3B}{T^2}$, for T

Solve. [11.4c]

22. $x^4 - 13x^2 + 36 = 0$

23. $15x^{-2} - 2x^{-1} - 1 = 0$

24. $(x^2 - 4)^2 - (x^2 - 4) - 6 = 0$

25. $x - 13\sqrt{x} + 36 = 0$

For each quadratic function in Exercises 26–28, find and label **(a)** the vertex, **(b)** the line of symmetry, and **(c)** the maximum or minimum value. Then **(d)** graph the function. [11.5c], [11.6a]

26. $f(x) = -\frac{1}{2}(x - 1)^2 + 3$

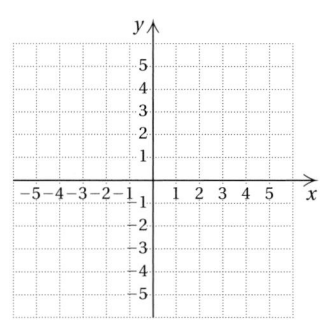

Vertex: (____, ____)
Line of symmetry: $x = $ ____
_____ value: ____

27. $f(x) = x^2 - x + 6$

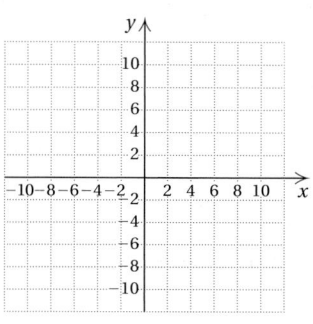

Vertex: (____, ____)
Line of symmetry: $x = $ ____
_____ value: ____

28. $f(x) = -3x^2 - 12x - 8$

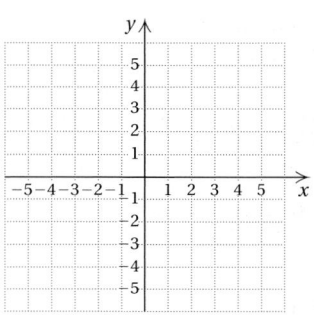

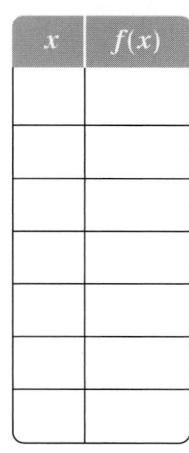

Vertex: (____, ____)
Line of symmetry: $x = $ ____
_____ value: ____

Find the x- and y-intercepts. [11.6b]

29. $f(x) = x^2 - 9x + 14$

30. $g(x) = x^2 - 4x - 3$

31. What is the minimum product of two numbers whose difference is 22? What numbers yield this product? [11.7a]

32. Find a quadratic function that fits the data points $(0, -2), (1, 3)$, and $(3, 7)$. [11.7b]

33. *Live Births by Age.* The average number of live births per 1000 women rises and falls according to age, as seen in the following bar graph. [11.7b]

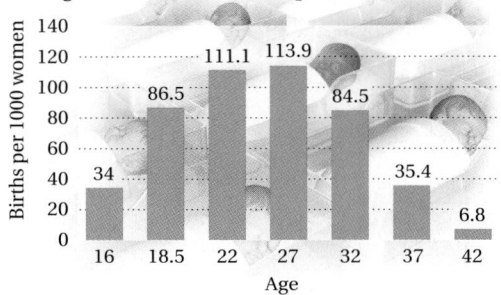

Average Number of Live Births per 1000 Women

SOURCE: Centers for Disease Control and Prevention

a) Use the data points $(16, 34), (27, 113.9)$, and $(37, 35.4)$ to fit a quadratic function to the data.

b) Use the quadratic function to estimate the number of live births per 1000 women of age 30.

Solve. [11.8a, b]

34. $(x + 2)(x - 1)(x - 2) > 0$

35. $\dfrac{(x + 4)(x - 1)}{(x + 2)} < 0$

36. Determine the nature of the solutions of $x^2 - 10x + 25 = 0$. [11.4a]

  A. Infinite number of solutions

  B. One real solution

  C. Two real solutions

  D. No real solutions

37. Solve: $2x^2 - 6x + 5 = 0$. [11.2a]

  A. $\dfrac{3}{2} \pm \dfrac{\sqrt{19}}{2}$ **B.** $3 \pm i$

  C. $3 \pm \sqrt{19}$ **D.** $\dfrac{3}{2} \pm \dfrac{i}{2}$

Synthesis

38. A quadratic function has x-intercepts $(-3, 0)$ and $(5, 0)$ and y-intercept $(0, -7)$. Find an equation for the function. What is its maximum or minimum value? [11.7a, b]

39. Find h and k such that $3x^2 - hx + 4k = 0$, the sum of the solutions is 20, and the product of the solutions is 80. [11.2a], [11.7b]

40. The average of two numbers is 171. One of the numbers is the square root of the other. Find the numbers. [11.3a]

Understanding Through Discussion and Writing

1. Does the graph of every quadratic function have a y-intercept? Why or why not? [11.6b]

2. Explain how the leading coefficient of a quadratic function can be used to determine whether a maximum or minimum function value exists. [11.7a]

3. Explain, without plotting points, why the graph of $f(x) = (x + 3)^2 - 4$ looks like the graph of $f(x) = x^2$ translated 3 units to the left and 4 units down. [11.5c]

4. Describe a method that could be used to create quadratic inequalities that have no solution. [11.8a]

5. Is it possible for the graph of a quadratic function to have only one x-intercept if the vertex is off the x-axis? Why or why not? [11.6b]

6. Explain how the x-intercepts of a quadratic function can be used to help find the vertex of the function. What piece of information would still be missing? [11.6a, b]

CHAPTER

11

Test

For Extra Help

CHAPTER
Test Prep
VIDEOS

Step-by-step test solutions are found on the Chapter Test Prep Videos available via the Video Resources
on DVD, in **MyMathLab**, and on You**Tube** (search "BittingerIntroInter" and click on "Channels").

1. a) Solve: $3x^2 - 4 = 0$.
   b) Find the x-intercepts of $f(x) = 3x^2 - 4$.

Solve.

2. $x^2 + x + 1 = 0$

3. $x - 8\sqrt{x} + 7 = 0$

4. $4x(x - 2) - 3x(x + 1) = -18$

5. $4x^4 - 17x^2 + 15 = 0$

6. $x^2 + 4x = 2$. Give exact solutions and approximate solutions to three decimal places.

7. $\dfrac{1}{4 - x} + \dfrac{1}{2 + x} = \dfrac{3}{4}$

8. Solve $x^2 - 4x + 1 = 0$ by completing the square. Show your work.

9. *Free-Falling Objects.* The Peachtree Plaza in Atlanta, Georgia, is 723 ft tall. Use the function $s(t) = 16t^2$ to approximate how long it would take an object to fall from the top.

10. *Marine Travel.* The Columbia River flows at a rate of 2 mph for the length of a popular boating route. In order for a motorized dinghy to travel 3 mi upriver and then return in a total of 4 hr, how fast must the boat be able to travel in still water?

11. *Memory Board.* A computer-parts company wants to make a rectangular memory board that has a perimeter of 28 cm. What dimensions will allow the board to have the maximum area?

12. *Hang Time.* Use the function $V(T) = 48T^2$. Nate Robinson of the New York Knicks has a vertical leap of 43 in. What is his hang time?

13. Determine the nature of the solutions of the equation $x^2 + 5x + 17 = 0$.

14. Write a quadratic equation having the solutions $\sqrt{3}$ and $3\sqrt{3}$.

15. Solve $V = 48T^2$ for T.

For the quadratic functions in Exercises 16 and 17, find and label **(a)** the vertex, **(b)** the line of symmetry, and **(c)** the maximum or minimum value. Then **(d)** graph the function.

16. $f(x) = -x^2 - 2x$

x	$f(x)$

Vertex: (____ , ____)
Line of symmetry: $x =$ ____
_____ value: ____

17. $f(x) = 4x^2 - 24x + 41$

x	$f(x)$

Vertex: (____ , ____)
Line of symmetry: $x =$ ____
_____ value: ____

18. Find the x- and y-intercepts:
$$f(x) = -x^2 + 4x - 1.$$

19. What is the minimum product of two numbers whose difference is 8? What numbers yield this product?

20. Find the quadratic function that fits the data points $(0, 0)$, $(3, 0)$, and $(5, 2)$.

21. *Foreign Adoptions.*   The graph at right shows the number of foreign adoptions to the United States for various years. It appears that the graph might be fit by a quadratic function.

a) Use the data points $(0, 18.5)$, $(4, 22.9)$, and $(8, 17.4)$ to fit a quadratic function $A(x) = ax^2 + bx + c$ to the data, where A is the number of foreign adoptions to the United States x years since 2000 and $x = 0$ corresponds to 2000.

b) Use the quadratic function to estimate the number of adoptions in 2009.

Foreign Adoptions to the United States

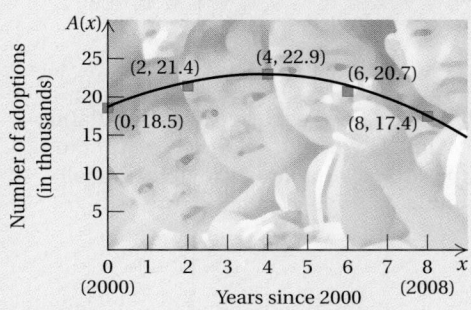

SOURCE: Intercountry Adoption, Office of Children's Issues, U.S. Department of State

Solve.

22. $x^2 < 6x + 7$

23. $\dfrac{x - 5}{x + 3} < 0$

24. $\dfrac{(x - 2)}{(x + 3)(x - 1)} \geq 0$

25. Write a quadratic equation whose solutions are $\dfrac{i}{2}$ and $-\dfrac{i}{2}$.

A. $4x^2 - 4ix - 1 = 0$
B. $x^2 - \dfrac{1}{4} = 0$
C. $4x^2 + 1 = 0$
D. $x^2 - ix + 1 = 0$

Synthesis

26. A quadratic function has x-intercepts $(-2, 0)$ and $(7, 0)$ and y-intercept $(0, 8)$. Find an equation for the function. What is its maximum or minimum value?

27. One solution of $kx^2 + 3x - k = 0$ is -2. Find the other solution.

Cumulative Review

1. *Golf Courses.* Most golf courses have a hole such as the one shown here, where the safe way to the hole is to hit straight out on a first shot (the distance a) and then make subsequent shots at a right angle to cover the distance b. Golfers are often lured, however, into taking a shortcut over trees, houses, or lakes. If a golfer makes a hole in one on this hole, how long is the shot?

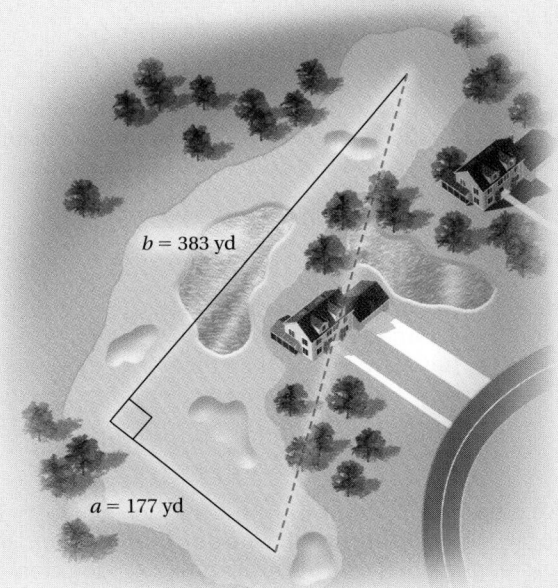

$b = 383$ yd

$a = 177$ yd

Simplify.

2. $(4 + 8x^2 - 5x) - (-2x^2 + 3x - 2)$

3. $(2x^2 - x + 3)(x - 4)$

4. $\dfrac{a^2 - 16}{5a - 15} \cdot \dfrac{2a - 6}{a + 4}$

5. $\dfrac{y}{y^2 - y - 42} \div \dfrac{y^2}{y - 7}$

6. $\dfrac{2}{m + 1} + \dfrac{3}{m - 5} - \dfrac{m^2 - 1}{m^2 - 4m - 5}$

7. $(9x^3 + 5x^2 + 2) \div (x + 2)$

8. $\dfrac{\dfrac{1}{x} - \dfrac{1}{y}}{x + y}$

9. $\sqrt{0.36}$

10. $\sqrt{9x^2 - 36x + 36}$

11. $6\sqrt{45} - 3\sqrt{20}$

12. $\dfrac{2\sqrt{3} - 4\sqrt{2}}{\sqrt{2} - 3\sqrt{6}}$

13. $(8^{2/3})^4$

14. $(3 + 2i)(5 - i)$

15. $\dfrac{6 - 2i}{3i}$

Factor.

16. $2t^2 - 7t - 30$

17. $a^2 + 3a - 54$

18. $-3a^3 + 12a^2$

19. $64a^2 - 9b^2$

20. $3a^2 - 36a + 108$

21. $\dfrac{1}{27}a^3 - 1$

22. $24a^3 + 18a^2 - 20a - 15$

23. $(x + 1)(x - 1) + (x + 1)(x + 2)$

Solve.

24. $3(4x - 5) + 6 = 3 - (x + 1)$

25. $F = \dfrac{mv^2}{r}$, for r

26. $5 - 3(2x + 1) \le 8x - 3$

27. $3x - 2 < -6 \ or \ x + 3 > 9$

28. $|4x - 1| \leq 14$

29. $5x + 10y = -10,$
$\quad -2x - 3y = 5$

30. $2x + y - z = 9,$
$\quad 4x - 2y + z = -9,$
$\quad 2x - y + 2z = -12$

31. $10x^2 + 28x - 6 = 0$

32. $\dfrac{2}{n} - \dfrac{7}{n} = 3$

33. $\dfrac{1}{2x - 1} = \dfrac{3}{5x}$

34. $A = \dfrac{mh}{m + a},$ for m

35. $\sqrt{2x - 1} = 6$

36. $\sqrt{x - 2} + 1 = \sqrt{2x - 6}$

37. $16(t - 1) = t(t + 8)$

38. $x^2 - 3x + 16 = 0$

39. $\dfrac{18}{x + 1} - \dfrac{12}{x} = \dfrac{1}{3}$

40. $P = \sqrt{a^2 - b^2},$ for a

41. $\dfrac{(x + 3)(x + 2)}{(x - 1)(x + 1)} < 0$

42. Solve: $4x^2 - 25 > 0.$

Graph.

43. $x + y = 2$

44. $y \geq 6x - 5$

45. $x < -3$

46. $3x - y > 6,$
$\quad 4x + y \leq 3$

47. $f(x) = x^2 - 1$

48. $f(x) = -2x^2 + 3$

49. Find an equation of the line with slope $\frac{1}{2}$ and through the point $(-4, 2)$.

50. Find an equation of the line parallel to the line $3x + y = 4$ and through the point $(0, 1)$.

51. *Marine Travel.* The Connecticut River flows at a rate of 4 km/h for the length of a popular scenic route. In order for a cruiser to travel 60 km upriver and then return in a total of 8 hr, how fast must the boat be able to travel in still water?

52. *Architecture.* An architect is designing a rectangular family room with a perimeter of 56 ft. What dimensions will yield the maximum area? What is the maximum area?

53. The perimeter of a hexagon with all six sides the same length is the same as the perimeter of a square. One side of the hexagon is 3 less than the side of the square. Find the perimeter of each polygon.

54. Two pipes can fill a tank in $1\frac{1}{2}$ hr. One pipe requires 4 hr longer running alone to fill the tank than the other. How long would it take the faster pipe, working alone, to fill the tank?

55. Complete the square: $f(x) = 5x^2 - 20x + 15.$
 A. $f(x) = 5(x - 2)^2 - 5$ **B.** $f(x) = 5(x + 2)^2 + 15$
 C. $f(x) = 5(x + 2)^2 + 6$ **D.** $f(x) = 5(x + 2)^2 + 11$

56. How many times does the graph of $f(x) = x^4 - 6x^2 - 16$ cross the x-axis?
 A. 1   **B.** 2
 C. 3   **D.** 4

Synthesis

57. Solve: $\dfrac{2x + 1}{x} = 3 + 7\sqrt{\dfrac{2x + 1}{x}}.$

58. Factor: $\dfrac{a^3}{8} + \dfrac{8b^3}{729}.$

Exponential and Logarithmic Functions

Real-World Application

Twenty-one women competed in the Summer Olympic Games in Paris in 1900, the first year in which women participated in the Games. Female participation grew exponentially through the years, reaching a total of 4746 competitors in Beijing in 2008. We let $t =$ the number of years since 1900. Then $t = 0$ corresponds to 1900 and $t = 108$ corresponds to 2008. Use the data points $(0, 21)$ and $(108, 4746)$ to find the exponential growth rate and then the exponential growth function. Use the function to predict the number of female competitors in 2016 and to determine the year in which there were about 2552 female competitors.

Sources: The Complete Book of the Olympics, David Wallechinsky; www.olympic.org/uk

This problem appears as Example 6 in Section 12.7.

Exponential Functions

OBJECTIVES

a Graph exponential equations and functions.

b Graph exponential equations in which x and y have been interchanged.

c Solve applied problems involving applications of exponential functions and their graphs.

SKILL TO REVIEW
Objective 3.2a: Graph linear equations of the type $y = mx + b$.

Graph.

1. $y = x + 3$

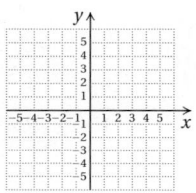

2. $y = \frac{1}{2}x - 2$

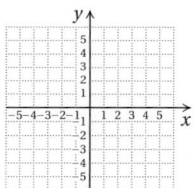

The rapidly rising graph shown below approximates the graph of an *exponential function*. We will consider such functions and some of their applications.

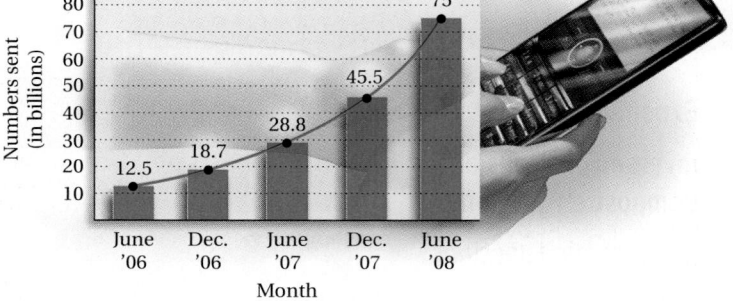

Text Messages Sent in the United States

SOURCE: CTIA, The Wireless Association

a Graphing Exponential Functions

In Chapter 10, we gave meaning to exponential expressions with rational-number exponents such as

$$8^{1/4}, \quad 3^{-3/4}, \quad 7^{2.34}, \quad 5^{1.73}.$$

For example, $5^{1.73}$, or $5^{173/100}$, or $\sqrt[100]{5^{173}}$, means to raise 5 to the 173rd power and then take the 100th root. We now develop the meaning of exponential expressions with irrational exponents. Examples of expressions with irrational exponents are

$$5^{\sqrt{3}}, \quad 7^{\pi}, \quad 9^{-\sqrt{2}}.$$

Since we can approximate irrational numbers with decimal approximations, we can also approximate expressions with irrational exponents. For example, consider $5^{\sqrt{3}}$. We know that $5^{\sqrt{3}} \approx 5^{1.73} = \sqrt[100]{5^{173}}$. As rational values of r get close to $\sqrt{3}$, 5^r gets close to some real number. Note the following:

r closes in on $\sqrt{3}$.	5^r closes in on some real number p.
r	5^r
$1 < \sqrt{3} < 2$	$5 = 5^1 < p < 5^2 = 25$
$1.7 < \sqrt{3} < 1.8$	$15.426 \approx 5^{1.7} < p < 5^{1.8} \approx 18.119$
$1.73 < \sqrt{3} < 1.74$	$16.189 \approx 5^{1.73} < p < 5^{1.74} \approx 16.452$
$1.732 < \sqrt{3} < 1.733$	$16.241 \approx 5^{1.732} < p < 5^{1.733} \approx 16.267$

As r closes in on $\sqrt{3}$, 5^r closes in on some real number p. We define $5^{\sqrt{3}}$ to be that number p. To seven decimal places, we have

$$5^{\sqrt{3}} \approx 16.2424508.$$

Answers

Skill to Review:

1.

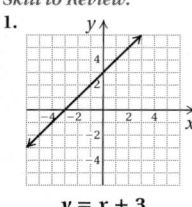

$y = x + 3$

2.

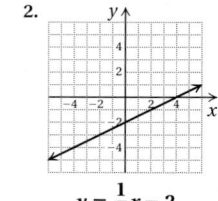

$y = \frac{1}{2}x - 2$

Any positive irrational exponent can be defined in a similar way. Negative irrational exponents are then defined in the same way as negative integer exponents. Then the expression a^x has meaning for any real number x. The general laws of exponents still hold, but we will not prove that here.

We now define exponential functions.

EXPONENTIAL FUNCTION

The function $f(x) = a^x$, where a is a positive constant different from 1, is called an **exponential function**, base a.

We restrict the base a to being positive to avoid the possibility of taking even roots of negative numbers such as the square root of -1, $(-1)^{1/2}$, which is not a real number. We restrict the base from being 1 because for $a = 1$, $t(x) = 1^x = 1$, which is a constant. The following are examples of exponential functions:

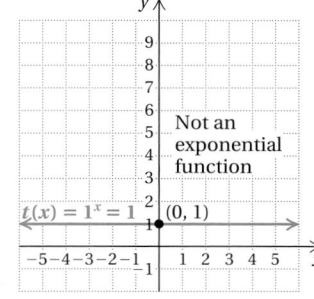

$$f(x) = 2^x, \qquad f(x) = \left(\tfrac{1}{2}\right)^x, \qquad f(x) = (0.4)^x.$$

Note that in contrast to polynomial functions like $f(x) = x^2$ and $f(x) = x^3$, the variable is *in the exponent*. Let's consider graphs of exponential functions.

EXAMPLE 1   Graph the exponential function $f(x) = 2^x$.

We compute some function values and list the results in a table. It is a good idea to begin by letting $x = 0$.

$f(0) = 2^0 = 1;$

$f(1) = 2^1 = 2;$

$f(2) = 2^2 = 4;$

$f(3) = 2^3 = 8;$

$f(-1) = 2^{-1} = \dfrac{1}{2^1} = \dfrac{1}{2};$

$f(-2) = 2^{-2} = \dfrac{1}{2^2} = \dfrac{1}{4};$

$f(-3) = 2^{-3} = \dfrac{1}{2^3} = \dfrac{1}{8}$

x	$f(x)$
0	1
1	2
2	4
3	8
-1	$\frac{1}{2}$
-2	$\frac{1}{4}$
-3	$\frac{1}{8}$

Next, we plot these points and connect them with a smooth curve.

In graphing, be sure to plot enough points to determine how steeply the curve rises.

The curve comes very close to the x-axis, but does not touch or cross it.

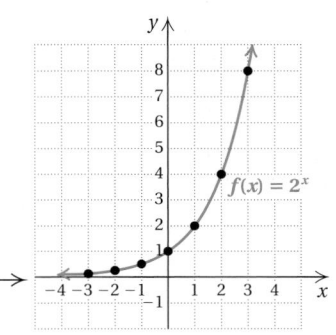

1. Graph: $f(x) = 3^x$. Complete this table of solutions. Then plot the points from the table and connect them with a smooth curve.

x	$f(x)$
0	
1	
2	
3	
−1	
−2	
−3	

2. Graph: $f(x) = \left(\dfrac{1}{3}\right)^x$. Complete this table of solutions. Then plot the points from the table and connect them with a smooth curve.

x	$f(x)$
0	
1	
2	
3	
−1	
−2	
−3	

Note that as x increases, the function values increase indefinitely. As x decreases, the function values decrease, getting very close to 0. The x-axis, or the line $y = 0$, is an *asymptote*, meaning here that as x gets very small, the curve comes very close to but never touches the axis.

> Do Exercise 1.

EXAMPLE 2 Graph the exponential function $f(x) = \left(\dfrac{1}{2}\right)^x$.

We compute some function values and list the results in a table. Before we do so, note that

$$f(x) = \left(\tfrac{1}{2}\right)^x = (2^{-1})^x = 2^{-x}.$$

Then we have

$$f(0) = 2^{-0} = 1;$$

$$f(1) = 2^{-1} = \frac{1}{2^1} = \frac{1}{2};$$

$$f(2) = 2^{-2} = \frac{1}{2^2} = \frac{1}{4};$$

$$f(3) = 2^{-3} = \frac{1}{2^3} = \frac{1}{8};$$

$$f(-1) = 2^{-(-1)} = 2^1 = 2;$$

$$f(-2) = 2^{-(-2)} = 2^2 = 4;$$

$$f(-3) = 2^{-(-3)} = 2^3 = 8.$$

x	$f(x)$
0	1
1	$\frac{1}{2}$
2	$\frac{1}{4}$
3	$\frac{1}{8}$
−1	2
−2	4
−3	8

Next, we plot these points and draw the curve. Note that this graph is a reflection across the y-axis of the graph in Example 1. The line $y = 0$ is again an asymptote.

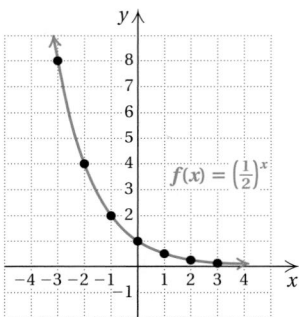

> Do Exercise 2.

The preceding examples illustrate exponential functions with various bases. Let's list some of their characteristics. Keep in mind that the definition of an exponential function, $f(x) = a^x$, requires that the base be positive and different from 1.

Answers

1.

x	$f(x)$
0	1
1	3
2	9
3	27
−1	$\frac{1}{3}$
−2	$\frac{1}{9}$
−3	$\frac{1}{27}$

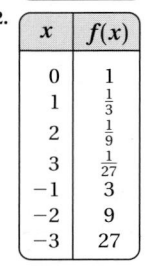

$f(x) = 3^x$

2.

x	$f(x)$
0	1
1	$\frac{1}{3}$
2	$\frac{1}{9}$
3	$\frac{1}{27}$
−1	3
−2	9
−3	27

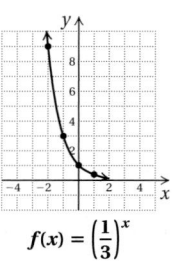

$f(x) = \left(\dfrac{1}{3}\right)^x$

When $a > 1$, the function $f(x) = a^x$ increases from left to right. The greater the value of a, the steeper the curve. As x gets smaller and smaller, the curve gets closer to the line $y = 0$: It is an asymptote.

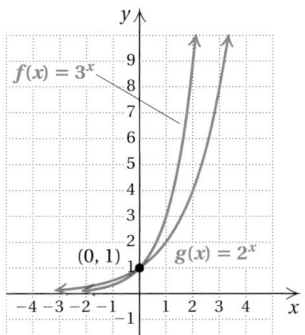

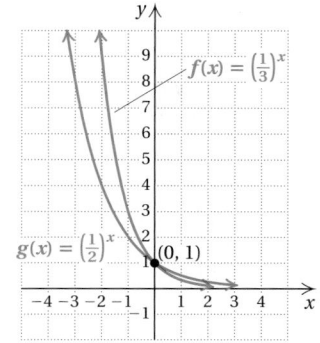

When $0 < a < 1$, the function $f(x) = a^x$ decreases from left to right. As a approaches 1, the curve becomes less steep. As x gets larger and larger, the curve gets closer to the line $y = 0$: It is an asymptote.

> ### y-INTERCEPT OF AN EXPONENTIAL FUNCTION
>
> All functions $f(x) = a^x$ go through the point $(0, 1)$. That is, the y-intercept is $(0, 1)$.

Do Exercises 3 and 4.

EXAMPLE 3 Graph: $f(x) = 2^{x-2}$.

We construct a table of values. Then we plot the points and connect them with a smooth curve. Be sure to note that $x - 2$ is the *exponent*.

$$f(0) = 2^{0-2} = 2^{-2} = \frac{1}{2^2} = \frac{1}{4};$$

$$f(1) = 2^{1-2} = 2^{-1} = \frac{1}{2^1} = \frac{1}{2};$$

$$f(2) = 2^{2-2} = 2^0 = 1;$$

$$f(3) = 2^{3-2} = 2^1 = 2;$$

$$f(4) = 2^{4-2} = 2^2 = 4;$$

$$f(-1) = 2^{-1-2} = 2^{-3} = \frac{1}{2^3} = \frac{1}{8};$$

$$f(-2) = 2^{-2-2} = 2^{-4} = \frac{1}{2^4} = \frac{1}{16}$$

x	$f(x)$
0	$\frac{1}{4}$
1	$\frac{1}{2}$
2	1
3	2
4	4
-1	$\frac{1}{8}$
-2	$\frac{1}{16}$

The graph has the same shape as the graph of $g(x) = 2^x$, but it is translated 2 units to the right.

The y-intercept of $g(x) = 2^x$ is $(0, 1)$. The y-intercept of $f(x) = 2^{x-2}$ is $\left(0, \frac{1}{4}\right)$. The line $y = 0$ is still an asymptote.

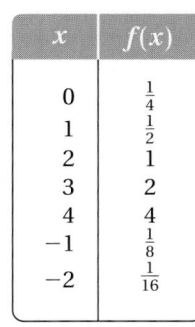

Graph.

3. $f(x) = 4^x$

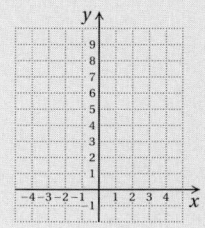

4. $f(x) = \left(\frac{1}{4}\right)^x$

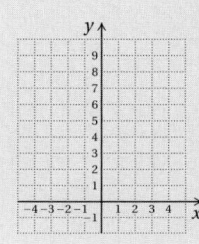

Answers

3.
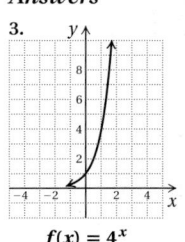
$f(x) = 4^x$

4.
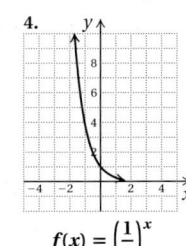
$f(x) = \left(\frac{1}{4}\right)^x$

5. Graph: $f(x) = 2^{x+2}$.

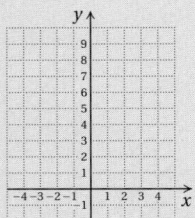

Do Exercise 5.

EXAMPLE 4   Graph: $f(x) = 2^x - 3$.

We construct a table of values. Then we plot the points and connect them with a smooth curve. Note that the only expression in the exponent is x.

$f(0) = 2^0 - 3 = 1 - 3 = -2;$

$f(1) = 2^1 - 3 = 2 - 3 = -1;$

$f(2) = 2^2 - 3 = 4 - 3 = 1;$

$f(3) = 2^3 - 3 = 8 - 3 = 5;$

$f(4) = 2^4 - 3 = 16 - 3 = 13;$

$f(-1) = 2^{-1} - 3 = \frac{1}{2} - 3 = -\frac{5}{2};$

$f(-2) = 2^{-2} - 3 = \frac{1}{4} - 3 = -\frac{11}{4}$

x	$f(x)$
0	-2
1	-1
2	1
3	5
4	13
-1	$-\frac{5}{2}$
-2	$-\frac{11}{4}$

6. Graph: $f(x) = 2^x - 4$.

x	$f(x)$
0	
1	
2	
3	
4	
-1	
-2	

The graph has the same shape as the graph of $g(x) = 2^x$, but it is translated 3 units down. The y-intercept is $(0, -2)$. The line $y = -3$ is an asymptote. The curve gets closer to this line as x gets smaller and smaller.

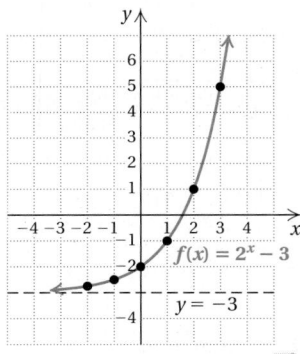

Do Exercise 6.

Calculator Corner

Graphing Exponential Functions   We can use a graphing calculator to graph exponential functions. It might be necessary to try several sets of window dimensions in order to find the ones that give a good view of the curve.

To graph $f(x) = 3^x - 1$, we enter the equation as y_1 by pressing ③ ⌃ (X,T,θ,n) ⊖ ①. We can begin graphing with the standard window $[-10, 10, -10, 10]$ by pressing (ZOOM) ⑥. Although this window gives a good view of the curve, we might want to adjust it to show more of the curve in the first quadrant. Changing the dimensions to $[-10, 10, -5, 15]$ accomplishes this.

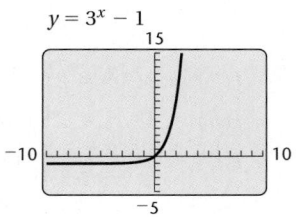

$y = 3^x - 1$

Exercises:

1. Use a graphing calculator to graph the functions in Examples 1–4.

2. Use a graphing calculator to graph the functions in Margin Exercises 1–6.

Answers

5.

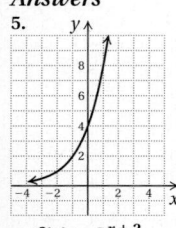

$f(x) = 2^{x+2}$

6.

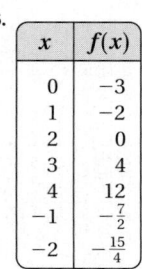

x	$f(x)$
0	-3
1	-2
2	0
3	4
4	12
-1	$-\frac{7}{2}$
-2	$-\frac{15}{4}$

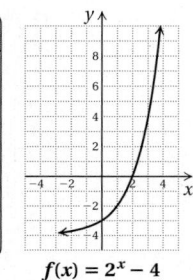

$f(x) = 2^x - 4$

b) Equations with x and y Interchanged

It will be helpful in later work to be able to graph an equation in which the x and the y in $y = a^x$ are interchanged.

EXAMPLE 5   Graph: $x = 2^y$.

Note that x is alone on one side of the equation. We can find ordered pairs that are solutions more easily by choosing values for y and then computing the x-values.

For $y = 0, x = 2^0 = 1$.

For $y = 1, x = 2^1 = 2$.

For $y = 2, x = 2^2 = 4$.

For $y = 3, x = 2^3 = 8$.

For $y = -1, x = 2^{-1} = \dfrac{1}{2^1} = \dfrac{1}{2}$.

For $y = -2, x = 2^{-2} = \dfrac{1}{2^2} = \dfrac{1}{4}$.

For $y = -3, x = 2^{-3} = \dfrac{1}{2^3} = \dfrac{1}{8}$.

x	y
1	0
2	1
4	2
8	3
$\frac{1}{2}$	−1
$\frac{1}{4}$	−2
$\frac{1}{8}$	−3

(1) Choose values for y.
(2) Compute values for x.

We plot the points and connect them with a smooth curve. What happens as y-values become smaller?

This curve does not touch or cross the y-axis.

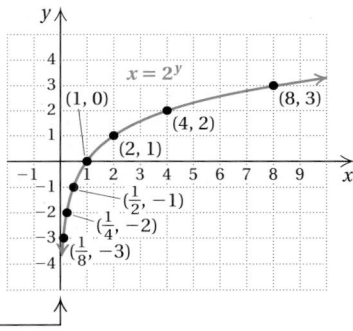

Note that this curve $x = 2^y$ has the same shape as the graph of $y = 2^x$, except that it is reflected, or flipped, across the line $y = x$, as shown below.

$y = 2^x$			$x = 2^y$	
x	**y**		**x**	**y**
0	1		1	0
1	2		2	1
2	4		4	2
3	8		8	3
−1	$\frac{1}{2}$		$\frac{1}{2}$	−1
−2	$\frac{1}{4}$		$\frac{1}{4}$	−2
−3	$\frac{1}{8}$		$\frac{1}{8}$	−3

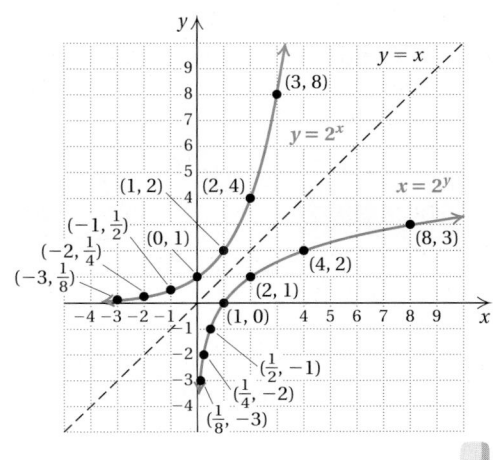

Do Exercise 7.

7. Graph: $x = 3^y$.

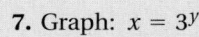
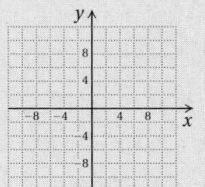

Answer

7.

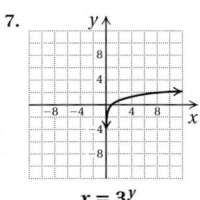

$x = 3^y$

(c) Applications of Exponential Functions

When interest is paid on interest, we call it **compound interest**. This is the type of interest paid on investments and loans. Suppose you have $100,000 in a savings account at an interest rate of 4%. This means that in 1 year, the account will contain the original $100,000 plus 4% of $100,000. Thus the total in the account after 1 year will be

$100,000 plus $100,000 × 0.04.

This can also be expressed as

$100,000 + $100,000 × 0.04 = $100,000 × 1 + $100,000 × 0.04

$= $100,000(1 + 0.04)$ Factoring out $100,000 using the distributive law

$= $100,000(1.04)$

$= $104,000.$

Now suppose that the total of $104,000 remains in the account for another year. At the end of the second year, the account will contain the $104,000 plus 4% of $104,000. The total in the account will be

$104,000 plus $104,000 × 0.04,

or

$104,000(1.04) = [\$100,000(1.04)](1.04) = \$100,000(1.04)^2$

$= $108,160.$

Note that in the second year, interest is earned on the first year's interest as well as the original amount. When this happens, we say that the interest is **compounded annually**. If the original amount of $100,000 earned only simple interest for 2 years, the interest would be

$100,000 × 0.04 × 2, or $8000,

and the amount in the account would be

$100,000 + $8000 = $108,000,

less than the $108,160 when interest is compounded annually.

> **8. Interest Compounded Annually.** Find the amount in an account after 1 year and after 2 years if $40,000 is invested at 2%, compounded annually.

> Do Exercise 8.

The following table shows how the computation continues over 4 years.

$100,000 IN AN ACCOUNT

YEAR	WITH INTEREST COMPOUNDED ANNUALLY	WITH SIMPLE INTEREST
Beginning of 1st year End of 1st year	$100,000 $100,000(1.04)^1 = $104,000	 $104,000
Beginning of 2nd year End of 2nd year	$104,000 $100,000(1.04)^2 = $108,160	 $108,000
Beginning of 3rd year End of 3rd year	$108,160 $100,000(1.04)^3 = $112,486.40	 $112,000
Beginning of 4th year End of 4th year	$112,486.40 $100,000(1.04)^4 ≈ $116,985.86	 $116,000

Answer

8. $40,800; $41,616

We can express interest compounded annually using an exponential function.

EXAMPLE 6 *Interest Compounded Annually.* The amount of money A that a principal P will grow to after t years at interest rate r, compounded annually, is given by the formula

$$A = P(1 + r)^t.$$

Suppose that $100,000 is invested at 4% interest, compounded annually.

a) Find a function for the amount in the account after t years.

b) Find the amount of money in the account at $t = 0$, $t = 4$, $t = 8$, and $t = 10$.

c) Graph the function.

a) If $P = \$100,000$ and $r = 4\% = 0.04$, we can substitute these values and form the following function:

$$A(t) = \$100,000(1 + 0.04)^t = \$100,000(1.04)^t.$$

b) To find the function values, you might find a calculator with a power key helpful.

$$A(0) = \$100,000(1.04)^0 = \$100,000;$$
$$A(4) = \$100,000(1.04)^4 \approx \$116,985.86;$$
$$A(8) = \$100,000(1.04)^8 \approx \$136,856.91;$$
$$A(10) = \$100,000(1.04)^{10} \approx \$148,024.43$$

c) We use the function values computed in (b) with others, if we wish, to draw the graph as follows. Note that the axes are scaled differently because of the large numbers.

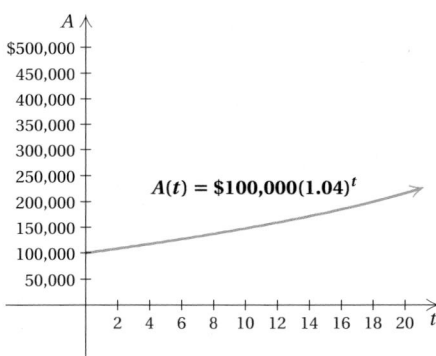

Do Exercise 9.

Suppose the principal of $100,000 we just considered were **compounded semiannually**—that is, every half year. Interest would then be calculated twice a year at a rate of 4% ÷ 2, or 2%, each time. The computations are as follows:

After the first $\frac{1}{2}$ year, the account will contain 102% of $100,000:

$$\$100,000 \times 1.02 = \$102,000.$$

After a second $\frac{1}{2}$ year (1 full year), the account will contain 102% of $102,000:

$$\$102,000 \times 1.02 = \$100,000 \times (1.02)^2 = \$104,040.$$

After a third $\frac{1}{2}$ year $\left(1\frac{1}{2}\text{ full years}\right)$, the account will contain 102% of $104,040:

$$\$104,040 \times 1.02 = \$100,000 \times (1.02)^3 = \$106,120.80.$$

9. Interest Compounded Annually. Suppose that $40,000 is invested at 5% interest, compounded annually.

a) Find a function for the amount in the account after t years.

b) Find the amount of money in the account at $t = 0$, $t = 4$, $t = 8$, and $t = 10$.

c) Graph the function.

10. A couple invests $7000 in an account paying 3.4%, compounded quarterly. Find the amount in the account after $5\frac{1}{2}$ years.

Answers

9. (a) $A(t) = \$40,000(1.05)^t$;
(b) $40,000; $48,620.25; $59,098.22; $65,155.79
(c)

$A(t) = \$40,000(1.05)^t$

10. $8432.72

The Compound-Interest Formula If $1000 is invested at 3%, compounded quarterly, how much is in the account at the end of 2 years?

We use the compound-interest formula, substituting 1000 for P, 0.03 for r, 4 for n (compounding quarterly), and 2 for t. Then we get

$$A = P\left(1 + \frac{r}{n}\right)^{n \cdot t}$$

$$= 1000\left(1 + \frac{0.03}{4}\right)^{4 \cdot 2}.$$

To do this computation on a calculator, we press ① ⓪ ⓪ ⓪ ⦗ ① ＋ ⦘ ⊙ ⓪ ③ ÷ ④ ⦘ ⌃ ⦗ ④ × ② ⦘ ENTER. The result is approximately $1061.60.

Exercises:

1. If $1000 is invested at 2%, compounded semiannually, how much is in the account at the end of 2 years?

2. If $1000 is invested at 2.4%, compounded monthly, how much is in the account at the end of 2 years?

3. If $20,000 is invested at 4.2%, compounded quarterly, how much is in the account at the end of 10 years?

4. If $10,000 is invested at 5.4%, how much is in the account at the end of 1 year, if interest is compounded **(a)** annually? **(b)** semiannually? **(c)** quarterly? **(d)** daily? **(e)** hourly?

After a fourth $\frac{1}{2}$ year (2 full years), the account will contain 102% of $106,120.80:

$$\$106{,}120.80 \times 1.02 = \$100{,}000 \times (1.02)^4$$

$$\approx \$108{,}243.22. \qquad \text{Rounded to the nearest cent}$$

Comparing these results with those in the table on p. 880, we can see that by having more compounding periods, we increase the amount in the account.

We have illustrated the following result.

COMPOUND-INTEREST FORMULA

If a principal P has been invested at interest rate r, compounded n times a year, in t years it will grow to an amount A given by

$$A = P \cdot \left(1 + \frac{r}{n}\right)^{n \cdot t}.$$

EXAMPLE 7 The Ibsens invest $4000 in an account paying $2\frac{5}{8}\%$, compounded quarterly. Find the amount in the account after $2\frac{1}{2}$ years.

The compounding is quarterly—that is, four times a year—so in $2\frac{1}{2}$ years, there are ten $\frac{1}{4}$-year periods. We substitute $4000 for P, $2\frac{5}{8}\%$, or 0.02625, for r, 4 for n, and $2\frac{1}{2}$, or $\frac{5}{2}$, for t and compute A:

$$A = P \cdot \left(1 + \frac{r}{n}\right)^{n \cdot t}$$

$$= 4000 \cdot \left(1 + \frac{2\frac{5}{8}\%}{4}\right)^{4 \cdot \frac{5}{2}}$$

$$= 4000 \cdot \left(1 + \frac{0.02625}{4}\right)^{10}$$

$$= 4000(1.0065625)^{10} \qquad \text{Using a calculator}$$

$$\approx \$4270.39.$$

The amount in the account after $2\frac{1}{2}$ years is $4270.39.

Do Exercise 10 on the preceding page.

22. $x = \left(\dfrac{1}{3}\right)^y$

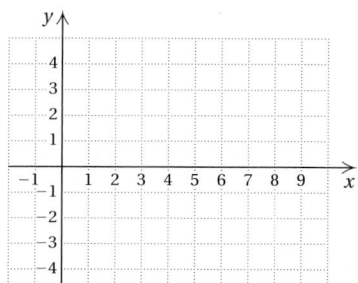

23. $x = 5^y$

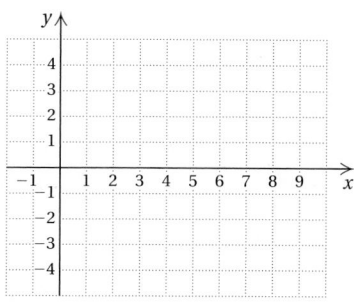

24. $x = \left(\dfrac{2}{3}\right)^y$

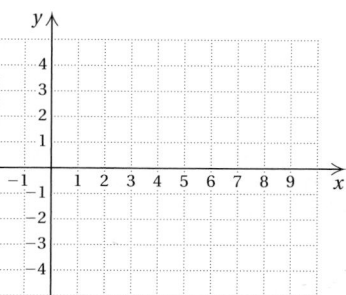

Graph both equations using the same set of axes.

25. $y = 2^x,\ x = 2^y$

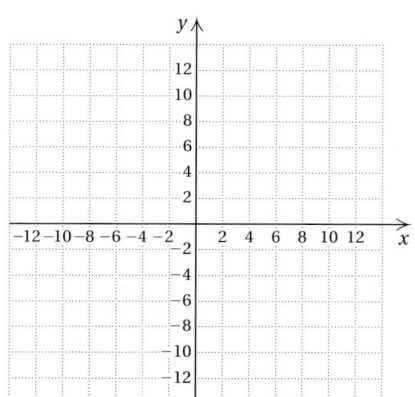

26. $y = \left(\dfrac{1}{2}\right)^x,\ x = \left(\dfrac{1}{2}\right)^y$

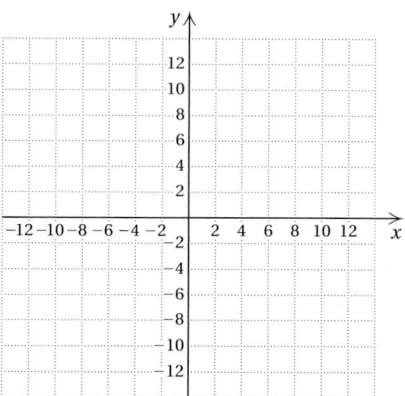

C Solve.

27. *Interest Compounded Annually.* Suppose that $50,000 is invested at 2% interest, compounded annually.

   a) Find a function A for the amount in the account after t years.

   b) Complete the following table of function values.

t	$A(t)$
0	
1	
2	
4	
8	
10	
20	

   c) Graph the function.

28. *Interest Compounded Annually.* Suppose that $50,000 is invested at 3% interest, compounded annually.

   a) Find a function A for the amount in the account after t years.

   b) Complete the following table of function values.

t	$A(t)$
0	
1	
2	
4	
8	
10	
20	

   c) Graph the function.

29. *Interest Compounded Semiannually.* Jesse deposits $2000 in an account paying 2.6%, compounded semi-annually. Find the amount in the account after 3 years.

30. *Interest Compounded Semiannually.* Rory deposits $3500 in an account paying 3.2%, compounded semi-annually. Find the amount in the account after 2 years.

31. *Interest Compounded Quarterly.* The Jansens invest $4500 in an account paying 3.6%, compounded quarterly. Find the amount in the account after $4\frac{1}{2}$ years.

32. *Interest Compounded Quarterly.* The Gemmers invest $4000 in an account paying 2.8%, compounded quarterly. Find the amount in the account after $3\frac{1}{2}$ years.

33. *Wind Power.* Wind power is the conversion of wind energy into a useful form, such as electricity, using wind turbines. The wind power W installed in the United States, in megawatts (MW), t years after 2005 can be approximated by

$$W(t) = 8733.5(1.406)^t,$$

where $t = 0$ corresponds to 2005.
Source: American Wind Energy Association

a) What was the wind power capacity in the United States in 2006? in 2008? in 2010?
b) Graph the function.

34. *Eco-Friendly Schools.* As energy costs rise, the greening of America's schools is part of a larger trend toward more energy-efficient construction. The number of schools C seeking eco-friendly certification t years after 2002 can be approximated by

$$C(t) = 16.4(1.69)^t,$$

where $t = 0$ corresponds to 2002.
Source: U.S. Green Building Council

a) How many schools sought eco-friendly certification in 2005? in 2008? in 2009?
b) Graph the function.

35. *Flat-Panel TVs.* Lower-than-expected demand for flat-panel TVs has spurred manufacturers to cut prices in recent years. The average price P of a flat-panel TV t years after 2004 can be approximated by

$$P(t) = 5105(0.698)^t,$$

where $t = 0$ corresponds to 2004.
Source: Pacific Media Associates

a) What was the average price of a flat-panel TV in 2004? in 2006? in 2008?
b) Graph the function.

36. *Salvage Value.* An office machine is purchased for $5200. Its value each year is about 80% of the value the preceding year. Its value after t years is given by the exponential function

$$V(t) = \$5200(0.8)^t.$$

a) Find the value of the machine after 0 year, 1 year, 2 years, 5 years, and 10 years.
b) Graph the function.

37. *Recycling Aluminum Cans.* Although Americans throw 1500 aluminum cans in the trash every second of every day, 51.5% of the aluminum in the cans is recycled. If a beverage company distributes 500,000 cans, the amount of aluminum still in use after t years can be made into N cans, where

$$N(t) = 500{,}000(0.515)^t.$$

Source: The Container Recycling Institute

a) How many cans can be made from the original 500,000 cans after 1 year? after 3 years? after 7 years?

b) Graph the function.

38. *Growth of Bacteria.* Bladder infections are often caused when the bacteria *Escherichia coli* reach the human bladder. Suppose that 3000 of the bacteria are present at time $t = 0$. Then t minutes later, the number of bacteria present will be

$$N(t) = 3000(2)^{t/20}.$$

Source: Chris Hayes, "Detecting a Human Health Risk: *E. coli*," *Laboratory Medicine* 29, no. 6, June 1998: 347–355

a) How many bacteria will be present after 10 min? 20 min? 30 min? 40 min? 60 min?

b) Graph the function.

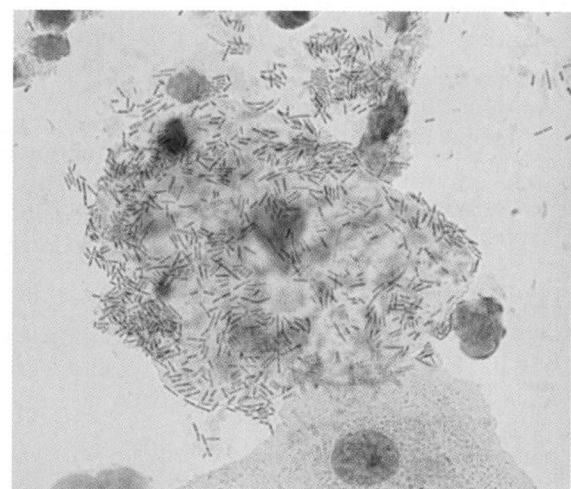

Skill Maintenance

39. Multiply and simplify: $x^{-5} \cdot x^3$. [4.1d, f]

40. Simplify: $(x^{-3})^4$. [4.1f], [4.2a]

Simplify. [4.1b]

41. 9^0

42. $\left(\frac{2}{3}\right)^0$

43. $\left(\frac{2}{3}\right)^1$

44. 2.7^1

Divide and simplify. [4.1e, f]

45. $\dfrac{x^{-3}}{x^4}$

46. $\dfrac{x}{x^{11}}$

47. $\dfrac{x}{x^0}$

48. $\dfrac{x^{-3}}{x^{-4}}$

Synthesis

49. Simplify: $\left(5^{\sqrt{2}}\right)^{2\sqrt{2}}$.

50. Which is larger: $\pi^{\sqrt{2}}$ or $\left(\sqrt{2}\right)^{\pi}$?

Graph.

51. $y = 2^x + 2^{-x}$

52. $y = |2^x - 2|$

53. $y = \left|\left(\frac{1}{2}\right)^x - 1\right|$

54. $y = 2^{-x^2}$

Graph both equations using the same set of axes.

55. $y = 3^{-(x-1)}, \; x = 3^{-(y-1)}$

56. $y = 1^x, \; x = 1^y$

57. ⌁ Use a graphing calculator to graph each of the equations in Exercises 51–54.

12.2

Inverse Functions and Composite Functions

OBJECTIVES

a Find the inverse of a relation if it is described as a set of ordered pairs or as an equation.

b Given a function, determine whether it is one-to-one and has an inverse that is a function.

c Find a formula for the inverse of a function, if it exists, and graph inverse relations and functions.

d Find the composition of functions and express certain functions as a composition of functions.

e Determine whether a function is an inverse by checking its composition with the original function.

When we go from an output of a function back to its input or inputs, we get an *inverse relation*. When that relation is a function, we have an *inverse function*. We now study inverse functions and how to find formulas when the original function has a formula. We do so to understand the relationships among the special functions that we study in this chapter.

a Inverses

A set of ordered pairs is called a **relation**. When we consider the graph of a function, we are thinking of a set of ordered pairs. Thus a function can be thought of as a special kind of relation, in which to each first coordinate there corresponds one and only one second coordinate.

Consider the relation h given as follows:

$$h = \{(-7, 4), (3, -1), (-6, 5), (0, 2)\}.$$

Suppose we *interchange* the first and second coordinates. The relation we obtain is called the **inverse** of the relation h and is given as follows:

Inverse of $h = \{(4, -7), (-1, 3), (5, -6), (2, 0)\}.$

INVERSE RELATION

Interchanging the coordinates of the ordered pairs in a relation produces the **inverse relation**.

EXAMPLE 1 Consider the relation g given by

$$g = \{(2, 4), (-1, 3), (-2, 0)\}.$$

In the figure below, the relation g is shown in red. The inverse of the relation is

$$\{(4, 2), (3, -1), (0, -2)\}$$

and is shown in blue.

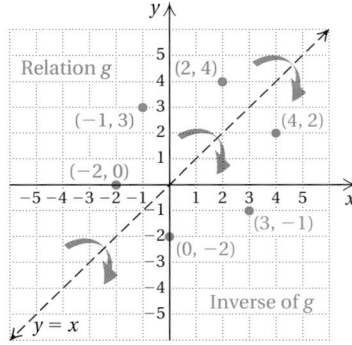

1. Consider the relation g given by $g = \{(2, 5), (-1, 4), (-2, 1)\}.$ The graph of the relation is shown below in red. Find the inverse and draw its graph in blue.

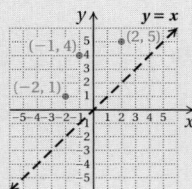

Answer

1. Inverse of $g = \{(5, 2), (4, -1), (1, -2)\}$

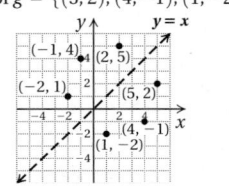

Do Exercise 1.

INVERSE RELATION

If a relation is defined by an equation, interchanging the variables produces an equation of the **inverse relation**.

EXAMPLE 2   Find an equation of the inverse of the relation

$$y = 3x - 4.$$

Then graph both the relation and its inverse.

We interchange x and y and obtain an equation of the inverse:

$$x = 3y - 4.$$

Relation: $y = 3x - 4$ ⟶ *Inverse*: $x = 3y - 4$

x	y
0	-4
1	-1
2	2
3	5

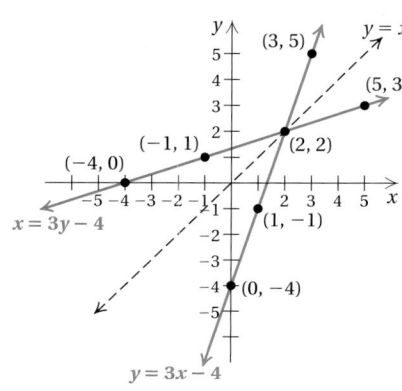

x	y
-4	0
-1	1
2	2
5	3

Note in Example 2 that the relation $y = 3x - 4$ is a function and its inverse relation $x = 3y - 4$ is also a function. Each graph passes the vertical-line test. (See Section 7.1.)

EXAMPLE 3   Find an equation of the inverse of the relation

$$y = 6x - x^2.$$

Then graph both the original relation and its inverse.

We interchange x and y and obtain an equation of the inverse:

$$x = 6y - y^2.$$

Relation: $y = 6x - x^2$ ⟶ *Inverse*: $x = 6y - y^2$

x	y
-1	-7
0	0
1	5
3	9
5	5

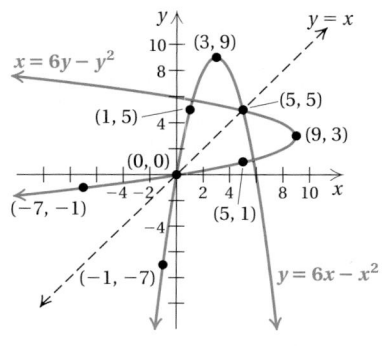

x	y
-7	-1
0	0
5	1
9	3
5	5

2. Find an equation of the inverse relation. Then complete the table and graph both the original relation and its inverse.

Relation:

$y = 6 - 2x$

x	y
0	6
2	2
3	0
5	-4

Inverse:

_____

x	y
	0
	2
	3
	5

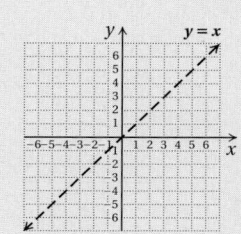

Answers

2. *Inverse:*

   $x = 6 - 2y$

x	y
6	0
2	2
0	3
-4	5

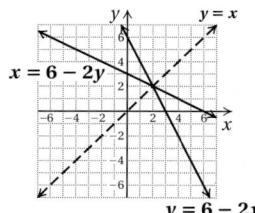

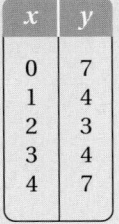
Note in Example 3 that the relation $y = 6x - x^2$ is a function because it passes the vertical-line test. However, its inverse relation $x = 6y - y^2$ is not a function because its graph fails the vertical-line test. Therefore, the inverse of a function is *not* always a function.

Do Exercises 2 and 3. (Exercise 2 is on the preceding page.)

b Inverses and One-To-One Functions

Let's consider the following two functions.

NUMBER (Domain)	CUBE (Range)
$-3 \longrightarrow$	-27
$-2 \longrightarrow$	-8
$-1 \longrightarrow$	-1
$0 \longrightarrow$	0
$1 \longrightarrow$	1
$2 \longrightarrow$	8
$3 \longrightarrow$	27

YEAR (Domain)	FIRST-CLASS POSTAGE COST, IN CENTS (Range)
1999	33
2000	
2001	34
2002	37
2006	39
2007	41
2008	42
2009	44

SOURCE: U.S. Postal Service

Suppose we reverse the arrows. Are these inverse relations functions?

CUBE ROOT (Range)	NUMBER (Domain)
$-3 \longleftarrow$	-27
$-2 \longleftarrow$	-8
$-1 \longleftarrow$	-1
$0 \longleftarrow$	0
$1 \longleftarrow$	1
$2 \longleftarrow$	8
$3 \longleftarrow$	27

YEAR (Range)	FIRST-CLASS POSTAGE COST, IN CENTS (Domain)
1999	33
2000	
2001	34
2002	37
2006	39
2007	41
2008	42
2009	44

We see that the inverse of the cubing function is a function. The inverse of the postage function is not a function, however, because the input 33 has *two* outputs, 1999 and 2000. Recall that for a function, each input has exactly one output. However, it can happen that the same output comes from two or more different inputs. If this is the case, the inverse cannot be a function. When this possibility is excluded, the inverse is also a function.

In the cubing function, different inputs have different outputs. Thus its inverse is also a function. The cubing function is what is called a **one-to-one function**. If the inverse of a function f is also a function, it is named f^{-1} (read "f-inverse").

Caution!

The -1 in f^{-1} is *not* an exponent and f^{-1} does *not* represent a reciprocal!

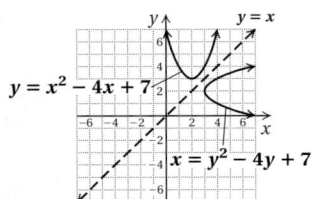

ONE-TO-ONE FUNCTION AND INVERSES

A function f is **one-to-one** if different inputs have different outputs—that is,

    if $a \neq b$,   then $f(a) \neq f(b)$.   Or,

A function f is **one-to-one** if when the outputs are the same, the inputs are the same—that is,

    if $f(a) = f(b)$,   then $a = b$.

If a function is one-to-one, then its inverse is a function.

The domain of a one-to-one function f is the range of the inverse f^{-1}.

The range of a one-to-one function f is the domain of the inverse f^{-1}.

How can we tell graphically whether a function is one-to-one and thus has an inverse that is a function?

EXAMPLE 4 The graph of the exponential function $f(x) = 2^x$, or $y = 2^x$, is shown on the left below. The graph of the inverse $x = 2^y$ is shown on the right. How can we tell by examining only the graph on the left whether it has an inverse that is a function?

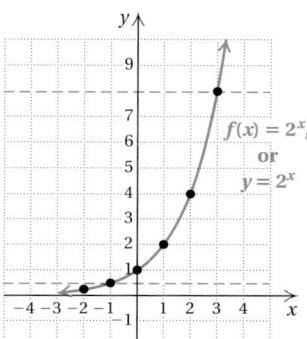
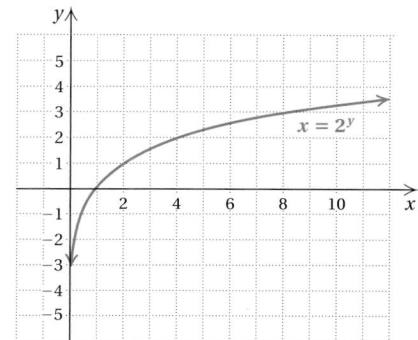

We see that the graph on the right passes the vertical-line test, so we know it is the graph of a function. However, if we look only at the graph on the left, we think as follows:

A function is one-to-one if different inputs have different outputs. In other words, no two x-values will have the same y-value. For this function, we cannot find two x-values that have the same y-value. Note also that no horizontal line can be drawn that will cross the graph more than once. The function is thus one-to-one and its inverse is a function.

THE HORIZONTAL-LINE TEST

If it is possible for a horizontal line to intersect the graph of a function more than once, then the function is not one-to-one and therefore its inverse is not a function.

Determine whether the function is one-to-one and thus has an inverse that is also a function.

4. $f(x) = 4 - x$

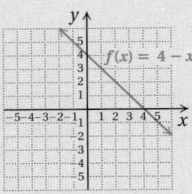

5. $f(x) = x^2 - 1$

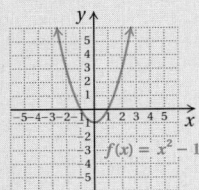

6. $f(x) = 4^x$
(Sketch this graph yourself.)

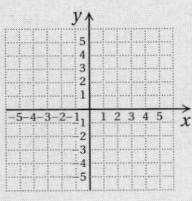

7. $f(x) = |x| - 3$
(Sketch this graph yourself.)

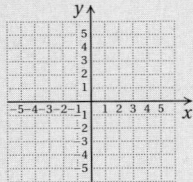

A graph is that of a function if no vertical line crosses the graph more than once. A function has an inverse that is also a function if no horizontal line crosses the graph more than once.

EXAMPLE 5 Determine whether the function $f(x) = x^2$ is one-to-one and has an inverse that is also a function.

The graph of $f(x) = x^2$, or $y = x^2$, is shown on the left below. There are many horizontal lines that cross the graph more than once, so this function is not one-to-one and does not have an inverse that is a function.

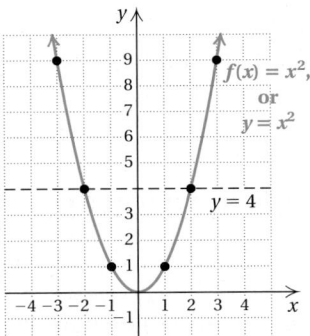
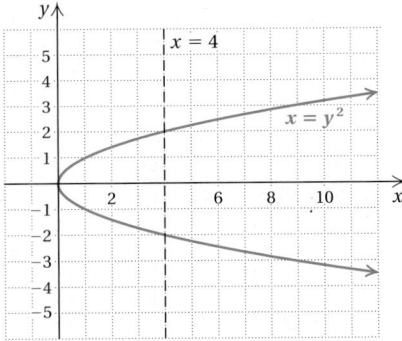

The inverse of the function $y = x^2$ is the relation $x = y^2$. The graph of $x = y^2$ is shown on the right above. It fails the vertical-line test and is not a function.

Do Exercises 4–7.

c Inverse Formulas and Graphs

Suppose that a function is described by a formula. If it has an inverse that is a function, how do we find a formula for the inverse function? If for any equation with two variables such as x and y we interchange the variables, we obtain an equation of the inverse relation. We proceed as follows to find a formula for f^{-1}.

If a function f is one-to-one, a formula for its inverse f^{-1} can be found as follows:

1. Replace $f(x)$ with y.
2. Interchange x and y. (This gives the inverse relation.)
3. Solve for y.
4. Replace y with $f^{-1}(x)$.

EXAMPLE 6 Given $f(x) = x + 1$:

a) Determine whether the function is one-to-one.

b) If it is one-to-one, find a formula for $f^{-1}(x)$.

c) Graph the inverse function, if it exists.

Answers

4. Yes **5.** No **6.** Yes **7.** No

a) The graph of $f(x) = x + 1$ is shown below. It passes the horizontal-line test, so it is one-to-one. Thus its inverse is a function.

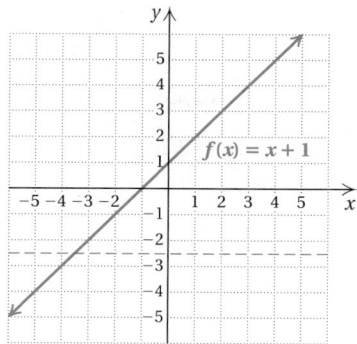

b) 1. Replace $f(x)$ with y: $\qquad y = x + 1$.

    2. Interchange x and y: $\qquad x = y + 1$.   This gives the inverse relation.

    3. Solve for y: $\qquad x - 1 = y$.

    4. Replace y with $f^{-1}(x)$: $\quad f^{-1}(x) = x - 1$.

c) We graph $f^{-1}(x) = x - 1$, or $y = x - 1$. The graph is shown below.

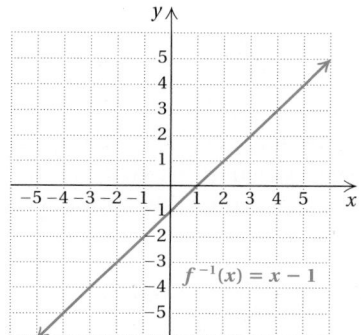

EXAMPLE 7   Given $f(x) = 2x - 3$:

a) Determine whether the function is one-to-one.

b) If it is one-to-one, find a formula for $f^{-1}(x)$.

c) Graph the inverse function, if it exists.

a) The graph of $f(x) = 2x - 3$ is shown below. It passes the horizontal-line test and is one-to-one.

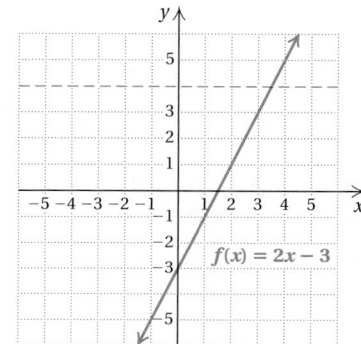

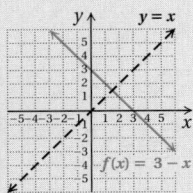
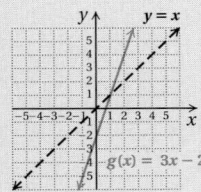
b) **1.** Replace $f(x)$ with y: $y = 2x - 3$.

    2. Interchange x and y: $x = 2y - 3$.

    3. Solve for y: $x + 3 = 2y$

$$\frac{x + 3}{2} = y.$$

    4. Replace y with $f^{-1}(x)$: $f^{-1}(x) = \dfrac{x + 3}{2}$.

c) We graph

$$f^{-1}(x) = \frac{x + 3}{2}, \quad \text{or}$$

$$y = \frac{1}{2}x + \frac{3}{2}.$$

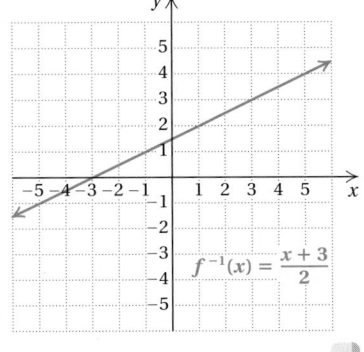

Do Exercises 8 and 9.

Let's now consider inverses of functions in terms of a function machine. Suppose that a one-to-one function f is programmed into a machine. If the machine has a reverse switch, when the switch is thrown, the machine performs the inverse function f^{-1}. Inputs then enter at the opposite end, and the entire process is reversed.

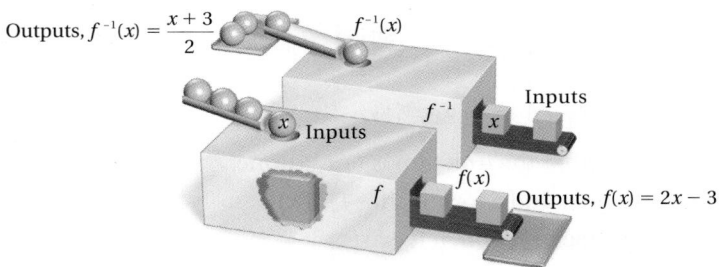

Consider $f(x) = 2x - 3$ and $f^{-1}(x) = (x + 3)/2$ from Example 7. For the input 5,

$$f(5) = 2 \cdot 5 - 3 = 10 - 3 = 7.$$

The output is 7. Now we use 7 for the input in the inverse:

$$f^{-1}(7) = \frac{7 + 3}{2} = \frac{10}{2} = 5.$$

The function f takes 5 to 7. The inverse function f^{-1} takes the number 7 back to 5.

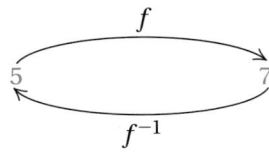

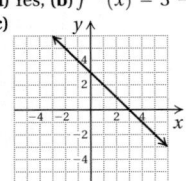
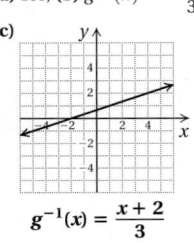

How do the graphs of a function and its inverse compare?

EXAMPLE 8 Graph $f(x) = 2x - 3$ and $f^{-1}(x) = (x + 3)/2$ using the same set of axes. Then compare.

The graph of each function follows. Note that the graph of f^{-1} can be drawn by reflecting the graph of f across the line $y = x$. That is, if we graph $f(x) = 2x - 3$ in wet ink and fold the paper along the line $y = x$, the graph of $f^{-1}(x) = (x + 3)/2$ will appear as the impression made by f.

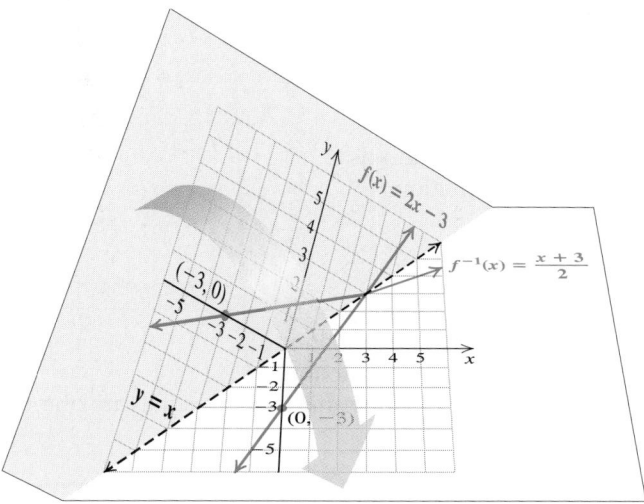

When x and y are interchanged to find a formula for the inverse, we are, in effect, flipping the graph of $f(x) = 2x - 3$ over the line $y = x$. For example, when the coordinates of the y-intercept of the graph of f, $(0, -3)$, are reversed, we get the x-intercept of the graph of f^{-1}, $(-3, 0)$.

> The graph of f^{-1} is a reflection of the graph of f across the line $y = x$.

Do Exercise 10.

10. Graph $g(x) = 3x - 2$ and $g^{-1}(x) = (x + 2)/3$ using the same set of axes.

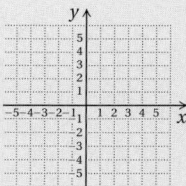

EXAMPLE 9 Consider $g(x) = x^3 + 2$.

a) Determine whether the function is one-to-one.

b) If it is one-to-one, find a formula for its inverse.

c) Graph the inverse, if it exists.

a) The graph of $g(x) = x^3 + 2$ is shown at right in red. It passes the horizontal-line test and thus is one-to-one.

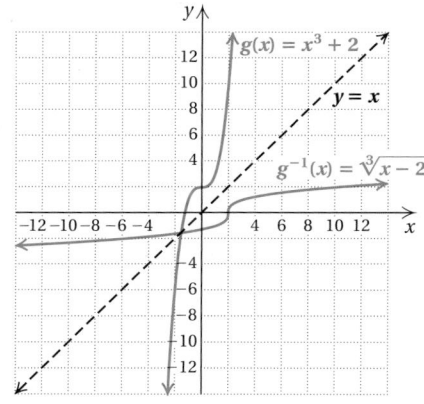

Answer

10.

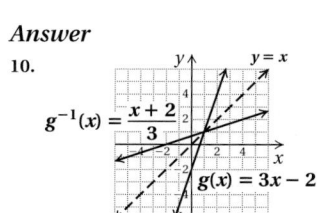

$$g^{-1}(x) = \frac{x + 2}{3}$$

$$g(x) = 3x - 2$$

11. Given $f(x) = x^3 + 1$:

a) Determine whether the function is one-to-one.

b) If it is one-to-one, find a formula for its inverse.

c) Graph the function and its inverse using the same set of axes.

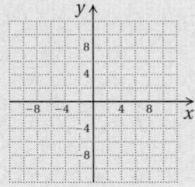

b) 1. Replace $g(x)$ with y: $\qquad y = x^3 + 2.$

2. Interchange x and y: $\qquad x = y^3 + 2.$

3. Solve for y: $\qquad\qquad x - 2 = y^3$

$\qquad\qquad\qquad\qquad \sqrt[3]{x - 2} = y.$ $\quad$ Since a number has only one cube root, we can solve for y.

4. Replace y with $g^{-1}(x)$: $\quad g^{-1}(x) = \sqrt[3]{x - 2}.$

c) To find the graph, we reflect the graph of $g(x) = x^3 + 2$ across the line $y = x$, as we did in Example 8. It can also be found by substituting into $g^{-1}(x) = \sqrt[3]{x - 2}$ and plotting points. The graphs of g and g^{-1} are shown together on the preceding page.

> Do Exercise 11.

We can now see why we exclude 1 as a base for an exponential function. Consider

$$f(x) = a^x = 1^x = 1.$$

The graph of f is the horizontal line $y = 1$. The graph is not one-to-one. The function does not have an inverse that is a function. All other positive bases yield exponential functions that are one-to-one.

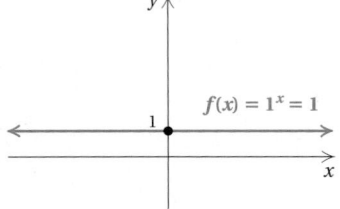

Calculator Corner

Graphing an Inverse Function The DRAWINV operation can be used to graph a function and its inverse on the same screen. A formula for the inverse function need not be found in order to do this. The graphing calculator must be set in FUNC mode when this operation is used.

To graph $f(x) = 2x - 3$ and $f^{-1}(x)$ using the same set of axes, we first clear any existing equations on the equation-editor screen and then enter $y_1 = 2x - 3$. Now, we press **2ND** **DRAW** **8** to select the DRAWINV operation. (DRAW is the second operation associated with the **PRGM** key.) We press **VARS** ▷ **1** **1** to indicate that we want to graph the inverse of y_1. Then we press **ENTER** to see the graph of the function and its inverse. The graphs are shown here in a squared window.

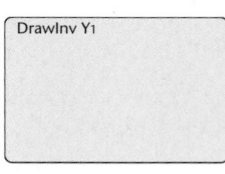

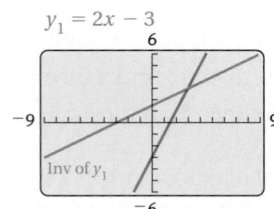

Exercises: Use the DRAWINV operation on a graphing calculator to graph each function with its inverse on the same screen.

1. $f(x) = x - 5$

2. $f(x) = \frac{2}{3}x$

3. $f(x) = x^2 + 2$

4. $f(x) = x^3 - 3$

Answer

11. **(a)** Yes; **(b)** $f^{-1}(x) = \sqrt[3]{x - 1}$;
(c)

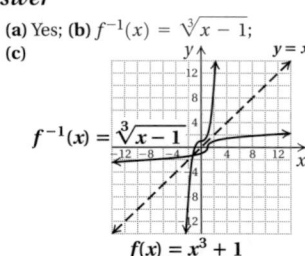

d Composite Functions

In the real world, functions frequently occur in which some quantity depends on a variable that, in turn, depends on another variable. For instance, the number of employees hired by a firm may depend on the firm's profits, which may in turn depend on the number of items the firm produces. Such functions are called **composite functions**.

For example, the function g that gives a correspondence between women's shoe sizes in the United States and those in Italy is given by $g(x) = 2x + 24$, where x is the U.S. size and $g(x)$ is the Italian size. Thus a U.S. size 4 corresponds to a shoe size of $g(4) = 2 \cdot 4 + 24$, or 32, in Italy.

There is also a function that gives a correspondence between women's shoe sizes in Italy and those in Britain. The function is given by $f(x) = \frac{1}{2}x - 14$, where x is the Italian size and $f(x)$ is the corresponding British size. Thus an Italian size 32 corresponds to a British size $f(32) = \frac{1}{2}(32) - 14$, or 2.

It seems reasonable to conclude that a shoe size of 4 in the United States corresponds to a size of 2 in Britain and that some function h describes this correspondence. Can we find a formula for h? If we look at the following tables, we might guess that such a formula is $h(x) = x - 2$, and that is indeed correct. But, for more complicated formulas, we would need to use algebra.

g		f	
$g(x) = 2x + 24$		$f(x) = \frac{1}{2}x - 14$	

U.S.	Italy	Britain
4	32	2
5	34	3
6	36	4
7	38	5
8	40	6

h

$$h(x) = ?$$

A shoe size x in the United States corresponds to a shoe size $g(x)$ in Italy, where

$$g(x) = 2x + 24.$$

Now $2x + 24$ is a shoe size in Italy. If we replace x in $f(x) = \frac{1}{2}x - 14$ with $g(x)$, or $2x + 24$, we can find the corresponding shoe size in Britain:

$$f(x) = \tfrac{1}{2}x - 14$$
$$f(g(x)) = f(2x + 24) = \tfrac{1}{2}[2x + 24] - 14$$
$$= x + 12 - 14 = x - 2.$$

This gives a formula for h:

$$h(x) = x - 2.$$

Thus a shoe size of 4 in the United States corresponds to a shoe size of $h(4) = 4 - 2$, or 2, in Britain. The function h is the **composition** of f and g, symbolized by $f \circ g$. To find $(f \circ g)(x)$, we substitute $g(x)$ for x in $f(x)$.

COMPOSITE FUNCTION

The **composite function** $f \circ g$, the **composition** of f and g, is defined as

$$(f \circ g)(x) = f(g(x)).$$

We can visualize the composition of functions as follows.

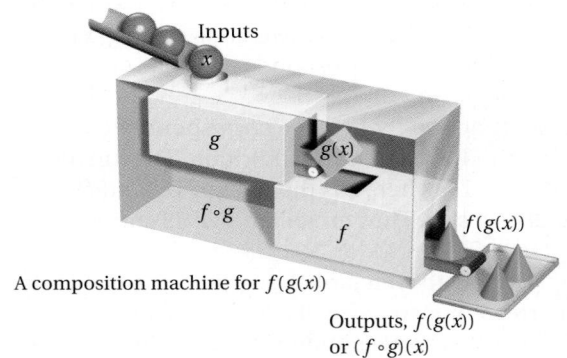

Inputs
x

g

$g(x)$

$f \circ g$

f

$f(g(x))$

A composition machine for $f(g(x))$

Outputs, $f(g(x))$
or $(f \circ g)(x)$

EXAMPLE 10   Given $f(x) = 3x$ and $g(x) = 1 + x^2$:

a) Find $(f \circ g)(5)$ and $(g \circ f)(5)$.

b) Find $(f \circ g)(x)$ and $(g \circ f)(x)$.

We consider each function separately:

$$f(x) = 3x \qquad \text{This function multiplies each input by 3.}$$

and $g(x) = 1 + x^2$.   This function adds 1 to the square of each input.

a) $(f \circ g)(5) = f(g(5)) = f(1 + 5^2) = f(26) = 3(26) = 78;$

$(g \circ f)(5) = g(f(5)) = g(3 \cdot 5) = g(15) = 1 + 15^2 = 1 + 225 = 226$

b) $(f \circ g)(x) = f(g(x))$

$\qquad\qquad = f(1 + x^2)$ Substituting $1 + x^2$ for $g(x)$

$\qquad\qquad = 3(1 + x^2)$

$\qquad\qquad = 3 + 3x^2;$

$(g \circ f)(x) = g(f(x))$

$\qquad\qquad = g(3x)$ Substituting $3x$ for $f(x)$

$\qquad\qquad = 1 + (3x)^2$

$\qquad\qquad = 1 + 9x^2$

We can check the values in part (a) with the formulas found in part (b):

$(f \circ g)(x) = 3 + 3x^2$	$(g \circ f)(x) = 1 + 9x^2$
$(f \circ g)(5) = 3 + 3 \cdot 5^2$	$(g \circ f)(5) = 1 + 9 \cdot 5^2$
$\qquad = 3 + 3 \cdot 25$	$\qquad = 1 + 9 \cdot 25$
$\qquad = 3 + 75$	$\qquad = 1 + 225$
$\qquad = 78;$	$\qquad = 226.$

Example 10 shows that $(f \circ g)(5) \neq (g \circ f)(5)$ and, in general, $(f \circ g)(x) \neq (g \circ f)(x)$.

12. Given $f(x) = x + 5$ and $g(x) = x^2 - 1$, find $(f \circ g)(x)$ and $(g \circ f)(x)$.

Do Exercise 12.

Answer

12. $x^2 + 4; \ x^2 + 10x + 24$

EXAMPLE 11 Given $f(x) = \sqrt{x}$ and $g(x) = x - 1$, find $(f \circ g)(x)$ and $(g \circ f)(x)$.

$$(f \circ g)(x) = f(g(x)) = f(x - 1) = \sqrt{x - 1};$$
$$(g \circ f)(x) = g(f(x)) = g(\sqrt{x}) = \sqrt{x} - 1$$

Do Exercise 13.

13. Given $f(x) = 4x + 5$ and $g(x) = \sqrt[3]{x}$, find $(f \circ g)(x)$ and $(g \circ f)(x)$.

It is important to be able to recognize how a function can be expressed, or "broken down," as a composition. Such a situation can occur in a study of calculus.

EXAMPLE 12 Find $f(x)$ and $g(x)$ such that $h(x) = (f \circ g)(x)$:

$$h(x) = (7x + 3)^2.$$

This is $7x + 3$ to the 2nd power. Two functions that can be used for the composition are $f(x) = x^2$ and $g(x) = 7x + 3$. We can check by forming the composition:

$$h(x) = (f \circ g)(x) = f(g(x)) = f(7x + 3) = (7x + 3)^2.$$

This is the most "obvious" answer to the question. There can be other less obvious answers. For example, if

$$f(x) = (x - 1)^2 \quad \text{and} \quad g(x) = 7x + 4,$$

then $h(x) = (f \circ g)(x) = f(g(x)) = f(7x + 4) = (7x + 4 - 1)^2 = (7x + 3)^2.$

14. Find $f(x)$ and $g(x)$ such that $h(x) = (f \circ g)(x)$. Answers may vary.

a) $h(x) = \sqrt[3]{x^2 + 1}$

b) $h(x) = \dfrac{1}{(x + 5)^4}$

Do Exercise 14.

(e) Inverse Functions and Composition

Suppose that we used some input x for the function f and found its output, $f(x)$. The function f^{-1} would then take that output back to x. Similarly, if we began with an input x for the function f^{-1} and found its output, $f^{-1}(x)$, the original function f would then take that output back to x.

If a function f is one-to-one, then f^{-1} is the unique function for which

$$(f^{-1} \circ f)(x) = x \quad \text{and} \quad (f \circ f^{-1})(x) = x.$$

15. Let $f(x) = \frac{2}{3}x - 4$. Use composition to show that
$$f^{-1}(x) = \frac{3x + 12}{2}.$$

EXAMPLE 13 Let $f(x) = 2x - 3$. Use composition to show that

$$f^{-1}(x) = \frac{x + 3}{2}. \qquad \text{(See Example 7.)}$$

We find $(f^{-1} \circ f)(x)$ and $(f \circ f^{-1})(x)$ and check to see that each is x.

$$\begin{aligned} (f^{-1} \circ f)(x) &= f^{-1}(f(x)) \\ &= f^{-1}(2x - 3) \\ &= \frac{(2x - 3) + 3}{2} \\ &= \frac{2x}{2} \\ &= x; \end{aligned}$$

$$\begin{aligned} (f \circ f^{-1})(x) &= f(f^{-1}(x)) \\ &= f\left(\frac{x + 3}{2}\right) \\ &= 2 \cdot \frac{x + 3}{2} - 3 \\ &= x + 3 - 3 \\ &= x \end{aligned}$$

Answers

13. $4\sqrt[3]{x} + 5$; $\sqrt[3]{4x + 5}$

14. **(a)** $f(x) = \sqrt[3]{x}$; $g(x) = x^2 + 1$;

(b) $f(x) = \dfrac{1}{x^4}$; $g(x) = x + 5$

15. $(f^{-1} \circ f)(x) = f^{-1}(f(x)) = f^{-1}\left(\frac{2}{3}x - 4\right)$

$$= \frac{3\left(\frac{2}{3}x - 4\right) + 12}{2}$$

$$= \frac{2x - 12 + 12}{2}$$

$$= \frac{2x}{2} = x;$$

$(f \circ f^{-1})(x) = f(f^{-1}(x)) = f\left(\frac{3x + 12}{2}\right)$

$$= \frac{2}{3}\left(\frac{3x + 12}{2}\right) - 4$$

$$= \frac{6x + 24}{6} - 4$$

$$= x + 4 - 4 = x$$

Do Exercise 15.

a Find the inverse of each relation. Graph the original relation in red and then graph the inverse relation in blue.

1. $\{(1, 2), (6, -3), (-3, -5)\}$

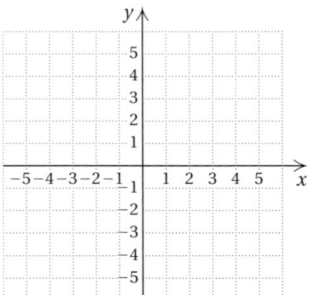

2. $\{(3, -1), (5, 2), (5, -3), (2, 0)\}$

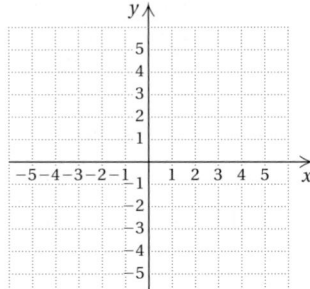

Find an equation of the inverse of the relation. Then complete the second table and graph both the original relation and its inverse.

3. $y = 2x + 6$

x	y
−1	4
0	6
1	8
2	10
3	12

x	y
4	
6	
8	
10	
12	

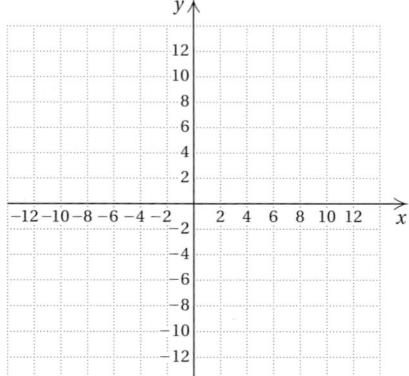

4. $y = \frac{1}{2}x^2 - 8$

x	y
−4	0
−2	−6
0	−8
2	−6
4	0

x	y
0	
−6	
−8	
−6	
0	

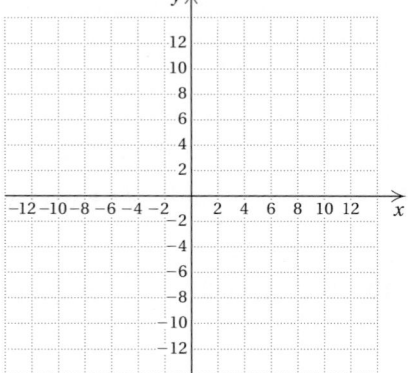

b Determine whether each function is one-to-one.

5. $f(x) = x - 5$

6. $f(x) = 3 - 6x$

7. $f(x) = x^2 - 2$

8. $f(x) = 4 - x^2$

9. $f(x) = |x| - 3$

10. $f(x) = |x - 2|$

11. $f(x) = 3^x$

12. $f(x) = \left(\frac{1}{2}\right)^x$

C   Determine whether each function is one-to-one. If it is, find a formula for its inverse.

13. $f(x) = 5x - 2$

14. $f(x) = 4 + 7x$

15. $f(x) = \dfrac{-2}{x}$

16. $f(x) = \dfrac{1}{x}$

17. $f(x) = \frac{4}{3}x + 7$

18. $f(x) = -\frac{7}{8}x + 2$

19. $f(x) = \dfrac{2}{x + 5}$

20. $f(x) = \dfrac{1}{x - 8}$

21. $f(x) = 5$

22. $f(x) = -2$

23. $f(x) = \dfrac{2x + 1}{5x + 3}$

24. $f(x) = \dfrac{2x - 1}{5x + 3}$

25. $f(x) = x^3 - 1$

26. $f(x) = x^3 + 5$

27. $f(x) = \sqrt[3]{x}$

28. $f(x) = \sqrt[3]{x - 4}$

Graph each function and its inverse using the same set of axes.

29. $f(x) = \frac{1}{2}x - 3$,
     $f^{-1}(x) = $ _____

x	$f(x)$
-4	
0	
2	
4	

x	$f^{-1}(x)$
-5	
-3	
-2	
-1	

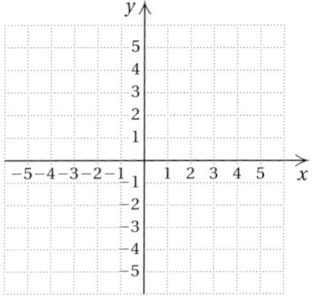

30. $g(x) = x + 4$,
     $g^{-1}(x) = $ _____

x	$g(x)$
-1	
0	
3	
5	

x	$g^{-1}(x)$
-5	
-4	
-1	
1	

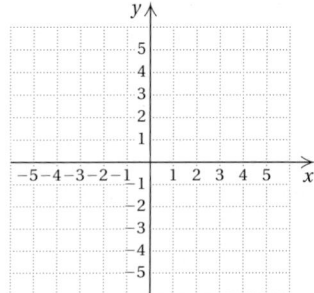

31. $f(x) = x^3$,
$f^{-1}(x) =$ _____

x	$f(x)$
0	
1	
2	
3	
−1	
−2	
−3	

x	$f^{-1}(x)$

32. $f(x) = x^3 - 1$,
$f^{-1}(x) =$ _____

x	$f(x)$
0	
1	
2	
3	
−1	
−2	
−3	

x	$f^{-1}(x)$

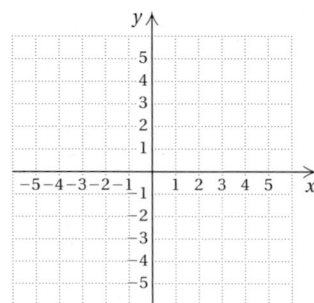

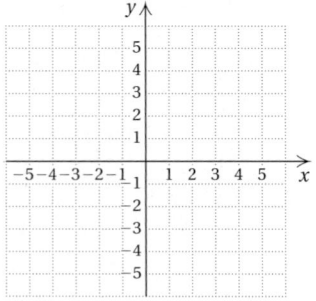

d   Find $(f \circ g)(x)$ and $(g \circ f)(x)$.

33. $f(x) = 2x - 3$,
    $g(x) = 6 - 4x$

34. $f(x) = 9 - 6x$,
    $g(x) = 0.37x + 4$

35. $f(x) = 3x^2 + 2$,
    $g(x) = 2x - 1$

36. $f(x) = 4x + 3$,
    $g(x) = 2x^2 - 5$

37. $f(x) = 4x^2 - 1$,
    $g(x) = \dfrac{2}{x}$

38. $f(x) = \dfrac{3}{x}$,
    $g(x) = 2x^2 + 3$

39. $f(x) = x^2 + 5$,
    $g(x) = x^2 - 5$

40. $f(x) = \dfrac{1}{x^2}$,
    $g(x) = x - 1$

Find $f(x)$ and $g(x)$ such that $h(x) = (f \circ g)(x)$. Answers may vary.

41. $h(x) = (5 - 3x)^2$

42. $h(x) = 4(3x - 1)^2 + 9$

43. $h(x) = \sqrt{5x + 2}$

44. $h(x) = (3x^2 - 7)^5$

45. $h(x) = \dfrac{1}{x - 1}$

46. $h(x) = \dfrac{3}{x} + 4$

47. $h(x) = \dfrac{1}{\sqrt{7x + 2}}$

48. $h(x) = \sqrt{x - 7} - 3$

49. $h(x) = \left(\sqrt{x} + 5\right)^4$

50. $h(x) = \dfrac{x^3 + 1}{x^3 - 1}$

e For each function, use composition to show that the inverse is correct.

51. $f(x) = \frac{4}{5}x,$
$f^{-1}(x) = \frac{5}{4}x$

52. $f(x) = x - 3,$
$f^{-1}(x) = x + 3$

53. $f(x) = \frac{x + 7}{2},$
$f^{-1}(x) = 2x - 7$

54. $f(x) = \frac{3}{4}x - 1,$
$f^{-1}(x) = \frac{4x + 4}{3}$

55. $f(x) = \frac{1 - x}{x},$
$f^{-1}(x) = \frac{1}{x + 1}$

56. $f(x) = x^3 - 5,$
$f^{-1}(x) = \sqrt[3]{x + 5}$

Find the inverse of the given function by thinking about the operations of the function and then reversing, or undoing, them. Then use composition to show whether the inverse is correct.

Function	*Inverse*	*Function*	*Inverse*
57. $f(x) = 3x$	$f^{-1}(x) = \underline{\hspace{1.5cm}}$	**58.** $f(x) = \frac{1}{4}x + 7$	$f^{-1}(x) = \underline{\hspace{1.5cm}}$
59. $f(x) = -x$	$f^{-1}(x) = \underline{\hspace{1.5cm}}$	**60.** $f(x) = \sqrt[3]{x} - 5$	$f^{-1}(x) = \underline{\hspace{1.5cm}}$
61. $f(x) = \sqrt[3]{x - 5}$	$f^{-1}(x) = \underline{\hspace{1.5cm}}$	**62.** $f(x) = x^{-1}$	$f^{-1}(x) = \underline{\hspace{1.5cm}}$

63. *Dress Sizes in the United States and France.* A size-6 dress in the United States is size 38 in France. A function that converts dress sizes in the United States to those in France is

$$f(x) = x + 32.$$

a) Find the dress sizes in France that correspond to sizes of 8, 10, 14, and 18 in the United States.
b) Determine whether this function has an inverse that is a function. If so, find a formula for the inverse.
c) Use the inverse function to find dress sizes in the United States that correspond to sizes of 40, 42, 46, and 50 in France.

64. *Dress Sizes in the United States and Italy.* A size-6 dress in the United States is size 36 in Italy. A function that converts dress sizes in the United States to those in Italy is

$$f(x) = 2(x + 12).$$

a) Find the dress sizes in Italy that correspond to sizes of 8, 10, 14, and 18 in the United States.
b) Determine whether this function has an inverse that is a function. If so, find a formula for the inverse.
c) Use the inverse function to find dress sizes in the United States that correspond to sizes of 40, 44, 52, and 60 in Italy.

Skill Maintenance

Use rational exponents to simplify. [10.2d]

65. $\sqrt[6]{a^2}$

66. $\sqrt[6]{x^4}$

67. $\sqrt{a^4b^6}$

68. $\sqrt[3]{8t^6}$

69. $\sqrt[8]{81}$

70. $\sqrt[4]{32}$

71. $\sqrt[12]{64x^6y^6}$

72. $\sqrt[8]{p^4t^2}$

73. $\sqrt[5]{32a^{15}b^{40}}$

74. $\sqrt[3]{1000x^9y^{18}}$

75. $\sqrt[4]{81a^8b^8}$

76. $\sqrt[3]{27p^3q^9}$

Synthesis

In Exercises 77–80, use a graphing calculator to help determine whether or not the given functions are inverses of each other.

77. $f(x) = 0.75x^2 + 2$; $g(x) = \sqrt{\dfrac{4(x-2)}{3}}$

78. $f(x) = 1.4x^3 + 3.2$; $g(x) = \sqrt[3]{\dfrac{x-3.2}{1.4}}$

79. $f(x) = \sqrt{2.5x + 9.25}$; $g(x) = 0.4x^2 - 3.7$, $x \geq 0$

80. $f(x) = 0.8x^{1/2} + 5.23$; $g(x) = 1.25(x^2 - 5.23)$, $x \geq 0$

81. Use a graphing calculator to help match each function in Column A with its inverse from Column B.

Column A

(1) $y = 5x^3 + 10$

(2) $y = (5x + 10)^3$

(3) $y = 5(x + 10)^3$

(4) $y = (5x)^3 + 10$

Column B

A. $y = \dfrac{\sqrt[3]{x} - 10}{5}$

B. $y = \sqrt[3]{\dfrac{x}{5}} - 10$

C. $y = \sqrt[3]{\dfrac{x - 10}{5}}$

D. $y = \dfrac{\sqrt[3]{x - 10}}{5}$

In Exercises 82 and 83, graph the inverse of f.

82.

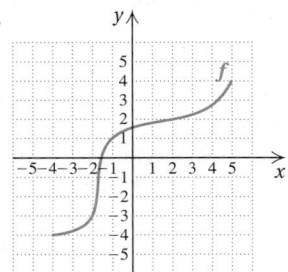

83.

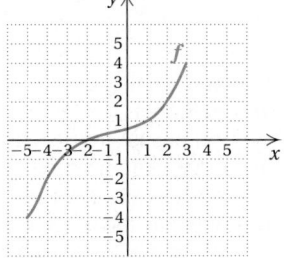

84. Examine the following table. Does it appear that f and g could be inverses of each other? Why or why not?

x	$f(x)$	$g(x)$
6	6	6
7	6.5	8
8	7	10
9	7.5	12
10	8	14
11	8.5	16
12	9	18

85. Assume in Exercise 84 that f and g are both linear functions. Find equations for $f(x)$ and $g(x)$. Are f and g inverses of each other?

12.3 Logarithmic Functions

We are now ready to study inverses of exponential functions. These functions have many applications and are referred to as *logarithm*, or *logarithmic*, *functions*.

a Graphing Logarithmic Functions

Consider the exponential function $f(x) = 2^x$. Like all exponential functions, f is one-to-one. Can a formula for f^{-1} be found? To answer this, we use the method of Section 12.2:

1. Replace $f(x)$ with y: $y = 2^x$.

2. Interchange x and y: $x = 2^y$.

3. Solve for y: $y =$ the power to which we raise 2 to get x.

4. Replace y with $f^{-1}(x)$: $f^{-1}(x) =$ the power to which we raise 2 to get x.

We now define a new symbol to replace the words "the power to which we raise 2 to get x."

MEANING OF LOGARITHMS

$\log_2 x$, read "the logarithm, base 2, of x," or "log, base 2, of x," means "the power to which we raise 2 to get x."

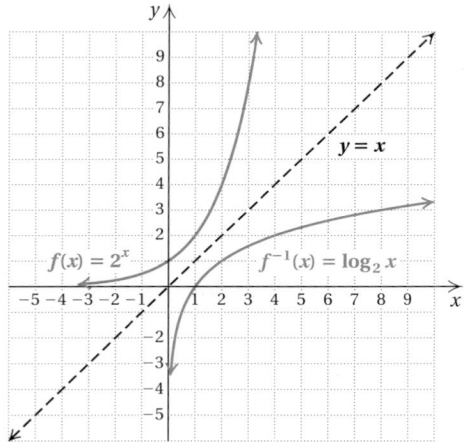

$f(x) = 2^x$ $y = x$ $f^{-1}(x) = \log_2 x$

Thus if $f(x) = 2^x$, then $f^{-1}(x) = \log_2 x$. Note that $f^{-1}(8) = \log_2 8 = 3$, because 3 is the *power to which we raise 2 to get 8*; that is, $2^3 = 8$.

Although expressions like $\log_2 13$ can be only approximated, remember that $\log_2 13$ represents the **power** *to which we raise 2 to get 13*. That is, $2^{\log_2 13} = 13$.

Do Exercise 1.

OBJECTIVES

a Graph logarithmic functions.

b Convert from exponential equations to logarithmic equations and from logarithmic equations to exponential equations.

c Solve logarithmic equations.

d Find common logarithms on a calculator.

1. Write the meaning of $\log_2 64$. Then find $\log_2 64$.

Answer

1. $\log_2 64$ is the power to which we raise 2 to get 64; 6

For any exponential function $f(x) = a^x$, the inverse is called a **logarithmic function, base a**. The graph of the inverse can, of course, be drawn by reflecting the graph of $f(x) = a^x$ across the line $y = x$. It will be helpful to remember that the inverse of $f(x) = a^x$ is given by $f^{-1}(x) = \log_a x$. Normally, we use a number a that is greater than 1 for the logarithm base.

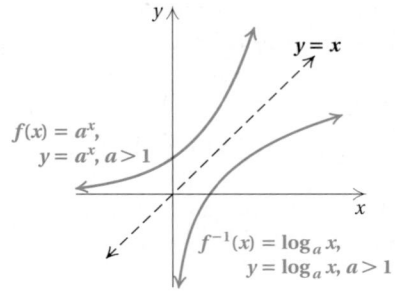

LOGARITHMS

The inverse of $f(x) = a^x$ is given by

$$f^{-1}(x) = \log_a x.$$

We read "$\log_a x$" as "the logarithm, base a, of x." We define $y = \log_a x$ as that number y such that $a^y = x$, where $x > 0$ and a is a positive constant other than 1.

It is helpful in dealing with logarithmic functions to remember that the logarithm of a number is an **exponent**. For instance, $\log_a x$ is the exponent y in $x = a^y$. Keep thinking, "The logarithm, base a, of a number x is the power to which a must be raised in order to get x."

EXPONENTIAL FUNCTION	LOGARITHMIC FUNCTION
$y = a^x$	$x = a^y$
$f(x) = a^x$	$f^{-1}(x) = \log_a x$
$a > 0, a \neq 1$	$a > 0, a \neq 1$
Domain = The set of real numbers	Range = The set of real numbers
Range = The set of positive numbers	Domain = The set of positive numbers

Why do we exclude 1 from being a logarithm base? See the graph below. If we allow 1 as a logarithm base, the graph of the relation $y = \log_1 x$, or $x = 1^y = 1$, is a vertical line, which is not a function and therefore not a logarithmic function.

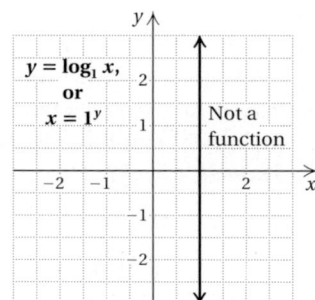

EXAMPLE 1 Graph: $y = f(x) = \log_5 x$.

The equation $y = \log_5 x$ is equivalent to $5^y = x$. We can find ordered pairs that are solutions by choosing values for y and computing the corresponding x-values.

For $y = 0$, $x = 5^0 = 1$.

For $y = 1$, $x = 5^1 = 5$.

For $y = 2$, $x = 5^2 = 25$.

For $y = 3$, $x = 5^3 = 125$.

For $y = -1$, $x = 5^{-1} = \dfrac{1}{5}$.

For $y = -2$, $x = 5^{-2} = \dfrac{1}{25}$.

x, or 5^y	y
1	0
5	1
25	2
125	3
$\frac{1}{5}$	-1
$\frac{1}{25}$	-2

(1) Select y.
(2) Compute x.

The table shows the following:

$\left.\begin{array}{l} \log_5 1 = 0; \\ \log_5 5 = 1; \\ \log_5 25 = 2; \\ \log_5 125 = 3; \\ \log_5 \frac{1}{5} = -1; \\ \log_5 \frac{1}{25} = -2. \end{array}\right\}$ These can all be checked using the equations above.

We plot the ordered pairs and connect them with a smooth curve. The graph of $y = 5^x$ has been shown only for reference.

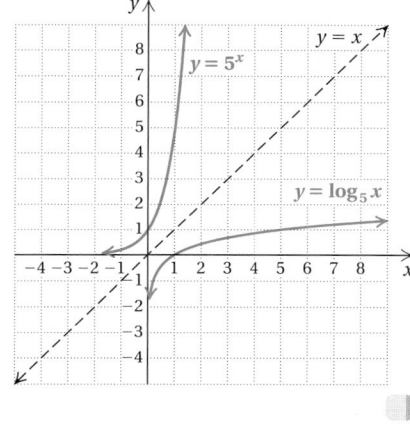

2. Graph: $y = f(x) = \log_3 x$.

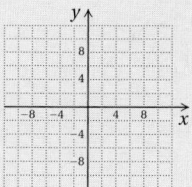

Do Exercise 2.

(b) Converting Between Exponential Equations and Logarithmic Equations

We use the definition of logarithms to convert from exponential equations to logarithmic equations.

CONVERTING BETWEEN EXPONENTIAL EQUATIONS AND LOGARITHMIC EQUATIONS

$$y = \log_a x \longrightarrow a^y = x; \qquad a^y = x \longrightarrow y = \log_a x$$

Be sure to memorize this relationship! It is probably the most important definition in the chapter. Often this definition will be a justification for a proof or a procedure that we are considering.

Answer

2.

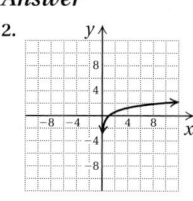

$y = f(x) = \log_3 x$

EXAMPLES Convert to a logarithmic equation.

Convert to a logarithmic equation.

3. $6^0 = 1$

4. $10^{-3} = 0.001$

5. $16^{0.25} = 2$

6. $m^T = P$

2. $8 = 2^x \longrightarrow x = \log_2 8$ The exponent is the logarithm.

                  The base remains the same.

3. $y^{-1} = 4 \longrightarrow -1 = \log_y 4$

4. $a^b = c \longrightarrow b = \log_a c$

Do Exercises 3–6.

We also use the definition of logarithms to convert from logarithmic equations to exponential equations.

EXAMPLES Convert to an exponential equation.

5. $y = \log_3 5 \longrightarrow 3^y = 5$ The logarithm is the exponent.

                  The base does not change.

Convert to an exponential equation.

7. $\log_2 32 = 5$

8. $\log_{10} 1000 = 3$

9. $\log_a Q = 7$

10. $\log_t M = x$

6. $-2 = \log_a 7 \longrightarrow a^{-2} = 7$

7. $a = \log_b d \longrightarrow b^a = d$

Do Exercises 7–10.

(c) Solving Certain Logarithmic Equations

Certain equations involving logarithms can be solved by first converting to exponential equations. We will solve more complicated equations later.

EXAMPLE 8 Solve: $\log_2 x = -3$.

$$\log_2 x = -3$$
$$2^{-3} = x \qquad \text{Converting to an exponential equation}$$
$$\frac{1}{2^3} = x$$
$$\frac{1}{8} = x$$

Check: $\log_2 \frac{1}{8}$ is the exponent to which we raise 2 to get $\frac{1}{8}$. Since $2^{-3} = \frac{1}{8}$, we know that $\frac{1}{8}$ checks and is the solution.

EXAMPLE 9 Solve: $\log_x 16 = 2$.

$$\log_x 16 = 2$$
$$x^2 = 16 \qquad \text{Converting to an exponential equation}$$
$$x = 4 \quad or \quad x = -4 \qquad \text{Using the principle of square roots}$$

Check: $\log_4 16 = 2$ because $4^2 = 16$. Thus, 4 is a solution. Since all logarithm bases must be positive, $\log_{-4} 16$ is not defined. Therefore, -4 is not a solution.

Solve.

11. $\log_{10} x = 4$

12. $\log_x 81 = 4$

13. $\log_2 x = -2$

Do Exercises 11–13.

Answers

3. $0 = \log_6 1$ **4.** $-3 = \log_{10} 0.001$
5. $0.25 = \log_{16} 2$ **6.** $T = \log_m P$
7. $2^5 = 32$ **8.** $10^3 = 1000$ **9.** $a^7 = Q$
10. $t^x = M$ **11.** $10{,}000$ **12.** 3 **13.** $\frac{1}{4}$

To think of finding logarithms as solving equations may help in some cases.

EXAMPLE 10 Find $\log_{10} 1000$.

METHOD 1: Let $\log_{10} 1000 = x$. Then

$$10^x = 1000 \qquad \text{Converting to an exponential equation}$$
$$10^x = 10^3$$
$$x = 3. \qquad \text{The exponents are the same.}$$

Therefore, $\log_{10} 1000 = 3$.

METHOD 2: Think of the meaning of $\log_{10} 1000$. It is the exponent to which we raise 10 to get 1000. That exponent is 3. Therefore, $\log_{10} 1000 = 3$.

EXAMPLE 11 Find $\log_{10} 0.01$.

METHOD 1: Let $\log_{10} 0.01 = x$. Then

$$10^x = 0.01 \qquad \text{Converting to an exponential equation}$$
$$10^x = \frac{1}{100}$$
$$10^x = 10^{-2}$$
$$x = -2. \qquad \text{The exponents are the same.}$$

Therefore, $\log_{10} 0.01 = -2$.

METHOD 2: $\log_{10} 0.01$ is the exponent to which we raise 10 to get 0.01. Noting that

$$0.01 = \frac{1}{100} = \frac{1}{10^2} = 10^{-2},$$

we see that the exponent is -2. Therefore, $\log_{10} 0.01 = -2$.

EXAMPLE 12 Find $\log_5 1$.

METHOD 1: Let $\log_5 1 = x$. Then

$$5^x = 1 \qquad \text{Converting to an exponential equation}$$
$$5^x = 5^0$$
$$x = 0. \qquad \text{The exponents are the same.}$$

Therefore, $\log_5 1 = 0$.

METHOD 2: $\log_5 1$ is the exponent to which we raise 5 to get 1. That exponent is 0. Therefore, $\log_5 1 = 0$.

Do Exercises 14-16.

Find each of the following.

14. $\log_{10} 10{,}000$

15. $\log_{10} 0.0001$

16. $\log_7 1$

THE LOGARITHM OF 1

For any base a,

$$\log_a 1 = 0.$$

The logarithm, base a, of 1 is always 0.

Answers

14. 4 **15.** -4 **16.** 0

The proof follows from the fact that $a^0 = 1$. This is equivalent to the logarithmic equation $\log_a 1 = 0$.

Another property follows similarly. We know that $a^1 = a$ for any real number a. In particular, it holds for any positive number a. This is equivalent to the logarithmic equation $\log_a a = 1$.

> **THE LOGARITHM, BASE a, OF a**
>
> For any base a,
> $$\log_a a = 1.$$

EXAMPLE 13 Simplify: $\log_m 1$ and $\log_t t$.

$$\log_m 1 = 0; \qquad \log_t t = 1$$

Do Exercises 17–20.

(d) Finding Common Logarithms on a Calculator

Base-10 logarithms are called **common logarithms**. Before calculators became so widely available, common logarithms were used extensively to do complicated calculations. In fact, that is why logarithms were invented. The abbreviation **log**, with no base written, is used for the common logarithm, base-10. Thus,

$$\log 29 \quad \text{means} \quad \log_{10} 29.$$

> Be sure to memorize $\log a = \log_{10} a$.

We can approximate $\log 29$. Note the following:

$$\log 100 = \log_{10} 100 = \mathbf{2};$$
$$\log 29 = \mathbf{?};$$
$$\log 10 = \log_{10} 10 = \mathbf{1}.$$

It seems reasonable to conclude that $\log 29$ is between 1 and 2.

On a scientific or graphing calculator, the key for common logarithms is generally marked **LOG**. We find that

$$\log 29 \approx 1.462397998 \approx 1.4624,$$

rounded to four decimal places. This also tells us that $10^{1.4624} \approx 29$.

On some scientific calculators, the keystrokes for doing such a calculation might be

(2)(9) **LOG** (=).   The display would then read   1.462398.

Using a graphing calculator, the keystrokes might be

LOG (2)(9) **ENTER**.   The display would then read   1.462397998.

EXAMPLES Find the common logarithm, to four decimal places, on a scientific or graphing calculator.

Function Value	Readout	Rounded
14. $\log 287{,}523$	5.458672591	5.4587
15. $\log 0.000486$	-3.313363731	-3.3134
16. $\log(-5)$	NONREAL ANS	Does not exist as a real number

In Example 16, log (-5) does not exist as a real number because there is no real-number power to which we can raise 10 to get -5. The number 10 raised to any power is nonnegative. The logarithm of a negative number does not exist as a real number (though it can be defined as a complex number).

Do Exercises 21–24 on the preceding page.

We can use common logarithms to express any positive number as a power of 10. We simply find the common logarithm of the number on a calculator. Considering very large or very small numbers as powers of 10 might be a helpful way to compare those numbers.

EXAMPLE 17 Complete the following table to express each number in the first column as a power of 10. Round each exponent to the nearest ten-thousandth.

We simply find the common logarithm of the number using a calculator.

NUMBER	EXPRESSED AS A POWER OF 10
4	$4 \approx 10^{0.6021}$
625	$625 \approx 10^{2.7959}$
134,567	$134{,}567 \approx 10^{5.1289}$
0.00567	$0.00567 \approx 10^{-2.2464}$
0.000374859	$0.000374859 \approx 10^{-3.4261}$
186,000	$186{,}000 \approx 10^{5.2695}$
186,000,000	$186{,}000{,}000 \approx 10^{8.2695}$

Do Exercise 25.

The inverse of a logarithmic function is an exponential function. Thus, if $f(x) = \log x$, then $f^{-1}(x) = 10^x$. Because of this, on many calculators, the **LOG** key doubles as the ⟨10ˣ⟩ key after a **2ND** or $\boxed{\text{SHIFT}}$ key has been pressed. To find $10^{5.4587}$ on a scientific calculator, we might enter 5.4587 and press ⟨10ˣ⟩. On many graphing calculators, we press **2ND** ⟨10ˣ⟩, followed by 5.4587. In either case, we get the approximation

$$10^{5.4587} \approx 287{,}541.1465.$$

Compare this computation to Example 14. Note that, apart from the rounding error, $10^{5.4587}$ takes us back to about 287,523.

Do Exercise 26.

Using the scientific keys on a calculator would allow us to construct a graph of $f(x) = \log_{10} x = \log x$ by finding function values directly, rather than converting to exponential form as we did in Example 1.

x	$f(x)$
0.5	-0.3010
1	0
2	0.3010
3	0.4771
5	0.6990
9	0.9542
10	1

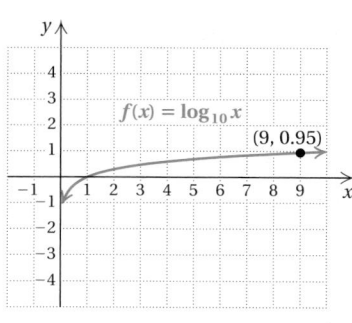

25. Complete the following table to express each number in the first column as a power of 10. Round each exponent to the nearest ten-thousandth.

NUMBER	EXPRESSED AS A POWER OF 10
8	
947	
634,567	
0.00708	
0.000778899	
18,600,000	
1860	

26. Find $10^{4.8934}$ using a calculator. (Compare your computation to that of Margin Exercise 21.)

Answers

25. $10^{0.9031}$; $10^{2.9763}$; $10^{5.8025}$; $10^{-2.1500}$; $10^{-3.1085}$; $10^{7.2695}$; $10^{3.2695}$ **26.** 78,234.8042

a Graph.

1. $f(x) = \log_2 x$, or $y = \log_2 x$
$y = \log_2 x \implies x = $ _____

x, or 2^y	y
	0
	1
	2
	3
	-1
	-2
	-3

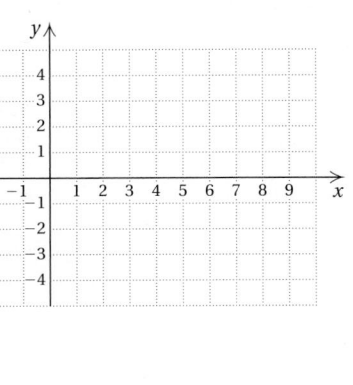

2. $f(x) = \log_{10} x$, or $y = \log_{10} x$
$y = \log_{10} x \implies x = $ _____

x, or 10^y	y
	0
	1
	2
	3
	-1
	-2
	-3

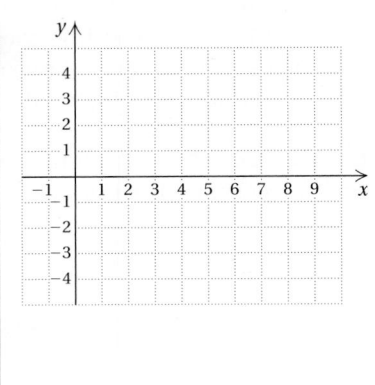

3. $f(x) = \log_{1/3} x$

x	y

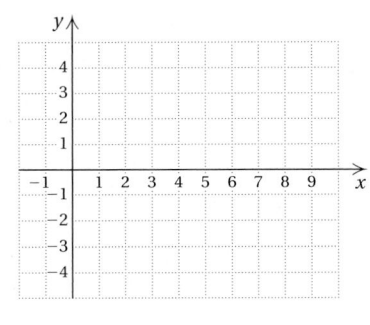

4. $f(x) = \log_{1/2} x$

x	y

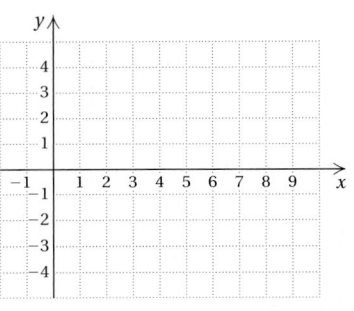

Graph both functions using the same set of axes.

5. $f(x) = 3^x$, $f^{-1}(x) = \log_3 x$

6. $f(x) = 4^x$, $f^{-1}(x) = \log_4 x$

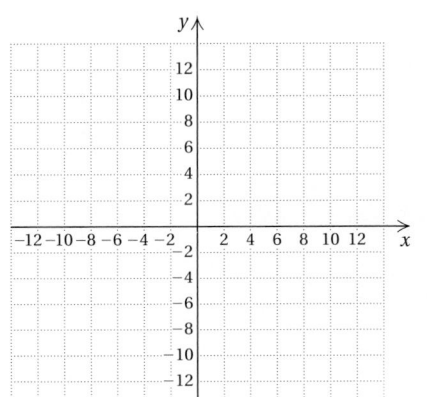

Convert to a logarithmic equation.

7. $10^3 = 1000$

8. $10^2 = 100$

9. $5^{-3} = \dfrac{1}{125}$

10. $4^{-5} = \dfrac{1}{1024}$

11. $8^{1/3} = 2$

12. $16^{1/4} = 2$

13. $10^{0.3010} = 2$

14. $10^{0.4771} = 3$

15. $e^2 = t$

16. $p^k = 3$

17. $Q^t = x$

18. $P^m = V$

19. $e^2 = 7.3891$

20. $e^3 = 20.0855$

21. $e^{-2} = 0.1353$

22. $e^{-4} = 0.0183$

Convert to an exponential equation.

23. $w = \log_4 10$

24. $t = \log_5 9$

25. $\log_6 36 = 2$

26. $\log_7 7 = 1$

27. $\log_{10} 0.01 = -2$

28. $\log_{10} 0.001 = -3$

29. $\log_{10} 8 = 0.9031$

30. $\log_{10} 2 = 0.3010$

31. $\log_e 100 = 4.6052$

32. $\log_e 10 = 2.3026$

33. $\log_t Q = k$

34. $\log_m P = a$

c Solve.

35. $\log_3 x = 2$

36. $\log_4 x = 3$

37. $\log_x 16 = 2$

38. $\log_x 64 = 3$

39. $\log_2 16 = x$

40. $\log_5 25 = x$

41. $\log_3 27 = x$

42. $\log_4 16 = x$

43. $\log_x 25 = 1$

44. $\log_x 9 = 1$

45. $\log_3 x = 0$

46. $\log_2 x = 0$

47. $\log_2 x = -1$

48. $\log_3 x = -2$

49. $\log_8 x = \dfrac{1}{3}$

50. $\log_{32} x = \dfrac{1}{5}$

Find each of the following.

51. $\log_{10} 100$

52. $\log_{10} 100{,}000$

53. $\log_{10} 0.1$

54. $\log_{10} 0.001$

55. $\log_{10} 1$

56. $\log_{10} 10$

57. $\log_5 625$

58. $\log_2 64$

59. $\log_7 49$

60. $\log_5 125$

61. $\log_2 8$

62. $\log_8 64$

63. $\log_9 \dfrac{1}{81}$

64. $\log_5 \dfrac{1}{125}$

65. $\log_8 1$

66. $\log_6 6$

67. $\log_e e$

68. $\log_e 1$

69. $\log_{27} 9$

70. $\log_8 2$

d Find the common logarithm, to four decimal places, on a calculator.

71. $\log 78{,}889.2$

72. $\log 9{,}043{,}788$

73. $\log 0.67$

74. $\log 0.0067$

75. $\log(-97)$

76. $\log 0$

77. $\log\left(\dfrac{289}{32.7}\right)$

78. $\log\left(\dfrac{23}{86.2}\right)$

79. Complete the following table to express each number in the first column as a power of 10. Round each exponent to the nearest ten-thousandth.

NUMBER	EXPRESSED AS A POWER OF 10
6	
84	
987,606	
0.00987606	
98,760.6	
70,000,000	
7000	

80. Complete the following table to express each number in the first column as a power of 10. Round each exponent to the nearest ten-thousandth.

NUMBER	EXPRESSED AS A POWER OF 10
7	
314	
31.4	
31,400,000	
0.000314	
3.14	
0.0314	

Skill Maintenance

In each of Exercises 81–88, fill in the blank with the correct term from the given list. Some of the choices may not be used.

81. The _____ of a complex number $a + bi$ is $a - bi$.
[10.8e]

82. If a situation gives rise to a linear function $f(x) = kx$, where k is a positive constant, the situation is an example of _____ variation.   [6.9a]

83. In the term $-2x^4$, -2 is the _____ and 4 is the _____.   [4.1a], [4.3d]

84. The expression $b^2 - 4ac$ in the _____ formula is called the _____.   [11.4a]

85. A system of equations that has no solution is called a(n) _____ system.   [8.1a]

86. Graphs of quadratic functions are called _____.
[11.5a]

87. For the graph of $f(x) = (x - 2)^2 + 4$, the line $x = 2$ is called the _____.   [11.5a]

88. A complex number is any number that can be named _____ , where a and b are any real numbers and $i = \sqrt{-1}$.   [10.8a]

direct

indirect

$b^2 - 4ac$

$a + bi$

discriminant

radical

quadratic

parabolas

polynomials

line of symmetry

conjugate

exponent

coefficient

consistent

inconsistent

Synthesis

Graph.

89. $f(x) = \log_3 |x + 1|$

90. $f(x) = \log_2 (x - 1)$

Solve.

91. $\log_{125} x = \frac{2}{3}$

92. $|\log_3 x| = 3$

93. $\log_{128} x = \frac{5}{7}$

94. $\log_4 (3x - 2) = 2$

95. $\log_8 (2x + 1) = -1$

96. $\log_{10} (x^2 + 21x) = 2$

Simplify.

97. $\log_{1/4} \frac{1}{64}$

98. $\log_{81} 3 \cdot \log_3 81$

99. $\log_{10} (\log_4 (\log_3 81))$

100. $\log_2 (\log_2 (\log_4 256))$

101. $\log_{1/5} 25$

12.4

Properties of Logarithmic Functions

OBJECTIVES

a Express the logarithm of a product as a sum of logarithms, and conversely.

b Express the logarithm of a power as a product.

c Express the logarithm of a quotient as a difference of logarithms, and conversely.

d Convert from logarithms of products, quotients, and powers to expressions in terms of individual logarithms, and conversely.

e Simplify expressions of the type $\log_a a^k$.

SKILL TO REVIEW
Objective 10.2c: Use the laws of exponents with rational exponents.

Simplify.

1. **(a)** $10^3 \cdot 10^{-5}$; **(b)** $\dfrac{2^6}{2^2}$

2. $(7^{3/4})^{5/6}$

Express as a sum of logarithms.
1. $\log_5 (25 \cdot 5)$

2. $\log_b (PQ)$

Express as a single logarithm.
3. $\log_3 7 + \log_3 5$

4. $\log_a J + \log_a A + \log_a M$

Answers

Skill to Review:
1. **(a)** 10^{-2}, or $\dfrac{1}{100}$, or 0.01; **(b)** 2^4, or 16
2. $7^{5/8}$

Margin Exercises:
1. $\log_5 25 + \log_5 5$ **2.** $\log_b P + \log_b Q$
3. $\log_3 35$ **4.** $\log_a (JAM)$

The ability to manipulate logarithmic expressions is important in many applications and in more advanced mathematics. We now establish some basic properties that are useful in manipulating logarithmic expressions.

a Logarithms of Products

> **PROPERTY 1: THE PRODUCT RULE**
>
> For any positive numbers M and N,
>
> $$\log_a (M \cdot N) = \log_a M + \log_a N.$$
>
> (The logarithm of a product is the sum of the logarithms of the factors. The number a can be any logarithm base.)

EXAMPLE 1   Express as a sum of logarithms: $\log_2 (4 \cdot 16)$.

$$\log_2 (4 \cdot 16) = \log_2 4 + \log_2 16 \qquad \text{By Property 1}$$

EXAMPLE 2   Express as a single logarithm: $\log_{10} 0.01 + \log_{10} 1000$.

$$\log_{10} 0.01 + \log_{10} 1000 = \log_{10} (0.01 \times 1000) \qquad \text{By Property 1}$$
$$= \log_{10} 10$$

Do Margin Exercises 1–4.

A PROOF OF PROPERTY 1 (*OPTIONAL*):   We let $\log_a M = x$ and $\log_a N = y$. Converting to exponential equations, we have $a^x = M$ and $a^y = N$. Then we multiply to obtain

$$M \cdot N = a^x \cdot a^y = a^{x+y}.$$

Converting $M \cdot N = a^{x+y}$ back to a logarithmic equation, we get

$$\log_a (M \cdot N) = x + y.$$

Remembering what x and y represent, we get

$$\log_a (M \cdot N) = \log_a M + \log_a N.$$

b Logarithms of Powers

> **PROPERTY 2: THE POWER RULE**
>
> For any positive number M and any real number k,
>
> $$\log_a M^k = k \cdot \log_a M.$$
>
> (The logarithm of a power of M is the exponent times the logarithm of M. The number a can be any logarithm base.)

EXAMPLES Express as a product.

3. $\log_a 9^{-5} = -5\log_a 9$ By Property 2

4. $\log_a \sqrt[4]{5} = \log_a 5^{1/4}$ Writing exponential notation

$\quad\quad = \frac{1}{4}\log_a 5$ By Property 2

Do Exercises 5 and 6.

A PROOF OF PROPERTY 2 (*OPTIONAL*): We let $x = \log_a M$. Then we convert to an exponential equation to get $a^x = M$. Raising both sides to the kth power, we obtain

$$(a^x)^k = M^k, \quad \text{or} \quad a^{xk} = M^k.$$

Converting back to a logarithmic equation with base a, we get $\log_a M^k = xk$. But $x = \log_a M$, so

$$\log_a M^k = (\log_a M)k = k \cdot \log_a M.$$

(c) Logarithms of Quotients

PROPERTY 3: THE QUOTIENT RULE

For any positive numbers M and N,

$$\log_a \frac{M}{N} = \log_a M - \log_a N.$$

(The logarithm of a quotient is the logarithm of the numerator minus the logarithm of the denominator. The number a can be any logarithm base.)

EXAMPLE 5 Express as a difference of logarithms: $\log_t \frac{6}{U}$.

$$\log_t \frac{6}{U} = \log_t 6 - \log_t U \quad \text{By Property 3}$$

EXAMPLE 6 Express as a single logarithm: $\log_b 17 - \log_b 27$.

$$\log_b 17 - \log_b 27 = \log_b \frac{17}{27} \quad \text{By Property 3}$$

EXAMPLE 7 Express as a single logarithm: $\log_{10} 10{,}000 - \log_{10} 100$.

$$\log_{10} 10{,}000 - \log_{10} 100 = \log_{10} \frac{10{,}000}{100} = \log_{10} 100$$

Do Exercises 7 and 8.

A PROOF OF PROPERTY 3 (*OPTIONAL*): The proof makes use of Property 1 and Property 2.

$$\log_a \frac{M}{N} = \log_a M \cdot \frac{1}{N} = \log_a MN^{-1} \quad \frac{1}{N} = N^{-1}$$

$$= \log_a M + \log_a N^{-1} \quad \text{By Property 1}$$

$$= \log_a M + (-1)\log_a N \quad \text{By Property 2}$$

$$= \log_a M - \log_a N$$

Express as a product.

5. $\log_7 4^5$ **6.** $\log_a \sqrt{5}$

Calculator Corner

Properties of Logarithms Use a table or a graph to determine whether each of the following is correct.

1. $\log(5x) = \log 5 \cdot \log x$

2. $\log(5x) = \log 5 + \log x$

3. $\log x^2 = \log x \cdot \log x$

4. $\log x^2 = 2\log x$

5. $\log\left(\frac{x}{3}\right) = \frac{\log x}{\log 3}$

6. $\log\left(\frac{x}{3}\right) = \log x - \log 3$

7. $\log(x + 2) = \log x + \log 2$

8. $\log(x + 2) = \log x \cdot \log 2$

7. Express as a difference of logarithms:

$$\log_b \frac{P}{Q}.$$

8. Express as a single logarithm:

$$\log_2 125 - \log_2 25.$$

Answers

5. $5\log_7 4$ **6.** $\frac{1}{2}\log_a 5$ **7.** $\log_b P - \log_b Q$

8. $\log_2 5$

(d) Using the Properties Together

EXAMPLES Express in terms of logarithms of w, x, y, and z.

8. $\log_a \dfrac{x^2 y^3}{z^4} = \log_a (x^2 y^3) - \log_a z^4$ Using Property 3

$= \log_a x^2 + \log_a y^3 - \log_a z^4$ Using Property 1

$= 2 \log_a x + 3 \log_a y - 4 \log_a z$ Using Property 2

9. $\log_a \sqrt[4]{\dfrac{xy}{z^3}} = \log_a \left(\dfrac{xy}{z^3} \right)^{1/4}$ Writing exponential notation

$= \tfrac{1}{4} \log_a \dfrac{xy}{z^3}$ Using Property 2

$= \tfrac{1}{4} (\log_a xy - \log_a z^3)$ Using Property 3 (note the parentheses)

$= \tfrac{1}{4} (\log_a x + \log_a y - 3 \log_a z)$ Using Properties 1 and 2

$= \tfrac{1}{4} \log_a x + \tfrac{1}{4} \log_a y - \tfrac{3}{4} \log_a z$ Distributive law

10. $\log_b \dfrac{xy}{w^3 z^4} = \log_b xy - \log_b w^3 z^4$ Using Property 3

$= (\log_b x + \log_b y) - (\log_b w^3 + \log_b z^4)$ Using Property 1

$= \log_b x + \log_b y - \log_b w^3 - \log_b z^4$ Removing parentheses

$= \log_b x + \log_b y - 3 \log_b w - 4 \log_b z$ Using Property 2

Do Exercises 9–11.

EXAMPLES Express as a single logarithm.

11. $\dfrac{1}{2} \log_a x - 7 \log_a y + \log_a z$

$= \log_a x^{1/2} - \log_a y^7 + \log_a z$ Using Property 2

$= \log_a \dfrac{\sqrt{x}}{y^7} + \log_a z$ Using Property 3

$= \log_a \dfrac{z\sqrt{x}}{y^7}$ Using Property 1

12. $\log_a \dfrac{b}{\sqrt{x}} + \log_a \sqrt{bx}$

$= \log_a b - \log_a \sqrt{x} + \log_a \sqrt{bx}$ Using Property 3

$= \log_a b - \tfrac{1}{2} \log_a x + \tfrac{1}{2} \log_a (bx)$ Using Property 2

$= \log_a b - \tfrac{1}{2} \log_a x + \tfrac{1}{2} (\log_a b + \log_a x)$ Using Property 1

$= \log_a b - \tfrac{1}{2} \log_a x + \tfrac{1}{2} \log_a b + \tfrac{1}{2} \log_a x$

$= \tfrac{3}{2} \log_a b$ Collecting like terms

$= \log_a b^{3/2}$ Using Property 2

Example 12 could also be done as follows:

$\log_a \dfrac{b}{\sqrt{x}} + \log_a \sqrt{bx} = \log_a \dfrac{b}{\sqrt{x}} \sqrt{bx}$ Using Property 1

$= \log_a \dfrac{b}{\sqrt{x}} \cdot \sqrt{b} \cdot \sqrt{x}$

$= \log_a b\sqrt{b}, \text{ or } \log_a b^{3/2}.$

Express in terms of logarithms of w, x, y, and z.

9. $\log_a \sqrt{\dfrac{z^3}{xy}}$

10. $\log_a \dfrac{x^2}{y^3 z}$

11. $\log_a \dfrac{x^3 y^4}{z^5 w^9}$

Express as a single logarithm.

12. $5 \log_a x - \log_a y + \dfrac{1}{4} \log_a z$

13. $\log_a \dfrac{\sqrt{x}}{b} - \log_a \sqrt{bx}$

Answers

9. $\tfrac{3}{2} \log_a z - \tfrac{1}{2} \log_a x - \tfrac{1}{2} \log_a y$

10. $2 \log_a x - 3 \log_a y - \log_a z$

11. $3 \log_a x + 4 \log_a y - 5 \log_a z - 9 \log_a w$

12. $\log_a \dfrac{x^5 z^{1/4}}{y}$, or $\log_a \dfrac{x^5 \sqrt[4]{z}}{y}$

13. $\log_a \dfrac{1}{b\sqrt{b}}$, or $\log_a b^{-3/2}$

Do Exercises 12 and 13 on the preceding page.

EXAMPLES Given $\log_a 2 = 0.301$ and $\log_a 3 = 0.477$, find each of the following.

13. $\log_a 6 = \log_a (2 \cdot 3) = \log_a 2 + \log_a 3$ Property 1
$$= 0.301 + 0.477 = 0.778$$

14. $\log_a \dfrac{2}{3} = \log_a 2 - \log_a 3$ Property 3
$$= 0.301 - 0.477 = -0.176$$

15. $\log_a 81 = \log_a 3^4 = 4 \log_a 3$ Property 2
$$= 4(0.477) = 1.908$$

16. $\log_a \dfrac{1}{3} = \log_a 1 - \log_a 3$ Property 3
$$= 0 - 0.477 = -0.477$$

17. $\log_a \sqrt{a} = \log_a a^{1/2} = \dfrac{1}{2} \log_a a = \dfrac{1}{2} \cdot 1 = \dfrac{1}{2}$ Property 2

18. $\log_a 2a = \log_a 2 + \log_a a$ Property 1
$$= 0.301 + 1 = 1.301$$

19. $\log_a 5$ There is no way to find this using these properties ($\log_a 5 \neq \log_a 2 + \log_a 3$).

20. $\dfrac{\log_a 3}{\log_a 2} = \dfrac{0.477}{0.301} \approx 1.58$ We simply divide the logarithms, not using any property.

Do Exercises 14–21.

Caution!

Keep in mind that, in general,

$\log_a (M + N) \neq \log_a M + \log_a N$,

$\log_a (M - N) \neq \log_a M - \log_a N$,

$\log_a (MN) \neq (\log_a M)(\log_a N)$,

and

$\log_a (M/N) \neq (\log_a M) \div (\log_a N)$.

Given
$$\log_a 2 = 0.301,$$
$$\log_a 5 = 0.699,$$
find each of the following.

14. $\log_a 4$ **15.** $\log_a 10$

16. $\log_a \dfrac{2}{5}$ **17.** $\log_a \dfrac{5}{2}$

18. $\log_a \dfrac{1}{5}$ **19.** $\log_a \sqrt{a^3}$

20. $\log_a 5a$ **21.** $\log_a 16$

(e) The Logarithm of the Base to a Power

PROPERTY 4

For any base a,
$$\log_a a^k = k.$$

(The logarithm, base a, of a to a power is the power.)

A PROOF OF PROPERTY 4 (*OPTIONAL*): The proof involves Property 2 and the fact that $\log_a a = 1$:

$\log_a a^k = k(\log_a a)$ Using Property 2
$ = k \cdot 1$ Using $\log_a a = 1$
$ = k$.

EXAMPLES Simplify.

21. $\log_3 3^7 = 7$

22. $\log_{10} 10^{5.6} = 5.6$

23. $\log_e e^{-t} = -t$

Do Exercises 22–24.

Simplify.

22. $\log_2 2^6$ **23.** $\log_{10} 10^{3.2}$

24. $\log_e e^{12}$

Answers

14. 0.602   **15.** 1   **16.** -0.398 **17.** 0.398
18. -0.699 **19.** $\dfrac{3}{2}$ **20.** 1.699   **21.** 1.204
22. 6   **23.** 3.2   **24.** 12

a Express as a sum of logarithms.

1. $\log_2 (32 \cdot 8)$

2. $\log_3 (27 \cdot 81)$

3. $\log_4 (64 \cdot 16)$

4. $\log_5 (25 \cdot 125)$

5. $\log_a Qx$

6. $\log_r 8Z$

Express as a single logarithm.

7. $\log_b 3 + \log_b 84$

8. $\log_a 75 + \log_a 5$

9. $\log_c K + \log_c y$

10. $\log_t H + \log_t M$

b Express as a product.

11. $\log_c y^4$

12. $\log_a x^3$

13. $\log_b t^6$

14. $\log_{10} y^7$

15. $\log_b C^{-3}$

16. $\log_c M^{-5}$

c Express as a difference of logarithms.

17. $\log_a \dfrac{67}{5}$

18. $\log_t \dfrac{T}{7}$

19. $\log_b \dfrac{2}{5}$

20. $\log_a \dfrac{z}{y}$

Express as a single logarithm.

21. $\log_c 22 - \log_c 3$

22. $\log_d 54 - \log_d 9$

d Express in terms of logarithms of a single variable or a number.

23. $\log_a x^2 y^3 z$

24. $\log_a 5xy^4 z^3$

25. $\log_b \dfrac{xy^2}{z^3}$

26. $\log_b \dfrac{p^2 q^5}{m^4 n^7}$

27. $\log_c \sqrt[3]{\dfrac{x^4}{y^3 z^2}}$

28. $\log_a \sqrt{\dfrac{x^6}{p^5 q^8}}$

29. $\log_a \sqrt[4]{\dfrac{m^8 n^{12}}{a^3 b^5}}$

30. $\log_a \sqrt{\dfrac{a^6 b^8}{a^2 b^5}}$

Express as a single logarithm and, if possible, simplify.

31. $\dfrac{2}{3} \log_a x - \dfrac{1}{2} \log_a y$

32. $\dfrac{1}{2} \log_a x + 3 \log_a y - 2 \log_a x$

33. $\log_a 2x + 3(\log_a x - \log_a y)$

34. $\log_a x^2 - 2 \log_a \sqrt{x}$

35. $\log_a \dfrac{a}{\sqrt{x}} - \log_a \sqrt{ax}$

36. $\log_a (x^2 - 4) - \log_a (x - 2)$

Given $\log_b 3 = 1.099$ and $\log_b 5 = 1.609$, find each of the following.

37. $\log_b 15$ **38.** $\log_b 8$ **39.** $\log_b \dfrac{5}{3}$ **40.** $\log_b \dfrac{3}{5}$

41. $\log_b \dfrac{1}{5}$ **42.** $\log_b \dfrac{1}{3}$ **43.** $\log_b \sqrt{b}$ **44.** $\log_b \sqrt{b^3}$

45. $\log_b 5b$ **46.** $\log_b 3b$ **47.** $\log_b 2$ **48.** $\log_b 75$

 Simplify.

49. $\log_e e^t$ **50.** $\log_w w^8$ **51.** $\log_p p^5$ **52.** $\log_Y Y^{-4}$

Solve for x.

53. $\log_2 2^7 = x$ **54.** $\log_9 9^4 = x$ **55.** $\log_e e^x = -7$ **56.** $\log_a a^x = 2.7$

Skill Maintenance

Compute and simplify. Express answers in the form $a + bi$, where $i^2 = -1$.   [10.8b, c, d, e]

57. i^{29} **58.** i^{34} **59.** $(2 + i)(2 - i)$ **60.** $\dfrac{2 + i}{2 - i}$

61. $(7 - 8i) - (-16 + 10i)$ **62.** $2i^2 \cdot 5i^3$ **63.** $(8 + 3i)(-5 - 2i)$ **64.** $(2 - i)^2$

Synthesis

65. Use the TABLE and GRAPH features to show that $\log x^2 \neq (\log x)(\log x)$.

66. Use the TABLE and GRAPH features to show that $\dfrac{\log x}{\log 4} \neq \log x - \log 4$.

Express as a single logarithm and, if possible, simplify.

67. $\log_a (x^8 - y^8) - \log_a (x^2 + y^2)$ **68.** $\log_a (x + y) + \log_a (x^2 - xy + y^2)$

Express as a sum or a difference of logarithms.

69. $\log_a \sqrt{1 - s^2}$ **70.** $\log_a \dfrac{c - d}{\sqrt{c^2 - d^2}}$

Determine whether each is true or false.

71. $\dfrac{\log_a P}{\log_a Q} = \log_a \dfrac{P}{Q}$ **72.** $\dfrac{\log_a P}{\log_a Q} = \log_a P - \log_a Q$ **73.** $\log_a 3x = \log_a 3 + \log_a x$

74. $\log_a 3x = 3 \log_a x$ **75.** $\log_a (P + Q) = \log_a P + \log_a Q$ **76.** $\log_a x^2 = 2 \log_a x$

Mid-Chapter Review

Concept Reinforcement

Determine whether each statement is true or false.

_____ **1.** The graph of an exponential function never crosses the x-axis. [12.1a]

_____ **2.** A function f is one-to-one if different inputs have different outputs. [12.2b]

_____ **3.** $\log_a 0 = 1$ [12.3b]

_____ **4.** $\log_a \dfrac{m}{n} = \log_a m - \log_a n$ [12.4c]

Guided Solutions

Fill in each box with the number and/or symbol that creates a correct statement or solution.

5. Solve: $\log_5 x = 3$. [12.3c]

$\log_5 x = 3$

$\boxed{}^{\boxed{}} = x$ Converting to an exponential equation

$\boxed{} = x$ Simplifying

6. Given $\log_a 2 = 0.648$ and $\log_a 9 = 2.046$, find:
(a) $\log_a 18$; **(b)** $\log_a \frac{1}{2}$. [12.4d]

a) $\log_a 18 = \log_a (\boxed{} \cdot \boxed{})$

$ = \log_a 2 \,\boxed{}\, \log_a \boxed{}$

$ = 0.648 \,\boxed{}\,\boxed{}$

$ = \boxed{}$

b) $\log_a \frac{1}{2} = \log_a 1 \,\boxed{}\, \log_a \boxed{}$

$\phantom{\log_a \frac{1}{2}} = \boxed{}\,\boxed{}\,0.648$

$\phantom{\log_a \frac{1}{2}} = \boxed{}$

Mixed Review

Graph. [12.1a], [12.3a]

7. $f(x) = 3^{x-1}$

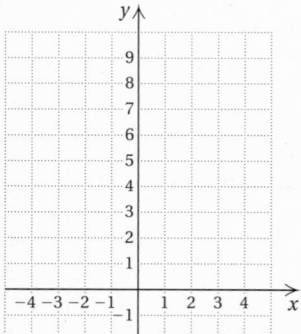

8. $f(x) = \left(\dfrac{3}{4}\right)^x$

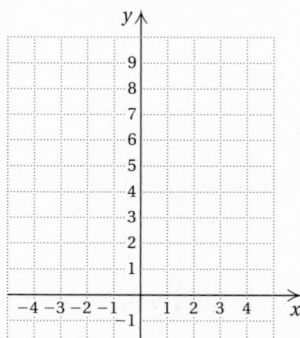

9. $f(x) = \log_4 x$

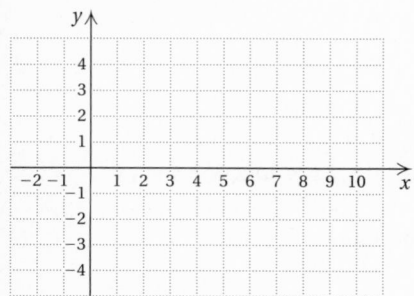

10. $f(x) = \log_{1/4} x$

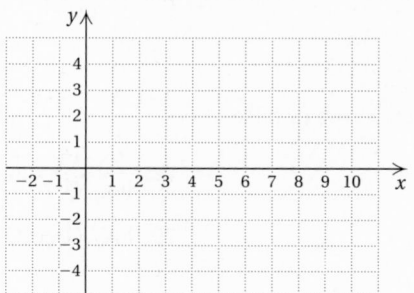

11. *Interest Compounded Annually.* Lucas invests $500 at 4% interest, compounded annually. [12.1c]

a) Find a function A for the amount in the account after t years.

b) Find the amount in the account at $t = 0$, $t = 4$, and $t = 10$.

12. *Interest Compounded Quarterly.* The Currys invest $1500 in an account paying 3.5% interest, compounded quarterly. Find the amount in the account after $1\frac{1}{2}$ years. [12.1c]

Determine whether each function is one-to-one. If it is, find a formula for its inverse. [12.2c]

13. $f(x) = 3x + 1$

14. $f(x) = x^3 + 2$

Find $(f \circ g)(x)$ and $(g \circ f)(x)$. [12.2d]

15. $f(x) = 2x - 5$, $g(x) = 3 - x$

16. $f(x) = x^2 + 1$, $g(x) = 3x - 1$

Find $f(x)$ and $g(x)$ such that $h(x) = (f \circ g)(x)$. Answers may vary. [12.2d]

17. $h(x) = \dfrac{3}{x + 4}$

18. $h(x) = \sqrt{6x - 7}$

For each function, use composition to show that the inverse is correct. [12.2e]

19. $f(x) = \dfrac{x}{3}$, $f^{-1}(x) = 3x$

20. $f(x) = \sqrt[3]{x + 4}$, $f^{-1}(x) = x^3 - 4$

Convert to a logarithmic equation. [12.3b]

21. $7^3 = 343$

22. $3^{-4} = \dfrac{1}{81}$

Convert to an exponential equation. [12.3b]

23. $\log_6 12 = t$

24. $\log_n T = m$

Solve. [12.3c]

25. $\log_4 64 = x$

26. $\log_x \dfrac{1}{4} = -2$

Find each of the following. [12.3c]

27. $\log_7 49$

28. $\log_2 32$

Use a calculator to find the logarithm, to four decimal places. [12.3d]

29. $\log 243.7$

30. $\log 0.23$

Express in terms of logarithms of x, y, and z. [12.4d]

31. $\log_b \dfrac{2xy^2}{z^3}$

32. $\log_a \sqrt[3]{\dfrac{x^2 y^5}{z^4}}$

Express as a single logarithm and, if possible, simplify. [12.4d]

33. $\log_a x - 2 \log_a y + \dfrac{1}{2} \log_a z$

34. $\log_m (b^2 - 16) - \log_m (b + 4)$

Simplify. [12.3c], [12.4e]

35. $\log_8 1$

36. $\log_3 3$

37. $\log_a a^{-3}$

38. $\log_c c^5$

Understanding Through Discussion and Writing

39. The function $V(t) = 750(1.2)^t$ is used to predict the value V of a certain rare stamp t years from 1999. Do not calculate $V^{-1}(t)$ but explain how V^{-1} could be used. [12.2c]

40. Explain in your own words what is meant by $\log_a b = c$. [12.3b]

41. Find a way to express $\log_a (x/5)$ as a difference of logarithms without using the quotient rule. Explain your work. [12.4a, b]

42. A student incorrectly reasons that

$$\log_b \frac{1}{x} = \log_b \frac{x}{x \cdot x}$$
$$= \log_b x - \log_b x + \log_b x$$
$$= \log_b x.$$

What mistake has the student made? Explain what the answer should be. [12.4a, c]

12.5

Natural Logarithmic Functions

OBJECTIVES

a Find logarithms or powers, base e, using a calculator.

b Use the change-of-base formula to find logarithms with bases other than e or 10.

c Graph exponential and logarithmic functions, base e.

SKILL TO REVIEW
Objective 12.3d: Find common logarithms on a calculator.

Find the common logarithm, to four places, on a calculator.

1. $\log \dfrac{8}{3}$ **2.** $\dfrac{\log 8}{\log 3}$

n	$A(n) = \left(1 + \dfrac{1}{n}\right)^n$
1 (compounded annually)	$2.00
2 (compounded semiannually)	$2.25
3	$2.370370
4 (compounded quarterly)	$2.441406
5	$2.488320
100	$2.704814
365 (compounded daily)	$2.714567
8760 (compounded hourly)	$2.718127

Any positive number other than 1 can serve as the base of a logarithmic function. Common, or base-10, logarithms, which were introduced in Section 12.3, are useful because they have the same base as our "commonly" used decimal system of naming numbers.

Today, another base is widely used. It is an irrational number named e. We now consider e and **natural logarithms**, or logarithms base e.

a The Base e and Natural Logarithms

When interest is computed n times per year, the compound-interest formula is

$$A = P\left(1 + \frac{r}{n}\right)^{nt},$$

where A is the amount that an initial investment P will grow to after t years at interest rate r. Suppose that $1 could be invested at 100% interest for 1 year. (In reality, no financial institution would pay such an interest rate.) The preceding formula becomes a function A defined in terms of the number of compounding periods n:

$$A(n) = \left(1 + \frac{1}{n}\right)^n. \quad \text{(See Section 12.1.)}$$

Let's find some function values, using a calculator and rounding to six decimal places. The numbers in the table at left approach a very important number called e. It is an irrational number, so its decimal representation neither terminates nor repeats.

THE NUMBER e

$e \approx 2.7182818284\ldots$

Logarithms, base e, are called **natural logarithms**, or **Naperian logarithms**, in honor of John Napier (1550–1617), a Scotsman who invented logarithms. The abbreviation **ln** is commonly used with natural logarithms. Thus,

$\ln 29$ means $\log_e 29$.   Be sure to memorize $\ln a = \log_e a$.

We usually read "ln 29" as "the natural log of 29," or simply "el en of 29."

On a calculator, the key for natural logarithms is generally marked **LN**. Using that key, we find that

$\ln 29 \approx 3.36729583 \approx 3.3673,$

rounded to four decimal places. This also tells us that $e^{3.3673} \approx 29$.

On some scientific calculators, the keystrokes for doing such a calculation might be

Answers
Skill to Review:
1. 0.4260   **2.** 1.8928

If we were to use a graphing calculator, the keystrokes might be

 .

EXAMPLES Find the natural logarithm, to four decimal places, on a calculator.

Function Value	Readout	Rounded
1. $\ln 287{,}523$	12.56905814	12.5691
2. $\ln 0.000486$	-7.629301934	-7.6293
3. $\ln(-5)$	NONREAL ANS	Does not exist as a real number
4. $\ln(e)$	1	1
5. $\ln 1$	0	0

Do Exercises 1–5.

The inverse of a logarithmic function is an exponential function. Thus, if $f(x) = \ln x$, then $f^{-1}(x) = e^x$. Because of this, on many calculators, the **LN** key doubles as the ⓔˣ key after a **2ND** or SHIFT key has been pressed.

EXAMPLE 6 Find $e^{12.5691}$ using a calculator.

On a scientific calculator, we might enter 12.5691 and press ⓔˣ. On a graphing calculator, we might press **2ND** ⓔˣ, followed by 12.5691 **ENTER**. In either case, we get the approximation

$$e^{12.5691} \approx 287{,}535.0371.$$

Compare this computation to Example 1. Note that, apart from the rounding error, $e^{12.5691}$ takes us back to about 287,523.

EXAMPLE 7 Find $e^{-1.524}$ using a calculator.

On a scientific calculator, we might enter −1.524 and press ⓔˣ. On a graphing calculator, we might press **2ND** ⓔˣ, followed by −1.524 **ENTER**. In either case, we get the approximation

$$e^{-1.524} \approx 0.2178.$$

Do Exercises 6 and 7.

b Changing Logarithm Bases

Most calculators give the values of both common logarithms and natural logarithms. To find a logarithm with some other base, we can use the following conversion formula.

> ### THE CHANGE-OF-BASE FORMULA
>
> For any logarithm bases a and b and any positive number M,
>
> $$\log_b M = \frac{\log_a M}{\log_a b}.$$

Find the natural logarithm, to four decimal places, on a calculator.

1. $\ln 78{,}235.4$

2. $\ln 0.0000309$

3. $\ln(-3)$

4. $\ln 0$

5. $\ln 10$

6. Find $e^{11.2675}$ using a calculator. (Compare this computation to that of Margin Exercise 1.)

7. Find e^{-2} using a calculator.

Answers

1. 11.2675   **2.** −10.3848   **3.** Does not exist as a real number   **4.** Does not exist
5. 2.3026   **6.** 78,237.1596   **7.** 0.1353

The Change-of-Base Formula   To find a logarithm with a base other than 10 or e, we use the change-of-base formula, $\log_b M = \dfrac{\log_a M}{\log_a b}$, where a and b are any logarithm bases and M is any positive number. For example, we can find $\log_5 8$ using common logarithms.

   We let $a = 10$, $b = 5$, and $M = 8$ and substitute in the change-of-base formula. We press (LOG) (8) (▸) (÷) (LOG) (5) (▸) (ENTER). Note that the parentheses must be closed in the numerator to enter the expression correctly. We also close the parentheses in the denominator for completeness. The result is about 1.2920. We could have let $a = e$ and used natural logarithms to find $\log_5 8$ as well.

> log(8)/log(5)
>   1.292029674

A PROOF OF THE CHANGE-OF-BASE FORMULA (*OPTIONAL*):   We let $x = \log_b M$. Then, writing an equivalent exponential equation, we have $b^x = M$. Next, we take the logarithm base a on both sides. This gives us

$$\log_a b^x = \log_a M.$$

By Property 2, the Power Rule,

$$x \log_a b = \log_a M,$$

and solving for x, we obtain

$$x = \frac{\log_a M}{\log_a b}.$$

But $x = \log_b M$, so we have

$$\log_b M = \frac{\log_a M}{\log_a b},$$

which is the change-of-base formula.

EXAMPLE 8   Find $\log_4 7$ using common logarithms.

   We let $a = 10$, $b = 4$, and $M = 7$. Then we substitute into the change-of-base formula:

$$\log_b M = \frac{\log_a M}{\log_a b}$$

$$\log_4 7 = \frac{\log_{10} 7}{\log_{10} 4} \qquad \text{Substituting 10 for } a, \\ 4 \text{ for } b, \text{ and } 7 \text{ for } M$$

$$= \frac{\log 7}{\log 4}$$

$$\approx 1.4037.$$

To check, we use a calculator with a power key $\boxed{y^x}$ or ⌢ to verify that

$$4^{1.4037} \approx 7.$$

   We can also use base e for a conversion.

EXAMPLE 9   Find $\log_4 7$ using natural logarithms.

$$\log_b M = \frac{\log_a M}{\log_a b}$$

$$\log_4 7 = \frac{\log_e 7}{\log_e 4} \qquad \text{Substituting } e \text{ for } a, \\ 4 \text{ for } b, \text{ and } 7 \text{ for } M$$

$$= \frac{\ln 7}{\ln 4}$$

$$\approx 1.4037 \qquad \text{Note that this is the same answer as that for Example 8.}$$

8. a) Find $\log_6 7$ using common logarithms.

   b) Find $\log_6 7$ using natural logarithms.

9. Find $\log_2 46$ using natural logarithms.

EXAMPLE 10   Find $\log_5 29$ using natural logarithms.

   Substituting e for a, 5 for b, and 29 for M, we have

$$\log_5 29 = \frac{\log_e 29}{\log_e 5} \qquad \text{Using the change-of-base formula}$$

$$= \frac{\ln 29}{\ln 5} \approx 2.0922.$$

Answers

8. (a) 1.0860; **(b)** 1.0860   **9.** 5.5236

Do Exercises 8 and 9.

c Graphs of Exponential and Logarithmic Functions, Base e

EXAMPLE 11 Graph $f(x) = e^x$ and $g(x) = e^{-x}$.

We use a calculator with an $\boxed{e^x}$ key to find approximate values of e^x and e^{-x}. Using these values, we can graph the functions.

x	e^x	e^{-x}
0	1	1
1	2.7	0.4
2	7.4	0.1
−1	0.4	2.7
−2	0.1	7.4

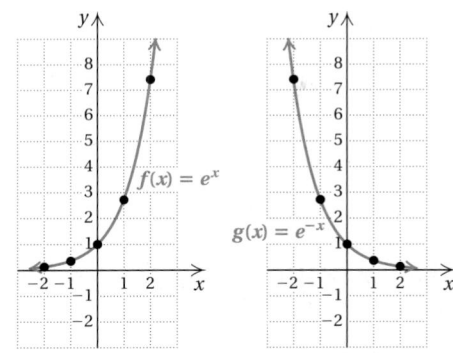

Note that each graph is the image of the other reflected across the y-axis.

EXAMPLE 12 Graph: $f(x) = e^{-0.5x}$.

We find some solutions with a calculator, plot them, and then draw the graph. For example, $f(2) = e^{-0.5(2)} = e^{-1} \approx 0.4$.

x	$e^{-0.5x}$
0	1
1	0.6
2	0.4
3	0.2
−1	1.6
−2	2.7
−3	4.5

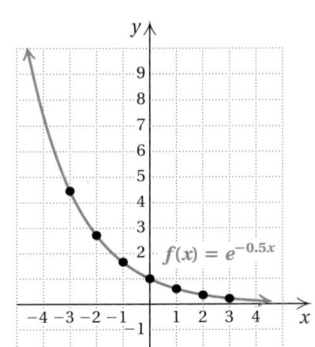

Do Exercises 10 and 11.

EXAMPLE 13 Graph: $g(x) = \ln x$.

We find some solutions with a calculator and then draw the graph. As expected, the graph is a reflection across the line $y = x$ of the graph of $y = e^x$.

x	$\ln x$
1	0
4	1.4
7	1.9
0.5	−0.7

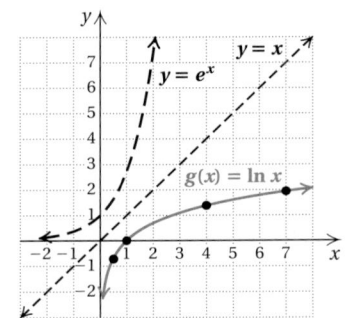

Graph.

10. $f(x) = e^{2x}$

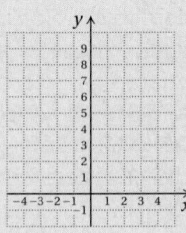

11. $g(x) = \frac{1}{2}e^{-x}$

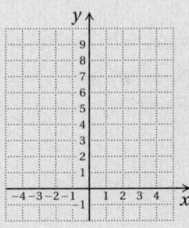

Answers

10. **11.**

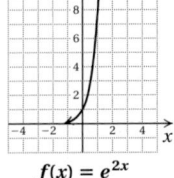

$f(x) = e^{2x}$ $g(x) = \frac{1}{2}e^{-x}$

Graph.

12. $f(x) = 2\ln x$

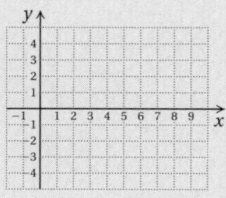

13. $g(x) = \ln(x - 2)$

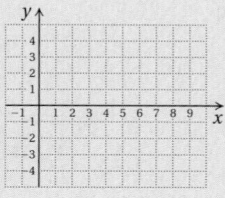

EXAMPLE 14 Graph: $f(x) = \ln(x + 3)$.

We find some solutions with a calculator, plot them, and then draw the graph.

x	$\ln(x + 3)$
0	1.1
1	1.4
2	1.6
3	1.8
4	1.9
−1	0.7
−2	0
−2.5	−0.7

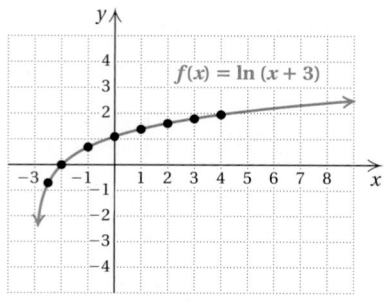

The graph of $y = \ln(x + 3)$ is the graph of $y = \ln x$ translated 3 units to the left.

Do Exercises 12 and 13.

Calculator Corner

Graphing Logarithmic Functions   We can graph logarithmic functions with base 10 or base e by entering the function on the equation-editor screen using the **LOG** or **LN** key. To graph a logarithmic function with a base other than 10 or e, we must first use the change-of-base formula to change the base to 10 or e.

In Example 1 of Section 12.3, we graphed the function $y = \log_5 x$ by finding a table of x- and y-values and plotting points. We will now graph this function on a graphing calculator by first changing the base to e. We let $a = e$, $b = 5$, and $M = x$ and substitute in the change-of-base formula. We enter $y_1 = \dfrac{\ln x}{\ln 5}$ on the equation-editor screen, select a window, and press **GRAPH**. The TI-84+ forces the use of parentheses with the ln function, so the parenthesis in the numerator must be closed: ln (x)/ln (5). The right parenthesis following the 5 is optional but we include it for completeness.

We could have let $a = 10$ and used base-10 logarithms to graph this function as well.

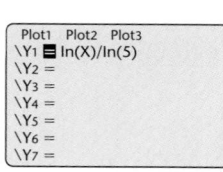

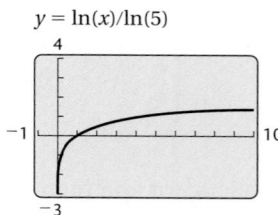

$y = \ln(x)/\ln(5)$

Exercises   Graph each of the following on a graphing calculator.

1. $y = \log_2 x$

2. $y = \log_3 x$

3. $y = \log_{1/2} x$

4. $y = \log_{2/3} x$

Answers

12.

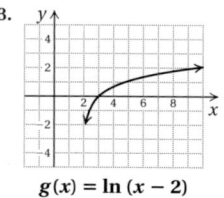

$f(x) = 2\ln x$

13.

$g(x) = \ln(x - 2)$

Visualizing for Success

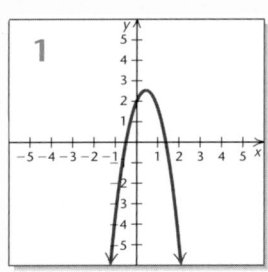

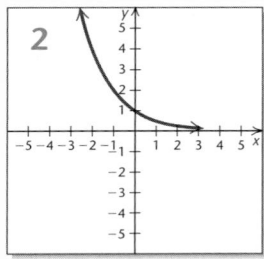

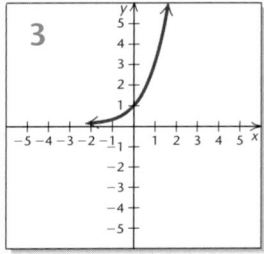

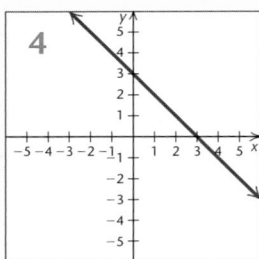

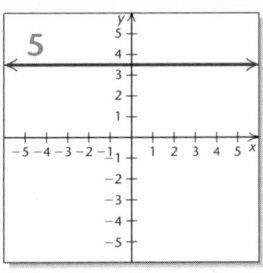

Match each graph with its function.

A. $f(x) = ax^2 + bx + c,$
  $a < 0, c < 0$

B. $f(x) = a^x, 0 < a < 1$

C. $f(x) = a^x, a < 0$

D. $f(x) = \log_a x, 0 < a < 1$

E. $f(x) = \log_a x, a < 0$

F. $f(x) = mx + b, m > 0, b < 0$

G. $f(x) = mx + b, m < 0, b > 0$

H. $f(x) = mx + b, m < 0, b < 0$

I. $f(x) = ax^2 + bx + c, a > 0, c < 0$

J. $f(x) = ax^2 + bx + c, a < 0, c > 0$

K. $f(x) = \log_a x, a > 1$

L. $f(x) = ax^2 + bx + c, a > 0, c > 0$

M. $f(x) = mx + b, m < 0, b = 0$

N. $f(x) = mx + b, m = 0, b > 0$

O. $f(x) = a^x, a > 1$

Answers on page A-40

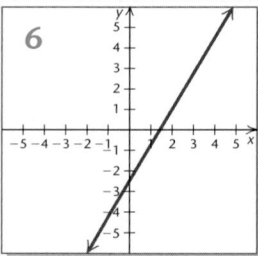

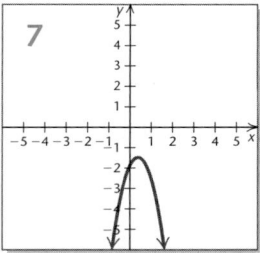

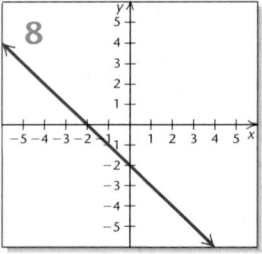

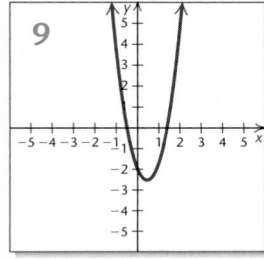

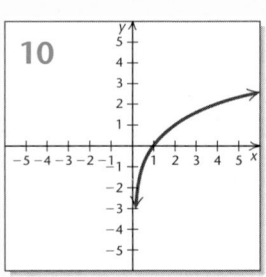

12.5 | Exercise Set

For Extra Help

MyMathLab · Math XL PRACTICE · WATCH · DOWNLOAD · READ · REVIEW

a Find each of the following logarithms or powers, base e, using a calculator. Round answers to four decimal places.

1. $\ln 2$ **2.** $\ln 5$ **3.** $\ln 62$ **4.** $\ln 30$ **5.** $\ln 4365$ **6.** $\ln 901.2$

7. $\ln 0.0062$ **8.** $\ln 0.00073$ **9.** $\ln 0.2$ **10.** $\ln 0.04$ **11.** $\ln 0$ **12.** $\ln(-4)$

13. $\ln\left(\dfrac{97.4}{558}\right)$ **14.** $\ln\left(\dfrac{786.2}{77.2}\right)$ **15.** $\ln e$ **16.** $\ln e^2$ **17.** $e^{2.71}$ **18.** $e^{3.06}$

19. $e^{-3.49}$ **20.** $e^{-2.64}$ **21.** $e^{4.7}$ **22.** $e^{1.23}$ **23.** $\ln e^5$ **24.** $e^{\ln 7}$

b Find each of the following logarithms using the change-of-base formula.

25. $\log_6 100$ **26.** $\log_3 100$ **27.** $\log_2 100$ **28.** $\log_7 100$ **29.** $\log_7 65$ **30.** $\log_5 42$

31. $\log_{0.5} 5$ **32.** $\log_{0.1} 3$ **33.** $\log_2 0.2$ **34.** $\log_2 0.08$ **35.** $\log_\pi 200$ **36.** $\log_\pi \pi$

c Graph.

37. $f(x) = e^x$ **38.** $f(x) = e^{0.5x}$ **39.** $f(x) = e^{-0.5x}$ **40.** $f(x) = e^{-x}$

x	$f(x)$
0	
1	
2	
3	
−1	
−2	
−3	

x	$f(x)$
0	
1	
2	
3	
−1	
−2	
−3	

x	$f(x)$

x	$f(x)$

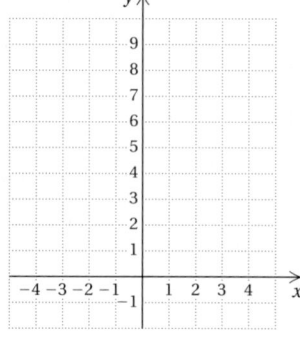

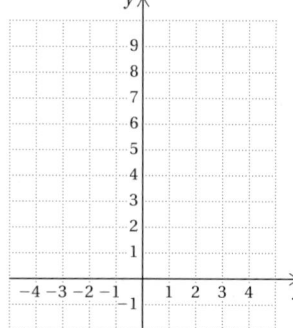

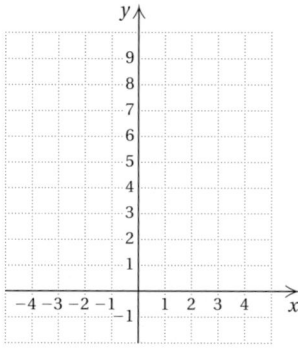

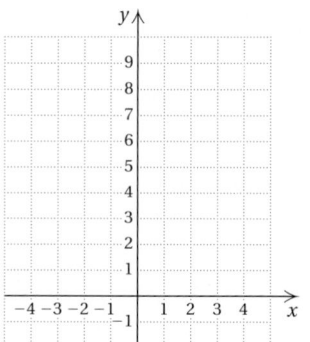

41. $f(x) = e^{x-1}$

x	$f(x)$

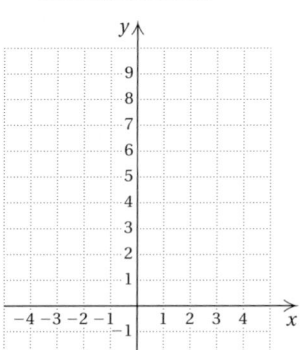

42. $f(x) = e^{-x} + 3$

x	$f(x)$

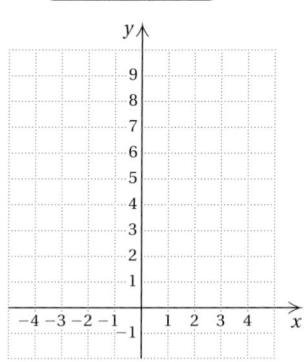

43. $f(x) = e^{x+2}$

x	$f(x)$

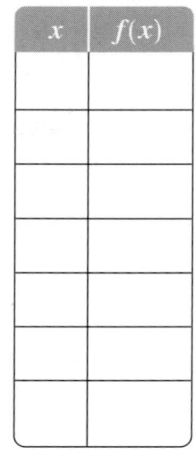

44. $f(x) = e^{x-2}$

x	$f(x)$

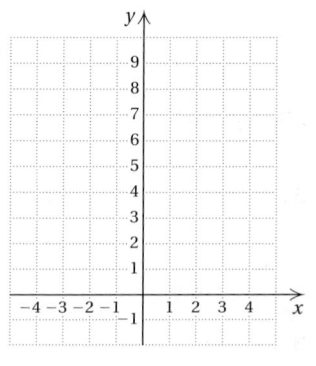

45. $f(x) = e^x - 1$

x	$f(x)$

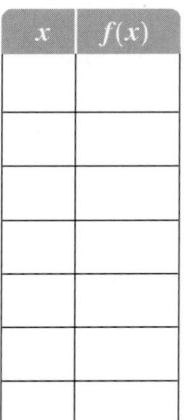

46. $f(x) = 2e^{0.5x}$

x	$f(x)$

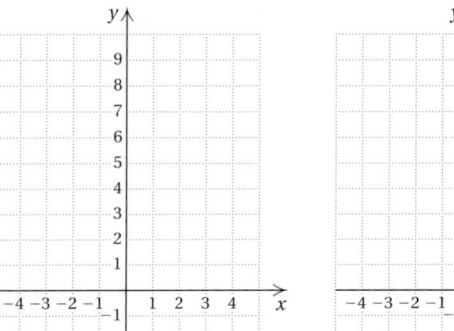
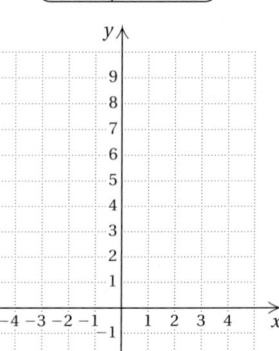

47. $f(x) = \ln(x + 2)$

x	$f(x)$
0	
1	
2	
3	
-0.5	
-1	
-1.5	

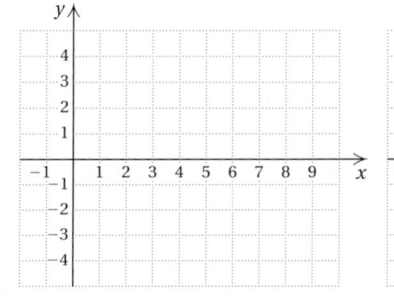

48. $f(x) = \ln(x + 1)$

x	$f(x)$
0	
1	
2	
3	
4	
-0.5	
-0.75	

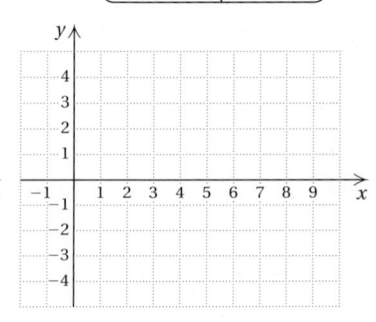

49. $f(x) = \ln(x - 3)$

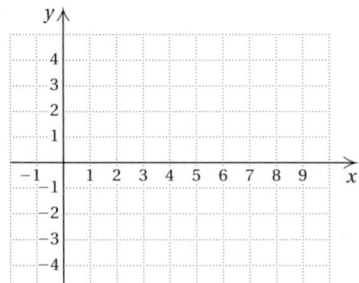

50. $f(x) = 2\ln(x - 2)$

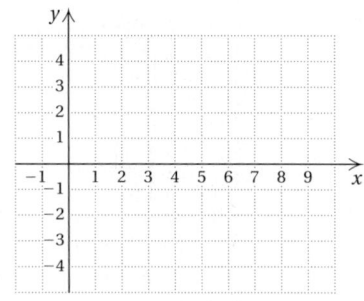

51. $f(x) = 2\ln x$

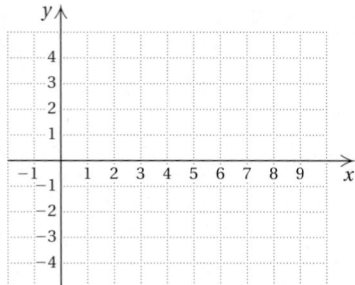

52. $f(x) = \ln x - 3$

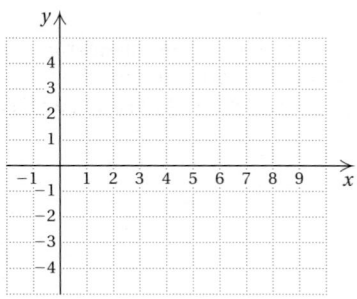

53. $f(x) = \frac{1}{2}\ln x + 1$

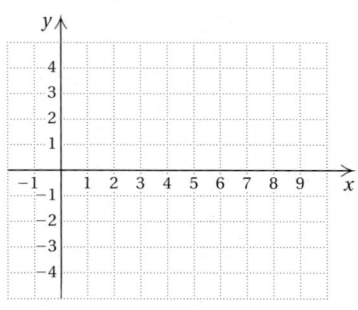

54. $f(x) = \ln x^2$

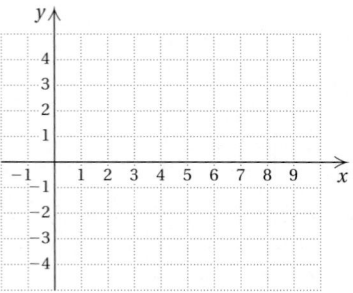

55. $f(x) = |\ln x|$

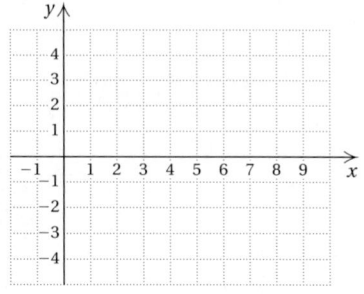

56. $f(x) = \ln|x|$

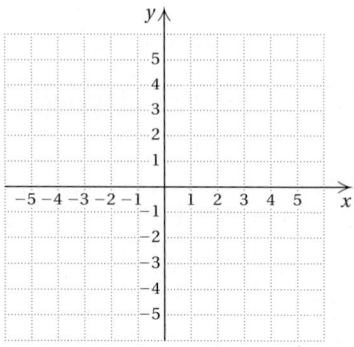

Skill Maintenance

Solve.　[11.4c]

57. $x^{1/2} - 6x^{1/4} + 8 = 0$　　　　**58.** $2y - 7\sqrt{y} + 3 = 0$　　　　**59.** $x - 18\sqrt{x} + 77 = 0$　　　　**60.** $x^4 - 25x^2 + 144 = 0$

Synthesis

Use the graph of the function to find the domain and the range.

61. $f(x) = 10x^2 e^{-x}$　　　　　　**62.** $f(x) = 7.4e^x \ln x$　　　　　　**63.** $f(x) = 100(1 - e^{-0.3x})$

Find the domain.

64. $f(x) = \log_3 x^2$　　　　　　　　　　**65.** $f(x) = \log(2x - 5)$

12.6

Solving Exponential and Logarithmic Equations

a Solving Exponential Equations

Equations with variables in exponents, such as $5^x = 12$ and $2^{7x} = 64$, are called **exponential equations**. Sometimes, as is the case with $2^{7x} = 64$, we can write each side as a power of the *same* number:

$$2^{7x} = 2^6.$$

Since the base is the same, 2, the exponents are the same. We can set them equal and solve:

$$7x = 6$$
$$x = \tfrac{6}{7}.$$

We use the following property, which is true because exponential functions are one-to-one.

THE PRINCIPLE OF EXPONENTIAL EQUALITY

For any $a > 0$, $a \neq 1$,

$$a^x = a^y \longrightarrow x = y.$$

(When powers are equal, the exponents are equal.)

EXAMPLE 1 Solve: $2^{3x-5} = 16$.

Note that $16 = 2^4$. Thus we can write each side as a power of the same number:

$$2^{3x-5} = 2^4.$$

Since the base is the same, 2, the exponents must be the same. Thus,

$$3x - 5 = 4$$
$$3x = 9$$
$$x = 3.$$

Check:
$$2^{3x-5} = 16$$
$$\overline{2^{3\cdot3-5} \; \overset{?}{\mid} \; 16}$$
$$2^{9-5}$$
$$2^4$$
$$16 \mid \qquad \text{TRUE}$$

The solution is 3.

Do Margin Exercises 1 and 2.

SKILL TO REVIEW
Objective 5.8b: Solve quadratic equations by factoring and then using the principle of zero products.

Solve.

1. $y^2 - y - 6 = 0$

2. $x^2 - 3x = 4$

Solve.

1. $3^{2x} = 9$

2. $4^{2x-3} = 64$

Answers

Skill to Review:
1. $-2, 3$ **2.** $-1, 4$

Margin Exercises:
1. 1   **2.** 3

✳ Algebraic-Graphical Connection

The solution, 3, of the equation $2^{3x-5} = 16$ in Example 1 is the x-coordinate of the point of intersection of the graphs of $y = 2^{3x-5}$ and $y = 16$, as we see in the graph on the left below.

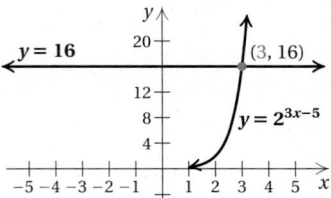
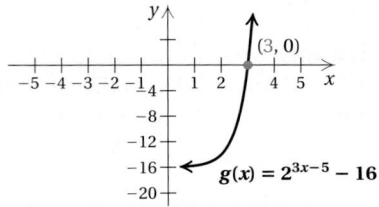

If we subtract 16 on both sides of $2^{3x-5} = 16$, we get $2^{3x-5} - 16 = 0$. The solution, 3, is then the x-coordinate of the x-intercept of the function $g(x) = 2^{3x-5} - 16$, as we see in the graph on the right above.

When it does not seem possible to write both sides of an equation as powers of the same base, we can use the following principle along with the properties developed in Section 12.4.

THE PRINCIPLE OF LOGARITHMIC EQUALITY

For any logarithm base a, and for $x, y > 0$,

$$\log_a x = \log_a y \longrightarrow x = y.$$

(If the logarithms, base a, of two expressions are the same, then the expressions are the same.)

Because calculators can generally find only common or natural logarithms (without resorting to the change-of-base formula), we usually take the common or natural logarithm on both sides of the equation.

The principle of logarithmic equality is useful anytime a variable appears as an exponent.

EXAMPLE 2 Solve: $5^x = 12$.

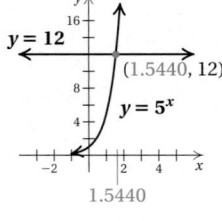

$5^x = 12$

$\log 5^x = \log 12$ Taking the common logarithm on both sides

$x \log 5 = \log 12$ Property 2

$x = \dfrac{\log 12}{\log 5}$ ⟵ --- *Caution!* ---

This is not $\log \frac{12}{5}$!

This is an exact answer. We cannot simplify further, but we can approximate using a calculator:

$$x = \frac{\log 12}{\log 5} \approx 1.5440.$$

We can also partially check this answer by finding $5^{1.5440}$ using a calculator:

$$5^{1.5440} \approx 12.00078587.$$

We get an answer close to 12, due to the rounding. This checks.

Do Exercise 3.

3. Solve: $7^x = 20$.

If the base is e, we can make our work easier by taking the logarithm, base e, on both sides.

EXAMPLE 3   Solve: $e^{0.06t} = 1500$.

We take the natural logarithm on both sides:

$$e^{0.06t} = 1500$$
$$\ln e^{0.06t} = \ln 1500 \qquad \text{Taking ln on both sides}$$
$$\log_e e^{0.06t} = \ln 1500 \qquad \text{Definition of natural logarithms}$$
$$0.06t = \ln 1500 \qquad \text{Here we use Property 4: } \log_a a^k = k.$$
$$t = \frac{\ln 1500}{0.06}.$$

We can approximate using a calculator:

$$t = \frac{\ln 1500}{0.06} \approx \frac{7.3132}{0.06} \approx 121.89.$$

We can also partially check this answer using a calculator.

Check: $$\begin{array}{c} e^{0.06t} = 1500 \\ \hline e^{0.06(121.89)} \; ? \; 1500 \\ e^{7.3134} \quad | \\ 1500.269444 \quad | \qquad \text{TRUE} \end{array}$$

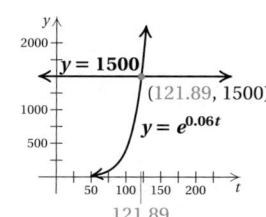

The solution is about 121.89.

Do Exercise 4.

Calculator Corner

Solving Exponential Equations   Use the INTERSECT feature to solve the equations in Examples 1–3. (See the Calculator Corner on p. 572 for the procedure.)

4. Solve: $e^{0.3t} = 80$.

(b) Solving Logarithmic Equations

Equations containing logarithmic expressions are called **logarithmic equations**. We solved some logarithmic equations in Section 12.3 by converting to equivalent exponential equations.

EXAMPLE 4   Solve: $\log_2 x = 3$.

We obtain an equivalent exponential equation:

$$x = 2^3$$
$$x = 8.$$

The solution is 8.

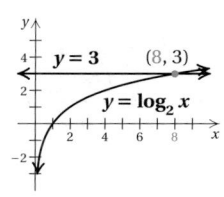

Do Exercise 5.

5. Solve: $\log_5 x = 2$.

Answers

3. 1.5395   **4.** 14.6068   **5.** 25

> To solve a logarithmic equation, first try to obtain a single logarithmic expression on one side and then write an equivalent exponential equation.

EXAMPLE 5 Solve: $\log_4 (8x - 6) = 3$.

We already have a single logarithmic expression, so we write an equivalent exponential equation:

$$8x - 6 = 4^3 \qquad \text{Writing an equivalent exponential equation}$$
$$8x - 6 = 64$$
$$8x = 70$$
$$x = \tfrac{70}{8}, \text{ or } \tfrac{35}{4}.$$

Check:

$$\frac{\log_4 (8x - 6) = 3}{\log_4 \left(8 \cdot \tfrac{35}{4} - 6\right) \; \overset{?}{\mid} \; 3}$$
$$\log_4 (70 - 6)$$
$$\log_4 64$$
$$3 \quad \mid \quad \text{TRUE}$$

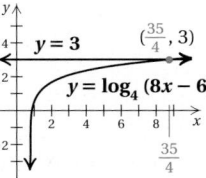

The solution is $\tfrac{35}{4}$.

6. Solve: $\log_3 (5x + 7) = 2$.

Do Exercise 6.

EXAMPLE 6 Solve: $\log x + \log (x - 3) = 1$.

Here we have common logarithms. It helps to first write in the 10's before we obtain a single logarithmic expression on the left.

$$\log_{10} x + \log_{10} (x - 3) = 1$$
$$\log_{10} [x(x - 3)] = 1 \qquad \text{Using Property 1 to obtain a single logarithm}$$
$$x(x - 3) = 10^1 \qquad \text{Writing an equivalent exponential expression}$$
$$x^2 - 3x = 10$$
$$x^2 - 3x - 10 = 0$$
$$(x + 2)(x - 5) = 0 \qquad \text{Factoring}$$
$$x + 2 = 0 \quad or \quad x - 5 = 0 \qquad \text{Using the principle of zero products}$$
$$x = -2 \quad or \qquad x = 5$$

Check: For -2:

$$\frac{\log x + \log (x - 3) = 1}{\log (-2) + \log (-2 - 3) \; \overset{?}{\mid} \; 1}$$

The number -2 does *not* check because negative numbers do not have logarithms.

For 5:

$$\frac{\log x + \log (x - 3) = 1}{\log 5 + \log (5 - 3) \; \overset{?}{\mid} \; 1}$$
$$\log 5 + \log 2$$
$$\log (5 \cdot 2)$$
$$\log 10$$
$$1 \quad \mid \quad \text{TRUE}$$

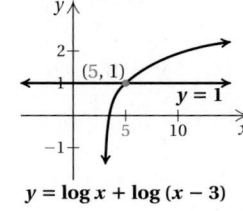

7. Solve: $\log x + \log (x + 3) = 1$.

The solution is 5.

Do Exercise 7.

EXAMPLE 7 Solve: $\log_2(x + 7) - \log_2(x - 7) = 3$.

$\log_2(x + 7) - \log_2(x - 7) = 3$

$$\log_2 \frac{x + 7}{x - 7} = 3 \qquad \text{Using Property 3 to obtain a single logarithm}$$

$$\frac{x + 7}{x - 7} = 2^3 \qquad \text{Writing an equivalent exponential expression}$$

$$\frac{x + 7}{x - 7} = 8$$

$$x + 7 = 8(x - 7) \qquad \text{Multiplying by the LCM, } x - 7$$

$$x + 7 = 8x - 56 \qquad \text{Using a distributive law}$$

$$63 = 7x$$

$$\frac{63}{7} = x$$

$$9 = x$$

Check:

$$\frac{\log_2(x + 7) - \log_2(x - 7) = 3}{\log_2(9 + 7) - \log_2(9 - 7) \;?\; 3}$$

$$\log_2 16 - \log_2 2$$

$$\log_2 \frac{16}{2}$$

$$\log_2 8$$

$$3 \quad | \quad \text{TRUE}$$

The solution is 9.

Do Exercise 8.

8. Solve:
$\log_3(2x - 1) - \log_3(x - 4) = 2.$

STUDY TIPS

BEGINNING TO STUDY FOR THE FINAL EXAM: THREE DAYS TO TWO WEEKS OF STUDY TIME

1. **Begin by browsing through each chapter, reviewing the highlighted or boxed information regarding important formulas in both the text and the Summary and Review.** There may be some formulas that you will need to memorize. Summarize them on an index card and quiz yourself frequently.

2. **Retake each chapter test that you took in class, assuming your instructor has returned it. You can also use the chapter tests in the book.** Restudy the objectives in the text that correspond to each question you missed.

3. **Work the Cumulative Review during the last couple of days before the final.** Skip any questions corresponding to objectives not covered. Again, restudy the objectives in the text that correspond to each question you missed.

4. **For remaining difficulties, see your instructor, go to a tutoring session, or participate in a study group.**

Answer

8. 5

a Solve.

1. $2^x = 8$ **2.** $3^x = 81$ **3.** $4^x = 256$ **4.** $5^x = 125$

5. $2^{2x} = 32$ **6.** $4^{3x} = 64$ **7.** $3^{5x} = 27$ **8.** $5^{7x} = 625$

9. $2^x = 11$ **10.** $2^x = 20$ **11.** $2^x = 43$ **12.** $2^x = 55$

13. $5^{4x-7} = 125$ **14.** $4^{3x+5} = 16$ **15.** $3^{x^2} \cdot 3^{4x} = \dfrac{1}{27}$ **16.** $3^{5x} \cdot 9^{x^2} = 27$

17. $4^x = 8$ **18.** $6^x = 10$ **19.** $e^t = 100$ **20.** $e^t = 1000$

21. $e^{-t} = 0.1$ **22.** $e^{-t} = 0.01$ **23.** $e^{-0.02t} = 0.06$ **24.** $e^{0.07t} = 2$

25. $2^x = 3^{x-1}$ **26.** $3^{x+2} = 5^{x-1}$ **27.** $(3.6)^x = 62$ **28.** $(5.2)^x = 70$

b Solve.

29. $\log_4 x = 4$ **30.** $\log_7 x = 3$ **31.** $\log_2 x = -5$ **32.** $\log_9 x = \dfrac{1}{2}$

33. $\log x = 1$ **34.** $\log x = 3$ **35.** $\log x = -2$ **36.** $\log x = -3$

37. $\ln x = 2$ **38.** $\ln x = 1$ **39.** $\ln x = -1$ **40.** $\ln x = -3$

41. $\log_3 (2x + 1) = 5$ **42.** $\log_2 (8 - 2x) = 6$ **43.** $\log x + \log (x - 9) = 1$

44. $\log x + \log (x + 9) = 1$

45. $\log x - \log (x + 3) = -1$

46. $\log (x + 9) - \log x = 1$

47. $\log_2 (x + 1) + \log_2 (x - 1) = 3$

48. $\log_2 x + \log_2 (x - 2) = 3$

49. $\log_4 (x + 6) - \log_4 x = 2$

50. $\log_4 (x + 3) - \log_4 (x - 5) = 2$

51. $\log_4 (x + 3) + \log_4 (x - 3) = 2$

52. $\log_5 (x + 4) + \log_5 (x - 4) = 2$

53. $\log_3 (2x - 6) - \log_3 (x + 4) = 2$

54. $\log_4 (2 + x) - \log_4 (3 - 5x) = 3$

Skill Maintenance

Solve.   [11.4c]

55. $x^4 + 400 = 104x^2$

56. $x^{2/3} + 2x^{1/3} = 8$

57. $(x^2 + 5x)^2 + 2(x^2 + 5x) = 24$

58. $10 = x^{-2} + 9x^{-1}$

59. Simplify: $(125x^3y^{-2}z^6)^{-2/3}$.   [10.2c]

60. Simplify: i^{79}.   [10.8d]

Synthesis

61. ⊞ Find the value of x for which the natural logarithm is the same as the common logarithm.

62. ⊞ Use a graphing calculator to check your answers to Exercises 4, 20, 36, and 54.

63. ⊞ Use a graphing calculator to solve each of the following equations.
  a) $e^{7x} = 14$ **b)** $8e^{0.5x} = 3$
  c) $xe^{3x-1} = 5$ **d)** $4 \ln (x + 3.4) = 2.5$

64. ⊞ Use the INTERSECT feature of a graphing calculator to find the points of intersection of the graphs of each pair of functions.
  a) $f(x) = e^{0.5x-7}$, $g(x) = 2x + 6$
  b) $f(x) = \ln 3x$, $g(x) = 3x - 8$
  c) $f(x) = \ln x^2$, $g(x) = -x^2$

Solve.

65. $2^{2x} + 128 = 24 \cdot 2^x$

66. $27^x = 81^{2x-3}$

67. $8^x = 16^{3x+9}$

68. $\log_x (\log_3 27) = 3$

69. $\log_6 (\log_2 x) = 0$

70. $x \log \frac{1}{8} = \log 8$

71. $\log_5 \sqrt{x^2 - 9} = 1$

72. $2^{x^2+4x} = \frac{1}{8}$

73. $\log (\log x) = 5$

74. $\log_5 |x| = 4$

75. $\log x^2 = (\log x)^2$

76. $\log_3 |5x - 7| = 2$

77. $\log_a a^{x^2+4x} = 21$

78. $\sqrt{x} \cdot \sqrt[3]{x} \cdot \sqrt[4]{x} \cdot \sqrt[5]{x} = 146$

79. $3^{2x} - 8 \cdot 3^x + 15 = 0$

80. If $x = (\log_{125} 5)^{\log_5 125}$, what is the value of $\log_3 x$?

12.7

Mathematical Modeling with Exponential and Logarithmic Functions

OBJECTIVES

a Solve applied problems involving logarithmic functions.

b Solve applied problems involving exponential functions.

SKILL TO REVIEW
Objective 12.6b: Solve logarithmic equations.

Solve.

1. $\log x = 2$ **2.** $\ln x = -2$

Exponential and logarithmic functions can now be added to our library of functions that can serve as models for many kinds of applications. Let's review some of their graphs.

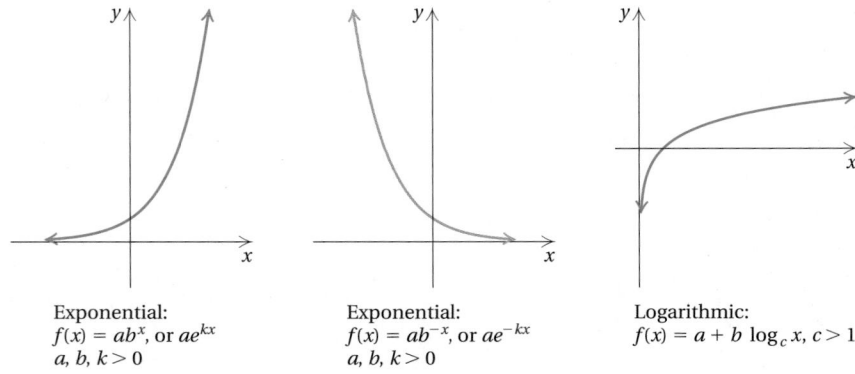

Exponential:
$f(x) = ab^x$, or ae^{kx}
$a, b, k > 0$

Exponential:
$f(x) = ab^{-x}$, or ae^{-kx}
$a, b, k > 0$

Logarithmic:
$f(x) = a + b \log_c x, c > 1$

a Applications of Logarithmic Functions

EXAMPLE 1 *Sound Levels.* To measure the "loudness" of any particular sound, the decibel scale is used. The loudness L, in decibels (dB), of a sound is given by

$$L = 10 \cdot \log \frac{I}{I_0},$$

where I is the intensity of the sound, in watts per square meter (W/m^2), and $I_0 = 10^{-12} \, W/m^2$. (I_0 is approximately the intensity of the softest sound that can be heard.)

a) An iPod can produce sounds of more than $10^{-0.5} \, W/m^2$, a volume that can damage the hearing of a person exposed to the sound for more than 28 sec. How loud, in decibels, is this sound level?

b) Audiologists and physicians recommend that earplugs be worn when one is exposed to sounds in excess of 90 dB. What is the intensity of such sounds?

Source: American Speech–Language-Hearing Association

a) To find the loudness, in decibels, we use the above formula:

$$L = 10 \cdot \log \frac{I}{I_0}$$

$$= 10 \cdot \log \frac{10^{-0.5}}{10^{-12}} \quad \text{Substituting}$$

$$= 10 \cdot \log 10^{11.5} \quad \text{Subtracting exponents}$$

$$= 10 \cdot 11.5 \quad \log 10^a = a$$

$$= 115.$$

The sound level is 115 decibels.

Answers

Skill to Review:
1. 100   **2.** $e^{-2} \approx 0.1353$

b) We substitute and solve for I:

$$L = 10 \cdot \log \frac{I}{I_0}$$

$$90 = 10 \cdot \log \frac{I}{10^{-12}} \qquad \text{Substituting}$$

$$9 = \log \frac{I}{10^{-12}} \qquad \text{Dividing by 10}$$

$$9 = \log I - \log 10^{-12} \qquad \text{Using Property 3}$$

$$9 = \log I - (-12) \qquad \log 10^a = a$$

$$-3 = \log I \qquad \text{Adding } -12$$

$$10^{-3} = I. \qquad \text{Converting to an exponential equation}$$

Earplugs are recommended for sounds with intensities that exceed $10^{-3}\,\text{W/m}^2$.

Do Exercises 1 and 2.

EXAMPLE 2 *Chemistry: pH of Liquids.* In chemistry, the pH of a liquid is defined as

$$\text{pH} = -\log[\text{H}^+],$$

where $[\text{H}^+]$ is the hydrogen ion concentration in moles per liter.

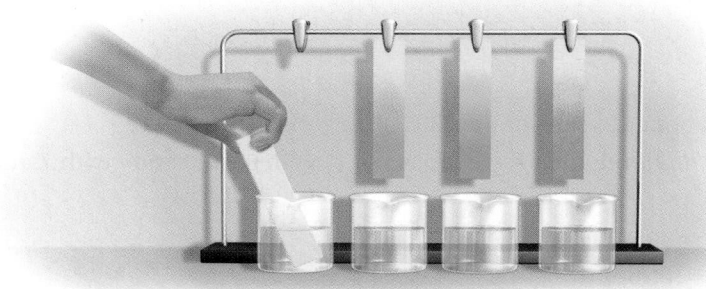

a) The hydrogen ion concentration of human blood is normally about 3.98×10^{-8} moles per liter. Find the pH.

b) The pH of seawater is about 8.3. Find the hydrogen ion concentration.

a) To find the pH of human blood, we use the above formula:

$$\text{pH} = -\log[\text{H}^+] = -\log[3.98 \times 10^{-8}]$$

$$\approx -(-7.400117) \qquad \text{Using a calculator}$$

$$\approx 7.4.$$

The pH of human blood is normally about 7.4.

b) We substitute and solve for $[\text{H}^+]$:

$$8.3 = -\log[\text{H}^+] \qquad \text{Using pH} = -\log[\text{H}^+]$$

$$-8.3 = \log[\text{H}^+] \qquad \text{Dividing by } -1$$

$$10^{-8.3} = [\text{H}^+] \qquad \text{Converting to an exponential equation}$$

$$5.01 \times 10^{-9} \approx [\text{H}^+]. \qquad \text{Using a calculator; writing scientific notation}$$

The hydrogen ion concentration of seawater is about 5.01×10^{-9} moles per liter.

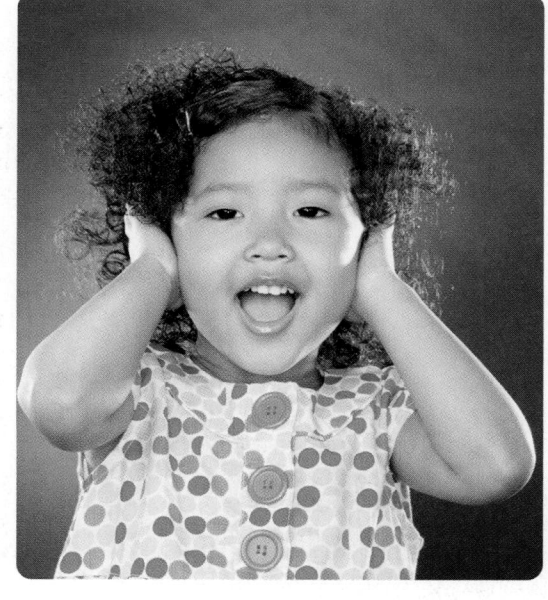

1. **Acoustics.** The intensity of sound in normal conversation is about $3.2 \times 10^{-6}\,\text{W/m}^2$. How high is this sound level in decibels?

2. **Audiology.** Overexposure to excessive sound levels can diminish one's hearing to the point where the softest sound that is audible is 28 dB. What is the intensity of such a sound?

Answers

1. About 65 decibels   2. $10^{-9.2}\,\text{W/m}^2$

3. Coffee. The hydrogen ion concentration of freshly brewed coffee is about 1.3×10^{-5} moles per liter. Find the pH.

4. Acidosis. When the pH of a patient's blood drops below 7.4, a condition called *acidosis* sets in. Acidosis can be fatal at a pH level of 7.0. What would the hydrogen ion concentration of the patient's blood be at that point?

Do Exercises 3 and 4.

b Applications of Exponential Functions

EXAMPLE 3 *Interest Compounded Annually.* Suppose that $30,000 is invested at 4% interest, compounded annually. In t years, it will grow to the amount A given by the function

$$A(t) = 30{,}000(1.04)^t.$$

(See Example 6 in Section 12.1.)

a) How long will it take to accumulate $150,000 in the account?

b) Let $T =$ the amount of time it takes for the $30,000 to double itself; T is called the **doubling time**. Find the doubling time.

a) We set $A(t) = 150{,}000$ and solve for t:

$$150{,}000 = 30{,}000(1.04)^t$$

$$\frac{150{,}000}{30{,}000} = (1.04)^t \qquad \text{Dividing by 30,000}$$

$$5 = (1.04)^t$$

$$\log 5 = \log (1.04)^t \qquad \text{Taking the common logarithm on both sides}$$

$$\log 5 = t \log 1.04 \qquad \text{Using Property 2}$$

$$\frac{\log 5}{\log 1.04} = t \qquad \text{Dividing by log 1.04}$$

$$41.04 \approx t. \qquad \text{Using a calculator}$$

It will take about 41 years for the $30,000 to grow to $150,000.

b) To find the *doubling time T,* we replace $A(t)$ with 60,000 and t with T and solve for T:

$$60{,}000 = 30{,}000(1.04)^T$$

$$2 = (1.04)^T \qquad \text{Dividing by 30,000}$$

$$\log 2 = \log (1.04)^T \qquad \text{Taking the common logarithm on both sides}$$

$$\log 2 = T \log 1.04 \qquad \text{Using Property 2}$$

$$T = \frac{\log 2}{\log 1.04} \approx 17.7. \qquad \text{Using a calculator}$$

The doubling time is about 17.7 years.

Do Exercise 5.

The function in Example 3 illustrates exponential growth. Populations often grow exponentially according to the following model.

5. Interest Compounded Annually. Suppose that $40,000 is invested at 4.3% interest, compounded annually.

a) After what amount of time will there be $250,000 in the account?

b) Find the doubling time.

Answers

3. About 4.9   **4.** 10^{-7} moles per liter
5. (a) 43.5 years; **(b)** 16.5 years

EXPONENTIAL GROWTH MODEL

An **exponential growth model** is a function of the form

$$P(t) = P_0 e^{kt}, \quad k > 0,$$

where P_0 is the population at time 0, $P(t)$ is the population at time t, and k is the **exponential growth rate** for the situation. The **doubling time** is the amount of time necessary for the population to double in size.

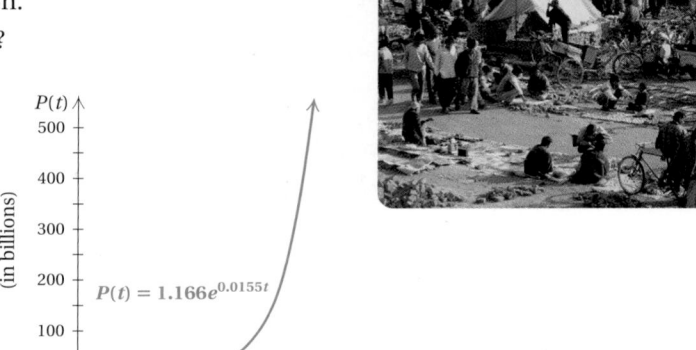

The exponential growth rate is the rate of growth of a population at any *instant* in time. Since the population is continually growing, the percent of total growth after one year will exceed the exponential growth rate.

EXAMPLE 4 *Population Growth in India.*   In 2009, India's population was 1.166 billion, and the exponential growth rate was 1.55% per year.
Source: Central Intelligence Agency

a) Find the exponential growth function.

b) What will the population be in 2015?

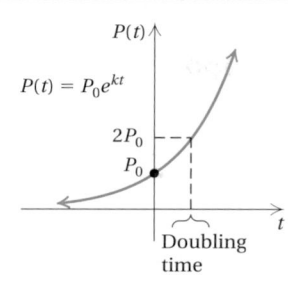

a) We are trying to find a model. The given information allows us to create one. At $t = 0$ (2009), the population was 1.166 billion. We substitute 1.166 for P_0 and 1.55%, or 0.0155, for k to obtain the exponential growth function:

$$P(t) = P_0 e^{kt}$$
$$P(t) = 1.166 e^{0.0155t}.$$

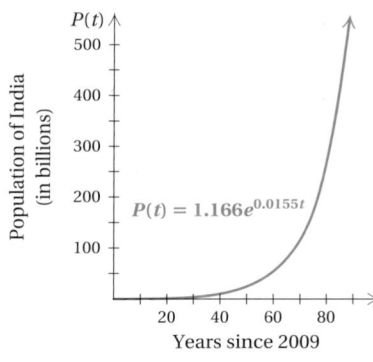

$P(t) = 1.166e^{0.0155t}$

Years since 2009

b) In 2015, we have $t = 6$. That is, 6 yr have passed since 2009. To find the population in 2015, we substitute 6 for t:

$$P(6) = 1.166e^{0.0155(6)} \quad \text{Substituting 6 for } t$$
$$\approx 1.280 \text{ billion.} \quad \text{Using a calculator}$$

The population of India will be about 1.280 billion in 2015.

Do Exercise 6.

6. Population Growth in India. What will the population of India be in 2020? in 2025?

EXAMPLE 5 *Interest Compounded Continuously.*   Suppose that an amount of money P_0 is invested in a savings account at interest rate k, compounded continuously. That is, suppose that interest is computed every "instant" and added to the amount in the account. The balance $P(t)$, after t years, is given by the exponential growth model

$$P(t) = P_0 e^{kt}.$$

a) Suppose that $30,000 is invested and grows to $34,855.03 in 5 years. Find the interest rate and then the exponential growth function.

Answer
6. 1.383 billion; 1.494 billion

b) What is the balance after 10 years?

c) What is the doubling time?

a) We have $P_0 = 30,000$. Thus the exponential growth function is

$$P(t) = 30,000e^{kt},$$

where k must still be determined. We know that $P(5) = 34,855.03$. We substitute and solve for k:

$$34,855.03 = 30,000e^{k(5)} = 30,000e^{5k}$$

$$\frac{34,855.03}{30,000} = e^{5k} \qquad \text{Dividing by 30,000}$$

$$1.161834 \approx e^{5k}$$

$$\ln 1.161834 = \ln e^{5k} \qquad \text{Taking the natural logarithm on both sides}$$

$$0.15 \approx 5k \qquad \text{Finding ln 1.161834 on a calculator and simplifying } \ln e^{5k}$$

$$\frac{0.15}{5} = 0.03 \approx k.$$

The interest rate is about 0.03, or 3%, compounded continuously. Note that since interest is being compounded continuously, the interest earned each year is more than 3%. The exponential growth function is

$$P(t) = 30,000e^{0.03t}.$$

b) We substitute 10 for t:

$$P(10) = 30,000e^{0.03(10)} \approx \$40,495.76.$$

The balance in the account after 10 years will be \$40,495.76.

c) To find the doubling time T, we replace $P(t)$ with 60,000 and solve for T:

$$60,000 = 30,000e^{0.03T}$$

$$2 = e^{0.03T} \qquad \text{Dividing by 30,000}$$

$$\ln 2 = \ln e^{0.03T} \qquad \text{Taking the natural logarithm on both sides}$$

$$\ln 2 = 0.03T$$

$$\frac{\ln 2}{0.03} = T \qquad \text{Dividing}$$

$$23.1 \approx T.$$

Thus the original investment of \$30,000 will double in about 23.1 years, as shown in the following graph of the growth function.

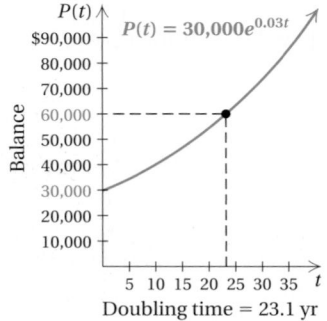

Doubling time = 23.1 yr

Do Exercise 7.

7. Interest Compounded Continuously.

a) Suppose that \$5000 is invested and grows to \$6356.25 in 4 years. Find the interest rate and then the exponential growth function.

b) What is the balance after 1 year? 2 years? 10 years?

c) What is the doubling time?

Answers

7. (a) $k = 6\%$, $P(t) = 5000e^{0.06t}$;
(b) \$5309.18; \$5637.48; \$9110.59;
(c) about 11.6 years

EXAMPLE 6 *Female Olympians.* Twenty-one women competed in the Summer Olympic Games in Paris in 1900, the first year in which women participated in the Games. Female participation grew exponentially through the years, reaching a total of 4746 competitors in Beijing in 2008, as shown in the graph below.

Sources: *The Complete Book of the Olympics,* David Wallechinsky; www.olympic.org/uk

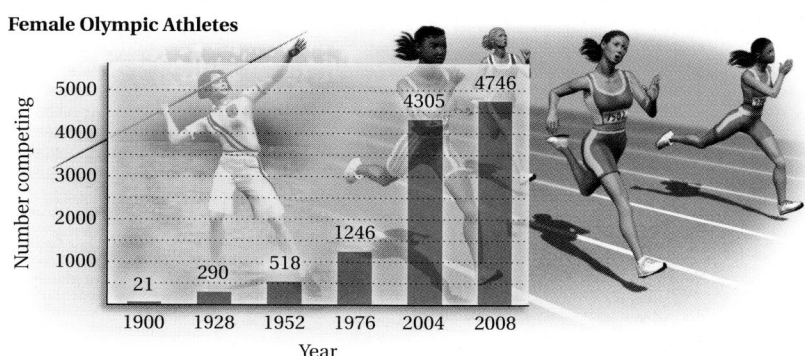

Female Olympic Athletes

a) We let t = the number of years since 1900. Then $t = 0$ corresponds to 1900 and $t = 108$ corresponds to 2008. Use the data points $(0, 21)$ and $(108, 4746)$ to find the exponential growth rate and then the exponential growth function.

b) Use the function found in part (a) to predict the number of female competitors in 2016.

c) Use the function to determine the year in which there were about 2552 female competitors.

a) We use the equation $P(t) = P_0e^{kt}$, where $P(t)$ is the number of women competing in the Summer Olympics t years after 1900. In 1900, at $t = 0$, there were 21 female competitors. Thus we substitute 21 for P_0:

$$P(t) = 21e^{kt}.$$

To find the exponential growth rate k, note that 108 yr later, in 2008, 4746 women competed. We substitute and solve for k:

$$\left. \begin{array}{l} P(108) = 21e^{k(108)} \\ 4746 = 21e^{k(108)} \end{array} \right\} \quad \text{Substituting}$$

$$226 = e^{108k} \qquad \text{Dividing by 21}$$

$$\ln 226 = \ln e^{108k} \qquad \text{Taking the natural logarithm on both sides}$$

$$5.4205 \approx 108k \qquad \ln e^a = a$$

$$0.05 \approx k.$$

The exponential growth rate is 0.05, or 5%, and the exponential growth function is $P(t) = 21e^{0.05t}$.

b) Since 2016 is 116 yr after 1900, we substitute 116 for t:

$$P(116) = 21e^{0.05(116)} \approx 6936.$$

There will be about 6936 female competitors in the Summer Olympic Games in 2016.

8. Recycling Cell Phones. The number of cell phones that are recycled has grown exponentially in recent years. The graph below shows the number of cell phones collected by the largest U.S. recycler.

Cell-Phone Recycling

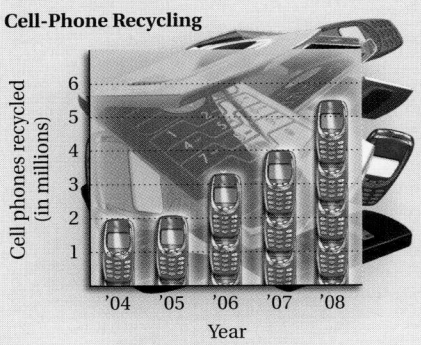

SOURCE: ReCellular.com

a) Let $P(t) = P_0 e^{kt}$, where $P(t)$ is the number of cell phones recycled, in millions, t years after 2004. Then $t = 0$ corresponds to 2004 and $t = 4$ corresponds to 2008. Use the data points $(0, 2)$ and $(4, 5.5)$ to find the exponential growth rate and then the exponential growth function.

b) Use the function found in part (a) to estimate the number of cell phones recycled by the largest recycler in 2010.

c) Assuming exponential growth continues at the same rate, predict the year in which 12 million cell phones will be recycled.

c) To determine when there were about 2552 female competitors in the Summer Olympics, we substitute 2552 for $P(t)$ and solve for t:

$$2552 = 21e^{0.05t}$$
$$121.5238 \approx e^{0.05t} \qquad \text{Dividing by 21}$$
$$\ln 121.5238 = \ln e^{0.05t} \qquad \text{Taking the natural logarithm on both sides}$$
$$4.8001 \approx 0.05t \qquad \ln e^a = a$$
$$96 \approx t.$$

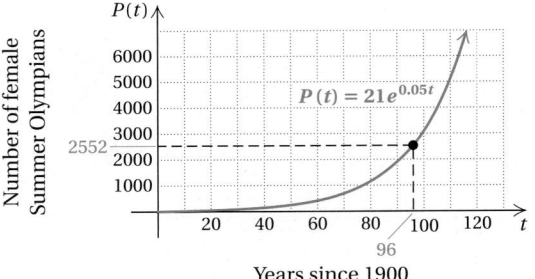

We see that, according to this model, 96 yr after 1900, or in 1996, there were about 2552 female competitors in the Summer Olympics.

Do Exercise 8.

In some real-life situations, a quantity or population is *decreasing* or *decaying* exponentially.

EXPONENTIAL DECAY MODEL

An **exponential decay model** is a function of the form

$$P(t) = P_0 e^{-kt}, \quad k > 0,$$

where P_0 is the quantity present at time 0, $P(t)$ is the amount present at time t, and k is the **decay rate**. The **half-life** is the amount of time necessary for half of the quantity to decay.

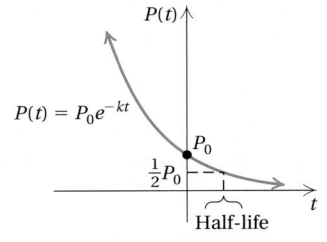

EXAMPLE 7 *Carbon Dating.* The radioactive element carbon-14 has a half-life of 5750 yr. The percentage of carbon-14 present in the remains of organic matter can be used to determine the age of that organic matter. In a cave in Spain, archaeologists have found charcoal samples that have lost 96.5% of their carbon-14. The age of these samples suggests that Neanderthals were in existence about 4000 yr longer than had been previously thought. What is the age of the samples?

Source: *Nature*, September 13, 2006

Answers

8. **(a)** $k \approx 0.253$; $P(t) = 2e^{0.253t}$; **(b)** about 9.1 million cell phones; **(c)** 2011

We first find k. To do so, we use the concept of half-life. When $t = 5750$ (the half-life), $P(t)$ will be half of P_0. Then

$$0.5P_0 = P_0 e^{-k(5750)}$$

$$0.5 = e^{-5750k} \qquad \text{Dividing by } P_0$$

$$\ln 0.5 = \ln e^{-5750k} \qquad \text{Taking the natural logarithm on both sides}$$

$$\ln 0.5 = -5750k$$

$$\frac{\ln 0.5}{-5750} = k$$

$$0.00012 \approx k.$$

Now we have a function for the decay of carbon-14:

$$P(t) = P_0 e^{-0.00012t}. \qquad \text{This completes the first part of our solution.}$$

(*Note*: This equation can be used for any subsequent carbon-dating problem.)

If the charcoal has lost 96.5% of its carbon-14 from an initial amount P_0, then $100\% - 96.5\%$, or 3.5%, of P_0 is still present. To find the age t of the charcoal, we solve the following equation for t:

$$3.5\% P_0 = P_0 e^{-0.00012t} \qquad \text{We want to find } t \text{ for which } P(t) = 0.035P_0.$$

$$0.035 = e^{-0.00012t} \qquad \text{Dividing by } P_0$$

$$\ln 0.035 = \ln e^{-0.00012t} \qquad \text{Taking the natural logarithm on both sides}$$

$$\ln 0.035 = -0.00012t \qquad \ln e^a = a$$

$$\frac{\ln 0.035}{-0.00012} \approx t \qquad \text{Dividing by } -0.00012$$

$$28{,}000 \approx t. \qquad \text{Rounding}$$

The charcoal samples are about 28,000 yr old.

Do Exercise 9.

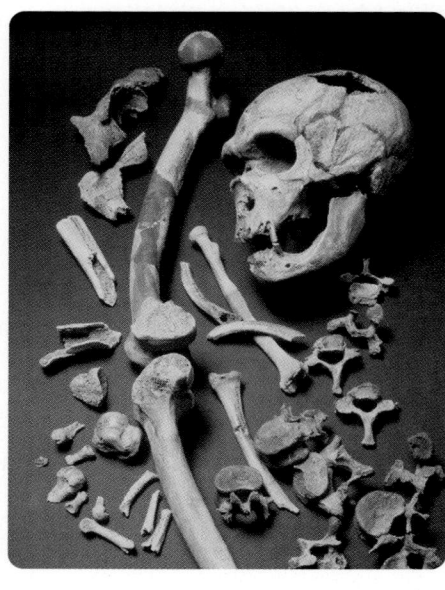

9. Carbon Dating. How old is an animal bone that has lost 30% of its carbon-14?

Answer

9. About 3000 yr

Translating for Success

1. Grain Flow. Grain flows through spout A four times as fast as through spout B. When grain flows through both spouts, a grain bin is filled in 8 hr. How many hours would it take to fill the bin if grain flows through spout B alone?

2. Rectangle Dimensions. The perimeter of a rectangle is 50 ft. The width of the rectangle is 10 ft shorter than the length. Find the length and the width.

3. Wire Cutting. A 1086-in. wire is cut into three pieces. The second piece is 8 in. longer than the first. The third is four-fifths as long as the first. How long is each piece?

4. iPod Sales. Global sales of iPods totaled 0.1 million in 2002 and were growing exponentially at a rate of 126% per year. Write an exponential growth function I for which $I(t)$ approximates the global sales of iPods t years after 2002.

5. Charitable Contributions. In 2010, Jeff donated $500 to charities. This was an 8% increase over his donations in 2008. How much did Jeff donate to charities in 2008?

The goal of these matching questions is to practice step (2), *Translate*, of the five-step problem-solving process. Translate each word problem to an equation or a system of equations and select a correct translation from equations A–O.

A. $I(t) = 0.1e^{1.26t}$

B. $40x = 50(x - 3)$

C. $x^2 + (x - 10)^2 = 50^2$

D. $\dfrac{8}{x} + \dfrac{8}{4x} = 1$

E. $x + 8\%x = 500$

F. $\dfrac{500}{x} + \dfrac{500}{x - 2} = 8$

G. $x + y = 90,$
$0.1x + 0.25y = 16.50$

H. $x + (x + 1) + (x + 2) = 39$

I. $x + (x + 8) + \dfrac{4}{5}x = 1086$

J. $x + (x + 2) + (x + 4) = 39$

K. $I(t) = 1.26e^{0.1t}$

L. $x^2 + (x + 8)^2 = 1086$

M. $2x + 2(x - 10) = 50$

N. $\dfrac{500}{x} = \dfrac{500}{x + 2} + 8$

O. $x + y = 90,$
$0.1x + 0.25y = 1650$

Answers on page A-40

---

6. Uniform Numbers. The numbers on three baseball uniforms are consecutive integers whose sum is 39. Find the integers.

7. Triangle Dimensions. The hypotenuse of a right triangle is 50 ft. The length of one leg is 10 ft shorter than the other. Find the lengths of the legs.

8. Coin Mixture. A collection of dimes and quarters is worth $16.50. There are 90 coins in all. How many of each coin are there?

9. Car Travel. Emma drove her car 500 mi to see her friend. The return trip was 2 hr faster at a speed that was 8 mph more. Find her return speed.

10. Train Travel. An Amtrak train leaves a station and travels east at 40 mph. Three hours later, a second train leaves on a parallel track traveling east at 50 mph. After what amount of time will the second train overtake the first?

For Extra Help

MyMathLab | Math XL PRACTICE | WATCH | DOWNLOAD | READ | REVIEW

a Solve.

Sound Levels. Use the decibel formula from Example 1 for Exercises 1–4.

1. *Sound of Cicadas.* The intensity of sound generated by a large swarm of cicadas can reach 10^{-3} W/m². What is this sound level, in decibels?

2. *Sound of an Alarm Clock.* The intensity of sound of an alarm clock is 10^{-4} W/m². What is this sound level, in decibels?

3. *Dishwasher Noise.* A top-of-the-line dishwasher, built to muffle noise, has a sound measurement of 45 dB. A less-expensive dishwasher can have a sound measurement of 60 dB. What is the intensity of each sound?

4. *Jackhammer Noise.* A jackhammer can generate sound measurements of 130 dB. What is the intensity of such sounds?

pH. Use the pH formula from Example 2 for Exercises 5–8.

5. *Milk.* The hydrogen ion concentration of milk is about 1.6×10^{-7} moles per liter. Find the pH.

6. *Mouthwash.* The hydrogen ion concentration of mouthwash is about 6.3×10^{-7} moles per liter. Find the pH.

7. *Alkalosis.* When the pH of a person's blood rises above 7.4, a condition called *alkalosis* sets in. Alkalosis can be fatal at a pH level above 7.8. What would the hydrogen ion concentration of the person's blood be at that point?

8. *Orange Juice.* The pH of orange juice is 3.2. What is its hydrogen ion concentration?

Walking Speed. In a study by psychologists Bornstein and Bornstein, it was found that the average walking speed w, in feet per second, of a person living in a city of population P, in thousands, is given by the function

$$w(P) = 0.37 \ln P + 0.05.$$

In Exercises 9–12, various cities and their populations are given. Find the walking speed of people in each city.

Source: *International Journal of Psychology*

9. Albuquerque, New Mexico: 518,271

10. Nashville, Tennessee: 590,867

11. Chicago, Illinois: 2,836,654

12. Philadelphia, Pennsylvania: 1,449,834

b Solve.

13. *Organic Food Sales.* A growing number of consumers are buying organic foods. Sales of organic food and beverages S in the United States, in billions of dollars, are approximated by the exponential function

$$S(t) = 13.9(1.19)^t,$$

where t is the number of years since 2005.

Source: Organic Trade Association

a) What were the sales of organic food and beverages in 2009?
b) In what year will sales total $56 billion?
c) What is the doubling time for sales of organic food and beverages?

14. *Spread of Rumor.* The number of people who hear a rumor increases exponentially. If 20 people start a rumor and if each person who hears the rumor repeats it to two people per day, the number of people N who have heard the rumor after t days is given by the function

$$N(t) = 20(3)^t.$$

a) How many people have heard the rumor after 5 days?
b) After what amount of time will 1000 people have heard the rumor?
c) What is the doubling time for the number of people who have heard the rumor?

15. *Internet Usage.* Internet users could soon experience a slowdown in speed as the use of interactive- and video-intensive services overwhelms the capacities of local cable, telephone, and wireless providers. Internet usage can be approximated by the exponential function

$$U(t) = 2469(2.47)^t,$$

where U is in petabytes per month and t is the number of years since 2006. [One petabyte (PB) $= 10^{15}$ bytes.]

Source: Nemertes Research

a) Find Internet usage in 2012.
b) In what year was Internet usage 91,900 PB per month?
c) What is the doubling time of Internet usage?

16. *Salvage Value.* A color photocopier is purchased for $4800. Its value each year is about 70% of its value in the preceding year. Its salvage value, in dollars, after t years is given by the exponential function

$$V(t) = 4800(0.7)^t.$$

a) Find the salvage value of the copier after 3 yr.
b) After what amount of time will the salvage value be $1200?
c) After what amount of time will the salvage value be half the original value?

Growth. Use the exponential growth model $P(t) = P_0 e^{kt}$ for Exercises 17–22.

17. *Interest Compounded Continuously.* Suppose that P_0 is invested in a savings account in which interest is compounded continuously at 3% per year.

a) Express $P(t)$ in terms of P_0 and 0.03.
b) Suppose that $5000 is invested. What is the balance after 1 year? 2 years? 10 years?
c) When will the investment of $5000 double itself?

18. *Interest Compounded Continuously.* Suppose that P_0 is invested in a savings account in which interest is compounded continuously at 5.4% per year.

a) Express $P(t)$ in terms of P_0 and 0.054.
b) Suppose that $10,000 is invested. What is the balance after 1 year? 2 years? 10 years?
c) When will the investment of $10,000 double itself?

19. *World Population Growth.* In 2009, the population of the world reached 6.8 billion, and the exponential growth rate was 1.188% per year.

Sources: U.S. Census Bureau; Central Intelligence Agency

a) Find the exponential growth function.
b) What will the world population be in 2014?
c) In what year will the world population reach 15 billion?
d) What is the doubling time of the world population?

20. *Population Growth of the United States.* In 2009, the population of the United States was 307 million, and the exponential growth rate was 0.975% per year.

Source: U.S Census Bureau

a) Find the exponential growth function.
b) What will the U.S. population be in 2015?
c) In what year will the U.S. population reach 335 million?
d) What is the doubling time of the U.S. population?

21. *Tax Preparation Cost.* As the U.S. tax code becomes increasingly complex, individuals and businesses are spending more each year on tax preparation. In 1990, $80 billion was spent on tax preparation. This cost was estimated to grow exponentially to $368 billion in 2010.

Source: Tax Foundation

a) Let $t = 0$ correspond to 1990 and $t = 20$ correspond to 2010. Then t is the number of years since 1990. Use the data points $(0, 80)$ and $(20, 368)$ to find the exponential growth rate and fit an exponential growth function $C(t) = C_0 e^{kt}$ to the data, where $C(t)$ is the amount spent on tax preparation t years after 1990.
b) Use the function found in part (a) to estimate the total cost of tax preparation in 2012.
c) When will the total cost reach $500 billion?

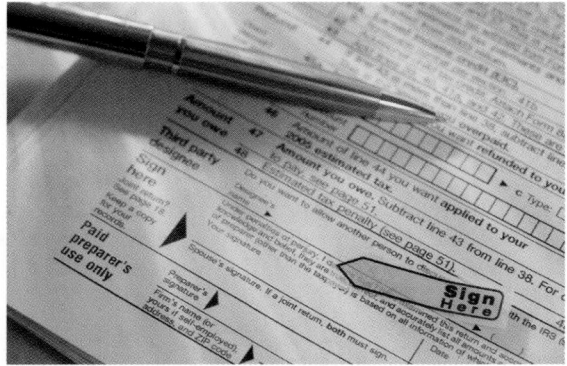

22. *First-Class Postage.* First-class postage (for the first ounce) was 34¢ in 2000 and 44¢ in 2009. Assume the cost increases according to an exponential growth function.

Source: U.S. Postal Service

a) Let $t = 0$ correspond to 2000 and $t = 9$ correspond to 2009. Then t is the number of years since 2000. Use the data points $(0, 34)$ and $(9, 44)$ to find the exponential growth rate and fit an exponential growth function $P(t) = P_0 e^{kt}$ to the data, where $P(t)$ is the cost of first-class postage, in cents, t years after 2000.
b) Use the function found in part (a) to predict the cost of first-class postage in 2016.
c) When will the cost of first-class postage be $1.00, or 100¢?

Carbon Dating. Use the carbon-14 decay function $P(t) = P_0 e^{-0.00012t}$ for Exercises 23 and 24.

23. *Carbon Dating.* When archaeologists found the Dead Sea scrolls, they determined that the linen wrapping had lost 22.3% of its carbon-14. How old was the linen wrapping?

24. *Carbon Dating.* In 1998, researchers found an ivory tusk that had lost 18% of its carbon-14. How old was the tusk?

Decay. Use the exponential decay function $P(t) = P_0 e^{-kt}$ for Exercises 25 and 26.

25. *Chemistry.* The decay rate of iodine-131 is 9.6% per day. What is the half-life?

26. *Chemistry.* The decay rate of krypton-85 is 6.3% per day. What is the half-life?

27. *Home Construction.* The chemical urea formaldehyde was found in some insulation used in houses built during the mid- to late 1960s. Unknown at the time was the fact that urea formaldehyde emitted toxic fumes as it decayed. The half-life of urea formaldehyde is 1 yr. What is its decay rate?

28. *Plumbing.* Lead pipes and solder are often found in older buildings. Unfortunately, as lead decays, toxic chemicals can get in the water resting in the pipes. The half-life of lead is 22 yr. What is its decay rate?

29. *Decline in Home Milk Deliveries.* The number of home milk deliveries has declined considerably over the years. In 1963, home deliveries accounted for 29.7% of milk distribution, but by 2005, this figure had dropped to 0.4%. Assume this percent is decreasing according to the exponential decay model.
Source: U.S. Department of Agriculture

a) Find the exponential decay rate, and write an exponential function D that represents the percent of milk distribution that consists of home deliveries t years after 1963, where $D_0 = 29.7$.

b) Estimate the percent of milk distribution accounted for by home deliveries in 1990.

c) In what year did home deliveries account for 1% of milk distribution?

30. *Covered Bridges.* There were as many as 15,000 covered bridges in the United States in the 1800s. Now their number is decreasing exponentially, partly as a result of vandalism. In 1965, there were 1156 covered bridges, but by 2007, only 750 covered bridges remained.
Source: National Society for the Preservation of Covered Bridges

a) Find the exponential decay rate, and write an exponential function B that represents the number of covered bridges t years after 1965.

b) Estimate the number of covered bridges in 2002.

c) In what year were there 900 covered bridges?

31. *Population Decline of Pittsburgh.* The population of the metropolitan Pittsburgh area declined from 2.431 million in 2000 to 2.356 million in 2007. Assume the population decreases according to the exponential decay model.
Source: U.S. Census Bureau

a) Find the exponential decay rate, and write an exponential function that represents the population of Pittsburgh t years after 2000.

b) Estimate the population of Pittsburgh in 2020.

c) In what year will the population of Pittsburgh reach 2.182 million?

32. *Solar Power.* Solar energy capacity is increasing exponentially worldwide. In 2005, 1460 megawatts (MW) of capacity had been installed. This capacity increased to 5948 MW in 2008.
Source: Solarbuzz Inc.

a) Find the exponential growth rate, and write a function that represents solar energy capacity t years after 2005.

b) Estimate the world's solar energy capacity in 2012.

c) In what year will solar energy capacity reach 100,000 MW?

33. *Value of a Sports Card.* Because he objected to smoking, and because his first baseball card was issued in cigarette packs, the great shortstop Honus Wagner halted production of his card before many were produced. One of these cards was sold in 1996 for $640,500 and again in 2007 for $2,800,000. Assume that the card's value increases exponentially.

a) Find the exponential growth rate, and write a function V that represents the value of the card t years after 1996.

b) Estimate the card's value in 2000.

c) What is the doubling time of the value of the card?

d) In what year did the value of the card first exceed $3,250,000?

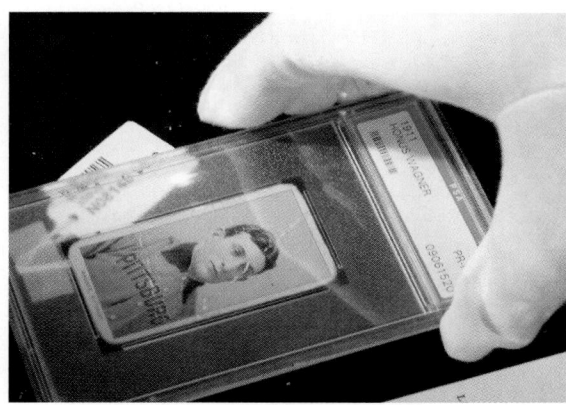

34. *Art Masterpieces.* The most ever paid for a painting is $104,168,000, paid in 2004 for Pablo Picasso's "Garçon à la Pipe." The same painting sold for $30,000 in 1950.
Source: BBC News, 5/6/04

a) Find the exponential growth rate, and write an exponential growth function V that represents the painting's value, in millions of dollars, t years after 1950.

b) Estimate the value of the painting in 2009.

c) What is the doubling time for the value of the painting?

d) How long after 1950 will the value of the painting be $1 billion?

Skill Maintenance

Compute and simplify. Express answers in the form $a + bi$, where $i^2 = -1$. [10.8c, d, e]

35. i^{46} **36.** i^{48} **37.** i^{53} **38.** i^{97} **39.** $i^{14} + i^{15}$

40. $i^{18} - i^{16}$ **41.** $\dfrac{8 - i}{8 + i}$ **42.** $\dfrac{2 + 3i}{5 - 4i}$ **43.** $(5 - 4i)(5 + 4i)$ **44.** $(-10 - 3i)^2$

Synthesis

Use a graphing calculator to solve each of the following equations.

45. $2^x = x^{10}$ **46.** $(\ln 2)x = 10 \ln x$ **47.** $x^2 = 2^x$ **48.** $x^3 = e^x$

49. *Sports Salaries.* In 2001, Derek Jeter of the New York Yankees signed a $189 million 10-yr contract that paid him $21 million in 2010. How much would the Yankee organization need to invest in 2001 at 5% interest, compounded continuously, in order to have the $21 million for Jeter in 2010?

50. *Nuclear Energy.* Plutonium-239 (Pu-239) is used in nuclear energy plants. The half-life of Pu-239 is 24,360 yr. How long will it take for a fuel rod of Pu-239 to lose 90% of its radioactivity?
Source: *Microsoft Encarta 97 Encyclopedia*

Key Terms and Properties

exponential function, p. 875
compound interest, p. 880
inverse relation, p. 888
one-to-one function, p. 890
composite function, p. 897
logarithmic function, p. 906

common logarithms, p. 910
natural logarithms, p. 924
exponential equation, p. 933
logarithmic equation, p. 935
doubling time, p. 942
exponential growth model, p. 943

exponential growth rate, p. 943
exponential decay model, p. 946
decay rate, p. 946
half-life, p. 946

Exponential Functions: $f(x) = a^x$, $f(x) = e^x$

Composition of Functions: $(f \circ g)(x) = f(g(x))$

Definition of Logarithms: $\log_a x$ is that number y such that $x = a^y$,
where $x > 0$ and a is a positive constant other than 1.

Properties of Logarithms:

$\log M = \log_{10} M,$ $\qquad\qquad \log_a 1 = 0,$ $\qquad\qquad \ln M = \log_e M,$ $\qquad \log_a a = 1,$

$\log_a MN = \log_a M + \log_a N,$ $\quad \log_a a^k = k,$ $\qquad\qquad \log_a M^k = k \cdot \log_a M,$ $\quad \log_b M = \dfrac{\log_a M}{\log_a b},$

$\log_a \dfrac{M}{N} = \log_a M - \log_a N,$ $\qquad e \approx 2.7182818284\ldots$

Growth: $P(t) = P_0 e^{kt}$

Decay: $P(t) = P_0 e^{-kt}$

Carbon Dating: $P(t) = P_0 e^{-0.00012t}$

Interest Compounded Annually: $\qquad\qquad A = P(1 + r)^t$

Interest Compounded n Times per Year: $A = P\left(1 + \dfrac{r}{n}\right)^{nt}$

Interest Compounded Continuously: $\qquad P(t) = P_0 e^{kt},$ where P_0 dollars are invested for t years at interest rate k

Concept Reinforcement

Determine whether each statement is true or false.

_____ **1.** The y-intercept of a function $f(x) = a^x$ is $(0, 1)$.   [12.1a]

_____ **2.** If it is possible for a horizontal line to intersect the graph of a function more than once, its inverse is a function.   [12.2b]

_____ **3.** A function and its inverse have the same domain.   [12.2b]

_____ **4.** A logarithm of a number is an exponent.   [12.3a]

_____ **5.** $\log_a 1 = 0,$ $a > 0$ [12.3c]

_____ **6.** If we find that $\log (78) \approx 1.8921$ on a calculator, we also know that $10^{1.8921} \approx 78$.   [12.3d]

_____ **7.** $\ln (35) = \ln 7 \cdot \ln 5$ [12.4a]

_____ **8.** The functions $f(x) = e^x$ and $g(x) = \ln x$ are inverses of each other.   [12.5a]

Important Concepts

Objective 12.1a Graph exponential equations and functions.

Example Graph: $f(x) = 4^x$.

We compute some function values and list the results in a table:

$f(-2) = 4^{-2} = \dfrac{1}{4^2} = \dfrac{1}{16}$;

$f(-1) = 4^{-1} = \dfrac{1}{4}$;

$f(0) = 4^0 = 1$;

$f(1) = 4^1 = 4$;

$f(2) = 4^2 = 16$.

x	$f(x)$
-2	$\dfrac{1}{16}$
-1	$\dfrac{1}{4}$
0	1
1	4
2	16

Now we plot the points $(x, f(x))$ and connect them with a smooth curve.

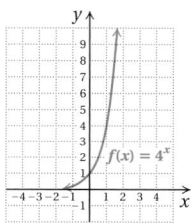

Practice Exercise

1. Graph: $f(x) = 2^x$.

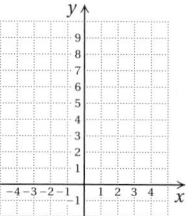

Objective 12.2b Given a function, determine whether it is one-to-one and has an inverse that is a function.

Example Determine whether the function $f(x) = x + 5$ is one-to-one and thus has an inverse that is also a function.

The graph of $f(x) = x + 5$ is shown below. No horizontal line crosses the graph more than once, so the function is one-to-one and has an inverse that is a function.

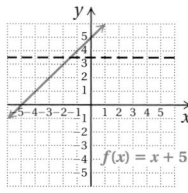

If there is a horizontal line that crosses the graph of a function more than once, the function is not one-to-one and does not have an inverse that is a function.

Practice Exercise

2. Determine whether the function $f(x) = 3^x$ is one-to-one.

Objective 12.2c Find a formula for the inverse of a function, if it exists, and graph inverse relations and functions.

Example Determine whether the function $f(x) = 3x - 1$ is one-to-one. If it is, find a formula for its inverse.

The graph of $f(x) = 3x - 1$ passes the horizontal-line test, so it is one-to-one. Now we find a formula for $f^{-1}(x)$.

1. Replace $f(x)$ with y: $y = 3x - 1$.

2. Interchange x and y: $x = 3y - 1$.

3. Solve for y: $x + 1 = 3y$

$$\frac{x + 1}{3} = y.$$

4. Replace y with $f^{-1}(x)$: $f^{-1}(x) = \dfrac{x + 1}{3}$.

Example Graph the one-to-one function $g(x) = x - 3$ and its inverse using the same set of axes.

We graph $g(x) = x - 3$ and then draw its reflection across the line $y = x$.

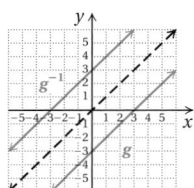

Practice Exercises

3. Determine whether the function $g(x) = 4 - x$ is one-to-one. If it is, find a formula for its inverse.

4. Graph the one-to-one function $f(x) = 2x + 1$ and its inverse using the same set of axes.

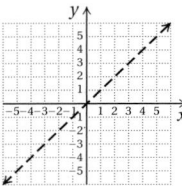

Objective 12.2d Find the composition of functions and express certain functions as a composition of functions.

Example Given $f(x) = x - 2$ and $g(x) = x^2$, find $(f \circ g)(x)$ and $(g \circ f)(x)$.

$(f \circ g)(x) = f(g(x))$
$\qquad = f(x^2) = x^2 - 2;$
$(g \circ f)(x) = g(f(x))$
$\qquad = g(x - 2) = (x - 2)^2$
$\qquad = x^2 - 4x + 4$

Example Find $f(x)$ and $g(x)$ such that $h(x) = (f \circ g)(x)$:
$$h(x) = \sqrt[3]{x} - 5.$$

Two functions that can be used are $f(x) = \sqrt[3]{x}$ and $g(x) = x - 5$. There are other correct answers.

Practice Exercises

5. Given $f(x) = 2x$ and $g(x) = 4x + 1$, find $(f \circ g)(x)$ and $(g \circ f)(x)$.

6. Find $f(x)$ and $g(x)$ such that $h(x) = (f \circ g)(x)$:
$$h(x) = \frac{1}{3x + 2}.$$

Objective 12.3a Graph logarithmic functions.

Example Graph: $y = f(x) = \log_4 x$.

The equation $y = \log_4 x$ is equivalent to $4^y = x$.

For $y = -2$, $x = 4^{-2} = \dfrac{1}{4^2} = \dfrac{1}{16}$.

For $y = -1$, $x = 4^{-1} = \dfrac{1}{4}$.

For $y = 0$, $x = 4^0 = 1$.

For $y = 1$, $x = 4^1 = 4$.

For $y = 2$, $x = 4^2 = 16$.

x	y
$\dfrac{1}{16}$	-2
$\dfrac{1}{4}$	-1
1	0
4	1
16	2

Now we plot these points and connect them with a smooth curve.

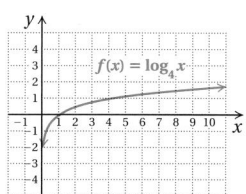

Practice Exercise

7. Graph: $y = \log_5 x$.

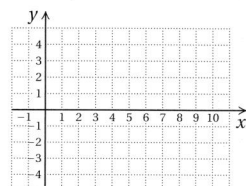

Objective 12.4d Convert from logarithms of products, quotients, and powers to expressions in terms of individual logarithms, and conversely.

Example Express

$$\log_a \frac{x^2 y}{z^3}$$

in terms of logarithms of x, y, and z.

$$\log_a \frac{x^2 y}{z^3} = \log_a (x^2 y) - \log_a z^3$$
$$= \log_a x^2 + \log_a y - \log_a z^3$$
$$= 2 \log_a x + \log_a y - 3 \log_a z$$

Example Express

$$4 \log_a x - \frac{1}{2} \log_a y$$

as a single logarithm.

$$4 \log_a x - \frac{1}{2} \log_a y = \log_a x^4 - \log_a y^{1/2}$$
$$= \log_a \frac{x^4}{y^{1/2}}, \text{ or } \log_a \frac{x^4}{\sqrt{y}}$$

Practice Exercises

8. Express $\log_a \sqrt[5]{\dfrac{x^3}{y^2}}$ in terms of logarithms of x and y.

9. Express $\dfrac{1}{2} \log_a x - 3 \log_a y$ as a single logarithm.

Objective 12.5c Graph exponential and logarithmic functions, base e.

Example Graph: $f(x) = e^{x-1}$.

x	$f(x)$
-1	0.1
0	0.4
1	1
2	2.7

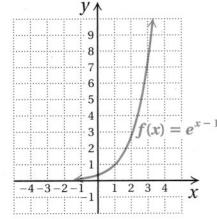

Example Graph: $g(x) = \ln x + 3$.

x	$g(x)$
0.5	2.3
1	3
3	4.1
5	4.6
8	5.1
10	5.3

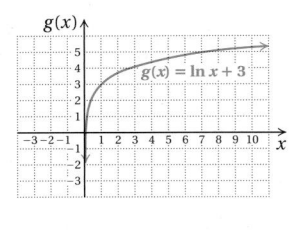

Practice Exercises

10. Graph: $f(x) = e^x - 1$.

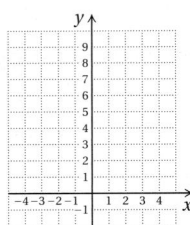

11. Graph: $f(x) = \ln (x + 3)$.

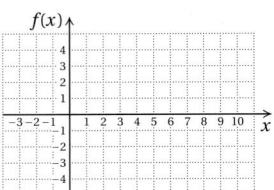

Objective 12.6a Solve exponential equations.

Example Solve: $3^{x-1} = 81$.

$$3^{x-1} = 81$$
$$3^{x-1} = 3^4$$

Since the bases are the same, the exponents must be the same:

$$x - 1 = 4$$
$$x = 5.$$

The solution is 5.

Practice Exercise

12. Solve: $2^{3x} = 16$.

Objective 12.6b Solve logarithmic equations.

Example Solve: $\log x + \log (x + 3) = 1$.

$$\log x + \log (x + 3) = 1$$
$$\log_{10} [x(x + 3)] = 1$$
$$x(x + 3) = 10^1$$
$$x^2 + 3x = 10$$
$$x^2 + 3x - 10 = 0$$
$$(x + 5)(x - 2) = 0$$
$$x + 5 = 0 \quad or \quad x - 2 = 0$$
$$x = -5 \quad or \quad x = 2$$

The number -5 does not check, but 2 does. The solution is 2.

Practice Exercise

13. Solve: $\log_3 (2x + 3) = 2$.

Review Exercises

1. Find the inverse of the relation
$\{(-4, 2), (5, -7), (-1, -2), (10, 11)\}$.   [12.2a]

Determine whether each function is one-to-one. If it is, find a formula for its inverse.   [12.2b, c]

2. $f(x) = 4 - x^2$

3. $g(x) = \dfrac{2x - 3}{7}$

4. $f(x) = 8x^3$

5. $f(x) = \dfrac{4}{3 - 2x}$

6. Graph the function $f(x) = x^3 + 1$ and its inverse using the same set of axes.   [12.2c]

Graph.

7. $f(x) = 3^{x-1}$ [12.1a]

x	$f(x)$
0	
1	
2	
3	
-1	
-2	
-3	

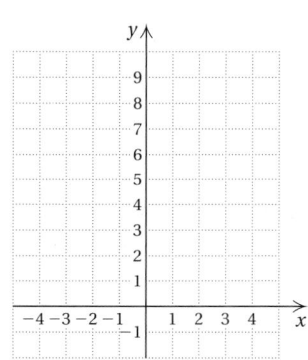

8. $f(x) = \log_3 x$, or $y = \log_3 x$ [12.3a]
$y = \log_3 x \rightarrow x = $ _____

x, or 3^y	y
	0
	1
	2
	3
	-1
	-2
	-3

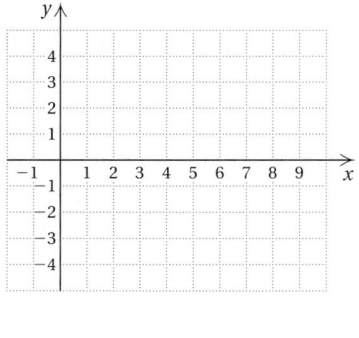

9. $f(x) = e^{x+1}$ [12.5c]

x	$f(x)$
0	
1	
2	
3	
-1	
-2	
-3	

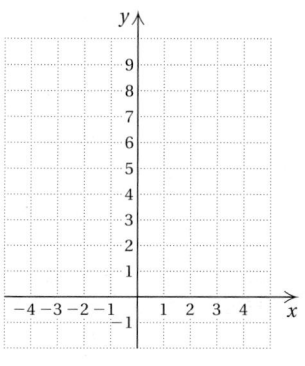

10. $f(x) = \ln(x - 1)$ [12.5c]

x	$f(x)$

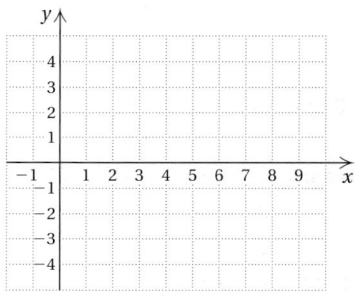

11. Find $(f \circ g)(x)$ and $(g \circ f)(x)$ if $f(x) = x^2$ and $g(x) = 3x - 5$.   [12.2d]

12. If $h(x) = \sqrt{4 - 7x}$, find $f(x)$ and $g(x)$ such that $h(x) = (f \circ g)(x)$.   [12.2d]

Convert to a logarithmic equation.   [12.3b]

13. $10^4 = 10,000$

14. $25^{1/2} = 5$

Convert to an exponential equation.   [12.3b]

15. $\log_4 16 = x$

16. $\log_{1/2} 8 = -3$

Find each of the following.   [12.3c]

17. $\log_3 9$

18. $\log_{10} \frac{1}{10}$

19. $\log_m m$

20. $\log_m 1$

Find the common logarithm, to four decimal places, using a calculator.   [12.3d]

21. $\log \left(\dfrac{78}{43{,}112} \right)$

22. $\log (-4)$

Express in terms of logarithms of x, y, and z.   [12.4d]

23. $\log_a x^4 y^2 z^3$

24. $\log \sqrt[4]{\dfrac{z^2}{x^3 y}}$

Express as a single logarithm.   [12.4d]

25. $\log_a 8 + \log_a 15$

26. $\frac{1}{2} \log a - \log b - 2 \log c$

Simplify.   [12.4e]

27. $\log_m m^{17}$

28. $\log_m m^{-7}$

Given $\log_a 2 = 1.8301$ and $\log_a 7 = 5.0999$, find each of the following.   [12.4d]

29. $\log_a 28$

30. $\log_a 3.5$

31. $\log_a \sqrt{7}$

32. $\log_a \frac{1}{4}$

Find each of the following, to four decimal places, using a calculator.   [12.5a]

33. $\ln 0.06774$

34. $e^{-0.98}$

35. $e^{2.91}$

36. $\ln 1$

37. $\ln 0$

38. $\ln e$

Find each logarithm using the change-of-base formula.
[12.5b]

39. $\log_5 2$

40. $\log_{12} 70$

Solve. Where appropriate, give approximations to four decimal places.   [12.6a, b]

41. $\log_3 x = -2$

42. $\log_x 32 = 5$

43. $\log x = -4$

44. $3 \ln x = -6$

45. $4^{2x-5} = 16$

46. $2^{x^2} \cdot 2^{4x} = 32$

47. $4^x = 8.3$

48. $e^{-0.1t} = 0.03$

49. $\log_4 16 = x$

50. $\log_4 x + \log_4 (x - 6) = 2$

51. $\log_2 (x + 3) - \log_2 (x - 3) = 4$

52. $\log_3 (x - 4) = 3 - \log_3 (x + 4)$

Solve.   [12.7a, b]

53. *Sound Level.*   The intensity of sound of a symphony orchestra playing at its peak can reach $10^{1.7}\,\text{W/m}^2$. How high is this sound level, in decibels? (Use $L = 10 \cdot \log (I/I_0)$ and $I_0 = 10^{-12}\,\text{W/m}^2$.)

54. e-filing. An increasing number of taxpayers are filing their federal income tax returns electronically. The number R, in millions, of returns e-filed t years after 2005 can be approximated by the exponential function

$$R(t) = 68.2(1.076)^t.$$

Source: Internal Revenue Service

a) Estimate the number of returns filed electronically in 2008, in 2010, and in 2012.
b) In what year will 131.9 million returns be e-filed?
c) What is the doubling time for the number of e-filed returns?
d) Graph the function.

55. Investment. In 2009, Lucy invested $40,000 in a mutual fund. By 2012, the value of her investment was $53,000. Assume that the value of her investment increased exponentially.

a) Find the value k, and write an exponential function that describes the value of Lucy's investment t years after 2009.
b) Predict the value of her investment in 2019.
c) In what year will the value of her investment first reach $85,000?

56. The population of a colony of bacteria doubled in 3 days. What was the exponential growth rate?

57. How long will it take $7600 to double itself if it is invested at 3.4%, compounded continuously?

58. How old is a skeleton that has lost 34% of its carbon-14? (Use $P(t) = P_0 e^{-0.00012t}$.)

59. What is the inverse of the function $f(x) = 5^x$, if it exists? [12.3a]

A. $f^{-1}(x) = x^5$ **B.** $f^{-1}(x) = \log_x 5$
C. $f^{-1}(x) = \log_5 x$ **D.** Does not exist

60. Solve: $\log(x^2 - 9) - \log(x + 3) = 1$. [12.6b]

A. 4   **B.** 5
C. 7   **D.** 13

Synthesis

Solve. [12.6a, b]

61. $\ln(\ln x) = 3$

62. $5^{x+y} = 25$, $2^{2x-y} = 64$

Understanding Through Discussion and Writing

1. Explain how the graph of $f(x) = e^x$ could be used to obtain the graph of $g(x) = 1 + \ln x$. [12.2c], [12.5a]

2. Christina first determines that the solution of $\log_3(x + 4) = 1$ is -1, but then rejects it. What mistake do you think she might be making? [12.6b]

3. An organization determines that the cost per person of chartering a bus is given by the function

$$C(x) = \frac{100 + 5x}{x},$$

where x is the number of people in the group and $C(x)$ is in dollars. Determine $C^{-1}(x)$ and explain how this inverse function could be used. [12.2c]

4. Explain how the equation $\ln x = 3$ could be solved using the graph of $f(x) = \ln x$. [12.6b]

5. Explain why you cannot take the logarithm of a negative number. [12.3a]

6. Write a problem for a classmate to solve in which data that seem to fit an exponential growth function are provided. Try to find data in a newspaper to make the problem as realistic as possible. [12.7b]

CHAPTER

12

Test

For Extra Help

CHAPTER
Test Prep
VIDEOS

Step-by-step test solutions are found on the Chapter Test Prep Videos available via the Video Resources on DVD, in *MyMathLab* ▌, and on You Tube (search "BittingerIntroInter" and click on "Channels").

Graph.

1. $f(x) = 2^{x+1}$

x	$f(x)$

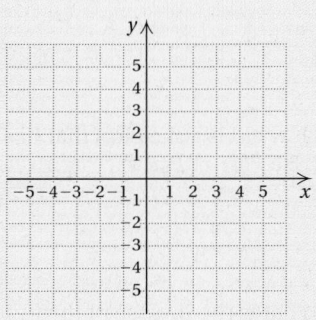

2. $y = \log_2 x$

x	y

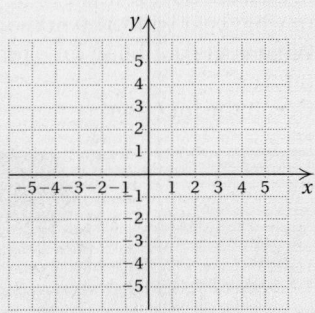

3. $f(x) = e^{x-2}$

x	$f(x)$

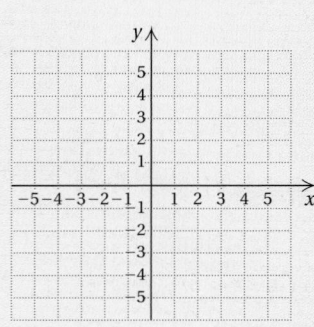

4. $f(x) = \ln(x - 4)$

x	$f(x)$

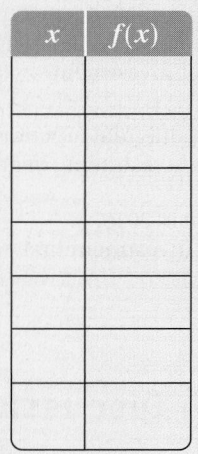
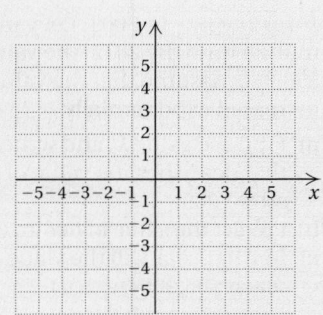

5. Find the inverse of the relation $\{(-4, 3), (5, -8), (-1, -3), (10, 12)\}$.
[12.2a]

Determine whether each function is one-to-one. If it is, find a formula for its inverse.

6. $f(x) = 4x - 3$

[12.2b, c]

7. $f(x) = (x + 1)^3$

[12.2b, c]

8. $f(x) = 2 - |x|$
[12.2b]

9. Find $(f \circ g)(x)$ and $(g \circ f)(x)$ if $f(x) = x + x^2$ and $g(x) = 5x - 2$.
[12.2d]

10. Convert to a logarithmic equation:
$$256^{1/2} = 16.$$

11. Convert to an exponential equation:
$$m = \log_7 49.$$

Find each of the following.

12. $\log_5 125$

13. $\log_t t^{23}$

14. $\log_p 1$

Find the common logarithm, to four decimal places, using a calculator.

15. $\log 0.0123$

16. $\log(-5)$

17. Express in terms of logarithms of a, b, and c:
$$\log \frac{a^3 b^{1/2}}{c^2}.$$

18. Express as a single logarithm:
$$\tfrac{1}{3}\log_a x - 3\log_a y + 2\log_a z.$$

Given $\log_a 2 = 0.301$, $\log_a 6 = 0.778$, and $\log_a 7 = 0.845$, find each of the following.

19. $\log_a \frac{2}{7}$

20. $\log_a 12$

Find each of the following, to four decimal places, using a calculator.

21. $\ln 807.39$

22. $e^{4.68}$

23. $\ln 1$

24. Find $\log_{18} 31$ using the change-of-base formula.

Solve. Where appropriate, give approximations to four decimal places.

25. $\log_x 25 = 2$

26. $\log_4 x = \frac{1}{2}$

27. $\log x = 4$

28. $\ln x = \frac{1}{4}$

29. $7^x = 1.2$

30. $\log(x^2 - 1) - \log(x - 1) = 1$

31. $\log_5 x + \log_5 (x + 4) = 1$

32. *Tomatoes.* What is the pH of tomatoes if the hydrogen ion concentration is 6.3×10^{-5} moles per liter? (Use $\text{pH} = -\log [\text{H}^+]$.)

33. *Cost of Health Care.* Spending on health care in the United States is projected to follow the exponential function

$$H(t) = 2.37(1.076)^t,$$

where H is the spending, in trillions of dollars, and t is the number of years since 2008.

Source: National Coalition on Health Care

a) Find the spending on health care in 2012.
b) In what year will spending on health care reach $5 trillion?
c) What is the doubling time of health-care spending?

34. *Interest Compounded Continuously.* Suppose a $1000 investment, compounded continuously, grows to $1150.27 in 5 years.

a) Find the interest rate and the exponential growth function.
b) What is the balance after 8 years?
c) When will the balance be $1439?
d) What is the doubling time?

35. The population of Masonville grew exponentially and doubled in 23 yr. What was the exponential growth rate?

36. How old is an animal bone that has lost 43% of its carbon-14? (Use $P(t) = P_0 e^{-0.00012t}$.)

37. Solve: $\log(3x - 1) + \log x = 1$.

A. There are one positive solution and one negative solution.
B. There is exactly one solution, and it is positive.
C. There is exactly one solution, and it is negative.
D. There is no solution.

Synthesis

38. Solve: $\log_3 |2x - 7| = 4$.

39. If $\log_a x = 2$, $\log_a y = 3$, and $\log_a z = 4$, find

$$\log_a \frac{\sqrt[3]{x^2 z}}{\sqrt[3]{y^2 z^{-1}}}.$$

Cumulative Review

Solve.

1. $\frac{1}{3}x - \frac{1}{5} \geq \frac{1}{5}x - \frac{1}{3}$

2. $|x| > 6.4$

3. $17 - |4 - x| = 2$

4. $\begin{aligned} 3x + y &= 4, \\ -6x - y &= -3 \end{aligned}$

5. $x^4 - 13x^2 + 36 = 0$

6. $2x^2 = x + 3$

7. $3x - \frac{6}{x} = 7$

8. $|x + 6| \leq 13$

9. $x(x + 10) = -21$

10. $2x^2 + x + 1 = 0$

11. $4(y - 1) + 6 = -8 - y$

12. $\frac{x + 1}{x - 2} > 0$

13. $\log_3 x = 2$

14. $x^2 - 1 \geq 0$

15. $\log_2 x + \log_2 (x + 7) = 3$

16. $\sqrt{x + 5} = x - 1$

17. $|x - 2| = |3x + 1|$

18. $7^x = 30$

19. $\begin{aligned} x - y + 2z &= 3, \\ -x \quad\ + z &= 4, \\ 2x + y - z &= -3 \end{aligned}$

20. $3 \leq 4x + 7 < 31$

21. $\frac{3}{x - 3} - \frac{x + 2}{x^2 + 2x - 15} = \frac{1}{x + 5}$

22. $P = \frac{3}{4}(M + 2N)$, for N

Solve.

23. *World Demand for Lumber.* The world is experiencing an exponential demand for lumber. The amount of timber N, in billions of cubic feet, consumed t years after 2000, can be approximated by

$$N(t) = 65(1.018)^t,$$

where $t = 0$ corresponds to 2000.

Sources: U. N. Food and Agricultural Organization; American Forest and Paper Association

a) How much timber is projected to be consumed in 2012? in 2015?
b) What is the doubling time?
c) Graph the function.

24. *Interest Compounded Annually.* Suppose that $50,000 is invested at 4% interest, compounded annually.

   a) Find a function A for the amount in the account after t years.

   b) Find the amount of money in the account at $t = 0$, $t = 4$, $t = 8$, and $t = 10$.

   c) Graph the function.

Simplify.

25. $(2x + 3)(x^2 - 2x - 1)$

26. $(3x^2 + x^3 - 1) - (2x^3 + x + 5)$

27. $\dfrac{2m^2 + 11m - 6}{m^3 + 1} \cdot \dfrac{m^2 - m + 1}{m + 6}$

28. $\dfrac{x}{x - 1} + \dfrac{2}{x + 1} - \dfrac{2x}{x^2 - 1}$

29. $\dfrac{1 - \dfrac{5}{x}}{x - 4 - \dfrac{5}{x}}$

30. $(x^4 + 3x^3 - x + 4) \div (x + 1)$

31. $\dfrac{\sqrt{75x^5y^2}}{\sqrt{3xy}}$

32. $4\sqrt{50} - 3\sqrt{18}$

33. $(16^{3/2})^{1/2}$

34. $\left(2 - i\sqrt{2}\right)\left(5 + 3i\sqrt{2}\right)$

35. $\dfrac{5 + i}{2 - 4i}$

36. $\left| \dfrac{2}{3} - \dfrac{4}{5} \right|$

37. $\dfrac{63x^2y^3}{-7x^{-4}y}$

38. $1000 \div 10^2 \cdot 25 \div 4$

39. $5x - 3[4(x - 2) - 2(x + 1)]$

40. Find the x- and y-intercepts of the graph of $7x - 14 = 28y$.

41. Find the slope, if it exists, of the line containing the given points.

   a) $(-3, 13), (8, -3)$

   b) $(-1, 2), \left(-1, \dfrac{2}{5}\right)$

42. Find an equation of the line containing the points $(1, 4)$ and $(-1, 0)$.

43. Find an equation of the line containing the point $(1, 2)$ and perpendicular to the line whose equation is $2x - y = 3$.

Graph.

44. $y - x < -2$

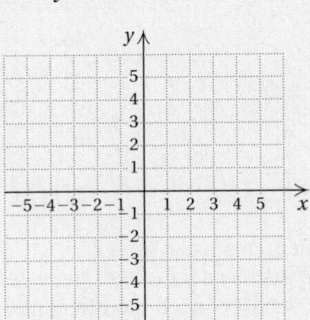

45. $y = -\dfrac{1}{2}x - 3$

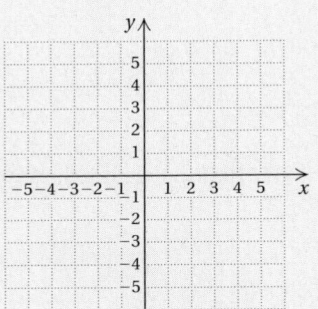

46. $4y - 3x = 12$

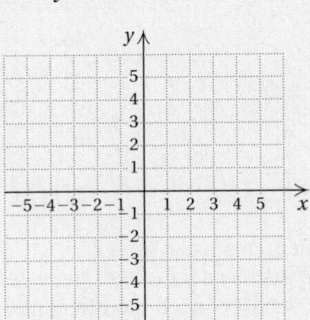

47. $y < -2$

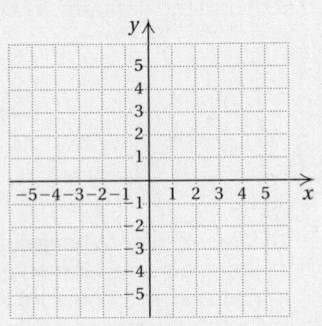

48. $h(x) = 2 + x - x^2$

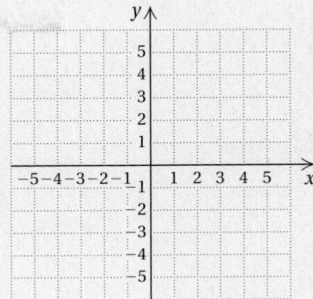

49. $x + y \leq 0,$
$\qquad x \geq -4,$
$\qquad y \geq -1$

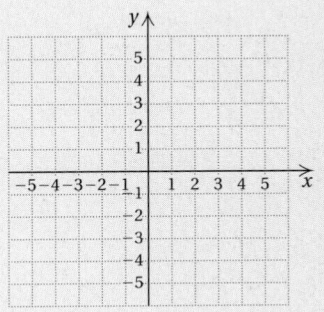

50. $x = 3.5$

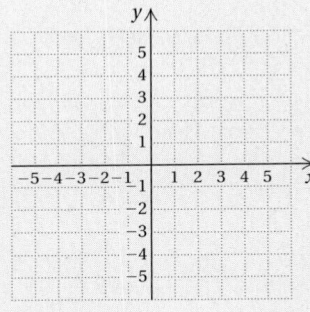

51. $f(x) = 2x^2 - 8x + 9$

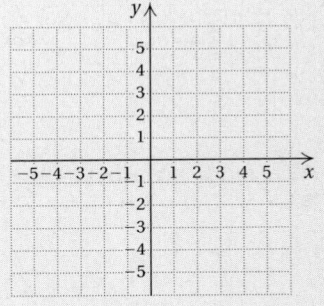

52. $f(x) = e^{-x}$

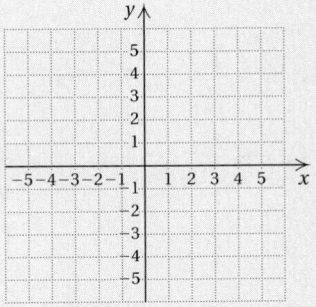

53. $f(x) = \log_2 x$

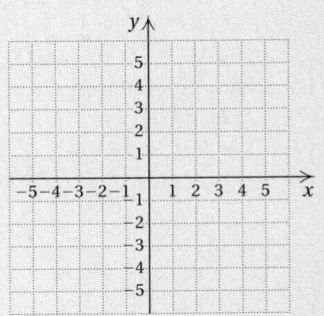

Factor.

54. $2x^4 - 12x^3 + x - 6$

55. $3a^2 - 12ab - 135b^2$

56. $x^2 - 17x + 72$

57. $81m^4 - n^4$

58. $16x^2 - 16x + 4$

59. $81a^3 - 24$

60. $10x^2 + 66x - 28$

61. $6x^3 + 27x^2 - 15x$

62. Find $f^{-1}(x)$ when $f(x) = 2x - 3$.

63. z varies directly as x and inversely as the cube of y, and $z = 5$ when $x = 4$ and $y = 2$. What is z when $x = 10$ and $y = 5$?

64. Given the function f described by $f(x) = x^3 - 2$, find $f(-2)$.

65. Given the function g described by $g(x) = -30$, find $g(0)$.

66. *Food and Drink Sales.* Total sales S of food and beverages in U.S. restaurants, in billions of dollars, can be modeled by the function

$$S(t) = 18t + 344.7,$$

where t is the number of years since 2000.

a) Find the total sales of food and drink in U.S. restaurants in 2005, in 2008, and in 2010.
b) Graph the function.
c) Find the y-intercept.
d) Find the slope.
e) Find the rate of change.

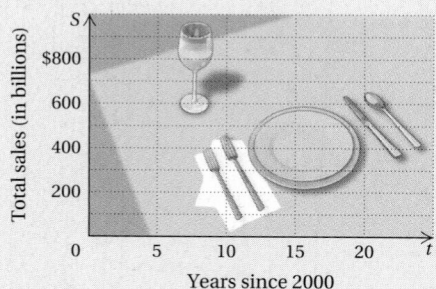

Years since 2000

67. Rationalize the denominator: $\dfrac{5 + \sqrt{a}}{3 - \sqrt{a}}$.

68. Find the domain: $f(x) = x^2 - 2x + 3$.

69. Find the domain: $h(x) = \dfrac{2 - x}{x + 2}$.

Solve.

70. *Book Club.* A book club offers two types of membership. Limited members pay a fee of $20 per year and can buy books for $20 each. Preferred members pay $40 per year and can buy books for $15 each. For what numbers of annual book purchases would it be less expensive to be a preferred member?

71. *Train Travel.* A passenger train travels at twice the speed of a freight train. The freight train leaves a station at 2 A.M. and travels north at 34 mph. The passenger train leaves the station at 11 A.M, traveling north on a parallel track. How far from the station will the passenger train overtake the freight train?

72. *Perimeters of Polygons.* A pentagon with all five sides the same size has a perimeter equal to that of an octagon in which all eight sides are the same size. One side of the pentagon is 2 less than three times one side of the octagon. What is the perimeter of each figure?

73. *Ammonia Solutions.* A chemist has two solutions of ammonia and water. Solution A is 6% ammonia and solution B is 2% ammonia. How many liters of each solution are needed in order to obtain 80 L of a solution that is 3.2% ammonia?

74. *Air Travel.* An airplane can fly 190 mi with the wind in the same time that it takes to fly 160 mi against the wind. The speed of the wind is 30 mph. How fast can the plane fly in still air?

75. *Work.* Christy can do a certain job in 21 min. Madeline can do the same job in 14 min. How long would it take to do the job if the two worked together?

76. *Centripetal Force.* The centripetal force F of an object moving in a circle varies directly as the square of the velocity v and inversely as the radius r of the circle. If $F = 8$ when $v = 1$ and $r = 10$, what is F when $v = 2$ and $r = 16$?

77. *Maximizing Area.* A farmer wants to fence in a rectangular area next to a river. (Note that no fence will be needed along the river.) What is the area of the largest region that can be fenced in with 100 ft of fencing?

78. *Carbon Dating.* Use the function $P(t) = P_0 e^{-0.00012t}$ to find the age of a bone that has lost 25% of its carbon-14.

79. *Beam Load.* The weight W that a horizontal beam can support varies inversely as the length L of the beam. If a 14-m beam can support 1440 kg, what weight can a 6-m beam support?

80. Fit a linear function to the data points $(2, -3)$ and $(5, -4)$.

81. Fit a quadratic function to the data points $(-2, 4)$, $(-5, -6)$, and $(1, -3)$.

82. Convert to a logarithmic equation: $10^6 = r$.

83. Convert to an exponential equation: $\log_3 Q = x$.

84. Express as a single logarithm:
$$\frac{1}{5}(7 \log_b x - \log_b y - 8 \log_b z).$$

85. Express in terms of logarithms of x, y, and z:
$$\log_b \left(\frac{xy^5}{z} \right)^{-6}.$$

86. What is the maximum product of two numbers whose sum is 26?

87. Determine whether the function $f(x) = 4 - x^2$ is one-to-one.

88. For the graph of function f shown here, determine (a) $f(2)$; (b) the domain; (c) all x-values such that $f(x) = -5$; and (d) the range.

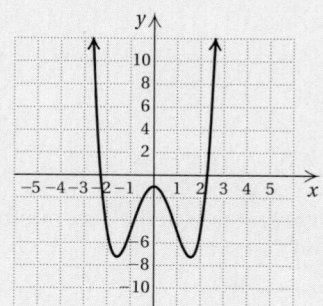

89. *Population Growth of Nevada.* In 2008, the population of Nevada was 2,600,167. It had grown from a population of 1,998,257 in 2000. Nevada was the fastest growing state in the United States. Assume the population growth increases according to an exponential growth function.
Source: U.S. Census Bureau

a) Let $t = 0$ correspond to 2000 and $t = 8$ correspond to 2008. Then t is the number of years since 2000. Use the data points $(0, 1,998,257)$ and $(8, 2,600,167)$ to find the exponential growth rate and fit an exponential growth function $P(t) = P_0 e^{kt}$ to the data, where $P(t)$ is the population of Nevada t years after 2000.

b) Use the function found in part (a) to predict the population of Nevada in 2015.

c) If growth continues at this rate, when will the population reach 3.5 million?

Synthesis

90. Solve: $\dfrac{9}{x} - \dfrac{9}{x + 12} = \dfrac{108}{x^2 + 12x}$.

91. Solve: $\log_2 (\log_3 x) = 2$.

92. Solve $ax^2 - bx = 0$ for x.

93. Factor: $\dfrac{a^3}{8} + \dfrac{8b^3}{729}$.

94. Simplify: $\left[\dfrac{1}{(-3)^{-2}} - (-3)^1 \right] \cdot [(-3)^2 + (-3)^{-2}]$.

Appendixes

A

Factoring and LCMs

OBJECTIVES

a Find all the factors of numbers and find prime factorizations of numbers.

b Find the LCM of two or more numbers using prime factorizations.

a Factors and Prime Factorizations

We begin our review with *factoring*, a necessary skill for addition and subtraction with fraction notation. Factoring is also an important skill in algebra.

The numbers we will be factoring are **natural numbers**:

1, 2, 3, 4, 5, and so on.

To **factor** a number means to express the number as a product. Consider the product $12 = 3 \cdot 4$. We say that 3 and 4 are **factors** of 12 and that $3 \cdot 4$ is a **factorization** of 12. Since $12 = 12 \cdot 1$, we also know that 12 and 1 are factors of 12 and that $12 \cdot 1$ is a factorization of 12.

EXAMPLE 1 Find all the factors of 12.

We first find all two-factor products:

$$12 = 1 \cdot 12, \quad 12 = 2 \cdot 6, \quad 12 = 3 \cdot 4.$$

The factors of 12 are 1, 2, 3, 4, 6, and 12.

EXAMPLE 2 Find all the factors of 150.

We first find some factorizations:

$$150 = 1 \cdot 150, \quad 150 = 2 \cdot 75, \quad 150 = 3 \cdot 50,$$
$$150 = 5 \cdot 30, \quad 150 = 6 \cdot 25, \quad 150 = 10 \cdot 15.$$

The factors of 150 are 1, 2, 3, 5, 6, 10, 15, 25, 30, 50, 75, and 150.

Note that the word "factor" is used both as a noun and as a verb. You **factor** when you express a number as a product. The numbers you multiply together to get the product are **factors**.

Find all the factors of each number.

1. 9

2. 16

3. 24

4. 180

Do Margin Exercises 1–4.

PRIME NUMBER

A natural number that has *exactly two different factors*, itself and 1, is called a **prime number**.

EXAMPLE 3 Which of these numbers are prime? 7, 4, 11, 18, 1

7 is prime. It has exactly two different factors, 1 and 7.

4 is *not* prime. It has three different factors, 1, 2, and 4.

11 is prime. It has exactly two different factors, 1 and 11.

18 is *not* prime. It has factors 1, 2, 3, 6, 9, and 18.

1 is *not* prime. It does not have two *different* factors.

Answers

1. 1, 3, 9   **2.** 1, 2, 4, 8, 16   **3.** 1, 2, 3, 4, 6, 8, 12, 24   **4.** 1, 2, 3, 4, 5, 6, 9, 10, 12, 15, 18, 20, 30, 36, 45, 60, 90, 180

In the margin at right is a table of the prime numbers from 2 to 157. There are more extensive tables, but these prime numbers will be the most helpful to you in this text.

Do Exercise 5.

If a natural number, other than 1, is not prime, we call it **composite**. Every composite number can be factored into a product of prime numbers. Such a factorization is called a **prime factorization**.

EXAMPLE 4 Find the prime factorization of 36.

We begin by factoring 36 any way we can. One way is like this:

$$36 = 4 \cdot 9.$$

The factors 4 and 9 are not prime, so we factor them:

$$36 = \underset{\downarrow}{4} \cdot \underset{\downarrow}{9}$$
$$= 2 \cdot 2 \cdot 3 \cdot 3.$$

The factors in the last factorization are all prime, so we now have the *prime factorization* of 36. Note that 1 is *not* part of this factorization because it is not prime.

Another way to find the prime factorization of 36 is like this:

$$36 = 2 \cdot 18 = 2 \cdot 3 \cdot 6 = 2 \cdot 3 \cdot 2 \cdot 3.$$

In effect, we begin factoring any way we can think of and keep factoring until all factors are prime. Using a **factor tree** might also be helpful.

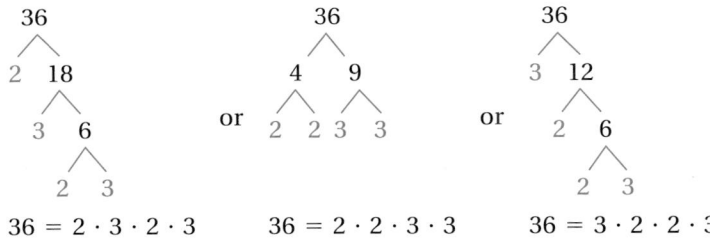

$$36 = 2 \cdot 3 \cdot 2 \cdot 3 \qquad 36 = 2 \cdot 2 \cdot 3 \cdot 3 \qquad 36 = 3 \cdot 2 \cdot 2 \cdot 3$$

No matter which way we begin, the result is the same: The prime factorization of 36 contains two factors of 2 and two factors of 3. Every composite number has a *unique* prime factorization.

EXAMPLE 5 Find the prime factorization of 60.

This time, we use the list of primes from the table. We go through the table until we find a prime that is a factor of 60. The first such prime is 2.

$$60 = 2 \cdot 30$$

We keep dividing by 2 until it is not possible to do so.

$$60 = 2 \cdot 2 \cdot 15$$

Now we go to the next prime in the table that is a factor of 60. It is 3.

$$60 = 2 \cdot 2 \cdot 3 \cdot 5$$

Each factor in $2 \cdot 2 \cdot 3 \cdot 5$ is a prime. Thus this is the prime factorization.

Do Exercises 6–9.

5. Which of these numbers are prime?

8, 6, 13, 14, 1

Find the prime factorization.

6. 48   **7.** 50

8. 770   **9.** 2340

Answers

5. 13   **6.** $2 \cdot 2 \cdot 2 \cdot 2 \cdot 3$
7. $2 \cdot 5 \cdot 5$ **8.** $2 \cdot 5 \cdot 7 \cdot 11$
9. $2 \cdot 2 \cdot 3 \cdot 3 \cdot 5 \cdot 13$

b Least Common Multiples

Least common multiples are used to add and subtract with fraction notation.

The **multiples** of a number all have that number as a factor. For example, the multiples of 2 are

$$2, \quad 4, \quad 6, \quad 8, \quad 10, \quad 12, \quad 14, \quad 16, \ldots.$$

We could name each of them in such a way as to show 2 as a factor. For example, $14 = 2 \cdot 7$.

The multiples of 3 all have 3 as a factor:

$$3, \quad 6, \quad 9, \quad 12, \quad 15, \quad 18, \ldots.$$

Two or more numbers always have many multiples in common. From lists of multiples, we can find common multiples.

EXAMPLE 6 Find the common multiples of 2 and 3.

We make lists of their multiples and circle the multiples that appear in both lists.

2, 4, ⑥, 8, 10, ⑫, 14, 16, ⑱, 20, 22, ㉔, 26, 28, ㉚, 32, 34, ㊱, . . . ;

3, ⑥, 9, ⑫, 15, ⑱, 21, ㉔, 27, ㉚, 33, ㊱, ...

The common multiples of 2 and 3 are

$$6, \quad 12, \quad 18, \quad 24, \quad 30, \quad 36, \ldots.$$

> **10.** Find the common multiples of 3 and 5 by making lists of multiples.
>
> **11.** Find the common multiples of 9 and 15 by making lists of multiples.

Do Exercises 10 and 11.

In Example 6, we found common multiples of 2 and 3. The *least*, or smallest, of those common multiples is 6. We abbreviate **least common multiple** as **LCM**.

There are several methods that work well for finding the LCM of several numbers. Some of these do not work well when we consider expressions with variables such as $4ab$ and $12abc$. We now review a method that will work in arithmetic *and in algebra as well*. To see how it works, let's look at the prime factorizations of 9 and 15 in order to find the LCM:

$$9 = 3 \cdot 3, \qquad 15 = 3 \cdot 5.$$

Any multiple of 9 must have *two* 3's as factors. Any multiple of 15 must have *one* 3 and *one* 5 as factors. The smallest multiple of 9 and 15 is

Two 3's; 9 is a factor
$$3 \cdot 3 \cdot 5 = 45.$$
One 3, one 5; 15 is a factor

The LCM must have all the factors of 9 and all the factors of 15, *but the factors are not repeated when they are common to both numbers.*

> To find the LCM of several numbers using prime factorizations:
>
> **a)** Write the prime factorization of each number.
>
> **b)** Form the LCM by writing the product of the different factors from step (a), using each factor the greatest number of times that it occurs in any *one* of the factorizations.

EXAMPLE 7   Find the LCM of 40 and 100.

a) We find the prime factorizations:

$$40 = 2 \cdot 2 \cdot 2 \cdot 5,$$
$$100 = 2 \cdot 2 \cdot 5 \cdot 5.$$

b) The different prime factors are 2 and 5. We write 2 as a factor three times (the greatest number of times that it occurs in any *one* factorization). We write 5 as a factor two times (the greatest number of times that it occurs in any *one* factorization).

The LCM is $2 \cdot 2 \cdot 2 \cdot 5 \cdot 5$, or 200.

Do Exercises 12 and 13.

Find each LCM by factoring.
12. 8 and 10   **13.** 18 and 27

EXAMPLE 8   Find the LCM of 27, 90, and 84.

a) We factor:

$$27 = 3 \cdot 3 \cdot 3,$$
$$90 = 2 \cdot 3 \cdot 3 \cdot 5,$$
$$84 = 2 \cdot 2 \cdot 3 \cdot 7.$$

b) We write 2 as a factor two times, 3 three times, 5 one time, and 7 one time.

The LCM is $2 \cdot 2 \cdot 3 \cdot 3 \cdot 3 \cdot 5 \cdot 7$, or 3780.

Do Exercise 14.

14. Find the LCM of 18, 24, and 30.

EXAMPLE 9   Find the LCM of 7 and 21.

Since 7 is prime, it has no prime factorization. It still, however, must be a factor of the LCM:

$$7 = 7,$$
$$21 = 3 \cdot 7.$$

The LCM is $7 \cdot 3$, or 21.

If one number is a factor of another, then the LCM is the larger of the two numbers.

Do Exercises 15 and 16.

Find each LCM.
15. 3, 18   **16.** 12, 24

EXAMPLE 10   Find the LCM of 8 and 9.

We have

$$8 = 2 \cdot 2 \cdot 2,$$
$$9 = 3 \cdot 3.$$

The LCM is $2 \cdot 2 \cdot 2 \cdot 3 \cdot 3$, or 72.

If two or more numbers have no common prime factor, then the LCM is the product of the numbers.

Do Exercises 17 and 18.

Find each LCM.
17. 4, 9   **18.** 5, 6, 7

Answers

12. 40   **13.** 54   **14.** 360   **15.** 18
16. 24   **17.** 36   **18.** 210

Always review the objectives before doing an exercise set. See page 972. Note how the objectives are keyed to the exercises.

a Find all the factors of each number.

1. 20   **2.** 36   **3.** 72   **4.** 81

Find the prime factorization of each number.

5. 15   **6.** 14   **7.** 22   **8.** 33   **9.** 9

10. 25   **11.** 49   **12.** 121   **13.** 18   **14.** 24

15. 40   **16.** 56   **17.** 90   **18.** 120   **19.** 210

20. 330   **21.** 91   **22.** 143   **23.** 119   **24.** 221

b Find the prime factorization of the numbers. Then find the LCM.

25. 4, 5   **26.** 18, 40   **27.** 24, 36   **28.** 24, 27   **29.** 3, 15

30. 20, 40   **31.** 30, 40   **32.** 50, 60   **33.** 13, 23   **34.** 17, 29

35. 18, 30   **36.** 45, 72   **37.** 30, 36   **38.** 30, 50   **39.** 24, 30

40. 60, 70   **41.** 12, 18   **42.** 18, 24   **43.** 12, 28   **44.** 35, 45

45. 2, 3, 5   **46.** 3, 5, 7   **47.** 24, 36, 12   **48.** 8, 16, 22

49. 5, 12, 15   **50.** 12, 18, 40   **51.** 6, 12, 18   **52.** 24, 35, 45

B Fraction Notation

We now review fraction notation and its use with addition, subtraction, multiplication, and division of *arithmetic numbers*.

a Equivalent Expressions and Fraction Notation

An example of **fraction notation** for a number is

$$\frac{2}{3} \begin{array}{l} \leftarrow \text{Numerator} \\ \leftarrow \text{Denominator} \end{array}$$

The top number is called the **numerator**, and the bottom number is called the **denominator**.

The **whole numbers** consist of the natural numbers and 0:

$$0, \quad 1, \quad 2, \quad 3, \quad 4, \quad 5, \ldots.$$

The **arithmetic numbers**, also called the **nonnegative rational numbers**, consist of the whole numbers and the fractions, such as $\frac{2}{3}$ and $\frac{9}{5}$.

> ### ARITHMETIC NUMBERS
>
> The **arithmetic numbers** are the whole numbers and the fractions, such as $8, \frac{3}{4}$, and $\frac{6}{5}$. All these numbers can be named with fraction notation $\frac{a}{b}$, where a and b are whole numbers and $b \neq 0$.

Note that all whole numbers can be named with fraction notation. For example, we can name the whole number 8 as $\frac{8}{1}$. We call 8 and $\frac{8}{1}$ **equivalent expressions**.

Being able to find an equivalent expression is critical to a study of algebra. Two simple but powerful properties of numbers that allow us to find equivalent expressions are the identity properties of 0 and 1.

> ### THE IDENTITY PROPERTY OF 0 (ADDITIVE IDENTITY)
>
> For any number a,
>
> $$a + 0 = a.$$
>
> (Adding 0 to any number gives that same number—for example, $12 + 0 = 12$.)

> ### THE IDENTITY PROPERTY OF 1 (MULTIPLICATIVE IDENTITY)
>
> For any number a,
>
> $$a \cdot 1 = a.$$
>
> $\left(\text{Multiplying any number by 1 gives that same number—for example, } \frac{3}{5} \cdot 1 = \frac{3}{5}.\right)$

Here are some ways to name the number 1:

$$\frac{5}{5}, \quad \frac{3}{3}, \quad \text{and} \quad \frac{26}{26}.$$

The following property allows us to find equivalent fraction expressions.

> **EQUIVALENT EXPRESSIONS FOR 1**
>
> For any number a, $a \neq 0$,
>
> $$\frac{a}{a} = 1.$$

We can use the identity property of 1 and the preceding result to find equivalent fraction expressions.

EXAMPLE 1 Write a fraction expression equivalent to $\frac{2}{3}$ with a denominator of 15.

Note that $15 = 3 \cdot 5$. We want fraction notation for $\frac{2}{3}$ that has a denominator of 15, but the denominator 3 is missing a factor of 5. We multiply by 1, using $\frac{5}{5}$ as an equivalent expression for 1. Recall from arithmetic that to multiply with fraction notation, we multiply numerators and we multiply denominators:

$$\frac{2}{3} = \frac{2}{3} \cdot 1 \qquad \text{Using the identity property of 1}$$

$$= \frac{2}{3} \cdot \frac{5}{5} \qquad \text{Using } \frac{5}{5} \text{ for 1}$$

$$= \frac{10}{15}. \qquad \text{Multiplying numerators and denominators}$$

Do Exercises 1–3.

b Simplifying Expressions

We know that $\frac{1}{2}, \frac{2}{4}, \frac{4}{8}$, and so on, all name the same number. Any arithmetic number can be named in many ways. The **simplest fraction notation** is the notation that has the smallest numerator and denominator. We call the process of finding the simplest fraction notation **simplifying**. We reverse the process of Example 1 by first factoring the numerator and the denominator. Then we factor the fraction expression and remove a factor of 1 using the identity property of 1.

EXAMPLE 2 Simplify: $\frac{10}{15}$.

$$\frac{10}{15} = \frac{2 \cdot 5}{3 \cdot 5} \qquad \text{Factoring the numerator and the denominator. In this case, each is the prime factorization.}$$

$$= \frac{2}{3} \cdot \frac{5}{5} \qquad \text{Factoring the fraction expression}$$

$$= \frac{2}{3} \cdot 1$$

$$= \frac{2}{3} \qquad \text{Using the identity property of 1 (removing a factor of 1)}$$

1. Write a fraction expression equivalent to $\frac{2}{3}$ with a denominator of 12.

2. Write a fraction expression equivalent to $\frac{3}{4}$ with a denominator of 28.

3. Multiply by 1 to find three different fraction expressions for $\frac{7}{8}$.

Answers

1. $\frac{8}{12}$ 2. $\frac{21}{28}$ 3. $\frac{14}{16}, \frac{21}{24}, \frac{28}{32}$; answers may vary

EXAMPLE 3 Simplify: $\dfrac{36}{24}$.

$$\dfrac{36}{24} = \dfrac{2 \cdot 3 \cdot 2 \cdot 3}{2 \cdot 2 \cdot 3 \cdot 2} \qquad \text{Factoring the numerator and the denominator}$$

$$= \dfrac{2 \cdot 3 \cdot 2}{2 \cdot 3 \cdot 2} \cdot \dfrac{3}{2} \qquad \text{Factoring the fraction expression}$$

$$= 1 \cdot \dfrac{3}{2}$$

$$= \dfrac{3}{2} \qquad \text{Removing a factor of 1}$$

It is always a good idea to check at the end to see if you have indeed factored out all the common factors of the numerator and the denominator.

Canceling

Canceling is a shortcut that you may have used to remove a factor of 1 when working with fraction notation. With *great* concern, we mention it as a possible way to speed up your work. You should use canceling only when removing common factors in numerators and denominators. Each common factor allows us to remove a factor of 1 in a product. **Canceling *cannot* be done when adding.** Our concern is that "canceling" be performed with care and understanding. Example 3 might have been done faster as follows:

$$\dfrac{36}{24} = \dfrac{2 \cdot \cancel{3} \cdot 2 \cdot 3}{2 \cdot 2 \cdot \cancel{3} \cdot 2} = \dfrac{3}{2}, \quad \text{or} \quad \dfrac{36}{24} = \dfrac{3 \cdot \cancel{12}}{2 \cdot \cancel{12}} = \dfrac{3}{2}, \quad \text{or} \quad \dfrac{\overset{3}{\cancel{\overset{18}{\cancel{36}}}}}{\underset{2}{\cancel{\underset{12}{\cancel{24}}}}} = \dfrac{3}{2}.$$

Caution!

The difficulty with canceling is that it is often applied incorrectly in situations like the following:

$$\dfrac{2 + 3}{\cancel{2}} = 3; \qquad \dfrac{\cancel{4} + 1}{\cancel{4} + 2} = \dfrac{1}{2}; \qquad \dfrac{1\cancel{5}}{\cancel{5}4} = \dfrac{1}{4}.$$

$$\quad\text{Wrong!} \qquad\qquad \text{Wrong!} \qquad\qquad \text{Wrong!}$$

The correct answers are

$$\dfrac{2 + 3}{2} = \dfrac{5}{2}; \qquad \dfrac{4 + 1}{4 + 2} = \dfrac{5}{6};$$

$$\dfrac{15}{54} = \dfrac{3 \cdot 5}{3 \cdot 18} = \dfrac{3}{3} \cdot \dfrac{5}{18} = \dfrac{5}{18}.$$

In each situation, the number canceled was not a factor of 1. Factors are parts of products. For example, in $2 \cdot 3$, 2 and 3 are factors, but in $2 + 3$, 2 and 3 are *not* factors.

Do Exercises 4–7.

Simplify.

4. $\dfrac{18}{45}$ **5.** $\dfrac{38}{18}$

6. $\dfrac{72}{27}$ **7.** $\dfrac{32}{56}$

Answers

4. $\dfrac{2}{5}$ **5.** $\dfrac{19}{9}$ **6.** $\dfrac{8}{3}$ **7.** $\dfrac{4}{7}$

We can always insert the number 1 as a factor. The identity property of 1 allows us to do that.

EXAMPLE 4 Simplify: $\dfrac{18}{72}$.

$$\frac{18}{72} = \frac{2 \cdot \cancel{9}}{8 \cdot \cancel{9}} = \frac{2}{8} = \frac{2 \cdot 1}{2 \cdot 4} = \frac{1}{4}, \quad \text{or} \quad \frac{18}{72} = \frac{1 \cdot \cancel{18}}{4 \cdot \cancel{18}} = \frac{1}{4}$$

EXAMPLE 5 Simplify: $\dfrac{72}{9}$.

$$\frac{72}{9} = \frac{8 \cdot 9}{1 \cdot 9} \qquad \text{Factoring and inserting a factor of 1 in the denominator}$$

$$= \frac{8 \cdot \cancel{9}}{1 \cdot \cancel{9}} \qquad \text{Removing a factor of 1: } \frac{9}{9} = 1$$

$$= \frac{8}{1} = 8 \qquad \text{Simplifying}$$

Do Exercises 8 and 9.

Do Exercises 8 and 9.

(c) Multiplication, Addition, Subtraction, and Division

After we have performed an operation of multiplication, addition, subtraction, or division, the answer may not be in simplified form. We simplify, if at all possible.

Multiplication

To multiply using fraction notation, we multiply the numerators to get the new numerator, and we multiply the denominators to get the new denominator.

> **MULTIPLYING FRACTIONS**
>
> To multiply fractions, multiply the numerators and multiply the denominators:
>
> $$\frac{a}{b} \cdot \frac{c}{d} = \frac{a \cdot c}{b \cdot d}.$$

EXAMPLE 6 Multiply and simplify: $\dfrac{5}{6} \cdot \dfrac{9}{25}$.

$$\frac{5}{6} \cdot \frac{9}{25} = \frac{5 \cdot 9}{6 \cdot 25} \qquad \text{Multiplying numerators and denominators}$$

$$= \frac{5 \cdot 3 \cdot 3}{2 \cdot 3 \cdot 5 \cdot 5} \qquad \text{Factoring the numerator and the denominator}$$

$$= \frac{\cancel{5} \cdot \cancel{3} \cdot 3}{2 \cdot \cancel{3} \cdot \cancel{5} \cdot 5} \qquad \text{Removing a factor of 1: } \frac{3 \cdot 5}{3 \cdot 5} = 1$$

$$= \frac{3}{10} \qquad \text{Simplifying}$$

Do Exercises 10 and 11.

Do Exercises 10 and 11.

Simplify.

8. $\dfrac{27}{54}$ **9.** $\dfrac{48}{12}$

Multiply and simplify.

10. $\dfrac{6}{5} \cdot \dfrac{25}{12}$ **11.** $\dfrac{3}{8} \cdot \dfrac{5}{3} \cdot \dfrac{7}{2}$

Answers

8. $\dfrac{1}{2}$ **9.** 4 **10.** $\dfrac{5}{2}$ **11.** $\dfrac{35}{16}$

Addition

When denominators are the same, we can add by adding the numerators and keeping the same denominator.

ADDING FRACTIONS WITH LIKE DENOMINATORS

To add fractions when denominators are the same, add the numerators and keep the same denominator:

$$\frac{a}{c} + \frac{b}{c} = \frac{a+b}{c}.$$

EXAMPLE 7 Add: $\dfrac{4}{8} + \dfrac{5}{8}$.

The common denominator is 8. We add the numerators and keep the common denominator:

$$\frac{4}{8} + \frac{5}{8} = \frac{4+5}{8} = \frac{9}{8}.$$

In arithmetic, we generally write $\frac{9}{8}$ as $1\frac{1}{8}$. (See a review of converting from a mixed numeral to fraction notation at right.) In algebra, you will find that *improper fraction* symbols such as $\frac{9}{8}$ are more useful and are quite *proper* for our purposes.

What do we do when denominators are different? We find a common denominator. We can do this by multiplying by 1. Consider adding $\frac{1}{6}$ and $\frac{3}{4}$. There are several common denominators that can be obtained. Let's look at two possibilities.

A. $\dfrac{1}{6} + \dfrac{3}{4} = \dfrac{1}{6} \cdot 1 + \dfrac{3}{4} \cdot 1$

$\qquad = \dfrac{1}{6} \cdot \dfrac{4}{4} + \dfrac{3}{4} \cdot \dfrac{6}{6}$

$\qquad = \dfrac{4}{24} + \dfrac{18}{24}$

$\qquad = \dfrac{22}{24}$

$\qquad = \dfrac{11}{12}$ \qquad Simplifying

B. $\dfrac{1}{6} + \dfrac{3}{4} = \dfrac{1}{6} \cdot 1 + \dfrac{3}{4} \cdot 1$

$\qquad = \dfrac{1}{6} \cdot \dfrac{2}{2} + \dfrac{3}{4} \cdot \dfrac{3}{3}$

$\qquad = \dfrac{2}{12} + \dfrac{9}{12}$

$\qquad = \dfrac{11}{12}$

We had to simplify in **A**. We didn't have to simplify in **B**. In **B**, we used the least common multiple of the denominators, 12. That number is called the **least common denominator**, or **LCD**. Using the LCD allows us to add fractions using the smallest numbers possible.

> To convert from a mixed numeral to fraction notation:
>
> $$\overset{\text{(b)}}{\curvearrowright}\ 3\underset{\text{(a)}}{\overset{5}{}}\underset{\curvearrowleft}{} = \frac{29}{8} \leftarrow \text{(c)}$$
>
> (a) Multiply the whole number by the denominator:
> $$3 \cdot 8 = 24.$$
>
> (b) Add the result to the numerator:
> $$24 + 5 = 29.$$
>
> (c) Keep the denominator.

ADDING FRACTIONS WITH DIFFERENT DENOMINATORS

To add fractions when denominators are different:

a) Find the least common multiple of the denominators. That number is the least common denominator, LCD.

b) Multiply by 1, using the appropriate notation n/n for each fraction to express fractions in terms of the LCD.

c) Add the numerators, keeping the same denominator.

d) Simplify, if possible.

EXAMPLE 8 Add and simplify: $\dfrac{3}{8} + \dfrac{5}{12}$.

The LCM of the denominators, 8 and 12, is 24. Thus the LCD is 24. We multiply each fraction by 1 to obtain the LCD:

$$\frac{3}{8} + \frac{5}{12} = \frac{3}{8} \cdot \frac{3}{3} + \frac{5}{12} \cdot \frac{2}{2}$$

Multiplying by 1. Since $3 \cdot 8 = 24$, we multiply the first number by $\frac{3}{3}$. Since $2 \cdot 12 = 24$, we multiply the second number by $\frac{2}{2}$.

$$= \frac{9}{24} + \frac{10}{24}$$

$$= \frac{9 + 10}{24}$$

Adding the numerators and keeping the same denominator

$$= \frac{19}{24}.$$

EXAMPLE 9 Add and simplify: $\dfrac{11}{30} + \dfrac{5}{18}$.

We first look for the LCM of 30 and 18. That number is then the LCD. We find the prime factorization of each denominator:

$$\frac{11}{30} + \frac{5}{18} = \frac{11}{5 \cdot 2 \cdot 3} + \frac{5}{2 \cdot 3 \cdot 3}.$$

The LCD is $5 \cdot 2 \cdot 3 \cdot 3$, or 90. To get the LCD in the first denominator, we need a factor of 3. To get the LCD in the second denominator, we need a factor of 5. We get these numbers by multiplying by 1:

$$\frac{11}{30} + \frac{5}{18} = \frac{11}{5 \cdot 2 \cdot 3} \cdot \frac{3}{3} + \frac{5}{2 \cdot 3 \cdot 3} \cdot \frac{5}{5}$$

Multiplying by 1

$$= \frac{33}{5 \cdot 2 \cdot 3 \cdot 3} + \frac{25}{2 \cdot 3 \cdot 3 \cdot 5}$$

The denominators are now the LCD.

$$= \frac{58}{5 \cdot 2 \cdot 3 \cdot 3}$$

Adding the numerators and keeping the LCD

$$= \frac{2 \cdot 29}{5 \cdot 2 \cdot 3 \cdot 3}$$

Factoring the numerator and removing a factor of 1

$$= \frac{29}{45}.$$

Simplifying

Do Exercises 12–15.

Subtraction

When subtracting, we also multiply by 1 to obtain the LCD. After we have made the denominators the same, we can subtract by subtracting the numerators and keeping the same denominator.

EXAMPLE 10 Subtract and simplify: $\dfrac{9}{8} - \dfrac{4}{5}$.

$$\frac{9}{8} - \frac{4}{5} = \frac{9}{8} \cdot \frac{5}{5} - \frac{4}{5} \cdot \frac{8}{8}$$

The LCD is 40.

$$= \frac{45}{40} - \frac{32}{40}$$

$$= \frac{45 - 32}{40} = \frac{13}{40}$$

Subtracting the numerators and keeping the same denominator

Add and simplify.

12. $\dfrac{4}{5} + \dfrac{3}{5}$

13. $\dfrac{5}{6} + \dfrac{7}{6}$

14. $\dfrac{5}{6} + \dfrac{7}{10}$

15. $\dfrac{13}{24} + \dfrac{7}{40}$

Answers

12. $\dfrac{7}{5}$ **13.** 2 **14.** $\dfrac{23}{15}$ **15.** $\dfrac{43}{60}$

EXAMPLE 11 Subtract and simplify: $\dfrac{7}{10} - \dfrac{1}{5}$.

$$\frac{7}{10} - \frac{1}{5} = \frac{7}{10} - \frac{1}{5} \cdot \frac{2}{2} \qquad \text{The LCD is 10; } \frac{7}{10} \text{ already has the LCD.}$$

$$= \frac{7}{10} - \frac{2}{10} = \frac{7-2}{10}$$

$$= \frac{5}{10}$$

$$= \frac{1 \cdot \cancel{5}}{2 \cdot \cancel{5}} = \frac{1}{2} \qquad \text{Removing a factor of 1: } \frac{5}{5} = 1$$

Do Exercises 16 and 17.

Subtract and simplify.

16. $\dfrac{7}{8} - \dfrac{2}{5}$ **17.** $\dfrac{5}{12} - \dfrac{2}{9}$

Reciprocals

Two numbers whose product is 1 are called **reciprocals**, or **multiplicative inverses**, of each other. All the arithmetic numbers, except zero, have reciprocals.

EXAMPLES

12. The reciprocal of $\frac{2}{3}$ is $\frac{3}{2}$ because $\frac{2}{3} \cdot \frac{3}{2} = \frac{6}{6} = 1$.

13. The reciprocal of 9 is $\frac{1}{9}$ because $9 \cdot \frac{1}{9} = \frac{9}{9} = 1$.

14. The reciprocal of $\frac{1}{4}$ is 4 because $\frac{1}{4} \cdot 4 = \frac{4}{4} = 1$.

Do Exercises 18–21.

Find each reciprocal.

18. $\dfrac{4}{11}$ **19.** $\dfrac{15}{7}$

20. 5   **21.** $\dfrac{1}{3}$

Reciprocals and Division

Reciprocals and the number 1 can be used to justify a fast way to divide arithmetic numbers. We multiply by 1, carefully choosing the expression for 1.

EXAMPLE 15 Divide $\dfrac{2}{3}$ by $\dfrac{7}{5}$.

This is a symbol for 1.

$$\frac{2}{3} \div \frac{7}{5} = \frac{\frac{2}{3}}{\frac{7}{5}} = \frac{\frac{2}{3}}{\frac{7}{5}} \cdot \frac{\frac{5}{7}}{\frac{5}{7}} \qquad \text{Multiplying by } \frac{\frac{5}{7}}{\frac{5}{7}}. \text{ We use } \frac{5}{7} \text{ because it is the reciprocal of } \frac{7}{5}.$$

$$= \frac{\frac{2}{3} \cdot \frac{5}{7}}{\frac{7}{5} \cdot \frac{5}{7}} \qquad \text{Multiplying numerators and denominators}$$

$$= \frac{\frac{10}{21}}{\frac{35}{35}} = \frac{\frac{10}{21}}{1} \qquad \frac{35}{35} = 1$$

$$= \frac{10}{21} \qquad \text{Simplifying}$$

After multiplying in Example 15, we had a denominator of $\frac{35}{35}$, or 1. That was because we used $\frac{5}{7}$, the reciprocal of the divisor, for both the numerator and the denominator of the symbol for 1.

Do Exercise 22.

22. Divide by multiplying by 1:

$$\frac{\frac{3}{5}}{\frac{4}{7}}.$$

Answers

16. $\dfrac{19}{40}$ **17.** $\dfrac{7}{36}$ **18.** $\dfrac{11}{4}$ **19.** $\dfrac{7}{15}$

20. $\dfrac{1}{5}$ **21.** 3   **22.** $\dfrac{21}{20}$

When multiplying by 1 to divide, we get a denominator of 1. What do we get in the numerator? In Example 15, we got $\frac{2}{3} \cdot \frac{5}{7}$. This is the product of $\frac{2}{3}$, the dividend, and $\frac{5}{7}$, the reciprocal of the divisor. This gives us a procedure for dividing fractions.

DIVIDING FRACTIONS

To divide fractions, multiply by the reciprocal of the divisor:

$$\frac{a}{b} \div \frac{c}{d} = \frac{a}{b} \cdot \frac{d}{c}.$$

EXAMPLE 16   Divide by multiplying by the reciprocal of the divisor: $\frac{1}{2} \div \frac{3}{5}$.

$$\frac{1}{2} \div \frac{3}{5} = \frac{1}{2} \cdot \frac{5}{3} \qquad \frac{5}{3} \text{ is the reciprocal of } \frac{3}{5}$$

$$= \frac{5}{6} \qquad \text{Multiplying}$$

After dividing, always simplify if possible.

EXAMPLE 17   Divide and simplify: $\frac{2}{3} \div \frac{4}{9}$.

$$\frac{2}{3} \div \frac{4}{9} = \frac{2}{3} \cdot \frac{9}{4} \qquad \frac{9}{4} \text{ is the reciprocal of } \frac{4}{9}$$

$$= \frac{2 \cdot 9}{3 \cdot 4} \qquad \text{Multiplying numerators and denominators}$$

$$= \frac{2 \cdot 3 \cdot 3}{3 \cdot 2 \cdot 2} \qquad \text{Removing a factor of 1: } \frac{2 \cdot 3}{2 \cdot 3} = 1$$

$$= \frac{3}{2}$$

Do Exercises 23–26.

Divide by multiplying by the reciprocal of the divisor. Then simplify.

23. $\dfrac{4}{3} \div \dfrac{7}{2}$ **24.** $\dfrac{5}{4} \div \dfrac{3}{2}$

25. $\dfrac{\frac{2}{9}}{\frac{5}{12}}$ **26.** $\dfrac{\frac{5}{6}}{\frac{45}{22}}$

EXAMPLE 18   Divide and simplify: $\dfrac{5}{6} \div 30$.

$$\frac{5}{6} \div 30 = \frac{5}{6} \div \frac{30}{1} = \frac{5}{6} \cdot \frac{1}{30} = \frac{5 \cdot 1}{6 \cdot 30} = \frac{5 \cdot 1}{6 \cdot 5 \cdot 6} = \frac{1}{6 \cdot 6} = \frac{1}{36}$$

$$\uparrow$$
$$\text{Removing a factor of 1: } \frac{5}{5} = 1$$

EXAMPLE 19   Divide and simplify: $24 \div \dfrac{3}{8}$.

$$24 \div \frac{3}{8} = \frac{24}{1} \div \frac{3}{8} = \frac{24}{1} \cdot \frac{8}{3} = \frac{24 \cdot 8}{1 \cdot 3} = \frac{3 \cdot 8 \cdot 8}{1 \cdot 3} = \frac{8 \cdot 8}{1} = 64$$

$$\uparrow$$
$$\text{Removing a factor of 1: } \frac{3}{3} = 1$$

Divide and simplify.

27. $\dfrac{7}{8} \div 56$ **28.** $36 \div \dfrac{4}{9}$

Do Exercises 27 and 28.

Answers

23. $\frac{8}{21}$ **24.** $\frac{5}{6}$ **25.** $\frac{8}{15}$

26. $\frac{11}{27}$ **27.** $\frac{1}{64}$ **28.** 81

a Write an equivalent expression for each of the following. Use the indicated name for 1.

1. $\dfrac{3}{4}\left(\text{Use }\dfrac{3}{3}\text{ for }1.\right)$
2. $\dfrac{5}{6}\left(\text{Use }\dfrac{10}{10}\text{ for }1.\right)$
3. $\dfrac{3}{5}\left(\text{Use }\dfrac{20}{20}\text{ for }1.\right)$
4. $\dfrac{8}{9}\left(\text{Use }\dfrac{4}{4}\text{ for }1.\right)$

Write an equivalent expression with the given denominator.

5. $\dfrac{7}{8}$ (Denominator: 24)
6. $\dfrac{2}{9}$ (Denominator: 54)

b Simplify.

7. $\dfrac{18}{27}$
8. $\dfrac{49}{56}$
9. $\dfrac{56}{14}$
10. $\dfrac{48}{27}$
11. $\dfrac{6}{42}$
12. $\dfrac{13}{104}$

13. $\dfrac{56}{7}$
14. $\dfrac{132}{11}$
15. $\dfrac{19}{76}$
16. $\dfrac{17}{51}$
17. $\dfrac{100}{20}$
18. $\dfrac{150}{25}$

19. $\dfrac{425}{525}$
20. $\dfrac{625}{325}$
21. $\dfrac{2600}{1400}$
22. $\dfrac{4800}{1600}$
23. $\dfrac{8 \cdot x}{6 \cdot x}$
24. $\dfrac{13 \cdot v}{39 \cdot v}$

c Compute and simplify.

25. $\dfrac{1}{3} \cdot \dfrac{1}{4}$
26. $\dfrac{15}{16} \cdot \dfrac{8}{5}$
27. $\dfrac{15}{4} \cdot \dfrac{3}{4}$
28. $\dfrac{10}{11} \cdot \dfrac{11}{10}$
29. $\dfrac{4}{9} + \dfrac{13}{18}$

30. $\dfrac{4}{5} + \dfrac{8}{15}$
31. $\dfrac{3}{10} + \dfrac{8}{15}$
32. $\dfrac{9}{8} + \dfrac{7}{12}$
33. $\dfrac{5}{4} - \dfrac{3}{4}$
34. $\dfrac{12}{5} - \dfrac{2}{5}$

35. $\dfrac{11}{12} - \dfrac{3}{8}$
36. $\dfrac{15}{16} - \dfrac{5}{12}$
37. $\dfrac{7}{6} \div \dfrac{3}{5}$
38. $\dfrac{7}{5} \div \dfrac{3}{4}$
39. $\dfrac{8}{9} \div \dfrac{4}{15}$

40. $\dfrac{3}{4} \div \dfrac{3}{7}$
41. $\dfrac{\frac{13}{12}}{\frac{39}{5}}$
42. $\dfrac{\frac{17}{6}}{\frac{3}{8}}$
43. $100 \div \dfrac{1}{5}$
44. $78 \div \dfrac{1}{6}$

45. $\dfrac{3}{4} \div 10$
46. $\dfrac{5}{6} \div 15$
47. $1000 - \dfrac{1}{100}$
48. $\dfrac{147}{50} - 2$

C

Exponential Notation and Order of Operations

OBJECTIVES

a Write exponential notation for a product.

b Evaluate exponential expressions.

c Simplify expressions using the rules for order of operations.

a Exponential Notation

Exponents provide a shorter way of writing products. An abbreviation for a product in which the factors are the same is called a **power**. An expression for a power is called **exponential notation**. For

$$\underbrace{10 \cdot 10 \cdot 10,}_{3 \text{ factors}} \quad \text{we write} \quad 10^3.$$

This is read "ten to the third power." We call the number 3 an **exponent** and we say that 10 is the **base**. For example,

$$a \cdot a \cdot a \cdot a = a^4.$$

This is the exponent.
This is the base.

An exponent of 2 or greater tells how many times the base is used as a factor.

EXPONENTIAL NOTATION

For any natural number n greater than or equal to 2,

$$b^n = \overbrace{b \cdot b \cdot b \cdot b \cdots b}^{n \text{ factors}}.$$

EXAMPLE 1 Write exponential notation for $10 \cdot 10 \cdot 10 \cdot 10 \cdot 10$.

$$10 \cdot 10 \cdot 10 \cdot 10 \cdot 10 = 10^5$$

Do Exercises 1–3.

Write exponential notation.
1. $4 \cdot 4 \cdot 4$

2. $6 \cdot 6 \cdot 6 \cdot 6 \cdot 6$

3. 1.08×1.08

b Evaluating Exponential Expressions

EXAMPLE 2 Evaluate: 3^4.

We have

$$3^4 = 3 \cdot 3 \cdot 3 \cdot 3 = 9 \cdot 9 = 81.$$

Do Exercises 4–7.

Evaluate.
4. 10^4

5. 8^3

6. $(1.1)^3$

7. $\left(\dfrac{2}{9}\right)^2$

c Order of Operations

What does $4 + 5 \times 2$ mean? If we add 4 and 5 and multiply the result by 2, we get 18. If we multiply 5 and 2 and add 4 to the result, we get 14. Since the results are different, we see that the order in which we carry out operations is important. To indicate which operation is to be done first, we use grouping symbols such as parentheses ( ), or brackets [ ], or braces { }. For example, $(3 \times 5) + 6 = 15 + 6 = 21$, but $3 \times (5 + 6) = 3 \times 11 = 33$.

Answers

1. 4^3 2. 6^5 3. 1.08^2 4. 10,000
5. 512   6. 1.331   7. $\dfrac{4}{81}$

Grouping symbols tell us what to do first. If there are no grouping symbols, there is a set of rules for the order in which operations should be done.

> **RULES FOR ORDER OF OPERATIONS**
>
> 1. Do all calculations within grouping symbols before operations outside.
> 2. Evaluate all exponential expressions.
> 3. Do all multiplications and divisions in order from left to right.
> 4. Do all additions and subtractions in order from left to right.

EXAMPLE 3 Calculate: $15 - 2 \times 5 + 3$.

$$
\begin{aligned}
15 - 2 \times 5 + 3 &= 15 - 10 + 3 & \text{Multiplying}\\
&= 5 + 3 & \text{Subtracting}\\
&= 8 & \text{Adding}
\end{aligned}
$$

Do Exercises 8 and 9.

Always calculate within parentheses first. When there are exponents and no parentheses, simplify powers first.

EXAMPLE 4 Calculate: $(3 \times 4)^2$.

$$
\begin{aligned}
(3 \times 4)^2 &= (12)^2 & \text{Working within parentheses first}\\
&= 144 & \text{Evaluating the exponential expression}
\end{aligned}
$$

EXAMPLE 5 Calculate: 3×4^2.

$$
\begin{aligned}
3 \times 4^2 &= 3 \times 16 & \text{Evaluating the exponential expression}\\
&= 48 & \text{Multiplying}
\end{aligned}
$$

Note that Examples 4 and 5 show that $(3 \times 4)^2 \neq 3 \times 4^2$.

EXAMPLE 6 Calculate: $7 + 3 \times 29 - 4^2$.

$$
\begin{aligned}
7 + 3 \times 29 - 4^2 &= 7 + 3 \times 29 - 16 & \text{There are no parentheses, so we find } 4^2 \text{ first.}\\
&= 7 + 87 - 16 & \text{Multiplying}\\
&= 94 - 16 & \text{Adding}\\
&= 78 & \text{Subtracting}
\end{aligned}
$$

Do Exercises 10–13.

EXAMPLE 7 Calculate: $100 \div 20 \div 2$.

$$
\begin{aligned}
100 \div 20 \div 2 &= 5 \div 2 & \text{Doing the divisions in order from left to right}\\
&= \frac{5}{2}, \text{ or } 2.5 & \text{Doing the second division}
\end{aligned}
$$

Calculate.

8. $16 - 3 \times 5 + 4$

9. $4 + 5 \times 2$

Calculate.

10. $18 - 4 \times 3 + 7$

11. $(2 \times 5)^3$

12. 2×5^3

13. $8 + 2 \times 5^3 - 4 \cdot 20$

Answers

8. 5   **9.** 14   **10.** 13
11. 1000   **12.** 250   **13.** 178

EXAMPLE 8 Calculate: $1000 \div \frac{1}{10} \cdot \frac{4}{5}$.

$$1000 \div \frac{1}{10} \cdot \frac{4}{5} = (1000 \cdot 10) \cdot \frac{4}{5} \qquad \text{Doing the division first}$$

$$= 10,000 \cdot \frac{4}{5} \qquad \text{Multiplying inside the parentheses}$$

$$= 8000 \qquad \text{Multiplying}$$

Do Exercises 14 and 15.

Sometimes combinations of grouping symbols are used. The rules still apply. We begin with the innermost grouping symbols and work to the outside.

EXAMPLE 9 Calculate: $5[4 + (8 - 2)]$.

$$5[4 + (8 - 2)] = 5[4 + 6] \qquad \text{Subtracting within the parentheses first}$$

$$= 5[10] \qquad \text{Adding inside the brackets}$$

$$= 50 \qquad \text{Multiplying}$$

A fraction bar can play the role of a grouping symbol.

EXAMPLE 10 Calculate: $\dfrac{12(9 - 7) + 4 \cdot 5}{3^3 - 2^4}$.

We do the calculations separately in the numerator and in the denominator, and then divide the results:

$$\frac{12(9 - 7) + 4 \cdot 5}{3^3 - 2^4} = \frac{12(2) + 4 \cdot 5}{27 - 16}$$

$$= \frac{24 + 20}{11}$$

$$= \frac{44}{11} = 4.$$

Do Exercises 16 and 17.

Calculate.

14. $51.2 \div 0.64 \div 40$

15. $1000 \cdot \dfrac{1}{10} \div \dfrac{4}{5}$

Calculate.

16. $4[(8 - 3) + 7]$

17. $\dfrac{13(10 - 6) + 4 \cdot 9}{5^2 - 3^2}$

Answers

14. 2 **15.** 125 **16.** 48 **17.** $\dfrac{11}{2}$

a Write exponential notation.

1. $5 \times 5 \times 5 \times 5$

2. $3 \times 3 \times 3 \times 3 \times 3$

3. $10 \cdot 10 \cdot 10$

4. $1 \cdot 1 \cdot 1$

5. $10 \times 10 \times 10 \times 10 \times 10 \times 10$

6. $18 \cdot 18$

b Evaluate.

7. 7^2

8. 4^3

9. 9^5

10. 12^4

11. 10^2

12. 1^5

13. 1^4

14. $(1.8)^2$

15. $(2.3)^2$

16. $(0.1)^3$

17. $(0.2)^3$

18. $(14.8)^2$

19. $(20.4)^2$

20. $\left(\dfrac{4}{5}\right)^2$

21. $\left(\dfrac{3}{8}\right)^2$

22. 2^4

23. 5^3

24. $(1.4)^3$

25. $1000 \times (1.02)^3$

26. $2000 \times (1.06)^2$

c Calculate.

27. $9 + 2 \times 8$

28. $14 + 6 \times 6$

29. $9(8) + 7(6)$

30. $30(5) + 2(2)$

31. $39 - 4 \times 2 + 2$

32. $14 - 2 \times 6 + 7$

33. $9 \div 3 + 16 \div 8$

34. $32 - 8 \div 4 - 2$

35. $7 + 10 - 10 \div 2$

36. $(5 \cdot 4)^2$

37. $(6 \cdot 3)^2$

38. $3 \cdot 2^3$

39. $4 \cdot 5^2$

40. $(7 + 3)^2$

41. $(8 + 2)^3$

42. $7 + 2^2$

43. $6 + 4^2$

44. $(5 - 2)^2$

45. $(3 - 2)^2$

46. $10 - 3^2$

47. $4^3 \div 8 - 4$

48. $20 + 4^3 \div 8 - 4$

49. $120 - 3^3 \cdot 4 \div 6$

50. $7 \times 3^4 + 18$

51. $6[9 + (3 + 4)]$

52. $8[(13 + 6) - 11]$

53. $8 + (7 + 9)$

54. $(8 + 7) + 9$

55. $15(4 + 2)$

56. $15 \cdot 4 + 15 \cdot 2$

57. $12 - (8 - 4)$

58. $(12 - 8) - 4$

59. $1000 \div 100 \div 10$

60. $256 \div 32 \div 4$

61. $2000 \div \dfrac{3}{50} \cdot \dfrac{3}{2}$

62. $400 \times 0.64 \div 3.2$

63. $75 \div 15 \cdot 4 \cdot 8 \div 32$

64. $84 \div 12 \cdot 10 \div 35 \cdot 8 \cdot 2 \div 16$

65. $16 \cdot 5 \div 80 \div 12 \cdot 36 \cdot 9$

66. $20 \cdot 45 \div 15 \div 15 \cdot 60 \div 12$

67. $\dfrac{80 - 6^2}{9^2 + 3^2}$

68. $\dfrac{5^2 + 4^3 - 3}{9^2 - 2^2 + 1^5}$

69. $\dfrac{3(6 + 7) - 5 \cdot 4}{6 \cdot 7 + 8(4 - 1)}$

70. $\dfrac{20(8 - 3) - 4(10 - 3)}{10(6 + 2) + 2(5 + 2)}$

71. $8 \cdot 2 - (12 - 0) \div 3 - (5 - 2)$

72. $95 - 2^3 \cdot 5 \div (24 - 4)$

Review of Factoring Polynomials

(a) Factoring Trinomials: $x^2 + bx + c$

Constant Term Positive

Recall the FOIL method* of multiplying two binomials:

$$(x + 3)(x + 5) = x^2 + 5x + 3x + 15$$
$$= x^2 + 8x + 15.$$

The product is a trinomial. In this example, the leading term has a coefficient of 1. The constant term is positive. To factor $x^2 + 8x + 15$, we think of FOIL in reverse.

EXAMPLE 1 Factor: $x^2 + 9x + 8$.

Think of FOIL in reverse. The first term of each factor is x. We are looking for numbers p and q such that

$$x^2 + 9x + 8 = (x + p)(x + q) = x^2 + (p + q)x + pq.$$

We look for two numbers p and q whose product is 8 and whose sum is 9. Since both 8 and 9 are positive, we need consider only positive factors.

PAIRS OF FACTORS	SUMS OF FACTORS
2, 4	6
1, 8	9 ←

The numbers we need are 1 and 8.

The factorization is $(x + 1)(x + 8)$. We can check by multiplying:

$$(x + 1)(x + 8) = x^2 + 9x + 8.$$

Do Margin Exercises 1 and 2.

EXAMPLE 2 Factor: $y^2 - 9y + 20$.

Since the constant term, 20, is positive and the coefficient of the middle term, -9, is negative, we look for a factorization of 20 in which both factors are negative. Their sum must be -9.

PAIRS OF FACTORS	SUMS OF FACTORS
-1, -20	-21
-2, -10	-12
-4, -5	-9 ←

The numbers we need are -4 and -5.

The factorization is $(y - 4)(y - 5)$.

Do Exercises 3 and 4.

---

* For a review of the *ac*-method, see Section 5.4.

OBJECTIVES

(a) Factor trinomials of the type $x^2 + bx + c$.

(b) Factor trinomials of the type $ax^2 + bx + c$, $a \neq 1$, by the FOIL method.

(c) Factor trinomial squares.

(d) Factor differences of squares.

(e) Factor sums and differences of cubes.

Factor. Check by multiplying.

1. $x^2 + 5x + 6$

2. $y^2 + 7y + 10$

Factor.

3. $m^2 - 8m + 12$

4. $24 - 11t + t^2$

Answers

1. $(x + 2)(x + 3)$ **2.** $(y + 2)(y + 5)$
3. $(m - 2)(m - 6)$ **4.** $(t - 3)(t - 8)$, or $(3 - t)(8 - t)$

Constant Term Negative

EXAMPLE 3   Factor: $x^3 - x^2 - 30x$.

Always look first for the largest common factor. This time x is the common factor. We first factor it out:

$$x^3 - x^2 - 30x = x(x^2 - x - 30).$$

Now consider $x^2 - x - 30$. Since the constant term, -30, is negative, we look for a factorization of -30 in which one factor is positive and one factor is negative. The sum of the factors must be -1, the coefficient of the middle term, so the negative factor must have the larger absolute value. Thus we consider only pairs of factors in which the negative factor has the larger absolute value.

PAIRS OF FACTORS	SUMS OF FACTORS
1, −30	−29
2, −15	−13
3, −10	−7
5, −6	−1 ←

The numbers we want are 5 and −6.

The factorization of $x^2 - x - 30$ is $(x + 5)(x - 6)$. But do not forget the common factor! The factorization of the original trinomial is

$$x(x + 5)(x - 6).$$

Do Exercises 5–7.

EXAMPLE 4   Factor: $x^2 + 17x - 110$.

Since the constant term, -110, is negative, we look for a factorization of -110 in which one factor is positive and one factor is negative. Their sum must be 17, so the positive factor must have the larger absolute value.

PAIRS OF FACTORS	SUMS OF FACTORS
−1, 110	109
−2, 55	53
−5, 22	17 ←
−10, 11	1

We consider only pairs of factors in which the positive term has the larger absolute value.

The numbers we need are −5 and 22.

The factorization is $(x - 5)(x + 22)$.

Do Exercises 8–10.

---

5. a) Factor: $x^2 - x - 20$.

  b) Explain why you would not consider these pairs of factors in factoring $x^2 - x - 20$.

PAIRS OF FACTORS	PRODUCTS OF FACTORS
1, 20	
2, 10	
4, 5	
−1, −20	
−2, −10	
−4, −5	

Factor.

6. $x^3 - 3x^2 - 54x$

7. $2x^3 - 2x^2 - 84x$

Factor.

8. $x^3 + 4x^2 - 12x$

9. $y^2 - 4y - 12$

10. $x^2 - 110 - x$

Answers

5. **(a)** $(x - 5)(x + 4)$; **(b)** The product of each pair is positive.   **6.** $x(x - 9)(x + 6)$
7. $2x(x - 7)(x + 6)$ **8.** $x(x + 6)(x - 2)$
9. $(y - 6)(y + 2)$ **10.** $(x + 10)(x - 11)$

b Factoring Trinomials: $ax^2 + bx + c, a \neq 1$

We first consider the **FOIL method** for factoring trinomials of the type $ax^2 + bx + c, a \neq 1$. Consider the following multiplication.

$$
\begin{array}{cccc}
\text{F} & \text{O} & \text{I} & \text{L} \\
\downarrow & \downarrow & \downarrow & \downarrow
\end{array}
$$

$$(3x + 2)(4x + 5) = 12x^2 + \underbrace{15x + 8x}_{} + 10$$

$$= 12x^2 + \quad 23x \quad + 10$$

To factor $12x^2 + 23x + 10$, we must reverse what we just did. We look for two binomials whose product is this trinomial. The product of the First terms must be $12x^2$. The product of the Outside terms plus the product of the Inside terms must be $23x$. The product of the Last terms must be 10. We know from the preceding discussion that the answer is $(3x + 2)(4x + 5)$. In general, however, finding such an answer involves trial and error. We use the following method.

THE FOIL METHOD

To factor trinomials of the type $ax^2 + bx + c, a \neq 1$, using the **FOIL** method:

1. Factor out the largest common factor.

2. Find two First terms whose product is ax^2:

$$(\square x + \quad)(\square x + \quad) = ax^2 + bx + c.$$
$$\text{FOIL}$$

3. Find two Last terms whose product is c:

$$(\quad x + \square)(\quad x + \square) = ax^2 + bx + c.$$
$$\text{FOIL}$$

4. Repeat steps (2) and (3) until a combination is found for which the sum of the Outside and Inside products is bx:

$$(\square x + \square)(\square x + \square) = ax^2 + bx + c.$$
$$\text{I} \qquad \text{FOIL}$$
$$\text{O}$$

5. Always check by multiplying.

EXAMPLE 5   Factor: $3x^2 + 10x - 8$.

1. First, we factor out the largest common factor, if any. There is none (other than 1 or -1).

2. Next, we factor the first term, $3x^2$. The only possibility is $3x \cdot x$. The desired factorization is then of the form $(3x + \square)(x + \square)$.

3. We then factor the last term, -8, which is negative. The possibilities are $(-8)(1), 8(-1), 2(-4),$ and $(-2)(4)$. They can be written in either order.

4. We look for combinations of factors from steps (2) and (3) such that the sum of the outside and the inside products is the middle term, $10x$:

$$\overset{\overset{3x}{\frown}}{(3x - 8)(x + 1)}_{\underset{-8x}{\smile}} = 3x^2 - 5x - 8; \qquad \overset{\overset{-3x}{\frown}}{(3x + 8)(x - 1)}_{\underset{8x}{\smile}} = 3x^2 + 5x - 8;$$
Wrong middle term · Wrong middle term

$$\overset{\overset{-12x}{\frown}}{(3x + 2)(x - 4)}_{\underset{2x}{\smile}} = 3x^2 - 10x - 8; \qquad \overset{\overset{12x}{\frown}}{(3x - 2)(x + 4)}_{\underset{-2x}{\smile}} = 3x^2 + 10x - 8.$$
Wrong middle term · Correct middle term!

There are four other possibilities that we could try, but we have a factorization: $(3x - 2)(x + 4)$.

5. *Check*: $(3x - 2)(x + 4) = 3x^2 + 10x - 8.$

Do Exercises 11 and 12.

EXAMPLE 6 Factor: $18x^6 - 57x^5 + 30x^4$.

1. First, we factor out the largest common factor, if any. The expression $3x^4$ is common to all terms, so we factor it out: $3x^4(6x^2 - 19x + 10)$.

2. Next, we factor the trinomial $6x^2 - 19x + 10$. We factor the first term, $6x^2$, and get $6x \cdot x$, or $3x \cdot 2x$. We then have these as possibilities for factorizations: $(3x + \square)(2x + \square)$ or $(6x + \square)(x + \square)$.

3. We then factor the last term, 10, which is positive. The possibilities are $(10)(1), (-10)(-1), (5)(2)$, and $(-5)(-2)$. They can be written in either order.

4. We look for combinations of factors from steps (2) and (3) such that the sum of the outside and the inside products is the middle term, $-19x$. The sign of the middle term is negative, but the sign of the last term, 10, is positive. Thus the signs of both factors of the last term, 10, must be negative. From our list of factors in step (3), we can use only $-10, -1$ and $-5, -2$ as possibilities. This reduces the possibilities for factorizations by half. We begin by using these factors with $(3x + \square)(2x + \square)$. Should we not find the correct factorization, we will consider $(6x + \square)(x + \square)$.

$$\overset{\overset{-3x}{\frown}}{(3x - 10)(2x - 1)}_{\underset{-20x}{\smile}} = 6x^2 - 23x + 10; \quad \overset{\overset{-30x}{\frown}}{(3x - 1)(2x - 10)}_{\underset{-2x}{\smile}} = 6x^2 - 32x + 10;$$
Wrong middle term · Wrong middle term

$$\overset{\overset{-6x}{\frown}}{(3x - 5)(2x - 2)}_{\underset{-10x}{\smile}} = 6x^2 - 16x + 10; \quad \overset{\overset{-15x}{\frown}}{(3x - 2)(2x - 5)}_{\underset{-4x}{\smile}} = 6x^2 - 19x + 10$$
Wrong middle term · Correct middle term!

We have a correct answer. We need not consider $(6x + \square)(x + \square)$.

The factorization of $6x^2 - 19x + 10$ is $(3x - 2)(2x - 5)$. But do not forget the common factor! We must include it in order to get a complete factorization of the original trinomial:

$$18x^6 - 57x^5 + 30x^4 = 3x^4(3x - 2)(2x - 5).$$

Factor by the FOIL method.

11. $3x^2 - 13x - 56$

12. $3x^2 + 5x + 2$

Answers

11. $(x - 7)(3x + 8)$ **12.** $(3x + 2)(x + 1)$

5. *Check*: $3x^4(3x - 2)(2x - 5) = 3x^4(6x^2 - 19x + 10)$
$$= 18x^6 - 57x^5 + 30x^4.$$

Do Exercises 13 and 14.

Factor.

13. $24y^2 - 46y + 10$

14. $20x^5 - 46x^4 + 24x^3$

Next, we consider some special factoring methods. When we recognize certain types of polynomials, we can factor more quickly using these special methods. Most of them are the reverse of the methods of special multiplication.

c Trinomial Squares

Consider the trinomial $x^2 + 6x + 9$. We look for factors of 9 whose sum is 6. We see that these factors are 3 and 3 and the factorization is

$$x^2 + 6x + 9 = (x + 3)(x + 3) = (x + 3)^2.$$

Note that the result is the square of a binomial. We also call $x^2 + 6x + 9$ a **trinomial square**, or **perfect-square trinomial**.

The factors of a trinomial square are two identical binomials. We use the following equations.

TRINOMIAL SQUARES

$A^2 + 2AB + B^2 = (A + B)^2$;
$A^2 - 2AB + B^2 = (A - B)^2$

EXAMPLE 7 Factor: $x^2 - 10x + 25$.

$$x^2 - 10x + 25 = (x - 5)^2$$

We find the square terms and write their square roots with a minus sign between them.

Note the sign!

EXAMPLE 8 Factor: $16y^2 + 49 + 56y$.

$$16y^2 + 49 + 56y = 16y^2 + 56y + 49$$

Rewriting in descending order

$$= (4y + 7)^2$$

We find the square terms and write their square roots with a plus sign between them.

Do Exercises 15 and 16.

Factor.

15. $x^2 + 14x + 49$

16. $9y^2 - 30y + 25$

d Differences of Squares

The following are *differences of squares*:

$$x^2 - 9, \quad 49 - 4y^2, \quad a^2 - 49b^2.$$

To factor a difference of two expressions that are squares, we can use a pattern for multiplying a sum and a difference that we used earlier.

Answers
13. $2(4y - 1)(3y - 5)$
14. $2x^3(2x - 3)(5x - 4)$
15. $(x + 7)^2$
16. $(3y - 5)^2$

> **FACTORING A DIFFERENCE OF SQUARES**
>
> $$A^2 - B^2 = (A + B)(A - B)$$

To factor a difference of squares $A^2 - B^2$, we find A and B, which are square roots of the expressions A^2 and B^2. We then use A and B to form two factors. One is the sum $A + B$, and the other is the difference $A - B$.

EXAMPLE 9 Factor: $x^2 - 9$.

$$\underset{\downarrow\quad\downarrow\quad\ \downarrow\quad\downarrow\quad\downarrow\quad\downarrow}{A^2 - B^2 = (A + B)(A - B)}$$

$$x^2 - 9 = x^2 - 3^2 = (x + 3)(x - 3)$$

EXAMPLE 10 Factor: $x^2 - \frac{1}{16}$.

$$x^2 - \tfrac{1}{16} = x^2 - \left(\tfrac{1}{4}\right)^2 = \left(x + \tfrac{1}{4}\right)\left(x - \tfrac{1}{4}\right)$$

Do Exercises 17–19.

Factor.

17. $y^2 - 4$

18. $49x^4 - 25y^{10}$

19. $m^2 - \dfrac{1}{9}$

(e) Sums or Differences of Cubes

We can factor the sum or the difference of two expressions that are cubes.
Consider the following products:

$$(A + B)(A^2 - AB + B^2) = A(A^2 - AB + B^2) + B(A^2 - AB + B^2)$$
$$= A^3 - A^2B + AB^2 + A^2B - AB^2 + B^3$$
$$= A^3 + B^3$$

and
$$(A - B)(A^2 + AB + B^2) = A(A^2 + AB + B^2) - B(A^2 + AB + B^2)$$
$$= A^3 + A^2B + AB^2 - A^2B - AB^2 - B^3$$
$$= A^3 - B^3.$$

The above equations (reversed) show how we can factor a sum or a difference of two cubes. Each factors as a product of a binomial and a trinomial.

> **SUM OR DIFFERENCE OF CUBES**
>
> $$A^3 + B^3 = (A + B)(A^2 - AB + B^2);$$
> $$A^3 - B^3 = (A - B)(A^2 + AB + B^2)$$

Note that what we are considering here is a sum or a difference of cubes. We are not cubing a binomial. For example, $(A + B)^3$ is *not* the same as $A^3 + B^3$. The table of cubes in the margin is helpful.

N	N^3
0.2	0.008
0.1	0.001
0	0
1	1
2	8
3	27
4	64
5	125
6	216
7	343
8	512
9	729
10	1000

EXAMPLE 11 Factor: $x^3 - 1000$.

We have

$$\underset{\quad\ \downarrow\qquad\ \downarrow}{A^3 - B^3}$$

$$x^3 - 1000 = x^3 - 10^3.$$

In one set of parentheses, we write the cube root of the first term, x. Then we write the cube root of the second term, -10. This gives us the expression $x - 10$:

$$(x - 10)(\qquad\qquad).$$

Answers

17. $(y + 2)(y - 2)$
18. $(7x^2 + 5y^5)(7x^2 - 5y^5)$
19. $\left(m + \tfrac{1}{3}\right)\left(m - \tfrac{1}{3}\right)$

To get the next factor, we think of $x - 10$ and do the following:

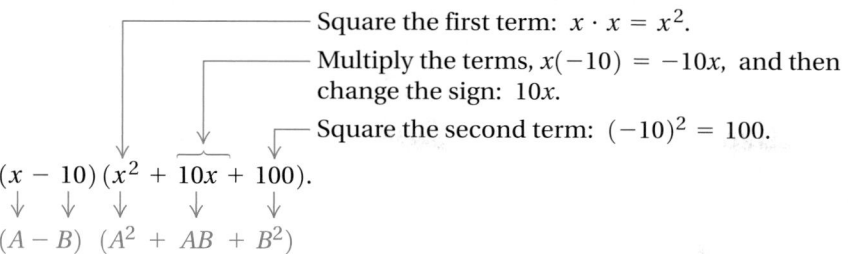

Square the first term: $x \cdot x = x^2$.

Multiply the terms, $x(-10) = -10x$, and then change the sign: $10x$.

Square the second term: $(-10)^2 = 100$.

$(x - 10)(x^2 + 10x + 100)$.

$(A - B) \quad (A^2 + AB + B^2)$

Note that we cannot factor $x^2 + 10x + 100$. It is not a trinomial square nor can it be factored by trial and error. Check this on your own.

Do Exercises 20 and 21.

Factor.

20. $w^3 - 27$

21. $1 - x^3$

EXAMPLE 12 Factor: $125 + 8y^3$.

We have

$$125 + 8y^3 = (5)^3 + (2y)^3.$$

In one set of parentheses, we write the cube root of the first term, 5. Then we write a plus sign, and then the cube root of the second term, $2y$:

$$(5 + 2y)(\qquad\qquad).$$

To get the next factor, we think of $5 + 2y$ and do the following:

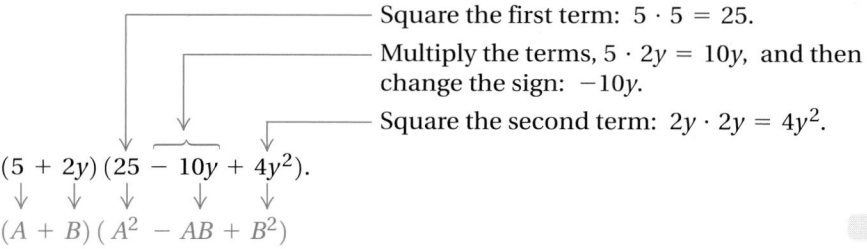

Square the first term: $5 \cdot 5 = 25$.

Multiply the terms, $5 \cdot 2y = 10y$, and then change the sign: $-10y$.

Square the second term: $2y \cdot 2y = 4y^2$.

$(5 + 2y)(25 - 10y + 4y^2)$.

$(A + B)(A^2 - AB + B^2)$

Do Exercises 22 and 23.

Factor.

22. $x^3 + 1000$

23. $343 + 64y^3$

FACTORING SUMMARY

Sum of cubes: $\quad A^3 + B^3 = (A + B)(A^2 - AB + B^2)$;

Difference of cubes: $\quad A^3 - B^3 = (A - B)(A^2 + AB + B^2)$;

Difference of squares: $\quad A^2 - B^2 = (A + B)(A - B)$;

Sum of squares: $\quad A^2 + B^2$ cannot be factored as the square of a binomial: $A^2 + B^2 \neq (A + B)^2$.

D **Exercise Set**

For Extra Help

MyMathLab

Math XL
PRACTICE

WATCH

DOWNLOAD

READ

REVIEW

a Factor.

1. $y^2 - 10y + 21$

2. $x^2 - 8x - 33$

3. $t^2 - 15 - 2t$

4. $2y^3 - 16y^2 + 32y$

5. $2a^3 - 20a^2 + 50a$

6. $p^2 + 3p - 54$

7. $m^2 + m - 72$

8. $12x + x^2 + 27$

9. $10y + y^2 + 24$

10. $y^2 - \dfrac{2}{3}y + \dfrac{1}{9}$

11. $p^2 + \dfrac{2}{5}p + \dfrac{1}{25}$

12. $t^2 - 4t + 3$

b Factor.

13. $3x^2 - 14x - 5$

14. $8x^2 - 6x - 9$

15. $10y^3 + y^2 - 21y$

16. $6x^3 + x^2 - 12x$

17. $3c^2 - 20c + 32$

18. $12b^2 - 8b + 1$

19. $35y^2 + 34y + 8$

20. $9a^2 + 18a + 8$

21. $4t + 10t^2 - 6$

22. $8x + 30x^2 - 6$

23. $8x^2 - 16 - 28x$

24. $18x^2 - 24 - 6x$

25. $12x^3 - 31x^2 + 20x$

26. $15x^3 - 19x^2 - 10x$

27. $14x^4 - 19x^3 - 3x^2$

28. $70x^4 - 68x^3 + 16x^2$

29. $3a^2 - a - 4$

30. $6a^2 - 7a - 10$

31. $9x^2 + 15x + 4$

32. $6y^2 - y - 2$

Factor.

33. $x^2 - 4x + 4$

34. $y^2 - 16y + 64$

35. $y^2 + 18y + 81$

36. $x^2 + 8x + 16$

37. $x^2 + 1 + 2x$

38. $x^2 + 1 - 2x$

39. $9y^2 + 12y + 4$

40. $25x^2 - 60x + 36$

41. $-18y^2 + y^3 + 81y$

42. $24a^2 + a^3 + 144a$

43. $12a^2 + 36a + 27$

44. $20y^2 + 100y + 125$

 Factor.

45. $x^2 - 16$

46. $y^2 - 9$

47. $p^2 - 49$

48. $m^2 - 64$

49. $4a^3 - 49a$

50. $9x^3 - 25x$

51. $\dfrac{1}{36} - z^2$

52. $\dfrac{1}{100} - y^2$

 Factor.

53. $z^3 + 27$

54. $a^3 + 8$

55. $x^3 - 1$

56. $c^3 - 64$

57. $8a^3 + 1$

58. $27x^3 + 1$

59. $8 - 27b^3$

60. $64 - 125x^3$

61. $2y^3 - 128$

62. $3z^3 - 3$

63. $24a^3 + 3$

64. $54x^3 + 2$

E Introductory Algebra Review

This text is appropriate for a two-semester course that combines the study of introductory and intermediate algebra. Students who take only the second-semester course (which generally begins with Chapter 7) often need a review of the topics covered in the first-semester course. This appendix is a guide for a review of the first six chapters of this text. Below is a syllabus of selected exercises that can be used as a condensed review of the main objectives in the first half of the text. For extra help, consult the *Student's Solutions Manual,* which contains fully worked-out solutions with step-by-step annotations for all the odd-numbered exercises in the exercise sets.

SECTION/OBJECTIVE	EXAMPLES	EXERCISES IN EXERCISE SET
1.3a	5–13	3, 11, 19, 25, 33, 39, 43
1.4a	6–12	5, 9, 15, 19, 21, 51, 55, 69, 81
1.5a	4–12, 17	17, 29, 33, 39, 53, 69
1.6c	19–23	49, 53, 55, 57
1.7c	14–17	47, 49, 59
1.7d	28–31	73, 89, 91
1.7e	32–38	99, 107, 111, 117
1.8b	7, 11, 12	15, 21
1.8c	16	29, 35
1.8d	20–23	41, 51, 61, 81
2.1b	6, 7	19, 39, 47
2.2a	2, 3, 6	3, 7, 31, 37
2.3a	2	5, 15
2.3b	7, 9	23, 43, 51
2.3c	11, 12, 13	71, 73, 83, 87
2.5a	4, 5, 6, 8	7, 15, 25, 35
2.6a	1, 2, 5, 8	1, 7, 15, 27
2.7e	13, 16	53, 61, 75
3.2a	2, 3, 5	5, 13, 21
3.3a	1	21, 25
3.3b	3, 4	43, 49
3.4a	1	9, 11
3.4c	5, 6, 7, 8	27, 39, 43
4.1b	2	15, 17
4.1c	5, 6	27, 35
4.1d, e, f	8, 10, 12, 15, 22–27	63, 71, 83, 87, 93, 101
4.2a, b	1–4, 10, 15, 17	3, 35, 41, 45
4.4a	3, 4	7, 17
4.4c	9, 10	33, 41, 49
4.5d	10	63
4.6a	2–6	9, 29
4.6b	11–13	43, 47
4.6c	16–18	63, 67, 71
4.7c	4–6	21, 27
4.7f	11, 12	41, 57
4.8b	7, 8, 10	25, 29, 33, 43

(continued)

SECTION/OBJECTIVE	EXAMPLES	EXERCISES IN EXERCISE SET
5.1b	9–12	19, 27
5.1c	14, 15, 18	43, 51
5.2a	1, 2, 3	1, 21, 27
5.3a	1, 2	3, 11, 25, 39, 71
5.4a	1	19, 35
5.5b	4, 6	13, 23, 33
5.5d	13, 17, 19	55, 59, 67, 81
5.6a	1, 2	1, 9, 23, 27
5.8b	4	25, 29, 37
5.9a	6	25
6.1d	11, 12	57, 63
6.2b	6, 8	13, 31
6.3c	4, 6	25, 31
6.4a	5, 6, 8	11, 17, 27, 59
6.5a	3, 4	9, 17, 39
6.7a	1, 3, 5	5, 11, 21, 35
6.8a	2	13
6.8b	3	27

Two other features of the text that can be used for review of the first six chapters are as follows.

- At the end of each chapter is a *Summary and Review* that provides an extensive set of review exercises. Reference codes beside each exercise or direction line preceding it allow the student to easily return to the objective being reviewed. Answers to all of these exercises appear at the back of the book.

- The *Cumulative Review* that follows every chapter starting with Chapter 2 can also be used for review. Each reviews material from all preceding chapters. At the back of the text are answers to all Cumulative Review exercises, together with section and objective references, so that students know exactly what material to study if they miss an exercise.

The extensive supplements package that accompanies this text also includes material appropriate for a structured review of the first six chapters. Consult the preface in the text for detailed descriptions of each of the following.

- Lecture videos and Chapter Test Prep videos on DVD
- InterAct Math Tutorial Website
- MathXL® Tutorials on CD
- MyMathLab

F

Handling Dimension Symbols

OBJECTIVES

a Perform calculations with dimension symbols.

b Make unit changes.

a Calculating with Dimension Symbols

In many applications, we add, subtract, multiply, and divide quantities having units, or dimensions, such as ft, km, sec, and hr. For example, to find average speed, we divide total distance by total time. What results is notation very much like a rational expression.

EXAMPLE 1 A car travels 150 km in 2 hr. What is its average speed?

$$\text{Speed} = \frac{150 \text{ km}}{2 \text{ hr}}, \text{ or } 75 \frac{\text{km}}{\text{hr}}$$

(The standard abbreviation for km/hr is km/h, but it does not suit our present discussion well.)

The symbol km/hr makes it look as though we are dividing kilometers by hours. It can be argued that we can divide only numbers. Nevertheless, we treat dimension symbols, such as km, ft, and hr, as if they were numerals or variables, obtaining correct results mechanically.

Do Exercise 1.

1. A truck travels 210 mi in 3 hr. What is its average speed?

EXAMPLES Compare the following.

2. $\dfrac{150x}{2y} = \dfrac{150}{2} \cdot \dfrac{x}{y} = 75\dfrac{x}{y}$ with $\dfrac{150 \text{ km}}{2 \text{ hr}} = \dfrac{150}{2}\dfrac{\text{km}}{\text{hr}} = 75\dfrac{\text{km}}{\text{hr}}$

3. $3x + 2x = (3 + 2)x = 5x$ with $3 \text{ ft} + 2 \text{ ft} = (3 + 2) \text{ ft} = 5 \text{ ft}$

4. $5x \cdot 3x = 15x^2$ with $5 \text{ ft} \cdot 3 \text{ ft} = 15 \text{ ft}^2$ (square feet)

Do Exercises 2-4.

Perform each calculation.

2. $\dfrac{100 \text{ m}}{4 \text{ sec}}$

3. $7 \text{ yd} + 9 \text{ yd}$

4. $24 \text{ in.} \cdot 3 \text{ in.}$

If 5 men work 8 hours, the total amount of labor is 40 man-hours.

EXAMPLE 5 Compare

$$5x \cdot 8y = 40xy \quad \text{with} \quad 5 \text{ men} \cdot 8 \text{ hours} = 40 \text{ man-hours}.$$

Do Exercise 5.

5. Calculate: 6 men · 11 hours.

Answers

1. $70\dfrac{\text{mi}}{\text{hr}}$ **2.** $25\dfrac{\text{m}}{\text{sec}}$ **3.** 16 yd **4.** 72 in^2
5. 66 man-hours

EXAMPLE 6 Compare

$$\frac{300x \cdot 240y}{15t} = 4800\frac{xy}{t} \quad \text{with} \quad \frac{300\text{ kW} \cdot 240\text{ hr}}{15\text{ da}} = 4800\frac{\text{kW-hr}}{\text{da}}.$$

If an electrical device uses 300 kW (kilowatts) for 240 hr over a period of 15 days, its rate of usage of energy is 4800 kilowatt-hours per day. The standard abbreviation for kilowatt-hours is kWh.

Do Exercise 6.

6. Calculate:
$$\frac{200\text{ kW} \cdot 140\text{ hr}}{35\text{ da}}.$$

(b) Making Unit Changes

We can treat dimension symbols much like numerals or variables, because we obtain correct results that way. We can change units by substituting or by multiplying by 1, as shown below.

EXAMPLE 7 Convert 3 ft to inches.

METHOD 1.   We have 3 ft. We know that 1 ft = 12 in., so we substitute 12 in. for ft:

$$3\text{ ft} = 3 \cdot 12\text{ in.} = 36\text{ in.}$$

METHOD 2.   We want to convert from "ft" to "in." We multiply by 1 using a symbol for 1 with "ft" in the denominator since we are converting from "ft," and with "in." in the numerator since we are converting to "in."

$$3\text{ ft} = 3\text{ ft} \cdot \frac{12\text{ in.}}{1\text{ ft}}$$

$$= \frac{3 \cdot 12}{1} \cdot \frac{\text{ft}}{\text{ft}} \cdot \text{in.} = 36\text{ in.}$$

Do Exercise 7.

7. Convert 7 ft to inches.

We can multiply by 1 several times to make successive conversions. In the following example, we convert mi/hr to ft/sec by converting successively from mi/hr to ft/hr to ft/min to ft/sec.

EXAMPLE 8 Convert 60 mi/hr to ft/sec.

$$60\frac{\text{mi}}{\text{hr}} = 60\frac{\text{mi}}{\text{hr}} \cdot \frac{5280\text{ ft}}{1\text{ mi}} \cdot \frac{1\text{ hr}}{60\text{ min}} \cdot \frac{1\text{ min}}{60\text{ sec}}$$

$$= \frac{60 \cdot 5280}{60 \cdot 60} \cdot \frac{\text{mi}}{\text{mi}} \cdot \frac{\text{hr}}{\text{hr}} \cdot \frac{\text{min}}{\text{min}} \cdot \frac{\text{ft}}{\text{sec}} = 88\frac{\text{ft}}{\text{sec}}.$$

Do Exercise 8.

8. Convert 90 mi/hr to ft/sec.

Answers

6. $800\frac{\text{kW-hr}}{\text{da}}$ **7.** 84 in.   **8.** $132\frac{\text{ft}}{\text{sec}}$

a Add the measures.

1. 45 ft + 23 ft

2. 55 km/hr + 27 km/hr

3. 17 g + 28 g

4. 3.4 lb + 5.2 lb

Find the average speeds, given total distance and total time.

5. 90 mi, 6 hr

6. 640 km, 20 hr

7. 9.9 m, 3 sec

8. 76 ft, 4 min

Perform the calculations.

9. $\dfrac{3 \text{ in.} \cdot 8 \text{ lb}}{6 \text{ sec}}$

10. $\dfrac{60 \text{ men} \cdot 8 \text{ hr}}{20 \text{ da}}$

11. $36 \text{ ft} \cdot \dfrac{1 \text{ yd}}{3 \text{ ft}}$

12. $55 \dfrac{\text{mi}}{\text{hr}} \cdot 4 \text{ hr}$

13. $5 \text{ ft}^3 + 11 \text{ ft}^3$

14. $\dfrac{3 \text{ lb}}{14 \text{ ft}} \cdot \dfrac{7 \text{ lb}}{6 \text{ ft}}$

15. Divide \$4850 by 5 days.

16. Divide \$25.60 by 8 hr.

b Make the unit changes.

17. Change 3.2 lb to oz (16 oz = 1 lb).

18. Change 6.2 km to m.

19. Change 35 mi/hr to ft/min.

20. Change \$375 per day to dollars per minute.

21. Change 8 ft to in.

22. Change 25 yd to ft.

23. How many years ago is 1 million sec ago? Let 365 days = 1 yr.

24. How many years ago is 1 billion sec ago?

25. How many years ago is 1 trillion sec ago?

26. Change 20 lb to oz.

27. Change $60 \dfrac{\text{lb}}{\text{ft}}$ to $\dfrac{\text{oz}}{\text{in.}}$.

28. Change $44 \dfrac{\text{ft}}{\text{sec}}$ to $\dfrac{\text{mi}}{\text{hr}}$.

29. Change 2 days to seconds.

30. Change 128 hr to days.

31. Change 216 in^2 to ft^2.

32. Change 1440 man-hours to man-days.

33. Change $80 \dfrac{\text{lb}}{\text{ft}^3}$ to $\dfrac{\text{ton}}{\text{yd}^3}$.

34. Change the speed of light, 186,000 mi/sec, to mi/yr.

G   Mean, Median, and Mode

a   Mean, Median, and Mode

One way to analyze data is to look for a single representative number, called a **center point** or **measure of central tendency**. Those most often used are the **mean** (or **average**), the **median**, and the **mode**.

Mean

> **MEAN, OR AVERAGE**
>
> The **mean**, or **average**, of a set of numbers is the sum of the numbers divided by the number of addends.

EXAMPLE 1 Consider the number of prescriptions, in millions, filled in supermarkets in 1997, 1999, 2001, 2003, 2005, and 2007:

$$269, \quad 357, \quad 418, \quad 462, \quad 465, \quad 478.$$

What is the mean, or average, of the numbers?

Source: IMS HEALTH and NACDS Economics Department

First, we add the numbers:

$$269 + 357 + 418 + 462 + 465 + 478 = 2449.$$

Then we divide by the number of addends, 6:

$$\frac{2449}{6} \approx 408. \quad \text{Rounding to the nearest one}$$

The mean, or average, number of prescriptions filled in super-markets in those six years is about 408 million.

Note that if the number of prescriptions had been the average (same) for each of the six years, we would have

$$408 + 408 + 408 + 408 + 408 + 408 = 2448 \approx 2449.$$

The number 408 is called the mean, or average, of the set of numbers.

> Do Exercises 1–3.

Median

The *median* is useful when we wish to de-emphasize extreme values. For example, suppose five workers in a technology company manufactured the following number of computers during one month's work:

Sarah: 88	Jen: 94	Matt: 92
Mark: 91	Pat: 66	

Let's first list the values in order from smallest to largest:

66 88 **91** 92 94.
          ↑
    Middle number

The middle number—in this case, 91—is the **median**.

OBJECTIVE

a Find the mean (average), the median, and the mode of a set of data and solve related applied problems.

Find the mean. Round to the nearest tenth.

1. 28, 103, 39

2. 85, 46, 105.7, 22.1

3. A student scored the following on five tests:

78, 95, 84, 100, 82.

What was the average score?

Answers
1. 56.7   **2.** 64.7   **3.** 87.8

Once a set of data has been arranged from smallest to largest, the **median** of the set of data is the middle number if there is an odd number of data numbers. If there is an even number of data numbers, then there are two middle numbers and the median is the *average* of the two middle numbers.

EXAMPLE 2 What is the median of the following set of yearly salaries?

$76,000, $58,000, $87,000, $32,500, $64,800, $62,500

We first rearrange the numbers in order from smallest to largest.

$32,500, $58,000, $62,500, $64,800, $76,000, $87,000

↑
Median

There is an even number of numbers. We look for the middle two, which are $62,500 and $64,800. In this case, the median is the average of $62,500 and $64,800:

$$\frac{\$62,500 + \$64,800}{2} = \$63,650.$$

Do Exercises 4–6.

Mode

The last center point we consider is called the *mode*. A number that occurs most often in a set of data can be considered a representative number or center point.

> **MODE**
>
> The **mode** of a set of data is the number or numbers that occur most often. If each number occurs the same number of times, there is *no* mode.

EXAMPLE 3 Find the mode of the following data:

23, 24, 27, 18, 19, 27

The number that occurs most often is 27. Thus the mode is 27.

EXAMPLE 4 Find the mode of the following data:

83, 84, 84, 84, 85, 86, 87, 87, 87, 88, 89, 90.

There are two numbers that occur most often, 84 and 87. Thus the modes are 84 and 87.

EXAMPLE 5 Find the mode of the following data:

115, 117, 211, 213, 219.

Each number occurs the same number of times. The set of data has *no* mode.

Do Exercises 7–10.

Find the median.

4. 17, 13, 18, 14, 19

5. 17, 18, 16, 19, 13, 14

6. 122, 102, 103, 91, 83, 81, 78, 119, 88

Find any modes that exist.

7. 33, 55, 55, 88, 55

8. 90, 54, 88, 87, 87, 54

9. 23.7, 27.5, 54.9, 17.2, 20.1

10. In conducting laboratory tests, Carole discovers bacteria in different lab dishes grew to the following areas, in square millimeters:

25, 19, 29, 24, 28.

a) What is the mean?

b) What is the median?

c) What is the mode?

Answers

4. 17   **5.** 16.5   **6.** 91   **7.** 55   **8.** 54, 87
9. No mode exists.   **10. (a)** 25 mm^2;
(b) 25 mm^2; **(c)** no mode exists.

a   For each set of numbers, find the mean (average), the median, and any modes that exist.

1. 17, 19, 29, 18, 14, 29

2. 13, 32, 25, 27, 13

3. 4.3, 7.4, 1.2, 5.7, 8.3

4. 13.4, 13.4, 12.6, 42.9

5. 234, 228, 234, 229, 234, 278

6. $29.95, $28.79, $30.95, $28.79

7. *Atlantic Storms and Hurricanes.*   The following bar graph shows the number of Atlantic storms or hurricanes that formed in various months from 1980 to 2007. What is the average number for the 9 months given? the median? the mode?

Atlantic Storms and Hurricanes

Tropical storm and hurricane formation in 1980–2007, by month

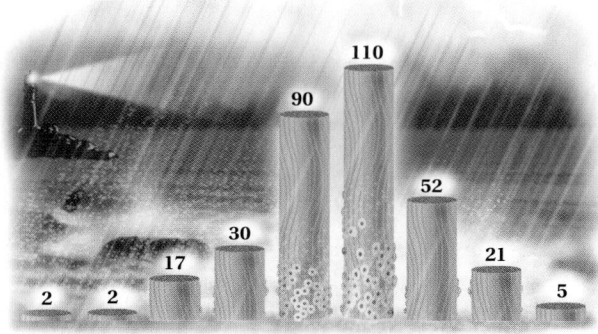

SOURCE: Colorado State University, Department of Atmospheric Science, Phil Klotzbach, Ph.D., Research Scientist

8. *Tornadoes.*   The following bar graph shows the average number of tornado deaths by month since 1950. What is the average number of tornado deaths for the 12 months? the median? the mode?

Average Number of Deaths by Tornado by Month

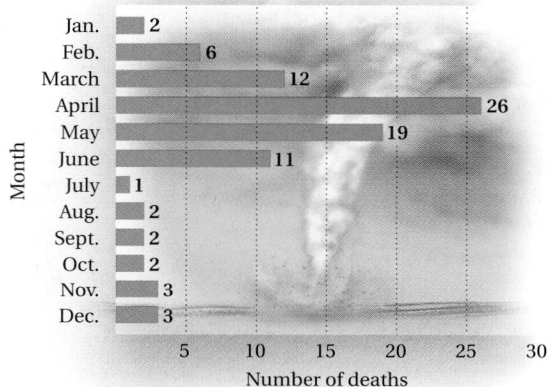

SOURCE: National Weather Service's Storm Prediction Center

9. *Brussels Sprouts.*   The following prices per stalk of Brussels sprouts were found at five supermarkets:

$3.99, $4.49, $4.99, $3.99, $3.49.

What was the average price per stalk? the median price? the mode?

10. *Cheddar Cheese Prices.*   The following prices per pound of sharp cheddar cheese were found at five supermarkets:

$5.99, $6.79, $5.99, $6.99, $6.79.

What was the average price per pound? the median price? the mode?

H

Synthetic Division

a Use synthetic division to divide a polynomial by a binomial of the type $x - a$.

a Synthetic Division

To divide a polynomial by a binomial of the type $x - a$, we can streamline the general procedure by a process called **synthetic division**.

Compare the following. In **A,** we perform a division. In **B,** we also divide but we do not write the variables.

A.

$$
\begin{array}{r}
4x^2 + 5x + 11 \\
x - 2\overline{)4x^3 - 3x^2 + x + 7} \\
\underline{4x^3 - 8x^2} \\
5x^2 + x \\
\underline{5x^2 - 10x} \\
11x + 7 \\
\underline{11x - 22} \\
29
\end{array}
$$

B.

$$
\begin{array}{r}
4 + 5 + 11 \\
1 - 2\overline{)4 - 3 + 1 + 7} \\
\underline{4 - 8} \\
5 + 1 \\
\underline{5 - 10} \\
11 + 7 \\
\underline{11 - 22} \\
29
\end{array}
$$

In **B,** there is still some duplication of writing. Also, since we can subtract by adding the opposite, we can use 2 instead of -2 and then add instead of subtracting.

C. *Synthetic Division*

a) $\quad \underline{2}\,|\,4 \quad -3 \quad 1 \quad 7$

$\qquad \overline{}$
$\qquad 4$

Write the 2, the opposite of -2 in the divisor $x - 2$, and the coefficients of the dividend.

Bring down the first coefficient.

b) $\quad \underline{2}\,|\,4 \quad -3 \quad 1 \quad 7$

$\qquad \qquad 8$

$\qquad \overline{4 \quad 5}$

Multiply 4 by 2 to get 8. Add 8 and -3.

c) $\quad \underline{2}\,|\,4 \quad -3 \quad 1 \quad 7$

$\qquad \qquad 8 \quad 10$

$\qquad \overline{4 \quad 5 \quad 11}$

Multiply 5 by 2 to get 10. Add 10 and 1.

d) $\quad \underline{2}\,|\,4 \quad -3 \quad 1 \quad 7$

$\qquad \qquad 8 \quad 10 \quad 22$

$\qquad \overline{4 \quad 5 \quad 11\,|\,29}$

Multiply 11 by 2 to get 22. Add 22 and 7.

$\qquad\qquad$ Quotient $\quad$ Remainder

The last number, 29, is the remainder. The other numbers are the coefficients of the quotient with that of the term of highest degree first, as follows. Note that the degree of the term of highest degree is 1 less than the degree of the dividend.

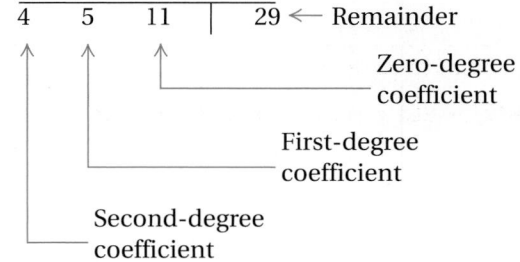

The answer is $4x^2 + 5x + 11$, R 29; or $4x^2 + 5x + 11 + \dfrac{29}{x-2}$.

EXAMPLE 1 Use synthetic division to divide:

$$(x^3 + 6x^2 - x - 30) \div (x - 2).$$

We have

$$
\begin{array}{r|rrr}
2 & 1 & 6 & -1 & -30 \\
  &   & 2 & 16 & 30 \\
\hline
  & 1 & 8 & 15 & 0
\end{array}
$$

The answer is $x^2 + 8x + 15$, R 0; or just $x^2 + 8x + 15$.

Do Exercise 1.

When there are missing terms, be sure to write 0's for their coefficients.

EXAMPLES Use synthetic division to divide.

2. $(2x^3 + 7x^2 - 5) \div (x + 3)$

There is no x-term, so we must write a 0 for its coefficient. Note that $x + 3 = x - (-3)$, so we write -3 at the left.

$$
\begin{array}{r|rrrr}
-3 & 2 & 7 & 0 & -5 \\
   &   & -6 & -3 & 9 \\
\hline
   & 2 & 1 & -3 & 4
\end{array}
$$

The answer is $2x^2 + x - 3$, R 4; or $2x^2 + x - 3 + \dfrac{4}{x+3}$.

3. $(x^3 + 4x^2 - x - 4) \div (x + 4)$

Note that $x + 4 = x - (-4)$, so we write -4 at the left.

$$
\begin{array}{r|rrrr}
-4 & 1 & 4 & -1 & -4 \\
   &   & -4 & 0 & 4 \\
\hline
   & 1 & 0 & -1 & 0
\end{array}
$$

The answer is $x^2 - 1$.

4. $(x^4 - 1) \div (x - 1)$

The divisor is $x - 1$, so we write 1 at the left.

$$
\begin{array}{r|rrrrr}
1 & 1 & 0 & 0 & 0 & -1 \\
  &   & 1 & 1 & 1 & 1 \\
\hline
  & 1 & 1 & 1 & 1 & 0
\end{array}
$$

The answer is $x^3 + x^2 + x + 1$.

5. $(8x^5 - 6x^3 + x - 8) \div (x + 2)$

Note that $x + 2 = x - (-2)$, so we write -2 at the left.

$$
\begin{array}{r|rrrrrr}
-2 & 8 & 0 & -6 & 0 & 1 & -8 \\
   &   & -16 & 32 & -52 & 104 & -210 \\
\hline
   & 8 & -16 & 26 & -52 & 105 & -218
\end{array}
$$

The answer is $8x^4 - 16x^3 + 26x^2 - 52x + 105$, R -218; or

$$8x^4 - 16x^3 + 26x^2 - 52x + 105 + \dfrac{-218}{x+2}.$$

Do Exercises 2 and 3.

It is important to remember that in order for synthetic division to work, the divisor must be of the form $x - a$, that is, a variable minus a constant. The coefficient of the variable must be 1.

1. Use synthetic division to divide:

$$(2x^3 - 4x^2 + 8x - 8) \div (x - 3).$$

Use synthetic division to divide.

2. $(x^3 - 2x^2 + 5x - 4) \div (x + 2)$

3. $(y^3 + 1) \div (y + 1)$

Answers

1. $2x^2 + 2x + 14$, R 34; or
$2x^2 + 2x + 14 + \dfrac{34}{x - 3}$

2. $x^2 - 4x + 13$, R -30; or
$x^2 - 4x + 13 + \dfrac{-30}{x + 2}$

3. $y^2 - y + 1$

Exercise Set

a Use synthetic division to divide.

1. $(x^3 - 2x^2 + 2x - 5) \div (x - 1)$

2. $(x^3 - 2x^2 + 2x - 5) \div (x + 1)$

3. $(a^2 + 11a - 19) \div (a + 4)$

4. $(a^2 + 11a - 19) \div (a - 4)$

5. $(x^3 - 7x^2 - 13x + 3) \div (x - 2)$

6. $(x^3 - 7x^2 - 13x + 3) \div (x + 2)$

7. $(3x^3 + 7x^2 - 4x + 3) \div (x + 3)$

8. $(3x^3 + 7x^2 - 4x + 3) \div (x - 3)$

9. $(y^3 - 3y + 10) \div (y - 2)$

10. $(x^3 - 2x^2 + 8) \div (x + 2)$

11. $(3x^4 - 25x^2 - 18) \div (x - 3)$

12. $(6y^4 + 15y^3 + 28y + 6) \div (y + 3)$

13. $(x^3 - 8) \div (x - 2)$

14. $(y^3 + 125) \div (y + 5)$

15. $(y^4 - 16) \div (y - 2)$

16. $(x^5 - 32) \div (x - 2)$

I Determinants and Cramer's Rule

In Chapter 8, you probably noticed that the elimination method concerns itself primarily with the coefficients and constants of the equations. Here we learn a method for solving a system of equations using just the coefficients and constants. This method involves *determinants*.

a Evaluating Determinants

The following symbolism represents a **second-order determinant**:

$$\begin{vmatrix} a_1 & b_1 \\ a_2 & b_2 \end{vmatrix}.$$

To evaluate a determinant, we do two multiplications and subtract.

EXAMPLE 1 Evaluate:

$$\begin{vmatrix} 2 & -5 \\ 6 & 7 \end{vmatrix}.$$

We multiply and subtract as follows:

$$\begin{vmatrix} 2 & -5 \\ 6 & 7 \end{vmatrix} = 2 \cdot 7 - 6 \cdot (-5) = 14 + 30 = 44.$$

Determinants are defined according to the pattern shown in Example 1.

> ### SECOND-ORDER DETERMINANT
>
> The determinant $\begin{vmatrix} a_1 & b_1 \\ a_2 & b_2 \end{vmatrix}$ is defined to mean $a_1b_2 - a_2b_1$.

The value of a determinant is a *number*. In Example 1, the value is 44.

Do Exercises 1 and 2.

b Third-Order Determinants

A **third-order determinant** is defined as follows.

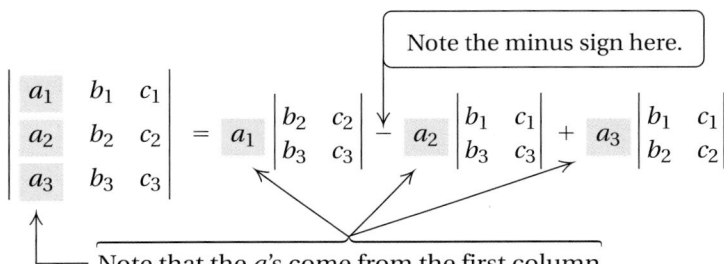

Note the minus sign here.

$$\begin{vmatrix} a_1 & b_1 & c_1 \\ a_2 & b_2 & c_2 \\ a_3 & b_3 & c_3 \end{vmatrix} = a_1 \begin{vmatrix} b_2 & c_2 \\ b_3 & c_3 \end{vmatrix} - a_2 \begin{vmatrix} b_1 & c_1 \\ b_3 & c_3 \end{vmatrix} + a_3 \begin{vmatrix} b_1 & c_1 \\ b_2 & c_2 \end{vmatrix}$$

Note that the a's come from the first column.

OBJECTIVES

a Evaluate second-order determinants.

b Evaluate third-order determinants.

c Solve systems of equations using Cramer's rule.

Evaluate.

1. $\begin{vmatrix} 3 & 2 \\ 4 & 1 \end{vmatrix}$

2. $\begin{vmatrix} 5 & -2 \\ -1 & -1 \end{vmatrix}$

Answers

1. -5 **2.** -7

Note that the second-order determinants on the right can be obtained by crossing out the row and the column in which each a occurs.

$$\text{For } a_1: \quad \begin{vmatrix} \cancel{a_1} & \cancel{b_1} & \cancel{c_1} \\ a_2 & b_2 & c_2 \\ a_3 & b_3 & c_3 \end{vmatrix} \qquad \text{For } a_2: \quad \begin{vmatrix} a_1 & b_1 & c_1 \\ \cancel{a_2} & \cancel{b_2} & \cancel{c_2} \\ a_3 & b_3 & c_3 \end{vmatrix}$$

$$\text{For } a_3: \quad \begin{vmatrix} a_1 & b_1 & c_1 \\ a_2 & b_2 & c_2 \\ \cancel{a_3} & \cancel{b_3} & \cancel{c_3} \end{vmatrix}$$

EXAMPLE 2 Evaluate this third-order determinant:

$$\begin{vmatrix} -1 & 0 & 1 \\ -5 & 1 & -1 \\ 4 & 8 & 1 \end{vmatrix} = -1 \begin{vmatrix} 1 & -1 \\ 8 & 1 \end{vmatrix} - (-5) \begin{vmatrix} 0 & 1 \\ 8 & 1 \end{vmatrix} + 4 \begin{vmatrix} 0 & 1 \\ 1 & -1 \end{vmatrix}.$$

We calculate as follows:

$$-1 \begin{vmatrix} 1 & -1 \\ 8 & 1 \end{vmatrix} - (-5) \begin{vmatrix} 0 & 1 \\ 8 & 1 \end{vmatrix} + 4 \begin{vmatrix} 0 & 1 \\ 1 & -1 \end{vmatrix}$$
$$= -1[1 \cdot 1 - 8(-1)] + 5(0 \cdot 1 - 8 \cdot 1) + 4[0 \cdot (-1) - 1 \cdot 1]$$
$$= -1(9) + 5(-8) + 4(-1)$$
$$= -9 - 40 - 4$$
$$= -53.$$

Do Exercises 3 and 4.

Evaluate.

3. $\begin{vmatrix} 2 & -1 & 1 \\ 1 & 2 & -1 \\ 3 & 4 & -3 \end{vmatrix}$

4. $\begin{vmatrix} 3 & 2 & 2 \\ -2 & 1 & 4 \\ 4 & -3 & 3 \end{vmatrix}$

c Solving Systems Using Determinants

Here is a system of two equations in two variables:

$$a_1x + b_1y = c_1,$$
$$a_2x + b_2y = c_2.$$

We form three determinants, which we call D, D_x, and D_y.

$$D = \begin{vmatrix} a_1 & b_1 \\ a_2 & b_2 \end{vmatrix} \qquad \text{In } D, \text{ we have the coefficients of } x \text{ and } y.$$

$$D_x = \begin{vmatrix} c_1 & b_1 \\ c_2 & b_2 \end{vmatrix} \qquad \text{To form } D_x, \text{ we replace the } x\text{-coefficients in } D \text{ with the constants on the right side of the equations.}$$

$$D_y = \begin{vmatrix} a_1 & c_1 \\ a_2 & c_2 \end{vmatrix} \qquad \text{To form } D_y, \text{ we replace the } y\text{-coefficients in } D \text{ with the constants on the right.}$$

It is important that the replacement be done *without changing the order of the columns*. Then the solution of the system can be found as follows. This is known as **Cramer's rule**.

Answers

3. -6 4. 93

CRAMER'S RULE

$$x = \frac{D_x}{D}, \qquad y = \frac{D_y}{D}$$

EXAMPLE 3   Solve using Cramer's rule:

$$3x - 2y = 7,$$
$$3x + 2y = 9.$$

We compute D, D_x, and D_y:

$$D = \begin{vmatrix} 3 & -2 \\ 3 & 2 \end{vmatrix} = 3 \cdot 2 - 3 \cdot (-2) = 6 + 6 = 12;$$

$$D_x = \begin{vmatrix} 7 & -2 \\ 9 & 2 \end{vmatrix} = 7 \cdot 2 - 9(-2) = 14 + 18 = 32;$$

$$D_y = \begin{vmatrix} 3 & 7 \\ 3 & 9 \end{vmatrix} = 3 \cdot 9 - 3 \cdot 7 = 27 - 21 = 6.$$

Then

$$x = \frac{D_x}{D} = \frac{32}{12}, \text{ or } \frac{8}{3} \quad \text{and} \quad y = \frac{D_y}{D} = \frac{6}{12} = \frac{1}{2}.$$

The solution is $\left(\frac{8}{3}, \frac{1}{2}\right)$.

Do Exercise 5.

5. Solve using Cramer's rule:
$$4x - 3y = 15,$$
$$x + 3y = 0.$$

Cramer's rule for three equations is very similar to that for two.

$$a_1 x + b_1 y + c_1 z = d_1$$
$$a_2 x + b_2 y + c_2 z = d_2$$
$$a_3 x + b_3 y + c_3 z = d_3$$

$$D = \begin{vmatrix} a_1 & b_1 & c_1 \\ a_2 & b_2 & c_2 \\ a_3 & b_3 & c_3 \end{vmatrix} \qquad D_x = \begin{vmatrix} d_1 & b_1 & c_1 \\ d_2 & b_2 & c_2 \\ d_3 & b_3 & c_3 \end{vmatrix}$$

$$D_y = \begin{vmatrix} a_1 & d_1 & c_1 \\ a_2 & d_2 & c_2 \\ a_3 & d_3 & c_3 \end{vmatrix}$$

D is again the determinant of the coefficients of x, y, and z. This time we have one more determinant, D_z. We get it by replacing the z-coefficients in D with the constants on the right:

$$D_z = \begin{vmatrix} a_1 & b_1 & d_1 \\ a_2 & b_2 & d_2 \\ a_3 & b_3 & d_3 \end{vmatrix}.$$

Answer

5. $(3, -1)$

The solution of the system is given by the following.

> ## CRAMER'S RULE
>
> $$x = \frac{D_x}{D}, \qquad y = \frac{D_y}{D}, \qquad z = \frac{D_z}{D}$$

EXAMPLE 4 Solve using Cramer's rule:

$$x - 3y + 7z = 13,$$
$$x + y + z = 1,$$
$$x - 2y + 3z = 4.$$

We compute D, D_x, D_y, and D_z:

$$D = \begin{vmatrix} 1 & -3 & 7 \\ 1 & 1 & 1 \\ 1 & -2 & 3 \end{vmatrix} = -10; \qquad D_x = \begin{vmatrix} 13 & -3 & 7 \\ 1 & 1 & 1 \\ 4 & -2 & 3 \end{vmatrix} = 20;$$

$$D_y = \begin{vmatrix} 1 & 13 & 7 \\ 1 & 1 & 1 \\ 1 & 4 & 3 \end{vmatrix} = -6; \qquad D_z = \begin{vmatrix} 1 & -3 & 13 \\ 1 & 1 & 1 \\ 1 & -2 & 4 \end{vmatrix} = -24.$$

Then

$$x = \frac{D_x}{D} = \frac{20}{-10} = -2;$$

$$y = \frac{D_y}{D} = \frac{-6}{-10} = \frac{3}{5};$$

$$z = \frac{D_z}{D} = \frac{-24}{-10} = \frac{12}{5}.$$

The solution is $\left(-2, \frac{3}{5}, \frac{12}{5}\right)$.

In Example 4, we would not have needed to evaluate D_z. Once we found x and y, we could have substituted them into one of the equations to find z. In practice, it is faster to use determinants to find only two of the numbers; then we find the third by substitution into an equation.

> Do Exercise 6.

In using Cramer's rule, we divide by D. If D were 0, we could not do so.

> ## INCONSISTENT SYSTEMS; DEPENDENT EQUATIONS
>
> If $D = 0$ and at least one of the other determinants is not 0, then the system does not have a solution, and we say that it is *inconsistent*.
>
> If $D = 0$ and all the other determinants are also 0, then there is an infinite set of solutions. In that case, we say that the equations in the system are *dependent*.

6. Solve using Cramer's rule:
$$x - 3y - 7z = 6,$$
$$2x + 3y + z = 9,$$
$$4x + y = 7.$$

Answer

6. $(1, 3, -2)$

a Evaluate.

1. $\begin{vmatrix} 3 & 7 \\ 2 & 8 \end{vmatrix}$

2. $\begin{vmatrix} 5 & 4 \\ 4 & -5 \end{vmatrix}$

3. $\begin{vmatrix} -3 & -6 \\ -5 & -10 \end{vmatrix}$

4. $\begin{vmatrix} 4 & 5 \\ -7 & 9 \end{vmatrix}$

5. $\begin{vmatrix} 8 & 2 \\ 12 & -3 \end{vmatrix}$

6. $\begin{vmatrix} 1 & 1 \\ 9 & 8 \end{vmatrix}$

7. $\begin{vmatrix} 2 & -7 \\ 0 & 0 \end{vmatrix}$

8. $\begin{vmatrix} 0 & -4 \\ 0 & -6 \end{vmatrix}$

b Evaluate.

9. $\begin{vmatrix} 0 & 2 & 0 \\ 3 & -1 & 1 \\ 1 & -2 & 2 \end{vmatrix}$

10. $\begin{vmatrix} 3 & 0 & -2 \\ 5 & 1 & 2 \\ 2 & 0 & -1 \end{vmatrix}$

11. $\begin{vmatrix} -1 & -2 & -3 \\ 3 & 4 & 2 \\ 0 & 1 & 2 \end{vmatrix}$

12. $\begin{vmatrix} 1 & 2 & 2 \\ 2 & 1 & 0 \\ 3 & 3 & 1 \end{vmatrix}$

13. $\begin{vmatrix} 3 & 2 & -2 \\ -2 & 1 & 4 \\ -4 & -3 & 3 \end{vmatrix}$

14. $\begin{vmatrix} 2 & -1 & 1 \\ 1 & 2 & -1 \\ 3 & 4 & -3 \end{vmatrix}$

15. $\begin{vmatrix} 3 & 2 & 4 \\ 1 & 1 & 1 \\ 1 & 1 & 1 \end{vmatrix}$

16. $\begin{vmatrix} -1 & 6 & -5 \\ 2 & 4 & 4 \\ 5 & 3 & 10 \end{vmatrix}$

c Solve using Cramer's rule.

17. $3x - 4y = 6,$
$5x + 9y = 10$

18. $5x + 8y = 1,$
$3x + 7y = 5$

19. $-2x + 4y = 3,$
$3x - 7y = 1$

20. $5x - 4y = -3,$
$7x + 2y = 6$

21. $4x + 2y = 11,$
$3x - y = 2$

22. $3x - 3y = 11,$
$9x - 2y = 5$

23. $x + 4y = 8,$
$3x + 5y = 3$

24. $x + 4y = 5,$
$-3x + 2y = 13$

25. $2x - 3y + 5z = 27,$
$x + 2y - z = -4,$
$5x - y + 4z = 27$

26. $x - y + 2z = -3,$
$x + 2y + 3z = 4,$
$2x + y + z = -3$

27. $r - 2s + 3t = 6,$
$2r - s - t = -3,$
$r + s + t = 6$

28. $a \quad - 3c = 6,$
$b + 2c = 2,$
$7a - 3b - 5c = 14$

29. $4x - y - 3z = 1,$
$8x + y - z = 5,$
$2x + y + 2z = 5$

30. $3x + 2y + 2z = 3,$
$x + 2y - z = 5,$
$2x - 4y + z = 0$

31. $p + q + r = 1,$
$p - 2q - 3r = 3,$
$4p + 5q + 6r = 4$

32. $x + 2y - 3z = 9,$
$2x - y + 2z = -8,$
$3x - y - 4z = 3$

J

Elimination Using Matrices

OBJECTIVE

a Solve systems of two or three equations using matrices.

The elimination method concerns itself primarily with the coefficients and constants of the equations. In what follows, we learn a method for solving systems using just the coefficients and the constants. This procedure involves what are called *matrices*.

In solving systems of equations, we perform computations with the constants. The variables play no important role until the end. Thus we can simplify writing a system by omitting the variables. For example, the system

$$3x + 4y = 5, \qquad \text{simplifies to} \qquad \begin{array}{ccc} 3 & 4 & 5 \\ 1 & -2 & 1 \end{array}$$

if we omit the variables, the operation of addition, and the equals signs. The result is a rectangular array of numbers. Such an array is called a **matrix** (plural, **matrices**). We ordinarily write brackets around matrices. The following are matrices.

$$\begin{bmatrix} 4 & 1 & 3 & 5 \\ 1 & 0 & 1 & 2 \\ 6 & 3 & -2 & 0 \end{bmatrix}, \quad \begin{bmatrix} 6 & 2 & 1 & 4 & 7 \\ 1 & 2 & 1 & 3 & 1 \\ 4 & 0 & -2 & 0 & -3 \end{bmatrix}, \quad \begin{bmatrix} 1 & 2 \\ 145 & 0 \\ -7 & 9 \\ 8 & 1 \\ 0 & 0 \end{bmatrix}.$$

The **rows** of a matrix are horizontal, and the **columns** are vertical.

$$\begin{bmatrix} 5 & -2 & 2 \\ 1 & 0 & 1 \\ 0 & 1 & 2 \end{bmatrix} \begin{array}{l} \longleftarrow \text{row 1} \\ \longleftarrow \text{row 2} \\ \longleftarrow \text{row 3} \end{array}$$

$$\begin{array}{ccc} \uparrow & \uparrow & \uparrow \\ \text{column 1} & \text{column 2} & \text{column 3} \end{array}$$

Let's now use matrices to solve systems of linear equations.

EXAMPLE 1 Solve the system

$$5x - 4y = -1,$$
$$-2x + 3y = 2.$$

We write a matrix using only the coefficients and the constants, keeping in mind that x corresponds to the first column and y to the second. A dashed line separates the coefficients from the constants at the end of each equation:

$$\begin{bmatrix} 5 & -4 & \vdots & -1 \\ -2 & 3 & \vdots & 2 \end{bmatrix}. \qquad \text{The individual numbers are called } elements \text{ or } entries.$$

Our goal is to transform this matrix into one of the form

$$\begin{bmatrix} a & b & \vdots & c \\ 0 & d & \vdots & e \end{bmatrix}.$$

The variables can then be reinserted to form equations from which we can complete the solution.

We do calculations that are similar to those that we would do if we wrote the entire equations. The first step, if possible, is to multiply and/or interchange the rows so that each number in the first column below the first number is a multiple of that number. In this case, we do so by multiplying Row 2 by 5. This corresponds to multiplying the second equation by 5.

$$\begin{bmatrix} 5 & -4 & \vdots & -1 \\ -10 & 15 & \vdots & 10 \end{bmatrix}$$ New Row 2 = 5(Row 2)

Next, we multiply the first row by 2 and add the result to the second row. This corresponds to multiplying the first equation by 2 and adding the result to the second equation. Although we write the calculations out here, we generally try to do them mentally:

$$2 \cdot 5 + (-10) = 0; \quad 2(-4) + 15 = 7; \quad 2(-1) + 10 = 8.$$

$$\begin{bmatrix} 5 & -4 & \vdots & -1 \\ 0 & 7 & \vdots & 8 \end{bmatrix}$$ New Row 2 = 2(Row 1) + (Row 2)

If we now reinsert the variables, we have

$$5x - 4y = -1, \quad \textbf{(1)}$$
$$7y = 8. \quad \textbf{(2)}$$

We can now proceed as before, solving equation (2) for y:

$$7y = 8 \quad \textbf{(2)}$$
$$y = \tfrac{8}{7}.$$

Next, we substitute $\tfrac{8}{7}$ for y back in equation (1). This procedure is called *back-substitution*.

$$5x - 4y = -1 \quad \textbf{(1)}$$
$$5x - 4 \cdot \tfrac{8}{7} = -1 \quad \text{Substituting } \tfrac{8}{7} \text{ for } y \text{ in equation (1)}$$
$$x = \tfrac{5}{7} \quad \text{Solving for } x$$

The solution is $\left(\tfrac{5}{7}, \tfrac{8}{7}\right)$.

Do Exercise 1.

1. Solve using matrices:
$$5x - 2y = -44,$$
$$2x + 5y = -6.$$

EXAMPLE 2 Solve the system

$$2x - y + 4z = -3,$$
$$x \quad\quad - 4z = 5,$$
$$6x - y + 2z = 10.$$

We first write a matrix, using only the coefficients and the constants. Where there are missing terms, we must write 0's:

$$\begin{bmatrix} 2 & -1 & 4 & \vdots & -3 \\ 1 & 0 & -4 & \vdots & 5 \\ 6 & -1 & 2 & \vdots & 10 \end{bmatrix}. \quad \begin{matrix} \textbf{(P1)} \\ \textbf{(P2)} \\ \textbf{(P3)} \end{matrix}$$ (P1), (P2), and (P3) designate the equations that are in the first, second, and third position, respectively.

Our goal is to find an equivalent matrix of the form

$$\begin{bmatrix} a & b & c & \vdots & d \\ 0 & e & f & \vdots & g \\ 0 & 0 & h & \vdots & i \end{bmatrix}.$$

A matrix of this form can be rewritten as a system of equations from which a solution can be found easily.

Answer

1. $(-8, 2)$

The first step, if possible, is to interchange the rows so that each number in the first column below the first number is a multiple of that number. In this case, we do so by interchanging Rows 1 and 2:

$$\begin{bmatrix} 1 & 0 & -4 & \vdots & 5 \\ 2 & -1 & 4 & \vdots & -3 \\ 6 & -1 & 2 & \vdots & 10 \end{bmatrix}.$$
This corresponds to interchanging the first two equations.

Next, we multiply the first row by -2 and add it to the second row:

$$\begin{bmatrix} 1 & 0 & -4 & \vdots & 5 \\ 0 & -1 & 12 & \vdots & -13 \\ 6 & -1 & 2 & \vdots & 10 \end{bmatrix}.$$
This corresponds to multiplying new equation (P1) by -2 and adding it to new equation (P2). The result replaces the former (P2). We perform the calculations mentally.

Now we multiply the first row by -6 and add it to the third row:

$$\begin{bmatrix} 1 & 0 & -4 & \vdots & 5 \\ 0 & -1 & 12 & \vdots & -13 \\ 0 & -1 & 26 & \vdots & -20 \end{bmatrix}.$$
This corresponds to multiplying equation (P1) by -6 and adding it to equation (P3).

Next, we multiply Row 2 by -1 and add it to the third row:

$$\begin{bmatrix} 1 & 0 & -4 & \vdots & 5 \\ 0 & -1 & 12 & \vdots & -13 \\ 0 & 0 & 14 & \vdots & -7 \end{bmatrix}.$$
This corresponds to multiplying equation (P2) by -1 and adding it to equation (P3).

Reinserting the variables gives us

$$
\begin{aligned}
x \qquad - 4z &= 5, & \textbf{(P1)} \\
-y + 12z &= -13, & \textbf{(P2)} \\
14z &= -7. & \textbf{(P3)}
\end{aligned}
$$

We now solve (P3) for z:

$$
\begin{aligned}
14z &= -7 & \textbf{(P3)} \\
z &= -\tfrac{7}{14} & \text{Solving for } z \\
z &= -\tfrac{1}{2}.
\end{aligned}
$$

Next, we back-substitute $-\frac{1}{2}$ for z in (P2) and solve for y:

$$
\begin{aligned}
-y + 12z &= -13 & \textbf{(P2)} \\
-y + 12\left(-\tfrac{1}{2}\right) &= -13 & \text{Substituting } -\tfrac{1}{2} \text{ for } z \text{ in equation (P2)} \\
-y - 6 &= -13 \\
-y &= -7 \\
y &= 7. & \text{Solving for } y
\end{aligned}
$$

Since there is no y-term in (P1), we need only substitute $-\frac{1}{2}$ for z in (P1) and solve for x:

$$
\begin{aligned}
x - 4z &= 5 & \textbf{(P1)} \\
x - 4\left(-\tfrac{1}{2}\right) &= 5 & \text{Substituting } -\tfrac{1}{2} \text{ for } z \text{ in equation (P1)} \\
x + 2 &= 5 \\
x &= 3. & \text{Solving for } x
\end{aligned}
$$

The solution is $\left(3, 7, -\frac{1}{2}\right)$.

Do Exercise 2.

2. Solve using matrices:

$$
\begin{aligned}
x - 2y + 3z &= 4, \\
2x - y + z &= -1, \\
4x + y + z &= 1.
\end{aligned}
$$

Answer

2. $(-1, 2, 3)$

All the operations used in the preceding example correspond to operations with the equations and produce equivalent systems of equations. We call the matrices **row-equivalent** and the operations that produce them **row-equivalent operations**.

> The best overall method of solving systems of equations is by row-equivalent matrices; graphing calculators and computers are programmed to use them. Matrices are part of a branch of mathematics known as linear algebra. They are also studied in more detail in many courses in finite mathematics.

ROW-EQUIVALENT OPERATIONS

Each of the following row-equivalent operations produces an equivalent matrix:

a) Interchanging any two rows.

b) Multiplying each element of a row by the same nonzero number.

c) Multiplying each element of a row by a nonzero number and adding the result to another row.

J Exercise Set

For Extra Help

MyMathLab Math XL  PRACTICE   WATCH   DOWNLOAD   READ   REVIEW

a Solve using matrices.

1. $4x + 2y = 11,$
$3x - y = 2$

2. $3x - 3y = 11,$
$9x - 2y = 5$

3. $x + 4y = 8,$
$3x + 5y = 3$

4. $x + 4y = 5,$
$-3x + 2y = 13$

5. $5x - 3y = -2,$
$4x + 2y = 5$

6. $3x + 4y = 7,$
$-5x + 2y = 10$

7. $2x - 3y = 50,$
$5x + y = 40$

8. $4x + 5y = -8,$
$7x + 9y = 11$

9. $4x - y - 3z = 1,$
$8x + y - z = 5,$
$2x + y + 2z = 5$

10. $3x + 2y + 2z = 3,$
$x + 2y - z = 5,$
$2x - 4y + z = 0$

11. $p + q + r = 1,$
$p - 2q - 3r = 3,$
$4p + 5q + 6r = 4$

12. $x + 2y - 3z = 9,$
$2x - y + 2z = -8,$
$3x - y - 4z = 3$

13. $x - y + 2z = 0,$
$x - 2y + 3z = -1,$
$2x - 2y + z = -3$

14. $4a + 9b = 8,$
$8a + 6c = -1,$
$6b + 6c = -1$

15. $3p + 2r = 11,$
$q - 7r = 4,$
$p - 6q = 1$

16. $m + n + t = 6,$
$m - n - t = 0,$
$m + 2n + t = 5$

K

The Algebra of Functions

OBJECTIVE

a Given two functions *f* and *g*, find their sum, difference, product, and quotient.

a The Sum, Difference, Product, and Quotient of Functions

Suppose that *a* is in the domain of two functions, *f* and *g*. The input *a* is paired with $f(a)$ by *f* and with $g(a)$ by *g*. The outputs can then be added to get $f(a) + g(a)$.

EXAMPLE 1 Let $f(x) = x + 4$ and $g(x) = x^2 + 1$. Find $f(2) + g(2)$.

We visualize two function machines. Because 2 is in the domain of each function, we can compute $f(2)$ and $g(2)$.

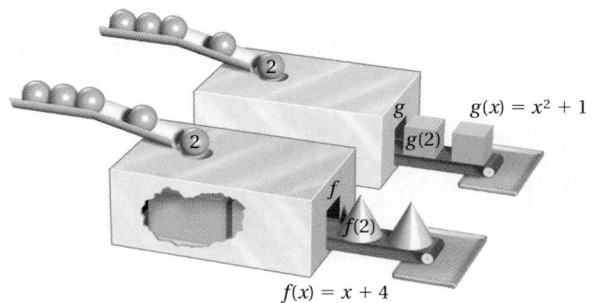

Since

$$f(2) = 2 + 4 = 6 \quad \text{and} \quad g(2) = 2^2 + 1 = 5,$$

we have

$$f(2) + g(2) = 6 + 5 = 11.$$

In Example 1, suppose that we were to write $f(x) + g(x)$ as $(x + 4) + (x^2 + 1)$, or $f(x) + g(x) = x^2 + x + 5$. This could then be regarded as a "new" function: $(f + g)(x) = x^2 + x + 5$. We can alternatively find $f(2) + g(2)$ with $(f + g)(x)$:

$$(f + g)(x) = x^2 + x + 5$$
$$(f + g)(2) = 2^2 + 2 + 5 \qquad \text{Substituting 2 for } x$$
$$= 4 + 2 + 5$$
$$= 11.$$

Similar notations exist for subtraction, multiplication, and division of functions.

THE SUM, DIFFERENCE, PRODUCT, AND QUOTIENT OF FUNCTIONS

For any functions *f* and *g*, we can form new functions defined as:

1. The **sum $f + g$**: $(f + g)(x) = f(x) + g(x)$;
2. The **difference $f - g$**: $(f - g)(x) = f(x) - g(x)$;
3. The **product fg**: $(f \cdot g)(x) = f(x) \cdot g(x)$;
4. The **quotient f/g**: $(f/g)(x) = f(x)/g(x)$, where $g(x) \neq 0$.

EXAMPLE 2 Given f and g described by $f(x) = x^2 - 5$ and $g(x) = x + 7$, find $(f + g)(x)$, $(f - g)(x)$, $(f \cdot g)(x)$, $(f/g)(x)$, and $(g \cdot g)(x)$.

$(f + g)(x) = f(x) + g(x) = (x^2 - 5) + (x + 7) = x^2 + x + 2;$

$(f - g)(x) = f(x) - g(x) = (x^2 - 5) - (x + 7) = x^2 - x - 12;$

$(f \cdot g)(x) = f(x) \cdot g(x) = (x^2 - 5)(x + 7) = x^3 + 7x^2 - 5x - 35;$

$(f/g)(x) = f(x)/g(x) = \dfrac{x^2 - 5}{x + 7};$

$(g \cdot g)(x) = g(x) \cdot g(x) = (x + 7)(x + 7) = x^2 + 14x + 49$

Note that the sum, difference, and product of polynomials are also polynomial functions, but the quotient may not be.

Do Exercise 1.

1. Given $f(x) = x^2 + 3$ and $g(x) = x^2 - 3$, find each of the following.

a) $(f + g)(x)$

b) $(f - g)(x)$

c) $(f \cdot g)(x)$

d) $(f/g)(x)$

e) $(f \cdot f)(x)$

EXAMPLE 3 For $f(x) = x^2 - x$ and $g(x) = x + 2$, find $(f + g)(3)$, $(f - g)(-1)$, $(f \cdot g)(5)$, and $(f/g)(-4)$.

We first find $(f + g)(x)$, $(f - g)(x)$, $(f \cdot g)(x)$, and $(f/g)(x)$.

$(f + g)(x) = f(x) + g(x) = x^2 - x + x + 2$
$\qquad\qquad = x^2 + 2;$

$(f - g)(x) = f(x) - g(x) = x^2 - x - (x + 2)$
$\qquad\qquad = x^2 - x - x - 2$
$\qquad\qquad = x^2 - 2x - 2;$

$(f \cdot g)(x) = f(x) \cdot g(x) = (x^2 - x)(x + 2)$
$\qquad\qquad = x^3 + 2x^2 - x^2 - 2x$
$\qquad\qquad = x^3 + x^2 - 2x;$

$(f/g)(x) = \dfrac{f(x)}{g(x)} = \dfrac{x^2 - x}{x + 2}.$

Then we substitute.

$(f + g)(3) = 3^2 + 2 \qquad$ Using $(f + g)(x) = x^2 + 2$
$\qquad\quad = 9 + 2 = 11;$

$(f - g)(-1) = (-1)^2 - 2(-1) - 2 \qquad$ Using $(f - g)(x) = x^2 - 2x - 2$
$\qquad\quad = 1 + 2 - 2 = 1;$

$(f \cdot g)(5) = 5^3 + 5^2 - 2 \cdot 5 \qquad$ Using $(f \cdot g)(x) = x^3 + x^2 - 2x$
$\qquad\quad = 125 + 25 - 10 = 140;$

$(f/g)(-4) = \dfrac{(-4)^2 - (-4)}{-4 + 2} \qquad$ Using $(f/g)(x) = (x^2 - x)/(x + 2)$

$\qquad\quad = \dfrac{16 + 4}{-2} = \dfrac{20}{-2} = -10$

Do Exercise 2.

2. Given $f(x) = x^2 + x$ and $g(x) = 2x - 3$, find each of the following.

a) $(f + g)(-2)$

b) $(f - g)(4)$

c) $(f \cdot g)(-3)$

d) $(f/g)(2)$

Answers

1. (a) $2x^2$; (b) 6; (c) $x^4 - 9$; (d) $\dfrac{x^2 + 3}{x^2 - 3}$;
(e) $x^4 + 6x^2 + 9$ **2.** (a) -5; (b) 15;
(c) -54; (d) 6

K **Exercise Set**

For Extra Help

MyMathLab

Math XL
PRACTICE

WATCH

DOWNLOAD

READ

REVIEW

a

Let $f(x) = -3x + 1$ and $g(x) = x^2 + 2$. Find each of the following.

1. $f(2) + g(2)$

2. $f(-1) + g(-1)$

3. $f(5) - g(5)$

4. $f(4) - g(4)$

5. $f(-1) \cdot g(-1)$

6. $f(-2) \cdot g(-2)$

7. $f(-4)/g(-4)$

8. $f(3)/g(3)$

9. $g(1) - f(1)$

10. $g(2)/f(2)$

11. $g(0)/f(0)$

12. $g(6) - f(6)$

Let $f(x) = x^2 - 3$ and $g(x) = 4 - x$. Find each of the following.

13. $(f + g)(x)$

14. $(f - g)(x)$

15. $(f + g)(-4)$

16. $(f + g)(-5)$

17. $(f - g)(3)$

18. $(f - g)(2)$

19. $(f \cdot g)(x)$

20. $(f/g)(x)$

21. $(f \cdot g)(-3)$

22. $(f \cdot g)(-4)$

23. $(f/g)(0)$

24. $(f/g)(1)$

25. $(f/g)(-2)$

26. $(f/g)(-1)$

For each pair of functions f and g, find $(f + g)(x)$, $(f - g)(x)$, $(f \cdot g)(x)$, and $(f/g)(x)$.

27. $f(x) = x^2$,
$g(x) = 3x - 4$

28. $f(x) = 5x - 1$,
$g(x) = 2x^2$

29. $f(x) = \dfrac{1}{x - 2}$,
$g(x) = 4x^3$

30. $f(x) = 3x^2$,
$g(x) = \dfrac{1}{x - 4}$

31. $f(x) = \dfrac{3}{x - 2}$,
$g(x) = \dfrac{5}{4 - x}$

32. $f(x) = \dfrac{5}{x - 3}$,
$g(x) = \dfrac{1}{x - 2}$

L   Distance, Midpoints, and Circles

In carpentry, surveying, engineering, and other fields, it is often necessary to determine distances and midpoints and to produce accurately drawn circles.

a The Distance Formula

Suppose that two points are on a horizontal line, and thus have the same second coordinate. We can find the distance between them by subtracting their first coordinates. This difference may be negative, depending on the order in which we subtract. So, to make sure we get a positive number, we take the absolute value of this difference. The distance between two points on a horizontal line (x_1, y_1) and (x_2, y_1) is thus $|x_2 - x_1|$. Similarly, the distance between two points on a vertical line (x_2, y_1) and (x_2, y_2) is $|y_2 - y_1|$.

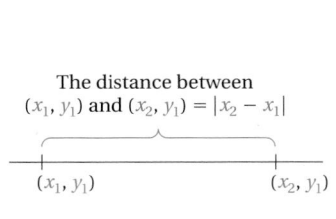

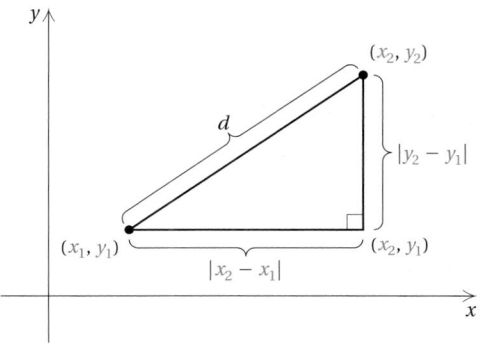

Now consider *any* two points (x_1, y_1) and (x_2, y_2). If $x_1 \neq x_2$ and $y_1 \neq y_2$, these points are vertices of a right triangle, as shown. The other vertex is then (x_2, y_1). The lengths of the legs are $|x_2 - x_1|$ and $|y_2 - y_1|$. We find d, the length of the hypotenuse, by using the Pythagorean equation:

$$d^2 = |x_2 - x_1|^2 + |y_2 - y_1|^2.$$

Since the square of a number is the same as the square of its opposite, we don't need these absolute-value signs. Thus,

$$d^2 = (x_2 - x_1)^2 + (y_2 - y_1)^2.$$

Taking the principal square root, we obtain the formula for the distance between two points.

THE DISTANCE FORMULA

The distance between any two points (x_1, y_1) and (x_2, y_2) is given by

$$d = \sqrt{(x_2 - x_1)^2 + (y_2 - y_1)^2}.$$

This formula holds even when the two points *are* on a vertical line or a horizontal line.

EXAMPLE 1   Find the distance between $(4, -3)$ and $(-5, 4)$. Give an exact answer and an approximation to three decimal places.

We substitute into the distance formula:

$$d = \sqrt{(-5 - 4)^2 + [4 - (-3)]^2} \quad \text{Substituting}$$
$$= \sqrt{(-9)^2 + 7^2}$$
$$= \sqrt{130} \approx 11.402. \quad \text{Using a calculator}$$

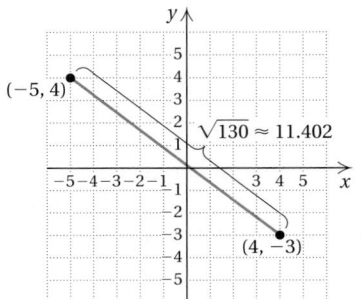

Do Exercises 1 and 2.

Find the distance between each
pair of points. Where appropriate,
give an approximation to three
decimal places.

1. $(2, 6)$ and $(-4, -2)$

2. $(-2, 1)$ and $(4, 2)$

(b) Midpoints of Segments

The distance formula can be used to derive a formula for finding the midpoint of a segment when the coordinates of the endpoints are known.

> **THE MIDPOINT FORMULA**
>
> If the endpoints of a segment are (x_1, y_1) and (x_2, y_2), then the coordinates of the midpoint are
>
> $$\left(\frac{x_1 + x_2}{2}, \frac{y_1 + y_2}{2} \right).$$
>
> (To locate the midpoint, determine the average of the x-coordinates and the average of the y-coordinates.)

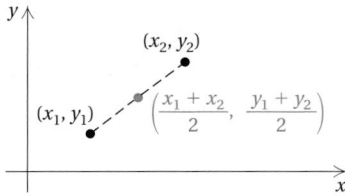

EXAMPLE 2   Find the midpoint of the segment with endpoints $(-2, 3)$ and $(4, -6)$.

Using the midpoint formula, we obtain

$$\left(\frac{-2 + 4}{2}, \frac{3 + (-6)}{2} \right), \quad \text{or} \quad \left(\frac{2}{2}, \frac{-3}{2} \right), \quad \text{or} \quad \left(1, -\frac{3}{2} \right).$$

Do Exercises 3 and 4.

Find the midpoint of the segment
with the given endpoints.

3. $(-3, 1)$ and $(6, -7)$

4. $(10, -7)$ and $(8, -3)$

Answers

1. 10   **2.** $\sqrt{37} \approx 6.083$ **3.** $\left(\frac{3}{2}, -3 \right)$

4. $(9, -5)$

(c) Circles

Another conic section, or curve, shown in the figure at the beginning of this section is a *circle*. A **circle** is defined as the set of all points in a plane that are a fixed distance from a point in that plane.

Let's find an equation for a circle. We call the center (h, k) and let the radius have length r. Suppose that (x, y) is any point on the circle. By the distance formula, we have

$$\sqrt{(x - h)^2 + (y - k)^2} = r.$$

Squaring both sides gives an equation of the circle in standard form:

$$(x - h)^2 + (y - k)^2 = r^2.$$

When $h = 0$ and $k = 0$, the circle is centered at the origin. Otherwise, we can think of that circle being translated $|h|$ units horizontally and $|k|$ units vertically from the origin.

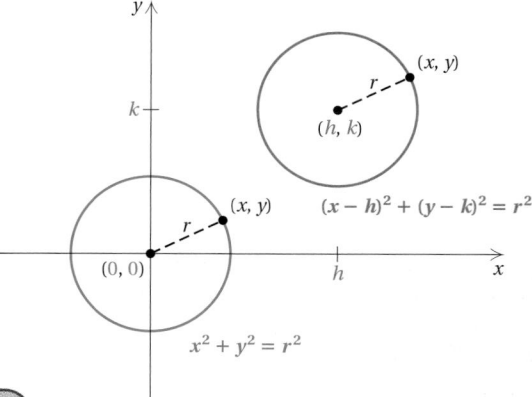

EQUATIONS OF CIRCLES

A circle centered at the origin with radius r has equation

$$x^2 + y^2 = r^2.$$

A circle with center (h, k) and radius r has equation

$$(x - h)^2 + (y - k)^2 = r^2. \quad \text{(Standard form)}$$

EXAMPLE 3 Find the center and the radius and graph this circle:

$$(x + 2)^2 + (y - 3)^2 = 16.$$

First, we find an equivalent equation in standard form:

$$[x - (-2)]^2 + (y - 3)^2 = 4^2.$$

Thus the center is $(-2, 3)$ and the radius is 4. We draw the graph, shown below, by locating the center and then using a compass, setting its radius at 4, to draw the circle.

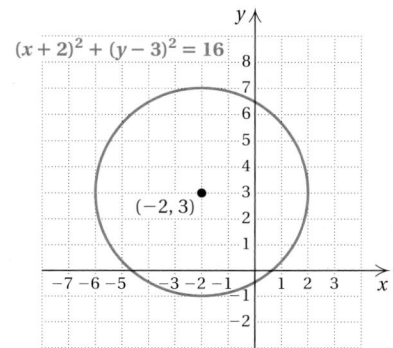

Do Exercises 5 and 6.

5. Find the center and the radius of the circle

$$(x - 5)^2 + \left(y + \tfrac{1}{2}\right)^2 = 9.$$

Then graph the circle.

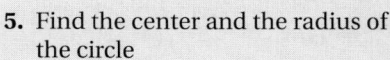

6. Find the center and the radius of the circle $x^2 + y^2 = 64$.

Answers

5.
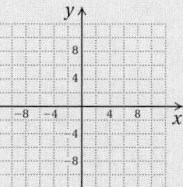
Center: $\left(5, -\tfrac{1}{2}\right)$
Radius: 3

6. $(0, 0); r = 8$

EXAMPLE 4 Write an equation of a circle with center $(9, -5)$ and radius $\sqrt{2}$.

We use standard form $(x - h)^2 + (y - k)^2 = r^2$ and substitute:

$$(x - 9)^2 + [y - (-5)]^2 = \left(\sqrt{2}\right)^2 \quad \text{Substituting}$$
$$(x - 9)^2 + (y + 5)^2 = 2. \quad \text{Simplifying}$$

7. Find an equation of a circle with center $(-3, 1)$ and radius 6.

Do Exercise 7.

With certain equations not in standard form, we can complete the square to show that the equations are equations of circles. We proceed in much the same manner as we did in Section 11.6.

EXAMPLE 5 Find the center and the radius and graph this circle:

$$x^2 + y^2 + 8x - 2y + 15 = 0.$$

First, we regroup the terms and then complete the square twice, once with $x^2 + 8x$ and once with $y^2 - 2y$:

$$x^2 + y^2 + 8x - 2y + 15 = 0$$
$$(x^2 + 8x) + (y^2 - 2y) = -15 \quad \text{Regrouping and subtracting 15}$$
$$(x^2 + 8x + 0) + (y^2 - 2y + 0) = -15 \quad \text{Adding 0}$$
$$(x^2 + 8x + 16 - 16) + (y^2 - 2y + 1 - 1) = -15 \quad \left(\tfrac{8}{2}\right)^2 = 4^2 = 16;$$
$$\left(\tfrac{-2}{2}\right)^2 = 1;$$
substituting $16 - 16$ and $1 - 1$ for 0

$$(x^2 + 8x + 16) + (y^2 - 2y + 1) - 16 - 1 = -15 \quad \text{Regrouping}$$
$$(x^2 + 8x + 16) + (y^2 - 2y + 1) = -15 + 16 + 1 \quad \text{Adding 16 and 1 on both sides}$$
$$(x + 4)^2 + (y - 1)^2 = 2 \quad \text{Factoring and simplifying}$$
$$[x - (-4)]^2 + (y - 1)^2 = \left(\sqrt{2}\right)^2. \quad \text{Writing standard form}$$

8. Find the center and the radius of the circle

$$x^2 + 2x + y^2 - 4y + 2 = 0.$$

Then graph the circle.

The center is $(-4, 1)$ and the radius is $\sqrt{2}$.

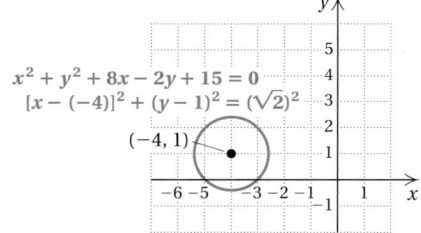

$x^2 + y^2 + 8x - 2y + 15 = 0$
$[x - (-4)]^2 + (y - 1)^2 = (\sqrt{2})^2$
$(-4, 1)$

Do Exercise 8.

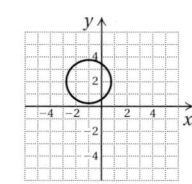

Exercise Set

a Find the distance between each pair of points. Where appropriate, give an approximation to three decimal places.

1. $(6, -4)$ and $(2, -7)$

2. $(1, 2)$ and $(-4, 14)$

3. $(0, -4)$ and $(5, -6)$

4. $(8, 3)$ and $(8, -3)$

5. $(9, 9)$ and $(-9, -9)$

6. $(2, 22)$ and $(-8, 1)$

7. $(2.8, -3.5)$ and $(-4.3, -3.5)$

8. $(6.1, 2)$ and $(5.6, -4.4)$

9. $\left(\dfrac{5}{7}, \dfrac{1}{14}\right)$ and $\left(\dfrac{1}{7}, \dfrac{11}{14}\right)$

10. $\left(0, \sqrt{7}\right)$ and $\left(\sqrt{6}, 0\right)$

11. $(-23, 10)$ and $(56, -17)$

12. $(34, -18)$ and $(-46, -38)$

13. (a, b) and $(0, 0)$

14. $(0, 0)$ and (p, q)

15. $\left(\sqrt{2}, -\sqrt{3}\right)$ and $\left(-\sqrt{7}, \sqrt{5}\right)$

16. $\left(\sqrt{8}, \sqrt{3}\right)$ and $\left(-\sqrt{5}, -\sqrt{6}\right)$

17. $(1000, -240)$ and $(-2000, 580)$

18. $(-3000, 560)$ and $(-430, -640)$

b Find the midpoint of the segment with the given endpoints.

19. $(-1, 9)$ and $(4, -2)$

20. $(5, 10)$ and $(2, -4)$

21. $(3, 5)$ and $(-3, 6)$

22. $(7, -3)$ and $(4, 11)$

23. $(-10, -13)$ and $(8, -4)$

24. $(6, -2)$ and $(-5, 12)$

25. $(-3.4, 8.1)$ and $(2.9, -8.7)$

26. $(4.1, 6.9)$ and $(5.2, -6.9)$

27. $\left(\dfrac{1}{6}, -\dfrac{3}{4}\right)$ and $\left(-\dfrac{1}{3}, \dfrac{5}{6}\right)$

28. $\left(-\dfrac{4}{5}, -\dfrac{2}{3}\right)$ and $\left(\dfrac{1}{8}, \dfrac{3}{4}\right)$

29. $\left(\sqrt{2}, -1\right)$ and $\left(\sqrt{3}, 4\right)$

30. $\left(9, 2\sqrt{3}\right)$ and $\left(-4, 5\sqrt{3}\right)$

C   Find the center and the radius of each circle. Then graph the circle.

31. $(x + 1)^2 + (y + 3)^2 = 4$

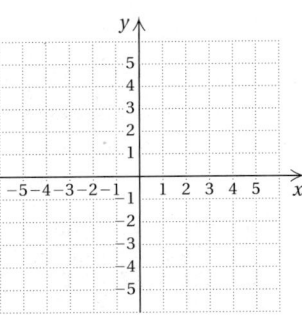

32. $(x - 2)^2 + (y + 3)^2 = 1$

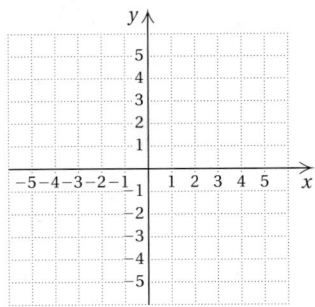

33. $(x - 3)^2 + y^2 = 2$

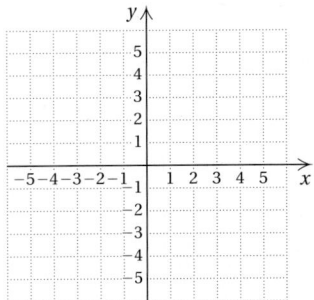

34. $x^2 + (y - 1)^2 = 3$

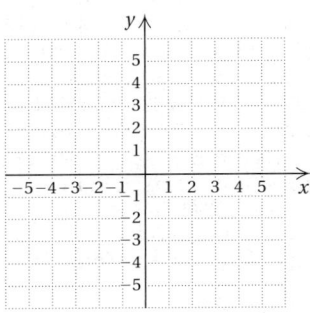

35. $x^2 + y^2 = 25$

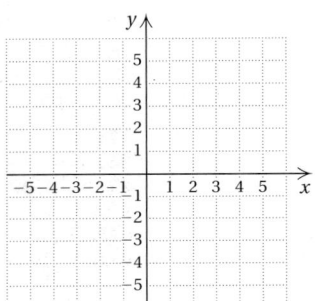

36. $x^2 + y^2 = 9$

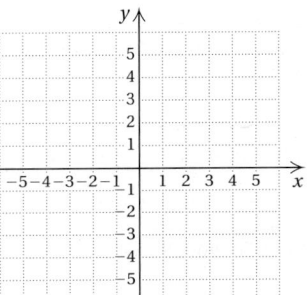

Find an equation of the circle having the given center and radius.

37. Center $(0, 0)$, radius 7

38. Center $(0, 0)$, radius 4

39. Center $(-5, 3)$, radius $\sqrt{7}$

40. Center $(4, 1)$, radius $3\sqrt{2}$

Find the center and the radius of each circle.

41. $x^2 + y^2 + 8x - 6y - 15 = 0$

42. $x^2 + y^2 + 6x - 4y - 15 = 0$

43. $x^2 + y^2 - 8x + 2y + 13 = 0$

44. $x^2 + y^2 + 6x + 4y + 12 = 0$

45. $x^2 + y^2 - 4x = 0$

46. $x^2 + y^2 + 10y - 75 = 0$

Answers

CHAPTER 1

Exercise Set 1.1, p. 6

1. 32 min; 69 min; 81 min **3.** 1935 m^2 **5.** 260 mi
7. 24 ft^2 **9.** 56 **11.** 8 **13.** 1 **15.** 6 **17.** 2
19. $b + 7$, or $7 + b$ **21.** $c - 12$ **23.** $q + 4$, or $4 + q$
25. $a + b$, or $b + a$ **27.** $x \div y$, or $\frac{x}{y}$, or x/y, or $x \cdot \frac{1}{y}$
29. $x + w$, or $w + x$ **31.** $n - m$ **33.** $x + y$, or $y + x$
35. $2z$ **37.** $3m$ **39.** $4a + 6$, or $6 + 4a$ **41.** $xy - 8$
43. $2t - 5$ **45.** $3n + 11$, or $11 + 3n$ **47.** $4x + 3y$, or
$3y + 4x$ **49.** 89%s, or 0.89s, where s is the salary
51. $s + 0.05s$ **53.** $65t$ miles **55.** $\$50 - x$ **57.** $\$8.50n$
59. $\frac{1}{4}$ **61.** 0

Calculator Corner, p. 13

1. 8.717797887 **2.** 17.80449381 **3.** 67.08203932
4. 35.4807407 **5.** 3.141592654 **6.** 91.10618695
7. 530.9291585 **8.** 138.8663978

Calculator Corner, p. 14

1. -0.75 **2.** -0.45 **3.** -0.125 **4.** -1.8 **5.** -0.675
6. -0.6875 **7.** -3.5 **8.** -0.76

Calculator Corner, p. 16

1. 5 **2.** 17 **3.** 0 **4.** 6.48 **5.** 12.7 **6.** 0.9 **7.** $\frac{5}{7}$ **8.** $\frac{4}{3}$

Exercise Set 1.2, p. 18

1. -282 **3.** 24; -2 **5.** 3,600,000,000; -460
7. Alley Cats: -34; Strikers: 34
9. $\frac{10}{3}$ **11.** -5.2
13. $-4\frac{2}{5}$ **15.** -0.875 **17.** $0.8\overline{3}$
19. $-1.1\overline{6}$ **21.** $0.\overline{6}$ **23.** 0.1 **25.** -0.5 **27.** 0.16
29. $>$ **31.** $<$ **33.** $<$ **35.** $<$ **37.** $>$ **39.** $<$ **41.** $>$
43. $<$ **45.** $<$ **47.** $>$ **49.** $<$ **51.** $<$ **53.** $x < -6$
55. $y \geq -10$ **57.** False **59.** True **61.** True
63. False **65.** 3 **67.** 10 **69.** 0 **71.** 30.4 **73.** $\frac{2}{3}$
75. 0 **77.** 2.65 **79.** $7\frac{4}{5}$ **80.** 105 **81.** 9
82. 13 **83.** 39 **84.** 3 **85.** 11 **86.** 3 **87.** 1
89. $-\frac{2}{3}, -\frac{2}{5}, -\frac{1}{3}, -\frac{2}{7}, -\frac{1}{7}, \frac{1}{3}, \frac{2}{5}, \frac{2}{3}, \frac{9}{8}$ **91.** $-100, -8\frac{7}{8}, -8\frac{5}{8}, -\frac{67}{8}$,
$-5, 0, 1^7, |3|, \frac{14}{4}, 4, |-6|, 7^1$ **93.** $\frac{1}{1}$

Exercise Set 1.3, p. 26

1. -7 **3.** -6 **5.** 0 **7.** -8 **9.** -7 **11.** -27
13. 0 **15.** -42 **17.** 0 **19.** 0 **21.** 3 **23.** -9

25. 7 **27.** 0 **29.** 35 **31.** -3.8 **33.** -8.1 **35.** $-\frac{1}{5}$
37. $-\frac{7}{9}$ **39.** $-\frac{3}{8}$ **41.** $-\frac{19}{24}$ **43.** $\frac{1}{24}$ **45.** $\frac{8}{15}$ **47.** $\frac{16}{45}$
49. 37 **51.** 50 **53.** -1409 **55.** -24 **57.** 26.9
59. -8 **61.** $\frac{13}{8}$ **63.** -43 **65.** $\frac{4}{3}$ **67.** 24 **69.** $\frac{3}{8}$
71. 13,796 ft **73.** $-3°$F **75.** $-\$20,300$ **77.** He owes $85.
79. -0.625 **80.** $0.\overline{3}$ **81.** $-0.08\overline{3}$ **82.** 0.65 **83.** 2.3
84. 0 **85.** $\frac{4}{5}$ **86.** 21.4 **87.** All positive numbers **89.** B

Exercise Set 1.4, p. 32

1. -7 **3.** -6 **5.** 0 **7.** -4 **9.** -7 **11.** -6 **13.** 0
15. 14 **17.** 11 **19.** -14 **21.** 5 **23.** -1 **25.** 18
27. -3 **29.** -21 **31.** 5 **33.** -8 **35.** 12 **37.** -23
39. -68 **41.** -73 **43.** 116 **45.** 0 **47.** -1 **49.** $\frac{1}{12}$
51. $-\frac{17}{12}$ **53.** $\frac{1}{8}$ **55.** 19.9 **57.** -8.6 **59.** -0.01
61. -193 **63.** 500 **65.** -2.8 **67.** -3.53 **69.** $-\frac{1}{2}$
71. $\frac{6}{7}$ **73.** $-\frac{41}{30}$ **75.** $-\frac{2}{15}$ **77.** $-\frac{1}{48}$ **79.** $-\frac{43}{60}$ **81.** 37
83. -62 **85.** -139 **87.** 6 **89.** 108.5 **91.** $\frac{1}{4}$ **93.** 2319 m
95. $347.94 **97.** 5676 ft **99.** 381 ft **101.** 1130°F
103. $y + 7$, or $7 + y$ **104.** $t - 41$ **105.** $a - h$
106. $6c$, or $c \cdot 6$ **107.** $r + s$, or $s + r$ **108.** $y - x$
109. False; $3 - 0 \neq 0 - 3$ **111.** True **113.** True

Mid-Chapter Review: Chapter 1, p. 36

1. True **2.** False **3.** True **4.** False
5. $-x = -(-4) = 4$;
$-(-x) = -(-(-4)) = -(4) = -4$
6. $5 - 13 = 5 + (-13) = -8$ **7.** $-6 - 7 = -6 + (-7) = -13$
8. 4 **9.** 11 **10.** $3y$ **11.** $n - 5$ **12.** 450; -79
13. -3.5 **14.** -0.8 **15.** $2.\overline{3}$ **16.** $<$
17. $>$ **18.** False **19.** True **20.** $5 > y$ **21.** $t \leq -3$
22. 15.6 **23.** 18 **24.** 0 **25.** $\frac{12}{5}$ **26.** 5.6 **27.** $-\frac{7}{4}$
28. 0 **29.** 49 **30.** 19 **31.** 2.3 **32.** -2 **33.** $-\frac{1}{8}$
34. 0 **35.** -17 **36.** $-\frac{11}{24}$ **37.** -8.1 **38.** -9 **39.** -2
40. -10.4 **41.** 16 **42.** $\frac{7}{20}$ **43.** -12 **44.** -4 **45.** $-\frac{4}{3}$
46. -1.8 **47.** 13 **48.** 9 **49.** -23 **50.** 75 **51.** 14
52. 33°C **53.** $54.80 **54.** Answers may vary. Three
examples are $\frac{6}{13}$, -23.8, and $\frac{43}{5}$. These are rational numbers

because they can be named in the form $\frac{a}{b}$, where a and b are

integers and b is not 0. They are not integers, however, because
they are neither whole numbers nor the opposites of whole
numbers. **55.** Answers may vary. Three examples are π, $-\sqrt{7}$,
and 0.31311311131111. . . . Irrational numbers cannot be
written as the quotient of two integers. Real numbers that are not
rational are irrational. Decimal notation for rational numbers
either terminates or repeats. Decimal notation for irrational

numbers neither terminates nor repeats. **56.** Answers may vary. If we think of the addition on the number line, we start at 0, move to the left to a negative number, and then move to the left again. This always brings us to a point on the negative portion of the number line. **57.** Yes; consider $m - (-n)$, where both m and n are positive. Then $m - (-n) = m + n$. Now $m + n$, the sum of two positive numbers, is positive.

Exercise Set 1.5, p. 42

1. -8 **3.** -48 **5.** -24 **7.** -72 **9.** 16 **11.** 42
13. -120 **15.** -238 **17.** 1200 **19.** 98 **21.** -72
23. -12.4 **25.** 30 **27.** 21.7 **29.** $-\frac{2}{5}$ **31.** $\frac{1}{12}$
33. -17.01 **35.** $-\frac{5}{12}$ **37.** 420 **39.** $\frac{2}{7}$ **41.** -60
43. 150 **45.** $-\frac{2}{45}$ **47.** 1911 **49.** 50.4 **51.** $\frac{10}{189}$ **53.** -960
55. 17.64 **57.** $-\frac{5}{784}$ **59.** 0 **61.** -720 **63.** $-30{,}240$
65. 1 **67.** $16, -16; 16, -16$ **69.** $441; -147$ **71.** $20; 20$
73. $-2; 2$ **75.** $-20\,\text{lb}$ **77.** -54°C **79.** 12.71
81. $-32\,\text{m}$ **83.** 38°F **85.** 2 **86.** $\frac{4}{15}$ **87.** $-\frac{1}{3}$ **88.** -4.3
89. 44 **90.** $-\frac{1}{12}$ **91.** True **92.** False **93.** False
94. False **95.** A
97.

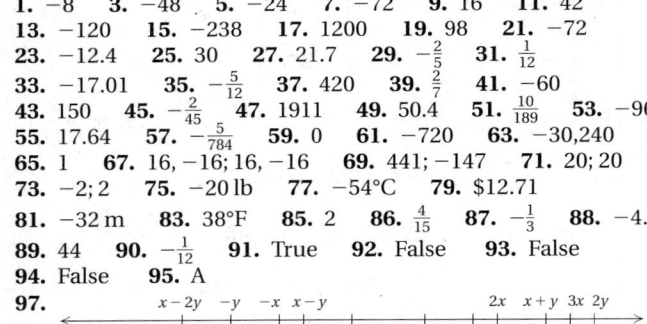

Calculator Corner, p. 50

1. -4 **2.** -0.3 **3.** -12 **4.** -9.5 **5.** -12 **6.** 2.7
7. -2 **8.** -5.7 **9.** -32 **10.** -1.8 **11.** 35
12. 14.44 **13.** -2 **14.** -0.8 **15.** 1.4 **16.** 4

Exercise Set 1.6, p. 51

1. -8 **3.** -14 **5.** -3 **7.** 3 **9.** -8 **11.** 2 **13.** -12
15. -8 **17.** Not defined **19.** 0 **21.** $\frac{7}{15}$ **23.** $-\frac{13}{47}$
25. $\frac{1}{13}$ **27.** $-\frac{1}{32}$ **29.** -7.1 **31.** 9 **33.** $4y$ **35.** $\frac{3b}{2a}$
37. $4 \cdot \left(\frac{1}{17}\right)$ **39.** $8 \cdot \left(-\frac{1}{13}\right)$ **41.** $13.9 \cdot \left(-\frac{1}{1.5}\right)$ **43.** $\frac{2}{3} \cdot \left(-\frac{5}{4}\right)$
45. $x \cdot y$ **47.** $(3x + 4)\left(\frac{1}{5}\right)$ **49.** $-\frac{9}{8}$ **51.** $\frac{5}{3}$ **53.** $\frac{9}{14}$
55. $\frac{9}{64}$ **57.** $-\frac{5}{4}$ **59.** $-\frac{27}{5}$ **61.** $\frac{11}{13}$ **63.** -2
65. -16.2 **67.** -2.5 **69.** -1.25 **71.** Not defined
73. 23.5% **75.** -3.3% **77.** $-\frac{1}{4}$ **78.** 5 **79.** -42
80. -48 **81.** 8.5 **82.** $-\frac{1}{8}$ **83.** -0.09 **84.** $0.91\overline{6}$
85. 3.75 **86.** $-3.\overline{3}$ **87.** $\frac{1}{-10.5}; -10.5$, the reciprocal of the reciprocal is the original number. **89.** Negative
91. Positive **93.** Negative

Exercise Set 1.7, p. 63

1. $\frac{3y}{5y}$ **3.** $\frac{10x}{15x}$ **5.** $\frac{2x}{x^2}$ **7.** $-\frac{3}{2}$ **9.** $-\frac{7}{6}$ **11.** $\frac{4s}{3}$ **13.** $8 + y$
15. nm **17.** $xy + 9$, or $9 + yx$ **19.** $c + ab$, or $ba + c$
21. $(a + b) + 2$ **23.** $8(xy)$ **25.** $a + (b + 3)$ **27.** $(3a)b$
29. $2 + (b + a), (2 + a) + b, (b + 2) + a$; answers may vary
31. $(5 + w) + v, (v + 5) + w, (w + v) + 5$; answers may vary
33. $(3x)y, y(x \cdot 3), 3(yx)$; answers may vary
35. $a(7b), b(7a), (7b)a$; answers may vary **37.** $2b + 10$
39. $7 + 7t$ **41.** $30x + 12$ **43.** $7x + 28 + 42y$
45. $7x - 21$ **47.** $-3x + 21$ **49.** $\frac{2}{3}b - 4$ **51.** $7.3x - 14.6$
53. $-\frac{3}{5}x + \frac{3}{5}y - 6$ **55.** $45x + 54y - 72$
57. $-4x + 12y + 8z$ **59.** $-3.72x + 9.92y - 3.41$
61. $4x, 3z$ **63.** $7x, 8y, -9z$ **65.** $2(x + 2)$ **67.** $5(6 + y)$
69. $7(2x + 3y)$ **71.** $7(2t - 1)$ **73.** $8(x - 3)$
75. $6(3a - 4b)$ **77.** $-4(y - 8)$, or $4(-y + 8)$
79. $5(x + 2 + 3y)$ **81.** $8(2m - 4n + 1)$
83. $4(3a + b - 6)$ **85.** $2(4x + 5y - 11)$ **87.** $a(x - 1)$
89. $a(x - y - z)$ **91.** $-6(3x - 2y - 1)$, or $6(-3x + 2y + 1)$
93. $\frac{1}{3}(2x - 5y + 1)$ **95.** $6(6x - y + 3z)$ **97.** $19a$ **99.** $9a$
101. $8x + 9z$ **103.** $7x + 15y^2$ **105.** $-19a + 88$
107. $4t + 6y - 4$ **109.** b **111.** $\frac{13}{4}y$ **113.** $8x$ **115.** $5n$
117. $-16y$ **119.** $17a - 12b - 1$ **121.** $4x + 2y$

123. $7x + y$ **125.** $0.8x + 0.5y$ **127.** $\frac{35}{6}a + \frac{3}{2}b - 42$
129. -26 **130.** -115 **131.** 160 **132.** 8 **133.** -5
134. 50 **135.** 180 **136.** $\frac{4}{13}$ **137.** True **138.** False
139. True **140.** True **141.** Not equivalent;
$3 \cdot 2 + 5 \neq 3 \cdot 5 + 2$ **143.** Equivalent; commutative law of addition **145.** $q(1 + r + rs + rst)$

Calculator Corner, p. 72

1. -11 **2.** 9 **3.** 114 **4.** 117,649 **5.** $-1{,}419{,}857$
6. $-1{,}124{,}864$ **7.** $-117{,}649$ **8.** $-1{,}419{,}857$ **9.** $-1{,}124{,}864$
10. -4 **11.** -2 **12.** 787

Exercise Set 1.8, p. 73

1. $-2x - 7$ **3.** $-8 + x$ **5.** $-4a + 3b - 7c$
7. $-6x + 8y - 5$ **9.** $-3x + 5y + 6$ **11.** $8x + 6y + 43$
13. $5x - 3$ **15.** $-3a + 9$ **17.** $5x - 6$ **19.** $-19x + 2y$
21. $9y - 25z$ **23.** $-7x + 10y$ **25.** $37a - 23b + 35c$
27. 7 **29.** -40 **31.** 19 **33.** $12x + 30$ **35.** $3x + 30$
37. $9x - 18$ **39.** $-4x - 64$ **41.** -7 **43.** -7 **45.** -16
47. -334 **49.** 14 **51.** 1880 **53.** 12 **55.** 8 **57.** -86
59. 37 **61.** -1 **63.** -10 **65.** -67 **67.** -7988
69. -3000 **71.** 60 **73.** 1 **75.** 10 **77.** $-\frac{13}{45}$ **79.** $-\frac{23}{18}$
81. -122 **83.** Integers **84.** Additive inverses
85. Commutative law **86.** Identity property of 1
87. Associative law **88.** Associative law **89.** Multiplicative inverses **90.** Identity property of 0
91. $6y - (-2x + 3a - c)$ **93.** $6m - (-3n + 5m - 4b)$
95. $-2x - f$ **97.** (a) 52; 52; 28.130169;
(b) $-24; -24; -108.307025$ **99.** -6

Summary and Review: Chapter 1, p. 77

Concept Reinforcement

1. True **2.** True **3.** False **4.** False

Important Concepts

1. 14 **2.** $<$ **3.** $\frac{5}{4}$ **4.** -8.5 **5.** -2 **6.** 56 **7.** -8
8. $\frac{9}{20}$ **9.** $\frac{5}{3}$ **10.** $5x + 15y - 20z$ **11.** $9(3x + y - 4z)$
12. $5a - 2b$ **13.** $4a - 4b$ **14.** -2

Review Exercises

1. 4 **2.** $19\%x$, or $0.19x$ **3.** $-45, 72$ **4.** 38 **5.** 126
6. -2.5 (number line: -6 to 6) **7.** $\frac{8}{9}$ (number line: -6 to 6)
8. $<$ **9.** $>$ **10.** $>$ **11.** $<$ **12.** $x > -3$ **13.** True
14. False **15.** -3.8 **16.** $\frac{3}{4}$ **17.** $\frac{8}{3}$ **18.** $-\frac{1}{7}$ **19.** 34
20. 5 **21.** -3 **22.** -4 **23.** -5 **24.** 1 **25.** $-\frac{7}{5}$
26. -7.9 **27.** 54 **28.** -9.18 **29.** $-\frac{2}{7}$ **30.** -210
31. -7 **32.** -3 **33.** $\frac{3}{4}$ **34.** 40.4 **35.** -2 **36.** 2
37. -2 **38.** 8-yd gain **39.** $-\$130$ **40.** 4.64
41. 18.95 **42.** $15x - 35$ **43.** $-8x + 10$ **44.** $4x + 15$
45. $-24 + 48x$ **46.** $2(x - 7)$ **47.** $-6(x - 1)$, or $6(-x + 1)$
48. $5(x + 2)$ **49.** $-3(x - 4y + 4)$, or $3(-x + 4y - 4)$
50. $7a - 3b$ **51.** $-2x + 5y$ **52.** $5x - y$ **53.** $-a + 8b$
54. $-3a + 9$ **55.** $-2b + 21$ **56.** 6 **57.** $12y - 34$
58. $5x + 24$ **59.** $-15x + 25$ **60.** D **61.** B **62.** $-\frac{5}{8}$
63. -2.1 **64.** 1000 **65.** $4a + 2b$

Understanding Through Discussion and Writing

1. The sum of each pair of opposites such as -50 and 50, -49 and 49, and so on is 0. The sum of these sums and the remaining integer, 0, is 0. **2.** The product of an even number of negative numbers is positive, and the product of an odd number of negative numbers is negative. Now $(-7)^8$ is the product of 8 factors of -7 so it is positive, and $(-7)^{11}$ is the product of 11 factors of -7 so it is negative. **3.** Consider $\frac{a}{b} = q$, where a and b are both negative numbers. Then $q \cdot b = a$, so q must be a positive number in order for the product to be negative.   **4.** Consider $\frac{a}{b} = q$, where a is a negative number and b is a positive number. Then $q \cdot b = a$, so q must be a negative number in order for the product to be negative.   **5.** We use the distributive law when we collect like terms even though we might not always write this step.   **6.** Jake expects the calculator to multiply 2 and 3 first and then divide 18 by that product. This procedure does not follow the rules for order of operations.

Test: Chapter 1, p. 83

1. [1.1a] 6 **2.** [1.1b] $x - 9$ **3.** [1.2d] $>$ **4.** [1.2d] $<$
5. [1.2d] $>$ **6.** [1.2d] $-2 > x$ **7.** [1.2d] True **8.** [1.2e] 7
9. [1.2e] $\frac{9}{4}$ **10.** [1.2e] 2.7 **11.** [1.3b] $-\frac{2}{3}$ **12.** [1.3b] 1.4
13. [1.6b] $-\frac{1}{2}$ **14.** [1.6b] $\frac{7}{4}$ **15.** [1.3b] 8 **16.** [1.4a] 7.8
17. [1.3a] -8 **18.** [1.3a] $\frac{7}{40}$ **19.** [1.4a] 10 **20.** [1.4a] -2.5
21. [1.4a] $\frac{7}{8}$ **22.** [1.5a] -48 **23.** [1.5a] $\frac{3}{16}$ **24.** [1.6a] -9
25. [1.6c] $\frac{3}{4}$ **26.** [1.6c] -9.728 **27.** [1.8d] -173
28. [1.8d] -5 **29.** [1.3c], [1.4b] Up 15 points **30.** [1.4b] 14°F
31. [1.5b] 16,080 **32.** [1.6d] -0.75°C each minute
33. [1.7c] $18 - 3x$ **34.** [1.7c] $-5y + 5$
35. [1.7d] $2(6 - 11x)$ **36.** [1.7d] $7(x + 3 + 2y)$
37. [1.4a] 12 **38.** [1.8b] $2x + 7$ **39.** [1.8b] $9a - 12b - 7$
40. [1.8c] $68y - 8$ **41.** [1.8d] -4 **42.** [1.8d] 448
43. [1.2d] B **44.** [1.2e], [1.8d] 15 **45.** [1.8c] $4a$
46. [1.7e] $4x + 4y$

CHAPTER 2

Exercise Set 2.1, p. 90

1. Yes **3.** No **5.** No **7.** Yes **9.** Yes **11.** No **13.** 4
15. -20 **17.** -14 **19.** -18 **21.** 15 **23.** -14 **25.** 2
27. 20 **29.** -6 **31.** $6\frac{1}{2}$ **33.** 19.9 **35.** $\frac{7}{3}$ **37.** $-\frac{7}{4}$
39. $\frac{41}{24}$ **41.** $-\frac{1}{20}$ **43.** 5.1 **45.** 12.4 **47.** -5 **49.** $1\frac{5}{6}$
51. $-\frac{10}{21}$ **53.** -11 **54.** 5 **55.** $-\frac{5}{12}$ **56.** $\frac{1}{3}$ **57.** $-\frac{3}{2}$
58. -5.2 **59.** $-\frac{1}{24}$ **60.** 172.72 **61.** $83 - x$ **62.** $65t$ miles
63. 342.246 **65.** $-\frac{26}{15}$ **67.** -10 **69.** All real numbers
71. $-\frac{5}{17}$ **73.** 13, -13

Exercise Set 2.2, p. 96

1. 6 **3.** 9 **5.** 12 **7.** -40 **9.** 1 **11.** -7 **13.** -6
15. 6 **17.** -63 **19.** -48 **21.** 36 **23.** -9 **25.** -21
27. $-\frac{3}{5}$ **29.** $-\frac{3}{2}$ **31.** $\frac{9}{2}$ **33.** 7 **35.** -7 **37.** 8 **39.** 15.9
41. -50 **43.** -14 **45.** $7x$ **46.** $-x + 5$ **47.** $8x + 11$
48. $-32y$ **49.** $x - 4$ **50.** $-5x - 23$ **51.** $-10y - 42$
52. $-22a + 4$ **53.** $8r$ miles **54.** $\frac{1}{2} b \cdot 10$ m², or $5b$ m²
55. -8655 **57.** No solution **59.** No solution

61. $\dfrac{b}{3a}$ **63.** $\dfrac{4b}{a}$

Calculator Corner, p. 101

1. Left to the student

Exercise Set 2.3, p. 105

1. 5 **3.** 8 **5.** 10 **7.** 14 **9.** -8 **11.** -8 **13.** -7
15. 12 **17.** 6 **19.** 4 **21.** 6 **23.** -3 **25.** 1 **27.** 6
29. -20 **31.** 7 **33.** 2 **35.** 5 **37.** 2 **39.** 10 **41.** 4
43. 0 **45.** -1 **47.** $-\frac{4}{3}$ **49.** $\frac{2}{5}$ **51.** -2 **53.** -4 **55.** $\frac{4}{5}$
57. $-\frac{28}{27}$ **59.** 6 **61.** 2 **63.** No solution **65.** All real numbers **67.** 6 **69.** 8 **71.** 1 **73.** All real numbers
75. No solution **77.** 17 **79.** $-\frac{5}{3}$ **81.** -3 **83.** 2
85. $\frac{4}{7}$ **87.** No solution **89.** All real numbers **91.** $-\frac{51}{31}$
93. -6.5 **94.** -75.14 **95.** $7(x - 3 - 2y)$
96. $8(y - 11x + 1)$ **97.** -160 **98.** $-17x + 18$

99. $91x - 242$ **100.** 0.25 **101.** $-\frac{5}{32}$ **103.** $\frac{52}{45}$

Exercise Set 2.4, p. 113

1. (a) 57,000 Btu's; (b) $a = \dfrac{B}{30}$ **3.** (a) 1.6 mi; (b) $t = 5M$
5. (a) 1423 students; (b) $n = 15f$ **7.** 10.5 calories per ounce
9. 42 games **11.** $x = \dfrac{y}{5}$ **13.** $c = \dfrac{a}{b}$ **15.** $m = n - 11$
17. $x = y + \dfrac{3}{5}$ **19.** $x = y - 13$ **21.** $x = y - b$
23. $x = 5 - y$ **25.** $x = a - y$ **27.** $y = \dfrac{5x}{8}$, or $\dfrac{5}{8}x$
29. $x = \dfrac{By}{A}$ **31.** $t = \dfrac{W - b}{m}$ **33.** $x = \dfrac{y - c}{b}$ **35.** $h = \dfrac{A}{b}$
37. $w = \dfrac{P - 2l}{2}$, or $\dfrac{1}{2}P - l$ **39.** $a = 2A - b$
41. $b = 3A - a - c$ **43.** $t = \dfrac{A - b}{a}$ **45.** $x = \dfrac{c - By}{A}$
47. $a = \dfrac{F}{m}$ **49.** $c^2 = \dfrac{E}{m}$ **51.** $t = \dfrac{3k}{v}$ **53.** 7
54. $-21a + 12b$ **55.** -13.2 **56.** $-\frac{3}{2}$ **57.** $-35\frac{1}{2}$ **58.** $-\frac{1}{6}$
59. -9.325 **60.** $3\frac{3}{4}$ **61.** $\frac{11}{8}$ **62.** -1 **63.** -3 **64.** $\frac{9}{7}$
65. (a) 1901 calories;
(b) $a = \dfrac{917 + 6w + 6h - K}{6}$;
$h = \dfrac{K - 917 - 6w + 6a}{6}$;
$w = \dfrac{K - 917 - 6h + 6a}{6}$
67. $b = \dfrac{Ha - 2}{H}$, or $a - \dfrac{2}{H}$; $a = \dfrac{2 + Hb}{H}$, or $\dfrac{2}{H} + b$
69. A quadruples. **71.** A increases by $2h$ units.

Mid-Chapter Review: Chapter 2, p. 117

1. False **2.** True **3.** True **4.** False
5.
$$x + 5 = -3$$
$$x + 5 - 5 = -3 - 5$$
$$x + 0 = -8$$
$$x = -8$$
6.
$$-6x = 42$$
$$\frac{-6x}{-6} = \frac{42}{-6}$$
$$1 \cdot x = -7$$
$$x = -7$$
7.
$$5y + z = t$$
$$5y + z - z = t - z$$
$$5y = t - z$$
$$\frac{5y}{5} = \frac{t - z}{5}$$
$$y = \frac{t - z}{5}$$

8. 6 **9.** -12 **10.** 7 **11.** -10 **12.** 20 **13.** 5 **14.** $\frac{3}{4}$
15. -1.4 **16.** 6 **17.** -17 **18.** -9 **19.** 17 **20.** 21
21. 18 **22.** -15 **23.** $-\frac{3}{2}$ **24.** 1 **25.** -3 **26.** $\frac{3}{2}$
27. -1 **28.** 3 **29.** -7 **30.** 4 **31.** 2 **32.** $\frac{9}{8}$ **33.** $-\frac{21}{5}$
34. 9 **35.** -2 **36.** 0 **37.** All real numbers
38. No solution **39.** $-\frac{13}{2}$ **40.** All real numbers **41.** $b = \dfrac{A}{4}$

42. $x = y + 1.5$ **43.** $m = s - n$ **44.** $t = \dfrac{9w}{4}$

45. $t = \dfrac{B + c}{a}$ **46.** $y = 2M - x - z$ **47.** Equivalent
expressions have the same value for all possible replacements for the variable(s). Equivalent equations have the same solution(s). **48.** The equations are not equivalent because they do not have the same solutions. Although 5 is a solution of both equations, -5 is a solution of $x^2 = 25$ but not of $x = 5$.
49. For an equation $x + a = b$, add the opposite of a

(or subtract a) on both sides of the equation.   **50.** The student probably added $\frac{1}{3}$ on the right side of the equation rather than adding $-\frac{1}{3}$ (or subtracting $\frac{1}{3}$) on the right side. The correct solution is -2.   **51.** For an equation $ax = b$, multiply by $1/a$ (or divide by a) on both sides of the equation.   **52.** Answers may vary. A walker who knows how far and how long she walks each day wants to know her average speed each day.

Exercise Set 2.5, p. 123

1. 20%   **3.** 150   **5.** 546   **7.** 24%   **9.** 2.5   **11.** 5%
13. 25%   **15.** 84   **17.** 24%   **19.** 16%   **21.** $46\frac{2}{3}$ **23.** 0.8
25. 5   **27.** 40   **29.** $16.1   **31.** $2.1   **33.** About 12%
35. $2.646 billion   **37.** $390   **39.** (a) 16%; (b) $29
41. (a) $3.75; (b) $28.75   **43.** (a) $30; (b) $34.50   **45.** About 85,821 acres   **47.** About 22.6%   **49.** 800%   **51.** 10%
53. About 144%   **55.** $a + c$ **56.** $7x - 9y$ **57.** -3.9
58. $-6\frac{1}{8}$ **59.** Division; subtraction   **60.** Exponential; division; subtraction   **61.** 6 ft 7 in.

Translating for Success, p. 138

1. B   **2.** H   **3.** G   **4.** N   **5.** J   **6.** C   **7.** L   **8.** E
9. F   **10.** D

Exercise Set 2.6, p. 139

1. 3113 manatees   **3.** 180 in.; 60 in.   **5.** $16.56   **7.** $699\frac{1}{3}$ mi.
9. 1204 and 1205   **11.** 41, 42, 43   **13.** 61, 63, 65   **15.** Length: 48 ft; width: 14 ft   **17.** $75   **19.** $85   **21.** 11 visits
23. 28°, 84°, 68°   **25.** 33°, 38°, 109°   **27.** $350   **29.** $852.94
31. 12 mi   **33.** $36   **35.** $25 and $50   **37.** -12 **39.** $-\frac{47}{40}$
40. $-\frac{17}{40}$ **41.** $-\frac{3}{10}$ **42.** $-\frac{32}{15}$ **43.** -10 **44.** 1.6
45. 409.6   **46.** -9.6 **47.** -41.6 **48.** 0.1   **49.** All real numbers   **50.** No solution   **51.** No solution   **52.** All real numbers   **53.** 120 apples   **55.** About 0.65 in.   **57.** $9.17, not $9.10

Exercise Set 2.7, p. 152

1. (a) Yes; (b) yes; (c) no; (d) yes; (e) yes
3. (a) No; (b) no; (c) no; (d) yes; (e) no
5.
7.
9.
11.
13.
15. $\{x|x > -5\}$;
17. $\{x|x \le -18\}$;
19. $\{y|y > -5\}$
21. $\{x|x > 2\}$ **23.** $\{x|x \le -3\}$ **25.** $\{x|x < 4\}$
27. $\{t|t > 14\}$ **29.** $\left\{y|y \le \frac{1}{4}\right\}$ **31.** $\left\{x|x > \frac{7}{12}\right\}$
33. $\{x|x < 7\}$;
35. $\{x|x < 3\}$;
37. $\left\{y|y \ge -\frac{2}{5}\right\}$ **39.** $\{x|x \ge -6\}$ **41.** $\{y|y \le 4\}$
43. $\left\{x|x > \frac{17}{3}\right\}$ **45.** $\left\{y|y < -\frac{1}{14}\right\}$ **47.** $\left\{x|x \le \frac{3}{10}\right\}$
49. $\{x|x < 8\}$ **51.** $\{x|x \le 6\}$ **53.** $\{x|x < -3\}$
55. $\{x|x > -3\}$ **57.** $\{x|x \le 7\}$ **59.** $\{x|x > -10\}$
61. $\{y|y < 2\}$ **63.** $\{y|y \ge 3\}$ **65.** $\{y|y > -2\}$
67. $\{x|x > -4\}$ **69.** $\{x|x \le 9\}$ **71.** $\{y|y \le -3\}$
73. $\{y|y < 6\}$ **75.** $\{m|m \ge 6\}$ **77.** $\left\{t|t < -\frac{5}{3}\right\}$
79. $\{r|r > -3\}$ **81.** $\left\{x|x \ge -\frac{57}{34}\right\}$ **83.** $\{x|x > -2\}$
85. -74 **86.** 4.8   **87.** $-\frac{5}{8}$ **88.** -1.11 **89.** -38 **90.** $-\frac{7}{8}$
91. -9.4 **92.** 1.11   **93.** 140   **94.** 41   **95.** $-2x - 23$
96. $37x - 1$ **97.** (a) Yes; (b) yes; (c) no; (d) no; (e) no; (f) yes; (g) yes   **99.** No solution

Exercise Set 2.8, p. 159

1. $n \ge 7$ **3.** $w > 2$ kg   **5.** 90 mph $< s <$ 110 mph
7. $w \le 20$ hr   **9.** $c \ge \$1.50$ **11.** $x > 8$ **13.** $y \le -4$
15. $n \ge 1300$ **17.** $W \le 500$ L   **19.** $3x + 2 < 13$
21. $\{x|x \ge 84\}$ **23.** $\{C|C < 1063°\}$ **25.** $\{Y|Y \ge 1935\}$
27. $\{L|L \ge 5$ in.$\}$ **29.** 15 or fewer copies   **31.** 5 min or more   **33.** 2 courses   **35.** 4 servings or more   **37.** Lengths greater than or equal to 92 ft; lengths less than or equal to 92 ft
39. Lengths less than 21.5 cm   **41.** The blue-book value is greater than or equal to $10,625.   **43.** It has at least 16 g of fat.
45. Dates at least 6 weeks after July 1   **47.** Heights greater than or equal to 4 ft   **49.** 21 calls or more   **51.** Even
52. Odd   **53.** Additive   **54.** Multiplicative   **55.** Equivalent
56. Addition principle   **57.** Multiplication principle; is reversed   **58.** Solution   **59.** Temperatures between $-15°C$ and $-9\frac{4}{9}°C$ **61.** They contain at least 7.5 g of fat per serving.

Summary and Review: Chapter 2, p. 164

Concept Reinforcement

1. True   **2.** True   **3.** False   **4.** True

Important Concepts

1. -12 **2.** All real numbers   **3.** No solution   **4.** $b = \dfrac{2A}{h}$
5.
6.
7. $\{y|y > -4\}$

Review Exercises

1. -22 **2.** 1   **3.** 25   **4.** 9.99   **5.** $\frac{1}{4}$ **6.** 7   **7.** -192
8. $-\frac{7}{3}$ **9.** $-\frac{15}{64}$ **10.** -8 **11.** 4   **12.** -5 **13.** $-\frac{1}{3}$ **14.** 3
15. 4   **16.** 16   **17.** All real numbers   **18.** 6   **19.** -3
20. 28   **21.** 4   **22.** No solution   **23.** Yes   **24.** No
25. Yes   **26.** $\left\{y|y \ge -\frac{1}{2}\right\}$ **27.** $\{x|x \ge 7\}$ **28.** $\{y|y > 2\}$
29. $\{y|y \le -4\}$ **30.** $\{x|x < -11\}$ **31.** $\{y|y > -7\}$
32. $\left\{x|x > -\frac{9}{11}\right\}$ **33.** $\left\{x|x \ge -\frac{1}{12}\right\}$
34.
35.
36.
37. $d = \dfrac{C}{\pi}$ **38.** $B = \dfrac{3V}{h}$
39. $a = 2A - b$ **40.** $x = \dfrac{y - b}{m}$ **41.** Length: 365 mi; width: 275 mi   **42.** 345, 346   **43.** $2117   **44.** 27 subscriptions
45. 35°, 85°, 60°   **46.** 15   **47.** 18.75%   **48.** 600
49. About 18%   **50.** $220   **51.** $53,400   **52.** $138.95
53. 86   **54.** $\{w|w > 17$ cm$\}$ **55.** C   **56.** A   **57.** 23, -23
58. 20, -20 **59.** $a = \dfrac{y - 3}{2 - b}$

Understanding Through Discussion and Writing

1. The end result is the same either way. If s is the original salary, the new salary after a 5% raise followed by an 8% raise is $1.08(1.05s)$. If the raises occur the other way around, the new salary is $1.05(1.08s)$. By the commutative and associative laws of multiplication, we see that these are equal. However, it would be better to receive the 8% raise first, because this increase yields a higher salary initially than a 5% raise.
2. No; Erin paid 75% of the original price and was offered credit for 125% of this amount, not to be used on sale items. Now, 125% of 75% is 93.75%, so Erin would have a credit of 93.75% of the original price. Since this credit can be applied only to non-sale items, she has less purchasing power than if the amount she paid were refunded and she could spend it on sale items.

3. The inequalities are equivalent by the multiplication principle for inequalities. If we multiply on both sides of one inequality by -1, the other inequality results.
4. For any pair of numbers, their relative position on the number line is reversed when both are multiplied by the same negative number. For example, -3 is to the left of 5 on the number line $(-3 < 5)$, but 12 is to the right of -20 $(-3(-4) > 5(-4))$.
5. Answers may vary. Fran is more than 3 years older than Todd.
6. Let n represent "a number." Then "five more than a number" translates to the *expression* $n + 5$, or $5 + n$, and "five is more than a number" translates to the *inequality* $5 > n$.

Test: Chapter 2, p. 169

1. [2.1b] 8　**2.** [2.1b] 26　**3.** [2.2a] -6　**4.** [2.2a] 49
5. [2.3b] -12　**6.** [2.3a] 2　**7.** [2.3a] -8　**8.** [2.1b] $-\frac{7}{20}$
9. [2.3c] 7　**10.** [2.3c] $\frac{5}{3}$　**11.** [2.3b] $\frac{5}{2}$
12. [2.3c] No solution　**13.** [2.3c] All real numbers
14. [2.7c] $\{x|x \le -4\}$　**15.** [2.7c] $\{x|x > -13\}$
16. [2.7d] $\{x|x \le 5\}$　**17.** [2.7d] $\{y|y \le -13\}$
18. [2.7d] $\{y|y \ge 8\}$　**19.** [2.7d] $\{x|x \le -\frac{1}{20}\}$
20. [2.7e] $\{x|x < -6\}$　**21.** [2.7e] $\{x|x \le -1\}$
22. [2.7b]

$$y \le 9$$
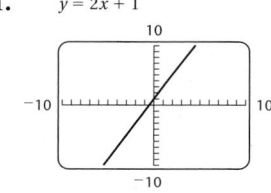

23. [2.7b, e]

$$x < 1$$

24. [2.7b]

$$-2 \le x \le 2$$

25. [2.5a] 18
26. [2.5a] 16.5%　**27.** [2.5a] 40,000　**28.** [2.5a] About 25.8%
29. [2.6a] Width: 7 cm; length: 11 cm　**30.** [2.5a] About $310 billion　**31.** [2.6a] 2509, 2510, 2511　**32.** [2.6a] $880
33. [2.6a] 3 m, 5 m　**34.** [2.8b] $\{l|l \ge 174 \text{ yd}\}$
35. [2.8b] $\{b|b \le \$105\}$　**36.** [2.8b] $\{c|c \le 143{,}750\}$
37. [2.4b] $r = \dfrac{A}{2\pi h}$　**38.** [2.4b] $x = \dfrac{y - b}{8}$
39. [2.5a] D　**40.** [2.4b] $d = \dfrac{1 - ca}{-c}$, or $\dfrac{ca - 1}{c}$
41. [1.2e], [2.3a] 15, -15　**42.** [2.6a] 60 tickets

Cumulative Review: Chapters 1–2, p. 171

1. [1.1a] $\frac{3}{2}$　**2.** [1.1a] $\frac{15}{4}$　**3.** [1.1a] 0　**4.** [1.1b] $2w - 4$
5. [1.2d] $>$　**6.** [1.2d] $>$　**7.** [1.2d] $<$　**8.** [1.3b], [1.6b] $-\frac{2}{5}, \frac{5}{2}$
9. [1.2e] 3　**10.** [1.2e] $\frac{3}{4}$　**11.** [1.2e] 0　**12.** [1.3a] -4.4
13. [1.4a] $-\frac{5}{2}$　**14.** [1.5a] $\frac{5}{6}$　**15.** [1.5a] -105　**16.** [1.6a] -9
17. [1.6c] -3　**18.** [1.6c] $\frac{32}{125}$　**19.** [1.7c] $15x + 25y + 10z$
20. [1.7c] $-12x - 8$　**21.** [1.7c] $-12y + 24x$
22. [1.7d] $2(32 + 9x + 12y)$　**23.** [1.7d] $8(2y - 7)$
24. [1.7d] $5(a - 3b + 5)$　**25.** [1.7e] $15b + 22y$
26. [1.7e] $4 + 9y + 6z$　**27.** [1.7e] $1 - 3a - 9d$
28. [1.7e] $-2.6x - 5.2y$　**29.** [1.8b] $3x - 1$　**30.** [1.8b] $-2x - y$
31. [1.8b] $-7x + 6$　**32.** [1.8b] $8x$　**33.** [1.8c] $5x - 13$
34. [2.1b] 4.5　**35.** [2.2a] $\frac{4}{25}$　**36.** [2.1b] 10.9　**37.** [2.1b] $3\frac{5}{6}$
38. [2.2a] -48　**39.** [2.2a] $-\frac{3}{8}$　**40.** [2.2a] -6.2
41. [2.3a] -3　**42.** [2.3b] $-\frac{12}{5}$　**43.** [2.3b] 8　**44.** [2.3c] 7
45. [2.3b] $-\frac{4}{5}$　**46.** [2.3b] $-\frac{10}{3}$　**47.** [2.3c] All real numbers
48. [2.3c] No solution　**49.** [2.7c] $\{x|x < 2\}$
50. [2.7e] $\{y|y < -3\}$　**51.** [2.7e] $\{y|y \ge 4\}$
52. [2.4b] $m = 65 - H$　**53.** [2.4b] $t = \dfrac{I}{Pr}$　**54.** [2.5a] 25.2
55. [2.5a] 45%　**56.** [2.5a] $363　**57.** [2.6a] $24.60
58. [2.6a] $45　**59.** [2.6a] $1050
60. [2.6a] 50 m, 53 m, 40 m　**61.** [2.8b] $\{s|s \ge 84\}$
62. [1.8d] C　**63.** [2.6a] $45,200　**64.** [2.6a] 30%
65. [1.2e], [2.3a] 4, -4　**66.** [2.3b] 3　**67.** [2.4b] $Q = \dfrac{2 - pm}{p}$

CHAPTER 3

Exercise Set 3.1, p. 178

1.
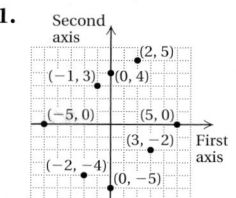

3. II　**5.** IV　**7.** III
9. On an axis, not in a quadrant
11. II　**13.** IV　**15.** I
17. Positive　**19.** II　**21.** I, IV
23. I, III

25. A: $(3, 3)$; B: $(0, -4)$; C: $(-5, 0)$; D: $(-1, -1)$; E: $(2, 0)$
27. No　**29.** No　**31.** Yes
33.
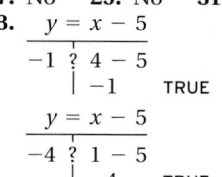
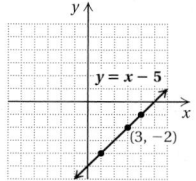

35.
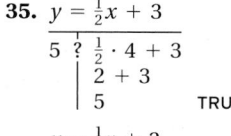
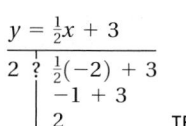
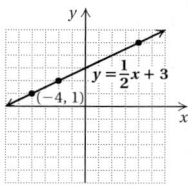

37.
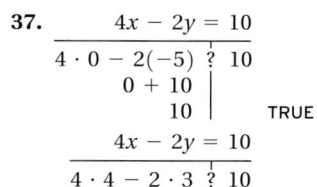
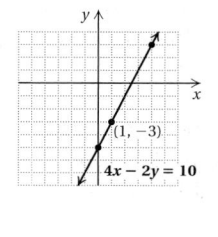

39. 8　**40.** $\dfrac{7}{4}$　**41.** All real numbers　**42.** No solution
43. $\dfrac{17}{10}$　**44.** $\dfrac{1}{3}$　**45.** $3.57　**46.** $48.60　**47.** 20%　**48.** $18
49. $45.15　**50.** $55　**51.** $(-1, -5)$　**55.** 26 linear units
53.
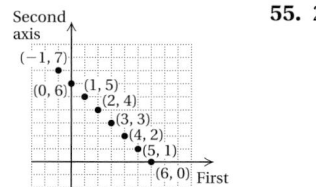

Calculator Corner, p. 181

1. Left to the student

Calculator Corner, p. 187

1.
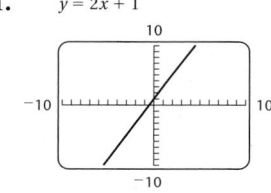
$y = 2x + 1$

2.
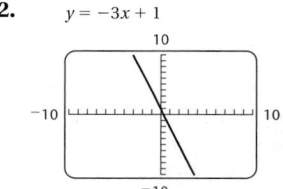
$y = -3x + 1$

3. $y = -5x + 3$

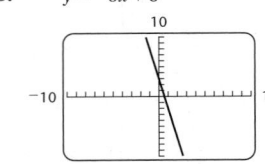

4. $y = 4x - 5$

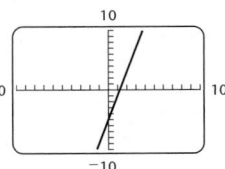

5. $y = \frac{4}{5}x + 2$

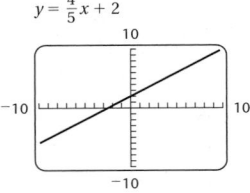

6. $y = -\frac{3}{5}x - 1$

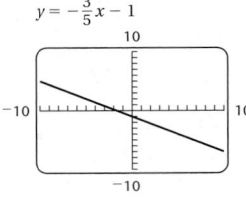

7. $y = 2.085x + 5.08$

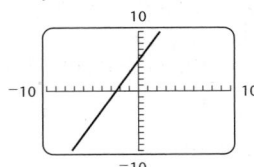

8. $y = -3.45x - 1.68$

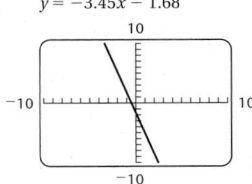

Exercise Set 3.2, p. 188

1.

x	y
−2	−1
−1	0
0	1
1	2
2	3

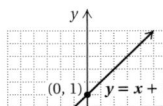

3.

x	y
−2	−2
−1	−1
0	0
1	1
2	2

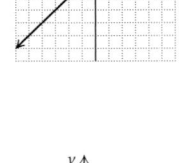

5.

x	y
−2	−1
0	0
4	2

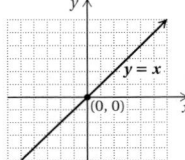

7.

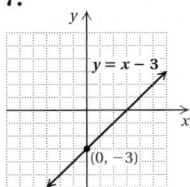

9.

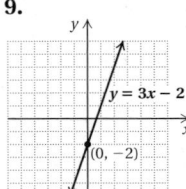

11.

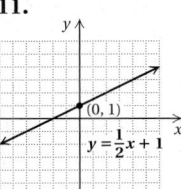

13.

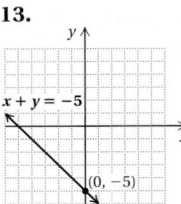

15.

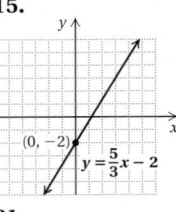

17.

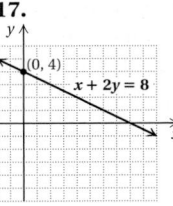

19.

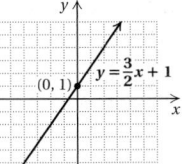

21.

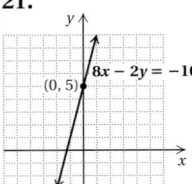

23.

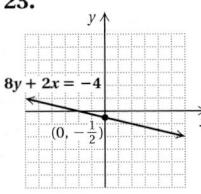

25. (a) 2002: \$52,620; 2007: \$44,130; 2010: \$39,036;
(b) about \$47,500;
(c) 9 yr after 2002, or in 2011

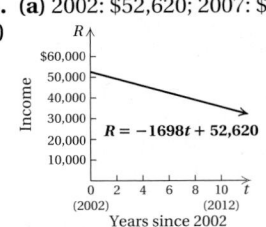

27. (a) 18.24 gal; 34.44 gal; 43.44 gal;
(b) about 31 gal;
(c) 11 yr after 2000, or in 2011

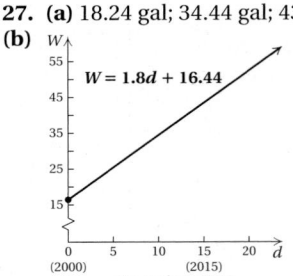

29. 12 **30.** 4.89 **31.** 0 **32.** $\frac{4}{5}$ **33.** 3.4 **34.** $\sqrt{2}$ **35.** $\frac{2}{3}$
36. $\frac{7}{8}$ **37.** 48 patients **38.** About 30.7% **39.** $-\frac{3}{25}$
40. 3 **41.** $-\frac{15}{16}$ **42.** −420 **43.** $\frac{32}{7}$ **44.** −9

Calculator Corner, p. 194

1. y-intercept: $(0, -15)$;
   x-intercept: $(-2, 0)$;

   $y = -7.5x - 15$

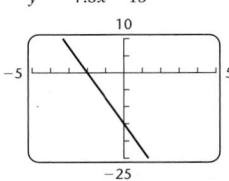

   Xscl = 1   Yscl = 5

2. y-intercept: $(0, 43)$;
   x-intercept: $(-20, 0)$;

   $y = 2.15x + 43$

   Xscl = 5   Yscl = 5

3. y-intercept: $(0, -30)$;
   x-intercept: $(25, 0)$;

   $y = (6x - 150)/5$

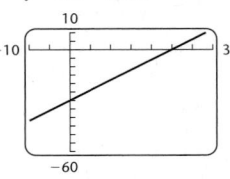

   Xscl = 5   Yscl = 5

4. y-intercept: $(0, -4)$;
   x-intercept: $(20, 0)$;

   $y = 0.2x - 4$

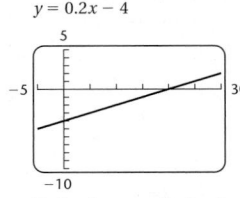

   Xscl = 5   Yscl = 1

5. y-intercept: $(0,-15)$;
x-intercept: $(10, 0)$;
$y = 1.5x - 15$

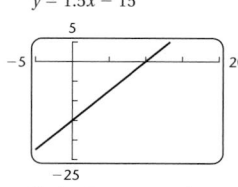
Xscl = 5   Yscl = 5

6. y-intercept: $\left(0, -\frac{1}{2}\right)$;
x-intercept: $\left(\frac{2}{5}, 0\right)$;
$y = (5x - 2)/4$

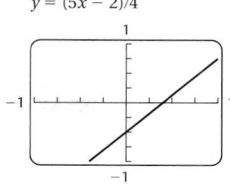
Xscl = 0.25   Yscl = 0.25

Visualizing for Success, p. 197

1. E **2.** C **3.** G **4.** A **5.** I **6.** D **7.** F **8.** J
9. B **10.** H

Exercise Set 3.3, p. 198

1. (a) $(0, 5)$; **(b)** $(2, 0)$ **3. (a)** $(0, -4)$; **(b)** $(3, 0)$
5. (a) $(0, 3)$; **(b)** $(5, 0)$ **7. (a)** $(0, -14)$; **(b)** $(4, 0)$
9. (a) $\left(0, \frac{10}{3}\right)$; **(b)** $\left(-\frac{5}{2}, 0\right)$ **11. (a)** $\left(0, -\frac{1}{3}\right)$; **(b)** $\left(\frac{1}{2}, 0\right)$

13.
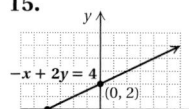

15.
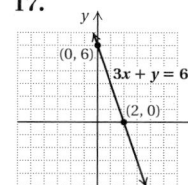

17.
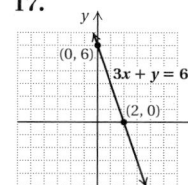

19.
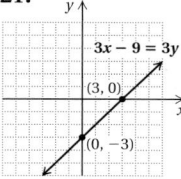

21.
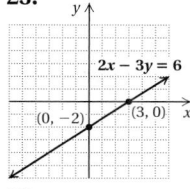

23.
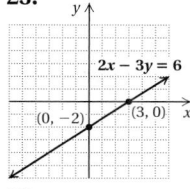

25.
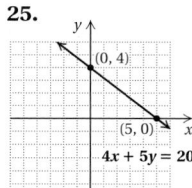

27.
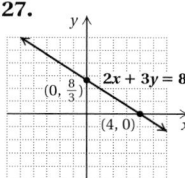

29.
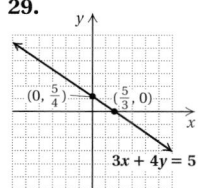

31.
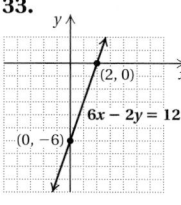

33.
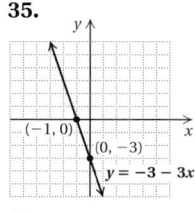

35.

37.
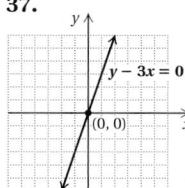

39.
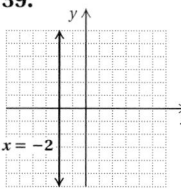

41.
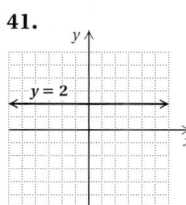

43.
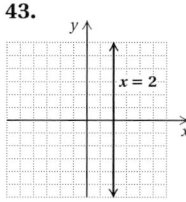

45.
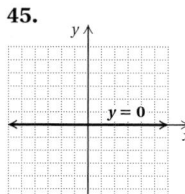

47.
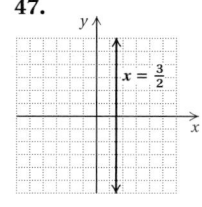

49.
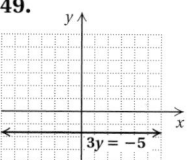

51.
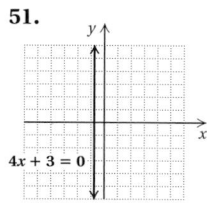

53.
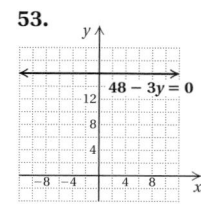

55. $y = -1$ **57.** $x = 4$ **59.** $\{x | x > -40\}$
60. $\{x | x \le -7\}$ **61.** $\{x | x < 1\}$ **62.** $\{x | x \ge 2\}$
63. $\{x | x \le 7\}$ **64.** $\{x | x > 1\}$ **65.** About 89,434
66. \$43,200 **67.** $y = -4$ **69.** $k = 12$

Mid-Chapter Review: Chapter 3, p. 203

1. True **2.** False **3.** True **4.** False
5. (a) The y-intercept is $(0, -3)$. **(b)** The x-intercept is $(-3, 0)$.
6. (a) The x-intercept is $(c, 0)$. **(b)** The y-intercept is $(0, d)$.
7. A: $(-1, 0)$; B: $(2, 5)$; C: $(-5, -4)$; D: $(6, -2)$; E: $(-4, 2)$
8. F: $(0, -3)$; G: $(5, 0)$; H: $(1, 5)$; I: $(-4, 4)$; J: $(3, -6)$
9. No **10.** Yes **11.** x-intercept: $(-6, 0)$; y-intercept: $(0, 9)$
12. x-intercept: $\left(\frac{1}{2}, 0\right)$; y-intercept: $\left(0, -\frac{1}{20}\right)$
13. x-intercept: $(40, 0)$; y-intercept: $(0, -2)$
14. x-intercept: $(-42, 0)$; y-intercept: $(0, 105)$
15.
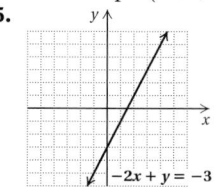

16.
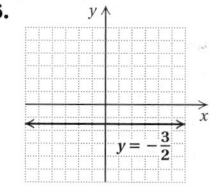

17.
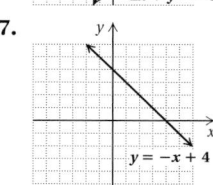

18.
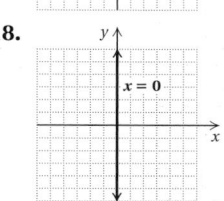

19. D **20.** C **21.** B **22.** E **23.** A **24.** No; an equation $x = a, a \ne 0$, does not have a y-intercept. **25.** Most would probably say that the second equation would be easier to graph because it has been solved for y. This makes it more efficient to find the y-value that corresponds to a given x-value. **26.** $A = 0$. If the line is horizontal, then the equation is of the form $y = a$ constant. Thus, Ax must be 0 and, hence, $A = 0$. **27.** Any ordered pair $(7, y)$ is a solution of $x = 7$. Thus all points on the graph are 7 units to the right of the y-axis, so they lie on a vertical line.

Calculator Corner, p. 208

1. This line will pass through the origin and slant up from left to right. This line will be steeper than $y = 10x$. **2.** This line will pass through the origin and slant up from left to right. This line will be less steep than $y = \frac{5}{32}x$. **3.** This line will pass through the origin and slant down from left to right. This line will be steeper than $y = -10x$. **4.** This line will pass through the origin and slant down from left to right. This line will be less steep than $y = -\frac{5}{32}x$.

Exercise Set 3.4, p. 211

1. $-\frac{3}{7}$ **3.** $\frac{2}{3}$ **5.** $\frac{3}{4}$ **7.** 0
9. $-\frac{4}{5}$;

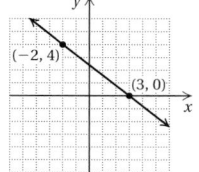

11. 3;

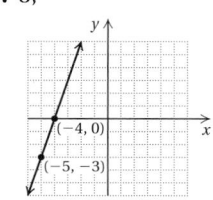

13. $-\frac{2}{3}$;

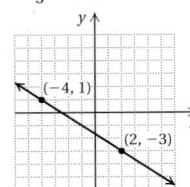

15. $\frac{7}{8}$;

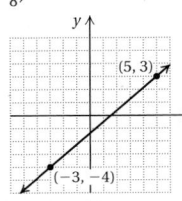

17. $\frac{2}{3}$ **19.** Not defined **21.** $-\frac{5}{13}$ **23.** 0 **25.** -10
27. 3.78 **29.** 3 **31.** $-\frac{1}{5}$ **33.** $-\frac{3}{2}$ **35.** Not defined
37. -1 **39.** 3 **41.** $\frac{5}{4}$ **43.** 0 **45.** $\frac{4}{3}$ **47.** $-\frac{21}{8}$ **49.** $\frac{12}{41}$
51. $\frac{28}{129}$ **53.** 3.0%; yes **55.** About $-2{,}170{,}000$ acres per year
57. About 82,900 people per year **59.** 19,000 tons per year
61. equivalent equations **62.** addition principle
63. multiplication principle **64.** Horizontal **65.** Vertical
66. slope **67.** x-intercept **68.** y-intercept
69. $y = -x + 5$ **71.** $y = x + 2$ **73.** 10

Summary and Review: Chapter 3, p. 216

Concept Reinforcement

1. True **2.** False **3.** True **4.** True **5.** False
6. False

Important Concepts

1. $F: (2, 4)$; $G: (-2, 0)$; $H: (-3, -5)$
2.

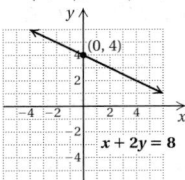

3.

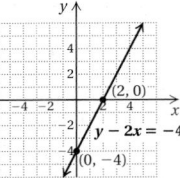

4.

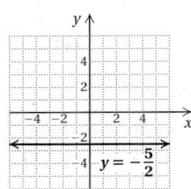

5.

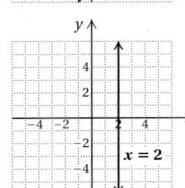

6. m is not defined. **7.** $\frac{3}{2}$ **8.** 0 **9.** m is not defined.
10. -2 **11.** 0 **12.** About 25,700 per year

Review Exercises

1. $(-5, -1)$ **2.** $(-2, 5)$ **3.** $(3, 0)$
4.–6.

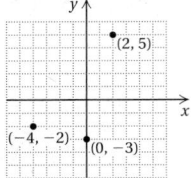

7. IV **8.** III **9.** I **10.** No
11. Yes

12.

$$2x - y = 3$$
$$\begin{array}{c|c} 2 \cdot 0 - (-3) & 3 \\ 0 + 3 & \\ 3 & \quad \text{TRUE} \end{array}$$

$$2x - y = 3$$
$$\begin{array}{c|c} 2 \cdot 2 - 1 & 3 \\ 4 - 1 & \\ 3 & \quad \text{TRUE} \end{array}$$

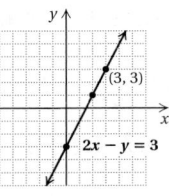

13.

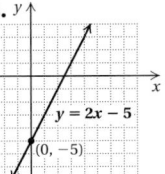

14.

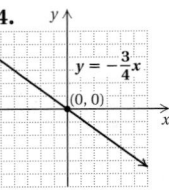

15.

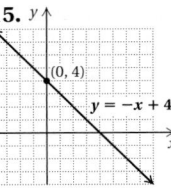

16.

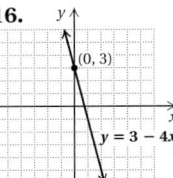

17.

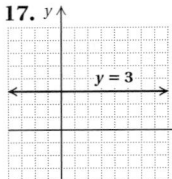

18.

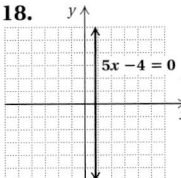

19.

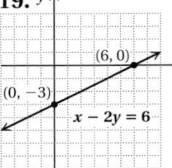

20.

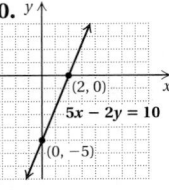

21. (a) $14\frac{1}{2}\,\text{ft}^3$, $16\,\text{ft}^3$, $20\frac{1}{2}\,\text{ft}^3$, $28\,\text{ft}^3$;
(b)

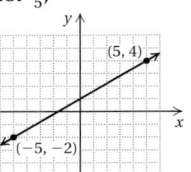

$20\,\text{ft}^3$;
(c) 6 residents

22. (a) 2.4 driveways per hour; **(b)** 25 minutes per driveway
23. 4 manicures per hour **24.** $\frac{1}{3}$ **25.** $-\frac{1}{3}$
26. $\frac{3}{5}$; **27.** -1;

28. 7% **29.** $-\frac{5}{8}$ **30.** $\frac{1}{2}$ **31.** Not defined **32.** 0
33. $\frac{1}{10}$ **34.** $\frac{3}{2}$ **35.** D **36.** C **37.** 45 square units;
28 linear units **38. (a)** $239.58\overline{3}$ ft per minute; **(b)** about
0.004 min per foot

Understanding Through Discussion and Writing

1. With slope $\frac{5}{3}$, for each horizontal change of 3 units, there is a
vertical change of 5 units. With slope $\frac{4}{3}$, for each horizontal
change of 3 units, there is a vertical change of 4 units. Since
$5 > 4$, the line with slope $\frac{5}{3}$ has a steeper slant. **2.** No; the
equation $y = b$, $b \neq 0$, does not have an x-intercept.
3. The y-intercept is the point at which the graph crosses
the y-axis. Since a point on the y-axis is neither left nor right
of the origin, the first or x-coordinate of the point is 0.
4. Any ordered pair $(x, -2)$ is a solution of $y = -2$. All points
on the graph are 2 units below the x-axis, so they lie on a
horizontal line.

Test: Chapter 3, p. 222

1. [3.1a] II **2.** [3.1a] III **3.** [3.1b] $(-5, 1)$
4. [3.1b] $(0, -4)$

5. [3.1c]

$$y - 2x = 5$$

$$\frac{-3 - 2(-4) \;?\; 5}{\begin{array}{c} -3 + 8 \\ 5 \end{array}} \quad \text{TRUE}$$

$$y - 2x = 5$$

$$\frac{3 - 2(-1) \;?\; 5}{\begin{array}{c} 3 + 2 \\ 5 \end{array}} \quad \text{TRUE}$$

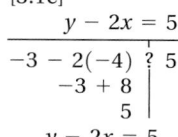

6. [3.2a] 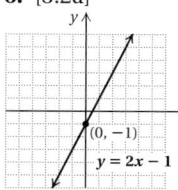 **7.** [3.2a] 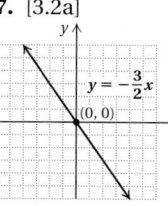 **8.** [3.3a]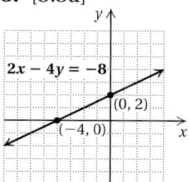

9. [3.3a] 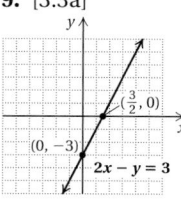 **10.** [3.3b] 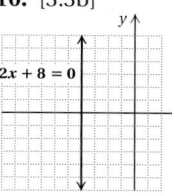 **11.** [3.3b]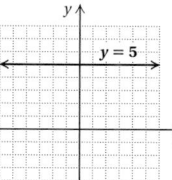

12. [3.4a] -2
13. [3.4a] $\frac{3}{8}$;

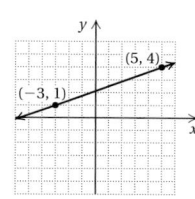

14. [3.4b] $\frac{2}{5}$ **15.** [3.4b] Not defined **16.** [3.4b] 0
17. [3.4b] -11 **18.** [3.4c] $-\frac{1}{20}$ **19.** [3.2b] **(a)** 1990: \$7800;
1996: \$12,000; 2005: \$18,300; 2010: \$21,800

(b) 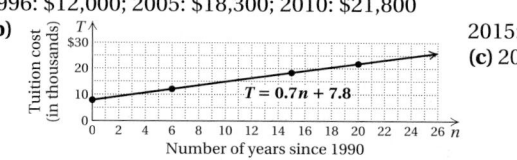 2015: \$25,000
(c) 2020

20. [3.4c] **(a)** 14.5 floors per minute; **(b)** $4\frac{4}{29}$ seconds per floor
21. [3.4c] 87.5 mph **22.** [3.3a], [3.4b] B **23.** [3.3b] $y = 3$
24. [3.1a] 25 square units; 20 linear units

Cumulative Review: Chapters 1–3, p. 225

1. [1.1a] $-\frac{4}{5}$ **2.** [1.7c] $-\frac{2}{3}x + 4y - 2$ **3.** [1.7d] $3(6w - 8 + 3y)$
4. [1.2c] $-0.\overline{7}$ **5.** [1.2e] $2\frac{1}{5}$ **6.** [1.3b] -8.17 **7.** [1.6b] $-\frac{7}{8}$
8. [1.7e] $-x - y$ **9.** [1.3a] -3 **10.** [1.8d] -6 **11.** [1.6c] $-\frac{2}{3}$
12. [1.8c] $11x + 9$ **13.** [1.5a] 2.64 **14.** [1.8d] -2
15. [2.2a] -81 **16.** [2.3c] No solution **17.** [2.3b] 3
18. [2.3c] All real numbers **19.** [2.7e] $\{x \mid x \le -\frac{11}{8}\}$
20. [2.1b] $\frac{4}{3}$ **21.** [3.4b] $\frac{3}{4}$ **22.** [3.4b] Not defined
23. [3.4b] 0 **24.** [2.4b] $s = \dfrac{7t}{z}$ **25.** [2.4b] $h = \dfrac{2A}{b + c}$

26. [3.1a] IV **27.** [3.3a] y-intercept: $(0, -3)$; x-intercept: $\left(\frac{21}{2}, 0\right)$
28. [2.7b]

29. [3.3a] 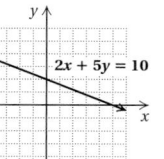 **30.** [3.3b] 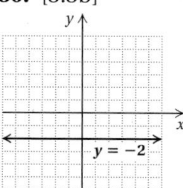 **31.** [3.2a]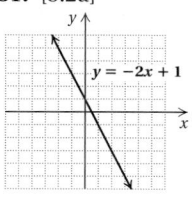

32. [3.2a] 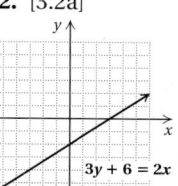 **33.** [3.2a] 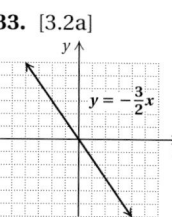 **34.** [3.3b]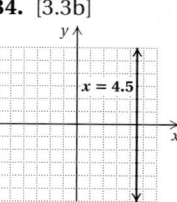

35. [3.4a] $-\frac{1}{4}$ **36.** [2.6a] 20.2 million people
37. [2.8b] $\{x \mid x \le 8\}$ **38.** [2.6a] First: 50 m; second: 53 m; third:
40 m **39.** [1.8d] A **40.** [3.4a] D **41.** [1.2e], [2.3a] $-4, 4$
42. [2.3b] 3 **43.** [2.4b] $Q = \dfrac{2 - pm}{p}$, or $\dfrac{2}{p} - m$

CHAPTER 4

Exercise Set 4.1, p. 234

1. $3 \cdot 3 \cdot 3 \cdot 3$ **3.** $(-1.1)(-1.1)(-1.1)(-1.1)(-1.1)$
5. $\left(\frac{2}{3}\right)\left(\frac{2}{3}\right)\left(\frac{2}{3}\right)\left(\frac{2}{3}\right)$ **7.** $(7p)(7p)$ **9.** $8 \cdot k \cdot k \cdot k$
11. $-6 \cdot y \cdot y \cdot y \cdot y$ **13.** 1 **15.** b **17.** 1 **19.** -7.03
21. 1 **23.** ab **25.** a **27.** 27 **29.** 19 **31.** -81
33. 256 **35.** 93 **37.** 136 **39.** 10; 4 **41.** 3629.84 ft^2
43. $\dfrac{1}{3^2} = \dfrac{1}{9}$ **45.** $\dfrac{1}{10^3} = \dfrac{1}{1000}$ **47.** $\dfrac{1}{a^3}$ **49.** $8^2 = 64$ **51.** y^4
53. $\dfrac{5}{z^4}$ **55.** $\dfrac{x}{y^2}$ **57.** 4^{-3} **59.** x^{-3} **61.** a^{-5} **63.** 2^7
65. 8^{14} **67.** x^5 **69.** 9^{38} **71.** $(3y)^{12}$ **73.** $(7y)^{17}$
75. 3^3 **77.** 1 **79.** x^{17} **81.** $\dfrac{1}{x^{13}}$ **83.** $\dfrac{1}{a^{10}}$ **85.** s^3t^7
87. 7^3 **89.** y^8 **91.** $\dfrac{1}{16^6}$ **93.** $\dfrac{1}{m^6}$ **95.** $\dfrac{1}{(8x)^4}$ **97.** 1
99. x^2 **101.** x^9 **103.** $\dfrac{1}{z^4}$ **105.** x^3 **107.** 1 **109.** a^3b^2
111. $5^2 = 25; 5^{-2} = \frac{1}{25}; \left(\frac{1}{5}\right)^2 = \frac{1}{25}; \left(\frac{1}{5}\right)^{-2} = 25; -5^2 = -25;$
$(-5)^2 = 25; -\left(-\frac{1}{5}\right)^2 = -\frac{1}{25}; \left(-\frac{1}{5}\right)^{-2} = 25$ **113.** 8 in.; 4 in.
114. 228, 229 **115.** 25,543.75 ft^2 **116.** $51°, 27°, 102°$
117. $\frac{23}{14}$ **118.** $\frac{11}{10}$ **119.** $4(x - 3 + 6y)$
120. $2(128 - a - 2b)$ **121.** No **123.** No **125.** y^{5x}
127. a^{4t} **129.** 1 **131.** $>$ **133.** $<$ **135.** $-\frac{1}{10,000}$
137. No; for example, $(3 + 4)^2 = 49$, but $3^2 + 4^2 = 25$.

Calculator Corner, p. 242

1. 1.3545×10^{-4} **2.** 3.2×10^5 **3.** 3×10^{-6} **4.** 8×10^{-26}

Exercise Set 4.2, p. 245

1. 2^6 **3.** $\dfrac{1}{5^6}$ **5.** x^{12} **7.** $\dfrac{1}{a^{18}}$ **9.** t^{18} **11.** $\dfrac{1}{t^{12}}$ **13.** x^8
15. a^3b^3 **17.** $\dfrac{1}{a^3b^3}$ **19.** $\dfrac{1}{m^3n^6}$ **21.** $16x^6$ **23.** $\dfrac{9}{x^8}$
25. $\dfrac{1}{x^{12}y^{15}}$ **27.** $x^{24}y^8$ **29.** $\dfrac{a^{10}}{b^{35}}$ **31.** $\dfrac{25t^6}{r^8}$ **33.** $\dfrac{b^{21}}{a^{15}c^6}$
35. $\dfrac{9x^6}{y^{16}z^6}$ **37.** $\dfrac{16x^6}{y^4}$ **39.** $a^{12}b^8$ **41.** $\dfrac{y^6}{4}$ **43.** $\dfrac{a^8}{b^{12}}$
45. $\dfrac{8}{y^6}$ **47.** $49x^6$ **49.** $\dfrac{x^6y^3}{z^3}$ **51.** $\dfrac{c^2d^6}{a^4b^2}$ **53.** 2.8×10^{10}
55. 9.07×10^{17} **57.** 3.04×10^{-6} **59.** 1.8×10^{-8}
61. 10^{11} **63.** 4.19854×10^8 **65.** \$2.4 \times 10^9$
67. 87,400,000 **69.** 0.00000005704 **71.** 10,000,000
73. 0.00001 **75.** 6×10^9 **77.** 3.38×10^4
79. 8.1477×10^{-13} **81.** 2.5×10^{13} **83.** 5.0×10^{-4}
85. 3.0×10^{-21} **87.** Approximately 1.325×10^{14} ft^3
89. The mass of Jupiter is 3.18×10^2 times the mass of Earth.

91. 1×10^{22} **93.** 7.5×10^{-7} m **95.** 4.375×10^2 days
97. $9(x - 4)$ **98.** $2(2x - y + 8)$ **99.** $3(s + t + 8)$
100. $-7(x + 2)$ **101.** $\frac{7}{4}$ **102.** 2 **103.** $-\frac{12}{7}$ **104.** $-\frac{11}{2}$
105. 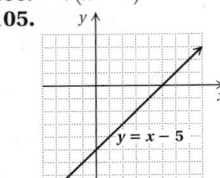 **106.**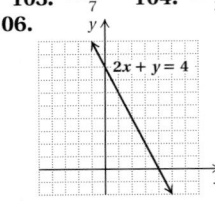

107. 2.478125×10^{-1} **109.** $\frac{1}{5}$ **111.** 3^{11} **113.** 7
115. $\frac{1}{0.4}$, or 2.5 **117.** False **119.** False **121.** True

Calculator Corner, p. 252

1. 3; 2.25; -27 **2.** 44; 0; 9.28

Exercise Set 4.3, p. 257

1. -18; 7 **3.** 19; 14 **5.** -12; -7 **7.** $\frac{13}{3}$; 5 **9.** 9; 1
11. 56; -2 **13.** 1112 ft **15.** \$18,750; \$24,000
17. $-4, 4, 5, 2.75, 1$ **19. (a)** 2728.4 billion kilowatt-hours;
3521.8 billion kilowatt-hours; 4315.2 billion kilowatt-hours;
5108.6 billion kilowatt-hours; 5902 billion kilowatt-hours;
(b) left to the student **21.** 9 words **23.** 6 **25.** 15
27. $2, -3x, x^2$ **29.** $-2x^4, \frac{1}{3}x^3, -x, 3$ **31.** $6x^2$ and $-3x^2$
33. $2x^4$ and $-3x^4$; $5x$ and $-7x$ **35.** $3x^5$ and $14x^5$; $-7x$
and $-2x$; 8 and -9 **37.** $-3, 6$ **39.** $5, \frac{3}{4}, 3$ **41.** $-5, 6,$
$-2.7, 1, -2$ **43.** $-3x$ **45.** $-8x$ **47.** $11x^3 + 4$ **49.** $x^3 - x$
51. $4b^5$ **53.** $\frac{3}{4}x^5 - 2x - 42$ **55.** x^4 **57.** $\frac{15}{16}x^3 - \frac{7}{6}x^2$
59. $x^5 + 6x^3 + 2x^2 + x + 1$ **61.** $15y^9 + 7y^8 + 5y^3 - y^2 + y$
63. $x^6 + x^4$ **65.** $13x^3 - 9x + 8$ **67.** $-5x^2 + 9x$
69. $12x^4 - 2x + \frac{1}{4}$ **71.** 1, 0; 1 **73.** 2, 1, 0; 2
75. 3, 2, 1, 0; 3 **77.** 2, 1, 6, 4; 6
79.

Term	Coefficient	Degree of the Term	Degree of the Polynomial
$-7x^4$	-7	4	
$6x^3$	6	3	
$-x^2$	-1	2	4
$8x$	8	1	
-2	-2	0	

81. x^2, x **83.** x^3, x^2, x^0 **85.** None missing
87. $x^3 + 0x^2 + 0x - 27$; x^3 $\qquad$ $- 27$
89. $x^4 + 0x^3 + 0x^2 - x + 0x^0$; x^4 $\qquad$ $- x$
91. None missing **93.** Trinomial **95.** None of these
97. Binomial **99.** Monomial **101.** 27 apples **102.** -19
103. $-\frac{17}{24}$ **104.** $\frac{5}{8}$ **105.** -2.6 **106.** $\frac{15}{2}$ **107.** $b = \frac{C + r}{a}$
108. 45%; 37.5%; 17.5% **109.** $3(x - 5y + 21)$ **111.** $3x^6$
113. 10 **115.** $-4, 4, 5, 2.75, 1$ **117.** 9

Calculator Corner, p. 266

1. Yes **2.** Yes **3.** No **4.** Yes **5.** No **6.** Yes

Exercise Set 4.4, p. 267

1. $-x + 5$ **3.** $x^2 - \frac{11}{2}x - 1$ **5.** $2x^2$ **7.** $5x^2 + 3x - 30$
9. $-2.2x^3 - 0.2x^2 - 3.8x + 23$ **11.** $6 + 12x^2$
13. $-\frac{1}{2}x^4 + \frac{2}{3}x^3 + x^2$ **15.** $0.01x^5 + x^4 - 0.2x^3 +$
$0.2x + 0.06$ **17.** $9x^8 + 8x^7 - 6x^4 + 8x^2 + 4$
19. $1.05x^4 + 0.36x^3 + 14.22x^2 + x + 0.97$ **21.** $5x$
23. $x^2 - \frac{3}{2}x + 2$ **25.** $-12x^4 + 3x^3 - 3$ **27.** $-3x + 7$
29. $-4x^2 + 3x - 2$ **31.** $4x^4 - 6x^2 - \frac{3}{4}x + 8$ **33.** $7x - 1$
35. $-x^2 - 7x + 5$ **37.** -18 **39.** $6x^4 + 3x^3 - 4x^2 + 3x - 4$
41. $4.6x^3 + 9.2x^2 - 3.8x - 23$ **43.** $\frac{3}{4}x^3 - \frac{1}{2}x$

45. $0.06x^3 - 0.05x^2 + 0.01x + 1$ **47.** $3x + 6$
49. $11x^4 + 12x^3 - 9x^2 - 8x - 9$ **51.** $x^4 - x^3 + x^2 - x$
53. $\frac{23}{2}a + 12$ **55.** $5x^2 + 4x$ **57.** $(r + 11)(r + 9)$;
$9r + 99 + 11r + r^2$, or $r^2 + 20r + 99$
59. $(x + 3)(x + 3)$, or $(x + 3)^2$; $x^2 + 3x + 9 + 3x$,
or $x^2 + 6x + 9$ **61.** $\pi r^2 - 25\pi$ **63.** $18z - 64$ **65.** 6
66. -19 **67.** $-\frac{7}{22}$ **68.** 5 **69.** 5 **70.** 1 **71.** $\frac{39}{2}$ **72.** $\frac{37}{2}$
73. $\{x | x \geq -10\}$ **74.** $\{x | x < 0\}$ **75.** $20w + 42$
77. $2x^2 + 20x$ **79.** $y^2 - 4y + 4$ **81.** $12y^2 - 23y + 21$
83. $-3y^4 - y^3 + 5y - 2$

Mid-Chapter Review: Chapter 4, p. 271

1. True **2.** False **3.** False **4.** True
5. $4w^3 + 6w - 8w^3 - 3w = (4 - 8)w^3 + (6 - 3)w =$
$-4w^3 + 3w$ **6.** $(3y^4 - y^2 + 11) - (y^4 - 4y^2 + 5) =$
$3y^4 - y^2 + 11 - y^4 + 4y^2 - 5 = 2y^4 + 3y^2 + 6$ **7.** z **8.** 1
9. -32 **10.** 1 **11.** 5^7 **12.** $(3a)^9$ **13.** $\frac{1}{x^3}$ **14.** 1
15. 7^4 **16.** $\frac{1}{x^2}$ **17.** w^8 **18.** $\frac{1}{y^4}$ **19.** 3^{15} **20.** $\frac{x^{18}}{y^{12}}$
21. $\frac{a^{24}}{5^6}$ **22.** $\frac{x^2 z^4}{4y^6}$ **23.** 2.543×10^7 **24.** 1.2×10^{-4}
25. 0.000036 **26.** 144,000,000 **27.** 6×10^3 **28.** 5×10^{-7}
29. 16; 1 **30.** -16; 9 **31.** $-2x^5 - 5x^2 + 4x + 2$
32. $8x^6 + 2x^3 - 8x^2$ **33.** 3, 1, 0; 3 **34.** 1, 4, 6; 6
35. Binomial **36.** Trinomial **37.** $8x^2 + 5$
38. $5x^3 - 2x^2 + 2x - 11$ **39.** $-4x - 10$
40. $-0.4x^2 - 3.4x + 9$ **41.** $3y + 3y^2$ **42.** The area of the
smaller square is x^2, and the area of the larger square is $(3x)^2$, or
$9x^2$, so the area of the larger square is nine times the area of the
smaller square. **43.** The volume of the smaller cube is x^3, and
the volume of the larger cube is $(2x)^3$, or $8x^3$, so the volume of
the larger cube is eight times the volume of the smaller cube.
44. Exponents are added when powers with like bases are
multiplied. Exponents are multiplied when a power is raised
to a power. **45.** $3^{-29} = \frac{1}{3^{29}}$ and $2^{-29} = \frac{1}{2^{29}}$. Since $3^{29} > 2^{29}$,
we have $\frac{1}{3^{29}} < \frac{1}{2^{29}}$. **46.** It is better to evaluate a polynomial after
like terms have been collected, because there are fewer terms
to evaluate. **47.** Yes; consider the following: $(x^2 + 4) +$
$(4x - 7) = x^2 + 4x - 3$.

Calculator Corner, p. 276

1. Correct **2.** Correct **3.** Not correct **4.** Not correct

Exercise Set 4.5, p. 277

1. $40x^2$ **3.** x^3 **5.** $32x^8$ **7.** $0.03x^{11}$ **9.** $\frac{1}{15}x^4$ **11.** 0
13. $-24x^{11}$ **15.** $-2x^2 + 10x$ **17.** $-5x^2 + 5x$
19. $x^5 + x^2$ **21.** $6x^3 - 18x^2 + 3x$ **23.** $-6x^4 - 6x^3$
25. $18y^6 + 24y^5$ **27.** $x^2 + 9x + 18$ **29.** $x^2 + 3x - 10$
31. $x^2 + 3x - 4$ **33.** $x^2 - 7x + 12$ **35.** $x^2 - 9$
37. $x^2 - 16$ **39.** $3x^2 + 11x + 10$ **41.** $25 - 15x + 2x^2$
43. $4x^2 + 20x + 25$ **45.** $x^2 - 6x + 9$ **47.** $x^2 - \frac{21}{10}x - 1$
49. $x^2 + 2.4x - 10.81$ **51.** $(x + 2)(x + 6)$, or $x^2 + 8x + 12$
53. $(x + 1)(x + 6)$, or $x^2 + 7x + 6$

55. **57.** **59.**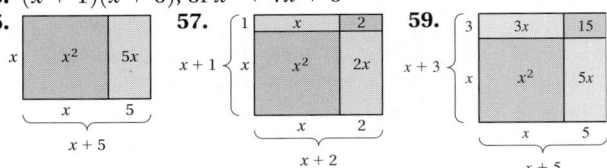

61. $x^3 - 1$ **63.** $4x^3 + 14x^2 + 8x + 1$
65. $3y^4 - 6y^3 - 7y^2 + 18y - 6$ **67.** $x^6 + 2x^5 - x^3$
69. $-10x^5 - 9x^4 + 7x^3 + 2x^2 - x$ **71.** $-1 - 2x - x^2 + x^4$
73. $6t^4 + t^3 - 16t^2 - 7t + 4$ **75.** $x^9 - x^5 + 2x^3 - x$

77. $x^4 - 1$ **79.** $x^4 + 8x^3 + 12x^2 + 9x + 4$
81. $2x^4 - 5x^3 + 5x^2 - \frac{19}{10}x + \frac{1}{5}$ **83.** $-\frac{3}{4}$ **84.** 6.4   **85.** 96
86. 32   **87.** $3(5x - 6y + 4)$ **88.** $4(4x - 6y + 9)$
89. $-3(3x + 15y - 5)$ **90.** $100(x - y + 10a)$
91.

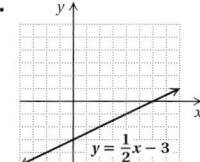
$y = \frac{1}{2}x - 3$

92. $\frac{23}{19}$ **93.** $75y^2 - 45y$
95. $V = (4x^3 - 48x^2 + 144x) \text{ in}^3; S = (-4x^2 + 144) \text{ in}^2$
97. 5   **99.** $(x^3 + 2x^2 - 210) \text{ m}^3$ **101.** 0   **103.** 0

Visualizing for Success, p. 286
1. E, F   **2.** B, O   **3.** K, S   **4.** G, R   **5.** D, M   **6.** J, P
7. C, L   **8.** N, Q   **9.** A, H   **10.** I, T

Exercise Set 4.6, p. 287
1. $x^3 + x^2 + 3x + 3$ **3.** $x^4 + x^3 + 2x + 2$ **5.** $y^2 - y - 6$
7. $9x^2 + 12x + 4$ **9.** $5x^2 + 4x - 12$ **11.** $9t^2 - 1$
13. $4x^2 - 6x + 2$ **15.** $p^2 - \frac{1}{16}$ **17.** $x^2 - 0.01$
19. $2x^3 + 2x^2 + 6x + 6$ **21.** $-2x^2 - 11x + 6$
23. $a^2 + 14a + 49$ **25.** $1 - x - 6x^2$ **27.** $\frac{9}{64}y^2 - \frac{5}{8}y + \frac{25}{36}$
29. $x^5 + 3x^3 - x^2 - 3$ **31.** $3x^6 - 2x^4 - 6x^2 + 4$
33. $13.16x^2 + 18.99x - 13.95$ **35.** $6x^7 + 18x^5 + 4x^2 + 12$
37. $8x^6 + 65x^3 + 8$ **39.** $4x^3 - 12x^2 + 3x - 9$
41. $4y^6 + 4y^5 + y^4 + y^3$ **43.** $x^2 - 16$ **45.** $4x^2 - 1$
47. $25m^2 - 4$ **49.** $4x^4 - 9$ **51.** $9x^8 - 16$ **53.** $x^{12} - x^4$
55. $x^8 - 9x^2$ **57.** $x^{24} - 9$ **59.** $4y^{16} - 9$ **61.** $\frac{25}{64}x^2 - 18.49$
63. $x^2 + 4x + 4$ **65.** $9x^4 + 6x^2 + 1$ **67.** $a^2 - a + \frac{1}{4}$
69. $9 + 6x + x^2$ **71.** $x^4 + 2x^2 + 1$ **73.** $4 - 12x^4 + 9x^8$
75. $25 + 60t^2 + 36t^4$ **77.** $x^2 - \frac{5}{4}x + \frac{25}{64}$ **79.** $9 - 12x^3 + 4x^6$
81. $4x^3 + 24x^2 - 12x$ **83.** $4x^4 - 2x^2 + \frac{1}{4}$ **85.** $9p^2 - 1$
87. $15t^5 - 3t^4 + 3t^3$ **89.** $36x^8 + 48x^4 + 16$
91. $12x^3 + 8x^2 + 15x + 10$ **93.** $64 - 96x^4 + 36x^8$
95. $t^3 - 1$ **97.** 25; 49   **99.** 56; 16   **101.** $a^2 + 2a + 1$
103. $t^2 + 10t + 24$ **105.** Lamps: 500 watts; air conditioner:
2000 watts; television: 50 watts   **106.** $\frac{28}{27}$ **107.** $-\frac{41}{7}$

108. $\frac{27}{4}$ **109.** $y = \dfrac{3x - 12}{2}$, or $y = \frac{3}{2}x - 6$

110. $a = \dfrac{5d + 4}{3}$, or $a = \frac{5}{3}d + \frac{4}{3}$ **111.** $30x^3 + 35x^2 - 15x$

113. $a^4 - 50a^2 + 625$ **115.** $81t^{16} - 72t^8 + 16$ **117.** -7
119. First row: 90, -432, -63; second row: 7, -18, -36, -14, 12,
-6, -21, -11; third row: 9, -2, -2, 10, -8, -8, -8, -10, 21;
fourth row: -19, -6 **121.** Yes   **123.** No

Exercise Set 4.7, p. 295
1. -1 **3.** -15 **5.** 240   **7.** -145 **9.** 3.715 L   **11.** 205.9 m
13. 44.46 in^2 **15.** 63.78125 in^2 **17.** Coefficients: 1, -2, 3, -5;
degrees: 4, 2, 2, 0; 4   **19.** Coefficients: 17, -3, -7; degrees: 5, 5,
0; 5   **21.** $-a - 2b$ **23.** $3x^2y - 2xy^2 + x^2$ **25.** $20au + 10av$
27. $8u^2v - 5uv^2$ **29.** $x^2 - 4xy + 3y^2$ **31.** $3r + 7$
33. $-b^2a^3 - 3b^3a^2 + 5ba + 3$ **35.** $ab^2 - a^2b$
37. $2ab - 2$ **39.** $-2a + 10b - 5c + 8d$
41. $6z^2 + 7zu - 3u^2$ **43.** $a^4b^2 - 7a^2b + 10$ **45.** $a^6 - b^2c^2$
47. $y^6x + y^4x + y^4 + 2y^2 + 1$ **49.** $12x^2y^2 + 2xy - 2$
51. $12 - c^2d^2 - c^4d^4$ **53.** $m^3 + m^2n - mn^2 - n^3$
55. $x^9y^9 - x^6y^6 + x^5y^5 - x^2y^2$ **57.** $x^2 + 2xh + h^2$
59. $9a^2 + 12ab + 4b^2$ **61.** $r^6t^4 - 8r^3t^2 + 16$
63. $p^8 + 2m^2n^2p^4 + m^4n^4$ **65.** $3a^3 - 12a^2b + 12ab^2$
67. $m^2 + 2mn + n^2 - 6m - 6n + 9$ **69.** $a^2 - b^2$
71. $4a^2 - b^2$ **73.** $c^4 - d^2$ **75.** $a^2b^2 - c^2d^4$
77. $x^2 + 2xy + y^2 - 9$ **79.** $x^2 - y^2 - 2yz - z^2$

81. $a^2 + 2ab + b^2 - c^2$
83. $3x^4 - 7x^2y + 3x^2 - 20y^2 + 22y - 6$ **85.** IV   **86.** III
87. I   **88.** II
89.

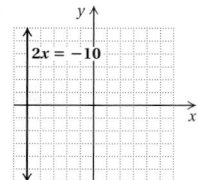
$2x = -10$

90.

$y = -4$

91.

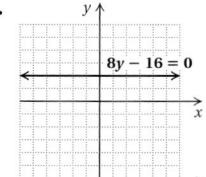
$8y - 16 = 0$

92.

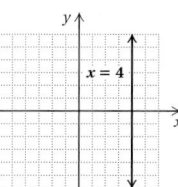
$x = 4$

93. $4xy - 4y^2$ **95.** $2xy + \pi x^2$
97. $2\pi nh + 2\pi mh + 2\pi n^2 - 2\pi m^2$ **99.** 16 gal
101. \$12,351.94

Exercise Set 4.8, p. 304
1. $3x^4$ **3.** $5x$ **5.** $18x^3$ **7.** $4a^3b$ **9.** $3x^4 - \frac{1}{2}x^3 + \frac{1}{8}x^2 - 2$
11. $1 - 2u - u^4$ **13.** $5t^2 + 8t - 2$ **15.** $-4x^4 + 4x^2 + 1$
17. $6x^2 - 10x + \frac{3}{2}$ **19.** $9x^2 - \frac{5}{2}x + 1$ **21.** $6x^2 + 13x + 4$

23. $3rs + r - 2s$ **25.** $x + 2$ **27.** $x - 5 + \dfrac{-50}{x - 5}$

29. $x - 2 + \dfrac{-2}{x + 6}$ **31.** $x - 3$ **33.** $x^4 - x^3 + x^2 - x + 1$

35. $2x^2 - 7x + 4$ **37.** $x^3 - 6$ **39.** $t^2 + 1$

41. $y^2 - 3y + 1 + \dfrac{-5}{y + 2}$ **43.** $3x^2 + x + 2 + \dfrac{10}{5x + 1}$

45. $6y^2 - 5 + \dfrac{-6}{2y + 7}$ **47.** Product   **48.** Monomial

49. Multiplication; equivalent   **50.** $x = a$ **51.** Trinomial
52. Quotient   **53.** Absolute value   **54.** Slope   **55.** $x^2 + 5$

57. $a + 3 + \dfrac{5}{5a^2 - 7a - 2}$ **59.** $2x^2 + x - 3$

61. $a^5 + a^4b + a^3b^2 + a^2b^3 + ab^4 + b^5$ **63.** -5 **65.** 1

Summary and Review: Chapter 4, p. 307
Concept Reinforcement
1. True   **2.** False   **3.** False   **4.** True

Important Concepts
1. z^8 **2.** a^2b^6 **3.** $\dfrac{y^6}{27x^{12}z^9}$ **4.** 7.63×10^5 **5.** 0.0003

6. 6×10^4 **7.** $2x^4 - 4x^2 - 3$ **8.** $3x^4 + x^3 - 2x^2 + 2$
9. $x^6 - 6x^4 + 11x^2 - 6$ **10.** $2y^2 + 11y + 12$ **11.** $x^2 - 25$
12. $9w^2 + 24w + 16$ **13.** $-2a^3b^2 - 5a^2b + ab^2 - 2ab$
14. $y^2 - 4y + \frac{8}{5}$ **15.** $x - 9 + \dfrac{48}{x + 5}$

Review Exercises
1. $\dfrac{1}{7^2}$ **2.** y^{11} **3.** $(3x)^{14}$ **4.** t^8 **5.** 4^3 **6.** $\dfrac{1}{a^3}$ **7.** 1

8. $9t^8$ **9.** $36x^8$ **10.** $\dfrac{y^3}{8x^3}$ **11.** t^{-5} **12.** $\dfrac{1}{y^4}$

13. 3.28×10^{-5} **14.** 8,300,000   **15.** 2.09×10^4
16. 5.12×10^{-5} **17.** 1.54468×10^{10} slices   **18.** 10
19. $-4y^5, 7y^2, -3y, -2$ **20.** x^2, x^0 **21.** 3, 2, 1, 0; 3
22. Binomial   **23.** None of these   **24.** Monomial

25. $-2x^2 - 3x + 2$ **26.** $10x^4 - 7x^2 - x - \frac{1}{2}$
27. $x^5 - 2x^4 + 6x^3 + 3x^2 - 9$
28. $-2x^5 - 6x^4 - 2x^3 - 2x^2 + 2$ **29.** $2x^2 - 4x$
30. $x^5 - 3x^3 - x^2 + 8$ **31.** Perimeter: $4w + 6$; area: $w^2 + 3w$
32. $(t + 3)(t + 4)$, $t^2 + 7t + 12$ **33.** $x^2 + \frac{7}{6}x + \frac{1}{3}$
34. $49x^2 + 14x + 1$ **35.** $12x^3 - 23x^2 + 13x - 2$
36. $9x^4 - 16$ **37.** $15x^7 - 40x^6 + 50x^5 + 10x^4$
38. $x^2 - 3x - 28$ **39.** $9y^4 - 12y^3 + 4y^2$ **40.** $2t^4 - 11t^2 - 21$
41. 49   **42.** Coefficients: 1, -7, 9, -8; degrees: 6, 2, 2, 0; 6
43. $-y + 9w - 5$ **44.** $m^6 - 2m^2n + 2m^2n^2 + 8n^2m - 6m^3$
45. $-9xy - 2y^2$ **46.** $11x^3y^2 - 8x^2y - 6x^2 - 6x + 6$
47. $p^3 - q^3$ **48.** $9a^8 - 2a^4b^3 + \frac{1}{9}b^6$ **49.** $5x^2 - \frac{1}{2}x + 3$
50. $3x^2 - 7x + 4 + \dfrac{1}{2x + 3}$ **51.** $0, 3.75, -3.75, 0$ **52.** B
53. D   **54.** $\frac{1}{2}x^2 - \frac{1}{2}y^2$ **55.** $400 - 4a^2$ **56.** $-28x^8$
57. $\frac{94}{13}$ **58.** $x^4 + x^3 + x^2 + x + 1$ **59.** 80 ft by 40 ft

Understanding Through Discussion and Writing

1. 578.6×10^{-7} is not in scientific notation because 578.6 is not a number greater than or equal to 1 and less than 10.
2. When evaluating polynomials, it is essential to know the order in which the operations are to be performed.
3. We label the figure as shown.

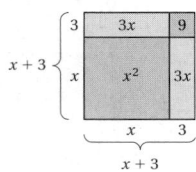

Then we see that the area of the figure is $(x + 3)^2$, or $x^2 + 3x + 3x + 9 \neq x^2 + 9$.   **4.** Emma did not divide *each* term of the polynomial by the divisor. The first term was divided by $3x$, but the second was not. Multiplying Emma's "quotient" by the divisor $3x$, we get $12x^3 - 18x^2 \neq 12x^3 - 6x$. This should convince her that a mistake has been made.   **5.** Yes; for example, $(x^2 + xy + 1) + (3x - xy + 2) = x^2 + 3x + 3$.
6. Yes; consider $a + b + c + d$. This is a polynomial in 4 variables but it has degree 1.

Test: Chapter 4, p. 313

1. [4.1d, f] $\dfrac{1}{6^5}$ **2.** [4.1d] x^9 **3.** [4.1d] $(4a)^{11}$ **4.** [4.1e] 3^3
5. [4.1e, f] $\dfrac{1}{x^5}$ **6.** [4.1b, e] 1   **7.** [4.2a] x^6 **8.** [4.2a, b] $-27y^6$
9. [4.2a, b] $16a^{12}b^4$ **10.** [4.2b] $\dfrac{a^3b^3}{c^3}$ **11.** [4.1d], [4.2a, b] $-216x^{21}$ **12.** [4.1d], [4.2a, b] $-24x^{21}$ **13.** [4.1d], [4.2a, b] $162x^{10}$ **14.** [4.1d], [4.2a, b] $324x^{10}$ **15.** [4.1f] $\dfrac{1}{5^3}$
16. [4.1f] y^{-8} **17.** [4.2c] 3.9×10^9 **18.** [4.2c] 0.00000005
19. [4.2d] 1.75×10^{17} **20.** [4.2d] 1.296×10^{22}
21. [4.2e] 1.5×10^4 files   **22.** [4.3a] -43 **23.** [4.3d] $\frac{1}{3}, -1, 7$
24. [4.3g] 3, 0, 1, 6; 6   **25.** [4.3i] Binomial   **26.** [4.3e] $5a^2 - 6$
27. [4.3e] $\frac{7}{4}y^2 - 4y$ **28.** [4.3f] $x^5 + 2x^3 + 4x^2 - 8x + 3$
29. [4.4a] $4x^5 + x^4 + 2x^3 - 8x^2 + 2x - 7$
30. [4.4a] $5x^4 + 5x^2 + x + 5$ **31.** [4.4c] $-4x^4 + x^3 - 8x - 3$
32. [4.4c] $-x^5 + 0.7x^3 - 0.8x^2 - 21$
33. [4.5b] $-12x^4 + 9x^3 + 15x^2$ **34.** [4.6c] $x^2 - \frac{2}{3}x + \frac{1}{9}$
35. [4.6b] $9x^2 - 100$ **36.** [4.6a] $3b^2 - 4b - 15$
37. [4.6a] $x^{14} - 4x^8 + 4x^6 - 16$ **38.** [4.6a] $48 + 34y - 5y^2$
39. [4.5d] $6x^3 - 7x^2 - 11x - 3$ **40.** [4.6c] $25t^2 + 20t + 4$
41. [4.7c] $-5x^3y - y^3 + xy^3 - x^2y^2 + 19$
42. [4.7e] $8a^2b^2 + 6ab - 4b^3 + 6ab^2 + ab^3$
43. [4.7f] $9x^{10} - 16y^{10}$ **44.** [4.8a] $4x^2 + 3x - 5$

45. [4.8b] $2x^2 - 4x - 2 + \dfrac{17}{3x + 2}$
46. [4.3a] 3, 1.5, -3.5, -5, -5.25
47. [4.4d] $(t + 2)(t + 2)$, $t^2 + 4t + 4$ **48.** [4.4d] B
49. [4.5b], [4.6a] $V = l^3 - 3l^2 + 2l$ **50.** [2.3b], [4.6b, c] $-\frac{61}{12}$

Cumulative Review: Chapters 1–4, p. 315

1. [1.1a] $\frac{5}{2}$ **2.** [4.3a] -4 **3.** [4.7a] -14 **4.** [1.2e] 4
5. [1.6b] $\frac{1}{5}$ **6.** [1.3a] $-\frac{11}{60}$ **7.** [1.4a] 4.2   **8.** [1.5a] 7.28
9. [1.6c] $-\frac{5}{12}$ **10.** [4.2d] 2.2×10^{22} **11.** [4.2d] 4×10^{-5}
12. [1.7a] -3 **13.** [1.8b] $-2y - 7$ **14.** [1.8c] $5x + 11$
15. [1.8d] -2 **16.** [4.4a] $2x^5 - 2x^4 + 3x^3 + 2$
17. [4.7d] $3x^2 + xy - 2y^2$ **18.** [4.4c] $x^3 + 5x^2 - x - 7$
19. [4.4c] $-\frac{1}{3}x^2 - \frac{3}{4}x$ **20.** [1.7c] $12x - 15y + 21$
21. [4.5a] $6x^8$ **22.** [4.5b] $2x^5 - 4x^4 + 8x^3 - 10x^2$
23. [4.5d] $3y^4 + 5y^3 - 10y - 12$
24. [4.7f] $2p^4 + 3p^3q + 2p^2q^2 - 2p^4q - p^3q^2 - p^2q^3 + pq^3$
25. [4.6a] $6x^2 + 13x + 6$ **26.** [4.6c] $9x^4 + 6x^2 + 1$
27. [4.6b] $t^2 - \frac{1}{4}$ **28.** [4.6b] $4y^4 - 25$
29. [4.6a] $4x^6 + 6x^4 - 6x^2 - 9$ **30.** [4.6c] $t^2 - 4t^3 + 4t^4$
31. [4.7f] $15p^2 - pq - 2q^2$ **32.** [4.8a] $6x^2 + 2x - 3$
33. [4.8b] $3x^2 - 2x - 7$ **34.** [2.1b] -1.2
35. [2.2a] -21 **36.** [2.3a] 9   **37.** [2.2a] $-\frac{20}{3}$
38. [2.3b] 2   **39.** [2.1b] $\frac{13}{8}$ **40.** [2.3c] $-\frac{17}{21}$ **41.** [2.3b] -17
42. [2.3b] 2   **43.** [2.7e] $\{x | x < 16\}$ **44.** [2.7e] $\{x | x \le -\frac{11}{8}\}$
45. [2.4b] $x = \dfrac{A - P}{Q}$ **46.** [2.5a] \$3.50
47. [4.4d] $(\pi r^2 - 18)$ ft²   **48.** [2.6a] 18 and 19
49. [2.6a] 20 ft, 24 ft   **50.** [2.6a] 10°   **51.** [4.1d, f] y^4
52. [4.1e, f] $\dfrac{1}{x}$ **53.** [4.2a, b] $-\dfrac{27x^9}{y^6}$ **54.** [4.1d, e, f] x^3
55. [3.3a]

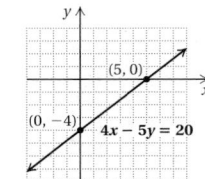

56. [4.1a, f] $3^2 = 9, 3^{-2} = \frac{1}{9}, \left(\frac{1}{3}\right)^2 = \frac{1}{9}, \left(\frac{1}{3}\right)^{-2} = 9, -3^2 = -9,$
$(-3)^2 = 9, \left(-\frac{1}{3}\right)^2 = \frac{1}{9}, \left(-\frac{1}{3}\right)^{-2} = 9$
57. [4.4d] $(4x - 4)$ in²
58. [4.1d], [4.2a, b], [4.4a] $12x^5 - 15x^4 - 27x^3 + 4x^2$
59. [4.4a], [4.6c] $5x^2 - 2x + 10$ **60.** [2.3b], [4.6a, c] $\frac{11}{7}$
61. [2.3b], [4.8b] 1   **62.** [1.2e], [2.3a] $-5, 5$
63. [2.3b], [4.6a], [4.8b] All real numbers except 5

CHAPTER 5

Exercise Set 5.1, p. 325

1. x **3.** x^2 **5.** 2   **7.** $17xy$ **9.** x **11.** x^2y^2
13. $x(x - 6)$ **15.** $2x(x + 3)$ **17.** $x^2(x + 6)$
19. $8x^2(x^2 - 3)$ **21.** $2(x^2 + x - 4)$
23. $17xy(x^4y^2 + 2x^2y + 3)$ **25.** $x^2(6x^2 - 10x + 3)$
27. $x^2y^2(x^3y^3 + x^2y + xy - 1)$
29. $2x^3(x^4 - x^3 - 32x^2 + 2)$
31. $0.8x(2x^3 - 3x^2 + 4x + 8)$
33. $\frac{1}{3}x^3(5x^3 + 4x^2 + x + 1)$ **35.** $(x + 3)(x^2 + 2)$
37. $(3z - 1)(4z^2 + 7)$ **39.** $(3x + 2)(2x^2 + 1)$
41. $(2a - 7)(5a^3 - 1)$ **43.** $(x + 3)(x^2 + 2)$
45. $(x + 3)(2x^2 + 1)$ **47.** $(2x - 3)(4x^2 + 3)$
49. $(3p - 4)(4p^2 + 1)$ **51.** $(x - 1)(5x^2 - 1)$
53. $(x + 8)(x^2 - 3)$ **55.** $(x - 4)(2x^2 - 9)$
57. $\{x | x > -24\}$ **58.** $\{x | x \le \frac{14}{5}\}$ **59.** 27

60. $p = 2A - q$ **61.** $y^2 + 12y + 35$ **62.** $y^2 + 14y + 49$
63. $y^2 - 49$ **64.** $y^2 - 14y + 49$

65.
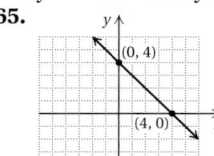

66.
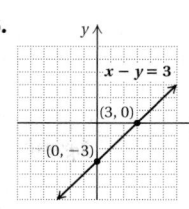

67.
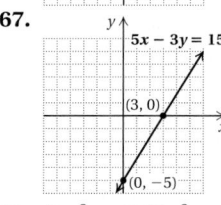

68.
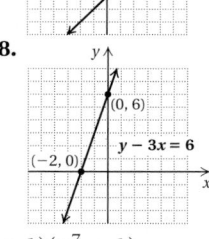

69. $(2x^2 + 3)(2x^3 + 3)$ **71.** $(x^5 + 1)(x^7 + 1)$
73. Not factorable by grouping

Exercise Set 5.2, p. 333

1.

Pairs of Factors	Sums of Factors
1, 15	16
−1, −15	−16
3, 5	8
−3, −5	−8

$(x + 3)(x + 5)$

3.

Pairs of Factors	Sums of Factors
1, 12	13
−1, −12	−13
2, 6	8
−2, −6	−8
3, 4	7
−3, −4	−7

$(x + 3)(x + 4)$

5.

Pairs of Factors	Sums of Factors
1, 9	10
−1, −9	−10
3, 3	6
−3, −3	−6

$(x - 3)^2$

7.

Pairs of Factors	Sums of Factors
−1, 14	13
1, −14	−13
−2, 7	5
2, −7	−5

$(x + 2)(x - 7)$

9.

Pairs of Factors	Sums of Factors
1, 4	5
−1, −4	−5
2, 2	4
−2, −2	−4

$(b + 1)(b + 4)$

11.

Pairs of Factors	Sums of Factors
$\frac{1}{3}, \frac{1}{3}$	$\frac{2}{3}$
$-\frac{1}{3}, -\frac{1}{3}$	$-\frac{2}{3}$
$1, \frac{1}{9}$	$\frac{10}{9}$
$-1, -\frac{1}{9}$	$-\frac{10}{9}$

$\left(x + \frac{1}{3}\right)^2$

13. $(d - 2)(d - 5)$ **15.** $(y - 1)(y - 10)$ **17.** Prime
19. $(x - 9)(x + 2)$ **21.** $x(x - 8)(x + 2)$
23. $y(y - 9)(y + 5)$ **25.** $(x - 11)(x + 9)$
27. $(c^2 + 8)(c^2 - 7)$ **29.** $(a^2 + 7)(a^2 - 5)$
31. $(x - 6)(x + 7)$ **33.** Prime   **35.** $(x + 10)^2$
37. $2z(z - 4)(z + 3)$ **39.** $3t^2(t^2 + t + 1)$
41. $x^2(x - 25)(x + 4)$ **43.** $(x - 24)(x + 3)$
45. $(x - 9)(x - 16)$ **47.** $(a + 12)(a - 11)$ **49.** $3(t + 1)^2$
51. $w^2(w - 4)^2$ **53.** $-1(x - 10)(x + 3)$, or
$(-x + 10)(x + 3)$, or $(x - 10)(-x - 3)$
55. $-1(a - 2)(a + 12)$, or $(-a + 2)(a + 12)$, or
$(a - 2)(-a - 12)$ **57.** $(x - 15)(x - 8)$
59. $-1(x + 12)(x - 9)$, or $(-x - 12)(x - 9)$, or
$(x + 12)(-x + 9)$ **61.** $(y - 0.4)(y + 0.2)$
63. $(p + 5q)(p - 2q)$ **65.** $-1(t + 14)(t - 6)$, or
$(-t - 14)(t - 6)$, or $(t + 14)(-t + 6)$ **67.** $(m + 4n)(m + n)$
69. $(s + 3t)(s - 5t)$ **71.** $6a^8(a + 2)(a - 7)$
73. $16x^3 - 48x^2 + 8x$ **74.** $28w^2 - 53w - 66$
75. $49w^2 + 84w + 36$ **76.** $16w^2 - 88w + 121$
77. $16w^2 - 121$ **78.** $y^3 - 3y^2 + 5y$
79. $6x^2 + 11xy - 35y^2$ **80.** $27x^{12}$ **81.** $\frac{8}{3}$ **82.** $-\frac{7}{2}$
83. 1743 arrests   **84.** 100°, 25°, 55°
85. 15, −15, 27, −27, 51, −51   **87.** $\left(x + \frac{1}{4}\right)\left(x - \frac{3}{4}\right)$
89. $(x + 5)\left(x - \frac{5}{7}\right)$ **91.** $(b^n + 5)(b^n + 2)$
93. $2x^2(4 - \pi)$

Calculator Corner, p. 338

1. Correct   **2.** Correct   **3.** Not correct   **4.** Not correct

Exercise Set 5.3, p. 342

1. $(2x + 1)(x - 4)$ **3.** $(5x + 9)(x - 2)$
5. $(3x + 1)(2x + 7)$ **7.** $(3x + 1)(x + 1)$
9. $(2x - 3)(2x + 5)$ **11.** $(2x + 1)(x - 1)$
13. $(3x - 2)(3x + 8)$ **15.** $(3x + 1)(x - 2)$
17. $(3x + 4)(4x + 5)$ **19.** $(7x - 1)(2x + 3)$
21. $(3x + 2)(3x + 4)$ **23.** $(3x - 7)^2$, or $(7 - 3x)^2$
25. $(24x - 1)(x + 2)$ **27.** $(5x - 11)(7x + 4)$
29. $-2(x - 5)(x + 2)$, or $2(-x + 5)(x + 2)$, or
$2(x - 5)(-x - 2)$ **31.** $4(3x - 2)(x + 3)$
33. $6(5x - 9)(x + 1)$ **35.** $2(3y + 5)(y - 1)$
37. $(3x - 1)(x - 1)$ **39.** $4(3x + 2)(x - 3)$
41. $(2x + 1)(x - 1)$ **43.** $(3x + 2)(3x - 8)$
45. $5(3x + 1)(x - 2)$ **47.** $p(3p + 4)(4p + 5)$
49. $-1(3x + 2)(3x - 8)$, or $(-3x - 2)(3x - 8)$, or
$(3x + 2)(-3x + 8)$ **51.** $-1(5x - 3)(3x - 2)$, or
$(-5x + 3)(3x - 2)$, or $(5x - 3)(-3x + 2)$
53. $x^2(7x - 1)(2x + 3)$ **55.** $3x(8x - 1)(7x - 1)$
57. $(5x^2 - 3)(3x^2 - 2)$ **59.** $(5t + 8)^2$
61. $2x(3x + 5)(x - 1)$ **63.** Prime   **65.** Prime
67. $(4m + 5n)(3m - 4n)$ **69.** $(2a + 3b)(3a - 5b)$
71. $(3a + 2b)(3a + 4b)$ **73.** $(5p + 2q)(7p + 4q)$
75. $6(3x - 4y)(x + y)$ **77.** $q = \dfrac{A + 7}{p}$

78. $x = \dfrac{y - b}{m}$ **79.** $y = \dfrac{6 - 3x}{2}$ **80.** $q = p + r - 2$

81. $\{x \mid x > 4\}$ **82.** $\left\{x \mid x \le \frac{8}{11}\right\}$
83.

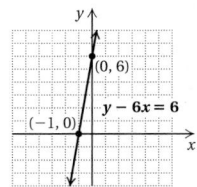

$y = \frac{2}{5}x - 1$

84. y^8 **85.** y-intercept: $(0, -4)$; x-intercept: $(16, 0)$
86. y-intercept: $(0, 4)$; x-intercept: $(16, 0)$
87. y-intercept: $(0, -5)$; x-intercept: $(6.5, 0)$
88. y-intercept: $\left(0, \frac{2}{3}\right)$; x-intercept: $\left(\frac{5}{8}, 0\right)$
89. y-intercept: $(0, 4)$; x-intercept: $\left(\frac{4}{5}, 0\right)$
90. y-intercept: $(0, -5)$; x-intercept: $\left(\frac{5}{2}, 0\right)$
91. $(2x^n + 1)(10x^n + 3)$ **93.** $(x^{3a} - 1)(3x^{3a} + 1)$
95.–103. Left to the student

Exercise Set 5.4, p. 347

1. $(x + 2)(x + 7)$ **3.** $(x - 4)(x - 1)$
5. $(3x + 2)(2x + 3)$ **7.** $(3x - 4)(x - 4)$
9. $(7x - 8)(5x + 3)$ **11.** $(2x + 3)(2x - 3)$
13. $(x^2 + 3)(2x^2 + 5)$ **15.** $(2x - 1)(x + 4)$
17. $(3x + 5)(x - 3)$ **19.** $(2x + 7)(3x + 1)$
21. $(3x - 1)(x - 1)$ **23.** $(2x + 3)(2x - 5)$
25. $(2x - 1)(x + 1)$ **27.** $(3x + 2)(3x - 8)$
29. $(3x - 1)(x + 2)$ **31.** $(3x - 4)(4x - 5)$
33. $(7x + 1)(2x - 3)$ **35.** $(3x + 2)(3x + 4)$
37. $(3x - 7)^2$, or $(7 - 3x)^2$ **39.** $(24x + 1)(x - 2)$
41. $-1(3a - 1)(3a + 5)$, or $(-3a + 1)(3a + 5)$, or $(3a - 1)(-3a - 5)$ **43.** $-2(x - 5)(x + 2)$, or $2(-x + 5)(x + 2)$, or $2(x - 5)(-x - 2)$
45. $4(3x - 2)(x + 3)$ **47.** $6(5x - 9)(x + 1)$
49. $2(3y + 5)(y - 1)$ **51.** $(3x - 1)(x - 1)$
53. $4(3x + 2)(x - 3)$ **55.** $(2x + 1)(x - 1)$
57. $(3x - 2)(3x + 8)$ **59.** $5(3x + 1)(x - 2)$
61. $p(3p + 4)(4p + 5)$ **63.** $-1(5x - 4)(x + 1)$, or $(-5x + 4)(x + 1)$, or $(5x - 4)(-x - 1)$
65. $-3(2t - 1)(t - 5)$, or $3(-2t + 1)(t - 5)$, or $3(2t - 1)(-t + 5)$ **67.** $x^2(7x - 1)(2x + 3)$
69. $3x(8x - 1)(7x - 1)$ **71.** $(5x^2 - 3)(3x^2 - 2)$
73. $(5t + 8)^2$ **75.** $2x(3x + 5)(x - 1)$ **77.** Prime
79. Prime   **81.** $(4m + 5n)(3m - 4n)$
83. $(2a + 3b)(3a - 5b)$ **85.** $(3a - 2b)(3a - 4b)$
87. $(5p + 2q)(7p + 4q)$ **89.** $6(3x - 4y)(x + y)$
91. $-6x(x - 5)(x + 2)$, or $6x(-x + 5)(x + 2)$, or $6x(x - 5)(-x - 2)$ **93.** $x^3(5x - 11)(7x + 4)$
95. $\{x \mid x < -100\}$ **96.** $\{x \mid x \ge 217\}$
97. $\{x \mid x \le 8\}$ **98.** $\{x \mid x < 2\}$ **99.** $\left\{x \mid x \ge \frac{20}{3}\right\}$
100. $\{x \mid x > 17\}$ **101.** $\left\{x \mid x > \frac{26}{7}\right\}$ **102.** $\left\{x \mid x \ge \frac{77}{17}\right\}$
103. About 6369 km, or 3949 mi   **104.** $40°$ **105.** $(3x^5 - 2)^2$
107. $(4x^5 + 1)^2$ **109.–117.** Left to the student

Mid-Chapter Review: Chapter 5, p. 351

1. True   **2.** False   **3.** True   **4.** False
5. $10y^3 - 18y^2 + 12y = 2y \cdot 5y^2 - 2y \cdot 9y + 2y \cdot 6$
$= 2y(5y^2 - 9y + 6)$
6. $a \cdot c = 2 \cdot (-6) = -12;$
$-x = -4x + 3x;$
$2x^2 - x - 6 = 2x^2 - 4x + 3x - 6$
$= 2x(x - 2) + 3(x - 2)$
$= (x - 2)(2x + 3)$
7. x **8.** x^2 **9.** $6x^3$ **10.** 4   **11.** $5x^2y$ **12.** x^2y^2
13. $x(x^2 - 8)$ **14.** $3x(x + 4)$ **15.** $2(y^2 + 4y - 2)$
16. $t^3(3t^3 - 5t - 2)$ **17.** $(x + 1)(x + 3)$ **18.** $(z - 2)^2$
19. $(x + 4)(x^2 + 3)$ **20.** $8y^3(y^2 - 6)$

21. $6xy(x^2 + 4xy - 7y^2)$ **22.** $(4t - 3)(t - 2)$
23. $(z - 1)(z + 5)$ **24.** $(z + 4)(2z^2 + 5)$
25. $(3p - 2)(p^2 - 3)$ **26.** $5x^3(2x^5 - 5x^3 - 3x^2 + 7)$
27. $(2w + 3)(w^2 - 3)$ **28.** $x^2(4x^2 - 5x + 3)$
29. $(6y - 5)(y + 2)$ **30.** $3(x - 3)(x + 2)$
31. $(3x + 2)(2x^2 + 1)$ **32.** $(w - 5)(w - 3)$
33. $(2x + 5)(4x^2 + 1)$ **34.** $(5z + 2)(2z - 5)$
35. $(2x + 1)(3x + 2)$ **36.** $(x - 6y)(x - 4y)$
37. $(2z + 1)(3z^2 + 1)$ **38.** $a^2b^3(ab^4 + a^2b^2 - 1 + a^3b^3)$
39. $(4y + 5z)(y - 3z)$ **40.** $3x(x + 2)(x + 5)$
41. $(x - 3)(x^2 - 2)$ **42.** $(3y + 1)^2$ **43.** $(y + 2)(y + 4)$
44. $3(2y + 5)(y + 3)$ **45.** $(x - 7)(x^2 + 4)$
46. $-1(y - 4)(y + 1)$, or $(-y + 4)(y + 1)$, or $(y - 4)(-y - 1)$
47. $4(2x + 3)(2x - 5)$ **48.** $(5a - 3b)(2a - b)$
49. $(2w - 5)(3w^2 - 5)$ **50.** $y(y + 6)(y + 3)$
51. $(4x + 3y)(x + 2y)$ **52.** $-1(3z - 2)(2z + 3)$, or $(-3z + 2)(2z + 3)$, or $(3z - 2)(-2z - 3)$
53. $(3t + 2)(4t^2 - 3)$ **54.** $(y - 4z)(y + 5z)$
55. $(3x - 4y)(3x + 2y)$ **56.** $(3z - 1)(z + 3)$
57. $(m - 8n)(m + 2n)$ **58.** $2(w - 3)^2$
59. $2t(3t - 2)(3t - 1)$ **60.** $(z + 3)(5z^2 + 1)$
61. $(t - 2)(t + 7)$ **62.** $(2t - 5)^2$ **63.** $(t - 2)(t + 6)$
64. $-1(2z + 3)(z - 4)$, or $(-2z - 3)(z - 4)$, or $(2z + 3)(-z + 4)$ **65.** $-1(y - 6)(y + 2)$, or $(-y + 6)(y + 2)$, or $(y - 6)(-y - 2)$ **66.** Find the product of two binomials. For example, $(2x^2 + 3)(x - 4) = 2x^3 - 8x^2 + 3x - 12$.
67. There is a finite number of pairs of numbers with the correct product, but there are infinitely many pairs with the correct sum.   **68.** Since both constants are negative, the middle term will be negative so $(x - 17)(x - 18)$ cannot be a factorization of $x^2 + 35x + 306$.   **69.** No; both $2x + 6$ and $2x + 8$ contain a factor of 2, so $2 \cdot 2$, or 4, must be factored out to reach the complete factorization. In other words, the largest common factor is 4, not 2.

Exercise Set 5.5, p. 359

1. Yes   **3.** No   **5.** No   **7.** Yes   **9.** $(x - 7)^2$
11. $(x + 8)^2$ **13.** $(x - 1)^2$ **15.** $(x + 2)^2$ **17.** $(y + 6)^2$
19. $(t - 4)^2$ **21.** $(q^2 - 3)^2$ **23.** $(4y + 7)^2$ **25.** $2(x - 1)^2$
27. $x(x - 9)^2$ **29.** $3(2q - 3)^2$ **31.** $(7 - 3x)^2$, or $(3x - 7)^2$
33. $5(y^2 + 1)^2$ **35.** $(1 + 2x^2)^2$ **37.** $(2p + 3q)^2$
39. $(a - 3b)^2$ **41.** $(9a - b)^2$ **43.** $4(3a + 4b)^2$ **45.** Yes
47. No   **49.** No   **51.** Yes   **53.** $(y + 2)(y - 2)$
55. $(p + 3)(p - 3)$ **57.** $(t + 7)(t - 7)$
59. $(a + b)(a - b)$ **61.** $(5t + m)(5t - m)$
63. $(10 + k)(10 - k)$ **65.** $(4a + 3)(4a - 3)$
67. $(2x + 5y)(2x - 5y)$ **69.** $2(2x + 7)(2x - 7)$
71. $x(6 + 7x)(6 - 7x)$ **73.** $\left(\frac{1}{4} + 7x^4\right)\left(\frac{1}{4} - 7x^4\right)$
75. $(0.3y + 0.02)(0.3y - 0.02)$ **77.** $(7a^2 + 9)(7a^2 - 9)$
79. $(a^2 + 4)(a + 2)(a - 2)$ **81.** $5(x^2 + 9)(x + 3)(x - 3)$
83. $(1 + y^4)(1 + y^2)(1 + y)(1 - y)$
85. $(x^6 + 4)(x^3 + 2)(x^3 - 2)$ **87.** $\left(y + \frac{1}{4}\right)\left(y - \frac{1}{4}\right)$
89. $\left(5 + \frac{1}{7}x\right)\left(5 - \frac{1}{7}x\right)$ **91.** $(4m^2 + t^2)(2m + t)(2m - t)$
93. -11 **94.** 400   **95.** $-\frac{5}{6}$ **96.** -0.9 **97.** 2   **98.** -160
99. $x^2 - 4xy + 4y^2$ **100.** $\frac{1}{2}\pi x^2 + 2xy$ **101.** y^{12}
102. $25a^4b^6$
103.

y-axis, x-axis graph with points $(0, 6)$, $(-1, 0)$; line labeled $y - 6x = 6$

104.

y-axis, x-axis graph with $3x - 5y = 15$; points $(5, 0)$, $(0, -3)$

105. Prime   **107.** $(x + 11)^2$ **109.** $2x(3x + 1)^2$

111. $(x^4 + 2^4)(x^2 + 2^2)(x + 2)(x - 2)$
113. $3x^3(x + 2)(x - 2)$ **115.** $2x\left(3x + \frac{2}{5}\right)\left(3x - \frac{2}{5}\right)$
117. $p(0.7 + p)(0.7 - p)$ **119.** $(0.8x + 1.1)(0.8x - 1.1)$
121. $x(x + 6)$ **123.** $\left(x + \frac{1}{x}\right)\left(x - \frac{1}{x}\right)$
125. $(9 + b^{2k})(3 - b^k)(3 + b^k)$ **127.** $(3b^n + 2)^2$
129. $(y + 4)^2$ **131.** 9   **133.** Not correct   **135.** Not correct

Exercise Set 5.6, p. 365
1. $(z + 3)(z^2 - 3z + 9)$ **3.** $(x - 1)(x^2 + x + 1)$
5. $(y + 5)(y^2 - 5y + 25)$ **7.** $(2a + 1)(4a^2 - 2a + 1)$
9. $(y - 2)(y^2 + 2y + 4)$ **11.** $(2 - 3b)(4 + 6b + 9b^2)$
13. $(4y + 1)(16y^2 - 4y + 1)$ **15.** $(2x + 3)(4x^2 - 6x + 9)$
17. $(a - b)(a^2 + ab + b^2)$ **19.** $\left(a + \frac{1}{2}\right)\left(a^2 - \frac{1}{2}a + \frac{1}{4}\right)$
21. $2(y - 4)(y^2 + 4y + 16)$ **23.** $3(2a + 1)(4a^2 - 2a + 1)$
25. $r(s + 4)(s^2 - 4s + 16)$ **27.** $5(x - 2z)(x^2 + 2xz + 4z^2)$
29. $(x + 0.1)(x^2 - 0.1x + 0.01)$
31. $8(2x^2 - t^2)(4x^4 + 2x^2t^2 + t^4)$
33. $2y(y - 4)(y^2 + 4y + 16)$
35. $(z - 1)(z^2 + z + 1)(z + 1)(z^2 - z + 1)$
37. $(t^2 + 4y^2)(t^4 - 4t^2y^2 + 16y^4)$
39. $(2w^3 - z^3)(4w^6 + 2w^3z^3 + z^6)$
41. $\left(\frac{1}{2}c + d\right)\left(\frac{1}{4}c^2 - \frac{1}{2}cd + d^2\right)$
43. $(0.1x - 0.2y)(0.01x^2 + 0.02xy + 0.04y^2)$
45. $\dfrac{343}{y^{15}}$ **46.** a^8b^{18} **47.** $\dfrac{16}{x^6}$ **48.** $4y^{10} - 9$
49. $w^2 - \frac{2}{3}w + \frac{1}{9}$ **50.** $x^2 + 0.4x - 0.05$
51. (a) $\pi h(R + r)(R - r)$; (b) 3,014,400 cm^3
53. $3(x^a + 2y^b)(x^{2a} - 2x^ay^b + 4y^{2b})$
55. $\frac{1}{3}\left(\frac{1}{2}xy + z\right)\left(\frac{1}{4}x^2y^2 - \frac{1}{2}xyz + z^2\right)$ **57.** $y(3x^2 + 3xy + y^2)$
59. $4(3a^2 + 4)$

Exercise Set 5.7, p. 373
1. $3(x + 8)(x - 8)$ **3.** $(a - 5)^2$ **5.** $(2x - 3)(x - 4)$
7. $x(x + 12)^2$ **9.** $(x + 3)(x + 2)(x - 2)$
11. $3(4x + 1)(4x - 1)$ **13.** $3x(3x - 5)(x + 3)$
15. Prime   **17.** $x(x^2 + 7)(x - 3)$ **19.** $x^3(x - 7)^2$
21. $-2(x - 2)(x + 5)$, or $2(-x + 2)(x + 5)$, or
$2(x - 2)(-x - 5)$ **23.** Prime
25. $4(x^2 + 4)(x + 2)(x - 2)$
27. $(1 + y^4)(1 + y^2)(1 + y)(1 - y)$ **29.** $x^3(x - 3)(x - 1)$
31. $\frac{1}{9}\left(\frac{1}{3}x^3 - 4\right)^2$ **33.** $\left(\frac{1}{10}m - \frac{1}{3}n\right)\left(\frac{1}{100}m^2 + \frac{1}{30}mn + \frac{1}{9}n^2\right)$
35. $9xy(xy - 4)$ **37.** $2\pi r(h + r)$ **39.** $(a + b)(2x + 1)$
41. $(x + 1)(x - 1 - y)$ **43.** $(n + 2)(n + p)$
45. $(2q - 1)(3q + p)$ **47.** $(2b - a)^2$, or $(a - 2b)^2$
49. $(4x + 3y)^2$ **51.** $(7m^2 - 8n)^2$ **53.** $(y^2 + 5z^2)^2$
55. $\left(\frac{1}{2}a + \frac{1}{3}b\right)^2$ **57.** $(a + b)(a - 2b)$
59. $(m + 20n)(m - 18n)$ **61.** $(mn - 8)(mn + 4)$
63. $r^3(rs - 2)(rs - 8)$ **65.** $a^3(a - b)(a + 5b)$
67. $\left(a + \frac{1}{5}b\right)\left(a - \frac{1}{5}b\right)$
69. $7(x + y)(x^2 - xy + y^2)(x - y)(x^2 + xy + y^2)$
71. $(4 + p^2q^2)(2 + pq)(2 - pq)$
73. $(1 + 4x^6y^6)(1 + 2x^3y^3)(1 - 2x^3y^3)$
75. $(q + 8)(q + 1)(q - 1)$ **77.** $ab(2ab + 1)(3ab - 2)$
79. $(m + 1)(m - 1)(m + 2)(m - 2)$ **81.** $(t + 1)^2(t - 1)^2$
82. $9x^2 - 25y^2$ **83.** $9x^2 - 30xy + 25y^2$
84. $9x^2 + 30xy + 25y^2$ **85.** $6x^2 + 11xy - 35y^2$
86. $25x^2 - 10xt + t^2$ **87.** $-\frac{14}{11}$ **88.** $\{x | x < 32\}$
89. $X = \dfrac{A + 7}{a + b}$ **91.** $(x - 5)(x + 2)(x - 2)$

93. $(3.5x - 1)^2$ **95.** $(5x + 4)(x + 1.8)$
97. $(y - 2)(y + 3)(y - 3)$ **99.** $(x - 1)(x + 2)(x - 2)$
101. Prime   **103.** $(y - 1)^3$ **105.** $(y + 4 + x)^2$

Calculator Corner, p. 382
1. Left to the student

Exercise Set 5.8, p. 383
1. $-4, -9$ **3.** $-3, 8$ **5.** $-12, 11$ **7.** $0, -3$ **9.** $0, -18$
11. $-\frac{5}{2}, -4$ **13.** $-\frac{1}{5}, 3$ **15.** $4, \frac{1}{4}$ **17.** $0, \frac{2}{3}$ **19.** $-\frac{1}{10}, \frac{1}{27}$
21. $\frac{1}{3}, -20$ **23.** $0, \frac{2}{3}, \frac{1}{2}$ **25.** $-5, -1$ **27.** $-9, 2$ **29.** $3, 5$
31. $0, 8$ **33.** $0, -18$ **35.** $-4, 4$ **37.** $-\frac{2}{3}, \frac{2}{3}$ **39.** -3
41. 4 **43.** $0, \frac{6}{5}$ **45.** $-1, \frac{5}{3}$ **47.** $-\frac{1}{4}, \frac{2}{3}$ **49.** $-1, \frac{2}{3}$
51. $-\frac{7}{10}, \frac{7}{10}$ **53.** $-2, 9$ **55.** $\frac{4}{5}, \frac{3}{2}$ **57.** $(-4, 0), (1, 0)$
59. $\left(-\frac{5}{2}, 0\right), (2, 0)$ **61.** $(-3, 0), (5, 0)$ **63.** $-1, 4$
65. $-1, 3$ **67.** $(a + b)^2$ **68.** $a^2 + b^2$ **69.** -16
70. -4.5 **71.** $-\frac{10}{3}$ **72.** $\frac{3}{10}$ **73.** $-5, 4$ **75.** $-3, 9$
77. $-\frac{1}{8}, \frac{1}{8}$ **79.** $-4, 4$ **81.** Answers may vary.
(a) $x^2 - x - 12 = 0$; (b) $x^2 + 7x + 12 = 0$; (c) $4x^2 - 4x + 1 = 0$;
(d) $x^2 - 25 = 0$; (e) $40x^3 - 14x^2 + x = 0$ **83.** $2.33, 6.77$
85. $0, 2.74$

Translating for Success, p. 392
1. O   **2.** M   **3.** K   **4.** I   **5.** G   **6.** E   **7.** C   **8.** A
9. H   **10.** B

Exercise Set 5.9, p. 393
1. Length: 42 in.; width: 14 in.   **3.** Length: 12 ft; width: 2 ft
5. Height: 4 cm; base: 14 cm   **7.** Base: 8 m; height: 16 m
9. 182 games   **11.** 12 teams   **13.** 4950 handshakes
15. 25 people   **17.** 20 people   **19.** 14 and 15
21. 12 and 14; -12 and -14 **23.** 15 and 17; -15 and -17
25. Hypotenuse: 17 ft; leg: 15 ft   **27.** 32 ft   **29.** 9 ft
31. Dining room: 12 ft by 12 ft; kitchen: 12 ft by 10 ft
33. 4 sec   **35.** 5 and 7   **37.** Factor   **38.** Factor
39. Product   **40.** Common factor   **41.** Trinomial
42. Quotient rule   **43.** y-intercept   **44.** Slope
45. 35 ft   **47.** 5 ft   **49.** 30 cm by 15 cm   **51.** 7 in.

Summary and Review: Chapter 5, p. 399
Concept Reinforcement
1. False   **2.** True   **3.** False   **4.** True

Important Concepts
1. $4xy$ **2.** $9x^2(3x^3 - x + 2)$ **3.** $(z - 3)(z^2 + 4)$
4. $(x + 2)(x + 4)$ **5.** $3(z - 4)(2z + 1)$
6. $(3y - 1)(2y + 3)$ **7.** $(2x + 1)^2$ **8.** $2(3x + 2)(3x - 2)$
9. $(3 - 5x)(9 + 15x + 25x^2)$
10. $\left(\frac{1}{2}q + 2a\right)\left(\frac{1}{4}q^2 - qa + 4a^2\right)$ **11.** $-5, 1$

Review Exercises
1. $5y^2$ **2.** $12x$ **3.** $5(1 + 2x^3)(1 - 2x^3)$ **4.** $x(x - 3)$
5. $(3x + 2)(3x - 2)$ **6.** $(x + 6)(x - 2)$ **7.** $(x + 7)^2$
8. $3x(2x^2 + 4x + 1)$ **9.** $(x + 1)(x^2 + 3)$
10. $(3x - 1)(2x - 1)$ **11.** $(x^2 + 9)(x + 3)(x - 3)$
12. $3x(3x - 5)(x + 3)$ **13.** $2(x + 5)(x - 5)$
14. $(x + 4)(x^3 - 2)$ **15.** $(4x^2 + 1)(2x + 1)(2x - 1)$
16. $4x^4(2x^2 - 8x + 1)$ **17.** $3(2x + 5)^2$ **18.** Prime
19. $x(x - 6)(x + 5)$ **20.** $(2x - 5)(4x^2 + 10x + 25)$
21. $(3x - 5)^2$ **22.** $2(3x + 4)(x - 6)$ **23.** $(x - 3)^2$
24. $(2x + 1)(x - 4)$ **25.** $2(3x - 1)^2$ **26.** $3(x + 3)(x - 3)$
27. $(x - 5)(x - 3)$ **28.** $(5x - 2)^2$ **29.** $(7b^5 - 2a^4)^2$
30. $(xy + 4)(xy - 3)$ **31.** $3(2a + 7b)^2$
32. $(m + 5)(m + t)$ **33.** $32(x^2 - 2y^2z^2)(x^2 + 2y^2z^2)$

34. $5(y + 2t)(y^2 - 2yt + 4t^2)$ **35.** $1, -3$ **36.** $-7, 5$
37. $-4, 0$ **38.** $\frac{2}{3}, 1$ **39.** $-8, 8$ **40.** $-2, 8$
41. $(-5, 0), (-4, 0)$ **42.** $\left(-\frac{3}{2}, 0\right), (5, 0)$
43. Height: 6 cm; base: 5 cm **44.** -18 and -16; 16 and 18
45. -19 and -17; 17 and 19 **46.** On the ground: 4 ft; on the
tree: 3 ft **47.** 6 km **48.** B **49.** A **50.** 2.5 cm **51.** $0, 2$
52. Length: 12 in.; width: 6 in. **53.** No solution **54.** $2, -3, \frac{5}{2}$
55. $-2, \frac{5}{4}, 3$ **56.** $x^2(\pi - 2)$

Understanding Through Discussion and Writing

1. Although $x^3 - 8x^2 + 15x$ can be factored as
$(x^2 - 5x)(x - 3)$, this is not a complete factorization of the
polynomial since $x^2 - 5x = x(x - 5)$. Gwen should always look
for a common factor first. **2.** Josh is correct, because answers
can easily be checked by multiplying.
3. For $x = -3$:
$$(x - 4)^2 = (-3 - 4)^2 = (-7)^2 = 49;$$
$$(4 - x)^2 = [4 - (-3)]^2 = 7^2 = 49.$$
For $x = 1$:
$$(x - 4)^2 = (1 - 4)^2 = (-3)^2 = 9;$$
$$(4 - x)^2 = (4 - 1)^2 = 3^2 = 9.$$
In general, $(x - 4)^2 = [-(-x + 4)]^2 = [-(4 - x)]^2 =$
$(-1)^2(4 - x)^2 = (4 - x)^2$.
4. The equation is not in the form $ab = 0$. The correct
procedure is
$$(x - 3)(x + 4) = 8$$
$$x^2 + x - 12 = 8$$
$$x^2 + x - 20 = 0$$
$$(x + 5)(x - 4) = 0$$
$$x + 5 = 0 \quad or \quad x - 4 = 0$$
$$x = -5 \quad or \quad x = 4.$$
The solutions are -5 and 4.
5. One solution of the equation is 0. Dividing both sides of the
equation by x, leaving the solution $x = 3$, is equivalent to
dividing by 0. **6.** She could use the measuring sticks to draw
a right angle as shown below. Then she could use the 3-ft and
4-ft sticks to extend one leg to 7 ft and the 4-ft and 5-ft sticks to
extend the other leg to 9 ft.

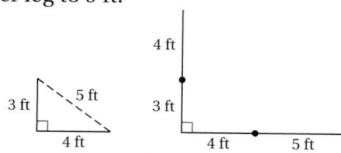

Next, she could draw another right angle with either the 7-ft
side or the 9-ft side as a side.

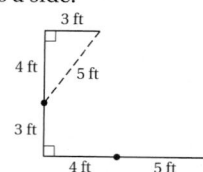

Then she could use the sticks to extend the other side to the
appropriate length. Finally, she would draw the remaining side
of the rectangle.

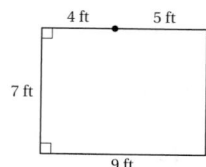

Test: Chapter 5, p. 405

1. [5.1a] $4x^3$ **2.** [5.2a] $(x - 5)(x - 2)$ **3.** [5.5b] $(x - 5)^2$
4. [5.1b] $2y^2(2y^2 - 4y + 3)$ **5.** [5.1c] $(x + 1)(x^2 + 2)$
6. [5.1b] $x(x - 5)$ **7.** [5.2a] $x(x + 3)(x - 1)$
8. [5.3a], [5.4a] $2(5x - 6)(x + 4)$
9. [5.5d] $(2x + 3)(2x - 3)$ **10.** [5.2a] $(x - 4)(x + 3)$
11. [5.3a], [5.4a] $3m(2m + 1)(m + 1)$
12. [5.5d] $3(w + 5)(w - 5)$ **13.** [5.5b] $5(3x + 2)^2$
14. [5.5d] $3(x^2 + 4)(x + 2)(x - 2)$ **15.** [5.5b] $(7x - 6)^2$
16. [5.3a], [5.4a] $(5x - 1)(x - 5)$ **17.** [5.1c] $(x + 2)(x^3 - 3)$
18. [5.5d] $5(4 + x^2)(2 + x)(2 - x)$
19. [5.3a], [5.4a] $3t(2t + 5)(t - 1)$
20. [5.3a], [5.4a] $(2x + 3)(2x - 5)$
21. [5.2a] $3(m + 2n)(m - 5n)$
22. [5.6a] $(10a - 3b)(100a^2 + 30ab + 9b^2)$
23. [5.8b] $0, 3$ **24.** [5.8b] $-4, 4$ **25.** [5.8b] $-4, 5$
26. [5.8b] $-5, \frac{3}{2}$ **27.** [5.8b] $-4, 7$ **28.** [5.8b] $(-5, 0), (7, 0)$
29. [5.8b] $\left(\frac{2}{3}, 0\right), (1, 0)$ **30.** [5.9a] Length: 8 m; width: 6 m
31. [5.9a] Height: 4 cm; base: 14 cm **32.** [5.9a] 5 ft
33. [5.5d] A **34.** [5.9a] Length: 15 m; width: 3 m
35. [5.2a] $(a - 4)(a + 8)$ **36.** [5.8b] $-\frac{8}{3}, 0, \frac{2}{5}$
37. [4.6b], [5.5d] D

Cumulative Review: Chapters 1–5, p. 407

1. [1.2d] $<$ **2.** [1.2d] $>$ **3.** [1.4a] 0.35 **4.** [1.6c] -1.57
5. [1.5a] $-\frac{1}{14}$ **6.** [1.6c] $-\frac{6}{5}$ **7.** [1.8c] $4x + 1$ **8.** [1.8d] -8
9. [4.2a, b] $\dfrac{8x^6}{y^3}$ **10.** [4.1d, e] $-\dfrac{1}{6x^3}$
11. [4.4a] $x^4 - 3x^3 - 3x^2 - 4$ **12.** [4.7e] $2x^2y^2 - x^2y - xy$
13. [4.8b] $x^2 + 3x + 2 + \dfrac{3}{x - 1}$ **14.** [4.6c] $4t^2 - 12t + 9$
15. [4.6b] $x^4 - 9$ **16.** [4.6a] $6x^2 + 4x - 16$
17. [4.5b] $2x^4 + 6x^3 + 8x^2$ **18.** [4.5d] $4y^3 + 4y^2 + 5y - 4$
19. [4.6b] $x^2 - \frac{4}{9}$ **20.** [5.2a] $(x + 4)(x - 2)$
21. [5.5d] $(2x + 5)(2x - 5)$ **22.** [5.1c] $(3x - 4)(x^2 + 1)$
23. [5.5b] $(x - 13)^2$ **24.** [5.5d] $3(5x + 6y)(5x - 6y)$
25. [5.3a], [5.4a] $(3x + 7)(2x - 9)$
26. [5.2a] $(x^2 - 3)(x^2 + 1)$
27. [5.7a] $2(2y - 3)(y - 1)(y + 1)$
28. [5.3a], [5.4a] $(3p - q)(2p + q)$
29. [5.3a], [5.4a] $2x(5x + 1)(x + 5)$ **30.** [5.5b] $x(7x - 3)^2$
31. [5.3a], [5.4a] Prime **32.** [5.1b] $3x(25x^2 + 9)$
33. [5.5d] $3(x^4 + 4y^4)(x^2 + 2y^2)(x^2 - 2y^2)$
34. [5.2a] $14(x + 2)(x + 1)$
35. [5.7a] $(x + 1)(x - 1)(2x^3 + 1)$ **36.** [2.3b] 15
37. [2.7e] $\{y | y < 6\}$ **38.** [5.8a] $15, -\frac{1}{4}$ **39.** [5.8a] $0, -37$
40. [5.8b] $5, -5, -1$ **41.** [5.8b] $6, -6$ **42.** [5.8b] $\frac{1}{3}$
43. [5.8b] $-10, -7$ **44.** [5.8b] $0, \frac{3}{2}$ **45.** [2.3a] 0.2
46. [5.8b] $-4, 5$ **47.** [2.7e] $\{x | x \le 20\}$
48. [2.3c] All real numbers **49.** [2.4b] $m = \dfrac{y - b}{x}$
50. [2.6a] 50, 52 **51.** [5.9a] -20 and -18; 18 and 20
52. [5.9a] Length: 6 ft; height: 3 ft **53.** [2.6a] 150 m by 350 m
54. [2.5a] \$6500 **55.** [5.9a] 17 m **56.** [2.6a] 30 m, 60 m, 10 m
57. [2.5a] \$29 **58.** [5.9a] Height: 18 cm; base: 16 cm
59. [3.3a]

(graph showing line $3x + 4y = -12$ through points $(-4, 0)$ and $(0, -3)$)

60. [2.7e], [4.6a] $\left\{x | x \ge -\frac{13}{3}\right\}$ **61.** [2.3b] 22
62. [5.8b] $-6, 4$ **63.** [5.7a] $(x - 3)(x - 2)(x + 1)$
64. [5.7a] $(2a + 3b + 3)(2a - 3b - 5)$ **65.** [5.5a] 25
66. [5.9a] 2 cm

CHAPTER 6

Exercise Set 6.1, p. 416

1. 0 3. 8 5. $-\frac{5}{2}$ 7. $-4, 7$ 9. $-5, 5$ 11. None
13. $\dfrac{(4x)(3x^2)}{(4x)(5y)}$ 15. $\dfrac{2x(x-1)}{2x(x+4)}$ 17. $\dfrac{(3-x)(-1)}{(4-x)(-1)}$
19. $\dfrac{(y+6)(y-7)}{(y+6)(y+2)}$ 21. $\dfrac{x^2}{4}$ 23. $\dfrac{8p^2q}{3}$ 25. $\dfrac{x-3}{x}$
27. $\dfrac{m+1}{2m+3}$ 29. $\dfrac{a-3}{a+2}$ 31. $\dfrac{a-3}{a-4}$ 33. $\dfrac{x+5}{x-5}$
35. $a+1$ 37. $\dfrac{x^2+1}{x+1}$ 39. $\frac{3}{2}$ 41. $\dfrac{6}{t-3}$ 43. $\dfrac{t+2}{2(t-4)}$
45. $\dfrac{t-2}{t+2}$ 47. -1 49. -1 51. -6 53. $-x-1$
55. $\dfrac{56x}{3}$ 57. $\dfrac{2}{dc^2}$ 59. 1 61. $\dfrac{(a+3)(a-3)}{a(a+4)}$
63. $\dfrac{2a}{a-2}$ 65. $\dfrac{(t+2)(t-2)}{(t+1)(t-1)}$ 67. $\dfrac{x+4}{x+2}$ 69. $\dfrac{5(a+6)}{a-1}$
71. 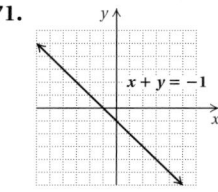 72.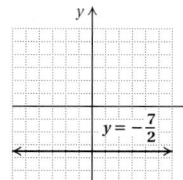
73. 18 and 20; -20 and -18 74. 3.125 L
75. $(x-8)(x+7)$ 76. $(a-8)^2$ 77. $x^3(x-7)(x+5)$
78. $(2y^2+1)(y-5)$ 79. $(2+t)(2-t)(4+t^2)$
80. $10(x+7)(x+1)$ 81. $(x-7)(x-2)$ 82. Prime
83. $(4x-5y)^2$ 84. $(a-7b)(a-2b)$ 85. $x+2y$
87. $\dfrac{(t-9)^2(t-1)}{(t^2+9)(t+1)}$ 89. $\dfrac{x-y}{x-5y}$
91. $\dfrac{5(2x+5)-25}{10} = \dfrac{10x+25-25}{10}$
$= \dfrac{10x}{10}$
$= x$

You get the same number you selected. To do a number trick, ask someone to select a number and then perform these operations. The person will probably be surprised that the result is the original number.

Exercise Set 6.2, p. 423

1. $\dfrac{x}{4}$ 3. $\dfrac{1}{x^2-y^2}$ 5. $a+b$ 7. $\dfrac{x^2-4x+7}{x^2+2x-5}$ 9. $\frac{3}{10}$
11. $\frac{1}{4}$ 13. $\dfrac{b}{a}$ 15. $\dfrac{(a+2)(a+3)}{(a-3)(a-1)}$ 17. $\dfrac{(x-1)^2}{x}$ 19. $\frac{1}{2}$
21. $\frac{15}{8}$ 23. $\frac{15}{4}$ 25. $\dfrac{a-5}{3(a-1)}$ 27. $\dfrac{(x+2)^2}{x}$ 29. $\frac{3}{2}$
31. $\dfrac{c+1}{c-1}$ 33. $\dfrac{y-3}{2y-1}$ 35. $\dfrac{x+1}{x-1}$ 37. $\{x|x \ge 77\}$
38. Height: 7 in.; base: 10 in. 39. $8x^3-11x^2-3x+12$
40. $-2p^2+4pq-4q^2$ 41. $\dfrac{4y^8}{x^6}$ 42. $\dfrac{125x^{18}}{y^{12}}$ 43. $\dfrac{4x^6}{y^{10}}$
44. $\dfrac{1}{a^{15}b^{20}}$ 45. $-\dfrac{1}{b^2}$ 47. $\dfrac{a+1}{5ab^2(a^2+4)}$

Exercise Set 6.3, p. 427

1. 108 3. 72 5. 126 7. 360 9. 500 11. $\frac{65}{72}$
13. $\frac{29}{120}$ 15. $\frac{23}{180}$ 17. $12x^3$ 19. $18x^2y^2$ 21. $6(y-3)$
23. $t(t+2)(t-2)$ 25. $(x+2)(x-2)(x+3)$
27. $t(t+2)^2(t-4)$ 29. $(a+1)(a-1)^2$

31. $(m-3)(m-2)^2$ 33. $(2+3x)(2-3x)$
35. $10v(v+4)(v+3)$ 37. $18x^3(x-2)^2(x+1)$
39. $6x^3(x+2)^2(x-2)$ 41. $120w^6$ 43. $(x-3)^2$
44. $2x(3x+2)$ 45. $(x+3)(x-3)$ 46. $(x+7)(x-3)$
47. $(x+3)^2$ 48. $(x-7)(x+3)$ 49. $120x^4; 8x^3; 960x^7$
50. $48x^6; 16x^5; 768x^{11}$ 51. $20x^2; 10x; 200x^3$
52. $48ab^3; 4ab; 192a^2b^4$ 53. $120x^3; 2x^2; 240x^5$
54. $a^{15}; a^5; a^{20}$ 55. The product of the LCM and the GCF is the product of the two expressions.

Exercise Set 6.4, p. 433

1. 1 3. $\dfrac{6}{3+x}$ 5. $\dfrac{-4x+11}{2x-1}$ 7. $\dfrac{2x+5}{x^2}$ 9. $\dfrac{41}{24r}$
11. $\dfrac{2(2x+3y)}{x^2y^2}$ 13. $\dfrac{4+3t}{18t^3}$ 15. $\dfrac{x^2+4xy+y^2}{x^2y^2}$
17. $\dfrac{6x}{(x-2)(x+2)}$ 19. $\dfrac{11x+2}{3x(x+1)}$ 21. $\dfrac{x(x+6)}{(x+4)(x-4)}$
23. $\dfrac{6}{z+4}$ 25. $\dfrac{3x-1}{(x-1)^2}$ 27. $\dfrac{11a}{10(a-2)}$
29. $\dfrac{2(x^2+4x+8)}{x(x+4)}$ 31. $\dfrac{7a+6}{(a-2)(a+1)(a+3)}$
33. $\dfrac{2(x^2-2x+17)}{(x-5)(x+3)}$ 35. $\dfrac{3a+2}{(a+1)(a-1)}$
37. $\frac{1}{4}$ 39. $-\dfrac{1}{t}$ 41. $\dfrac{-x+7}{x-6}$, or $\dfrac{7-x}{x-6}$, or $\dfrac{x-7}{6-x}$
43. $y+3$ 45. $\dfrac{2(b-7)}{(b+4)(b-4)}$ 47. $a+b$ 49. $\dfrac{5x+2}{x-5}$
51. -1 53. $\dfrac{-x^2+9x-14}{(x-3)(x+3)}$ 55. $\dfrac{2(x+3y)}{(x+y)(x-y)}$
57. $\dfrac{a^2+7a+1}{(a+5)(a-5)}$ 59. $\dfrac{5t-12}{(t+3)(t-3)(t-2)}$ 61. x^2-1
62. $13y^3-14y^2+12y-73$ 63. $\dfrac{1}{8x^{12}y^9}$ 64. $\dfrac{x^6}{25y^2}$
65. $\dfrac{1}{x^{12}y^{21}}$ 66. $\dfrac{25}{x^4y^6}$ 67.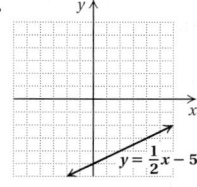
68. 69. 70.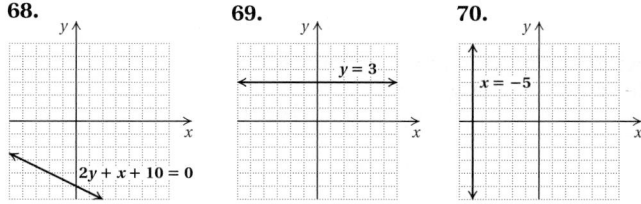
71. -8 72. $\frac{5}{6}$ 73. 3, 5 74. $-2, 9$ 75. Perimeter: $\dfrac{16y+28}{15}$; area: $\dfrac{y^2+2y-8}{15}$ 77. $\dfrac{(z+6)(2z-3)}{(z+2)(z-2)}$
79. $\dfrac{11z^4-22z^2+6}{(z^2+2)(z^2-2)(2z^2-3)}$

Exercise Set 6.5, p. 441

1. $\dfrac{4}{x}$ 3. 1 5. $\dfrac{1}{x-1}$ 7. $\dfrac{-a-4}{10}$ 9. $\dfrac{7z-12}{12z}$
11. $\dfrac{4x^2-13xt+9t^2}{3x^2t^2}$ 13. $\dfrac{2(x-20)}{(x+5)(x-5)}$ 15. $\dfrac{3-5t}{2t(t-1)}$
17. $\dfrac{2s-st-s^2}{(t+s)(t-s)}$ 19. $\dfrac{y-19}{4y}$ 21. $\dfrac{-2a^2}{(x+a)(x-a)}$

23. $\frac{8}{3}$ **25.** $\frac{13}{a}$ **27.** $\frac{8}{y-1}$ **29.** $\frac{x-2}{x-7}$

31. $\frac{4}{(a+5)(a-5)}$ **33.** $\frac{2(x-2)}{x-9}$ **35.** $\frac{3(3x+4)}{(x+3)(x-3)}$

37. $\frac{1}{2}$ **39.** $\frac{x-3}{(x+3)(x+1)}$ **41.** $\frac{18x+5}{x-1}$ **43.** 0

45. $\frac{-9}{2x-3}$ **47.** $\frac{20}{2y-1}$ **49.** $\frac{2a-3}{2-a}$ **51.** $\frac{z-3}{2z-1}$

53. $\frac{2}{x+y}$ **55.** x^5 **56.** $30x^{12}$ **57.** $\frac{b^{20}}{a^8}$ **58.** $18x^3$

59. $\frac{6}{x^3}$ **60.** $\frac{10}{x^3}$ **61.** $-\frac{11}{35}$ **62.** 10 **63.** $x^2-9x+18$

64. $(4-\pi)r^2$ **65.** $\frac{30}{(x-3)(x+4)}$

67. $\frac{x^2+xy-x^3+x^2y-xy^2+y^3}{(x^2+y^2)(x+y)^2(x-y)}$

69. Missing side: $\frac{-2a-15}{a-6}$; area: $\frac{-2a^3-15a^2+12a+90}{2(a-6)^2}$

Mid-Chapter Review: Chapter 6, p. 445

1. False   **2.** True   **3.** True   **4.** False   **5.** True

6. $\dfrac{x-1}{x-2}-\dfrac{x+1}{x+2}-\dfrac{x-6}{4-x^2}$

$=\dfrac{x-1}{x-2}-\dfrac{x+1}{x+2}-\dfrac{x-6}{4-x^2}\cdot\dfrac{-1}{-1}$

$=\dfrac{x-1}{x-2}-\dfrac{x+1}{x+2}-\dfrac{6-x}{x^2-4}$

$=\dfrac{x-1}{x-2}-\dfrac{x+1}{x+2}-\dfrac{6-x}{(x-2)(x+2)}$

$=\dfrac{x-1}{x-2}\cdot\dfrac{x+2}{x+2}-\dfrac{x+1}{x+2}\cdot\dfrac{x-2}{x-2}-\dfrac{6-x}{(x-2)(x+2)}$

$=\dfrac{x^2+x-2}{(x-2)(x+2)}-\dfrac{x^2-x-2}{(x-2)(x+2)}-\dfrac{6-x}{(x-2)(x+2)}$

$=\dfrac{x^2+x-2-x^2+x+2-6+x}{(x-2)(x+2)}$

$=\dfrac{3x-6}{(x-2)(x+2)}$

$=\dfrac{3(x-2)}{(x-2)(x+2)}=\dfrac{x-2}{x-2}\cdot\dfrac{3}{x+2}=\dfrac{3}{x+2}$

7. None   **8.** 3, 8   **9.** $\frac{7}{2}$ **10.** $\frac{x-1}{x-3}$ **11.** $\frac{2(y+4)}{y-1}$

12. -1 **13.** $\frac{1}{-x+3}$, or $\frac{1}{3-x}$ **14.** $10x^3(x-10)^2(x+10)$

15. $\frac{a+1}{a-3}$ **16.** $\frac{y}{(y-2)(y-3)}$ **17.** $x+11$ **18.** $\frac{1}{x-y}$

19. $\frac{a^2+5ab-b^2}{a^2b^2}$ **20.** $\frac{2(3x^2-4x+6)}{x(x+2)(x-2)}$ **21.** E   **22.** A

23. D   **24.** B   **25.** F   **26.** C   **27.** If the numbers have a common factor, then their product contains that factor more than the greatest number of times it occurs in any one factorization. In this case, their product is not their least common multiple.   **28.** Yes; consider the product $\frac{a}{b}\cdot\frac{c}{d}=\frac{ac}{bd}$. The reciprocal of the product is $\frac{bd}{ac}$. This is equal to the product of the reciprocals of the two original factors: $\frac{bd}{ac}=\frac{b}{a}\cdot\frac{d}{c}$.

29. Although multiplying the denominators of the expressions being added results in a common denominator, it is often not the *least* common denominator. Using a common denominator other than the LCD makes the expressions more complicated, requires additional simplification after the addition has been performed, and leaves more room for error.   **30.** Their sum is 0. Another explanation is that

$$-\left(\frac{1}{3-x}\right)=\frac{1}{-(3-x)}=\frac{1}{x-3}.$$

31. $\frac{x+3}{x-5}$ is undefined for $x=5$, $\frac{x-7}{x+1}$ is undefined for $x=-1$, and $\frac{x+1}{x-7}$ (the reciprocal of $\frac{x-7}{x+1}$) is undefined for $x=7$.   **32.** The binomial is a factor of the trinomial.

Exercise Set 6.6, p. 451

1. $\frac{25}{4}$ **3.** $\frac{1}{3}$ **5.** -6 **7.** $\frac{1+3x}{1-5x}$ **9.** $\frac{2x+1}{x}$ **11.** 8

13. $x-8$ **15.** $\frac{y}{y-1}$ **17.** $-\frac{1}{a}$ **19.** $\frac{ab}{b-a}$

21. $\frac{p^2+q^2}{q+p}$ **23.** $\frac{2a(a+2)}{5-3a^2}$ **25.** $\frac{15(4-a^3)}{14a^2(9+2a)}$

27. $\frac{ac}{bd}$ **29.** 1   **31.** $\frac{4x+1}{5x+3}$ **33.** $\{x|x\le 96\}$

34. $\left\{b|b>\frac{22}{9}\right\}$ **35.** $\{x|x<-3\}$ **36.** $\left\{m|m\ge\frac{22}{25}\right\}$

37. $4x^4+3x^3+2x-7$ **38.** 0   **39.** $(p-5)^2$

40. $(p+5)^2$ **41.** $50(p^2-2)$ **42.** $5(p+2)(p-10)$

43. 14 yd   **44.** 12 ft, 5 ft   **45.** $\frac{(x-1)(3x-2)}{5x-3}$ **47.** $\frac{5x+3}{3x+2}$

Calculator Corner, p. 456

1.–2. Left to the student

Study Tips, p. 457

1. Rational expression   **2.** Solutions   **3.** Rational expression   **4.** Rational expression   **5.** Rational expression   **6.** Solutions   **7.** Rational expression   **8.** Solutions   **9.** Solutions   **10.** Solutions   **11.** Rational expression   **12.** Solutions   **13.** Rational expression

Exercise Set 6.7, p. 458

1. $\frac{6}{5}$ **3.** $\frac{40}{29}$ **5.** $\frac{47}{2}$ **7.** -6 **9.** $\frac{24}{7}$ **11.** $-4,-1$
13. $-4,4$ **15.** 3   **17.** $\frac{14}{3}$ **19.** 5   **21.** 5   **23.** $\frac{5}{2}$ **25.** -2
27. $-\frac{13}{2}$ **29.** $\frac{17}{2}$ **31.** No solution   **33.** -5 **35.** $\frac{5}{3}$
37. $\frac{1}{2}$ **39.** No solution   **41.** No solution   **43.** 4
45. No solution   **47.** $-2,2$ **49.** 7   **51.** Quotient
52. Product   **53.** Reciprocals   **54.** Factoring
55. Greatest   **56.** Not   **57.** Subtract   **58.** Additive inverses   **59.** $-\frac{1}{6}$ **61.** Left to the student

Translating for Success, p. 470

1. K   **2.** E   **3.** C   **4.** N   **5.** D   **6.** O   **7.** F   **8.** H
9. B   **10.** A

Exercise Set 6.8, p. 471

1. $2\frac{2}{9}$ hr   **3.** $25\frac{5}{7}$ min   **5.** $3\frac{15}{16}$ hr   **7.** $22\frac{2}{9}$ min   **9.** $3\frac{3}{4}$ min
11. Sarah: 30 km/h; Rick: 70 km/h   **13.** Passenger: 80 mph; freight: 66 mph   **15.** 20 mph   **17.** Hank: 14 km/h; Kelly: 19 km/h; Ralph: 5 km/h; Bonnie: 8 km/h   **21.** 3 hr
23. $\frac{10}{3}$ students/teacher   **25.** 2.3 km/h   **27.** 66 g   **29.** 1.92 g
31. 1.75 lb   **33.** $1\frac{11}{39}$ kg   **35.** (a) 0.269; (b) 168 hits; (c) 188 hits   **37.** 22 in.; 55.8 cm   **39.** $7\frac{1}{4}$; 57.9 cm
41. $7\frac{1}{2}$; $23\frac{3}{5}$ in.   **43.** 287 trout   **45.** 200 duds
47. (a) 4.8 tons; (b) 48 lb   **49.** $\frac{21}{2}$ **51.** $\frac{8}{3}$ **53.** $\frac{35}{3}$ **55.** 15 ft
57. 0   **58.** -2 **59.** x^{11} **60.** x **61.** $\frac{1}{x^{11}}$ **62.** $\frac{1}{x}$

63.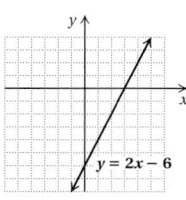

64.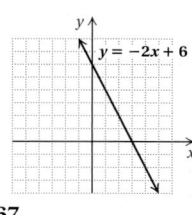

65.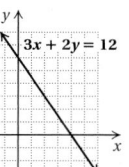

66.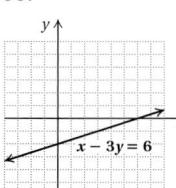

67.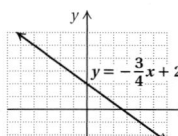

68.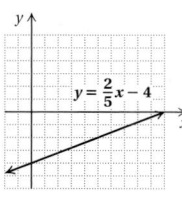

69. Ann: 6 hr; Betty: 12 hr   **71.** $27\frac{3}{11}$ min

73. $t = \dfrac{ab}{b + a}$

Exercise Set 6.9, p. 485

1. $5; y = 5x$ **3.** $\frac{2}{15}; y = \frac{2}{15}x$ **5.** $\frac{9}{4}; y = \frac{9}{4}x$
7. 175 semi trucks   **9.** 135,209,760 cans   **11.** 90 g
13. 40 kg   **15.** $98; y = \dfrac{98}{x}$ **17.** $36; y = \dfrac{36}{x}$
19. $0.05; y = \dfrac{0.05}{x}$ **21.** 3.5 hr   **23.** $\frac{2}{9}$ ampere   **25.** 960 lb
27. $5\frac{5}{7}$ hr   **29.** $y = 15x^2$ **31.** $y = \dfrac{0.0015}{x^2}$ **33.** $y = xz$
35. $y = \frac{3}{10}xz^2$ **37.** $y = \dfrac{xz}{5wp}$ **39.** 2.5 m   **41.** 199.4 lb
43. 95 earned runs   **45.** 729 gal   **47.** $(x - 8)(x + 7)$
48. $(a - 8)^2$ **49.** $x^3(x - 7)(x + 5)$ **50.** $(2y^2 + 1)(y - 5)$
51. $(2 - t)(2 + t)(4 + t^2)$ **52.** $10(x + 7)(x + 1)$
53. $(x - 7)(x - 2)$ **54.** Not factorable   **55.** $(4x - 5y)^2$
56. $(a - 7b)(a - 2b)$ **57.** $3(x - y)(x^2 + xy + y^2)$
58. $(w + t)(w^2 - wt + t^2)(w - t)(w^2 + wt + t^2)$
59. (a) Inversely; **(b)** neither; **(c)** directly; **(d)** directly   **61.** $7.20

Summary and Review: Chapter 6, p. 489

Concept Reinforcement
1. True   **2.** False   **3.** True   **4.** True   **5.** False

Important Concepts
1. $\dfrac{x - 1}{2(x + 5)}$ **2.** $\dfrac{y + 5}{5(y + 3)}$ **3.** $\dfrac{b + 7}{b + 8}$ **4.** -1
5. $\dfrac{x^2 - 4x - 10}{(x + 2)(x + 1)(x - 1)}$ **6.** $\dfrac{3(2y - 5)}{5(9 - y)}$ **7.** 1
8. $y = 150x; y = 300$ **9.** $y = \dfrac{225}{x}; y = 22.5$

Review Exercises
1. 0   **2.** 6   **3.** $-6, 6$ **4.** $-6, 5$ **5.** -2 **6.** None
7. $\dfrac{x - 2}{x + 1}$ **8.** $\dfrac{7x + 3}{x - 3}$ **9.** $\dfrac{y - 5}{y + 5}$ **10.** $\dfrac{a - 6}{5}$
11. $\dfrac{6}{2t - 1}$ **12.** $-20t$ **13.** $\dfrac{2x(x - 1)}{x + 1}$ **14.** $30x^2y^2$
15. $4(a - 2)$ **16.** $(y - 2)(y + 2)(y + 1)$ **17.** $\dfrac{-3(x - 6)}{x + 7}$
18. -1 **19.** $\dfrac{2a}{a - 1}$ **20.** $d + c$ **21.** $\dfrac{4}{x - 4}$ **22.** $\dfrac{x + 5}{2x}$

23. $\dfrac{2x + 3}{x - 2}$ **24.** $\dfrac{-x^2 + x + 26}{(x - 5)(x + 5)(x + 1)}$ **25.** $\dfrac{2(x - 2)}{x + 2}$
26. $\dfrac{z}{1 - z}$ **27.** $c - d$ **28.** 8   **29.** $-5, 3$ **30.** $5\frac{1}{7}$ hr
31. 95 mph, 175 mph   **32.** 240 km/h, 280 km/h
33. 160 defective calculators   **34. (a)** $\frac{12}{13}$ c; **(b)** $4\frac{1}{5}$ c; **(c)** $9\frac{1}{3}$ c
35. 10,000 blue whales   **36.** 6   **37.** $y = 4x$ **38.** $y = \dfrac{2500}{x}$
39. $y = 6xz$ **40.** $y = \dfrac{2x}{z}$ **41.** 20 min
42. About 77.7   **43.** 500 watts   **44.** C   **45.** A
46. $\dfrac{5(a + 3)^2}{a}$ **47.** They are equivalent proportions.

Understanding Through Discussion and Writing
1. No; when we are adding, no sign changes are required so the result is the same regardless of use of parentheses. When we are subtracting, however, the sign of each term of the expression being subtracted must be changed and parentheses are needed to make sure this is done.   **2.** Graph each side of the equation and determine the number of points of intersection of the graphs.   **3.** Canceling removes a factor of 1, allowing us to rewrite $a \cdot 1$ as a.   **4.** Inverse variation; the greater the average gain per play, the smaller the number of plays required.
5. Form a rational expression that has factors of $x + 3$ and $x - 4$ in the denominator.   **6.** If we multiply both sides of a rational equation by a variable expression in order to clear fractions, it is possible that the variable expression is equal to 0. Thus an equivalent equation might not be produced.

Test: Chapter 6, p. 495

1. [6.1a] 0   **2.** [6.1a] -8 **3.** [6.1a] $-7, 7$ **4.** [6.1a] 1, 2
5. [6.1a] 1   **6.** [6.1a] None   **7.** [6.1c] $\dfrac{3x + 7}{x + 3}$
8. [6.1d] $\dfrac{a + 5}{2}$ **9.** [6.2b] $\dfrac{(5x + 1)(x + 1)}{3x(x + 2)}$
10. [6.3a] $(y - 3)(y + 3)(y + 7)$ **11.** [6.4a] $\dfrac{23 - 3x}{x^3}$
12. [6.5a] $\dfrac{2(4 - t)}{t^2 + 1}$ **13.** [6.4a] $\dfrac{-3}{x - 3}$ **14.** [6.5a] $\dfrac{2x - 5}{x - 3}$
15. [6.4a] $\dfrac{8t - 3}{t(t - 1)}$ **16.** [6.5a] $\dfrac{-x^2 - 7x - 15}{(x + 4)(x - 4)(x + 1)}$
17. [6.5b] $\dfrac{x^2 + 2x - 7}{(x - 1)^2(x + 1)}$ **18.** [6.6a] $\dfrac{3y + 1}{y}$
19. [6.7a] 12   **20.** [6.7a] $-3, 5$ **21.** [6.9a] $y = 2x; 50$
22. [6.9a] $y = 0.5x; 12.5$ **23.** [6.9c] $y = \dfrac{18}{x}; \frac{9}{50}$
24. [6.9c] $y = \dfrac{22}{x}; \frac{11}{50}$ **25.** [6.9e] $Q = 2.5xy$ **26.** [6.9b] 240 km
27. [6.9d] $1\frac{1}{5}$ hr   **28.** [6.8b] 16 defective spark plugs
29. [6.8b] 50 zebras   **30.** [6.8a] 12 min   **31.** [6.8a] Craig: 65 km/h; Marilyn: 45 km/h   **32.** [6.8b] 15   **33.** [6.7a] D
34. [6.8a] Rema: 4 hr; Reggie: 10 hr   **35.** [6.6a] $\dfrac{3a + 2}{2a + 1}$

Cumulative Review: Chapters 1–6, p. 497

1. [1.2e] 3.5   **2.** [4.3g] 3, 2, 1, 0; 3   **3.** [2.5a] About 275 mg
4. [2.5a] About 7.3%   **5.** [2.5a] $2500   **6.** [6.8a] 35 mph, 25 mph   **7.** [5.9a] 14 ft   **8. (a)** [6.9b] $M = 0.4B$;
(b) [6.9b] 76.8 lb   **9.** [4.3e] $2x^3 - 3x^2 - 2$
10. [1.8c] $\frac{3}{8}x + 1$ **11.** [4.1e], [4.2a, b] $\dfrac{9}{4x^8}$
12. [6.6a] $\dfrac{4(2x - 3)}{17x}$ **13.** [4.7e] $-2xy^2 - 4x^2y^2 + xy^3$

14. [4.4a] $2x^5 + 6x^4 + 2x^3 - 10x^2 + 3x - 9$

15. [6.1d] $\dfrac{2}{3(y+2)}$ **16.** [6.2b] 2   **17.** [6.4a] $x + 4$

18. [6.5a] $\dfrac{2(x-3)}{(x+2)(x-2)}$ **19.** [4.6a] $a^2 - 9$

20. [4.6c] $36x^2 - 60x + 25$ **21.** [4.6b] $4x^6 - 1$
22. [5.3a], [5.4a] $(9a - 2)(a + 6)$ **23.** [5.5b] $(3x - 5y)^2$
24. [5.5d] $(7x - 1)(7x + 1)$ **25.** [2.3c] 3   **26.** [5.8b] $-4, \frac{1}{2}$
27. [5.8b] 0, 10   **28.** [2.7e] $\{x \mid x \geq -26\}$ **29.** [6.7a] 2

30. [2.4b] $a = \dfrac{t}{x+y}$ **31.** [3.4a] Not defined

32. [3.4a] $-\frac{3}{7}$ **33.** [3.4a] $-\dfrac{9}{4}$ **34.** [3.4a] 0

35. [3.3a] y-intercept: $(0, -2)$;   **36.** [3.3a] y-intercept: $\left(0, -\frac{1}{8}\right)$;
    x-intercept: $(-6, 0)$ x-intercept: $\left(\frac{3}{8}, 0\right)$

37. [3.3a] y-intercept: $(0, 25)$;   **38.** [3.3a] y-intercept: none;
    x-intercept: none x-intercept: $\left(-\frac{1}{4}, 0\right)$

39. [3.3b] 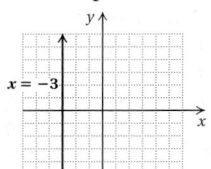   **40.** [3.3b]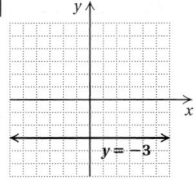

41. [3.3a] 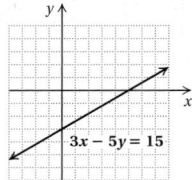   **42.** [3.3a]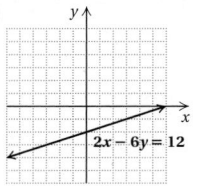

43. [3.2a] 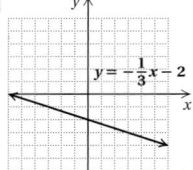   **44.** [3.2a], [3.3a]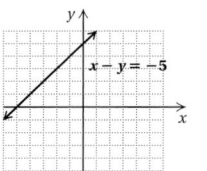

45. [6.1a],[6.6a] $0, 3, \frac{5}{2}$

CHAPTER 7
Calculator Corner, p. 503
1. -13.3 **2.** -14.4 **3.** 14   **4.** 34

Calculator Corner, p. 505
1. $y = x - 4$ 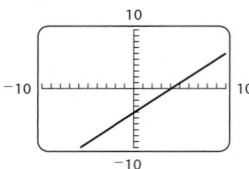   **2.** $y = -2x - 3$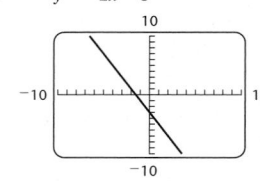

3. $y = 1 - x^2$ 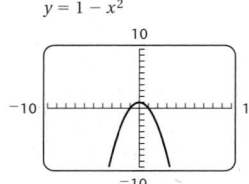   **4.** $y = 3x^2 - 4x + 1$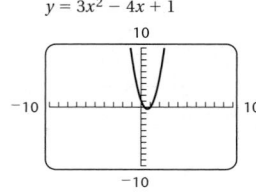

5. $y = x^3$ 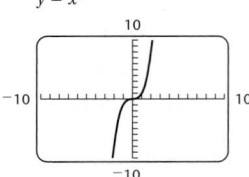   **6.** $y = |x + 3|$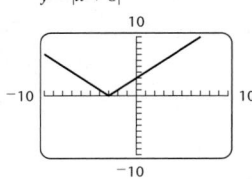

Exercise Set 7.1, p. 508
1. Yes   **3.** Yes   **5.** No   **7.** No   **9.** Yes   **11.** No
13. Yes   **15.** (a) 9; (b) 12; (c) 2; (d) 5; (e) 7.4; (f) $5\frac{2}{3}$
17. (a) -21; (b) 15; (c) 2; (d) 0; (e) $18a$; (f) $3a + 3$ **19.** (a) 7;
(b) -17; (c) 6; (d) 4; (e) $3a - 2$; (f) $3a + 3h + 4$ **21.** (a) 0;
(b) 5; (c) 2; (d) 170; (e) 65; (f) $32a^2 - 12a$ **23.** (a) 1; (b) 3;
(c) 3; (d) 11; (e) $|a - 1| + 1$; (f) $|a + h| + 1$ **25.** (a) 0; (b) -1;
(c) 8; (d) 1000; (e) -125; (f) $-27a^3$ **27.** 2003: about 61 yr; 2009:
about 63 yr   **29.** $1\frac{20}{33}$ atm; $1\frac{10}{11}$ atm; $4\frac{1}{33}$ atm   **31.** 1.792 cm;
2.8 cm; 11.2 cm

33. 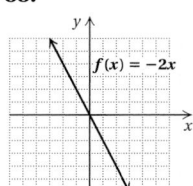   **35.** 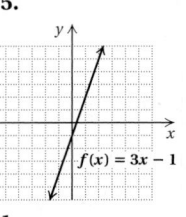   **37.**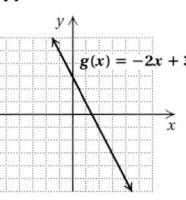

39. 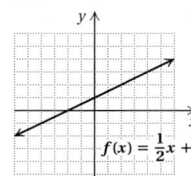   **41.** 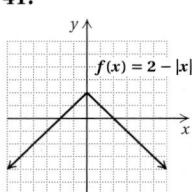   **43.**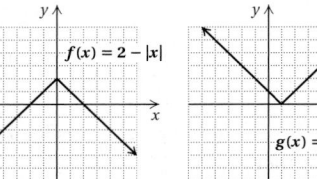

45. 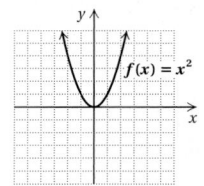   **47.**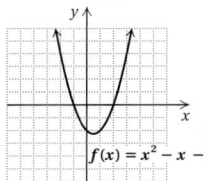

49. 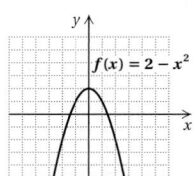   **51.**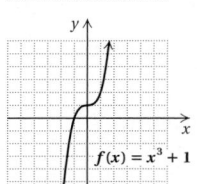

53. Yes   **55.** Yes   **57.** No   **59.** No   **61.** About 1150
stations   **63.** About 1.9 billion images   **65.** Quadrants
66. Relation   **67.** Function; domain; range; domain; range
68. Graph   **69.** Inputs   **70.** Solutions   **71.** Addition
principle   **72.** Vertical-line test   **73.** $g(-2) = 39$
75. 26; 99   **77.** $g(x) = \frac{15}{4}x - \frac{13}{4}$

Exercise Set 7.2, p. 517
1. (a) 3; (b) $\{-4, -3, -2, -1, 0, 1, 2\}$; (c) $-2, 0$; (d) $\{1, 2, 3, 4\}$
3. (a) $2\frac{1}{2}$; (b) $\{x \mid -3 \leq x \leq 5\}$; (c) $2\frac{1}{4}$; (d) $\{y \mid 1 \leq y \leq 4\}$
5. (a) $2\frac{1}{4}$; (b) $\{x \mid -4 \leq x \leq 3\}$; (c) 0; (d) $\{y \mid -5 \leq y \leq 4\}$
7. (a) 1; (b) all real numbers; (c) 3; (d) all real numbers
9. (a) 1; (b) all real numbers; (c) $-2, 2$; (d) $\{y \mid y \geq 0\}$
11. (a) -1; (b) $\{x \mid -6 \leq x \leq 5\}$; (c) $-4, 0, 3$; (d) $\{y \mid -2 \leq y \leq 2\}$
13. $\{x \mid x$ is a real number $and \ x \neq -3\}$

15. All real numbers   **17.** All real numbers
19. $\{x|x \text{ is a real number } and \ x \neq \frac{14}{5}\}$ **21.** All real numbers
23. $\{x|x \text{ is a real number } and \ x \neq \frac{7}{4}\}$ **25.** $\{x|x \text{ is a real}$
number $and \ x \neq 1\}$ **27.** All real numbers   **29.** All real
numbers   **31.** $\{x|x \text{ is a real number } and \ x \neq \frac{5}{2}\}$
33. All real numbers   **35.** $\{x|x \text{ is a real number } and \ x \neq -\frac{5}{4}\}$
37. $-8; 0; -2$ **39.** $a - 1$ **40.** $\dfrac{2(y + 2)}{7(y + 7)}$ **41.** $\dfrac{5}{x + 2}$
42. $t - 4$ **43.** $w + 1$, R 2; or $w + 1 + \dfrac{2}{w + 3}$
44. $x^4 + x^3 - 2x^2 + x - 2$, R $2x + 3$; or
$x^4 + x^3 - 2x^2 + x - 2 + \dfrac{2x + 3}{x^2 - 1}$ **45.** $14x^2 + 57x - 27$
46. $a^2 - 1$ **47.** $81y^2 + 180y + 100$ **48.** $8w^2 - w - \dfrac{1}{4}$
49. $\{y|y \text{ is a real number and } y \neq 0\}$; $\{y|y \geq 2\}$; $\{y|y \geq -4\}$;
$\{y|y \geq 0\}$ **51.** All real numbers

Mid-Chapter Review: Chapter 7, p. 519
1. True   **2.** False   **3.** True   **4.** True   **5.** False
6.

x	$f(x)$
0	1
2	-2
-2	4
4	-5

7.

x	$f(x)$
-2	0
-2 or 3	0
0	-6
2	-4
-1	-4
1	-6

8. Yes   **9.** No   **10.** Domain: $\{x|-3 \leq x \leq 3\}$;
range: $\{y|-2 \leq y \leq 1\}$ **11.** -3 **12.** -7 **13.** 8   **14.** 9
15. 9000   **16.** 0   **17.** Yes   **18.** No   **19.** Yes   **20.** $\{x|x \text{ is}$
a real number $and \ x \neq 4\}$ **21.** All real numbers
22. $\{x|x \text{ is a real number } and \ x \neq -2\}$ **23.** All real numbers
24.   **25.**   **26.**

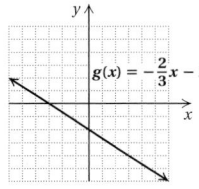 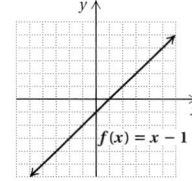 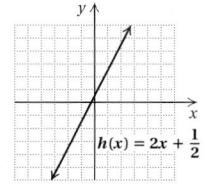

27.   **28.**   **29.**

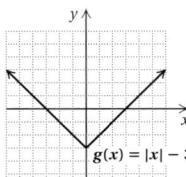 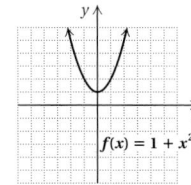 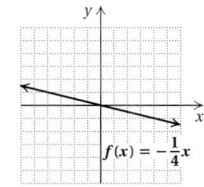

30. No; since each input has exactly one output, the number
of outputs cannot exceed the number of inputs.   **31.** When
$x < 0$, then $y < 0$ and the graph contains points in quadrant III.
When $0 < x < 30$, then $y < 0$ and the graph contains points in
quadrant IV. When $x > 30$, then $y > 0$ and the graph contains
points in quadrant I. Thus the graph passes through three
quadrants.   **32.** The output -3 corresponds to the input 2.
The number -3 in the range is paired with the number 2 in the
domain. The point $(2, -3)$ is on the graph of the function.
33. The domain of a function is the set of all inputs, and the
range is the set of all outputs.

Calculator Corner, p. 521
1. The graph of $y_2 = x + 4$ is the same as the graph of $y_1 = x$,
but it is moved up 4 units.   **2.** The graph of $y_3 = x - 3$ is the
same as the graph of $y_1 = x$, but it is moved down 3 units.

3. The graph of $y = x + 8$ will be the same as the graph of
$y_1 = x$, but it will be moved up 8 units. The graph of $y = x - 5$
will be the same as the graph of $y_1 = x$, but it will be moved
down 5 units.

Calculator Corner, p. 525
1. The graph of $y = 10x$ will slant up from left to right. It will be
steeper than the other graphs.   **2.** The graph of $y = 0.005x$
will slant up from left to right. It will be less steep than the other
graphs.   **3.** The graph of $y = -10x$ will slant down from left to
right. It will be steeper than the other graphs.   **4.** The graph of
$y = -0.005x$ will slant down from left to right. It will be less
steep than the other graphs.

Exercise Set 7.3, p. 530
1. $m = 4$; y-intercept: $(0, 5)$ **3.** $m = -2$; y-intercept: $(0, -6)$
5. $m = -\frac{3}{8}$; y-intercept: $\left(0, -\frac{1}{5}\right)$ **7.** $m = 0.5$;
y-intercept: $(0, -9)$ **9.** $m = \frac{2}{3}$; y-intercept: $\left(0, -\frac{8}{3}\right)$
11. $m = 3$; y-intercept: $(0, -2)$ **13.** $m = -8$;
y-intercept: $(0, 12)$ **15.** $m = 0$; y-intercept: $\left(0, \frac{4}{17}\right)$
17. $m = -\frac{1}{2}$ **19.** $m = \frac{1}{3}$ **21.** $m = 2$ **23.** $m = \frac{2}{3}$
25. $m = -\frac{1}{3}$ **27.** $\frac{2}{25}$, or 8%   **29.** $\frac{13}{41}$, or about 31.7%
31. The rate of change is -2.55 deaths per year.   **33.** The rate
of change is $-\$900$ per year.   **35.** The rate of change is 4313.4
servicemen per year.   **37.** -1323 **38.** $45x + 54$
39. $350x - 60y + 120$ **40.** 25   **41.** Square: 15 yd;
triangle: 20 yd   **42.** $(2 - 5x)(4 + 10x + 25x^2)$
43. $(c - d)(c^2 + cd + d^2)(c + d)(c^2 - cd + d^2)$
44. $7(2x - 1)(4x^2 + 2x + 1)$ **45.** $a - 10$, R -4; or
$a - 10 + \dfrac{-4}{a - 1}$

Calculator Corner, p. 533
1. $y = -3.2x - 16$

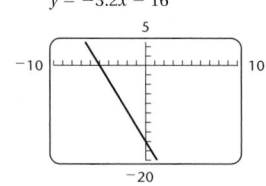

Xscl = 1, Yscl = 2

2. $y = 4.25x + 85$

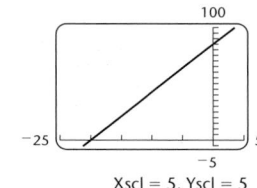

Xscl = 5, Yscl = 5

3. $y = (-6x + 90)/5$

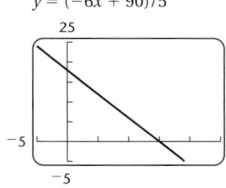

Xscl = 5, Yscl = 5

4. $y = (5x - 30)/6$

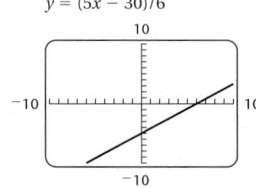

5. $y = (-8x + 9)/3$

6. $y = 0.4x - 5$

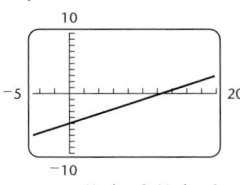

Xscl = 2, Yscl = 1

7. $y = 1.2x - 12$

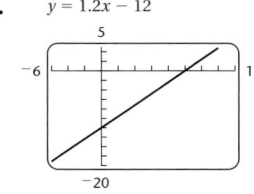

8. $y = (4x - 2)/5$

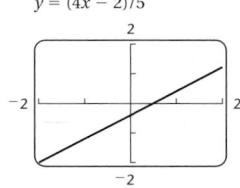

Xscl = 2, Yscl = 2

Visualizing for Success, p. 539

1. D   **2.** I   **3.** H   **4.** C   **5.** F   **6.** A   **7.** G   **8.** B
9. E   **10.** J

Exercise Set 7.4, p. 540

1.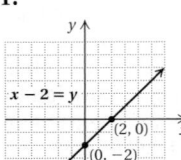

3.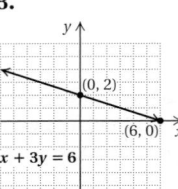

5.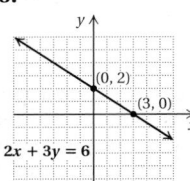

7.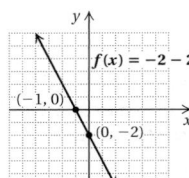

9.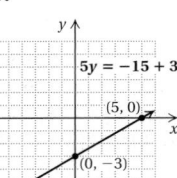

11.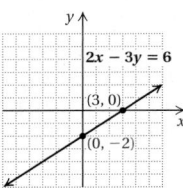

13.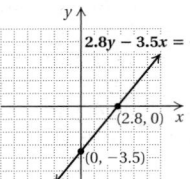

15.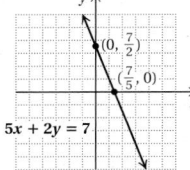

17.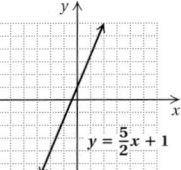

19.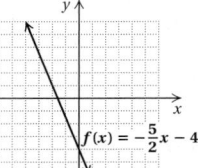

21.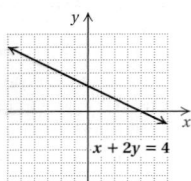

23.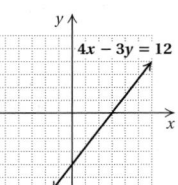

25.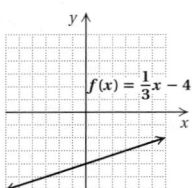

27.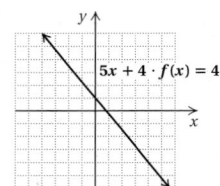

29. Not defined

31. $m = 0$

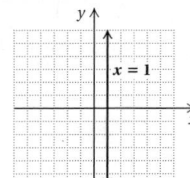

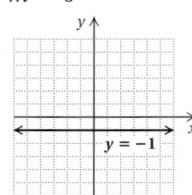

33. $m = 0$

35. $m = 0$

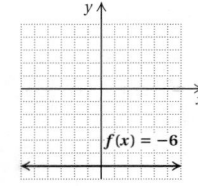

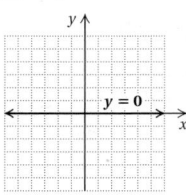

37. $m = 0$

39. Not defined

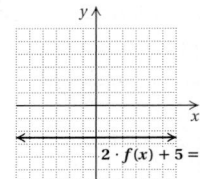

41. Yes   **43.** No   **45.** Yes   **47.** Yes   **49.** Yes   **51.** No
53. No   **55.** Yes   **57.** 5.3×10^{10} **58.** 4.7×10^{-5}
59. 1.8×10^{-2} **60.** 9.9902×10^{7} **61.** 0.0000213
62. $901{,}000{,}000$ **63.** $20{,}000$ **64.** 0.085677 **65.** $3(3x - 5y)$
66. $3a(4 + 7b)$ **67.** $7p(3 - q + 2)$ **68.** $64(x - 2y + 4)$
69. $y = 3$ **71.** $a = 2$ **73.** $y = \frac{2}{15}x + \frac{2}{5}$ **75.** $y = 0$; yes
77. $m = -\frac{3}{4}$ **79.** (a) II; (b) IV; (c) I; (d) III

Exercise Set 7.5, p. 552

1. $y = -8x + 4$ **3.** $y = 2.3x - 1$ **5.** $f(x) = -\frac{7}{3}x - 5$
7. $f(x) = \frac{2}{3}x + \frac{5}{8}$ **9.** $y = 5x - 17$ **11.** $y = -3x + 33$
13. $y = x - 6$ **15.** $y = -2x + 16$ **17.** $y = -7$
19. $y = \frac{2}{3}x - \frac{8}{3}$ **21.** $y = \frac{1}{2}x + \frac{7}{2}$ **23.** $y = x$
25. $y = \frac{7}{4}x + 7$ **27.** $y = \frac{3}{2}x$ **29.** $y = \frac{1}{6}x$
31. $y = 13x - \frac{15}{4}$ **33.** $y = -\frac{1}{2}x + \frac{17}{2}$ **35.** $y = \frac{5}{7}x - \frac{17}{7}$
37. $y = \frac{1}{3}x + 4$ **39.** $y = \frac{1}{2}x + 4$ **41.** $y = \frac{4}{3}x - 6$
43. $y = \frac{5}{2}x + 9$ **45.** (a) $C(t) = 40t + 85$;
(b)   **(c)** $345

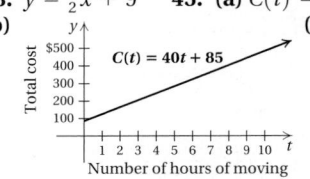

47. (a) $V(t) = 750 - 25t$;
(b)   **(c)** $425

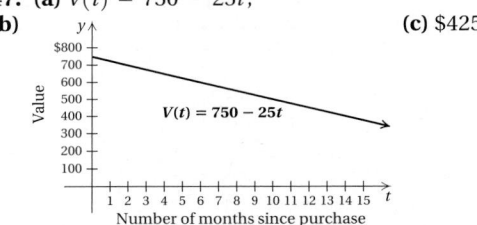

49. (a) $W(x) = 379.6x + 2862$; (b) 4001 cases; 12,352 cases
51. (a) $D(x) = -231.88x + 24{,}026$; (b) 21,939 dealerships;
(c) about 26 yr after 1991, or in 2017
53. (a) $M(t) = 0.236t + 71.8$; (b) about 75.8 yr   **55.** -1
56. $b + 1$ **57.** $\dfrac{x - 3}{2(x - 5)}$ **58.** $\dfrac{4}{y - 9}$ **59.** $\dfrac{2}{7}$ **60.** 0
61. Not defined   **62.** -2

Summary and Review: Chapter 7, p. 555

Concept Reinforcement

1. False   **2.** True   **3.** False

Important Concepts

1. No   **2.** $g(0) = -2; g(-2) = -3; g(6) = 1$
3.   **4.** Yes   **5.** Domain: $\{x | -4 \le x \le 5\}$;
range: $\{y | -2 \le y \le 4\}$ **6.** $\{x | x$ is a
real number $and\ x \ne -3\}$ **7.** -2
8. Slope: $-\frac{1}{2}$; y-intercept: $(0, 2)$

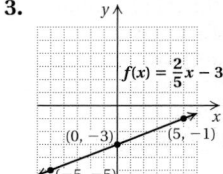

9.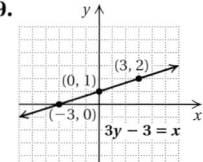

10.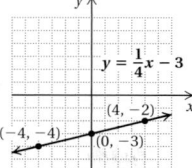

11.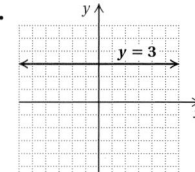

12.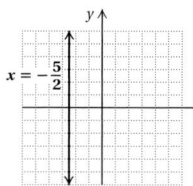

13. Parallel   **14.** Perpendicular   **15.** $y = -8x + 0.3$
16. $y = -4x - 1$ **17.** $y = -\frac{5}{3}x + \frac{11}{3}$ **18.** $y = \frac{4}{3}x - \frac{23}{3}$
19. $y = -\frac{3}{4}x - \frac{7}{2}$

Review Exercises

1. No   **2.** Yes   **3.** $g(0) = 5; g(-1) = 7$
4. $f(0) = 7; f(-1) = 12$ **5.** About $6810

6.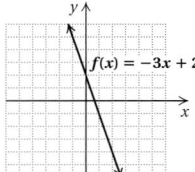

7.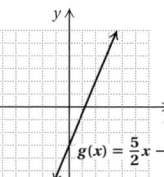

8.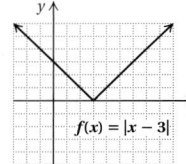

9.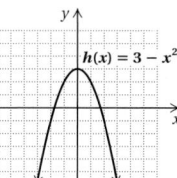

10. Yes   **11.** No   **12.** (a) $f(2) = 3$; (b) $\{x | -2 \le x \le 4\}$;
(c) -1; **(d)** $\{y | 1 \le y \le 5\}$ **13.** $\{x | x \text{ is a real number } and \, x \ne 4\}$
14. All real numbers   **15.** Slope: -3; y-intercept: $(0, 2)$
16. Slope: $-\frac{1}{2}$; y-intercept: $(0, 2)$ **17.** $m = \frac{11}{3}$
18. 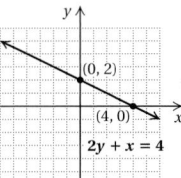   **19.** 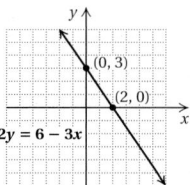   **20.**

21. 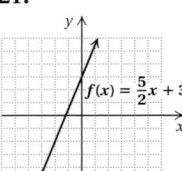   **22.** 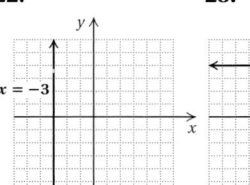   **23.**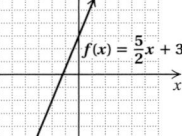

24. Perpendicular   **25.** Parallel   **26.** Parallel
27. Perpendicular   **28.** $f(x) = 4.7x - 23$ **29.** $y = -3x + 4$
30. $y = -\frac{3}{2}x$ **31.** $y = -\frac{5}{7}x + 9$ **32.** $y = \frac{1}{3}x + \frac{1}{3}$
33. **(a)** $R(x) = -0.064x + 46.8$; **(b)** about 44.37 sec; 44.24 sec
34. C   **35.** A   **36.** $f(x) = 3.09x + 3.75$

Understanding Through Discussion and Writing

1. A line's x- and y-intercepts are the same only when the line passes through the origin. The equation for such a line is of the form $y = mx$.   **2.** The concept of slope is useful in describing

how a line slants. A line with positive slope slants up from left to right. A line with negative slope slants down from left to right. The larger the absolute value of the slope, the steeper the slant.
3. Find the slope–intercept form of the equation:

$$4x + 5y = 12$$
$$5y = -4x + 12$$
$$y = -\frac{4}{5}x + \frac{12}{5}.$$

This form of the equation indicates that the line has a negative slope and thus should slant down from left to right. The student may have graphed $y = \frac{4}{5}x + \frac{12}{5}$.   **4.** For $R(t) = 50t + 35$, $m = 50$ and $b = 35$; 50 signifies that the cost per hour of a repair is $50; 35 signifies that the minimum cost of a repair job is $35.
5. $m = \dfrac{\text{change in } y}{\text{change in } x}$

As we move from one point to another on a vertical line, the y-coordinate changes but the x-coordinate does not. Thus the change in y is a nonzero number whereas the change in x is 0. Since division by 0 is undefined, the slope of a vertical line is undefined.

As we move from one point to another on a horizontal line, the y-coordinate does not change but the x-coordinate does. Thus the change in y is 0 whereas the change in x is a nonzero number, so the slope is 0.   **6.** Using algebra, we find that the slope–intercept form of the equation is $y = \frac{5}{2}x - \frac{3}{2}$. This indicates that the y-intercept is $\left(0, -\frac{3}{2}\right)$, so a mistake has been made. It appears that the student graphed $y = \frac{5}{2}x + \frac{3}{2}$.

Test: Chapter 7, p. 564

1. [7.1a] Yes   **2.** [7.1a] No   **3.** [7.1b] $-4; 2$ **4.** [7.1b] $7; 8$
5. [7.1b] $-6; -6$ **6.** [7.1b] $3; 0$
7. [7.1c]   **8.** [7.1c]   **9.** [7.1c]

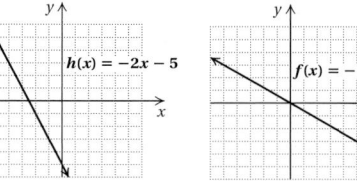

10. [7.1c]   **11.** [7.4c]   **12.** [7.4c]

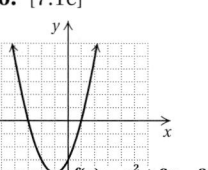   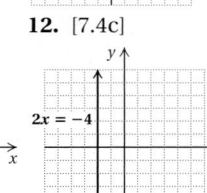

13. [7.1e] **(a)** 8.666 yr; **(b)** 1998   **14.** [7.1d] Yes   **15.** [7.1d] No
16. [7.2a] $\left\{x | x \text{ is a real number } and \, x \ne -\frac{3}{2}\right\}$ **17.** [7.2a] All
real numbers   **18.** [7.2a] **(a)** 1; **(b)** $\{x | -3 \le x \le 4\}$; **(c)** -3;
(d) $\{y | -1 \le y \le 2\}$ **19.** [7.3b] Slope: $-\frac{3}{5}$; y-intercept: $(0, 12)$
20. [7.3b] Slope: $-\frac{2}{5}$; y-intercept: $\left(0, -\frac{7}{5}\right)$ **21.** [7.3b] $m = \frac{5}{8}$
22. [7.3b] $m = 0$ **23.** [7.3c] m (or rate of change) $= \frac{4}{5}$ km/min
24. [7.4a]   **25.** [7.4b]

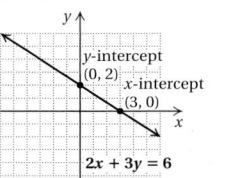

26. [7.4d] Parallel   **27.** [7.4d] Perpendicular
28. [7.5a] $y = -3x + 4.8$ **29.** [7.5a] $f(x) = 5.2x - \frac{5}{8}$
30. [7.5b] $y = -4x + 2$ **31.** [7.5c] $y = -\frac{3}{2}x$

32. [7.5d] $y = \frac{1}{2}x - 3$ **33.** [7.5d] $y = 3x - 1$
34. [7.5e] **(a)** $A(x) = 0.122x + 23.2$; **(b)** 27.84 yr; 28.69 yr
35. [7.5b] B   **36.** [7.5d] $\frac{24}{5}$ **37.** [7.1b] $f(x) = 3$; answers may vary

Cumulative Review: Chapters 1-7, p. 567

1. [1.8d] -6.8 **2.** [4.2d] 3.12×10^{-2} **3.** [1.6c] $-\frac{3}{4}$
4. [4.1d, e] $\frac{1}{8}$ **5.** [6.1c] $\frac{x + 3}{2x - 1}$ **6.** [6.1c] $\frac{t - 4}{t + 4}$
7. [6.6a] $\frac{x^2(x + 1)}{x + 4}$ **8.** [4.6a] $2 - 10x^2 + 12x^4$
9. [4.7f] $4a^4b^2 - 20a^3b^3 + 25a^2b^4$
10. [4.6b], [4.7f] $9x^4 - 16y^2$ **11.** [4.5b] $-2x^3 + 4x^4 - 6x^5$
12. [4.5d] $8x^3 + 1$ **13.** [4.6b] $64 - \frac{1}{9}x^2$
14. [4.4c] $-y^3 - 2y^2 - 2y + 7$
15. [4.8b] $x^2 - x - 1 + \frac{-2}{2x - 1}$ **16.** [6.4a] $\frac{-5x - 28}{5(x - 5)}$
17. [6.5a] $\frac{4x - 1}{x - 2}$ **18.** [6.1d] $\frac{y}{(y - 1)^2}$ **19.** [6.2b] $\frac{3(x + 1)}{2x}$
20. [5.1b] $3x^2(2x^3 - 12x + 3)$
21. [5.5d] $(4y^2 + 9)(2y + 3)(2y - 3)$
22. [5.3a], [5.4a] $(3x - 2)(x + 4)$
23. [5.5b] $(2x^2 - 3y)^2$ **24.** [5.3a] $3m(m + 5)(m - 3)$
25. [5.7a] $(x + 1)^2(x - 1)$ **26.** [2.3c] -9 **27.** [5.8a] $0, \frac{5}{2}$
28. [2.7e] $\{x | x \le 20\}$ **29.** [2.3c] 0.3 **30.** [5.8b] $13, -13$
31. [5.8b] $\frac{5}{3}, 3$ **32.** [6.7a] -1 **33.** [6.7a] No solution
34. [2.3c] All real numbers   **35.** [2.7e] $\left\{ y | y \ge -\frac{5}{2} \right\}$
36. [2.3c] No solution   **37.** [2.4b] $x = \frac{N + t}{r}$ **38.** [2.6a] \$35
39. [6.8a] $6\frac{2}{3}$ hr   **40.** [5.9a] Hypotenuse: 13 in.; leg: 5 in.
41. [6.8b] 75 chips   **42.** [5.9a] Height: 9 ft; base: 4 ft
43. [6.9d] 72 ft: 360   **44.** [6.9a] $y = 0.2x$ **45.** [7.1b] 6; 9
46. [7.1b] 11; 3   **47.** [3.4a] 0   **48.** [7.3b] $-\frac{2}{3}$; $(0, 2)$
49. [7.5c] $y = -\frac{10}{7}x - \frac{8}{7}$ **50.** [7.5a] $y = 6x - 3$
51. [3.3b]   **52.** [3.3a]

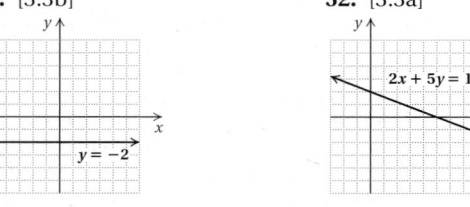

53. [7.1c]   **54.** [3.2a]

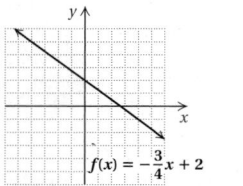

55. [3.3b]   **56.** [3.3a]

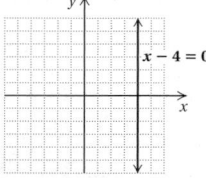

57. [2.7e] No solution   **58.** [1.8d], [6.2b], [6.5a] 0
59. [7.5d] $-\frac{3}{10}$

CHAPTER 8

Calculator Corner, p. 572

1. $(2, 3)$ **2.** $(-4, -1)$ **3.** $(-1, 5)$ **4.** $(3, -1)$

Exercise Set 8.1, p. 576

1. $(3, 1)$; consistent; independent   **3.** $(1, -2)$; consistent; independent   **5.** $(4, -2)$; consistent; independent   **7.** $(2, 1)$; consistent; independent   **9.** $\left(\frac{5}{2}, -2\right)$; consistent; independent
11. $(3, -2)$; consistent; independent   **13.** No solution; inconsistent; independent   **15.** Infinitely many solutions; consistent; dependent   **17.** $(4, -5)$; consistent; independent
19. $(2, -3)$; consistent; independent   **21.** Consistent; independent; F   **23.** Consistent; dependent; B
25. Inconsistent; independent; D   **27.** $y = \frac{3}{5}x + \frac{22}{5}$
28. $y = \frac{3}{8}x + \frac{9}{4}$ **29.** $y = -\frac{3}{2}x$ **30.** $y = \frac{4}{3}x - 14$
31. $(2.23, 1.14)$ **33.** $(3, 3), (-5, 5)$

Calculator Corner, p. 580

Left to the student

Exercise Set 8.2, p. 583

1. $(2, -3)$ **3.** $\left(\frac{21}{5}, \frac{12}{5}\right)$ **5.** $(2, -2)$ **7.** $(-2, -6)$ **9.** $(-2, 1)$
11. $\left(\frac{1}{2}, \frac{1}{2}\right)$ **13.** $\left(\frac{19}{8}, \frac{1}{8}\right)$ **15.** No solution   **17.** Length: 40 ft; width: 20 ft   **19.** 48° and 132°   **21.** Wins: 23; ties: 14
23. 1.3   **24.** $-15y - 39$ **25.** $p = \frac{7A}{q}$ **26.** $\frac{7}{3}$ **27.** -23
28. $\frac{29}{22}$ **29.** $m = -\frac{1}{2}$; $b = \frac{5}{2}$ **31.** Length: 57.6 in.; width: 20.4 in.

Exercise Set 8.3, p. 591

1. $(1, 2)$ **3.** $(-1, 3)$ **5.** $(-1, -2)$ **7.** $(5, 2)$ **9.** Infinitely many solutions   **11.** $\left(\frac{1}{2}, -\frac{1}{2}\right)$ **13.** $(4, 6)$ **15.** No solution
17. $(10, -8)$ **19.** $(12, 15)$ **21.** $(10, 8)$ **23.** $(-4, 6)$
25. $(10, -5)$ **27.** $(140, 60)$ **29.** 36 and 27   **31.** 18 and -15
33. 48° and 42°   **35.** Two-point shots: 21; free-throws: 6
37. Lanterns: 4; grills: 8   **39.** 1   **40.** 5   **41.** 3   **42.** 291
43. 15   **44.** $12a^2 - 2a + 1$ **45.** 53   **46.** 8.92
47. $\{x | x$ is a real number *and* $x \ne -7\}$ **48.** Domain: all real numbers; range: $\{y | y \le 5\}$ **49.** $y = -\frac{3}{5}x - 7$
50. $y = x + 12$ **51.** $(23.12, -12.04)$ **53.** $A = 2, B = 4$
55. $p = 2, q = -\frac{1}{3}$

Translating for Success, p. 603

1. G   **2.** E   **3.** D   **4.** A   **5.** J   **6.** B   **7.** C   **8.** I
9. F   **10.** H

Exercise Set 8.4, p. 604

1. 32 brushes at \$8.50; 13 brushes at \$9.75   **3.** Humulin: 21 vials; Novolin: 29 vials   **5.** 30-sec: 4; 60-sec: 8   **7.** 5 lb of each
9. 25% acid: 4 L; 50% acid: 6 L   **11.** 10 silk neckties
13. \$7500 at 6%; \$4500 at 9%   **15.** Whole milk: $169\frac{3}{13}$ lb; cream: $30\frac{10}{13}$ lb   **17.** \$5 bills: 7; \$1 bills: 15   **19.** \$7400 at 5.5%; \$10,600 at 4%   **21.** 375 mi   **23.** 14 km/h   **25.** 144 mi
27. 2 hr   **29.** $1\frac{1}{3}$ hr   **31.** About 1489 mi   **33.** -7 **34.** -11
35. -3 **36.** 33   **37.** -15 **38.** $8a - 7$ **39.** -23
40. 0.2   **41.** -4 **42.** -17 **43.** $-12h - 7$ **44.** 3993
45. $4\frac{4}{7}$ L   **47.** City: 261 mi; highway: 204 mi
49. Brown: 0.8 gal; neutral: 0.2 gal

Mid-Chapter Review: Chapter 8, p. 608

1. False   **2.** False   **3.** True   **4.** True
5. $x + 2(x - 6) = 3$
$x + 2x - 12 = 3$
$3x - 12 = 3$
$3x = 15$
$x = 5$

$$y = 5 - 6$$
$$y = -1$$

The solution is $(5, -1)$.

6.
$$6x - 4y = 10$$
$$\underline{2x + 4y = 14}$$
$$8x \qquad = 24$$
$$x = 3$$

$$2 \cdot 3 + 4y = 14$$
$$6 + 4y = 14$$
$$4y = 8$$
$$y = 2$$

The solution is $(3, 2)$.

7. $(5, -1)$, consistent; independent   **8.** $(0, 3)$; consistent; independent   **9.** Infinitely many solutions; consistent; dependent   **10.** No solution; inconsistent; independent
11. $(8, 6)$ **12.** $(2, -3)$ **13.** $(-3, 5)$ **14.** $(-1, -2)$
15. $(2, -2)$ **16.** $(5, -4)$ **17.** $(-1, -2)$ **18.** $(3, 1)$
19. No solution   **20.** Infinitely many solutions   **21.** $(10, -12)$
22. $(-9, 8)$ **23.** Length: 12 ft; width: 10 ft   **24.** $2100 at 2%; $2900 at 3%   **25.** 20% acid: 56 L; 50% acid: 28 L   **26.** 26 mph
27. *Graphically*: **1.** Graph $y = \frac{3}{4}x + 2$ and $y = \frac{2}{5}x - 5$ and find the point of intersection. The first coordinate of this point is the solution of the original equation. **2.** Rewrite the equation as $\frac{7}{20}x + 7 = 0$. Then graph $y = \frac{7}{20}x + 7$ and find the x-intercept. The first coordinate of this point is the solution of the original equation.
Algebraically: **1.** Use the addition and multiplication principles for equations. **2.** Multiply by 20 to clear the fractions and then use the addition and multiplication principles for equations.
28. (a) Answers may vary.
$$x + y = 1,$$
$$x - y = 7$$
(b) Answers may vary.
$$x + 2y = 5,$$
$$3x + 6y = 10$$
(c) Answers may vary.
$$x - 2y = 3,$$
$$3x - 6y = 9$$

29. Answers may vary. Form a linear expression in two variables and set it equal to two different constants. See Exercises 10 and 19 in this review for examples.   **30.** Answers may vary. Let any linear equation be one equation in the system. Multiply by a constant on both sides of that equation to get the second equation in the system. See Exercises 9 and 20 in this review for examples.

Exercise Set 8.5, p. 615

1. $(1, 2, -1)$ **3.** $(2, 0, 1)$ **5.** $(3, 1, 2)$ **7.** $(-3, -4, 2)$
9. $(2, 4, 1)$ **11.** $(-3, 0, 4)$ **13.** $(2, 2, 4)$ **15.** $\left(\frac{1}{2}, 4, -6\right)$
17. $(-2, 3, -1)$ **19.** $\left(\frac{1}{2}, \frac{1}{3}, \frac{1}{6}\right)$ **21.** $(3, -5, 8)$ **23.** $(15, 33, 9)$
25. $(4, 1, -2)$ **27.** $(17, 9, 79)$ **28.** $a = \dfrac{F}{3b}$
29. $a = \dfrac{Q - 4b}{4}$, or $\dfrac{Q}{4} - b$ **30.** $d = \dfrac{tc - 2F}{t}$, or $c - \dfrac{2F}{t}$
31. $c = \dfrac{2F + td}{t}$, or $\dfrac{2F}{t} + d$ **32.** $y = \dfrac{c - Ax}{B}$
33. $y = \dfrac{Ax - c}{B}$ **34.** Slope: $-\frac{2}{3}$; y-intercept: $\left(0, -\frac{5}{4}\right)$
35. Slope: -4; y-intercept: $(0, 5)$ **36.** Slope: $\frac{2}{5}$; y-intercept: $(0, -2)$ **37.** Slope: 1.09375; y-intercept: $(0, -3.125)$
39. $(1, -2, 4, -1)$

1. Reading: 502; math: 515; writing: 494   **3.** $32°, 96°, 52°$
5. $-7, 20, 42$ **7.** Automatic transmission: $865; power door locks: $520; air conditioning: $375   **9.** Small: 10; medium: 16; large: 8   **11.** First fund: $45,000; second fund: $10,000; third fund: $25,000   **13.** Dog: $200; cat: $81; bird: $9
15. Roast beef: 2; baked potato: 1; broccoli: 2   **17.** A: 1500 lenses; B: 1900 lenses; C: 2300 lenses   **19.** Par-3: 6 holes; par-4: 8 holes; par-5: 4 holes   **21.** Two-pointers: 32; three-pointers: 5; foul shots: 13   **23.** At most   **24.** At least   **25.** Linear
26. Negative   **27.** Consistent   **28.** Perpendicular
29. y-intercept   **30.** Horizontal   **31.** $180°$ **33.** 464

Summary and Review: Chapter 8, p. 625

Concept Reinforcement

1. False   **2.** True   **3.** True   **4.** False

Important Concepts

1. $(4, -1)$; consistent; independent   **2.** $(-1, 4)$ **3.** $(-2, 3)$
4. $8700 at 6%; $14,300 at 5%   **5.** $(3, -5, 1)$

Review Exercises

1. $(-2, 1)$; consistent; independent   **2.** Infinitely many solutions; consistent; dependent   **3.** No solution; inconsistent; independent   **4.** $(1, -1)$ **5.** No solution
6. $\left(\frac{2}{5}, -\frac{4}{5}\right)$ **7.** $(6, -3)$ **8.** $(2, 2)$ **9.** $(5, -3)$
10. Infinitely many solutions   **11.** 150 mph   **12.** CD: $18; DVD: $34   **13.** 5 L of each   **14.** $5\frac{1}{2}$ hr   **15.** $(10, 4, -8)$
16. $(-1, 3, -2)$ **17.** $(2, 0, 4)$ **18.** $\left(2, \frac{1}{3}, -\frac{2}{3}\right)$
19. $90°, 67\frac{1}{2}°, 22\frac{1}{2}°$ **20.** $20 bills: 5; $5 bills: 15; $1 bills: 19
21. C   **22.** C   **23.** A   **24.** $(0, 2)$ and $(1, 3)$

Understanding Through Discussion and Writing

1. Answers may vary. One day, a florist sold a total of 23 hanging baskets and flats of petunias. Hanging baskets cost $10.95 each and flats of petunias cost $12.95 each. The sales totaled $269.85. How many of each were sold?   **2.** We know that Eldon, Dana, and Casey can weld 74 linear feet per hour when working together. We also know that Eldon and Dana together can weld 44 linear feet per hour, which leads to the conclusion that Casey can weld $74 - 44$ or 30 linear feet per hour alone. We also know that Eldon and Casey together can weld 50 linear feet per hour. This, along with the earlier conclusion that Casey can weld 30 linear feet per hour alone, leads to two conclusions: Eldon can weld $50 - 30$, or 20 linear feet per hour alone, and Dana can weld $74 - 50$, or 24 linear feet per hour alone.   **3.** Let $x =$ the number of adults in the audience, $y =$ the number of senior citizens, and $z =$ the number of children. The total attendance is 100, so we have equation (1), $x + y + z = 100$. The amount taken in was $100, so equation (2) is $10x + 3y + 0.5z = 100$. There is no other information that can be translated to an equation. Clearing decimals in equation (2) and then eliminating z gives us equation (3), $95x + 25y = 500$. Dividing by 5 on both sides, we have equation (4), $19x + 5y = 100$. Since we have only two equations, it is not possible to eliminate z from another pair of equations. However, in $19x + 5y = 100$, note that 5 is a factor of both $5y$ and 100. Therefore, 5 must also be a factor of $19x$, and hence of x, since 5 is not a factor of 19. Then for some positive integer n, $x = 5n$. (We require n to be positive, since the number of adults clearly cannot be negative and must also be nonzero since the exercise states that the audience consists of adults, senior citizens, and children.) We have

$$19 \cdot 5n + 5y = 100$$
$$19n + y = 20. \qquad \text{Dividing by 5}$$

Since n and y must both be positive, $n = 1$. (If $n > 1$, then $19n + y > 20$.) Then $x = 5 \cdot 1$, or 5.

$$19 \cdot 5 + 5y = 100 \qquad \text{Substituting in (4)}$$
$$y = 1$$

$$5 + 1 + z = 100 \qquad \text{Substituting in (1)}$$
$$z = 94$$

There were 5 adults, 1 senior citizen, and 94 children in the audience.

Test: Chapter 8, p. 631

1. [8.1a] $(-2, 1)$; consistent; independent **2.** [8.1a] No solution; inconsistent; independent **3.** [8.1a] Infinitely many solutions; consistent; dependent **4.** [8.2a] $(2, -3)$ **5.** [8.2a] Infinitely many solutions **6.** [8.2a] $(-4, 5)$ **7.** [8.2a] $(8, -6)$ **8.** [8.3a] $(-1, 1)$ **9.** [8.3a] $\left(-\frac{3}{2}, -\frac{1}{2}\right)$ **10.** [8.3a] No solution **11.** [8.3a] $(10, -20)$ **12.** [8.2b] Length: 93 ft; width: 51 ft **13.** [8.4b] 120 km/h **14.** [8.3b], [8.4a] Buckets: 17; dinners: 11 **15.** [8.4a] 20% solution: 12 L; 45% solution: 8 L **16.** [8.5a] $\left(2, -\frac{1}{2}, -1\right)$ **17.** [8.6a] 3.5 hr **18.** [8.6a] B **19.** [8.3a] $m = 7$; $b = 10$

Cumulative Review: Chapters 1-8; p. 633

1. [4.7f] $9x^8 - 4y^{10}$ **2.** [4.6c] $x^4 + 8x^2 + 16$
3. [4.6a] $8x^2 - \frac{1}{8}$ **4.** [6.5a] $\frac{2(2x + 1)}{2x - 1}$
5. [4.4 a, c] $-3x^3 + 8x^2 - 5x$ **6.** [6.1d] $\frac{2(x - 5)}{3(x - 1)}$
7. [6.2b] $\frac{(x + 1)(x - 3)}{2(x + 3)}$ **8.** [4.8b] $3x^2 + 4x + 9 + \frac{13}{x - 2}$
9. [5.5d] $3(1 + 2x^4)(1 - 2x^4)$ **10.** [5.1b] $4t(3 - t - 12t^3)$
11. [5.3a], [5.4a] $2(3x - 2)(x - 4)$
12. [5.1c], [5.7a] $(2x + 1)(2x - 1)(x + 1)$
13. [5.5b] $(4x^2 - 7)^2$ **14.** [5.2a] $(x + 15)(x - 12)$
15. [7.3b] Slope: $\frac{4}{5}$; y-intercept: $(0, 4)$
16. [7.5b] $y = -3x + 17$ **17.** [7.5d] $y = \frac{1}{3}x + 4$
18. [7.4d] Perpendicular **19.** [5.8b] $-17, 0$ **20.** [2.3b] $\frac{2}{5}$
21. [6.7a] $-\frac{12}{5}$ **22.** [5.8b] $-5, 6$ **23.** [2.7e] $\left\{x | x \le -\frac{9}{2}\right\}$
24. [6.7a] $\frac{1}{3}$ **25.** [2.4b] $p = \frac{4A}{r + q}$ **26.** [8.2a] $\left(\frac{8}{5}, -\frac{1}{5}\right)$
27. [8.3a] $(1, -1)$ **28.** [8.3a] $(-1, 3)$ **29.** [8.5a] $(2, 0, -1)$
30. [7.4c]
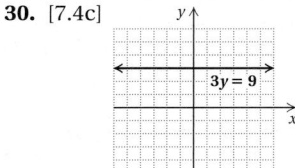
31. [7.1c]
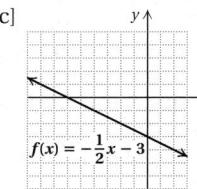
32. [7.4b]
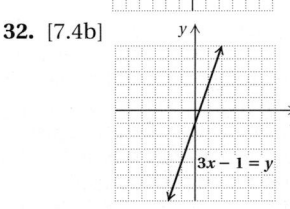
33. [7.4a]
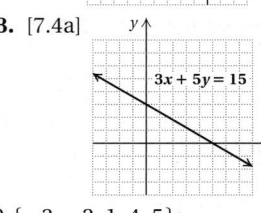
34. [7.2a] **(a)** $\{-5, -3, -1, 1, 3\}$; **(b)** $\{-3, -2, 1, 4, 5\}$;
(c) -2; **(d)** 3 **35.** [7.2a] $\left\{x | x \text{ is a real number } and \ x \ne \frac{1}{2}\right\}$
36. [7.1b] -1; 1; -17 **37.** [8.4a] 15%: 21 L; 25%: 9 L

38. [8.6a] $120 **39.** [6.8b] 20 defective resistors
40. [5.9a] Length: 15 m; width: 12 m **41.** [6.9d] 0.4 ft
42. [2.6a] 38°, 76°, 66° **43.** [7.5e] $151,000
44. [8.3a] $m = -\frac{5}{9}$; $b = -\frac{2}{9}$

CHAPTER 9

Translating for Success, p. 645

1. F **2.** I **3.** C **4.** E **5.** D **6.** J **7.** O **8.** M
9. B **10.** L

Exercise Set 9.1, p. 646

1. No, no, no, yes **3.** No, yes, yes, no, no **5.** $(-\infty, 5)$
7. $[-3, 3]$ **9.** $(-8, -4)$ **11.** $(-2, 5)$ **13.** $(-\sqrt{2}, \infty)$
15. $\{x | x > -1\}$, or $(-1, \infty)$ **17.** $\{y | y < 6\}$, or $(-\infty, 6)$

19. $\{a | a \le -22\}$, or $(-\infty, -22]$

21. $\{t | t \ge -4\}$, or $[-4, \infty)$

23. $\{y | y > -6\}$, or $(-6, \infty)$ **25.** $\{x | x \le 9\}$, or $(-\infty, 9]$

27. $\{x | x \ge 3\}$, or $[3, \infty)$

29. $\{x | x < -60\}$, or $(-\infty, -60)$ **31.** $\{x | x > 3\}$, or $(3, \infty)$

33. $\{x | x \le 0.9\}$, or $(-\infty, 0.9]$ **35.** $\left\{x | x \le \frac{5}{6}\right\}$, or $\left(-\infty, \frac{5}{6}\right]$
37. $\{x | x < 6\}$, or $(-\infty, 6)$ **39.** $\{y | y \le -3\}$, or $(-\infty, -3]$
41. $\left\{y | y > \frac{2}{3}\right\}$, or $\left(\frac{2}{3}, \infty\right)$ **43.** $\{x | x \ge 11.25\}$, or $[11.25, \infty)$
45. $\left\{x | x \le \frac{1}{2}\right\}$, or $\left(-\infty, \frac{1}{2}\right]$ **47.** $\left\{y | y \le -\frac{75}{2}\right\}$, or $\left(-\infty, -\frac{75}{2}\right]$
49. $\left\{x | x > -\frac{2}{17}\right\}$, or $\left(-\frac{2}{17}, \infty\right)$ **51.** $\left\{m | m > \frac{7}{3}\right\}$, or $\left(\frac{7}{3}, \infty\right)$
53. $\{r | r < -3\}$, or $(-\infty, -3)$ **55.** $\{x | x \ge 2\}$, or $[2, \infty)$
57. $\{y | y < 5\}$, or $(-\infty, 5)$ **59.** $\left\{x | x \le \frac{4}{7}\right\}$, or $\left(-\infty, \frac{4}{7}\right]$
61. $\{x | x < 8\}$, or $(-\infty, 8)$ **63.** $\left\{x | x \ge \frac{13}{2}\right\}$, or $\left[\frac{13}{2}, \infty\right)$
65. $\left\{x | x < \frac{11}{18}\right\}$, or $\left(-\infty, \frac{11}{18}\right)$ **67.** $\left\{x | x \ge -\frac{51}{31}\right\}$, or $\left[-\frac{51}{31}, \infty\right)$
69. $\{a | a \le 2\}$, or $(-\infty, 2]$ **71.** $\{W | W < \text{(approximately)}$ 189.5 lb$\}$ **73.** $\{S | S \ge 84\}$ **75.** $\{B | B \ge \$11,500\}$
77. $\{S | S > \$7000\}$ **79.** $\{n | n > 25\}$ **81.** $\{p | p > 80\}$
83. $\{s | s > 8\}$ **85.** **(a)** 8.2 million gal; 9.4 million gal; 10.6 million gal; **(b)** years after 2012
87. $3x^2 + 20x - 32$ **88.** $6r^2 - 23rs - 4s^2$
89. $6a^2 + 7a - 55$ **90.** $t^2 - 7st - 18s^2$
91. $(2x - 9)^2$ **92.** $16(5y + 1)(5y - 1)$
93. $(3w - 2)(9w^2 + 6w + 4)$ **94.** $2(8 - 3x)(5 + x)$
95. $\{x | x \text{ is a real number } and \ x \ne -8\}$ **96.** All real numbers
97. All real numbers **98.** $\left\{x | x \text{ is a real number } and \ x \ne \frac{2}{3}\right\}$
99. **(a)** $\{p | p > 10\}$; **(b)** $\{p | p < 10\}$ **101.** True
103. All real numbers **105.** All real numbers

Exercise Set 9.2, p. 660

1. $\{9, 11\}$ **3.** $\{b\}$ **5.** $\{9, 10, 11, 13\}$ **7.** $\{a, b, c, d, f, g\}$
9. $\varnothing$ **11.** $\{3, 5, 7\}$
13. $(-4, 1]$
15. $(1, 6)$
17. $\{x | -4 \le x < 5\}$, or $[-4, 5)$;

19. $\{x|x \geq 2\}$, or $[2, \infty)$;　**21.** $\varnothing$

<————————[—————————→
　　　　　0　2

23. $\{x|-8 < x < 6\}$, or $(-8, 6)$

25. $\{x|-6 < x \leq 2\}$, or $(-6, 2]$

27. $\{x|-1 < x \leq 6\}$, or $(-1, 6]$

29. $\{y|-1 < y \leq 5\}$, or $(-1, 5]$

31. $\{x|-\frac{5}{3} \leq x \leq \frac{4}{3}\}$, or $\left[-\frac{5}{3}, \frac{4}{3}\right]$

33. $\{x|-\frac{7}{2} < x \leq \frac{11}{2}\}$, or $\left(-\frac{7}{2}, \frac{11}{2}\right]$

35. $\{x|10 < x \leq 14\}$, or $(10, 14]$

37. $\{x|-\frac{13}{3} \leq x \leq 9\}$, or $\left[-\frac{13}{3}, 9\right]$

39. <———————)———(———————→; $(-\infty, -2) \cup (1, \infty)$
　　　　　　-2　0　1

41. <———————]———(———————→; $(-\infty, -3] \cup (1, \infty)$
　　　　　　-3　　0　1

43. $\{x|x < -5 \; or \; x > -1\}$, or $(-\infty, -5) \cup (-1, \infty)$;

<———————)———(———————→
　　　-5　　-1　0

45. $\{x|x \leq \frac{5}{2} \; or \; x \geq 4\}$, or $\left(-\infty, \frac{5}{2}\right] \cup [4, \infty)$;

　　　　　　　$\frac{5}{2}$
<———————]——[———————→
　　　0　　　　4

47. $\{x|x \geq -3\}$, or $[-3, \infty)$;

<———————[—————————→
　　　-3　　0

49. $\{x|x \leq -\frac{5}{4} \; or \; x > -\frac{1}{2}\}$, or $\left(-\infty, -\frac{5}{4}\right] \cup \left(-\frac{1}{2}, \infty\right)$
51. All real numbers, or $(-\infty, \infty)$
53. $\{x|x < -4 \; or \; x > 2\}$, or $(-\infty, -4) \cup (2, \infty)$
55. $\{x|x < \frac{79}{4} \; or \; x > \frac{89}{4}\}$, or $\left(-\infty, \frac{79}{4}\right) \cup \left(\frac{89}{4}, \infty\right)$
57. $\{x|x \leq -\frac{13}{2} \; or \; x \geq \frac{29}{2}\}$, or $\left(-\infty, -\frac{13}{2}\right] \cup \left[\frac{29}{2}, \infty\right)$
59. $\{d|0 \text{ ft} \leq d \leq 198 \text{ ft}\}$　　**61.** Between 23 beats and 27 beats
63. $\{W|140.2 \text{ lb} \leq W \leq 188.8 \text{ lb}\}$
65. $\{d|250 \text{ mg} < d < 500 \text{ mg}\}$　　**67.** $(-1, 2)$　　**68.** $(-3, -5)$
69. $(4, -4)$　　**70.** $y = -11x + 29$　　**71.** $y = -4x + 7$
72. $y = -x + 2$　　**73.** $6a^2 + 7ab - 5b^2$　　**74.** $25y^2 + 35y + 6$
75. $21x^2 - 59x + 40$　　**76.** $13x^2 + 37xy - 6y^2$
77. $\{x|-4 < x \leq 1\}$, or $(-4, 1]$　　**79.** $\{x|\frac{2}{5} \leq x \leq \frac{4}{5}\}$,
or $\left[\frac{2}{5}, \frac{4}{5}\right]$　　**81.** $\{x|-\frac{1}{8} < x < \frac{1}{2}\}$, or $\left(-\frac{1}{8}, \frac{1}{2}\right)$
83. $\{x|10 < x \leq 18\}$, or $(10, 18]$　　**85.** True　　**87.** False
89. All real numbers; $\varnothing$

Mid-Chapter Review: Chapter 9, p. 664

1. True　　**2.** False　　**3.** True　　**4.** True
5. $8 - 5x \leq x + 20$　　　　　　　**6.** $-17 < 3 - x < 36$
　　　$-5x \leq x + 12$　　　　　　　　　　$-20 < -x < 33$
　　　$-6x \leq 12$　　　　　　　　　　　$20 > x > -33$
　　　$x \geq -2$
7. G　**8.** B　**9.** H　**10.** E　**11.** F　**12.** A　**13.** $\{0, 10\}$
14. $\{b, d, e, f, g, h\}$　　**15.** $\left\{\frac{1}{4}, \frac{3}{8}\right\}$　　**16.** $\varnothing$
17. $\{y|y \leq -2\}$; $(-\infty, -2]$　　**18.** $\{x|x \leq 4\}$; $(-\infty, 4]$
19. $\{x|x < -9 \; or \; x > 1\}$; $(-\infty, -9) \cup (1, \infty)$
20. $\{x|3 \leq x < 24\}$; $[3, 24)$　　**21.** $\{x|x < -10 \; or \; x > 1\}$;
$(-\infty, -10) \cup (1, \infty)$　　**22.** $\{t|t > 2\}$; $(2, \infty)$
23. $\{x|x \geq -3\}$; $[-3, \infty)$　　**24.** $\left\{y|y > \frac{10}{11}\right\}$; $\left(\frac{10}{11}, \infty\right)$
25. $\{x|-\frac{17}{2} < x < \frac{25}{2}\}$; $\left(-\frac{17}{2}, \frac{25}{2}\right)$　　**26.** $\{x|x \geq \frac{35}{6}\}$; $\left[\frac{35}{6}, \infty\right)$
27. $\{x|4 < x < 17\}$; $(4, 17)$　　**28.** $\{x|x \text{ is a real number}\}$;
$(-\infty, \infty)$　　**29.** $\{S|S \geq 86\}$　　**30.** $3000
31. When the signs of the quantities on either side of the inequality symbol are changed, their relative positions on the number line are reversed.　　**32.** **(1)** $-9(x + 2) = -9x - 18$, not $-9x + 2$. **(2)** This would be correct if (1) were correct except that the inequality symbol should not have been reversed.

(3) If (2) were correct, the right-hand side would be -5, not 8.
(4) The inequality symbol should be reversed. The correct solution is

$$7 - 9x + 6x < -9(x + 2) + 10x$$
$$7 - 9x + 6x < -9x - 18 + 10x$$
$$7 - 3x < x - 18$$
$$-4x < -25$$
$$x > \frac{25}{4}.$$

33. By definition, the notation $3 < x < 5$ indicates that $3 < x$ and $x < 5$. A solution of the disjunction $3 < x \; or \; x < 5$ must be in at least one of these sets but not necessarily in both, so the disjunction cannot be written as $3 < x < 5$.

Exercise Set 9.3, p. 673

1. $9|x|$　　**3.** $2x^2$　　**5.** $2x^2$　　**7.** $6|y|$　　**9.** $\frac{2}{|x|}$　　**11.** $\frac{x^2}{|y|}$

13. $4|x|$　　**15.** $\frac{y^2}{3}$　　**17.** 38　　**19.** 19　　**21.** 6.3　　**23.** 5

25. $\{-3, 3\}$　　**27.** $\varnothing$　　**29.** $\{0\}$　　**31.** $\{-9, 15\}$　　**33.** $\left\{-\frac{1}{2}, \frac{7}{2}\right\}$
35. $\left\{-\frac{5}{4}, \frac{23}{4}\right\}$　　**37.** $\{-11, 11\}$　　**39.** $\{-291, 291\}$
41. $\{-8, 8\}$　　**43.** $\{-7, 7\}$　　**45.** $\{-2, 2\}$　　**47.** $\{-7, 8\}$
49. $\{-12, 2\}$　　**51.** $\left\{-\frac{5}{2}, \frac{7}{2}\right\}$　　**53.** $\varnothing$　　**55.** $\left\{-\frac{13}{54}, -\frac{7}{54}\right\}$
57. $\left\{-\frac{11}{2}, \frac{3}{4}\right\}$　　**59.** $\left\{\frac{3}{2}\right\}$　　**61.** $\left\{5, -\frac{3}{5}\right\}$　　**63.** All real numbers
65. $\left\{-\frac{3}{2}\right\}$　　**67.** $\left\{\frac{24}{23}, 0\right\}$　　**69.** $\left\{32, \frac{8}{3}\right\}$　　**71.** $\{x|-3 < x < 3\}$,
or $(-3, 3)$　　**73.** $\{x|x \leq -2 \; or \; x \geq 2\}$, or $(-\infty, -2] \cup [2, \infty)$
75. $\{x|0 < x < 2\}$, or $(0, 2)$　　**77.** $\{x|-6 \leq x \leq -2\}$, or
$[-6, -2]$　　**79.** $\{x|-\frac{1}{2} \leq x \leq \frac{7}{2}\}$, or $\left[-\frac{1}{2}, \frac{7}{2}\right]$　　**81.** $\{y|y < -\frac{3}{2} \; or$
$y > \frac{17}{2}\}$, or $\left(-\infty, -\frac{3}{2}\right) \cup \left(\frac{17}{2}, \infty\right)$　　**83.** $\{x|x \leq -\frac{5}{4} \; or \; x \geq \frac{23}{4}\}$, or
$\left(-\infty, -\frac{5}{4}\right] \cup \left[\frac{23}{4}, \infty\right)$　　**85.** $\{y|-9 < y < 15\}$, or $(-9, 15)$
87. $\{x|-\frac{7}{2} \leq x \leq \frac{1}{2}\}$, or $\left[-\frac{7}{2}, \frac{1}{2}\right]$　　**89.** $\{y|y < -\frac{4}{3} \; or \; y > 4\}$,
$\left(-\infty, -\frac{4}{3}\right) \cup (4, \infty)$　　**91.** $\{x|x \leq -\frac{5}{4} \; or \; x \geq \frac{23}{4}\}$, or
$\left(-\infty, -\frac{5}{4}\right] \cup \left[\frac{23}{4}, \infty\right)$　　**93.** $\{x|-\frac{9}{2} < x < 6\}$, or $\left(-\frac{9}{2}, 6\right)$
95. $\{x|x \leq -\frac{25}{6} \; or \; x \geq \frac{23}{6}\}$, or $\left(-\infty, -\frac{25}{6}\right] \cup \left[\frac{23}{6}, \infty\right)$
97. $\{x|-5 < x < 19\}$, or $(-5, 19)$
99. $\{x|x \leq -\frac{2}{15} \; or \; x \geq \frac{14}{15}\}$, or $\left(-\infty, -\frac{2}{15}\right] \cup \left[\frac{14}{15}, \infty\right)$
101. $\{m|-12 \leq m \leq 2\}$, or $[-12, 2]$　　**103.** $\{x|\frac{1}{2} \leq x \leq \frac{5}{2}\}$, or
$\left[\frac{1}{2}, \frac{5}{2}\right]$　　**105.** $\{x|-1 \leq x \leq 2\}$, or $[-1, 2]$　　**107.** Union
108. Disjoint　　**109.** At least　　**110.** $[a, b]$　　**111.** Absolute
value　**112.** Equation　　**113.** Equivalent　　**114.** Inequality
115. $\left\{d|5\frac{1}{2} \text{ ft} \leq d \leq 6\frac{1}{2} \text{ ft}\right\}$　　**117.** $\{x|x \geq -5\}$, or $[-5, \infty)$
119. $\left\{1, -\frac{1}{4}\right\}$　　**121.** $\varnothing$　　**123.** All real numbers
125. $|x| < 3$　　**127.** $|x| \geq 6$　　**129.** $|x + 3| > 5$

Visualizing for Success, p. 686

1. D　**2.** B　**3.** E　**4.** C　**5.** I　**6.** G　**7.** F　**8.** H
9. A　**10.** J

Exercise Set 9.4, p. 687

1. Yes　**3.** Yes
5.　　　　　　　　**7.**　　　　　　　　**9.**

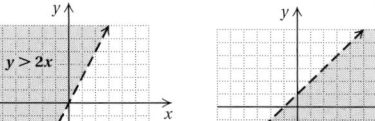

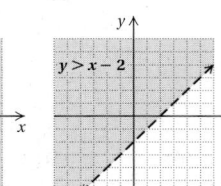

11.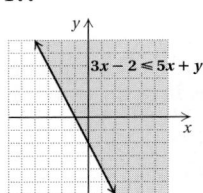
$x + y < 4$

13.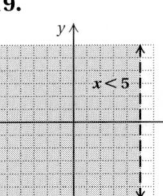
$3x + 4y \le 12$

15.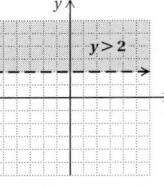
$2y - 3x > 6$

17.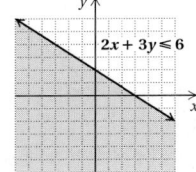
$3x - 2 \le 5x + y$

19.
$x < 5$

21.
$y > 2$

23.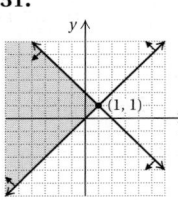
$2x + 3y \le 6$

25. F   **27.** B   **29.** C

31.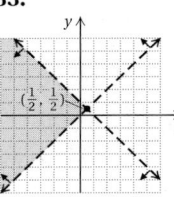
$(1, 1)$

33.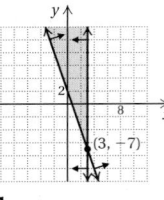
$\left(\frac{1}{2}, \frac{1}{2}\right)$

35.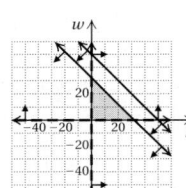
$(3, -7)$

37.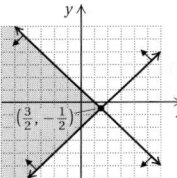
$\left(\frac{3}{2}, -\frac{1}{2}\right)$

39.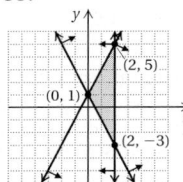
$(2, 5)$, $(0, 1)$, $(2, -3)$

41.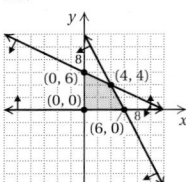
$(0, 6)$, $(4, 4)$, $(0, 0)$, $(6, 0)$

43. $\frac{10}{17}$ **44.** $-\frac{14}{13}$ **45.** -2 **46.** $\frac{29}{11}$ **47.** -12 **48.** $\frac{333}{245}$
49. 2   **50.** 3   **51.** 1   **52.** 8   **53.** 4   **54.** $|2 - 2a|$, or
$2|1 - a|$ **55.** 6   **56.** 0.2
57. $w > 0$,
   $h > 0$,
   $w + h + 30 \le 62$, or
   $w + h \ge 32$,
   $2w + 2h + 30 \le 130$, or
   $w + h \le 50$

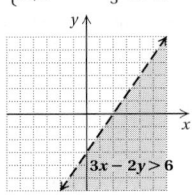

Summary and Review: Chapter 9, p. 691

Concept Reinforcement
1. True   **2.** False   **3.** False   **4.** False   **5.** True   **6.** False
7. True

Important Concepts
1. -2 is not a solution; 5 is a solution   **2. (a)** $(-\infty, -8)$;
(b) $[-7, 10)$; **(c)** $[3, \infty)$ **3.** $\{y|y < -2\}$, or $(-\infty, -2)$;

4. $\{z|-2 \le z < 1\}$, or $[-2, 1)$;

5. $\{z|z < -1 \text{ or } z \ge 1\}$, or $(-\infty, -1) \cup [1, \infty)$;

6. $\left\{-\frac{8}{5}, 2\right\}$ **7.** $\left\{3, -\frac{1}{2}\right\}$ **8. (a)** $\{x|-4 < x < 1\}$, or $(-4, 1)$;
(b) $\left\{x|x \le -\frac{10}{3} \text{ or } x \ge 2\right\}$, or $\left(-\infty, -\frac{10}{3}\right] \cup [2, \infty)$

9.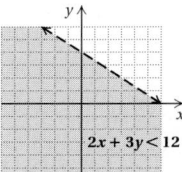
$3x - 2y > 6$

10.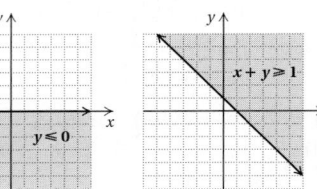
$(1, 3)$, $\left(1, -\frac{3}{2}\right)$, $(4, 0)$

Review Exercises
1. -3 is not a solution; 7 is a solution   **2.** $[-8, 9)$
3. $(-\infty, 40]$ **4.** ; $(-\infty, -2]$
5. ; $(1, \infty)$
6. $\{a|a \le -21\}$, or $(-\infty, -21]$ **7.** $\{y|y \ge -7\}$, or $[-7, \infty)$
8. $\{y|y > -4\}$, or $(-4, \infty)$ **9.** $\{y|y > -30\}$, or $(-30, \infty)$
10. $\{x|x > -3\}$, or $(-3, \infty)$ **11.** $\left\{y|y \le -\frac{6}{5}\right\}$, or $\left(-\infty, -\frac{6}{5}\right]$
12. $\{x|x < -3\}$, or $(-\infty, -3)$ **13.** $\{y|y > -10\}$, or $(-10, \infty)$
14. $\left\{x|x \le -\frac{5}{2}\right\}$, or $\left(-\infty, -\frac{5}{2}\right]$ **15.** $\left\{t|t > 4\frac{1}{4} \text{ hr}\right\}$
16. \$10,000   **17.** ; $[-2, 5)$
18. ; $(-\infty, -2] \cup (5, \infty)$
19. $\{1, 5, 9\}$ **20.** $\{1, 2, 3, 5, 6, 9\}$ **21.** $\varnothing$
22. $\{x|-7 < x \le 2\}$, or $(-7, 2]$ **23.** $\left\{x|-\frac{5}{4} < x < \frac{5}{2}\right\}$, or $\left(-\frac{5}{4}, \frac{5}{2}\right)$
24. $\{x|x < -3 \text{ or } x > 1\}$, or $(-\infty, -3) \cup (1, \infty)$
25. $\{x|x < -11 \text{ or } x \ge -6\}$, or $(-\infty, -11) \cup [-6, \infty)$
26. $\{x|x \le -6 \text{ or } x \ge 8\}$, or $(-\infty, -6] \cup [8, \infty)$
27. $\frac{3}{|x|}$ **28.** $\frac{2|x|}{y^2}$ **29.** $\frac{4}{|y|}$
30. 62   **31.** $\{-6, 6\}$ **32.** $\{-5, 9\}$ **33.** $\left\{-14, \frac{4}{3}\right\}$
34. $\varnothing$ **35.** $\left\{x|-\frac{17}{2} < x < \frac{7}{2}\right\}$, or $\left(-\frac{17}{2}, \frac{7}{2}\right)$
36. $\{x|x \le -3.5 \text{ or } x \ge 3.5\}$, or $(-\infty, -3.5] \cup [3.5, \infty)$
37. $\left\{x|x \le -\frac{11}{3} \text{ or } x \ge \frac{19}{3}\right\}$, or $\left(-\infty, -\frac{11}{3}\right] \cup \left[\frac{19}{3}, \infty\right)$ **38.** $\varnothing$
39.   **40.**   **41.**
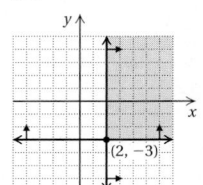   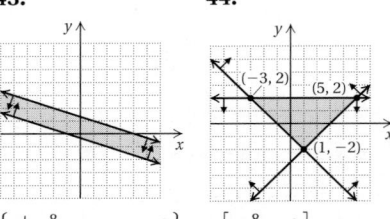
$2x + 3y < 12$ $y \le 0$ $x + y \ge 1$

42.   **43.**   **44.**
$(2, -3)$ $(-3, 2)$, $(5, 2)$, $(1, -2)$

45. D   **46.** C   **47.** $\left\{x|-\frac{8}{3} \le x \le -2\right\}$, or $\left[-\frac{8}{3}, -2\right]$

Understanding Through Discussion and Writing
1. When $b \ge c$, then the intervals overlap and
$[a, b] \cup [c, d] = [a, d]$.   **2.** The distance between x and -5 is
$|x - (-5)|$, or $|x + 5|$. Then the solutions of the inequality
$|x + 5| \le 2$ can be interpreted as "all those numbers x
whose distance from -5 is at most 2 units."   **3.** No; the
symbol $\ge$ does not always yield a graph in which the half-plane
above the line is shaded. For the inequality $-y \ge 3$, for example,
the half-plane below the line $y = -3$ is shaded.

4. The solutions of $|x| \geq 6$ are those numbers whose distance from 0 is greater than or equal to 6. In addition to the numbers in $[6, \infty)$, the distance of the numbers in $(-\infty, -6]$ from 0 is also greater than or equal to 6. Thus, $[6, \infty)$ is only part of the solution of the inequality.

Test: Chapter 9, p. 697

1. [9.1b] $(-3, 2]$ **2.** [9.1b] $(-4, \infty)$
3. [9.1c] ; $(-\infty, 6]$
4. [9.1c] ; $(-\infty, -2]$
5. [9.1c] $\{x \mid x \geq 10\}$, or $[10, \infty)$ **6.** [9.1c] $\{y \mid y > -50\}$, or $(-50, \infty)$ **7.** [9.1c] $\{a \mid a \leq \frac{11}{5}\}$, or $(-\infty, \frac{11}{5}]$
8. [9.1c] $\{y \mid y > 1\}$, or $(1, \infty)$ **9.** [9.1c] $\{x \mid x > \frac{5}{2}\}$, or $(\frac{5}{2}, \infty)$
10. [9.1c] $\{x \mid x \leq \frac{7}{4}\}$, or $(-\infty, \frac{7}{4}]$ **11.** [9.1d] $\{h \mid h > 2\frac{1}{10} \text{ hr}\}$
12. [9.2c] $\{d \mid 33 \text{ ft} \leq d \leq 231 \text{ ft}\}$
13. [9.2a] 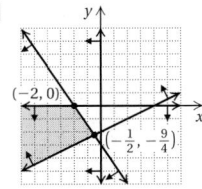 ; $[-3, 4]$
14. [9.2b] ; $(-\infty, -3) \cup (4, \infty)$
15. [9.2a] $\{x \mid x \geq 4\}$, or $[4, \infty)$ **16.** [9.2a] $\{x \mid -1 < x < 6\}$, or $(-1, 6)$ **17.** [9.2a] $\{x \mid -\frac{2}{5} < x \leq \frac{9}{5}\}$, or $(-\frac{2}{5}, \frac{9}{5}]$
18. [9.2b] $\{x \mid x < -4 \text{ or } x > -\frac{5}{2}\}$, or $(-\infty, -4) \cup (-\frac{5}{2}, \infty)$
19. [9.2b] All real numbers, or $(-\infty, \infty)$
20. [9.2b] $\{x \mid x < 3 \text{ or } x > 6\}$, or $(-\infty, 3) \cup (6, \infty)$
21. [9.3a] $\frac{7}{|x|}$ **22.** [9.3a] $2|x|$ **23.** [9.3b] 8.4
24. [9.2a] $\{3, 5\}$ **25.** [9.2b] $\{1, 3, 5, 7, 9, 11, 13\}$
26. [9.3c] $\{-9, 9\}$ **27.** [9.3c] $\{-6, 12\}$ **28.** [9.3d] $\{1\}$
29. [9.3c] $\varnothing$ **30.** [9.3e] $\{x \mid -0.875 < x < 1.375\}$, or $(-0.875, 1.375)$ **31.** [9.3e] $\{x \mid x < -3 \text{ or } x > 3\}$, or $(-\infty, -3) \cup (3, \infty)$ **32.** [9.3e] $\{x \mid -99 \leq x \leq 111\}$, or $[-99, 111]$ **33.** [9.3e] $\{x \mid x \leq -\frac{13}{5} \text{ or } x \geq \frac{7}{5}\}$, or $(-\infty, -\frac{13}{5}] \cup [\frac{7}{5}, \infty)$

34. [9.4b]

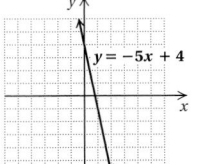

35. [9.4c]   **36.** [9.4c]

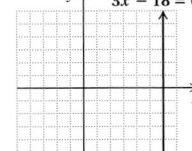

37. [9.3e] B   **38.** [9.3e] $\varnothing$ **39.** [9.2a] $\{x \mid \frac{1}{5} < x < \frac{4}{5}\}$, or $(\frac{1}{5}, \frac{4}{5})$

Cumulative Review: Chapters 1–9, p. 699

1. [3.2a]   **2.** [3.3b]

3. [9.4b]

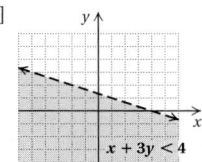

4. [9.4c]

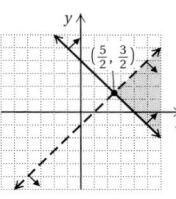

5. [7.1b] 11   **6.** [7.2a] $\{x \mid x$ is a real number and $x \neq -5$ and $x \neq 5\}$ **7.** [7.2a] Domain: $\{x \mid -5 < x < 5\}$; range: $\{y \mid -2 \leq y \leq 4\}$ **8.** [4.6c] $36m^2 - 12mn + n^2$
9. [4.6a] $15a^2 - 14ab - 8b^2$ **10.** [6.1d] $\frac{y - 2}{3}$
11. [6.2b] $\frac{3x - 5}{x + 4}$ **12.** [6.5a] $\frac{6x + 13}{20(x - 3)}$ **13.** [6.6a] $\frac{y^3 - 2y}{y^3 - 1}$
14. [4.4c] $16p^2 - 8p$ **15.** [6.5b] $\frac{4x + 1}{(x + 2)(x - 2)}$
16. [4.8b] $2x^2 - 11x + 23 + \frac{-49}{x + 2}$ **17.** [2.3c] $\frac{15}{2}$
18. [9.2a] $\{x \mid -3 < x < -\frac{3}{2}\}$, or $(-3, -\frac{3}{2})$
19. [6.7a] No solution   **20.** [2.4b] $a = \frac{bP}{3 - P}$
21. [2.4b] $C = \frac{5}{9}(F - 32)$ **22.** [9.3e] $\{x \mid x \leq -2.1 \text{ or } x \geq 2.1\}$, or $(-\infty, -2.1] \cup [2.1, \infty)$ **23.** [6.7a] -1 **24.** [5.8b] $\frac{1}{4}$
25. [5.8b] $-2, \frac{7}{2}$ **26.** [8.2a], [8.3a] Infinite number of solutions
27. [8.2a], [8.3a] $(-2, 1)$ **28.** [8.5a] $(3, 2, -1)$
29. [8.5a] $(\frac{5}{8}, \frac{1}{16}, -\frac{3}{4})$ **30.** [5.1b] $2x^2(2x + 9)$
31. [5.1c] $(2a - 1)(4a^2 - 3)$ **32.** [5.2a] $(x - 6)(x + 14)$
33. [5.3a], [5.4a] $(2x + 5)(3x - 2)$
34. [5.5d] $(4y + 9)(4y - 9)$ **35.** [5.5b] $(t - 8)^2$
36. [5.6a] $8(2x + 1)(4x^2 - 2x + 1)$
37. [5.6a] $(0.3b - 0.2c)(0.09b^2 + 0.06bc + 0.04c^2)$
38. [5.7a] $x^2(x^2 + 1)(x + 1)(x - 1)$
39. [5.3a], [5.4a] $(4x - 1)(5x + 3)$ **40.** [7.5b] $y = -\frac{1}{2}x - 1$
41. [7.5d] $y = \frac{1}{2}x - \frac{5}{2}$ **42.** [8.6a] Win: 38 games; lose: 30 games; tie: 13 games   **43.** [6.9b] About 202.3 lb
44. [6.7a] A   **45.** [5.8b] C   **46.** [6.8a] B
47. [8.6a] $a = 1, b = -5, c = 6$ **48.** [5.8b] $0, \frac{1}{4}, -\frac{1}{4}$
49. [6.7a] All real numbers except 9 and -5

CHAPTER 10

Exercise Set 10.1, p. 709

1. $4, -4$ **3.** $12, -12$ **5.** $20, -20$ **7.** $-\frac{7}{6}$ **9.** 14
11. 0.06   **13.** Does not exist as a real number   **15.** 18.628
17. 1.962   **19.** $y^2 + 16$ **21.** $\frac{x}{y - 1}$ **23.** $\sqrt{20} \approx 4.472$; 0; does not exist as a real number; does not exist as a real number
25. $\sqrt{11} \approx 3.317$; does not exist as a real number; $\sqrt{11} \approx 3.317$; 12   **27.** Domain $= \{x \mid x \geq 2\} = [2, \infty)$
29. 21 spaces; 25 spaces
31.   **33.**   **35.**

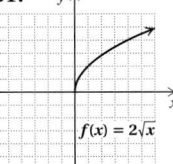
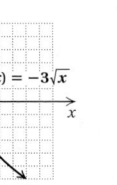
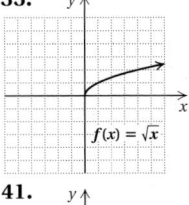

37.   **39.**   **41.**

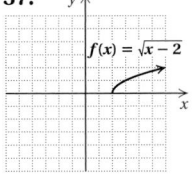
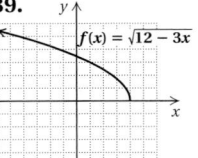
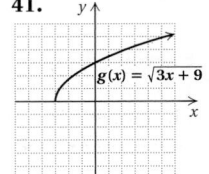

43. $4|x|$ **45.** $12|c|$ **47.** $|p+3|$ **49.** $|x-2|$ **51.** 3
53. $-4x$ **55.** -6 **57.** $0.7(x+1)$ **59.** $2;3;-2;-4$
61. $-1;\ -\sqrt[3]{-20}$, or $\sqrt[3]{20}\approx 2.714;\ -4;\ -10$ **63.** -5
65. -1 **67.** $-\frac{2}{3}$ **69.** $|x|$ **71.** $5|a|$ **73.** 6 **75.** $|a+b|$
77. y **79.** $x-2$ **81.** $-2,1$ **82.** $-1,0$ **83.** $-\frac{7}{2},\frac{7}{2}$
84. $4,9$ **85.** $-2,\frac{5}{3}$ **86.** $\frac{5}{2}$ **87.** $0,\frac{5}{2}$ **88.** $0,1$
89. $a^9b^6c^{15}$ **90.** $10a^{10}b^9$
91. Domain $= \{x|-3\le x<2\}=[-3,2)$ **93.** $1.7;2.2;3.2$
95. **(a)** Domain: $(-\infty,\infty)$; range: $(-\infty,\infty)$;
(b) domain: $(-\infty,\infty)$; range: $(-\infty,\infty)$;
(c) domain: $[-3,\infty)$; range: $(-\infty,2]$; **(d)** domain: $[0,\infty)$;
range: $[0,\infty)$; **(e)** domain: $[3,\infty)$; range: $[0,\infty)$

Calculator Corner, p. 715
1. 3.344 **2.** 3.281 **3.** 0.283 **4.** 11.053 **5.** 5.527×10^{-5}
6. 2

Exercise Set 10.2, p. 718
1. $\sqrt[7]{y}$ **3.** 2 **5.** $\sqrt[5]{a^3b^3}$ **7.** 8 **9.** 343 **11.** $17^{1/2}$
13. $18^{1/3}$ **15.** $(xy^2z)^{1/5}$ **17.** $(3mn)^{3/2}$ **19.** $(8x^2y)^{5/7}$
21. $\frac{1}{3}$ **23.** $\frac{1}{1000}$ **25.** $\frac{3}{x^{1/4}}$ **27.** $\frac{1}{(2rs)^{3/4}}$ **29.** $\frac{2a^{3/4}c^{2/3}}{b^{1/2}}$
31. $\left(\frac{8yz}{7x}\right)^{3/5}$ **33.** $x^{2/3}$ **35.** $\frac{x^4}{2^{1/3}y^{2/7}}$ **37.** $\frac{7x}{z^{1/3}}$
39. $\frac{5ac^{1/2}}{3}$ **41.** $5^{7/8}$ **43.** $7^{1/4}$ **45.** $4.9^{1/2}$ **47.** $6^{3/28}$
49. $a^{23/12}$ **51.** $a^{8/3}b^{5/2}$ **53.** $\frac{1}{x^{2/7}}$ **55.** $\frac{y^{1/3}}{x^{1/2}}$ **57.** $m^{3/5}n^2$
59. $\sqrt[3]{a}$ **61.** x^5 **63.** $\frac{1}{x^3}$ **65.** a^5b^5 **67.** $\sqrt{2}$ **69.** $\sqrt[3]{2x}$
71. x^2y^3 **73.** $2c^2d^3$ **75.** $\sqrt[12]{7^4\cdot 5^3}$ **77.** $\sqrt[20]{5^5\cdot 7^4}$
79. $\sqrt[6]{4x^5}$ **81.** a^6b^{12} **83.** $\sqrt[18]{m}$ **85.** $\sqrt[12]{x^4y^3z^2}$
87. $\sqrt[30]{\dfrac{d^{35}}{c^{99}}}$ **89.** $\left\{-\frac{4}{7},2\right\}$ **90.** $\{-40,40\}$
91. $\left\{-\frac{15}{2},\frac{5}{2}\right\}$ **92.** $\left\{-\frac{11}{8},\frac{3}{8}\right\}$ **93.** Left to the student

Exercise Set 10.3, p. 725
1. $2\sqrt{6}$ **3.** $3\sqrt{10}$ **5.** $5\sqrt[3]{2}$ **7.** $6x^2\sqrt{5}$ **9.** $3x^2\sqrt[3]{2x^2}$
11. $2t^2\sqrt[3]{10t^2}$ **13.** $2\sqrt[4]{5}$ **15.** $4a\sqrt{2b}$ **17.** $3x^2y^2\sqrt[4]{3y^2}$
19. $2xy^3\sqrt[5]{3x^2}$ **21.** $5\sqrt{2}$ **23.** $3\sqrt{10}$ **25.** 2 **27.** $30\sqrt{3}$
29. $3x^4\sqrt{2}$ **31.** $5bc^2\sqrt{2b}$ **33.** $a\sqrt[3]{10}$ **35.** $2y^3\sqrt[3]{2}$
37. $4\sqrt[4]{4}$ **39.** $4a^3b\sqrt{6ab}$ **41.** $\sqrt[6]{200}$ **43.** $\sqrt[4]{12}$
45. $a\sqrt[5]{a}$ **47.** $b\sqrt[10]{b^9}$ **49.** $xy\sqrt[6]{xy^5}$ **51.** $2ab\sqrt[4]{2a^3}$
53. $3\sqrt{2}$ **55.** $\sqrt{5}$ **57.** 3 **59.** $y\sqrt{7y}$ **61.** $2\sqrt[3]{a^2b}$
63. $4\sqrt{xy}$ **65.** $2x^2y^2$ **67.** $\frac{1}{\sqrt[6]{a}}$ **69.** $\sqrt[12]{a^5}$
71. $\sqrt[12]{x^2y^5}$ **73.** $\frac{5}{6}$ **75.** $\frac{4}{7}$ **77.** $\frac{5}{3}$ **79.** $\frac{7}{y}$ **81.** $\frac{5y\sqrt{y}}{x^2}$
83. $\frac{3y^3\sqrt[3]{3y^2}}{4}$ **85.** $\frac{3a\sqrt[3]{a}}{2b}$ **87.** $\frac{3x}{2}$ **89.** $\frac{2a^3}{bc^4}$ **91.** $\frac{2x\sqrt[5]{x^3}}{y^2}$
93. $\frac{w\sqrt[5]{w^2}}{z^2}$ **95.** $\frac{x^2\sqrt[6]{x}}{yz^2}$ **97.** $-10,9$ **98.** Height: 4 in.;
base: 6 in. **99.** 8 **100.** $\frac{15}{2}$ **101.** No solution
102. No solution **103.** **(a)** 1.62 sec; **(b)** 1.99 sec; **(c)** 2.20 sec
105. $2yz\sqrt{2z}$

Exercise Set 10.4, p. 731
1. $11\sqrt{5}$ **3.** $\sqrt[3]{7}$ **5.** $13\sqrt[3]{y}$ **7.** $-8\sqrt{6}$ **9.** $6\sqrt[3]{3}$
11. $21\sqrt{3}$ **13.** $38\sqrt{5}$ **15.** $122\sqrt{2}$ **17.** $9\sqrt[3]{2}$
19. $29\sqrt{2}$ **21.** $(1+6a)\sqrt{5a}$ **23.** $(2-x)\sqrt[3]{3x}$

25. $(21x+1)\sqrt{3x}$ **27.** $2+3\sqrt{2}$ **29.** $15\sqrt[3]{4}$
31. $(x+1)\sqrt[3]{6x}$ **33.** $3\sqrt{a-1}$ **35.** $(x+3)\sqrt{x-1}$
37. $4\sqrt{5}-10$ **39.** $\sqrt{6}-\sqrt{21}$ **41.** $-12+6\sqrt{3}$
43. $2\sqrt{15}-6\sqrt{3}$ **45.** -6 **47.** $6y-12\sqrt[3]{y^2}$ **49.** $3a\sqrt[3]{2}$
51. 1 **53.** -12 **55.** 44 **57.** 1 **59.** 3 **61.** -19
63. $a-b$ **65.** $1+\sqrt{5}$ **67.** $7+3\sqrt{3}$ **69.** -6
71. $a+\sqrt{3a}+\sqrt{2a}+\sqrt{6}$ **73.** $2\sqrt[3]{9}-3\sqrt[3]{6}-2\sqrt[3]{4}$
75. $7+4\sqrt{3}$ **77.** $\sqrt[5]{72}+3-\sqrt[5]{24}-\sqrt[5]{81}$
79. $\dfrac{x(x^2+4)}{(x+4)(x+3)}$ **80.** $\dfrac{(a+2)(a+4)}{a}$ **81.** $a-2$
82. $\dfrac{(y-3)(y-3)}{y+3}$ **83.** $\dfrac{4(3x-1)}{3(4x+1)}$ **84.** $\dfrac{x}{x+1}$ **85.** $\dfrac{pq}{q+p}$
86. $\dfrac{a^2b^2}{b^2-ab+a^2}$ **87.** $-\frac{29}{3},5$ **88.** $\left\{x|-\frac{29}{3}<x<5\right\}$,
or $\left(-\frac{29}{3},5\right)$ **89.** $\left\{x|x\le -\frac{29}{3}\ or\ x\ge 5\right\}$, or $\left(-\infty,-\frac{29}{3}\right]\cup[5,\infty)$
90. $-12,-\frac{2}{5}$ **91.** Domain $= (-\infty,\infty)$ **93.** 6
95. $14+2\sqrt{15}-6\sqrt{2}-2\sqrt{30}$ **97.** $3\sqrt[3]{3}+2\sqrt[3]{9}-8$

Mid-Chapter Review: Chapter 10, p. 735
1. False **2.** True **3.** False **4.** True
5. $\sqrt{6}\sqrt{10} = \sqrt{6\cdot 10} = \sqrt{2\cdot 3\cdot 2\cdot 5} = 2\sqrt{15}$
6. $5\sqrt{32} - 3\sqrt{18} = 5\sqrt{16\cdot 2} - 3\sqrt{9\cdot 2} = 5\cdot 4\sqrt{2} - 3\cdot 3\sqrt{2} = 20\sqrt{2} - 9\sqrt{2} = 11\sqrt{2}$ **7.** 9 **8.** -12 **9.** $\frac{4}{5}$
10. Does not exist as a real number **11.** 3; does not exist as a real number **12.** Domain $= \{x|x\le 4\} = (-\infty,4]$
13. **14.**

$f(x) = -2\sqrt{x}$

$g(x) = \sqrt{x+1}$

15. $6|z|$ **16.** $|x-4|$ **17.** -4 **18.** $-3a$ **19.** 2 **20.** $|y|$
21. 5 **22.** $\sqrt[4]{a^3b}$ **23.** $16^{1/5}$ **24.** $(6m^2n)^{1/3}$ **25.** $\frac{1}{3^{3/8}}$
26. $7^{4/5}$ **27.** $\dfrac{x^{3/2}}{y^{4/3}}$ **28.** $\dfrac{1}{n^{3/4}}$ **29.** $\sqrt[3]{4}$ **30.** $\sqrt{ab}$
31. $\sqrt[6]{y^5}$ **32.** $\sqrt[15]{a^{10}b^9}$ **33.** $5\sqrt{3}$ **34.** $2xy\sqrt[3]{3y^2}$
35. $2\sqrt[3]{5}$ **36.** $\dfrac{7a^2\sqrt{a}}{b^4}$ **37.** $11\sqrt{7}$ **38.** $(9x-24)\sqrt{2x}$
39. $2\sqrt{3}-15$ **40.** $3-4\sqrt{x}+x$ **41.** $m-n$
42. $11+4\sqrt{7}$ **43.** $-42+\sqrt{15}$ **44.** Yes; since x^2 is nonnegative for any value of x, the nth root of x^2 exists regardless of whether n is even or odd. Thus the nth root of x^2 always exists. **45.** Formulate an expression containing a radical term with an even index and a radicand R such that the solution of the inequality $R \ge 0$ is $\{x|x \le 5\}$. One expression is $\sqrt{5-x}$. Other expressions could be formulated as $a\sqrt[k]{b(5-x)} + c$, where $a \ne 0, b > 0$, and k is an even integer.
46. Since $x^6 \ge 0$ and $x^2 \ge 0$ for any value of x, then $\sqrt[3]{x^6} = x^2$. However, $x^3 \ge 0$ only for $x \ge 0$, so $\sqrt{x^6} = x^3$ only when $x \ge 0$.
47. No; for example, $\dfrac{\sqrt{8}}{\sqrt{2}} = \sqrt{\dfrac{8}{2}} = \sqrt{4} = 2$.

Exercise Set 10.5, p. 740
1. $\dfrac{\sqrt{15}}{3}$ **3.** $\dfrac{\sqrt{22}}{2}$ **5.** $\dfrac{2\sqrt{15}}{35}$ **7.** $\dfrac{2\sqrt[3]{6}}{3}$ **9.** $\dfrac{\sqrt[3]{75ac^2}}{5c}$
11. $\dfrac{y\sqrt[3]{9yx^2}}{3x^2}$ **13.** $\dfrac{\sqrt[4]{s^3t^3}}{st}$ **15.** $\dfrac{\sqrt{15x}}{10}$ **17.** $\dfrac{\sqrt[3]{100xy}}{5x^2y}$
19. $\dfrac{\sqrt[4]{2xy}}{2x^2y}$ **21.** $\dfrac{54+9\sqrt{10}}{26}$ **23.** $-2\sqrt{35}+2\sqrt{21}$

25. $\dfrac{18\sqrt{6} + 6\sqrt{15}}{13}$ 27. $\dfrac{3\sqrt{2} - 3\sqrt{5} + \sqrt{10} - 5}{-3}$

29. $\dfrac{3 + \sqrt{21} - \sqrt{6} - \sqrt{14}}{-4}$ 31. $\dfrac{\sqrt{15} + 20 - 6\sqrt{2} - 8\sqrt{30}}{-77}$

33. $\dfrac{6 - 5\sqrt{a} + a}{9 - a}$ 35. $\dfrac{6 + 5\sqrt{x} - 6x}{9 - 4x}$ 37. $\dfrac{3\sqrt{6} + 4}{2}$

39. $\dfrac{x - 2\sqrt{xy} + y}{x - y}$ 41. 30   42. $-\frac{19}{5}$ 43. 1   44. $\dfrac{x - 2}{x + 3}$

45. Left to the student   47. $-\dfrac{3\sqrt{a^2 - 3}}{a^2 - 3}$

Calculator Corner, p. 744

1. Left to the student   2. Left to the student

Exercise Set 10.6, p. 748

1. $\frac{19}{2}$ 3. $\frac{49}{6}$ 5. 57   7. $\frac{92}{5}$ 9. -1 11. No solution
13. 3   15. 19   17. -6 19. $\frac{1}{64}$ 21. 9   23. 15
25. 2, 5   27. 6   29. 5   31. 9   33. 7   35. $\frac{80}{9}$
37. 2, 6   39. -1 41. No solution   43. 3   45. About 44.1 mi
47. About 680 ft   49. About 117 ft   51. 151.25 ft; 281.25 ft
53. About 25°F   55. About 0.81 ft   57. About 3.9 ft
59. $4\frac{4}{9}$ hr   60. Jeff: $1\frac{1}{3}$ hr; Grace: 4 hr   61. 2808 mi
62. 84 hr   63. 0, -2.8 64. 0, $\frac{5}{3}$ 65. $-8, 8$ 66. $-3, \frac{7}{2}$
67. 11   68. -7 69. 3   70. $-9\frac{1}{9}$ 71. Left to the student
73. 6912   75. 0   77. $-6, -3$ 79. 2   81. $0, \frac{125}{4}$
83. 2   85. $\frac{1}{2}$ 87. 3

Translating for Success, p. 755

1. J   2. B   3. O   4. M   5. K   6. I   7. G   8. E
9. F   10. A

Exercise Set 10.7, p. 756

1. $\sqrt{34}$; 5.831   3. $\sqrt{450}$; 21.213   5. 5   7. $\sqrt{43}$; 6.557
9. $\sqrt{12}$; 3.464   11. $\sqrt{n-1}$ 13. $\sqrt{116}$ ft; 10.770 ft
15. 7.1 ft   17. 50 ft   19. $\sqrt{10{,}561}$ ft; 102.767 ft
21. $s + s\sqrt{2}$ 23. $\sqrt{181}$ cm; 13.454 cm   25. $(3, 0), (-3, 0)$
27. $\sqrt{340} + 8$ ft; 26.439 ft   29. $\sqrt{420.125}$ in.; 20.497 in.
31. Flash: $67\frac{2}{3}$ mph; Crawler: $53\frac{2}{3}$ mph   32. $3\frac{3}{4}$ mph
33. $-7, \frac{3}{2}$ 34. 3, 8   35. 1   36. $-2, 2$ 37. 13   38. 7
39. 26 packets   41. $\sqrt{75}$ cm

Calculator Corner, p. 765

1. $-2 - 9i$ 2. $20 + 17i$ 3. $-47 - 161i$ 4. $-\frac{151}{290} + \frac{73}{290}i$
5. -20 6. -28.373 7. $-\frac{16}{25} - \frac{1}{50}i$ 8. 81   9. $117 + 118i$
10. $-\frac{14}{169} + \frac{34}{169}i$

Exercise Set 10.8, p. 767

1. $i\sqrt{35}$, or $\sqrt{35}i$ 3. $4i$ 5. $-2i\sqrt{3}$, or $-2\sqrt{3}i$
7. $i\sqrt{3}$, or $\sqrt{3}i$ 9. $9i$ 11. $7i\sqrt{2}$, or $7\sqrt{2}i$
13. $-7i$ 15. $4 - 2\sqrt{15}i$, or $4 - 2i\sqrt{15}$ 17. $\left(2 + 2\sqrt{3}\right)i$
19. $12 - 4i$ 21. $9 - 5i$ 23. $7 + 4i$ 25. $-4 - 4i$
27. $-1 + i$ 29. $11 + 6i$ 31. -18 33. $-\sqrt{14}$ 35. 21
37. $-6 + 24i$ 39. $1 + 5i$ 41. $18 + 14i$ 43. $38 + 9i$
45. $2 - 46i$ 47. $5 - 12i$ 49. $-24 + 10i$ 51. $-5 - 12i$
53. $-i$ 55. 1   57. -1 59. i 61. -1 63. $-125i$
65. 8   67. $1 - 23i$ 69. 0   71. 0   73. 1   75. $5 - 8i$
77. $2 - \dfrac{\sqrt{6}}{2}i$ 79. $\frac{9}{10} + \frac{13}{10}i$ 81. $-i$ 83. $-\frac{3}{7} - \frac{8}{7}i$
85. $\frac{6}{5} - \frac{2}{5}i$ 87. $-\frac{8}{41} + \frac{10}{41}i$ 89. $-\frac{4}{3}i$ 91. $-\frac{1}{2} - \frac{1}{4}i$
93. $\frac{3}{5} + \frac{4}{5}i$
95.
$$x^2 - 2x + 5 = 0$$
$$\begin{array}{c|c} (1 - 2i)^2 - 2(1 - 2i) + 5 \ ? \ 0 & \\ 1 - 4i + 4i^2 - 2 + 4i + 5 & \\ 1 - 4i - 4 - 2 + 4i + 5 & \\ 0 & \text{TRUE} \end{array}$$
Yes

97.
$$x^2 - 4x - 5 = 0$$
$$\begin{array}{c|c} (2 + i)^2 - 4(2 + i) - 5 \ ? \ 0 & \\ 4 + 4i + i^2 - 8 - 4i - 5 & \\ 4 + 4i - 1 - 8 - 4i - 5 & \\ -10 & \text{FALSE} \end{array}$$
No

99. Rational   100. Difference of squares   101. Coordinates
102. Positive   103. Proportion   104. Trinomial square
105. Negative   106. Zero products   107. $-4 - 8i; -2 + 4i$;
$8 - 6i$ 109. $-3 - 4i$ 111. $-88i$ 113. 8   115. $\frac{3}{5} + \frac{9}{5}i$
117. 1

Summary and Review: Chapter 10, p. 771

Concept Reinforcement

1. True   2. False   3. False   4. False   5. True   6. True

Important Concepts

1. $6|y|$ 2. $|a + 2|$ 3. $\sqrt[5]{z^3}$ 4. $(6ab)^{5/2}$ 5. $\dfrac{1}{9^{3/2}} = \dfrac{1}{27}$
6. $\sqrt[4]{a^3b}$ 7. $5y\sqrt{6}$ 8. $2\sqrt{a}$ 9. $2\sqrt{3}$ 10. $25 - 10\sqrt{x} + x$
11. 5   12. 6   13. $-21 - 20i$ 14. $\frac{4}{5} - \frac{7}{5}i$

Review Exercises

1. 27.893   2. 6.378   3. $f(0), f(-1),$ and $f(1)$ do not exist as
real numbers; $f\left(\frac{41}{3}\right) = 5$ 4. Domain $= \left\{x \mid x \geq \frac{16}{3}\right\}$, or $\left[\frac{16}{3}, \infty\right)$
5. $9|a|$ 6. $7|z|$ 7. $|6 - b|$ 8. $|x + 3|$ 9. -10 10. $-\frac{1}{3}$
11. $2; -2; 3$ 12. $|x|$ 13. 3   14. $\sqrt[5]{a}$ 15. 512   16. $31^{1/2}$
17. $(a^2b^3)^{1/5}$ 18. $\frac{1}{7}$ 19. $\dfrac{1}{4x^{2/3}y^{2/3}}$ 20. $\dfrac{5b^{1/2}}{a^{3/4}c^{2/3}}$
21. $\dfrac{3a}{t^{1/4}}$ 22. $\dfrac{1}{x^{2/5}}$ 23. $7^{1/6}$ 24. x^7 25. $3x^2$
26. $\sqrt[12]{x^4y^3}$ 27. $\sqrt[12]{x^7}$ 28. $7\sqrt{5}$ 29. $-3\sqrt[3]{4}$
30. $5b^2\sqrt[3]{2a^2}$ 31. $\frac{7}{6}$ 32. $\dfrac{4x^2}{3}$ 33. $\dfrac{2x^2}{3y^3}$ 34. $\sqrt{15xy}$
35. $3a\sqrt[3]{a^2b^2}$ 36. $\sqrt[15]{a^5b^9}$ 37. $y\sqrt[3]{6}$ 38. $\frac{5}{2}\sqrt{x}$
39. $\sqrt[12]{x^5}$ 40. $7\sqrt[3]{x}$ 41. $3\sqrt{3}$ 42. $15\sqrt{2}$
43. $(2x + y^2)\sqrt[3]{x}$ 44. $-43 - 2\sqrt{10}$ 45. $8 - 2\sqrt{7}$
46. $9 - \sqrt[3]{4}$ 47. $\dfrac{2\sqrt{6}}{3}$ 48. $\dfrac{2\sqrt{a} - 2\sqrt{b}}{a - b}$ 49. 4   50. 13
51. 1   52. About 4166 rpm   53. 4480 rpm   54. 9 cm
55. $\sqrt{24}$ ft; 4.899 ft   56. 25   57. $\sqrt{46}$; 6.782
58. $\left(5 + 2\sqrt{2}\right)i$ 59. $-2 - 9i$ 60. $1 + i$ 61. 29   62. i
63. $9 - 12i$ 64. $\frac{2}{5} + \frac{3}{5}i$ 65. $\frac{1}{10} - \frac{7}{10}i$
66.   67. D   68. -1 69. 3

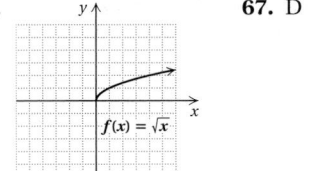
$f(x) = \sqrt{x}$

Understanding Through Discussion and Writing

1. $f(x) = (x + 5)^{1/2}(x + 7)^{-1/2}$. Consider $(x + 5)^{1/2}$.
Since the exponent is $\frac{1}{2}$, $x + 5$ must be nonnegative. Then
$x + 5 \geq 0$, or $x \geq -5$. Consider $(x + 7)^{-1/2}$. Since the exponent
is $-\frac{1}{2}$, $x + 7$ must be positive. Then $x + 7 > 0$, or $x > -7$. Then
the domain of $f = \{x \mid x \geq -5 \text{ and } x > -7\}$, or $\{x \mid x \geq -5\}$.
2. Since $\sqrt{x}$ exists only for $\{x \mid x \geq 0\}$, this is the domain of
$y = \sqrt{x} \cdot \sqrt{x}$.   3. The distributive law is used to collect
radical expressions with the same indices and radicands just as
it is used to collect monomials with the same variables and
exponents.   4. No; when n is odd, it is true that if $a^n = b^n$,

then $a = b$. **5.** Use a calculator to show that $\dfrac{5 + \sqrt{2}}{\sqrt{18}} \neq 2$.

Explain that we multiply by 1 to rationalize a denominator. In this case, we would write 1 as $\sqrt{2}/\sqrt{2}$. **6.** When two radical expressions are conjugates, their product contains no radicals. Similarly, the product of a complex number and its conjugate does not contain i.

Test: Chapter 10, p. 777

1. [10.1a] 12.166 **2.** [10.1a] 2; does not exist as a real number
3. [10.1a] Domain $= \{x | x \leq 2\}$, or $(-\infty, 2]$ **4.** [10.1b] $3|q|$
5. [10.1b] $|x + 5|$ **6.** [10.1c] $-\frac{1}{10}$ **7.** [10.1d] x **8.** [10.1d] 4
9. [10.2a] $\sqrt[3]{a^2}$ **10.** [10.2a] 8 **11.** [10.2a] $37^{1/2}$
12. [10.2a] $(5xy^2)^{5/2}$ **13.** [10.2b] $\frac{1}{10}$ **14.** [10.2b] $\dfrac{8a^{3/4}}{b^{3/2}c^{2/5}}$
15. [10.2c] $\dfrac{x^{8/5}}{y^{9/5}}$ **16.** [10.2c] $\dfrac{1}{2.9^{31/24}}$ **17.** [10.2d] $\sqrt[4]{x}$
18. [10.2d] $2x\sqrt{x}$ **19.** [10.2d] $\sqrt[15]{a^6b^5}$ **20.** [10.2d] $\sqrt[12]{8y^7}$
21. [10.3a] $2\sqrt{37}$ **22.** [10.3a] $2\sqrt[4]{5}$ **23.** [10.3a] $2a^3b^4\sqrt[3]{3a^2b}$
24. [10.3b] $\dfrac{2x\sqrt[3]{2x^2}}{y^2}$ **25.** [10.3b] $\dfrac{5x}{6y^2}$ **26.** [10.3a] $\sqrt[3]{10xy^2}$
27. [10.3a] $xy\sqrt[4]{x}$ **28.** [10.3b] $\sqrt[5]{x^2y^2}$ **29.** [10.3b] $2\sqrt{a}$
30. [10.4a] $38\sqrt{2}$ **31.** [10.4b] -20 **32.** [10.4b] $9 + 6\sqrt{x} + x$
33. [10.5b] $\dfrac{13 + 8\sqrt{2}}{-41}$ **34.** [10.6a] 35 **35.** [10.6b] 7
36. [10.6a] 5 **37.** [10.7a] 7 ft **38.** [10.6c] 3600 ft
39. [10.7a] $\sqrt{98}$; 9.899 **40.** [10.7a] 2 **41.** [10.8a] $11i$
42. [10.8b] $7 + 5i$ **43.** [10.8c] $37 + 9i$ **44.** [10.8d] $-i$
45. [10.8e] $-\frac{77}{50} + \frac{7}{25}i$ **46.** [10.8f] No **47.** [10.6a] A
48. [10.8c, e] $-\frac{17}{4}i$ **49.** [10.6b] 3

Cumulative Review: Chapters 1–10, p. 779

1. [4.4a] $-3x^3 + 9x^2 + 3x - 3$ **2.** [4.6c] $4x^4 - 4x^2y + y^2$
3. [4.5d] $15x^4 - x^3 - 9x^2 + 5x - 2$ **4.** [6.1d] $\dfrac{(x + 4)(x - 7)}{x + 7}$
5. [6.2b] $\dfrac{y - 6}{y - 9}$ **6.** [6.5b] $\dfrac{-2x + 4}{(x + 2)(x - 3)}$, or $\dfrac{-2(x - 2)}{(x + 2)(x - 3)}$
7. [4.8b] $y^2 + y - 2 + \dfrac{-1}{y + 2}$ **8.** [10.1c] $-2x$
9. [10.1b], [10.3a] $4(x - 1)$ **10.** [10.4a] $57\sqrt{3}$
11. [10.3a] $4xy^2\sqrt{y}$ **12.** [10.5b] $\sqrt{30} + \sqrt{15}$
13. [10.1d], [10.3b] $\dfrac{m^2n^4}{2}$ **14.** [10.2c] $6^{8/9}$
15. [10.8b] $3 + 5i$ **16.** [10.8e] $\frac{7}{61} - \frac{16}{61}i$ **17.** [2.3a] 2
18. [2.4b] $c = 8M + 3$ **19.** [9.1c] $\{a | a > -7\}$, or $(-7, \infty)$
20. [9.2a] $\{x | -10 < x < 13\}$, or $(-10, 13)$ **21.** [9.3c] $\frac{4}{3}, \frac{8}{3}$
22. [5.8b] $\frac{25}{7}, -\frac{25}{7}$ **23.** [8.3a] $(5, 3)$ **24.** [8.5a] $(-1, 0, 4)$
25. [6.7a] -5 **26.** [6.7a] $\frac{1}{3}$ **27.** [2.4a], [6.7a] $R = \dfrac{nE - nrI}{I}$
28. [10.6a] 6 **29.** [10.6b] $-\frac{1}{4}$ **30.** [10.6a] 5
31. [7.1c] **32.** [7.4a] **33.** [9.4b]

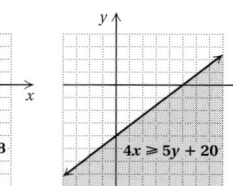

34. [9.4c] **35.** [7.1c] **36.** [7.1c]

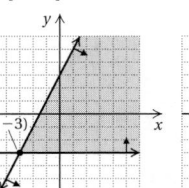 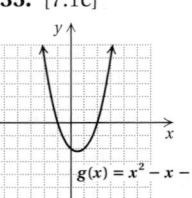 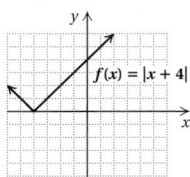

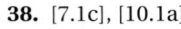

37. [7.1c] **38.** [7.1c], [10.1a]

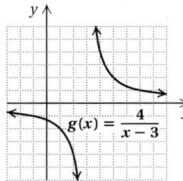 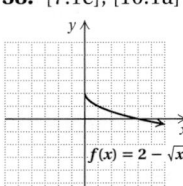

39. [5.1b] $6xy^2(2x - 5y)$ **40.** [5.3a], [5.4a] $(3x + 4)(x - 7)$
41. [5.2a] $(y + 11)(y - 12)$ **42.** [5.6a] $(3y + 2)(9y^2 - 6y + 4)$
43. [5.5d] $(2x + 25)(2x - 25)$ **44.** [7.2a] Domain: $[-5, 5]$;
range: $[-3, 4]$ **45.** [7.2a] Domain: $(-\infty, \infty)$; range: $[-5, \infty)$
46. [7.3b] Slope: $\frac{3}{2}$; y-intercept: $(0, -4)$ **47.** [7.5d] $y = -\frac{1}{3}x + \frac{13}{3}$
48. [6.9d] 125 ft; 1000 ft^2 **49.** [6.8a] 1 hr **50.** [6.9f] 64 L
51. [10.2a] D **52.** [6.8a] A **53.** [4.8b] A **54.** [10.6a] B
55. [10.6b] $-\frac{8}{9}$

CHAPTER 11

Calculator Corner, p. 786

The calculator returns an ERROR message because the graph of $y = 4x^2 + 9$ has no x-intercepts. This indicates that the equation $4x^2 + 9 = 0$ has no real-number solutions.

Exercise Set 11.1, p. 792

1. (a) $\sqrt{5}, -\sqrt{5}$, or $\pm\sqrt{5}$; (b) $\left(-\sqrt{5}, 0\right), \left(\sqrt{5}, 0\right)$ **3.** (a) $\frac{5}{3}i$,
$-\frac{5}{3}i$, or $\pm\frac{5}{3}i$; (b) no x-intercepts **5.** $\pm\dfrac{\sqrt{6}}{2}$; ± 1.225 **7.** $5, -9$
9. $8, 0$ **11.** $11 \pm \sqrt{7}$; $13.646, 8.354$ **13.** $7 \pm 2i$ **15.** $18, 0$
17. $\dfrac{3}{2} \pm \dfrac{\sqrt{14}}{2}$; $3.371, -0.371$ **19.** $5, -11$ **21.** $9, 5$
23. $-2 \pm \sqrt{6}$ **25.** $11 \pm 2\sqrt{33}$ **27.** $-\dfrac{1}{2} \pm \dfrac{\sqrt{5}}{2}$
29. $\dfrac{5}{2} \pm \dfrac{\sqrt{53}}{2}$ **31.** $-\dfrac{3}{4} \pm \dfrac{\sqrt{57}}{4}$ **33.** $\dfrac{9}{4} \pm \dfrac{\sqrt{105}}{4}$ **35.** $2, -8$
37. $-11 \pm \sqrt{19}$ **39.** $5 \pm \sqrt{29}$ **41.** (a) $-\dfrac{7}{2} \pm \dfrac{\sqrt{57}}{2}$;
(b) $\left(-\dfrac{7}{2} - \dfrac{\sqrt{57}}{2}, 0\right), \left(-\dfrac{7}{2} + \dfrac{\sqrt{57}}{2}, 0\right)$ **43.** (a) $\dfrac{5}{4} \pm \dfrac{\sqrt{39}}{4}i$;
(b) no x-intercepts **45.** $\dfrac{3}{4} \pm \dfrac{\sqrt{17}}{4}$ **47.** $\dfrac{3}{4} \pm \dfrac{\sqrt{145}}{4}$
49. $\dfrac{2}{3} \pm \dfrac{\sqrt{7}}{3}$ **51.** $-\dfrac{1}{2} \pm \dfrac{\sqrt{7}}{2}i$ **53.** $2 \pm 3i$ **55.** About
0.816 sec **57.** About 6.8 sec **59.** About 5.9 sec **61.** About
7.1 sec **63.** (a) $B(t) = 0.022t + 2.6$, where t is the number of years since 1930; (b) about 4.4 million; (c) 2016
64. **65.**

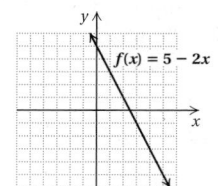

66.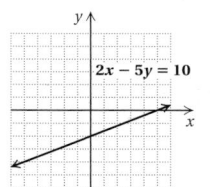

67.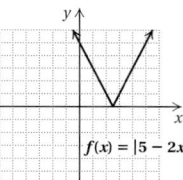

68. $2\sqrt{22}$ **69.** $\dfrac{\sqrt{10}}{5}$ **70.** 4 **71.** 5 **72.** 4

73. No solution **75.** Left to the student **77.** $16, -16$
79. $0, \frac{7}{2}, -8, -\frac{10}{3}$

Calculator Corner, p. 798
1.–3. Left to the student

Calculator Corner, p. 800
1. $-3, 0.8$ **2.** $-1.5, 5$ **3.** $3, 8$ **4.** $2, 4$

Exercise Set 11.2, p. 801
1. $-4 \pm \sqrt{14}$ **3.** $\dfrac{-4 \pm \sqrt{13}}{3}$ **5.** $\dfrac{1}{2} \pm \dfrac{\sqrt{3}}{2}i$ **7.** $2 \pm 3i$
9. $\dfrac{-3 \pm \sqrt{41}}{2}$ **11.** $-1 \pm 2i$ **13. (a)** $0, -1$; **(b)** $(0, 0), (-1, 0)$
15. (a) $\dfrac{3 \pm \sqrt{229}}{22}$; **(b)** $\left(\dfrac{3 + \sqrt{229}}{22}, 0\right), \left(\dfrac{3 - \sqrt{229}}{22}, 0\right)$
17. (a) $\frac{2}{5}$; **(b)** $\left(\frac{2}{5}, 0\right)$ **19.** $-1, -2$ **21.** $5, 10$ **23.** $\dfrac{17 \pm \sqrt{249}}{10}$
25. $2 \pm i$ **27.** $\frac{2}{3}, \frac{3}{2}$ **29.** $2 \pm \sqrt{10}$ **31.** $\frac{3}{4}, -2$ **33.** $\frac{1}{2} \pm \frac{3}{2}i$
35. $1, -\dfrac{1}{2} \pm \dfrac{\sqrt{3}}{2}i$ **37.** $-3 \pm \sqrt{5}; -0.764, -5.236$
39. $3 \pm \sqrt{5}; 5.236, 0.764$ **41.** $\dfrac{3 \pm \sqrt{65}}{4}; 2.766, -1.266$
43. $\dfrac{4 \pm \sqrt{31}}{5}; 1.914, -0.314$ **45.** 2 **46.** 3 **47.** 10 **48.** 8
49. No solution **50.** No solution **51.** $\frac{15}{4}$ **52.** $\frac{17}{3}$
53. Left to the student; $-0.797, 0.570$ **55.** $\dfrac{1 \pm \sqrt{1 + 8\sqrt{5}}}{4}$
57. $\dfrac{-i \pm i\sqrt{1 + 4i}}{2}$ **59.** $\dfrac{-1 \pm 3\sqrt{5}}{6}$ **61.** $3 \pm \sqrt{13}$

Translating for Success, p. 809
1. B **2.** G **3.** F **4.** L **5.** N **6.** C **7.** J **8.** E
9. K **10.** A

Exercise Set 11.3, p. 810
1. Length: 9 ft; width: 2 ft **3.** Length: 18 yd; width: 9 yd
5. Height: 16 m; base: 7 m **7.** Length: $\dfrac{51 + \sqrt{122{,}399}}{2}$ ft;
width: $\dfrac{\sqrt{122{,}399} - 51}{2}$ ft **9.** 2 in. **11.** 6 ft, 8 ft
13. 28 and 29 **15.** Length: $2 + \sqrt{14}$ ft ≈ 5.742 ft; width:
$\sqrt{14} - 2$ ft ≈ 1.742 ft **17.** $\dfrac{17 - \sqrt{109}}{2}$ in. ≈ 3.280 in.
19. $7 + \sqrt{239}$ ft ≈ 22.460 ft; $\sqrt{239} - 7$ ft ≈ 8.460 ft
21. First part: 60 mph; second part: 50 mph **23.** 40 mph
25. Cessna: 150 mph; Beechcraft: 200 mph; or Cessna:
200 mph; Beechcraft: 250 mph **27.** To Hillsboro: 10 mph;
return trip: 4 mph **29.** About 11 mph
31. $s = \sqrt{\dfrac{A}{6}}$ **33.** $r = \sqrt{\dfrac{Gm_1m_2}{F}}$ **35.** $c = \sqrt{\dfrac{E}{m}}$
37. $b = \sqrt{c^2 - a^2}$ **39.** $k = \dfrac{3 + \sqrt{9 + 8N}}{2}$
41. $r = \dfrac{-\pi h + \sqrt{\pi^2 h^2 + 2\pi A}}{2\pi}$ **43.** $g = \dfrac{4\pi^2 L}{T^2}$

45. $H = \sqrt{\dfrac{703W}{I}}$ **47.** $v = \dfrac{c\sqrt{m^2 - (m_0)^2}}{m}$ **49.** $\dfrac{1}{x - 2}$
50. $\dfrac{(x + 1)(x^2 + 2)}{(x - 1)(x^2 + x + 1)}$ **51.** $\dfrac{-x}{(x + 3)(x - 1)}$ **52.** $3x^2\sqrt{x}$
53. $2i\sqrt{5}$ **54.** $\dfrac{3(x + 1)}{3x + 1}$ **55.** $\dfrac{4b}{a(3b^2 - 4a)}$ **57.** $\pm\sqrt{2}$
59. $A(S) = \dfrac{\pi S}{6}$ **61.** $l = \dfrac{w + w\sqrt{5}}{2}$

Exercise Set 11.4, p. 820
1. One real **3.** Two nonreal **5.** Two real **7.** One real
9. Two nonreal **11.** Two real **13.** Two real
15. One real **17.** $x^2 - 16 = 0$ **19.** $x^2 + 16 = 0$
21. $x^2 - 16x + 64 = 0$ **23.** $25x^2 - 20x - 12 = 0$
25. $12x^2 - (4k + 3m)x + km = 0$ **27.** $x^2 - \sqrt{3}x - 6 = 0$
29. $x^2 + 36 = 0$ **31.** $\pm\sqrt{3}$ **33.** $1, 81$ **35.** $-1, 1, 5, 7$
37. $-\frac{1}{4}, \frac{1}{9}$ **39.** 1 **41.** $-1, 1, 4, 6$ **43.** $\pm 2, \pm 5$ **45.** $-1, 2$
47. $\pm\dfrac{\sqrt{15}}{3}, \pm\dfrac{\sqrt{6}}{2}$ **49.** $-1, 125$ **51.** $-\frac{11}{6}, -\frac{1}{6}$ **53.** $-\frac{3}{2}$
55. $\dfrac{9 \pm \sqrt{89}}{2}, -1 \pm \sqrt{3}$ **57.** $\left(\frac{4}{25}, 0\right)$ **59.** $(4, 0), (-1, 0),$
$\left(\dfrac{3 + \sqrt{33}}{2}, 0\right), \left(\dfrac{3 - \sqrt{33}}{2}, 0\right)$ **61.** $(-8, 0), (1, 0)$
63. Kenyan: 30 lb; Peruvian: 20 lb **64.** Solution A: 4 L;
solution B: 8 L **65.** $4x$ **66.** $3x^2$ **67.** $3a\sqrt[4]{2a}$ **68.** 4
69.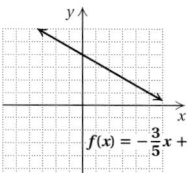
70.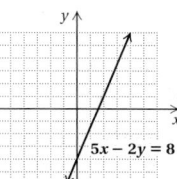

71.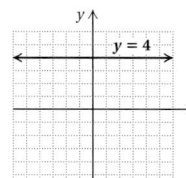
72.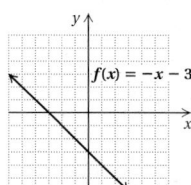

73. Left to the student **75. (a)** $-\frac{3}{5}$; **(b)** $-\frac{1}{3}$
77. $x^2 - \sqrt{3}x + 8 = 0$ **79.** $a = 1, b = 2, c = -3$ **81.** $\frac{100}{99}$
83. 259 **85.** $1, 3$

Mid-Chapter Review: Chapter 11, p. 824
1. False **2.** True **3.** True **4.** False
5.
$$5x^2 + 3x = 4$$
$$\frac{1}{5}(5x^2 + 3x) = \frac{1}{5} \cdot 4$$
$$x^2 + \frac{3}{5}x = \frac{4}{5}$$
$$x^2 + \frac{3}{5}x + \frac{9}{100} = \frac{4}{5} + \frac{9}{100}$$
$$\left(x + \frac{3}{10}\right)^2 = \frac{89}{100}$$
$$x + \frac{3}{10} = \sqrt{\frac{89}{100}} \quad or \quad x + \frac{3}{10} = -\sqrt{\frac{89}{100}}$$
$$x + \frac{3}{10} = \frac{\sqrt{89}}{10} \quad or \quad x + \frac{3}{10} = -\frac{\sqrt{89}}{10}$$
$$x = -\frac{3}{10} + \frac{\sqrt{89}}{10} \quad or \quad x = -\frac{3}{10} - \frac{\sqrt{89}}{10}$$

The solutions are $-\dfrac{3}{10} \pm \dfrac{\sqrt{89}}{10}$.

6.

$$5x^2 + 3x = 4$$
$$5x^2 + 3x - 4 = 0$$
$$5x^2 + 3x + (-4) = 0$$
$$a = 5, \quad b = 3, \quad c = -4$$
$$x = \frac{-b \pm \sqrt{b^2 - 4ac}}{2a}$$
$$x = \frac{-3 \pm \sqrt{3^2 - 4 \cdot 5 \cdot (-4)}}{2 \cdot 5}$$
$$x = \frac{-3 \pm \sqrt{9 + 80}}{10}$$
$$x = \frac{-3 \pm \sqrt{89}}{10}$$
$$x = -\frac{3}{10} \pm \frac{\sqrt{89}}{10}$$

7. $-2 \pm \sqrt{3}$ **8.** $-3, \frac{1}{2}$ **9.** $-5 \pm \sqrt{31}$ **10.** $\frac{1}{2} \pm \frac{\sqrt{21}}{2}$

11. One real solution; one x-intercept **12.** Two real solutions; two x-intercepts **13.** Two nonreal solutions; no x-intercepts **14.** Two nonreal solutions; no x-intercepts **15.** Two real solutions; two x-intercepts **16.** Two real solutions; two x-intercepts **17.** $x^2 - 9x - 10 = 0$ **18.** $x^2 - 169 = 0$ **19.** $x^2 - 2\sqrt{5}x - 15 = 0$ **20.** $x^2 + 16 = 0$ **21.** $x^2 + 12x + 36 = 0$ **22.** $21x^2 + 22x - 8 = 0$

23. 60 mph **24.** $s = \sqrt{\dfrac{R}{a}}$ **25.** $-\frac{4}{3}, 1$ **26.** $\pm\sqrt{3}, \pm\sqrt{5}$

27. $\dfrac{15 \pm \sqrt{145}}{8}$ **28.** $-1, -\frac{2}{7}$ **29.** $-1, 0$ **30.** $-11, 5$

31. $\pm\frac{4}{7}i$ **32.** $\pm\sqrt{6}, \pm 2i$ **33.** $\dfrac{-5 \pm \sqrt{73}}{2}$ **34.** $-6 \pm i$

35. $\dfrac{5 \pm \sqrt{11}}{2}$ **36.** $\dfrac{7 \pm \sqrt{13}}{6}$ **37.** $\dfrac{1 \pm \sqrt{2}}{2}$ **38.** $-1 \pm 4i$
39. $8 \pm \sqrt{3}$ **40.** $3 \pm \sqrt{10}i$ **41.** $4 \pm \sqrt{26}$ **42.** 9
43. Given the solutions of a quadratic equation, it is possible to find an equation equivalent to the original equation but not necessarily expressed in the same form as the original equation. For example, we can find a quadratic equation with solutions -2 and 4:

$$[x - (-2)](x - 4) = 0$$
$$(x + 2)(x - 4) = 0$$
$$x^2 - 2x - 8 = 0.$$

Now $x^2 - 2x - 8 = 0$ has solutions -2 and 4. However, the original equation might have been in another form, such as $2x(x - 3) - x(x - 4) = 8$. **44.** Given the quadratic equation $ax^2 + bx + c = 0$, we find $x = \dfrac{-b + \sqrt{b^2 - 4ac}}{2a}$ or

$x = \dfrac{-b - \sqrt{b^2 - 4ac}}{2a}$ using the quadratic formula.
Then we have $ax^2 + bx + c =$
$\left(x - \dfrac{-b + \sqrt{b^2 - 4ac}}{2a}\right)\left(x - \dfrac{-b - \sqrt{b^2 - 4ac}}{2a}\right).$
Consider $5x^2 + 8x - 3$. First, we use the quadratic formula to solve $5x^2 + 8x - 3 = 0$:

$$x = \frac{-8 \pm \sqrt{8^2 - 4 \cdot 5 \cdot (-3)}}{2 \cdot 5}$$
$$x = \frac{-8 \pm \sqrt{124}}{10} = \frac{-8 \pm 2\sqrt{31}}{10}$$
$$x = \frac{-4 \pm \sqrt{31}}{5}.$$

Then $5x^2 + 8x - 3 = \left(x - \dfrac{-4 - \sqrt{31}}{5}\right)\left(x - \dfrac{-4 + \sqrt{31}}{5}\right).$

45. Set the product
$(x - 1)(x - 2)(x - 3)(x - 4)(x - 5)(x - 6)(x - 7)$
equal to 0. **46.** Write an equation of the form
$a(3x^2 + 1)^2 + b(3x^2 + 1) + c = 0$, where $a \neq 0$. To ensure that this equation has real-number solutions, select a, b, and c so that $b^2 - 4ac \geq 0$ and $3x^2 + 1 \geq 0$.

Exercise Set 11.5, p. 832

1.

x	$f(x)$
0	0
1	4
2	16
-1	4
-2	16

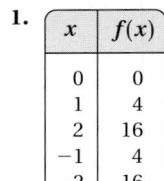

3.

x	$f(x)$
0	0
1	$\frac{1}{3}$
2	$\frac{4}{3}$
-1	$\frac{1}{3}$
-2	$\frac{4}{3}$

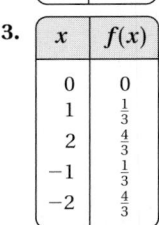

5.

x	$f(x)$
-3	0
-2	1
-1	4
-4	1
-5	4

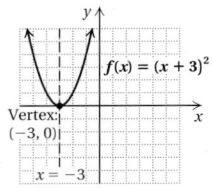

7.
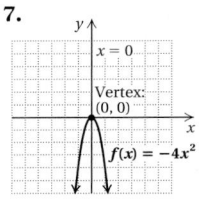

9.
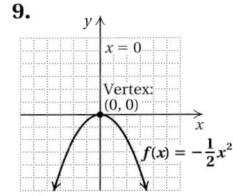

11.
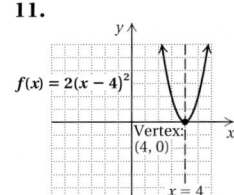

13.

x	$f(x)$
-2	0
-3	-2
-1	-2
-4	-8
0	-8

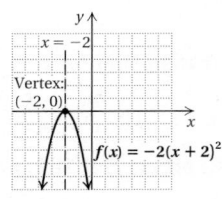

15.
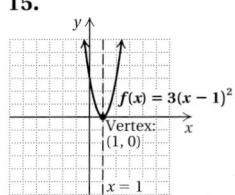

17.
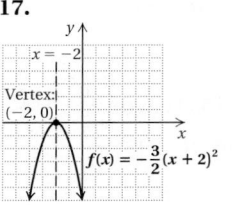

19.
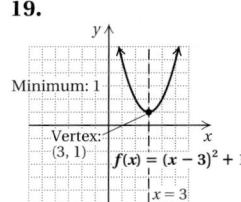

21.
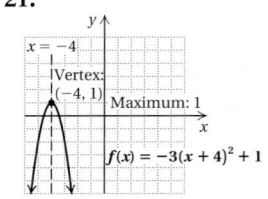

23.

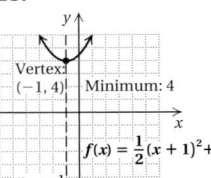

Vertex: $(-1, 4)$; Minimum: 4; $f(x) = \frac{1}{2}(x+1)^2 + 4$; $x = -1$

25.

$x = -1$; Maximum: -2; Vertex: $(-1, -2)$; $f(x) = -(x+1)^2 - 2$

27. $5xy^2\sqrt[4]{x}$ **28.** $12a^2b^2$

Visualizing for Success, p. 840

1. F   **2.** H   **3.** A   **4.** I   **5.** C   **6.** J   **7.** G   **8.** B
9. E   **10.** D

Exercise Set 11.6, p. 841

1.

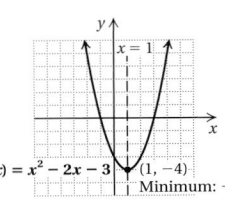

$x = 1$; $f(x) = x^2 - 2x - 3$; $(1, -4)$; Minimum: -4

3.

$x = -2$; $(-2, 2)$; Maximum: 2; $f(x) = -x^2 - 4x - 2$

5.

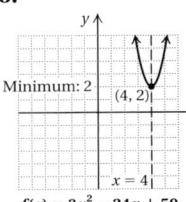

Minimum: 2; $(4, 2)$; $x = 4$; $f(x) = 3x^2 - 24x + 50$

7.

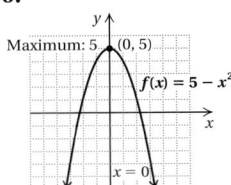

$\left(-\frac{1}{2}, \frac{7}{2}\right)$; Maximum: $\frac{7}{2}$; $x = -\frac{1}{2}$; $f(x) = -2x^2 - 2x + 3$

9.

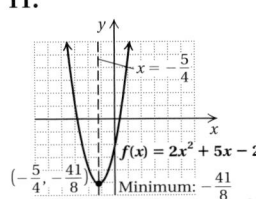

Maximum: 5; $(0, 5)$; $f(x) = 5 - x^2$; $x = 0$

11.

$x = -\frac{5}{4}$; $f(x) = 2x^2 + 5x - 2$; $\left(-\frac{5}{4}, -\frac{41}{8}\right)$; Minimum: $-\frac{41}{8}$

13. y-intercept: $(0, 1)$; x-intercepts: $\left(3 + 2\sqrt{2}, 0\right), \left(3 - 2\sqrt{2}, 0\right)$
15. y-intercept: $(0, 20)$; x-intercepts: $(5, 0), (-4, 0)$
17. y-intercept: $(0, 9)$; x-intercept: $\left(-\frac{3}{2}, 0\right)$ **19.** y-intercept: $(0, 8)$; x-intercepts: none   **21.** $D = 15w$ **22.** $C = \frac{89}{6}t$
23. 250; $y = \dfrac{250}{x}$ **24.** 250; $y = \dfrac{250}{x}$ **25.** $\frac{125}{2}$; $y = \frac{125}{2}x$
26. $\frac{2}{125}$; $y = \frac{2}{125}x$ **27. (a)** Minimum: -6.954;
(b) maximum: 7.014
29.

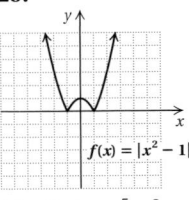

$f(x) = |x^2 - 1|$

31.

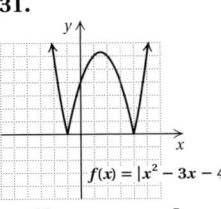

$f(x) = |x^2 - 3x - 4|$

33. $f(x) = \frac{5}{16}x^2 - \frac{15}{8}x - \frac{35}{16}$, or $f(x) = \frac{5}{16}(x - 3)^2 - 5$
35.

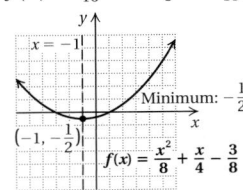

$x = -1$; Minimum: $-\frac{1}{2}$; $\left(-1, -\frac{1}{2}\right)$; $f(x) = \frac{x^2}{8} + \frac{x}{4} - \frac{3}{8}$

Calculator Corner, p. 846

1. Minimum: 1   **2.** Minimum: 4.875   **3.** Maximum: 6
4. Maximum: 0.5625

Exercise Set 11.7, p. 851

1. 180 ft by 180 ft   **3.** 3.5 in.   **5.** 3.5 hundred, or 350
7. 200 ft^2; 10 ft by 20 ft   **9.** 11 days after the concert was
announced; about 62 tickets   **11.** $P(x) = -x^2 + 980x - 3000$;
$\$237,100$ at $x = 490$ **13.** 121; 11 and 11   **15.** -4; 2 and -2
17. 36; -6 and -6 **19.** $f(x) = mx + b$
21. $f(x) = ax^2 + bx + c, a > 0$ **23.** Polynomial, neither
quadratic nor linear   **25.** $f(x) = ax^2 + bx + c, a < 0$
27. $f(x) = 2x^2 + 3x - 1$ **29.** $f(x) = -\frac{1}{4}x^2 + 3x - 5$
31. (a) $A(s) = \frac{3}{16}s^2 - \frac{135}{4}s + 1750$; **(b)** about 531 per 200,000,000
kilometers driven   **33.** $D(x) = -0.008x^2 + 0.8x$; 15 ft
35. Radical; radicand   **36.** Dependent   **37.** Sum
38. At least one   **39.** Inverse   **40.** Independent
41. Descending   **42.** x-intercept   **43.** $b = 19$ cm,
$h = 19$ cm; $A = 180.5$ cm^2

Calculator Corner, p. 856

1. $\{x|x < -4 \text{ or } x > 1\}$, or $(-\infty, -4) \cup (1, \infty)$
2. $\{x|-2 < x < 3\}$, or $(-2, 3)$
3. $\{x|x \le -2 \text{ or } 0 \le x \le 0.5\}$, or $(-\infty, -2] \cup [0, 0.5]$
4. $\{x|-4 \le x \le 0 \text{ or } x \ge 4\}$, or $[-4, 0] \cup [4, \infty)$

Exercise Set 11.8, p. 861

1. $\{x|x < -2 \text{ or } x > 6\}$, or $(-\infty, -2) \cup (6, \infty)$
3. $\{x|-2 \le x \le 2\}$, or $[-2, 2]$ **5.** $\{x|-1 \le x \le 4\}$, or $[-1, 4]$
7. $\{x|-1 < x < 2\}$, or $(-1, 2)$ **9.** All real numbers, or
$(-\infty, \infty)$ **11.** $\{x|2 < x < 4\}$, or $(2, 4)$
13. $\{x|x < -2 \text{ or } 0 < x < 2\}$, or $(-\infty, -2) \cup (0, 2)$
15. $\{x|-9 < x < -1 \text{ or } x > 4\}$, or $(-9, -1) \cup (4, \infty)$
17. $\{x|x < -3 \text{ or } -2 < x < 1\}$, or $(-\infty, -3) \cup (-2, 1)$
19. $\{x|x < 6\}$, or $(-\infty, 6)$ **21.** $\{x|x < -1 \text{ or } x > 3\}$, or
$(-\infty, -1) \cup (3, \infty)$ **23.** $\left\{x|-\frac{2}{3} \le x < 3\right\}$, or $\left[-\frac{2}{3}, 3\right)$
25. $\left\{x|2 < x < \frac{5}{2}\right\}$, or $\left(2, \frac{5}{2}\right)$ **27.** $\{x|x < -1 \text{ or } 2 < x < 5\}$, or
$(-\infty, -1) \cup (2, 5)$ **29.** $\{x|-3 \le x < 0\}$, or $[-3, 0)$
31. $\{x|1 < x < 2\}$, or $(1, 2)$ **33.** $\{x|x < -4 \text{ or } 1 < x < 3\}$,
or $(-\infty, -4) \cup (1, 3)$ **35.** $\left\{x|0 < x < \frac{1}{3}\right\}$, or $\left(0, \frac{1}{3}\right)$
37. $\{x|x < -3 \text{ or } -2 < x < 1 \text{ or } x > 4\}$, or
$(-\infty, -3) \cup (-2, 1) \cup (4, \infty)$ **39.** $\frac{5}{3}$ **40.** $\dfrac{5}{2a}$ **41.** $\dfrac{4a}{b^2}\sqrt{a}$
42. $\dfrac{3c}{7d}\sqrt[3]{c^2}$ **43.** $\sqrt{2}$ **44.** $17\sqrt{5}$ **45.** $(10a + 7)\sqrt[3]{2a}$
46. $3\sqrt{10} - 4\sqrt{5}$ **47.** Left to the student
49. $\{x|1 - \sqrt{3} \le x \le 1 + \sqrt{3}\}$, or $[1 - \sqrt{3}, 1 + \sqrt{3}]$
51. All real numbers except 0, or $(-\infty, 0) \cup (0, \infty)$
53. $\left\{x|x < \frac{1}{4} \text{ or } x > \frac{5}{2}\right\}$, or $\left(-\infty, \frac{1}{4}\right) \cup \left(\frac{5}{2}, \infty\right)$
55. (a) $\{t|0 < t < 2\}$, or $(0, 2)$; **(b)** $\{t|t > 10\}$, or $(10, \infty)$

Summary and Review: Chapter 11, p. 863

Concept Reinforcement

1. False   **2.** True   **3.** False

Important Concepts

1. $2 \pm 3i$ **2.** $6 \pm \sqrt{5}$ **3.** $5 \pm \sqrt{2}$, or 6.414 and 3.586
4. (a) Two real solutions; **(b)** two nonreal solutions
5. $5x^2 - 13x - 6 = 0$ **6.** $\pm\sqrt{2}, \pm 3$ **7.** Vertex: $(-1, -2)$;
line of symmetry: $x = -1$; maximum: -2;

$x = -1$; $(-1, -2)$; Maximum: -2

8. y-intercept: $(0, 4)$; x-intercepts: $\left(3 - \sqrt{5}, 0\right)$ and $\left(3 + \sqrt{5}, 0\right)$
9. $\{x | x < 4 \text{ or } x > 10\}$, or $(-\infty, 4) \cup (10, \infty)$
10. $\{x | 5 < x \le 11\}$, or $(5, 11]$

Review Exercises

1. (a) $\pm\dfrac{\sqrt{14}}{2}$; **(b)** $\left(-\dfrac{\sqrt{14}}{2}, 0\right), \left(\dfrac{\sqrt{14}}{2}, 0\right)$ **2.** $0, -\dfrac{5}{14}$ **3.** $3, 9$

4. $-\dfrac{3}{8} \pm \dfrac{\sqrt{7}}{8}i$ **5.** $\dfrac{7}{2} \pm \dfrac{\sqrt{3}}{2}i$ **6.** $3, 5$ **7.** $-2 \pm \sqrt{3}$;

$-0.268, -3.732$ **8.** $4, -2$ **9.** $4 \pm 4\sqrt{2}$ **10.** $\dfrac{1 \pm \sqrt{481}}{15}$

11. $-3 \pm \sqrt{7}$ **12.** 0.901 sec **13.** Length: 14 cm; width: 9 cm
14. 1 in. **15.** First part: 50 mph; second part: 40 mph
16. Two real **17.** Two nonreal **18.** $25x^2 + 10x - 3 = 0$
19. $x^2 + 8x + 16 = 0$ **20.** $p = \dfrac{9\pi^2}{N^2}$ **21.** $T = \sqrt{\dfrac{3B}{2A}}$
22. $2, -2, 3, -3$ **23.** $3, -5$ **24.** $\pm\sqrt{7}, \pm\sqrt{2}$ **25.** $81, 16$
26. (a) $(1, 3)$; **(b)** $x = 1$; **(c)** maximum: 3;
(d)

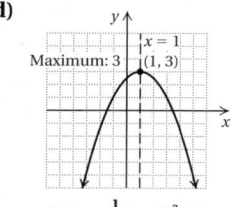

$f(x) = -\frac{1}{2}(x - 1)^2 + 3$

27. (a) $\left(\frac{1}{2}, \frac{23}{4}\right)$; **(b)** $x = \frac{1}{2}$; **(c)** minimum: $\frac{23}{4}$;
(d)

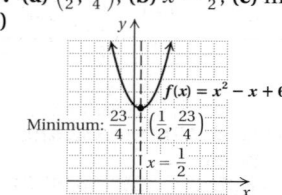

28. (a) $(-2, 4)$; **(b)** $x = -2$; **(c)** maximum: 4;
(d)

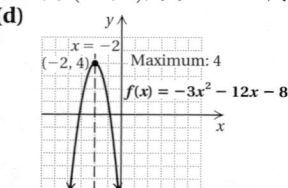

29. y-intercept: $(0, 14)$; x-intercepts: $(2, 0), (7, 0)$
30. y-intercept: $(0, -3)$; x-intercepts: $\left(2 - \sqrt{7}, 0\right)$ and $\left(2 + \sqrt{7}, 0\right)$ **31.** -121; 11 and -11
32. $f(x) = -x^2 + 6x - 2$ **33. (a)** $N(x) = -0.720x^2 + 38.211x - 393.127$; **(b)** about 105 live births
34. $\{x | -2 < x < 1 \text{ or } x > 2\}$, or $(-2, 1) \cup (2, \infty)$
35. $\{x | x < -4 \text{ or } -2 < x < 1\}$, or $(-\infty, -4) \cup (-2, 1)$
36. B **37.** D **38.** $f(x) = \frac{7}{15}x^2 - \frac{14}{15}x - 7$; minimum: $-\frac{112}{15}$
39. $h = 60, k = 60$ **40.** 18 and 324

Understanding Through Discussion and Writing

1. Yes; for any quadratic function $f(x) = ax^2 + bx + c$, $f(0) = c$, so the graph of every quadratic function has a y-intercept, $(0, c)$.
2. If the leading coefficient is positive, the graph of the function opens up and hence has a minimum value. If the leading coefficient is negative, the graph of the function opens down and hence has a maximum value. **3.** When an input of $y = (x + 3)^2$ is 3 less than (or 3 units to the left of) an input of $y = x^2$, the outputs are the same. In addition, for any input, the output of $f(x) = (x + 3)^2 - 4$ is 4 less than (or 4 units down from) the output of $f(x) = (x + 3)^2$. Thus the graph of $f(x) = (x + 3)^2 - 4$ looks like the graph of $f(x) = x^2$ translated 3 units to the left and 4 units down. **4.** Find a quadratic

function $f(x)$ whose graph lies entirely above the x-axis or a quadratic function $g(x)$ whose graph lies entirely below the x-axis. Then write $f(x) < 0, f(x) \le 0, g(x) > 0$, or $g(x) \ge 0$. For example, the quadratic inequalities $x^2 + 1 < 0$ and $-x^2 - 5 \ge 0$ have no solution. **5.** No; if the vertex is off the x-axis, then due to symmetry, the graph has either no x-intercept or two x-intercepts. **6.** The x-coordinate of the vertex lies halfway between the x-coordinates of the x-intercepts. The function must be evaluated for this value of x in order to determine the maximum or minimum value.

Test: Chapter 11, p. 869

1. [11.1a] **(a)** $\pm\dfrac{2\sqrt{3}}{3}$; **(b)** $\left(\dfrac{2\sqrt{3}}{3}, 0\right), \left(-\dfrac{2\sqrt{3}}{3}, 0\right)$
2. [11.2a] $-\dfrac{1}{2} \pm \dfrac{\sqrt{3}}{2}i$ **3.** [11.4c] $49, 1$ **4.** [11.2a] $9, 2$
5. [11.4c] $\pm\dfrac{\sqrt{5}}{2}, \pm\sqrt{3}$ **6.** [11.2a] $-2 \pm \sqrt{6}$; $0.449, -4.449$
7. [11.2a] $0, 2$ **8.** [11.1b] $2 \pm \sqrt{3}$ **9.** [11.1c] About 6.7 sec
10. [11.3a] About 2.89 mph **11.** [11.3a] 7 cm by 7 cm
12. [11.1c] About 0.946 sec **13.** [11.4a] Two nonreal
14. [11.4b] $x^2 - 4\sqrt{3}x + 9 = 0$
15. [11.3b] $T = \sqrt{\dfrac{V}{48}}$, or $\dfrac{\sqrt{3V}}{12}$
16. [11.6a] **(a)** $(-1, 1)$; **(b)** $x = -1$; **(c)** maximum: 1;
(d)

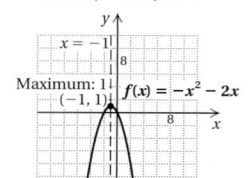

17. [11.6a] **(a)** $(3, 5)$; **(b)** $x = 3$; **(c)** minimum: 5;
(d)

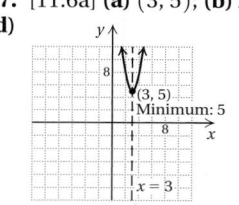

$f(x) = 4x^2 - 24x + 41$

18. [11.6b] y-intercept: $(0, -1)$; x-intercepts: $\left(2 - \sqrt{3}, 0\right)$, $\left(2 + \sqrt{3}, 0\right)$ **19.** [11.7a] -16; 4 and -4
20. [11.7b] $f(x) = \frac{1}{5}x^2 - \frac{3}{5}x$
21. [11.7b] **(a)** $A(x) = -0.3x^2 + 2.3x + 18.5$; **(b)** about 14.9 thousand adoptions **22.** [11.8a] $\{x | -1 < x < 7\}$, or $(-1, 7)$
23. [11.8b] $\{x | -3 < x < 5\}$, or $(-3, 5)$
24. [11.8b] $\{x | -3 < x < 1 \text{ or } x \ge 2\}$, or $(-3, 1) \cup [2, \infty)$
25. [11.4b] C **26.** [11.6a, b] $f(x) = -\frac{4}{7}x^2 + \frac{20}{7}x + 8$; maximum: $\frac{81}{7}$ **27.** [11.2a] $\frac{1}{2}$

Cumulative Review: Chapters 1–11, p. 871

1. [10.7a] About 422 yd **2.** [4.4c] $10x^2 - 8x + 6$
3. [4.5d] $2x^3 - 9x^2 + 7x - 12$ **4.** [6.1d] $\dfrac{2(a - 4)}{5}$
5. [6.2b] $\dfrac{1}{y^2 + 6y}$ **6.** [6.5b] $\dfrac{-(m - 3)(m - 2)}{(m + 1)(m - 5)}$
7. [4.8b] $9x^2 - 13x + 26 + \dfrac{-50}{x + 2}$ **8.** [6.6a] $\dfrac{y - x}{xy(x + y)}$
9. [10.1b] 0.6 **10.** [10.1b] $3(x - 2)$ **11.** [10.4a] $12\sqrt{5}$
12. [10.5b] $\dfrac{\sqrt{6} + 9\sqrt{2} - 12\sqrt{3} - 4}{-26}$ **13.** [10.2d] 256
14. [10.8c] $17 + 7i$ **15.** [10.8e] $-\frac{2}{3} - 2i$
16. [5.3a], [5.4a] $(2t + 5)(t - 6)$ **17.** [5.2a] $(a + 9)(a - 6)$
18. [5.1b] $-3a^2(a - 4)$ **19.** [5.5d] $(8a + 3b)(8a - 3b)$
20. [5.5b] $3(a - 6)^2$ **21.** [5.6a] $\left(\frac{1}{3}a - 1\right)\left(\frac{1}{9}a^2 + \frac{1}{3}a + 1\right)$
22. [5.1c] $(4a + 3)(6a^2 - 5)$ **23.** [5.1b] $(x + 1)(2x + 1)$

24. [2.3c] $\frac{11}{13}$ **25.** [2.4a] $r = \dfrac{mv^2}{F}$ **26.** [9.1c] $\left\{x \middle| x \geq \frac{5}{14}\right\}$, or $\left[\frac{5}{14}, \infty\right)$ **27.** [9.2b] $\left\{x \middle| x < -\frac{4}{3} \text{ or } x > 6\right\}$, or $\left(-\infty, -\frac{4}{3}\right) \cup (6, \infty)$ **28.** [9.3e] $\left\{x \middle| -\frac{13}{4} \leq x \leq \frac{15}{4}\right\}$, or $\left[-\frac{13}{4}, \frac{15}{4}\right]$

29. [8.3a] $(-4, 1)$ **30.** [8.5a] $\left(\frac{1}{2}, 3, -5\right)$

31. [5.8b] $\frac{1}{5}$, -3 **32.** [6.7a] $-\frac{5}{3}$ **33.** [6.7a] 3

34. [2.4a], [6.7a] $m = \dfrac{aA}{h - A}$ **35.** [10.6a] $\frac{37}{2}$ **36.** [10.6b] 11

37. [5.8b] 4 **38.** [11.2a] $\dfrac{3}{2} \pm \dfrac{\sqrt{55}}{2}i$ **39.** [11.2a] $\dfrac{17 \pm \sqrt{145}}{2}$

40. [11.3b] $a = \sqrt{P^2 + b^2}$

41. [11.8b] $\{x \mid -3 < x < -2 \text{ or } -1 < x < 1\}$, or $(-3, -2) \cup (-1, 1)$

42. [11.8a] $\left\{x \middle| x < -\frac{5}{2} \text{ or } x > \frac{5}{2}\right\}$, or $\left(-\infty, -\frac{5}{2}\right) \cup \left(\frac{5}{2}, \infty\right)$

43. [3.2a]   **44.** [9.4b]   **45.** [9.4b]

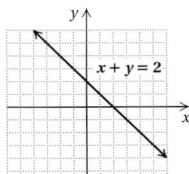 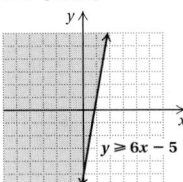 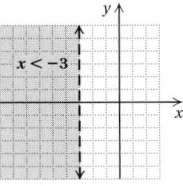

46. [9.4c]   **47.** [11.6a]   **48.** [11.6a]

 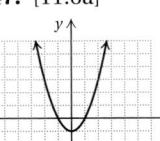 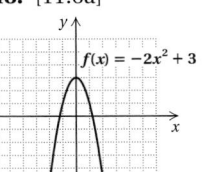

49. [7.5b] $y = \frac{1}{2}x + 4$ **50.** [7.5d] $y = -3x + 1$
51. [11.3a] 16 km/h   **52.** [11.7a] 14 ft by 14 ft; 196 ft^2
53. [8.2b], [8.3b] 36 **54.** [6.8a] 2 hr   **55.** [11.1b] A
56. [11.4c] B   **57.** [11.4c] $\dfrac{2}{51 + 7\sqrt{61}}$, or $\dfrac{-51 + 7\sqrt{61}}{194}$

58. [5.6a] $\left(\dfrac{a}{2} + \dfrac{2b}{9}\right)\left(\dfrac{a^2}{4} - \dfrac{ab}{9} + \dfrac{4b^2}{81}\right)$

CHAPTER 12

Calculator Corner, p. 878

1. Left to the student   **2.** Left to the student

Calculator Corner, p. 882

1. $1040.60   **2.** $1049.12   **3.** $30,372.65   **4.** **(a)** $10,540; **(b)** $10,547.29; **(c)** $10,551.03; **(d)** $10,554.80; **(e)** $10,554.84

Exercise Set 12.1, p. 883

1.

x	$f(x)$
0	1
1	2
2	4
3	8
-1	$\frac{1}{2}$
-2	$\frac{1}{4}$
-3	$\frac{1}{8}$

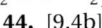

3.

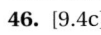

5.

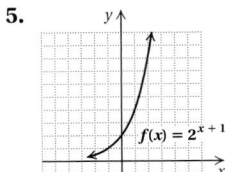

7.

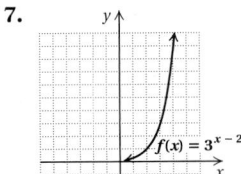

9.

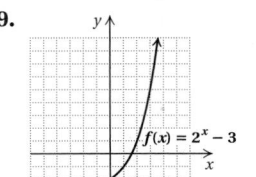

11.

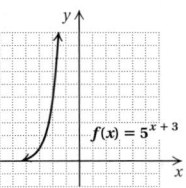

13.

x	$f(x)$
0	1
1	$\frac{1}{2}$
2	$\frac{1}{4}$
3	$\frac{1}{8}$
-1	2
-2	4
-3	8

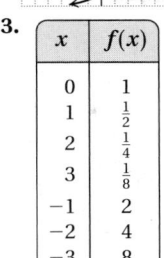

15.

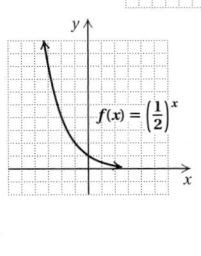

17.

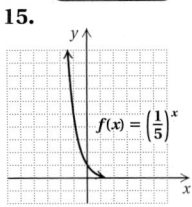

19.

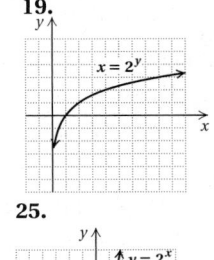

21.

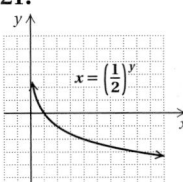

23.

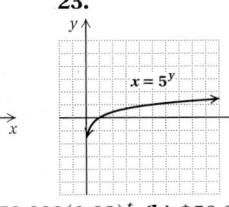

25.

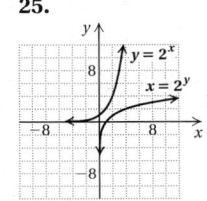

27. **(a)** $A(t) = \$50,000(1.02)^t$; **(b)** $50,000; $51,000; $52,020; $54,121.61; $58,582.97; $60,949.72; $74,297.37;   **29.** $2161.16   **31.** $5287.54
(c)

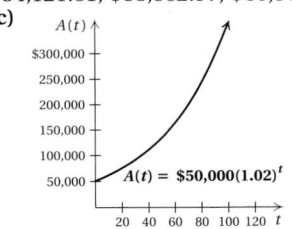

33. **(a)** 12,279 MW; 24,274 MW; 47,986 MW;
(b)

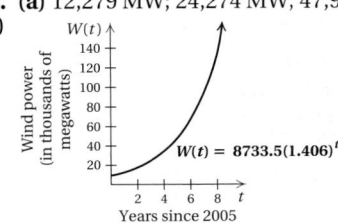

35. **(a)** $5105; $2487; $1212; **(b)**

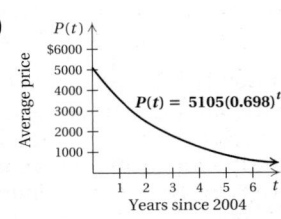

37. **(a)** 257,500 cans; 68,295 cans; 4804 cans; **(b)**

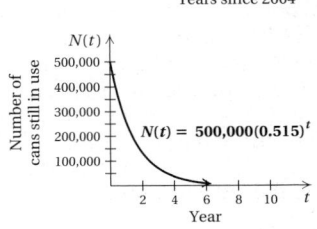

39. $\dfrac{1}{x^2}$ **40.** $\dfrac{1}{x^{12}}$ **41.** 1 **42.** 1 **43.** $\frac{2}{3}$ **44.** 2.7

45. $\dfrac{1}{x^7}$ **46.** $\dfrac{1}{x^{10}}$ **47.** x **48.** x **49.** 5^4, or 625

51.

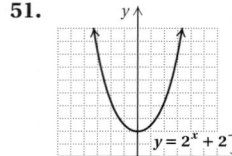

53.

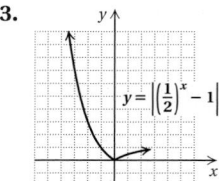

55.

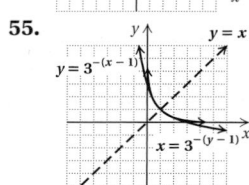

57. Left to the student

Calculator Corner, p. 896

1.

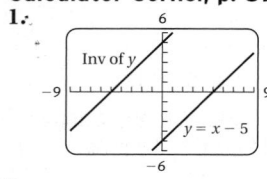

2.

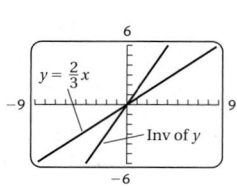

3.

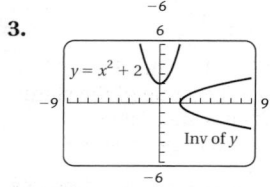

4.

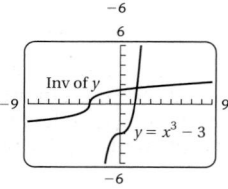

Exercise Set 12.2, p. 900

1. Inverse: $\{(2, 1), (-3, 6), (-5, -3)\}$

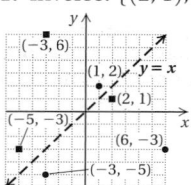

3. Inverse: $x = 2y + 6$

x	y
4	-1
6	0
8	1
10	2
12	3

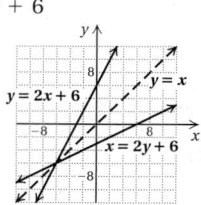

5. Yes **7.** No **9.** No **11.** Yes **13.** $f^{-1}(x) = \dfrac{x + 2}{5}$

15. $f^{-1}(x) = \dfrac{-2}{x}$ **17.** $f^{-1}(x) = \frac{3}{4}(x - 7)$

19. $f^{-1}(x) = \dfrac{2}{x} - 5$ **21.** Not one-to-one

23. $f^{-1}(x) = \dfrac{1 - 3x}{5x - 2}$ **25.** $f^{-1}(x) = \sqrt[3]{x + 1}$

27. $f^{-1}(x) = x^3$
29. $f^{-1}(x) = 2x + 6$

x	$f(x)$
-4	-5
0	-3
2	-2
4	-1

x	$f^{-1}(x)$
-5	-4
-3	0
-2	2
-1	4

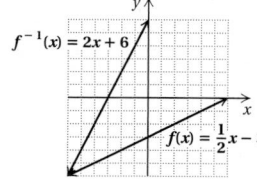

A-38 Answers

31. $f^{-1}(x) = \sqrt[3]{x}$

x	$f(x)$
0	0
1	1
2	8
3	27
-1	-1
-2	-8
-3	-27

x	$f^{-1}(x)$
0	0
1	1
8	2
27	3
-1	-1
-8	-2
-27	-3

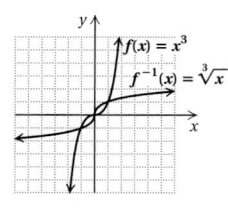

33. $-8x + 9; -8x + 18$ **35.** $12x^2 - 12x + 5; 6x^2 + 3$

37. $\dfrac{16}{x^2} - 1; \dfrac{2}{4x^2 - 1}$ **39.** $x^4 - 10x^2 + 30; x^4 + 10x^2 + 20$

41. $f(x) = x^2, g(x) = 5 - 3x$ **43.** $f(x) = \sqrt{x}, g(x) = 5x + 2$

45. $f(x) = \dfrac{1}{x}, g(x) = x - 1$ **47.** $f(x) = \dfrac{1}{\sqrt{x}}, g(x) = 7x + 2$

49. $f(x) = x^4, g(x) = \sqrt{x} + 5$

51. $(f^{-1} \circ f)(x) = f^{-1}(f(x)) = f^{-1}\left(\frac{4}{5}x\right) = \frac{5}{4}\left(\frac{4}{5}x\right) = x;$
$(f \circ f^{-1})(x) = f(f^{-1}(x)) = f\left(\frac{5}{4}x\right) = \frac{4}{5}\left(\frac{5}{4}x\right) = x$

53. $(f^{-1} \circ f)(x) = f^{-1}(f(x)) = f^{-1}\left(\dfrac{x + 7}{2}\right)$

$\quad = 2\left(\dfrac{x + 7}{2}\right) - 7 = x + 7 - 7 = x;$

$\quad (f \circ f^{-1})(x) = f(f^{-1}(x)) = f(2x - 7)$

$\quad = \dfrac{2x - 7 + 7}{2} = \dfrac{2x}{2} = x$

55. $(f^{-1} \circ f)(x) = f^{-1}(f(x)) = f^{-1}\left(\dfrac{1 - x}{x}\right)$

$\quad = \dfrac{1}{\dfrac{1 - x}{x} + 1} = \dfrac{1}{\dfrac{1}{x}} = x;$

$\quad (f \circ f^{-1})(x) = f(f^{-1}(x)) = f\left(\dfrac{1}{x + 1}\right)$

$\quad = \dfrac{1 - \dfrac{1}{x + 1}}{\dfrac{1}{x + 1}} = \dfrac{\dfrac{x}{x + 1}}{\dfrac{1}{x + 1}} = x$

57. $f^{-1}(x) = \frac{1}{3}x$ **59.** $f^{-1}(x) = -x$ **61.** $f^{-1}(x) = x^3 + 5$

63. **(a)** 40, 42, 46, 50; **(b)** $f^{-1}(x) = x - 32$; **(c)** 8, 10, 14, 18

65. $\sqrt[3]{a}$ **66.** $\sqrt[3]{x^2}$ **67.** a^2b^3 **68.** $2t^2$ **69.** $\sqrt{3}$ **70.** $2\sqrt[4]{2}$

71. $\sqrt{2xy}$ **72.** $\sqrt[4]{p^2t}$ **73.** $2a^3b^8$ **74.** $10x^3y^6$ **75.** $3a^2b^2$

76. $3pq^3$ **77.** No **79.** Yes **81.** (1) C; (2) A; (3) B; (4) D

83.

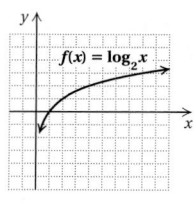

85. $f(x) = \frac{1}{2}x + 3$;
$g(x) = 2x - 6$; yes

Exercise Set 12.3, p. 912

1. $x = 2^y$

x, or 2^y	y
1	0
2	1
4	2
8	3
$\frac{1}{2}$	-1
$\frac{1}{4}$	-2
$\frac{1}{8}$	-3

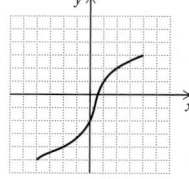

3. 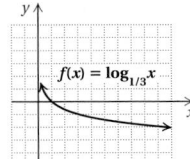   **5.**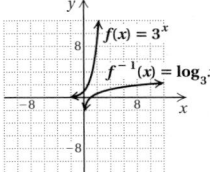

7. $3 = \log_{10} 1000$ **9.** $-3 = \log_5 \frac{1}{125}$ **11.** $\frac{1}{3} = \log_8 2$

13. $0.3010 = \log_{10} 2$ **15.** $2 = \log_e t$ **17.** $t = \log_Q x$

19. $2 = \log_e 7.3891$ **21.** $-2 = \log_e 0.1353$ **23.** $4^w = 10$

25. $6^2 = 36$ **27.** $10^{-2} = 0.01$ **29.** $10^{0.9031} = 8$

31. $e^{4.6052} = 100$ **33.** $t^k = Q$ **35.** 9   **37.** 4   **39.** 4

41. 3   **43.** 25   **45.** 1   **47.** $\frac{1}{2}$ **49.** 2   **51.** 2   **53.** -1

55. 0   **57.** 4   **59.** 2   **61.** 3   **63.** -2 **65.** 0   **67.** 1

69. $\frac{2}{3}$ **71.** 4.8970   **73.** -0.1739 **75.** Does not exist as a real number   **77.** 0.9464   **79.** $6 = 10^{0.7782}$; $84 = 10^{1.9243}$; $987,606 = 10^{5.9946}$; $0.00987606 = 10^{-2.0054}$; $98,760.6 = 10^{4.9946}$; $70,000,000 = 10^{7.8451}$; $7000 = 10^{3.8451}$ **81.** Conjugate

82. Direct   **83.** Coefficient; exponent   **84.** Quadratic; discriminant   **85.** Inconsistent   **86.** Parabolas

87. Line of symmetry   **88.** $a + bi$

89. 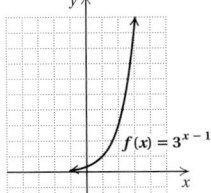   **91.** 25   **93.** 32
95. $-\frac{7}{16}$ **97.** 3
99. 0   **101.** -2

Calculator Corner, p. 917

1. Not correct   **2.** Correct   **3.** Not correct   **4.** Correct
5. Not correct   **6.** Correct   **7.** Not correct   **8.** Not correct

Exercise Set 12.4, p. 920

1. $\log_2 32 + \log_2 8$ **3.** $\log_4 64 + \log_4 16$ **5.** $\log_a Q + \log_a x$

7. $\log_b 252$ **9.** $\log_c Ky$ **11.** $4 \log_c y$ **13.** $6 \log_b t$

15. $-3 \log_b C$ **17.** $\log_a 67 - \log_a 5$ **19.** $\log_b 2 - \log_b 5$

21. $\log_c \frac{22}{3}$ **23.** $2 \log_a x + 3 \log_a y + \log_a z$

25. $\log_b x + 2 \log_b y - 3 \log_b z$ **27.** $\frac{4}{3} \log_c x - \log_c y - \frac{2}{3} \log_c z$

29. $2 \log_a m + 3 \log_a n - \frac{3}{4} - \frac{5}{4} \log_a b$ **31.** $\log_a \frac{x^{2/3}}{y^{1/2}}$, or $\log_a \frac{\sqrt[3]{x^2}}{\sqrt{y}}$ **33.** $\log_a \frac{2x^4}{y^3}$ **35.** $\log_a \frac{\sqrt{a}}{x}$ **37.** 2.708

39. 0.51   **41.** -1.609 **43.** $\frac{1}{2}$ **45.** 2.609   **47.** Cannot be found using the properties of logarithms   **49.** t **51.** 5

53. 7   **55.** -7 **57.** i **58.** -1 **59.** 5   **60.** $\frac{3}{5} + \frac{4}{5}i$

61. $23 - 18i$ **62.** $10i$ **63.** $-34 - 31i$ **64.** $3 - 4i$

65. Left to the student   **67.** $\log_a(x^6 - x^4 y^2 + x^2 y^4 - y^6)$

69. $\frac{1}{2} \log_a(1 - s) + \frac{1}{2} \log_a(1 + s)$ **71.** False   **73.** True

75. False

Mid-Chapter Review: Chapter 12, p. 922

1. False   **2.** True   **3.** False   **4.** True

5. $\log_5 x = 3$
$\quad 5^3 = x$
$\quad 125 = x$

6. (a) $\log_a 18 = \log_a(2 \cdot 9) = \log_a 2 + \log_a 9 = 0.648 + 2.046 = 2.694$; **(b)** $\log_a \frac{1}{2} = \log_a 1 - \log_a 2 = 0 - 0.648 = -0.648$

7. 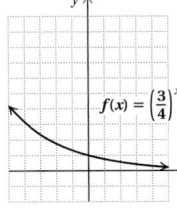   **8.** (graph)

9. 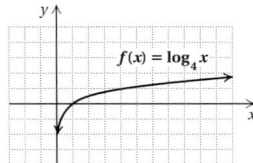   **10.**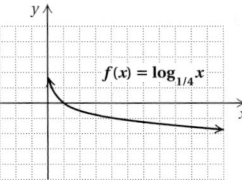

11. **(a)** $A(t) = \$500(1.04)^t$; **(b)** $\$500$; $\$584.93$; $\$740.12$

12. $\$1580.49$ **13.** $f^{-1}(x) = \frac{x - 1}{3}$ **14.** $f^{-1}(x) = \sqrt[3]{x - 2}$

15. $1 - 2x$; $8 - 2x$ **16.** $9x^2 - 6x + 2$; $3x^2 + 2$

17. $f(x) = \frac{3}{x}$; $g(x) = x + 4$ **18.** $f(x) = \sqrt{x}$; $g(x) = 6x - 7$

19. $(f^{-1} \circ f)(x) = f^{-1}(f(x)) = f^{-1}\left(\frac{x}{3}\right) = 3\left(\frac{x}{3}\right) = x$;

$\quad (f \circ f^{-1})(x) = f(f^{-1}(x)) = f(3x) = \frac{3x}{3} = x$

20. $(f^{-1} \circ f)(x) = f^{-1}(f(x)) = f^{-1}\left(\sqrt[3]{x + 4}\right)$
$\quad = \left(\sqrt[3]{x + 4}\right)^3 - 4 = x + 4 - 4 = x$;
$\quad (f \circ f^{-1})(x) = f(f^{-1}(x)) = f(x^3 - 4)$
$\quad = \sqrt[3]{x^3 - 4 + 4} = \sqrt[3]{x^3} = x$

21. $3 = \log_7 343$ **22.** $-4 = \log_3 \frac{1}{81}$ **23.** $6^t = 12$

24. $n^m = T$ **25.** 3   **26.** 2   **27.** 2   **28.** 5   **29.** 2.3869

30. -0.6383 **31.** $\log_b 2 + \log_b x + 2 \log_b y - 3 \log_b z$

32. $\frac{2}{3} \log_a x + \frac{5}{3} \log_a y - \frac{4}{3} \log_a z$ **33.** $\log_a \frac{x\sqrt{z}}{y^2}$

34. $\log_m(b - 4)$ **35.** 0   **36.** 1   **37.** -3 **38.** 5

39. $V^{-1}(t)$ could be used to predict when the value of the stamp will be t, where $V^{-1}(t)$ is the number of years after 1999.

40. $\log_a b$ is the number to which a is raised to get c. Since $\log_a b = c$, then $a^c = b$.   **41.** Express $\frac{x}{5}$ as $x \cdot 5^{-1}$ and then use

the product rule and the power rule to get $\log_a\left(\frac{x}{5}\right) =$

$\log_a(x \cdot 5^{-1}) = \log_a x + \log_a 5^{-1} = \log_a x + (-1) \log_a 5 = \log_a x - \log_a 5$.   **42.** The student didn't subtract the logarithm of the entire denominator after using the quotient rule. The correct procedure is as follows:

$$\log_b \frac{1}{x} = \log_b \frac{x}{xx}$$
$$= \log_b x - \log_b xx$$
$$= \log_b x - (\log_b x + \log_b x)$$
$$= \log_b x - \log_b x - \log_b x$$
$$= -\log_b x.$$

(Note that $-\log_b x$ is equivalent to $\log_b 1 - \log_b x$.)

Calculator Corner, p. 928

1. $y = \log_2 x$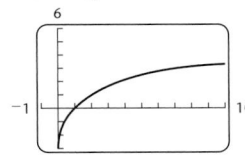

2. $y = \log_3 x$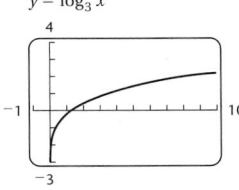

3. $y = \log_{1/2} x$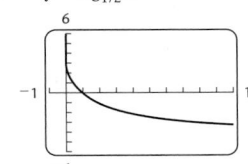

4. $y = \log_{2/3} x$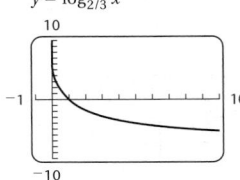

1. J **2.** B **3.** O **4.** G **5.** N **6.** F **7.** A **8.** H
9. I **10.** K

Exercise Set 12.5, p. 930

1. 0.6931 **3.** 4.1271 **5.** 8.3814 **7.** −5.0832 **9.** −1.6094
11. Does not exist **13.** −1.7455 **15.** 1 **17.** 15.0293
19. 0.0305 **21.** 109.9472 **23.** 5 **25.** 2.5702 **27.** 6.6439
29. 2.1452 **31.** −2.3219 **33.** −2.3219 **35.** 4.6284

37.

x	f(x)
0	1
1	2.7
2	7.4
3	20.1
−1	0.4
−2	0.1
−3	0.05

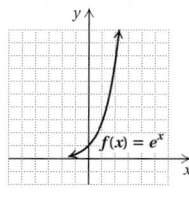

39. 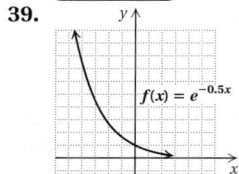   **41.**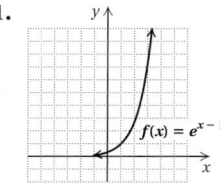

43. 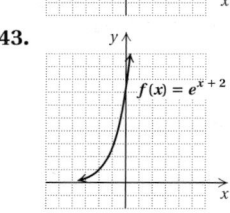   **45.**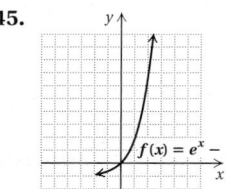

47.

x	f(x)
0	0.7
1	1.1
2	1.4
3	1.6
−0.5	0.4
−1	0
−1.5	−0.7

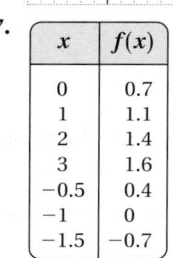

49. 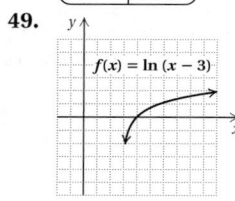   **51.**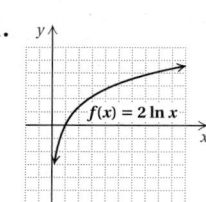

53. 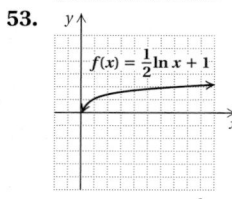   **55.**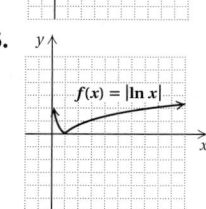

57. 16, 256 **58.** $\frac{1}{4}$, 9 **59.** 49, 121 **60.** ±3, ±4
61. Domain: $(-\infty, \infty)$; range: $[0, \infty)$
63. Domain: $(-\infty, \infty)$; range: $(-\infty, 100)$ **65.** $\left(\frac{5}{2}, \infty\right)$

Calculator Corner, p. 935

Left to the student

Exercise Set 12.6, p. 938

1. 3 **3.** 4 **5.** $\frac{5}{2}$ **7.** $\frac{3}{5}$ **9.** 3.4594 **11.** 5.4263 **13.** $\frac{5}{2}$
15. −3, −1 **17.** $\frac{3}{2}$ **19.** 4.6052 **21.** 2.3026 **23.** 140.6705

25. 2.7095 **27.** 3.2220 **29.** 256 **31.** $\frac{1}{32}$ **33.** 10 **35.** $\frac{1}{100}$
37. $e^2 \approx 7.3891$ **39.** $\frac{1}{e} \approx 0.3679$ **41.** 121 **43.** 10 **45.** $\frac{1}{3}$
47. 3 **49.** $\frac{2}{5}$ **51.** 5 **53.** No solution **55.** ±10, ±2
56. −64, 8 **57.** −2, −3, $\dfrac{-5 \pm \sqrt{41}}{2}$ **58.** $-\frac{1}{10}$, 1 **59.** $\dfrac{y^{4/3}}{25x^2z^4}$
60. −i **61.** 1 **63.** (a) 0.3770; (b) −1.9617; (c) 0.9036;
(d) −1.5318 **65.** 3, 4 **67.** −4 **69.** 2 **71.** $\pm\sqrt{34}$
73. $10^{100,000}$ **75.** 1, 100 **77.** 3, −7 **79.** 1, $\dfrac{\log 5}{\log 3} \approx 1.465$

Translating for Success, p. 948

1. D **2.** M **3.** I **4.** A **5.** E **6.** H **7.** C **8.** G
9. N **10.** B

Exercise Set 12.7, p. 949

1. 90 dB **3.** $10^{-7.5}$ W/m^2, or about 3.2×10^{-8} W/m^2;
-10^{-6} W/m^2 **5.** About 6.8 **7.** 1.58×10^{-8} moles per liter
9. 2.36 ft/sec **11.** 2.99 ft/sec **13.** (a) $27.87 billion;
(b) 2013; (c) about 4 yr **15.** About 560,664 PB per month;
(b) 2010; (c) about 0.77 yr **17.** (a) $P(t) = P_0 e^{0.03t}$;
(b) $5152.27; $5309.18; $6749.29; (c) in 23.1 yr
19. $P(t) = 6.8e^{0.01188t}$; (b) 7.2 billion; (c) 2076; (d) 58.3 yr
21. (a) $k \approx 0.076$; $C(t) = 80e^{0.076t}$; (b) about $426 billion;
(c) 2014 **23.** About 2103 yr **25.** About 7.2 days
27. 69.3% per year **29.** (a) $k \approx 0.103$; $D(t) = 29.7e^{-0.103t}$;
(b) about 1.84%; (c) 1996 **31.** (a) $k \approx 0.004$;
$P(t) = 2.431e^{-0.004t}$; (b) 2.244 million; (c) 2027
33. (a) $k \approx 0.134$; $V(t) = 640,500e^{0.134t}$; (b) $1,094,715;
(c) 5.2 yr; (d) 2008 **35.** −1 **36.** 1 **37.** i **38.** i
39. −1 − i **40.** −2 **41.** $\frac{63}{65} - \frac{16}{65}i$ **42.** $-\frac{2}{41} + \frac{23}{41}i$
43. 41 **44.** 91 + 60i **45.** −0.937, 1.078, 58.770
47. −0.767, 2, 4 **49.** $13.4 million

Summary and Review: Chapter 12, p. 954

Concept Reinforcement
1. True **2.** False **3.** False **4.** True **5.** True **6.** True
7. False **8.** True

Important Concepts
1. 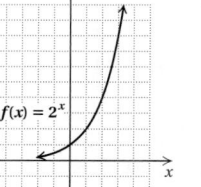   **2.** Yes **3.** $g^{-1}(x) = 4 - x$

4. 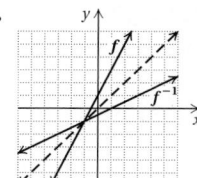   **5.** $8x + 2$; $8x + 1$

6. $f(x) = \dfrac{1}{x}$, $g(x) = 3x + 2$; answers may vary

7.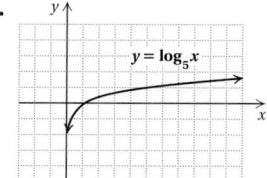

8. $\frac{3}{5}\log_a x - \frac{2}{5}\log_a y$ **9.** $\log_a \dfrac{\sqrt{x}}{y^3}$, or $\log_a \dfrac{x^{1/2}}{y^3}$

10.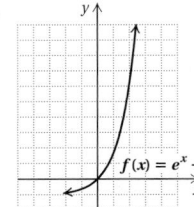

11.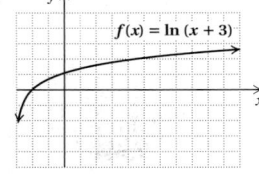

12. $\frac{4}{3}$ **13.** 3

Review Exercises

1. $\{(2, -4), (-7, 5), (-2, -1), (11, 10)\}$ **2.** Not one-to-one

3. $g^{-1}(x) = \dfrac{7x + 3}{2}$ **4.** $f^{-1}(x) = \frac{1}{2}\sqrt[3]{x}$ **5.** $f^{-1}(x) = \dfrac{3x - 4}{2x}$

6.

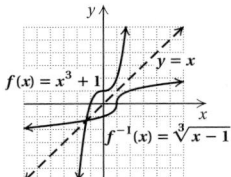

7.

x	$f(x)$
0	$\frac{1}{3}$
1	1
2	3
3	9
-1	$\frac{1}{9}$
-2	$\frac{1}{27}$
-3	$\frac{1}{81}$

$f(x) = 3^{x-1}$

8. 3^y

x, or 3^y	y
1	0
3	1
9	2
27	3
$\frac{1}{3}$	-1
$\frac{1}{9}$	-2
$\frac{1}{27}$	-3

$y = \log_3 x$

9.

x	$f(x)$
0	2.7
1	7.4
2	20.1
3	54.6
-1	1
-2	0.4
-3	0.1

$f(x) = e^{x+1}$

10.

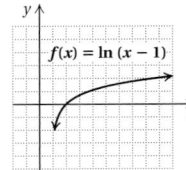

$f(x) = \ln (x - 1)$

11. $(f \circ g)(x) = 9x^2 - 30x + 25;\ (g \circ f)(x) = 3x^2 - 5$
12. $f(x) = \sqrt{x}, g(x) = 4 - 7x$; answers may vary
13. $4 = \log 10,000$ **14.** $\frac{1}{2} = \log_{25} 5$ **15.** $4^x = 16$
16. $\left(\frac{1}{2}\right)^{-3} = 8$ **17.** 2 **18.** -1 **19.** 1 **20.** 0
21. -2.7425 **22.** Does not exist as a real number
23. $4\log_a x + 2\log_a y + 3\log_a z$ **24.** $\frac{1}{2}\log z - \frac{3}{4}\log x - \frac{1}{4}\log y$
25. $\log_a 120$ **26.** $\log \dfrac{a^{1/2}}{bc^2}$ **27.** 17 **28.** -7 **29.** 8.7601
30. 3.2698 **31.** 2.54995 **32.** -3.6602 **33.** -2.6921

34. 0.3753 **35.** 18.3568 **36.** 0 **37.** Does not exist
38. 1 **39.** 0.4307 **40.** 1.7097 **41.** $\frac{1}{9}$ **42.** 2 **43.** $\dfrac{1}{10,000}$
44. $e^{-2} \approx 0.1353$ **45.** $\frac{7}{2}$ **46.** 1, -5 **47.** $\dfrac{\log 8.3}{\log 4} \approx 1.5266$
48. $\dfrac{\ln 0.03}{-0.1} \approx 35.0656$ **49.** 2 **50.** 8 **51.** $\frac{17}{5}$ **52.** $\sqrt{43}$
53. 137 dB **54.** (a) 85.0 million returns; 98.4 million returns; 113.9 million returns; (b) 2014; (c) about 9.5 yr
(d)

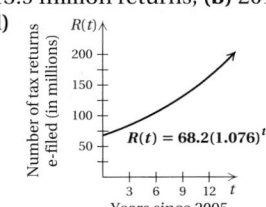

$R(t) = 68.2(1.076)^t$

55. (a) $k \approx 0.094$; $V(t) = 40,000e^{0.094t}$; (b) \$102,399; (c) 2017
56. $k \approx 0.231$ **57.** About 20.4 yr **58.** About 3463 yr
59. C **60.** D **61.** e^{e^3} **62.** $\left(\frac{8}{3}, -\frac{2}{3}\right)$

Understanding Through Discussion and Writing

1. Reflect the graph of $f(x) = e^x$ across the line $y = x$ and then translate it up one unit. **2.** Christina mistakenly thinks that, because negative numbers do not have logarithms, negative numbers cannot be solutions of logarithmic equations.

3.
$$C(x) = \frac{100 + 5x}{x}$$
$$y = \frac{100 + 5x}{x} \qquad \text{Replace } C(x) \text{ with } y.$$
$$x = \frac{100 + 5y}{y} \qquad \text{Interchange variables.}$$
$$y = \frac{100}{x - 5}; \qquad \text{Solve for } y.$$
$$C^{-1}(x) = \frac{100}{x - 5} \qquad \text{Replace } y \text{ with } C^{-1}(x).$$

$C^{-1}(x)$ gives the number of people in the group, where x is the cost per person, in dollars.
4. To solve $\ln x = 3$, graph $f(x) = \ln x$ and $g(x) = 3$ on the same set of axes. The solution is the first coordinate of the point of intersection of the two graphs. **5.** You cannot take the logarithm of a negative number because logarithm bases are positive and there is no real-number power to which a positive number can be raised to yield a negative number. **6.** Answers will vary.

Test: Chapter 12, p. 962

1. [12.1a]

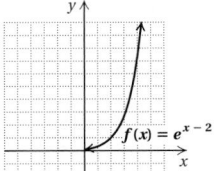

$f(x) = 2^{x+1}$

2. [12.3a]

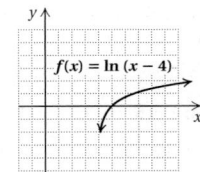

$y = \log_2 x$

3. [12.5c]

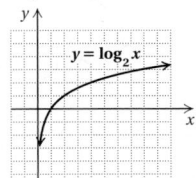

$f(x) = e^{x-2}$

4. [12.5c]

$f(x) = \ln (x - 4)$

5. [12.2a] $\{(3, -4), (-8, 5), (-3, -1), (12, 10)\}$
6. [12.2b, c] $f^{-1}(x) = \dfrac{x + 3}{4}$ **7.** [12.2b, c] $f^{-1}(x) = \sqrt[3]{x} - 1$
8. [12.2b] Not one-to-one

9. [12.2d] $(f \circ g)(x) = 25x^2 - 15x + 2$, $(g \circ f)(x) = 5x^2 + 5x - 2$ **10.** [12.3b] $\log_{256} 16 = \frac{1}{2}$
11. [12.3b] $7^m = 49$ **12.** [12.3c] 3   **13.** [12.4e] 23
14. [12.3c] 0   **15.** [12.3d] -1.9101 **16.** [12.3d] Does not exist as a real number   **17.** [12.4d] $3 \log a + \frac{1}{2} \log b - 2 \log c$
18. [12.4d] $\log_a \dfrac{x^{1/3}z^2}{y^3}$ **19.** [12.4d] -0.544
20. [12.4d] 1.079   **21.** [12.5a] 6.6938   **22.** [12.5a] 107.7701
23. [12.5a] 0   **24.** [12.5b] 1.1881   **25.** [12.6b] 5
26. [12.6b] 2   **27.** [12.6b] 10,000   **28.** [12.6b] $e^{1/4} \approx 1.2840$
29. [12.6a] $\dfrac{\log 1.2}{\log 7} \approx 0.0937$ **30.** [12.6b] 9
31. [12.6b] 1   **32.** [12.7a] 4.2   **33.** [12.7b] **(a)** $3.18 trillion; **(b)** 2018; **(c)** about 9.5 yr
34. [12.7b] **(a)** $k \approx 0.028$, or 2.8%; $P(t) = 1000e^{0.028t}$; **(b)** $1251.07; **(c)** after 13 yr; **(d)** about 24.8 yr
35. [12.7b] About 3%   **36.** [12.7b] About 4684 yr
37. [12.6b] B   **38.** [12.6b] $44, -37$ **39.** [12.4d] 2

Cumulative Review: Chapters 1–12, p. 965

1. [9.1c] $\{x | x \geq -1\}$, or $[-1, \infty)$
2. [9.3e] $\{x | x < -6.4 \text{ or } x > 6.4\}$, or $(-\infty, -6.4) \cup (6.4, \infty)$
3. [9.3c] $\{-11, 19\}$ **4.** [8.2a], [8.3a] $\left(-\dfrac{1}{3}, 5\right)$
5. [11.4c] $-3, -2, 2, 3$ **6.** [5.8 b] $-1, \dfrac{3}{2}$ **7.** [6.7a] $-\dfrac{2}{3}, 3$
8. [9.3e] $\{x | -19 \leq x \leq 7\}$, or $[-19, 7]$ **9.** [5.8b] $-7, -3$
10. [11.2a] $-\dfrac{1}{4} \pm i \dfrac{\sqrt{7}}{4}$ **11.** [2.3c] -2
12. [11.8b] $\{x | x < -1 \text{ or } x > 2\}$, or $(-\infty, -1) \cup (2, \infty)$
13. [12.6b] 9   **14.** [11.8a] $\{x | x \leq -1 \text{ or } x \geq 1\}$, or $(-\infty, -1] \cup [1, \infty)$ **15.** [12.6b] 1   **16.** [10.6a] 4
17. [9.3d] $\left\{-\dfrac{3}{2}, \dfrac{1}{4}\right\}$ **18.** [12.6a] 1.748   **19.** [8.5a] $(-1, 2, 3)$
20. [9.2a] $\{x | -1 \leq x < 6\}$, or $[-1, 6)$ **21.** [6.7a] -16
22. [2.4b] $N = \dfrac{4P - 3M}{6}$ **23.** [12.1c], [12.7b] **(a)** About 80.52 billion ft^3; about 84.94 billion ft^3, **(b)** [12.1c], [12.7b] about 39 yr;
(c)

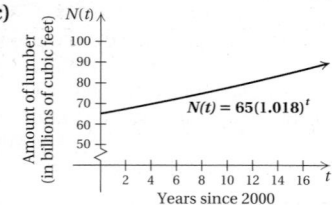

24. [12.7b] **(a)** $A(t) = \$50,000(1.04)^t$; **(b)** $50,000; $58,492.93; $68,428.45; $74,012.21; **(c)**

25. [4.5d] $2x^3 - x^2 - 8x - 3$ **26.** [4.4c] $-x^3 + 3x^2 - x - 6$
27. [6.1d] $\dfrac{2m - 1}{m + 1}$ **28.** [6.5b] $\dfrac{x + 2}{x + 1}$ **29.** [6.6a] $\dfrac{1}{x + 1}$
30. [4.8b] $x^3 + 2x^2 - 2x + 1 + \dfrac{3}{x + 1}$ **31.** [10.3b] $5x^2\sqrt{y}$
32. [10.4a] $11\sqrt{2}$ **33.** [10.2d] 8   **34.** [10.8c] $16 + i\sqrt{2}$
35. [10.8e] $\dfrac{3}{10} + \dfrac{11}{10}i$ **36.** [1.2e] $\dfrac{2}{15}$ **37.** [4.1e, f] $-9x^6y^2$
38. [1.8d] 62.5   **39.** [1.8c] $-x + 30$
40. [3.3a] x-intercept: $(2, 0)$; y-intercept: $\left(0, -\dfrac{1}{2}\right)$

41. **(a)** [3.4a] $-\dfrac{16}{11}$; **(b)** [3.4a] not defined

42. [7.5c] $y = 2x + 2$ **43.** [7.5d] $y = -\dfrac{1}{2}x + \dfrac{5}{2}$

44. [9.4b]   **45.** [3.2a]   **46.** [3.3a], [7.4a]

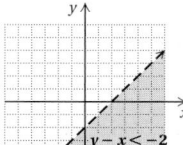   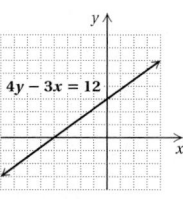

47. [9.4b]   **48.** [7.1c]   **49.** [9.4c]

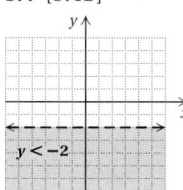   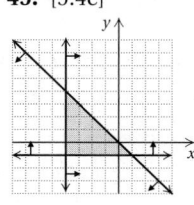

50. [3.3b], [7.4c]   **51.** [11.6a]   **52.** [12.5c]

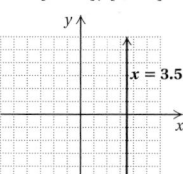   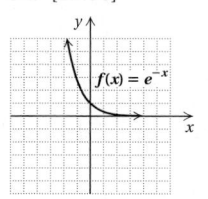

53. [12.3a]   **54.** [5.1c] $(x - 6)(2x^3 + 1)$

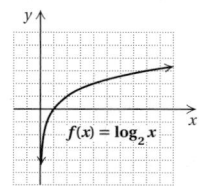

55. [5.1b], [5.2a] $3(a - 9b)(a + 5b)$
56. [5.2a] $(x - 8)(x - 9)$
57. [5.5d], [5.7a] $(9m^2 + n^2)(3m + n)(3m - n)$
58. [5.5b] $4(2x - 1)^2$ **59.** [5.6a] $3(3a - 2)(9a^2 + 6a + 4)$
60. [5.3a], [5.4a] $2(5x - 2)(x + 7)$
61. [5.1b], [5.3a] $3x(2x - 1)(x + 5)$
62. [12.2c] $f^{-1}(x) = \dfrac{1}{2}(x + 3)$ **63.** [6.9e] $\dfrac{4}{5}$ **64.** [7.1b] -10
65. [7.1b] -30 **66.** [7.3c], [7.5e] **(a)** $434.7 billion, $488.7 billion, $524.7 billion; **(b)**

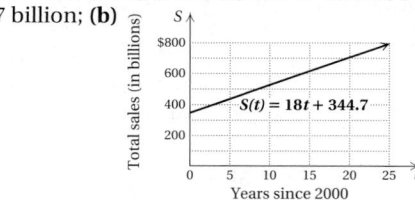

(c) $(0, 344.7)$; **(d)** 18; **(e)** an increase of $18 billion per year
67. [10.5b] $\dfrac{15 + 8\sqrt{a} + a}{9 - a}$ **68.** [7.2a] All real numbers
69. [7.2a] $\{x | x$ is a real number $and \ x \neq -2\}$, or $(-\infty, -2) \cup (-2, \infty)$ **70.** [9.1d] More than 4
71. [6.8a] 612 mi   **72.** [2.6a] $11\dfrac{3}{7}$ **73.** [8.4a] 24 L of A; 56 L of B   **74.** [6.8a] 350 mph   **75.** [6.8a] $8\dfrac{2}{5}$ min
76. [6.9f] 20   **77.** [11.7a] 1250 ft^2 **78.** [12.7b] 2397 yr
79. [6.9d] 3360 kg   **80.** [7.5c] $f(x) = -\dfrac{1}{3}x - \dfrac{7}{3}$
81. [11.7b] $f(x) = -\dfrac{17}{18}x^2 - \dfrac{59}{18}x + \dfrac{11}{9}$ **82.** [12.3b] $\log r = 6$

83. [12.3b] $3^x = Q$ **84.** [12.4d] $\log_b \left(\dfrac{x^7}{yz^8}\right)^{1/5}$, or $\log_b \dfrac{x^{7/5}}{y^{1/5}z^{8/5}}$

85. [12.4d] $-6 \log_b x - 30 \log_b y + 6 \log_b z$ **86.** [11.7a] 169
87. [12.2b] No **88.** [7.2 a] **(a)** -5; **(b)** all real numbers, or $(-\infty, \infty)$; **(c)** $-2, -1, 1, 2$; **(d)** $\{x | x \geq -7\}$, or $[-7, \infty)$
89. [12.7 b] **(a)** $P(t) = 1,998,257\, e^{0.03t}$; **(b)** 3,133,891;
(c) about 2019 **90.** [6.7a] All real numbers except 0 and -12

91. [12.6b] 81 **92.** [5.8b] $0, \dfrac{b}{a}$

93. [5.6a] $\left(\dfrac{a}{2} + \dfrac{2b}{9}\right)\left(\dfrac{a^2}{4} - \dfrac{ab}{9} + \dfrac{4b^2}{81}\right)$ **94.** [1.8d], [4.1f] $\dfrac{328}{3}$

APPENDIXES

Exercise Set A, p. 976

1. 1, 2, 4, 5, 10, 20 **3.** 1, 2, 3, 4, 6, 8, 9, 12, 18, 24, 36, 72
5. $3 \cdot 5$ **7.** $2 \cdot 11$ **9.** $3 \cdot 3$ **11.** $7 \cdot 7$ **13.** $2 \cdot 3 \cdot 3$
15. $2 \cdot 2 \cdot 2 \cdot 5$ **17.** $2 \cdot 3 \cdot 3 \cdot 5$ **19.** $2 \cdot 3 \cdot 5 \cdot 7$
21. $7 \cdot 13$ **23.** $7 \cdot 17$ **25.** $2 \cdot 2$; 5; 20 **27.** $2 \cdot 2 \cdot 2 \cdot 3$;
$2 \cdot 2 \cdot 3 \cdot 3$; 72 **29.** 3; $3 \cdot 5$; 15 **31.** $2 \cdot 3 \cdot 5$; $2 \cdot 2 \cdot 2 \cdot 5$; 120
33. 13; 23; 299 **35.** $2 \cdot 3 \cdot 3$; $2 \cdot 3 \cdot 5$; 90 **37.** $2 \cdot 3 \cdot 5$;
$2 \cdot 2 \cdot 3 \cdot 3$; 180 **39.** $2 \cdot 2 \cdot 2 \cdot 3$; $2 \cdot 3 \cdot 5$; 120 **41.** $2 \cdot 2 \cdot 3$;
$2 \cdot 3 \cdot 3$; 36 **43.** $2 \cdot 2 \cdot 3$; $2 \cdot 2 \cdot 7$; 84 **45.** 2; 3; 5; 30
47. $2 \cdot 2 \cdot 2 \cdot 3$; $2 \cdot 2 \cdot 3 \cdot 3$; $2 \cdot 2 \cdot 3$; 72 **49.** 5; $2 \cdot 2 \cdot 3$;
$3 \cdot 5$; 60 **51.** $2 \cdot 3$; $2 \cdot 2 \cdot 3$; $2 \cdot 3 \cdot 3$; 36

Exercise Set B, p. 985

1. $\frac{9}{12}$ **3.** $\frac{60}{100}$ **5.** $\frac{21}{24}$ **7.** $\frac{2}{3}$ **9.** 4 **11.** $\frac{1}{7}$ **13.** 8 **15.** $\frac{1}{4}$
17. 5 **19.** $\frac{17}{21}$ **21.** $\frac{13}{7}$ **23.** $\frac{4}{3}$ **25.** $\frac{1}{12}$ **27.** $\frac{45}{16}$ **29.** $\frac{7}{6}$
31. $\frac{5}{6}$ **33.** $\frac{1}{2}$ **35.** $\frac{13}{24}$ **37.** $\frac{35}{18}$ **39.** $\frac{10}{3}$ **41.** $\frac{5}{36}$
43. 500 **45.** $\frac{3}{40}$ **47.** $\frac{99,999}{100}$

Exercise Set C, p. 989

1. 5^4 **3.** 10^3 **5.** 10^6 **7.** 49 **9.** 59,049 **11.** 100
13. 1 **15.** 5.29 **17.** 0.008 **19.** 416.16 **21.** $\frac{9}{64}$ **23.** 125
25. 1061.208 **27.** 25 **29.** 114 **31.** 33 **33.** 5 **35.** 12
37. 324 **39.** 100 **41.** 1000 **43.** 22 **45.** 1 **47.** 4
49. 102 **51.** 96 **53.** 24 **55.** 90 **57.** 8 **59.** 1
61. 50,000 **63.** 5 **65.** 27 **67.** $\frac{22}{45}$ **69.** $\frac{19}{66}$ **71.** 9

Exercise Set D, p. 998

1. $(y - 7)(y - 3)$ **3.** $(t - 5)(t + 3)$ **5.** $2a(a - 5)(a - 5)$
7. $(m + 9)(m - 8)$ **9.** $(y + 6)(y + 4)$ **11.** $\left(p + \frac{1}{5}\right)\left(p + \frac{1}{5}\right)$
13. $(3x + 1)(x - 5)$ **15.** $y(5y - 7)(2y + 3)$
17. $(3c - 8)(c - 4)$ **19.** $(5y + 2)(7y + 4)$
21. $2(5t - 3)(t + 1)$ **23.** $4(2x + 1)(x - 4)$
25. $x(3x - 4)(4x - 5)$ **27.** $x^2(7x + 1)(2x - 3)$
29. $(3a - 4)(a + 1)$ **31.** $(3x + 1)(3x + 4)$ **33.** $(x - 2)^2$
35. $(y + 9)^2$ **37.** $(x + 1)^2$ **39.** $(3y + 2)^2$ **41.** $y(y - 9)^2$
43. $3(2a + 3)^2$ **45.** $(x + 4)(x - 4)$ **47.** $(p + 7)(p - 7)$
49. $a(2a + 7)(2a - 7)$ **51.** $\left(\frac{1}{6} + z\right)\left(\frac{1}{6} - z\right)$
53. $(z + 3)(z^2 - 3z + 9)$ **55.** $(x - 1)(x^2 + x + 1)$
57. $(2a + 1)(4a^2 - 2a + 1)$ **59.** $(2 - 3b)(4 + 6b + 9b^2)$
61. $2(y - 4)(y^2 + 4y + 16)$ **63.** $3(2a + 1)(4a^2 - 2a + 1)$

Exercise Set F, p. 1004

1. 68 ft **3.** 45 g **5.** $15 \dfrac{\text{mi}}{\text{hr}}$ **7.** $3.3 \dfrac{\text{m}}{\text{sec}}$ **9.** $4 \dfrac{\text{in.-lb}}{\text{sec}}$

11. 12 yd **13.** 16 ft^3 **15.** $\dfrac{\$970}{\text{day}}$ **17.** 51.2 oz

19. 3080 ft/min **21.** 96 in. **23.** Approximately 0.03 yr

25. Approximately 31,710 yr **27.** $80 \dfrac{\text{oz}}{\text{in.}}$ **29.** 172,800 sec

31. $\frac{3}{2}$ ft^2 **33.** $1.08 \dfrac{\text{ton}}{\text{yd}^3}$

Exercise Set G, p. 1007

1. Mean: 21; median: 18.5; mode: 29 **3.** Mean: 5.38; median:
5.7; no mode exists **5.** Mean: 239.5; median: 234; mode: 234
7. Mean: 36.5; median: 90; mode: 2 **9.** Mean: $4.19; median:
$3.99; mode: $3.99

Exercise Set H, p. 1010

1. $x^2 - x + 1$, R -4; or $x^2 - x + 1 + \dfrac{-4}{x - 1}$

3. $a + 7$, R -47; or $a + 7 + \dfrac{-47}{a + 4}$

5. $x^2 - 5x - 23$, R -43; or $x^2 - 5x - 23 + \dfrac{-43}{x - 2}$

7. $3x^2 - 2x + 2$, R -3; or $3x^2 - 2x + 2 + \dfrac{-3}{x + 3}$

9. $y^2 + 2y + 1$, R 12; or $y^2 + 2y + 1 + \dfrac{12}{y - 2}$

11. $3x^3 + 9x^2 + 2x + 6$ **13.** $x^2 + 2x + 4$
15. $y^3 + 2y^2 + 4y + 8$

Exercise Set I, p. 1015

1. 10 **3.** 0 **5.** -48 **7.** 0 **9.** -10 **11.** -3 **13.** 5
15. 0 **17.** $(2, 0)$ **19.** $\left(-\frac{25}{2}, -\frac{11}{2}\right)$ **21.** $\left(\frac{3}{2}, \frac{5}{2}\right)$ **23.** $(-4, 3)$
25. $(2, -1, 4)$ **27.** $(1, 2, 3)$ **29.** $\left(\frac{3}{2}, -4, 3\right)$ **31.** $(2, -2, 1)$

Exercise Set J, p. 1019

1. $\left(\frac{3}{2}, \frac{5}{2}\right)$ **3.** $(-4, 3)$ **5.** $\left(\frac{1}{2}, \frac{3}{2}\right)$ **7.** $(10, -10)$
9. $\left(\frac{3}{2}, -4, 3\right)$ **11.** $(2, -2, 1)$ **13.** $(0, 2, 1)$ **15.** $\left(4, \frac{1}{2}, -\frac{1}{2}\right)$

Exercise Set K, p. 1022

1. 1 **3.** -41 **5.** 12 **7.** $\frac{13}{18}$ **9.** 5 **11.** 2
13. $x^2 - x + 1$ **15.** 21 **17.** 5 **19.** $-x^3 + 4x^2 + 3x - 12$
21. 42 **23.** $-\frac{3}{4}$ **25.** $\frac{1}{6}$

27. $x^2 + 3x - 4$; $x^2 - 3x + 4$; $3x^3 - 4x^2$; $\dfrac{x^2}{3x - 4}$

29. $\dfrac{1}{x - 2} + 4x^3$; $\dfrac{1}{x - 2} - 4x^3$; $\dfrac{4x^3}{x - 2}$; $\dfrac{1}{4x^3(x - 2)}$

31. $\dfrac{3}{x - 2} + \dfrac{5}{4 - x}$; $\dfrac{3}{x - 2} - \dfrac{5}{4 - x}$; $\dfrac{15}{(x - 2)(4 - x)}$; $\dfrac{3(4 - x)}{5(x - 2)}$

Exercise Set L, p. 1027

1. 5 **3.** $\sqrt{29} \approx 5.385$ **5.** $\sqrt{648} \approx 25.456$ **7.** 7.1
9. $\dfrac{\sqrt{41}}{7} \approx 0.915$ **11.** $\sqrt{6970} \approx 83.487$ **13.** $\sqrt{a^2 + b^2}$
15. $\sqrt{17 + 2\sqrt{14} + 2\sqrt{15}} \approx 5.677$
17. $\sqrt{9,672,400} \approx 3110.048$ **19.** $\left(\frac{3}{2}, \frac{7}{2}\right)$ **21.** $\left(0, \frac{11}{2}\right)$
23. $\left(-1, -\frac{17}{2}\right)$ **25.** $(-0.25, -0.3)$ **27.** $\left(-\frac{1}{12}, \frac{1}{24}\right)$
29. $\left(\dfrac{\sqrt{2} + \sqrt{3}}{2}, \dfrac{3}{2}\right)$ **31.**

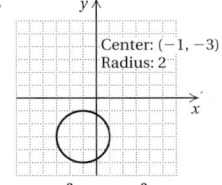

Center: $(-1, -3)$
Radius: 2

$(x + 1)^2 + (y + 3)^2 = 4$

33.

Center: $(3, 0)$
Radius: $\sqrt{2}$

$(x - 3)^2 + y^2 = 2$

35.

$x^2 + y^2 = 25$

Center: $(0, 0)$
Radius: 5

37. $x^2 + y^2 = 49$ **39.** $(x + 5)^2 + (y - 3)^2 = 7$
41. $(-4, 3)$, $r = 2\sqrt{10}$ **43.** $(4, -1)$, $r = 2$ **45.** $(2, 0)$, $r = 2$

Glossary

A

Abscissa   The first coordinate in an ordered pair of numbers

Absolute value   The distance that a number is from 0 on the number line

ac-method   A method for factoring trinomials of the type $ax^2 + bx + c, a \neq 1$, involving the product, ac, of the leading coefficient a and the last term c

Additive identity   The number 0

Additive inverse   A number's opposite; two numbers are additive inverses of each other if their sum is 0

Algebraic expression   An expression consisting of variables, constants, numerals, operation signs and/or grouping symbols

Area   The number of square units that fill a plane region

Arithmetic numbers   The whole numbers and the positive fractions; all these numbers can be named with fraction notation $\frac{a}{b}$, where a and b are whole numbers and $b \neq 0$

Ascending order   When a polynomial is written with the terms arranged according to degree from least to greatest, it is said to be in ascending order.

Associative law of addition   The statement that when three numbers are added, regrouping the addends gives the same sum

Associative law of multiplication   The statement that when three numbers are multiplied, regrouping the factors gives the same product

Average   A center point of a set of numbers found by adding the numbers and dividing by the number of items of data; also called the *mean*

Axes   Two perpendicular number lines used to identify points in a plane

Axis of symmetry   A line that can be drawn through a graph such that the part of the graph on one side of the line is an exact reflection of the part on the opposite side; also called *line of symmetry*

B

Base   In exponential notation, the number being raised to a power

Binomial   A polynomial composed of two terms

C

Circle   The set of all points in a plane that are a fixed distance r, called the radius, from a fixed point (h, k), called the center

Circumference   The distance around a circle

Coefficient   The numerical multiplier of a variable

Common logarithm   A logarithm with base 10

Commutative law of addition   The statement that when two numbers are added, changing the order in which the numbers are added does not affect the sum

Commutative law of multiplication   The statement that when two numbers are multiplied, changing the order in which the numbers are multiplied does not affect the product

Complementary angles   Angles whose sum is 90°

Completing the square   Adding a particular constant to an expression so that the resulting sum is a perfect square

Complex fraction expression   A rational expression that has one or more rational expressions within its numerator and/or denominator

Complex number   Any number that can be named as $a + bi$, where a and b are any real numbers

Complex number i   The square root of -1; that is, $i = \sqrt{-1}$ and $i^2 = -1$

Complex rational expression   A rational expression that has one or more rational expressions within its numerator and/or denominator

Complex-number system   A number system that contains the real-number system and is designed so that negative numbers have defined square roots

Composite function   A function in which a quantity depends on a variable that, in turn, depends on another variable

Composite number   A natural number, other than 1, that is not prime

Compound inequality   A statement in which two or more inequalities are joined by the word *and* or the word *or*

Compound interest   Interest computed on the sum of an original principal and the interest previously accrued by that principal

Conjugate of a complex number   The conjugate of a complex number $a + bi$ is $a - bi$ and the conjugate of $a - bi$ is $a + bi$.

Conjugates of radical terms Pairs of radical terms, like $\sqrt{a} + \sqrt{b}$ and $\sqrt{a} - \sqrt{b}$ or $c + \sqrt{d}$ and $c - \sqrt{d}$, whose product does not have a radical term

Conjunction A statement in which two or more sentences are joined by the word *and*

Consecutive even integers Even integers that are two units apart

Consecutive integers Integers that are one unit apart

Consecutive odd integers Odd integers that are two units apart

Consistent system of equations A system of equations that has at least one solution

Constant A known number

Constant function A function given by an equation of the form $y = b$, or $f(x) = b$, where b is a real number

Constant of proportionality The constant in an equation of direct or inverse variation

Coordinates The numbers in an ordered pair

Cube root The number c is the cube root of a, written $\sqrt[3]{a}$, if the third power of c is a.

D

Decay rate The variable k in the exponential decay model $P(t) = P_0 e^{-kt}$

Degree of a polynomial The degree of the term of highest degree in a polynomial

Degree of a term The sum of the exponents of the variables

Denominator The bottom number in a fraction

Dependent equations The equations in a system are dependent if one equation can be removed without changing the solution set. There are infinitely many solutions to the equations.

Descending order When a polynomial is written with the terms arranged according to degree from greatest to least, it is said to be in descending order.

Diameter A segment that passes through the center of a circle and has its endpoints on the circle

Difference of cubes Any expression that can be written in the form $A^3 - B^3$

Difference of squares Any expression that can be written in the form $A^2 - B^2$

Direct variation A situation that gives rise to a linear function $f(x) = kx$, or $y = kx$, where k is a positive constant

Discriminant The expression $b^2 - 4ac$, from the quadratic formula

Disjoint sets Two sets with an empty intersection

Disjunction A statement in which two or more sentences are joined by the word *or*

Distributive law of multiplication over addition The statement that multiplying a factor by the sum of two numbers gives the same result as multiplying the factor by each of the two numbers and then adding

Distributive law of multiplication over subtraction The statement that multiplying a factor by the difference of two numbers gives the same result as multiplying the factor by each of the two numbers and then subtracting

Domain The set of all first coordinates of the ordered pairs in a function

Doubling time The time necessary for a population to double in size

E

Elimination method An algebraic method that uses the addition principle to solve a system of equations

Empty set The set without members

Equation A number sentence that says that the expressions on either side of the equals sign, =, represent the same number

Equation of direct variation An equation described by $y = kx$, with k a positive constant, used to represent direct variation

Equation of inverse variation An equation described by $y = k/x$, with k a positive constant, used to represent inverse variation

Equivalent equations Equations with the same solutions

Equivalent expressions Expressions that have the same value for all allowable replacements

Equivalent inequalities Inequalities that have the same solution set

Evaluate To substitute a value for each occurrence of a variable in an expression

Even root When the number k in $\sqrt[k]{}$ is an even number, we say that we are taking an even root.

Exponent In expressions of the form a^n, the number n is an exponent. For n a natural number, a^n represents n factors of a.

Exponential decay model A decrease in quantity over time that can be modeled by an exponential function of the form $P(t) = P_0 e^{-kt}$, $k > 0$

Exponential equation An equation in which a variable appears as an exponent

Exponential function The function $f(x) = a^x$, where a is a positive constant different from 1

Exponential growth model An increase in quantity over time that can be modeled by an exponential function of the form $P(t) = P_0 e^{kt}$, $k > 0$

Exponential growth rate The variable k in the exponential growth model $P(t) = P_0 e^{kt}$

Exponential notation A representation of a number using a base raised to a power

F

Factor *Verb*: To write an equivalent expression that is a product. *Noun*: A multiplier

Factorization of a polynomial An expression that names the polynomial as a product

FOIL To multiply two binomials by multiplying the First terms, the Outside terms, the Inside terms, and then the Last terms

Formula An equation that uses numbers or letters to represent a relationship between two or more quantities

Fraction equation An equation containing one or more rational expressions; also called a *rational equation*

Fraction expression A quotient, or ratio, of polynomials; also called a *rational expression*

Fraction notation A number written using a numerator and a denominator

Function A correspondence that assigns to each member of a set called the domain *exactly one* member of a set called the range

G

Grade The measure of a road's steepness

Graph A picture or diagram of the data in a table; a line, curve, or collection of points that represents all the solutions of an equation

Greatest common factor (GCF) The common factor of a polynomial with the largest possible coefficient and the largest possible exponent(s)

H

Half-life The amount of time necessary for half of a quantity to decay

Hypotenuse In a right triangle, the side opposite the 90° angle

I

Identity property of 1 The statement that the product of a number and 1 is always the original number

Identity property of 0 The statement that the sum of a number and 0 is always the original number

Imaginary number A number that can be named bi, where b is some real number and $b \neq 0$

Inconsistent system of equations A system of equations for which there is no solution

Independent equations Equations that are not dependent and for which there exists either one or no solution

Index In the expression $\sqrt[k]{a}$, the number k is called the index.

Inequality A mathematical sentence using $<, >, \leq, \geq,$ or $\neq$

Input A member of the domain of a function

Integers The whole numbers and their opposites

Intercept The point at which a graph intersects the x- or y-axis

Intersection of sets A and B The set of all members that are common to A and B

Interval notation The use of a pair of numbers inside parentheses and brackets to represent the set of all numbers between those two numbers

Inverse relation The relation formed by interchanging the coordinates of the ordered pairs in a relation

Inverse variation A situation that gives rise to a function $f(x) = k/x$, or $y = k/x$, where k is a positive constant

Irrational number A real number that cannot be named as a ratio of two integers

J

Joint variation A situation that gives rise to an equation of the form $y = kxz$, where k is a positive constant

L

Leading coefficient The coefficient of the term of highest degree in a polynomial

Leading term The term of highest degree in a polynomial

Least common denominator (LCD) The least common multiple of the denominators

Least common multiple (LCM) The smallest number that is a multiple of both numbers

Legs In a right triangle, the two sides that form the right angle

Like radicals Radicals having the same index and radicand

Like terms Terms that have exactly the same variable factors

Line of symmetry A line that can be drawn through a graph such that the part of the graph on one side of the line is an exact reflection of the part on the opposite side; also called *axis of symmetry*

Linear equation Any equation that can be written in the form $y = mx + b$ or $Ax + By = C$, where x and y are variables

Linear equation in three variables An equation equivalent to one of the type $Ax + By + Cz = D$

Linear function A function that can be described by an equation of the form $y = mx + b$, where x and y are variables

Linear inequality An inequality whose related equation is a linear equation

Logarithmic equation An equation containing a logarithmic expression

Logarithmic function, base a The inverse of an exponential function $f(x) = a^x$

M

Mathematical model A model in which the essential parts of a problem are described in mathematical language

Matrix A rectangular array of numbers

Maximum The largest function value (output) achieved by a function

Mean A center point of a set of numbers found by adding the numbers and dividing by the number of items of data; also called the *average*

Median In a set of data listed in order from smallest to largest, the middle number if there is an odd number of data items, or the average of the two middle numbers if there is an even number of data items

Minimum The smallest function value (output) achieved by a function

Mode The number or numbers that occur most often in a set of data

Monomial An expression of the type ax^n, where a is a real number constant and n is a nonnegative integer

Motion problem A problem that deals with distance, speed (or rate), and time

Multiple A product of a number and some natural number

Multiplication property of 0 The statement that the product of 0 and any real number is 0

Multiplicative identity The number 1

Multiplicative inverses Reciprocals; two numbers whose product is 1

N

Natural logarithm A logarithm with base e

Natural numbers The counting numbers: 1, 2, 3, 4, 5, …

Negative integers   The integers to the left of zero on the number line

Nonnegative rational numbers   The whole numbers and the positive fractions; all these numbers can be named with fraction notation $\frac{a}{b}$, where a and b are whole numbers and $b \neq 0$

Numerator   The top number in a fraction

O

Odd root   When the number k in $\sqrt[k]{}$ is an odd number, we say we are taking an odd root.

One-to-one function   A function for which different inputs have different outputs

Opposite   The opposite, or additive inverse, of a number a is denoted $-a$. Opposites are the same distance from 0 on the number line but on different sides of 0.

Opposite of a polynomial   To find the opposite of a polynomial, replace each term with its opposite — that is, change the sign of every term.

Ordered pair   A pair of numbers of the form (h, k) for which the order in which the numbers are listed is important

Ordinate   The second coordinate in an ordered pair of numbers

Origin   The point on a graph where the two axes intersect

Output   A member of the range of a function

P

Parabola   A graph of a quadratic function

Parallel lines   Lines in the same plane that never intersect. Two lines are parallel if they have the same slope and different y-intercepts.

Parallelogram   A four-sided polygon with two pairs of parallel sides

Percent notation   A representation of a number as parts per 100

Perfect square   A rational number p for which there exists a number a for which $a^2 = p$

Perfect-square trinomial   A trinomial that is the square of a binomial

Perimeter   The sum of the lengths of the sides of a polygon

Perpendicular lines   Lines that form a right angle

Pi (π)   The number that results when the circumference of a circle is divided by its diameter; $\pi \approx 3.14$ or $22/7$

Point–slope equation   An equation of the form $y - y_1 = m(x - x_1)$, where m is the slope and (x_1, y_1) is a point on the line

Polynomial   A monomial or a combination of sums and/or differences of monomials

Polynomial equation   An equation in which two polynomials are set equal to each other

Positive integers   The natural numbers, or the integers to the right of zero on the number line

Prime factorization   A factorization of a composite number as a product of prime numbers

Prime number   A natural number that has *exactly two different factors:* itself and 1

Prime polynomial   A polynomial that cannot be factored using only integer coefficients

Principal square root   The nonnegative square root of a number

Principle of zero products   The statement that an equation $ab = 0$ is true if and only if $a = 0$ is true or $b = 0$ is true or both are true

Proportion   An equation stating that two ratios are equal

Proportional numbers   Two pairs of numbers having the same ratio

Pythagorean theorem   In any right triangle, if a and b are the lengths of the legs and c is the length of the hypotenuse, then $a^2 + b^2 = c^2$.

Q

Quadrants   The four regions into which the axes divide a plane

Quadratic equation   An equation equivalent to an equation of the type $ax^2 + bx + c = 0$, where $a \neq 0$

Quadratic formula   The solutions of $ax^2 + bx + c = 0$ are given by the equation $x = \dfrac{-b \pm \sqrt{b^2 - 4ac}}{2a}$.

Quadratic inequality   A second-degree polynomial inequality in one variable

R

Radical   The symbol $\sqrt{}$

Radical equation   An equation in which a variable appears in one or more radicands

Radical expression   An algebraic expression written with a radical

Radicand   The expression written under the radical

Radius   A segment with one endpoint on the center of a circle and the other endpoint on the circle

Range   The set of all second coordinates of the ordered pairs in a function

Rate   The ratio of two different kinds of measure

Ratio   The quotient of two quantities

Rational equation   An equation containing one or more rational expressions; also called *fraction equation*

Rational expression   A quotient, or ratio, of polynomials; also called *fraction expression*

Rational inequality   An inequality containing a rational expression

Rational number   A number that can be written in the form a/b, where a and b are integers and $b \neq 0$

Rationalizing the denominator   A procedure for finding an equivalent expression without a radical in the denominator

Real numbers   All rational and irrational numbers; the set of all numbers corresponding to points on the number line

Reciprocal   A multiplicative inverse; two numbers are reciprocals if their product is 1

Rectangle   A four-sided polygon with four right angles

Relation   A correspondence between a first set, the domain, and a second set, the range, such that each member of the domain corresponds to *at least one* member of the range

Repeating decimal   A decimal in which a number pattern repeats indefinitely

Right triangle   A triangle that includes a 90° angle

Rise   The change in the second coordinate between two points on a line

Roster notation   A way of naming sets by listing all the elements in the set

Rounding   Approximating the value of a number; used when estimating

Run   The change in the first coordinate between two points on a line

S

Scientific notation   A representation of a number of the form $M \times 10^n$, where n is an integer, $1 \le M < 10$, and M is expressed in decimal notation

Second-order determinant   The determinant of a two-by-two matrix $\begin{bmatrix} a_1 & b_1 \\ a_2 & b_2 \end{bmatrix}$ is denoted $\begin{vmatrix} a_1 & b_1 \\ a_2 & b_2 \end{vmatrix}$ and is defined to mean $a_1 b_2 - a_2 b_1$.

Set   A collection of objects

Set-builder notation   The naming of a set by describing basic characteristics of the elements in the set

Similar triangles   Triangles in which corresponding angles have the same measure and the lengths of corresponding sides are proportional

Simplest fraction notation   A fraction written with the smallest numerator and denominator

Simplify   To rewrite an expression in an equivalent, abbreviated form

Slope   The ratio of the rise to the run for any two points on a line

Slope–intercept equation   An equation of the form $y = mx + b$, where x and y are variables; the slope is m and the y-intercept is $(0, b)$

Solution   A replacement for the variable that makes an equation or inequality true

Solution set   The set of all solutions of an equation, an inequality, or a system of equations or inequalities

Solution of a system of linear inequalities   An ordered pair (x, y) that is a solution of *both* inequalities

Solution of a system of three equations   An ordered triple (x, y, z) that makes *all three* equations true

Solution of a system of two equations   An ordered pair (x, y) that makes *both* equations true

Solve   To find all solutions of an equation, an inequality, or a system of equations or inequalities; to find the solution(s) of a problem

Speed   The ratio of distance traveled to the time required to travel that distance

Square   A four-sided polygon with four right angles and all sides of equal length

Square of a number   A number multiplied by itself

Square root   The number c is a square root of a if $c^2 = a$.

Standard form of a quadratic equation   A quadratic equation in the form $ax^2 + bx + c = 0$, where $a \ne 0$

Subsets   Sets that are a part of other sets

Substitute   To replace a variable with a number

Substitution method   A nongraphical method for solving systems of equations

Sum of cubes   An expression that can be written in the form $A^3 + B^3$

Sum of squares   An expression that can be written in the form $A^2 + B^2$

Supplementary angles   Angles whose sum is 180°

Synthetic division   A simplified process for dividing a polynomial by a binomial of the type $x - a$

System of equations   A set of two or more equations that are to be solved simultaneously

System of linear inequalities   A set of two or more inequalities that are to be solved simultaneously

T

Term   A number, a variable, or a product or a quotient of numbers and/or variables

Terminating decimal   A decimal that can be written using a finite number of decimal places

Third-order determinant   The determinant of a three-by-three matrix $\begin{bmatrix} a_1 & b_1 & c_1 \\ a_2 & b_2 & c_2 \\ a_3 & b_3 & c_3 \end{bmatrix}$ is denoted $\begin{vmatrix} a_1 & b_1 & c_1 \\ a_2 & b_2 & c_2 \\ a_3 & b_3 & c_3 \end{vmatrix}$ and is defined to mean $a_1 \begin{vmatrix} b_2 & c_2 \\ b_3 & c_3 \end{vmatrix} - a_2 \begin{vmatrix} b_1 & c_1 \\ b_3 & c_3 \end{vmatrix} + a_3 \begin{vmatrix} b_1 & c_1 \\ b_2 & c_2 \end{vmatrix}.$

Trinomial   A polynomial that is composed of three terms

Trinomial square   The square of a binomial expressed as three terms

U

Union of sets A and B   The set of all elements belonging to A and/or B

V

Value   The numerical result after a number has been substituted into an expression

Variable   A letter that represents an unknown number

Variation constant   The constant in an equation of direct or inverse variation

Vertex   The point at which the graph of a quadratic equation crosses its axis of symmetry

Vertical-line test   The statement that a graph represents a function if it is impossible to draw a vertical line that intersects the graph more than once

W

Whole numbers   The natural numbers and 0: 0, 1, 2, 3, …

X

x-intercept   The point at which a graph crosses the x-axis

Y

y-intercept   The point at which a graph crosses the y-axis

Photo Credits

1, © Roger Ressmeyer/CORBIS 4, © Fancy/Veer/Corbis 18 (left), Raymond Gehman/National Geographic/Getty Images 18 (right), Melissa Farlow/National Geographic/Getty Images 27, © Roger Ressmeyer/CORBIS 31, Purestock/Getty Images 81, Thinkstock/Getty Images 85,© age fotostock/SuperStock 121, © Glow Images/SuperStock 122, Medioimages/Photodisc/Getty Images 124 (left), Jochen Sand, Digital Vision/Getty Images 124 (right), © Don Mason/Blend Images/Corbis 125, © Tetra Images/Tetra Images/Corbis 126 (left), © Fancy/Veer/Corbis 126 (right), © age fotostock/SuperStock 133, Corbis/Jupiter Images 135, AFP/Getty Images 136, Visual Ideas/ Camilo Morales/Jupiter Images 139 (left), © age fotostock/SuperStock 139 (right), © Photodisc/SuperStock 142, © Glow Images/Super-Stock 157, © Aladdin Color, Inc./Corbis 160, © Charles O'Rear/CORBIS 161, Tetra Images/Jupiter Images 162 (left), Copyright © A. Ramey/PhotoEd 162 (right), © Mark Karrass/Corbis 168, Rob Van Petten, Digital Vision/Getty Images 173, © David Sailors/CORBIS 210, Digital Vision/Getty Images 214 (left), © David Sailors/CORBIS 214 (right), © D. Robert & Lorri Franz/CORBIS 215 (left), Copyright © Jim West/PhotoEdit 215 (right), © Photodisc/SuperStock 221, Federal Highway Administration and Washington Infrastructure Services, Inc. 226, Peter Dazeley, Photographer's Choice/Getty Images 227, © Larry W. Smith/epa/Corbis 240, Francesca Yorke © Dorling Kindersley 241 (top), NASA 241 (bottom), Eyewire Collection 244 (left), Phototake 244 (right), Alfred Pasieka/ Peter Arnold Inc. 246, © Larry W. Smith/epa/Corbis 247 (left), Copyright © Dennis MacDonald/PhotoEdit 247 (right), NASA 248, © Creatas/SuperStock 295, Time&Life Pictures/Getty Images 317, © Stockbyte/SuperStock 350, GSFC/NASA 388, © SW Productions/Brand X/Corbis 391, Fredrik Nyman, Johner Images Royalty-Free/Getty Images 394 (top), Shaffer Smith Photo LLC/Jupiter Images 394 (bottom), Jupiter Images 395 (left), © Corbis 395 (right), Hans Neleman, Taxi/Getty Images 409, © age fotostock/ SuperStock 429, © Tetra Images/SuperStock 462, Alamy Images 466, Associated Press 467 (top), Joel Sartore, National Geographic/Getty Images 467 (bottom), Pacific Stock 468, © age fotostock/SuperStock 473, © Tim Pannell/Corbis 474 (left), PEARSON STUDENT PHOTO ARCHIVE 474 (right), Tom Turpin, Purdue University 475 (left), © MATTHEW CAVANAUGH/epa/Corbis 475 (right), © Noah K. Murray/Star Ledger/ Corbis 476, Axiom Photographic Agency/Getty Images 478, Medioimages/Photodisc/Getty Images 480, Comstock/Getty Images 481 (top), © PAULO WHITAKER/Reuters/Corbis 481 (bottom), Heather Charles/The Indianapolis Star 484, © Radius/SuperStock 485 (left), © moodboard/Corbis 485 (right), © Pixtal/SuperStock 486, C. Flanigan/Contributor, Film Magic/Getty Images 487, ballyscanlon, Photodisc/Getty Images 493 (left), PhotoEdit, Inc. 493 (right), PhotoLink/Getty Images 497, Monty Rakusen/Jupiter Images 499, Dave Smith, BeadsByTheBead 507, Photodisc/Getty Images 509 (left), Stock Connection 509 (right), Ron Edmonds, Associated Press 510, © Glowimages/Corbis 545, © Chad Ehlers—Stock Connection/Science Faction/Corbis 549, Dave Smith, BeadsByTheBead 550, © Molly Riley/Reuters/Corbis 551, © Royalty-Free/Corbis 553, Digital Vision/Getty Images 568, © mood-board/Corbis 569, Courtesy of the Press-Register 2009 © All rights reserved. Reprinted with permission. 570, Sharon Dominick/Getty Images 581, Getty Images Inc.—Stone Allstock 590, Courtesy of the Press-Register 2009 © All rights reserved. Reprinted with permission. 593, Comstock/Corbis 594, Image by © Charles Gupton/CORBIS 604, © age fotostock/SuperStock 605 (left), © Sean Justice/Corbis 605 (right), David Young-Wolff/PhotoEdit, Inc. 618, © moodboard/Corbis 620 (left), Charles Gupton/Corbis/Stock Market 620 (right), Copyright © David Young-Wolff/PhotoEdit 621, Copyright © Myrleen Ferguson Cate/PhotoEdit 622 (left), © Jim Craigmyle/Corbis 622 (right), © Construction Photography/Corbis 623 (left), © Don Hammond/Design Pics/Corbis 623 (right), © Ron Chapple/ Corbis 643, © David Butow/CORBIS SABA 644, © Fancy/Veer/Corbis 650, imagebroker.net/photolibrary 651, © Photodisc/ SuperStock 652, Copyright © Spencer Grant/PhotoEdit 662, © Régis Bossu/Sygma/Corbis 690, Getty Images, Inc.—Allsport Photography 700 (top), Dan Sams, Taxi/Getty Images 700 (bottom), © Radius Images/Corbis 701, WireImage/Getty 709, © Royalty-Free/Corbis 748, WireImage 751 (top), Photolibrary.com 751 (bottom left), © Galen Rowell/CORBIS 751 (bottom right), Getty Images Entertainment/ Getty Images 752, © Franz Faltermaier/Westend61/Corbis 759, Image99/Jupiter Images 780, Comstock/Jupiter Images 781, © Flirt/ SuperStock 791 (left), Sports Illustrated/Getty Images 791 (right), Getty Images Sport/Getty Images 794 (left), Guy Christian/Photo Library 794 (right), AFP/Getty Images 810, © Comstock/Corbis 812 (left), Thinkstock/Getty Images 812 (right), © HENNY RAY ABRAMS/Reuters/Corbis 814, © Tony Gentile/Reuters/Corbis 844, © Angela Drury/Corbis 849, © Corbis 873, AFP/Getty Images 886 (left), Cultura/Lilian Henglein/Getty Images 886 (right), Heather Wines/Gannett 887 (left), PhotoEdit, Inc. 887 (right), MANFRED KAGE/Peter Arnold, Inc. 897, photos_alyson/Getty Images 940, Asia Images Group/Getty Images 941, Aldon/ Stockbyte/Getty Images 943, Photodisc/Getty Images 945, AFP/Getty Images 947, Photo Researchers, Inc. 949 (top left), Todd Gipstein, National Geographic/Getty Images 949 (top right), Lilli Day, Photodisc/Getty Images 949 (bottom), © PhotoAlto/SuperStock 950, Anne Rippy, Digital Vision/Getty Images 951, Tetra Images/Getty Images 952 (left), Copyright © Mark Richards/ PhotoEdit 952 (right), Stockbyte/Getty Images 953 (left, right), AFP/Getty Images 960, Ambient Images 961, Copyright © David Young-Wolff/PhotoEdit

Index

Geometric Formulas

PLANE GEOMETRY

Rectangle
Area: $A = l \cdot w$
Perimeter: $P = 2 \cdot l + 2 \cdot w$

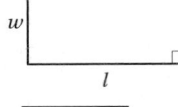

Square
Area: $A = s^2$
Perimeter: $P = 4 \cdot s$

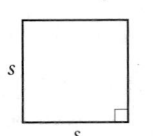

Triangle
Area: $A = \frac{1}{2} \cdot b \cdot h$

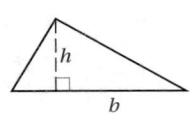

Sum of Angle Measures
$A + B + C = 180°$

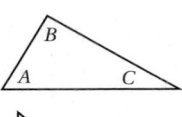

Right Triangle
Pythagorean Theorem:
$a^2 + b^2 = c^2$

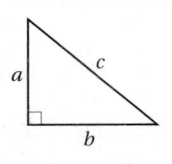

Parallelogram
Area: $A = b \cdot h$

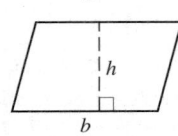

Trapezoid
Area: $A = \frac{1}{2} \cdot h \cdot (a + b)$

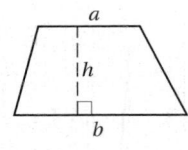

Circle
Area: $A = \pi \cdot r^2$
Circumference:
$C = \pi \cdot d = 2 \cdot \pi \cdot r$
$\left(\frac{22}{7} \text{ and } 3.14 \text{ are different} \right.$
$\left. \text{approximations for } \pi\right)$

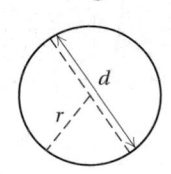

SOLID GEOMETRY

Rectangular Solid
Volume: $V = l \cdot w \cdot h$

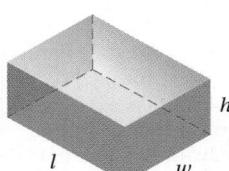

Cube
Volume: $V = s^3$

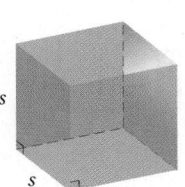

Right Circular Cylinder
Volume: $V = \pi \cdot r^2 \cdot h$
Surface Area:
$S = 2 \cdot \pi \cdot r \cdot h + 2 \cdot \pi \cdot r^2$

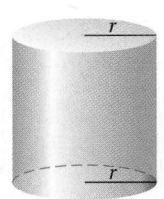

Right Circular Cone
Volume: $V = \frac{1}{3} \cdot \pi \cdot r^2 \cdot h$
Surface Area: $S = \pi \cdot r^2 + \pi \cdot r \cdot s$

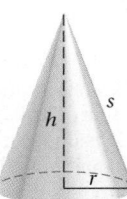

Sphere
Volume: $V = \frac{4}{3} \cdot \pi \cdot r^3$
Surface Area: $S = 4 \cdot \pi \cdot r^2$

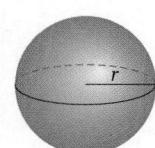